Extrinsic Geometric Flows

GRADUATE STUDIES IN MATHEMATICS 206

Extrinsic Geometric Flows

Ben Andrews
Bennett Chow
Christine Guenther
Mat Langford

AMERICAN MATHEMATICAL SOCIETY
Providence, Rhode Island

2010 *Mathematics Subject Classification.* Primary 53C44, 58J35, 53A07, 52A20, 35K20.

For additional information and updates on this book, visit
www.ams.org/bookpages/gsm-206

Library of Congress Cataloging-in-Publication Data

Names: Andrews, Ben, author. | Chow, Bennett, author. | Guenther, Christine (Christine Marie), 1966- author. | Langford, Mat (Mathew), 1987- author.
Title: Extrinsic geometric flows / Ben Andrews, Bennett Chow, Christine Guenther, Mat Langford.
Description: Providence, Rhode Island : American Mathematical Society, [2020] | Series: Graduate studies in mathematics, 1065-7339 ; 206 | Includes bibliographical references and index.
Identifiers: LCCN 2019059835 | ISBN 9781470455965 (v. 206 ; hardcover) | ISBN 9781470456863 (v. 206 ; ebook) | ISBN 9781470464578 (v. 206; softcover)
Subjects: LCSH: Global differential geometry. | Differential equations, Parabolic. | Flows (Differentiable dynamical systems). | Curvature. | Geometric analysis. | AMS: Differential geometry – Global differential geometry. | Global analysis, analysis on manifolds – Partial differential equations on manifolds; differential operators. | Differential geometry – Classical differential geometry – Higher-dimensional and -codimensional. | Convex and discrete geometry – General convexity – Convex sets in n dimensions (including convex hypersurfaces). | Partial differential equations – Parabolic equations and systems – Initial-boundary value problems for second-order parabolic equations.
Classification: LCC QA670 .A53 2020 | DDC 516.3/62–dc23
LC record available at https://lccn.loc.gov/2019059835

Printed in the United States of America.
Reprinted by the American Mathematical Society, 2021.

∞ The paper used in this book is acid-free and falls within the guidelines established to ensure permanence and durability.
Visit the AMS home page at `https://www.ams.org/`

11 10 9 8 7 6 5 4 3 2   26 25 24 23 22 21

Contents

Preface xiii

A Guide for the Reader xv

The heat equation (Chapter 1) xv

Curve shortening flow (Chapters 2–4) xv

Mean curvature flow (Chapters 5–14) xvi

Gauß curvature flows (Chapters 15–17) xix

Fully nonlinear curvature flows (Chapters 18–20) xx

Acknowledgments xx

Suggested Course Outlines xxiii

Notation and Symbols xxv

Chapter 1. The Heat Equation 1

§1.1. Introduction 1

§1.2. Gradient flow 3

§1.3. Invariance properties 3

§1.4. The maximum principle 8

§1.5. Well-posedness 12

§1.6. Asymptotic behavior 14

§1.7. The Bernstein method 17

§1.8. The Harnack inequality 17

§1.9. Further monotonicity formulae 19

§1.10. Sharp gradient estimates 21

§1.11. Notes and commentary 29
§1.12. Exercises 33

Chapter 2. Introduction to Curve Shortening 37
§2.1. Basic geometric theory of planar curves 38
§2.2. Curve shortening flow 42
§2.3. Graphs of functions 45
§2.4. The support function 48
§2.5. Short-time existence 50
§2.6. Smoothing 51
§2.7. Global existence 54
§2.8. Notes and commentary 59
§2.9. Exercises 59

Chapter 3. The Gage–Hamilton–Grayson Theorem 63
§3.1. The avoidance principle 64
§3.2. Preserving embeddedness 66
§3.3. Huisken's distance comparison estimate 68
§3.4. A curvature bound by distance comparison 74
§3.5. Grayson's theorem 82
§3.6. Singularities of immersed solutions 88
§3.7. Notes and commentary 90
§3.8. Exercises 92

Chapter 4. Self-Similar and Ancient Solutions 95
§4.1. Invariance properties 95
§4.2. Self-similar solutions 96
§4.3. Monotonicity formulae 102
§4.4. Ancient solutions 110
§4.5. Classification of convex ancient solutions on S^1 116
§4.6. Notes and commentary 122
§4.7. Exercises 123

Chapter 5. Hypersurfaces in Euclidean Space 125
§5.1. Parametrized hypersurfaces 125
§5.2. Alternative representations of hypersurfaces 143
§5.3. Dynamical properties 152
§5.4. Curvature flows 164

§5.5. Notes and commentary 169
§5.6. Exercises 169

Chapter 6. Introduction to Mean Curvature Flow 173
§6.1. The mean curvature flow 173
§6.2. Invariance properties and self-similar solutions 176
§6.3. Evolution equations 179
§6.4. Short-time existence 184
§6.5. The maximum principle 189
§6.6. The avoidance principle 192
§6.7. Preserving embeddedness 196
§6.8. Long-time existence 197
§6.9. Weak solutions 206
§6.10. Notes and commentary 215
§6.11. Exercises 219

Chapter 7. Mean Curvature Flow of Entire Graphs 223
§7.1. Introduction 223
§7.2. Preliminary calculations 224
§7.3. The Dirichlet problem 227
§7.4. A priori height and gradient estimates 228
§7.5. Local a priori estimates for the curvature 232
§7.6. Proof of Theorem 7.1 238
§7.7. Convergence to self-similarly expanding solutions 239
§7.8. Self-similarly shrinking entire graphs 240
§7.9. Notes and commentary 240
§7.10. Exercises 241

Chapter 8. Huisken's Theorem 243
§8.1. Pinching is preserved 244
§8.2. Pinching improves: The roundness estimate 246
§8.3. A gradient estimate for the curvature 256
§8.4. Huisken's theorem 259
§8.5. Regularity of the arrival time 266
§8.6. Huisken's theorem via width pinching 267
§8.7. Notes and commentary 274
§8.8. Exercises 278

Chapter 9. Mean Convex Mean Curvature Flow 281
§9.1. Singularity formation 281
§9.2. Preserving pinching conditions 284
§9.3. Pinching improves: Convexity and cylindrical estimates 294
§9.4. A natural class of initial data 301
§9.5. A gradient estimate for the curvature 303
§9.6. Notes and commentary 308
§9.7. Exercises 309

Chapter 10. Monotonicity Formulae 311
§10.1. Huisken's monotonicity formula 311
§10.2. Hamilton's Harnack estimate 319
§10.3. Notes and commentary 338
§10.4. Exercises 342

Chapter 11. Singularity Analysis 345
§11.1. Local uniform convergence of mean curvature flows 345
§11.2. Neck detection 354
§11.3. The Brakke–White regularity theorem 363
§11.4. Huisken's theorem revisited 366
§11.5. The structure of singularities 371
§11.6. Notes and commentary 389
§11.7. Exercises 394

Chapter 12. Noncollapsing 395
§12.1. The inscribed and exscribed curvatures 395
§12.2. Differential inequalities for the inscribed and exscribed curvatures 402
§12.3. The Gage–Hamilton and Huisken theorems via noncollapsing 412
§12.4. The Haslhofer–Kleiner curvature estimate 415
§12.5. Notes and commentary 421
§12.6. Exercises 422

Chapter 13. Self-Similar Solutions 425
§13.1. Shrinkers — an introduction 425
§13.2. The Gaußian area functional 426
§13.3. Mean convex shrinkers 431

§13.4. Compact embedded self-shrinking surfaces 443
§13.5. Translators — an introduction 452
§13.6. The Dirichlet problem for graphical translators 454
§13.7. Cylindrical translators 455
§13.8. Rotational translators 456
§13.9. The convexity estimates of Spruck, Sun, and Xiao 462
§13.10. Asymptotics 468
§13.11. X.-J. Wang's dichotomy 469
§13.12. Rigidity of the bowl soliton 470
§13.13. Flying wings 477
§13.14. Bowloids 490
§13.15. Notes and commentary 492
§13.16. Exercises 499

Chapter 14. Ancient Solutions 503
§14.1. Rigidity of the shrinking sphere 504
§14.2. A convexity estimate 509
§14.3. A gradient estimate for the curvature 511
§14.4. Asymptotics 513
§14.5. X.-J. Wang's dichotomy 516
§14.6. Ancient solutions to curve shortening flow revisited 525
§14.7. Ancient ovaloids 531
§14.8. Ancient pancakes 533
§14.9. Notes and commentary 536
§14.10. Exercises 540

Chapter 15. Gauß Curvature Flows 543
§15.1. Invariance properties and self-similar solutions 545
§15.2. Basic evolution equations 546
§15.3. Chou's long-time existence theorem 548
§15.4. Differential Harnack estimates 558
§15.5. Firey's conjecture 560
§15.6. Variational structure and entropy formulae 570
§15.7. Notes and commentary 578
§15.8. Exercises 578

Chapter 16. The Affine Normal Flow 581
§16.1. Affine invariance 582
§16.2. Affine-renormalized solutions 586
§16.3. Convergence and the limit flow 590
§16.4. Self-similarly shrinking solutions are ellipsoids 590
§16.5. Convergence without affine corrections 593
§16.6. Notes and commentary 601
§16.7. Exercises 602

Chapter 17. Flows by Superaffine Powers of the Gauß Curvature 607
§17.1. Bounds on diameter, speed, and inradius 607
§17.2. Convergence to a shrinking self-similar solution 613
§17.3. Shrinking self-similar solutions are round 618
§17.4. Notes and commentary 633
§17.5. Exercises 635

Chapter 18. Fully Nonlinear Curvature Flows 639
§18.1. Introduction 639
§18.2. Symmetric functions and their differentiability properties 641
§18.3. Examples 650
§18.4. Short-time existence 655
§18.5. The avoidance principle 658
§18.6. Differential Harnack estimates 660
§18.7. Entropy estimates 664
§18.8. Alexandrov reflection 670
§18.9. Notes and commentary 682
§18.10. Exercises 683

Chapter 19. Flows of Mean Curvature Type 687
§19.1. Convex hypersurfaces contract to round points 687
§19.2. Evolving nonconvex hypersurfaces 698
§19.3. Notes and commentary 708
§19.4. Exercises 709

Chapter 20. Flows of Inverse-Mean Curvature Type 711
§20.1. Convex hypersurfaces expand to round infinity 711
§20.2. Notes and commentary 723
§20.3. Exercises 724

Bibliography 727

Index 753

Preface

Geometric flows are evolution equations at the intersection of differential equations and geometry. The field of study is characterized by the deformation of geometric objects by geometric attributes such as curvature, and the equations that arise are, in an appropriate "gauge", nonlinear parabolic differential equations. These equations have extensive applications to physical and geometric problems arising in industry, materials science, computer vision and image processing, physics, and pure mathematics. The flows may be viewed from the perspectives of geometry and partial differential equations, or perhaps more aptly as a synthesis of both.

Geometric flows come in many flavors — there are, for instance, **extrinsic flows** (e.g., the curve shortening, mean curvature, Lagrangian mean curvature, Gauß curvature, inverse-mean curvature, and Willmore flows), **intrinsic flows** (e.g., the Ricci, Kähler–Ricci, Yamabe, Calabi, Chern–Ricci, cross curvature, and renormalization group flows), **flows of maps** (e.g., the heat equation on Riemannian manifolds, the harmonic map heat flow, and diffeomorphism-preserving flows), **flows of connections** (e.g., the Yang–Mills flow), as well as flows of further geometric structures (e.g., the G_2-flow), and even **discrete curvature flows** (e.g., discrete surface Ricci and discrete Yamabe flows).

Extrinsic flows are characterized by a submanifold evolving in an ambient space with velocity determined by its extrinsic curvature. The goal of this book is to give an extensive introduction to a few of the most prominent extrinsic flows, namely, the curve shortening flow, the mean curvature flow, the Gauß curvature flow, the inverse-mean curvature flow, and fully nonlinear flows of mean curvature and inverse-mean curvature type. We highlight techniques and behaviors that frequently arise in the study of these (and other) flows. To illustrate the broad applicability of the techniques developed, we also consider general classes of fully nonlinear curvature flows. We do not consider "higher-order" flows, such as the Willmore flow.

This book is written at the level of a graduate student who has had a basic course in differential geometry and has some familiarity with partial differential equations. It is intended also to be useful as a reference for specialists. In general, we provide detailed proofs, although for some more specialized results we may present only the main ideas; in such cases, we provide references for complete proofs. The coverage is not comprehensive; rather, selected topics are chosen that establish the foundation for further study. A brief survey of additional topics, with extensive references, can be found in the notes and commentary at the end of each chapter.

A unifying theme throughout these notes is *monotonicity formulae.* Such formulae are fundamental in the qualitative study of partial differential equations, geometric analysis, and geometric evolution equations. The reason is simple: A monotone quantity not only gives some control on a solution but usually also shows that the solution is improving in some sense (unless it is already in equilibrium). On the other hand, for many nonlinear evolution equations, singularities can occur. This is true even in the case of the diffusive, or heat-type, equations that we shall study. Often, the control provided by monotonicity formulae may be bootstrapped to obtain global existence and convergence for certain classes of initial data (such as closed convex hypersurfaces) or to analyze and classify singularities. The classification of singularities is a major aspect of formulating a way to flow past them. In this respect, similarity solutions are central. Similarity solutions are those solutions whose shapes do not change in time. So they do *not* improve; indeed, they are already in some sense the best possible solutions! As we shall see, monotonicity formulae may often be used to show that the shape of singularities approaches that of self-similar solutions.

A Guide for the Reader

The heat equation (Chapter 1)

This part consists of a single chapter, in which we briefly discuss the model for all geometric evolution equations — the classical heat equation on Euclidean space. We investigate its invariance properties and corresponding invariant solutions (a.k.a. "solitons"). We introduce the maximum principle, the Poincaré inequality, Bernstein estimates, the Harnack inequality, and entropy monotonicity formulae, all of which will arise in different contexts in later chapters. We characterize the long-time behavior of periodic solutions and introduce "multipoint maximum principle" methods, another theme which appears repeatedly throughout the book.

Curve shortening flow (Chapters 2–4)

Chapter 2 introduces the curve shortening flow of planar curves. We derive equations for the evolution of geometric quantities such as curvature, length, area, and the support function. We investigate special solutions such as the shrinking circle, the Grim Reaper, and the paperclip. We obtain Bernstein-type derivative estimates for the curvature and use them to characterize the singular time. Short-time existence and uniqueness of solutions are not proved here since they follow from the corresponding results for mean curvature flow proved in Chapter 6.

In Chapter 3, we prove the Gage–Hamilton and Grayson convergence theorems for the curve shortening flow of embedded planar curves. Our approach is to develop successive refinements of maximum principle arguments involving the distance function. The most basic is the avoidance principle, a modification of which implies that embeddedness is preserved under the flow via a chord-arc estimate. A little more work yields Huisken's distance comparison estimate and then the curvature bound by distance comparison by Bryan and the first author. This estimate, combined with straightforward bootstrapping arguments, leads quickly to the convergence result.

Finally, we introduce, in Chapter 4, self-similar and ancient solutions to the curve shortening flow. We discuss the classification by Abresch and Langer, Epstein and Weinstein, and Halldorsson of properly immersed self-similar solutions. We present various monotonicity formulae: Gage's isoperimetric ratio monotonicity, Hamilton's differential Harnack and entropy estimates, and Huisken's monotonicity formula. We examine the hairclip and paperclip ancient solutions and discuss the classication, by Daskalopoulos, Hamilton, and Šešum, of closed, convex ancient solutions. In a later chapter, we present a different route to this classification (which also applies in the noncompact setting).

Mean curvature flow (Chapters 5–14)

In the introductory chapter to this part, Chapter 5, we develop the basic theory of hypersurfaces in Euclidean space, including the fundamental Gauß, Codazzi, and Simons identities and variation formulae for surface areas and enclosed volumes. Our point of view is primarily that of parametrized hypersurfaces, although we also develop the geometry of graphical, level set, starshaped, and convex hypersurfaces. A convex hypersurface may be parametrized by its Gauß map and is characterized by its support function. We develop the "time-dependent" geometry of evolving hypersurfaces in two equivalent ways — via evolving orthonormal frames and pullback bundles — and briefly introduce general curvature flows of hypersurfaces.

In Chapter 6, we begin our discussion of mean curvature flow, which is the gradient flow for the area functional. We briefly discuss its invariance properties and self-similar solutions, including shrinking spheres and cylinders, which appear later in the analysis of singularities. We compute the evolution of various geometric quantities and prove short-time existence and uniqueness of solutions. We introduce the maximum principle for scalar and tensor heat-type equations and use it to prove the avoidance principle and the preservation of embeddedness and to characterize the singular time. Finally, we briefly survey the various notions of "weak solutions" to the mean curvature flow, including measure-theoretic solutions, level-set/viscosity solutions, shadow solutions, piecewise smooth solutions, and flows with surgery.

We then present, in Chapter 7, the important work of Ecker and Huisken on the mean curvature flow of entire graphs. We present their global existence theorem, which provides an entire graphical solution to the mean curvature flow for all positive time from any entire initial datum which is locally Lipschitz continuous. We first develop their local estimates for the second fundamental form and its covariant derivatives.

In Chapter 8, we present Huisken's theorem, which states that the mean curvature flow evolves closed, convex hypersurfaces to round points. We start by proving that pinching of the second fundamental form is preserved under the flow using the tensor maximum principle. The scalar maximum principle shows that a certain measure of the "roundness" of the hypersurface is preserved. A Stampacchia iteration argument, making use of the Michael–Simon Sobolev inequality, is then used to show that the roundness tends to perfection at high curvature scales. Convergence to a round point then follows from estimates for the covariant derivatives of the second fundamental form and straightforward bootstrapping arguments. We finish the chapter with a second proof of Huisken's theorem via a width pinching estimate and a compactness argument. Two further proofs will be presented in later chapters.

Chapter 9 concerns the mean curvature flow of *mean convex* hypersurfaces, that is, hypersurfaces with positive mean curvature. The main interest is singularity formation. To motivate this, we begin with a brief discussion of the neckpinch and degenerate neckpinch singularities. We present a general theory which shows that convex pinching conditions for the second fundamental form are preserved, including a rigidity statement: Nonnegative pinching strictly improves unless the solution splits off a line. We prove the Huisken–Sinestrari "convexity estimate" for mean convex solutions and "cylindrical estimates" for "m-convex" solutions using Stampacchia iteration. In the case of convex solutions, the latter estimate recovers Huisken's roundness estimate. Using a general setup, *all* of these estimates are proved at once, substantially shortening the original proofs. We then prove the Huisken–Sinestrari gradient estimate for the curvature using the maximum principle and develop some of its applications.

In Chapter 10, we present two fundamental monotonicity formulae for the mean curvature flow: Huisken's monotonicity formula and Hamilton's differential Harnack estimate. We first encountered these inequalities in the context of curve shortening flow. We also present important applications.

In Chapter 11, we study singularity formation for mean convex solutions. We consider sequences of space-time points approaching a singularity and the associated "blow-up sequences" and ancient limits. In order to take limits, we need an appropriate compactness theory. We achieve this by making use of the Cheeger–Gromov compactness theorem for sequences of Riemannian manifolds to obtain an analogue for sequences of hypersurfaces, which leads to a suitable compactness result for mean curvature flows. Our first application of the compactness theorem is in the analysis of regions of high curvature; namely, we prove the Huisken–Sinestrari "neck detection" lemma, which shows that a long, thin, round "neck" is necessarily forming at the first singular time for a 2-convex solution. As is typical, this follows by reductio ad absurdum using the compactness theorem (and the convexity estimates proved in Chapter 9). We then use Huisken's monotonicity formula and the compactness theory to prove Brakke's regularity theorem in the case of smooth flows (à la White).

We then address singularity formation. We begin by presenting Hamilton's proof of Huisken's theorem, which involves a compactness argument making use of Huisken's monotonicity formula and Hamilton's Harnack inequality. This argument is then extended to the mean convex setting, providing a classification of singularity models. We show that type-I singularity models are necessarily shrinking cylinders and type-II singularity models are necessarily translating self-similar solutions. We also discuss the notions of tangent flows and limit flows, including some of their established and conjectured properties.

Chapter 12 introduces the noncollapsing theory of W. Sheng and X.-J. Wang, as developed by the first author and others. The crucial objects are the inscribed and exscribed curvatures, which lead to the notions of interior and exterior noncollapsing, respectively. The inscribed and exscribed curvatures are extrema of a certain smooth "2-point" function. Using this function as a barrier, we prove that noncollapsing is preserved under the mean curvature flow. Related to these calculations are Simons-type inequalities for the maximum and minimum principal curvatures as well as the inscribed and exscribed curvatures. Noncollapsing leads to a *fourth* proof of Huisken's theorem which (unlike the previous proofs) also applies to the curve shortening flow, providing another proof of the Gage–Hamilton theorem. A further application of noncollapsing is the local curvature estimate of Haslhofer and Kleiner, which is used to obtain a new proof (by contradiction) of the convexity and cylindrical estimates discussed previously.

Chapter 13 investigates, in greater depth, self-similar and ancient solutions to the mean curvature flow. We first consider shrinking self-similar solutions. After deriving basic equations, we cover the classification by Huisken and Colding–Minicozzi of properly immersed shrinking self-similar solutions. We then present some topological classification results for shrinkers in $\mathbb{R}^3$ — Brendle's theorem that compact embedded genus zero shrinkers in $\mathbb{R}^3$ are round spheres and a result of Mramor and S. Wang which shows that shrinkers in $\mathbb{R}^3$ are unknotted. Next, we investigate translating self-similar solutions. After deriving basic equations, we construct several examples, including the oblique Grim hyperplanes, the bowl soliton, and the translating catenoid. We present a theorem of Spruck and Xiao, which shows that mean convex translators in $\mathbb{R}^3$ are necessarily convex and a theorem of Spruck and Sun which shows that solutions to the translator Dirichlet problem over convex domains are convex. We present rigidity results for the bowl soliton due to X.-J. Wang and R. Haslhofer and state X.-J. Wang's dichotomy (proved in the subsequent chapter): any convex translator which is not entire necessarily lies in a slab region. We construct the "flying wings" of Bourni, the fourth author, and Tinaglia, and the "bowloids" of Hoffman, Martín, Ilmanen, and White. A uniqueness result of Hoffman et al. completes the classification of mean convex translators in $\mathbb{R}^3$.

In Chapter 14, we consider ancient solutions. We present the characterization of the shrinking sphere among closed, convex ancient solutions due to Huisken–Sinestrari and Haslhofer–Hershkovits and prove a convexity estimate for mean convex ancient solutions, which follows a similar argument. We then present a gradient estimate for "entire" ancient solutions due also to Huisken and Sinestrari followed by a classification of the asymptotics of convex ancient solutions. We then present the proof of X.-J. Wang's remarkable dichotomy — every convex ancient solution which is not entire necessarily lies in a stationary slab region — and present some classification results for "ancient ovaloids" (closed, entire examples) and "ancient pancakes" (closed examples in slabs), including a different proof of the characterization of the shrinking sphere mentioned above.

Gauß curvature flows (Chapters 15–17)

We begin our investigation of Gauß curvature flows, in Chapter 15, by analyzing invariance properties and self-similar solutions. We present Chou's characterization of the singular time and the first author's proof of the Firey conjecture. We also discuss the second author's differential Harnack inequality and complete the chapter with an investigation of the variational structures and related entropies associated with the Gauß curvature flows.

In Chapter 16, we study the $\frac{1}{n+2}$-Gauß curvature flow. This flow is exceptional in that it is invariant under the full group of *affine* transformations of $\mathbb{R}^{n+1}$; indeed, it is tangential-diffeomorphism equivalent to the "affine normal flow". As a result, solutions cannot be expected to converge always to "round" points. Instead, we shall see that uniformly convex hypersurfaces converge to *ellipsoids* after appropriately renormalizing. Various methods are used to prove this, including Chou's estimate, the differential Harnack inequality, and the variational structure. However, in contrast to the other convergence results in this book, it is important that we exploit the full affine invariance of the equation.

In Chapter 17, we present the proof of the convergence to a shrinking self-similar solution for flows by powers $\alpha > \frac{1}{n+2}$ of the Gauß curvature due to P. Guan, L. Ni, and the first author, and the classification by Brendle, K. Choi, and Daskalopoulos of these shrinking self-similar solutions (they are round spheres). Combined, these results prove the general Firey conjecture.

Fully nonlinear curvature flows (Chapters 18–20)

We start by introducing, in Chapter 18, a general notion of evolution by curvature and deduce some general properties such as short-time existence of solutions and the avoidance principle, differential Harnack inequalities, entropy monotonicity formulae, and the Alexandrov reflection method.

We then consider, in Chapter 19, contracting flows by speeds that are 1-homogeneous in the principal curvatures, which tend to share phenomena with the mean curvature flow. Indeed, we show that, under certain structure conditions, convex hypersurfaces shrink to round points. We also discuss certain convexity, cylindrical, and noncollapsing estimates.

In Chapter 20 we discuss expanding curvature flows by speeds that are 1-homogeneous in the principal radii. We show that for a large class of such flows of compact convex hypersurfaces, solutions exist for infinite time, expand to infinity, and converge after rescaling to round spheres.

Throughout the book, we have included photos of a cross section of mathematicians who have made important contributions in the field of extrinsic geometric flows. The inclusion of these photos is in the spirit of encouragement and inspiration. We hope that they will enliven the results and make the book more interesting to read.[1]

Acknowledgments

We would like to give special thanks to AMS acquisitions editor Ina Mette and AMS publisher Sergei Gelfand for their continual encouragement, help, and support. We would also like to thank AMS acquisitions editor Eriko Hironaka and the Graduate Studies in Mathematics Editorial Committee for their help and support. We would like to thank Arlene O'Sean for her expert copy editing of the book. We would like to thank Marcia Almeida, Brian Bartling, and Peter Sykes of the AMS for their assistance in publishing the book.

[1]Results discussed in this book are often multiauthored, whereas the photos usually appear individually and in different parts of the book, reflecting the fact that many authors have multiple contributions.

The authors wish to acknowledge the valuable comments and suggestions of many people, including Simon Brendle, Kyeongsu Choi, Tobias Colding, Peng Lu, Stephen Lynch, Zilu Ma, Christos Mantoulidis, William Minicozzi, Alex Mramor, Julian Scheuer, Siksha Sivaramakrishna, Liming Sun, Ryan Unger, Bo Yang, and Yongjia Zhang, who provided useful feedback on earlier versions of the text. We are especially grateful to Tom Ilmanen for providing a number of important insights and conjectures regarding mean curvature flow singularity formation in Section 11.5.5. We are indebted to Greg Anderson for his elegant proof of Lemma 17.21 and to Theodora Bourni for explaining to us the proof of Theorems 13.43 and 14.13. The final product also benefited greatly from the suggestions of the anonymous reviewers (without whom Chapter 20 would not exist, for example).

During the preparation of this volume, Ben Andrews was partially supported by Laureate grant FL150100126 and Discovery grants DP120102462, DP120100097, DP150100375 of the Australian Research Council. Christine Guenther was partially supported by Simons Grant #283083.

Ben would like to thank Gerhard Huisken, Klaus Ecker, Leon Simon, Neil Trudinger, Rick Schoen, and S.-T. Yau for inspiration and encouragement over many years; his coauthors for persisting in what turned out to be a very long-term project; and Bean, Mark, Matt, Ambrose, Felix, Kylie, Rufus, Oliver, and Llewellyn for more than words can describe.

Bennett would like to thank Ed Dunne of the AMS for all of his support throughout the years on the previous Ricci flow books project. Bennett would especially like to thank Peng Lu for vast contributions to his expository development and for Peng's continued support and encouragement. Bennett is indebted to his coauthors Mat, Ben, and Chris for the fantastic work they have done to bring this book to fruition. Bennett would like to thank Gang Tian for his invitation to visit BICMR. Special thanks to Richard Hamilton and Mike Freedman for their encouragement. Bennett would like to express very special thanks to his wife, Jingwei Xia, his brother, Peter, his daughters, Michelle and Isabelle, and his stepdaughter, Gloriana, for their support and encouragement. Bennett dedicates this book to his parents, Yutze Chow and Wanlin Wu, and to the memory of his sister, Eleanor.

Chris thanks her coauthors for their dedication and expertise; it was a privilege to work with them. She wishes to express her appreciation to the many mathematicians whose work appears in these pages. She would like to thank Jim Isenberg, whose knowledge, humor, and encouragement continue to be invaluable. She is grateful to her sisters, Lisa and Karin, and to her parents, Ronald and MaryAnn, for their unfailing support of this and all of her endeavors. Chris dedicates this book to her husband, Manuel, her son, Miguel, and her daughter, Isabel.

Mat would like to thank his coauthors for their trust, encouragement, guidance, and friendship; his mentors, Ben Andrews, Julie Clutterbuck, Klaus Ecker, and James McCoy, for the generosity with which they have shared their wisdom; his collaborators, Theodora Bourni, Stephen Lynch, Alex Mramor, Huy The Nguyen, Julian Scheuer, and Giuseppe Tinaglia, for their patience and for making it so much fun; the many mathematicians without whose insights this book would not exist; and his hero, Neil Trudinger, for setting the gold standard in mathematical writing.

Ben Andrews
Australian National University

Bennett Chow
University of California San Diego

Christine Guenther
Pacific University

Mat Langford
University of Tennessee, Knoxville

Suggested Course Outlines

This book offers multiple pathways for a one- or two-semester course for students familiar with differential geometry of Euclidean hypersurfaces and some parabolic PDE.

Chapters 1 and 5 present background material (on the heat equation and hypersurface geometry, respectively) and can thus be omitted if not required.

(1) CURVE SHORTENING FLOW: Chapters 1 through 4 would be suitable for an undergraduate course on curve shortening flow.

(2) INTRODUCTION TO GEOMETRIC FLOWS: A selection of material from Chapters 1–3, 5, and 6, Sections 8.1, 8.2–8.4, Chapters 15–16, Section 19.1, and Chapter 20 would be suitable for a broad introduction to extrinsic geometric flows. Chapter 3 can be substituted with Sections 4.1–4.3. Sections 8.2–8.4 can be substituted with Section 8.6, or with Chapter 10 and Sections 11.1 and 11.4.

(3) MEAN CURVATURE FLOW: Chapters 5–12 and selected topics from Chapters 13 and 14 would be suitable for a course on mean curvature flow.

(4) GAUß CURVATURE FLOW: Chapters 2–4 and 15–17 would be suitable for a course on Gauß curvature flow.

(5) SELF-SIMILAR AND ANCIENT SOLUTIONS: An appropriate selection of material from Chapters 1, 2, 4, 6, 10, 11, and 13–17 would be suitable for a course on self-similar and ancient solutions to geometric flows.

Notation and Symbols

$\doteqdot$	defined to be equal to
$\times$	multiplication, if a formula does not fit on one line
$\cdot$	dot product or multiplication
$\otimes$	tensor product
$\odot$	symmetrized tensor product
$.^{\top}$	tangential projection
$.^{\perp}$	normal projection
∂_i	i-th coordinate basis element or partial derivative with respect to x^i : $\frac{\partial}{\partial x^i}$
∂_t	canonical vector field on $M^n \times (\alpha, \omega)$ or partial derivative with respect to "time", t: $\frac{\partial}{\partial t}$
∇	covariant derivative or gradient
∇_a	covariant derivative as directional derivative of functions on the frame bundle
∇_t	covariant time derivative (abstract or as a vector field on the frame bundle)
$\overline{\nabla}$	covariant derivative or gradient with respect to the standard metric on S^n
$\nabla\nabla$ or ∇^2	(covariant) Hessian
Δ	Laplacian (rough Laplacian when acting on tensors)
Δ_f	f-Laplacian
$\langle \cdot, \cdot \rangle$	Riemannian metric or inner product
$\partial\Omega$	boundary of a domain Ω
$\alpha^{\natural}$	dual vector field to the 1-form α

A	area enclosed by a plane curve
$\mathrm{A}_p = d\mathrm{G}\,\vert_p$	shape operator
$A_\#$	least shadow area
Area	area of a surface or volume of a hypersurface
$B_r(p)$	extrinsic ball of radius r centered at p
$\mathbb{C}$	set of complex numbers
$C^\infty(M)$	space of smooth functions on M
$C_c^\infty(M)$	space of C^∞ functions on M with compact support
$\mathcal{C}_m^n(R,\alpha)$	the class of $(m+1)$-convex hypersurfaces of $\mathbb{R}^{n+1}$ satisfying certain geometric bounds
const.	constant
d	distance
D	Euclidean covariant derivative
${}^X\!D$	pullback of Euclidean connection along X
dA	area form of a boundary
$d\mu$	Riemannian area element
dm	weighted area $\mathrm{e}^{-f}\,d\mu$
ds	arc length element
dt	canonical 1-form on $M^n \times (\alpha,\omega)$
diam	diameter
div	divergence
e	Euler's number
$\{e_i\}$	standard Euclidean basis
$E_r(x,t)$	heat ball of radius r based at (x,t)
End	bundle of selfadjoint endomorphisms of (TM,g)
exp	exponential map or exponential function
$\mathcal{F}$	entropy or Perelman's energy functional
$F(M)$	frame bundle of M
$\Gamma(E)$	space of smooth sections of E
Γ_{ij}^k	Christoffel symbols
G	Gauß map
g	Riemannian metric, including pullback metric
$g\,(t)$	time-dependent metric
$\bar{g}$	standard metric on S^n
H, $\vec{H}$	mean curvature, mean curvature vector
H_m	m-th mean curvature

H_{-1}	harmonic mean curvature
$\mathscr{H}^n$	n-dimensional Hausdorff measure
I	a "time" interval
I or g	first fundamental form or pullback metric
II or h	second fundamental form or its pullback
$\vec{\mathrm{II}}$ or $\vec{h}$	vector second fundamental form or its pullback
id	identity
inj	injectivity radius
int or $^\circ$	interior
J	counterclockwise rotation by $\pi/2$
κ	curvature of a plane curve
κ_i	principal curvature of a hypersurface
κ_1, κ_n	smallest, largest principal curvature
$\vec{\kappa}$	n-tuple of principal curvatures
K	Gauß curvature
k	extrinsic ball curvature
$\overline{k}$, $\underline{k}$	inscribed, exscribed curvature
L	various linear operators
L	length of a plane curve or Weingarten map
$\vec{\mathrm{L}}$	normal vector-valued Weingarten tensor
$\mathcal{L}$	Lie derivative or the operator $\alpha K^\alpha(\mathrm{II}^{-1})^{ij}\nabla_i\nabla_j$
log	natural logarithm
M^n	n-manifold (domain of a parametrized hypersurface $X : M^n \to \mathbb{R}^{n+1}$)
$\mathcal{M}^n$	immersed n-manifold (image of a parametrized hypersurface $X : M^n \to \mathbb{R}^{n+1}$)
μ	Riemannian measure
$\widetilde{M}$	universal covering of M
N, $\vec{n}$, ν	unit outward normal
MW	mean width
$\mathbb{N}$	the (positive) natural numbers
$N\mathcal{M}$, NM	normal bundle of $\mathcal{M}$, M
$\mathfrak{N}$, NM	normal bundle of $M^n \times (\alpha, \omega)$
$n\omega_n$ or σ_{n-1}	volume of the unit Euclidean $(n-1)$-sphere
ω_n	volume of the unit Euclidean n-ball
ODE	ordinary differential equation(s)

$OF(M)$	orthonormal frame bundle of M
OU	Ornstein–Uhlenbeck operator
ρ_y	reflection of S^n in the y-direction
ρ_Y	left action of $\mathrm{GL}(n, \mathbb{R})$ on $F(M)$
$P_r(x,t)$	parabolic ball of radius r based at (x,t)
PDE	partial differential equation(s)
$\mathbb{R}$ ($\mathbb{R}_+$)	the set of (positive) real numbers
$\mathbb{R}^n$	n-dimensional Euclidean space
r_i	principal radius of a hypersurface
R, Rc, Rm	scalar, Ricci, and Riemann curvature tensors
σ	support function
$\mathfrak{S}$, TM	spatial tangent bundle of $M \times (\alpha, \omega)$
S^n	unit radius n-dimensional sphere
$\mathrm{S}(n)$ or $\mathrm{S}^{n \times n}$	normed linear space of symmetric (real) $n \times n$ matrices (equipped with the Hilbert–Schmidt norm)
S^n_{y+} (S^n_{y-})	upper (lower) hemispheres in the y-direction
$S\mathcal{M}$, SM	unit tangent bundle of $\mathcal{M}$, M
$\mathrm{SO}(n)$	(real) special orthogonal group
$\mathfrak{so}(n)$	Lie algebra of $\mathrm{SO}(n)$
supp	support of a function
θ	normal angle of a plane curve
T	unit tangent vector to a curve
$T\mathcal{M}$, TM	tangent bundle of $\mathcal{M}$, M
$T^*\mathcal{M}$, T^*M	cotangent bundle of $\mathcal{M}$, M
T^n	n-dimensional torus
tr or trace	trace
Vol	volume enclosed by a hypersurface
w	width function
X	typically a parametrized hypersurface
X_t	parametrized hypersurface at time t
y^j_a	coordinates on the frame bundle
$\mathbb{Z}$	ring of integers

Chapter 1

The Heat Equation

The simplest example of a geometric flow equation is the classical scalar heat flow. Most of the analytical properties of the flows we shall explore in this book admit simpler analogues in this most basic case; as such, the heat equation serves as a useful model for geometric flows more generally.

1.1. Introduction

In 1807, motivated by experiments with various materials, Jean-Baptiste Joseph Fourier proposed a theory of heat based on the following postulates:

(1) **Fourier's law**: *The rate of flow of heat energy (heat flux $\vec{q}$) is proportional to the negative temperature gradient. That is, in appropriate units,*

$$\vec{q} = -\nabla u \,,$$

where $u = u(x,t)$ is the temperature at the point x in a Euclidean domain Ω at time t.

(2) **Conservation of energy**: *The heat energy $Q(\Omega, t)$ of Ω can only be gained or lost via flux through its boundary. That is,*

$$Q(\Omega, t_2) - Q(\Omega, t_1) = -\int_{t_1}^{t_2}\int_{\partial\Omega} \langle \vec{q}, \mathsf{N}\rangle \, dA \, dt \,,$$

where N and dA are the outward unit normal and area element of $\partial\Omega$, respectively. From these postulates, he derived

$$\int_\Omega \partial_t u \, dV = \frac{d}{dt} Q(\Omega, t) = -\int_{\partial\Omega} \langle \vec{q}, \mathsf{N}\rangle \, dA = \int_{\partial\Omega} \langle \nabla u, \mathsf{N}\rangle \, dA = \int_\Omega \Delta u \, dV \,,$$

where $\partial_t \doteqdot \frac{\partial}{\partial t}$, $\Delta u \doteqdot \operatorname{div}(\nabla u)$ is the Laplacian of u, dV is the volume element of Ω, and dA is the area element of $\partial\Omega$. Since the same argument applies to every subdomain of Ω, we actually obtain (assuming that u is sufficiently smooth) the pointwise equation

$$(\partial_t - \Delta)u = 0\,. \tag{1.1}$$

Solutions to (1.1) are sometimes referred to as **caloric functions**.

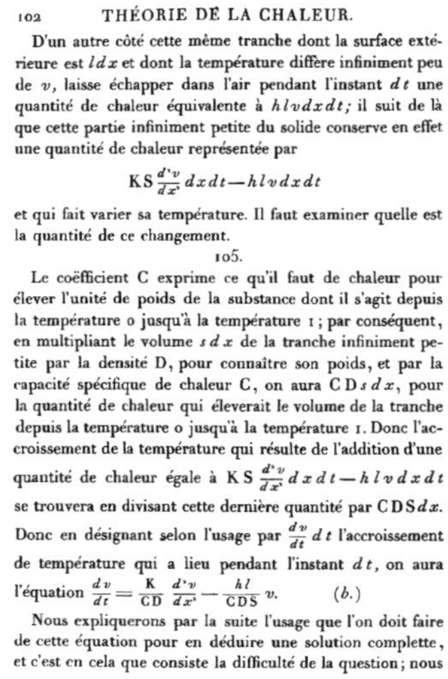

102 THÉORIE DE LA CHALEUR.

D'un autre côté cette même tranche dont la surface extérieure est $l\,dx$ et dont la température diffère infiniment peu de v, laisse échapper dans l'air pendant l'instant dt une quantité de chaleur équivalente à $hlv\,dx\,dt$; il suit de là que cette partie infiniment petite du solide conserve en effet une quantité de chaleur représentée par

$$\mathrm{KS}\frac{d^2v}{dx^2}dx\,dt - hlv\,dx\,dt$$

et qui fait varier sa température. Il faut examiner quelle est la quantité de ce changement.

105.

Le coëfficient C exprime ce qu'il faut de chaleur pour élever l'unité de poids de la substance dont il s'agit depuis la température 0 jusqu'à la température 1; par conséquent, en multipliant le volume $s\,dx$ de la tranche infiniment petite par la densité D, pour connaître son poids, et par la capacité spécifique de chaleur C, on aura $\mathrm{CD}s\,dx$, pour la quantité de chaleur qui éleverait le volume de la tranche depuis la température 0 jusqu'à la température 1. Donc l'accroissement de la température qui résulte de l'addition d'une quantité de chaleur égale à $\mathrm{KS}\frac{d^2v}{dx^2}dx\,dt - hlv\,dx\,dt$ se trouvera en divisant cette dernière quantité par $\mathrm{CDS}dx$. Donc en désignant selon l'usage par $\frac{dv}{dt}dt$ l'accroissement de température qui a lieu pendant l'instant dt, on aura l'équation $\frac{dv}{dt} = \frac{\mathrm{K}}{\mathrm{CD}}\frac{d^2v}{dx^2} - \frac{hl}{\mathrm{CDS}}v$. (b.)

Nous expliquerons par la suite l'usage que l'on doit faire de cette équation pour en déduire une solution complette, et c'est en cela que consiste la difficulté de la question; nous

Figure 1.1. An engraved portrait of Jean-Baptiste Joseph Fourier (1768–1830) and an excerpt from his treatise *Théorie de la Chaleur* in which the heat equation for a 1-dimensional rod is derived. (The reaction term accounts for heat loss to the surrounding air.)

The heat equation exhibits a multitude of remarkable properties, many of which will arise repeatedly in our later geometric setting. In order to get a feel for these properties with a minimum of technical hurdles, we shall usually consider, whenever boundary conditions need to be specified, **periodic** solutions of (1.1), that is, solutions $u : \mathbb{R}^n \times [0,\infty) \to \mathbb{R}$ satisfying

$$u(x,t) = u(x+e_i,t) \;\text{ for all }\; (x,t) \in \mathbb{R}^n \times [0,\infty) \;\text{ and }\; i = 1,\ldots,n\,,$$

where $\{e_i\}_{i=1}^n$ is the standard basis for $\mathbb{R}^n$. This is equivalent to considering solutions $u : T^n \times [0,\infty) \to \mathbb{R}$ on the n-dimensional torus, $T^n \doteqdot \mathbb{R}^n/\mathbb{Z}^n$.

Much of the discussion will hold equally well for solutions defined on smoothly bounded connected open sets $\Omega \subset \mathbb{R}^n$ which are smooth up to the boundary and satisfy the **Neumann** boundary condition:

$$\langle \nabla u, \mathsf{N} \rangle \equiv 0 \;\text{ on }\; \partial\Omega \times [0,T)\,.$$

By Fourier's law, this corresponds to a physical system for which there is no gain or loss of heat through the boundary.

The reader is invited to investigate which aspects change for solutions satisfying the **Dirichlet** boundary condition:

$$u \equiv 0 \quad \text{on} \quad \partial\Omega \times [0, T).$$

This corresponds to a physical system for which the external temperature remains constant (in this case, zero).

1.2. Gradient flow

The equilibrium states of the heat equation are given by the **harmonic functions** u, satisfying

$$-\Delta u = 0.$$

Recall that a smooth function $u : \Omega \to \mathbb{R}$ is harmonic if and only if it is a stationary point of the **Dirichlet energy**,

$$E(u) \doteqdot \frac{1}{2} \int_\Omega |\nabla u|^2 \, d\mu.$$

Indeed, if $U \in C^\infty(\Omega \times (-\varepsilon, \varepsilon))$ is a smooth variation of u which agrees with u outside a compact set, then, setting $\phi(x) \doteqdot \frac{d}{dt}\big|_{t=0} U(x,t)$, we find

$$\frac{d}{dt}\bigg|_{t=0} E(U) \doteqdot \int_\Omega \langle \nabla\phi, \nabla u \rangle \, d\mu = -\int_\Omega \phi \, \Delta u \, d\mu.$$

This proves that each harmonic function is a stationary point of E. To go the other way, consider, for any smooth function ϕ which is compactly supported in Ω, the variation $U(x,t) \doteqdot u(x) + t\phi(x)$.

This calculation also shows that the heat equation is the **L^2-gradient flow** of the Dirichlet energy: By the Cauchy–Schwarz inequality, E decreases most rapidly when ϕ is proportional to Δu. This suggests that solutions of the heat equation will tend towards harmonic equilibria.

1.3. Invariance properties

In this section we consider solutions of the heat equation defined on arbitrary open subsets of $\mathbb{R}^n \times \mathbb{R}$. It is not hard to check that the equation is invariant under the following local group actions, $u \mapsto u_\varepsilon$, in the sense that, for appropriate values of ε, performing the operation to a solution yields a new solution (possibly defined over a new space-time domain):

(1) Time translation: $u_\varepsilon(x,t) \doteqdot u(x, t-\varepsilon)$.

(2) Spatial translations: $u_\varepsilon(x,t) \doteqdot u(x - \varepsilon V, t)$, $V \in \mathbb{R}^n$.

(3) Scaling: $u_\varepsilon(x,t) \doteqdot \mathrm{e}^{-\varepsilon} u(x,t)$.

(4) Space-time dilation: $u_\varepsilon(x,t) \doteqdot u\left(\mathrm{e}^{-\varepsilon} x, \mathrm{e}^{-2\varepsilon} t\right)$.

(5) Rotations: $u_\varepsilon(x,t) \doteqdot u(e^{\varepsilon J} \cdot x, t)$, $J \in \mathfrak{so}(n)$, where $\mathfrak{so}(n)$ is the Lie algebra of $\mathrm{SO}(n)$ (i.e., the skew-symmetric $n \times n$ matrices).

(6) Superpositions: $u_\varepsilon(x,t) \doteqdot u(x,t) + \varepsilon v(x,t)$, where $(\partial_t - \Delta)v = 0$.

There are two further — much less obvious — symmetries, both of which mix all three (dependent and independent) variables. These are:

(7) Galilean boosts: $u_\varepsilon(x,t) \doteqdot e^{-(x-t\varepsilon V)\cdot\varepsilon V} u(x - 2t\varepsilon V, t)\,, \quad V \in \mathbb{R}^n$.

(8) The Appell transform [**535**]:

$$u_\varepsilon(x,t) \doteqdot (1+\varepsilon t)^{-\frac{n}{2}} \exp\left(\frac{-\varepsilon|x|^2}{4(1+\varepsilon t)}\right) u\left(\frac{x}{1+\varepsilon t}, \frac{t}{1+\varepsilon t}\right).$$

Although this latter pair of symmetries may appear quite mysterious, they can be derived systematically by the *method of prolongation.* We refer the reader to the book of Olver for a systematic exploration of local symmetry groups of partial differential equations [**431**].

Note that ε need not be positive or small. Indeed, in all the transformations except for the last, which is genuinely local, ε can take any value.

There heat equation also possesses symmetries which do not arise from the action of local Lie groups. For example, it is also preserved by convolutions: If $u : \mathbb{R}^n \times I \to \mathbb{R}$ is a solution, then so is the convolution

$$v(x,t) \doteqdot \int_{\mathbb{R}^n} u(x-y,t) v_0(y)\, dy$$

for any $v_0 : \mathbb{R}^n \to \mathbb{R}$, so long as the integral is finite and differentiation under the integral is allowed (for example, when v_0 is bounded and u and its derivatives are integrable).

1.3.1. Generating solutions from symmetries. A lot of useful information can be extracted from the presence of these symmetries. For example, we immediately see that if we start with a given solution u, then we can generate a large family new solutions. If we start with a nonzero constant solution, for instance, and apply a Galilean boost or an Appell transformation, then we obtain the **traveling wave** solution,

$$w(x,t) \doteqdot e^{-x_1+t}, \tag{1.2}$$

where x_1 is the first coordinate of x, or the **fundamental solution**,

$$\rho(x,t) \doteqdot (4\pi t)^{-\frac{n}{2}} e^{-\frac{|x|^2}{4t}}, \tag{1.3}$$

respectively. The profile of the traveling wave solution moves in the e_1-direction with unit speed:

$$w(x,t) = w(x - te_1, 0).$$

The fundamental solution can be interpreted as the solution to the heat equation starting with a unit amount of energy all localized at the origin. Indeed, $\rho(\cdot, t)$ converges in the distributional sense to the Dirac distribution as $t \searrow 0$.

More generally, given any solution u and an auxiliary solution v, we obtain the 8-parameter family of solutions

$$u_{\varepsilon_1\varepsilon_2\varepsilon_3\varepsilon_4\varepsilon_5\varepsilon_6\varepsilon_7\varepsilon_8} \doteqdot (4(1+\varepsilon_8 t))^{-\frac{n}{2}} \exp\left(\varepsilon_3 - \frac{\left(e^{\varepsilon_5 J}\cdot x - t\varepsilon_7 V\right)\cdot \varepsilon_7 V + \frac{\varepsilon_8}{4}|x|^2}{1+\varepsilon_8 t}\right)$$
$$\times\, u\left(\frac{e^{-\varepsilon_4}(e^{\varepsilon_5 J}\cdot x - 2t\varepsilon_7 V)}{1+\varepsilon_8 t} - \varepsilon_2, \frac{e^{-2\varepsilon_4} t}{1+\varepsilon_8 t} - \varepsilon_1\right) + \varepsilon_6 v\,.$$

Taking convolutions of these examples yields the new solutions

$$P[\phi] \doteqdot (4\pi t)^{-\frac{n}{2}} \int_{\mathbb{R}^n} e^{-\frac{|x-y|^2}{4t}} \phi(y)\, dy$$

and

$$L[\phi] \doteqdot \int_{\mathbb{R}^n} e^{x\cdot y + t|y|^2} \phi(y)\, dy\,.$$

The first convolution is the classical **Poisson transform** and the second is the **Laplace transform**. It is also well known that solutions to the heat equation are generated by the **Fourier transform**,

$$F[\phi] \doteqdot \int_{\mathbb{R}^n} e^{ix\cdot y - t|y|^2} \phi(y)\, dy\,.$$

Note that the (complex-valued) **plane wave** function

$$v(x,t) \doteqdot e^{ix\cdot y - t|y|^2}$$

satisfies the heat equation for each $y \in \mathbb{R}^n$.

Since the heat equation is linear, there is another useful way to generate special solutions from symmetries: If $\{u_\varepsilon\}_{\varepsilon\in(-\varepsilon_0,\varepsilon_0)}$ is a 1-parameter family of solutions of the heat equation, then the variation field $\frac{d}{d\varepsilon}\big|_{\varepsilon=0} u_\varepsilon$ is also a solution. In particular, starting with a given solution u, we can generate families of new solutions associated with each of the symmetries above (some of them trivial):

(1) Time translation: $\frac{d}{d\varepsilon}\Big|_{\varepsilon=0} u_\varepsilon = -\partial_t u = -\Delta u$.

(2) Spatial translations: $\frac{d}{d\varepsilon}\Big|_{\varepsilon=0} u_\varepsilon = -\nabla_V u$, $V \in \mathbb{R}^n$.

(3) Scaling: $\frac{d}{d\varepsilon}\Big|_{\varepsilon=0} u_\varepsilon = -u$.

(4) Space-time dilation: $\frac{d}{d\varepsilon}\Big|_{\varepsilon=0} u_\varepsilon = -x\cdot\nabla u - 2t\,\partial_t u$.

(5) Rotations: $\frac{d}{d\varepsilon}\Big|_{\varepsilon=0} u_\varepsilon = \nabla u \cdot Jx$, $J \in \mathfrak{so}(n)$.

(6) Superpositions: $\left.\frac{d}{d\varepsilon}\right|_{\varepsilon=0} u_\varepsilon = v.$

(7) Galilean boosts: $\left.\frac{d}{d\varepsilon}\right|_{\varepsilon=0} u_\varepsilon = -\left(ux + 2t\nabla u\right) \cdot V.$

(8) The Appell transform: $\left.\frac{d}{d\varepsilon}\right|_{\varepsilon=0} u_\varepsilon = -\frac{1}{4}\left(|x|^2 + 2nt\right)u - tx\nabla u - t^2\partial_t u.$

Example 1.1. Applying the generator of the Appell transform to the constant solution, $u \equiv 4$, we see that the function u defined by

$$u(x,t) \doteqdot |x|^2 + 2nt$$

is a solution to the heat equation.

In the same way, we can obtain further solutions by looking at the higher-order variation fields.

1.3.2. Invariant solutions. A solution to the heat equation is an **invariant solution** (a.k.a. a **self-similar solution**) if it is left unchanged by a local symmetry of the equation. That is, if $u(x,t) \equiv u_\varepsilon(x,t)$ for some local symmetry $u \mapsto u_\varepsilon = \mathrm{e}^{\varepsilon g} \cdot u$. Such a solution necessarily solves the equation

$$0 = \left.\frac{d}{d\varepsilon}\right|_{\varepsilon=0} u_\varepsilon = g \cdot u\,. \tag{1.4}$$

In addition to the invariant solutions corresponding to the generators listed above, we mention the following special classes of invariant solutions.

(1) Traveling waves:

$$u(x,t) = u(x - \varepsilon V, t - \varepsilon) \quad \leftrightarrow \quad 0 = \Delta u + \nabla_V u\,.$$

If u is a traveling wave solution, then, denoting by ∇_V the directional derivative in the direction V,

$$u(x,t) = u(x - tV, 0) \doteqdot u_0(x - tV)\,,$$

where u_0 satisfies

$$\Delta u_0 + \nabla_V u_0 = 0\,. \tag{1.5}$$

Conversely, if u_0 satisfies (1.5), then $u(x,t) \doteqdot u_0(x - tV)$ satisfies the heat equation. The solution defined in (1.2) is a traveling wave solution.

(2) Similarity solutions:

$$u(x,t) = \mathrm{e}^{a\varepsilon} u\left(\mathrm{e}^{-\varepsilon} x, \mathrm{e}^{-2\varepsilon} t\right) \quad \leftrightarrow \quad 0 = au - x \cdot \nabla u - 2t\Delta u$$

for some $a \in \mathbb{R}$. If u is a similarity solution, then it is defined either for all $t < 0$ or all $t > 0$ and

$$u(x,t) = (\pm t)^{\frac{a}{2}} u\left(\frac{x}{\sqrt{\pm t}}, \pm 1\right) \doteqdot (\pm t)^{\frac{a}{2}} u_{\pm 1}\left(\frac{x}{\sqrt{\pm t}}\right),$$

where $u_{\pm 1}$ satisfies

(1.6)
$$\pm \Delta u_{\pm 1} + \frac{1}{2} x \cdot \nabla u_{\pm 1} = \frac{a}{2} u_{\pm 1} \,.$$

Conversely, if $u_{\pm 1}$ satisfies (1.6), then $u(x,t) \doteqdot (\pm t)^{\frac{a}{2}} u_{\pm 1}\left(\frac{x}{\sqrt{\pm t}}\right)$ satisfies the heat equation.

(3) Vortices:

$$u(x,t) = u(\mathrm{e}^{\varepsilon J} \cdot x, t - \varepsilon) \quad \leftrightarrow \quad \nabla u \cdot Jx = \Delta u \,.$$

Vortices are defined for all $t \in \mathbb{R}$ and satisfy

$$u(x,t) = u(\mathrm{e}^{tJ} \cdot x, 0) \doteqdot u_0(\mathrm{e}^{tJ} \cdot x) \,,$$

where u_0 satisfies

(1.7)
$$\Delta u_0 = \nabla u_0 \cdot Jx \,.$$

Conversely, if u_0 satisfies (1.7), then $u(x,t) \doteqdot u_0(\mathrm{e}^{tJ} \cdot x)$ satisfies the heat equation.

Of course, a more systematic approach to invariant solutions can be taken. For a comprehensive discussion of invariant solutions to partial differential equations, we refer the reader to [**431**].

Example 1.2. Let $u : \mathbb{R} \times (0,\infty) \to \mathbb{R}$ be the fundamental solution. The functions $v : \mathbb{R} \times (0,\infty) \to \mathbb{R}$ and $w : \mathbb{R} \times (0,\infty) \to \mathbb{R}$ defined by

$$v(x,t) \doteqdot \int_{-\infty}^{x} u(\xi,t)\, d\xi \quad \text{and} \quad w(x,t) \doteqdot \int_{-\infty}^{x} (2v(\xi,t) - 1)\, d\xi \,,$$

respectively, also satisfy the heat equation. (The fundamental solution u evolves out of the Dirac distribution at time $t = 0$, while the solutions v and w evolve out of the Heaviside distribution and the absolute value function $x \mapsto |x|$, respectively.)

The fundamental solution u is a similarity solution with $a = -1$. Indeed,

$$u(\lambda x, \lambda^2 t) = (4\pi\lambda^2 t)^{-\frac{1}{2}} \mathrm{e}^{-\frac{|\lambda x|^2}{4\lambda^2 t}} = \lambda^{-1} u(x,t) \,.$$

The solution v is a similarity solution with $a = 0$. Indeed,

$$\begin{aligned} v(\lambda x, \lambda^2 t) = \int_0^{\lambda x} u(\xi, \lambda^2 t)\, d\xi &= \lambda \int_0^x u(\lambda\eta, \lambda^2 t)\, d\eta \\ &= \int_0^x u(\eta, t)\, d\eta = v(x,t) \,. \end{aligned}$$

The solution w is a similarity solution with $a = 1$. Indeed,

$$\begin{aligned} w(\lambda x, \lambda^2 t) = \int_0^{\lambda x} (v(\xi, \lambda^2 t) - 1)\, d\xi &= \lambda \int_0^{x} (v(\lambda\eta, \lambda^2 t) - 1)\, d\eta \\ &= \lambda \int_0^{x} (v(\eta, t) - 1)\, d\eta = \lambda w(x,t)\,. \end{aligned}$$

1.4. The maximum principle

Suppose that u satisfies the heat equation. At any given time, the Hessian $\nabla^2 u$ is nonpositive definite at a point of maximal temperature and hence its trace, the Laplacian Δu, is also nonpositive at that point. By the heat equation, we conclude that the time derivative $\partial_t u$ is nonpositive, so the temperature seeks to decrease. Arguing similarly for the minimum, we deduce that

$$\max_{T^n \times \{t_2\}} u \le \max_{T^n \times \{t_1\}} u \text{ and } \min_{T^n \times \{t_2\}} u \ge \min_{T^n \times \{t_1\}} u\,, \quad \text{whenever } t_2 \ge t_1\,.$$

This basic idea applies very generally to heat-type equations and is referred to as the **maximum principle**. However, to make the above argument work rigorously, we actually need a *strict* inequality for the time derivative (consider the behavior of the function $t \mapsto t^3$ at the origin), but this is not hard to arrange: First note that, by translating in time if necessary, it suffices to assume that $t_1 = 0$. Consider for any $\varepsilon > 0$ the function $u_\varepsilon \div u - \varepsilon t$. We claim that

$$\sup_{T^n \times \mathbb{R}} u_\varepsilon \le \max_{T^n \times \{0\}} u_\varepsilon = \max_{T^n \times \{0\}} u\,. \tag{1.8}$$

Indeed, if this is not the case, then there must be some $\delta > 0$, some $x \in T^n$, and some $t > 0$ such that

$$u_\varepsilon(x,t) = \max_{T^n \times \{0\}} u + \delta\,.$$

Since $\delta > 0$ and u_ε is continuous, the first time at which this happens exists and is positive. Assuming t is the first such time, we see that $u_\varepsilon(x, \cdot)$ must be nondecreasing at t and hence

$$\Delta u_\varepsilon(x,t) = \partial_t u_\varepsilon(x,t) + \varepsilon \ge \varepsilon > 0\,.$$

On the other hand, since the point x is a spatial maximum of $u_\varepsilon(\cdot, t)$, the Laplacian of $u_\varepsilon(\cdot, t)$ must be nonpositive there, a contradiction. So (1.8) does indeed hold. Taking $\varepsilon \to 0$ yields the claim. The proof for the minimum is similar.

Pushing the idea even further, we obtain the following more general comparison result. Let $\mathrm{S}(n)$ denote the space of symmetric $n \times n$ real matrices and recall that T^n is the n-dimensional torus.

Theorem 1.3. *Suppose that* $u \in C^\infty(T^n \times (0,T)) \cap C^0(T^n \times [0,T))$ *satisfies*

$$(\partial_t - a^{ij}\nabla_i\nabla_j - b^k\nabla_k - c)u \leq 0$$

on $T^n \times (0,T)$ *for some nonnegative definite family of symmetric matrices* $a : T^n \times (0,T) \to \mathrm{S}(n)$*, some vector field* $b : T^n \times (0,T) \to \mathbb{R}^n$*, and some locally bounded function* $c : T^n \times [0,T) \to \mathbb{R}$*. If* $u(\cdot,0) \leq 0$*, then* $u(\cdot,t) \leq 0$ *for all* $t \in [0,T)$*.*

Proof. Given $\varepsilon > 0$ and $s \in (0,T)$, consider $u_{\varepsilon,s}(x,t) \doteqdot u(x,t) - \varepsilon \mathrm{e}^{(C+1)t}$, where $C \doteqdot \max_{T^n \times [0,s]} c$. We claim that $u_{\varepsilon,s} < 0$ in $T^n \times [0,s]$. Suppose, to the contrary, that $u_{\varepsilon,s}(x_0,t_0) \geq 0$ for some point $(x_0,t_0) \in T^n \times [0,s]$. Since $u_{\varepsilon,s}(\cdot,0) < 0$, there exists a positive earliest such time, which we take to be t_0, in which case $u(x_0,t_0)$. At the point (x_0,t_0),

$$\begin{aligned} 0 \leq (\partial_t - a^{ij}\nabla_i\nabla_j - b^k\nabla_k)u_{\varepsilon,s} &= cu - \varepsilon(C+1)\mathrm{e}^{(C+1)t} \\ &= \varepsilon \mathrm{e}^{(C+1)t}c - \varepsilon(C+1)\mathrm{e}^{(C+1)t} \\ &\leq -\varepsilon \mathrm{e}^{(C+1)t} < 0\,, \end{aligned}$$

which is absurd. We conclude that $u_{\varepsilon,s} < 0$ in $T^n \times [0,s]$. But $\varepsilon > 0$ and $s \in (0,T)$ were arbitrary. Taking $\varepsilon \to 0$ and then $s \to T$ yields the claim. ☐

Replacing u by $-u$ yields a corresponding minimum comparison for nonnegative supersolutions. We also note the following useful ODE comparison principle for equations with nonlinear reaction term.

Corollary 1.4. *Suppose the function* $u \in C^\infty(T^n \times (0,T)) \cap C^0(T^n \times [0,T))$ *satisfies*

$$(\partial_t - a^{ij}\nabla_i\nabla_j - b^k\nabla_k)u \leq F(u)$$

on $T^n \times (0,T)$ *for some nonnegative definite family of symmetric matrices* $a : T^n \times (0,\infty) \to \mathrm{S}(n)$*, some vector field* $b : T^n \times (0,\infty) \to \mathbb{R}^n$*, and some locally Lipschitz* $F : \mathbb{R} \to \mathbb{R}$*. If* $u(\cdot,0) \leq \phi_0 \in \mathbb{R}$*, then* $u(\cdot,t) \leq \phi(t)$ *for all* $t \in [0,T)$*, where* ϕ *is the solution to the associated* ODE*,*

$$\frac{d\phi}{dt} = F(\phi)\,, \quad \phi(0) = \phi_0\,.$$

Proof. Fix $s \in (0,T)$. Since F is locally Lipschitz, there exists some $L < \infty$ such that

$$\begin{aligned} (\partial_t - a^{ij}\nabla_i\nabla_j - b^k\nabla_k)(u - \phi) &\leq F(u) - F(\phi) \\ &\leq L|u - \phi| = L\,\mathrm{sign}(u-\phi)(u-\phi) \end{aligned}$$

in $T^n \times (0,s]$, where $\mathrm{sign}(u-\phi)$ is the sign of the expression $u - \phi$. The claim now follows, within $T^n \times [0,s]$, from Theorem 1.3. Taking $s \to T$ completes the proof. ☐

Replacing u by $-u$ and $F(u)$ by $-F(-u)$ yields a corresponding lower comparison principle for supersolutions.

The **strong maximum principle** states that, in fact, the maximum temperature is actually strictly decreasing, except when it is already constant (in space and time). This is one manifestation of *infinite propagation speed* for parabolic equations. For the heat equation on a bounded domain $\Omega \subset \mathbb{R}^n$, the strong maximum principle can be obtained relatively easily from the **mean value property**, which states that

$$u(x_0, t_0) = \frac{1}{r^n} \iint_{E_r(x_0,t_0)} u(x,t) \frac{|x-x_0|^2}{(t-t_0)^2}\, dx\, dt\,, \tag{1.9}$$

where the **heat ball** $E_r(x,t)$ of radius $r > 0$ centered at (x,t) is defined by

$$E_r(x,t) \doteqdot \{(y,s) \in \mathbb{R}^n \times \mathbb{R} : s < t\,,\ \rho(x-y, t-s) \geq r^{-n}\}\,.$$

For more general parabolic equations it is somewhat harder to prove. The following result is essentially proved in [**368**, Chapter II].

Theorem 1.5 (Strong maximum principle). *Given a connected open set $\Omega \subset \mathbb{R}^n$, suppose that $u \in C^\infty(\Omega^n \times (0,T))$ is nonpositive and satisfies*

$$(\partial_t - a^{ij}\nabla_i\nabla_j - b^k\nabla_k - c)u \leq 0$$

for some positive definite family of symmetric matrices $a : \Omega\times(0,T) \to \mathrm{S}(n)$, some vector field $b : \Omega \times (0,T) \to \mathbb{R}^n$, and some function $c : \Omega \times (0,T) \to \mathbb{R}$. If $u(x_0,t_0) = 0$ for some $(x_0,t_0) \in \Omega \times (0,T)$, then $u(x,t) = 0$ for all $(x,t) \in \Omega \times (0,t_0]$.

There is also a boundary version of the strong maximum principle, known as the **Hopf lemma** or **Hopf boundary point lemma** (see [**368**, Chapter II]).

Lemma 1.6 (The Hopf lemma). *Given $\alpha > 0$, $\gamma < \infty$, and $\nu < \infty$, suppose (for some $R > 0$, $\lambda > 0$, and $(x_0,t_0) \in \mathbb{R}^n \times \mathbb{R}$) that $(a,b,c) : PF_{\alpha,\lambda,R}(x_0,t_0) \to \mathrm{S}_+(n) \times \mathbb{R}^n \times \mathbb{R}$ satisfies*

$$\lambda\delta^{ij} \leq a^{ij} \leq \gamma\lambda\delta^{ij} \quad \text{and} \quad \left(\frac{|b|}{\lambda}\right)^2 + \frac{|c|}{\lambda} \leq \frac{\nu}{R^2}\,,$$

*where the **parabolic frustum** $PF_{\alpha,\lambda,R}(x_0,t_0)$ is defined by*

$$PF_{\alpha,\lambda,R}(x_0,t_0) \doteqdot \left\{(x,t) \in \mathbb{R}^n \times (-\infty, t_0] : |x-x_0|^2 + \frac{\lambda}{\alpha}(t_0 - t) \leq R^2\right\}.$$

Let $u \in P^2(PF_{\alpha,\lambda,R}(x_0,t_0))$ be a positive supersolution to

$$u_t = a^{ij}u_{ij} + b^k u_k + cu \quad \text{in} \quad PF_{\alpha,\lambda,R}(x_0,t_0)\,.$$

Suppose that $u(x_1, t_0) = 0$ *for some* $x_1 \in \partial B_R(x_0)$. *Then*

$$D^+_{\vec{\nu}} u(x_1, t_0) \doteqdot \limsup_{h \nearrow 0} \frac{u(x_1 + h\vec{\nu}, t_0) - u(x_1, t_0)}{h} < 0 ,$$

where $\vec{\nu} \doteqdot (x_1 - x_0)/|x_1 - x_0|$.

The fundamental solution can be exploited to obtain a maximum principle for solutions on $\mathbb{R}^n$ which grow at most exponentially in space.

Theorem 1.7. *Let* $u \in C^\infty(\mathbb{R}^n \times (0, T]) \cap C^0(\mathbb{R}^n \times [0, T])$ *be a subsolution to the heat equation. Suppose that*

$$u(x, t) \leq a e^{a|x|^2} \text{ for all } (x, t) \in \mathbb{R}^n \times [0, T]$$

for some $a < \infty$. *Then*

$$\sup_{\mathbb{R}^n \times \{T\}} u \leq \sup_{\mathbb{R}^n \times \{0\}} u . \tag{1.10}$$

Proof. We follow [**216**, Theorem 6] (see also [**321**, Chapter 7]). It suffices to prove the claim in case $4aT < 1$. Choose $\varepsilon > 0$ and $\gamma > 0$ such that

$$4(a + \gamma)(T + \varepsilon) < 1 .$$

Given $y \in \mathbb{R}^n$ and $\mu > 0$, set

$$v(x, t) \doteqdot u(x, t) - \mu(T + \varepsilon - t)^{-\frac{n}{2}} e^{\frac{|x-y|^2}{4(T+\varepsilon-t)}} .$$

Since the second term is just a multiple of a space-time translation of the fundamental solution, v is a subsolution of the heat equation. The maximum principle then implies, for any $r > 0$, that

$$\max_{B_r(y) \times [0,T]} v = \max_{B_r(y) \times \{0\} \cup \partial B_r(y) \times [0,T]} v .$$

Since, for $x \in \partial B_r(y)$,

$$\begin{aligned} v(x, t) &= u(x, t) - \mu(T + \varepsilon - t)^{-\frac{n}{2}} e^{\frac{r^2}{4(T+\varepsilon-t)}} \\ &\leq a e^{a|x|^2} - \mu(T + \varepsilon - t)^{-\frac{n}{2}} e^{\frac{r^2}{4(T+\varepsilon-t)}} \\ &\leq a e^{a(|y|+r)^2} - \mu(T + \varepsilon - t)^{-\frac{n}{2}} e^{\frac{r^2}{4(T+\varepsilon)}} \\ &\leq a e^{a(|y|+r)^2} - \mu(T + \varepsilon - t)^{-\frac{n}{2}} e^{(a+\gamma)r^2} \\ &= e^{ar^2} \left(a e^{a(|y|^2+2r|y|)} - \mu(T + \varepsilon - t)^{-\frac{n}{2}} e^{\gamma r^2} \right) , \end{aligned}$$

we can choose r sufficiently large that

$$\max_{B_r(y) \times [0,T]} v = \max_{B_r(y) \times \{0\}} v \leq \max_{B_r(y) \times \{0\}} u .$$

The claim follows since $y \in \mathbb{R}^n$ and $\mu > 0$ were arbitrary. ∎

By constructing supersolutions of the form

$$v(x,t) = \mathrm{e}^{at}(|x|^2+1)^q$$

for any $q > 1$, Daskalopoulos and Huisken obtain the following result for subsolutions to more general linear parabolic equations [**193**].

Figure 1.2. Toti Daskalopoulos.

Theorem 1.8 (Daskalopoulos and Huisken). *Let $u \in C^\infty(\mathbb{R}^n \times (0,T]) \cap C^0(\mathbb{R}^n \times [0,T])$ be a subsolution to the linear parabolic equation*

$$\partial_t u = (a^{ij}\nabla_i\nabla_j + b^k\nabla_k + c)u$$

with measurable coefficients satisfying

$$\lambda^{-1}|\xi|^2 \leq a(x,t)^{ij}\xi_i\xi_j \leq \Lambda|\xi|^2\,, \quad |b(x,t)|^2 \leq \Lambda(|x|^2+1)\,, \quad \textit{and} \quad |c(x,t)| \leq \Lambda$$

for all $(x,t) \in \mathbb{R}^n \times [0,T]$ for some $\Lambda < \infty$.

Suppose that

$$u(x,t) \leq C(|x|^2+1)^p \ \ \textit{for all} \ \ (x,t) \in \mathbb{R}^n \times [0,T]$$

for some $C < \infty$ and $p > 0$. Then

$$\sup_{\mathbb{R}^n \times \{t\}} u \leq \sup_{\mathbb{R}^n \times \{0\}} u\,. \tag{1.11}$$

1.5. Well-posedness

Up to now, we have ignored the subtle question of the *existence* of solutions (with specified initial-boundary conditions). For solutions of the heat equation on all of $\mathbb{R}^n$, we can make use of the **heat kernel**, $k \in C^\infty(\mathbb{R}^n \times \mathbb{R}^n \times (0,\infty))$, which is defined by

$$k(x,y,t) \doteq \rho(x-y,t) = (4\pi t)^{-\frac{n}{2}}\mathrm{e}^{-\frac{|x-y|^2}{4t}}\,. \tag{1.12}$$

Theorem 1.9. *If $\varphi : \mathbb{R}^n \to \mathbb{R}$ is bounded and continuous*[1]*, then the function $u \in C^\infty(\mathbb{R}^n \times (0,\infty)) \cap C^0(\mathbb{R}^n \times [0,\infty))$ defined by*

$$u(x,t) \doteqdot \int_{\mathbb{R}^n} k(x,y,t)\varphi(y)\,dy \tag{1.13}$$

satisfies the heat equation and converges pointwise to φ as $t \searrow 0$.

Let $v \in C^\infty(\mathbb{R}^n \times (0,\infty)) \cap C^0(\mathbb{R}^n \times [0,\infty))$ be a second solution to the heat equation which converges pointwise to φ as $t \searrow 0$. If

$$|v(x,t)| \le a e^{a|x|^2} \text{ for all } (x,t) \in \mathbb{R}^n \times (0,\infty)$$

for some $a < \infty$, then $v \equiv u$.

Proof. The first claim is a computation which we leave to the reader (see [**321**, Chapter 7]). The second is a consequence of the (noncompact) maximum principle (Theorem 1.7). □

The uniqueness of the solution (1.13) fails if the growth bound is removed. Indeed, Tychonoff constructed infinitely many distinct solutions with zero initial condition. These solutions are of the form

$$u_\alpha(x,t) = \sum_{j=0}^{\infty} \frac{g_\alpha^{(j)}(t) x^{2j}}{(2j)!}$$

for each $\alpha > 1$, where

$$g_\alpha(t) \doteqdot e^{-t^{-\alpha}}.$$

See [**321**, Chapter 7]. On the other hand, by a theorem of Widder [**536**], the formula (1.13) yields the only nonnegative solution if the initial condition is nonnegative.

For solutions over bounded domains $\Omega \subset \mathbb{R}^n$, uniqueness fails even among bounded solutions if appropriate boundary conditions are not specified: Compare the solutions v and w from Example 1.2 with the solutions $v'(x,t) \equiv 1$ and $w'(x,t) \equiv x$, respectively: Although v coincides with v' (and w coincides with w') everywhere in $(0,\infty)$ at time $t = 0$, the solutions differ at positive times. Another example is the fundamental solution u extended by $u(x,t) \equiv 0$ for $t < 0$. This defines a smooth solution to the heat equation on $\mathbb{R}^2 \setminus \{(0,0)\}$ which is identically zero for $t \le 0$ and positive for $t > 0$. This also illustrates that solutions to the heat equation are not necessarily analytic. On the other hand, the maximum principle immediately implies that two solutions which agree on the parabolic boundary $\overline{\Omega} \times \{0\} \cup \partial\Omega \times [0,T)$ necessarily agree in $\overline{\Omega} \times [0,T)$.

[1] If φ is measurable and satisfies $|\varphi(x)| \le M e^{a|x|^2}$ for all $x \in \mathbb{R}^n$ for some $a < \infty$ and $M < \infty$, then (1.13) yields a smooth solution on the time interval $t \in (0, \frac{1}{4a})$.

Proving the *existence* of solutions (with appropriately specified initial-boundary conditions) to the heat equation over smoothly bounded domains $\Omega \subset \mathbb{R}^n$ is a far more subtle issue, although there are various methods available. One well-known approach relies on the spectral theory of the Laplacian on Ω (see [**137**]), which allows us to represent the initial condition u_0 as

$$u_0(x) = \sum_i a_i e_i(x)\,,$$

where $\{e_i\}_{i=1}^n$ is an L^2-orthonormal basis of smooth eigenfunctions e_i of the Laplacian satisfying the corresponding (appropriately specified) boundary condition. The solution to the heat equation with initial condition u_0 (and the specified boundary condition) is then given by

$$u(x,t) = \sum_i a_i \mathrm{e}^{-\lambda_i t} e_i(x)\,,$$

where λ_i is the eigenvalue corresponding to e_i. Once solutions to the heat equation are obtained, the inhomogeneous heat equation can be solved using Duhamel's principle (see [**216**, §2.3]). More general (inhomogeneous) linear parabolic equations can then be solved via the method of continuity once appropriate a priori estimates (in Hölder spaces, say) have been established.

Existence of solutions to *nonlinear* parabolic equations can be established (for a short-time) by solving the linearization of the problem and applying the inverse function theorem. Once again, this requires a priori estimates.

We refer the reader to the books of Cannon [**135**], Evans [**216**], Friedman [**226**], Krylov [**340–342**], Ladyženskaja, Solonnikov, and Ural′ceva [**343**], and Lieberman [**368**] for a comprehensive treatment of the existence theory for linear and nonlinear parabolic equations following various approaches.

1.6. Asymptotic behavior

Suppose that $u \in C^\infty(T^n \times (0,\infty)) \cap C^0(T^n \times [0,\infty))$ satisfies the heat equation. Observe that the average temperature,

$$\overline{u}(t) \doteqdot \frac{1}{|T^n|}\int_{T^n} u(x,t)\,dx = \int_{T^n} u(x,t)\,dx\,,$$

remains constant ($|T^n| = 1$ denotes the volume of T^n). Indeed, by the divergence theorem

$$\frac{d}{dt}\overline{u} = \frac{d}{dt}\int_{T^n} u\,dx = \int_{T^n} \partial_t u\,dx = \int_{T^n} \Delta u\,dx = \int_{T^n} \operatorname{div}(\nabla u)\,dx = 0\,.$$

The gradient flow property and the maximum principle both suggest that the temperature will tend to even out towards the average as $t \to \infty$. This is indeed the case. To prove it, we will need the **Poincaré inequality**.

Theorem 1.10 (Poincaré inequality). *Given any $n \in \mathbb{N}$ and any connected, open, bounded $\Omega \subset \mathbb{R}^n$ with Lipschitz boundary*[2], *there exists a constant $\delta > 0$ such that every $u \in C^1(\Omega)$ satisfies*

$$\delta \int_\Omega |u - \overline{u}|^2 dx \leq \operatorname{diam}(\Omega)^2 \int_\Omega |\nabla u|^2 dx \,, \tag{1.14}$$

where $\overline{u}$ is the average value of u over Ω and $\operatorname{diam}(\Omega) = \sup_{x,y\in\Omega} |x - y|$.

Since $\overline{u}$ is constant in time for a solution $u : T^n \times [0, \infty) \to \mathbb{R}$ of the heat equation, the Poincaré inequality yields

$$\frac{d}{dt}\int_{T^n} |u - \overline{u}|^2 dx = -2\int_{T^n} |\nabla u|^2 dx \leq -2\delta \int_{T^n} |u - \overline{u}|^2 dx \,.$$

Integrating, we deduce that u decays in L^2 to its average value $\overline{u}$ exponentially in time:

$$\|u - \overline{u}\|_{L^2} \doteqdot \left(\int_{T^n} |u - \overline{u}|^2 dx\right)^{\frac{1}{2}} \leq D_{-1} \mathrm{e}^{-\delta t} \,,$$

where D_{-1} is a constant. This weak convergence can be "bootstrapped" to obtain convergence in C^∞ using **interpolation inequalities**. We claim that

$$\max_{T^n} |u - \overline{u}| \leq C \max\left\{\left(\int_{T^n} |u - \overline{u}|^2\right)^{\frac{1}{2+n}} \max_{T^n} |\nabla u|^{\frac{n}{2+n}}, \left(\int_{T^n} |u - \overline{u}|^2\right)^{\frac{1}{2}}\right\},$$

where $C < \infty$ is a constant which depends only on n. To prove this, choose $\overline{x}$ so that $|u(\overline{x}) - \overline{u}| = \max_{T^n} |u - \overline{u}|$. Then, by Taylor's theorem,

$$|u(y) - \overline{u}| \geq |u(\overline{x}) - \overline{u}| - |\overline{x} - y| \sup_{T^n} |\nabla u| \geq \frac{1}{2}|u(\overline{x}) - \overline{u}|$$

for $y \in B_{\overline{r}}(\overline{x})$, where $\overline{r} \doteqdot \min\left\{\frac{\max_{T^n} |u-\overline{u}|}{2\max_{T^n} |\nabla u|}, 1\right\}$. Thus,

$$\begin{aligned}\int_{T^n} |u - \overline{u}|^2 dx &\geq \int_{B_{\overline{r}}(\overline{x})} |u - \overline{u}|^2 dx \\ &\geq \frac{1}{4} \max_{T^n} |u - \overline{u}|^2 |B_{\overline{r}}(\overline{x})| \\ &\geq \frac{\omega_n}{4} \max_{T^n} |u - \overline{u}|^2 \min\left\{1, \left(\frac{\max_{T^n} |u - \overline{u}|}{2\max_\Omega |\nabla u|}\right)^n\right\},\end{aligned}$$

where $\omega_n \doteqdot |B_1(0)|$. The claim follows. Thus, in order to obtain convergence in C^0 (exponentially in time), it suffices to bound the derivative of u. This

[2]That is, the boundary is locally representable as the graph of a Lipschitz function.

can be achieved using the maximum principle: Observe that

$$\begin{aligned}(\partial_t - \Delta)|\nabla u|^2 &= 2\left\langle (\partial_t - \Delta)\nabla u, \nabla u\right\rangle - 2|\nabla^2 u|^2\\ &= 2\left\langle \nabla(\partial_t - \Delta)u, \nabla u\right\rangle - 2|\nabla^2 u|^2\\ &= -2|\nabla^2 u|^2\\ &\le 0\,.\end{aligned}$$

So the maximum principle yields

$$\|\nabla u\|_\infty^2(t) \doteqdot \max_{T^n}|\nabla u(\cdot,t)|^2 \le C_1 \doteqdot \max_{T^n\times\{0\}}|\nabla u|^2\,.$$

It follows that u decays to its average $\overline{u}$ in C^0 exponentially in time.

To obtain convergence of all derivatives of u to zero, we shall need some further interpolation inequalities. First observe that

$$\|\nabla u\|_\infty \le \|u-\overline{u}\|_\infty^{\frac{1}{2}}\|\nabla^2 u\|_\infty^{\frac{1}{2}}\,.$$

Indeed, by Taylor's theorem,

$$u(x+hv) \ge u(x) + h\nabla_v u(x) - \frac{h^2}{2}\|\nabla^2 u\|_\infty$$

for any unit vector v. Taking v proportional to ∇u and $h > 0$ yields

$$|\nabla u(x)| \le \frac{u(x+hv)-u(x)}{h} + \frac{h}{2}\|\nabla^2 u\|_\infty\,,$$

so that

$$\|\nabla u\|_\infty \le \frac{2}{h}\|u-\overline{u}\|_\infty + \frac{h}{2}\|\nabla^2 u\|_\infty\,.$$

This is true for any $h > 0$. If $\|\nabla^2 u\|_\infty$ is zero, we can take h tending to infinity to obtain the claim. Otherwise, we can take $h \doteqdot 2\sqrt{\frac{\|u\|_\infty}{\|\nabla^2 u\|_\infty}}$.

Replacing $u-\overline{u}$ by $\nabla^{k-1}u$, we obtain

$$\|\nabla^k u\|_\infty \le c_k\|\nabla^{k-1}u\|_\infty^{\frac{1}{2}}\|\nabla^{k+1}u\|_\infty^{\frac{1}{2}}$$

for each $k \ge 2$, where c_k is a constant which depends only on n and k.

Exponential decay of all derivatives of u can now be obtained via induction and the estimates

$$|\nabla^k u|^2 \le C_k \doteqdot \max_{T^n\times\{0\}}|\nabla^k u|^2\,,$$

which follow from the maximum principle since

$$(\partial_t - \Delta)|\nabla^k u|^2 = -2|\nabla^{k+1}u|^2 \le 0\,.$$

1.7. The Bernstein method

Suppose that $u \in C^\infty(T^n \times (0,\infty)) \cap C^0(T^n \times [0,\infty))$ satisfies the heat equation. Consider

$$Q_m \doteqdot \sum_{\ell=0}^{m} a_\ell t^\ell |\nabla^\ell u|^2 ,$$

where $a_\ell \doteqdot \frac{2^\ell}{\ell!}$. Commuting derivatives and using the fact that u satisfies the heat equation, we obtain

$$\begin{aligned}(\partial_t - \Delta) Q_m &= \sum_{\ell=1}^{m} \ell a_\ell t^{\ell-1} |\nabla^\ell u|^2 - 2\sum_{\ell=0}^{m} a_\ell t^\ell |\nabla^{\ell+1} u|^2 \\ &= \sum_{\ell=1}^{m} (\ell a_\ell - 2a_{\ell-1})\, t^{\ell-1} |\nabla^\ell u|^2 - 2a_m t^m |\nabla^{m+1} u|^2 \\ &\leq 0 .\end{aligned}$$

We conclude from the maximum principle that

$$\sum_{\ell=0}^{m} a_\ell t^\ell |\nabla^\ell u|^2 \leq \max_{T^n \times \{0\}} u^2 .$$

So this gives convergence in C^∞ to a constant as $t \to \infty$ without the need for the interpolation inequality. It also illustrates the powerful smoothing property of the heat equation: Even if we start the heat flow with merely bounded initial data, the solution becomes immediately smooth, with derivatives decaying rapidly in time.

1.8. The Harnack inequality

Suppose that at some instant in time we have heat concentrated near a point, so that the heat distribution approximates the Dirac delta function. As time evolves the heat diffuses to the surrounding region, so we should have a lower bound for the heat at a space-time point in terms of the elapsed time and distance from the origin. Qualitatively, the lower bound should decrease as time increases and as the distance increases. This behavior of heat flow is captured in the following quantitative estimate [**405**].

Theorem 1.11 (Harnack inequality). *Assume that $u \in C^\infty(T^n \times (0,T)) \cap C^0(T^n \times [0,T))$ is a positive solution to the heat equation. Then*

$$\frac{u(x_2,t_2)}{u(x_1,t_1)} \geq \left(\frac{t_2}{t_1}\right)^{-\frac{n}{2}} \exp\left(-\frac{|x_2 - x_1|^2}{4(t_2 - t_1)}\right) \tag{1.15}$$

for all $x_1, x_2 \in T^n$ and all $t_1 < t_2 \in (0,T)$.

We will present a robust approach to the Harnack inequality developed by Li and Yau in their classical work on the heat equation on Riemannian manifolds [**364**] (cf. Exercise 1.19).

Observe that the fundamental solution $\rho : \mathbb{R}^n \times (0,\infty) \to \mathbb{R}$ satisfies the identity

$$\Delta \log \rho + \frac{n}{2t} = 0\,.$$

Proposition 1.12 (Differential Harnack inequality of Li and Yau [**364**]). *Let $u \in C^\infty(T^n \times (0,T))$ be a positive solution to the heat equation. Then*

$$\Delta \log u + \frac{n}{2t} \geq 0\,. \tag{1.16}$$

Proof. First note that

$$(\partial_t - \Delta)\log u = \frac{|\nabla u|^2}{u^2} = |\nabla \log u|^2\,.$$

We will prove the claim by applying the maximum principle to the function

$$Q \doteqdot \Delta \log u\,.$$

Observe that

$$\begin{aligned}(\partial_t - \Delta)Q &= (\partial_t - \Delta)\Delta \log u\\ &= \Delta(\partial_t - \Delta)\log u\\ &= \Delta|\nabla \log u|^2\\ &= 2\,\langle \nabla \log u, \nabla Q\rangle + 2|\nabla^2 \log u|^2\\ &\geq 2\,\langle \nabla \log u, \nabla Q\rangle + \frac{2}{n}Q^2\,.\end{aligned} \tag{1.17}$$

Setting

$$P \doteqdot 2tQ + n\,,$$

we obtain

$$\begin{aligned}(\partial_t - \Delta)P &\geq 2t\left(2\,\langle \nabla \log u, \nabla Q\rangle + \frac{2}{n}Q^2\right) + 2Q\\ &= 2\,\langle \nabla \log u, \nabla P\rangle + \frac{2Q}{n}P\,.\end{aligned}$$

Since P is initially positive, we conclude from the maximum principle that it can never reach zero. That is,

$$\Delta \log u + \frac{n}{2t} \geq 0\,.$$

□

In fact, a stronger estimate (referred to as the **matrix Harnack inequality**) holds:

$$\nabla^2 \log u + \frac{\mathrm{I}}{2t} \geq 0\,,$$

where I is the Euclidean metric. See Exercise 1.14.

By integrating, we obtain the classical Harnack inequality.

Proof of the Harnack estimate. Given any two space-time points (x_1,t_1) and (x_2,t_2) with $t_1 < t_2$, let $\gamma : [t_1,t_2] \to T^n$ be a path from x_1 to x_2. Using the Cauchy–Schwarz inequality and the differential Harnack inequality, we can estimate

$$\begin{aligned}\frac{d}{dt}\log u(\gamma(t),t) &= \partial_t \log u + \frac{\nabla_{\gamma'} u}{u} \\ &\geq \frac{|\nabla u|^2}{u^2} - \frac{n}{2t} - \frac{|\nabla u|}{u}|\gamma'| \\ &\geq -\frac{n}{2t} - \frac{|\gamma'|^2}{4}\,.\end{aligned}$$

Integrating between t_1 and t_2 then yields

$$\log\frac{u(x_2,t_2)}{u(x_1,t_1)} \geq -\frac{n}{2}\log\frac{t_2}{t_1} - \frac{1}{4}\int_{t_1}^{t_2} |\gamma'(t)|^2\,dt\,.$$

Taking γ to be the optimal path (a straight line segment) we conclude that

$$\frac{u(x_2,t_2)}{u(x_1,t_1)} \geq \left(\frac{t_2}{t_1}\right)^{-\frac{n}{2}} \exp\left(-\frac{|x_2-x_1|^2}{4(t_2-t_1)}\right)\,. \qquad \square$$

1.9. Further monotonicity formulae

1.9.1. The Nash entropy.

Let $u : T^n \times [0,T) \to \mathbb{R}$ be a positive solution to the heat equation on the torus. The **Nash entropy** (or **Boltzmann entropy**) is defined by

$$E\,(u) \doteqdot -\int_{T^n} u \log u\, d\theta\,.$$

Observe that

$$\begin{aligned}\frac{dE}{dt} &= -\int_{T^n} (u_t \log u + u_t)\, d\theta \\ &= -\int_{T^n} \Delta u \log u\, d\theta \\ &= \int_{T^n} \frac{|\nabla u|^2}{u} d\theta.\end{aligned}$$

In particular, E is nondecreasing.

1.9.2. Weighted monotonicity formulae. Now let u be a solution to the heat equation on $\mathbb{R}^n$ and ϕ be a solution to the L^2-**adjoint heat equation**. That is,

$$(\partial_t + \Delta)\phi = 0\,.$$

Then, assuming all integrals are finite and all integrations by parts are legal,

$$\frac{d}{dt}\int_{\mathbb{R}^n} u\phi\,dx = \int_{\mathbb{R}^n}(\partial_t u\phi + u\partial_t\phi)\,dx = \int_{\mathbb{R}^n}(\Delta u\phi - u\Delta\phi)\,dx = 0\,.$$

That is, the mass of u is constant with respect to the weighted volume, $\phi\,dx$.

Next observe, assuming again that everything is legal, that

$$\frac{d}{dt}\int_{\mathbb{R}^n} u^2\phi\,dx = \int_{\mathbb{R}^n}\left(2u\Delta u\phi - u^2\Delta\phi\right)dx = -2\int_{\mathbb{R}^n}|\nabla u|^2\,\phi\,dx \le 0\,.$$

So the weighted L^2-norm of u is strictly decreasing unless u is constant. Similarly,

$$\begin{aligned}\frac{d}{dt}\int_{\mathbb{R}^n}|\nabla^m u|^2\phi\,dx &= \int_{\mathbb{R}^n}\left(2\left\langle\nabla^m u, \Delta\nabla^m u\right\rangle\phi - |\nabla^m u|^2\Delta\phi\right)dx\\ &= -2\int_{\mathbb{R}^n}\left|\nabla^{m+1}u\right|^2\phi\,dx \le 0\,.\end{aligned}$$

1.9.3. Semilinear heat equations. Consider, for $p > 1$, the semilinear reaction diffusion equation

$$(\partial_t - \Delta)u = |u|^{p-1}u \tag{1.18}$$

on $\mathbb{R}^n\times[0,T)$. Equation (1.18) is the gradient flow of the energy with energy density

$$e(u) = \frac{1}{2}|\nabla u|^2 - \frac{1}{p+1}|u|^{p+1}\,.$$

Unlike solutions of the heat equation, solutions of (1.18) only exist, in general, for a finite time. Indeed, for positive solutions with periodic boundary conditions, say, a straightforward application of the ODE comparison principle yields

$$\left(\underline{u}_0^{1-p} - 2t\right)^{-\frac{1}{2}} \le u \le \left(\overline{u}_0^{1-p} - 2t\right)^{-\frac{1}{2}},$$

where $\underline{u}_0 \doteqdot \min_{T^n\times\{0\}} u$ and $\overline{u}_0 \doteqdot \max_{T^n\times\{0\}}$. So the latest time on which the solution is defined is certainly no greater than $\frac{1}{2}\underline{u}_0^{1-p}$ (in general it will be less). For convenience, we will assume that the solution is defined on the time interval $[-T, 0)$, which can be arranged by a translation in time.

Observe that (1.18) is invariant under the scaling $u \mapsto u_\varepsilon$, where

$$u_\varepsilon(x,t) = \mathrm{e}^{\frac{2}{p-1}\varepsilon}u\left(\mathrm{e}^\varepsilon x, \mathrm{e}^{2\varepsilon}t\right)\,.$$

Note also that

$$e(u_\varepsilon)(x,t) = \mathrm{e}^{2\varepsilon\frac{p+1}{p-1}}e(u)\left(\mathrm{e}^\varepsilon x, \mathrm{e}^{2\varepsilon}t\right).$$

We shall call a solution u **scale invariant** if $u_\varepsilon(x,t) = u(x,t)$. A scale-invariant solution u satisfies (cf. Section 1.3.2)

$$\partial_t u + \frac{1}{2t}x \cdot \nabla u + \frac{1}{(p-1)t}u = 0\,. \tag{1.19}$$

Conversely, if u_{-1} satisfies

$$0 = \Delta u_{-1} + |u_{-1}|^{p-1}u_{-1} + \frac{1}{2}x \cdot \nabla u_{-1} + \frac{1}{p-1}u_{-1},$$

then

$$u(x,t) \doteqdot (-t)^{\frac{1}{p-1}}u_{-1}\left(\frac{x}{\sqrt{-t}}\right)$$

is a scale-invariant solution to (1.18).

Consider the modified backward heat kernel

$$\Phi_\gamma(x,t) = \frac{1}{(-4\pi t)^{\frac{n-\gamma}{2}}}e^{\frac{|x|^2}{4t}}\,.$$

Setting $\gamma = 2(p+1)/(p-1)$, we find that

$$\frac{d}{dt}\int_{\mathbb{R}^n} e(u)\Phi_\gamma\, dx = 0\,.$$

The following monotonicity formula was discovered by Giga and Kohn [**244**] (see also [**206**]).

Theorem 1.13 (Giga and Kohn). *If u satisfies* (1.18) *on the time interval* $(-T,0)$, *then*

$$\frac{d}{dt}\int_{\mathbb{R}^n}\left(e(u) - \frac{1}{2(p-1)t}u^2\right)\Phi_\gamma\, dx$$
$$= -\int_{\mathbb{R}^n}\left(\partial_t u + \frac{1}{2t}x \cdot \nabla u + \frac{1}{(p-1)t}u\right)^2 \Phi_\gamma\, dx\,.$$

So the energy

$$F(u) \doteqdot \int_{\mathbb{R}^n}\left(e(u) - \frac{1}{2(p-1)t}u^2\right)\Phi_\gamma\, dx$$

is strictly decreasing unless u is scale invariant.

1.10. Sharp gradient estimates

Here, we discuss an approach to obtaining estimates for the *modulus of continuity* of the heat equation. The argument is based on the maximum principle; a novel feature is that we apply it to a function involving not one but *two* points in the domain of the solution. Such techniques go back to the work of Kružkov [**338**]. The inequalities we present were developed in the thesis of Julie Clutterbuck [**174**] and in the joint work [**47, 48**] and were used to obtain sharp modulus of continuity and gradient estimates for solutions to certain quasilinear parabolic equations.

Figure 1.3. Julie Clutterbuck.

To illustrate the methods, consider a solution u to the 1-dimensional L-periodic heat equation:

$$\begin{cases} u_t = u'' \text{ in } \mathbb{R} \times [0, \infty), \\ u(x + L, \cdot) = u(x, \cdot) \text{ for } x \in \mathbb{R}. \end{cases}$$

Define its (spatial) **modulus of continuity** $\omega : [0, L/2] \times [0, \infty) \to \mathbb{R}$ by

$$\omega(s, t) \doteqdot \sup \left\{ \frac{u(y, t) - u(x, t)}{2} : \frac{y - x}{2} = s \right\}. \tag{1.20}$$

Proposition 1.14. *The modulus of continuity of a solution to the 1-dimensional L-periodic heat equation is a* ***viscosity subsolution*** *to the 1-dimensional Dirichlet heat equation.*

Do not be put off by the "viscosity subsolution" terminology: This is a very simple way to define a subsolution in a situation where the function may not be smooth (which could be the case for the modulus of continuity since it is defined as a supremum). It requires that any smooth function $\varphi : (s_0 - r, s_0 + r) \times (t_0 - r^2, t_0] \to \mathbb{R}$, $r > 0$, which **supports** ω **from above** at $(s_0, t_0) \in [0, L/2] \times (0, \infty)$, in the sense that

$$\varphi \geq \omega \text{ in } (s_0 - r, s_0 + r) \times (t_0 - r^2, t_0] \text{ with } \varphi(s_0, t_0) = \omega(s_0, t_0),$$

satisfies

$$\varphi_t(s_0, t_0) \leq \varphi''(s_0, t_0).$$

Viscosity supersolutions are defined analogously.

Proof of Proposition 1.14. Clearly, $\omega(0,t) = \omega(L/2,t) = 0$. Given (s_0,t_0), suppose that $\varphi \in C^\infty((s_0-r,s_0+r)\times(t_0-r^2,t_0])$ supports ω from above at (s_0,t_0) and consider the function $Z:\mathbb{R}\times\mathbb{R}\times[0,\infty)\to\mathbb{R}$ defined by

$$Z(x,y,t) \doteqdot u(y,t)-u(x,t)-2\varphi\left(\frac{y-x}{2},t\right). \tag{1.21}$$

Choose (x_0,y_0) so that $2\omega(s_0,t_0)=u(y_0,t_0)-u(x_0,t_0)$. Then

$$Z\le 0 \text{ in } (x_0-r/2,x_0+r/2)\times(y_0-r/2,y_0+r/2)\times(t_0-r^2/4,t_0]$$

with

$$Z(x_0,y_0,t_0)=0.$$

Since Z is smooth, this implies that

$$0\le \partial_t Z|_{(x_0,y_0,t_0)} = u_t(y_0,t_0)-u_t(x_0,t_0)-2\varphi_t(s_0,t_0),$$

$$0 = DZ|_{(x_0,y_0,t_0)} = \left(-u'(x_0,t_0)+\varphi'(s_0,t_0),u'(y_0,t_0)-\varphi'(s_0,t_0)\right),$$

and

$$0\ge D^2Z|_{(x_0,y_0,t_0)} = \begin{pmatrix} -u''(x_0,t_0)-\varphi''(s_0,t_0) & -\varphi''(s_0,t_0)\\ \varphi''(s_0,t_0) & u''(y_0,t_0)-\varphi''(s_0,t_0)\end{pmatrix}.$$

Thus,

$$\begin{aligned} 0 &\le \partial_t Z|_{(x_0,y_0,t_0)} - \left.\frac{d^2}{dh^2}\right|_{h=0} Z(x_0+h,y_0-h,t_0) \\ &= \partial_t Z|_{(x_0,y_0,t_0)} - (\partial_x-\partial_y)^2 Z|_{(x_0,y_0,t_0)} \\ &= -2(\varphi_t-\varphi'')|_{(s_0,t_0)}, \end{aligned} \tag{1.22}$$

which is the claim. □

Note that, had we simply used the Laplacian of Z in (1.22), we would only have obtained the rough estimate

$$2\varphi_t(s_0,t_0)\le\varphi''(s_0,t_0).$$

The variation chosen corresponds to a variation of the parameter $s=\frac{y-x}{2}$.

As a corollary, we obtain a sharp a priori estimate for the gradient of a solution. Recall that the **error function** $\operatorname{erf}:\mathbb{R}\to\mathbb{R}$ is defined by

$$\operatorname{erf}(x)\doteqdot\frac{2}{\sqrt{\pi}}\int_0^x e^{-y^2}\,dy.$$

Corollary 1.15. *The modulus of continuity ω of a solution u to the L-periodic heat equation satisfies*

$$\omega(s,t)\le M\operatorname{erf}\left(\frac{s}{\sqrt{4t}}\right), \tag{1.23}$$

where

$$M \doteqdot \max_{s\in[0,\frac{L}{2}]} \omega(s,0) = \max_{x,y\in\mathbb{R}} \frac{u(x,0)-u(y,0)}{2}\,.$$

In particular, we have the gradient estimate

$$|u'(x,t)| \le \frac{M}{\sqrt{\pi t}}\,. \tag{1.24}$$

Proof. Observe that the function $v(s,t) \doteqdot \operatorname{erf}\left(\frac{s}{\sqrt{4t}}\right)$ satisfies the heat equation with

$$\begin{aligned} \lim_{t\searrow 0} v(s,t) &= 0 \quad \text{for} \quad s\in(0,\infty)\,, \\ v(0,t) &\equiv 0 \quad \text{for} \quad t\in(0,\infty)\,, \quad \text{and} \\ v'(s,t) &> 0 \quad \text{for} \quad (s,t)\in\mathbb{R}\times(0,\infty)\,. \end{aligned}$$

We claim that

$$\omega(s,t) - Mv(s,t) - \varepsilon \mathrm{e}^t \le 0$$

for all $s\in[0,\frac{L}{2}]$, $t>0$, and $\varepsilon>0$. Suppose that this is not true. Then, since

$$\omega(s,0) - Mv(s,0) - \varepsilon\mathrm{e}^0 \le -\varepsilon < 0\,,$$

we can find $\varepsilon>0$, $t_0>0$, and $s_0\in[0,\frac{L}{2}]$ such that

$$\omega(s,t) - Mv(s,t) - \varepsilon\mathrm{e}^t < 0$$

for all $(s,t)\in[0,\frac{L}{2}]\times[0,T_0)$, while

$$\omega(s_0,t_0) - Mv(s_0,t_0) - \varepsilon\mathrm{e}^{t_0} = 0\,.$$

That is, the function φ defined by

$$\varphi(s,t) \doteqdot Mv(s,t) + \varepsilon\mathrm{e}^t$$

supports ω from above at (s_0,t_0). Since $\omega(0,t) = \omega(\frac{L}{2},t) = 0$, we certainly have $s_0\in(0,\frac{L}{2})$. But then

$$0 \ge (\varphi_t - \varphi'')|_{(s_0,t_0)} = \varepsilon\mathrm{e}^{t_0} > 0\,,$$

which is absurd. Taking $\varepsilon\to 0$ yields the estimate for ω. The gradient estimate follows easily:

$$\begin{aligned} |u'(x,t)| &\le \sup_{y\ne x} \frac{|u(x,t)-u(y,t)|}{|x-y|} \\ &\le M \sup_{y\ne x} \frac{|v(x,t)-v(y,t)|}{|x-y|} \\ &= Mv'(0,t) \\ &= \frac{M}{\sqrt{\pi t}}\,. \end{aligned}$$

□

Note that these modulus of continuity and gradient estimates are sharp since equality holds on the error function solution.

The argument also works in higher dimensions and for different boundary conditions: Consider, for example, a solution u to the Neumann heat equation on a bounded, strictly convex domain $\Omega \subset \mathbb{R}^n$ of diameter $D \doteqdot \operatorname{diam}(\Omega)$. Define its (spatial) **modulus of continuity** $\omega : [0, D/2] \times [0, \infty) \to \mathbb{R}$ by

$$\omega(s,t) \doteqdot \sup\left\{\frac{u(y,t)-u(x,t)}{2} : \frac{|y-x|}{2} = s\right\}. \tag{1.25}$$

Proposition 1.16. *The modulus of continuity of a solution to the Neumann heat equation over a smoothly bounded, strictly convex domain of diameter D is a viscosity subsolution to the* 1*-dimensional heat equation with Dirichlet–Neumann boundary conditions with respect to spatially nondecreasing barriers.*

Proof. Given (s_0, t_0), suppose that $\varphi \in C^\infty((s_0 - r, s_0 + r) \times (t_0 - r^2, t_0])$ supports ω from above at (s_0, t_0) and choose (x_0, y_0) so that

$$2\omega(s_0, t_0) = u(y_0, t_0) - u(x_0, t_0)\,.$$

If $s_0 = D/2$, then the points x_0 and y_0 are of maximal distance. Thus, $x_0, y_0 \in \partial\Omega$ and the outward pointing normal N to $\partial\Omega$ satisfies

$$\mathsf{N}(y_0) = \frac{y_0 - x_0}{|y_0 - x_0|} = -\mathsf{N}(x_0)\,.$$

If φ is nondecreasing in s, then

$$\begin{aligned}
0 &\le \varphi'(D/2, t_0)\\
&= \limsup_{h \searrow 0} \frac{\varphi(D/2, t_0) - \varphi(D/2 - h, t_0)}{h}\\
&\le \limsup_{h \searrow 0} \frac{\omega(D/2, t_0) - \omega(D/2 - h, t_0)}{h}\\
&\le \limsup_{h \searrow 0} \frac{u(y_0, t_0) - u(x_0, t_0) - [u(y_0 - h\mathsf{N}(y_0)) - u(x_0 - h\mathsf{N}(x_0))]}{h}\\
&= D_{\mathsf{N}} u(y_0, t_0) - D_{\mathsf{N}} u(x_0, t_0)\\
&= 0\,.
\end{aligned}$$

On the other hand, $\omega(0, t) = 0$. So ω satisfies mixed Dirichlet–Neumann boundary conditions (in the viscosity sense).

Now suppose that $s_0 < D/2$ and consider the function $Z : \Omega \times \Omega \times [0, \infty) \to \mathbb{R}$ defined by

$$Z(x, y, t) \doteqdot u(y, t) - u(x, t) - 2\varphi\left(\frac{|y-x|}{2}, t\right). \tag{1.26}$$

Then

$$Z \le 0 \text{ in } B_{r/2}(x_0) \times B_{r/2}(y_0) \times (t_0 - r^2/4, t_0] \text{ with } Z(x_0, y_0, t_0) = 0\,.$$

Since Z is smooth, this implies that

$$0 \le \partial_t Z|_{(x_0,y_0,t_0)} = u_t(y_0, t_0) - u_t(x_0, t_0) - 2\varphi_t(s_0, t_0)\,.$$

Note that, if $y_0 \in \partial\Omega$, then, since Ω is convex,

$$0 \le DZ|_{(x_0,y_0,t_0)}(0, \mathsf{N}(y_0)) = D_{\mathsf{N}}u(y_0, t_0) - \varphi'(s_0, t_0)\frac{y_0 - x_0}{|y_0 - x_0|} \cdot \mathsf{N}(y_0) \le 0\,.$$

So

$$DZ|_{(x_0,y_0,t_0)}(0, \mathsf{N}(y_0)) = 0\,.$$

Similarly, if $x_0 \in \partial\Omega$, then

$$DZ|_{(x_0,y_0,t_0)}(\mathsf{N}(x_0), 0) = 0\,.$$

Thus, (x_0, y_0) is a critical point of Z even if one or both of the points are in $\partial\Omega$, and hence

$$0 \ge D^2 Z|_{(x_0,y_0,t_0)}\,.$$

Choose an orthonormal basis $\{e_i\}_{i=1}^n$ for $\mathbb{R}^n$ such that $e_n = \frac{y_0 - x_0}{|y_0 - x_0|}$. Then

$$0 \ge D^2 Z|_{(x_0,y_0,t_0)}(e_i, e_i) = u_{ii}(y_0, t_0) - u_{ii}(x_0, t_0)$$

for each $i = 1, \ldots, n-1$ and

$$0 \ge \left.\frac{d^2}{dh^2}\right|_{h=0} Z(x_0 + he_n, y_0 - he_n, t_0) = u_{nn}(y_0, t_0) - u_{nn}(x_0, t_0) - 2\varphi''(s_0, t_0)\,.$$

We conclude that

$$\begin{aligned} 0 \le \partial_t Z|_{(x_0,y_0,t_0)} &- \sum_{i=1}^{n-1} D^2 Z|_{(x_0,y_0,t_0)}(e_i, e_i) \\ &- \left.\frac{d^2}{dh^2}\right|_{h=0} Z(x_0 + he_n, y_0 - he_n, t_0) \\ = &-2(\varphi_t - \varphi'')|_{(s_0,t_0)} \end{aligned} \tag{1.27}$$

as claimed. □

Once again, it was crucial that we made the correct choice in the second variation of Z in (1.27). Our choice was motivated by the 1-dimensional case: Suppose that $\Omega = A \times [-D/2, D/2]$, where A is a convex subset of $\mathbb{R}^{n-1}$ and $u(x, y, t) = f(y, t)$. Then u solves the Neumann heat equation in Ω if and only if f solves the Neumann heat equation in $[-D/2, D/2]$. If $f(\cdot, t)$ is odd, increasing, and concave for $y > 0$, then $\omega(s, t) = f(s, t)$ (for $0 < s < D/2$) and hence ω satisfies the 1-dimensional heat equation. The variation corresponding to ∂_s is given by the e_n direction in (1.27).

Comparing the solution with the error function solution again gives a sharp gradient estimate (since the gradient of the latter is positive).

Corollary 1.17. *The modulus of continuity ω of a solution u to the Neumann heat equation over a smoothly bounded, strictly convex domain $\Omega \subset \mathbb{R}^{n+1}$ of diameter D satisfies*

$$\omega(s,t) \le M\operatorname{erf}\left(\frac{s}{\sqrt{4t}}\right), \tag{1.28}$$

where

$$M \doteqdot \max_{s\in[0,\frac{D}{2}]} \omega(s,0) = \max_{x,y\in\mathbb{R}} \frac{u(x,0)-u(y,0)}{2}.$$

In particular, we have the gradient estimate

$$|\nabla u(x,t)| \le \frac{M}{\sqrt{\pi t}}. \tag{1.29}$$

Moduli of continuity estimates for solutions to the heat equation also have a beautiful connection to eigenvalue problems for the Laplacian [**50**]. Let us illustrate this connection with a simple proof of the Payne–Weinberger inequality (first proved in [**437**]), which is a consequence of the following observation.

Corollary 1.18. *The modulus of continuity ω of a solution to the Neumann heat equation over a bounded, strictly convex domain of diameter D satisfies*

$$\omega(s,t) \le C\mathrm{e}^{-\frac{\pi^2}{D^2}t}\sin\left(\frac{\pi s}{D}\right) \tag{1.30}$$

for all $(s,t)\in[0,D/2]\times[0,\infty)$, where

$$C \doteqdot \sup_{(0,D/2)} \frac{\omega(s,0)}{\sin\left(\frac{\pi s}{D}\right)}.$$

Note that

$$\sup_{s\in(0,D/2)} \frac{\omega(\cdot,0)}{\sin\left(\frac{\pi s}{D}\right)} \le D \sup_{x\neq y\in\Omega} \frac{u(y,0)-u(x,0)}{|y-x|}$$

since $\frac{\sin z}{z} \ge \frac{2}{\pi}$ for $z\in(0,\frac{\pi}{2})$.

Proof of Corollary 1.18. We need to show that the maximum principle works for viscosity subsolutions. For each $t\ge 0$, set

$$\psi(t) \doteqdot \sup_{s\in(0,D/2)} \frac{\omega(s,t)}{f(s,t)},$$

where

$$f(s,t) \doteqdot \mathrm{e}^{-\frac{\pi^2}{D^2}t}\sin\left(\frac{\pi s}{D}\right).$$

We claim that

$$\omega(s,t) - \psi(0)f(s,t) - \varepsilon(1+s+t) \le 0$$

for all $t > 0$, $s \in [0, D/2]$, and $\varepsilon > 0$. Suppose that this is not true. Then, since

$$\omega(s,0) - \psi(0)f(s,0) - \varepsilon(1+s) \le -\varepsilon(1+s) < 0\,,$$

we can find $\varepsilon > 0$, $t_0 > 0$, and $s_0 \in [0, D/2]$ such that

$$\omega(s,t) - \psi(0)f(s,t) - \varepsilon(1+s+t) < 0$$

for all $(s,t) \in [0, D/2] \times [0, t_0)$, while

$$\omega(s_0,t_0) - \psi(0)f(s_0,t_0) - \varepsilon(1+s_0+t_0) = 0\,.$$

That is, the function φ defined by $\varphi(s,t) \doteqdot \psi(0)f(s,t)+\varepsilon(1+s+t)$ supports ω from above at (s_0, t_0). Since it is nondecreasing in s, it is an admissible barrier for Proposition 1.16. Clearly, $s_0 > 0$. If $s_0 = D/2$, then

$$0 \ge \psi(0)f'(D/2,t) + \varepsilon = \varepsilon > 0\,,$$

which is absurd. So $s_0 < D/2$. Since f satisfies the 1-dimensional heat equation, we conclude that

$$0 \ge (\varphi_t - \varphi'')|_{(s_0,t_0)} = \varepsilon > 0\,,$$

which is also absurd. Thus,

$$\omega(s,t) - \psi(0)f(s,t) - \varepsilon(1+s+t) \le 0$$

for all $t > 0$, $s \in [0, D/2]$, and $\varepsilon > 0$. Taking $\varepsilon \to 0$ yields

$$\omega(s,t) \le \psi(0)f(s,t)$$

for all $t > 0$, $s \in [0, D/2]$ and hence $\psi(t) \le \psi(0)$ for all $t > 0$, which is the claim. □

Theorem 1.19. *The first nontrivial eigenvalue λ_1 of the Neumann Laplacian on a bounded convex domain $\Omega \subset \mathbb{R}^n$ of diameter D satisfies*

$$\lambda_1 \ge \frac{\pi^2}{D^2}\,. \tag{1.31}$$

Equality holds in the limit of cylindrical domains $\Omega = A \times [-D/2, D/2]$, wherein the diameter of A approaches zero.

Proof. Let $u_1(x)$ be the eigenfunction corresponding to λ_1. Then

$$u(x,t) \doteqdot \mathrm{e}^{-\lambda_1 t} u_1(x)$$

satisfies the Neumann heat equation on Ω. Let $\omega : [0, D/2] \times [0, \infty) \to \mathbb{R}$ be its modulus of continuity. By Corollary 1.18,

$$\omega(s,t) \le C\mathrm{e}^{-\frac{\pi^2}{D^2}t} \sin\left(\frac{\pi s}{D}\right)$$

for all $(s,t) \in [0, D/2] \times [0,\infty)$. Taking the supremum over s then yields

$$\operatorname{osc}(u_1)\mathrm{e}^{-\lambda_1 t} = \operatorname{osc}(u(\cdot,t)) \le C\mathrm{e}^{-\frac{\pi^2}{D^2}t}.$$

Were λ_1 less than $\frac{\pi^2}{D^2}$, this would be impossible for sufficiently large t. We leave the characterization of the equality case as an exercise. □

Similar arguments were used to prove the fundamental gap conjecture in [**49**].

The arguments presented here can also be modified to obtain *interior* gradient estimates [**174**]. Similar interior gradient estimates for the heat equation on $\mathbb{R}^n$ (or any complete Riemannian manifold with nonnegative Ricci curvature) were obtained by P. Souplet and Q. Zhang [**492**] by applying the maximum principle to the function $|\nabla \ln(\ln M - \ln u)|$ (modified by appropriate cutoff functions); cf. [**364**].

1.11. Notes and commentary

1.11.1. The Omori–Yau maximum principle. Note that compactness of the torus was crucial in proving the weak maximum principle since we needed to find a point where the spatial maximum was attained (this is not the case for the strong maximum principle, which is local in nature). The following result, usually referred to as the **Omori–Yau maximum principle**, makes similar arguments possible even in the noncompact setting.

Theorem 1.20 (Omori [**433**], Yau [**538**]). *Let (M,g) be a Riemannian manifold. If the Ricci curvature of (M,g) is bounded from below,* $\mathrm{Rc} \ge Kg$ *for some $K \in \mathbb{R}$, then given any bounded $f \in C^2(M)$ there exists a sequence of points $x_k \in M$ such that*

$$\lim_{k\to\infty} f(x_k) = \sup_M f\,, \quad \lim_{k\to\infty} |\nabla f(x_k)| = 0, \quad \textit{and} \quad \limsup_{k\to\infty} \Delta f(x_k) \le 0\,.$$

If the sectional curvature of (M,g) is bounded from below, $\sec \ge K$ *for some $K \in \mathbb{R}$, then given any bounded $f \in C^2(M)$ there exists a sequence of points $x_k \in M$ such that*

$$\lim_{k\to\infty} f(x_k) = \sup_M f\,, \quad \lim_{k\to\infty} |\nabla f(x_k)| = 0, \quad \textit{and} \quad \limsup_{k\to\infty} \max_{|v|=1} \nabla_v\nabla_v f(x_k) \le 0\,.$$

Of course, both hypotheses are satisfied by $\mathbb{R}^n$.

1.11.2. Sturm's theorem. Parabolic equations in only one spatial variable satisfy some nice properties which are not available in higher dimensions. One of these is the parabolic version of **Sturm's theorem** [**500**].

Consider solutions $u : [0,1] \times [0,T] \to \mathbb{R}$ to the 1-dimensional linear heat-type equation

$$(a(x,t)\partial_x^2 + b(x,t)\partial_x + c(x,t))u(x,t) = 0\,. \tag{1.32}$$

The following result shows that, under very general hypotheses, the zero set of u is nonincreasing in time.

Theorem 1.21 (Angenent [**64**]). *Let $u : [0,1] \times [0,T] \to \mathbb{R}$ be a bounded solution of* (1.32) *satisfying the periodic boundary condition $u(0,t) = u(1,t)$ for all t. Suppose that the functions a, a^{-1}, a_t, a_x, a_{xx}, b, b_t, b_x, and c are all bounded. Define*

$$z(t) \doteqdot |\{x \in [0,1] : u(x,t) = 0\}|\,.$$

If u is not identically zero, then $z(t)$ is finite for all $t > 0$ and nonincreasing in t.

The theorem is certainly not true in higher dimensions: It is easy to construct examples where the zero set can increase and even become infinite.

Sturm's theorem is a useful tool in the study of curve shortening flow (see for instance Sections 3.6 and 4.3.6).

1.11.3. The differential Harnack inequality. There is an elegant proof of the (Euclidean) matrix Harnack inequality, due to Helmensdorfer and Topping [**283**], which makes use of the representation formula for bounded solutions (Theorem 1.9) and the fact that the sum of log-convex functions is log-convex.

Claim 1.22. Let $u \in C^\infty(\mathbb{R}^n \times (0,\infty))$ be a bounded, positive solution to the heat equation. Then the function $x \mapsto \frac{u(x,t)}{\rho(x,t)}$ is log-convex for every $t > 0$.

Proof. Given $\varepsilon > 0$, set $v_\varepsilon(x,t) \doteqdot u(x,t+\varepsilon)$. Given $t > 0$, define

$$F(x,t) \doteqdot \frac{v_\varepsilon(x,t)}{\rho(x,t)}\,.$$

Then, by Theorem 1.9,

$$F(x,t) = \int_{\mathbb{R}^n} v_\varepsilon(y,0)G(x,y,t)\,dy\,,$$

where

$$G(x,y,t) \doteqdot \frac{\rho(x-y,t)}{\rho(x,t)}\,.$$

Since $G(\cdot, y, t)$ is log-convex for each y and t, Hölder's inequality yields

$$\begin{aligned} F(sw + (1-s)z, t) &= \int_{\mathbb{R}^n} v_\varepsilon(y,0)G(sw + (1-s)z, y, t)\,dy \\ &\leq \int_{\mathbb{R}^n} v_\varepsilon(y,0)G^s(w,y,t)G^{1-s}(z,y,t)\,dy \\ &\leq \left(\int_{\mathbb{R}^n} v_\varepsilon(y,0)G(w,y,t)dy\right)^s \left(\int_{\mathbb{R}^n} v_\varepsilon(y,0)G(z,y,t)dy\right)^{1-s} \\ &= F(w,t)^s F(z,t)^{1-s}. \end{aligned}$$

The claim follows since $\varepsilon > 0$ was arbitrary. □

Differentiating yields

$$0 \leq \nabla^2 \log \frac{u}{\rho} = \nabla^2 \log u + \frac{\mathrm{I}}{2t}.$$

1.11.4. Ancient solutions. Unlike most other physical processes, heat-type phenomena are not time-symmetric; parabolic equations are well-posed *forwards* in time — given appropriate initial-boundary conditions, a solution can be found for a (perhaps short) *forwards* time interval (cf. Theorem 1.9) — and diffusion tends to drive solutions towards equilibria (cf. Section 1.6). In particular, we should not expect to be able to find many solutions defined on large *backwards* time intervals. A solution to a parabolic equation which is defined for times $t \in (-\infty, T)$, $T \in (-\infty, \infty]$, is called an **ancient** solution. Ancient solutions to parabolic equations can be thought of as the parabolic analogue of entire solutions to the corresponding elliptic equation.

The following Liouville-type theorem for the heat equation was obtained by P. Souplet and Q. Zhang as a consequence of their sharp interior gradient estimates [**492**].

Theorem 1.23. *Let u be an ancient solution to the heat equation on $\mathbb{R}^n$ (or any complete Riemannian manifold with nonnegative Ricci curvature).*

- *If u satisfies $u(x,t) \leq o(|x| + \sqrt{-t})$ as $|x| + \sqrt{-t} \to \infty$, then u is constant.*
- *If u is positive and satisfies $u(x,t) \leq \mathrm{e}^{o(|x|+\sqrt{-t})}$ as $|x|+\sqrt{-t} \to \infty$, then u is constant.*

Note that the hypotheses are necessary: Consider the affine and traveling wave solutions $(x,t) \mapsto ax_1 + b$ and $(x,t) \mapsto \mathrm{e}^{x_1+t}$, respectively.

Consider now solutions $u : \mathbb{R}^n \times (t_0, T) \to \mathbb{R}$ to the *semilinear* heat equation (1.18) for subcritical exponents $1 < p < \frac{n+2}{n-2}$. Positive solutions blow up in finite time (cf. Exercise 1.10). After appropriately rescaling, singularities are modeled by self-similar solutions [**244**].

F. Merle and H. Zaag [**399**], building on the work of Y. Giga and R. Kohn [**244**], obtained the following Liouville-type result.

Theorem 1.24. *Let $u : \mathbb{R}^n \times (-\infty, T) \to \mathbb{R}$ be a (maximal) positive ancient solution to the semilinear heat equation* (1.18). *If*

$$u(x,t) = O((T-t)^{-\frac{1}{p-1}}),$$

then

$$u = ((p-1)(T-t))^{-\frac{1}{p-1}}.$$

T. Colding and W. P. Minicozzi II proved that (on any Riemannian manifold with polynomial volume growth) the dimension of the space of ancient solutions having given polynomial growth is bounded by the degree of growth times the dimension of the space of harmonic functions with the same growth [**187**].

1.11.5. The heat equation on Riemannian manifolds. Consider now, instead of Euclidean space, $\mathbb{R}^n$, a smooth, connected, complete Riemannian n-manifold[3], (M^n, g). The Riemannian metric g gives rise to gradient and divergence operators, and hence to a natural analogue of the Laplacian — the Laplace–Beltrami operator, $\Delta \cdot \doteqdot \operatorname{div}(\nabla \cdot)$. Much of what we have discussed so far in the context of the heat equation on subsets of Euclidean space applies equally well to solutions of the heat equation

$$(\partial_t - \Delta)u = 0$$

on (M^n, g). A notable exception, however, is the presence of the symmetries described in Section 1.3 and the consequences thereof. Of course, we can still — rather trivially — translate in time, scale the solution, and superimpose solutions to obtain new solutions but, unless (M^n, g) itself admits symmetries (e.g., some number of conformal Killing fields), translation, rotation, and scaling in space are no longer available.

The heat flow on (M^n, g) is the gradient flow of the Dirichlet energy,

$$E(u) \doteqdot \frac{1}{2}\int_M |\nabla u|^2 \, d\mu,$$

with respect to the L^2-structure induced by the Riemannian metric (recall that, in a local coordinate chart $x : U \to \mathbb{R}^n$ on M, $d\mu = \sqrt{\det((x^{-1})^* g)}\, dx$).

The maximum principle and the ODE comparison principle continue to hold in the geometric setting, with more or less the same form and proof (see also Chapter 6). The strong maximum principle, being in essence a local property, also continues to hold.

[3]For treatments of Riemannian geometry, we encourage the reader to consult one of the many beautiful books on the subject, such as [**138, 198, 361, 442**].

It is also still possible to construct a fundamental solution, which arises as the solution to the heat equation with initial condition given by a Dirac distribution (combined with appropriate boundary conditions) — one first solves the initial value problem over a precompact domain and then takes a limit over an exhaustion of M [**137**].

The long-term behavior of solutions will depend on the geometry of (M^n, g). Indeed, the optimal constant in the Poincaré inequality is determined by the rate of diffusion for the heat equation and vice versa.

Bernstein's method is also available in the geometric setting but is somewhat complicated by the fact that commuting the Laplacian past a gradient gives rise to Ricci curvature terms, which must be controlled. This will be an important point later in the book (for instance in Sections 2.6 and 6.8 and Chapter 7). Similarly, the Harnack inequality and its differential version are available so long as the Ricci curvature can be controlled (cf. Sections 4.3.2 and 10.2).

The moduli of continuity estimates also pass to the Riemannian setting, by making use of the Riemannian distance function and appropriate comparison theorems. See [**42, 50**].

1.12. Exercises

Exercise 1.1. Check that the heat equation is indeed invariant under each of the local group actions (1)–(8) introduced in Section 1.3.

Exercise 1.2. Suppose that $u_i : \mathbb{R} \times (0, \infty) \to \mathbb{R}$ satisfies the heat equation for $i = 1, \ldots, n$. Show that the function $u : \mathbb{R}^n \times (0, \infty) \to \mathbb{R}$ defined by

$$u(x, t) \doteqdot u_1(x_1, t) \cdots u_n(x_n, t)$$

satisfies the heat equation.

Exercise 1.3. Prove that

$$\int_{\mathbb{R}^n} \rho(x, t)\, dx \equiv 1\,,$$

where ρ is the fundamental solution to the heat equation on $\mathbb{R}^n$. Prove that

$$\lim_{t \searrow 0} \int_{\mathbb{R}^n} \rho(x, t)\varphi(x)\, dx = \varphi(0)$$

for all smooth functions $\varphi : \mathbb{R}^n \to \mathbb{R}$ with compact support.

Exercise 1.4. Find all traveling wave solutions of the heat equation on $\mathbb{R}$.

Exercise 1.5. Show that the fundamental solution to the heat equation on $\mathbb{R}^n$ is invariant under all Galilean boosts. Find all solutions $u : \mathbb{R} \times (0, \infty)$ of the heat equation on $\mathbb{R}$ which are invariant under the Galilean boosts.

Exercise 1.6. Suppose that $\varphi : \mathbb{R}^n \to \mathbb{R}$ is a bounded continuous function. Prove that the function $u \in C^\infty(\mathbb{R}^n \times (0,\infty)) \cap C^0(\mathbb{R}^n \times [0,\infty))$ defined by

$$u(x,t) \doteqdot \int_{\mathbb{R}^n} k(x,y,t)\varphi(y)\, dy\,,$$

where k is the heat kernel (defined in (1.12)), satisfies the heat equation and converges pointwise to φ as $t \searrow 0$.

Exercise 1.7. Consider the **heat** (a.k.a. **caloric**) **polynomials**

$$P_m(x,t) \doteqdot \sum_{\ell=0}^{\lfloor m/2 \rfloor} \frac{m!}{\ell!(m-2\ell)!} x^{m-2\ell} t^\ell\,.$$

Show that

$$P_{m+1} = (x + 2t\partial_x)P_m\,.$$

Deduce that P_m is a solution to the heat equation on $\mathbb{R}$. HINT: $P_{m+1} = xP_m + 2mtP_{m-1}$.

Exercise 1.8. Let $u \in C^\infty(T^n \times (0,T)) \cap C^0(T^n \times [0,T))$ be a subsolution to the heat equation with drift term $b : T^n \times [0,T) \to \mathbb{R}^n$. That is,

$$(\partial_t - \Delta - b^i\nabla_i)u \le 0\,.$$

Show that

$$\max_{T^n \times \{t\}} u \le \max_{T^n \times \{t_0\}} u$$

for all $t > t_0 \in [0,T)$.

Exercise 1.9. Theorem 1.3 and Corollary 1.4 also hold for subsolutions $u \in C^\infty(\Omega \times (0,T)) \cap C^0(\overline{\Omega} \times [0,T))$ defined in bounded open sets $\Omega \subset \mathbb{R}^n$, under the additional hypotheses that $u \le 0$ on $\partial\Omega \times (0,T)$ and $u \le \phi_0$ on $\partial\Omega \times (0,T)$, respectively. Identify where in the respective proofs these hypotheses are needed. Provide a "physical" argument for why they should be necessary.

Exercise 1.10. Suppose that $u : T^n \times [0,T) \to \mathbb{R}$ is a smooth, positive solution to the reaction-diffusion equation

$$(\partial_t - \Delta)u = u^p$$

where $p > 1$. Show that

$$\left(\underline{u}_0^{1-p} - (p-1)t\right)^{-\frac{1}{p-1}} \le u \le \left(\overline{u}_0^{1-p} - (p-1)t\right)^{-\frac{1}{p-1}}\,,$$

where $\underline{u}_0 \doteqdot \min_{T^n \times \{0\}} u$ and $\overline{u}_0 \doteqdot \max_{T^n \times \{0\}}$. Deduce that T is finite. Suppose that $\limsup_{t \to T} \max_{T^n \times \{t\}} u = \infty$. Prove that

$$\min_{T^n \times \{t\}} u \le ((p-1)(T-t))^{-\frac{1}{p-1}} \le \max_{T^n \times \{t\}} u\,.$$

Exercise 1.11. Prove the 1-dimensional Poincaré inequality:

$$\int_0^1 |u(x) - \overline{u}|\, dx \le \int_0^1 |u'(x)|^2\, dx$$

for every $u \in C^1((0,1))$.

Exercise 1.12. Show that the fundamental solution $\rho : \mathbb{R}^n \times (0,\infty) \to \mathbb{R}$ (defined in (1.3)) satisfies

$$\Delta \log \rho + \frac{n}{2t} = 0 .$$

Exercise 1.13. Show that the fundamental solution $\rho : \mathbb{R}^n \times (0,\infty) \to \mathbb{R}$ satisfies

$$\frac{\rho(x_2,t_2)}{\rho(x_1,t_1)} \ge \left(\frac{t_1}{t_2}\right)^{\frac{n}{2}} \exp\left(-\frac{|x_2 - x_1|^2}{4(t_2 - t_1)}\right)$$

for all $x_1, x_2 \in \mathbb{R}^n$ and $t_1 < t_2 \in (0,\infty)$, with equality if and only if $\frac{x_1}{t_1} = \frac{x_2}{t_2}$.

Exercise 1.14. Modify the proof of Theorem 1.11 to obtain the **matrix Harnack inequality**,

$$\nabla^2 \log u + \frac{\mathrm{I}}{2t} \ge 0$$

for positive periodic solutions $u : T^n \times [0,\infty) \to \mathbb{R}$ to the heat equation, where I is the Euclidean inner product and $T^n \doteqdot \mathbb{R}^n/\mathbb{Z}^n$ is the torus. HINT: Consider $Q \doteqdot \nabla_V \nabla_V \log u$ for any $V \in S^n$.

Exercise 1.15. Prove that if $u \in C^\infty(T^n \times (-\infty, 0))$ is a positive ancient solution to the heat equation, then

$$\Delta \log u \ge 0 .$$

Exercise 1.16. Let $u : T^n \times (-\infty, 0) \to \mathbb{R}$ be a bounded ancient solution to the heat equation on the n-torus, T^n. Prove that u is constant.

Exercise 1.17. Prove versions of Theorem 1.3 and Corollary 1.4 for subsolutions defined on compact Riemannian manifolds.

Exercise 1.18. Let (M, g) be a compact Riemannian manifold satisfying $\mathrm{Rc} \ge 0$ and let $u : M \times [0,T) \to \mathbb{R}$ be a solution to the heat equation. Show that

$$(\partial_t - \Delta)|\nabla u|^2 \le -2|\nabla^2 u|^2 ,$$

where $\nabla^2 u$ is the covariant Hessian of u. Deduce a Bernstein-type decay estimate for $|\nabla u|^2$.

Exercise 1.19. Let (M, g) be a compact Riemannian manifold satisfying $\mathrm{Rc} \ge 0$ and let $u : M \times [0,T) \to \mathbb{R}$ be a solution to the heat equation. Prove that

$$\Delta \log u + \frac{n}{2t} \ge 0 .$$

Deduce that

$$\frac{u(x_2,t_2)}{u(x_1,t_1)} \geq \left(\frac{t_2}{t_1}\right)^{-\frac{n}{2}} \exp\left(-\frac{|x_2-x_1|^2}{4(t_2-t_1)}\right)$$

for all $x_1, x_2 \in M$ and $t_1, t_2 \in (0,T)$ with $t_1 < t_2$.

Exercise 1.20. Let (M,g) be a compact Riemannian manifold satisfying $\mathrm{Rc} \geq 0$ and let $u : M \times [0,T) \to \mathbb{R}$ be a solution to the heat equation. Prove that

$$\nabla^2 \log u + \frac{n}{2t} g \geq 0 \,. \tag{1.33}$$

Exercise 1.21. Let (M,g) be a compact Riemannian manifold satisfying $\mathrm{Rc} \geq 0$. Show that the heat kernel $k : M \times M \times (0,\infty) \to \mathbb{R}$ satisfies

$$k(x,y,t) \geq (4\pi t)^{-\frac{n}{2}} \exp\left(-\frac{d(x,y)^2}{4t}\right) .$$

HINT: You will need the fact that

$$k(x,x,t) = (4\pi t)^{-\frac{n}{2}} (1 + O(t))$$

as $t \to 0$. See [**137**, Chapters VI and VIII].

Exercise 1.22. Let (M,g) be a compact Riemannian manifold satisfying $\mathrm{Rc} \geq -kg$ for some $k \geq 0$ and let ϕ be a solution to the adjoint heat equation,

$$(\partial_t + \Delta)\phi = 0 \,.$$

Prove that

$$\frac{d}{dt}\left(\mathrm{e}^{-2kt} \int_M |\nabla u|^2 \phi \, d\mu\right) \leq -2\mathrm{e}^{-2kt} \int_M |\nabla^2 u|^2 \phi \, d\mu$$

and

$$\frac{d}{dt} \int_M \left(|\nabla u|^2 + ku^2\right) \phi \, d\mu \leq -2 \int_M |\nabla^2 u|^2 \phi \, d\mu$$

for any solution u of the heat equation on (M,g).

Chapter 2

Introduction to Curve Shortening

The curve shortening flow has its origins in material science. It was explicitly described in 1956 by Mullins to model the motion of idealized grain boundaries [**410**], and there is related work by von Neumann in 1952 [**514**]. Renewed interest in the curve shortening flow resulted from the works of Gage and Hamilton in 1986 on convex planar curves [**228**] and Grayson in 1987 on embedded planar curves [**247**]. Later, Grayson [**248**] used curve shortening to prove the theorem of Lusternik and Shnirelmann [**375**] on the existence of at least three simple closed geodesics on a Riemannian 2-sphere. More recently, Sigurd Angenent has used it to prove the existence of closed geodesics on surfaces with certain prescribed flat knot types [**69**] (see also [**65, 67, 231, 232**]). The curve shortening flow even shows up in Perelman's proof of the geometrization conjecture, where it was used to prove that the Ricci flow with surgery terminates after a finite time [**440**]. It continues to be influential for both theory and applications.

Another motivation for the curve shortening flow is that a number of techniques in the field of geometric flows exhibit themselves in the context of curve shortening flow in an elegant and less technical way.

After reviewing the geometric theory of planar curves, we introduce the curve shortening flow and derive equations for associated quantities such as the curvature. After considering graphs and the support function, we discuss smoothing properties of the flow, including short-time and long-time existence theorems.

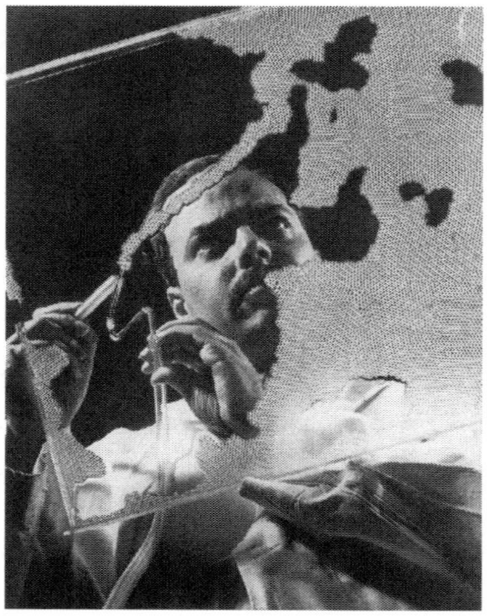

I. INTRODUCTION

IT has been shown by Beck[1] that the grain boundaries of a recrystallized metal, when annealed, migrate toward their centers of curvature. The concomitant reduction in the area of the boundaries, all having positive free energies when referred to an equivalent amount of crystal, provides the driving force for this motion. Thus, it is easily shown[2] that a boundary of mean curvature K and free energy per unit area σ is urged toward its nearest center of curvature with a pressure given by $p=K\sigma$. Such pressures and the motions they produce are now recognized to be responsible for normal grain growth.[3]

Smoluchowski[4] and Turnbull[5] have shown that a pressure p, of the type discussed, produces an unbalance in the fluxes of atoms crossing a boundary. This, in turn, causes the boundary to move in the direction favored by the pressure with a speed S given by $S=pM=K\sigma M$, where $M=Ae^{-Q/kT}$ is the speed per unit pressure or mobility. In the ideal case of a pure metal whose boundaries do not undergo any progressive change in their structure or mode of motion, M and σ remain unchanged during isothermal annealing. Hence, the boundary speed at any point will be proportional to the mean curvature at that point. The observed

Figure 2.1. William W. Mullins and an excerpt from his paper *Two-dimensional motion of idealized grain boundaries*, in which curve shortening flow is developed as a model for the motion of grain boundaries.

Here, as in the rest of the book, we shall, unless otherwise stated, assume that all objects under consideration are smooth, i.e., C^∞. This assumption is made more out of convenience than necessity and we refer the reader to the notes and commentary at the end of this chapter for references to results with weaker regularity hypotheses.

2.1. Basic geometric theory of planar curves

We consider parametrized immersed planar curves, described by smooth maps $X : I \subset \mathbb{R} \to \mathbb{R}^2$ satisfying the "immersion" condition: $|X'(u)| \neq 0$ for every $u \in I$, where $X' = \frac{dX}{du}$. Here I is a (possibly infinite) interval in the real line. In particular we are interested in properties of such curves which are invariant under both reparametrization of the domain $\mathbb{R}$ and rigid motions of the codomain $\mathbb{R}^2$.

2.1.1. Immersed, embedded, and closed curves. A parametrized curve X is called **embedded** if X is a homeomorphism to its image set (that is, X is injective and maps closed sets to closed sets). A particular case which will be important to us is that of closed curves, which we understand for the purposes of this chapter to mean curves for which the immersion X is defined on all of $\mathbb{R}$ and is periodic with some period, so that there exists some $A > 0$ such that $X(x + A) = X(x)$ for all $x \in I$. In this case we say X is embedded if X is injective modulo periodicity, so that $X(x) = X(y)$ implies $y - x = kA$ for some $k \in \mathbb{Z}$. In this case we also say that the curve is a **simple closed curve**.

For a closed curve given by a periodic immersion $X:\ \mathbb{R}\to\mathbb{R}^2$, we can induce a map (which we also call X) from the quotient $\mathbb{R}/(A\mathbb{Z})$ to $\mathbb{R}^2$ by defining $X([x]) = X(x)$ for each x, where $[x] = \{y\in\mathbb{R} :\ y - x\in A\mathbb{Z}\}$ is the equivalence class of x. Then the condition for embeddedness is simply that X is injective on $\mathbb{R}/(A\mathbb{Z})$.

2.1.2. Arc length, tangency, and normalcy. Invariance under reparametrization of the curve can be guaranteed by finding a canonical parametrization. This is provided by the arc length parametrization, which is uniquely defined (up to translations and reversal of direction) by the condition that the derivative of the map X have length 1. It will be convenient to restrict our attention largely to changes of parametrization which do not reverse direction, but we will remark on the effects of reversing direction as we proceed.

The arc length parametrization immediately gives us an important invariant of the embedding: the **unit tangent vector** T, which is simply the derivative of X with respect to an arc length parameter. Note that translation of the arc length parameter does not affect T, while reversal of direction changes T to $-\mathsf{T}$. By convention we will refer to arc length parameters as s and will use primes to denote derivatives with respect to an arbitrary (increasing) parameter. Let $\frac{d}{ds} = \frac{1}{|X'|}\frac{d}{du}$. We see that

$$\mathsf{T} = \frac{dX}{ds} = \frac{X'}{|X'|}. \tag{2.1}$$

Let $J:\ \mathbb{R}^2\to\mathbb{R}^2$ be the counterclockwise rotation through angle $\pi/2$. Then we can define a **unit normal vector** N to the curve in a canonical way by choosing $\mathsf{N} = -J\mathsf{T}$ (our choice of convention here means that a simple closed curve with counterclockwise parametrization has an outward-pointing unit normal). Again, reversal of parametrization reverses the direction of N.

2.1.3. Curvature. Differentiating the unit tangent and normal vectors[1] gives the final invariant we need, the **curvature** κ: Since T has unit length, the derivative of T with respect to an arc length parameter is orthogonal to T and hence is a multiple of N. We define the curvature by this relation and are led to the **Frenet–Serret equations** for a plane curve:

$$\frac{d}{ds}\mathsf{T} = -\kappa\mathsf{N}, \tag{2.2a}$$

$$\frac{d}{ds}\mathsf{N} = \kappa\mathsf{T}. \tag{2.2b}$$

[1]Note that T and N are themselves vector-valued functions from $I\subset\mathbb{R}$ to $\mathbb{R}^2$ and can be differentiated by differentiating each component function.

Since reversal of direction of parametrization reverses s and T and N, it also reverses the sign of κ. However we note that the product $\kappa\mathsf{N}$ is unaffected. This is called the **curvature vector** of the curve. By (2.2a) and (2.2b), we have

$$\kappa = \left\langle \frac{d\mathsf{N}}{ds}, \mathsf{T} \right\rangle = -\left\langle \frac{d\mathsf{T}}{ds}, \mathsf{N} \right\rangle. \tag{2.3}$$

2.1.4. Round circles. The **round circle** of radius r and center p is given as a parametrized curve by the map $X(\theta) = (r\cos\theta, r\sin\theta)$. Since $\frac{\partial X}{\partial\theta} = r(-\sin\theta, \cos\theta)$ has length r, an arc length parameter is given by $s = \theta r$. Accordingly, an arc length parametrization of the curve is given by $X(s) = r\left(\cos\left(\frac{s}{r}\right), \sin\left(\frac{s}{r}\right)\right)$.

If, for example, X parametrizes a round circle of radius r, then choosing N to be the *outward* normal yields $\kappa \equiv \frac{1}{r}$. Indeed, a parametrization of the circle of radius r centered at the origin $X : S^1 = \mathbb{R}/2\pi\mathbb{Z} \to \mathbb{R}^2$ is induced by its lift $\tilde{X} : \mathbb{R} \to \mathbb{R}^2$ defined by

$$\tilde{X}(\theta) = (r\cos\theta, r\sin\theta).$$

The vector field $\frac{d}{d\theta}$ on $\mathbb{R}$ descends to a vector field on S^1 also denoted by $\frac{d}{d\theta}$. Then the "lifted arc length parameter" $\tilde{s} : \mathbb{R} \to \mathbb{R}$ is defined by

$$\tilde{s}(\theta) = r\theta.$$

The vector field $\frac{d}{d\tilde{s}} = \frac{1}{r}\frac{d}{d\theta}$ on $\mathbb{R}$ induces the vector field $\frac{d}{ds} = \frac{1}{r}\frac{d}{d\theta}$ on S^1. The unit tangent vector field of the lifted curve $\tilde{X}$ is

$$\tilde{\mathsf{T}}(\theta) = \frac{1}{r}\frac{d\tilde{X}}{d\theta}(\theta) = (-\sin\theta, \cos\theta).$$

Sacrificing some formal accuracy (i.e., ignoring the fact that θ is not a global coordinate), we simply write $\mathsf{T} : S^1 \to S^1$ as $\mathsf{T}(\theta) = (-\sin\theta, \cos\theta)$. Then

$$\frac{d\mathsf{T}}{ds} = \frac{1}{r}\frac{d}{d\theta}(-\sin\theta, \cos\theta) = -\frac{1}{r}(\cos\theta, \sin\theta).$$

Choosing $\mathsf{N} : S^1 \to S^1$ to be the outward unit normal to X, we have

$$\mathsf{N}(\theta) = (\cos\theta, \sin\theta). \tag{2.4}$$

Thus, for the circle of radius r, we have

$$\kappa = -\left\langle \frac{d\mathsf{T}}{ds}, \mathsf{N} \right\rangle = \frac{1}{r}. \tag{2.5}$$

2.1.5. The normal angle. Now we return to the general case of an immersed planar curve X. Take M^1 to be either an interval I or diffeomorphic to S^1. Let the function $\theta : M^1 \to \mathbb{R}$ be a choice, for each $u \in M^1$, of the angle that the unit normal $\mathsf{N}(u)$ makes with the positive x-axis. We call θ the **normal angle** (a.k.a. **turning angle**) of the curve. Then

$$\mathsf{N}(u) = (\cos\theta(u), \sin\theta(u)). \tag{2.6}$$

Note that, for each $u \in M^1$, the possible choices for the normal angle $\theta(u)$ differ by integer multiples of 2π.

Fix $u_0 \in M^1$ and specify a choice of $\theta(u_0)$. If $M^1 = I$ and if we require the normal angle θ to be smooth (equivalently, continuous), then θ is uniquely determined. Otherwise, if $M^1 \cong S^1$, then we need to remove a point $u_1 \neq u_0$ from M^1 to obtain a uniquely determined smooth $\theta : M^1 - \{u_0\} \to \mathbb{R}$. In any case, globally, $\theta : M^1 \to \mathbb{R}/2\pi\mathbb{Z}$ is well-defined. The differential $d\theta$ is a well-defined 1-form on M^1 independent of the choice of $\theta(u_0)$.

Alternatively, let $p : \widetilde{M}^1 \to M^1$ be the **universal covering** of M^1. Given $\tilde{u}_0 \in \widetilde{M}^1$ and $\tilde{\theta}_0 \in \mathbb{R}$ such that $\mathsf{N}(p(\tilde{u}_0)) = (\cos\tilde{\theta}_0, \sin\tilde{\theta}_0)$, there exists a unique smooth function $\tilde{\theta} : \widetilde{M}^1 \to \mathbb{R}$ such that

$$\mathsf{N}(p(\tilde{u})) = (\cos\tilde{\theta}(\tilde{u}), \sin\tilde{\theta}(\tilde{u})) \quad \text{for } \tilde{u} \in \widetilde{M}^1, \tag{2.7a}$$

$$\tilde{\theta}(\tilde{u}_0) = \tilde{\theta}_0. \tag{2.7b}$$

We leave it to the reader to check the relations between $\tilde{\theta}$ and the θ in the previous paragraph.

For convenience, we shall make the assumption that the immersed curve X is parametrized and the unit normal is chosen so that the frame field $\{\mathsf{N}, \mathsf{T}\}$ is positively oriented with respect to the standard orientation of $\mathbb{R}^2$; that is, it has the same orientation as $\{\frac{\partial}{\partial x}, \frac{\partial}{\partial y}\}$. In this case we say that the curve X is **positively oriented**. Then (2.6) implies

$$\mathsf{T}(u) = (-\sin\theta(u), \cos\theta(u)). \tag{2.8}$$

By (2.3) we have

$$\begin{aligned}\kappa &= \left\langle \frac{d}{ds}(\cos\theta, \sin\theta), (-\sin\theta, \cos\theta) \right\rangle \\ &= \frac{d\theta}{ds}.\end{aligned} \tag{2.9}$$

We shall also use the notation $\theta_s \doteqdot \frac{d\theta}{ds}$, so that $\kappa = \theta_s$. If $\kappa \neq 0$ on all of M^1, then the vector field $\frac{\partial}{\partial\theta}$ is defined on M^1. Otherwise, it is defined at the points where $\kappa \neq 0$.

2.1.6. Arc length element, rotation index. The **arc length element** ds is the smooth 1-form on M^1 dual to the vector field $\frac{d}{ds}$ on M^1; i.e., it is uniquely characterized by $ds(\frac{d}{ds}) = 1$. We have

$$d\theta = \kappa\, ds \tag{2.10}$$

as 1-forms on M^1.

If $M^1 \cong S^1$ and X is a positively oriented immersed curve, then

$$\frac{1}{2\pi}\int_{M^1} \kappa\, ds = \frac{1}{2\pi}\int_{M^1} d\theta \doteqdot w \in \mathbb{Z} \tag{2.11}$$

and w is called the **rotation index**. To see that w is an integer, choose $u_0 \in M^1$ and $\tilde{u}_0 \in \widetilde{M}^1$ with $p(\tilde{u}_0) = u_0$ and choose $\tilde{\theta}_0 \in \mathbb{R}$ such that $\mathsf{N}(u_0) = (\cos\tilde{\theta}_0, \sin\tilde{\theta}_0)$. Let $\tilde{\theta} : \widetilde{M}^1 \to \mathbb{R}$ be as in (2.7a) and (2.7b). Then $p^*(d\theta) = d\tilde{\theta}$. Let $\tilde{u}_1 > \tilde{u}_0$ be the smallest number such that $p([\tilde{u}_0, \tilde{u}_1]) = M^1$. Then $p(\tilde{u}_1) = p(\tilde{u}_0)$ and $\tilde{u}_1 = \tilde{u}_0 + 2\pi w$. The theorem of turning tangents (a.k.a. the Umlaufsatz) states that the rotation index of an embedded closed curve is equal to $+1$ or -1, depending on whether T is pointing counterclockwise or clockwise, respectively. Equivalently,

$$\int_{M^1} \kappa\, ds = 2\pi \tag{2.12}$$

for a simple closed curve parametrized counterclockwise.

2.2. Curve shortening flow

Now let $X : M^1 \times [0,T) \to \mathbb{R}^2$ be a smooth map, which is an immersion for each t (we say in this situation that X is a smooth 1-parameter family of immersions).

We say that $X(t)$ is a solution to the **curve shortening flow** if

$$\frac{\partial X}{\partial t}(u,t) = -\kappa(u,t)\, \mathsf{N}(u,t). \tag{2.13}$$

We shall also use the abbreviation $X_t \doteqdot \frac{\partial X}{\partial t}$. By (2.2a) and $X_s = \mathsf{T}$, the curve shortening flow may be written as a heat-type equation:

$$X_t = X_{ss}. \tag{2.14}$$

Since the arc length derivative $\frac{d}{ds}$ depends on time, X_{ss} does not depend linearly on X. In particular, the curve shortening flow is not a linear equation.

Example 2.1. Given $r_0 > 0$, the shrinking round circle solution to curve shortening flow with the initial circle having radius r_0 is the map $X : S^1 \times [0, r_0/\sqrt{2}) \to \mathbb{R}^2$ defined by

$$X(\theta, t) = (r(t)\cos\theta, r(t)\sin\theta),$$

where $r(t) = \sqrt{r_0^2 - 2t}$. Note that $\kappa(\theta, t) = (r_0^2 - 2t)^{-1/2}$.

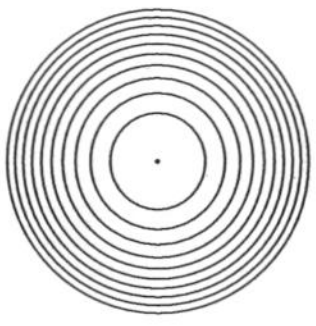

Figure 2.2. The unit circle shrinking to a point; time step $\Delta t = 0.05$.

We wish to compute the evolution of various geometric quantities associated to a curve evolving by curve shortening flow. First we compute a general formula. For a function f, let $f_t \doteqdot \frac{\partial f}{\partial t}$. By Clairaut's theorem on the equality of mixed partial derivatives, for any C^2 function $f : M^1 \times [0, T) \to \mathbb{R}$ we have $\frac{\partial^2 f}{\partial t \partial u} = \frac{\partial^2 f}{\partial u \partial t}$; i.e., $f_{ut} = f_{tu}$ for short. On the other hand, this formula does not hold in general when we replace u by an arc length parameter s since in general s will depend on time. Instead, we have the following.

Lemma 2.2. *Let X be a curve evolving by*

$$\frac{\partial X}{\partial t}(u, t) = -\phi(u, t)\, \mathsf{N}(u, t), \tag{2.15}$$

where $\phi : M^1 \times [0, T) \to \mathbb{R}$. For any C^2 function $f : M^1 \times [0, T) \to \mathbb{R}$ we have

$$f_{st} = f_{ts} + \phi\, \kappa\, f_s.$$

In other words, the commutator of the differential operators $\frac{\partial}{\partial t}$ and $\frac{\partial}{\partial s}$ is

$$\left[\frac{\partial}{\partial t}, \frac{\partial}{\partial s}\right] \doteqdot \frac{\partial^2}{\partial t \partial s} - \frac{\partial^2}{\partial s \partial t} = \phi\, \kappa \frac{\partial}{\partial s}.$$

Proof. Using $\frac{\partial}{\partial s} = \left|\frac{\partial X}{\partial u}\right|^{-1} \frac{\partial}{\partial u}$, we compute that

$$\begin{aligned} f_{st} &= (|X_u|^{-1} f_u)_t \\ &= -|X_u|^{-3} \langle X_{tu}, X_u \rangle f_u + |X_u|^{-1} f_{tu} \\ &= |X_u|^{-3} \langle \phi\, \mathsf{N}_u, X_u \rangle f_u + f_{ts} \\ &= \phi \langle \mathsf{N}_s, T \rangle f_s + f_{ts} \\ &= \phi\, \kappa\, f_s + f_{ts} \end{aligned}$$

since $\langle \mathsf{N}, X_u \rangle = 0$ and $\langle \mathsf{N}_s, \mathsf{T} \rangle = \kappa$. □

Corollary 2.3 (Commutator of $\frac{\partial}{\partial t}$ and $\frac{\partial}{\partial s}$). *Under curve shortening flow,*

$$f_{st} = f_{ts} + \kappa^2 f_s \,. \tag{2.16}$$

The **length** of a curve is

$$\mathrm{L} \doteqdot \int_{M^1} ds \,. \tag{2.17}$$

By (2.53) below (in Exercise 2.3), under (2.15) the length satisfies

$$\frac{d\,\mathrm{L}}{dt} = \int_{M^1} (ds)_t = -\int_{M^1} \kappa\,\phi\, ds \,. \tag{2.18}$$

In particular, under curve shortening flow we have

$$\frac{\partial}{\partial t}(ds) = -\kappa^2 ds \quad \text{and} \quad \frac{d\,\mathrm{L}}{dt} = -\int_{M^1} \kappa^2 ds \,. \tag{2.19}$$

Consider the space $\mathcal{C}$ of smooth immersed plane curves modulo orientation-preserving reparametrizations. Given any smooth 1-parameter family of immersions $X(t) : M^1 \to \mathbb{R}^2$, $M^1 \cong S^1$, we may uniquely project its variation vector field $\partial_t X$ to the normal vector field $(\partial_t X)^\perp = \partial_t X - \langle \partial_t X, \mathsf{T} \rangle\, \mathsf{T}$. Then $(\partial_t X)^\perp$ is the variation vector field of suitable reparametrizations of the $X(t)$. Because of this, the formal tangent space $T_{[X]}\,\mathcal{C}$ at the equivalence class $[X]$ of a curve $X : M^1 \to \mathbb{R}^2$ may be identified with the space of smooth functions $C^\infty(M^1)$, where $f \in C^\infty(M^1)$ corresponds to the normal variation $f\,\mathsf{N} \in T_{[X]}\,\mathcal{C}$.

Define the L^2 **metric** on $T_{[X]}\,\mathcal{C}$ by

$$\langle f_1, f_2 \rangle_{L^2} \doteqdot \int_{M^1} f_1 f_2 \, ds \,.$$

By (2.18), under the variation

$$\partial_t X = f\,\mathsf{N} \,,$$

we have

$$\frac{d\,\mathrm{L}}{dt} = \int_{M^1} \kappa\, f\, ds = \langle \kappa, f \rangle_{L^2} \,. \tag{2.20}$$

In other words, the L^2-gradient of the length functional is given by $\operatorname{grad} \mathrm{L} = \kappa$, which is identified with the vector field $\kappa\,\mathsf{N}$.

Lemma 2.4. *The negative gradient flow of the length functional is given by* $\partial_t X = -\kappa\,\mathsf{N}$*, which is the curve shortening flow.*

The same formula as (2.16) holds for Euclidean vector-valued functions, so that we have

$$X_{st} = X_{ts} + \kappa^2 X_s \,. \tag{2.21}$$

That is, under curve shortening flow, the unit tangent vector field evolves by

$$\mathsf{T}_t = (-\kappa\,\mathsf{N})_s + \kappa^2\mathsf{T} \tag{2.22}$$
$$= -\kappa_s\mathsf{N}$$

since $\mathsf{N}_s = \kappa\mathsf{T}$ by (2.2b). From this and $\frac{d}{dt}\langle\mathsf{N},\mathsf{T}\rangle = 0$ we easily derive that the unit normal evolves by

$$\mathsf{N}_t = \kappa_s\mathsf{T} \tag{2.23}$$

and that the normal angle evolves by

$$\theta_t = \kappa_s\,. \tag{2.24}$$

Lemma 2.5 (Evolution of κ under curve shortening flow). *Under the curve shortening flow, the curvature evolves by the heat-type equation*:

$$\kappa_t = \kappa_{ss} + \kappa^3\,. \tag{2.25}$$

Proof. Using $\theta_s = \kappa$, $\theta_t = \kappa_s$, and (2.16), we compute

$$\kappa_t = \theta_{st} = \theta_{ts} + \kappa^2\theta_s = \kappa_{ss} + \kappa^3.$$ □

2.3. Graphs of functions

In this section we consider solutions to curve shortening flow whose images are the graphs of functions.

Let $f : I \to \mathbb{R}$ be a smooth function defined on an interval. The graph of the function $y = f(x)$ is parametrized by the map $X : I \to \mathbb{R}^2$ defined by $X(x) \doteqdot (x, y)$, where $y = f(x)$. We have $X_x(x) = (1, y_x)$, so that the unit tangent vector field to X is

$$\mathsf{T} = \frac{(1, y_x)}{\left(1 + (y_x)^2\right)^{1/2}},$$

where I has the standard orientation with $\frac{d}{dx}$ being positively oriented. The unit normal N of X, with $\{\mathsf{N},\mathsf{T}\}$ being positively oriented, is

$$\mathsf{N} = \frac{(y_x, -1)}{\left(1 + (y_x)^2\right)^{1/2}}. \tag{2.26}$$

The curvature of X is

$$\kappa = \frac{\langle\mathsf{N}_x, \mathsf{T}\rangle}{\left(1 + (y_x)^2\right)^{1/2}} \tag{2.27}$$
$$= \frac{y_{xx}}{\left(1 + (y_x)^2\right)^{3/2}}.$$

By (2.26), the normal angle satisfies

$$\tan\theta(x) = -\frac{1}{y_x(x)}. \tag{2.28}$$

Example 2.6. If $y = -\sqrt{r^2 - x^2}$ for $-r < x < r$ (the lower hemicircle of radius r), then

$$\begin{aligned} y_x &= x\left(r^2 - x^2\right)^{-1/2}, \\ y_{xx} &= r^2\left(r^2 - x^2\right)^{-3/2}, \\ 1 + (y_x)^2 &= r^2\left(r^2 - x^2\right)^{-1}, \end{aligned}$$

so that $\kappa \equiv r^{-1}$, agreeing with (2.5).

Now we can formulate the equation for a 1-parameter family of graphs to be (the reparametrization of) a solution to curve shortening flow. Suppose that curves $\mathcal{M}_t \subset \mathbb{R}^2$ are the graphs of functions $y = f(x,t)$. Then

$$X(x,t) \doteqdot (x, f(x,t))$$

are parametrizations of $\mathcal{M}_t$. We have that X is a reparametrization of a solution to curve shortening flow if and only if

$$\frac{\partial X}{\partial t} = -\kappa\,\mathsf{N} + \phi\frac{\partial X}{\partial x} \tag{2.29}$$

for some function ϕ. Here, $\phi\frac{\partial X}{\partial x}$ represents an arbitrary time-dependent tangent vector field. Now $\frac{\partial X}{\partial t} = (0, y_t)$, $\mathsf{N} = \frac{(y_x,-1)}{\left(1+(y_x)^2\right)^{1/2}}$, and $\frac{\partial X}{\partial x} = (1, y_x)$, so that (2.29) becomes

$$(0, y_t) = \kappa\frac{(-y_x, 1)}{\left(1 + (y_x)^2\right)^{1/2}} + \phi(1, y_x).$$

The equality of the first components implies that

$$\phi = \frac{\kappa\, y_x}{\left(1 + (y_x)^2\right)^{1/2}},$$

so that the equality of the second components yields that the reparametrized curve shortening flow is equivalent to

$$\begin{aligned} y_t &= \frac{\kappa}{\left(1 + (y_x)^2\right)^{1/2}} + \phi y_x \\ &= \kappa\left(1 + (y_x)^2\right)^{1/2} \\ &= \frac{y_{xx}}{1 + (y_x)^2}. \end{aligned}$$

Lemma 2.7 (Curve shortening flow for graphs). *A 1-parameter family of graphs $y = f(x,t)$ solves the reparametrized curve shortening flow if and only if*

$$y_t = \frac{y_{xx}}{1 + (y_x)^2}. \tag{2.30}$$

2.3.1. Grim Reaper solution. An important special solution of curve shortening flow is the **Grim Reaper** solution, defined implicitly by

$$y = t + \log \sec x, \tag{2.31}$$

for $x \in (-\pi/2, \pi/2)$ and for all $t \in \mathbb{R}$. Note that this solution translates in the y-direction at constant (unit) speed. Using

$$y_t = 1, \quad y_x = \tan x, \quad y_{xx} = \sec^2 x,$$

we easily verify that (2.30) holds; that is, the family of curves defined by (2.31) indeed satisfies curve shortening flow. Since this solution is defined for all time $t \in \mathbb{R}$, we call it an **eternal solution**. It earns its name since, like the swing of Death's scythe, it "kills" every solution in the strip $\left(-\frac{\pi}{2}, \frac{\pi}{2}\right) \times \mathbb{R}$, in accordance with the avoidance principle (see Section 3.1).

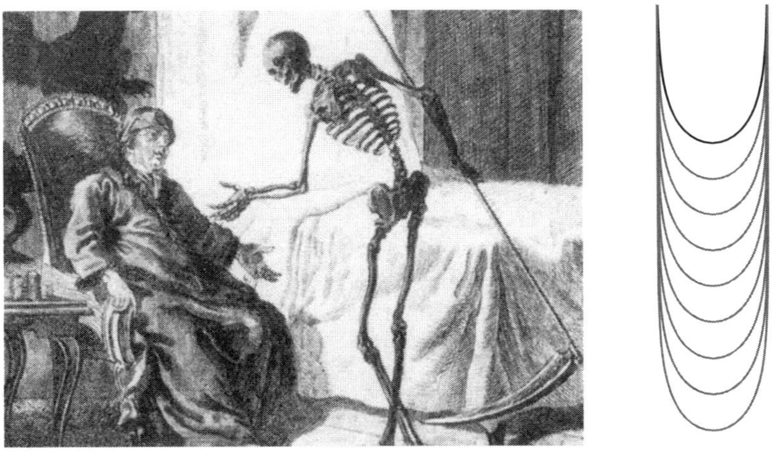

Figure 2.3. L: The Grim Reaper collects a soul. R: The Grim Reaper curve translating vertically under curve shortening flow.

If we let θ denote the angle that the unit tangent vector T makes with the positive x-axis (which modulo 2π is the normal angle plus $\pi/2$), then we have

$$\kappa = \cos \theta. \tag{2.32}$$

Indeed, from (2.27) we compute that

$$\begin{aligned}\kappa &= \frac{\sec^2 x}{\left(1+\tan^2 x\right)^{3/2}} \\ &= \cos x\,;\end{aligned} \tag{2.33}$$

moreover, $y_x = \tan x$ implies that $x = \theta$. Note that κ in (2.32) satisfies the ODE

$$\kappa_{\theta\theta} + \kappa = 0\,. \tag{2.34}$$

2.4. The support function

Let $X : M^1 \to \mathbb{R}^2$ be an immersed planar curve. The **support function** $\sigma : M^1 \to \mathbb{R}$ is defined by

$$\sigma\left(u\right) \doteqdot \left\langle X\left(u\right), \mathsf{N}\left(u\right)\right\rangle . \tag{2.35}$$

Geometrically, $\sigma\left(u\right)$ is equal to the signed distance to the origin of the tangent line passing through $X\left(u\right)$.

Let M^1 be closed, i.e., a circle. We have that

$$\begin{aligned}\int_{M^1} \sigma\,\kappa\,ds &= \int_{M^1} \left\langle X, \kappa\mathsf{N}\right\rangle ds \\ &= -\int_{M^1} \left\langle X, X_{ss}\right\rangle ds \\ &= \int_{M^1} \left|X_s\right|^2 ds \\ &= \mathrm{L}\end{aligned} \tag{2.36}$$

is the length, where we integrated by parts to obtain the third equality.

2.4.1. Evolution of the area.

2.4.1. Evolution of the area. Now assume that X is a counterclockwise parametrization of an embedded closed curve $\mathcal{M} = X(M^1)$ and let A denote the **area** enclosed by $\mathcal{M}$. With $X(s) \doteqdot (x(s), y(s))$, we have $\mathsf{T} = (x_s, y_s)$ and $\mathsf{N} = (y_s, -x_s)$. By Green's theorem, we have

$$\begin{aligned}\mathrm{A} &= \frac{1}{2}\int_{\mathcal{M}} \left(x\,dy - y\,dx\right) \\ &= \frac{1}{2}\int_{M^1} \left(x, y\right)\cdot\left(y_s, -x_s\right) ds \\ &= \frac{1}{2}\int_{M^1} \left\langle X, \mathsf{N}\right\rangle ds,\end{aligned} \tag{2.37}$$

where $\cdot$ denotes the Euclidean inner product. Hence the evolution of the area is given by

$$\begin{aligned}\frac{d\,\mathrm{A}}{dt} &= \frac{1}{2}\int_{M^1} \left(\left(\langle X_t, \mathsf{N}\rangle + \langle X, \mathsf{N}_t\rangle\right) ds + \langle X, \mathsf{N}\rangle\,(ds)_t\right) \\ &= \frac{1}{2}\int_{M^1} \left(-\kappa + \kappa_s \langle X, \mathsf{T}\rangle - \langle X, \mathsf{N}\rangle\,\kappa^2\right) ds\end{aligned}$$

since $\mathsf{N}_t = \kappa_s \mathsf{T}$. Integrating by parts yields

$$\begin{aligned}\int_{M^1} \kappa_s \langle X, \mathsf{T}\rangle\, ds &= -\int_{M^1} \kappa\left(\langle X_s, \mathsf{T}\rangle + \langle X, \mathsf{T}_s\rangle\right) ds \\ &= -\int_{M^1} \kappa\left(1 - \kappa\langle X, \mathsf{N}\rangle\right) ds\end{aligned}$$

since $\mathsf{T}_s = -\kappa\,\mathsf{N}$. Therefore, recalling (2.11),

Lemma 2.8 (Area evolution). *For an embedded solution $X : M^1 \times (0, T) \to \mathbb{R}^2$, $M^1 \cong S^1$, to the curve shortening flow, the area $\mathrm{A}(t)$ enclosed by $\mathcal{M}$ evolves according to*

$$\frac{d\,\mathrm{A}}{dt} = -\int_{M^1} \kappa\, ds = -2\pi\,. \tag{2.38}$$

2.4.2. Characterization for convex curves to be closed. Now assume that M^1 is a finite closed interval and the curve $X : M^1 \to \mathbb{R}^2$ satisfies $\kappa > 0$ with respect to some smooth choice of N; i.e., X is **convex**. Assume also that X is parametrized by a smooth choice of angle $\theta \in [a, b]$. Recall that $\mathsf{T}(\theta) = (-\sin\theta, \cos\theta)$. Since $\frac{dX}{d\theta} = \frac{1}{\kappa}\frac{dX}{ds} = \frac{1}{\kappa}\mathsf{T}$, we have for any $\theta_0, \theta \in [a, b]$,

$$\begin{aligned}X(\theta) &= X(\theta_0) + \int_{\theta_0}^{\theta} \frac{dX}{d\theta} d\theta \\ &= X(\theta_0) + \int_{\theta_0}^{\theta} \frac{(-\sin\theta, \cos\theta)}{\kappa(\theta)} d\theta\,.\end{aligned}$$

In other words, writing $X \doteqdot (x, y)$, we have

$$x(\theta) = x(\theta_0) - \int_{\theta_0}^{\theta} \frac{\sin\theta}{\kappa(\theta)} d\theta, \tag{2.39a}$$

$$y(\theta) = y(\theta_0) + \int_{\theta_0}^{\theta} \frac{\cos\theta}{\kappa(\theta)} d\theta\,. \tag{2.39b}$$

Thus a convex curve $X : [a, b] \to \mathbb{R}^2$ is closed, i.e., $X(b) = X(a)$, if and only if

$$\int_a^b \frac{\sin\theta}{\kappa(\theta)} d\theta = 0, \qquad \int_a^b \frac{\cos\theta}{\kappa(\theta)} d\theta = 0. \tag{2.40}$$

We shall now show that a smooth positive function $k : S^1 \to \mathbb{R}$ is the curvature of a smooth embedded closed convex curve X if and only if (2.40) is satisfied:

$$\int_{S^1} \frac{\sin\theta}{k(\theta)} d\theta = 0, \qquad \int_{S^1} \frac{\cos\theta}{k(\theta)} d\theta = 0. \tag{2.41}$$

Indeed, if X is a smooth embedded closed convex curve, then by (2.39a) and (2.39b) and $X(2\pi) = X(0)$, we have (2.41) with $k(\theta) = \kappa(\theta)$. Conversely, if a smooth positive function $k : S^1 \to \mathbb{R}$ satisfies (2.41), then we define $\tilde{X} = (\tilde{x}, \tilde{y}) : \mathbb{R} \to \mathbb{R}^2$ by

$$\begin{aligned} \tilde{x}(\theta) &= -\int_0^\theta \frac{\sin\theta}{k(\theta)} d\theta, \\ \tilde{y}(\theta) &= \int_0^\theta \frac{\cos\theta}{k(\theta)} d\theta \end{aligned}$$

for $\theta \in \mathbb{R}$. Then $\tilde{x}(2\pi + \theta) = \tilde{x}(\theta)$ and $\tilde{y}(2\pi + \theta) = \tilde{y}(\theta)$, so that $\tilde{X}$ descends to a map $X = (x, y) : S^1 \to \mathbb{R}^2$ with

$$\begin{aligned} \frac{dx}{d\theta}(\theta) &= -\frac{\sin\theta}{k(\theta)}, \\ \frac{dy}{d\theta}(\theta) &= \frac{\cos\theta}{k(\theta)}. \end{aligned}$$

That is,

$$\frac{1}{\kappa(\theta)} \mathsf{T} = \frac{dX}{d\theta}(\theta) = \frac{1}{k(\theta)} \mathsf{T}.$$

Hence $\kappa(\theta) = k(\theta)$.

2.5. Short-time existence

Our analysis of the curve shortening flow requires the following short-time existence theorem:

Theorem 2.9 (Short-time existence for curve shortening flow). *Let M^1 be a closed 1-manifold and let $X_0 : M^1 \to \mathbb{R}^2$ be a smooth immersion. Then there exist $\varepsilon > 0$ and a smooth solution $X(t) : M^1 \to \mathbb{R}^2$ to the curve shortening flow defined for $t \in [0, \varepsilon)$ and satisfying $X(0) = X_0$.*

We will not prove this result here. The more general result for the mean curvature flow is proved in Chapter 6 (see Proposition 6.8). For now, we mention only some of the issues which arise: The equation (2.13) is not strictly parabolic since in a general parametrization it becomes

$$\frac{\partial X}{\partial t} = \frac{1}{|X'|^2} \pi_{\mathsf{N}}(X''),$$

where $X' = \frac{dX}{du}$, $X'' = \frac{d^2X}{du^2}$, and $\pi_{\mathsf{N}}(v) = (v \cdot \mathsf{N})\mathsf{N}$ is the orthogonal projection onto the subspace generated by the unit normal N. Since there is no dependence on the component of X'' tangent to the curve, the symbol is degenerate and the flow is not parabolic. This means that standard existence methods for partial differential equations cannot be applied. As we shall see in Section 6.4, the degeneracy is a direct consequence of the invariance of the flow under reparametrization. There are at least three approaches to overcoming this degeneracy: The first is to write the evolving curves as graphs over some fixed curve (such as the initial curve, or a nearby smooth curve if the initial curve is not sufficiently regular). This produces a strictly parabolic evolution equation for a scalar function. One can then produce from the graphical solution a solution of the original curve shortening flow. The second approach is to apply the Nash–Moser inverse function theorem. For this approach see Section 2 of Gage and Hamilton [**228**]. The third approach is DeTurck's trick, which is employed in the proof of Proposition 6.8.

2.6. Smoothing

One of the most important properties of geometric heat equations (and parabolic equations more generally) is **smoothing**: Even for highly irregular initial data, solutions become smooth (infinitely many times differentiable) at any positive time. In this section we will show how to prove this (assuming only a bound on the curvature) for curve shortening flow using maximum principle estimates. A very similar procedure can be applied to other flows, such as the higher-dimensional mean curvature flow and the Ricci flow.

We start with the first derivative of curvature estimate. The method of proof, following the **Bernstein technique**, is to apply the maximum principle to a suitable first derivative quantity.

Lemma 2.10. *Suppose that we have a solution to curve shortening flow on a closed curve with*

$$|\kappa| \leq K \quad \text{on } M^1 \times [0, t_0), \tag{2.42}$$

where $t_0 \in (0, K^{-2}]$ and $K < \infty$. Then on $M^1 \times (0, t_0)$

$$|\kappa_s| \leq \frac{4K}{\sqrt{t}}. \tag{2.43}$$

Proof. First note that from $\kappa_t = \kappa_{ss} + \kappa^3$ we may derive

$$\begin{aligned} \left(\kappa^2\right)_t &= 2\kappa\left(\kappa_{ss} + \kappa^3\right) \\ &= \left(\kappa^2\right)_{ss} - 2\left(\kappa_s\right)^2 + 2\kappa^4. \end{aligned} \tag{2.44}$$

Next, differentiating (2.25) with respect to s while using the commutator formula (2.16) yields

$$(2.45)\qquad \begin{aligned}(\kappa_s)_t &= \kappa_{ts} + \kappa^2\kappa_s \\ &= (\kappa_s)_{ss} + 4\kappa^2\kappa_s .\end{aligned}$$

Thus

$$(2.46)\qquad \begin{aligned}((\kappa_s)^2)_t &= 2\kappa_s\left((\kappa_s)_{ss} + 4\kappa^2\kappa_s\right) \\ &= ((\kappa_s)^2)_{ss} - 2(\kappa_{ss})^2 + 8\kappa^2(\kappa_s)^2 .\end{aligned}$$

From (2.44) and (2.46) we compute for any constant $b > 0$ that

$$(t(\kappa_s)^2 + b\kappa^2)_t - (t(\kappa_s)^2 + b\kappa^2)_{ss} = -2t(\kappa_{ss})^2 + \left(8t\kappa^2 + 1 - 2b\right)(\kappa_s)^2 + 2b\kappa^4.$$

Choosing $b = 5$ and by (2.42) and $t_0 \leq K^{-2}$, we obtain the heat-type inequality

$$(2.47)\qquad (t(\kappa_s)^2 + 5\kappa^2)_t - (t(\kappa_s)^2 + 5\kappa^2)_{ss} \leq -2t(\kappa_{ss})^2 + 10K^4$$

on $M^1 \times [0, t_0)$. At $t = 0$ we have $t(\kappa_s)^2 + 5\kappa^2 \leq 5K^2$. Thus, by the maximum principle we conclude that

$$t(\kappa_s)^2 + 5\kappa^2 \leq 15K^2$$

on $M^1 \times [0, t_0)$, which implies (2.43). □

Next we consider the higher derivative estimates for curve shortening flow. Given a function f on $M^1 \times [0, t_0)$, let $f^{(m)} \doteqdot \frac{\partial^m f}{\partial s^m}$ for $m \geq 0$. Define

$$Z_m \doteqdot \kappa_t^{(m)} - \kappa_{ss}^{(m)}.$$

Note that $Z_0 = \kappa^3$. Then

$$(2.48)\qquad \begin{aligned}Z_{m+1} &= \kappa_t^{(m+1)} - \kappa_{ss}^{(m+1)} \\ &= \kappa_{st}^{(m)} - \kappa_{ss}^{(m+1)} \\ &= \kappa_{ts}^{(m)} - \kappa_{ss}^{(m+1)} + \kappa^2\kappa_s^{(m)} \\ &= (Z_m)_s + \kappa^2\kappa^{(m+1)}.\end{aligned}$$

Using (2.48), we calculate for example

$$\begin{aligned}Z_1 &= 4\kappa^2\kappa^{(1)}, \\ Z_2 &= 5\kappa^2\kappa^{(2)} + 8\kappa(\kappa^{(1)})^2, \\ Z_3 &= 6\kappa^2\kappa^{(3)} + 26\kappa\kappa^{(1)}\kappa^{(2)} + 8(\kappa^{(1)})^3.\end{aligned}$$

We claim that (see Exercise 2.9)

$$(2.49)\qquad Z_m = (m+3)\,\kappa^2\kappa^{(m)} + \sum c_{i_1,i_2,i_3}\kappa^{(i_1)}\kappa^{(i_2)}\kappa^{(i_3)}$$

for each $m \geq 1$, where c_{i_1,i_2,i_3} are nonnegative integers depending only on i_1, i_2, i_3 and where the sum is over $0 \leq i_1 \leq i_2 \leq i_3 \leq m-1$ with $i_1+i_2+i_3 = m$.

In the rest of this section constants depend only on their indices.

Theorem 2.11 (Derivative estimates for curve shortening flow). *For each $m \geq 1$ there exists $C_m < \infty$ such that if $|\kappa| \leq K$ on $M^1 \times [0, t_0)$, where $t_0 \in (0, K^{-2}]$ and $K < \infty$, then*

$$|\kappa^{(m)}| \leq \frac{C_m K}{t^{m/2}}$$

on $M^1 \times (0, t_0)$.

Proof. Consider the function

$$\Phi_m \doteqdot \sum_{i=0}^{m} a_i t^i (\kappa^{(i)})^2$$

for some to-be-determined constants $\{a_m\}_{m=0}^{\infty}$. Observe that

$$\Phi_{m+1} = a_{m+1} t^{m+1} (\kappa^{(m+1)})^2 + \Phi_m$$

and hence

$$\begin{aligned}(\Phi_{m+1})_t - (\Phi_{m+1})_{ss} \\ &= a_{m+1} t^m \Big((m+1)(\kappa^{(m+1)})^2 + t\big[(\kappa^{(m+1)})_t^2 - (\kappa^{(m+1)})_{ss}^2 \big] \Big) \\ &\quad + (\Phi_m)_t - (\Phi_m)_{ss}\end{aligned}$$

for each $m \geq 0$.

We claim that there exist positive constants $\{a_m\}_{i=0}^{\infty}$ and $\{A_m\}_{m=1}^{\infty}$ depending only on $m \in \mathbb{N}$ such that

$$(\Phi_m)_t - (\Phi_m)_{ss} \leq -2 a_m t^m (\kappa^{(m+1)})^2 + A_m K^4 \tag{2.50}$$

on $M^1 \times [0, t_0)$ for each $m \in \mathbb{N}$.

First set $a_0 = 5$ and $a_1 = 1$, so that $\Phi_1 = t\,(\kappa_s)^2 + 5\kappa^2$. Then, by (2.47), Φ_1 satisfies (2.50) with $A_1 = 10$. So suppose that

$$(\Phi_i)_t - (\Phi_i)_{ss} \leq -2 a_i t^i (\kappa^{(i+1)})^2 + A_i K^4$$

for each $i = 0, \ldots, m$. The ODE comparison principle then yields

$$t^{i/2} \left| \kappa^{(i)} \right| \leq C_i K$$

on $M^1 \times [0, t_0)$ for each $i = 1, \ldots, m$. In particular, by (2.49),

$$\begin{aligned}(\kappa^{(m+1)})^2_t - (\kappa^{(m+1)})^2_{ss} &= 2(m+4)\kappa^2(\kappa^{(m+1)})^2 - 2(\kappa^{(m+2)})^2 \\ &\quad + 2\kappa^{(m+1)} \sum_{\substack{0 \le i_1 \le i_2 \le i_3 \le m, \\ i_1+i_2+i_3=m+1}} c_{i_1,i_2,i_3} \kappa^{(i_1)} \kappa^{(i_2)} \kappa^{(i_3)} \\ &\le 2(m+4)\kappa^2(\kappa^{(m+1)})^2 - 2(\kappa^{(m+2)})^2 \\ &\quad + 2B_m K^3 t^{-\frac{m+1}{2}} |\kappa^{(m+1)}| \\ &\le 2(m+4)\kappa^2(\kappa^{(m+1)})^2 - 2(\kappa^{(m+2)})^2 \\ &\quad + B_m^2 K^6 t^{-m} + t^{-1}(\kappa^{(m+1)})^2,\end{aligned}$$

where $B_m \doteqdot \sum_{i_1+i_2+i_3=m} c_{i_1,i_2,i_3} C_{i_1} C_{i_2} C_{i_3}$ and we used the Cauchy–Schwarz inequality.

Thus,

$$\begin{aligned}(\Phi_{m+1})_t - (\Phi_{m+1})_{ss} &\le \left(\left(m + 2 + 2\,(m+4)\,t\kappa^2\right) a_{m+1} - 2a_m\right) t^m (\kappa^{(m+1)})^2 \\ &\quad + \left(B_m^2 K^2 t + A_m\right) K^4 - 2a_{m+1} t^{m+1} (\kappa^{(m+2)})^2 .\end{aligned}$$

Choosing $a_{m+1} = \frac{2a_m}{3(m+4)}$ therefore yields

$$(\Phi_{m+1})_t - (\Phi_{m+1})_{ss} \le \; - 2a_{m+1} t^{m+1} (\kappa^{(m+2)})^2 + A_{m+1} K^4,$$

where $A_{m+1} = B_m^2 a_{m+1} + A_m$. That is, (2.50) holds with m replaced by $m+1$. Since $\Phi_{m+1} \le a_0 K^2$ at $t = 0$, the ODE comparison principle yields

$$a_{m+1} t^{m+1} (\kappa^{(m+1)})^2 \le \Phi_{m+1} \le a_0 K^2 + A_{m+1} K^4 t \le (a_0 + A_{m+1})\, K^2,$$

so that

$$t^{(m+1)/2} \left|\kappa^{(m+1)}\right| \le C_{m+1} K$$

on $M^1 \times [0, t_0)$, where $C_{m+1} = \sqrt{\frac{a_0 + A_{m+1}}{a_{m+1}}}$. □

The derivative estimates we proved are "coarse" in the sense that the optimality of the constants is not essential for applications and that one can adjust to some extent the constants in the quantities we considered.

2.7. Global existence

The following result, essentially due to Gage and Hamilton [**228**], is the basic **long-time existence theorem** for curve shortening flow.

It is a special case of the corresponding result for mean curvature flow.

Figure 2.4. Michael Gage. Photo courtesy of Sandra Cherin.

Theorem 2.12. *Let $X : M^1 \times [0,T) \to \mathbb{R}^2$ be a smooth solution of curve shortening flow. If T is the maximal time of existence, then*

$$\limsup_{t\to T} \sup_{M^1\times\{t\}} |\kappa| = \infty.$$

Proof. We will prove the contrapositive: If $\sup_{M^1\times[0,T)} |\kappa| \doteqdot K < \infty$, then there exists $\delta > 0$ such that X extends smoothly to $M^1 \times [0, T + \delta)$.

To achieve this, we convert the curvature derivative bounds of Theorem 2.11 into bounds for the parametrization map X. We may then apply the Arzelà–Ascoli theorem to obtain a smooth limit immersion as $t \to T$ and thereby extend the solution via the short-time existence theorem (Theorem 2.9). We begin with a very simple supremum bound:

$$|X(x,t) - X(x,0)| \le \int_0^t \left|\frac{\partial X}{\partial t}(x,\tau)\right| d\tau \le K|t| \le KT. \tag{2.51}$$

Note that the derivatives of the map X are not invariant under reparametrization, so we fix a parameter x (for convenience, the arc length parameter for the curve when $t = 0$) and control all of the derivatives of X with respect to this fixed parametrization.

A crucial step is to bound the first derivatives of X both above and below (the lower bound ensures that the limiting map is an immersion).

Lemma 2.13. *For any $x \in M^1$ and $t \in [0, T)$,*

$$\exp\{-K^2T\} \le |X'(x,t)| \le 1.$$

Proof. We compute

$$\begin{aligned}\frac{\partial}{\partial t}|X'| &= \frac{X'}{|X'|}\cdot\left(\frac{\partial X}{\partial t}\right)' \\ &= \mathsf{T}\cdot|X'|\frac{\partial}{\partial s}(-\kappa\mathsf{N}) \\ &= |X'|\mathsf{T}\cdot\left(-\kappa_s\mathsf{N}-\kappa^2\mathsf{T}\right) \\ &= -\kappa^2|X'|,\end{aligned} \tag{2.52}$$

where we used equation (2.2b) and the relation $\frac{\partial}{\partial s}(\cdot) = \frac{1}{|X'|}(\cdot)'$. This gives $0 \geq \frac{\partial}{\partial t}\log|X'| \geq -K^2$, and by integration

$$0 \geq \log|X'(x,t)| - \log|X'(x,0)| \geq -K^2 t \geq -K^2 T.$$

The result follows since by our choice of coordinate we have $|X'(x,0)| = 1$ for all x. □

Next, we bound the higher derivatives of X. In the following we denote by $X^{(k)}$ the k-th derivative of X with respect to the fixed spatial parameter, so that $f^{(0)} = f$ and $\left(f^{(k)}\right)' = f^{(k+1)}$ for any function f. To establish bounds on $X^{(k)}$ we first derive an evolution equation. This would be very messy to write down precisely, but all we need is a general form:

Lemma 2.14. *For each $k \geq 1$,*

$$\mathsf{N}\cdot X^{(k+1)} = \sum_{\substack{p_1+2p_2+\cdots+kp_k=k+1,\\ 1+q_0+2q_1+\cdots+kq_{k-1}=p_1+\cdots+p_k}} A^{p_1,\ldots,p_k}_{q_0,\ldots,q_{k-1}} \prod_{i=1}^{k}\left(\mathsf{T}\cdot X^{(i)}\right)^{p_i}\prod_{j=0}^{k-1}\left(\partial_s^j\kappa\right)^{q_j}.$$

Under curve shortening flow,

$$\begin{aligned}\frac{\partial X^{(k)}}{\partial t} = &\sum_{\substack{p_1+2p_2+\cdots+kp_k=k,\\ q_0+2q_1+\cdots+(k+1)q_k=1+p_1+\cdots+p_{k+1}}} C^{p_1,\ldots,p_k}_{q_0,\ldots,q_k} \prod_{i=1}^{k}\left(\mathsf{T}\cdot X^{(i)}\right)^{p_i}\prod_{j=0}^{k}\left(\partial_s^j\kappa\right)^{q_j}\mathsf{N} \\ &+ \sum_{\substack{p_1+2p_2+\cdots+kp_k=k,\\ q_0+2q_1+\cdots+(k+1)q_k=1+p_1+\cdots+p_{k+1}}} D^{p_1,\ldots,p_k}_{q_0,\ldots,q_k} \prod_{i=1}^{k}\left(\mathsf{T}\cdot X^{(i)}\right)^{p_i}\prod_{j=0}^{k}\left(\partial_s^j\kappa\right)^{q_j}\mathsf{T},\end{aligned}$$

where $A^{p_0,\ldots,p_k}_{q_0,\ldots,q_{k-1}}$, $C^{p_0,\ldots,p_k}_{q_0,\ldots,q_k}$, and $D^{p_0,\ldots,p_k}_{q_0,\ldots,q_k}$ are constants.

Remark 2.15. These expressions, while convenient for the proof, look rather ugly. The reader will find it instructive to write out the meaning of the expressions in the first few cases. For example, when $k = 1$ the only possibility in the expression for $\mathsf{N}\cdot X''$ is $p_1 = 2$ and $q_0 = 1$, yielding $\mathsf{N}\cdot X''=A_1^2|X'|^2\kappa$ (in fact by the definition of curvature we have $\mathsf{N}\cdot X''=-|X'|^2\kappa$). For small values of k the expressions become

$$\begin{aligned}\mathsf{N}\cdot X''' &= A^{3,0}_{2,0}|X'|^3\kappa^2+A^{3,0}_{0,1}|X'|^3\kappa_s+A^{1,1}_{1,0}|X'|\mathsf{T}\cdot X''\kappa,\\ \mathsf{N}\cdot X^{(4)} &= A^{4,0,0}_{3,0,0}|X'|^4\kappa^3+A^{4,0,0}_{1,1,0}|X'|^4\kappa\kappa_s+A^{4,0,0}_{0,0,1}|X'|^4\kappa_{ss}\\ &\quad+A^{2,1,0}_{2,0,0}|X'|^2\mathsf{T}\cdot X''\kappa^2+A^{2,1,0}_{0,1,0}|X'|^2\mathsf{T}\cdot X''\kappa_s\\ &\quad+A^{1,0,1}_{1,0,0}|X'|\mathsf{T}\cdot X'''\kappa+A^{0,2,0}_{1,0,0}(\mathsf{T}\cdot X'')^2\kappa.\end{aligned}$$

One can compute these exactly, obtaining

$$\mathsf{N}\cdot X''' = -3|X'|\mathsf{T}\cdot X''\kappa-|X'|^3\kappa_s$$

and

$$\mathsf{N}\cdot X^{(4)}=-|X'|^4(3\kappa^3+\kappa_{ss})-6|X'|^2\mathsf{T}\cdot X''\kappa_s-3(\mathsf{T}\cdot X'')^2\kappa-4|X'|\mathsf{T}\cdot X'''\kappa.$$

Similarly in the evolution equations we get the expressions

$$\frac{\partial X'}{\partial t}=(C^1_{2,0}|X'|\kappa^2+C^1_{0,1}|X'|\kappa_s)\mathsf{N}+(D^1_{2,0}|X'|\kappa^2+D^1_{0,1}|X'|\kappa_s)\mathsf{T}$$

and

$$\begin{aligned}\frac{\partial X''}{\partial t}&=|X'|^2\left(C^{2,0}_{3,0,0}\kappa^3+C^{2,0}_{1,1,0}\kappa\kappa_s+C^{2,0}_{0,0,1}\kappa_{ss}\right)\mathsf{N}\\ &\quad+\mathsf{T}\cdot X''\left(C^{0,1}_{2,0,0}\kappa^2+C^{0,1}_{0,1,0}\kappa_s\right)\mathsf{N}\\ &\quad+|X'|^2\left(D^{2,0}_{3,0,0}\kappa^3+D^{2,0}_{1,1,0}\kappa\kappa_s+D^{2,0}_{0,0,1}\kappa_{ss}\right)\mathsf{T}\\ &\quad+\mathsf{T}\cdot X''\left(D^{0,1}_{2,0,0}\kappa^2+D^{0,1}_{0,1,0}\kappa_s\right)\mathsf{T}.\end{aligned}$$

Computing these exactly we find

$$\frac{\partial X'}{\partial t}=-|X'|\kappa_s\mathsf{N}-|X'|\kappa^2\mathsf{T}$$

and

$$\frac{\partial X''}{\partial t}=\left(|X'|^2\kappa^3-|X'|^2\kappa_{ss}-\mathsf{T}\cdot X''\kappa_s\right)\mathsf{N}-\left(3|X'|^2\kappa\kappa_s+\mathsf{T}\cdot X''\kappa^2\right)\mathsf{T}.$$

We see from these that the expressions in the lemma are rather coarse as not all of the allowed terms arise, but they are sufficient for our purposes. In fact, inspecting the proof below we could have used a much weaker form: $\frac{\partial X^{(k)}}{\partial t}$ has the form $P\cdot X^{(k)}+Q$, where $P=-\kappa_s\mathsf{N}-\kappa^2\mathsf{T}$ and Q is a polynomial in $\mathsf{T}\cdot X^{(j)}$ for $j<k$ and $\partial_s^j\kappa$ for $j\le k$.

Proof. The proof is a straightforward induction on k: In the case $k=1$ both expressions are clearly true (we verified these in the remark above). Now suppose we have established both expressions for $k=1,\dots,m$. Then

$$\mathsf{N}\cdot X^{(m+2)}=(\mathsf{N}\cdot X^{(m+1)})'-\mathsf{N}'\cdot X^{(m+1)}=(\mathsf{N}\cdot X^{(m+1)})'-|X'|\kappa\mathsf{T}\cdot X^{(m+1)},$$

where we used the relation $f' = |X'| f_s$ for any function f and the Frenet–Serret equation. In the first term we substitute the expression for $\mathsf{N} \cdot X^{(m+1)}$ coming from the inductive hypothesis. Differentiating a factor $\mathsf{T} \cdot X^{(i)}$ gives $\mathsf{T} \cdot X^{(i+1)} - |X'|\kappa \mathsf{N} \cdot X^{(i)}$. In the second of these terms we substitute the expression for $\mathsf{N} \cdot X^{(i)}$ coming from the inductive hypothesis. We leave it to the reader to check that all of the resulting terms are of the form allowed in the claimed expression for $\mathsf{N} \cdot X^{(m+2)}$. Similarly, differentiating a term $\partial_s^j \kappa$ gives $|X'| \partial_s^{j+1} \kappa$, which again is of the form allowed, and we have completed the induction. The argument for the evolution equation is similar. □

We will now prove bounds on $X^{(k)}$:

Lemma 2.16. *For each k there exist constants A_k and B_k such that $|X^{(k)}| \leq A_k$ and $\left|\frac{\partial}{\partial t} X^{(k)}\right| \leq B_k$.*

Proof. We prove the two statements simultaneously by induction on k. The statement is clearly true for $k = 0$ by the curvature bound and equation (2.51). We suppose the bounds on $X^{(k)}$ hold for $k = 1, \dots, m$ and consider the evolution equation for $X^{(m+1)}$ from Lemma 2.14: This gives (using the bounds from the inductive hypothesis)

$$\left| \frac{\partial}{\partial t} X^{(m+1)} \right| \leq C_1 \left| X^{(m+1)} \right| + C_2,$$

from which bounds on $|X^{(m+1)}|$ (and hence also the time derivative) follow since the time interval is finite. This completes the induction and proves the lemma. □

We now complete the proof of convergence as t approaches T: We use the fact that $C^k(\mathbb{R}/(A\mathbb{Z}), \mathbb{R}^2)$ is complete. The bound on the time derivative gives

$$\|X(\cdot, t_2) - X(\cdot, t_1)\|_{C^k} \leq C|t_2 - t_1|$$

for any t_1 and t_2 in $[0, t_0]$. In particular, for any sequence t_n approaching T, the sequence $X(\cdot, t_n)$ is Cauchy in C^k and hence converges to a limit $X(\cdot, T)$. But then we also have

$$\|X(\cdot, t) - X(\cdot, T)\|_{C^k} \leq C(T - t),$$

proving that $X(\cdot, t)$ converges to $X(\cdot, T)$ in C^k.

Finally, we apply the short-time existence result (Theorem 2.9) to prove the theorem. We leave it to the reader to check that if we extend the solution to the time interval $[T, T+\delta)$ in this way, then the resulting map X is smooth and is a solution of curve shortening flow on $M^1 \times [0, T + \delta)$. □

2.8. Notes and commentary

Two of the earliest works on curve shortening flow are by von Neumann [**514**] and Mullins [**410**]. In the latter paper, the curve shortening flow equation and the Grim Reaper solution (a terminology later coined by Matt Grayson) appear explicitly.

There is a book on curve shortening flow by Kai-Seng Chou and Xi-Ping Zhu [**156**].

The curve shortening flow is well-defined for curves in spaces of higher dimension. Curve shortening flow of space curves (i.e., curves in $\mathbb{R}^3$) was studied by Altschuler [**19**] and Altschuler and Grayson [**21**]. D. Altschuler, S. Altschuler, S. Angenent, and L. Wu constructed a "zoo" of soliton solutions[2] to the curve shortening flow in $\mathbb{R}^n$ [**18**].

The curve shortening flow for "networks" of curves in the plane was introduced by C. Mantegazza, M. Novaga, and V. M. Tortorelli [**385**]. For related work, see [**83–85, 120, 225, 312, 379, 384, 393, 452, 453, 464**] and the survey [**381**].

2.9. Exercises

Exercise 2.1. Let $X : \mathbb{R} \to \mathbb{R}^2$ be a periodic curve with period $A > 0$. Show that the induced map $X : \mathbb{R}/(A\mathbb{Z}) \to \mathbb{R}^2$ is a homeomorphism to its image set if we place the quotient topology on $\mathbb{R}/(A\mathbb{Z})$. HINT: A much more general result holds: The inverse map of a continuous bijection from a compact topological space is always continuous.

Exercise 2.2. Given $k \in \mathbb{N}$ and $r \in (0,1)$, consider the curve $X_{k,r} : \mathbb{R}/2k\pi\mathbb{Z} \to \mathbb{R}^2$ defined by

$$X_{k,r}(u) \doteqdot \left(-\cos u + kr\cos\left(\tfrac{u}{k}\right), \sin u - kr\sin\left(\tfrac{u}{k}\right)\right).$$

(a) Show that the rotation index of $X_{k,r}$ is equal to k.

(b) Show that the curvature $\kappa_{k,r}$ of $X_{k,r}$ is equal to

$$\kappa_{k,r} = \frac{k + r^2 - (k+1)r\cos\left(\frac{k-1}{k}u\right)}{k\left(1 + r^2 - 2r\cos\left(\frac{k-1}{k}u\right)\right)^{\frac{3}{2}}}.$$

Exercise 2.3. Let X be a curve evolving by (2.15). Show that the time derivative of $\frac{\partial}{\partial s}$, as a vector field on M^1, is given by

$$\frac{\partial}{\partial t}\left(\frac{\partial}{\partial s}\right) = \phi\kappa\frac{\partial}{\partial s}.$$

[2] Animations of some of their examples can be viewed at `https://www.youtube.com/user/solitons2012/videos?view=1`.

Conclude from this that the time derivative of the 1-form ds is

$$\frac{\partial}{\partial t}(ds) = -\phi\,\kappa\, ds\,. \tag{2.53}$$

Exercise 2.4. Show that under curve shortening flow

$$\frac{\partial}{\partial t}|X|^2 = \frac{\partial^2}{\partial s^2}|X|^2 - 2, \tag{2.54a}$$

$$\frac{\partial}{\partial t}\langle X, V\rangle = \frac{\partial^2}{\partial s^2}\langle X, V\rangle \tag{2.54b}$$

for any fixed vector $V \in \mathbb{R}^2$.

Exercise 2.5. Show that under curve shortening flow the support function $\sigma = \langle X, \mathsf{N}\rangle$ satisfies

$$\sigma_t = \sigma_{ss} + \kappa^2\sigma - 2\kappa\,. \tag{2.55}$$

Derive evolution equations for $|X|^2 + 2t$ and $\kappa + \sigma$.

Exercise 2.6. Let $v \doteqdot \langle \mathsf{N}, V\rangle$, where $V \in \mathbb{R}^2$ is fixed. Show that

$$v_t = v_{ss} + \kappa^2 v\,. \tag{2.56}$$

Let $w = v^{-1}$. Show that

$$w_t = w_{ss} - 2w^{-1}(w_s)^2 - \kappa^2 w\,, \tag{2.57}$$

$$(\kappa^2 w^2)_t = (\kappa^2 w^2)_{ss} - 2\frac{w_s}{w}(\kappa^2 w^2)_s - 2((\kappa w)_s)^2, \tag{2.58}$$

and

$$\left(\frac{\partial}{\partial t} - \frac{\partial^2}{\partial s^2} + 2\frac{w_s}{w}\frac{\partial}{\partial s}\right)\left(\left(2t\kappa^2 + 1\right)w^2\right) \le 0\,. \tag{2.59}$$

Exercise 2.7. Find a family of diffeomorphisms $y : \left(-\frac{\pi}{2}, \frac{\pi}{2}\right) \times (-\infty, \infty) \to \left(-\frac{\pi}{2}, \frac{\pi}{2}\right)$ such that the parametrization

$$(y, t) \mapsto (x(y,t), t - \log\cos(x(y,t)))$$

of the Grim Reaper satisfies the parametrized curve shortening flow equation.

Exercise 2.8. The Grim Reaper solution (2.31) satisfies the implicit equation $\mathrm{e}^{-y} = \mathrm{e}^{-t}\cos x$. Show that the paperclip, given implicitly by $\cosh y = \mathrm{e}^{-t}\cos x$ (see [**331**] and [**119**, (14)]), and the hairclip, given implicitly by $\sinh y = \mathrm{e}^{-t}\cos x$, are solutions to the curve shortening flow. We further discuss these solutions in Section 4.4.

Exercise 2.9. Prove, by induction on m, that

$$(\kappa^{(m)})_t - (\kappa^{(m)})_{ss} = (m+3)\,\kappa^2\kappa^{(m)} + \sum c_{i_1,i_2,i_3}\kappa^{(i_1)}\kappa^{(i_2)}\kappa^{(i_3)} \tag{2.60}$$

for each $m \geq 1$, where the c_{i_1,i_2,i_3} are nonnegative integers depending only on i_1, i_2, i_3 and where the sum is over indices i_1, i_2, and i_3 satisfying $0 \leq i_1 \leq i_2 \leq i_3 \leq m-1$ with $i_1 + i_2 + i_3 = m$.

Chapter 3

The Gage–Hamilton–Grayson Theorem

The main goal of this chapter is to prove the Gage–Hamilton–Grayson Theorem, which states that every embedded planar curve contracts to a round point under curve shortening flow. In contrast to the "coarse" derivative estimates of the previous chapter, in this chapter the estimates are "fine" — they provide precise comparisons with model cases such as round circles.

Our main tool is the maximum principle. We first show that the distance between two solutions of the flow is nondecreasing in time. An extension of the same argument shows that an embedded curve remains embedded under the flow. Huisken's distance comparison estimate is a refinement of these methods. Further refinements, due to Paul Bryan and the first author, yield a curvature estimate which leads directly to the theorem.

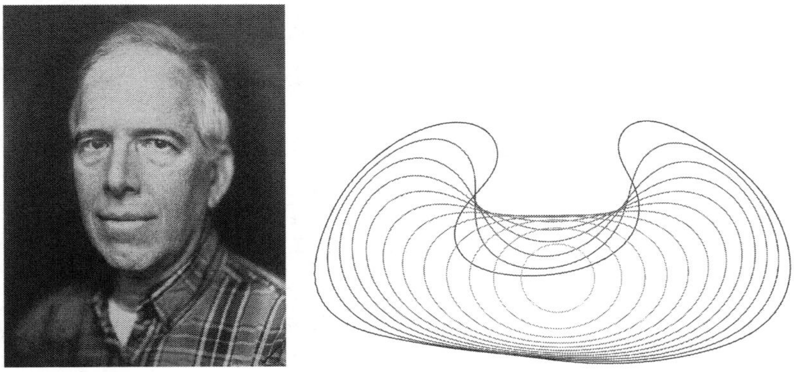

Figure 3.1. L: Matt Grayson. Photo by Geoffrey Berliner / Penumbra Foundation. R: A simple closed curve shrinking to a round point.[1]

[1] We are grateful to Anthony Carapetis for permission to include this drawing created using his interactive curve shortening flow visualization `https://a.carapetis.com/csf`.

3.1. The avoidance principle

We begin with the following beautiful geometric principle.

Theorem 3.1. *Let $X_i : M_i^1 \times [0,T) \to \mathbb{R}^2$ be solutions to curve shortening flow satisfying $X_1(M_1^1, 0) \cap X_2(M_2^1, 0) = \emptyset$. Then $X_1(M_1^1, t) \cap X_2(M_2^1, t) = \emptyset$ for each $t \in [0,T)$.*

Proof. In fact we will prove something stronger: The length of the shortest line segment joining the two curves is not decreasing in time. There is a simple geometric reason behind this: If we take the shortest such line segment and rotate the picture to make this vertical, then the curvature at the "top" point must be more in the upward direction than the curvature at the "bottom" point (otherwise nearby vertical segments have shorter length), and so the curves are moving apart at the endpoints (see (3.5) below).

To achieve this, we employ the maximum principle but, unlike our previous maximum principle arguments, we will work with a function defined not just on an evolving curve but on a product of *two* curves, that is, a function of a pair of points $(x, y) \in M_1 \times M_2$. This is a useful technique which we will exploit later in other situations.

Define a function $d : M_1 \times M_2 \times [0,T) \to \mathbb{R}$ by

$$d(x, y, t) = |X_2(y, t) - X_1(x, t)| . \tag{3.1}$$

Since M_1 and M_2 are compact, the initial minimal distance is positive:

$$d_0 \doteqdot \inf\{d(x, y, 0) : (x, y) \in M_1 \times M_2\} > 0 .$$

We will prove that $de^{\varepsilon(1+t)} > d_0$ for every $\varepsilon > 0$. Suppose that this is not the case. Then, since $\inf\{d(x, y, t)e^{\varepsilon(1+t)} : (x, y) \in M_1 \times M_2\}$ is continuous in t and positive when $t = 0$, there must be a positive first time $t_0 \in (0, T)$ such that $\inf\{d(x, y, t_0)e^{\varepsilon(1+t_0)} : (x, y) \in M_1 \times M_2\} = d_0$. Since $M_1 \times M_2$ is compact, we can then find $(x_0, y_0) \in M_1 \times M_2$ such that $d(x_0, y_0, t_0) = d_0$. We will derive a contradiction by considering derivatives of d at the point (x_0, y_0, t_0). Specifically, we have the inequalities $\frac{\partial}{\partial t}\left(de^{\varepsilon(1+t)}\right)\Big|_{(x_0,y_0,t_0)} \leq 0$, the first derivatives on $M_1 \times M_2$ vanish, and the matrix of second derivatives is nonnegative definite at this point. Our computations are made simpler by exploiting the parametrization-invariance of the curve shortening flow: We choose our parametrization of M_i to be an arc length parametrization induced by the immersion $X_i(\cdot, t_0)$ (this still leaves us freedom to reverse the direction of parametrization, a fact which we will exploit below).

The vanishing of the first spatial derivatives gives the following identities at (x_0, y_0, t_0):

$$0 = \frac{\partial d}{\partial x} = -\langle w, \mathsf{T}_1 \rangle, \quad 0 = \frac{\partial d}{\partial y} = \langle w, \mathsf{T}_2 \rangle, \tag{3.2}$$

where $w \doteqdot \frac{X_2 - X_1}{|X_2 - X_1|}$ is the unit vector pointing from $X_1(x_0, t_0)$ to $X_2(y_0, t_0)$ and T_1 and T_2 are the unit tangent vectors to the parametrized curves $X_1(\cdot, t_0)$ and $X_2(\cdot, t_0)$ at x_0 and y_0, respectively. It follows that both T_1 and T_2 are orthogonal to the unit vector w. By reversing the direction of parametrization of M_i if necessary, we can ensure that $\mathsf{N}_2 = \mathsf{N}_1 = w$ (hence also $\mathsf{T}_1(x_0, t_0) = \mathsf{T}_2(y_0, t_0)$). Finally, we compute the second derivatives

$$\frac{\partial^2 d}{\partial x^2} = \frac{\langle \mathsf{T}_1 - \langle w, \mathsf{T}_1 \rangle w, \mathsf{T}_1 \rangle}{|X_1 - X_2|} - \langle w, \kappa_1 \mathsf{N}_1 \rangle, \tag{3.3a}$$

$$\frac{\partial^2 d}{\partial y^2} = \frac{\langle \mathsf{T}_2 - \langle w, \mathsf{T}_2 \rangle w, \mathsf{T}_2 \rangle}{|X_1 - X_2|} + \langle w, \kappa_2 \mathsf{N}_2 \rangle, \tag{3.3b}$$

$$\frac{\partial^2 d}{\partial x \partial y} = -\frac{\langle \mathsf{T}_1 - \langle w, \mathsf{T}_1 \rangle w, \mathsf{T}_2 \rangle}{|X_1 - X_2|}, \tag{3.3c}$$

where κ_1 and κ_2 are the curvatures of $X_1(\cdot, t_0)$ and $X_2(\cdot, t_0)$ at x_0 and y_0, respectively. Here we used the Frenet–Serret equation (2.2a) to differentiate the unit tangent vectors arising in (3.2).

Observe that equations (3.3a)–(3.3c) hold at any point (x, y, t). Using the identities $\mathsf{T}_1 = \mathsf{T}_2$ and $\mathsf{N}_i = w$, at (x_0, y_0, t_0) these become

$$\frac{\partial^2 d}{\partial x^2} = \frac{1}{d} - \kappa_1, \quad \frac{\partial^2 d}{\partial y^2} = \frac{1}{d} + \kappa_2, \quad \frac{\partial^2 d}{\partial x \partial y} = -\frac{1}{d}. \tag{3.4}$$

In particular the nonnegativity of the Hessian implies that

$$0 \leq \left(\frac{\partial}{\partial x} + \frac{\partial}{\partial y} \right)^2 d = \kappa_2 - \kappa_1. \tag{3.5}$$

The inequality from the time derivative is as follows:

$$\begin{aligned} 0 &\geq \mathrm{e}^{-\varepsilon(1+t_0)} \frac{\partial}{\partial t} \left(d\mathrm{e}^{\varepsilon(1+t)} \right) \Big|_{(x_0, y_0, t_0)} \\ &= \varepsilon d + \left\langle \frac{X_2 - X_1}{|X_2 - X_1|}, -\kappa_1 \mathsf{N}_1 + \kappa_2 \mathsf{N}_2 \right\rangle \\ &> \langle w, -\kappa_1 \mathsf{N}_1 + \kappa_2 \mathsf{N}_2 \rangle \\ &= \kappa_2 - \kappa_1 \geq 0, \end{aligned}$$

where in the last step we applied the inequality (3.5). This contradicts our assumption that $d\mathrm{e}^{\varepsilon(1+t)}$ decreases to d_0 at some time, so we have proved that $d\mathrm{e}^{\varepsilon(1+t)} > d_0$ for each $\varepsilon > 0$, and hence $d \geq d_0$ on $M_1 \times M_2 \times [0, t_0]$. $\square$

Remark 3.2. In the proof there is an important point which will reappear several times in the discussion below. Since X_2 and X_1 both evolve according to the curve shortening flow, it is not surprising that the quantity d we construct from them also satisfies a heat-type equation on $M_1 \times M_2$; however, it is important that we do not try to make it too much like the usual heat equation. Essentially we showed in the proof above that

$$\frac{\partial d}{\partial t} = \left(\frac{\partial}{\partial s_1} + \frac{\partial}{\partial s_2}\right)^2 d \quad \text{(modulo gradient terms)}$$

(see also Exercise 3.3 below), where s_1 and s_2 are the arc length parameters on M_1 and M_2. This is not the heat equation on $M_1 \times M_2$ in the product metric, but rather a degenerate heat equation with nontrivial diffusion coefficient in only one tangent direction $\frac{\partial}{\partial s_1} + \frac{\partial}{\partial s_2}$. By (3.3a)–(3.3c) and related to (3.4), we can of course write d as a solution of something more like a standard heat equation,

$$\frac{\partial d}{\partial t} = \left(\frac{\partial^2}{\partial s_1^2} + \frac{\partial^2}{\partial s_2^2}\right) d - \frac{2}{d} \quad \text{(modulo gradient terms)}, \tag{3.6}$$

but then the maximum principle cannot be applied to keep d greater than d_0. To put this another way, the nonnegativity of the Hessian of d as a function on $M_1 \times M_2$ contains more information than just the nonnegativity of the Laplacian, and it is crucial in the argument that this be exploited.

Corollary 3.3. *If $X_0 : \mathbb{R}/(A\mathbb{Z}) \to \mathbb{R}^2$ is an immersion, then the maximal time of existence for the curve shortening flow with initial data X_0 is finite.*

Proof. Let $X : \mathbb{R}/(A\mathbb{Z}) \to \mathbb{R}^2$ be the solution of curve shortening flow with initial data X_0. Since $\mathbb{R}/(A\mathbb{Z})$ is compact, the image of X_0 is contained in some large ball $B_R(0)$ in $\mathbb{R}^2$. In particular, X_0 is disjoint from the circle $\tilde{X}_0(\theta) = (R\cos\theta, R\sin\theta)$. By the avoidance principle this remains true for positive times, and so $X(\mathbb{R}/(A\mathbb{Z}), t)$ is enclosed by the circle given by the image of the map $\tilde{X}(\theta, t) = \sqrt{R^2 - 2t}\,(\cos\theta, \sin\theta)$. The enclosed region disappears at time $\frac{R^2}{2}$, so the time of existence can be no greater than this. □

3.2. Preserving embeddedness

Next we prove a result which is closely related to the avoidance principle: the preservation of embeddedness under the curve shortening flow. While the avoidance principle states that two curves which are initially disjoint cannot cross each other, the preservation of embeddedness says that a curve which does not initially cross itself will not develop self-intersections.

Theorem 3.4. *Let $X : M^1 \times [0, t_0] \to \mathbb{R}^2$, $M^1 \cong S^1$, be a smooth solution to the curve shortening flow, and suppose that $X(\cdot, 0)$ is an embedding. Then $X(\cdot, t)$ is an embedding for each $t \in [0, t_0]$.*

Proof. Since M^1 is compact, it suffices to prove that $X(\cdot, t)$ is injective for each $t \in [0, t_0]$. We would like to proceed as in the proof of Theorem 3.1, but there is an important complication. Define the extrinsic distance function $d : M^1 \times M^1 \times [0, t_0] \to \mathbb{R}$ by

$$d(x, y, t) = |X(y, t) - X(x, t)|.$$

Then d is zero on the "diagonal" subset $\{(x, x) : x \in M^1\}$, so we cannot hope to show that d remains positive on $M^1 \times M^1$. Instead, we will first use the fact that the curvature is bounded to control d on a neighborhood of the diagonal and then apply the maximum principle to show that d remains positive on the remaining region.

Lemma 3.5. *If $X : M^1 \to \mathbb{R}^2$, $M^1 \cong S^1$, is an immersion with $|\kappa(x)| \leq K$ for all $x \in M^1$, then*

$$|X(y) - X(x)| \geq \frac{2}{K} \sin\left(\frac{K\ell(x, y)}{2}\right) \tag{3.7}$$

for all x, y with $\ell(x, y) \leq \frac{\pi}{K}$, where $\ell(x, y)$ is the arc length from x to y.

Proof. Let x and y be two points on the curve joined by a curve segment of length $\ell \leq \frac{\pi}{K}$. Let s be an arc length parameter with $s(x) = -\ell/2$ and $s(y) = \ell/2$. For any $s \in [-\ell/2, \ell/2]$ we have $|\theta(s) - \theta(0)| \leq \left|\int_0^s \kappa \, ds\right| \leq K|s| \leq \frac{K\ell}{2} \leq \frac{\pi}{2}$. We compute

$$\begin{aligned}
\langle X(y) - X(x), \mathsf{T}(0)\rangle &= \int_{-\ell/2}^{\ell/2} \langle \mathsf{T}(s), \mathsf{T}(0)\rangle \, ds \\
&= \int_{-\ell/2}^{\ell/2} \cos(|\theta(s) - \theta(0)|) \, ds \\
&\geq \int_{-\ell/2}^{\ell/2} \cos(K|s|) \, ds \\
&= \frac{2}{K} \sin\left(\frac{K\ell}{2}\right)
\end{aligned}$$

since $\cos(x)$ is decreasing on $[0, \pi/2]$. This completes the proof of the lemma since $\langle X(y) - X(x), \mathsf{T}(0)\rangle \leq |X(y) - X(x)|$. $\square$

Remark 3.6. Inequality (3.7) is an example of a **chord-arc estimate**. For any two points on a curve, the intrinsic distance $\ell(x,y)$ bounds (sharply) the extrinsic distance $|X(y) - X(x)|$. A chord-arc estimate is a bound in the reverse direction.

Now we can complete the proof that $X(\cdot,t)$ is injective. Since $M^1 \times [0,t_0]$ is compact, there exists a constant K such that $|\kappa(x,t)| \leq K$ for all $x \in M^1$ and $t \in [0,t_0]$. We have $|X(y,t) - X(x,t)| \geq \frac{2}{K} \sin\left(\frac{K\ell(x,y,t)}{2}\right) > 0$ whenever $0 < \ell(x,y,t) \leq \frac{\pi}{K}$ by Lemma 3.5. Now we apply the maximum principle to d on the set $\mathcal{S} = \{(x,y,t) \in M^1 \times M^1 \times [0,t_0] : \ell(x,y,t) \geq \frac{\pi}{K}\}$. On the spatial boundary of $\mathcal{S}$ we have $\inf\{d(x,y,t) : \ell(x,y,t) = \frac{\pi}{K}\} \geq \frac{2}{K}$ by Lemma 3.5. It follows from the result of Exercise 3.3 that

$$\begin{aligned}\frac{\partial d}{\partial t} &= \left(\frac{\partial^2}{\partial s_y^2} + \frac{\partial^2}{\partial s_x^2} + 2\mathsf{T}_x \cdot \mathsf{T}_y \frac{\partial^2}{\partial s_y \partial s_x}\right) d \\ &\quad - \frac{1}{d}\left(\left(\frac{\partial d}{\partial s_y}\right)^2 + \left(\frac{\partial d}{\partial s_x}\right)^2 - 2\mathsf{T}_x \cdot \mathsf{T}_y \frac{\partial d}{\partial s_y}\frac{\partial d}{\partial s_x}\right)\end{aligned} \tag{3.8}$$

on $\mathcal{S}$. The maximum principle then yields

$$d(x,y,t) \geq \min\left\{\inf\left\{d(x,y,0) : \ell(x,y,0) \geq \frac{\pi}{K}\right\}, \frac{2}{K}\right\}.$$

Since $X(\cdot,0)$ is injective, we have $\inf\{d(x,y,0) : \ell(x,y,0) \geq \frac{\pi}{K}\} > 0$. This establishes that $d(x,y,t) > 0$ for all $y \neq x$ and $t \in [0,t_0]$, so $X(\cdot,t)$ is injective for each $t \in [0,t_0]$ and the theorem is proved. □

3.3. Huisken's distance comparison estimate

Next we consider a refinement of the above argument due to Gerhard Huisken [**298**]. As in the argument above we derive a positive lower bound on $d(x,y,t) = |X(y,t) - X(x,t)|$ for $y \neq x$, but this time the lower bound will not depend explicitly on a bound on curvature. Instead we will show that d can be bounded from below in terms of the arc length $\ell(x,y,t)$ and the total length of the curve.

The motivation for the precise estimate proved by Huisken comes from a consideration of the shrinking circle solution. Why is this important? If an estimate is to be useful for analyzing singularities, it should be scale invariant. But the shrinking circle solution is itself scale invariant (self-similar), so any scale-invariant quantity is constant in time. If the maximum principle applies to preserve an inequality, then the strong maximum principle says that the inequality becomes strict unless the quantity is constant along the directions where the diffusion coefficient is nontrivial. For this reason it makes sense to look for quantities that are constant on circles (to consider quantities which are constant on self-similar solutions follows Hamilton's philosophy).

There is essentially a unique identity relating the distance d to the arc length ℓ and the total length L for a pair of points on a circle: If θ is the angle subtended at the center by the two points, then we have $\ell/\mathrm{L} = \theta/2\pi$ and $d/\mathrm{L} = 2\sin(\theta/2)/2\pi$. Rearranging this we find that

$$d = \frac{\mathrm{L}}{\pi}\sin\left(\frac{\pi\ell}{\mathrm{L}}\right). \tag{3.9}$$

Huisken proved the following chord-arc estimate using a 2-point function maximum principle argument.

Theorem 3.7. *Let $X: M^1 \times [0,T) \to \mathbb{R}^2$, $M^1 \cong S^1$, be a smooth solution to the curve shortening flow, such that $X(\cdot,0)$ is an embedding. Define a function $Z: \left((M^1 \times M^1) \setminus \{(x,x): x \in M^1\}\right) \times [0,T) \to \mathbb{R}$ by*

$$Z(x,y,t) = \frac{\mathrm{L}(t)}{d(x,y,t)}\sin\left(\frac{\pi\ell(x,y,t)}{\mathrm{L}(t)}\right), \tag{3.10}$$

where $d(x,y,t) = |X(y,t) - X(x,t)|$, $\ell(x,y,t)$ is the arc length from x to y at time t, and L is the total length of the curve $X(M^1,t)$. Then $\sup\{Z(x,y,t): x,y \in M^1\}$ is nonincreasing in t, strictly unless the curve is a round circle.

Note that this result implies in particular that embeddedness is preserved, with estimates that do not depend on any explicit curvature bound.

Proof. We apply the maximum principle to Z as a function on $M^1 \times M^1 \times [0,T)$. To understand the behavior of Z near the diagonal $\{(x,x): x \in M^1\}$, we begin with a simple observation.

Lemma 3.8. *If $X(\cdot,t)$ is embedded, then Z is smooth and extends to a continuous function on $M^1 \times M^1 \times [0,T)$ with $Z(x,x,t) = \pi$.*

Proof. Embeddedness implies that $d(x,y,t) \neq 0$ when $x \neq y$ in M^1. Since d^2 is smooth and d is positive, d is also, and Z is smooth on $\left(M^1 \times M^1\right) \setminus \{(x,x): x \in M^1\}$. We proceed to show continuity on the diagonal. Clearly the arc length cannot be less than the chord distance, so we have $d \leq \ell$. By Lemma 3.5 we also have $d \geq \frac{2}{K}\sin\left(\frac{K\ell(x,y)}{2}\right)$ for $\ell \leq \pi/K$, where $K = \sup_{M^1}|\kappa|$. Since $x - \frac{1}{6}x^3 \leq \sin(x) \leq x$, we have

$$Z = \frac{\mathrm{L}}{d}\sin\left(\frac{\pi\ell}{\mathrm{L}}\right) \leq \frac{\mathrm{L}\left(\frac{\pi\ell}{\mathrm{L}}\right)}{\frac{2}{K}\sin\left(\frac{K\ell}{2}\right)} \leq \frac{\pi\ell}{\frac{2}{K}\left(\frac{K\ell}{2} - \frac{1}{6}\left(\frac{K\ell}{2}\right)^3\right)} = \frac{\pi}{1 - \frac{K^2}{24}\ell^2}$$

for $\ell < \pi/K$, while

$$Z \geq \frac{\mathrm{L}\left(\frac{\pi\ell}{\mathrm{L}} - \frac{1}{6}\left(\frac{\pi\ell}{\mathrm{L}}\right)^3\right)}{\ell} = \pi\left(1 - \frac{\pi^2}{6\,\mathrm{L}^2}\ell^2\right).$$

It follows that $|Z(x,y,t) - \pi| \leq C\ell(x,y,t)^2$ for $\ell < r$, where C and r depend only on K and L. $\square$

This is complemented by the following result.

Lemma 3.9. $\sup\{Z(x,y,t) : x, y \in M^1\} > \pi$ *if* $X(M^1,t)$ *is not a round circle.*

Proof. Suppose that $X(M^1,t)$ is not a round circle, and let K be the maximum absolute value of the curvature κ at time t. Then

$$K > \frac{1}{\mathrm{L}}\int_{M^1} \kappa\, ds = \frac{2\pi}{\mathrm{L}}, \tag{3.11}$$

where the strict inequality holds because the curve is not circular and the curvature is therefore not constant. The second equality is the theorem of turning tangents. Choose an arc length parameter s such that $s = 0$ is a point where the maximum curvature is attained. We use the following Taylor expansion for X, which can be computed using the definition of T and the Frenet–Serret formulae (2.2a) and (2.2b)

$$\begin{aligned} X(s,t) = X(0,t) + s\mathsf{T} - \frac{s^2}{2}\kappa\mathsf{N} - \frac{s^3}{6}(\kappa_s\mathsf{N} + \kappa^2\mathsf{T}) \\ + \frac{s^4}{24}\left((\kappa^3 - \kappa_{ss})\mathsf{N} - 3\kappa\kappa_s\mathsf{T}\right) + O(s^5) \end{aligned} \tag{3.12}$$

as $s \to 0$, where T, N, κ, κ_s, and κ_{ss} are all evaluated at $(0,t)$. From this we find

$$X\left(\frac{s}{2},t\right) - X\left(-\frac{s}{2},t\right) = s\mathsf{T} - \frac{s^3}{24}\left(\kappa_s\mathsf{N} + \kappa^2\mathsf{T}\right) + O(s^5) \tag{3.13}$$

as $s \to 0$, so that

$$\left|X\left(\frac{s}{2},t\right) - X\left(-\frac{s}{2},t\right)\right|^2 = s^2 - \frac{s^4}{12}\kappa^2 + O(s^6), \tag{3.14}$$

and therefore

$$d\left(\frac{s}{2},-\frac{s}{2},t\right) = \left|X\left(\frac{s}{2},t\right) - X\left(-\frac{s}{2},t\right)\right| = s - \frac{s^3}{24}\kappa^2 + O(s^5). \tag{3.15}$$

We also have $\sin\left(\frac{\pi s}{\mathrm{L}}\right) = \frac{\pi s}{\mathrm{L}} - \frac{1}{6}\left(\frac{\pi s}{\mathrm{L}}\right)^3 + O(s^5)$ from the Taylor expansion of the sine function. Combining these we find

$$Z\left(\frac{s}{2}, -\frac{s}{2}, t\right) = \frac{\pi\left(1 - \frac{\pi^2 s^2}{6\,\mathrm{L}^2} + O(s^4)\right)}{1 - \frac{s^2}{24}\kappa^2 + O(s^4)} = \pi\left(1 + \frac{1}{24}\left(\kappa^2 - \frac{4\pi^2}{\mathrm{L}^2}\right)s^2 + O(s^4)\right). \tag{3.16}$$

Since we are working at the point of maximum absolute value of curvature, we have $\kappa^2 = K^2 > \frac{4\pi^2}{\mathrm{L}^2}$ by equation (3.11), and it follows that $Z > \pi$ for sufficiently small positive values of s. □

Of course, if the initial curve is a round circle, then the solution is the shrinking round circle solution and Z remains constant. So we proceed by assuming that the curve is not circular and prove that Z is strictly decreasing. More precisely, we prove the equivalent statement that for any $t_0 \in [0, T)$ there does not exist $t_1 > t_0$ such that

$$\sup\{Z(x, y, t_1) : \; x, y \in M^1\} \geq \sup\{Z(x, y, t_0) : \; x, y \in M^1\}.$$

For this it suffices to prove that there does not exist $t_1 > t_0$ such that

$$\sup\{Z(x, y, t_1) : \; x, y \in M^1\} = \sup\{Z(x, y, t) : \; x, y \in M^1, \; t \in [t_0, t_1]\}.$$

Suppose that such times t_0 and t_1 exist. Then there exist x_0 and y_0 in M^1 such that $Z(x_0, y_0, t_1) = \sup\{Z(x, y, t) : \; x, y \in M^1, \; t \in [t_0, t_1]\}$. By Lemma 3.8 and Lemma 3.9 we must have $x_0 \neq y_0$. At this point we have $\frac{\partial Z}{\partial t} \geq 0$ and the matrix of second derivatives of Z is nonpositive. We derive a contradiction from these two inequalities.

We first compute the time derivative. Observing that $\frac{d}{dt}\mathrm{L} = -\int_{M^1} \kappa^2\, ds$ and $\frac{\partial}{\partial t}\ell(x, y, t) = -\int_x^y \kappa^2\, ds$ by (2.19), we find that

$$d\frac{\partial Z}{\partial t} = Z\,\langle w, \kappa_y \mathsf{N}_y - \kappa_x \mathsf{N}_x\rangle - \left(\sin\left(\frac{\pi\ell}{\mathrm{L}}\right) - \frac{\pi\ell}{\mathrm{L}}\cos\left(\frac{\pi\ell}{\mathrm{L}}\right)\right)\int_{M^1}\kappa^2\, ds - \pi\cos\left(\frac{\pi\ell}{\mathrm{L}}\right)\int_x^y \kappa^2\, ds. \tag{3.17}$$

Since we can assume without loss of generality that $0 \leq \ell \leq \mathrm{L}/2$, we have $\sin\left(\frac{\pi\ell}{\mathrm{L}}\right) - \frac{\pi\ell}{\mathrm{L}}\cos\left(\frac{\pi\ell}{\mathrm{L}}\right) \geq 0$ and $\cos\left(\frac{\pi\ell}{\mathrm{L}}\right) \geq 0$, so the terms after the first are nonpositive. To understand the first term we use the spatial second variation. First, we compute the first derivatives, working at a point (x, y, t) with respect to arc length parametrization at time t. We orient the curve so that the parameter increases from x to y along the shorter of the two arcs of the curve, so that $\frac{\partial \ell}{\partial x} = -1$ and $\frac{\partial \ell}{\partial y} = 1$. This gives

$$d\frac{\partial Z}{\partial x} = Z\langle w, \mathsf{T}_x\rangle - \pi\cos\left(\frac{\pi\ell}{\mathrm{L}}\right), \tag{3.18a}$$

$$d\frac{\partial Z}{\partial y} = -Z\langle w, \mathsf{T}_y\rangle + \pi\cos\left(\frac{\pi\ell}{\mathrm{L}}\right). \tag{3.18b}$$

From this we find the following expressions for the second derivatives (at an extremum where the first derivatives vanish):

$$d\frac{\partial^2 Z}{\partial x^2} = -Z\langle w, \kappa_x \mathsf{N}_x\rangle - \frac{Z}{d}\left(1 - \langle w, \mathsf{T}_x\rangle^2\right) - \frac{\pi^2}{L}\sin\left(\frac{\pi\ell}{\mathrm{L}}\right), \tag{3.19a}$$

$$d\frac{\partial^2 Z}{\partial y^2} = Z\langle w, \kappa_y \mathsf{N}_y\rangle - \frac{Z}{d}\left(1 - \langle w, \mathsf{T}_y\rangle^2\right) - \frac{\pi^2}{L}\sin\left(\frac{\pi\ell}{\mathrm{L}}\right), \tag{3.19b}$$

$$d\frac{\partial^2 Z}{\partial x \partial y} = \frac{Z}{d}\left(\langle \mathsf{T}_y, \mathsf{T}_x\rangle - \langle w, \mathsf{T}_y\rangle\langle w, \mathsf{T}_x\rangle\right) + \frac{\pi^2}{L}\sin\left(\frac{\pi\ell}{\mathrm{L}}\right). \tag{3.19c}$$

The vanishing of the first derivatives in identities (3.18a) and (3.18b) tells us that

$$\langle w, \mathsf{T}_x\rangle = \langle w, \mathsf{T}_y\rangle = \frac{\pi}{Z}\cos\left(\frac{\pi\ell}{\mathrm{L}}\right).$$

This leads to two possibilities: Either the two tangent directions T_x and T_y are parallel or they are bisected by the direction w. Suppose first that the two tangent directions T_x and T_y are parallel (as in Figure 3.2). Then

$$\begin{aligned} d\left(\frac{\partial}{\partial t}Z - \left(\frac{\partial}{\partial x} + \frac{\partial}{\partial y}\right)^2 Z\right) &= -\left(\sin\left(\frac{\pi\ell}{\mathrm{L}}\right) - \frac{\pi\ell}{\mathrm{L}}\cos\left(\frac{\pi\ell}{\mathrm{L}}\right)\right)\int_M \kappa^2\, ds \\ &\quad - \pi\cos\left(\frac{\pi\ell}{\mathrm{L}}\right)\int_x^y \kappa^2\, ds \\ &< 0. \end{aligned} \tag{3.20}$$

This contradicts the fact that we are at a point where a new supremum of Z is attained. So suppose that the two tangent directions T_x and T_y are bisected by the direction w (as in Figure 3.3). Let θ be the angle from w to T_x, so that in this case the angle from T_x to T_y is 2θ. Then we compute

$$\begin{aligned} d\left(\frac{\partial}{\partial t}Z - \left(\frac{\partial}{\partial x} - \frac{\partial}{\partial y}\right)^2 Z\right) &= -\frac{Z}{d}\left(2\sin^2\theta + 2(\cos(2\theta) - \cos^2\theta)\right) \\ &\quad + \frac{4\pi^2}{L}\sin\left(\frac{\pi\ell}{\mathrm{L}}\right) - \pi\cos\left(\frac{\pi\ell}{\mathrm{L}}\right)\int_x^y \kappa^2\, ds \\ &\quad - \left(\sin\left(\frac{\pi\ell}{\mathrm{L}}\right) - \frac{\pi\ell}{\mathrm{L}}\cos\left(\frac{\pi\ell}{\mathrm{L}}\right)\right)\int_M \kappa^2\, ds. \end{aligned} \tag{3.21}$$

The terms in the first set of parentheses simplify to zero. The remainder we estimate as follows: The coefficients of each of the two integrals of κ^2 are nonnegative, so we can estimate each of the integrals using the Cauchy–Schwarz inequality and the turning-angle identity (2.12). The integral over the whole curve gives

$$\int_M \kappa^2\, ds > \frac{1}{\mathrm{L}}\left(\int_M \kappa\, ds\right)^2 = \frac{4\pi^2}{\mathrm{L}}, \tag{3.22}$$

where the strict inequality holds because κ is not constant if the curve is not a round circle. The integral over the segment from x to y gives

$$\int_x^y \kappa^2\, ds \geq \frac{1}{\ell}\left(\int_x^y \kappa\, ds\right)^2 = \frac{4\theta^2}{\ell}. \tag{3.23}$$

Substituting these, we arrive at the inequality

$$d\left(\frac{\partial}{\partial t}Z - \left(\frac{\partial}{\partial x} - \frac{\partial}{\partial y}\right)^2 Z\right) < -\frac{4\pi}{\ell}\cos\left(\frac{\pi\ell}{\mathrm{L}}\right)\left(\theta^2 - \left(\frac{\pi\ell}{\mathrm{L}}\right)^2\right). \tag{3.24}$$

To complete the argument we need to compare θ to the ratio $\pi\ell/\mathrm{L}$. The required estimate comes from first order identity (3.18a), which says that $Z\cos(\theta) = \pi\cos\left(\frac{\pi\ell}{\mathrm{L}}\right)$, together with the fact that $Z > \pi$ at the supremum by Lemma 3.9. These imply that $0 \leq \cos\theta \leq \cos\left(\frac{\pi\ell}{\mathrm{L}}\right)$, and since the cosine function is decreasing on the interval $[0, \pi]$, we deduce that $\theta \geq \frac{\pi\ell}{\mathrm{L}}$. The last set of parentheses is therefore nonnegative and the time derivative of Z is strictly negative at the maximum point. This is a contradiction, and so we have ruled out both cases and the theorem is proved. □

Remark 3.10. The reader may have noted that the "heat equation" we derive for Z differs in the two cases. There is a simple geometric reason for this, which we will try to elaborate here. Our choice of the second-order operator is governed by the need to get the best inequality possible. We want to show that Z is decreasing, which we do by showing that the time derivative of Z minus a suitable second-order operator applied to Z is nonpositive. To give ourselves the best chance of proving this, we choose the operator to give values as large as possible at the maximum point, and this corresponds to having the operator applied to d taking values as small as possible.

Note that d is a convex function, and the second derivatives are strictly positive except in radial directions. This means that the second derivative of d is minimized when we keep the vector w parallel. In the first case of the proof the tangent lines are parallel and traversed in the same direction (see Figure 3.2), so w is constant if we move x and y in direction $\partial_x + \partial_y$ (that is, x moves in direction ∂_x and y moves in direction ∂_y).

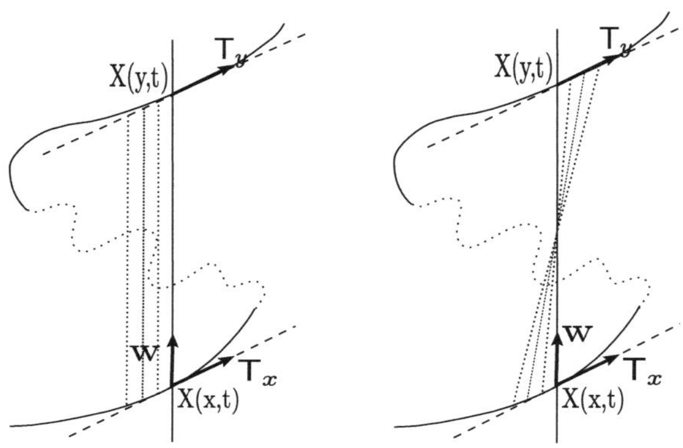

Figure 3.2. Case 1: $\mathsf{T}_x = \mathsf{T}_y$. Moving (x, y) in direction $\partial_y + \partial_x$ keeps w constant, while moving in direction $\partial_y - \partial_x$ rotates w and gives a positive second derivative of d.

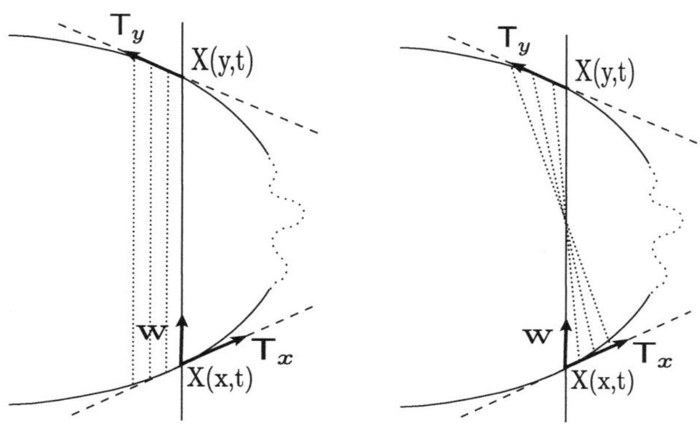

Figure 3.3. Case 2: $\mathsf{T}_x \neq \mathsf{T}_y$. Moving (x, y) in direction $\partial_y - \partial_x$ keeps w constant, while moving in direction $\partial_y + \partial_x$ rotates w and gives a positive second derivative of d.

In the second case of the proof the corresponding tangent lines meet the w direction at equal angles, as in Figure 3.3. In this case w remains constant if we move x and y in direction $\partial_y - \partial_x$, so the optimal operator is the second derivative in this direction.

3.4. A curvature bound by distance comparison

Now we come to one final refinement of the argument. In our first proof that embeddedness is preserved, we assumed a curvature bound and deduced "injectivity" bounds depending on a bound for curvature. Huisken's estimate improves this by removing the dependence on curvature, producing lower bounds for d in terms of ℓ which depend only on initial lower bounds and the total length of the curve. We want to prove Grayson's theorem, which says that if the initial curve is embedded, then the evolving curve shrinks to a point, becoming circular in the process. In particular the solution continues to exist as long as the length remains positive.

Huisken's estimate is not enough by itself to deduce that the solution continues to exist while the length remains positive. To apply the global existence theorem to deduce this, we would need to prove that a curvature bound holds as long as the length remains positive. We will later see how an analysis of blow-up limits would allow us to derive Grayson's theorem from Huisken's estimate. But first we will see how to improve Huisken's argument to derive a curvature bound directly, giving a rather easy proof of Grayson's theorem.

The idea is as follows. If the curvature κ is large at a point x on a smooth curve, then it looks locally like a part of circle of radius $1/\kappa$. In particular we would expect that the distance d is close to $\frac{2}{\kappa}\sin\left(\frac{\kappa\ell}{2}\right) \simeq \ell - \frac{\kappa^2\ell^3}{24}$ for pairs of points close to x. Conversely, if we can prove that $d(x,y) \geq \varphi(\ell(x,y))$ where $\varphi(x) \geq x - Cx^3 + o(x^3)$ as $x \to 0$, then we must have $\kappa^2 \leq 24C$ at every point of the curve. So the challenge is to prove a lower bound on d of this kind. Note that the bound coming from Huisken's estimate is not strong enough. That estimate gives $d \geq \varphi(\ell) = \frac{c\mathrm{L}}{\pi}\sin\left(\frac{\pi\ell}{\mathrm{L}}\right)$ for some $c \in (0,1)$, so we have $\varphi(x) \simeq cx$ as $x \to 0$.

First, let us establish the implication more carefully.

Definition 3.11. Let $X : M^1 \to \mathbb{R}^2$ be a smooth embedding. The **chord-arc profile** of X is the function $\psi_X : [0,\infty) \to \mathbb{R}$ defined by

$$\psi_X(z) = \inf\{|X(x) - X(y)| : \ell(x,y) = z\}.$$

Proposition 3.12. *Let $X : M^1 \to \mathbb{R}^2$ be a smooth embedding and let $K = \sup_{M^1}|\kappa|$. Then*

$$\psi_X(z) = z - \frac{K^2}{24}z^3 + O(z^5) \quad \textit{as } z \to 0.$$

Proof. By Lemma 3.5 we have $d(x,y) \geq \frac{2}{K}\sin\left(\frac{K\ell}{2}\right)$ for all (x,y) with $\ell(x,y) \leq \frac{\pi}{K}$, so $\psi_X(z) \geq \frac{2}{K}\sin\left(\frac{Kz}{2}\right) = z - \frac{K^2}{24}z^3 + O(z^5)$ as $z \to 0$. The proof of Lemma 3.9 gives the reverse inequality: We showed there (in equation (3.15)) that for any small z there exist points (x,y) with $\ell(x,y) = z$ and

$$d(x,y) \leq z - \frac{K^2}{24}z^3 + Cz^5.$$

Therefore by definition we have also $\psi_X(z) \leq z - \frac{K^2}{24}z^3 + O(z^5)$ as $z \to 0$. □

It follows that we can read off the maximum curvature from the chord-arc profile and that a strong enough lower bound for the chord-arc profile (of the form $\psi_X(z) \geq z - Cz^3$ for small z) implies a curvature bound (precisely, $\kappa^2 \leq 24C$).

The next problem is: How can we prove good lower bounds for the chord-arc profile? Let us see what conditions are required for a lower bound on the chord-arc profile to be preserved. To take into account the fact that the curve is expected to shrink to a point, it makes sense to consider a relative chord-arc profile, which compares the chord distance as a proportion of total length to the arc length as a proportion of total length (the relative chord-arc profile of a curve X is therefore just the chord-arc profile of the curve obtained by rescaling X to have total length 1). Define

$$Z(x,y,t) = d(x,y,t) - \mathrm{L}(t)\,\varphi\left(\frac{\ell(x,y,t)}{\mathrm{L}(t)}, t\right), \tag{3.25}$$

where φ is a smooth function yet to be determined. The idea is to find conditions on φ which will imply that positivity of Z is preserved. We will begin somewhat formally, by considering what the implications are for φ when a new interior zero minimum point (x,y,t) of Z is attained with $y \neq x$. At such a point we compute the following (working in arc length parametrization):

$$\frac{\partial Z}{\partial x} = \langle w, -\mathsf{T}_x\rangle + \varphi', \quad \frac{\partial Z}{\partial y} = \langle w, \mathsf{T}_y\rangle - \varphi'. \tag{3.26}$$

Thus, as before, the cosine of the angle between T_x and $w = \frac{X(y,t)-X(x,t)}{|X(y,t)-X(x,t)|}$ and the angle between T_y and w both equal φ'. The second derivatives give

$$\frac{\partial^2 Z}{\partial x^2} = \langle w, \kappa_x \mathsf{N}_x\rangle + \frac{1}{d}\langle \mathsf{T}_x - \langle w, \mathsf{T}_x\rangle\, w, \mathsf{T}_x\rangle - \frac{\varphi''}{\mathrm{L}}, \tag{3.27a}$$

$$\frac{\partial^2 Z}{\partial y^2} = -\langle w, \kappa_y \mathsf{N}_y\rangle + \frac{1}{d}\langle \mathsf{T}_y - \langle w, \mathsf{T}_y\rangle\, w, \mathsf{T}_y\rangle - \frac{\varphi''}{\mathrm{L}}, \tag{3.27b}$$

$$\frac{\partial^2 Z}{\partial x \partial y} = -\frac{1}{d}\langle \mathsf{T}_x - \langle w, \mathsf{T}_x\rangle\, w, \mathsf{T}_y\rangle + \frac{\varphi''}{\mathrm{L}}. \tag{3.27c}$$

The time derivative of Z is as follows:
(3.28)

$$\frac{\partial Z}{\partial t} = \langle w, \kappa_x \mathsf{N}_x - \kappa_y \mathsf{N}_y\rangle + \left(\varphi - \frac{\ell}{\mathrm{L}}\varphi'\right)\int_M \kappa^2\, ds + \varphi' \int_x^y \kappa^2\, ds - \mathrm{L}\frac{\partial \varphi}{\partial t}.$$

The vanishing of the first derivatives implies as in Huisken's result (Theorem 3.7) that there are two possible cases: Either $\mathsf{T}_x = \mathsf{T}_y$ or the direction w bisects T_x and T_y. We can eliminate one of these possibilities as long as φ is concave.

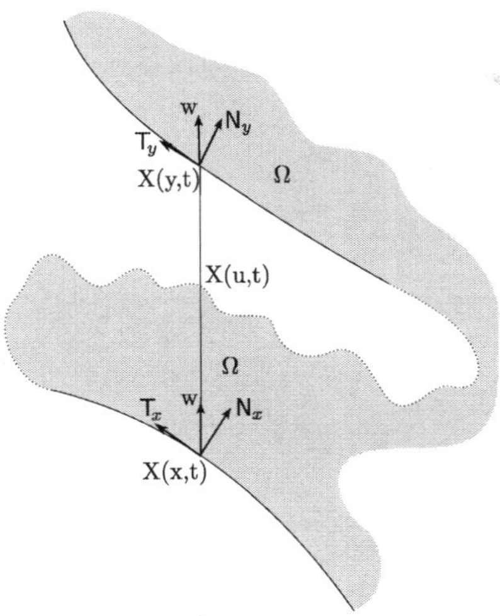

Figure 3.4. In Case 2 there must be a point $X(u,t)$ on the curve along the line joining $X(x,t)$ to $X(y,t)$, and it follows (if $\varphi(\cdot,t)$ is concave) that (x,y) does not minimize $Z(\cdot,\cdot,t)$.

Lemma 3.13. *Suppose that $\varphi(\cdot,t)$ is symmetric about $z=\frac{1}{2}$ and strictly concave with $|\varphi'(\cdot,t)|<1$ on $(0,1)$, and let $(x,y)\in(M^1\times M^1)\setminus\{(x,x): x\in M^1\}$ be a critical point of Z with $\mathsf{T}_x=\mathsf{T}_y$ and $Z(x,y,t)=0$. Then there exists $(u,v)\in M^1\times M^1$ such that $Z(u,v,t)<0$.*

Proof. By the Jordan curve theorem, the curve $X(\cdot,t)$ separates $\mathbb{R}^2$ into precisely two open connected regions. Note that by equation (3.26) we have $\langle w,\mathsf{T}_x\rangle=\varphi'\in(-1,1)$, and so T_x and w are linearly independent. It follows that for small $a>0$, the point $X(x,t)+aw$ lies in one of the two connected components of $\mathbb{R}^2\setminus X(M^1)$. Call this component Ω. Since $\varphi(\cdot,t)$ is symmetric about $z=\frac{1}{2}$, we can assume (by reversing parametrization if necessary) that Ω also contains the points $X(x,t)+a\mathsf{N}(x,t)$ for small $a>0$. But then by connectedness, Ω also contains the points $X(z,t)+a\mathsf{N}(z,t)$ for any $z\in M^1$, and in particular the points $X(y,t)+a\mathsf{N}(y,t)$. But since $\mathsf{T}_y=\mathsf{T}_x$, we have $\mathsf{N}_y=\mathsf{N}_x$, so Ω contains the points $X(y,t)+aw$ for small $a>0$, hence not the points $X(y,t)-aw$ for small $a>0$.

Write $d=d(x,y,t)$, and let $s_*=\sup\{s\in(0,d): X(x,t)+sw\in\Omega\}$. Then the above argument shows that $0<s_*<d$ and $X(x,t)+s_*w\in\partial\Omega=X(M^1)$. Therefore, there exists $u\in M^1$ such that $X(u,t)=X(x,t)+s_*w$. In particular, we have $d(x,u,t)+d(u,y,t)=d(x,y,t)$, and we also have $\ell(x,u,t)+\ell(u,y,t)=\ell(x,y,t)$. Therefore by strict concavity we have $\varphi(\frac{\ell(x,u,t)}{\mathrm{L}},t)+\varphi(\frac{\ell(u,y,t)}{\mathrm{L}},t)>\varphi(\frac{\ell(x,y,t)}{\mathrm{L}})$. By assumption we have that $\varphi(\frac{\ell(x,y,t)}{\mathrm{L}},t)=\frac{d(x,y,t)}{\mathrm{L}}=\frac{d(x,u,t)}{\mathrm{L}}+\frac{d(u,y,t)}{\mathrm{L}}$, so we deduce that $\varphi(\frac{\ell(x,u,t)}{\mathrm{L}},t)+\varphi(\frac{\ell(u,y,t)}{\mathrm{L}},t)>\frac{d(x,u,t)}{\mathrm{L}}+\frac{d(u,y,t)}{\mathrm{L}}$ and hence either

$\varphi(\frac{\ell(x,u,t)}{\mathrm{L}}, t) > \frac{d(x,u,t)}{\mathrm{L}}$ or $\varphi(\frac{\ell(u,y,t)}{\mathrm{L}}, t) > \frac{d(u,y,t)}{\mathrm{L}}$. That is, either $Z(x,u,t) < 0$ or $Z(u,y,t) < 0$. $\square$

Since by hypothesis $Z(\cdot,\cdot,t) \geq 0$, it follows that the only possibility is the case where T_x and T_y are bisected by w. As before, we let θ be the angle between T_x and w, which then equals the angle between w and T_y.

Now we compute

$$\left(\frac{\partial}{\partial t} - \left(\frac{\partial}{\partial x} - \frac{\partial}{\partial y}\right)^2\right) Z = -\mathrm{L}\frac{\partial \varphi}{\partial t} + 4\frac{\varphi''}{\mathrm{L}} + \varphi' \int_x^y \kappa^2\, ds + \left(\varphi - \frac{\ell}{\mathrm{L}}\varphi'\right)\int_{M^1} \kappa^2\, ds. \tag{3.29}$$

To proceed as we did in the proof of Theorem 3.7 we must assume that the coefficients of the two integrals of κ^2 are nonpositive, so that φ must be increasing on $[0, \frac{1}{2}]$ and we must have $\varphi(z,t) - z\varphi'(z,t) \geq 0$. These follow from the assumption of concavity that we made previously:

Lemma 3.14. *If $\varphi(\cdot, t)$ is strictly concave and positive on $[0,1]$ and $\varphi(z,t) = \varphi(1-z)$, then for all $z \in [0, 1/2)$,*

$$\varphi' > 0 \quad \textit{and} \quad \varphi - z\varphi' > 0.$$

Proof. By symmetry φ has a critical point at $1/2$ and hence, by strict concavity, $\varphi > 0$ for $z < 1/2$. The function $u(z) = \varphi - z\varphi'$ is nonnegative at zero and satisfies $u'(z) = -z\varphi'' > 0$, so $u > 0$ on $(0, 1]$. $\square$

We can use the Cauchy–Schwarz inequality to estimate the integrals, yielding

$$\left(\frac{\partial}{\partial t} - \left(\frac{\partial}{\partial x} - \frac{\partial}{\partial y}\right)^2\right) Z > -\mathrm{L}\frac{\partial \varphi}{\partial t} + 4\frac{\varphi''}{\mathrm{L}} + \frac{4\theta^2}{\ell}\varphi' + \frac{4\pi^2}{\mathrm{L}}\left(\varphi - \frac{\ell}{\mathrm{L}}\varphi'\right). \tag{3.30}$$

The strict inequality holds because we have a strict inequality in the Cauchy–Schwarz inequality if κ is not constant (that is, if the curve is not a circle). Finally, the first derivative condition allows us to relate θ to φ. This says that $\cos\theta = \langle w, T_x\rangle = \varphi'$, or $\theta = \arccos(\varphi')$. This gives

$$\left(\frac{\partial}{\partial t} - \left(\frac{\partial}{\partial x} - \frac{\partial}{\partial y}\right)^2\right) Z > -\mathrm{L}\frac{\partial \varphi}{\partial t} + 4\frac{\varphi''}{\mathrm{L}} + \frac{4\left(\arccos(\varphi')\right)^2}{\ell}\varphi' + \frac{4\pi^2}{\mathrm{L}}\left(\varphi - \frac{\ell}{\mathrm{L}}\varphi'\right). \tag{3.31}$$

The important thing to note is that the right-hand side of this equation depends only on φ (and ℓ and L — note that φ is itself a function of ℓ/L

and t). To put this another way, we can rule out the possibility of an interior maximum of Z if φ satisfies the differential inequality

$$\mathrm{L}^2\frac{\partial\varphi}{\partial t}(z,t) \le 4\varphi'' + \frac{4(\arccos(\varphi'))^2\varphi'}{z} + 4\pi^2(\varphi - z\varphi'). \tag{3.32}$$

The appearance of the total length L at first sight looks like a problem since this depends on time itself. However, we can remove this factor simply by redefining our "time variable"; that is, we define a new parameter τ by the equation $\tau = \int_0^t \frac{1}{\mathrm{L}^2}\,dt$. The inverse trigonometric function can also be removed, using the following estimate:

Lemma 3.15. *The function $x \mapsto (\arccos(x))^2$ is convex on $[0,1]$, and $(\arccos(x))^2 \ge \pi^2 z^2 + 2\frac{\pi z}{\sin(\pi z)}(\cos(\pi z) - x)$ for $x, z \in [0,1]$.*

Proof. If $y = (\arccos(x))^2$, a direct computation yields

$$y' = -\frac{2\sqrt{y}}{\sin(\sqrt{y})} \quad \text{and} \quad y'' = \frac{2\cos(\sqrt{y})}{\sin^3(\sqrt{y})}(\tan(\sqrt{y}) - \sqrt{y}) > 0.$$

So y is convex. Therefore $y(x) \ge y(\cos(\pi z)) + y'(\cos \pi z)(x - \cos(\pi z))$, and the result follows since $y(\cos(\pi z)) = \pi^2 z^2$ and $y'(\cos(\pi z)) = -\frac{2\pi z}{\sin(\pi z)}$. □

We give a comment on the form of the result above, which might otherwise look rather mysterious. In seeking to estimate the terms which arise, we must keep an important principle in mind: If we want to prove a result which is strong enough to imply that the curves become circular, then in particular we must be able to prove that all inequalities we use hold exactly in the case of the shrinking circle. In the case of a circle the chord-arc profile is exactly $\varphi = \frac{1}{\pi}\sin(\pi x)$, as derived in equation (3.9). In this case we have $\varphi' = \cos(\pi z)$. For this reason, it is natural to look for an estimate which holds exactly in the case $x = \cos(\pi z)$, which we obtain in the lemma by using the linear approximation about this point. Now we can put everything together to get the following result:

Proposition 3.16. *Let $X : \mathcal{M}^1 \times [0,T) \to \mathbb{R}^2$ be an embedded solution of curve shortening flow and let $\varphi : [0,1] \times [0,\infty) \to \mathbb{R}$ be a smooth nonnegative function which is symmetric about $z = \frac{1}{2}$, strictly concave in its first argument, and satisfies $|\varphi'| < 1$. Suppose that φ satisfies the inequality*

$$\frac{\partial\varphi}{\partial\tau} \le 4\left(\varphi'' + \pi^2\varphi\right) + \frac{8\pi\varphi'}{\tan(\pi z)} - \frac{8\pi(\varphi')^2}{\sin(\pi x)} \tag{3.33}$$

at any point with $|\varphi'| \le 1$. If

$$d(x,y,0) \ge \mathrm{L}(0)\varphi\left(\frac{\ell(x,y,0)}{\mathrm{L}(0)}, 0\right)$$

for all $x, y \in \mathcal{M}^1$*, then*

$$d(x,y,t) \geq \mathrm{L}(t)\varphi\left(\frac{\ell(x,y,t)}{\mathrm{L}(t)}, \tau(t)\right)$$

for all $x, y \in \mathcal{M}^1$ *and all* $t \in [0,T)$*, where* $\tau(t) = \int_0^t \frac{1}{\mathrm{L}(t)^2}\,dt$.

This leaves us with the problem of how to find a function φ that satisfies the conditions of Proposition 3.16. Equation (3.33) is not a simple equation, so it is quite remarkable that we can find a simple, explicit function for which equality holds. To find this we seek a "similarity" solution of a quite general form; namely we try the ansatz $\varphi(z,\tau) = A(t)\Psi(B(\tau)C(z))$, where A, B, C, and Ψ are smooth functions. We want a solution which always has $\varphi'(0,\tau) = 1$, in order to derive a curvature bound using Proposition 3.12. Since $\varphi'(0,\tau) = A(\tau)B(\tau)\Psi'(0)C'(0)$, we look for a solution with $\Psi'(0) = 1$, $C'(0) = 1$, and $A(\tau)B(\tau) = 1$. Substituting this form in equation (3.33), we obtain

$$\begin{aligned} A'\Psi - \frac{CA'}{A}\Psi' &= \frac{4}{A}\Psi''(C')^2 + 4\Psi'C'' + 4\pi^2 A\Psi \\ &\quad + \frac{8\pi\Psi'C'}{\tan(\pi z)}\left(1 - \frac{C'\Psi'}{\cos(\pi z)}\right). \end{aligned} \tag{3.34}$$

We cancel the Ψ terms by choosing $A = \mathrm{e}^{4\pi^2\tau}$. We also simplify the last set of parentheses by making the choice $C = \frac{1}{\pi}\sin(\pi z)$, so that the factor $C'/\cos(\pi z)$ disappears. Miraculously the second term on the left then cancels exactly with the second term on the right, and the equation becomes

$$0 = 4\mathrm{e}^{-4\pi^2\tau}\cos^2(\pi z)\left(\Psi'' + 2\frac{\Psi'(1-\Psi')}{\xi}\right), \tag{3.35}$$

where $\xi = \frac{1}{\pi}\mathrm{e}^{-4\pi^2\tau}\sin(\pi z)$ is the argument of Ψ. Thus the problem has been reduced to finding the solutions of the first-order ODE $\Psi'' + 2\frac{\Psi'(1-\Psi')}{\xi} = 0$ with the initial conditions $\Psi(0) = 0$ and $\Psi'(0) = 1$. These are $\Psi(z) = z$, $\Psi(z) = \frac{1}{a}\arctan(az)$, or $\Psi(z) = \frac{1}{a}\mathrm{arctanh}(az)$. The first of these yields the circle chord-arc profile $\varphi = \frac{1}{\pi}\sin(\pi z)$. The third has $\varphi' > 1$ and so is not useful for our purposes, but the second family is exactly what we need.

Corollary 3.17. *If* $X : M^1 \times [0,T) \to \mathbb{R}^2$ *is a smooth solution of curve shortening flow with initial bound*

$$d(x,y,0) \geq \frac{\mathrm{L}(0)}{a}\arctan\left(\frac{a}{\pi}\sin\left(\frac{\pi\ell(x,y,0)}{\mathrm{L}(0)}\right)\right) \tag{3.36}$$

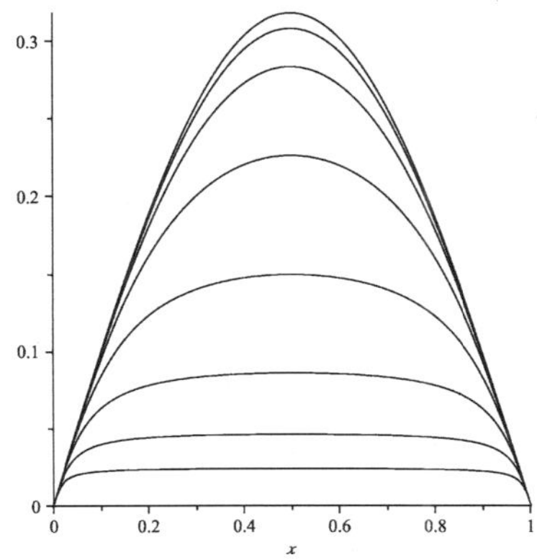

Figure 3.5. Plots of the chord-arc comparisons $\frac{1}{a}\arctan\left(\frac{a}{\pi}\sin(\pi x)\right)$ for varying values of a. As a increases to infinity the function approaches zero, while as a approaches zero the function approaches the chord-arc profile of the circle $\frac{1}{\pi}\sin(\pi x)$.

for some $a > 0$ and all $x, y \in M^1$, then we obtain the global bound

$$d(x,y,t) \geq \frac{\mathrm{L}(t)\,\mathrm{e}^{4\pi^2\tau(t)}}{a}\arctan\left(\frac{a}{\pi\mathrm{e}^{4\pi^2\tau(t)}}\sin\left(\frac{\pi\ell(x,y,t)}{\mathrm{L}(t)}\right)\right) \tag{3.37}$$

for all $x, y \in M^1$ and $t \in [0, T)$, where $\tau(t) = \int_0^t \frac{1}{\mathrm{L}(t)^2}\,dt$.

We leave it to the reader to check that this function φ is concave and has gradient less than 1 on $(0,1)$ as required in our argument.

Importantly, this lower bound on the chord-arc profile is indeed strong enough to deduce a curvature bound using Proposition 3.12. Indeed, the first few terms in the Taylor expansion about zero give the following:

$$\frac{1}{a}\arctan\left(\frac{a}{\pi}\sin(\pi z)\right) = z - \left(\frac{a^2}{3} + \frac{\pi^2}{6}\right)z^3 + O(z^5)\,. \tag{3.38}$$

Comparing with Proposition 3.12 we see that a bound of the form $d \geq \frac{\mathrm{L}}{a}\arctan(a/\pi\sin(\pi\ell/\mathrm{L}))$ implies a curvature bound

$$\kappa^2 \leq (4\pi^2 + 8a^2)/\mathrm{L}^2\,.$$

In particular the bound (3.37) implies a curvature bound

$$\kappa^2 \leq \left(\frac{2\pi}{\mathrm{L}}\right)^2 + \frac{8a^2\mathrm{e}^{-8\pi^2\tau}}{\mathrm{L}^2}\,.$$

Corollary 3.18. *If $X : M^1 \times [0,T) \to \mathbb{R}^2$ is a smooth solution of curve shortening flow satisfying* (3.36) *for some $a > 0$ and all $x, y \in M^1$, then*

$$\kappa^2(x,t) \le \left(\frac{2\pi}{\mathrm{L}(t)}\right)^2 \left(1 + \frac{2a^2}{\pi^2}\mathrm{e}^{-8\pi^2\tau(t)}\right) \tag{3.39}$$

for all $x \in M^1$ and all $t \in [0,T)$.

We need one more observation to make this useful, which we leave to the reader in Exercise 3.5 at the end of the chapter.

3.5. Grayson's theorem

We are now in a position to prove the most famous result concerning curve shortening flow:

Theorem 3.19 (Grayson's theorem). *Let $X_0 : M^1 \to \mathbb{R}^2$ be a smooth embedding of a compact connected 1-manifold M^1. Then the solution of the curve shortening flow with initial data X_0 exists on a maximal time interval $[0,T)$ and converges to a point $p \in \mathbb{R}^2$ as $t \to T$. The rescaled embeddings*

$$\tilde{X}(\cdot,t) = \frac{X(\cdot,t) - p}{\sqrt{2(T-t)}}$$

converge in C^∞ to a limit embedding $\tilde{X}_T$ with image equal to the unit circle about the origin.

Proof. The idea is to show that as the final time is approached, the length approaches zero and $\tau(t)$ approaches infinity. The curvature bound (3.39) then shows that the maximum curvature approaches $2\pi/\mathrm{L}(t)$ near the final time. Since the theorem of turning tangents implies the average curvature is exactly $2\pi/\mathrm{L}(t)$, this means the curvature must be approaching a constant and the curve is becoming circular. There is further work required to prove the strong convergence statement of the theorem.

By Exercise 3.5 there exists $a > 0$ such that (3.36) holds for all $x, y \in M^1$. By Corollary 3.18 we have the curvature bound

$$\kappa^2(x,t) \le \left(\frac{2\pi}{\mathrm{L}(t)}\right)^2 \left(1 + \frac{2a^2}{\pi^2}\mathrm{e}^{-8\pi^2\tau(t)}\right), \tag{3.40}$$

where $\tau(t) \ge 0$ for each $t \in [0,T)$.

Lemma 3.20. $\lim_{t\to T} \mathrm{L}(t) = 0$.

Proof. We know that $\mathrm{L}(t)$ is decreasing since $\frac{\partial}{\partial t} ds = -\kappa^2\, ds$ by (2.53). If $\mathrm{L}(t)$ does not approach zero, then we have $\mathrm{L}(t) \ge L_0 > 0$ for all $t \in [0,T)$ and hence $\kappa^2 \le C/L_0^2$ remains bounded. This contradicts the global existence theorem (Theorem 2.12). □

Lemma 3.21. *$\mathrm{L}(t)$ is comparable to $\sqrt{T-t}$:*

$$2\pi\sqrt{2(T-t)} \leq \mathrm{L}(t) \leq 2\pi\sqrt{2\left(1+\frac{2a^2}{\pi^2}\right)(T-t)}$$

for all $t \in [0,T)$.

Proof. These inequalities follow from the evolution equation

$$\frac{d}{dt}\mathrm{L} = -\int_{M^1} \kappa^2\, ds$$

and the inequalities

$$\int_{M^1} \kappa^2\, ds \geq \frac{\left(\int \kappa\, ds\right)^2}{\mathrm{L}} = \frac{4\pi^2}{\mathrm{L}}$$

and

$$\int_{M^1} \kappa^2\, ds \leq \frac{4\pi^2(1+\frac{2a^2}{\pi^2})}{\mathrm{L}(t)}. \qquad \square$$

Substituting the inequalities in Lemma 3.21 into the identity $\tau = \int_0^t \frac{1}{\mathrm{L}^2}\, dt$ we conclude the following:

Lemma 3.22. *$\tau(t)$ approaches infinity as $t \to T$, and*

$$-\frac{1}{8\pi^2+16a^2}\log\left(1-\frac{t}{T}\right) \leq \tau(t) \leq -\frac{1}{8\pi^2}\log\left(1-\frac{t}{T}\right).$$

Now the curvature bound (3.40) becomes

$$\kappa^2 \leq \left(\frac{2\pi}{\mathrm{L}(t)}\right)^2\left(1+\frac{2a^2}{\pi^2}\left(1-\frac{t}{T}\right)^{\frac{1}{1+2a^2/\pi^2}}\right).$$

Using this in the evolution equation for $\mathrm{L}(t)$ yields an improvement of Lemma 3.21:

Lemma 3.23. *Set $C \doteqdot \frac{2a^2}{\pi^2}T^{-\frac{1}{1+2a^2/\pi^2}}$. Then*

$$2\pi\sqrt{2(T-t)} \leq \mathrm{L}(t) \leq 2\pi\sqrt{2(T-t)\left(1+C(T-t)^{\frac{1}{1+2a^2/\pi^2}}\right)}.$$

This implies in particular that the isoperimetric ratio $\frac{\mathrm{L}^2}{4\pi\mathrm{A}}$ approaches 1 as the final time is approached since the enclosed area is given by (see Exercise 3.7)

$$\mathrm{A}(t) = 2\pi(T-t).$$

We also obtain better estimates for κ and L.

Corollary 3.24. *There exists a constant $C < \infty$ such that*

$$\kappa^2 \le \left(\frac{2\pi}{\mathrm{L}}\right)^2 (1 + C(T-t))$$

and

$$\mathrm{L} \le 2\pi\sqrt{2(T-t)}(1 + C(T-t))\,.$$

Proof. See Exercise 3.8. □

In order to extract a limit of the rescaled embeddings $\tilde{X}_t$ as t approaches T, we will require bounds on $\tilde{X}_t$ and all of its derivatives. The first step is to control the derivatives of curvature.

Lemma 3.25. *There exist constants $\tilde{C}_k$ for each $k > 0$ such that*

$$\left|\frac{\partial^k \kappa}{\partial s^k}\right|^2 \le \frac{\tilde{C}_k}{(T-t)^{1+k}}.$$

Proof. We apply the smoothing estimates from Theorem 2.11. The curvature bound (3.40) gives that $|\kappa|$ is bounded by a multiple of L^{-1}, which by Lemma 3.21 is comparable to $(T-t)^{1/2}$, so there exists C_0 such that

$$|\kappa| \le C_0 (T-t)^{-1/2}.$$

In particular, on the time interval $[T-b, T - \frac{C_0^2}{1+C_0^2}b]$ of length $\frac{b}{1+C_0^2}$ we have $|\kappa| \le \sqrt{\frac{1+C_0^2}{b}}$. Therefore we can apply Theorem 2.11 to deduce

$$\left|\frac{\partial^k \kappa}{\partial s^k}\right| \le C_k \sqrt{\frac{1+C_0^2}{b}}(t - T + b)^{-k/2}.$$

For given t, choosing $b = \min\left\{T, \frac{1+C_0^2}{C_0^2}(T-t)\right\}$ gives

$$\left|\frac{\partial^k \kappa}{\partial s^k}\right| \le C_k \max\left\{\sqrt{\frac{1+C_0^2}{T}}t^{-k/2}, \left(\frac{C_0^2}{T-t}\right)^{\frac{k+1}{2}}\right\}. \tag{3.41}$$

We also have estimates for small t from the short-time existence theorem, and the result follows. □

Note that for each k, the expression $(T-t)^{\frac{k+1}{2}}\frac{\partial^k \kappa}{\partial s^k}$ is just the k-th derivative of curvature with respect to arc length on the rescaled curve, so Lemma 3.25 simply states that all derivatives of curvature remain bounded on the rescaled curves. Next we deduce that the curvature of the rescaled curves approaches 1.

Lemma 3.26. *There exists* $C > 0$ *such that*

$$\left|\kappa\sqrt{2(T-t)} - 1\right| \leq C\sqrt{T-t}.$$

Proof. First we prove convergence in an integral sense. By applying the curvature bound from Corollary 3.24 (in the first term on the right), the theorem of turning tangents (in the second), and the length estimate from Corollary 3.24 (in both terms on the second line), we find that

$$\begin{aligned}\frac{1}{\mathrm{L}}\int_{M^1}\left|\kappa\sqrt{2(T-t)}-1\right|^2 ds &= \frac{2(T-t)}{\mathrm{L}}\int_{M^1}\kappa^2\,ds - \frac{2\sqrt{2(T-t)}}{\mathrm{L}}\int_{M^1}\kappa\,ds + 1\\ &\leq \frac{8\pi^2}{\mathrm{L}^2}(T-t+C(T-t)^2) - \frac{4\pi}{\mathrm{L}}\sqrt{2(T-t)} + 1\\ &\leq C(T-t)\,. \end{aligned} \tag{3.42}$$

Since we have bounds on the derivatives of curvature from Lemma 3.25, we can convert the integral bounds into pointwise bounds. The idea is that if there is some point where the difference $|\kappa\sqrt{2(T-t)}-1|$ is large, then the derivative estimates imply that the difference is large on a neighborhood, so that the integral must also be large and we deduce that (see Exercise 3.9)

$$\left|\kappa\sqrt{2(T-t)} - 1\right| \leq C(T-t)^{1/2}. \tag{3.43}$$

This is a simple example of an **interpolation inequality**, which is a common tool in the analysis of partial differential equations. We remark that by using interpolation inequalities involving higher derivatives, one can improve the right-hand side of the estimate to $C_\gamma(T-t)^\gamma$ for any $\gamma < 1$. See Exercise 3.11. □

By a similar argument we can show that all of the derivatives of the curvature of the rescaled curves decay to zero.

Lemma 3.27. *For each* $k \in \mathbb{N}$ *there exists* $\bar{C}_k < \infty$ *such that*

$$(T-t)^{\frac{1+k}{2}}\left|\frac{\partial^k\kappa}{\partial s^k}\right| \leq \bar{C}_k(T-t)^{1/4}.$$

Proof. This is again an exercise in interpolation. For a smooth function $f : M^1 \to \mathbb{R}$, if the second derivative is bounded and the absolute value of f is very small, then the first derivative must also be quite small. The reason is that if the derivative were large somewhere, then the second derivative bound would imply that the derivative was large on a whole neighborhood, and so the function itself could not be small everywhere. A simple proof can be given using the Taylor approximation of degree 2 (following Exercise 15 in Chapter 5 of Rudin [**449**]). Choosing x to be a point where the maximum derivative is attained, we have

$$f(x+h) = f(x) + f'(x)h + \frac{h^2}{2} f''(z)$$

for some z. Rearranging this gives

$$f'(x) = \frac{f(x+h) - f(x)}{h} - \frac{h}{2} f''(z),$$

from which we deduce

$$M_1 \le \frac{2}{h} M_0 + \frac{h}{2} M_2,$$

where $M_k = \sup_{M^1} \left|\partial_s^k f\right|$ for each k. Optimizing over h gives

$$M_1 \le 2 M_0^{1/2} M_2^{1/2}. \tag{3.44}$$

Applying this with f replaced by the k-th derivative of f gives

$$M_{k+1} \le 2 M_k^{1/2} M_{k+1}^{1/2}. \tag{3.45}$$

Using this we can control the size of any derivative of f if we have control on a higher derivative and a lower derivative. Indeed, (3.45) implies that

$$M_j \le 2^{(k-j)(j-i)} M_i^{\frac{k-j}{k-i}} M_k^{\frac{j-i}{k-i}} \tag{3.46}$$

for any triple of integers i, j, and k satisfying $0 \le i < j < k$ (see Exercise 3.10). For a given k we use equation (3.46) applied to $f = \kappa\sqrt{2(T-t)} - 1$ to interpolate the k-th derivative of κ between the result of Lemma 3.26 and the bound on a very high derivative of κ from Lemma 3.25. For concreteness we can interpolate using the $2k$-th derivative:

$$\begin{aligned} |\partial_s^k \kappa| &\le 2^{\frac{k^2}{2}} \sup_{M^1} \left|\kappa\sqrt{2(T-t)} - 1\right|^{\frac{1}{2}} \sup_{M^1} |\partial_s^{2k}\kappa|^{\frac{1}{2}} (2(T-t))^{-\frac{1}{4}} \\ &\le C(T-t)^{\frac{1}{4}} \tilde{C}_{2k}^{\frac{1}{2}} (T-t)^{-\frac{1+2k}{4}} (T-t)^{-\frac{1}{4}} \\ &= C(T-t)^{-\frac{k}{2}-\frac{1}{4}}. \end{aligned} \tag{3.47}$$

This implies

$$(T-t)^{\frac{1+k}{2}} |\partial_s^k \kappa| \le C(T-t)^{\frac{1}{4}}, \tag{3.48}$$

as required. We remark that by interpolating with higher derivatives both here and in Lemma 3.26 we could improve the bound on the right to $C(T-t)^\gamma$ for any $\gamma < 1$. □

Now we can prove that the curves converge to a point p:

Lemma 3.28. *There exists $p \in \mathbb{R}^2$ such that $X(\cdot, t)$ converges uniformly to p as t approaches T.*

Proof. Fix any $z_0 \in M^1$. Then

$$|X(z_0,t_2) - X(z_0,t_1)| \leq \int_{t_1}^{t_2} |\kappa(x_0,t)|\,dt \leq C\sqrt{T-t_1}$$

for all $t_1 < t_2 < T$ since $|\kappa| \leq C(T-T)^{-1/2}$ by Lemma 3.26. In particular $X(z_0,t)$ is Cauchy in $\mathbb{R}^2$ as $t \to T$ and hence converges to some $p \in \mathbb{R}^2$, with $|X(z_0,t) - p| \leq C\sqrt{T-t}$. For any $z \in M^1$ we have $|X(z,t) - X(z_0,t)| \leq \frac{L(t)}{2} \leq C\sqrt{T-t}$ by Corollary 3.24, so $|X(z,t) - p| \leq C\sqrt{T-t}$, proving the lemma. □

With the point p identified, we can now define the rescaled embeddings $\tilde{X}(\cdot,t)$ given in the statement of Grayson's theorem. Using the estimates of Lemma 3.27 we can control all of the derivatives of $\tilde{X}$:

Lemma 3.29. *There exists $C_- > 0$ such that*

$$|\tilde{X}'| \geq C_-,$$

and for each $k \geq 1$ there exists C_k such that

$$|\tilde{X}^{(k)}| \leq C_k \quad \text{and} \quad \left|\frac{\partial}{\partial t}\tilde{X}^{(k)}\right| \leq C_k(T-t)^{-3/4}.$$

Proof. This is a straightforward adaptation of the proofs of Lemmas 2.13 and 2.16. We have $\tilde{X}' = \frac{X'}{\sqrt{2(T-t)}}$, and so

$$\begin{aligned}\frac{\partial}{\partial t}\tilde{X}' &= -\,|\tilde{X}'|\partial_s(\kappa\mathsf{N}) + \frac{X'}{(2(T-t))^{3/2}} \\ &= -\,|\tilde{X}'|\left(\kappa_s\mathsf{N} + \left(\kappa^2 - \frac{1}{2(T-t)}\right)\mathsf{T}\right).\end{aligned} \tag{3.49}$$

From this we find that

$$\frac{\partial}{\partial t}\log|\tilde{X}'| = -\left(\kappa^2 - \frac{1}{2(T-t)}\right).$$

The right-hand side is bounded by a constant by Lemma 3.26, so $\log|\tilde{X}'|$ is uniformly bounded (since the time of existence is finite).

To control the higher derivatives we adapt the result of Lemma 2.14. First, noting that $\partial_{\tilde{s}} = \frac{1}{\sqrt{2(T-t)}}\partial_s$ where $\tilde{s}$ is the arc length parameter on $\tilde{X}$, we rewrite equation (3.49) as

$$\frac{\partial}{\partial t}\tilde{X}' = -\frac{|\tilde{X}'|}{2(T-t)}\left(\tilde{\kappa}_{\tilde{s}}\mathsf{N} + (\tilde{\kappa}^2 - 1)\,\mathsf{T}\right).$$

On differentiating this we find that $\frac{\partial \tilde{X}^{(k)}}{\partial t}$ is equal to $\frac{1}{2(T-t)}$ times a polynomial in $\tilde{\kappa},\dots,\partial_{\tilde{s}}\tilde{\kappa},\dots,\partial_{\tilde{s}}^k\tilde{\kappa}$ and $\mathsf{T}\cdot\tilde{X}^{(i)}$ for $i \leq k$, with every term containing either a factor $\partial_{\tilde{s}}^i\tilde{\kappa}$ for some $i > 1$ or a factor $\tilde{\kappa}^2 - 1$, and every term containing at most one factor $\mathsf{T}\cdot X^{(k)}$. We proceed by induction on k. Assuming that we have already obtained the result for $k = 1, \ldots, n$, we observe that since $|\tilde{\kappa}^2 - 1| \le C\sqrt{T-t}$ and $|\partial_{\tilde{s}}^i\tilde{\kappa}| \le C(T-t)^{1/4}$ for each $i \ge 1$, we have

$$\left|\frac{\partial}{\partial t}\tilde{X}^{(n+1)}\right| \le C(T-t)^{-3/4}\left(\left|\tilde{X}^{(n+1)}\right| + 1\right). \tag{3.50}$$

It follows that $\tilde{X}^{(n+1)}$ is bounded since $(T-t)^{-3/4}$ is integrable on $[0, T)$. Substituting this bound into equation (3.50) gives

$$\left|\frac{\partial}{\partial t}\tilde{X}^{(n+1)}\right| \le C(T-t)^{-3/4}. \tag{3.51}$$

This completes the induction and the proof. □

Notice that we have controlled the derivatives of $\tilde{X}$ but not yet $\tilde{X}$ itself. However it is immediate that the derivative $\tilde{X}'$ converges smoothly to a nonzero limit $Y(\cdot, T)$, with $|\tilde{X}'(\cdot, t) - Y(\cdot)| \le C(T-t)^{1/4}$. It follows that $\mathsf{N} = J\left(\frac{\tilde{X}'}{|\tilde{X}'|}\right)$ converges smoothly to $\mathsf{N}_T = J(Y/|Y|)$, with $|\mathsf{N}(\cdot,t) - \mathsf{N}_T(\cdot)| \le C(T-t)^{1/4}$. We can use this to show that $\tilde{X}$ converges, as follows (this is the only place that we use the choice of the point p): We compute

$$\begin{aligned}\left|\partial_t(X - \sqrt{(2(T-t)}\mathsf{N})\right| &= \left|\left(\frac{1}{\sqrt{2(T-t)}} - \kappa\right)\mathsf{N} - \sqrt{2(T-t)}\kappa_s\mathsf{T}\right| \\ &\le C(T-t)^{-1/4}\end{aligned}$$

by Lemma 3.26 and the case $k = 1$ of Lemma 3.27. Since $X - \sqrt{2(T-t)}\mathsf{N}$ converges to p as $t \to T$, we conclude that

$$\left|X - p - \sqrt{2(T-t)}\mathsf{N}\right| \le C(T-t)^{3/4}.$$

Dividing through by $\sqrt{2(T-t)}$ gives

$$|\tilde{X}(\cdot,t) - \mathsf{N}(\cdot,t)| \le C(T-t)^{1/4},$$

and hence $\tilde{X}$ converges uniformly to N_T. Since $\tilde{X}'$ converges in C^∞, it follows that the convergence of $\tilde{X}$ to N_T is in C^∞, and we conclude the proof by observing that N_T is a smooth diffeomorphism to S^1 since $\mathsf{N}' = |X'|\kappa$ which converges to the nonzero limit $|Y|$. □

3.6. Singularities of immersed solutions

We conclude the chapter by noting that Grayson's theorem fails when embeddedness of the initial curve is relaxed. Indeed, it is already clear from the Whitney–Graustein theorem[2] that the conclusion necessarily fails for any closed curve with rotation index not equal to one. When the rotation index is zero (as is the case, e.g., for the "figure-8" curve), this also eliminates the possibility that the solution limits to a multiple covering of a round S^1.

It is instructive to consider an explicit example (due to Angenent [**66**]). Consider the evolution $\gamma : M^1 \times [0,T) \to \mathbb{R}^2$ by curve shortening of an immersion $\gamma_0 : M^1 \to \mathbb{R}^2$ with rotation index 2 and a single point of self-intersection.

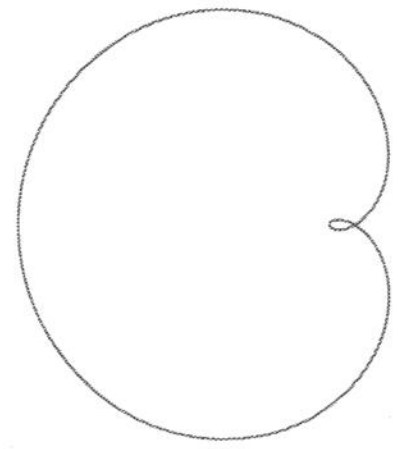

Figure 3.6. A closed planar curve which forms a cusp singularity under curve shortening flow.

We claim that the number of self-intersections of γ_t is 1 for all $t \in [0,T)$. Indeed, by the Whitney–Graustein theorem, this number certainly cannot decrease to zero. On the other hand, given any $(x_0,t_0) \in M^1 \times (0,T)$, we can find a ball B_r of radius $r > 0$ about $\gamma(x_0,t_0)$ such that, for each $t \in (t_0 - r^2, t_0]$, the set $\gamma_t^{-1}(B_r)$ consists of $|\gamma_{t_0}^{-1}(\{x_0\})|$ components, any two of which can be represented by graphs of the time t slice of smooth functions $u_1 : B_r \cap L \times (t_0 - r^2, t_0] \to \mathbb{R}$ and $u_2 : B_r \cap L \times (t_0 - r^2, t_0] \to \mathbb{R}$ over some line $L \subset \mathbb{R}^2$ (which, after a rotation, we take to be the x-axis) satisfying the graphical curve shortening flow. Their difference, $w \doteqdot u_2 - u_1$, satisfies a linear parabolic equation of the form

$$\partial_t w(x,t) = a(x,t)\partial_x^2 w(x,t) + b(x,t)\partial_x w(x,t)\,.$$

Sturm's theorem (Theorem 1.21) then implies that the number of zeroes of w does not increase. We conclude that the number of self-intersections of Γ_t cannot increase and is therefore always 1. In particular, $\Gamma_t \doteqdot \gamma_t(M^1)$ always consists of two simple closed curves, Γ_t^1 and Γ_t^2. Denote the areas enclosed by Γ_t^1 and Γ_t^2 by $A_1(t)$ and $A_2(t)$, respectively. Then

$$-A_i'(t) = \int_{\Gamma_t^i} \kappa_i \, ds_t^i = \begin{cases} 3\pi - \alpha(t) & \text{if } \ i = 1\,, \\ \pi + \alpha(t) & \text{if } \ i = 2\,, \end{cases}$$

[2]The **rotation index** of a closed, immersed curve is preserved under regular homotopies [**534**] (see also [**484**]).

where κ_i and ds_i are the curvature and length element of Γ_i and $\alpha(t) \in [0, \pi]$ is the angle between the tangent lines at the point of self-intersection. In particular,

$$\pi < -A_2' < 2\pi < -A_1' < 3\pi .$$

So if $A_1(0) > 3A_2(0)$, then the area enclosed by Γ_t^2 must disappear before time $A_2(0)/\pi$, while the area enclosed by Γ_t^1 remains strictly positive at this time. So Γ_t becomes singular before it shrinks to a point.

3.7. Notes and commentary

That the curve shortening flow evolves closed *convex* curves to round points was proved by Gage and Hamilton [**228**], with earlier work by Gage [**229**], [**230**]. The space of convex curves (or convex bodies) in $\mathbb{R}^2$ has some interesting structure which can be exploited to understand curve flows, and the argument of Gage and Hamilton made use of this structure in several ways: Gage [**229**] proved that the isoperimetric ratio of the curves improves with time in the convex case. Gage and Hamilton proved a Li–Yau-type differential Harnack estimate and also found a monotone "entropy" quantity. These were both used in the proof of convergence, and similar ideas play a central role in our understanding of higher-dimensional generalizations of the curve shortening flow and Ricci flow. We discuss some of these techniques for curve shortening flow in the next chapter and expand on them later in a more general context.

Grayson [**247**] extended the Gage–Hamilton theorem to general closed embedded curves, proving that in this situation the evolving curves eventually become convex. Grayson's proof made strong use of "zero-counting" arguments, which in this case imply that the number of inflection points of the curve cannot increase with time. He made use of this to prove a key geometric lemma (the "δ-whisker lemma") which can be seen as a precursor of the distance comparison estimates we discussed above. The argument involves a careful accounting of the behavior and interaction between "nice" subarcs of the curve and requires analysis of various possible cases.

There are now a multitude of approaches to Grayson's theorem using the idea of controlling or proving monotonicity of an isoperimetric ratio, all of which greatly simplify Grayson's original argument. We presented in Section 3.3 the approach using chord-arc distance comparison due to Huisken [**298**]. A similar argument was provided by Hamilton [**270**], who showed how to control the geometry by bounding the isoperimetric profile of the region enclosed by the evolving curve. Both of these methods can be combined with a blow-up argument and some classification of singularities to deduce Grayson's theorem. We discuss the techniques required for this in the next chapter. In the presentation we adopted here, based on a refinement of Huisken's distance comparison due to Bryan and the first author [**45**], no separate analysis of the convex case is needed and no analysis of singularities is required. A similar extension of Hamilton's isoperimetric profile estimate was found later [**44**], with the additional feature that the isoperimetric profile can be compared to that of a suitable model solution of curve shortening flow (such as the paperclip solution), so that one does not have to construct "by hand" a solution of the differential inequality as we did here in equations (3.34) and (3.35) — an exact solution is provided by the model solution.

While these approaches to the proof allow for a streamlined exposition and a quick route to a famous theorem, the drawback is that many beautiful and useful techniques have not been discussed along the way. To counter such accusations, we will devote the next chapter to the elaboration of some of the other results and methods which have been developed for curve shortening flow, many of which later appear and play a crucial role in the analysis of other geometric evolution equations.

In Section 3.6, we saw that immersed curves can suffer cusp-like singularities before contracting to a point. In Chapter 11, we shall see that such singularities look, after appropriately rescaling, like the translating Grim Reaper solution. Finer asymptotics for certain cusp-forming solutions have been obtained by Angenent [**66**] and Angenent and Velázquez [**74**] (see also [**24**]). In particular, the curvature of these solutions blows up like

$$O\left((T-t)^{-\frac{1}{2}}\right) < \max_{\mathcal{M}^1\times\{t\}} \kappa = (1+o(1))\sqrt{\frac{\log\log\left(\frac{1}{T-t}\right)}{T-t}} < O\left((T-t)^{-\frac{1}{2}-\varepsilon}\right)$$

for any $\varepsilon > 0$, and the asymptotic shape is a cusp of the form

$$y = \left(\frac{\pi}{4}+o(1)\right)\frac{x}{\log\log\left(\frac{1}{x}\right)}\,.$$

The curve shortening flow of curves on surfaces was studied by Grayson [**248**], Angenent [**65**], [**67**], and Oaks [**428**]. Applications of curve shortening flow to closed geodesics on surfaces were given by Angenent [**69**]. Oaks [**428**] and Chou and Zhu [**155**] gave some extensions of Grayson's theorem to anisotropic analogues of curve shortening flow. Angenent, Sapiro, and Tannenbaum [**73, 455**] proved a result similar to Grayson's theorem for an affine-geometric curvature flow of nonconvex curves in the plane (in which the speed is proportional to the cube root of the curvature at each point). This flow will be studied in higher dimensions in Chapter 16. Chou and Zhu [**156**] studied flows where the speed is proportional to other powers of curvature.

Flows of plane curves by nonzero powers α of curvature were studied by the first author in [**36**]. When $\alpha > 1/3$, the only embedded shrinking self-similar solutions are round circles centered at the origin [**36**, Theorem 1.5]. When $1/3 < \alpha < 1$, solutions converge to round points [**36**, Corollary 1.6]. When $\alpha \in (0, 1/3)$, the isoperimetric ratio becomes unbounded as the final time is approached for generic symmetric initial data.

Lauer [**355**] considered the curve shortening flow with highly irregular initial data (essentially an arbitrary connected, locally connected compact subset of the plane), proving in particular the existence of a unique smooth solution starting from any Jordan curve with finite length.

Huisken and Polden [**300**] also considered the curve shortening flow in noncompact situations, assuming some control on the curves near infinity.

Convergence results for flows by negative powers of curvature have been obtained by Urbas [**509**], [**510**], Gerhardt [**236**], and Andrews [**32**]. Expanding flows by nonhomogeneous functions of curvature were studied by the second author and Tsai [**171**], [**504**], and the second author, Liou, and Tsai [**168**]. Further flows of curves were studied by Y.-L. Ni and M.-J. Zhu [**424**]. Yagisita [**537**] studied expanding flows of starshaped curves with speed $1-\kappa$ and initial curvature less than 1.

3.8. Exercises

Exercise 3.1. Define $f : M^1 \times [0,T) \to \mathbb{R}$ by $f(x,t) = |X(x,t)|^2$. Show that $\frac{\partial f}{\partial t} = \frac{\partial^2 f}{\partial s^2} - 2$. Use this to give another proof of Corollary 3.3.

Exercise 3.2. Define $g : M^1 \times M^1 \to \mathbb{R}$ by $g(x_1, x_2, t) = |X(x_1,t) - X(x_2,t)|^2$. Show that $\frac{\partial g}{\partial t} = \left(\frac{\partial^2}{\partial s_1^2} + \frac{\partial^2}{\partial s_2^2}\right) g - 4$. Observe that this implies (3.6).

Exercise 3.3. Show that d satisfies the following degenerate heat-type equation on $M_1 \times M_2$:

$$\frac{\partial d}{\partial t} = \left(\frac{\partial^2}{\partial s_1^2} + \frac{\partial^2}{\partial s_2^2} + 2\mathsf{T}_1 \cdot \mathsf{T}_2 \frac{\partial^2}{\partial s_1 \partial s_2}\right) d - \frac{1}{d}\left(\left(\frac{\partial d}{\partial s_1}\right)^2 + \left(\frac{\partial d}{\partial s_2}\right)^2 - 2\mathsf{T}_1 \cdot \mathsf{T}_2 \frac{\partial d}{\partial s_1}\frac{\partial d}{\partial s_2}\right)$$

where s_1 and s_2 are arc length parameters on M_1 and M_2, respectively. Theorem 3.1 then follows by applying the maximum principle.

Exercise 3.4. Derive the Taylor expansion (3.12).

Exercise 3.5. Show that for any smooth initial embedding $X_0 : M^1 \to \mathbb{R}^2$ from a compact 1-manifold to $\mathbb{R}^2$, there exists a constant $a > 0$ such that the estimate (3.36) holds.

Exercise 3.6. Why is it necessary to rule out the "Case 2" situation using Lemma 3.13, rather than proceeding as we did in the Huisken estimate in Section 3.3?

Exercise 3.7. Let $X : M^1 \times [0,T) \to \mathbb{R}^2$, $M^1 \cong S^1$, be a maximal embedded solution of curve shortening flow. Prove, using the fact that $L(t) \to 0$ as $t \to T$, that the area $\mathrm{A}(t)$ enclosed by $\mathcal{M}_t$ is given by $\mathrm{A}(t) = 2\pi(T-t)$.

Exercise 3.8. Use the result of Lemma 3.23 and the definition of τ to prove $\kappa^2 \le \left(\frac{2\pi}{\mathrm{L}}\right)^2(1+C(T-t))$ and $\mathrm{L} \le 2\pi\sqrt{2(T-t)}(1+C(T-t))$ for some C.

Exercise 3.9. Let f be a smooth function on a simple closed curve of length L, with $M_1 = \sup_{M^1}|f_s|$, $\|f\|_2 = \left(\int_{M^1} f^2\,ds\right)^{1/2}$, and $M_0 = \sup_{M^1}|f|$. Using the gradient bound find an interval on which $|f| \ge \frac{1}{2}M_0$, and use this to obtain a lower bound on $\|f\|_2$. Deduce that

$$M_0 \le \max\{2^{2/3}\|f\|_2^{2/3}M_1^{1/3}, 2\|f\|_2/\,\mathrm{L}\}.$$

Exercise 3.10. Using the inequality (3.45), show that

$$M_j \le 2^{(k-j)(j-i)}M_i^{\frac{k-j}{k-i}}M_k^{\frac{j-i}{k-i}} \quad \text{for integers } 0 \le i < j < k.$$

Exercise 3.11. Using interpolation inequalities involving higher derivatives, improve the estimate of Lemma 3.26 by showing that

$$\left|\kappa\sqrt{2(T-t)} - 1\right| \le C_\gamma(T-t)^\gamma$$

for any $\gamma < 1$, where C_γ is a constant which may depend on γ.

Chapter 4

Self-Similar and Ancient Solutions

This chapter concerns special solutions to the curve shortening flow. We first describe some invariance properties and corresponding self-similar solutions, which are solutions whose shapes do not change, although their sizes may change. We discuss the classification of self-similar solutions. We then describe various monotonicity formulae, including the Harnack inequality and monotonicity of the isoperimetric ratio, the Nash entropy, and the Gaußian density. We describe two explicit ancient solutions to the curve shortening flow before discussing the classification of convex, embedded ancient solutions to curve shortening flow on closed curves due to Daskalopoulos, Hamilton, and Šešum.

4.1. Invariance properties

The curve shortening flow (in the plane) possesses some useful invariance properties, which shed light on the behavior of solutions and can be exploited in various ways. Given a solution $X : M^1 \times I \to \mathbb{R}^2$ of curve shortening flow, the following local symmetries, $X \mapsto X_\varepsilon$, result, for appropriate values of ε, in new solutions X_ε (possibly defined over new space-time domains).

(1) Reparametrization: $X_\varepsilon(x,t) \doteqdot X(\phi(x,\varepsilon),t)$, where ϕ is the flow of some $V \in \Gamma(TM^1)$.

(2) Time translation: $X_\varepsilon(x,t) \doteqdot X(x,t+\varepsilon)$.

(3) Space translation: $X_\varepsilon(x,t) \doteqdot X(x,t) + \varepsilon V$, $V \in \mathbb{R}^2$.

(4) Rotation: $X_\varepsilon(x,t) \doteqdot \mathrm{e}^{\varepsilon J} \cdot X(x,t)$, $J \in \mathfrak{so}(2)$.

(5) Space-time dilation: $X_\varepsilon(x,t) \doteqdot \mathrm{e}^{\varepsilon} X(x,\mathrm{e}^{-2\varepsilon}t)$.

A solution $X : M^1 \times I \to \mathbb{R}^2$ to curve shortening flow is an **invariant solution** if it is invariant under a 1-parameter family of local symmetries.

Example 4.1.

(1) Given $a, b \in \mathbb{R}^2$, the stationary line

$$X(x,t) \doteqdot a + xb\,, \quad (x,t) \in \mathbb{R} \times (-\infty, \infty), \tag{4.1}$$

is invariant under time translations:

$$X(x, t+\varepsilon) = X(x,t)\,.$$

If $a = 0$, then X is also invariant under space-time dilation modulo reparametrization:

$$\mathrm{e}^{\varepsilon} X(\phi(x,\varepsilon), \mathrm{e}^{-2\varepsilon} t) = X(x,t)\,,$$

where $\phi(x,\varepsilon) \doteqdot \mathrm{e}^{-\varepsilon} x$.

(2) The shrinking circle,

$$X(x,t) \doteqdot \sqrt{-2t}\, x\,, \quad (x,t) \in S^1 \times (-\infty, 0)\,, \tag{4.2}$$

is invariant under space-time dilation:

$$\mathrm{e}^{\varepsilon} X(x, \mathrm{e}^{-2\varepsilon} t) = X(x,t)\,.$$

(3) The Grim Reaper,

$$X(x,t) \doteqdot \left(\arctan\left(\mathrm{e}^{-t}\tan x\right), \log\sqrt{\tan^2 x + \mathrm{e}^{2t}}\right), \tag{4.3}$$

$(x,t) \in \left(-\frac{\pi}{2}, \frac{\pi}{2}\right) \times (-\infty, \infty)$, is invariant under translation modulo reparametrization:

$$X(\phi(x,\varepsilon), t+\varepsilon) - \varepsilon e_2 = X(x,t)\,,$$

where

$$\phi(x,\varepsilon) \doteqdot \arctan\left(\mathrm{e}^{\varepsilon}\tan x\right)\,.$$

4.2. Self-similar solutions

An important class of invariant solutions are the *self-similar* solutions. Since we will be dealing with rotations, it is convenient to identify $\mathbb{R}^2$ with the complex plane, $\mathbb{C}$. A solution $X : M^1 \times I \to \mathbb{C}$ to curve shortening flow is **self-similar** if there exist smooth functions $a : (-\varepsilon_0, \varepsilon_0) \to \mathbb{R}$, $b : (-\varepsilon_0, \varepsilon_0) \to \mathbb{R}$, $c : (-\varepsilon_0, \varepsilon_0) \to \mathbb{C}$, and $s : (-\varepsilon_0, \varepsilon_0) \to \mathbb{R}$ with $a(0) = 0$, $b(0) = 0$, and $c(0) = 0$, and a smooth 1-parameter family of diffeomorphisms $\phi : M^1 \times (-\varepsilon_0, \varepsilon_0) \to M^1$ with $\phi(\cdot, 0) = 1$ such that

$$\mathrm{e}^{-a(\varepsilon) - ib(\varepsilon)} X\left(\phi(u,\varepsilon), \mathrm{e}^{2a(\varepsilon)} t + s(\varepsilon)\right) - c(\varepsilon) = X(u,t)\,. \tag{4.4}$$

We also assume that $\frac{da}{d\varepsilon}(0)$ and $\frac{ds}{d\varepsilon}(0)$ are not both zero — aside from ruling out the trivial transformation, this also ensures that the function

$$\tau(\varepsilon) \doteqdot e^{2a(\varepsilon)}t_0 + s(\varepsilon)$$

is locally invertible for every $t_0 \in I \setminus \{-\frac{S}{2A}\}$ (or every $t_0 \in I$ if $\frac{da}{d\varepsilon}(0) = 0$). Choosing such a time $t_0 \in I$ and denoting by ε a local inverse for τ, the solution is then determined locally by its time t_0 slice via

$$X(v,t) = e^{a(\varepsilon(t))+ib(\varepsilon(t))}\left(c(\varepsilon(t)) + X\left(\phi^{-1}(v,\varepsilon(t)), t_0\right)\right).$$

Differentiating (4.4) with respect to ε at $\varepsilon = 0$ and setting $t = t_0 \in I$, we find that the t_0 time slice must satisfy

$$(A+iB)\,X + C = \frac{\partial X}{\partial u}V - (2At_0 + S)\kappa\mathsf{N},$$

where $A \doteqdot \frac{da}{d\varepsilon}(0)$, $B \doteqdot \frac{db}{d\varepsilon}(0)$, $C \doteqdot \frac{dC}{d\varepsilon}(0)$, $S \doteqdot \frac{ds}{d\varepsilon}(0)$, and $V \doteqdot \frac{d\phi}{d\varepsilon}(\cdot,0)$. Assuming that $\{\mathsf{N},\mathsf{T}\}$ is positively oriented, so that $i\mathsf{N} = \mathsf{T}$, we find, in particular, that

$$A\langle X,\mathsf{N}\rangle + B\langle X,\mathsf{T}\rangle + \langle C,\mathsf{N}\rangle = -(2At_0 + S)\kappa. \tag{4.5}$$

A complete classification of solutions to (4.5) is possible [**258**]: Consider first the case $A = B = 0$. We claim that X is, modulo rigid motions and dilation, either a straight line or the Grim Reaper curve.

Lemma 4.2. *Suppose that M^1 is connected and that $X : M^1 \to \mathbb{R}^2$ is a proper immersion satisfying*

$$\langle C,\mathsf{N}\rangle = S\kappa$$

for some $C \in \mathbb{R}^2$ and some $S \in \mathbb{R}$, not both zero. Then X is, modulo rigid motions and dilations, either a straight line or the Grim Reaper curve.

Proof. If either C or S is zero, then it is easy to see that X must be a straight line. If $C \neq 0$, then X is the image under a rotation and a dilation of a curve satisfying

$$\cos\theta = -\langle e_2,\mathsf{N}\rangle = \kappa,$$

where θ is the angle between T and the x-axis. Choose a point $p_0 \in M^1$ such that $\kappa(p_0) \neq 0$ and let $U \subset M^1$ be the largest connected neighborhood of p_0 on which $\kappa \neq 0$. Then, parametrizing $X|_U$ with respect to θ, we find

$$\begin{aligned} X_{\theta\theta} &= \frac{1}{\kappa}\left(\frac{1}{\kappa}X_s\right)_s \\ &= -\frac{1}{\kappa}\left(\frac{\kappa_s}{\kappa^2}X_s - \frac{1}{\kappa}X_{ss}\right) \\ &= -\frac{1}{\kappa}\left(\frac{\kappa_\theta}{\kappa}\mathsf{T} + \mathsf{N}\right) \\ &= \frac{1}{\cos\theta}\left(\tan\theta\mathsf{T} - \mathsf{N}\right) \\ &= \left(0, \sec^2\theta\right). \end{aligned}$$

Integrating and recalling that $X_\theta = \kappa^{-1}\mathsf{T} = (1, \tan\theta)$ yields

$$X(\theta) = (\theta + x_0, y_0 - \log\cos\theta).$$

It follows that $U = M^1$ and X is, modulo a translation, the Grim Reaper. $\square$

On the other hand, if $A + iB \neq 0$, then the translated and dilated immersion

$$\widetilde{X} \doteqdot (At_0 + S)^{-\frac{1}{2}}\left(X + \frac{C}{A+iB}\right)$$

satisfies

$$A\left\langle \widetilde{X}, \widetilde{\mathsf{N}}\right\rangle + B\left\langle \widetilde{X}, \widetilde{\mathsf{T}}\right\rangle = -\widetilde{\kappa}.$$

So it suffices to consider the case that $C = 0$, $A + iB \neq 0$, and $At_0 + S = 1$. That is, it suffices to consider solutions $X : M^1 \to \mathbb{C}$ to

$$A\langle X, \mathsf{N}\rangle + B\langle X, \mathsf{T}\rangle = -\kappa. \tag{4.6}$$

Depending on the signs of A and B, we shall see that one has rotating, expanding and shrinking solutions, as well as their combinations. Instructive illustrations of each of these behaviors can be found in [**258**]. To begin, we derive an associated system of ODE (given in (4.10)) and use it to prove that for any value of A and B there exists an immersed curve that satisfies (4.6).

Using $\frac{d}{ds}\mathsf{N} = \kappa\mathsf{T}$, we calculate

$$\begin{aligned} \frac{d}{ds}\langle X, \mathsf{T}\rangle &= \left\langle \frac{d}{ds}X, \mathsf{T}\right\rangle + \left\langle X, \frac{d}{ds}\mathsf{T}\right\rangle \\ &= 1 + \left\langle X, \frac{d}{ds}(-i\mathsf{N})\right\rangle \\ &= 1 + \langle X, -i\kappa\mathsf{T}\rangle \\ &= 1 - \kappa\langle X, \mathsf{N}\rangle \end{aligned} \tag{4.7}$$

and

$$\frac{d}{ds}\langle X, \mathsf{N}\rangle = \left\langle X, \frac{d}{ds}\mathsf{N}\right\rangle = \kappa\langle X, \mathsf{T}\rangle . \tag{4.8}$$

Consider the functions

$$x = A\langle X, \mathsf{N}\rangle + B\langle X, \mathsf{T}\rangle \quad \text{and} \quad y = -B\langle X, \mathsf{N}\rangle + A\langle X, \mathsf{T}\rangle . \tag{4.9}$$

Then

$$x' = B + \kappa y \quad \text{and} \quad y' = A - \kappa x ,$$

where the prime denotes differentiation with respect to arc length. But $x = -\kappa$, so

$$x' = B - xy, \qquad y' = A + x^2. \tag{4.10}$$

We are ready to prove the existence theorem:

Theorem 4.3. *For each value of A and B there exists an immersed curve X satisfying the self-similar solution equation* (4.6).

Proof. There is a unique solution to system (4.10) with initial values $x(0) = x_0$, $y(0) = y_0$. By the inequality

$$\frac{d}{ds}\sqrt{x^2+y^2} = \frac{Bx+Ay}{\sqrt{x^2+y^2}} \leq \sqrt{A^2+B^2},$$

the solution is defined on all of $\mathbb{R}$. Define the curve $X : \mathbb{R} \to \mathbb{C}$ by

$$X(s) \doteqdot \frac{x(s)+iy(s)}{B+iA}\mathrm{e}^{i\theta(s)} ,$$

where

$$\theta(s) = -\int_0^s x(\sigma)d\sigma .$$

Then

$$\frac{dX}{ds} = \mathrm{e}^{i\theta}\frac{i\theta'(x+iy)+x'+iy'}{B+iA} = \mathrm{e}^{i\theta} .$$

So X is parametrized by arc length and its tangent is $\mathsf{T} = \mathrm{e}^{i\theta}$. It follows that its curvature is $\kappa = -x$. Finally, it satisfies (4.6) since

$$\begin{aligned} A\langle X, \mathsf{N}\rangle + B\langle X, \mathsf{T}\rangle &= \left\langle A\mathsf{N} + B\mathsf{T}, \frac{x+iy}{B+iA}\right\rangle \mathrm{e}^{i\theta} \\ &= \left\langle A\mathsf{N} + B\mathsf{T}, \frac{Bx+Ay-i(Ax-By)}{A^2+B^2}\mathrm{e}^{i\theta}\right\rangle \\ &= \left\langle A\mathsf{N} + B\mathsf{T}, \frac{Bx+Ay}{A^2+B^2}\mathsf{T} + \frac{Ax-By}{A^2+B^2}\mathsf{N}\right\rangle \\ &= x . \end{aligned}$$

□

Closed immersed shrinking self-similar solutions to curve shortening flow were classified by Abresch and Langer [**2**]; see also Epstein and Weinstein [**212**]. The general classification of immersed self-similar solutions of curve shortening flow was completed by Halldorsson in [**258**].

Theorem 4.4 (Classification of self-similar solutions to curve shortening flow [**2, 212, 258**]). *Let $X : M^1 \times I \to \mathbb{C}$ be a properly immersed self-similar solution to curve shortening flow. If X is not a stationary line, then it is one of the following*[1].

(1) (Shrinking nonrotating) $A < 0$, $B = 0$. *Each of the curves is contained in an annulus around the origin and consists of a series of identical excursions between the two boundaries of the annulus. They form a 1-dimensional family, parametrized by distance to the origin. The family consists of both closed and open curves.*

(a) Closed curves: *These are the* **Abresch–Langer curves**. *In addition to the circle, they include, for each pair of relatively prime positive integers (p, q) satisfying $\frac{1}{2} < \frac{p}{q} < \frac{\sqrt{2}}{2}$, a closed curve with rotation index p which touches each boundary of the annulus q times. In the embedded case, one has only the round circle.*

(b) Open curves: *These curves consist of infinitely many paths that go between the inner and outer boundaries of the annulus and include curves whose image is dense in the annulus.*

(2) (Steady nonrotating) $A = B = 0$. *This class contains only the Grim Reaper solution.*

(3) (Expanding nonrotating) $A > 0$, $B = 0$. *These curves are convex, embedded, and asymptotic to a wedge (with angle $\in (0, \pi)$). They form a 1-dimensional family parametrized by distance to the origin.*

(4) (Shrinking rotating) $A < 0$, $B \neq 0$. *This is a 2-dimensional family of curves, each of which has one end asymptotic to a circle. The other end can wrap around the circle or spiral toward infinity.*

(5) (Steady rotating) $A = 0$, $B \neq 0$. *This is a 1-dimensional family of embedded curves, each of which has two arms which spiral to infinity. The symmetric curve is the "yin-yang" spiral. The tip of the spiral looks like the Grim Reaper as the distance to the origin increases.*

(6) (Expanding rotating) $A > 0$, $B \neq 0$. *This is a 2-dimensional family of embedded curves, each of which has two arms that spiral to infinity.*

---

[1] We warn the reader that our conventions for A and B (recall (4.6)) and the functions x and y (recall (4.9)) differ from those of Halldorsson [**258**].

We refer the reader to [**258**] for the complete proof but sketch some calculations in each case to give an idea of how the behaviors arise.

In case (1), by rescaling we may assume that $A = -1$. The system is

$$x' = -xy, \qquad y' = -1 + x^2,$$

so $x(s) = x_0 e^{-\int_0^s y(\sigma)d\sigma}$. Thus, the curvature does not change sign and, by reparametrizing if necessary, we can assume that $x_0 > 0$ and hence $x > 0$. Observe that $(1, 0)$ is a fixed point of the system, corresponding to the unit circle. It can further be shown that x and y are periodic. One has the explicit solution $x = x_0 e^{\frac{x_0^2 - r^2}{2}}$, where $r \doteqdot |X| = \sqrt{\frac{x^2+y^2}{A^2+B^2}} = \sqrt{x^2 + y^2}$ and x_0 is either $x_{\min}$ or $x_{\max}$. Then $r_{\min} = x_{\min}$ and $r_{\max} = x_{\max}$ indicating the excursions between annuli of radius $r_{\min}$ and $r_{\max}$.

Case (2) is covered by Lemma 4.2.

In case (3), we rescale such that $A = 1$ and obtain the system

$$x' = -xy, \qquad y' = 1 + x^2.$$

By the symmetries of the equation, it is clear that x is even and y is odd; hence X is even. One has the explicit solution $x = x_{\max} e^{\frac{r^2 - x_{\max}^2}{2}}$ and, integrating $x(s)$, one can see that the total curvature is no greater than π. We conclude that X is convex, embedded, and the graph of an even function. By a result of Ishimura [**313**], it is asymptotic to a cone.

In case (4), without loss of generality $B = 1$. Writing $A = -\alpha^2$, we have

$$x' = 1 - xy, \qquad y' = -\alpha^2 + x^2.$$

The circle $(\alpha, \frac{1}{\alpha})$ is an asymptotically attracting fixed point of the associated system. One can have either $\lim_{s\to\pm\infty} r = \frac{1}{\alpha}$, or $\lim_{s\to-\infty} r = \frac{1}{\alpha}$ and $\lim_{s\to\infty} r = \infty$, corresponding to the two cases of an end wrapping around the circle or spiraling toward infinity.

In case (5), a simple calculation yields

$$\begin{aligned} \frac{d}{ds}\left(r^2 + \frac{2}{B}\theta\right) &= \frac{2xx'(s) + 2yy'(s)}{B^2} - \frac{2}{B}x(s) \\ &= \frac{2x(xy + A) + 2y(-x^2) - 2Bx}{B^2} \\ &= 0. \end{aligned}$$

So $r^2 + \frac{2}{B}\theta$ is constant, and the curves are embedded.

In case (6), one can prove that the curves are embedded as well. Without loss of generality, $B > 0$. If $x(s) = 0$, then $x'(s) = B > 0$, and so it has at most one zero and is negative before and positive afterwards. One sees immediately that $\lim_{s\to\pm\infty} y = \pm\infty$ since $y' \geq A > 0$. This implies that

$\lim_{s\to\pm\infty} r = \infty$. In fact, r has one extremum which is a global minimum since

$$\frac{d}{ds}r^2 = 2\langle X, \mathsf{T}\rangle$$

and $\langle X, \mathsf{T}\rangle = 0$ implies $\frac{d}{ds}\langle X, \mathsf{T}\rangle = A\langle X, \mathsf{N}\rangle^2 + 1 > 0$.

4.3. Monotonicity formulae

In this section we discuss isoperimetric ratio monotonicity, the differential Harnack estimate, entropy monotonicity, and Huisken's monotonicity formula. As an earlier and alternative approach to the one presented in this book, one may prove the Gage–Hamilton–Grayson Theorem based in part on these formulae. These monotonicity formulae are more elementary than the advanced formulae discussed in the previous chapter.

4.3.1. Isoperimetric ratio. The **isoperimetric inequality** in the plane says that for any embedded closed curve we have

$$\frac{\mathrm{L}^2}{\mathrm{A}} \geq 4\pi, \tag{4.11}$$

with equality if and only if the curve is a round circle. See Osserman [**435**] for an introduction to and a survey of isoperimetric inequalities.

Lemma 4.5 (Gage's inequality). *For any embedded closed convex curve,*

$$\int_{M^1} \kappa^2 ds \geq \frac{\pi\,\mathrm{L}}{\mathrm{A}}. \tag{4.12}$$

Proof. The idea is to show that there exists a choice of origin such that the support function σ (defined in Section 2.4) satisfies the inequality

$$\int_{M^1} \sigma^2 ds \leq \frac{\mathrm{L}\,\mathrm{A}}{\pi} \tag{4.13}$$

(see Gage [**229**] for the details of the proof of this). Assuming this inequality, we deduce from (2.36) that

$$\mathrm{L}^2 = \left(\int_{M^1} \sigma\,\kappa\, ds\right)^2 \leq \int_{M^1} \sigma^2 ds \int_{M^1} \kappa^2 ds \leq \frac{\mathrm{L}\,\mathrm{A}}{\pi}\int_{M^1} \kappa^2 ds,$$

which yields (4.12). □

Proposition 4.6 (Monotonicity of the isoperimetric ratio under curve shortening flow). *For any embedded closed convex curve evolving by curve shortening flow, we have*

$$\frac{d}{dt}\left(\frac{\mathrm{L}^2}{\mathrm{A}}\right) \leq 0. \tag{4.14}$$

Proof. Using (2.19) and (2.38), we compute that

$$\frac{d}{dt}\left(\frac{\mathrm{L}^2}{\mathrm{A}}\right) = \frac{\mathrm{L}}{\mathrm{A}^2}\left(2\,\mathrm{A}\,\frac{d\,\mathrm{L}}{dt} - \mathrm{L}\,\frac{d\,\mathrm{A}}{dt}\right) = \frac{\mathrm{L}}{\mathrm{A}^2}\left(-2\,\mathrm{A}\int_{M^1}\kappa^2 ds + 2\pi\,\mathrm{L}\right) \leq 0,$$

where we used (4.12) to obtain the inequality. □

4.3.2. Differential Harnack estimate. A general approach to understanding the qualitative behavior of a solution to a geometric evolution equation is to consider a suitable quantity associated to the solution, such as one involving derivatives of the solution, and to bound this quantity by computing its evolution and then applying the maximum principle. We employed this approach in the previous chapter. In the same vein, differential Harnack estimates are inequalities for gradient and Laplacian-type quantities, which usually vanish on self-similar solutions to the equation and enable one to compare the solution at different points in space and time.

Let $X : M^1 \times [0,T) \to \mathbb{R}^2$ be a solution to the curve shortening flow with $\kappa > 0$, where $M^1 \cong S^1$. Then for each $t \in [0,T)$ the angle parameter

$$\theta(t) : M^1 \to \mathbb{R}/\left(2\pi w\mathbb{Z}\right) \cong S^1$$

is a well-defined diffeomorphism, where w is the turning index of X. So we may consider $\kappa \doteqdot \kappa \circ \theta^{-1}$ as a function on $\mathbb{R}/\left(2\pi w\mathbb{Z}\right) \times [0,T)$.

As before, let $\frac{\partial}{\partial t}$ denote the partial derivative with respect to t in the original coordinates (u,t). This is different than the partial derivative with respect to t in the coordinates (θ,t), which we denote instead by $\frac{\partial}{\partial \tau}$ (we change t to τ in our notation to alert the reader of this difference). Note that

$$\frac{\partial \theta}{\partial \tau} = 0, \tag{4.15}$$

and for any function $f : \mathbb{R}/\left(2\pi w\mathbb{Z}\right) \times [0,T) \to \mathbb{R}$ we have

$$f_{\tau\theta} = f_{\theta\tau}. \tag{4.16}$$

Lemma 4.7 (Time derivative under parametrization by θ). *For any convex solution to the curve shortening flow, we have*

$$\frac{\partial}{\partial t} = \frac{\partial}{\partial \tau} + \kappa^{-1}\kappa_s\frac{\partial}{\partial s}. \tag{4.17}$$

Proof. We may write

$$\frac{\partial}{\partial t} = \frac{\partial}{\partial \tau} + \varphi\frac{\partial}{\partial s}, \tag{4.18}$$

where φ is a function which we now determine. Applying (4.18) to θ, we have

$$\kappa_s = \varphi \frac{\partial \theta}{\partial s} = \varphi \kappa$$

since $\theta_\tau = 0$, $\theta_t = \kappa_s$, and $\theta_s = \kappa$. Hence $\varphi = \kappa^{-1}\kappa_s$. □

Lemma 4.8 (Evolution of $\kappa(\theta,\tau)$ under curve shortening flow). *As a function of θ, the curvature evolves by*

$$\kappa_\tau = \kappa^2 (\kappa_{\theta\theta} + \kappa) . \tag{4.19}$$

Proof. By Lemma 4.7,

$$\kappa_t = \kappa_\tau + \kappa^{-1} (\kappa_s)^2 .$$

Since $\frac{\partial}{\partial s} = \kappa \frac{\partial}{\partial \theta}$, we have

$$\kappa_{ss} = \kappa (\kappa \kappa_\theta)_\theta = \kappa^2 \kappa_{\theta\theta} + \kappa (\kappa_\theta)^2 .$$

Combining these two equations with Lemma 2.5 yields

$$\begin{aligned} \kappa_\tau + \kappa^{-1} (\kappa_s)^2 &= \kappa_t \\ &= \kappa_{ss} + \kappa^3 \\ &= \kappa^2 \kappa_{\theta\theta} + \kappa (\kappa_\theta)^2 + \kappa^3, \end{aligned}$$

and the lemma follows canceling terms using $(\kappa_s)^2 = \kappa^2 (\kappa_\theta)^2$. □

Remark 4.9. Note that if κ is a solution to (4.19), then for any $c > 0$,

$$\kappa_c(\tau) \doteqdot c^{-1} \kappa \left(c^{-2} \tau\right)$$

is also a solution to (4.19). This is the curvature of the rescaled solution to curve shortening flow

$$X_c(u,t) \doteqdot cX\left(u, c^{-2} t\right) .$$

Given a solution to curve shortening flow with $\kappa > 0$, we define

$$Q \doteqdot \kappa (\kappa_{\theta\theta} + \kappa) = \frac{\kappa_\tau}{\kappa}, \tag{4.20}$$

where the equality holds by (4.19). Note that by (2.34), Q vanishes on the Grim Reaper solution. We compute that

$$\begin{aligned} Q_\tau &= \kappa_\tau (\kappa_{\theta\theta} + \kappa) + \kappa (\kappa_{\tau\theta\theta} + \kappa_\tau) \\ &= Q^2 + \kappa ((\kappa Q)_{\theta\theta} + \kappa Q) \\ &= \kappa^2 Q_{\theta\theta} + 2\kappa\kappa_\theta Q_\theta + 2Q^2 . \end{aligned} \tag{4.21}$$

Notice the similarity of this evolution equation with the formula (1.17) which is used to prove the Li–Yau inequality for the heat equation. Such formulae are thematic in proving differential Harnack estimates for geometric flows.

Applying the maximum principle, we obtain the following.

Theorem 4.10 (Harnack estimate). *On any immersed solution $X : M^1 \times [\alpha, T) \to \mathbb{R}^2$, $M^1 \cong S^1$, to curve shortening flow with $\kappa > 0$,*

$$\kappa\left(\kappa_{\theta\theta} + \kappa\right) + \frac{1}{2\left(t - \alpha\right)} \geq 0 \tag{4.22}$$

for $t \in (\alpha, T)$.

Proof. Recalling (4.21), the ODE comparison principle yields

$$Q\left(\theta, t\right) \geq q\left(t\right),$$

where $q\left(t\right) \doteqdot -\frac{1}{2(t-\alpha)}$ is the solution to

$$\begin{aligned} \frac{dq}{dt} &= 2q^2, \\ \lim_{t \searrow \alpha} q\left(t\right) &= -\infty. \end{aligned}$$

□

A **singular solution** to curve shortening flow is a solution $X : M^1 \times [0, T) \to \mathbb{R}^2$ for which $T < \infty$ and

$$\sup_{M^1 \times [0,T)} \left|\kappa\right|\left(u, t\right) = \infty.$$

An **ancient solution** is a solution defined on a time interval of the form $(-\infty, \omega)$, where $\omega \in (-\infty, \infty]$. As we see later in this chapter, such solutions arise as the rescalings of singular solutions.

Corollary 4.11 (Harnack for ancient solutions). *For an ancient solution with $\kappa > 0$ and defined on the time interval $(-\infty, 0)$, we have*

$$\frac{\kappa_\tau}{\kappa^2} = \kappa_{\theta\theta} + \kappa \geq 0. \tag{4.23}$$

Proof. Since the solution is ancient, for any $\alpha \in (-\infty, 0)$ the solution exists on $[\alpha, 0)$, so that (4.22) holds. The corollary follows from taking the limit as $\alpha \to -\infty$. □

Using $\frac{\partial}{\partial\theta} = \kappa^{-1}\frac{\partial}{\partial s}$, we see that in terms of the arc length parameter the Harnack estimate (4.22) is equivalent to

$$\left(\log \kappa\right)_{ss} + \kappa^2 + \frac{1}{2\left(t - \alpha\right)} \geq 0. \tag{4.24}$$

4.3.3. Entropy monotonicity. By analogy with the Nash entropy, which is monotone on a solution to the heat equation (see Section 1.9.1, the **entropy** of an embedded convex curve is defined by

$$E \doteqdot -\int_{M^1} \kappa \log \kappa \, ds = -\int_{M^1} \log \kappa \, d\theta \,. \tag{4.25}$$

This invariant is well defined since $\kappa > 0$. Under curve shortening flow, the time derivative of the entropy is related to the differential Harnack inequality by the identity

$$N \doteqdot -\frac{dE}{dt} = \int_{M^1} \frac{\kappa_\tau}{\kappa} \, d\theta = \int_{M^1} Q \, d\theta \,, \tag{4.26}$$

where Q is defined by (4.20).

Recalling the calculation (4.21), we find

$$\begin{aligned} \frac{dN}{dt} &= \int_{M^1} Q_\tau \, d\theta \\ &= \int_{M^1} \left(\kappa^2 Q_{\theta\theta} + 2\kappa\kappa_\theta Q_\theta + 2Q^2\right) d\theta \\ &= 2\int_{M^1} Q^2 \, d\theta \,, \end{aligned} \tag{4.27}$$

where we integrated by parts to obtain the last equality. Hence

$$\frac{dN}{dt} \geq \frac{1}{\pi}\left(\int_{M^1} Q \, d\theta\right)^2 = \frac{1}{\pi} N^2. \tag{4.28}$$

Let A_0 denote the area enclosed by the initial curve. Then the area enclosed by the curve at time t is $\mathrm{A}(t) = \mathrm{A}_0 - 2\pi t$. By the results of the previous chapter, we have that the solution to curve shortening flow exists for $t \in [0, \frac{\mathrm{A}_0}{2\pi})$. For the original proof, see Theorem 4.3.1 (curvature bound) and Theorem 4.5 in [**228**].

We may use the ordinary differential inequality for N to estimate the entropy E. By (4.28), we have $\frac{dN}{d(-t)} \leq -\frac{1}{\pi} N^2$. The solution to the ODE

$$\frac{dn}{d(-t)} = -\frac{1}{\pi} n^2, \qquad \lim_{t \to \frac{\mathrm{A}_0}{2\pi}} n(t) = +\infty$$

is $n(t) = \frac{\pi}{\frac{\mathrm{A}_0}{2\pi} - t} = \frac{2\pi^2}{\mathrm{A}(t)}$. By the ODE comparison theorem, we have

$$-\frac{dE}{dt}(t) = N(t) \leq n(t) = \frac{2\pi^2}{\mathrm{A}(t)}. \tag{4.29}$$

Now

$$\frac{d}{dt} \log \mathrm{A}(t) = -\frac{2\pi}{\mathrm{A}(t)}. \tag{4.30}$$

Define the **normalized entropy** to be

$$\tilde{E} \doteqdot -\int_{M^1} \log\left(\mathrm{A}^{1/2}\kappa\right) d\theta, \tag{4.31}$$

where we have inserted the area A to make the quantity scale invariant. Then (4.29) and (4.30) imply the following.

Lemma 4.12 (Normalized entropy monotonicity). *Under curve shortening flow we have*

$$\frac{d\tilde{E}}{dt}(t) = \frac{dE}{dt}(t) + \frac{2\pi^2}{\mathrm{A}(t)} \geq 0. \tag{4.32}$$

4.3.4. A sketch of the Gage–Hamilton proof that convex embedded curves converge to round points. In Section 4.3 of Gage and Hamilton [**228**], a curvature estimate for curve shortening flow of convex embedded plane curves is established. They show that the curvature is bounded as long as the enclosed area is positive. Define the **maximum sustained curvature** of a curve by

$$\kappa^* = \sup\{\, k \mid \kappa(\theta) \geq k \text{ on some interval } [\theta_0, \theta_0 + \pi]\,\}. \tag{4.33}$$

This provides a link between the isoperimetric ratio and the entropy:

(a) We have the estimate $\kappa^* < \frac{L}{A}$. By the isoperimetric estimate $L(t) \leq C\sqrt{A(t)}$; this implies

$$\kappa^*(t) < \frac{C}{\sqrt{A(t)}}. \tag{4.34}$$

(b) If $\kappa^*(t) \leq M$ for $t \in [0, T)$, then the entropy satisfies the bound

$$E(t) \leq E(0) + 2ML(0) + 2\pi M^2 T. \tag{4.35}$$

(c) If $E(t)$ is bounded on $[0, T)$, then $\kappa(t)$ is bounded on $[0, T)$.

Definition 4.13. Let $\mathcal{M}$ be a closed embedded plane curve. The **inradius** $\rho_-(\mathcal{M})$ of $\mathcal{M}$ is the radius of the largest ball enclosed by $\mathcal{M}$. The **circumradius** $\rho_+(\mathcal{M})$ of $\mathcal{M}$ is the radius of the smallest ball which contains $\mathcal{M}$.

By the estimates above and by the Bernstein estimates for the derivatives of curvature, the solution to curve shortening flow exists until the area tends to zero. By work of Gage, it then follows that the isoperimetric ratio L^2/A tends to 4π, so that the ratio of the circumradius to the inradius tends to 1. One then shows by geometric methods that the ratio of the maximum curvature to the minimum curvature also tends to 1. By estimating the L^2-norm of derivatives of the curvature, one proves that the derivatives of the curvature tend to 0 exponentially fast for normalized curve shortening flow. Smooth convergence to a round circle follows.

4.3.5. Huisken's monotonicity formula. The 1-dimensional special case of Huisken's monotonicity formula for the mean curvature flow is the following. See Exercise 4.3 for a proof of this special case. We give a proof of this result in general dimensions in Chapter 10.

Theorem 4.14 (Huisken's monotonicity formula for curve shortening flow). *Let $X(t) : M^1 \to \mathbb{R}^2$, $t \in [0, T)$, be a smooth closed immersed curve evolving by the curve shortening flow. Then*

$$\frac{d}{dt}\int_{M^1} \Phi(X,\tau)\, ds = -\int_{M^1} \left(\kappa - \frac{\langle X, \mathsf{N}\rangle}{2\tau}\right)^2 \Phi(X,\tau)\, ds, \tag{4.36}$$

where $\Phi(X,\tau) \doteqdot (4\pi\tau)^{-1/2}\, \mathrm{e}^{-\frac{|X|^2}{4\tau}}$ is the backward heat kernel and $\tau \doteqdot T-t$.

4.3.6. Monotonicity via Sturm's theorem. Let $X : M^1 \times [0,T) \to \mathbb{R}^2$ be an immersed solution to the curve shortening flow with $\kappa > 0$ and rotation index $w \in \mathbb{N}$ (defined by (2.11)). Let $\mathbb{S}^1_w = \mathbb{R}/(2\pi w\mathbb{Z})$. Using the normal angle parametrization (2.6), we have that the curvature $\kappa : \mathbb{S}^1_w \times [0,T) \to \mathbb{R}^2$ satisfies (4.19). Define a constant, depending only on the initial curve, by

$$K \doteqdot \max_{\theta\in\mathbb{S}^1_w} (\kappa_\theta^2 + \kappa^2)^{1/2}(\theta, 0)\,.$$

By Sturm's theorem, we can prove the following Harnack-style estimate (compare with (4.22)).

Lemma 4.15. *For each $(\theta, t) \in \mathbb{S}^1_w \times (0,T)$ we have*

$$(\kappa_{\theta\theta} + \kappa)(\theta,t) > 0 \quad \text{or} \quad (\kappa_\theta^2 + \kappa^2)^{1/2}(\theta,t) \le K\,. \tag{4.37}$$

Hence, wherever $\kappa > K$, we have $\partial_t\kappa > 0$, where the time derivative is with respect to the normal angle parametrization.

Proof. Suppose that (θ_0, t_0) is such that

$$J \doteqdot (\kappa_\theta^2 + \kappa^2)^{1/2}(\theta_0, t_0) > K\,.$$

Then there exists (a unique) $\theta_1 \in \left(-\frac{\pi}{2}, \frac{\pi}{2}\right)$ such that

$$\kappa(\theta_0, t_0) = J\cos\theta_1 \quad \text{and} \quad \kappa_\theta(\theta_0, t_0) = J\sin\theta_1\,.$$

Define the "Grim Reaper" comparison function (see (2.32))

$$\kappa^\star(\theta) \doteqdot J\cos(\theta_1 + \theta_0 - \theta)\,, \tag{4.38}$$

which is a static solution of (4.19) with $\kappa^\star(\theta_0) = \kappa(\theta_0, t_0)$. Consider the difference

$$f(\theta,t) = \kappa(\theta,t) - \kappa^\star(\theta)\,,$$

which is defined for $\theta \in I$, $t \in [0, T)$, where $I \doteqdot (\theta_1 + \theta_0 - \frac{\pi}{2}, \theta_1 + \theta_0 + \frac{\pi}{2})$. We have that $f(\theta_0, t_0) = 0$ and $f_\theta(\theta_0, t_0) = 0$. Observe that

$$(\kappa^\star)^2(\theta) + (\kappa_\theta^\star)^2(\theta) = J^2 > K^2 \geq \kappa^2(\theta) + \kappa_\theta^2(\theta) .$$

Hence, if $\theta \in (\theta_1 + \theta_0 - \frac{\pi}{2}, \theta_1 + \theta_0)$ is a zero of $f(\theta, 0)$, then

$$(\kappa_\theta(\theta, 0) - \kappa_\theta^\star(\theta))(\kappa_\theta(\theta, 0) + \kappa_\theta^\star(\theta)) < 0 .$$

Since $\kappa_\theta^\star(\theta) > 0$, this implies $f_\theta(\theta, 0) < 0$ at each zero of $f(\theta, 0)$ on the interval $(\theta_1 + \theta_0 - \frac{\pi}{2}, \theta_1 + \theta_0)$. Similarly, one shows that $f_\theta(\theta, 0) > 0$ at each zero of $f(\theta, 0)$ on the interval $(\theta_1 + \theta_0, \theta_1 + \theta_0 + \frac{\pi}{2})$. One concludes from this that $f(\theta, 0)$ has exactly two zeros, both with multiplicity one, on the interval I. By Sturm's theorem, Theorem 1.21, we conclude for each $t \in (0, T)$ that $f(\theta, t)$ has at most two zeros counting multiplicity on the interval I. On the other hand, $f(\theta, t_0)$ has a zero of multiplicity two at $\theta = \theta_0 \in I$. Thus $f(\theta, t_0)$ has no other zeros on I. Since $f(\theta, t_0) = \kappa(\theta, t_0) > 0$ for $\theta \in \partial I$, we have that $f(\theta, t_0) > 0$ for $\theta \in I \backslash \{\theta_0\}$. Therefore $f_{\theta\theta}(\theta_0, t_0) \geq 0$; in fact, $f_{\theta\theta}(\theta_0, t_0) > 0$ since otherwise the multiplicity at θ_0 would be at least three. We conclude that $(\kappa_{\theta\theta} + \kappa)(\theta_0, t_0) > 0$. □

Define the maximum curvature function

$$\overline{\kappa}(t) = \max_{\theta \in \mathbb{S}_w^1} \kappa(\theta, t) . \tag{4.39}$$

For each $t \in (0, T)$, choose a point at which κ achieves its maximum: $\theta_t \in \mathbb{S}_w^1$ such that $\kappa(\theta_t, t) = \overline{\kappa}(t)$. For sufficiently large curvature points, we have a comparison with the curvature of a scaled Grim Reaper.

Lemma 4.16. *For any t with $\overline{\kappa}(t) > K$, we have*

$$\kappa(\theta, t) > \overline{\kappa}(t) \cos(\theta - \theta_t) \tag{4.40}$$

for all $\theta \in \mathbb{S}_w^1$ such that $|\theta - \theta_t| < \cos^{-1}\left(\frac{K}{\overline{\kappa}(t)}\right)$.

Proof. Integrating by parts while using $\kappa(\theta_t, t) = \overline{\kappa}(t)$ and $\kappa_\theta(\theta_t, t) = 0$, we see that

$$\kappa(\theta, t) = \overline{\kappa}(t) \cos(\theta - \theta_t) + \int_{\theta_t}^{\theta} \sin(\theta - \tilde{\theta})(\kappa_{\theta\theta} + \kappa)(\tilde{\theta}, t) d\tilde{\theta} . \tag{4.41}$$

Since $(\kappa_\theta^2 + \kappa^2)^{1/2}(\theta_t, t) = \overline{\kappa}(t) > K$, by Lemma 4.15 we have

$$(\kappa_{\theta\theta} + \kappa)(\theta, t) > 0$$

at $\theta = \theta_t$ and hence for θ near θ_t. Thus, by (4.41) we have that

$$\kappa(\theta, t) > \overline{\kappa}(t) \cos(\theta - \theta_t)$$

for θ close to and at least θ_t. Suppose for a contradiction there exists $\theta_0 \in \left(\theta_t, \theta_t + \cos^{-1}\left(\frac{K}{\overline{\kappa}(t)}\right)\right)$ such that $\theta = \theta_0$ satisfies $\kappa(\theta, t) \leq \overline{\kappa}(t)\cos(\theta - \theta_t)$. Let θ_0 be the smallest θ with this property. Then for $\theta \in [\theta_t, \theta_0)$ we have

$$\kappa(\theta, t) > \overline{\kappa}(t)\cos(\theta - \theta_t) \geq \overline{\kappa}(t)\,\frac{K}{\overline{\kappa}(t)} = K\,.$$

Hence, by Lemma 4.15 again, we have

$$(\kappa_{\theta\theta} + \kappa)(\theta, t) > 0$$

for $\theta \in [\theta_t, \theta_0)$, which implies that $\kappa(\theta_0, t) > \overline{\kappa}(t)\cos(\theta_0 - \theta_t)$, a contradiction. By the corresponding considerations for θ at most θ_t, the lemma follows. □

Using the results above, Angenent (see Theorem C in [**66**]) proved that if an immersed solution of curve shortening flow has $\kappa > 0$ and is type-II, then there exists a sequence of space-time points (θ_j, t_j) with $t_j \to T$ such that

$$\lim_{j\to\infty} \frac{\kappa(\theta + \theta_j, t_j)}{\overline{\kappa}(t_j)} = \cos\theta \quad \text{for} \quad -\frac{\pi}{2} < \theta < \frac{\pi}{2}\,,$$

where the convergence is in C^∞ on compact subsets of $\left(-\frac{\pi}{2}, \frac{\pi}{2}\right)$. Thus the Grim Reaper is a singularity model of the flow.

4.4. Ancient solutions

The self-similarly shrinking, rotating, and translating solutions to curve shortening constructed in Section 4.2 provide examples of ancient solutions to the flow. But these solutions are very special and the question naturally arises whether there exist ancient solutions which do not arise through self-similarities of the equation. In this section, we construct two explicit ancient solutions to curve shortening flow which do not evolve by self-similarities.

4.4.1. The hairclip solution. Consider the graph of the function

$$y\,(x, t) = \log\left(\cos x + \sqrt{\mathrm{e}^{2t} + \cos^2 x}\right) - t \tag{4.42}$$

for $x \in \mathbb{R}$ and $t \in (-\infty, \infty)$.

We have $y\,(x + 2\pi, t) = y\,(x, t) = y\,(-x, t)$ and $y(x + \pi, t) = y(\pi - x, t) = -y(x, t)$. Note, for example, that $y\,(x, t) > 0$ for $x \in (-\frac{\pi}{2}, \frac{\pi}{2})$ while $y(\pm\frac{\pi}{2}, t) = 0$. We compute that

$$y_x\,(x, t) = -\frac{\sin x}{\sqrt{\mathrm{e}^{2t} + \cos^2 x}}\,,$$

$$y_{xx}\,(x, t) = -\frac{\left(\mathrm{e}^{2t} + 1\right)\cos x}{\left(\mathrm{e}^{2t} + \cos^2 x\right)^{3/2}}\,.$$

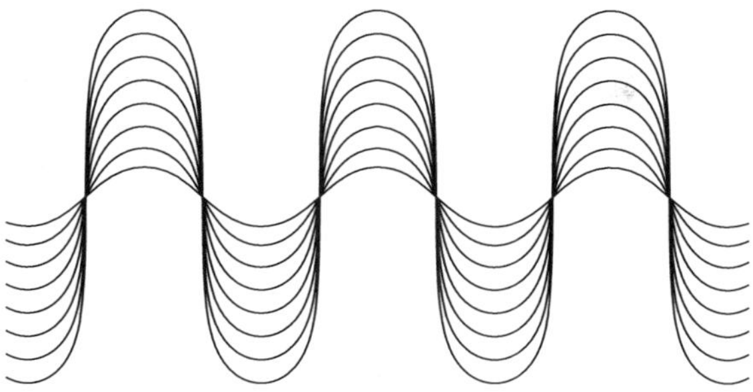

Figure 4.1. Time slices of the hairclip solution.

The curve shortening flow equation (2.30) for a graph holds since

$$y_t\,(x,t) = -\frac{\cos x}{\left(\mathrm{e}^{2t} + \cos^2 x\right)^{1/2}} = \frac{y_{xx}}{1 + (y_x)^2}\,(x,t)\,.$$

We have the backward in time limit

$$\lim_{t\to-\infty}\,(y\,(x,t) + t - \log 2) = \log\,(\cos x)$$

for $x \in \mathbb{R}$ such that $\cos x > 0$, which is the Grim Reaper solution (2.31) (at $t = 0$). Where $\cos x < 0$, we have

$$\lim_{t\to-\infty}\,(y\,(x,t) - t + \log 2) = -\log\,(|\cos x|)\,.$$

We have $\lim_{t\to\infty} y(x,t) = 0$ uniformly in $x \in \mathbb{R}$.

Note also that by (2.27), the curvature is

$$\kappa = -\frac{\cos x}{\sqrt{\mathrm{e}^{2t} + 1}}.$$

From (2.26), the (downward pointing) unit normal is

$$\mathsf{N} = -\frac{\left(\sin x, \sqrt{\mathrm{e}^{2t} + \cos^2 x}\right)}{\sqrt{\mathrm{e}^{2t} + 1}}.$$

By (2.28), the normal angle $\theta\,(x)$ satisfies

$$\tan\theta\,(x) = \frac{\sqrt{\mathrm{e}^{2t} + \cos^2 x}}{\sin x}$$

and

$$\theta\,(x) = \cos^{-1}\left(\frac{\sin x}{\sqrt{\mathrm{e}^{2t} + 1}}\right) - \pi\,.$$

4.4.2. The paperclip solution. Besides a straight line, a round shrinking circle, the Grim Reaper, and the graph (4.42), there is yet another explicit ancient embedded solution to curve shortening flow, which we now describe.

In view of condition (2.41), this convex solution on S^1 may be constructed, via determining its curvature $\kappa(\theta, \tau)$, as follows. Let

$$p \doteqdot \kappa^2, \tag{4.43}$$

which is called the **pressure function**. Using (4.19), we calculate that

$$\begin{aligned} p_\tau &= 2\kappa\kappa_\tau \\ &= 2\kappa^3(\kappa_{\theta\theta} + \kappa) \\ &= p\,p_{\theta\theta} - \frac{1}{2}(p_\theta)^2 + 2p^2 \end{aligned} \tag{4.44}$$

since $p_{\theta\theta} = 2\kappa\kappa_{\theta\theta} + 2(\kappa_\theta)^2$. Define

$$\alpha \doteqdot p_\theta. \tag{4.45}$$

Differentiating (4.44) implies

$$\begin{aligned} \alpha_\tau &= p_{\tau\theta} \\ &= \left(p p_{\theta\theta} - \frac{1}{2}(p_\theta)^2 + 2p^2\right)_\theta \\ &= p p_{\theta\theta\theta} + 4p p_\theta \\ &= p(\alpha_{\theta\theta} + 4\alpha). \end{aligned} \tag{4.46}$$

We search for a solution to equation (4.44) of the form

$$p(\theta, \tau) = a(\theta) + b(\tau), \tag{4.47a}$$

$$\int_{S^1} a(\theta)\, d\theta = 0 \tag{4.47b}$$

for $\theta \in S^1$ and τ in some maximal interval to be determined. We may think of this ansatz, which is equivalent to

$$\alpha_\tau = p_{\theta\tau} = 0 \tag{4.48}$$

(while normalizing a), as saying that the curvature squared is "weakly time dependent".

Assume that $a(\theta) \not\equiv 0$, i.e., the curve is not a round circle. By (4.46), we have

$$\alpha_{\theta\theta} + 4\alpha \equiv 0.$$

Since $(p_{\theta\theta} + 4p)_\theta = 0$ and $\int_{S^1} a(\theta)\, d\theta = 0$, we have that

$$a_{\theta\theta} + 4a = 0. \tag{4.49}$$

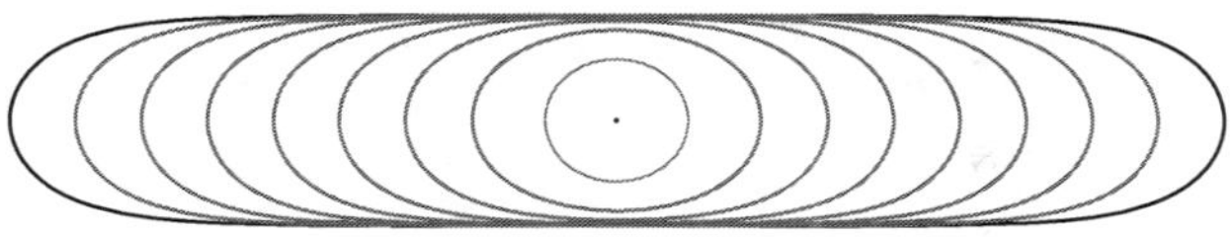

Figure 4.2. Time slices of the paperclip solution.

Multiplying this by $2a_\theta$ and integrating yields

$$(a_\theta)^2 + 4a^2 = 4A^2 \tag{4.50}$$

for some constant $A > 0$. Equation (4.44) then becomes

$$\begin{aligned} b_\tau &= (a+b)\,a_{\theta\theta} - \frac{1}{2}(a_\theta)^2 + 2(a+b)^2 \\ &= 2b^2 - 2A^2. \end{aligned} \tag{4.51}$$

Up to a translation in time, (4.51) yields

$$b(\tau) = A\coth(-2A\tau),$$

which is defined for $\tau \in (-\infty, 0)$. By rotating the solution in the plane, (4.50) implies that

$$a(\theta) = A\cos 2\theta.$$

By (4.47a), this solution satisfies

$$\kappa^2(\theta,\tau) = p(\theta,\tau) = A(\cos 2\theta + \coth(-2A\tau)) > 0 \tag{4.52}$$

for $\theta \in S^1$ and $\tau < 0$; i.e.,

$$\kappa(\theta,\tau) = \sqrt{A(\cos 2\theta + \coth(-2A\tau))}. \tag{4.53}$$

Clearly we have the symmetries

$$\kappa(\theta,\tau) = \kappa(\pi-\theta,\tau) = \kappa(\theta+\pi,\tau) = \kappa(-\theta,\tau).$$

Let

$$S(\theta) \doteqdot \frac{\sin\theta}{\kappa(\theta)}, \qquad C(\theta) \doteqdot \frac{\cos\theta}{\kappa(\theta)}.$$

Then

$$S(\theta+\pi) = -S(\theta), \qquad C(\theta+\pi) = -C(\theta).$$

From this we easily deduce that condition (2.41), for κ in (4.52) to be the curvature of a smooth embedded closed convex curve X, holds:

$$\int_0^{2\pi} S(\theta)\,d\theta = 0, \qquad \int_0^{2\pi} C(\theta)\,d\theta = 0.$$

We call this solution to curve shortening flow the **paperclip**[2]. It is also called the **Angenent oval** in honor of Sigurd Angenent [**68**, p. 25], although it was discovered independently by multiple people [**373, 412**].

[2]A cuter name, suggested by Sigurd Angenent, is **jousting Grim Reapers**.

Figure 4.3. Sigurd Angenent. Author: Jürgen Pöschel. Source: Archives of the Mathematisches Forschungsinstitut Oberwolfach.

Remark 4.17. The parameter $A > 0$ corresponds to rescalings of the solution in the $A = 1$ case (take $c = A^{-1/2}$ in Remark 4.9).

Given any $\tau < 0$, the maximum of the curvature

$$\kappa_{\max}(\tau) = \sqrt{A\left(1 + \coth(-2A\tau)\right)} \tag{4.54}$$

is attained at $\theta = 0$ and $\theta = \pi$, whereas the minimum

$$\kappa_{\min}(\tau) = \sqrt{A \coth(-2A\tau)} \tag{4.55}$$

is attained at $\theta = -\frac{\pi}{2}$ and $\theta = \frac{\pi}{2}$. Now assume for convenience that $A = 1$. Via (2.39b), the y-component of a solution to curve shortening flow with curvature (4.53) is given by

$$\begin{aligned} y(\theta, \tau) &= \int_0^\theta \frac{\cos\theta}{\sqrt{\cos 2\theta + \coth(-2\tau)}} \, d\theta \\ &= \frac{1}{\sqrt{2}} \sin^{-1}\left(\sqrt{\frac{2}{\coth(-2\tau) + 1}} \sin\theta\right). \end{aligned} \tag{4.56}$$

We have $y(\pi - \theta, \tau) = y(\theta, \tau)$ and $y(\theta + \pi, \tau) = -y(\theta, \tau) = y(-\theta, \tau)$. In particular, $y(0, \tau) = 0 = y(\pi, \tau)$. Note that $y(\theta, \tau)$ attains its maximum value $\frac{1}{\sqrt{2}} \sin^{-1}\left(\sqrt{\frac{2}{\coth(-2\tau)+1}}\right)$ at $\theta = \frac{\pi}{2}$ and $y(\theta, \tau)$ attains its minimum value $-\frac{1}{\sqrt{2}} \sin^{-1}\left(\sqrt{\frac{2}{\coth(-2\tau)+1}}\right)$ at $\theta = -\frac{\pi}{2}$.

By (2.39a), the x-component is

$$\begin{aligned} x(\theta,\tau) &= x_0(\tau) - \int_0^\theta \frac{\sin\theta}{\sqrt{\cos 2\theta + \coth(-2\tau)}}\,d\theta \\ &= x_0(\tau) + \frac{1}{\sqrt{2}}\log\left(4\cos\theta + 2\sqrt{2}\sqrt{\cos 2\theta + \coth(-2\tau)}\right) \\ &\quad - \frac{1}{\sqrt{2}}\log\left(4 + 2\sqrt{2}\sqrt{1+\coth(-2\tau)}\right). \end{aligned}$$

Thus the solution is centered with respect to x if $x(0) + x(\pi) = 0$; i.e.,

$$\begin{aligned} x_0(\tau) &= -\frac{1}{2\sqrt{2}}\log\left(8\left(\coth(-2\tau) - 1\right)\right) \\ &\quad + \frac{1}{\sqrt{2}}\log\left(4 + 2\sqrt{2}\sqrt{1+\coth(-2\tau)}\right) \\ &= \frac{1}{2\sqrt{2}}\log\left(\frac{4 + 2\sqrt{2}\sqrt{1+\coth(-2\tau)}}{2\sqrt{2}\sqrt{1+\coth(-2\tau)} - 4}\right). \end{aligned}$$

Assuming this, we have

$$\begin{aligned} x(\theta,\tau) &= \frac{1}{\sqrt{2}}\log\left(4\cos\theta + 2\sqrt{2}\sqrt{\cos 2\theta + \coth(-2\tau)}\right) \\ &\quad - \frac{1}{2\sqrt{2}}\log\left(8\left(\coth(-2\tau) - 1\right)\right) \\ &= \frac{1}{\sqrt{2}}\log\left(\frac{4\cos\theta + 2\sqrt{2}\sqrt{\cos 2\theta + \coth(-2\tau)}}{\sqrt{8\left(\coth(-2\tau) - 1\right)}}\right). \end{aligned} \tag{4.57}$$

We have $x(\pi-\theta,\tau) = -x(\theta,\tau) = x(\theta+\pi,\tau)$ and $x(-\theta,\tau) = x(\theta,\tau)$. In particular, $x\left(\frac{\pi}{2},\tau\right) = 0 = x\left(\frac{3\pi}{2},\tau\right)$.

We compute directly from (4.56) that

$$\frac{\partial y}{\partial \tau}(\theta,\tau) = \frac{-\coth(-2\tau) + 1}{\sqrt{\cos 2\theta + \coth(-2\tau)}}\sin\theta, \tag{4.58a}$$

$$\frac{\partial y}{\partial \theta}(\theta,\tau) = \frac{\cos\theta}{\sqrt{\cos 2\theta + \coth(-2\tau)}} \tag{4.58b}$$

and from (4.57) that

$$\frac{\partial x}{\partial \tau}(\theta,\tau) = -\frac{\coth(-2\tau) + 1}{\sqrt{\cos 2\theta + \coth(-2\tau)}}\cos\theta, \tag{4.59a}$$

$$\frac{\partial x}{\partial \theta}(\theta,\tau) = -\frac{\sin\theta}{\sqrt{\cos 2\theta + \coth(-2\tau)}}. \tag{4.59b}$$

Thus $X(\theta,\tau)$ is a reparametrization of curve shortening flow since (4.58a) and (4.58b) and (4.59a) and (4.59b) imply

$$\frac{\partial X}{\partial \tau}+\kappa \mathsf{N}=\sin 2\theta \frac{\partial X}{\partial \theta}.$$

Remark 4.18. Define the reflections $R_x:\mathbb{R}^2\to\mathbb{R}^2$ and $R_y:\mathbb{R}^2\to\mathbb{R}^2$ by

$$R_x(x,y)=(-x,y) \quad \text{and} \quad R_y(x,y)=(x,-y).$$

From the symmetries of $x(\theta,\tau)$ and $y(\theta,\tau)$, we conclude that $X(\theta,\tau)$ satisfies

$$\begin{aligned} X(\pi-\theta,\tau) &= (R_x\circ X)(\theta,\tau), \\ X(-\theta,\tau) &= (R_y\circ X)(\theta,\tau), \\ X(\theta+\pi,\tau) &= (R_x\circ R_y\circ X)(\theta,\tau). \end{aligned}$$

By the Gage–Hamilton theorem, as $t\to 0$, the paperclip shrinks to a round point. In fact, by (4.54) and (4.55), we have

$$\lim_{\tau\to 0}\frac{\kappa_{\max}(\tau)}{\kappa_{\min}(\tau)}=\lim_{\tau\to 0}\sqrt{1+\tanh(-2\tau)}=1.$$

Remark 4.19 (Grim Reaper as a backward limit)**.** Note that

$$\lim_{\tau\to-\infty}\kappa(\theta,\tau)=\sqrt{\cos 2\theta+1}=\sqrt{2}\left|\cos\theta\right|. \tag{4.60}$$

This exhibits the fact that the limit as $\tau\to-\infty$ at either end of the oval is the Grim Reaper (without rescaling).

4.5. Classification of convex ancient solutions on S^1

We now present the proof[3], by Daskalopoulos, Hamilton, and Šešum [**192**], that the paperclips and the shrinking circles are the only convex ancient solutions to curve shortening flow on S^1.

Recall that $\alpha \doteqdot p_\theta=(\kappa^2)_\theta$. Define the functional

$$I(\alpha)\doteqdot\int_{S^1}\left((\alpha_\theta)^2-4\alpha^2\right)d\theta. \tag{4.61}$$

For the paperclip, by (4.53) we have $\alpha_\tau=0=\alpha_{\theta\theta}+4\alpha$ and $I(\alpha)=0$.

[3]We shall present a different proof of this result, which extends to the noncompact case, in Section 14.6.

Figure 4.4. Natasa Šešum.

Lemma 4.20 (Monotonicity of I). *Under curve shortening flow, we have*

$$\frac{dI}{dt} = -2\int_{S^1} \frac{(\alpha_\tau)^2}{p} d\theta \leq 0. \tag{4.62}$$

Proof. Integrating by parts and using (4.46), we compute that

$$\begin{aligned} \frac{dI}{dt} &= \int_{S^1} (2\alpha_\theta \alpha_{\tau\theta} - 8\alpha\, \alpha_\tau)\, d\theta \\ &= -2\int_{S^1} \alpha_\tau (\alpha_{\theta\theta} + 4\alpha)\, d\theta \\ &= -2\int_{S^1} \frac{(\alpha_\tau)^2}{p} d\theta. \end{aligned} \tag{4.63}$$ □

The paperclip is characterized by I being constant.

Corollary 4.21. *If an embedded solution to curve shortening flow satisfies $\kappa > 0$ and $\frac{dI}{dt} \equiv 0$, then it is the paperclip.*

Proof. By (4.62), we have $\alpha_\tau \equiv 0$; i.e., (4.48) holds. By the derivation in the previous section for solutions of this form, we conclude that the solution is a paperclip. □

Note that, by integrating by parts on (4.61), we obtain

$$I(\alpha) = \int_{S^1} -\alpha (\alpha_{\theta\theta} + 4\alpha)\, d\theta = -\int_{S^1} \frac{\alpha\, \alpha_\tau}{p} d\theta. \tag{4.64}$$

Define the functional

$$J(t) \doteqdot \int_{S^1} \frac{\alpha^2}{2p} d\theta = \int_{S^1} \frac{(p_\theta)^2}{2p} d\theta \geq 0.$$

Using (4.44) and (4.46), we compute

$$\begin{aligned} \frac{dJ}{dt} &= \int_{S^1} \left(\frac{\alpha\, \alpha_\mathcal{T}}{p} - \frac{\alpha^2 p_\mathcal{T}}{2p^2} \right) d\theta \\ &= -I(\alpha) - \int_{S^1} \frac{\alpha^2 p_\mathcal{T}}{2p^2}\, d\theta \\ &\leq -I(\alpha) \end{aligned} \tag{4.65}$$

by (4.64) and since $p_\mathcal{T} = 2\kappa\, \kappa_\mathcal{T} \geq 0$ by the Harnack estimate (4.23).

Lemma 4.22 (I as $t \to -\infty$). *The monotone quantity $I(\alpha(t))$ satisfies*

$$\lim_{t \to -\infty} I(\alpha(t)) \leq 0. \tag{4.66}$$

Proof. By (4.65), for any $t_1 < t_2$, we have

$$-\int_{t_1}^{t_2} I(\alpha(t))\, dt \geq J(t_2) - J(t_1) \geq -J(t_1). \tag{4.67}$$

On the other hand, by the Harnack estimate and (4.44),

$$\begin{aligned} 0 &\leq \int_{S^1} \frac{p_\mathcal{T}}{p}\, d\theta \\ &= \int_{S^1} \frac{1}{p} \left(p\, p_{\theta\theta} - \frac{1}{2}(p_\theta)^2 + 2p^2 \right) d\theta \\ &= \int_{S^1} \left(-\frac{(p_\theta)^2}{2p} + 2p \right) d\theta. \end{aligned}$$

Thus

$$J = \int_{S^1} \frac{(p_\theta)^2}{2p}\, d\theta \leq 2 \int_{S^1} p\, d\theta.$$

Combine this with

$$\int_{t_1}^{t_2} I(\alpha(t))\, dt \leq J(t_1) \leq 2 \int_{S^1} p(\theta, t_1)\, d\theta \leq 2 \int_{S^1} p(\theta, t_2)\, d\theta,$$

where we again used $p_\mathcal{T} \geq 0$. The lemma follows from fixing $t_2 \in (-\infty, 0)$, taking $t_1 \to -\infty$, and using the monotonicity (4.62). □

Combining the lemma and (4.63), we see that

$$I(\alpha(t)) \leq 0 \tag{4.68}$$

for all $t \in (-\infty, 0)$. For ancient solutions, the monotonicity of I becomes constancy.

Proposition 4.23 (I vanishes for ancient solutions of curve shortening flow). *For any convex embedded ancient solution to curve shortening flow on a closed curve,*

$$\lim_{t\to 0} I\left(\alpha\left(t\right)\right) \geq 0. \tag{4.69}$$

This implies

$$I\left(\alpha\left(t\right)\right) \equiv 0 \quad \textit{for } t \in \left(-\infty, 0\right).$$

Proof. First observe that combining (4.69) with

$$\frac{d}{dt} I\left(\alpha\left(t\right)\right) \leq 0 \quad \text{and} \quad \lim_{t\to-\infty} I\left(\alpha\left(t\right)\right) \leq 0$$

yields $I\left(\alpha\left(t\right)\right) \equiv 0$ for $t \in \left(-\infty, 0\right)$.

The rest of the proof is devoted to establishing (4.69). Note that

$$\int_{S^1} \alpha\left(\theta, \tau\right) d\theta = \int_{S^1} p_\theta\left(\theta, \tau\right) d\theta = 0.$$

Let

$$\alpha_{\mathrm{c}}\left(\tau\right) \doteqdot \frac{1}{\pi} \int_{S^1} \alpha\left(\theta, \tau\right) \cos\theta \, d\theta \quad \text{and} \quad \alpha_{\mathrm{s}}\left(\tau\right) \doteqdot \frac{1}{\pi} \int_{S^1} \alpha\left(\theta, \tau\right) \sin\theta \, d\theta.$$

Define

$$\beta\left(\theta, \tau\right) \doteqdot \alpha\left(\theta, \tau\right) - \alpha_{\mathrm{c}}\left(\tau\right) \cos\theta - \alpha_{\mathrm{s}}\left(\tau\right) \sin\theta,$$

so that

$$\left\langle \beta, 1\right\rangle_{L^2} = 0, \tag{4.70a}$$

$$\left\langle \beta, \cos\theta\right\rangle_{L^2} = 0, \tag{4.70b}$$

$$\left\langle \beta, \sin\theta\right\rangle_{L^2} = 0. \tag{4.70c}$$

By (4.70a)–(4.70c) and expanding β as a Fourier series, we see that

$$\int_{S^1} \left(\left(\beta_\theta\right)^2 - 4\beta^2\right) d\theta \geq 0.$$

From this, we obtain

$$\begin{aligned} I\left(\alpha\right) &= \int_{S^1} \left(\left(\alpha_\theta\right)^2 - 4\alpha^2\right) d\theta \\ &= \int_{S^1} \left(\left(\beta_\theta - \alpha_{\mathrm{c}} \sin\theta + \alpha_{\mathrm{s}} \cos\theta\right)^2 - 4\left(\beta + \alpha_{\mathrm{c}} \cos\theta + \alpha_{\mathrm{s}} \sin\theta\right)^2\right) d\theta \\ &= \int_{S^1} \left(\left(\beta_\theta\right)^2 - 4\beta^2\right) d\theta - 3\pi\left(\alpha_{\mathrm{c}}^2 + \alpha_{\mathrm{s}}^2\right) \\ &\geq -3\pi\left(\alpha_{\mathrm{c}}^2 + \alpha_{\mathrm{s}}^2\right). \end{aligned} \tag{4.71}$$

Hence the proposition will follow from showing that

$$\lim_{\tau\to 0} \alpha_{\mathrm{c}}\left(\tau\right) = \lim_{\tau\to 0} \alpha_{\mathrm{s}}\left(\tau\right) = 0. \tag{4.72}$$

Since $\alpha = p_\theta$, by integrating by parts, we have

$$\alpha_{\mathrm{c}}(\tau) = \frac{1}{\pi}\int_{S^1} p(\theta,\tau)\sin\theta\, d\theta,$$

$$\alpha_{\mathrm{s}}(\tau) = -\frac{1}{\pi}\int_{S^1} p(\theta,\tau)\cos\theta\, d\theta.$$

Note that for the paperclip, we have $\alpha_{\mathrm{c}}(\tau) \equiv \alpha_{\mathrm{s}}(\tau) \equiv 0$. Let

$$w \doteqdot p - \bar{p}, \quad \text{where } \bar{p} \doteqdot \frac{\int p\, d\theta}{\int d\theta}.$$

Observe that $\bar{p} \sim (-2\tau)^{-1}$ since $\kappa \sim (-2\tau)^{-1/2}$ (by Gage and Hamilton's theorem, the curve shrinks to a round point at time 0). Then

$$\int_{S^1} w\sin\theta\, d\theta = \pi\alpha_{\mathrm{c}}(\tau) \quad \text{and} \quad \int_{S^1} w\cos\theta\, d\theta = \pi\alpha_{\mathrm{s}}(\tau).$$

We have

$$\begin{aligned}\frac{1}{\kappa} &= \frac{1}{\sqrt{p}}\\ &= (w+\bar{p})^{-1/2}\\ &= \bar{p}^{-1/2}\left(\frac{w}{\bar{p}}+1\right)^{-1/2}\\ &= \bar{p}^{-1/2}\left(1-\frac{w}{2\bar{p}}+O\left(\left(\tfrac{w}{\bar{p}}\right)^2\right)\right).\end{aligned}$$

Substituting this into (2.40), we obtain

$$\begin{aligned}0 &= \bar{p}^{1/2}\int_{M^1}\frac{\cos\theta}{\kappa(\theta)}\, d\theta\\ &= \int_{M^1}\left(1-\frac{w}{2\bar{p}}+O\left(\left(\tfrac{w}{\bar{p}}\right)^2\right)\right)\cos\theta\, d\theta\\ &= -\frac{1}{2}\bar{p}^{-1}\int_{M^1} w\cos\theta\, d\theta + O\left(\left(\tfrac{w}{\bar{p}}\right)^2\right).\end{aligned}$$

Hence

$$\int_{M^1} w\cos\theta\, d\theta = O\left(\tfrac{w^2}{\bar{p}}\right). \tag{4.73}$$

On the other hand, by Lemma 4.24 below, we have

$$w = O\left((-\tau)^{-\lambda}\right) \tag{4.74}$$

for any $\lambda \in (0,1)$. Since $\bar{p} \sim (-2\tau)^{-1}$, we find that

$$O\left(\tfrac{w^2}{\bar{p}}\right) \leq (-\tau)^{1-2\lambda},$$

which tends to 0 as $\tau \to 0$ by taking $\lambda \in (0, \frac{1}{2})$. We conclude that $\alpha_{\mathrm{s}}(\tau) \to 0$ as $\tau \to 0$. Similarly, by replacing $\cos\theta$ by $\sin\theta$ in the above argument, we have $\alpha_{\mathrm{c}}(\tau) \to 0$ as $\tau \to 0$. □

As needed above, we prove the following regarding the rate of convergence of the curvature.

Lemma 4.24 (w grows more slowly than κ^2). *As before, let $w = p - \bar{p}$. For any $\lambda \in (0,1)$ there exists $C < \infty$ such that*

$$w = O\left(\left(-\tau\right)^{-\lambda}\right).$$

Proof. We rescale the solution by

$$\tilde{X}(\theta,\tau) \doteqdot (-2\tau)^{-1/2} X(\theta,\tau),$$
$$\tilde{\tau} \doteqdot -\frac{1}{2}\log(-\tau).$$

Then

$$\tilde{\kappa}(\theta,\tau) = (-2\tau)^{1/2}\kappa(\theta,\tau),$$
$$\frac{d\tilde{\tau}}{d\tau} = -\frac{1}{2\tau}.$$

From this and (4.19), we compute that

$$\begin{aligned}\frac{\partial\tilde{\kappa}}{\partial\tilde{\tau}} &= \left(\frac{d\tilde{\tau}}{d\tau}\right)^{-1}\frac{d}{d\tau}\left((-2\tau)^{1/2}\kappa\right)\\ &= -2\tau\left(-(-2\tau)^{-1/2}\kappa + (-2\tau)^{1/2}\kappa_\tau\right)\\ &= -2t\left(-(-2\tau)^{-1/2}\kappa + (-2\tau)^{1/2}\kappa^2\left(\kappa_{\theta\theta}+\kappa\right)\right)\\ &= \tilde{\kappa}^2\tilde{\kappa}_{\theta\theta} + \tilde{\kappa}^3 - \tilde{\kappa}.\end{aligned}$$

From this we see that a steady-state solution is given by $\tilde{\kappa} \equiv 1$. Now, by Corollary 5.7.12 in Gage and Hamilton [**228**], for any $\mu \in (0,1)$, there exists $C < \infty$ such that

$$\|\tilde{\kappa}_\theta(\cdot,\tilde{\tau})\|_{L^\infty(S^1)} \le C\mathrm{e}^{-2\mu\tilde{\tau}},$$

so that

$$\|\tilde{\kappa}(\theta,\tilde{\tau}) - 1\|_{L^\infty(S^1)} \le C\mathrm{e}^{-2\mu\tilde{\tau}}.$$

This implies that

$$\left\|\kappa(\theta,\tau) - (-2\tau)^{-1/2}\right\|_{L^\infty(S^1)} \le \frac{C}{\sqrt{2}}(-\tau)^{\mu-1/2}.$$

Hence

$$\begin{aligned}
\|w(\theta,\tau)\|_{L^\infty(S^1)} &\leq \kappa^2(\tau)_{\max} - \kappa^2(\tau)_{\min} \\
&\leq 2\left\|\kappa^2(\theta,\tau) - (-2\tau)^{-1}\right\|_{L^\infty(S^1)} \\
&\leq 2\left\|\kappa(\theta,\tau) - (-2\tau)^{-\frac{1}{2}}\right\|_{L^\infty(S^1)}\left\|\kappa(\theta,\tau) + (-2\tau)^{-\frac{1}{2}}\right\|_{L^\infty(S^1)} \\
&\leq \sqrt{2}C(-\tau)^{\mu-1/2}\left\|\kappa(\theta,\tau) + (-2\tau)^{-1/2}\right\|_{L^\infty(S^1)} \\
&\leq C'(-\tau)^{\mu-1}
\end{aligned}$$

since $\kappa(\cdot,\tau) \sim (-2\tau)^{-1/2}$. In other words, for any $\lambda \in (0,1)$, there exists $C < \infty$ such that

$$\|w(\theta,\tau)\|_{L^\infty(S^1)} \leq C(-\tau)^{-\lambda}. \qquad \square$$

By combining the results above, we conclude the following.

Theorem 4.25 (Convex ancient solutions to curve shortening flow)**.** *A non-round convex embedded ancient solution to curve shortening flow on a closed curve must be a paperclip.*

Proof. The claim follows from Proposition 4.23 and Corollary 4.21. $\square$

4.6. Notes and commentary

Bryan and Louie [**126**] showed that the only closed embedded ancient solutions to the curve shortening flow on the round 2-sphere are equators or shrinking round circles.

Curve flows by powers α of the curvature were studied in [**36**]. It is shown that for $\alpha \in (1/8,\infty)$ the only embedded self-similar solutions are round circles, except for ellipses which occur for $\alpha = 1/3$. For $\alpha \in (0,1/8)$, besides round circles, for each k with $3 \leq k \leq 1+\alpha^{-1}$, there exists an embedded k-fold symmetric self-similar solution which converges to the regular k-sided polygon as $\alpha \to 0$ and to the round circle as $\alpha \to \frac{1}{k^2-1}$. In Nien and Tsai [**425**] translating self-similar solutions is studied for flows by powers of curvature.

T. K.-K. Au [**80**] showed that the Abresch–Langer solutions serve, in a certain explicit sense, as "saddle points" between the class of solutions which limit (after rescaling) to nonsingular curves and those which limit (after rescaling) to singular curves, proving a conjecture of Abresch and Langer [**2**].

Angenent and You have constructed a large number of "ancient trombone" solutions to curve shortening flow whose total curvature is uniformly bounded, by gluing together an arbitrary chain of given Grim Reapers along their common asymptotes [**76**] (see Section 14.9).

4.7. Exercises

Exercise 4.1. Show that the paperclip described in Chapter 4 is the same as the paperclip described in Chapter 2.

Exercise 4.2. Give an alternate proof of (4.24) using (2.25) and (2.16).

Exercise 4.3 (Pointwise version of monotonicity)**.** Show that

$$\partial_t \Phi + \partial_s^2 \Phi = \left(\kappa^2 - \left(\kappa - (2\tau)^{-1} \langle X, \mathsf{N} \rangle\right)^2\right) \Phi .$$

Note that (4.36) follows from integrating this formula and $\frac{\partial}{\partial t} ds = -\kappa^2 ds$.

Exercise 4.4. Show that for any $k \in \mathbb{N}$ the solution (4.42) to curve shortening flow induces (descends to) a solution on the cylinder $(\mathbb{R}/2\pi k) \times \mathbb{R}$.

Exercise 4.5. Define $\gamma \doteqdot \alpha_{\theta\theta} + 4\alpha = \frac{\alpha_\tau}{p}$. Show that

$$I(\alpha) = -\int_{S^1} \alpha\gamma\, d\theta, \quad \int_{S^1} \alpha\gamma_\tau\, d\theta = \int_{S^1} p\gamma^2\, d\theta, \quad \text{and} \quad \int_{S^1} \alpha_\tau\gamma\, d\theta = \int_{S^1} p\gamma^2\, d\theta .$$

This yields another proof of (4.62).

Chapter 5

Hypersurfaces in Euclidean Space

In this third part, we investigate the mean curvature flow, which is the simplest generalization of the curve shortening flow to higher dimensions. Before beginning our study, it will be useful to recall the basic theory of hypersurfaces in Euclidean spaces.

Unless otherwise indicated (e.g., when we consider varifolds), all objects in this chapter (and throughout the book) are assumed to be of class C^∞.

5.1. Parametrized hypersurfaces

Let M^n be a smooth n-dimensional manifold, and let

$$X : M^n \to \mathbb{R}^{n+1}$$

be a smooth immersion (that is, a smooth map whose differential $dX : TM^n \to T\mathbb{R}^{n+1}$ is everywhere injective). We will often require that X is **proper**, which means that the preimage $X^{-1}(K)$ of any compact subset $K \subset \mathbb{R}^{n+1}$ is a compact subset of M^n. If X is an embedding (i.e., X is a diffeomorphism onto its image), then its image $\mathcal{M}^n \doteqdot X(M^n)$ has the structure of a smooth n-dimensional submanifold of $\mathbb{R}^{n+1}$. When X is a proper embedding, we say that $\mathcal{M}^n$ is **properly embedded**.

We can use the natural geometric structures of $\mathbb{R}^{n+1}$ to induce analogous structures on M^n. When X is an embedding, we may interpret the induced geometric structures as defined either on $\mathcal{M}^n$ or equivalently, by "pulling back" along X, on M^n. If X is merely an immersion, then the former point of view holds locally (since every immersion is locally an embedding), but in general not globally.

We will usually denote points of M^n by x, y, or z and points of $\mathcal{M}^n$ by p, q, or r. Given a coordinate system $(U, \{x^i\}_{i=1}^n)$ on M^n, we have local bases $\{\frac{\partial}{\partial x^i}\}_{i=1}^n$ for the tangent space TM on $U \subset M^n$ and $\{\frac{\partial X}{\partial x^i}\}_{i=1}^n$ on $X(U) \subset \mathcal{M}$. We will denote $\partial_i = \frac{\partial}{\partial x^i}$ and

$$\partial_i X = \frac{\partial X}{\partial x^i} \doteqdot dX(\partial_i)\,, \tag{5.1}$$

where $dX : TM^n \to T\mathbb{R}^{n+1}$ is the derivative of X.

Denote by $\langle\,\cdot\,,\,\cdot\,\rangle : \mathbb{R}^{n+1} \times \mathbb{R}^{n+1} \to \mathbb{R}$ the standard **Euclidean inner product** on $\mathbb{R}^{n+1}$, which we recall acts on pairs of vectors $u = \left(u^1, \ldots, u^{n+1}\right)$ and $v = \left(v^1, \ldots, v^{n+1}\right)$ via

$$\langle u, v\rangle \doteqdot \sum_{i=1}^{n+1} u^i v^i\,.$$

The Euclidean inner product induces an inner product $\langle\cdot,\cdot\rangle_p$ on each tangent space $T_p\mathbb{R}^{n+1}$ via the canonical identification $T_p\mathbb{R}^{n+1} \cong \mathbb{R}^{n+1}$. That is,

$$\langle u, v\rangle_p \doteqdot \langle \widetilde{u}, \widetilde{v}\rangle\ ,$$

where $\widetilde{u}, \widetilde{v} \in \mathbb{R}^{n+1}$ are defined with respect to the identity chart $y : \mathbb{R}^{n+1} \to \mathbb{R}^{n+1}$ by $[(y, \widetilde{u})] = u$ and $[(y, \widetilde{v})] = v$. It varies smoothly with respect to p in the sense that, given smooth vector fields $U, V \in \Gamma(T\mathbb{R}^{n+1})$, the induced function

$$p \mapsto \langle U_p, V_p\rangle_p$$

is smooth.

A smoothly varying family $g(\,\cdot\,,\,\cdot\,)$ of inner products $g_p(\,\cdot\,,\,\cdot\,)$ on the tangent spaces T_pM^n of a smooth manifold M^n is called a **Riemannian metric** on M, and a smooth manifold equipped with a Riemannian metric is called a **Riemannian manifold**. Treatments of Riemannian geometry can be found in the books [**138, 198, 361, 442**].

The Euclidean metric rather trivially induces a metric (which we also denote by $\langle\,\cdot\,,\,\cdot\,\rangle$) on the restriction bundle $T\mathbb{R}^{n+1}|_{\mathcal{M}}$ by restriction. Slightly less trivially, it also induces a metric ${}^X\langle\,\cdot\,,\,\cdot\,\rangle$ on the pullback bundle $X^*T\mathbb{R}^{n+1}$ over M^n by pulling back along X:

$${}^X\langle (x,u)\,,\,(x,v)\rangle_x \doteqdot \langle dX_x(u), dX_x(v)\rangle_{X(x)} \tag{5.2}$$

for each pair of pulled-back tangent vectors $(x,u), (x,v) \in (X^*T\mathbb{R}^{n+1})_x$.

We define the **first fundamental form** I of an embedded submanifold $\mathcal{M} \hookrightarrow \mathbb{R}^{n+1}$ to be the restriction of the Euclidean metric to the tangent spaces $T_p\mathcal{M}^n \hookrightarrow T_p\mathbb{R}^{n+1}$, $p \in \mathcal{M}^n$. That is,

$$\mathrm{I}_p(u,v) \doteqdot \langle u,v\rangle_p \tag{5.3}$$

for each pair of tangent vectors $u,v \in T_p\mathcal{M}^n$. Observe that I is bilinear, symmetric, and positive definite, and therefore defines an inner product on each tangent space $T_p\mathcal{M}^n$. It is also smooth in the sense that the function

$$p \mapsto \mathrm{I}_p(U_p, V_p)$$

is smooth whenever $U,V \in \Gamma(T\mathcal{M}^n)$ are smooth vector fields, so it defines a Riemannian metric on $\mathcal{M}^n$.

Since $dX_x : T_xM^n \to T_{X(x)}\mathcal{M}^n$ is an isomorphism, we may also think of the first fundamental form as a Riemannian metric g on M^n by pulling back to TM^n. That is,

$$g_x(u,v) \doteqdot \langle dX_x(u), dX_x(v)\rangle_p = \mathrm{I}_p(dX_x(u), dX_x(v)) \tag{5.4}$$

for each pair of tangent vectors $u,v \in T_xM^n$. In local coordinates $\{x^i\}_{i=1}^n$,

$$g_{ij} = \langle \partial_i X, \partial_j X\rangle\,. \tag{5.5}$$

The norms induced by I and g will be denoted by $|\cdot|$. Note that g is well-defined even when $X : M^n \to \mathbb{R}^{n+1}$ is merely an immersion.

The metric I (respectively, g) induces an isomorphism $\cdot^\flat : T\mathcal{M}^n \to T^*\mathcal{M}^n$ (respectively, $\cdot^\flat : TM^n \to T^*M^n$) via $v^\flat(u) \doteqdot \mathrm{I}(u,v)$ (respectively, $v^\flat(u) \doteqdot g(u,v)$). Its inverse is denoted by $\cdot^\sharp$. These **musical isomorphisms** induce a metric I on $T^*\mathcal{M}^n$ (respectively, g on T^*M^n) by $\mathrm{I}(\alpha,\beta) \doteqdot \mathrm{I}(\alpha^\sharp,\beta^\sharp)$ (respectively, $g(\alpha,\beta) \doteqdot g(\alpha^\sharp,\beta^\sharp)$).

Given any basis $\{e_i\}_{i=1}^n$ for $T_p\mathcal{M}^n$ (respectively, T_xM^n) we can define the dual basis $\{\vartheta^i\}_{i=1}^n$ for $T^*\mathcal{M}^n$ (respectively, $T_x^*M^n$) by

$$\vartheta^j(e_i) = \delta_i^j \doteqdot \begin{cases} 1 & \text{if } i = j\,, \\ 0 & \text{otherwise}\,. \end{cases}$$

With respect to dual bases, the matrix of components $\mathrm{I}^{ij} \doteqdot I(\vartheta^i,\vartheta^j)$ of the first fundamental form on $T^*\mathcal{M}^n$ (respectively, $g^{ij} \doteqdot g(\vartheta^i,\vartheta^j)$ on T^*M^n) is inverse to the matrix of components $\mathrm{I}_{ij} \doteqdot I(e_i,e_j)$ of the first fundamental form on $T\mathcal{M}^n$ (respectively, $g_{ij} \doteqdot g(e_i,e_j)$ on TM^n).

Similarly, $\langle\cdot,\cdot\rangle|_{\mathcal{M}^n}$ (respectively, ${}^X\langle\cdot\,,\cdot\rangle$) induces an isomorphism between $T\mathbb{R}^{n+1}|_{\mathcal{M}^n}$ and $T^*\mathbb{R}^{n+1}|_{\mathcal{M}^n}$ (respectively, $X^*T\mathbb{R}^{n+1}$ and $X^*T^*\mathbb{R}^{n+1}$). Together, I and $\langle\cdot,\cdot\rangle|_{\mathcal{M}^n}$ (respectively, g and ${}^X\langle\cdot\,,\cdot\rangle$) induce a metric on any tensor bundle obtained from $T\mathcal{M}^n$ and $T\mathbb{R}^{n+1}|_{\mathcal{M}^n}$ (respectively, TM^n and $X^*T\mathbb{R}^{n+1}$) by finitely many tensor products and direct sum operations by distributing over the direct sum and commuting with the tensor product.

Example 5.1. The squared norm of $dX \in T^*M^n \otimes X^*T\mathbb{R}^{n+1}$ is

$$|dX|^2 = \langle dX, dX\rangle = g^{ij}\,\langle dX(\partial_i), dX(\partial_j)\rangle = g^{ij}g_{ij} = n\,, \tag{5.6}$$

where we employed local coordinates $\{x^i\}_{i=1}^n$.

Given a vector $u \in \mathbb{R}^{n+1}$, denote by D_u the **directional derivative** in the direction u. This acts on smooth functions $f : \mathbb{R}^{n+1} \to \mathbb{R}$ by

$$f \mapsto D_u f \doteqdot u^i \frac{\partial f}{\partial y^i}$$

and on smooth vector fields $V : \mathbb{R}^{n+1} \to \mathbb{R}^{n+1}$ by

$$V \mapsto D_u V \doteqdot u^i\,\frac{\partial V^j}{\partial y^i}\frac{\partial}{\partial y^j}\,,$$

where $\left\{y^j\right\}_{j=1}^{n+1}$ are the standard rectilinear Euclidean coordinates on $\mathbb{R}^{n+1}$ and $\{\frac{\partial}{\partial y^j}\}_{j=1}^{n+1}$ are the standard basis vectors.

The directional derivative induces a differential operator associated with $T\mathbb{R}^{n+1}$ via the canonical identification $T_p\mathbb{R}^{n+1} \cong \mathbb{R}^{n+1}$. Given a tangent vector $u \in T_p\mathbb{R}^{n+1}$ and a vector field $V \in \Gamma(T\mathbb{R}^{n+1})$, the **Euclidean covariant derivative** of V in the direction u is defined by

$$\widetilde{D_u V} \doteqdot D_{\widetilde{u}}\widetilde{V}\,, \tag{5.7}$$

where $\widetilde{\cdot} : T_p\mathbb{R}^{n+1} \to \mathbb{R}^{n+1}$ is the canonical identification and $D_{\widetilde{u}}\widetilde{V}$ is the directional derivative of $\widetilde{V}$ in the direction $\widetilde{u}$.

There is another way to define the Euclidean covariant derivative, which we now describe. The canonical identification $T_p\mathbb{R}^{n+1} \cong \mathbb{R}^{n+1}$ gives rise to isomorphisms $\tau_{qp} : T_q\mathbb{R}^{n+1} \to T_p\mathbb{R}^{n+1}$ for $q \in \mathbb{R}^{n+1}$ called **parallel translation** maps. Given any vector field $V \in \Gamma(T\mathbb{R}^{n+1})$, we can define a "vector field" $\widetilde{V} : \mathbb{R}^{n+1} \to T_p\mathbb{R}^{n+1}$ by parallel translating V to $T_p\mathbb{R}^{n+1}$,

$$\widetilde{V}_q \doteqdot \tau_{qp}(V_q) = V_q^j\,\frac{\partial}{\partial y^j}\bigg|_p\,,$$

where $\{\frac{\partial}{\partial y^j}\big|_p\}_{j=1}^{n+1}$ is the canonical orthonormal basis at p. Given a tangent vector $u \in T_p\mathbb{R}^{n+1}$, choose a curve $\gamma : I \to \mathbb{R}^{n+1}$ with $0 \in I$, $\gamma(0) = p$, and $\gamma'(0) = u$. Since $\widetilde{V}$ is a smooth map into a fixed linear space, we can differentiate it along γ. Since parallel translation of $\{\frac{\partial}{\partial y^j}\big|_p\}_{j=1}^{n+1}$ does not depend on γ, we obtain

$$\frac{d}{ds}\bigg|_{s=0}(\widetilde{V}\circ\gamma) = \frac{d}{ds}\bigg|_{s=0}(V^j\circ\gamma)\,\frac{\partial}{\partial y^j}\bigg|_p = (uV^j)\,\frac{\partial}{\partial y^j}\bigg|_p\,.$$

So this definition of the Euclidean covariant derivative coincides with the previous one.

Observe that for any tangent vector $u \in T_p\mathbb{R}^{n+1}$ and vector field $V \in \Gamma(T\mathbb{R}^{n+1})$, $D_uV \in T_p\mathbb{R}^{n+1}$. Furthermore, for any tangent vector $u \in T_pM$, any pair of vector fields $V, W \in \Gamma(T\mathbb{R}^{n+1})$, and any real number $\lambda \in \mathbb{R}$,

$$D_u(V + \lambda W) = D_uV + \lambda D_uW.$$

For any pair of tangent vectors $u, v \in T_pM$, any vector field $W \in \Gamma(T\mathbb{R}^{n+1})$, and any real number $\lambda \in \mathbb{R}$,

$$D_{u+\lambda v}W = D_uW + \lambda D_vW,$$

and for any tangent vector $u \in T_pM$, any vector field $V \in \Gamma(T\mathbb{R}^{n+1})$, and any smooth function $f \in C^\infty(\mathbb{R}^{n+1})$,

$$D_u(fV) = (uf)V + f(p)D_uV.$$

This last identity is called the **Leibniz rule** (or **product rule**) as it resembles the traditional Leibniz (product) rule. Finally, observe that the function

$$p \mapsto D_UV(p) \doteqdot D_{U_p}V$$

is smooth for any pair of smooth vector fields $U, V \in \Gamma(T\mathbb{R}^{n+1})$.

A **connection** on M is a real bilinear map $\nabla : TM \times \Gamma(TM) \to TM$ which (1) respects the bundle structure, in the sense that $\pi(\nabla_uV) = \pi(u)$), (2) is smooth, in the sense that $\nabla_UV \in \Gamma(TM)$ for each pair $U, V \in \Gamma(TM)$, where $\nabla_UV(p) \doteqdot \nabla_{U_p}V \doteqdot \nabla(U_p, V)$, and (3) satisfies the Leibniz rule. Given $u \in TM$, the induced derivation $\nabla_u : \Gamma(TM) \to TM$ is called the **covariant derivative** in the direction u. The covariant derivative of any homogeneous tensor field can be defined by identifying it with the directional derivative when acting on functions and asserting the Leibniz rule with respect to tensor products and that it commutes with contractions.

The covariant differential $\nabla Y : u \mapsto \nabla_uY$ is uniquely determined by the restriction of Y to an open neighborhood of p. In fact, in order to differentiate a vector field Y in a direction $u \in T_pM$, we only need to know the values of Y along a curve through p with tangent u at p.

Observe that the Euclidean connection D is $\langle \cdot, \cdot \rangle$**-compatible** (or **metric compatible**):

$$u \langle V, W \rangle = \langle D_uV, W_p \rangle + \langle V_p, D_uW \rangle$$

for any tangent vector $u \in T_p\mathbb{R}^{n+1}$ and any pair of vector fields $V, W \in \Gamma(T\mathbb{R}^{n+1})$, and **symmetric**:

$$D_uV - D_vU = [U, V]_p$$

for any pair of tangent vectors $u, v \in T_p\mathbb{R}^{n+1}$ and respective extensions $U, V \in \Gamma(T\mathbb{R}^{n+1})$, where $[\cdot, \cdot] : \Gamma(T\mathbb{R}^{n+1}) \times \Gamma(T\mathbb{R}^{n+1}) \to \Gamma(T\mathbb{R}^{n+1})$ denotes the Lie bracket. Recall that symmetry and metric compatibility uniquely define a connection on any Riemannian manifold, called the **Levi-Civita connection**.

The Euclidean connection D defines the **restriction connection** (which we also denote by D) on the restriction $T\mathbb{R}^{n+1}|_{\mathcal{M}}$ of $T\mathbb{R}^{n+1}$ to $\mathcal{M}$ via

$$D_u V \doteqdot D_u \widetilde{V} \tag{5.8}$$

for any $u \in T\mathcal{M} \hookrightarrow T\mathbb{R}^{n+1}$ and any $V \in \Gamma(T\mathbb{R}^{n+1}|_{\mathcal{M}})$, where $\widetilde{V}$ is any extension of V to $\Gamma(T\mathbb{R}^{n+1})$.

The **Euclidean pullback connection** (which we again denote by D) on the pullback bundle $X^*T\mathbb{R}^{n+1}$ is defined by pulling back along X. That is,

$$D_u(X^*V) \doteqdot D_{dX(u)}V \tag{5.9}$$

for any tangent vector $u \in TM$ and any vector field $V \in \Gamma(T\mathbb{R}^{n+1})$. Although we have only defined the pullback connection on pulled-back sections, it can be extended to all sections of $X^*T\mathbb{R}^{n+1}$ by asserting the Leibniz rule since any section of $X^*T\mathbb{R}^{n+1}$ is locally a $C^\infty(M)$-linear combination of pulled-back sections.

Observe that in the embedded case the restriction bundle $T\mathbb{R}^{n+1}|_{\mathcal{M}}$ splits as an orthogonal sum $T\mathcal{M} \oplus^{\perp} N\mathcal{M}$, where $N\mathcal{M}$ is the **normal bundle**,

$$N\mathcal{M} \doteqdot \{(p, \nu) \in T\mathbb{R}^{n+1}|_{\mathcal{M}} : \langle u, \nu \rangle_p = 0 \ \text{ for all } \ u \in T_p\mathcal{M}\}\,. \tag{5.10}$$

Define the **tangential** and **normal projections** $\cdot^{\top} : T\mathbb{R}^n|_{\mathcal{M}} \to T\mathcal{M}$ and $\cdot^{\perp} : T\mathbb{R}^n|_{\mathcal{M}} \to N\mathcal{M}$, respectively, by

$$u^{\top} \doteqdot u - \langle u, \nu \rangle \, \nu \quad \text{and} \quad u^{\perp} \doteqdot \langle u, \nu \rangle \, \nu\,, \tag{5.11}$$

where ν is either of the two unit vectors in $N_p\mathcal{M}$, $p \doteqdot \pi(u)$. Pulling back to M yields the ${}^X\langle \cdot\,, \cdot \rangle$-orthogonal decomposition $X^*T\mathbb{R}^{n+1} \cong TM \oplus^{\perp} NM$, where

$$NM \doteqdot \{(x, \nu) \in X^*T\mathbb{R}^{n+1} : {}^X\langle dX_x(u)\,, \nu \rangle_x = 0 \ \text{ for all } \ u \in T_xM\} \tag{5.12}$$

is the **normal bundle** of $X : M \to \mathbb{R}^{n+1}$. The **tangential** and **normal projections**

$$\cdot^{\top} : X^*T\mathbb{R}^{n+1} \to TM \quad \text{and} \quad \cdot^{\perp} : X^*T\mathbb{R}^{n+1} \to NM$$

are defined similarly as before (see (5.11)).

Example 5.2. Let $\mathcal{M}^n \hookrightarrow \mathbb{R}^{n+1}$ be an embedded submanifold. Denote by $X : \mathcal{M}^n \to \mathbb{R}^{n+1}$ the inclusion map. Since $dX_p : T_p\mathcal{M}^n \to T_p\mathbb{R}^{n+1}|_{\mathcal{M}^n}$ is the inclusion map, we have

$$u\left(\frac{1}{2}|X|^2\right) = \langle dX(u), X \rangle = \langle u, X \rangle\ .$$

Thus, identifying X with a vector field $X \in \Gamma(T\mathbb{R}^{n+1}|_{\mathcal{M}^n})$,

$$X^\top = \operatorname{grad}\left(\frac{1}{2}|X|^2\right). \tag{5.13}$$

Pulling back to M^n, we can identify X with a vector field $X \in \Gamma(X^*T\mathbb{R}^{n+1})$. We then obtain the same identity, with $\cdot^\top$ and grad replaced by their "pulled-back" versions.

Note that the covariant derivative of a tangent vector field $V \in \Gamma(T\mathcal{M}^n)$ in the direction $u \in T\mathcal{M}^n$ does not define a covariant derivative on $\mathcal{M}^n$ since the result may not be tangential. This can be remedied by simply taking the tangential projection of the result.

Lemma 5.3. *If $\mathcal{M}^n \hookrightarrow \mathbb{R}^{n+1}$ is an embedded hypersurface, then the tangential projection of the Euclidean connection D,*

$$\nabla_u V \doteqdot (D_u V)^\top, \tag{5.14}$$

defines a connection $\nabla : T\mathcal{M}^n \times \Gamma(T\mathcal{M}^n) \to T\mathcal{M}^n$ on $\mathcal{M}^n$. It is metric compatible and symmetric; i.e., ∇ is the Levi-Civita connection of I.

Proof. Bilinearity and smoothness are immediate from the linearity of $\cdot^\top$. So to confirm that ∇ is a connection, we need only check the Leibniz rule. Given $u \in T_p\mathcal{M}^n$, $f \in C^\infty(\mathcal{M}^n)$, and $V \in \Gamma(T\mathcal{M}^n)$,

$$\begin{aligned}\nabla_u fV &= (D_u fV)^\top \\ &= ((uf)V_p + f(p)D_uV)^\top \\ &= (uf)V_p + f(p)(D_uV)^\top \\ &= (uf)V_p + f(p)\nabla_u V.\end{aligned}$$

Next observe that, for any tangent vector $u \in T_p\mathcal{M}^n$ and tangent vector fields $V, W \in \Gamma(T\mathcal{M}^n)$,

$$\begin{aligned}u\langle V, W\rangle &= \langle D_uV, W_p\rangle + \langle V_p, D_uW\rangle \\ &= \left\langle (D_uV)^\top, W_p\right\rangle + \left\langle V_p, (D_uW)^\top\right\rangle \\ &= \langle \nabla_u V, W_p\rangle + \langle V_p, \nabla_u W\rangle.\end{aligned}$$

So ∇ is metric compatible.

Finally, given any tangent vectors $u, v \in T_p\mathcal{M}^n$ and respective extensions $U, V \in \Gamma(T\mathcal{M}^n)$,

$$\nabla_u V - \nabla_v U = (D_uV)^\top - (D_vU)^\top = (D_uV - D_vU)^\top = [U, V]_p^\top.$$

But $[U, V]$ is already tangential. $\square$

Pulling back to M, we obtain a connection $\nabla : TM^n \times \Gamma(TM^n) \to TM^n$ via

$$dX(\nabla_u V) \doteqdot (D_u [dX(V)])^\top , \tag{5.15}$$

where D is the Euclidean pullback connection on $X^* T\mathbb{R}^{n+1}$. By essentially the same arguments, ∇ is a torsion-free, metric compatible connection on M, so the pullback connection ∇ is the Levi-Civita connection of g.

Recall that, with respect to local coordinates $x : U \to \mathbb{R}^n$ for M^n, the connection coefficients (a.k.a. Christoffel symbols) $\Gamma_{ij}^k : U \to \mathbb{R}$ are defined by

$$\Gamma_{ij}^k \partial_k \doteqdot \nabla_i \partial_j . \tag{5.16}$$

Applying the definitions (5.2) and (5.15), we obtain

$$\Gamma_{ij}^k = g^{k\ell} \langle \partial_i \partial_j X, \partial_\ell X \rangle , \tag{5.17}$$

where $\partial_j X \in \Gamma(X^* T\mathbb{R}^{n+1})$ and we have defined

$$\partial_i \partial_j X \doteqdot \frac{\partial^2 X}{\partial x^i \partial x^j} \doteqdot D_i (\partial_j X) .$$

Let S be any tensor field which is a section of finitely many tensor products of the bundles $T\mathcal{M}^n$ and $T\mathbb{R}^{n+1}|_{\mathcal{M}^n}$ (respectively, TM^n and $X^*T\mathbb{R}^{n+1}$) and their duals. We can define its covariant derivative $\nabla_u S$ as follows: First, we assert that $\nabla_u S$ is the directional derivative of S if S is a function, the covariant derivative of S induced by the restriction connection on $T\mathbb{R}^{n+1}|_{\mathcal{M}^n}$ (respectively, the pullback connection on $X^*T\mathbb{R}^{n+1}$) if $S \in \Gamma(T\mathbb{R}^{n+1}|_{\mathcal{M}^n})$ (respectively, $S \in \Gamma(X^*T\mathbb{R}^{n+1})$ and the covariant derivative of S induced by the induced connection on $T\mathcal{M}^n$ (respectively, TM^n) if $S \in \Gamma(T\mathcal{M}^n)$ (respectively, $S \in \Gamma(TM^n)$). The definition is then completed by asserting that the ∇_u distribute over addition, satisfy the Leibniz rule with respect to tensor products, and commute with contractions. The **covariant differential** of a tensor field S is the tensor field ∇S of one higher covariant degree defined by $\nabla S(u) \doteqdot \nabla_u S$ for any tangent vector u. We sometimes denote dX by ∇X.

Example 5.4. Given $u, v \in T_x M^n$, the derivative of dX satisfies

$$\begin{aligned} \nabla^2 X(u,v) &\doteqdot (\nabla_u dX)(v) \\ &= D_u(dX(v)) - dX(\nabla_u V) \\ &= (D_u [dX(V)])^\perp , \end{aligned} \tag{5.18}$$

where $V \in \Gamma(TM)$ is any extension of v.

The *normal* part of the Euclidean covariant derivative along $\mathcal{M}$ is a tensor, called the (**vector**) **second fundamental form** of $\mathcal{M}$.

Lemma 5.5. *Let $u, v \in T\mathcal{M}$, and let $V \in \Gamma(T\mathcal{M})$ be an extension of v. Then the smooth map $\vec{\mathrm{II}} : T\mathcal{M} \times T\mathcal{M} \to N\mathcal{M}$ defined by*

$$\vec{\mathrm{II}}(u, v) \doteqdot (D_u V)^{\perp} \tag{5.19}$$

is well-defined and bilinear. It also happens to be symmetric.

Proof. Let $\nu \in \Gamma(N\mathcal{M})$ be a normal vector field such that $|\nu| \equiv 1$ in a neighborhood of p. Then, given $u, v \in T_p\mathcal{M}$ and any extension $V \in \Gamma(T\mathcal{M})$ of v,

$$(D_u V)^{\perp} = \langle D_u V, \nu_p \rangle \, \nu_p = - \langle v, D_u \nu \rangle \, \nu_p .$$

It follows that $\vec{\mathrm{II}}$ is well-defined and bilinear. Now choose an extension $U \in \Gamma(T\mathcal{M})$ of u. Then

$$0 = [U, V]_p^{\perp} = (D_u V - D_v U)^{\perp} = \vec{\mathrm{II}}(u, v) - \vec{\mathrm{II}}(v, u) .$$

Thus, $\vec{\mathrm{II}}(u, v)$ is symmetric. □

Pulling back to M we obtain the symmetric, bilinear (**vector**) **second fundamental form** $\vec{h} : TM \times TM \to NM$ of $X : M \to \mathbb{R}^{n+1}$,

$$\vec{h}(u, v) \doteqdot (D_u V)^{\perp} , \tag{5.20}$$

where $V \in \Gamma(TM)$ is any extension of v.

Choosing a unit normal $\nu \in N_x M = N_p\mathcal{M}$, where $p \doteqdot X(x)$, we can write

$$\vec{h}_x(u, v) = -h_x(u, v)\nu = -\mathrm{II}_p(u, v)\nu = \vec{\mathrm{II}}_p(u, v) . \tag{5.21}$$

By (5.18)

$$\nabla^2 X = \vec{h} . \tag{5.22}$$

or, in components with respect to local bases $\{e_i\}_{i=1}^n$ for $T_x M$ and ν for $N_x M$,

$$\nabla_i \nabla_j X = -h_{ij} \, \nu . \tag{5.23}$$

To each symmetric bilinear form corresponds a selfadjoint endomorphism. The **Weingarten tensor** (or **Weingarten map**) $\mathrm{L}_p : T_p\mathcal{M} \to T_p\mathcal{M}$ of $\mathcal{M}$ (corresponding to the choice of unit normal ν at p) is the linear map defined by

$$\mathrm{L}_p(u) \doteqdot D_u N \tag{5.24}$$

for some extension $N \in \Gamma(N\mathcal{M})$ of ν satisfying $|N| \equiv 1$ in a neighborhood of p. It is well-defined because

$$0 = \frac{1}{2} D_u |N|^2 = \langle D_u N, \nu \rangle$$

and

$$\langle D_u N, v\rangle = -\langle \nu, D_u V\rangle = \mathrm{II}(u,v) \tag{5.25}$$

for any tangent vectors $u, v \in T_p\mathcal{M}$, any extension $V \in \Gamma(T\mathcal{M})$ of v, and any extension $N \in \Gamma(N\mathcal{M})$ of ν satisfying $|N| \equiv 1$ in a neighborhood of p.

Pulling back the map L_p to M^n, we obtain the **Weingarten tensor** $L_x : T_xM \to T_xM$ of $X : M^n \to \mathbb{R}^{n+1}$,

$$L_x(u) \doteqdot D_u N\,, \tag{5.26}$$

where $N \in \Gamma(NM)$ is some extension of ν satisfying $|N| \equiv 1$ in a neighborhood of p. Then

$$h_x(u,v) = g(L_x(u), v) = \langle \mathrm{L}_p(u), v\rangle = \mathrm{II}_p(u,v)\,.$$

That is, L corresponds to II and L to h under the canonical isomorphism induced by the respective metrics. Note that L_p is $\langle \cdot\,,\,\cdot\rangle_p$-selfadjoint and L_x is g_x-selfadjoint. With respect to local coordinates $\{x^i\}_{i=1}^n$ on M^n, we have

$$D_i N = \mathrm{L}_i{}^k \partial_k X = \mathrm{II}_{ij}\, g^{jk}\partial_k X\,. \tag{5.27}$$

If $\mathcal{M}$ is orientable, then we can choose a global unit normal field $\mathsf{N} \in \Gamma\mathcal{M}$, in which case the tensor fields $h \in \Gamma(T^*M \otimes T^*M)$, $L \in \Gamma(T^*M \otimes TM)$, $\mathrm{II} \in \Gamma(T^*\mathcal{M} \otimes T^*\mathcal{M})$, and $\mathrm{L} \in \Gamma(T^*\mathcal{M} \otimes T\mathcal{M})$ are all well-defined globally (the former two even when X is merely an immersion). We can identify $\mathsf{N} : \mathcal{M} \to N\mathcal{M} \subset T\mathbb{R}^{n+1}$ with a map

$$\mathrm{G} : \mathcal{M} \to S^n \subset \mathbb{R}^{n+1}\,, \tag{5.28}$$

known as the **Gauß map** of $\mathcal{M}$, via the canonical identification $T_p\mathbb{R}^{n+1} \cong \mathbb{R}^{n+1}$.

Observe that the tangent spaces $T_p\mathcal{M}$ and $T_{\mathrm{G}(p)}S^n$ coincide under the identification $T_p\mathbb{R}^{n+1} \cong \mathbb{R}^{n+1}$. Indeed, since the Gauß map of the sphere is the identity,

$$\begin{aligned} T_p\mathcal{M} &= \{v \in T_p\mathbb{R}^{n+1} : \langle v, \mathsf{N}_p\rangle_p = 0\} \\ &\cong \{v \in \mathbb{R}^{n+1} : \langle v, \mathrm{G}(p)\rangle = 0\} \\ &\cong \{v \in T_{\mathrm{G}(p)}\mathbb{R}^{n+1} : \left\langle v, \mathsf{N}_{\mathrm{G}(p)}\right\rangle_{\mathrm{G}(p)} = 0\} \\ &= T_{\mathrm{G}(p)}S^n\,, \end{aligned}$$

where $\mathsf{N}_{\mathrm{G}(p)}$ is the outward normal to S^n at $\mathrm{G}(p)$. The derivative

$$\mathrm{A}_p \doteqdot d\,\mathrm{G}\,|_p : T_p\mathcal{M} \to T_{\mathrm{G}(p)}S^n$$

of the Gauß map is called the **shape operator** of $\mathcal{M}$. Up to the identification of $T_p\mathcal{M}$ and $T_{\mathrm{G}(p)}S^n$, we have $\mathrm{A} = \mathrm{L}$. So the Weingarten tensor measures the rate at which the tangent hyperplanes of $\mathcal{M}$ are turning in $\mathbb{R}^{n+1}$.

The (ordered) eigenvalues

$$\kappa_1 \leq \kappa_2 \leq \cdots \leq \kappa_n$$

of the Weingarten tensor are called the **principal curvatures** of $X : M \to \mathbb{R}^{n+1}$. They switch sign according to the chosen orientation at each point. The corresponding unit eigenvectors are called **principal directions**. Since the Weingarten tensor is selfadjoint, the principal curvatures are real and there exists, at each point, an orthonormal basis $\{e_i\}_{i=1}^n$ for the tangent space consisting of principal directions. We refer to an orthonormal basis of principal directions at a point as a **principal basis** and a local or global orthonormal frame field (not necessarily continuous) of principal directions as a **principal frame field**.

Various important curvature invariants can be constructed from the second fundamental form. For example, the **mean curvature**, H, is the trace of the Weingarten tensor. That is,

$$H \doteqdot \mathrm{tr}(\mathrm{L}) = \kappa_1 + \cdots + \kappa_n = \mathrm{tr}(L)\,. \tag{5.29}$$

The mean curvature measures the local variation of area under small perturbations of the hypersurface. (We prove this more explicitly below, in Lemma 5.25). It is equal to the average of the second fundamental form over the tangent sphere and can also be related to the scalar part of the divergence of dX with respect to the induced covariant derivative and metric on $T^*M \otimes X^*TM$. We shall call a hypersurface $X : M^n \to \mathbb{R}^{n+1}$ **mean convex** if it admits a unit normal field with respect to which its mean curvature is nonnegative and **strictly mean convex** if it admits a unit normal field with respect to which its mean curvature is positive.

Recall that the **divergence** $\operatorname{div} V$ of a vector field V is the trace of its covariant differential; that is,

$$\operatorname{div} V \doteqdot \mathrm{tr}\,(u \mapsto \nabla_u V)\,.$$

Similarly, given a tensor field $S \in \Gamma(TM \otimes E)$, we can form the divergence

$$\operatorname{div} S \doteqdot \mathrm{tr}\,(u \mapsto \nabla_u S) \in E\,. \tag{5.30}$$

We can also form the divergence of a tensor field $S \in \Gamma(T^*M \otimes E)$ by identifying T^*M with TM prior to tracing. In particular, by (5.23),

$$\operatorname{div}(dX) = \mathrm{tr}\,(\nabla(dX)) = -g^{ij}h_{ij}\nu = \vec{H}\,.$$

Recall that the **Laplacian** Δf of a smooth function f is the divergence of its gradient. That is,

$$\Delta f \doteqdot \operatorname{div}(\operatorname{grad} f)\,, \tag{5.31}$$

where $\operatorname{grad} f$, the **gradient** of f, is the tangent vector field dual to the differential df of f. Thus, $\vec{H}$ can be interpreted as the Laplacian of X:

$$\Delta X \doteqdot \operatorname{div}(dX) = \operatorname{tr}(\nabla dX) = \vec{H}\,. \tag{5.32}$$

The **Gauß curvature**, K, is the determinant of the Weingarten tensor. That is,

$$K \doteqdot \det(\mathrm{L}) = \kappa_1 \cdots \kappa_n = \det(L)\,. \tag{5.33}$$

By Gauß's Theorema Egregium, the Gauß curvature is intrinsic to an immersed surface $X : M^2 \to \mathbb{R}^3$: It is invariant under isometries of the induced metric g.

More generally, the **m-th mean curvature** H_m, defined for $1 \leq m \leq n$, is the m-th elementary symmetric polynomial in the principal curvatures:

$$H_m \doteqdot \sum_{1\leq i_1 < i_2 < \cdots < i_m \leq n} \kappa_{i_1} \cdots \kappa_{i_m}\,. \tag{5.34}$$

In particular, H_1 is the mean curvature and H_n is the Gauß curvature.

The **harmonic mean curvature** H_{-1} is defined by

$$H_{-1} \doteqdot \frac{1}{\kappa_1^{-1} + \cdots + \kappa_n^{-1}} = \frac{H_n}{H_{n-1}}\,. \tag{5.35}$$

Example 5.6 (Round Euclidean spheres)**.** Denote by S_r^n the n-sphere of radius r centered at the origin in $\mathbb{R}^{n+1}$ and let $\mathsf{N}(x) \doteqdot \frac{x}{|x|}$ denote its unit outward normal vector field. The second fundamental form and mean curvature of S_r^n are given by

$$\mathrm{II} = \frac{1}{r}\mathrm{I}_{S_r^n} \tag{5.36}$$

and

$$H = \frac{n}{r},$$

respectively. Indeed, if $V, W \in TS_r^n$, then

$$\begin{aligned}
\mathrm{II}\,(V,W) &= \left\langle D_V\left(\frac{x}{|x|}\right), W\right\rangle \\
&= \left\langle \frac{V}{|x|} + V\left(\frac{1}{|x|}\right)x, W\right\rangle \\
&= \frac{1}{r}\langle V, W\rangle
\end{aligned}$$

since $|x| = r$ and $\langle x, W\rangle = 0$ (or $V\left(\frac{1}{|x|}\right) = 0$).

Let A be a subset of $\mathbb{R}^{n+1}$, let m be a nonnegative integer, and let ω_m denote the volume of the m-dimensional unit Euclidean ball. For a subset B of $\mathbb{R}^{n+1}$, let $\operatorname{diam}(B)$ denote the diameter of B and let $\operatorname{rad}(B) = \frac{1}{2}\operatorname{diam}(B)$. Given $\delta > 0$, define

$$\mathscr{H}^m_\delta(A) = \inf_{\mathscr{B}} \sum_i \omega_m \operatorname{rad}(B_i)^m,$$

where the infimum is taken over all countable covers $\mathscr{B} = \{B_i\}$ of A satisfying $\operatorname{diam}(B_i) \leq \delta$ for each $B_i \in \mathscr{B}$. Observe that $\mathscr{H}^m_\delta(A)$ is a nonincreasing function of δ. The **m-dimensional Hausdorff measure** of A is defined by

$$\mathscr{H}^m(A) = \lim_{\delta \to 0} \mathscr{H}^m_\delta(A) = \sup_{\delta > 0} \mathscr{H}^m_\delta(A) \in [0, \infty]. \tag{5.37}$$

Observe that $\mathscr{H}^0$ is the counting measure. The m-dimensional Hausdorff measure may be defined for all nonnegative real numbers m by defining $\omega_m = \pi^{m/2}\Gamma(\frac{m}{2}+1)$, where Γ is the gamma function. Also, $\mathscr{H}^m$ is a *Borel regular* outer measure on $\mathbb{R}^{n+1}$; that is, Borel sets are $\mathscr{H}^m$-measurable and each subset of $\mathbb{R}^{n+1}$ is contained in a Borel set of equal $\mathscr{H}^m$-measure. See, for example, Frank Morgan's book [**403**].

The **area** $\operatorname{Area}(K)$ of a compact subset K of an embedded hypersurface $\mathcal{M}^n \hookrightarrow \mathbb{R}^{n+1}$ is defined as the n-dimensional Hausdorff measure on $\mathbb{R}^{n+1}$ of K. That is,

$$\operatorname{Area}(K) \doteqdot \mathscr{H}^n(K).$$

Another way to measure area in a Riemannian manifold (M^n, g) is given by its **Riemannian measure** μ. Given any compact subset $K \subset M^n$, $\mu(K)$ is defined, with respect to a locally finite atlas $\{(U_i, \varphi_i)\}_{i\in\mathcal{I}}$ for M^n and subordinate partition of unity $\{\rho_i\}_{i\in\mathcal{I}}$, by

$$\mu(K) \doteqdot \sum_{i\in\mathcal{I}} \int_{U_i} (\rho_i \chi_K) \circ \varphi^{-1} \sqrt{\det g_{\varphi_i}} d\mathscr{H}^n,$$

where χ_K is the indicator function of K and g_{φ_i} is the component matrix of g with respect to the chart φ_i. By the change of variables formula for integration, the choice of locally finite atlas and subordinate partition of unity is irrelevant.

It turns out that when $X : M^n \to \mathbb{R}^{n+1}$ is an embedding the pullback of the Lebesgue measure $\mathscr{H}^n$ along X coincides with the Riemannian measure μ on (M^n, g):

$$\mu(X^{-1}(K)) = \mathscr{H}^n(K).$$

On the other hand, the two may not coincide when $X : M \to \mathbb{R}^{n+1}$ is merely an immersion.

Lemma 5.7. *Let $X : M^n \to \mathbb{R}^{n+1}$ be a compact immersed hypersurface. Then*

$$\text{Area}(M) = -\frac{1}{n}\int_M \langle X, \vec{H}\rangle \, d\mu\,. \tag{5.38}$$

Proof. Observe that

$$\begin{aligned}\frac{1}{2}\Delta|X|^2 &= \operatorname{div}\langle dX, X\rangle \\ &= |dX|^2 + \langle \operatorname{div}(dX), X\rangle \\ &= n + \big\langle X, \vec{H}\big\rangle .\end{aligned} \tag{5.39}$$

The divergence theorem then yields

$$\begin{aligned} n\,\text{Area}(M) &= \int_M n\, d\mu \\ &= \int_M \left(\frac{1}{2}\Delta|X|^2 - \big\langle X, \vec{H}\big\rangle\right) d\mu \\ &= -\int_M \big\langle X, \vec{H}\big\rangle \, d\mu\,. \end{aligned}$$

□

By the Jordan–Brouwer separation theorem [**255**, Chapter 2, §5], any compact embedded hypersurface $\mathcal{M}^n \hookrightarrow \mathbb{R}^{n+1}$ separates $\mathbb{R}^{n+1}$ into two disjoint open sets, exactly one of which is bounded. The **volume** $\text{Vol}(\mathcal{M})$ enclosed by $\mathcal{M}$ is defined as the volume $|\Omega|$ (measured with respect to the Lebesgue measure) of the bounded component $\Omega \subset \mathbb{R}^{n+1}$ of $\mathbb{R}^{n+1} \setminus \mathcal{M}^n$.

Lemma 5.8. *Let $\mathcal{M} \hookrightarrow \mathbb{R}^{n+1}$ be a compact embedded hypersurface. Then*

$$\text{Vol}(\mathcal{M}) = |\Omega| = \frac{1}{n+1}\int_{\mathcal{M}} \langle X, \mathsf{N}\rangle \, d\mu\,, \tag{5.40}$$

where N is the outer unit normal field to $\mathcal{M}$.

Proof. Denote by $X : \Omega \hookrightarrow \mathbb{R}^{n+1}$ the inclusion mapping. Then by the divergence theorem,

$$\begin{aligned} (n+1)\,\text{Vol}(\mathcal{M}) &= \int_\Omega (n+1)\, d\mathscr{H}^{n+1} \\ &= \int_\Omega \frac{1}{2}\Delta|X|^2 \, d\mathscr{H}^{n+1} \\ &= \int_{\mathcal{M}} \langle X, \mathsf{N}\rangle \, d\mu\,, \end{aligned}$$

where $\mathscr{H}^{n+1}$ is the Lebesgue measure on $\mathbb{R}^{n+1}$. □

When the hypersurface is a curve in the plane, we will refer to **length** instead of area and **enclosed area** instead of enclosed volume.

We next derive two fundamental formulae for hypersurfaces which relate their intrinsic and extrinsic curvatures to the ambient curvature (which is zero for Euclidean space). They are special cases of analogous formulae which hold for any smoothly immersed submanifold of any Riemannian manifold.

Proposition 5.9 (Fundamental structure equations). *Let* $X : M \to \mathbb{R}^{n+1}$ *be an embedded hypersurface. Given any* $p \in \mathcal{M}$*, tangent vectors* $u, v, w \in T_p\mathcal{M}$*, and a unit normal vector field* $\mathsf{N} \in \Gamma(N\mathcal{M})$*,*

$$\mathrm{Rm}(u,v)w = \mathrm{II}(u,w)\,\mathrm{L}(v) - \mathrm{II}(v,w)\,\mathrm{L}(u) \tag{5.41}$$

and

$$\nabla_u \mathrm{L}(v) = \nabla_v \mathrm{L}(u)\,, \tag{5.42}$$

where Rm *is the Riemann curvature tensor of the induced connection* ∇ *on* $\mathcal{M}$ *and where* II *and* L *are the second fundamental form and Weingarten tensor fields, respectively, associated to* N*.*

Proof. Fix $u, v, w \in T_p\mathcal{M}$. Given extensions $U, V, W \in \Gamma(TM)$ of u, v, w, respectively, Rm is defined by

$$\mathrm{Rm}(u,v)w = \nabla_v(\nabla_U W) - \nabla_u(\nabla_V W) - \nabla_{[V,U]_p} W. \tag{5.43}$$

Thus,

$$\begin{aligned}
\mathrm{Rm}(u,v)w &= \nabla_v(D_U W)^\top - \nabla_u(D_V W)^\top - (D_{[V,U]_p} W)^\top \\
&= \nabla_v(D_U W + \mathrm{II}(U,W)\mathsf{N}) - \nabla_u(D_V W + \mathrm{II}(V,W)\mathsf{N}) \\
&\quad - (D_{[V,U]_p} W)^\top \\
&= \left(D_v(D_U W) + \mathrm{II}(u,w)\,\mathrm{L}(v)\right)^\top \\
&\quad - \left(D_u(D_V W) + \mathrm{II}(v,w)\,\mathrm{L}(u)\right)^\top - \left(D_{[V,U]_p} W\right)^\top \\
&= \mathrm{II}(u,w)\,\mathrm{L}(v) - \mathrm{II}(v,w)\,\mathrm{L}(u)\,.
\end{aligned}$$

Similarly,

$$\begin{aligned}
\nabla_u \mathrm{L}(v) &= \nabla_u(\mathrm{L}(V)) - \mathrm{L}(\nabla_u V) \\
&= (D_u(D_V \mathsf{N}))^\top - \mathrm{L}(\nabla_u V) \\
&= (D_v(D_U \mathsf{N}) + D_{[U,V]_p}\mathsf{N})^\top - \mathrm{L}(\nabla_u V) \\
&= (D_v(D_U \mathsf{N}))^\top + \mathrm{L}([U,V]_p - \nabla_u V) \\
&= \nabla_v(\mathrm{L}(U)) - \mathrm{L}(\nabla_v U) = \nabla_v \mathrm{L}(u)\,.
\end{aligned}$$

□

The first equation (5.41) is known as the **Gauß equation**. In local coordinates, it takes the form

$$\mathrm{Rm}_{ijkl} = \mathrm{II}_{ik}\,\mathrm{II}_{jl} - \mathrm{II}_{jk}\,\mathrm{II}_{il}\,, \tag{5.44}$$

where

$$\mathrm{Rm}_{ijkl} \doteqdot \mathrm{I}(\mathrm{Rm}(\partial_i X, \partial_j X)\partial_k X, \partial_l X).$$

Taking the trace yields

$$\mathrm{Rc}(u,v) = H\,\mathrm{II}(u,v) - \mathrm{II}^2(u,v)\,, \tag{5.45}$$

where Rc is the Ricci curvature and

$$\mathrm{II}^2(u,v) \doteqdot \mathrm{I}(\mathrm{L}^2(u),v) = \mathrm{I}(\mathrm{L}(u),\mathrm{L}(v)).$$

In local coordinates,

$$\mathrm{Rc}_{ij} = g^{kl}\,\mathrm{Rm}_{ikjl} = H\,\mathrm{II}_{ij} - \mathrm{II}^2_{ij} = H\,\mathrm{II}_{ij} - \mathrm{I}^{kl}\,\mathrm{II}_{ik}\,\mathrm{II}_{lj}\ .$$

Tracing again yields

$$\mathrm{R} = H^2 - |\mathrm{II}|^2 = \sum_{i\neq j} \kappa_i\kappa_j\,, \tag{5.46}$$

where $\mathrm{R} \doteqdot \mathrm{tr}(\mathrm{Rc})$ is the scalar curvature.

The second equation (5.42) is known as the **Codazzi equation**. Since II is symmetric, the Codazzi equation implies that $\nabla\,\mathrm{II}$ is totally symmetric. In local coordinates, it takes the form

$$\nabla_k\,\mathrm{L}_i{}^j = \nabla_i\,\mathrm{L}_k{}^j$$

or

$$\nabla_k\,\mathrm{II}_{ij} = \nabla_i\,\mathrm{II}_{kj}\ . \tag{5.47}$$

Taking its trace yields

$$\mathrm{div}(\mathrm{L}) \doteqdot \mathrm{tr}(\nabla\,\mathrm{L}) = \nabla\,\mathrm{tr}(\mathrm{L}) = dH$$

(where the first trace is taken with respect to the first and third components), or, in local coordinates,

$$\mathrm{div}(\mathrm{L})_j \doteqdot \nabla_i\,\mathrm{L}_j{}^i = \nabla_j\,\mathrm{L}_i{}^i = \nabla_i H\,. \tag{5.48}$$

Essentially the same arguments yield analogous equations for h and L, namely,

$$\mathrm{Rm}(u,v)w = h(u,w)L(v) - h(v,w)L(u)$$

and

$$\nabla_u L(v) = \nabla_v L(u)\,,$$

and corresponding equations for their traces, where Rm is now the curvature of the connection ∇ induced on M.

The Gauß and Codazzi equations imply a remarkable formula which relates the second fundamental form to its second covariant derivatives, known as **Simons's equation**.

Proposition 5.10. *Given $p \in \mathcal{M}$, tangent vectors $u, v, w, z \in T_p\mathcal{M}$, and a unit normal vector field $\mathsf{N} \in \Gamma(N\mathcal{M})$,*

$$\begin{aligned}\nabla_u \nabla_v \mathrm{II}(w,z) - \nabla_w \nabla_z \mathrm{II}(u,v) =& \mathrm{II}(u,v)\,\mathrm{II}^2(w,z) - \mathrm{II}(v,w)\,\mathrm{II}^2(u,z)\\ &+ \mathrm{II}(u,z)\,\mathrm{II}^2(v,w) - \mathrm{II}(w,z)\,\mathrm{II}^2(u,v)\,. \end{aligned} \tag{5.49}$$

In particular,

$$\nabla_{(u}\nabla_{v)}\mathrm{II}(w,z) - \nabla_{(w}\nabla_{z)}\mathrm{II}(u,v) = \mathrm{II}(u,v)\mathrm{II}^2(w,z) - \mathrm{II}(w,z)\mathrm{II}^2(u,v), \tag{5.50}$$

where the brackets indicate the symmetric part:

$$\nabla_{(u}\nabla_{v)}\,\mathrm{II} \doteqdot \frac{1}{2}\left(\nabla_u\nabla_v\,\mathrm{II} + \nabla_v\nabla_u\,\mathrm{II}\right)\,. \tag{5.51}$$

Proof. The Gauß and Codazzi equations and the symmetry of II yield

$$\begin{aligned}\nabla_u\nabla_v\,\mathrm{II}(w,z) &= \nabla_u\nabla_w\,\mathrm{II}(v,z)\\ &= \nabla_w\nabla_u\,\mathrm{II}(v,z) + (\mathrm{Rm}(w,u)\,\mathrm{II})(v,z)\\ &= \nabla_w\nabla_u\,\mathrm{II}(z,v) - \mathrm{II}(\mathrm{Rm}(w,u)v,z) - \mathrm{II}(v,\mathrm{Rm}(w,u)z)\\ &= \nabla_w\nabla_z\,\mathrm{II}(u,v) - \mathrm{II}\left(\mathrm{II}(w,v)\,\mathrm{L}(u) - \mathrm{II}(u,v)\,\mathrm{L}(w), z\right)\\ &\quad - \mathrm{II}\left(v, \mathrm{II}(w,z)\,\mathrm{L}(u) - \mathrm{II}(u,z)\,\mathrm{L}(w)\right)\\ &= \nabla_w\nabla_z\,\mathrm{II}(u,v) - \mathrm{II}(w,v)\,\mathrm{II}^2(u,z) + \mathrm{II}(u,v)\,\mathrm{II}^2(w,z)\\ &\quad - \mathrm{II}(w,z)\,\mathrm{II}^2(u,v) + \mathrm{II}(u,z)\,\mathrm{II}^2(v,w)\,. \end{aligned}$$

□

In local coordinates, Simons's formulae become

$$\nabla_i\nabla_j\,\mathrm{II}_{kl} - \nabla_k\nabla_l\,\mathrm{II}_{ij} = \mathrm{II}_{ij}\,\mathrm{II}^2_{kl} - \mathrm{II}_{jk}\,\mathrm{II}^2_{il} + \mathrm{II}_{il}\,\mathrm{II}^2_{jk} - \mathrm{II}_{kl}\,\mathrm{II}^2_{ij} \tag{5.52}$$

and

$$\nabla_{(i}\nabla_{j)}\,\mathrm{II}_{kl} - \nabla_{(k}\nabla_{l)}\,\mathrm{II}_{ij} = \mathrm{II}_{ij}\,\mathrm{II}^2_{kl} - \mathrm{II}_{kl}\,\mathrm{II}^2_{ij}\,. \tag{5.53}$$

Pulling back to M, the same calculations yield

$$\begin{aligned}\nabla_u\nabla_v h(w,z) - \nabla_w\nabla_z h(u,v) =& h(u,v)h^2(w,z) - h(w,v)h^2(u,z)\\ &+ h(u,z)h^2(v,w) - h(w,z)h^2(u,v)\end{aligned}$$

and

$$\nabla_{(u}\nabla_{v)}h(w,z) - \nabla_{(w}\nabla_{z)}h(u,v) = h(u,v)h^2(w,z) - h(w,z)h^2(u,v)\,,$$

where

$$h^2(u,v) \doteqdot g(L^2(u),v) = g(L(u),L(v))\,.$$

Simons's equation is a rather miraculous identity: It shows that every hypersurface satisfies a hidden structure equation relating second derivatives of curvature to the curvature itself. It is particularly powerful when the hypersurface satisfies some additional relation, such as the minimal surface equation or the mean curvature flow. The following useful corollary provides an illustration of its effectiveness.

Corollary 5.11. *Let $X : M^n \to \mathbb{R}^{n+1}$, $n \geq 2$, be an immersed hypersurface. Suppose that $\mathrm{II} > 0$ and $\nabla\,\mathrm{II} \equiv 0$. Then $X(M^n)$ is* ***umbilic****. In particular, if X is proper, then the components of $X(M^n)$ are round spheres.*

Proof. Consider the 4-tensor C defined by

$$\mathrm{C}(u,v,w,z) \doteqdot \mathrm{II}(u,v)\,\mathrm{II}^2(w,z) - \mathrm{II}(w,z)\,\mathrm{II}^2(u,v)\,. \tag{5.54}$$

Observe that, with respect to a principal basis $\{e_i\}_{i=1}^n$ for T_xM at any $x \in M$,

$$\mathrm{C}_{ijkl} = \kappa_i\kappa_k(\kappa_k - \kappa_i)\delta_{ij}\delta_{kl}$$

and hence

$$|\mathrm{C}|^2 = \sum_{i,k=1}^n \kappa_i^2\kappa_k^2(\kappa_k - \kappa_i)^2\,. \tag{5.55}$$

If $\nabla\mathrm{II} \equiv 0$, then, by Simons's equation (5.50), $\mathrm{C} \equiv 0$. If $\mathrm{II} > 0$, then $\kappa_i > 0$ for each i and we conclude from (5.55) that $\kappa_i = \kappa_j$ for each i, j. That is, the hypersurface is umbilic. The conclusion then follows from Exercise 5.5. □

A slightly more sophisticated argument yields the following result. Here we assume constant mean curvature instead of parallel second fundamental form.

Proposition 5.12. *Let $X : M^n \to \mathbb{R}^{n+1}$, $n \geq 2$, be a compact immersed hypersurface. Suppose that $\mathrm{II} > 0$ and $\nabla H \equiv 0$. Then $X(M^n)$ is umbilic. In particular, the components of $X(M^n)$ are round spheres.*

Proof. Since H is locally constant, Simons's equation yields

$$\frac{1}{2}\Delta|\mathrm{II}|^2 = |\nabla\mathrm{II}|^2 + \mathrm{I}(\Delta\mathrm{II},\mathrm{II}) = |\nabla\mathrm{II}|^2 + \sum_{i,k}\left(\kappa_i^3\kappa_k - \kappa_i^2\kappa_k^2\right).$$

We claim that the final term is nonnegative. Indeed,

$$\begin{aligned}\sum_{i,k}\left(\kappa_i^3\kappa_k - \kappa_i^2\kappa_k^2\right) &= \sum_{i<k}\kappa_i^2\kappa_k\left(\kappa_i - \kappa_k\right) + \sum_{k<i}\kappa_i^2\kappa_k\left(\kappa_i - \kappa_k\right)\\ &= \sum_{i<k}\left(\kappa_i^2\kappa_k - \kappa_k^2\kappa_i\right)\left(\kappa_i - \kappa_k\right)\\ &= \sum_{i<k}\kappa_i\kappa_k\left(\kappa_i - \kappa_k\right)^2 \geq 0\,.\end{aligned}$$

In particular,

$$-\Delta|\mathrm{II}|^2 \leq 0\,.$$

Since M is compact, the maximum principle now implies that $|\mathrm{II}|$ is locally constant. But then $\Delta|\mathrm{II}|^2 \equiv 0$ and hence

$$0 = |\nabla \mathrm{II}|^2 + \sum_{i<k} \kappa_i \kappa_k \left(\kappa_i - \kappa_k\right)^2 \geq 0\,.$$

The claim follows. □

A further useful identity, known as the **Ricci identity**, provides a formula for commuting the Laplacian past a gradient.

Proposition 5.13. *Let $\mathcal{M}^n \subset \mathbb{R}^{n+1}$ be a hypersurface. Given any $u \in T\mathcal{M}^n$ and any $f \in C^\infty(\mathcal{M})$,*

$$\Delta \nabla_u f - \nabla_u \Delta f = \mathrm{Rc}(u, \nabla f) = H\, \mathrm{II}(u, \nabla f) - \mathrm{II}^2(u, \nabla f)\,. \tag{5.56}$$

Proof. Computing in local coordinates and at a point where $g_{ij} = \delta_{ij}$, while using the Einstein convention of summing over repeated indices, we have that

$$\begin{aligned}
\Delta \nabla_i f - \nabla_i \Delta f &= \nabla_j \nabla_j \nabla_i f - \nabla_i \nabla_j \nabla_j f \\
&= (\nabla_j \nabla_i - \nabla_i \nabla_j) \nabla_j f \\
&= \mathrm{Rm}_{jijk} \nabla_k f \\
&= \mathrm{Rc}_{ik} \nabla_k f \\
&= H\, \mathrm{II}_{ik} \nabla_k f - \mathrm{II}^2_{ik} \nabla_k f\,,
\end{aligned}$$

where we applied the contracted Gauß equation (5.45). □

Henceforth, we will conflate the extrinsic and instrinsic points of view. We will use the notation $\langle \cdot, \cdot \rangle$ or g for the induced metric, II for the second fundamental form, and L for the Weingarten tensor.

5.2. Alternative representations of hypersurfaces

In this section we discuss useful parametrizations of hypersurfaces in certain special cases, namely graphs of functions, starshaped hypersurfaces, and convex hypersurfaces.

5.2.1. Graphs of functions. The graph $\mathcal{M}^n = \operatorname{graph} f \doteqdot \{(x, f(x)) : x \in \Omega\}$ of a smooth function $f : \Omega \to \mathbb{R}$, $\Omega \underset{\text{open}}{\subset} \mathbb{R}^n$, is a smooth hypersurface. We call such a hypersurface a **graph**. Conversely, by the implicit function theorem, every smooth hypersurface can be locally represented as the graph of a smooth function. A graph, graph f, is called **entire** if $f : \Omega \to \mathbb{R}$, $\Omega \subset \mathbb{R}^n$ is an **entire function**, that is, if $\Omega = \mathbb{R}^n$. We briefly discussed 1-dimensional graphs in the context of the curve shortening flow in Chapter 2.

A graph hypersurface $\mathcal{M}^n = \operatorname{graph} f \doteqdot \{(x, f(x)) : x \in \Omega\}$ may be parametrized by the embedding $X : \Omega \to \mathbb{R}^{n+1}$ defined by

$$X(x) \doteqdot (x, f(x)). \tag{5.57}$$

Differentiating (5.57) in the x^i-direction yields

$$\frac{\partial X}{\partial x^i} = \left(e_i, \frac{\partial f}{\partial x^i}\right), \tag{5.58}$$

where $\{e_i\}_{i=1}^n$ is the standard basis for $\mathbb{R}^n$. Then $\{\frac{\partial X}{\partial x^1}, \ldots, \frac{\partial X}{\partial x^n}\}$ is a basis for each tangent space of $\mathcal{M}$. Denote by $\nabla f = (\frac{\partial f}{\partial x^1}, \ldots, \frac{\partial f}{\partial x^n})$ the Euclidean gradient of f. By (5.58), the unit vector field

$$\mathsf{N} \doteqdot \frac{(\nabla f, -1)}{\sqrt{1 + |\nabla f|^2}} \tag{5.59}$$

is normal to $\mathcal{M}^n$. Note that this normal points "downward": $\langle \mathsf{N}, e_{n+1} \rangle < 0$.

Lemma 5.14 (Graph of a function). *The second fundamental form and mean curvature of a graph $\mathcal{M} = \{(x, f(x)) : x \in \Omega\}$ are given by*

$$\mathrm{II}_{ij} = \frac{1}{\sqrt{1 + |\nabla f|^2}} \frac{\partial^2 f}{\partial x^i \partial x^j}, \tag{5.60}$$

$$H = \operatorname{div}\left(\frac{\nabla f}{\sqrt{1 + |\nabla f|^2}}\right), \tag{5.61}$$

where div *is the Euclidean divergence.*

Proof. We compute, using (5.58) and (5.59), that

$$\begin{aligned}
\mathrm{II}_{ij} &= \left\langle \frac{\partial \mathsf{N}}{\partial x^i}, \frac{\partial X}{\partial x^j} \right\rangle \\
&= \left\langle \frac{\partial}{\partial x^i}\left(\frac{(\nabla f, -1)}{\sqrt{1 + |\nabla f|^2}}\right), \left(e_j, \frac{\partial f}{\partial x^j}\right) \right\rangle \\
&= \frac{\partial}{\partial x^i}\left(\frac{1}{\sqrt{1 + |\nabla f|^2}} \frac{\partial f}{\partial x^j}\right) - \frac{\partial}{\partial x^i}\left(\frac{1}{\sqrt{1 + |\nabla f|^2}}\right) \frac{\partial f}{\partial x^j},
\end{aligned}$$

and formula (5.60) follows. On the other hand, the metric is

$$\mathrm{I}_{ij} = \left\langle \frac{\partial X}{\partial x^i}, \frac{\partial X}{\partial x^j} \right\rangle = \delta_{ij} + \frac{\partial f}{\partial x^i}\frac{\partial f}{\partial x^j}, \tag{5.62}$$

which has inverse

$$\mathrm{I}^{ij} = \delta_{ij} - \frac{\frac{\partial f}{\partial x^i}\frac{\partial f}{\partial x^j}}{1+|\nabla f|^2}, \tag{5.63}$$

so that $\mathrm{I}_{ij}\,\mathrm{I}^{jk} = \delta_i^k$. Thus the mean curvature is equal to

$$\begin{aligned} H &= \mathrm{I}^{ij}\,\mathrm{II}_{ij} \\ &= \left(\delta_{ij} - \frac{\frac{\partial f}{\partial x^i}\frac{\partial f}{\partial x^j}}{1+|\nabla f|^2}\right)\frac{1}{\sqrt{1+|\nabla f|^2}}\frac{\partial^2 f}{\partial x^i \partial x^j} \\ &= \frac{\partial}{\partial x^i}\left(\frac{1}{\sqrt{1+|\nabla f|^2}}\frac{\partial f}{\partial x^i}\right) \end{aligned} \tag{5.64}$$

and formula (5.61) follows. □

5.2.2. Level sets at regular values. By the implicit function theorem, the level set $\Sigma \doteqdot \{X \in \Omega : u(X) = 0\}$ of a smooth function $u : \Omega \to \mathbb{R}$, $\Omega \underset{\text{open}}{\subset} \mathbb{R}^{n+1}$, is a smooth hypersurface in a neighborhood of a **regular point** $X_0 \in \Sigma$ (i.e., $Du_{X_0} \neq 0$). In particular, Σ is smooth if 0 is a **regular value** (i.e., every point of the level set Σ is regular). The intrinsic geometry of (the regular part of) Σ is readily determined from the relation $u(X) = 0$. First, we determine the tangent space: Let $X : U \to \Sigma$, $U \underset{\text{open}}{\subset} \mathbb{R}^n$, be a local parametrization for Σ. Then

$$0 = u(X(x)) \quad \text{for all } x \in U. \tag{5.65}$$

Differentiating (5.65) in the x^i-direction yields

$$0 = Du(\partial_i X). \tag{5.66}$$

Thus the gradient $Du|_\Sigma$ is normal to Σ. We orient Σ by equipping it with the unit normal field

$$\mathsf{N} \doteqdot -\left.\frac{Du}{|Du|}\right|_\Sigma.$$

To determine the curvature of Σ, we differentiate (5.66) in the x^j-direction and apply (5.66) to obtain

$$0 = D^2u(\partial_i X, \partial_j X) + |Du|\,\mathrm{II}_{ij}.$$

That is,

$$\mathrm{II}_{ij} = -\frac{D^2u(\partial_i X, \partial_j X)}{|Du|}. \tag{5.67}$$

In particular, the mean curvature of Σ is given by

$$H = -\frac{1}{|Du|}\left(\Delta u - \frac{D^2u(Du,Du)}{|Du|^2}\right).$$

The choice of orientation ensures that $\mathrm{II} \geq 0$ when u is a concave function.

5.2.3. Starshaped hypersurfaces. A hypersurface $\mathcal{M}^n \subset \mathbb{R}^{n+1}$ is called **starshaped** if it may be written as a radial graph over a round sphere, that is, if

$$\mathcal{M} = \{p + r(z)\, z : z \in S^n\} \tag{5.68}$$

for some $p \in \mathbb{R}^{n+1}$ and some **radial function** $r : S^n \to (0,\infty)$.

Assuming $p = 0$, define the parametrization $X : S^n \to \mathcal{M} \subset \mathbb{R}^{n+1}$ by

$$X(z) \doteqdot r(z)\, z. \tag{5.69}$$

Given local coordinates $\left\{\theta^i\right\}_{i=1}^n$ on S^n, the components of the first and second fundamental forms of the starshaped hypersurface $\mathcal{M}$ are given by

$$g_{ij} \doteqdot g\left(\frac{\partial X}{\partial \theta^i}, \frac{\partial X}{\partial \theta^j}\right), \tag{5.70}$$

$$\mathrm{II}_{ij} \doteqdot \mathrm{II}\left(\frac{\partial X}{\partial \theta^i}, \frac{\partial X}{\partial \theta^j}\right). \tag{5.71}$$

Using the formula

$$\frac{\partial X}{\partial \theta^i} = r\frac{\partial}{\partial \theta^i} + \frac{\partial r}{\partial \theta^i} z,$$

which is obtained by applying $\frac{\partial}{\partial \theta^i}$ to (5.69), we compute that

$$\begin{aligned} g_{ij} &= \left\langle r\frac{\partial}{\partial \theta^i} + \frac{\partial r}{\partial \theta^i} z, r\frac{\partial}{\partial \theta^j} + \frac{\partial r}{\partial \theta^j} z\right\rangle \\ &= r^2 \overline{g}_{ij} + \overline{\nabla}_i r \overline{\nabla}_j r, \end{aligned} \tag{5.72}$$

where $\overline{\nabla}_i r \doteqdot \frac{\partial r}{\partial \theta^i}$, $\overline{g}$ is the standard metric on S^n, and

$$\overline{g}_{ij} \doteqdot \left\langle \frac{\partial}{\partial \theta^i}, \frac{\partial}{\partial \theta^j}\right\rangle = \overline{g}\left(\overline{\nabla}_i z, \overline{\nabla}_j z\right). \tag{5.73}$$

Similarly to the formula (5.63), we find that

$$g^{ij} = \frac{1}{r^2}\left(\overline{g}^{ij} - \frac{\overline{\nabla}^i r \overline{\nabla}^j r}{r^2 + \left|\overline{\nabla} r\right|^2}\right), \tag{5.74}$$

where $\overline{\nabla}^i r \doteqdot \overline{g}^{ik}\overline{\nabla}_k r$ and where $\left|\overline{\nabla} r\right|^2 \doteqdot \overline{g}^{ij}\overline{\nabla}_i r \overline{\nabla}_j r$.

We claim that the vector field

$$\mathsf{N}(z) = \frac{1}{\sqrt{r^2 + \left|\overline{\nabla} r\right|^2}}\left(rz - \overline{g}^{ij}\overline{\nabla}_i r \overline{\nabla}_j z\right) \tag{5.75}$$

is a unit normal field for $\mathcal{M}^n$. Indeed, $|\mathsf{N}(z)| = 1$ since

$$\left|rz - \overline{g}^{ij}\overline{\nabla}_i r\overline{\nabla}_j z\right|^2 = r^2 + \overline{g}^{ij}\overline{\nabla}_i r\overline{g}^{k\ell}\overline{\nabla}_k r\left\langle\overline{\nabla}_j z, \overline{\nabla}_\ell z\right\rangle = r^2 + \left|\overline{\nabla} r\right|^2,$$

where we used $\left\langle\overline{\nabla}_j z, z\right\rangle = \frac{1}{2}\overline{\nabla}_j |z|^2 \equiv 0$ and $\left\langle\overline{\nabla}_j z, \overline{\nabla}_\ell z\right\rangle = \overline{g}_{j\ell}$. The fact that $\mathsf{N}(z)$ is normal to M follows from the calculation

$$\begin{aligned}\left\langle rz - \overline{g}^{ij}\overline{\nabla}_i r\overline{\nabla}_j z, \overline{\nabla}_k X\right\rangle &= \left\langle rz - \overline{g}^{ij}\overline{\nabla}_i r\overline{\nabla}_j z, \left(\overline{\nabla}_k r\right) z + r\overline{\nabla}_k z\right\rangle \\ &= r\overline{\nabla}_k r - r\overline{g}^{ij}\overline{\nabla}_i r\left\langle\overline{\nabla}_j z, \overline{\nabla}_k z\right\rangle \\ &= 0\end{aligned}$$

for all $k = 1, \ldots, n$.

Observe also that $\mathsf{N}(z)$ is outward pointing for all $z \in S^n$: Choose $z_0 \in S^n$ such that $r(z_0) = \max_{z\in S^n} r(z)$. Then $\overline{\nabla} r(z_0) = 0$ and hence by (5.75),

$$\mathsf{N}(z_0) = z_0,$$

which is clearly outward pointing. Since $\mathsf{N}(z)$ is unit and smooth, it must be outward pointing for every $z \in S^n$.

The second fundamental form of a starshaped hypersurface has the following expression.

Lemma 5.15. *If $\mathcal{M} \subset \mathbb{R}^{n+1}$ is a starshaped hypersurface, then*

$$\mathrm{II}_{ij} = \frac{-r\overline{\nabla}_i\overline{\nabla}_j r + 2\overline{\nabla}_i r\overline{\nabla}_j r + r^2\overline{g}_{ij}}{\sqrt{r^2 + \left|\overline{\nabla} r\right|^2}}. \tag{5.76}$$

Proof. Recalling (5.19) and (5.23), we have

$$\mathrm{II}_{ij} = \left\langle\overline{\nabla}_j X, \overline{\nabla}_i \mathsf{N}\right\rangle = -\left\langle\overline{\nabla}_i\overline{\nabla}_j X, \mathsf{N}\right\rangle$$

and, by (5.75),

$$\mathsf{N} = \frac{1}{\sqrt{r^2 + \left|\overline{\nabla} r\right|^2}}\left(rz - \overline{g}^{k\ell}\overline{\nabla}_k r\overline{\nabla}_\ell z\right).$$

On the other hand, differentiating (5.69) yields

$$\overline{\nabla}_j X = z\overline{\nabla}_j r + r\overline{\nabla}_j z$$

and

$$\overline{\nabla}_i\overline{\nabla}_j X = z\overline{\nabla}_i\overline{\nabla}_j r + r\overline{\nabla}_i\overline{\nabla}_j z + \overline{\nabla}_i z\overline{\nabla}_j r + \overline{\nabla}_i r\overline{\nabla}_j z.$$

Thus

$$\begin{aligned}\left\langle\overline{\nabla}_i\overline{\nabla}_j X, z\right\rangle &= \overline{\nabla}_i\overline{\nabla}_j r + r\left\langle\overline{\nabla}_i\overline{\nabla}_j z, z\right\rangle \\ &= \overline{\nabla}_i\overline{\nabla}_j r - r\left\langle\overline{\nabla}_j z, \overline{\nabla}_i z\right\rangle \\ &= \overline{\nabla}_i\overline{\nabla}_j r - r\overline{g}_{ij}\end{aligned}$$

and

$$\left\langle \overline{\nabla}_i \overline{\nabla}_j X, \overline{\nabla}_\ell z \right\rangle = \overline{g}_{i\ell} \overline{\nabla}_j r + \overline{g}_{j\ell} \overline{\nabla}_i r$$

since $\overline{\nabla}_i \overline{\nabla}_j z = \overline{g}_{ij} z$ implies $\left\langle \overline{\nabla}_i \overline{\nabla}_j z, \overline{\nabla}_\ell z \right\rangle = 0$. We conclude that

$$\begin{aligned} \mathrm{II}_{ij} &= -\frac{1}{\sqrt{r^2 + \left|\overline{\nabla} r\right|^2}} \left(r \left\langle \overline{\nabla}_i \overline{\nabla}_j X, z \right\rangle - \overline{g}^{k\ell} \overline{\nabla}_k r \left\langle \overline{\nabla}_i \overline{\nabla}_j X, \overline{\nabla}_\ell z \right\rangle \right) \\ &= -\frac{1}{\sqrt{r^2 + \left|\overline{\nabla} r\right|^2}} \left(r \left(\overline{\nabla}_i \overline{\nabla}_j r - r \overline{g}_{ij} \right) - \overline{g}^{k\ell} \overline{\nabla}_k r \left(\overline{g}_{i\ell} \overline{\nabla}_j r + \overline{g}_{j\ell} \overline{\nabla}_i r \right) \right) \\ &= \frac{1}{\sqrt{r^2 + \left|\overline{\nabla} r\right|^2}} \left(-r \overline{\nabla}_i \overline{\nabla}_j r + r^2 \overline{g}_{ij} + 2 \overline{\nabla}_i r \overline{\nabla}_j r \right). \qquad \square \end{aligned}$$

Consequently,

Corollary 5.16. *The mean curvature of the starshaped hypersurface $\mathcal{M}$ is given by*

$$H = \frac{1}{r\sqrt{r^2 + \left|\overline{\nabla} r\right|^2}} \left(-\overline{\Delta} r + \frac{\overline{\nabla}^i r \overline{\nabla}^j r \overline{\nabla}_i \overline{\nabla}_j r}{r^2 + \left|\overline{\nabla} r\right|^2} + \frac{r \left|\overline{\nabla} r\right|^2}{r^2 + \left|\overline{\nabla} r\right|^2} + nr \right). \tag{5.77}$$

Proof. The formula is obtained by expanding the expression

$$H = \frac{1}{r\sqrt{r^2 + \left|\overline{\nabla} r\right|^2}} \left(\overline{g}^{ij} - \frac{\overline{\nabla}^i r \overline{\nabla}^j r}{r^2 + \left|\overline{\nabla} r\right|^2} \right) \left(-\overline{\nabla}_i \overline{\nabla}_j r + 2 \frac{\overline{\nabla}_i r \overline{\nabla}_j r}{r} + r \overline{g}_{ij} \right). \qquad \square$$

5.2.4. Convex hypersurfaces. We begin by recalling some basic definitions and results from convex geometry. Let $\Omega \subset \mathbb{R}^{n+1}$ be a **convex body**; that is, Ω is convex and has nonempty interior. A closed half-space H is said to **support** Ω (at p) if $\Omega \subset H$ and there is a point $p \in \partial\Omega \cap \partial H$. The boundary of a supporting half-space is called a **supporting hyperplane**. Through each boundary point of a convex body Ω there passes at least one supporting hyperplane, and the closure of Ω is the intersection of its supporting half-spaces [**447**, Theorem 11.5 and Corollary 11.6.1].

It is easy to see that the boundary of every convex body in $\mathbb{R}^{n+1}$ is the image of a continuous embedding of S^n. Conversely, a continuously embedded hypersurface $\mathcal{M}^n \subset \mathbb{R}^{n+1}$ is called a **convex hypersurface** if it bounds a convex body. A convex hypersurface $\mathcal{M}^n$ is **strictly convex** if each $p \in \mathcal{M}^n$ admits a unique supporting hyperplane. A continuously immersed hypersurface $X : M^n \to \mathbb{R}^{n+1}$ is called **locally (strictly) convex** if each point $x \in M^n$ admits a neighborhood $U \subset M^n$ such that $X(U)$ lies in a (strictly) convex hypersurface.

Lemma 5.17. *A smooth immersion $X : M^n \to \mathbb{R}^{n+1}$ is locally convex if and only if its second fundamental form is everywhere nonnegative definite. If its second fundamental form is positive definite at a point $x \in M^n$, then $X : M^n \to \mathbb{R}^{n+1}$ is strictly locally convex at x.*

Proof. By the implicit function theorem, any immersion can be represented locally as a graph over its tangent hyperplanes. The claims then follow from (5.60) and corresponding local convexity properties of smooth convex functions. □

The converse of the final statement is of course not true: The graph of the function $x \mapsto |x|^4$ is strictly convex, even though its second fundamental form is the zero form at $x = 0$. Moreover, a locally convex immersion need not be convex. We have already encountered smoothly immersed curves with positive curvature which do not bound convex bodies in the plane. Cylinders over such curves yield higher-dimensional counterexamples. The following theorem of van Heijenoort [**512**] (see also the prior work of Hadamard [**257**] and Stoker [**497**] and the stronger result of Sacksteder [**450**]) demonstrates a striking difference between curves and higher-dimensional hypersurfaces.

Theorem 5.18. *Suppose that $X : M^n \to \mathbb{R}^{n+1}$, $n \geq 2$, is a smooth, locally convex, proper immersion. If M^n is locally strictly convex at at least one point, then $\mathcal{M}^n \doteqdot X(M^n)$ is convex.*

Observe that any smooth immersion $X : M^n \to \mathbb{R}^{n+1}$ of a *compact* hypersurface M^n admits a locally strictly convex point. Indeed, consider the function $u : M^n \to \mathbb{R}$ defined by

$$u(x) \doteqdot \frac{1}{2}|X(x)|^2 .$$

At a point x where u attains its maximum,

$$0 = \nabla_i u = \langle \partial_i X, X \rangle \ , \quad \text{which implies} \quad X = |X| \mathsf{N}$$

for one of the two choices of unit normal N at x. Thus,

$$0 \geq \nabla_i \nabla_j u = g_{ij} - \mathrm{II}_{ij} \langle \mathsf{N}, X \rangle = g_{ij} - \mathrm{II}_{ij} |X|$$

at x. So $X : M^n \to \mathbb{R}^{n+1}$ is locally strictly convex at x. This proves the following corollary.

Corollary 5.19. *Suppose that $X : M^n \to \mathbb{R}^{n+1}$, $n \geq 2$, is a smooth immersion of a compact manifold M^n. If M^n has nonnegative definite second fundamental form, then $\mathcal{M}^n \doteqdot X(M^n)$ is convex.*

We refer to an immersion $X : M^n \to \mathbb{R}^{n+1}$ as **locally uniformly convex** if its second fundamental form is positive definite and **uniformly convex** if it is convex and its second fundamental form satisfies $\mathrm{II} \geq \varepsilon \, \mathrm{I}$ for some

$\varepsilon > 0$. On a smooth, locally uniformly convex hypersurface $\mathcal{M}^n \hookrightarrow \mathbb{R}^{n+1}$, the shape operator $\mathrm{A} = d\,\mathrm{G}$ is positive definite, and hence the Gauß map $\mathrm{G} : \mathcal{M}^n \to S^n$ is a local diffeomorphism. If $\mathcal{M}^n$ bounds a convex body, then G is injective and hence a diffeomorphism onto its image. Denote by $\overline{g}$ the standard metric on S^n and by $\overline{\nabla}$ the corresponding covariant derivative.

Lemma 5.20. *Let $\mathcal{M}^n \hookrightarrow \mathbb{R}^{n+1}$ be a convex, locally uniformly convex hypersurface. Then*

$$\mathrm{G}^*\,\overline{g} = \mathrm{II}^2\,. \tag{5.78}$$

Proof. By definition,

$$(\mathrm{G}^*\,\overline{g})_{ij} = (\mathrm{G}^*\,\overline{g})\left(\frac{\partial X}{\partial x^i}, \frac{\partial X}{\partial x^j}\right) = \overline{g}\left(\frac{\partial \mathsf{N}}{\partial x^i}, \frac{\partial \mathsf{N}}{\partial x^j}\right).$$

By (5.27), $\frac{\partial \mathsf{N}}{\partial x^i} = \mathrm{II}_{ij}\, g^{jk}\frac{\partial X}{\partial x^k}$, so that

$$(\mathrm{G}^*\,\overline{g})_{ij} = \left\langle \mathrm{II}_{ik}\, g^{k\ell}\frac{\partial X}{\partial x^\ell}, \mathrm{II}_{jp}\, g^{pq}\frac{\partial X}{\partial x^q}\right\rangle = \mathrm{II}_{ik}\, g^{kp}\,\mathrm{II}_{jp} \doteqdot \mathrm{II}^2_{ij}\,. \qquad \square$$

The **support function** $\sigma : S^n \to \mathbb{R}$ of a convex, locally uniformly convex hypersurface $\mathcal{M}^n \hookrightarrow \mathbb{R}^{n+1}$ is defined by (cf. (2.35))

$$\sigma(z) \doteqdot \left\langle \mathrm{G}^{-1}(z), z\right\rangle. \tag{5.79}$$

Here, $z \in S^n$ is the unit outward normal of $\mathcal{M}$ at the position vector $\mathrm{G}^{-1}(z)$.

Lemma 5.21. *Let $\mathcal{M}^n \hookrightarrow \mathbb{R}^{n+1}$ be a convex, locally uniformly convex hypersurface with support function σ. Then*

$$\mathrm{G}^{-1}(z) = \sigma(z) z + \overline{\nabla}\sigma(z). \tag{5.80}$$

In particular, the gradient of the support function is the tangential component of the position vector. Furthermore, the second fundamental form and the support function are related through the equation

$$\mathrm{G}^*\left(\overline{\nabla}^2\sigma + \sigma\overline{g}\right) = \mathrm{II}\,. \tag{5.81}$$

Equivalently,

$$\overline{\nabla}^2\sigma + \sigma\overline{g} = (\mathrm{G}^{-1})^*\,\mathrm{II}\,. \tag{5.82}$$

Proof. Denote by D the Euclidean covariant derivative and consider the embedding $X \doteqdot \mathrm{G}^{-1} : S^n \to \mathbb{R}^{n+1}$. Then

$$\sigma(z) = \langle X(z), \mathsf{N}(z)\rangle\,,$$

where $\mathsf{N}(z) = z$ is the outward unit normal field. Differentiating σ in the direction of a tangent vector field $W \in \Gamma(TS^n)$ then yields

$$\begin{aligned} W\sigma &= \langle D_W X, \mathsf{N}\rangle + \langle X, D_W \mathsf{N}\rangle \\ &= \langle X, W\rangle \end{aligned}$$

since $D_W \mathsf{N} = W$ is tangent to S^n. So the gradient $\overline{\nabla}\sigma$ of σ is given by

$$\overline{\nabla}\sigma = X^\top, \tag{5.83}$$

where $^\top : \mathbb{R}^{n+1} \to TS^n$ is the tangential projection map. This proves the identity (5.80). Differentiating again, we find that

$$\begin{aligned}\overline{\nabla}^2_{V,W}\sigma &\doteqdot V(W\sigma) - \left(\overline{\nabla}_V W\right)\sigma \\ &= \langle D_V X, W\rangle + \langle X, D_V W\rangle - \left\langle X, \overline{\nabla}_V W\right\rangle \\ &= \langle D_V X, W\rangle - \overline{g}(V,W)\langle X, \mathsf{N}\rangle\end{aligned}$$

since $D_V W - \overline{\nabla}_V W = -\overline{g}(V,W)\mathsf{N}$ (the difference of the Euclidean covariant derivative D and the induced covariant derivative $\overline{\nabla}$ on S^n is the second fundamental form of the unit sphere S^n, which is the same as the first fundamental form of S^n). Hence

$$\overline{\nabla}^2_{V,W}\sigma + \sigma\overline{g}(V,W) = \langle D_V X, W\rangle .$$

On the other hand,

$$\begin{aligned}\left(\left(\mathrm{G}^{-1}\right)^*\mathrm{II}\right)(V,W) &\doteqdot \mathrm{II}\left(\left(\mathrm{G}^{-1}\right)_* V, \left(\mathrm{G}^{-1}\right)_* W\right) \\ &= \left\langle W, \left(\mathrm{G}^{-1}\right)_* V\right\rangle \\ &= \langle W, D_V X\rangle\end{aligned}$$

since

$$\mathrm{II}\left(Y, \left(\mathrm{G}^{-1}\right)_* W\right) = \left\langle D_{\left(\mathrm{G}^{-1}\right)_* W}\mathsf{N}, Y\right\rangle = \langle W, Y\rangle$$

and $\left(\mathrm{G}^{-1}\right)_* V = D_V X$. The lemma follows. □

In what follows it will often be convenient to define the tensor $A = \overline{\nabla}^2\sigma + \sigma\overline{g} \in T^*S^n \otimes T^*S^n$. We will also use the notation $A[\sigma]$ when the dependence on the support function σ needs to be made explicit, and more generally we write $A[f]$ for the same expression with σ replaced by f, for any smooth function f:

$$A[f] \doteqdot \overline{\nabla}^2 f + f\overline{g}. \tag{5.84}$$

It will also be convenient to write

$$X[f] \doteqdot fz + \overline{\nabla} f \tag{5.85}$$

for any smooth function f. These two definitions are related by the identity

$$\partial_i X[f] = A[f]_{ip}\overline{g}^{pq}\partial_q z. \tag{5.86}$$

It follows from (5.78) and (5.81) that the eigenvalues $r_1, \ldots, r_n$ of $A[\sigma]$ with respect to $\overline{g}$ are the same as the eigenvalues of II with respect to II^2, which are $\kappa_1^{-1}, \ldots, \kappa_n^{-1}$. That is, $r_i = \kappa_i^{-1}$ are the **principal radii**. In particular, the inverses of the Gauß curvature and the harmonic mean curvature are given by

$$K^{-1} = \det_{\overline{g}}\left(\overline{\nabla}^2\sigma + \sigma\overline{g}\right) \tag{5.87}$$

and

$$\begin{aligned}(H_{-1})^{-1} &= \kappa_1^{-1} + \cdots + \kappa_n^{-1} \\ &= \operatorname{tr}_{\overline{g}}\left(\overline{\nabla}^2\sigma + \sigma\overline{g}\right) \\ &= \overline{\Delta}\sigma + n\sigma\,,\end{aligned} \tag{5.88}$$

respectively.

Recall that the Gauß curvature of a compact Riemannian surface (M^2, g) is related to the topology of M^2 by the **Gauß-Bonnet theorem**, which states that

$$\int_{M^2} K\, d\mu = \chi(M^2)\,,$$

where $\chi(M^2)$ is the **Euler characteristic** of M^2 (equal to $2 - 2g(M^2)$, where $g(M^2)$ is the genus of M^2). In higher dimensions, we obtain a similar formula for locally uniformly convex hypersurfaces of $\mathbb{R}^{n+1}$.

Proposition 5.22. *Let $X : M^n \to \mathbb{R}^{n+1}$ be a convex, locally uniformly convex embedding. Then*

$$\int_{M^n} K\, d\mu = \operatorname{Area}(S^n)\,, \tag{5.89}$$

where K is the Gauß curvature and S^n is the round sphere of unit radius.

Proof. Since the Gauß map $\mathrm{G} : M^n \to S^n$ is a diffeomorphism, the change of variables formula yields

$$\int_{M^n} K\, d\mu = \int_{M^n} \det(d\mathrm{G})\, d\mu = \int_{S^n} d\mu_{S^n} = \operatorname{Area}(S^n)\,. \qquad \square$$

5.3. Dynamical properties

It will be useful to recount some general dynamical properties of Euclidean hypersurfaces.[1] Let $X : M^n \times I \to \mathbb{R}^{n+1}$ be a smooth 1-parameter family of smooth immersions $X(\cdot, t) : M^n \to \mathbb{R}^{n+1}$, where I is an open interval.

---

[1]We will develop an abstract formulation in Section 5.3.1. The reader preferring a more concrete (coordinate-based) approach may refer to Section 5.4.2 below.

5.3.1. Geometry on the pullback bundle. Denote by ∂_t the **canonical vector field** on $M^n \times I$, which is defined via its action on smooth functions $f : M^n \times I \to \mathbb{R}$ by

$$\partial_t f(x,t) \doteqdot \frac{d}{ds}\Big|_{s=0} f(x,t+s)\,.$$

Observe that the tangent bundle $T(M^n \times I)$ splits into a direct sum of the **spatial tangent bundle** $\mathfrak{S} \doteqdot \{\xi \in T(M^n \times I) : dt(\xi) = 0\}$ with the line bundle $\mathbb{R}\,\partial_t$, where dt is the **canonical** 1**-form**, which is the 1-form dual to ∂_t (equivalently, the differential of the **time function**, $(x,t) \mapsto t$). We will refer to sections of the spatial tangent bundle as **time-dependent (tangential) vector fields**.

We subsequently abuse notation by conflating the tangent bundle TM^n (a bundle over M^n) and the spatial tangent bundle $\mathfrak{S}$ (a bundle over $M^n \times I$) but for clarity, we continue to use the notation $\mathfrak{S}$ in this section. The spatial tangent bundle admits an induced geometry akin to the standard Riemannian submanifold geometry of a "stationary" immersion but possesses additional information about the evolution of the immersions $X_t \doteqdot X(\cdot, t)$.

We begin by equipping the pullback bundle $X^*T\mathbb{R}^{n+1}$ over $M^n \times I$ with the pullback metric ${}^X\langle \cdot, \cdot \rangle$, which is defined on each fiber by

$$ {}^X\big\langle (x,t,u),(x,t,v)\big\rangle = \big\langle (X(x,t),u),(X(x,t),v)\big\rangle\,, \tag{5.90}$$

where $\langle \cdot, \cdot \rangle$ is the metric on $T\mathbb{R}^{n+1}$. Here (x,t,u) and (x,t,v) are elements of $(X^*T\mathbb{R}^{n+1})_{(x,t)}$, so that, by definition of the pullback bundle, $(X(x,t),u)$ and $(X(x,t),v)$ are elements of $T_{X(x,t)}\mathbb{R}^{n+1}$.

The pullback bundle also inherits the pullback connection, ${}^XD$, defined by

$$ {}^XD_\xi(X^*V) = D_{dX(\xi)}V \tag{5.91}$$

for every tangent vector $\xi \in T(M^n \times I)$ and every pulled-back section $X^*V \in \Gamma(X^*T\mathbb{R}^{n+1})$. This extends to all sections of $\Gamma(X^*T\mathbb{R}^{n+1})$ via the Leibniz rule, as we can always form a local basis of pulled-back sections.

The push-forward $dX(\mathfrak{S})$ of the spatial tangent bundle is a subbundle of $X^*T\mathbb{R}^{n+1}$ of rank n. Its orthogonal complement in $X^*T\mathbb{R}^{n+1}$ with respect to ${}^X\langle \cdot, \cdot \rangle$ is a subbundle of rank 1, which we denote by $\mathfrak{N}$ and refer to as the **normal bundle** of X. Subsequently, we will abuse notation by denoting $N\mathcal{M} = \mathfrak{N}$, but for this section we continue to use $\mathfrak{N}$. In any case, we have the ${}^X\langle \cdot, \cdot \rangle$-orthogonal decomposition $X^*T\mathbb{R}^{n+1} = dX(\mathfrak{S}) \oplus^{\perp} \mathfrak{N}$. We refer to the ${}^X\langle \cdot, \cdot \rangle$-orthogonal projections

$$\cdot^{\top} : X^*T\mathbb{R}^{n+1} \to dX(\mathfrak{S}) \quad \text{and} \quad \cdot^{\perp} : X^*T\mathbb{R}^{n+1} \to \mathfrak{N}$$

as, respectively, the **tangential** and **normal projection**.

The spatial tangent bundle $\mathfrak{S}$ of $M^n \times I$ inherits an **induced metric** $g \in \Gamma(\mathfrak{S}^* \otimes \mathfrak{S}^*)$ and an **induced connection** ∇ from the pullback metric and connection. These are defined by

$$g\,(u,v) \doteqdot {}^X\langle X_*u\,,\, X_*v\rangle\,, \quad u,v \in \mathfrak{S}_{(x,t)},\ (x,t) \in M^n \times I\,,$$

and

$$dX(\nabla_\xi V) \doteqdot \left({}^X\!D_\xi dX(V)\right)^\top, \quad V \in \Gamma(\mathfrak{S})\,,\ \xi \in T(M^n \times I)\,,$$

respectively.

For each $t \in I$, the restrictions of $\mathfrak{S}$ and $\mathfrak{N}$ to $M^n \times \{t\}$ are canonically isomorphic to the tangent and normal bundles, respectively, of the immersion X_t. Under this identification, g is the induced metric of the immersion X_t and, in spatial directions $\xi \in \mathfrak{S}$, ∇ is the induced connection on X_t (which we recall is the Levi-Civita connection of the induced metric). In fact, by the same straightforward computation as in the stationary setting, we find that ∇ satisfies the Levi-Civita conditions:

$$\nabla_u g = 0\,, \quad u \in \mathfrak{S}\,, \tag{5.92}$$

and

$$\nabla_U V - \nabla_V U - [U,V] = 0\,, \quad U,V \in \Gamma(\mathfrak{S})\,. \tag{5.93}$$

Note that the connection on $\mathfrak{S}^*\otimes\mathfrak{S}^*$ is defined, as usual, by commuting with contractions and obeying the Leibniz rule with respect to tensor products, so that

$$\xi g(U,V) = \nabla_\xi g(U,V) + g(\nabla_\xi U, V) + g(U, \nabla_\xi V)$$

for all $\xi \in T(M^n \times I)$ and all $U, V \in \Gamma(\mathfrak{S})$.

The normal part of the pullback connection, restricted to $\mathfrak{S}$, is a (normal bundle-valued) symmetric 2-tensor, which we call the **vector second fundamental form** of X and denote by $\vec{\mathrm{II}}$. More precisely, $\vec{\mathrm{II}} \in \Gamma(\mathfrak{S}^*\otimes\mathfrak{S}^*\otimes\mathfrak{N})$ is defined by

$$\vec{\mathrm{II}}(u,v) \doteqdot \left({}^X\!D_u dX(V)\right)^\perp, \tag{5.94}$$

for any $u, v \in \mathfrak{S}_{(x,t)}$, where V is any extension of v. That this is actually a well-defined, symmetric tensor follows similarly as in the stationary setting.

Since $\vec{\mathrm{II}}$ is a symmetric 2-tensor (taking values in the normal bundle), for every normal vector there is an associated g-selfadjoint endomorphism. This is encoded in the **vector Weingarten tensor** $\vec{\mathrm{L}} : \Gamma(\mathfrak{N}) \to \Gamma(\mathfrak{S}^*\otimes\mathfrak{S})$. Just as in the stationary setting, $\vec{\mathrm{L}}$ is given explicitly by (see Proposition 5.23)

$$dX(\vec{\mathrm{L}}_\nu(u)) = {}^X\!D_u N \tag{5.95}$$

for any $\nu \in \mathfrak{N}_{(x,t)}$ and $u \in \mathfrak{S}_{(x,t)}$, where N is any extension of ν satisfying $|N| \equiv 1$ in a neighborhood of (x,t).

If M^n is orientable, then we may choose a smooth global unit normal field $\mathsf{N} \in \Gamma(\mathfrak{N})$, with respect to which we may define the **second fundamental form**, II, by

$$\vec{\mathrm{II}}(u,v) = -\,\mathrm{II}(u,v)\nu$$

and the **Weingarten tensor**, L, by

$$dX(\mathrm{L}(u)) = {}^X\!D_u N\,.$$

The eigenvalues of L are the **principal curvatures** of X (with respect to ν), which we denote by $\kappa_1, \ldots, \kappa_n$, and the corresponding unit eigenvectors are the corresponding **principal directions**.

The above constructions yield a time-dependent intrinsic and extrinsic geometry for X which at each time $t \in I$ reduces to the intrinsic and extrinsic geometry of the immersion X_t. So we do not appear to have gained anything as yet; however, we have only considered the information coming from the spatial tangent bundle. There is additional information to be obtained from the "temporal" direction. This information gives an invariant characterization of the evolution of geometric structures induced by X_t.

First, we exhaust the temporal information contained in the pullback metric by assuming that the velocity of X is in its normal direction; that is, we assume that $\partial_t X = -F\mathsf{N}$ for some scalar F, where we denote

$$\partial_t X(x,t) \doteqdot dX(\partial_t) = \frac{d}{dt}X(x,t)\,,$$

recalling that ∂_t (in the middle expression) is the canonical vector field on $M^n \times I$. Note that this can be achieved for any time-dependent immersion by introducing a "time-dependent diffeomorphism" of M^n: Let $X : M^n \times I \to \mathbb{R}^{n+1}$ be a time-dependent immersion, and let $\phi : M^n \times I \to M^n$ be the flow of the negative of the tangential part of the velocity of X; that is,

$$\frac{d\phi}{dt} = -T \circ \phi\,,$$

where $T = dX^{-1}\left((\partial_t X)^\top\right)$. Then the velocity of the time-dependent immersion Y defined by $Y(y,t) \doteqdot X\left(\phi(x,t),t\right)$ is

$$Y_*\partial_t = dX_\phi\left(\frac{d\phi}{dt}\right) + \partial_t X_\phi = \left(\partial_t X_\phi\right)^\perp\,.$$

Thus, modulo a time-dependent diffeomorphism of M^n, we lose no information by assuming that the velocity is in the normal direction.

Next consider the temporal Lie derivative of the metric. We claim that

$$\mathcal{L}_t g = -2F\,\mathrm{II}\,,$$

where $\mathcal{L}_t \doteqdot \mathcal{L}_{\partial_t}$. Indeed, computing in a coordinate basis, we find

$$\begin{aligned}\mathcal{L}_t g_{ij} &= \partial_t g_{ij} \\ &= \left\langle {}^X\!D_i \partial_t X, \partial_j X\right\rangle + \left\langle \partial_i X, {}^X\!D_j \partial_t X\right\rangle \\ &= -\left\langle {}^X\!D_i (F\nu), \partial_j X\right\rangle - \left\langle \partial_i X, {}^X\!D_j (F\nu)\right\rangle \\ &= -2F\,\mathrm{II}_{ij}\,,\end{aligned} \tag{5.96}$$

where ${}^X\!D_i \doteqdot {}^X\!D_{\partial_i}$.

Finally, we consider the temporal information in the pullback connection. Given $V \in \mathfrak{S}$, we have

$$\begin{aligned}{}^X\!D_t dX(V) &= {}^X\!D_V \partial_t X + dX([\partial_t, V]) \\ &= -{}^X\!D_V(F\nu) + dX([\partial_t, V]) \\ &= -dF(V)\nu - F dX(\mathrm{L}(V)) + dX([\partial_t, V])\,,\end{aligned} \tag{5.97}$$

where ${}^X\!D_t \doteqdot {}^X\!D_{\partial_t}$. The tangential part of (5.97) yields the **temporal component of the connection** ∇ (cf. [**52**, Theorem 5.1]):

$$\nabla_t V \doteqdot \mathcal{L}_t V - F\,\mathrm{L}(V)\,. \tag{5.98}$$

It follows that

$$\begin{aligned}\nabla_t g(u,v) &= \mathcal{L}_t g(u,v) + g((\mathcal{L}_t - \nabla_t)u, v) + g(u, (\mathcal{L}_t - \nabla_t)v) \\ &= 0\,.\end{aligned} \tag{5.99}$$

The normal component of (5.97) yields the **temporal second fundamental form**:

$$\vec{\mathrm{II}}_t(u) \doteqdot \left({}^X\!D_t dX(u)\right)^\perp = -dF(u)\,\nu\,. \tag{5.100}$$

In this way, we may consider $\vec{\mathrm{II}}$ as a section of $T^*(M^n \times I) \otimes \mathfrak{S}^* \otimes \mathfrak{N}$. Similarly, we may consider $\vec{\mathrm{L}}$ as a section of $\mathfrak{N}^* \otimes T^*(M^n \times I) \otimes \mathfrak{S}$ by setting

$$\vec{\mathrm{L}}_\nu(\partial_t|_{(x,t)}) \doteqdot {}^X\!D_{\partial_t|_{(x,t)}} N\,, \tag{5.101}$$

for any $\nu \in \mathfrak{N}_{(x,t)}$, where N is any extension of ν satisfying $|N| \equiv 1$ in a neighborhood of (x,t).

To avoid confusion, when we consider the temporal component of the time-dependent connection ∇, we will always be explicit; that is, we reserve the notation ∇V for the *spatial* covariant differential.

We now state the structure equations for the time-dependent immersion X. In the spatial directions, these reduce to the standard structure equations for a stationary immersion but yield additional "evolution" identities for the temporal direction.

Proposition 5.23 (Structure equations for time-dependent hypersurfaces). *Let $X : M^n \times I \to \mathbb{R}^{n+1}$ be an oriented*[2] *time-dependent immersion equipped with unit normal field $\mathsf{N} \in \Gamma(S\mathfrak{N})$ and velocity*

$$\partial_t X = -F\mathsf{N}\,. \tag{5.102}$$

(1) *The **Weingarten equation**: For any $\xi \in T(M^n \times I)$ and any $u \in \mathfrak{S}$,*

$$g\left(\mathrm{L}(\xi), u\right) = -\,\mathrm{II}(\xi, u)\,. \tag{5.103}$$

This reduces to the standard Weingarten equation when $\xi \in \mathfrak{S}$. Setting $\xi = \partial_t$ yields the additional temporal Weingarten equation

$$^{X}D_t\mathsf{N} = dX(\operatorname{grad} F)\,. \tag{5.104}$$

(2) *The **Gauß equation**: For any $\xi \in T(M^n \times I)$ and any $u, v \in \mathfrak{S}$,*

$$\mathrm{Rm}(\xi, u)v = \mathrm{II}(\xi, v)\,\mathrm{L}(u) - \mathrm{II}(u, v)\,\mathrm{L}(\xi)\,, \tag{5.105}$$

where $\mathrm{Rm}(\xi, u)v \doteqdot \nabla_u(\nabla_\xi V) - \nabla_\xi(\nabla_u V) - \nabla_{[u,\xi]}V$, $V \in \Gamma(\mathfrak{S})$ any extension of v, is the curvature of $(\mathfrak{S}, \nabla)$. This reduces to the standard Gauß equation (5.41) when $\xi \in \mathfrak{S}$. Setting $\xi = \partial_t$ yields the additional temporal Gauß equation

$$\mathrm{Rm}(\partial_t, u)v = \nabla_v F\,\mathrm{L}(u) - \mathrm{II}(u, v)\operatorname{grad} F\,. \tag{5.106}$$

(3) *The **Codazzi equation**: For any $\xi \in T(M^n \times I)$ and any $u \in \mathfrak{S}$,*

$$\nabla_\xi\left(\mathrm{L}(u)\right) - \nabla_u\left(\mathrm{L}(\xi)\right) = \mathrm{L}\left([\xi, u]\right)\,. \tag{5.107}$$

This reduces to the standard Codazzi equation (5.42) when $\xi \in \mathfrak{S}$. Setting $\xi = \partial_t$ yields the additional temporal Codazzi equation

$$\nabla_t\,\mathrm{L} = \nabla\operatorname{grad} F + F\,\mathrm{L}^2\,. \tag{5.108}$$

Proof. The identities follow from similar calculations as in the stationary setting. The spatial and temporal equations follow immediately. □

Note that the trace of (5.106) yields

$$\mathrm{Rc}(\partial_t, v) = H\nabla_v F - \mathrm{II}(\operatorname{grad} F, v) \tag{5.109}$$

and the trace of (5.108) yields a variational formula for the mean curvature:

$$\partial_t H = \Delta F + |\mathrm{II}|^2 F\,. \tag{5.110}$$

Note that the commutation formula for space and time derivatives is not the usual commutation formula for covariant derivatives since the space-time

[2]This assumption is not strictly necessary; we have merely assumed it for clarity of exposition.

connection is not torsion free. Indeed, given a tensor field S over the spatial tangent bundle,

$$\begin{aligned}\nabla^2_{\partial_t,u}S &= \nabla_t(\nabla_u S) - \nabla_{\nabla_t u}S \\ &= \nabla_u(\nabla_t S) - \mathrm{Rm}(\partial_t,u)S + F\nabla_{\mathrm{L}(u)}S \\ &= \nabla^2_{u,\partial_t}S - \mathrm{Rm}(\partial_t,u)S + F\nabla_{\mathrm{L}(u)}S\,.\end{aligned} \tag{5.111}$$

So we gain the extra term $F\nabla_{\mathrm{L}(u)}S$.

Henceforth, the (vector) second fundamental form ($\vec{\mathrm{II}}$) II and (vector) Weingarten tensor ($\vec{\mathrm{L}}$) L will refer only to the "spatial" tensors introduced in (5.94) and (5.103), and we shall use the definitions (5.100) and (5.101) explicitly when we require the temporal components.

5.3.2. Evolving orthonormal frames. In the previous section, we introduced the time-dependent connection as a way of invariantly defining time derivatives of tensors. Here we describe a second method, which yields equivalent equations.

We begin with an elementary observation. Consider a time-dependent immersion $X : M^n \times [0,T) \to \mathbb{R}^{n+1}$ with normal velocity $-F\mathsf{N}$. Given any point $(x_0,t_0) \in M^n \times I$, choose an orthonormal basis $\{e_a^0\}_{a=1}^n$ at (x_0,t_0). We may extend $\{e_a^0\}_{a=1}^n$ to a neighborhood of (x_0,t_0) by first parallel translating in space along radial geodesics emanating from x_0, with respect to $g(t_0)$, and then extending in the time direction by solving the ODE

$$\frac{d}{dt}e_a = F\,\mathrm{II}(e_a) \tag{5.112}$$

for each $a = 1,\dots,n$. The resulting local frame field $\{e_a(t)\}_{a=1}^n$ remains orthonormal with respect to $g(t)$:

$$\begin{aligned}\frac{d}{dt}g_{ab} &= \left(\frac{d}{dt}g\right)_{ab} + g\left(\frac{d}{dt}e_a, e_b\right) + g\left(e_a, \frac{d}{dt}e_b\right) \\ &= -\,2F\,\mathrm{II}_{ab} + F\,\mathrm{II}_{ab} + F\,\mathrm{II}_{ab} \\ &= 0\,.\end{aligned} \tag{5.113}$$

We may formulate this more abstractly. Let E be any vector bundle over M^n that is isomorphic to TM^n. If we evolve any initial bundle isometry $\iota_0 : (E, \iota_0^*(g_0)) \to (TM^n, g(0))$ by the equation

$$\begin{aligned}\partial_t\,\iota(t) &= -F\,h \circ \iota(t)\,, \\ \iota(0) &= \iota_0\,,\end{aligned} \tag{5.114}$$

then the maps $\iota(t) : (E, \iota_0^*(g_0)) \to (TM^n, g(t))$ remain isometries.

There is yet another point of view, in which we view tensors as functions on the frame bundle $F(M)$ of M. For example, a 2-tensor such as the second fundamental form is identified with the function

$$\mathrm{II} : F(M) \to \mathbb{R}^{n^2}$$

that is defined by

$$\mathrm{II}\,(Y) \doteqdot \mathrm{II}\,(Y_a, Y_b)\,, \tag{5.115}$$

where $Y = \{Y_a\}_{a=1}^n \in F(M^n)$ is a frame. Observe that a covariant derivative ∇ on M^n defines a horizontal subspace of $TF\,(M^n)$, where the horizontal vectors at a point $Y \in F(M^n)$ are the vectors that are tangent to parallel sections of $F\,(M^n)$ along a path in M^n.

To understand this better, let $U \subset M^n$ be a coordinate neighborhood with local coordinates $\{x^i\}$ and define $\{y_a^j\,(Y)\}$ by $Y_a \doteqdot y_a^j \frac{\partial}{\partial x^j}$ for frames $Y \in F(M^n)|_U$. This determines a coordinate system $\{\pi \circ x^i, y_a^j\}$ on the open set $F(M^n)|_U$. The corresponding coordinate basis for $TF(M^n)$ is then $\left\{\frac{\partial}{\partial x^i}, \frac{\partial}{\partial y_a^j}\right\}$, where we write $\frac{\partial}{\partial x^i} \doteqdot \frac{\partial}{\partial (x^i \circ \pi)}$. Now let $\sigma(t)$ be a path in M^n with $\sigma(0) = x$ and $\sigma'(0) = Y_a(x)$, and consider the path

$$\gamma(t) = (\sigma(t), P_{\sigma(t)} Y_1, \ldots, P_{\sigma(t)} Y_n)\,,$$

where $P_{\sigma(t)}$ is parallel transport with respect to ∇ along the path $\sigma(t)$. Then the vector $\nabla_a \in T_Y F(M^n)$ is the (unique) horizontal lift of Y_a, defined by

$$\nabla_a = \left.\frac{d}{dt}\right|_{t=0} \gamma(t). \tag{5.116}$$

We have defined n global horizontal vector fields ∇_a on $F(M^n)$ and denote

$$\nabla = (\nabla_1, \ldots, \nabla_n)\,.$$

Covariant differentiation of a tensor can now be achieved by taking its directional derivative (as a function on the frame bundle) with respect to the vector fields ∇_a. For example, the tensor $\nabla\,\mathrm{II}$ is defined by

$$(\nabla\,\mathrm{II})(Y) \doteqdot \nabla(\mathrm{II}(Y))\,;$$

that is,

$$(\nabla_a\,\mathrm{II})_{bc} \doteqdot (\nabla\,\mathrm{II})(Y_a, Y_b, Y_c) \doteqdot \nabla_a(\mathrm{II}(Y_b, Y_c))\,. \tag{5.117}$$

We will now obtain a local expression for ∇_a. In coordinates, the components of a 1-form are given by $V_c = V\,(Y_c) = y_c^i V\left(\frac{\partial}{\partial x^i}\right) = y_c^i V_i$. Note that V_c is a function on $F(M^n)$, y_c^i is a function on $F(M^n)|_U$, but V_j is a function on U, and so to be precise we should write $V_c = y_c^i(V_i \circ \pi)$, where $V_i \circ \pi$ is constant on the fibers. We have

$$(\nabla_a V)_c = y_a^i y_c^j (\nabla_i V)_j = y_a^i y_c^j \left(\frac{\partial}{\partial x^i} V_j - \Gamma_{ij}^k V_k\right).$$

Since

$$\frac{\partial}{\partial x^i} V_c = y_c^j \frac{\partial}{\partial x^i} V_j$$

and

$$\frac{\partial}{\partial y_b^k} V_c = \frac{\partial}{\partial y_b^k} y_c^j V_j = \delta_c^b V_k ,$$

this implies that

$$(\nabla_a V)_c = y_a^i \left(\frac{\partial}{\partial x^i} - \Gamma_{ij}^k y_b^i \frac{\partial}{\partial y_b^k} \right) V_c . \tag{5.118}$$

Similar calculations yield local expression for the covariant derivative of any tensor field.

We next define vector fields tangent to the fibers. Given a frame Y at x, consider the path $\zeta : \mathbb{R} \to F(M^n)$ defined by

$$\zeta(t) = (x, Y_1, \ldots, Y_{a-1}, Y_a + tY_b, Y_{a+1}, \ldots, Y_n) .$$

In coordinates,

$$\zeta'(0) = (0, \ldots, 0, \underbrace{y_b^1, \ldots, y_b^n}_{a\text{-th position}}, 0, \ldots, 0) = y_b^i \frac{\partial}{\partial y_a^i} .$$

We define

$$\Lambda_b^a = y_b^i \frac{\partial}{\partial y_a^i} , \tag{5.119}$$

which represents the skewing of the a-axis into the b-axis for $a \neq b$, or the stretching of the axis when $a = b$.

It is easy to calculate the action of Λ_b^a on tensors. For example, for any function f we have that $\Lambda_c^b f = 0$. For a 1-form V,

$$\Lambda_b^a V_c = y_b^i \frac{\partial}{\partial y_a^i} (y_c^j V_j) = \delta_c^a V_b .$$

The vector fields ∇_a and Λ_a^b form an orthonormal basis on $F(M^n)$ with respect to the induced Riemannian metric.

Now assume that $X : M^n \times I \to \mathbb{R}^{n+1}$ evolves with velocity $\partial_t X = -F\mathsf{N}$. To understand the evolution of tensors as functions on the orthonormal frame bundle we first consider the (**general linear**) **frame bundle** before restricting to the orthonormal frame bundle. Note that now the frame bundle $F(M^n)$, a principal $\mathrm{GL}(n, \mathbb{R})$-bundle, and the orthonormal frame bundle $OF(M)$, a principal $\mathrm{O}(n)$-bundle, are "time-independent" — they are bundles over the space-time manifold $M^n \times I$. Given a frame $Y \in F(M)$ consider the map

$$\rho_Y : \mathrm{GL}(n, \mathbb{R}) \to F(M)$$

defined by the left action:

$$\rho_Y(G) \doteqdot G(Y) \quad \text{for } G \in \mathrm{GL}(n,\mathbb{R}). \tag{5.120}$$

We have a basis $\{e(a,b)\}_{a,b=1}^{n}$ of $\mathfrak{gl}(n,\mathbb{R})$ defined by

$$e(a,b)_c^d \doteqdot \delta_c^a \delta_b^d. \tag{5.121}$$

With this we can define a natural basis of vertical vector fields $\{\Lambda_b^a\}$ on $F(M)$ by

$$\Lambda_b^a(Y) \doteqdot (\rho_Y)_* e(a,b) \qquad \text{for all } Y \in F(M). \tag{5.122}$$

In local coordinates we have, by (5.119), $\Lambda_b^a = y_b^i \frac{\partial}{\partial y_a^i}$.

The vector fields Λ_b^a operate on 1-forms (as directional derivatives of functions on $F(M)$) by

$$\Lambda_b^a V_c = \delta_c^a V_b. \tag{5.123}$$

That is,

$$\Lambda_b^a V = \Bigg(0,\ldots,0,\underbrace{V_b}_{a\text{-th position}},0,\ldots,0\Bigg),$$

and the Λ_d^c operate on 2-tensors by

$$\Lambda_d^c V_{ab} = \delta_a^c V_{db} + \delta_b^c V_{ad}. \tag{5.124}$$

We also have horizontal vector fields defined by (5.116) and given in local coordinates by (5.118):

$$\nabla_a = y_a^i \left(\frac{\partial}{\partial x^i} - \Gamma_{ij}^k y_b^i \frac{\partial}{\partial y_b^k}\right). \tag{5.125}$$

Since $OF(M) \subset F(M)$ depends on time, we replace the usual time derivative ∂_t, which is a vector field on $F(M)\times[0,T)$, by a new time derivative ∇_t which stays tangent to the orthonormal frame bundle; i.e., ∇_t is tangent to $\bigcup_{t\in[0,T)} OF_{g(t)}(M) \times \{t\} \subset F(M) \times [0,T)$. More precisely, we choose ∇_t to be a vector field on $F(M)$ such that $\nabla_t - \partial_t$ is space-like (i.e., tangent to the slices $F(M)\times\{t\}$), vertical, and satisfies $\nabla_t g_{ab} = 0$. That is, ∇_t is a vector field of the form

$$\nabla_t = \partial_t + \alpha_c^d \Lambda_d^c, \tag{5.126}$$

where the α_c^d are functions on the frame bundle to be determined by the condition

$$\nabla_t g_{ab} = 0. \tag{5.127}$$

Using this and (5.124), we calculate

$$\begin{aligned} 0 &= \nabla_t g_{ab} \\ &= \partial_t g_{ab} + \alpha_c^d \Lambda_d^c g_{ab} \\ &= -2F\,\mathrm{II}_{ab} + \alpha_a^d g_{bd} + \alpha_b^d g_{ad}\,, \end{aligned}$$

which implies that

$$0 = -2F\,\mathrm{II}_{ab} + \alpha_a^b + \alpha_b^a\,.$$

So, defining $\alpha_a^b \doteqdot F\,\mathrm{II}_{ac}\,g^{cb}$, we arrive at

$$\nabla_t \doteqdot \partial_t + F\,\mathrm{II}_{ab}\,g^{bc}\Lambda_c^a\,. \tag{5.128}$$

Since $\nabla_t g_{ab} = 0$, we think of ∇_t as the covariant derivative in the time direction and we call ∇_t the **covariant time derivative**. We leave it as an exercise to the reader to check that the two notions of covariant time derivative coincide.

The following formula for commuting covariant space and time derivatives coincides with the formula (5.111) from the previous section.

Proposition 5.24 (Commutator of space and time derivatives). *The covariant space and time derivatives satisfy the following commutator formula*:

$$[\nabla_t, \nabla_a] = (\nabla_b F\,\mathrm{II}_{ac} - \nabla_c F\,\mathrm{II}_{ab})\Lambda_c^b + F\,\mathrm{II}_{ad}\,\nabla_d\,. \tag{5.129}$$

Proof. Observe that

$$[\Lambda_b^a, \nabla_c] = \Lambda_b^a \nabla_c - \nabla_c \Lambda_b^a = \delta_c^a \nabla_b \tag{5.130}$$

and, by the Levi-Civita formula (see Exercise 5.11),

$$\frac{\partial}{\partial t}\Gamma_{ij}{}^k = -g^{k\ell}\left[\nabla_i (F\,\mathrm{II})_{j\ell} + \nabla_j (F\,\mathrm{II})_{i\ell} - \nabla_\ell (F\,\mathrm{II})_{ij}\right]. \tag{5.131}$$

Using (5.118) and (5.131), we obtain

$$\begin{aligned} \left[\frac{\partial}{\partial t}, \nabla_a\right] &= -y_a^i \left(\frac{\partial}{\partial t}\Gamma_{ij}{}^k\right) y_b^j \frac{\partial}{\partial y_b^k} \\ &= g^{cd}(\nabla_a(F\,\mathrm{II}_{bc}) + \nabla_b(F\,\mathrm{II}_{ac}) - \nabla_c(F\,\mathrm{II}_{ab}))\Lambda_d^b. \end{aligned}$$

Thus, recalling (5.130), we have

$$\begin{aligned} [\nabla_t, \nabla_a] =& \left(\frac{\partial}{\partial t} + F\,\mathrm{II}_{cd}\,g^{de}\Lambda_e^c\right)\nabla_a - \nabla_a\left(\frac{\partial}{\partial t} + F\,\mathrm{II}_{cd}\,g^{de}\Lambda_e^c\right) \\ =& g^{cd}(\nabla_a(F\,\mathrm{II}_{bc}) + \nabla_b(F\,\mathrm{II}_{ac}) - \nabla_c(F\,\mathrm{II}_{ab}))\Lambda_d^b + F\,\mathrm{II}_{cd}\,g^{de}\delta_a^c\nabla_e \\ & - \nabla_a(F\,\mathrm{II}_{cd}\,g^{de})\Lambda_e^c \\ =& (\nabla_b(F\,\mathrm{II}_{ac}) - \nabla_c(F\,\mathrm{II}_{ab}))\Lambda_c^b + F\,\mathrm{II}_{ad}\,\nabla_d, \end{aligned}$$

which by the Codazzi equation yields (5.129). □

5.3.3. First variation of area and volume. Finally, we make note of the first variation formulae for the area and enclosed volume.

Lemma 5.25. *Let $X : M^n \times I \to \mathbb{R}^{n+1}$ be a smooth family of smooth immersions with velocity $\partial_t X = \vec{F}$. Then*

$$\frac{d}{dt}\operatorname{Area}(K) = -\int_K \langle \vec{F}, \vec{H}\rangle \, d\mu \tag{5.132}$$

for any compact subset $K \subset M^n$. Similarly,

$$\frac{d}{dt}\int_{M^n} f d\mu = \int_{M^n} \left(\partial_t f - \langle \vec{F}, \vec{H}\rangle f\right) d\mu \tag{5.133}$$

for any smooth function $f : M^n \times I \to \mathbb{R}$ with compact support.

If X is a family of embeddings and M^n is compact, then

$$\frac{d}{dt}\operatorname{Vol}(\mathcal{M}_t) = -\int_{M^n} F \, d\mu_t \,, \tag{5.134}$$

where, with respect to the outward unit normal field N, $F \doteqdot -\langle \vec{F}, \mathsf{N}\rangle$, $\mathcal{M}_t \doteqdot X_t(\mathcal{M})$, and μ_t is the Riemannian measure induced by X_t.

Proof. First, in any local coordinate system $x : U \to \mathbb{R}^n$ for M, by using (5.96) we compute that

$$\begin{aligned}
\partial_t \sqrt{\det g_x} &= \frac{1}{2\sqrt{\det g_x}} \partial_t \det g_x \\
&= \frac{1}{2\sqrt{\det g_x}} \det g_x \operatorname{tr}\left(g_x^{-1} \cdot \partial_t g_x\right) \\
&= -\sqrt{\det g_x}\, \langle \vec{F}, \vec{H}\rangle \,,
\end{aligned}$$

where g_x is the component matrix of g with respect to the coordinate basis induced by x. Thus, given any compact $K \subset M$,

$$\frac{d}{dt}\operatorname{Area}(K) = -\int_K \langle \vec{F}, \vec{H}\rangle \, d\mu \,.$$

If $f : M \times I \to \mathbb{R}$ is a smooth function with compact support, the same argument yields

$$\frac{d}{dt}\int_M f d\mu = \int_M \left(\partial_t f - \langle \vec{F}, \vec{H}\rangle f\right) d\mu \,.$$

Next, recall that

$$\operatorname{Vol}(\mathcal{M}_t) = \frac{1}{n+1}\int_M \langle X_t, \mathsf{N}_t\rangle \, d\mu_t$$

whenever $X_t : M \to \mathbb{R}^{n+1}$ is a family of embeddings. Thus,

$$\begin{aligned}\frac{d}{dt}\operatorname{Vol}(\mathcal{M}_t) &= \frac{1}{n+1}\int_M \left(\langle \partial_t X, \mathsf{N}\rangle + \langle X, D_t\mathsf{N}\rangle - \langle X, \mathsf{N}\rangle HF\right) d\mu_t \\ &= \frac{1}{n+1}\int_M \left(-F + \langle X, \operatorname{grad} F\rangle - \langle X, \mathsf{N}\rangle HF\right) d\mu_t\,,\end{aligned}$$

where $H \doteqdot -\langle \vec{H}, \mathsf{N}\rangle$. Observe that

$$\begin{aligned}\operatorname{div}_g(FX)^\top &= \operatorname{tr}_{T\mathcal{M}}\left(D(FX - F\langle X, \mathsf{N}\rangle \mathsf{N})\right) \\ &= \langle X, \operatorname{grad} F\rangle + nF - FH\langle X, \mathsf{N}\rangle\,.\end{aligned}$$

The divergence theorem then yields

$$\frac{d}{dt}\operatorname{Vol}(\mathcal{M}_t) = -\int_M F\, d\mu_t\,. \qquad \square$$

5.4. Curvature flows

Roughly speaking, a smooth 1-parameter family $X : M^n \times I \to \mathbb{R}^{n+1}$ of immersions $X_t \doteqdot X(\cdot, t)$ is said to **flow by curvature** if its normal speed $F = -\langle \partial_t X, \mathsf{N}\rangle$ is equal to a function of its curvature, that is, if after a time-dependent reparametrization X satisfies the **curvature flow**

$$\partial_t X(x,t) = -F(x,t)\,\mathsf{N}(x,t)\,, \tag{5.135}$$

where F is equal to a function of the curvature of X. One way to make this precise is to require that

$$F(x,t) = f([\mathrm{II}_{(x,t)}]) \tag{5.136}$$

for some function $f : \mathrm{S}(n) \to \mathbb{R}$, where $\mathrm{S}(n)$ is the linear space of symmetric $n \times n$-matrices and $[\mathrm{II}_{(x,t)}]$ is the component matrix of II with respect to some $g_{(x,t)}$-orthonormal basis for $T_{(x,t)}M^n$. In order to ensure that the resulting flow speed is independent of the choice of orthonormal basis, we are forced to consider speeds f which are $\mathrm{SO}(n)$**-invariant**, in the sense that

$$f(O \cdot A \cdot O^t) = f(A)$$

for any $A \in \mathrm{S}(n)$ and any $O \in \mathrm{SO}(n)$, where, defining the components O_i^k of O by $O(e_i) = O_i^k e_k$,

$$(O \cdot A \cdot O^t)_{ij} = O_i^k O_j^l A_{kl}\,.$$

Since every symmetric matrix can be diagonalized by conjugation with a special orthogonal one, this is equivalent to requiring that

$$F(x,t) = f(\vec{\kappa}(x,t)) \tag{5.137}$$

for some function $f : \mathbb{R}^n \to \mathbb{R}$ which is P_n**-invariant**, in the sense that it is symmetric under permutations $\sigma \in P_n$ of its variables, where $\vec{\kappa}$ is the

n-tuple of principal curvatures and P_n is the group of permutations of the set $\{1, \ldots, n\}$.

We discuss the general setting in greater depth in Chapter 18; for now, let us merely mention some explicit examples.

Example 5.26. Consider a smooth family $X : M \times I \to \mathbb{R}^{n+1}$ of immersions satisfying the curvature flow (5.135).

(1) If f is the trace function, then F becomes the mean curvature H and we call (5.135) the **mean curvature flow**. The mean curvature flow will be the topic of Chapters 6 to 14.

(2) If f is the determinant function, then F becomes the Gauß curvature K and we call (5.135) the **Gauß curvature flow**. The Gauß curvature flow and the related **α-Gauß curvature flows**, where $F = K^\alpha$, $\alpha > 0$, of uniformly convex hypersurfaces are the topic of Chapters 15 to 17.

(3) If f is minus one over the trace, then $F = -H^{-1}$ and we call (5.135) the **inverse-mean curvature flow**, which is a particular example of the flows studied in Chapter 20.

(4) If f is the harmonic mean, then F becomes the harmonic mean curvature $F = \left(\kappa_1^{-1} + \cdots + \kappa_n^{-1}\right)^{-1}$ and we call (5.135) the **harmonic mean curvature flow**. The harmonic mean curvature flow has natural connections with convex hypersurfaces; cf. Anada [**23**] and the first author [**27**]. We will study it, and a family of related flows, in Chapter 19.

The following lemma summarizes the variational equations of Section 5.3.

Lemma 5.27. *Let $X : M^n \times I \to \mathbb{R}^{n+1}$ be a smooth family of immersions solving the equation $\partial_t X = -F\mathsf{N}$, where $F : M^n \times I \to \mathbb{R}$ is a function of curvature satisfying* (5.137). *Then*

★

$$\nabla_t g = 0\,, \tag{5.138}$$

$$D_t\mathsf{N} = \nabla F\,, \tag{5.139}$$

$$\nabla_t \mathrm{II} = \nabla^2 F + \mathrm{II}^2\, F\,, \tag{5.140}$$

$$\partial_t H = \Delta F + |\mathrm{II}|^2\, F\,, \tag{5.141}$$

$$\partial_t K = K \operatorname{tr}\left(L^{-1} \cdot \nabla^2 F\right) + KHF\,, \tag{5.142}$$

$$\frac{d}{dt}\int_C d\mu = -\int_C FH\, d\mu \;\; \textit{for any } C \underset{\text{compact}}{\subset} M^n\,. \tag{5.143}$$

Proof. Equation (5.138) follows from the definition of the covariant time derivative (recall (5.99) or (5.128)). Equation (5.139) is the temporal component of the Weingarten equation (5.104). Equation (5.140) is the temporal component of the Codazzi equation (5.108). Equation (5.141) follows immediately from (5.138) and (5.140). Equation (5.142) follows from (5.140) and the derivative formula for the determinant. The final equation is the first variation formula for area (5.132). □

5.4.1. The linearized flow.

An important equation related to the curvature flow (5.135) is the linear equation

$$\partial_t u = \dot{F}(\nabla^2 u + \mathrm{II}^2\, u)\,, \tag{5.144}$$

where the tensor $\dot{F}$ is defined at each point with respect to some (and hence any) orthonormal basis $\{e_i\}_{i=1}^n$ by

$$\dot{F} \doteqdot \dot{f}^{ij}_{[\mathrm{II}]} e_i \otimes e_j\,, \tag{5.145}$$

where the $\dot{f}^{ij}$ are the partial derivatives of f, defined by

$$\dot{f}^{ij}_A B_{ij} = \left.\frac{d}{ds}\right|_{s=0} f(A+sB) \tag{5.146}$$

for any $A, B \in \mathrm{S}(n)$.

We will refer to (5.144) as the **linearized flow**, the **linearization** of (5.135), or the **Jacobi equation** for (5.135). It arises as the equation satisfied by the normal variation of a smooth family of solutions to (5.135).

Lemma 5.28. *Let $X : M \times I \times (-\varepsilon_0, \varepsilon_0) \to \mathbb{R}^{n+1}$ be a smooth family of solutions to the curvature flow* (5.135) *with $X|_{\varepsilon=0} = X_0$. The normal component,*

$$v \doteqdot \langle \partial_\varepsilon X, \mathsf{N}\rangle|_{\varepsilon=0}\,,$$

of the variation is a solution to the linearized flow (5.144).

Proof. Fix any point $(x_0, t_0) \in M \times I$ and let $\{e_i\}_{i=1}^n$ be a local orthonormal frame which forms a principal basis at $(x_0, t_0, 0)$. Observe that

$$\nabla_i v = \left(\langle {}^X\!D_\varepsilon dX(e_i), \mathsf{N}\rangle + \langle \partial_\varepsilon X, {}^X\!D_i \mathsf{N}\rangle\right)\big|_{\varepsilon=0}\,,$$

$$\begin{aligned}\nabla_i\nabla_j v = &-\left.\frac{\partial}{\partial\varepsilon}\right|_{\varepsilon=0} \mathrm{II}_{ij} + \langle \partial_\varepsilon X, \nabla\, \mathrm{II}_{ij}\rangle|_{\varepsilon=0} - {\mathrm{II}^2}_{ij}\big|_{\varepsilon=0}\, v \\ &+ \left\langle {}^X\!D_\varepsilon dX(e_i), {}^X\!D_j \mathsf{N}\right\rangle\big|_{\varepsilon=0} + \left\langle {}^X\!D_\varepsilon dX(e_j), {}^X\!D_i \mathsf{N}\right\rangle\big|_{\varepsilon=0}\,,\end{aligned}$$

and

$$\begin{aligned}\frac{\partial}{\partial t}v &= -\left.\frac{\partial}{\partial \varepsilon}\right|_{\varepsilon=0} F + \langle \partial_\varepsilon X, \nabla F\rangle|_{\varepsilon=0} \\ &= -\dot{F}^{ij}\left.\frac{\partial}{\partial \varepsilon}\right|_{\varepsilon=0} \mathrm{II}_{ij} + \left.\left\langle \partial_\varepsilon X, \dot{F}^{ij}\nabla \mathrm{II}_{ij}\right\rangle\right|_{\varepsilon=0}.\end{aligned}$$

At the point $(x_0, t_0, 0)$,

$$\begin{aligned}\left\langle {}^X\!D_\varepsilon dX(e_i), {}^X\!D_j \mathsf{N}\right\rangle\big|_{\varepsilon=0} &= \kappa_j \left\langle {}^X\!D_\varepsilon dX(e_i)\big|_{\varepsilon=0}, dX_0(e_j)\right\rangle \\ &= -\kappa_j \left\langle dX_0(e_i), {}^X\!D_\varepsilon dX_0(e_j)\big|_{\varepsilon=0}\right\rangle.\end{aligned}$$

So the second line in the equation for $\nabla^2 v$ is skew-symmetric in i and j at $(x_0, t_0, 0)$. Since $\dot{f}^{ij}$ is symmetric in i and j, we conclude that

$$\partial_t v = \dot{F}^{ij}\left(\nabla_i\nabla_j v + \mathrm{II}^2_{ij}\, v\right)$$

at $(x_0, t_0, 0)$. The claim follows since (x_0, t_0) was arbitrary. □

Corollary 5.29. *Let $X : M^n \times I \to \mathbb{R}^{n+1}$ be a solution to the curvature flow* (5.135). *Then*

$$\partial_t F = \dot{F}(\nabla^2 F + \mathrm{II}^2\, F)\,. \tag{5.147}$$

Proof. Consider the 1-parameter family of solutions to (5.135) X_ε defined by $X_\varepsilon(x,t) \doteqdot X(x, t-\varepsilon)$. Then

$$\langle \partial_\varepsilon|_{\varepsilon=0} X, \mathsf{N}\rangle = F\,.$$

□

5.4.2. Local coordinate calculations for flows. Let $X : M^n \to \mathbb{R}^{n+1}$ be a parametrized hypersurface. Recall from (5.5) that the metric components are $g_{ij} = \langle \partial_i X, \partial_j X\rangle$. The Christoffel symbols are given by (5.17) as

$$\Gamma_{ij}^k = g^{k\ell}\langle \partial_i\partial_j X, \partial_\ell X\rangle\,.$$

Observe that an alternative way to derive this equation is to simply use the standard formula

$$\Gamma_{ij}^k = \frac{1}{2} g^{k\ell}\left(\partial_i g_{j\ell} + \partial_j g_{i\ell} - \partial_\ell g_{ij}\right)$$

and to cancel terms. By (5.23), we have

$$\vec{h}_{ij} = -h_{ij}\,\mathsf{N} = \nabla_i\nabla_j X\,,$$

so that

$$h_{ij} = -\langle \nabla_i\nabla_j X, \mathsf{N}\rangle = -\langle \partial_i\partial_j X, \mathsf{N}\rangle, \tag{5.148}$$

whose traces are $\vec{H} = -H\mathsf{N} = \Delta X$ and $H = -\langle \Delta X, \mathsf{N}\rangle$. For the second equality in (5.148) as well as throughout our calculations below, we use

$\langle \partial_k X, \mathsf{N} \rangle = 0$. The Euclidean derivative of the normal with respect to a basis tangent vector is

$$\partial_i \mathsf{N} = D_i \mathsf{N} = D_{\partial_i X} \mathsf{N} = \mathrm{II}_{ij} g^{jk} \partial_k X \,,$$

which is tangential. For any tangent vector V,

$$V = g^{ij} \langle V, \partial_i X \rangle \, \partial_j X$$

(an easy way to see that this is the correct formula is to take the inner product of both sides with $\partial_k X$). In particular, the tangential component of the position vector is given by

$$X^\top = g^{ij} \langle X^\top, \partial_i X \rangle \partial_j X = g^{ij} \langle X, \partial_i X \rangle \partial_j X = \frac{1}{2} \nabla |X|^2.$$

Now let $X : M^n \times I \to \mathbb{R}^{n+1}$ be a family of parametrized hypersurfaces evolving by (5.135): $\partial_t X = -F\mathsf{N}$. The metric then evolves by

$$\begin{aligned} \partial_t g_{ij} &= \langle \partial_i(\partial_t X), \partial_j X \rangle + \langle \partial_j(\partial_t X), \partial_i X \rangle \\ &= \langle \partial_i(-F\mathsf{N}), \partial_j X \rangle + \langle \partial_j(-F\mathsf{N}), \partial_i X \rangle \\ &= -F\langle \partial_i \mathsf{N}, \partial_j X \rangle - F\langle \partial_j \mathsf{N}, \partial_i X \rangle \\ &= -2F\mathrm{II}_{ij} \,, \end{aligned} \tag{5.149}$$

which is equivalent to both (5.96) and (5.138).

Next, we compute

$$\langle \partial_t \mathsf{N}, \partial_i X \rangle = -\langle \mathsf{N}, \partial_i(\partial_t X) \rangle = -\langle \mathsf{N}, \partial_i(-F\mathsf{N}) \rangle = \partial_i F \,,$$

where we used $\langle \mathsf{N}, \partial_i \mathsf{N} \rangle = 0$. Since $\langle \partial_t \mathsf{N}, \mathsf{N} \rangle = 0$, we obtain

$$\partial_t \mathsf{N} = \nabla F \,,$$

which is equivalent to both (5.104) and (5.139).

The evolution of the second fundamental form is given by

$$\begin{aligned} \partial_t \mathrm{II}_{ij} &= -\langle \partial_i \partial_j (\partial_t X), \mathsf{N} \rangle - \langle \partial_i \partial_j X, \partial_t \mathsf{N} \rangle \\ &= \langle \partial_i \partial_j (F\mathsf{N}), \mathsf{N} \rangle - \langle \partial_i \partial_j X, g^{k\ell} \nabla_k F \partial_\ell X \rangle \\ &= \partial_i \partial_j F + \langle F \partial_i \partial_j \mathsf{N}, \mathsf{N} \rangle - \Gamma_{ij}^k \nabla_k F \\ &= \nabla_i \nabla_j F - F\mathrm{II}_{ik} g^{k\ell} \mathrm{II}_{\ell j} \,, \end{aligned} \tag{5.150}$$

where we used

$$\langle \partial_i \partial_j \mathsf{N}, \mathsf{N} \rangle = \langle \partial_i (\mathrm{II}_{jk} g^{k\ell} \partial_\ell X), \mathsf{N} \rangle = \mathrm{II}_{jk} g^{k\ell} \langle \partial_i \partial_\ell X, \mathsf{N} \rangle = -\mathrm{II}_{ik} g^{k\ell} \mathrm{II}_{\ell j} \,.$$

This is equivalent to both (5.108) and (5.140).

Finally, taking the trace, we obtain

$$\partial_t H = \partial_t g^{ij} \, \mathrm{II}_{ij} = g^{ij} \partial_t \mathrm{II}_{ij} - g^{ik} g^{j\ell} \partial_t g_{k\ell} \mathrm{II}_{ij} = \Delta F + |\mathrm{II}|^2 F \,,$$

which is the same as both (5.110) and (5.141).

5.5. Notes and commentary

An alternate way of making geometric calculations, besides using local coordinates, is to use an *orthonormal frame adapted to a hypersurface* (for this point of view see, for example, Hicks [**285**] or [**170**, Chapter 1, §7.4].

The "evolving orthonormal frame" formalism discussed in Section 5.3.2 is closely related to the "Uhlenbeck trick" used in the Ricci flow, which originates in the papers [**260, 262**] of Hamilton. It was first used in the context of mean curvature flow by Hamilton [**269**]. An exposition is given in [**52**, Chapter 5], where the space-time connection approach is also developed. The space-time connection was developed in the context of mean curvature flow by the first author and Baker [**43**].

5.6. Exercises

Exercise 5.1 (Tangential projection map). Show that the tangential projection defined by (5.11) is given by

$$W^{\top} = \langle W, \partial_i X\rangle \, g^{ij} \partial_j X \,. \tag{5.151}$$

Note by taking $W = X$ that this gives another verification of (5.13).

Exercise 5.2 (Saddle surface). Sketch the parametrized saddle surface defined by $(x_1, x_2) \mapsto (x_1, x_2, x_1 x_2)$, and calculate its second fundamental form and mean and Gauß curvatures. Verify the Gauß formula $2K = H^2 - |\text{II}|^2$.

Exercise 5.3 (Catenoid). (See, e.g., Osserman [**436**] or Boothby [**94**, p. 374].) The catenoid is the surface in $\mathbb{R}^3$ obtained by rotating the curve $x = \cosh z$ about the z-axis. One can parametrize the surface by $(z, \theta) \mapsto (\cosh z \cos\theta, \cosh z \sin\theta, z)$, where $-\infty < z < \infty$ and $0 < \theta < 2\pi$.

(a) Sketch the surface, and show that it is a minimal surface.

(b) The helicoid is a minimal surface in $\mathbb{R}^3$ that can be parametrized by $(u, v) \mapsto (u \cos v, u \sin v, v)$, where $0 < u < \infty, -\infty < v < \infty$. Sketch the helicoid, and then show that it is locally isometric to the catenoid via the isometry $v = \theta$, $u = \sinh z$.

Exercise 5.4 (Scherk's surface). Define the function $f : \Omega \to \mathbb{R}$ by

$$f\left(x^1, x^2\right) = \log\left(\frac{\cos x^1}{\cos x^2}\right),$$

where $\Omega \doteqdot \{(x^1, x^2) \in \mathbb{R}^2 : \cos x^1 \cdot \cos x^2 > 0\}$.

Show that $\mathcal{M}^n \doteqdot \operatorname{graph} f$ is a minimal surface.

Exercise 5.5. Consider an immersion $X : M^n \to \mathbb{R}^{n+1}$, $n \geq 2$.

(1) Suppose that $p \in M$ is an **umbilic point**; i.e.,

$$\mathrm{II}_p = \alpha_p g_p \, .$$

Show that the constant of proportionality α_p is equal to the average $\frac{1}{n}H(p)$ of the principal curvatures at p.

(2) Suppose that X is **totally umbilic**; i.e.,

$$\mathrm{II}_p \equiv \frac{1}{n} H g_p \, .$$

Show that $\nabla \mathrm{II} \equiv 0$ and hence, in particular, α is locally constant.

(3) Deduce that connected, totally umbilic hypersurfaces of $\mathbb{R}^{n+1}$ are open subsets of hyperplanes or round spheres.

Exercise 5.6. Show that for any starshaped hypersurface and $z \in S^n$,

$$\langle \mathsf{N}\,(z)\,, z\rangle = \frac{r}{\sqrt{r^2 + |\overline{\nabla} r|^2}} > 0 \, .$$

Exercise 5.7. Let $X : M^n \to \mathbb{R}^{n+1}$ be a connected embedded hypersurface.

(1) Suppose that $\mathrm{II} \equiv 0$. Show that $\mathcal{M} \doteqdot X(M^n)$ lies in a hyperplane. Hint: First show that N is constant.

(2) Suppose that $\mathrm{II} \equiv \kappa g$, where κ is a nonzero real constant. Show that $\mathcal{M}$ lies in an $(n-1)$-sphere of radius $|\kappa|^{-1}$. Hint: Define $Y : M^n \to \mathbb{R}^{n+1}$ by $Y = X - \kappa^{-1}\mathsf{N}$.

Exercise 5.8. Let $X : M^n \to \mathbb{R}^{n+1}$ be a closed immersed hypersurface. Show that there exists $p \in M$ such that $\mathrm{II}(p) > 0$. Deduce that there is no closed minimal surface in $\mathbb{R}^3$.

Exercise 5.9. Let $\mathcal{M}^2$ be a connected properly embedded surface in $\mathbb{R}^3$ and let $\mathrm{G} : \mathcal{M} \to \mathcal{S}^2$ be its Gauß map. Suppose that G is a conformal map, i.e., $\lambda\langle X, Y\rangle = \langle d\,\mathrm{G}(X), d\,\mathrm{G}(Y)\rangle = \langle \mathrm{L}(X), \mathrm{L}(Y)\rangle$ for some $\lambda : \mathcal{M} \to \mathbb{R}$, where L is the Weingarten map. Prove that $\mathcal{M}$ is a sphere or a minimal surface. Hint: Use the identity $\mathrm{L}^2 - H\,\mathrm{L} + K\,\mathrm{id} = 0$.

Exercise 5.10 (Bernstein and Mettler [**87**]). Let $\mathcal{M}^2 \subset \mathbb{R}^3$ be a minimal surface with $K < 0$.

(1) Show that $2|\mathrm{II}|^2|\nabla \mathrm{II}|^2 = |\nabla|\mathrm{II}|^2|^2$.

(2) Show that $\Delta \mathrm{II} + |\mathrm{II}|^2\mathrm{II} = 0$. Hint: Trace Simons's identity (5.49).

(3) Show that $\Delta \log|K| + 4|K| = 0$. Hint: $2K = H^2 - |\mathrm{II}|^2$.

Exercise 5.11. Let $X : M^n \times I \to \mathbb{R}^{n+1}$ be a smooth family of smooth hypersurfaces evolving with velocity

$$\partial_t X(x,t) = -F(x,t)\mathsf{N}(x,t)\,.$$

Prove that

$$\partial_t \Gamma_{ij}^k = -g^{k\ell}\left[\nabla_i (F\,\mathrm{II})_{j\ell} + \nabla_j (F\,\mathrm{II})_{i\ell} - \nabla_\ell (F\,\mathrm{II})_{ij}\right]\,.$$

Exercise 5.12. Let $X : M^n \to \mathbb{R}^{n+1}$ be a hypersurface and $f : \mathrm{S}(n) \to \mathbb{R}$ an $\mathrm{SO}(n)$-invariant function. Check that the corresponding tensor field $\dot{F}$ given by (5.136) and (5.145) is well-defined.

Exercise 5.13. Following the local coordinate calculations in Section 5.4.2, rederive equation (5.142) for $\partial_t K$. HINT: Use

$$\partial_t K = K\left((\mathrm{II}^{-1})^{ij}\partial_t \mathrm{II}_{ij} - g^{ij}\partial_t g_{ij}\right)\,.$$

Chapter 6

Introduction to Mean Curvature Flow

In the next several chapters we introduce the **mean curvature flow** of hypersurfaces in Euclidean space. We begin, in this introductory chapter, by setting the foundation with basic definitions, results, and properties of the flow.

The mean curvature flow has extensive geometric and physical applications, with a long history stemming from material science as early as the 1920s. The first comprehensive mathematical treatment — by Ken Brakke in 1978 — is from the point of view of geometric measure theory. Subsequently, the theory of classical solutions to mean curvature flow, including singularity formation, was developed by Gerhard Huisken and others [**291**], [**208**], [**296**], [**301**], et al. The mean curvature flow of curves on surfaces is the curve shortening flow, which we have treated in Chapters 2 through 4.

The mean curvature flow is (after choosing an appropriate "gauge") a parabolic equation, with the minimal surface equation as its static case and elliptic counterpart. Its linear analogue is the heat equation, with its elliptic counterpart being the Laplace equation. Not surprisingly, ideas that originally arose in other geometric equations of parabolic and elliptic type, including the minimal surface equation, the heat equation on Riemannian manifolds, the harmonic map equation and flow, and the Ricci flow, influence many of the results in this part. Conversely, results on mean curvature flow influence the study of other geometric evolution equations.

6.1. The mean curvature flow

A smooth 1-parameter family of immersions $X : M^n \times I \to \mathbb{R}^{n+1}$ satisfies **mean curvature flow** if

$$\partial_t X(x,t) = \vec{H}(x,t) \quad \text{for all} \quad (x,t) \in M^n \times [0,T)\,, \tag{6.1}$$

where $\vec{H}$ denotes the mean curvature vector of the family. We say that $X : M \times [0, T) \to \mathbb{R}^{n+1}$ is a solution of **reparametrized mean curvature flow** if

$$(\partial_t X(x,t))^\perp = \vec{H}(x,t)\,. \tag{6.2}$$

In this case, the vector field

$$V(x,t) \doteqdot (\partial_t X(x,t))^\perp - \partial_t X(x,t)$$

is tangential. Defining the 1-parameter family of self-diffeomorphisms $\varphi_t : M \to M$ by

$$\partial_t \varphi_t(x) = dX_t^{-1}(V(\varphi_t(x),t))\,, \quad \varphi_0 = \mathrm{id}_M\,,$$

we have that $\hat{X}_t \doteqdot X_t \circ \varphi_t$ is a solution to the mean curvature flow.

We also say that a smooth 1-parameter family $\{\mathcal{M}_t^n\}_{t\in I}$ of immersed hypersurfaces $\mathcal{M}_t^n \subset \mathbb{R}^{n+1}$ **evolves by mean curvature** if it admits a family $X : M^n \times I \to \mathbb{R}^{n+1}$ of parametrizations $X(\cdot,t)$ satisfying the mean curvature flow (equivalently, the reparametrized mean curvature flow).

6.1.1. Explicit solutions to the mean curvature flow. We begin by describing some important special solutions.

I. (Shrinking round spheres) Denote by $S_r^n \subset \mathbb{R}^{n+1}$ the n-sphere of radius $r > 0$ centered at the origin. Set $\mathcal{M}_0 = S_{r_0}^n$, where $r_0 > 0$. We seek a solution to mean curvature flow, starting at $\mathcal{M}_0$. It is reasonable to expect that the solution remains round, so we make the ansatz

$$\mathcal{M}_t = S_{r(t)}^n\,.$$

If we parametrize these spheres by the embeddings $X_t : S_1^n \hookrightarrow \mathbb{R}^{n+1}$, where

$$X_t(x) \doteqdot X(x,t) \doteqdot r(t)\,x\,,$$

then the mean curvature and unit outward normal of $\mathcal{M}_t$ are given by $H(x,t) = \frac{n}{r(t)}$ and $\mathsf{N}(x,t) = x$, respectively. Thus,

$$\frac{dr}{dt}(t)\,x = \partial_t X(x,t) = -\,H(x,t)\,\mathsf{N}(x,t) = -\frac{nx}{r(t)}\,.$$

So X_t satisfies mean curvature flow if and only if

$$\frac{dr}{dt}(t) = -\frac{n}{r(t)}\,.$$

The solution to this ODE with initial value $r(0) = r_0$ is

$$r(t) = \sqrt{r_0^2 - 2nt}\,. \tag{6.3}$$

Thus, the **shrinking sphere**

$$X_t(x) = \sqrt{r_0^2 - 2nt}\,x\,, \quad (x,t) \in S_1^n \times [0, r_0^2/2n), \tag{6.4}$$

is a solution of mean curvature flow. In fact, it is easy to see that this extends to a solution defined for times $t \in (-\infty, r_0^2/2n)$. Since the radius approaches zero as $t \to T \doteqdot r_0^2/2n$ (and hence its curvature becomes infinite), the solution has no smooth extension beyond time T.

The round shrinking sphere is the model for the singularities of compact convex hypersurfaces, which we study in Chapter 8.

II. (Minimal submanifolds) Recall that a **minimal hypersurface** $\mathcal{M}^n \subset \mathbb{R}^{n+1}$ is a hypersurface satisfying $\vec{H} \equiv 0$. Given a minimal hypersurface $\mathcal{M}_0^n$, the family $t \mapsto \mathcal{M}_t \doteqdot \mathcal{M}_0$, $t \in (-\infty, \infty)$, is clearly a solution to mean curvature flow. There are many examples of complete minimal hypersurfaces; for a classical book on this subject, see Osserman [**436**]. For a modern treatment, see Colding and Minicozzi [**179**].

III. (Cylinders over solutions) Suppose that $\{\mathcal{M}_t^{n-m}\}_{t\in I}$ is a solution to mean curvature flow in $\mathbb{R}^{n-m+1}$. Then the family of product hypersurfaces

$$\mathcal{M}_t^n \doteqdot \mathcal{M}_t^{n-m+1} \times \mathbb{R}^m \subset \mathbb{R}^{n+1} \tag{6.5}$$

is a solution to mean curvature flow in $\mathbb{R}^{n+1}$. In particular, the many special solutions to the curve shortening flow studied in Chapter 4, such as the translating Grim Reaper solution $\{\Gamma_t\}_{t\in(-\infty,\infty)}$, give rise to "cylinder solutions" in higher dimensions, such as the **Grim hyperplanes** $\{\Gamma_t \times \mathbb{R}^{n-1}\}_{t\in(-\infty,\infty)}$. Another class of important examples is the **shrinking cylinders** $\{S^{n-m}_{\sqrt{-2(n-m)t}} \times \mathbb{R}^m\}_{t\in(-\infty,0)}$.

IV. (Graphs) We shall see in Chapter 7 that a family $\{\mathcal{M}_t\}_{t\in I}$ of graphs

$$\mathcal{M}_t = \{(x, u(x,t)) : x \in \Omega_t \subset \mathbb{R}^n\} \subset \mathbb{R}^{n+1}$$

of smooth functions $u(\cdot, t) : \Omega_t \to \mathbb{R}$ satisfies mean curvature flow if and only if

$$\partial_t u = \sqrt{1 + |\nabla u|^2} \operatorname{div}\left(\frac{\nabla u}{\sqrt{1 + |\nabla u|^2}}\right). \tag{6.6}$$

V. (Self-similar solutions) We saw in Chapter 4 that the curve shortening flow admits a large number of self-similar solutions, which evolve by scaling, rotation, or translation, or combinations thereof. This is also the case for the mean curvature flow. These solutions are introduced in the following section and studied in greater depth in Chapter 13. In contrast to the curve shortening flow, many basic questions about higher-dimensional self-similar solutions remain open.

6.2. Invariance properties and self-similar solutions

The mean curvature flow shares similar invariance properties with the curve shortening flow. Given a solution $X : M^n \times I \to \mathbb{R}^{n+1}$ of mean curvature flow, the following local symmetries, $X \mapsto X_\varepsilon$, result, for appropriate values of ε, in new solutions X_ε (possibly defined over new space-time domains).

(1) Reparametrization:

$$X_\varepsilon(x,t) \doteqdot X(\phi(x,\varepsilon),t)\,, \quad \frac{d}{d\varepsilon}\phi = V \circ \phi,\ V \in \Gamma(TM^n).$$

(2) Time translation:

$$X_\varepsilon(x,t) \doteqdot X(x,t+\varepsilon).$$

(3) Space translation:

$$X_\varepsilon(x,t) \doteqdot X(x,t) + \varepsilon V\,, \quad V \in \mathbb{R}^{n+1}.$$

(4) Rotation:

$$X_\varepsilon(x,t) \doteqdot \mathrm{e}^{\varepsilon J} \cdot X(x,t)\,, \quad J \in \mathfrak{so}(n+1).$$

(5) Space-time dilation:

$$X_\varepsilon(x,t) \doteqdot \mathrm{e}^{\varepsilon} X(x,\mathrm{e}^{-2\varepsilon}t).$$

A solution $X : M^n \times I \to \mathbb{R}^{n+1}$ to mean curvature flow is an **invariant solution** if it is invariant under a 1-parameter family of local symmetries. An important class of invariant solutions is the *self-similar* solutions. A solution $X : M^n \times I \to \mathbb{R}^{n+1}$ to mean curvature flow is **self-similar** if it is invariant under a 1-parameter family of local symmetries generated by reparametrization, space translation, time translation, rotation, and space-time dilation.

6.2.1. Translators. A solution $X : M^n \times I \to \mathbb{R}^{n+1}$ to mean curvature flow is a **translating self-similar solution** if there is some $e \in \mathbb{R}^{n+1}$ and some $V \in \Gamma(TM)$ such that

$$X(\varphi(x,\varepsilon),t-\varepsilon) - \varepsilon e = X(x,t) \tag{6.7}$$

for all $(x,t) \in M^n \times I$ and $\varepsilon \in \mathbb{R}$ such that $t - \varepsilon \in I$, where φ is the flow of V. Fixing any $t_0 \in I$ and setting $\varepsilon = t_0 - t$, we find that

$$X(x,t) = X(\varphi(x,t-t_0),t_0) - (t-t_0)e\,.$$

Differentiating (6.7) with respect to ε at $\varepsilon = 0$, we find that

$$dX(V) - \vec{H} = -e.$$

In particular,

$$H = -\langle e, \mathsf{N}\rangle\,. \tag{6.8}$$

A hypersurface satisfying (6.8) is called a **translator**.

Conversely, if $X_0 : M^n \to \mathbb{R}^{n+1}$ is a translator, then the family $X : M^n \times \mathbb{R} \to \mathbb{R}^{n+1}$ defined by

$$X(x,t) \doteqdot X_0(\varphi(x,t)) + te\,,$$

where φ is the flow of the vector field $V \doteqdot -dX_0^{-1}(e^\top)$, is a translating self-similar solution to mean curvature flow. Indeed,

$$\partial_t X = dX_0(V) + e = e - e^\top = e^\perp = \vec{H}_0 = \vec{H}\,.$$

We have already seen the Grim Reaper example $\{\Gamma_t^1\}_{t\in(-\infty,\infty)}$ in dimension one. Higher-dimensional examples include the **Grim hyperplanes** $\{\Gamma_t^n\}_{t\in(-\infty,\infty)}$, which are defined as

$$\Gamma_t^n = \Gamma_t^1 \times \mathbb{R}^{n-1}\,.$$

6.2.2. Shrinkers and expanders. A solution $X : M^n \times I \to \mathbb{R}^{n+1}$ to mean curvature flow is a **homothetic self-similar solution** if there is some $V \in \Gamma(TM)$ such that

$$\mathrm{e}^{\varepsilon} X(\varphi(x,\varepsilon), \mathrm{e}^{-2\varepsilon} t) = X(x,t) \tag{6.9}$$

for all $(x,t) \in M^n \times I$ and $\varepsilon \in \mathbb{R}$ such that $t - \varepsilon \in I$, where φ is the flow of V. The solution is called a **shrinking self-similar solution** if $t < 0$ for each $t \in I$ and an **expanding self-similar solution** if $0 < t$ for each $t \in I$.

Fixing any nonzero $t_0 \in I$ and setting $\varepsilon = \frac{1}{2}\log(t/t_0)$, we find that

$$X(x,t) = \sqrt{t/t_0}X\left(\varphi(x, \tfrac{1}{2}\log(t/t_0)), t_0\right)\,. \tag{6.10}$$

Differentiating (6.9) with respect to ε at $\varepsilon = 0$, we find that

$$X + dX(V) - 2t\vec{H} = 0\,.$$

In particular,

$$H = \frac{1}{-2t}\langle X, \mathsf{N}\rangle\,.$$

A hypersurface satisfying

$$H = \frac{1}{2}\langle X, \mathsf{N}\rangle \tag{6.11}$$

is called a **shrinker** and a hypersurface satisfying

$$H = -\frac{1}{2}\langle X, \mathsf{N}\rangle \tag{6.12}$$

is called an **expander**.

If $X_0 : M^n \to \mathbb{R}^{n+1}$ is a shrinker, then the family $X : M^n \times (-\infty, 0) \to \mathbb{R}^{n+1}$ defined by

$$X(x,t) \doteqdot \sqrt{-t}X_0(\varphi(x, \tfrac{1}{2}\log(-t)))\,,$$

where φ is the flow of the vector field $V \doteqdot -dX_0^{-1}(X_0^\top)$, is a shrinking self-similar solution to mean curvature flow. Indeed,

$$\partial_t X = \frac{1}{2\sqrt{-t}}X_0 + \frac{1}{2\sqrt{-t}}dX_0(V) = \frac{1}{2\sqrt{-t}}X_0^\perp = \frac{1}{\sqrt{-t}}\vec{H}_0 = \vec{H}\,.$$

Similarly, if $X_0 : M^n \to \mathbb{R}^{n+1}$ is an expander, then the family $X : M^n \times (-\infty, 0) \to \mathbb{R}^{n+1}$ defined by

$$X(x,t) \doteqdot \sqrt{t}X_0(\varphi(x, \tfrac{1}{2}\log(t)))\,,$$

where φ is the flow of the vector field $V \doteqdot -dX_0^{-1}(X_0^\top)$, is an expanding self-similar solution to mean curvature flow. Examples of shrinkers include the shrinking round cylinders, $\left\{\mathbb{R}^m \times S^{n-m}_{\sqrt{-2(n-m)t}}\right\}_{t\in(-\infty,0)}$.

6.2.3. Rotators. A solution $X : M^n \times I \to \mathbb{R}^{n+1}$ to mean curvature flow is a **rotating self-similar solution** if there is some $V \in \Gamma(TM)$ and some $J \in \mathfrak{so}(n+1)$ such that

$$\mathrm{e}^{\varepsilon J} \cdot X(\varphi(x,\varepsilon), t-\varepsilon) = X(x,t) \tag{6.13}$$

for all $(x,t) \in M^n \times I$ and $\varepsilon \in \mathbb{R}$ such that $t - \varepsilon \in I$, where φ is the flow of V.

Fixing any nonzero $t_0 \in I$ and setting $\varepsilon = t_0 - t$, we find that

$$X(x,t) = \mathrm{e}^{(t-t_0)J} \cdot X\left(\varphi(x, t-t_0), t_0\right)\,.$$

Differentiating (6.13) with respect to ε at $\varepsilon = 0$, we find that

$$J \cdot X + dX(V) - \vec{H} = 0.$$

In particular,

$$H = -\langle J\cdot X, \mathsf{N}\rangle\,. \tag{6.14}$$

A hypersurface satisfying (6.14) is called a **rotator**.

If $X_0 : M^n \to \mathbb{R}^{n+1}$ is a rotator, then the family $X : M^n \times (-\infty, 0) \to \mathbb{R}^{n+1}$ defined by

$$X(x,t) \doteqdot \mathrm{e}^{tJ} \cdot X_0(\varphi(x,t))\,,$$

where φ is the flow of the vector field $V \doteqdot -dX_0^{-1}(JX_0^\top)$, is a rotating self-similar solution to mean curvature flow. Indeed,

$$\partial_t X = \mathrm{e}^{tJ} \cdot (J\cdot X_0 + dX_0(V)) = \mathrm{e}^{tJ} \cdot (J \cdot X_0)^\perp = \mathrm{e}^{tJ}\cdot \vec{H}_0 = \vec{H}\,.$$

6.3. Evolution equations

The following equations, which follow quite directly from the variational equations of Section 5.3, govern the evolution of the fundamental geometric structures of the evolving hypersurfaces under mean curvature flow.

Lemma 6.1 (Evolution equations under mean curvature flow). *Let $X : M^n \times I \to \mathbb{R}^{n+1}$ be a solution to mean curvature flow. The evolution of the first fundamental form (induced metric), normal, second fundamental form, mean curvature, and area are governed by*

$$\nabla_t g = 0, \tag{6.15}$$

$$\partial_t \mathsf{N} = \nabla H, \tag{6.16}$$

$$\nabla_t \mathrm{II} = \Delta \mathrm{II} + |\mathrm{II}|^2 \, \mathrm{II}, \tag{6.17}$$

$$\partial_t H = \Delta H + |\mathrm{II}|^2 \, H, \tag{6.18}$$

$$\frac{d}{dt}\int_K d\mu = -\int_K H^2 \, d\mu \text{ for any } K \underset{compact}{\subset} M^n, \tag{6.19}$$

and, if the time slices $\mathcal{M}_t^n = X(M^n, t)$ bound open sets Ω_t,

$$\frac{d}{dt}\int_{\Omega_t \cap K} d\mathscr{H}^{n+1} = -\int_{\mathcal{M}_t^n \cap K} H^2 \, d\mathscr{H}^n \text{ for any } K \underset{compact}{\subset} \mathbb{R}^{n+1}, \tag{6.20}$$

where ∇_t is the covariant time derivative (defined either by (5.98) or by (5.126)).

Proof. Equation (6.15) follows immediately from the definition of ∇_t; equation (6.16) is the temporal Weingarten equation for mean curvature flow (5.104); equation (6.18) follows by tracing the temporal Codazzi equation (5.108) (or by tracing (6.17) or by applying Lemma 5.28 to the variation $X_\varepsilon(x,t) \doteqdot X(x, t-\varepsilon)$), and equation (6.19) follows from the first variation formula for the area (5.132). To prove (6.17), we recall the temporal Codazzi equation (5.108), which for mean curvature flow implies

$$\nabla_t \mathrm{II} = \nabla^2 H + H \, \mathrm{II}^2 , \tag{6.21}$$

and Simons's equation (5.49), which after tracing yields

$$\nabla^2 H - \Delta \, \mathrm{II} = |\mathrm{II}|^2 \, \mathrm{II} - H \, \mathrm{II}^2 .$$

The remaining two identities follow immediately from Lemma 5.25. □

One benefit of introducing the covariant time derivative is that it simplifies many calculations involving tensors and their contractions. For example, (6.15) and (6.17) immediately imply that

$$\partial_t |\mathrm{II}|^2 = \Delta |\mathrm{II}|^2 - 2|\nabla \mathrm{II}|^2 + 2|\mathrm{II}|^4 . \tag{6.22}$$

On the other hand, if we compute in local coordinates and use the ordinary time derivative as in Section 5.4.2, then we find that the metric and second fundamental form evolve in local coordinates according to

$$\partial_t g_{ij} = -2H\,\mathrm{II}_{ij}, \tag{6.23}$$

$$\partial_t g^{ij} = 2H\,\mathrm{II}^{ij}, \quad \text{and} \tag{6.24}$$

$$\partial_t \mathrm{II}_{ij} = \Delta \mathrm{II}_{ij} + |\mathrm{II}|^2 \mathrm{II}_{ij} - 2H\,\mathrm{II}^2_{ij}, \tag{6.25}$$

and we arrive at (6.22) only after a more complicated calculation.

The following formulae for commuting the heat operator past gradients and Laplacians will also be useful in this regard.

Proposition 6.2 (Commutator of space and time derivatives)**.** *Suppose that $X : M^n \times I \to \mathbb{R}^{n+1}$ satisfies mean curvature flow. For any $f \in C^\infty(M \times I)$,*

$$(\nabla_t - \Delta)\nabla f = \nabla(\nabla_t - \Delta) f + \mathrm{II}^2(\nabla f), \tag{6.26}$$

and

$$(\partial_t - \Delta)\Delta f = \Delta(\partial_t - \Delta) f + 2H \left\langle \mathrm{II}, \nabla^2 f \right\rangle + 2\,\mathrm{II}(\nabla H, \nabla f). \tag{6.27}$$

Proof. We present two proofs. In the first proof we interpret ∇_t as a vector field on the frame bundle (see (5.126)). To derive (6.26), we use (5.129) acting on the function f, where $\Lambda_c^b f = 0$, to compute that

$$\begin{aligned}(\nabla_t - \Delta)\nabla_a f - \nabla_a(\nabla_t - \Delta) f &= (\nabla_t\nabla_a - \nabla_a\nabla_t)\, f + \nabla_a\Delta f - \Delta\nabla_a f \\ &= H\,\mathrm{II}_{ac}\nabla_c f - \mathrm{Rc}_{ac}\nabla_c f \\ &= H\,\mathrm{II}_{ac}\nabla_c f - (H\,\mathrm{II}_{ad} - \mathrm{II}_{ca}\mathrm{II}_{cd})\nabla_d f \\ &= \mathrm{II}_{ac}\mathrm{II}_{cd}\nabla_d f.\end{aligned}$$

Here, we used the Ricci identity (5.56) and the contracted Gauß equation (5.45).

Next, to derive (6.27), we first observe the following commutator formula. By (5.129) with $F = H$ and by $\Lambda_c^b V_e = \delta_e^b V_c$ from (5.123),

$$\begin{aligned}[\nabla_t, \nabla_a] V_e &= (\nabla_b H\,\mathrm{II}_{ac} - \nabla_c H\,\mathrm{II}_{ab})\Lambda_c^b V_e + H\,\mathrm{II}_{ad}\nabla_d V_e \\ &= (\nabla_e H\,\mathrm{II}_{ac} - \nabla_c H\,\mathrm{II}_{ae}) V_c + H\,\mathrm{II}_{ad}\nabla_d V_e.\end{aligned} \tag{6.28}$$

By taking the divergence of (6.26), we also have

$$\Delta(\nabla_t - \Delta) f = \nabla_a(\nabla_t - \Delta)\nabla_a f - \nabla_a(\mathrm{II}^2_{ab}\nabla_b f). \tag{6.29}$$

Applying both

$$(\nabla_t\nabla_a - \nabla_a\nabla_t)\nabla_a f = (\nabla_a H\,\mathrm{II}_{ac} - \nabla_c H H)\nabla_c f + H\,\mathrm{II}_{ad}\,\nabla_d\nabla_a f$$

from (6.28) and

$$\nabla_a\Delta\nabla_a f = \nabla_a\nabla_b\nabla_b\nabla_a f = \nabla_b\nabla_a\nabla_b\nabla_a f = \Delta\Delta f + \nabla_a(\mathrm{Rc}_{ab}\,\nabla_b f)$$

to (6.29) yields

(6.30)
$$\begin{aligned}\Delta(\nabla_t - \Delta)f - (\nabla_t - \Delta)\Delta f = &- \nabla_a((\mathrm{Rc}_{ab} + \mathrm{II}^2_{ab})\nabla_b f) - H\,\mathrm{II}_{ab}\,\nabla_b\nabla_a f\\ &- (\nabla_a H\,\mathrm{II}_{ab} - \nabla_b H H)\nabla_b f.\end{aligned}$$

Substituting the contracted Gauß equation (5.45), we obtain

$$\begin{aligned}\Delta(\nabla_t - \Delta)f - (\nabla_t - \Delta)\Delta f &= - \nabla_a(H\,\mathrm{II}_{ab}\,\nabla_b f) - H\,\mathrm{II}_{ab}\,\nabla_b\nabla_a f\\ &\quad - (\nabla_a H\,\mathrm{II}_{ab} - \nabla_b H H)\nabla_b f\\ &= - 2H\,\mathrm{II}_{ab}\,\nabla_b\nabla_a f - 2\nabla_a H\,\mathrm{II}_{ab}\,\nabla_b f.\end{aligned} \tag{6.31}$$

Since $\nabla_t = \partial_t$ when acting on functions, this proves (6.27).

Alternatively, if we interpret ∇_t in terms of the pullback connection, then since the curvature operator vanishes on functions, the commutator formula (5.111) (with $F = H$) yields

$$\begin{aligned}\nabla_t\nabla_k f &= \nabla_k\nabla_t f + H\,\mathrm{II}(\nabla f)\\ &= \nabla_k\partial_t f + H\,\mathrm{II}(\nabla f)\,.\end{aligned}$$

The Ricci identity (5.56) then yields

$$(\nabla_t - \Delta)\nabla f = \nabla(\partial_t - \Delta f) + \mathrm{II}^2(\nabla f, \cdot)\,.$$

Similar arguments, also using (5.111) as well as (5.109), yield

$$\begin{aligned}\partial_t\Delta f &= g^{ij}\nabla_t\nabla_i\nabla_j f\\ &= g^{ij}\big(\nabla_i\nabla_t\nabla_j f - \mathrm{Rm}(\partial_t, \partial_i)\nabla_j f + H\nabla_{\mathrm{II}(\partial_i)}\nabla_j f\big)\\ &= g^{ij}\big(\nabla_i\nabla_j\nabla_t f + \nabla_i(H\,\mathrm{II}(\nabla f))(\partial_j) - \mathrm{Rm}(\partial_t, \partial_i)\nabla_j f + H\nabla_{\mathrm{II}(\partial_i)}\nabla_j f\big)\\ &= \Delta\nabla_t f + \mathrm{II}(\nabla H, \nabla f) + Hg(\nabla f, \nabla H) - \mathrm{Rc}(\partial_t, \nabla f) + 2H\left\langle \mathrm{II}, \nabla^2 f\right\rangle\\ &= \Delta\partial_t f + 2\,\mathrm{II}(\nabla H, \nabla f) + 2H\left\langle \mathrm{II}, \nabla^2 f\right\rangle.\end{aligned}$$

Thus,

$$(\partial_t - \Delta)\Delta f = \Delta(\partial_t - \Delta)f + 2\,\mathrm{II}(\nabla H, \nabla f) + 2H\left\langle \mathrm{II}, \nabla^2 f\right\rangle,$$

which is (6.27). □

Remark 6.3. Another way to derive (6.27) is as follows. By taking the time derivative of $\Delta = g^{ij}\nabla_i\nabla_j = g^{ij}(\partial_i\partial_j - \Gamma_{ij}^k\partial_k)$, we obtain

$$\begin{aligned}\partial_t\Delta f =& \Delta\partial_t f + \partial_t(g^{ij})\nabla_i\nabla_j f - g^{ij}\partial_t(\Gamma_{ij}^k)\nabla_k f \\ =& \Delta\partial_t f + 2H\,\mathrm{II}_{ij}\,\nabla_i\nabla_j f \\ &+ g^{ij}(\nabla_i(H\,\mathrm{II}_j^k) + \nabla_j(H\,\mathrm{II}_i^k) - \nabla^k(H\,\mathrm{II}_{ij}))\nabla_k f \\ =& \Delta\partial_t f + 2H\langle \mathrm{II}, \nabla^2 f\rangle + 2\,\mathrm{II}(\nabla H, \nabla f),\end{aligned}$$

where we used Exercise 5.11.

6.3.1. The gradient flow of the area functional. One nice feature of the mean curvature flow is that it is a **gradient flow**. Let $\mathfrak{Hyp}$ denote the space of smoothly immersed hypersurfaces, $X : M^n \to \mathbb{R}^{n+1}$. Since any smooth deformation of a hypersurface is, up to a reparametrization, a normal deformation, we may identify the formal tangent space $T_X\mathfrak{Hyp}$ with the space of smooth functions on M^n via the identification $f \leftrightarrow f\mathsf{N}$. Using this identification, we may define the L^2-metric on $\mathfrak{Hyp}$ by the following inner product on each tangent space $T_{\mathcal{M}}\mathfrak{Hyp}$:

$$\langle f_1, f_2\rangle_{L^2} \doteqdot \int_{M^n} f_1 f_2\, d\mu,$$

where μ is the Riemannian measure induced by the immersion.

Proposition 6.4. *The mean curvature flow is the negative gradient flow of the area functional.*

Proof. By (5.132), under a compactly supported variation $\frac{\partial}{\partial s}X = -f\mathsf{N}$,

$$\frac{d}{ds}\int_{M^n} d\mu = -\int_{M^n} fH d\mu = -\langle f, H\rangle_{L^2}.$$

That is, the L^2-gradient of the area functional is equal to the negative of the mean curvature vector, $H\mathsf{N} = -\vec{H}$. □

6.3.2. Normalized mean curvature flow. The mean curvature flow that we have defined is sometimes called the **unnormalized** mean curvature flow. Due to the formation of singularities, it is often convenient to modify the flow in order to preserve some geometric quantity. In [**291**], for example, the flow is modified so that the area is preserved, preventing the solution from collapsing in finite time. One can do this simply by a rescaling in space and time. We seek ψ so that if

$$\widetilde{X}(x,t) = \psi(t)X(x,t), \tag{6.32}$$

$\psi > 0$, then the associated flow has constant area. In order to get this to work, we also need to reparametrize in time. Since

$$\tilde{g}_{ij} = \left\langle \frac{\partial \widetilde{X}}{\partial x^i}, \frac{\partial \widetilde{X}}{\partial x^j} \right\rangle = \psi^2 g_{ij},$$

we have

$$d\tilde{\mu} = \sqrt{\det(\tilde{g})} = \sqrt{\psi^{2n}\det(g)} = \psi^n d\mu,$$

and so the requirement $\frac{d}{dt}\int_{M^n} d\tilde{\mu} = 0$ implies the condition

$$\int_{M^n} n\psi^{n-1}\psi' d\mu + \psi^n \frac{d}{dt} d\mu = 0,$$

or by the evolution equation for the area functional (6.19),

$$n\psi^{n-1}\psi' \int_{M^n} d\mu = \psi^n \int_{M^n} H^2 d\mu.$$

Defining

$$h^2(t) \doteqdot \frac{\int_{M^n} H^2(x,t)\, d\mu(t)}{\int_{M^n} d\mu(t)},$$

we obtain

$$\psi(t) = \exp\left(\frac{1}{n}\int_0^t h^2(\tau)\, d\tau\right).$$

If we also reparametrize in time by

$$\tilde{t} = \int_0^t \psi^2(\tau)\, d\tau,$$

then using $\widetilde{H} = \frac{1}{\psi} H$, we have the associated **area-normalized** mean curvature flow

$$\frac{\partial}{\partial \tilde{t}} \widetilde{X} = -\widetilde{H}\mathsf{N} + \frac{1}{n}\tilde{h}^2 \widetilde{X}. \tag{6.33}$$

An alternate flow that has been studied is the **volume-normalized** mean curvature flow. Given a closed, embedded hypersurface, one can evolve it in such a way that the enclosed volume remains constant. Indeed, letting $h = \frac{\int_{M^n} H d\mu}{\int_{M^n} d\mu}$, we can see that the flow

$$\frac{\partial}{\partial t} X(p,t) = (h - H)\,\mathsf{N}$$

has this property. By Lemma 5.25, the change in volume is the integral (over the hypersurface) of the normal velocity; hence

$$\begin{aligned}\frac{d}{dt}\operatorname{Vol}(t) &= \int_{M^n}(h-H)\,d\mu\\ &= \frac{\int_{M^n} H\,d\mu}{\int_{M^n} d\mu}\int_{M^n} d\mu - \int_{M^n} H\,d\mu\\ &= 0.\end{aligned}$$

Whereas the mean curvature flow arises as a dynamic model for soap films, the volume-preserving mean curvature flow can be seen as a dynamic model for soap bubbles — the mean curvature term arises from surface tension and the volume constraint arises from the pressure differential between the inside and outside of the bubble.

Although the area- and volume-normalized mean curvature flow equations appear at first glance to be quite analogous (one fixes area and the other volume), they actually behave very differently: The area-normalized equation is merely a homothetic rescaling of the standard mean curvature flow, whereas the volume-normalized flow is forced by a global term which does *not* correspond to rescalings. Moreover, the latter has the remarkable property that it improves the isoperimetric ratio of the solution.

Proposition 6.5. *Let $\{\mathcal{M}_t^n\}_{t\in I}$ be a family of compact, embedded hypersurfaces evolving by the volume-normalized mean curvature flow. Then the isoperimetric ratio*

$$I(t) \doteqdot \frac{\operatorname{Area}(\mathcal{M}_t^n)^{n+1}}{\operatorname{Vol}(\mathcal{M}_t^n)^n}$$

is strictly decreasing at each time $t \in I$, unless $\mathcal{M}_t^n$ is a round sphere.

Proof. By Lemma 5.25 and Hölder's inequality,

$$\begin{aligned}\frac{d}{dt}\operatorname{Area}(\mathcal{M}_t^n) &= \int_{\mathcal{M}_t^n}(h-H)H\,d\mathscr{H}^n\\ &= \frac{\left(\int_{\mathcal{M}_t^n} H\,d\mathscr{H}^n\right)^2}{\int_{\mathcal{M}_t^n} d\mathscr{H}^n} - \int_{\mathcal{M}_t^n} H^2\,d\mathscr{H}^n\\ &\le 0\end{aligned}$$

with equality only if H is constant. The claim follows. □

6.4. Short-time existence

In this section we prove the short-time existence of solutions to mean curvature flow on closed hypersurfaces, following a discussion of the diffeomorphism invariance of geometric hypersurface flows.

6.4.1. Invariance under diffeomorphisms. If the speed f is a function only of the principal curvatures, then the curvature flow

$$\partial_t X = -f\mathsf{N} \tag{6.34}$$

depends only on the geometry of the hypersurface and not on how it is parametrized. In particular, given a diffeomorphism $\phi : M^n \to M^n$, if X_t is a solution to (6.34), then $X_t \circ \phi$ is also a solution to (6.34). It follows that the diffusion term is degenerate in directions tangential to the solution, which suggests that mean curvature flow is only *weakly* parabolic. As we see in Proposition 6.8, we need to overcome this degeneracy when proving short-time existence.

On the other hand, if a graph $\mathcal{M} = \{(x, u(x)) : x \in \mathbb{R}^n\}$ evolves by the mean curvature flow, then using (5.59) and (5.61), we obtain

$$\partial_t u = -H \langle \mathsf{N}, e_{n+1} \rangle = \sqrt{1 + |\nabla u|^2} \operatorname{div}_{\mathbb{R}^n} \left(\frac{\operatorname{grad}_{\mathbb{R}^n} u}{\sqrt{1 + |\nabla u|^2}} \right) \tag{6.35}$$

$$= \frac{1}{(1 + |\nabla u|^2)} \left(\left(1 + |\nabla u|^2\right) \delta_{ij} - \frac{\partial u}{\partial x^i} \frac{\partial u}{\partial x^j} \right) \frac{\partial^2 u}{\partial x^i \partial x^j}, \tag{6.36}$$

where the last equality is obtained from (5.64). The above equation is *strictly* parabolic. That is, considering a hypersurface locally as the graph of a function breaks the diffeomorphism invariance of the mean curvature flow.

Now we consider the class of flows which are diffeomorphism (i.e., geometrically) equivalent to the flow $\partial_t X = -f\mathsf{N}$. We obtain these flows simply by adding a tangential component to the velocity:

$$\partial_t X = -f\mathsf{N} + Y, \tag{6.37}$$

where Y is tangential, i.e., $Y \in T\mathcal{M}_t^n$. In this case the evolution of the unit normal is given by the following.

Lemma 6.6. *Under the evolution equation* (6.37), *we have*

$$\partial_t \mathsf{N} = \nabla f + \mathrm{II}(Y).$$

Proof. Again $\partial_t \mathsf{N}$ is tangential since $\langle \partial_t \mathsf{N}, \mathsf{N} \rangle = \frac{1}{2}\partial_t |\mathsf{N}|^2 = 0$. On the other hand,

$$\begin{aligned}
\left\langle \partial_t \mathsf{N}, \frac{\partial X}{\partial x^i} \right\rangle &= -\left\langle \mathsf{N}, \partial_t \left(\frac{\partial X}{\partial x^i} \right) \right\rangle = -\left\langle \mathsf{N}, \frac{\partial}{\partial x^i} (\partial_t X) \right\rangle \\
&= -\left\langle \mathsf{N}, \frac{\partial}{\partial x^i} (-f\mathsf{N} + Y) \right\rangle \\
&= \frac{\partial f}{\partial x^i} - \left\langle \mathsf{N}, \frac{\partial Y}{\partial x^i} \right\rangle \\
&= \frac{\partial f}{\partial x^i} + \mathrm{II}\left(\frac{\partial X}{\partial x^i}, Y \right).
\end{aligned}$$

The lemma now follows from $\partial_t \mathsf{N} = g^{ij} \left\langle \partial_t \mathsf{N}, \frac{\partial X}{\partial x^i} \right\rangle \frac{\partial X}{\partial x^j}$. □

Of particular note is the diffeomorphism equivalent flow which produces a time-independent normal vector.

Corollary 6.7. *If the $\mathcal{M}_t$ are locally uniformly convex and $Y = -\mathrm{II}^{-1} (\nabla f)$, then $\partial_t \mathsf{N} = 0$.*

Let X_t be a locally uniformly convex solution to

$$\partial_t X = -f\mathsf{N} - \mathrm{II}^{-1} (\nabla f). \tag{6.38}$$

If the $\mathcal{M}_t$ are compact so that the Gauß map $G_t : \mathcal{M}_t \to S^n$ is a diffeomorphism and if $X_0 = G_0^{-1}$, then we have $X_t = G_t^{-1}$. That is, if $\mathcal{M}_0$ is initially parametrized by the inverse of the Gauß map, then under (6.38) $\mathcal{M}_t$ remains parametrized by the inverse of the Gauß map. In the study of flows of convex hypersurfaces it is often useful to parametrize the hypersurfaces by the inverse of the Gauß map. See, e.g., [**28, 29, 506**].

6.4.2. Short-time existence.

The study of mean curvature flow is made possible by the following result.

Proposition 6.8 (Short-time existence). *Let $X_0 : M^n \to \mathbb{R}^m$ be a smooth immersion of a closed manifold M^n. There exist $\varepsilon > 0$ and a smooth solution $X : M^n \times [0, \varepsilon) \to \mathbb{R}^m$ to the mean curvature flow, with $X(\cdot, 0) = X_0$. The solution is unique: If $Y : M^n \times [0, \varepsilon) \to \mathbb{R}^m$ is a second solution and $Y(\cdot, 0) \equiv X_0$, then $Y \equiv X$.*

Proof. By formula (5.32), the mean curvature flow equation for an immersion $X_t : M^n \to \mathbb{R}^m$ is

$$\begin{aligned}
\partial_t X &= \Delta X, \\
X(0) &= X_0,
\end{aligned}$$

where $\Delta = \Delta_{g(t),g_{\mathbb{R}^m}}$ and $g(t) \doteqdot X_t^* g_{\mathbb{R}^m}$. As we have seen, this equation is only weakly parabolic due to the diffeomorphism invariance of the equation. In fact, the dependence of g on the map X changes the symbol of the operator $\Phi : X \mapsto \Delta X$. Indeed,

$$\Delta X = g^{ij}\left(\frac{\partial^2 X}{\partial x^i \partial x^j} - \Gamma_{ij}^k \frac{\partial X}{\partial x^k}\right). \tag{6.39}$$

The point is that, by (5.17),

$$\Gamma_{ij}^k = g^{k\ell}\left\langle \frac{\partial^2 X}{\partial x^i \partial x^j}, \frac{\partial X}{\partial x^\ell}\right\rangle;$$

i.e., the Christoffel symbols are second order in X, and

$$\begin{aligned}\Delta X &= g^{ij}\left(\frac{\partial^2 X}{\partial x^i \partial x^j} - g^{k\ell}\left\langle \frac{\partial^2 X}{\partial x^i \partial x^j}, \frac{\partial X}{\partial x^\ell}\right\rangle \frac{\partial X}{\partial x^k}\right)\\ &= g^{ij}\left(\frac{\partial^2 X}{\partial x^i \partial x^j} - \left(\frac{\partial^2 X}{\partial x^i \partial x^j}\right)^\top\right)\\ &= g^{ij}\left(\frac{\partial^2 X}{\partial x^i \partial x^j}\right)^\perp.\end{aligned}$$

In particular, we have a degeneracy of the Laplace operator in the tangential directions, which exactly corresponds to the diffeomorphism invariance of curvature flows discussed in Section 6.4.1 (note that tangential vector fields correspond to infinitesimal diffeomorphisms of the submanifold). Indeed, we see that the symbol $\sigma D\Phi_X(\zeta) : \mathbb{R}^{n+1} \to \mathbb{R}^{n+1}$ of the linearization $D\Phi$ of the operator Φ is given by

$$\begin{aligned}\sigma D\Phi_X(\zeta)(V) &= g^{ij}\zeta_i\zeta_j\left(V - g^{k\ell}\left\langle V, \frac{\partial X}{\partial x^\ell}\right\rangle \frac{\partial X}{\partial x^k}\right)\\ &= |\zeta|^2 V^\perp\end{aligned}$$

for $\zeta \in T^*M$ and $V \in \mathbb{R}^{n+1}$. Hence, for $\zeta \in T_x^*M \backslash \{0\}$, we have the nontrivial kernel $\ker D\Phi_X(\zeta) = T_{X(x)}\mathcal{M}$.

We use a version of "DeTurck's trick" to solve the degeneracy problem. Fix a background metric $\bar{g}$ on M^n (the initial one, say) and define the tangent vector field

$$V \doteqdot g^{ij}\left(\Gamma_{ij}^k - \bar{\Gamma}_{ij}^k\right)\frac{\partial X}{\partial x^k} \in \Gamma(X^* T\mathbb{R}^m), \tag{6.40}$$

where the $\bar{\Gamma}_{ij}^k$ are the Christoffel symbols of $\bar{g}$. Note that this is indeed a (globally defined) tangent vector field since the difference between any two connections is a tensor. Then (6.39) may be rewritten as

$$-H\mathsf{N} = \Delta X = g^{ij}\left(\frac{\partial^2 X}{\partial x^i \partial x^j} - \bar{\Gamma}_{ij}^k \frac{\partial X}{\partial x^k}\right) - V.$$

So consider the modified mean curvature flow

$$\partial_t X = -H\mathsf{N} + V = g^{ij}\left(\frac{\partial^2 X}{\partial x^i \partial x^j} - \bar{\Gamma}_{ij}^k \frac{\partial X}{\partial x^k}\right). \tag{6.41}$$

The difference between this equation and the original mean curvature flow is that the Christoffel symbols of the induced metric g are replaced by the Christoffel symbols of the fixed metric $\bar{g}$. Since V is tangential, this corresponds to reparametrizing the flow. So, geometrically, the modified mean curvature flow and the original mean curvature flow are the same. The modified mean curvature flow (6.41) is evidently a strictly parabolic second-order quasilinear PDE, so that, by standard theory, given any immersed smooth initial hypersurface $X_0 : M^n \to \mathbb{R}^m$, there exists a solution X to (6.41) with initial parametrization X_0 on a short-time interval $[0, \varepsilon)$ with $\varepsilon > 0$. Note that, for a short time, X remains an immersion and, if it starts as an embedding, then it also remains an embedding for a short time.

To obtain a solution to the parametrized mean curvature flow, we solve the ODE

$$\begin{aligned} \frac{d}{dt}\varphi_t &= -X_t^*\left[V(t)\right] \circ \varphi_t, \\ \varphi_0 &= \mathrm{id}_M, \end{aligned} \tag{6.42}$$

where $V(t)$ is defined by (6.40) with $g = g(t)$ (note $V(t) \in \Gamma(X_t^* T\mathbb{R}^m)$, whereas $X_t^*\left[V(t)\right] \in \Gamma(TM)$). So

$$\hat{X}_t \doteqdot X_t \circ \varphi_t : M^n \to \mathbb{R}^m$$

is a solution to the parametrized mean curvature flow with $\hat{X}_0 = X_0$. Indeed,

$$\begin{aligned} \partial_t \hat{X}_t &= (\partial_t X_t) \circ \varphi_t + D_{\frac{d}{dt}\varphi_t}(X_t \circ \varphi_t) \\ &= (-H\mathsf{N} + V(t)) \circ \varphi_t - D_{-X_t^*[V(t)]\circ\varphi_t}(X_t \circ \varphi_t) \\ &= -\hat{H}\hat{\mathsf{N}}. \end{aligned}$$

Note that, since φ_t is a diffeomorphism, $\hat{X}_t$ is also an immersion/embedding if X_t is. This proves the existence of a solution to the parametrized mean curvature flow initial value problem.

We now prove uniqueness. Suppose that $\hat{X}_1$ and $\hat{X}_2$ are solutions to the mean curvature flow on $M^n \times [0, \varepsilon)$ with $\hat{X}_1(\cdot, 0) = \hat{X}_2(\cdot, 0) = X_0$. Let $\hat{g}_i(t)$ be the pullback metrics via $\hat{X}_i(\cdot, t)$ on M^n, $i = 1, 2$. Let (taking ε smaller if necessary) $\varphi_i : M^n \times [0, \varepsilon) \to M$, $i = 1, 2$, be solutions to the harmonic map flow

$$\begin{aligned} \partial_t \varphi_i &= \Delta_{\hat{g}_i, \bar{g}}\, \varphi_i, \\ \varphi_i(\cdot, 0) &= \mathrm{id}_M. \end{aligned}$$

Since the harmonic map flow is strictly parabolic, solutions exist for a short time. We then have that $X_i(\cdot,t) \doteqdot \hat{X}_i(\cdot,t)\circ\varphi_i(\cdot,t)^{-1}$ are solutions to the modified mean curvature flow (6.41). Since $X_1(\cdot,0) = X_2(\cdot,0) = X_0$, by the uniqueness of solutions to strictly parabolic systems of equations, we have that $X_1(\cdot,t) = X_2(\cdot,t)$ for as long as they are both defined. This in turn implies that $\varphi_i(\cdot,t)$ are solutions to the same ODE (6.42) with $\varphi_1(\cdot,0) = \varphi_2(\cdot,0) = \mathrm{id}_M$. So, by ODE uniqueness theory, we have that $\varphi_1(\cdot,t) = \varphi_2(\cdot,t)$ and hence

$$\hat{X}_1(\cdot,t) = X_1(\cdot,t)\circ\varphi_1(\cdot,t) = X_2(\cdot,t)\circ\varphi_2(\cdot,t) = \hat{X}_2(\cdot,t)\,.$$

This completes the proof of uniqueness. □

The proof can be adapted in a straightforward way to the case of a Riemannian target manifold.

Remark 6.9. There are various ways to "break the diffeomorphism invariance" to prove short-time existence. For example, for a locally uniformly convex hypersurface one may use the support function. This is related to parametrizing the hypersurface by the inverse of the Gauß map. This approach is taken up Chapter 15. For starshaped hypersurfaces, one can work with the radial graph height. Another approach, which works in the general setting, is to seek solutions which are graphical over the initial immersion for a short time. This approach is taken up in Chapter 18.

6.5. The maximum principle

The maximum principle was an essential tool in the study of the curve shortening flow and is one of the most important tools for the analysis of the mean curvature flow as well. The philosophy is the same as for linear heat equations developed in Section 1.4.

6.5.1. Maximum principle for scalars.

The following simple lemma allows us to apply the maximum principle to functions satisfying linear heat-type inequalities on smoothly time-dependent Riemannian manifolds.

Lemma 6.10. *Let (M^n, g) be a Riemannian manifold equipped with its Levi-Civita connection ∇. If $f \in C^2(M^n)$ attains a local maximum at $x_0 \in M^n$, then*

$$0 = \nabla f(x_0) \quad \textit{and} \quad \nabla^2 f(x_0) \le 0\,.$$

Proof. Fix $v \in T_{x_0}M$ and let $\gamma : (-s_0, s_0) \to M^n$ be the geodesic through $x_0 = \gamma(0)$ satisfying $\gamma'(0) = v$. Since $f\circ\gamma$ attains a local maximum at 0, we obtain

$$0 = \left.\frac{d}{ds}\right|_{s=0} (f\circ\gamma) = \nabla_v f$$

and

$$0 \geq \left.\frac{d^2}{ds^2}\right|_{s=0} (f \circ \gamma) = \nabla_v \nabla_v f\,.$$

The claims follow. □

In particular, we obtain the following useful result.

Lemma 6.11. *Let $X : M^n \times I \to \mathbb{R}^{n+1}$ be a 1-parameter family of immersions of a compact*[1] *manifold M^n. Suppose that $u \in C^\infty(M^n \times (0,T)) \cap C^0(M^n \times [0,T))$ satisfies*

$$\partial_t u \leq \Delta u + \nabla_b u + cu$$

for some time-dependent vector field b and some locally bounded function $c : M^n \times [0,T) \to \mathbb{R}$, where the Laplacian Δ and covariant derivative ∇ are taken with respect to the induced metric. If $\max_{M^n \times \{0\}} u \leq 0$, then

$$\max_{M^n \times \{t\}} u \leq 0 \text{ for all } t \in [0,T]\,. \tag{6.43}$$

If $c \equiv 0$, then

$$\max_{M^n \times [0,T]} u = \max_{M^n \times \{0\}} u\,. \tag{6.44}$$

Proof. Using Lemma 6.10, the proof is the same as that of the classical maximum principle (see Section 1.4) and is left as an exercise (see Exercise 6.7). □

Of course, the same argument applies with the inequalities reversed, leading to a **minimum principle**.

We also obtain a useful ODE comparison principle.

Lemma 6.12 (ODE comparison principle). *Let $X : M^n \times I \to \mathbb{R}^{n+1}$ be a 1-parameter family of immersions of a compact manifold M^n. Suppose that $u \in C^\infty(M^n \times (0,T)) \cap C^0(M^n \times [0,T))$ satisfies*

$$\partial_t u \leq \Delta u + \nabla_b u + F(u)\,, \tag{6.45}$$

for some time-dependent vector field b and some locally Lipschitz function $F : \mathbb{R} \to \mathbb{R}$, where the Laplacian Δ and covariant derivative ∇ are taken with respect to the induced metric. If $u \leq \phi_0$ at $t = 0$ for some $\phi_0 \in \mathbb{R}$, then $u(x,t) \leq \phi(t)$ for all $x \in M^n$ and $0 \leq t < T$, where ϕ is the solution to the ODE

$$\begin{cases} \dfrac{d\phi}{dt} = F(\phi) & \text{in } (0,T), \\ \phi(0) = \phi_0\,. \end{cases} \tag{6.46}$$

[1] We will obtain a maximum principle for subsolutions to the heat equation along *noncompact* solutions to mean curvature flow in Proposition 10.7. See also [**163**, Chapter 12].

Proof. The proof is again the same as that of the classical version and is left as an exercise. □

Again, one can reverse the inequalities to obtain the corresponding ODE comparison from below.

The strong maximum principle also passes to our current setting.

Theorem 6.13. *Let $X : M^n \times I \to \mathbb{R}^{n+1}$ be a 1-parameter family of immersions of a connected manifold M^n. Suppose that $u \in C^\infty(M^n \times (0,T))$ is nonpositive and satisfies*

$$\partial_t u \leq \Delta u + \nabla_b u + cu \tag{6.47}$$

for some time-dependent vector field b and some function $c : M^n \times (0,T) \to \mathbb{R}$, where the Laplacian Δ and covariant derivative ∇ are taken with respect to the induced metric. If $u(x_0, t_0) = 0$ for some $(x_0, t_0) \in M^n \times (0,T)$, then $u(x,t) = 0$ for all $(x,t) \in M^n \times (0, t_0]$.

Proof. In local coordinates $\{x^i\}_{i=1}^n$ for a connected coordinate patch $U \subset M^n$ about x_0, u satisfies

$$\partial_t u \leq g^{ij} u_{ij} + (b^k + g^{ij}\Gamma_{ij}{}^k) u_k + cu\,.$$

The classical strong maximum principle (Theorem 1.5) then implies that $u \equiv 0$ in $U \times (0, t_0]$. Since M^n is connected, the claim follows from a standard "open-closed" argument. □

6.5.2. A maximum principle for tensors. A further useful tool is Hamilton's maximum principle for symmetric 2-tensors [**259**, Theorem 9.1] (see also [**260**, Theorems 4.3 and 8.3] for a more general result and [**167**, Theorem 4.6] for an exposition), of which the following result is a special case.

Theorem 6.14 (Tensor maximum principle). *Let $X : M^n \times [0,T) \to \mathbb{R}^{n+1}$ be a smooth 1-parameter family of immersions of a compact manifold M^n. Suppose that $S \in \Gamma(T^*M^n \odot T^*M^n)$ satisfies*

$$(\nabla_t - \Delta) S_{(x,t)}(v,v) \geq F(x,t) S_{(x,t)}(v,v) \ \ \textit{for all} \ \ (x,t,v) \in TM^n$$

for some locally bounded function $F : M^n \times [0,T) \to \mathbb{R}$. If $S_{(x,0)} \geq 0$ for all $x \in M^n$, then $S_{(x,t)} \geq 0$ for all $(x,t) \in M^n \times [0,T)$.

Proof. Fix $\varepsilon > 0$ and $\sigma \in (0,T)$. We claim that the tensor

$$S^{\varepsilon,\sigma} \doteqdot S + \varepsilon e^{(C_\sigma + 1)t} g$$

is positive definite in $M^n \times [0,\sigma]$, where $C_\sigma \doteqdot \max_{M^n \times [0,\sigma]} F$. By hypothesis, $S^{\varepsilon,\sigma}_{(x,0)} > 0$ for all $x \in M^n$. So suppose, contrary to the claim, that there exist $x_0 \in M^n$, $t_0 \in (0,\sigma]$, and $V_0 \in T_{x_0}M^n \setminus \{0\}$ such that $S^{\varepsilon,\sigma}_{(x,t)} > 0$ for each $(x,t)\in M^n\times[0,t_0)$ but $S^{\varepsilon,\sigma}_{(x_0,t_0)}(V_0,V_0)=0$. Extend V_0 locally in space by solving

$$\nabla_{\gamma'}V\equiv 0$$

along radial g_{t_0}-geodesics γ emanating from x_0 and then extend the resulting local vector field in the time direction by solving

$$\nabla_t V\equiv 0\,.$$

Then $\nabla V(x_0,t_0)=0$ and $\nabla_t V(x_0,t_0)=0$. We claim that we also have $\Delta V(x_0,t_0)=0$. To see this, let $\{e_i\}_{i=1}^n$ be an orthonormal frame at x_0 and parallel translate it along geodesics emanating from x_0, all of this with respect to g_{t_0}. We then may compute using $e_i=\gamma_i'$ along γ_i with $\gamma_i'(0)=e_i$ that

$$\Delta V(x_0,t_0)=\sum_{i=1}^n\left(\nabla_{e_i}(\nabla_{e_i}V)-\nabla_{\nabla_{e_i}e_i}V\right)(x_0,t_0)=0\,.$$

Now set

$$s_{\varepsilon,\sigma}(x,t)\doteqdot S^{\varepsilon,\sigma}_{(x,t)}(V_{(x,t)},V_{(x,t)})$$

for (x,t) near (x_0,t_0). Then $s_{\varepsilon,\sigma}(x,t)\geq 0$ for (x,t) in a small parabolic neighborhood $B_r(x_0,t_0)\times(t_0-r^2,t_0]$ of (x_0,t_0) and $s_{\varepsilon,\sigma}(x_0,t_0)=0$, and hence

$$\begin{aligned}
0&\geq(\partial_t-\Delta)s_{\varepsilon,\sigma}|_{(x_0,t_0)}\\
&=(\nabla_t-\Delta)S^{\varepsilon,\sigma}|_{(x_0,t_0)}(V_0,V_0)\\
&\geq F(x_0,t_0)S_{(x_0,t_0)}(V_0,V_0)+\varepsilon(C_\sigma+1)\mathrm{e}^{(C_\sigma+1)t}g_{(x_0,t_0)}(V_0,V_0)\\
&=-\varepsilon\mathrm{e}^{(C_\sigma+1)t_0}g_{(x_0,t_0)}(V_0,V_0)F(x_0,t_0)+\varepsilon(C_\sigma+1)\mathrm{e}^{(C_\sigma+1)t_0}g_{(x_0,t_0)}(V_0,V_0)\\
&\geq\varepsilon\mathrm{e}^{(C_\sigma+1)t_0}g_{(x_0,t_0)}(V_0,V_0)>0\,,
\end{aligned}$$

a contradiction. So $S^{\varepsilon,\sigma}$ indeed remains positive definite in $[0,\sigma]$. The claim follows since $\varepsilon>0$ and $\sigma\in(0,T)$ are arbitrary. □

Remark 6.15. In each of the results of this section, there was nothing special about the fact that $(M^n,g(\cdot,t))$, $t\in I$, was obtained from a family of hypersurface immersions — the arguments apply equally well to any smoothly time-dependent Riemannian manifold (see [**163**, Chapter 12]). Moreover, the Laplacian can be replaced by any (appropriately defined) time-dependent linear elliptic operator.

6.6. The avoidance principle

An important property of mean curvature flow, which we have also seen for the curve shortening flow, is the **avoidance principle**, which asserts that initially disjoint hypersurfaces remain disjoint under the flow.

Let $X_i : M_i^n \times [0,T) \to \mathbb{R}^{n+1}$, $i = 1,2$, be properly immersed solutions to mean curvature flow such that at least one of M_1^n or M_2^n is compact. Suppose that $X_1(M_1^n, 0) \cap X_2(M_2^n, 0) = \emptyset$. Given respective (local) unit normal fields N_i, denote by II_i and H_i the corresponding second fundamental forms and mean curvatures, respectively. Denote by g_i the respective induced metrics and by Δ_i the corresponding Laplacians. Define the distance d : $M_1^n \times M_2^n \times [0,T) \to \mathbb{R}$ between the two solutions by

$$d\,(x,y,t) = |X_2\,(y,t) - X_1\,(x,t)|\,. \tag{6.48}$$

By assumption, $d_0 \doteqdot \inf_{(x,y)\in M_1^n\times M_2^n} d\,(x,y,0) > 0$. We will prove that d remains positive.

Theorem 6.16. *If X_1 and X_2 are solutions to mean curvature flow on closed manifolds with $d_0 > 0$, then $d\,(x,y,t) \geq d_0$ for all x, y, t. In particular, if $X_1(x,0) \neq X_2(y,0)$ for all $x \in M_1^n$ and $y \in M_2^n$, then $X_1(x,t) \neq X_2(y,t)$ for all $x \in M_1^n$, $y \in M_2^n$, and $t \in [0,T)$.*

Proof. Fix $\varepsilon > 0$ and suppose that $\mathrm{e}^{\varepsilon(1+t)}d(x,y,t)$ is not everywhere strictly greater than d_0. Then there exists a first time $t_0 > 0$ such that $\mathrm{e}^{\varepsilon(1+t)}d(\cdot,\cdot,t)$ reaches d_0. That is,

$$\mathrm{e}^{\varepsilon(1+t)}d(\cdot,\cdot,t) > 0 \quad \text{for} \quad t < t_0$$

and

$$\mathrm{e}^{\varepsilon(1+t_0)}d\,(x_0,y_0,t_0) = d_0 \quad \text{for some} \quad (x_0,y_0) \in M_1^n \times M_2^n\,.$$

Then

$$\partial_t \left(\mathrm{e}^{\varepsilon(1+t)}d\right)\Big|_{(x_0,y_0,t_0)} \leq 0. \tag{6.49}$$

Denote by D the Euclidean directional derivative, by ∇_i the covariant derivative induced on M_i^n by X_i, and by ∇ the covariant derivative on $M_1^n \times M_2^n$ induced by ∇_1 and ∇_2. Since (x_0,y_0) is a local minimum of $d(\cdot,\cdot,t_0)$, we find

$$\nabla^2 d|_{(x_0,y_0,t_0)} \geq 0\,. \tag{6.50}$$

Let

$$w\,(x,y,t) \doteqdot \frac{X_2\,(y,t) - X_1\,(x,t)}{d\,(x,y,t)}\,. \tag{6.51}$$

Let $U, V \in T_x M_1^n \cong T_{X_1(x,t)}(\mathcal{M}_1)_t$ and $W, Y \in T_y M_2^n \cong T_{X_2(y,t)}(\mathcal{M}_2)_t$. We have

$$U(d) = -\frac{1}{d}\,\langle U, X_2\,(y,t) - X_1\,(x,t)\rangle = -\,\langle U, w\rangle \tag{6.52}$$

and

$$W(d) = \langle W, w\rangle\,. \tag{6.53}$$

Hence $D_V w = -\frac{V}{d} - \frac{w}{d}V(d) = -\frac{1}{d}(V - \langle V, w\rangle w)$. Extend U, V to vector fields in a neighborhood of x and extend W, Y to vector fields in a neighborhood of y. We compute that

$$\begin{aligned} V(U(d)) &= -\langle D_V U, w\rangle - \langle U, D_V w\rangle \\ &= -\langle \nabla_V U, w\rangle + \mathrm{II}_1(V, U)\langle \mathsf{N}_1, w\rangle + \left\langle U, \frac{1}{d}(V - \langle V, w\rangle w)\right\rangle \end{aligned}$$

since $D_V U = \nabla_V U - \mathrm{II}_1(V, U)\,\mathsf{N}_1$ for any vector field U tangent to M_1^n near x. Hence, for $U, V \in T_{X_1(x,t)}(\mathcal{M}_1)_t$,

$$\left(\nabla^2 d\right)(V, U) = \mathrm{II}_1(V, U)\langle \mathsf{N}_1, w\rangle + \frac{1}{d}\left(\langle U, V\rangle - \langle U, w\rangle\langle V, w\rangle\right).$$

Taking the trace of this, i.e., summing over an orthonormal frame at x, we obtain

(6.54) $$\Delta_1 d = H_1\langle \mathsf{N}_1, w\rangle + \frac{1}{d}\left(n - \left|w^{T_1}\right|^2\right),$$

where w^{T_1} denotes the projection of w onto $T_{X_1(x,t)}(\mathcal{M}_1)_t$.

Similarly, for $W, Y \in T_{X_2(y,t)}(\mathcal{M}_2)_t$ we have $D_Y w = \frac{1}{d}(Y - \langle Y, w\rangle w)$,

$$\left(\nabla^2 d\right)(Y, W) = -\mathrm{II}_2(Y, W)\langle \mathsf{N}_2, w\rangle + \frac{1}{d}\left(\langle W, Y\rangle - \langle W, w\rangle\langle Y, w\rangle\right),$$

and

(6.55) $$\Delta_2 d = -H_2\langle \mathsf{N}_2, w\rangle + \frac{1}{d}\left(n - \left|w^{T_2}\right|^2\right),$$

where w^{T_2} denotes the projection of w onto $T_{X_2(y,t)}(\mathcal{M}_2)_t$.

By (6.52) and (6.53), if $(\bar{x}, \bar{y}, \bar{t})$ is a point at which $\nabla_1 d = 0$ and $\nabla_2 d = 0$, then $w^{T_1} = 0$ and $w^{T_2} = 0$. At $(\bar{x}, \bar{y}, \bar{t})$ we may choose the orientation of the normals so that $\mathsf{N}_1 = \mathsf{N}_2 = w$. Hence at $(\bar{x}, \bar{y}, \bar{t})$ we have

$$\Delta_1 d = H_1 + \frac{n}{d}, \qquad \Delta_2 d = -H_2 + \frac{n}{d}.$$

Next we compute that for $U \in T_{X_1(x,t)}(\mathcal{M}_1)_t$ and $W \in T_{X_2(y,t)}(\mathcal{M}_2)_t$,

(6.56) $$\begin{aligned} \left(\nabla^2 d\right)(U, W) &= U(W(d)) \\ &= \langle W, D_U w\rangle \\ &= -\frac{1}{d}\langle W, U - \langle U, w\rangle w\rangle \\ &= -\frac{1}{d}\left(\langle W, U\rangle - \langle U, w\rangle\langle W, w\rangle\right). \end{aligned}$$

Since $T_{X_1(\bar{x},\bar{t})}(\mathcal{M}_1)_{\bar{t}} = T_{X_2(\bar{y},\bar{t})}(\mathcal{M}_2)_{\bar{t}} \doteqdot E$ and $w^{T_1} = w^{T_2} = 0$, we can trace this to obtain

$$\sum_{i=1}^{n} e_i^1(e_i^2 d) = -\frac{n}{d},$$

where $\{e_i\}_{i=1}^n$ is an orthonormal frame for E and where the superscript 1 or 2 indicates whether e_i is considered as tangent to $\mathcal{M}_1$ or $\mathcal{M}_2$. We conclude that at $(\bar{x}, \bar{y}, \bar{t})$, and in particular at (x_0, y_0, t_0), we have

$$\begin{aligned} \sum_{i=1}^{n} (e_i^1 + e_i^2) \left((e_i^1 + e_i^2) (d) \right) &= \Delta_1 d + \Delta_2 d + 2 \sum_{i=1}^{n} e_i^1 \left(e_i^2 d \right) \\ &= H_1 - H_2 \,. \end{aligned} \tag{6.57}$$

From (6.50) we have at (x_0, y_0, t_0) that $\sum_{i=1}^n (e_i^1 + e_i^2) \left((e_i^1 + e_i^2) (d) \right) \geq 0$. Hence

$$H_1 (x_0, t_0) - H_2 (y_0, t_0) \geq 0 \,. \tag{6.58}$$

On the other hand, we compute that

$$\begin{aligned} \partial_t d (x, y, t) &= \frac{1}{d} \langle X_1(x, t) - X_2(y, t), -H_1 \mathsf{N}_1 + H_2 \mathsf{N}_2 \rangle \\ &= \langle w, -H_1 \mathsf{N}_1 + H_2 \mathsf{N}_2 \rangle \,. \end{aligned} \tag{6.59}$$

By (6.49), at (x_0, y_0, t_0) we have

$$\begin{aligned} 0 &\geq \partial_t \left(\mathrm{e}^{\varepsilon(1+t)} d \right) |_{(x_0, y_0, t_0)} \\ &= \varepsilon \mathrm{e}^{\varepsilon(1+t_0)} d + \mathrm{e}^{\varepsilon(1+t_0)} \langle w, -H_1 \mathsf{N}_1 + H_2 \mathsf{N}_2 \rangle \\ &= \varepsilon \mathrm{e}^{\varepsilon(1+t_0)} d + \mathrm{e}^{\varepsilon(1+t_0)} (H_1 - H_2) \\ &\geq \varepsilon d_0 \,, \end{aligned} \tag{6.60}$$

which is a contradiction. □

From the proof, one sees that d satisfies a nice weakly parabolic evolution equation as follows. Firstly, from (6.54) and (6.55), at an arbitrary point (x, y, t) we have the strictly parabolic equation

$$(\partial_t d - \Delta_1 d - \Delta_2 d) (x, y, t) = \frac{1}{d} \left(-2n + \left| w^{T_2} \right|^2 + \left| w^{T_1} \right|^2 \right) . \tag{6.61}$$

We may rewrite this using (6.52) and (6.53), which say that $\nabla_1 d = -w^{T_1}$ and $\nabla_2 d = w^{T_2}$, respectively.

Secondly, from (6.56),

$$-U (W(d)) = \frac{1}{d} \left(\langle W, U \rangle - \langle U, w \rangle \langle W, w \rangle \right) . \tag{6.62}$$

Let $\{e_i\}_{i=1}^{n-1}$ be orthonormal in $T_{X_1(x,t)} (\mathcal{M}_1)_t \cap T_{X_2(y,t)} (\mathcal{M}_2)_t$ and choose e_n^1 and e_n^2 later. Then taking the trace of (6.62) with respect to these frames yields the weakly parabolic equation

(6.63)

$$\left(\frac{\partial d}{\partial t} - \Delta_1 d - \Delta_2 d - 2\sum_{i=1}^{n} e_i^1\left(e_i^2 d\right)\right)(x,y,t) = \frac{1}{d}\left(-2 + \left|w^{T_2}\right|^2 + \left|w^{T_1}\right|^2\right) + \frac{2}{d}\left(\langle e_n^1, e_n^2\rangle - \langle e_n^1, w\rangle\langle e_n^2, w\rangle\right).$$

Remark 6.17. Assume that

$$w \in \operatorname{span}\{e_n^1, e_n^2\} = \operatorname{span}\{\mathsf{N}_1, \mathsf{N}_2\} = \operatorname{span}\{e_n^1, \mathsf{N}_1\} = \operatorname{span}\{e_n^2, \mathsf{N}_2\}.$$

Then $w^{T_1} = \langle e_n^1, w\rangle\, e_n^1$ and $w^{T_2} = \langle e_n^2, w\rangle\, e_n^2$, so that $\left|w^{T_1}\right|^2 = \langle e_n^1, w\rangle^2$, $\left|w^{T_2}\right|^2 = \langle e_n^2, w\rangle^2$, and

$$\left(\frac{\partial d}{\partial t} - \Delta_1 d - \Delta_2 d - 2\sum_{i=1}^{n} e_i^1\left(e_i^2 d\right)\right) = \frac{1}{d}\left(\langle e_n^1 - e_n^2, w\rangle^2 - \left|e_n^1 - e_n^2\right|^2\right).$$

The right-hand side is nonpositive and, moreover, vanishes if and only if w and $e_n^1 - e_n^2$ are linearly dependent. At a point (x,y,t) where $T_{X_1(x,t)}(\mathcal{M}_1)_t = T_{X_2(y,t)}(\mathcal{M}_2)_t$ we may choose e_n^1 and e_n^2 so that $e_n^1 = e_n^2$, so in this case the right-hand side is zero. Suppose that $P \doteqdot T_x\mathcal{M}_1 \cap T_y\mathcal{M}_2$ has $\dim P = n-1$. Then $w \in \operatorname{span}\{e_n^1, e_n^2\}$ if and only if $w \perp P$.

6.7. Preserving embeddedness

As we saw for curve shortening flow, the proof of the avoidance principle can be modified to prove that embeddedness is preserved.

Theorem 6.18. *Let $X : M^n \times [0, t_0] \to \mathbb{R}^2$, M^n compact, be a smooth solution to the mean curvature flow, and suppose that $X(\cdot, 0)$ is an embedding. Then $X(\cdot, t)$ is an embedding for each $t \in [0, t_0]$.*

Proof. It suffices to prove that $X(\cdot, t)$ is injective for each $t \in [0, t_0]$.

Lemma 6.19. *If $X : M^n \to \mathbb{R}^{n+1}$, M^n compact, is an immersion satisfying $|\mathrm{II}_x| \le K$ for all $x \in M^n$, then*

$$|X(y) - X(x)| \ge \frac{2}{K}\sin\left(\frac{K\ell(x,y)}{2}\right) \tag{6.64}$$

for all x, y with $\ell(x,y) \le \frac{\pi}{K}$, where $\ell(x,y)$ is the intrinsic distance on M^n from x to y.

Proof. The proof is similar to that of Lemma 3.5 for the curve shortening flow. Let x and y be two points of M^n joined by a unit speed, minimizing geodesic segment $\gamma : [-\ell/2, \ell/2] \to M^n$ of intrinsic length $\ell \le \frac{\pi}{K}$ and let $\theta(s)$ denote the angle between $(X \circ \gamma)'(s)$ and $(X \circ \gamma)'(0)$. Then

$$\begin{aligned}\left|\frac{d}{ds}\left\langle (X\circ\gamma)'(s), (X\circ\gamma)'(0)\right\rangle\right| &= \left|\left\langle \mathrm{II}_{\gamma(s)}(\gamma'(s),\gamma'(s))\mathsf{N}(\gamma(s)), (X\circ\gamma)'(0)\right\rangle\right| \\ &\le K\end{aligned}$$

and hence,

$$|\theta(s)| \le K|s| \le \frac{K\ell}{2} \le \frac{K\pi}{2}.$$

Thus,

$$\begin{aligned}|X(x) - X(y)| &\ge \left\langle X(x) - X(y), (X\circ\gamma)'(0)\right\rangle \\ &= \int_{-\ell/2}^{\ell/2} \left\langle (X\circ\gamma)'(s), (X\circ\gamma)'(0)\right\rangle ds \\ &= \int_{-\ell/2}^{\ell/2} \cos(\theta(s))\, ds \\ &\ge \int_{-\ell/2}^{\ell/2} \cos(Ks)\, ds \\ &= \frac{2}{K}\sin\left(\frac{K\ell}{2}\right).\end{aligned}$$

□

Now we can complete the proof that $X(\cdot, t)$ is injective. Since $M^n \times [0, t_0]$ is compact, there exists a constant K such that $|\mathrm{II}_{(x,t)}| \le K$ for all $x \in M^n$ and $t \in [0, t_0]$. Then $|X(y,t) - X(x,t)| \ge \frac{2}{K}\sin\left(\frac{K\ell(x,y,t)}{2}\right) > 0$ whenever $0 < \ell(x,y,t) \le \frac{\pi}{K}$ by Lemma 6.19. Now we apply the maximum principle to d on the set $\mathcal{S} = \{(x,y,t) \in M^n \times M^n \times [0,t_0] : \ell(x,y,t) \ge \frac{\pi}{K}\}$. On the spatial boundary of $\mathcal{S}$, we have $\inf\{d(x,y,t) : \ell(x,y,t) = \frac{\pi}{K}\} \ge \frac{2}{K}$ by Lemma 6.19. The argument of Theorem 6.16 then shows that d can never decrease below its initial minimum in the spatial interior of $\mathcal{S}$, and we conclude that

$$d(x,y,t) \ge \min\left\{\inf\left\{d(x,y,0) : \ell(x,y,0) \ge \pi K^{-1}\right\}, 2K^{-1}\right\} > 0$$

in $\mathcal{S}$. So $X(\cdot, t)$ is injective for each $t \in [0, t_0]$ and the theorem is proved. □

6.8. Long-time existence

A proper solution $X : M^n \times [0,T) \to \mathbb{R}^{n+1}$, where $0 < T \le \infty$, to the mean curvature flow is a **maximal solution** if it has no smooth extension to a larger time interval, that is, if every smooth solution $Y : M^n \times [0,S) \to \mathbb{R}^{n+1}$ which agrees with X on $M^n \times ([0,T) \cap [0,S))$ satisfies $S \le T$. If $X : M^n \times [0,T) \to \mathbb{R}^{n+1}$ is a maximal solution, then T is its **maximal**

time. A proper solution $X : M^n \times [0,T) \to \mathbb{R}^{n+1}$ is a **singular solution** if $T < \infty$ and $\sup_{M^n\times[0,T)} |\mathrm{II}| = \infty$.

Just as for the curve shortening flow (Theorem 2.12), the maximal time is characterized by curvature blow-up when M^n is compact.

Theorem 6.20 (Long-time existence). *Let M^n be a compact manifold and $X : M^n \times [0,T) \to \mathbb{R}^{n+1}$ a smooth solution to mean curvature flow. If X is maximal and $T < \infty$, then*

$$\sup_{M^n\times[0,T)} |\mathrm{II}| = \infty\,.$$

The proof, a reductio ad absurdum, is similar to that of the corresponding result for curve shortening flow (Theorem 2.12). Recall that the main thrust of that argument was to show, using a Bernstein-type argument, that if II remains bounded up to time T, then we can also bound the higher derivatives of II. These bounds could then be converted into bounds for the immersion and its derivatives, allowing us to extend the solution up to time T using the Arzelà–Ascoli theorem and then beyond time T using the short-time existence theorem, yielding a contradiction to T being maximal. The remainder of this section will be devoted to filling in the details of this argument.

To obtain the derivative estimates, we first need an evolution equation for the norm of the m-th covariant derivative of the second fundamental form [**291**, Theorem 7.1]. Once again, it suffices to extract the general structure of the evolution equation rather than an explicit formula.

Definition 6.21 (Coarse tensor product). Given tensors A and B, we denote by $A * B$ any tensor obtained from the summing of constant multiples (depending only on the ranks of A and B) of contractions (possibly using the metric g or its inverse g^{-1}) of $A \otimes B$. The (associative) higher-order products $A * B * C * \cdots$ are defined similarly. We denote by A^{*p} any p-fold product $A * \cdots * A$. For example, the Gauß equation says that $\mathrm{Rm} = \mathrm{II} * \mathrm{II}$. Differentiating this yields $\nabla \mathrm{Rm} = \mathrm{II} * \nabla\mathrm{II}$.

Lemma 6.22 (Evolution of derivatives of II). *For every $m \in \mathbb{N} \cup \{0\}$*

$$(\nabla_t - \Delta)\nabla^m\mathrm{II} = \sum_{i+j+k=m} \nabla^i\mathrm{II} * \nabla^j\mathrm{II} * \nabla^k\mathrm{II}\,, \tag{6.65}$$

where ∇_t is the covariant time derivative, and hence

$$(\partial_t - \Delta)\,|\nabla^m\mathrm{II}|^2 = \sum_{i+j+k=m} \nabla^i\mathrm{II} * \nabla^j\mathrm{II} * \nabla^k\mathrm{II} * \nabla^m\mathrm{II} - 2\left|\nabla^{m+1}\mathrm{II}\right|^2. \tag{6.66}$$

Proof. We proceed by induction. Recall from (6.17) that

$$(\nabla_t - \Delta)\mathrm{II} = |\mathrm{II}|^2\,\mathrm{II} = \mathrm{II} * \mathrm{II} * \mathrm{II},$$

so (6.65) holds when $m = 0$. Suppose that it holds up to order $m \in \mathbb{N}$. Recall the commutation formulae

$$(\nabla_t\nabla - \nabla\nabla_t)T = \mathrm{II} * \nabla\,\mathrm{II} * T + \mathrm{II} * \mathrm{II} * \nabla T$$

and

$$\begin{aligned}(\nabla\Delta - \Delta\nabla)T &= \nabla\,\mathrm{Rm} * T + \mathrm{Rm} * \nabla T\\ &= \mathrm{II} * \nabla\,\mathrm{II} * T + \mathrm{II} * \mathrm{II} * \nabla T\end{aligned}$$

for a given tensor field T. The first identity follows from (5.111) and (5.106) and the second follows by applying the Gauß equation to the standard formula for commuting the gradient and Laplacian. Applying the induction hypothesis then yields

$$\begin{aligned}(\nabla_t - \Delta)\nabla^{m+1}\,\mathrm{II} &= \nabla(\nabla_t - \Delta)\nabla^m\,\mathrm{II} + \mathrm{II} * \nabla\,\mathrm{II} * \nabla^m\,\mathrm{II} + \mathrm{II} * \mathrm{II} * \nabla^{m+1}\,\mathrm{II}\\ &= \nabla \sum_{i+j+k=m} \nabla^i\,\mathrm{II} * \nabla^j\,\mathrm{II} * \nabla^k\,\mathrm{II} + \mathrm{II} * \nabla\,\mathrm{II} * \nabla^m\,\mathrm{II}\\ &\quad + \mathrm{II} * \mathrm{II} * \nabla^{m+1}\,\mathrm{II}\\ &= \sum_{i+j+k=m+1} \nabla^i\,\mathrm{II} * \nabla^j\,\mathrm{II} * \nabla^k\,\mathrm{II}\,.\end{aligned}$$

This completes the proof of (6.65). Equation (6.66) then follows immediately from

$$(\partial_t - \Delta)\left|\nabla^m\,\mathrm{II}\right|^2 = 2\langle(\partial_t - \Delta)\nabla^m\,\mathrm{II}, \nabla^m\,\mathrm{II}\rangle - 2\left|\nabla^{m+1}\,\mathrm{II}\right|^2. \qquad \square$$

Applying the maximum principle to the evolution equations for the derivatives of II, one can obtain estimates for the covariant derivatives of II assuming only a bound for II. We start with a coarser estimate and we then improve it.

Lemma 6.23. *Let M^n be a compact manifold and $X : M^n \times [0,T) \to \mathbb{R}^{n+1}$ a solution to mean curvature flow. Suppose that $T < \infty$ and $C_0 \doteqdot \sup_{M^n\times[0,T)} |\mathrm{II}| < \infty$. For each $m \in \mathbb{N}$ there exists a constant $C_m < \infty$, which depends only on n, m, T, C_0, and $\max_{M^n\times\{0\}} |\nabla^\ell \mathrm{II}|$, $\ell = 1, \ldots, m$, such that*

$$\sup_{M^n\times[0,T)} |\nabla^m \mathrm{II}| \le C_m\,. \tag{6.67}$$

Proof. Recalling (6.66), we may estimate

$$(\partial_t - \Delta)\left|\nabla^m \mathrm{II}\right|^2 \leq -2\left|\nabla^{m+1}\mathrm{II}\right|^2 + C_{n,m}\sum_{i+j+k=m}\left|\nabla^i\mathrm{II}\right|\left|\nabla^j\mathrm{II}\right|\left|\nabla^k\mathrm{II}\right|\left|\nabla^m\mathrm{II}\right|, \tag{6.68}$$

where $C_{n,m} < \infty$ is a constant which depends only on n and m. For example,

$$(\partial_t - \Delta)\left|\mathrm{II}\right|^2 = -2\left|\nabla\mathrm{II}\right|^2 + 2\left|\mathrm{II}\right|^4,$$

which was proved directly in Lemma 6.1, or

$$(\partial_t - \Delta)\left|\nabla\mathrm{II}\right|^2 \leq -2\left|\nabla^2\mathrm{II}\right|^2 + C_{n,1}\left|\mathrm{II}\right|^2\left|\nabla\mathrm{II}\right|^2.$$

Since $|\mathrm{II}| \leq C_0$, applying the maximum principle to this equation yields

$$|\nabla\mathrm{II}| \leq \left(\max_{M^n\times\{0\}}|\nabla\mathrm{II}|\right)\mathrm{e}^{C_{n,1}C_0^2T} \quad \text{on } M^n\times[0,T).$$

This proves (6.67) for $m = 1$.

We prove estimates for the higher derivatives of II by induction on m. Suppose for each $\ell \in \{1, \ldots, m-1\}$ that

$$\sup_{\mathcal{M}\times[0,T)}\left|\nabla^\ell\mathrm{II}\right|^2 \leq C_\ell$$

for some constant $C_\ell < \infty$. Recalling (6.68), we find

$$\begin{aligned}(\partial_t - \Delta)\left|\nabla^{m-1}\mathrm{II}\right|^2 &\leq C_{n,m-1}\sum_{i+j+k=m-1}\left|\nabla^i\mathrm{II}\right|\left|\nabla^j\mathrm{II}\right|\left|\nabla^k\mathrm{II}\right|\left|\nabla^{m-1}\mathrm{II}\right| - 2\left|\nabla^m\mathrm{II}\right|^2\\ &\leq c_{m-1} - 2\left|\nabla^m\mathrm{II}\right|^2,\end{aligned}$$

where c_{m-1} depends on n, m, and the constants $C_0, \ldots, C_{m-1}$. Similarly,

$$\begin{aligned}(\partial_t - \Delta)\left|\nabla^m\mathrm{II}\right|^2 &\leq C_{n,m}\sum_{i+j+k=m}\left|\nabla^i\mathrm{II}\right|\left|\nabla^j\mathrm{II}\right|\left|\nabla^k\mathrm{II}\right|\left|\nabla^m\mathrm{II}\right| - 2\left|\nabla^{m+1}\mathrm{II}\right|^2\\ &\leq C_{n,m}\left(\left|\mathrm{II}\right|^2\left|\nabla^m\mathrm{II}\right|^2 + \sum_{\substack{i+j+k=m\\ i,j,k\leq m-1}}\left|\nabla^i\mathrm{II}\right|\left|\nabla^j\mathrm{II}\right|\left|\nabla^k\mathrm{II}\right|\left|\nabla^m\mathrm{II}\right|\right)\\ &\leq C_{n,m}C_0^2\left|\nabla^m\mathrm{II}\right|^2 + c_m\left|\nabla^m\mathrm{II}\right|\\ &\leq 2C_{n,m}C_0^2\left|\nabla^m\mathrm{II}\right|^2 + \frac{c_m}{4C_{n,m}C_0^2},\end{aligned}$$

where c_m depends on n, m, and the constants $C_0, \ldots, C_{m-1}$. Thus,

$$(\partial_t - \Delta)\left(\left|\nabla^m\mathrm{II}\right|^2 + C_{n,m}C_0^2\left|\nabla^{m-1}\mathrm{II}\right|^2\right) \leq \frac{c_m}{4C_{n,m}C_0^2} + c_{m-1}C_{n,m}C_0^2.$$

So, by the comparison principle, $|\nabla^m \mathrm{II}|^2 + C_{n,m}C_0^2 \left|\nabla^{m-1}\mathrm{II}\right|^2$ grows at most linearly in time and we conclude that

$$|\nabla^m \mathrm{II}|^2 \leq C_m$$

for a constant $C_m < \infty$ which depends on n, m, T, $C_0, \ldots, C_{m-1}$, and $\max_{M^n \times \{0\}} |\nabla^m \mathrm{II}|$. The claim now follows by induction. $\square$

The bounds for the derivatives of II provided by Lemma 6.23 depend on the initial hypersurface and, in particular, on initial bounds for the derivatives of II. By applying the Bernstein technique as in Theorem 2.11, we can obtain rapidly decaying bounds for the derivatives of the curvature which depend only on a bound for the second fundamental form.

Theorem 6.24 (Rapid smoothing). *Let $n \geq 1$. Given any constant $C_0 < \infty$ and any $m \in \mathbb{N}$, there exists a constant $C_m < \infty$ with the following property: If $X : M^n \times [0, r^2] \to \mathbb{R}^{n+1}$ is a compact solution of mean curvature flow satisfying $|\mathrm{II}|^2 \leq C_0 r^{-2}$ on $M^n \times [0, r^2]$, then it also satisfies*

$$t^m |\nabla^m \mathrm{II}|^2 \leq C_m r^{-2} \quad \text{on } M^n \times [0, r^2]. \tag{6.69}$$

Proof. We prove the theorem by induction on m. By hypothesis, (6.69) is true for $m = 0$. Set

$$Q_m \doteqdot \sum_{\ell=0}^{m} a_\ell t^\ell |\nabla^\ell \mathrm{II}|^2 ,$$

where $a_0 = 1$ and the remaining coefficients $a_i > 0$ will be chosen explicitly in a moment. By (6.66),

$$\begin{aligned}(\partial_t - \Delta) Q_m = {} & \sum_{\ell=1}^{m} \ell a_\ell t^{\ell-1} |\nabla^\ell \mathrm{II}|^2 - 2\sum_{\ell=0}^{m} a_\ell t^\ell |\nabla^{\ell+1} \mathrm{II}|^2 \\ & + \sum_{\ell=0}^{m} \sum_{i+j+k=\ell} a_\ell t^\ell \nabla^i \mathrm{II} * \nabla^j \mathrm{II} * \nabla^k \mathrm{II} * \nabla^\ell \mathrm{II} .\end{aligned}$$

Suppose that $Q_\ell \leq C_\ell r^{-2}$ on $M^n \times [0, r^2]$ for each $\ell = 0, \ldots, m-1$, where C_ℓ depends only on C_0 and ℓ. Then $r t^{\frac{i}{2}} |\nabla^i \mathrm{II}| \leq \sqrt{\frac{C_i}{a_i}}$ for each $i = 1, \ldots, m-1$ and we can find a constant c depending only on n and m

such that

$$\begin{aligned}(\partial_t-\Delta)Q_m \le & -2a_m t^m|\nabla^{m+1}\mathrm{II}|^2+c|\mathrm{II}|^4\\ & +\sum_{\ell=1}^{m}(\ell a_\ell-2a_{\ell-1})\,t^{\ell-1}|\nabla^\ell\mathrm{II}|^2\\ & +c\sum_{\ell=1}^{m}t^\ell a_\ell\left(|\mathrm{II}|^2|\nabla^\ell\mathrm{II}|^2+2A_\ell^{\frac12}r^{-3}t^{-\frac{\ell}{2}}|\nabla^\ell\mathrm{II}|\right)\\ \le & -2a_m t^m|\nabla^{m+1}\mathrm{II}|^2+cr^{-4}\left(C_0^2+\sum_{\ell=1}^{m}A_\ell^{\frac23}a_\ell\right)\\ & +\sum_{\ell=1}^{m}\left((C_0+A_\ell^{\frac13})ca_\ell r^{-2}t+\ell a_\ell-2a_{\ell-1}\right)t^{\ell-1}|\nabla^\ell\mathrm{II}|^2,\end{aligned}$$

where

$$A_\ell \doteqdot \frac14\sum_{\substack{i+j+k=\ell\\ i,j,k<\ell}}\frac{C_iC_jC_k}{a_ia_ja_k}.$$

If we choose

$$a_\ell \doteqdot \frac{2a_{\ell-1}}{(C_0+A_\ell^{\frac13})c+\ell},$$

then

$$(\partial_t-\Delta)Q_m\le cr^{-4}\left(C_0^2+\sum_{\ell=1}^{m}A_\ell\right)\doteqdot C_mr^{-4}$$

and the maximum principle yields

$$Q_m\le C_mr^{-2}$$

in $M^n\times[0,r^2]$. □

In order to prove convergence of the immersions $X(\cdot,t)$ as $t\to T$, we need to work in an appropriate space. Fix any (time-independent) metric $\langle\cdot,\cdot\rangle$ and metric connection D on TM^n. For concreteness, let us take the induced metric and Levi-Civita connection of the initial immersion. Given an immersion $X:M^n\to\mathbb{R}^{n+1}$, recall that $dX\doteqdot X_*$ is a section of the bundle $TM^n\otimes X^*T\mathbb{R}^{n+1}$. Denote by $X^*\langle\cdot,\cdot\rangle$ and ${}^X\!D$ the pullback to $X^*T\mathbb{R}^{n+1}$ of the Euclidean metric and connection. We can extend the metrics $\langle\cdot,\cdot\rangle$ and $X^*\langle\cdot,\cdot\rangle$ and connections D and ${}^X\!D$ in the usual way to obtain a metric $\langle\cdot,\cdot\rangle$ and connection D on $(T^*M^n)^\ell\otimes X^*T\mathbb{R}^{n+1}$ for any $\ell\in\mathbb{N}$. Denote by $\|\cdot\|$ the norm induced by $\langle\cdot,\cdot\rangle$.

Note that the difference of the connections $\Lambda\doteqdot D-\nabla$ (acting on sections of TM^n) is a tensor and can be identified with a section of the bundle

$T^*M^n \otimes T^*M^n \otimes TM^n$. We claim that

$$D^m T = \sum_{i_0+2i_1+\cdots+mi_{m-1}+k=m} \Lambda^{i_0} * (D\Lambda)^{i_1} * \cdots * \left(D^{m-1}\Lambda\right)^{i_{m-1}} * \nabla^k T \tag{6.70}$$

for each $m \in \mathbb{N}\cup\{0\}$ and tensor field $T \in \Gamma((TM^n)^\ell \otimes X^*\mathbb{R}^{n+1})$. The identity is clearly true when $m = 0$. To prove it to $(m+1)$-st order, assuming it holds up to m-th order, simply differentiate (6.70),

$$\begin{aligned} D^{m+1}T = &\sum_{i_0+2i_1+\cdots+mi_{m-1}+k=m} \Big(\Lambda^{i_0-1} * (D\Lambda)^{i_1+1} * \cdots * \left(D^{m-1}\Lambda\right)^{i_{m-1}} \\ &+ \cdots + \Lambda^{i_0} * (D\Lambda)^{i_1} * \cdots * \left(D^{m-1}\Lambda\right)^{i_{m-1}-1} * D^m\Lambda\Big) * \nabla^k T \\ &+ \sum_{i_0+2i_1+\cdots+mi_{m-1}+k=m} \Lambda^{i_0} * (D\Lambda)^{i_1} * \cdots * \left(D^{m-1}\Lambda\right)^{i_{m-1}} \\ &\qquad * \left(\nabla^{k+1}T + \Lambda * \nabla^k T\right), \end{aligned}$$

and check that each of the resulting terms is of the desired form.

Lemma 6.25. *For every $m \in \mathbb{N}$*

$$\begin{aligned} D^{m+1}X = &- X_* D^{m-1}\Lambda + \sum_{\sum_{j=0}^{m-2}(j+1)i_j+\ell=m-1} \prod_{k=0}^{m-2} (D^k\Lambda)^{i_k} * \nabla^\ell \,\mathrm{II} * \mathsf{N} \\ &+ \sum_{\sum_{j=0}^{m-2}(j+1)i_j+\ell=m-1} \sum_{p=1}^{\ell} \prod_{k=0}^{m-2} (D^k\Lambda)^{i_k} * \nabla^{\ell-p}\,\mathrm{II} * \nabla^{p-1}\,\mathrm{II} * X_* \end{aligned} \tag{6.71}$$

and

$$\begin{aligned} D_t D^{m+1}X = &\sum_{\sum_{j=0}^{m-1}(j+1)i_j+\ell=m} \prod_{k=0}^{m-1} (D^k\Lambda)^{i_k} * \nabla^{\ell+1}\,\mathrm{II} * \mathsf{N} \\ &+ \sum_{\sum_{j=0}^{m-1}(j+1)i_j+\ell=m} \sum_{p=1}^{\ell+1} \prod_{k=0}^{m-1} (D^k\Lambda)^{i_k} * \nabla^{\ell-p+1}\,\mathrm{II} * \nabla^{p-1}\,\mathrm{II} * X_* . \end{aligned} \tag{6.72}$$

Example 6.26. Let us compute the first few derivatives explicitly. Observe first that

$$D^2X(u,v) = {}^X D_u X_* v - X_* D_u v = -X_*\Lambda(u,v) - \mathrm{II}(u,v)\mathsf{N}$$

for any pair of vector fields u and v. So

$$D^2 X = -X_*\Lambda - \mathrm{II} \otimes \mathsf{N}\,. \tag{6.73}$$

Differentiating (6.73), we obtain

$$\begin{aligned} D^3X &= X_*D\Lambda - D\,\mathrm{II}\otimes\mathsf{N} - \mathrm{II}\otimes X_*\,\mathrm{L} \\ &= X_*D\Lambda - (\nabla\,\mathrm{II}+\Lambda(\cdot,\mathrm{II}))\otimes\mathsf{N} - \mathrm{II}\otimes X_*\,\mathrm{L} \\ &= X_*D\Lambda - (\nabla\,\mathrm{II}+\Lambda*\mathrm{II})*\mathsf{N} + \mathrm{II}*\mathrm{II}*X_*\,. \end{aligned}$$

Differentiating again yields

$$\begin{aligned} D^4X &= X_*D^2\Lambda - (D\nabla\,\mathrm{II}+D\Lambda(\cdot,\mathrm{II})+\Lambda(\cdot,D\,\mathrm{II}))\otimes\mathsf{N} \\ &\quad - (\nabla\,\mathrm{II}+\Lambda(\cdot,\mathrm{II}))\otimes X_*\,\mathrm{L} - D\,\mathrm{II}\otimes X_*\,\mathrm{L} - \mathrm{II}\otimes X_*D\,\mathrm{L} \\ &= X_*D^2\Lambda - \big(\nabla^2\,\mathrm{II}+2\Lambda(\cdot,\nabla\,\mathrm{II})+\Lambda^2(\cdot,\mathrm{II})+D\Lambda(\cdot,\mathrm{II})\big)\otimes\mathsf{N} \\ &\quad - 2\,(\nabla\,\mathrm{II}+\Lambda(\cdot,\mathrm{II}))\otimes X_*\,\mathrm{L} - \mathrm{II}\otimes X_*(\nabla\,\mathrm{L}+\Lambda(\cdot,\mathrm{L})) \\ &= X_*D^2\Lambda - \big(\nabla^2\,\mathrm{II}+\Lambda*\nabla\,\mathrm{II}+\Lambda^2*\mathrm{II}+D\Lambda*\mathrm{II}\big)*\mathsf{N} \\ &\quad + \Lambda*\mathrm{II}*\mathrm{II}*X_* + \mathrm{II}*\nabla\,\mathrm{II}*X_*\,. \end{aligned}$$

Proof of Lemma 6.25. Applying the identity (6.70) to (6.73) yields

$$D^{m+1}X = -X_*D^{m-1}\Lambda + \sum_{\sum_{j=0}^{m-2}(j+1)i_j+\ell=m-1}\ \prod_{k=0}^{m-1}(D^k\Lambda)^{i_k}*\nabla^\ell(\mathrm{II}\otimes\mathsf{N})\,.$$

The first identity now follows from the iterated Leibniz rule,

$$\begin{aligned} \nabla^k(\mathrm{II}\otimes\mathsf{N}) &= \sum_{i=0}^{k}\binom{k}{i}\nabla^{k-i}\,\mathrm{II}\otimes{}^XD^i\mathsf{N} \\ &= \nabla^k\,\mathrm{II}\otimes\mathsf{N} + \sum_{i=1}^{k}\binom{k}{i}\nabla^{k-i}\,\mathrm{II}\otimes X_*\nabla^{i-1}\,\mathrm{L}\,. \end{aligned}$$

The proof of the second identity is similar. □

We can now obtain Theorem 6.20 from the Arzelà–Ascoli theorem.

Proof of Theorem 6.20. Suppose, contrary to the claim, that

$$\widetilde{C}_0 \doteqdot \sup_{M^n\times[0,T)}|\mathrm{II}| < \infty\,.$$

The curvature bound implies a bound for the speed of the hypersurface. Integrating then yields a displacement bound,

$$|X(x,t)-X(x,0)| \le \int_0^t |H(x,\sigma)|d\sigma \le n\widetilde{C}_0t \le n\widetilde{C}_0T,$$

for all $x\in M^n$ and $t\in[0,T)$.

Next, we note that the induced (time-dependent) metric is uniformly equivalent to our fixed metric (the initial one) under the flow. Indeed, the evolution equation for the induced metric yields

$$|\partial_t \log g(v,v)| = \left|2H\frac{\mathrm{II}(v,v)}{g(v,v)}\right| \leq 2n\widetilde{C}_0^2$$

for any nonzero $v \in TM^n$. Integrating then yields

$$\lambda \leq \frac{g_{(x,t)}(v,v)}{g_{(x,0)}(v,v)} \leq \lambda^{-1} \tag{6.74}$$

for all $(x,t) \in M^n \times [0,T)$, where $\lambda \doteqdot \mathrm{e}^{-2n\widetilde{C}_0^2 T} > 0$. So we are free to work with either metric.

Lemma 6.23 provides bounds $\widetilde{C}_m$ for $|\nabla^m \mathrm{II}|$ for each $m \in \mathbb{N}$. Using Lemma 6.25, we can convert these bounds into bounds for $D^m X$ for each $m \in \mathbb{N}$; we claim that

$$|D^m X| + |D_t D^m X| \leq A_m \tag{6.75}$$

for each $m \in \mathbb{N} \cup \{0\}$ for some constant $A_m < \infty$ that depends on m, n, and the curvature derivative bounds $\widetilde{C}_\ell$ for $\ell \leq m$. We proceed by induction. The argument above proves the claim when $m = 0$. So suppose that (6.75) holds up to order $m \in \mathbb{N} \cup \{0\}$. Then (6.71) implies (by an induction argument) that

$$|D^{m-1}\Lambda| \leq |D^{m+1}X| + B_m\,,$$

where B_m depends on n, m, and A_m, and (6.71) then yields

$$|D_t D^{m+1} X| \leq B'_m(|D^{m+1}X| + 1)\,.$$

It follows that

$$\partial_t |D^{m+1}X|^2 \leq B''_m(|D^{m+1}X|^2 + 1)\,.$$

Integrating yields a bound for $|D^{m+1}X|$ and hence also for $|D_t D^{m+1}X|$, proving (6.75).

We can now invoke the Arzelà–Ascoli theorem (see Proposition 6.39) to deduce that $X(\,\cdot\,,t)$ converges to X_T in $C^\infty(M^n, \mathbb{R}^{n+1})$. The estimate (6.74) implies that X_T is an immersion, so we can evolve it by mean curvature flow for a short time, $\delta > 0$ say, to obtain a map $\widetilde{X} : M^n \times [0, T+\delta) \to \mathbb{R}^{n+1}$ which is smooth away from $t = T$, smooth in space at $t = T$, and continuous in t. So it remains only to prove that X is smooth at time $t = T$. This is achieved by relating the mixed space-time derivatives to pure space derivatives. We claim that

$$D_t^m \mathrm{II} = \sum_{\sum_{j=0}^{2m}(j+1)i_j = 2m+1} \prod_{k=0}^{2m} (\nabla^k \mathrm{II})^{i_k}\,. \tag{6.76}$$

Clearly (6.76) holds when $m = 1$. So suppose it holds up to order $m \in \mathbb{N}$. Then

$$\begin{aligned} D_t^{m+1} \mathrm{II} &= D_t \sum_{\sum_{j=0}^{2m}(j+1)i_j = 2m+1} \mathrm{II}^{i_0} * \cdots * (\nabla^{2m} \mathrm{II})^{i_{2m}} \\ &= \sum_{k=0}^{2m} \sum_{\sum_{j=0}^{2m}(j+1)i_j = 2m+1} \mathrm{II}^{i_0} * \cdots * (\nabla^{k} \mathrm{II})^{i_k - 1} * \cdots * (\nabla^{2m} \mathrm{II})^{i_{2m}} \\ &\qquad * \left(\nabla^{k+2} \mathrm{II} + \sum_{p+q+r=k} \nabla^p \mathrm{II} * \nabla^q \mathrm{II} * \nabla^r \mathrm{II} \right). \end{aligned}$$

Each of these terms is of the desired form.

On the other hand,

$$\begin{aligned} D_t^{m+1} X = D_t^m (\mathrm{II} * \mathsf{N}) &= \sum_{j=0}^{m} D_t^{m-j} \mathrm{II} * D_t^j \mathsf{N} \\ &= D_t^m \mathrm{II} * \mathsf{N} + \sum_{j=1}^{m} D_t^{m-j} \mathrm{II} * D_t^{j-1} DH . \end{aligned} \tag{6.77}$$

Combining (6.76) and (6.77), we may now conclude, for each ℓ and $m \in \mathbb{N} \cup \{0\}$, that the left and right limits of $D^m D_t^\ell X(\cdot, t)$ in $C^\infty(M^n, \mathbb{R}^{n+1})$ as $t \nearrow T$ and $t \searrow T$ exist and agree. It follows that X is smooth at time T, completing the proof. □

6.9. Weak solutions

We have seen that, without additional hypotheses, smooth hypersurfaces generally develop singularities in finite time under mean curvature flow. There has been much interest in understanding the nature of the singular set, motivated in part by the desire to continue the flow past singularities (preferably in a canonical way). In this section, we provide a brief introduction to some of the main approaches to defining a weak version of the mean curvature flow.

6.9.1. The Brakke flow. In Ken Brakke's pioneering work on the mean curvature flow [**101**], he considered the evolution of **varifolds**, employing the tools of geometric measure theory. The main idea behind this approach is an integral formulation of the flow via test functions: Consider a smooth solution $\{\mathcal{M}_t\}_{t\in I}$ of the mean curvature flow which is, at each time, properly embedded in an open subset $U \subset \mathbb{R}^{n+1}$, and let $\phi : U \to \mathbb{R}$ be a C^1 function with compact support. Applying the chain rule and recalling the evolution equation (6.19) for the volume element, we observe that

$$\frac{d}{dt}\int_{\mathcal{M}} \phi \circ X \, d\mu = \int_{\mathcal{M}} (D\phi \cdot \vec{H} - \phi H^2)\, d\mu .$$

Figure 6.1. Ken Brakke.

Turning the idea around, we arrive at the following weak definition of mean curvature flow. Recall that on $\mathbb{R}^{n+1}$ a **Radon measure** is the same as a locally finite Borel regular measure.

Definition 6.27. A 1-parameter family $\{\mu_t\}_{t\in I}$ of Radon measures μ_t on $\mathbb{R}^{n+1}$ is called an **integral Brakke flow** if for almost every $t \in I$ there is an integral n-dimensional varifold V_t such that $\mu_t = \mu_{V_t}$, and

$$\frac{d}{dt}\int \phi \, d\mu \le \int (D\phi \cdot \vec{H} - \phi|\vec{H}|^2)\, d\mu \tag{6.78}$$

for all nonnegative C^1 functions $\phi : \mathbb{R}^{n+1} \to \mathbb{R}$ with compact support, where $\frac{d}{dt} f = \limsup_{\Delta t \to 0} \frac{f(t+\Delta t) - f(t)}{\Delta t}$ and $\vec{H}$ is the generalized mean curvature vector of the varifold V_t.

An **integral n-dimensional varifold** V in $\mathbb{R}^{n+1}$ is a pair $(\mathcal{M}, \vartheta)$, where the subset $\mathcal{M} \subset \mathbb{R}^{n+1}$ is **countably n-rectifiable** (i.e., contained in a countable union of C^1 hypersurfaces away from a set of zero $\mathscr{H}^n$ measure) and is $\mathscr{H}^n$-measurable and $\vartheta : \mathcal{M} \to \mathbb{N}$ is a positive, integer-valued, locally $\mathscr{H}^n$-measurable function on $\mathcal{M}$. Here ϑ is called the **multiplicity function**. The induced **weight measure** μ_V of V is the restriction of $\mathscr{H}^n$ to $\mathcal{M}$, weighted by ϑ. That is,

$$\mu_V(A) \doteqdot \int_{\mathcal{M}\cap A} \vartheta \, d\mathscr{H}^n \tag{6.79}$$

for any $\mathscr{H}^n$-measurable subset $A \subset \mathbb{R}^{n+1}$. The **support**, $\operatorname{spt} V$, of V is defined as the support, $\operatorname{spt} \mu_V$, of its weight measure μ_V; that is, $\operatorname{spt} V$

is the set of points $x \in \mathbb{R}^{n+1}$ for which $\mu_V(U) > 0$ on every open set U containing x. The **mass** (or **weight**) of V is

$$\mathbf{M}(V) \doteqdot \mu_V(\mathbb{R}^{n+1}). \tag{6.80}$$

We say that V has **multiplicity one** if $\vartheta = 1$ outside a set of zero $\mathscr{H}^n$ measure. We refer the reader to the books of Federer [**221**], Simon [**482**], and Morgan [**403**] for introductions to geometric measure theory. An excellent introduction to Brakke flows can be found in the book of Tonegawa [**502**].

Since Brakke's work, multiple slight variations of this definition have been introduced, for various technical reasons. Lahiri has proved that they are all, in a quite general setting, equivalent [**345**].

The main advantage of working with integral varifolds is their good compactness properties.

Theorem 6.28 (Brakke–Ilmanen compactness theorem for Brakke flows [**308**, Theorem 7.1] (cf. [**101**, Chapter 4])). *Let $\{\mu_{t,i}\}_{t\in I, i\in\mathbb{N}}$ be a sequence of integral Brakke flows satisfying the uniform local area bound*

$$\sup_{i\in\mathbb{N}}\sup_{t\in I}\mu_{t,i}(K) < \infty \text{ for every } K \underset{compact}{\subset} \mathbb{R}^{n+1}.$$

Then, after passing to a subsequence,

(i) *the measures $\mu_{i,t}$ converge weakly to a limit Radon measure μ_t for every $t \in I$,*

(ii) *the family $\{\mu_t\}_{t\in I}$ is an integral Brakke flow, and*

(iii) *for almost every $t \in I$, after passing to a further subsequence (depending on t), the varifolds $V_{i,t}$ converge to a limit varifold V_t in the sense of varifolds.*

We say that the subsequence converges ***in the sense of Brakke flows*** *to $\{\mu_t\}_{t\in I}$.*

We refer the reader to the monograph [**308**] for the proof of the compactness theorem for Brakke flows. We note only that it fails if the inequality (6.78) in the definition of Brakke flows is replaced by equality.

One of the most important results in the regularity theory of mean curvature flow is Brakke's regularity theorem [**101**, Section 6.12], which is in essence a parabolic analogue of Allard's famous regularity theorem [**15**] (see also [**482**, Chapter 5]). The following simplified version holds in the context of smooth flows (and hence also weak limits of sequences of smooth flows).

Theorem 6.29 (Brakke–White regularity theorem). *Given $n \in \mathbb{N}$, there exist constants $\varepsilon > 0$ and $C < \infty$ with the following property. If $\{\mathcal{M}^n\}_{t\in I}$*

is a mean curvature flow of smooth, properly embedded hypersurfaces $\mathcal{M}_t^n \subset \mathbb{R}^{n+1}$ *satisfying*

$$\sup_{(X,t)\in P(X_0,t_0,r)} \Theta(X,t,r) < 1+\varepsilon$$

for some $r>0$*, where* Θ *is the* ***Gaußian density ratio*** *of* $\{\mathcal{M}^n\}_{t\in I}$*, then*

$$\sup_{P(X_0,t_0,r/2)} |\mathrm{II}| \le Cr^{-1}\,. \tag{6.81}$$

We introduce the Gaußian density ratio and present White's proof [**532**] of the regularity theorem in the context of smooth mean curvature flows in Section 11.3. For a statement and proof in the context of general Brakke flows, which is much more difficult, we refer the reader to the work of Kasai and Tonegawa [**327**] and Lahiri [**344**], as well as the original monograph of Brakke [**101**]. An excellent exposition of the regularity theory for the mean curvature flow may be found in the book of Ecker [**205**].

6.9.2. The level set flow. A formulation of mean curvature flow based on the regular value theorem was suggested, and studied numerically, by Osher and Sethian [**434**] and Sethian [**478**] and developed analytically by Chen, Giga, and Goto [**143**] and by Evans and Spruck [**219**].

By the regular value theorem, if zero is a regular value of a smooth function $u:\mathbb{R}^{n+1}\to\mathbb{R}$, then the zero set $\mathcal{M}^n \doteqdot \{X\in\mathbb{R}^{n+1} : u(X)=0\}$ is a smooth hypersurface of $\mathbb{R}^{n+1}$. In particular, given a smooth 1-parameter family $u:\mathbb{R}^{n+1}\times I\to\mathbb{R}$, $I\subset\mathbb{R}$, of smooth functions $u(\cdot,t):\mathbb{R}^{n+1}\to\mathbb{R}$ for which 0 is a regular value, the level sets $\mathcal{M}_t^n \doteqdot \{X\in\mathbb{R}^{n+1} : u(X,t)=0\}$ are smoothly embedded hypersurfaces of $\mathbb{R}^{n+1}$. If we choose a smooth 1-parameter family $X : U\times I\to\mathbb{R}^{n+1}$ of smooth local parametrizations $X(\cdot,t):U\to\mathbb{R}^{n+1}$ for the hypersurfaces $\mathcal{M}_t$ defined near $X_0 = X(x_0,t_0)\in\mathcal{M}_{t_0}$, then the chain rule yields, on the one hand,

$$\partial_t u(X_0,t_0) = -Du|_{(X_0,t_0)}\cdot\partial_t X(x_0,t_0)\,.$$

On the other hand, recalling that the gradient Du of u is normal to its level sets, we see that the mean curvature of $\mathcal{M}_{t_0}$ at x_0 is given by

$$H(x_0,t_0) = -\operatorname{div}\left(\frac{Du}{|Du|}\right)(x_0,t_0)\,.$$

So the family $\{\mathcal{M}_t\}_{t\in I}$ evolves by mean curvature flow if and only if u satisfies the **level set flow**,

$$\partial_t u = |Du|\operatorname{div}\left(\frac{Du}{|Du|}\right). \tag{6.82}$$

The level set flow is a degenerate parabolic equation. Although the flow is not defined in the classical sense at critical points of u, Evans–Spruck [**219**], and Chen–Giga–Goto [**143**] were able (independently) to construct viscosity solutions $u : \mathbb{R}^{n+1} \times [0, \infty) \to \mathbb{R}$ of (6.82) evolving from any bounded continuous initial condition $u_0 : \mathbb{R}^{n+1} \to \mathbb{R}$. Solutions are *geometrically unique* in that the zero sets of two solutions will coincide whenever their initial zero sets coincide [**219**, Theorem 5.1]. Given a compact $\mathcal{M}_0 \subset \mathbb{R}^{n+1}$ that can be realized as the zero set of a bounded continuous function $u_0 : \mathbb{R}^{n+1} \to \mathbb{R}$, the family $\{\mathcal{M}_t\}_{t\in[0,\infty)}$ of closed sets $\mathcal{M}_t \doteqdot \{X \in \mathbb{R}^{n+1} : u(X,t) = 0\}$, where u is a solution of (6.82) with initial condition u_0, is called the **level set flow** of $\mathcal{M}_0$. The level set flow agrees with the classical smooth mean curvature flow of $\mathcal{M}_0$ whenever the latter exists [**219**, Theorem 6.1], so the level set flow provides a weak solution of the mean curvature flow which extends the classical one.

An interesting feature of the level set flow is the unpleasant phenomenon of **fattening**. Fattening is said to occur when the level sets become too large to be, in any meaningful way, hypersurfaces of $\mathbb{R}^{n+1}$. For example, the level sets could develop an interior or their Hausdorff dimension could increase. This is illustrated by the figure-8 curve, $\Gamma_0 \subset \mathbb{R}^2$. As a parametrized curve, it is immersed in the plane and admits a unique parametrized evolution by mean curvature up to its first singular time. However, as a level set, it is a singular curve and it turns out that its evolution by level set flow immediately develops an interior [**219**]. See Figure 6.2.

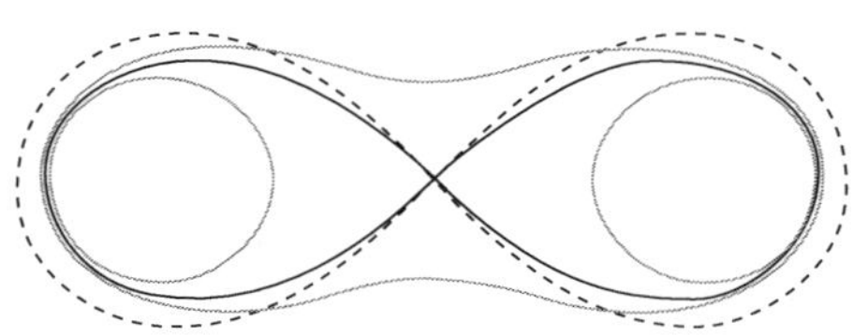

Figure 6.2. The level set flow of a figure-8 curve fattens. The dashed curve is the initial curve. The inner boundary of the level set flow consists of the two inner curves and the outer boundary is the outer curve. The classical flow (as an immersion) is the central figure-8 curve.

This can be interpreted as a form of nonuniqueness in the flow. Let F_0 be the union of the two closed disks bounded by the two loops of Γ_0 and let $\{F_t^+\}_{t\in[0,\infty)}$ and $\{F_t^-\}_{t\in[0,\infty)}$ be the level set flows of F_0 and $\overline{\mathbb{R}^2 \setminus F_0}$, respectively. Then the outer and inner boundaries

$$\Gamma_t^+ \doteqdot \partial F_t^+ \quad \text{and} \quad \Gamma_t^- \doteqdot \partial F_t^- ,$$

respectively, define (smooth for a short time) solutions of mean curvature flow which converge in the Hausdorff topology to Γ_0 as $t \to 0$. Moreover, neither of these two solutions coincides with the solution in the sense of parametrized immersions. An example of a smoothly embedded compact initial configuration in $\mathbb{R}^3$ whose level set flow fattens after a short time is described in [**530**, Section 4] and investigated numerically in [**71**].

The situation is somewhat simplified when considering *mean convex* solutions of the flow. For in that case the hypersurfaces $\mathcal{M}_t$ are always moving in the same direction, so we may consider the ansatz

$$\mathcal{M}_t = \{X \in \mathbb{R}^{n+1} : u(X) = t\}$$

for some function $u : \mathbb{R}^{n+1} \to \mathbb{R}$. Similar calculations as above then imply that u is a *translating solution* of the level set flow. That is,

$$-1 = |Du| \operatorname{div}\left(\frac{Du}{|Du|}\right), \tag{6.83}$$

so that $u(X,t) = u(X) - t$ satisfies the level set flow. The function u is called the **arrival time** of the solution $\{\mathcal{M}_t\}_{t\in I}$ since $u(X) = t$ is the time at which the solution "arrives" at the point X; that is, $X \in \mathcal{M}_t$.

Since we will be mostly concerned with mean convex mean curvature flow, we shall also refer to equation (6.83) as the level set flow. Lipschitz solutions of (6.83) can be found by the method of *elliptic regularization.*

Example 6.30. The function $u : \mathbb{R}^{n+1} \to \mathbb{R}$ defined by

$$u(x) \doteq -\tfrac{1}{2n}(x_1^2 + \cdots + x_{n+1}^2)$$

satisfies the level set flow equation (6.83). Its level sets $\{u(x) = t\}$ are spheres of radius $\sqrt{-2nt}$, so u is a representation of the shrinking sphere.

6.9.3. The viscosity approach.

6.9.3. The viscosity approach. The starting point for the viscosity approach to mean curvature flow, developed independently by Ilmanen [**307**] and Soner [**489**], is the avoidance principle.

Definition 6.31 ([**307, 489**])**.** A family $\{F_t\}_{t\in[0,T)}$ of closed subsets $F_t \subset \mathbb{R}^{n+1}$ is a **viscosity subsolution** of the mean curvature flow if for every smooth, compact, embedded solution $\{\mathcal{M}_t\}_{t\in[0,S)}$ of mean curvature flow

$$\mathcal{M}_0 \cap F_0 = \emptyset \quad \Longrightarrow \quad \mathcal{M}_t \cap F_t = \emptyset \text{ for all } t \in [0,T) \cap [0,S)\,.$$

Taking the closure of the union of all viscosity subsolutions with common initial condition F_0 yields the **maximal subsolution** evolving from F_0. The maximal subsolution coincides with the level set flow of F_0 whenever F_0 is the zero set of a Lipschitz function [**307, 489**].

6.9.4. The shadow flow of Sáez Trumper and Schnürer. M. Sáez Trumper and O. Schnürer [**451**] developed an ingenious weak notion of mean curvature flow based on the following simple observation.

Figure 6.3. M. Sáez Trumper. Photo courtesy of Paula Arenas.

Proposition 6.32. *Let $\{\mathcal{M}_t\}_{t\in I}$ be a smooth family of smooth hypersurfaces $\mathcal{M}_t \subset \mathbb{R}^{n+2}$ such that, for each $t \in I$, $\mathcal{M}_t$ is the graph of a smooth function $u_t : \Omega_t \subset \mathbb{R}^{n+1} \to \mathbb{R}$ satisfying $u_t(x) \to \infty$ as $x \to \partial\Omega_t$. If $\{\mathcal{M}_t\}_{t\in I}$ satisfies mean curvature flow, then so does its* ***shadow****, $\{\partial\Omega_t\}_{t\in I}$, so long as the latter is smooth.*

Proof. The translated graphs, $\operatorname{graph} u(\cdot,t) - se_{n+2}$, solve mean curvature flow and converge locally uniformly in the smooth topology to the cylinders $\Omega_t \times \mathbb{R}$ as $s \to \infty$. □

Turning the idea on its head, Sáez Trumper and Schnürer proved the following.

Theorem 6.33 (Existence of shadow flows [**451**, Theorem 1.1])**.** *Let $A \subset \mathbb{R}^{n+1}$ be a bounded open set and $u_0 : A \to \mathbb{R}$ a locally Lipschitz continuous function satisfying the maximality condition $u_0(x) \to \infty$ as $x \to \partial A$. There exist a relatively open set $\Omega = \bigcup_{t\in[0,\infty)} \Omega_t \times \{t\} \subset \mathbb{R}^{n+1} \times [0,\infty)$ with $\Omega_0 = A$ and a function $u \in C^0(\Omega) \cap C^\infty(\Omega \setminus (\Omega_0 \times \{0\}))$ with $u(\cdot,0) = u_0$ which solves graphical mean curvature flow in $\Omega \setminus (\Omega_0 \times \{0\})$ and satisfies $u(x,t) \to \infty$ as $(x,t) \to \partial\Omega$, where $\partial\Omega$ is the relative boundary of Ω in $\mathbb{R}^{n+1} \times [0,\infty)$.*

By Proposition 6.32, the boundaries of the sets Ω_t solve mean curvature flow in the classical sense until the first singular time. Beyond the first singular time the boundaries agree with the level set flow of $\partial\Omega_0$ until the latter fattens, so long as we interpret the boundaries in the measure-theoretic sense.

Theorem 6.34 ([**451**, Theorem 1.2])**.** *Let (A, u_0) and (Ω, u) be as in Theorem 6.33. Assume that the level set evolution of $\partial\Omega_0$ does not fatten. Then it coincides with $\{\partial^\mu\Omega_t\}_{t\in[0,\infty)}$, where*

$$\partial^\mu\Omega_t \doteqdot \{x \in \mathbb{R}^{n+2} : 0 < |\Omega_t \cap B_r(x)| < |B_r(x)| \text{ for all } r > 0\}$$

is the measure-theoretic boundary of Ω_t.

6.9.5. Mean curvature flow with surgery. The most powerful — and also the most restricted — approach to the continuation of the flow through singularities is the **mean curvature flow with surgery**. This is an ad hoc construction, whereby one stops the flow at some time before a singularity forms, "performs the surgery" — replacing parts of the hypersurface with simple topology which are in danger of becoming singular — and then continues the flow, repeating the process until the hypersurface is finally in some desired form. In order to achieve this goal, one needs to obtain a very precise description of almost singular regions. One would then like to prove that the surgeries do not accumulate and that the process terminates after finitely many of them.

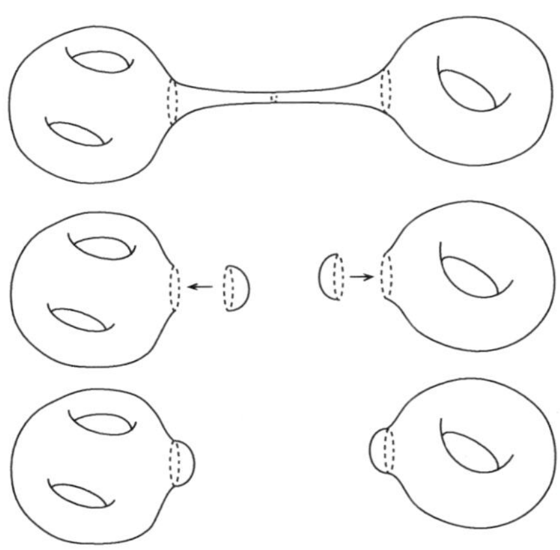

Figure 6.4. Performing an $(n-1)$-surgery to remove a neck region which is in danger of developing a singularity.

The basic observation of surgery theory is that the space $S^k \times S^{n-k-1}$ can be understood both as the boundary of $S^k \times D^{n-k}$ and of $D^{k+1} \times S^{n-k-1}$. For example, taking $k = n-1$, we see that the "neck" $S^{n-1} \times [-1,1]$ has the same boundary, $S^{n-1} \sqcup S^{n-1}$, as the "pair of spherical caps" $D^n \times \{-1,1\}$. In particular, given an n-manifold M^n and an embedding $\phi : S^k \times D^{n-k} \to M^n$, a new manifold

$$M^n_\sharp = \left[\left(M^n \setminus \operatorname{int} \phi(S^k \times D^{n-k})\right) \cup D^{k+1} \times S^{n-k-1}\right] / \sim$$

can be constructed by removing the interior of $\phi(S^k \times D^{n-k})$ and "attaching" $D^{k+1} \times S^{n-k-1}$ by identifying the boundaries of $M^n \setminus \operatorname{int} \phi(S^k \times D^{n-k})$ and $D^{k+1} \times S^{n-k-1}$. This process is referred to as a k-**surgery** on M^n. An example of an $(n-1)$-surgery is illustrated in Figure 6.4.

Surgery is closely related to handle attaching. An $(n+1)$-dimensional j-**handle** is a pair $(D^j \times D^{n+1-j}, D^j \times S^{n-j})$. In particular, an $(n+1)$-dimensional 0-handle is a pair (D^{n+1}, S^n) and an $(n+1)$-dimensional 1-handle is a pair $(D^1 \times D^n, D^1 \times S^{n-1})$. Given an $(n+1)$-dimensional manifold with boundary $(\Omega^{n+1}, \partial\Omega)$ and an embedding $S^k \times D^{n-k} \hookrightarrow \partial\Omega$, the $(n+1)$-dimensional manifold with boundary $(\Omega \cup (D^{k+1} \times D^{n-k}), (\partial\Omega \setminus S^k \times D^{n-k}) \cup (D^{k+1} \times S^{n-k-1}))$ is obtained by **attaching a** $(k+1)$-**handle**. If M'^n is the effect of a k-surgery on M^n, then attaching a corresponding $(k+1)$-handle to $M^n \times [0,1]$ via an embedding $S^k \times D^{n-k} \hookrightarrow M^n \times \{1\}$ yields a *cobordism* from M^n to M'^n, i.e., a compact $(n+1)$-manifold with boundary $(-M^n) \sqcup M'^n$. An $(n+1)$-dimensional r-**handlebody** is a finite union (by attaching) of $(n+1)$-dimensional j-handles, where $j \leq r$.

In the late 1980s, Hamilton proposed developing the mean curvature flow with surgery as a means to address the 4-dimensional Schoenflies conjecture.

Conjecture 6.35 (Schoenflies conjecture). *Let $\mathcal{M}^3 \subset \mathbb{R}^4$ be the image of a smooth embeddeding of S^3 into $\mathbb{R}^4$. There exists a diffeomorphism $\phi : \mathbb{R}^4 \to \mathbb{R}^4$ such that $\phi(\mathcal{M}^3) = S^3$.*

By the Jordan–Brouwer theorem, $\mathcal{M}^3$ separates $\mathbb{R}^4$ into two connected components — a bounded component, Ω, and an unbounded component, $\mathbb{R}^4 \setminus \overline{\Omega}$ — with common boundary $\mathcal{M}^3$. The Schoenflies conjecture asserts that Ω is diffeomorphic to B^4 and $\mathbb{R}^4 \setminus \overline{\Omega}$ is diffeomorphic to $\mathbb{R}^4 \setminus \overline{B}^4$. The analogous problem for smooth embeddings of S^n into $\mathbb{R}^{n+1}$, $n \neq 3$, was proved by Schoenflies ($n = 1$) [**470**], Alexander ($n = 2$) [**7**] (see [**278**] for a modern exposition), and Brown and Mazur ($n \geq 4$) [**121, 391, 392**].

Carrying out the mean curvature flow with surgery program is very difficult and has only been achieved in the restricted context of 2-convex hypersurfaces in $\mathbb{R}^{n+1}$. (Recall that a hypersurface is **2-convex** if at each point the sum of its smallest two principal curvatures is positive.) The reason for the success in this restricted setting is that, whenever a singularity forms, one can always find a nearby region which is close, after rescaling, to a long, round cylindrical segment $S^n_r \times [-L, L]$ (a "neck").

Theorem 6.36 (Huisken and Sinestrari [**303**]). *Starting from any smooth, compact, 2-convex immersed hypersurface in $\mathbb{R}^{n+1}$, $n \geq 3$, there exists a mean curvature flow with surgery which terminates after finitely many surgeries and decomposes the initial hypersurface via $(n-1)$-surgeries into finitely many topological spheres, S^n, and products $S^1 \times S^{n-1}$.*

Since an $(n-1)$-surgery is the inverse procedure of a connected sum, it is easy to reconstruct the diffeomorphism type of the initial hypersurface from the final pieces. Moreover, since mean curvature flow with surgery preserves embeddedness, it is also possible to control the bounded region.

Corollary 6.37 (Jordan–Schoenflies theorem for 2-convex hypersurfaces). *Every smooth, compact, 2-convex immersed hypersurface $M^n \to \mathbb{R}^{n+1}$, $n \geq 3$, is diffeomorphic either to the sphere, S^n, or to a finite connected sum of the product $S^1 \times S^{n-1}$. If M^n is embedded, then it bounds the corresponding $(n+1)$-dimensional 1-handlebody. In particular, if M^n is diffeomorphic to S^n, then it bounds a standard $(n+1)$-ball.*

The surgery approach of Huisken and Sinestrari was influenced by Hamilton's Ricci flow with surgery [**268**, **271**]. It was later extended to embedded, 2-dimensional, mean convex flows by Brendle and Huisken [**114**] and, independently, by Haslhofer and Kleiner [**277**] using "noncollapsing" methods introduced by Sheng and Wang [**481**] (which we study in Chapter 12). Alex Mramor refined these surgery arguments to obtain, in particular, a finiteness theorem for the space of closed, embedded, 2-convex hypersurfaces [**406**].

Mean curvature flow with surgery is a highly nonunique process: Each of the constructions mentioned is slightly different from the others and they all depend on various parameters, which control, for example, the surgery times and the quality and size of the neck regions to be removed. On the other hand, Head [**280**] and Lauer [**354**] proved, independently, that the mean curvature flow with surgery of Huisken and Sinestrari converges, in an appropriate sense, to the level set flow as the surgery parameters degenerate.

We shall discuss some of the machinery required for carrying out the surgery program in Chapters 9 through 12. See also Sections 11.5.6 and 11.5.7.

6.10. Notes and commentary

6.10.1. Maximum principles. Maximum principles for elliptic and parabolic systems were developed by Hans Weinberger [**524**] and others [**217**, **520**].

6.10.2. Constrained mean curvature flows. The volume-preserving mean curvature flow was introduced by G. Huisken in [**294**]. He proved that convex hypersurfaces remain convex and converge to round spheres in infinite time (cf. Chapter 8). J. McCoy [**395**] obtained a similar result for a large class of constrained mean curvature flows which preserve the "mixed volumes" (see Section 18.7). J. Escher and G. Simonett obtained convergence to a round sphere assuming that the initial data is already sufficiently close to a round sphere in an appropriate norm [**214**]. The behavior of nonconvex hypersurfaces under this flow is not nearly as nice since, in contrast to the mean curvature flow, other natural curvature conditions (such as mean convexity) are not preserved [**132**].

M. Athenassenas and S. Kandanaarachchi showed that axially symmetric hypersurfaces converge to a sphere (in the no boundary case) or to a hemisphere (in the free boundary case) under the volume-preserving mean curvature flow, so long as singularities do not develop along the axis of rotation [**79**] (see also the prior work of Athenassenas [**77, 78**]).

E. Cabezas-Rivas and V. Miquel proved analogous results for volume-preserving mean curvature flow in other ambient spaces [**129–131**].

M. Bertini, Cabezas-Rivas, and C. Sinestrari [**90, 133, 483**] obtained sphere-convergence results for volume constrained flows by powers of the k-th mean curvatures (see Section 18.3). See also [**91**].

F. Dittberner [**197**] obtained several nice results for constrained curve flows (see Section 6.3.2). In particular, she obtained an analogue of Huisken's distance comparison estimate (Theorem 3.7) for initial curves with total curvature at least $-\pi$ (and constructed counterexamples whenever this condition is violated). The area-preserving curve shortening flow with free boundary was studied by E. Mäder-Baumdicker [**377, 378**].

6.10.3. Well-posedness. In our presentation of the existence and uniqueness of solutions we did not discuss continuous dependence on initial data. Let $\{\Sigma_i\}_{i\in\mathbb{N}}$ be a sequence of smooth, compact hypersurfaces $\Sigma_i \subset \mathbb{R}^{n+1}$. Suppose that the hypersurfaces Σ_i converge to a limit hypersurface Σ. Then we may ask whether the corresponding mean curvature flows $\{\Sigma^i_t\}_{t\in[0,T_i)}$ converge to the flow $\{\Sigma_t\}_{t\in[0,T)}$ of Σ. The answer is yes, in the sense that $\liminf_{i\to\infty} T_i \geq T$ and $\{\Sigma^i_t\}_{t\in[0,T_i)\cap[0,T)} \to \{\Sigma_t\}_{t\in[0,T)}$ uniformly on compact time subintervals. This may be proved by representing each Σ_i as a normal graph over Σ (for i sufficiently large) and studying the scalar parabolic PDE satisfied by the normal graph heights of the solutions $\{\Sigma^i_t\}_{t\in[0,T_i)}$ (for small t). See [**382**, §1.5]. It is possible, however, that $\liminf_{i\to\infty} T_i > T$. This can be the case, for example, when the mean curvature flow of Σ suffers a degenerate neckpinch singularity (see Chapter 9): If the initial datum is perturbed slightly, the solution can avoid the degenerate neckpinch and shrink to a point at a much later time. On the other hand, Kevin Sonnanburg has proved that the maximal time $\Sigma \mapsto T_\Sigma$ is continuous at a mean convex hypersurface Σ whose mean curvature flow is type-I (see Chapter 11) [**490**].

6.10.4. Blow-up of the mean curvature. We saw, in Theorem 6.20, that the second fundamental form necessarily "blows up" at the final time for a compact solution of mean curvature flow.

Conjecture 6.38. *Let $X : M^n \times [0,T) \to \mathbb{R}^{n+1}$ be a smooth, maximal solution to mean curvature flow (with M^n compact and $n \leq 6$). Then*

$$\limsup_{t\to T} \sup_{M^n\times\{t\}} |H| = \infty .$$

The conjecture is trivially true when the initial hypersurface is mean convex (see Lemma 8.4 in Chapter 8). Cooper has proved that the intermediate quantity $|H\mathrm{II}|$ necessarily blows up [**189**] and Lin and Šešum showed that the conjecture is true when the trace-free second fundamental form (see (8.5)) of the initial hypersurface is small in the L^2 sense [**370**]. (In dimension 2, this is equivalent to smallness of the *Willmore energy*.) Le and Šešum [**358, 359**] proved that the mean curvature blows up at type-I singularities. In particular, for surfaces in $\mathbb{R}^3$, the mean curvature blows up at the first singular time provided that either Ilmanen's multiplicity one conjecture (Conjecture 11.35 in Chapter 11) holds or the Gaußian density is less than two. In a separate paper [**360**], they show that the second fundamental form stays bounded so long as it is bounded pointwise from below, $\mathrm{II} \geq -Bg$, and the mean curvature is bounded in $L^p(M^n \times [0,T))$ for some $p \geq n+2$ (a sharp condition in view of certain counterexamples).

6.10.5. The compact-open C^k topology. Let M and N be smooth manifolds. Given $k \in \mathbb{N} \cup \{0\}$, denote by $C^k(M,N)$ the set of all functions from M to N which are k times continuously differentiable. We equip $C^k(M,N)$ with the *compact-open C^k topology*. This is the topology generated by the neighborhood subbasis consisting of all the sets of the form

$$\mathcal{S}^k(f,\phi,U,\psi,V,K,\varepsilon) \doteqdot \Big\{ g \in C^k(M,N) : g(K) \subset V,$$

$$\max_{|\alpha|\leq k} \sup_{x\in\phi(K)} \left\| \partial_x^\alpha \left(\psi\circ g\circ\phi^{-1}\right)(x) - \partial_x^\alpha\left(\psi\circ f\circ\phi^{-1}\right)(x)\right\| < \varepsilon \Big\}, \tag{6.84}$$

where $f \in C^k(M,N)$, (ϕ,U) and (ψ,V) are charts for M and N, respectively, K is a compact subset of U with $f(K) \subset V$, and $\varepsilon > 0$ [**286**].

If M is compact, then given any $f \in C^k(M,N)$ we can find finitely many charts $\{(\phi_i,U_i)\}_{i=1}^r$ for M and $\{(\psi_i,V_i)\}_{i=1}^r$ for N and compact subsets $\{K_i\}_{i=1}^r$ of M such that $\bigcup_{i=1}^r K_i = M$ and, for each $i = 1,\ldots,r$, $K_i \subset U_i$ and $f(K_i) \subset V_i$. The sets

$$\mathcal{B}(f,\varepsilon) \doteqdot \textstyle\bigcap_{i=1}^r \mathcal{S}^k(f,\phi_i,U_i,\psi_i,V_i,K_i,\varepsilon) \tag{6.85}$$

form a neighborhood basis for $C^k(M,N)$. In particular, a sequence $\{f_j\}_{j\in\mathbb{N}}$ of maps $f_j \in C^k(M,N)$ converges in the compact-open C^k topology (or "in C^k" for short[2]) to a function $f \in C^k(M,N)$ if and only if given any $\varepsilon > 0$ there exists $N \in \mathbb{N}$ such that $f_j \in \mathcal{B}(f,\varepsilon)$ for all $j \geq N$.

The *compact-open C^∞ topology* $C^\infty(M,N) \doteqdot \bigcap_{k=0}^\infty C^k(M,N)$ is the smallest topology under which the inclusions $C^\infty(M,N) \hookrightarrow C^k(M,N)$ are

[2]When M is noncompact, we will also say that the sequence $\{f_j\}_{j\in\mathbb{N}}$ *converges in C^k uniformly on compact subsets of M*.

embeddings. In particular, a sequence $\{f_j\}_{j\in\mathbb{N}}$ of maps $f_j \in C^\infty(M,N)$ converges in C^∞ to a map $f \in C^\infty(M,N)$ if and only if it converges to f in C^k for each k.

Now suppose that (M,γ,∇) and (N,g,D) are Riemannian manifolds equipped with metric connections. Given a map $f \in C^k(M,N)$, we can define a connection ${}^fD$ and a nonnegative definite bilinear form $\langle\cdot,\cdot\rangle_f$ on $\Gamma((T^*M)^m \otimes f^*TN)$ for each $m \in \mathbb{N}$ by extending the metric γ and connection ∇ and the pullbacks of g and D in the usual way. For example, in local coordinates (x^i,U) for M near x and (y^α,V) for N near $f(x)$,

$$\begin{aligned}{}^fD^2 f(\partial_{x^j},\partial_{x^j}) &= f^*D_i(f_*\partial_{x^j}) - f_*\nabla_{x^i}\partial_{x^j}\\ &= f^*D_i\left(\frac{\partial f^\alpha}{\partial x^i}f^*\partial_{y^\alpha}\right) - \Gamma_{ij}^k\frac{\partial f^\alpha}{\partial x^k}f^*\partial_{y^\alpha}\\ &= \left(\frac{\partial^2 f^\gamma}{\partial x^i\partial x^j} + \frac{\partial f^\alpha}{\partial x^i}\frac{\partial f^\beta}{\partial x^j}f^*G_{\alpha\beta}^\gamma - \Gamma_{ij}^k\frac{\partial f^\gamma}{\partial x^k}\right)f^*\partial_{y^\gamma},\end{aligned}$$

where the Γ_{ij}^k are the coefficients of ∇ with respect to (x^i,U) and the $G_{\alpha\beta}^\gamma$ are the coefficients of D with respect to (y^α,V), and

$$\langle S,T\rangle_f = \gamma^{ik}\gamma^{jl}f^*g_{\alpha\beta}S_{ij}^\alpha T_{kl}^\beta$$

for any $S \in T^*M \otimes f^*TN$. More generally,

$$({}^fD^{m+1}f)_{i_1\ldots i_{m+1}}^\gamma = f_{i_1\ldots i_{m+1}}^\gamma + f_{i_{m+1}}^\alpha({}^fD^m f)_{i_1\ldots i_m}^\beta f^*G_{\alpha\beta}^\gamma - \sum_{j=1}^m \Gamma_{i_j i_{m+1}}^k({}^fD^m f)_{i_1\ldots i_{j-1}ki_{j+1}\ldots i_m}^\alpha, \tag{6.86}$$

where $f_{i_1\ldots i_\ell}^\alpha \doteq \frac{(\psi\circ f\circ\phi^{-1})^\alpha}{\partial x^{i_1}\cdots\partial x^{i_\ell}}$, and

$$\langle S,T\rangle = \gamma^{i_1j_1}\cdots\gamma^{i_mj_m}f^*g_{\alpha\beta}S_{i_1\ldots i_m}^\alpha S_{j_1\ldots j_m}^\beta \tag{6.87}$$

for any $S \in (T^*M)^m \otimes f^*TN$.

Proposition 6.39. *Let (M,γ,∇) and (N,g,D) be two smooth Riemannian manifolds equipped with metric connections, with M compact, and let $\{f_j\}_{j\in\mathbb{N}}$ be a sequence of functions $f_j \in C^{k+1}(M,N)$. Suppose that there is a compact subset $L \subset N$ and a constant $C < \infty$ such that $f_j(M) \subset L$ and*

$$\sup_{x\in M}\sum_{m=1}^{k+1}\|{}^{f_j}D^m f_j(x)\| < \infty$$

for every $j \in \mathbb{N}$. Then there is some $f \in C^k(M,N)$ and a subsequence of $\{f_j\}_{j\in\mathbb{N}}$ that converges to f in the compact-open C^k topology.

Proof. First choose finitely many compact subsets $K_1,\ldots,K_\mu$ of M and $L_1,\ldots,L_\nu$ of N whose interiors cover M and L, respectively, and for which charts $\varphi_1 : U_1 \to \mathbb{R}^m, \dots, \varphi_\mu : U_\mu \to \mathbb{R}^m$ and $\psi_1 : V_1 \to \mathbb{R}^n, \dots, \psi_\nu : V_\nu \to \mathbb{R}^n$ for M and N, respectively, can be chosen so that $K_1 \subset U_1, \dots, K_\mu \subset U_\mu$ and $L_1 \subset V_1, \dots, L_\nu \subset V_\mu$. Set $f_j^{pq} \doteqdot \psi_q \circ f_j \circ \varphi_p$. Since the components of the metrics and connections with respect to each coordinate chart are uniformly bounded in each of the corresponding compact sets, the hypotheses, equations (6.86), (6.87), and an induction argument imply that $\sup_{x \in K_p} \|D^m f_j^{pq}\|$ is bounded uniformly in j for each $p = 1, \dots, \mu$, $q = 1, \dots, \nu$, and $m = 1, \dots, k+1$. By the Arzelà–Ascoli theorem and a diagonal subsequence argument, we can pass to a subsequence along which f_j^{pq} converges uniformly in $C^k(\mathrm{int}(K_p), \mathrm{int}(K_q))$ to some limit $f^{pq} \in C^k(\mathrm{int}(K_p), \mathrm{int}(K_q))$ for every p and q. Since the sequences converge pointwise, their limits agree on the overlaps, and hence the limits paste together into a well-defined function $f \in C^k(M, N)$. It follows that the sequence $\{f_j\}_{j \in \mathbb{N}}$ converges to f in the compact-open topology. □

6.10.6. Regularity of the level set flow. Even though solutions of the level set flow (6.83) are a priori only continuous (solving (6.83) in the viscosity sense), Evans and Spruck proved that solutions are actually Lipschitz [**219**]. Huisken proved that solutions corresponding to the mean curvature flow of *convex* hypersurfaces are always C^2. In fact, Kohn and Serfaty proved, using the Gage–Hamilton theorem, that the solutions corresponding to convex curves in the plane are even C^3 [**334**]. On the other hand, Šešum has proved that third derivatives do not exist in general, even for the level set flow of convex hypersurfaces (of dimension at least two) [**477**]. Colding and Minicozzi have proved that, even though the arrival time may fail to be C^2, it is always twice differentiable everywhere, with uniformly bounded second derivative, and smooth away from its critical points (where (6.83) is degenerate) [**184**]. They also proved that it even behaves in certain ways like an analytic function [**185**].

6.10.7. Mean curvature flow of soap film clusters. A natural generalization of the flow of smooth hypersurfaces by mean curvature is the flow of surface clusters, where three hypersurfaces can meet under equal angles, forming a "hyperedge". These hyperedges can themselves then meet on lower-dimensional strata. We discussed the "network flow" of curves in the plane in the Notes and Commentary at the end of Chapter 2.

6.11. Exercises

Exercise 6.1. Prove that mean curvature flow is invariant under space-time translation, rotation, reparametrization, and space-time dilation (see Section 6.2).

Exercise 6.2. Let $X : M^n \times I \to \mathbb{R}^{n+1}$ be a solution to mean curvature flow. Show that each of the functions

$$u_1(x,t) \doteqdot \langle \mathsf{N}(x,t), e\rangle\,, \quad e \in \mathbb{R}^{n+1}\,,$$

$$u_2(x,t) \doteqdot \langle \mathsf{N}(x,t), J \cdot X(x,t)\rangle\,, \quad J \in \mathfrak{so}(n+1)\,,$$

$$u_3(x,t) \doteqdot \langle \mathsf{N}(x,t), X(x,t)\rangle + 2tH(x,t)$$

satisfies the Jacobi equation

$$(\partial_t - \Delta)u = |\mathrm{II}|^2 u\,.$$

Exercise 6.3. Generalize Lemma 5.27 to derive the evolution equations for a hypersurface flow into a Riemannian manifold $\mathcal{P}$. In particular, prove that if $X : M^n \times I \to \mathcal{P}^{n+1}$ satisfies $\partial_t X = -F\mathsf{N}$, then equations (5.140) and (5.141) become

$$\nabla_t \mathrm{II} = \nabla^2 F - F\,\mathrm{II}^2 + F\,\mathrm{Rm}_{\mathcal{P}}(\cdot, \mathsf{N}, \cdot, \mathsf{N}), \tag{6.88}$$

$$\partial_t H = \Delta F + F\left(|\mathrm{II}|^2 + \mathrm{Rc}_{\mathcal{P}}(\mathsf{N}, \mathsf{N})\right), \tag{6.89}$$

where $\mathrm{Rm}_{\mathcal{P}}$ and $\mathrm{Rc}_{\mathcal{P}}$ denote the Riemannian and Ricci curvatures of the target manifold $\mathcal{P}$.

Exercise 6.4 (Mean curvature flow coupled to Ricci flow)**.** Let $X : M^n \times I \to \mathcal{P}^{n+1}$ be a hypersurface evolving by the mean curvature flow, where the metric $g_{\mathcal{P}}(t)$ on $\mathcal{P}$ evolves by the Ricci flow:

$$\partial_t g_{\mathcal{P}} = -2\,\mathrm{Rc}\,(g_{\mathcal{P}})\,.$$

Compute $\partial_t \mathrm{II}_{ij}$. HINT: Note that the terms

$$-D_i\left(\mathrm{Rc}_{\mathcal{P}}\right)_{j\nu} - D_j\left(\mathrm{Rc}_{\mathcal{P}}\right)_{i\nu} + D_\nu\left(\mathrm{Rc}_{\mathcal{P}}\right)_{ij}$$

are canceled by new terms introduced by the Ricci flow. These terms represent the evolution of the Christoffel symbols under the Ricci flow.

Exercise 6.5. Show that if in (6.37) we take $Y = \nabla\left(-\frac{|X|^2}{2}\right) = -X^\top$, then we may rewrite (6.37) as

$$\partial_t\left(e^t X\right) = -\left(e^t f - \left\langle e^t X, \mathsf{N}\right\rangle\right)\mathsf{N}.$$

Exercise 6.6. Prove Proposition 6.8 for $\mathcal{M}_0^m \subset \mathcal{P}^n$, where $\mathcal{P}$ is a general Riemannian manifold. To do so replace $g_{\mathbb{R}^n}$ with $g_{\mathcal{P}^n}$, so that the mean curvature flow for a family of immersions $X : M^m \times I \to \mathcal{P}^n$ becomes

$$\partial_t X = \Delta_{g, g_{\mathcal{P}}} X\,,$$

where g is the metric induced by X.

Exercise 6.7. Prove Lemmas 6.11 and 6.12.

Exercise 6.8. Prove Lemma 6.23 for mean curvature flow in a general ambient manifold $\mathcal{P}$.

Exercise 6.9. Let M be a smooth, compact manifold, let N be a smooth manifold, and let k be a positive integer. Verify that the sets defined by (6.85) do indeed define a basis for the compact-open C^k topology on $C^k(M, N)$.

Exercise 6.10. Prove that the function $u : \mathbb{R}^{n+1} \to \mathbb{R}$ defined by $u(x) \doteqdot -\frac{1}{2n}|x|^2$ is a solution of the level set flow.

Exercise 6.11. Show that the level set flow of convex curves in $\mathbb{R}^2$ is equivalent to the equation

$$u_{xx}u_y^2 + u_{yy}u_x^2 - 2u_{xy}u_xu_y + u_x^2 + u_y^2 = 0\,. \tag{6.90}$$

Exercise 6.12. Show that the level set flow $u : \mathbb{R}^2 \to \mathbb{R}$ of a convex, (unit speed) translating solution of curve shortening flow satisfies the equations

$$u_y = 1 \text{ and } u_{xx} + u_x^2 + 1 = 0\,.$$

Find all solutions.

Exercise 6.13. Find all convex solutions $u : \Omega \subset \mathbb{R}^2 \to \mathbb{R}$ of the level set flow satisfying the ansatz

$$Du(x, y) = (F(x), G(y))\,.$$

Exercise 6.14. Show that the family of 1-sheeted hyperboloids

$$\mathrm{H}_t \doteqdot \{x \in \mathbb{R}^{n+1} : x_1^2 = x_2^2 + \cdots + x_{n+1}^2 - r^2 + 2nt\}, \quad t < T_r \doteqdot r^2/2n,$$

satisfies

$$-|Du|\operatorname{div}\left(\frac{Du}{|Du|}\right) \leq 1 \text{ in } \Omega_0 \setminus \{0\},$$

where $\Omega_t \doteqdot \{x \in \mathbb{R}^{n+1} : x_1^2 \leq x_2^2 + \cdots + x_{n+1}^2 - r^2 + 2nt\}$. Deduce that $\{\mathrm{H}_t\}_{t<T_r}$ satisfies the avoidance principle from the inside: Every smooth solution $\{\mathcal{M}_t\}_{t\in[t_0,T)}$ to mean curvature flow with $\mathcal{M}_0 \subset \Omega_0$ satisfies $\mathcal{M}_t \subset \Omega_t$ for all $t \in [t_0, T) \cap (-\infty, T_r)$. We call $\{\mathrm{H}_t\}_{t<t_r}$ a **subsolution** of mean curvature flow.

Chapter 7

Mean Curvature Flow of Entire Graphs

In this chapter we study the evolution of entire graphs under mean curvature flow, following the work of Klaus Ecker and Gerhard Huisken. Their analysis is based on a priori estimates on solutions and their derivatives, obtained via the maximum principle.

7.1. Introduction

In their papers [**208, 209**], Ecker and Huisken proved the global existence of smooth graphical solutions of mean curvature flow for Lipschitz continuous initial data. Moreover, they proved the convergence (after rescaling) to a self-similar solution for data that are "straight at infinity".[1] The asymptotic result is shown to be optimal through construction of a slowly oscillating initial surface that does not converge asymptotically. Furthermore, the only entire graphs that are shrinking self-similar solutions are hyperplanes.

We follow their arguments closely, diverging only in the proof of a sharp version of their gradient estimate (discovered independently by L. C. Evans and J. Spruck [**220**, Theorem 5.2] and T. Colding and W. Minicozzi [**180**]).

Theorem 7.1. *Let $\mathcal{M}_0 = X_0(\mathbb{R}^n)$ be a locally Lipschitz continuous entire graph. Then there exists a solution $X : \mathbb{R}^n \times [0,\infty) \to \mathbb{R}^{n+1}$ of mean curvature flow with $X \in C^0(\mathbb{R}^n \times [0,\infty)) \cap C^\infty(\mathbb{R}^n \times (0,\infty))$ and $X(0) = X_0$. Moreover, each $\mathcal{M}_t = X_t(\mathbb{R}^n)$ is an entire graph.*

[1]**Straight at infinity** means that there exists $\delta > 0$ such that $|\langle X, \mathsf{N}\rangle| \le \delta^{-1}(|X|+1)^{1-\delta}$ on $\mathcal{M}$. In particular, the normal component X grows more slowly than $|X|$.

There are several ingredients in the proof (which will be completed in the following sections). We first assume that X_0 is the graph of a $C^{2,\alpha}$ function u_0 in order to be able to apply known PDE existence results. Shrinking spherical barriers are then used to prove that the height $u(\cdot,t)$ in any ball is bounded by its maximum at time $t=0$ on a slightly larger ball (Proposition 7.7). We next derive a key gradient estimate (Theorem 7.8 below), in which $|\nabla u|$ is bounded locally in terms of u. Finally, we derive local a priori bounds on the second fundamental form and its derivatives (Theorems 7.13 and 7.14 below). Putting these together, we obtain a sequence of smooth local graphical solutions u over balls of radii R_k and take a limit as $R_k \to \infty$ to find an entire graphical solution $u \in C^0(\mathbb{R}^n \times [0,\infty)) \cap C^\infty(\mathbb{R}^n \times (0,\infty))$. Since the derivatives depend only on the initial height and gradient, an approximation argument implies a smooth solution for locally Lipschitz initial data.

7.2. Preliminary calculations

Let Ω be an open subset of $\mathbb{R}^n$, let $I \subset \mathbb{R}$ be an open interval, and let $u : \Omega \times I \to \mathbb{R}$ be a *graphical solution* to mean curvature flow, as defined in equation (6.35). That is, assume that u satisfies the equation

$$\partial_t u = \sqrt{1+|\nabla u|^2}\,\operatorname{div}\left(\frac{\nabla u}{\sqrt{1+|\nabla u|^2}}\right) \quad \text{in } \Omega \times I, \tag{7.1}$$

where ∇, div, and the norm are all with respect to the Euclidean metric on $\mathbb{R}^n$. Then the map $G : \Omega \times I \to \mathbb{R}^{n+1}$ given by $G(x,t) = (x, u(x,t))$ is a solution of the flow

$$(\partial_t G)^\perp = -H\mathsf{N},$$

which, as we have seen, is diffeomorphism equivalent to mean curvature flow.

We shall study such graphical solutions via the original parametric mean curvature flow

$$\partial_t X = -H\mathsf{N}\,. \tag{7.2}$$

In particular, given any graphical solution G, any time t_0 in the interior of I, and any open set U whose closure is in the interior of Ω, there exist an $\varepsilon > 0$ and a solution $X : U \times (t_0-\varepsilon, t_0+\varepsilon) \to \mathbb{R}^{n+1}$ to mean curvature flow which is diffeomorphism equivalent to G; that is, there exists a 1-parameter family $\Phi : U \times (t_0-\varepsilon, t_0+\varepsilon) \to \Omega$ of embeddings $\Phi(\cdot,t) : U \to \Omega$ such that $\Phi(\cdot,t_0)$ is the inclusion map and

$$X = \Phi^* G \tag{7.3}$$

in the sense that $X(x,t) = G(\Phi(x,t),t)$. Note that $DG(\partial_t\Phi) = -\Phi^*(\partial_t G)^\top$.

Given a fixed unit vector $\omega \in \mathbb{R}^{n+1} \setminus \{0\}$, let $\langle X, \omega \rangle$ be the **height** in the direction ω (over the hyperplane $\omega^\perp \doteqdot \{V \in \mathbb{R}^{n+1} : \langle V, \omega \rangle = 0\}$). Note that $\langle X, \omega \rangle$ is equal to u composed with a time-dependent diffeomorphism of Ω. The difference is that $u(y,t)$ is the height above a fixed point y in the hyperplane orthogonal to ω and $\langle X, \omega \rangle$ is the height of $X(x,t)$ for fixed $x \in \Omega$.

We first derive the evolution equations for the distance, support, and coordinate (height) functions which hold for arbitrary solutions of mean curvature flow. Denote $\mathcal{M}_t = X(\Omega, t)$ and observe that since $\langle \mathsf{N}, e_{n+1} \rangle > 0$ our choice of unit normal is

$$\mathsf{N} = \frac{(-\nabla u, 1)}{\sqrt{1 + |\nabla u|^2}}.$$

Lemma 7.2. *Under the (parametric) mean curvature flow,*

$$\partial_t |X|^2 = \Delta |X|^2 - 2n, \tag{7.4}$$

$$\partial_t \langle X, \mathsf{N} \rangle = \Delta \langle X, \mathsf{N} \rangle + |\mathrm{II}|^2 \langle X, \mathsf{N} \rangle - 2H, \tag{7.5}$$

$$\partial_t \langle X, \omega \rangle = \Delta \langle X, \omega \rangle, \tag{7.6}$$

where Δ is the Laplacian with respect to the induced metric.

Proof. (1) From (7.2) we compute

$$\partial_t |X|^2 = 2 \langle X, -H\mathsf{N} \rangle,$$

whereas by (5.39) we have

$$\Delta |X|^2 = -2H \langle X, \mathsf{N} \rangle + 2n.$$

Formula (7.4) follows immediately.

(2) We compute using (6.16) that the time derivative of $\langle X, \mathsf{N} \rangle$ is

$$\begin{aligned} \partial_t \langle X, \mathsf{N} \rangle &= \left\langle \frac{\partial X}{\partial t}, \mathsf{N} \right\rangle + \left\langle X, \frac{\partial \mathsf{N}}{\partial t} \right\rangle \\ &= -H + \langle X, \nabla H \rangle. \end{aligned}$$

On the other hand, by (5.27), the gradient of $\langle X, \mathsf{N} \rangle$ is determined by

$$\begin{aligned} \nabla_j \langle X, \mathsf{N} \rangle &= \left\langle \frac{\partial X}{\partial x^j}, \mathsf{N} \right\rangle + \left\langle X, \frac{\partial \mathsf{N}}{\partial x^j} \right\rangle \\ &= \mathrm{II}_{jk}\, g^{k\ell} \left\langle X, \frac{\partial X}{\partial x^\ell} \right\rangle = \frac{1}{2} \mathrm{II}_{jk}\, g^{k\ell} \nabla_\ell |X|^2. \end{aligned} \tag{7.7}$$

Invariantly, $\nabla \langle X, \mathsf{N}\rangle = \mathrm{II}\left(X^\top\right) = \frac{1}{2}\mathrm{II}(\nabla |X|^2)$, where $X^\top = X - \langle X, \mathsf{N}\rangle \mathsf{N}$. The Hessian of $|X|^2$ is determined by

$$\begin{aligned} \nabla_i \nabla_j |X|^2 &= 2\langle \nabla_i \nabla_j X, X\rangle + 2\langle \nabla_i X, \nabla_j X\rangle \\ &= -2\,\mathrm{II}_{ij}\langle X, \mathsf{N}\rangle + 2g_{ij}. \end{aligned}$$

So, by differentiating (7.7) and using the Codazzi identity, we have that the Laplacian of $\langle X, \mathsf{N}\rangle$ is

$$\begin{aligned} \Delta \langle X, \mathsf{N}\rangle &= g^{ij}\nabla_i \nabla_j \langle X, \mathsf{N}\rangle \\ &= \frac{1}{2} g^{ij}\nabla_i \,\mathrm{II}_{jk}\, g^{k\ell}\nabla_\ell |X|^2 + \frac{1}{2} g^{ij}\,\mathrm{II}_{jk}\, g^{k\ell}\nabla_i \nabla_\ell |X|^2 \\ &= g^{k\ell}\nabla_k H \left\langle X, \frac{\partial X}{\partial x^\ell}\right\rangle + g^{ij}\,\mathrm{II}_{jk}\, g^{k\ell}\left(-\,\mathrm{II}_{i\ell}\langle X, \mathsf{N}\rangle + g_{i\ell}\right) \\ &= \langle X, \nabla H\rangle - |\mathrm{II}|^2 \langle X, \mathsf{N}\rangle + H. \end{aligned}$$

Combining these equations yields (7.5).

(3) Formula (7.6) follows from taking the inner product of the equation $\partial_t X = \Delta X$ with ω. □

Corollary 7.3. *Under the mean curvature flow,*

$$(\partial_t - \Delta)\left(|X|^2 + 2nt\right) = 0, \tag{7.8}$$

$$(\partial_t - \Delta)\left(H + \langle X, \mathsf{N}\rangle\right) = |\mathrm{II}|^2 \left(H + \langle X, \mathsf{N}\rangle\right) - 2H. \tag{7.9}$$

Remark 7.4. In particular, (7.8) implies that if M^n is compact and if $X : M^n \to \mathbb{R}^{n+1}$ satisfies $|X| \leq C$ at $t = 0$, then $|X|^2 + 2nt \leq C^2$ for all $t \geq 0$; i.e.,

$$|X_t| \leq \sqrt{C^2 - 2nt} \tag{7.10}$$

(this also implies $t \leq \frac{C^2}{2n}$). The geometric interpretation of (7.10) is that this is simply a consequence of the avoidance principle applied to the solution X_t and the standard shrinking sphere $\bar{X}_t$ with $\left|\bar{X}_0\right| = C$, which satisfies (6.3), so $\left|\bar{X}_t\right| = \sqrt{C^2 - 2nt}$.

From (7.6), we deduce for $\langle X, \omega\rangle$ that

$$\begin{aligned} \partial_t\left(\langle X, \omega\rangle^2\right) &= \Delta\left(\langle X, \omega\rangle^2\right) - 2\left|\nabla \langle X, \omega\rangle\right|^2 \\ &= \Delta\left(\langle X, \omega\rangle^2\right) - 2\left|\omega^\top\right|^2 \end{aligned} \tag{7.11}$$

since $\nabla \langle X, \omega\rangle = \omega^\top$. Let

$$z = |X|^2 - \langle X, \omega\rangle^2 + 2nt. \tag{7.12}$$

By (7.8) and (7.11), we have

$$\partial_t z \geq \Delta z. \tag{7.13}$$

Now consider $\langle \mathsf{N}, \omega \rangle$. Note that $\cos^{-1} \langle \mathsf{N}, \omega \rangle$ measures the angle between the tangent space $T\mathcal{M}_t$ and the horizontal hyperplane $\omega^{\perp}$. We need to estimate this quantity from below so that we ensure that the solution stays a graph. We have

$$\partial_t \langle \mathsf{N}, \omega \rangle = \langle \nabla H, \omega \rangle = \left\langle \nabla H, \omega^{\top} \right\rangle$$

and

$$\nabla \langle \mathsf{N}, \omega \rangle = \mathrm{II}(\omega^{\top}), \tag{7.14}$$

$$\Delta \langle \mathsf{N}, \omega \rangle = \langle \nabla H, \omega^{\top} \rangle - |\mathrm{II}|^2 \langle \mathsf{N}, \omega \rangle . \tag{7.15}$$

To derive the last formula we used $\omega^{\top} = \omega - \langle \omega, \mathsf{N} \rangle \mathsf{N}$, which implies

$$\nabla_i \omega^{\top} = - \langle \omega, \mathsf{N} \rangle \nabla_i \mathsf{N} = - \langle \omega, \mathsf{N} \rangle \mathrm{II}(\partial_i),$$

so that $\mathrm{II}_{ij} \nabla_i \left(\omega^{\top}\right)_j = - |\mathrm{II}|^2 \langle \omega, \mathsf{N} \rangle = - |\mathrm{II}|^2 \langle \mathsf{N}, \omega \rangle$. Hence,

$$\partial_t \langle \mathsf{N}, \omega \rangle = \Delta \langle \mathsf{N}, \omega \rangle + |\mathrm{II}|^2 \langle \mathsf{N}, \omega \rangle . \tag{7.16}$$

By the maximum principle, if $\langle \mathsf{N}, \omega \rangle \geq c > 0$ at $t = 0$, then $\langle \mathsf{N}, \omega \rangle \geq c$ for all $t \geq 0$. The equation for $\langle \mathsf{N}, \omega \rangle$ also implies that

$$v \doteqdot \langle \mathsf{N}, \omega \rangle^{-1} \tag{7.17}$$

satisfies

$$\begin{aligned} \partial_t v &= - \langle \mathsf{N}, \omega \rangle^{-2} \partial_t \langle \mathsf{N}, \omega \rangle \\ &= -v^2 \left(\Delta \left(v^{-1}\right) + |\mathrm{II}|^2 v^{-1} \right) \\ &= \Delta v - 2v^{-1} |\nabla v|^2 - |\mathrm{II}|^2 v. \end{aligned} \tag{7.18}$$

By the maximum principle, if $v \leq C$ at $t = 0$, then $v \leq C$ for all $t \geq 0$.

Remark 7.5. Note that by definition, $v \geq |\omega|^{-1} = 1$.

7.3. The Dirichlet problem

Fix $R > 0$, and assume that $u_0 \in C^{2,\alpha}(\mathbb{R}^n)$. The following result provides smooth solutions, for all $t > 0$, to the Dirichlet problem for the graphical mean curvature flow over the ball $B_R \doteqdot B_R(0)$ of radius R centered at the origin in $\mathbb{R}^n$.

Theorem 7.6. *Let $G_0(x) = (x, u_0(x))$, $x \in \mathbb{R}^n$, be an entire graph with $u_0 \in C^{2,\alpha}(\mathbb{R}^n)$. For each fixed $R > 0$, the initial-boundary value problem for the graphical mean curvature flow,*

$$(\partial_t G(x,t))^{\perp} = -H(x,t)\mathsf{N}(x,t)\,, \quad (x,t) \in B_R \times (0,\infty)\,, \tag{7.19a}$$

$$G(x,0) = G_0(x), \quad x \in B_R\,, \tag{7.19b}$$

$$G(x,t) = G_0(x), \quad (x,t) \in \partial B_R \times [0,\infty)\,, \tag{7.19c}$$

where $G(x,t) = (x, u(x,t))$, admits a unique solution $u \in C^{2,\alpha}(\bar{B}_R \times [0,\infty)) \cap C^{\infty}(B_R \times (0,\infty))$.[2]

For the proof of this theorem, we refer the reader to Theorem 2.1 in Huisken [**295**], which holds in the more general case of weakly mean convex bounded domains with $C^{2,\alpha}$ boundary. See O. A. Ladyženskaja, V. A. Solonnikov, and N. N. Ural′ceva [**343**] or G. Lieberman [**368**] for the parabolic Schauder theory. We remark that (7.19a)–(7.19c) is the same as (7.1) together with initial-boundary conditions; i.e.,

$$\partial_t u = \sqrt{1+|\nabla u|^2}\,\operatorname{div}\left(\frac{\nabla u}{\sqrt{1+|\nabla u|^2}}\right) \quad \text{in } B_R \times (0,\infty)\,,$$

$$u(x,t) = u_0(x) \quad \text{on } (B_R \times \{0\}) \cup (\partial B_R \times (0,\infty))\,.$$

Huisken further proved that the solution converges as $t \to \infty$ to a minimal graph (i.e., a graphical hypersurface with $H \equiv 0$) over $\bar{B}_R$ with boundary values $u_0|_{\partial B_R}$. Recall from (7.3) that (7.19a) is diffeomorphism equivalent to the parametrized mean curvature flow (7.2).

7.4. A priori height and gradient estimates

We next derive a height bound, following T. Colding and W. Minicozzi [**180**]. We assume for ease of computation that the solution is locally a graph over a ball contained in $\mathbb{R}^n$ centered at the origin.

Let $\rho \geq \sqrt{2n+1}$ and assume that the graph of the time-dependent function

$$u : B_{\rho r} \times [0, r^2] \to \mathbb{R}$$

satisfies the mean curvature flow (7.1). For simplicity, assume first that $r = 1$. Consider the round sphere in $\mathbb{R}^{n+1}$ of radius ρ centered at the point $(0, y_0) \in \mathbb{R}^n \times \mathbb{R}$, where $y_0 \doteqdot \rho + \max_{x \in B_\rho} u(x,0) + \varepsilon$. By construction, this does not intersect the graph $G_0(x) = (x, u(x,0))$. If one evolves the round sphere by mean curvature flow, it generates concentric spheres S_t^n centered at $(0, y_0)$, with radii $r(t) = \sqrt{\rho^2 - 2nt}$. Now consider any time $0 \leq t \leq 1$. By the avoidance principle (Theorem 6.16), the graphical solution

[2]The $C^{2,\alpha}$-space is the standard parabolic Hölder space.

$G_t(x) = (x, u(x,t))$ and the spheres S_t^n must remain disjoint. The equation for a point (x, y) on S_t^n is

$$|x|^2 + (y - (\rho + \max_{B_\rho} u(\cdot, 0) + \varepsilon))^2 = \rho^2 - 2nt \geq \rho^2 - 2n,$$

so that on the part of the southern hemisphere of S_t^n over B_1 we have

$$y \leq \rho + \max_{B_\rho} u(\cdot, 0) + \varepsilon - (\rho^2 - (2n+1))^{1/2}.$$

By the avoidance principle and letting $\varepsilon \to 0$, this implies that

$$\max_{B_1 \times [0,1]} u(x,t) \leq \max_{B_\rho} u(\cdot, 0) + \rho - (\rho^2 - (2n+1))^{1/2}.$$

The above, together with the analogous argument for the minimum of $u(x,t)$ and scaling, yields the desired bound:

Proposition 7.7. *If $u : B_{\rho r} \times [0, r^2] \to \mathbb{R}$, where $\rho \geq \sqrt{2n+1}$, is a solution to the graphical mean curvature flow, then*

$$\begin{aligned} \max_{B_r \times [0,r^2]} |u(x,t)| &\leq r\left(\rho - (\rho^2 - (2n+1))^{1/2}\right) + \max_{B_{\rho r}} |u(\cdot, 0)| \\ &\leq \frac{(2n+1)r}{\rho} + \max_{B_{\rho r}} |u(\cdot, 0)|. \end{aligned} \tag{7.20}$$

We next derive the sharp gradient estimate for $|\nabla u|$ in terms of u discovered independently by Evans and Spruck [**220**, Theorem 5.2] and by Colding and Minicozzi [**180**] (which improves the original estimate of Ecker and Huisken [**209**, Theorem 2.3]).

Theorem 7.8. *Assume that $X_0(x) = (x, u(x, 0))$ is a graph over a ball $B_{\sqrt{2n+1}\,r}(0)$ and that the graph of $u : B_{\sqrt{2n+1}\,r}(0) \times [0, r^2] \to \mathbb{R}$ evolves by mean curvature flow. Then*

$$\log |\nabla u| \left(0, r^2/4n\right) \leq C(n) \left(1 + r^{-1} \|u(\cdot, 0)\|_\infty\right)^2, \tag{7.21}$$

with the convention that $\log 0 = -\infty$.

We follow the treatment in [**180**]. The result follows from three lemmas. It is convenient to work with $v \doteqdot \langle \mathsf{N}, \omega \rangle^{-1} = \sqrt{1 + |\nabla u|^2}$, instead of $|\nabla u|$.

Lemma 7.9 ([**180**, Lemma 1])**.** *Let $u : \Omega \times [0, T] \to \mathbb{R}$ be a graphical solution to mean curvature flow. Define $v = \sqrt{1 + |\nabla u|^2}$. If a function ϕ on $\Omega \times [0, T]$ satisfies $\phi \leq 0$ on $(\Omega \times \{0\}) \cup (\partial\Omega \times [0, T])$ and $\max_{\Omega \times [0,T]} \phi v > 0$, then*

$$(\partial_t - \Delta)\phi \geq 0$$

at any maximum point of ϕv, where the Laplacian is with respect to $g(t)$.

Proof. At a positive maximum point of ϕv, which must be in the interior since $\phi \le 0$ on the boundary, we know that

$$0 = \nabla(\phi v) = v\nabla\phi + \phi\nabla v \quad \text{and}$$

$$0 \le (\partial_t - \Delta)(\phi v) = \phi(\partial_t - \Delta)v + v(\partial_t - \Delta)\phi - 2\langle\nabla\phi, \nabla v\rangle .$$

Since $(\partial_t - \Delta)v \le -2|\nabla v|^2/v$ for the parametric mean curvature flow by (7.18), at a positive maximum point of ϕv we have

$$v^2(\partial_t - \Delta)\phi \ge -\phi v(\partial_t - \Delta)v + 2v\langle\nabla\phi, \nabla v\rangle \ge 0. \qquad \square$$

Recall from (5.62), (5.63), and (5.60) that

$$g_{ij} = \delta_{ij} + \partial_i u\, \partial_j u, \quad g^{ij} = \delta_{ij} - \frac{\partial_i u\, \partial_j u}{1+|\nabla u|^2}, \quad \mathrm{II}_{ij} = \frac{\partial_i\partial_j u}{\sqrt{1+|\nabla u|^2}}. \tag{7.22}$$

Set $y \doteqdot \langle X, \omega\rangle$, so that $y : \mathcal{M} \to \mathbb{R}$ is a reparametrization of $u : \mathbb{R}^n \to \mathbb{R}$. Let $|\cdot|_g$ and $|\cdot|_{\mathbb{R}^n}$ denote the norms with respect to the pulled-back metric and the Euclidean metric, respectively. By (7.22),

$$g^{ij} \ge \frac{1}{1+|\nabla u|^2_{\mathbb{R}^n}}\, \delta_{ij} .$$

Hence, for any 1-form α in $\mathbb{R}^n$, $|\alpha|_g \ge (1+|\nabla u|^2_{\mathbb{R}^n})^{-1/2}|\alpha|_{\mathbb{R}^n}$. In particular,

$$|\nabla y|_g \ge \frac{|\nabla u|_{\mathbb{R}^n}}{\sqrt{1+|\nabla u|^2_{\mathbb{R}^n}}}. \tag{7.23}$$

In the following, we shall often omit the norm subscripts in our notation.

The next lemma is a consequence of Lemma 7.2. Recall that $z \doteqdot |X|^2 - \langle X, \omega\rangle^2 + 2nt$ and that $y \doteqdot \langle X, \omega\rangle$ is the parametrized version of u.

Lemma 7.10 ([**180**, Lemma 2]). *Given $a \in \mathbb{R}$, set $\phi = \eta e^{ay^2/t}$, where $\eta \doteqdot 1 - z$. If $u : B_1 \times [0,1] \to \mathbb{R}$ is a graphical solution of mean curvature flow, then under the associated parametric flow*

$$(\partial_t - \Delta)\phi \le -\frac{e^{ay^2/t}}{t^2}\left(ay^2\eta + (4a^2y^2 + 2at)\eta\frac{|\nabla u|^2}{1+|\nabla u|^2} - 8|ay|t\frac{|\nabla u|}{1+|\nabla u|^2}\right),$$

where the Laplacian $\Delta = \Delta_g$ is with respect to the pulled-back metric and where the norms $|\cdot| = |\cdot|_{\mathbb{R}^n}$ are with respect to the Euclidean metric. Here we suppressed the pulling back by diffeomorphisms of the $|\nabla u|$ terms.

Proof. By (7.11), we have

$$(\partial_t - \Delta)\left(y^2\right) = -2|\nabla y|^2_g .$$

From this we compute that

$$\begin{aligned}(\partial_t - \Delta)\,\mathrm{e}^{ay^2/t} &= -\frac{ay^2}{t^2}\mathrm{e}^{ay^2/t} + \mathrm{e}^{ay^2/t}\left(\frac{a}{t}(\partial_t - \Delta)(y^2) - \frac{4a^2y^2}{t^2}|\nabla y|_g^2\right)\\ &= -\frac{\mathrm{e}^{ay^2/t}}{t^2}\left(ay^2 + (4a^2y^2 + 2at)|\nabla y|_g^2\right).\end{aligned}$$

By (7.13), we have $(\partial_t - \Delta)\,\eta \leq 0$. Hence

$$\begin{aligned}(\partial_t - \Delta)(\eta \mathrm{e}^{ay^2/t}) &\leq \eta(\partial_t - \Delta)\mathrm{e}^{ay^2/t} - 4\frac{ay}{t}\mathrm{e}^{ay^2/t}\langle \nabla\eta, \nabla y\rangle_g\\ &= -\frac{\mathrm{e}^{ay^2/t}}{t^2}(ay^2\eta + (4a^2y^2 + 2at)\eta|\nabla y|_g^2 - 4ayt\langle \nabla z, \nabla y\rangle_g).\end{aligned}$$

The lemma now follows from (7.23) and $|\nabla z|_{\mathbb{R}^n} = 2|X - \langle X, \omega\rangle\omega|_{\mathbb{R}^n} \leq 2$ (since we have assumed that the solution is defined in B_1). □

Lemma 7.11 ([**180**, Proposition 2]). *If the graph of $u : B_r \times [0, r^2] \to \mathbb{R}$ flows by mean curvature, then*

$$\log\left(1 + |\nabla u|_{\mathbb{R}^n}^2(0, r^2/4n)\right) \leq \log 40 + 16n\left(1 + 2r^{-1}\|u\|_\infty\right)^2.$$

Proof. By scaling (see Exercise 7.1), one only needs to prove this for $r = 1$. Let $a \leq -2$ be a constant to be chosen below. Translating $u \mapsto u + \|u\|_\infty + 1$, it suffices to consider the case where $u \geq 1$ and hence $\phi(x,t) = \eta \mathrm{e}^{au^2(x,t)/t}$ converges uniformly to zero as $t \to 0$. Observe also that $\phi \leq 0$ on $\partial B_1 \times [0,1]$. The previous two lemmas yield

$$ay^2\eta + (4a^2y^2 + 2at_0)\eta\frac{|\nabla u|^2}{1 + |\nabla u|^2} - 8|ay|t_0\frac{|\nabla u|}{1 + |\nabla u|^2} \leq 0$$

at a maximum point (x_0, t_0) of ϕv. Rearranging yields

$$(4a^2y^2 + ay^2 + 2at_0)(\eta|\nabla u|)^2 - 8|ay|t_0(\eta|\nabla u|) + ay^2\eta^2 \leq 0.$$

This quadratic inequality yields a bound for $|\nabla u|$ as follows. Since $y \geq 1$, $t_0 \leq 1$, and $a \leq -2$, we have $4a^2y^2 + ay^2 + 2at_0 \geq 2a^2y^2$. Thus,

$$2X^2 - 8t_0X + ay^2\eta^2 \leq 0,$$

where $X \doteqdot |ay|\eta|\nabla u|(x_0, t_0)$. This implies

$$X \leq \tfrac{1}{4}\left(8t_0 + \sqrt{(8t_0)^2 - 8ay^2\eta^2}\right) \leq 4 + \sqrt{|a|}\,y\,|\eta|,$$

which easily yields

$$\eta|\nabla u|(x_0, t_0) \leq 3.$$

Since $a < 0$ and $\eta \leq 1$, we conclude that

$$\begin{aligned}\max_{B_1\times[0,1]} \phi v = \phi v(x_0, t_0) &= \eta \mathrm{e}^{au^2(x_0,t_0)/t_0}\sqrt{1 + |\nabla u|^2(x_0, t_0)}\\ &\leq \sqrt{1 + (\eta|\nabla u|)^2(x_0, t_0)} \leq \sqrt{10}.\end{aligned}$$

Taking $a = -2$ and applying the estimate at $(0, 1/4n)$ (where $\eta = \frac{1}{2}$) yields

$$\log(1 + |\nabla u|^2)(0, 1/4n) \le \log 40 + 16nu^2(0, 1/4n). \qquad \square$$

Proof of Theorem 7.8. Lemmas 7.9, 7.10, and 7.11 together with inequality (7.20) prove the theorem. $\square$

Remark 7.12. Using Grim Reapers as barriers, Colding and Minicozzi showed that the inequality in Theorem 7.8 is sharp [**180**, Proposition 1].

7.5. Local a priori estimates for the curvature

Next, we need to localize the estimates for the second fundamental form and its derivatives obtained in Theorem 6.24.

Recall that a map is **proper** if the preimage of a compact set is compact. We say that a solution $X : M^n \times I \to \mathbb{R}^{n+1}$ of mean curvature flow is **properly defined** in a cylinder $U \times J$, where U is an open subset of $\mathbb{R}^{n+1}$ and J an interval, if $J \subset I$ and $X_t|_{X_t^{-1}(U)} : X_t^{-1}(U) \subset M^n \to U$ is a proper map for $t \in J$. We shall also denote by

$$P_R(x,t) \doteqdot B_R(x) \times (t - R^2, t]$$

the (**extrinsic**) **parabolic cylinder** of radius $R>0$ about $(x,t) \in \mathbb{R}^{n+1} \times \mathbb{R}$.

Figure 7.1. Klaus Ecker. Photo courtesy of Anja Gruber.

We first obtain a local bound for the curvature depending on a bound for the (local) graphical representation.

Theorem 7.13 ([**209**, Corollary 3.2(ii)], [**205**, Proposition 3.21]). *Suppose that $\{\mathcal{M}^n\}_{t \in I}$ is a solution of mean curvature flow which is properly defined in a parabolic cylinder $P_R(x_0, t_0)$. If*[3] $\sup_{P_R(x_0,t_0)} v^2 \le V$*, where*

[3]This implies that the mean curvature flow is a graphical when restricted to $P_R(x_0, t_0)$.

$v = \langle \mathsf{N}, \omega\rangle^{-1}$, $\omega \in S^n$, then

$$\sup_{P_{\theta R}(x_0,t_0)} |\mathrm{II}|^2 \le CR^{-2}, \tag{7.24}$$

where $C = c(1-\theta^2)^{-2}V^2$, with the constant c depending only on n.

Proof. First observe that, by Kato's inequality $|\nabla|\mathrm{II}|| \le |\nabla \mathrm{II}|$, we can estimate (6.22) by

$$(\partial_t - \Delta)|\mathrm{II}|^2 \le 2|\mathrm{II}|^4 - 2|\nabla|\mathrm{II}||^2. \tag{7.25}$$

The reaction term $|\mathrm{II}|^4 = |\mathrm{II}|^2 \cdot |\mathrm{II}|^2$ on the right-hand side is a bad term. We "improve" this equation by finding a suitable multiplicative factor. Define the function $\varphi(s) = \frac{s}{1-ks}$, where k is a positive constant to be chosen below. Write, for short,

$$\varphi = \varphi(v^2), \quad \varphi' = \varphi'(v^2), \quad \text{and} \quad \varphi'' = \varphi''(v^2),$$

where $v = \langle \mathsf{N}, \omega\rangle^{-1} \ge 1$. Note that $\varphi(v^2) \ge 1$ is well-defined provided $v^2 < k^{-1}$. From (7.25) and (7.18) we compute, for $1 - kv^2 > 0$,

$$\begin{aligned}
(\partial_t - \Delta)(|\mathrm{II}|^2\varphi) &= \varphi\,(\partial_t - \Delta)|\mathrm{II}|^2 + |\mathrm{II}|^2\varphi'\,(\partial_t - \Delta)(v^2) \\
&\quad -|\mathrm{II}|^2\varphi''\,|\nabla(v^2)|^2 - 2\nabla|\mathrm{II}|^2\cdot\nabla\varphi \\
&\le \varphi\,(2|\mathrm{II}|^4 - 2|\nabla|\mathrm{II}||^2) + |\mathrm{II}|^2\varphi'\,(-6|\nabla v|^2 - 2|\mathrm{II}|^2v^2) \\
&\quad -4|\mathrm{II}|^2\varphi''\,v^2|\nabla v|^2 - 4\varphi'\,v\,\nabla|\mathrm{II}|^2\cdot\nabla v.
\end{aligned}$$

Note that $\varphi' > 0$, so the second factor introduces useful terms. We may estimate the last term on the right-hand side, modulo a gradient term, by

$$\begin{aligned}
&-4\varphi'\,v\,\nabla|\mathrm{II}|^2\cdot\nabla v + 2\varphi^{-1}\varphi'\,v\nabla(|\mathrm{II}|^2\varphi)\cdot\nabla v \\
&\qquad = 4|\mathrm{II}|^2\varphi^{-1}(\varphi')^2v^2|\nabla v|^2 - 2\varphi'\,v\nabla|\mathrm{II}|^2\cdot\nabla v \\
&\qquad \le 6|\mathrm{II}|^2\varphi^{-1}(\varphi')^2v^2|\nabla v|^2 + 2\varphi\,|\nabla|\mathrm{II}||^2,
\end{aligned}$$

where we used the Peter–Paul inequality. Hence

$$\begin{aligned}
(\partial_t - \Delta)(|\mathrm{II}|^2\varphi) &\le -2\varphi^{-1}\varphi'\,v\nabla v\cdot\nabla(|\mathrm{II}|^2\varphi) + 2(\varphi - v^2\varphi')|\mathrm{II}|^4 \\
&\quad +|\mathrm{II}|^2|\nabla v|^2(-6\varphi' - 4v^2\varphi'' + 6v^2\varphi^{-1}(\varphi')^2).
\end{aligned}$$

Observe that φ satisfies the following identities:

$$\begin{aligned}
\varphi(s) - s\varphi'(s) &= -k\varphi^2(s), \\
-6\varphi' - 4s\varphi'' + 6s\varphi^{-1}(\varphi')^2 &= -\frac{2ks}{(1-ks)^3} = -\frac{2k}{(1-ks)^2}\varphi, \\
\varphi^{-1}(s)\varphi'(s) &= \frac{s}{1-ks} = \frac{1}{s^2}\varphi(s).
\end{aligned}$$

Therefore we obtain the evolution equation

$$(7.26)\qquad \begin{aligned}(\partial_t - \Delta)(|\mathrm{II}|^2\varphi) \le\ & -2\varphi\, v^{-3}\nabla v\cdot\nabla(|\mathrm{II}|^2\varphi) - 2k(|\mathrm{II}|^2\varphi)^2\\ & -\frac{2k}{(1-kv^2)^2}|\nabla v|^2\,|\mathrm{II}|^2\varphi.\end{aligned}$$

Next we localize the calculation above by introducing an appropriate cutoff function. Define $r(x,t) = |X(x,t)|^2 + 2nt$. By (7.8), under the mean curvature flow we have $(\partial_t - \Delta)r = 0$. By (5.13) we have

$$(7.27)\qquad |\nabla r|^2 = 4|X^\top|^2 \le 4|X|^2.$$

Define

$$\eta(x,t) = \eta(r(x,t)) = (R^2 - r(x,t))^2.$$

We consider η in the region where it is nonnegative, i.e., where $r \le R^2$. In the following, we conflate $\eta' = \eta'(r) = -2(R^2 - r)$ and $\eta'' = \eta''(r) = 2$. We compute that

$$(\partial_t - \Delta)\eta = \eta'\,(\partial_t - \Delta)r + \eta''\,|\nabla r|^2 = 2|\nabla r|^2.$$

By equation (7.26) and the above, we obtain

$$\begin{aligned}(\partial_t - \Delta)(|\mathrm{II}|^2\varphi\,\eta) \le\ & 2|\nabla r|^2|\mathrm{II}|^2\varphi - 2\varphi\, v^{-3}\nabla v\cdot\nabla(|\mathrm{II}|^2\varphi)\,\eta - 2k(|\mathrm{II}|^2\varphi)^2\,\eta\\ & -\frac{2k}{(1-kv^2)^2}|\nabla v|^2\,|\mathrm{II}|^2\varphi\,\eta - 2\nabla(|\mathrm{II}|^2\varphi)\cdot\nabla\eta.\end{aligned}$$

From the right-hand side we pick out the two terms we can estimate modulo a gradient term; namely

$$\begin{aligned}&-2\varphi\, v^{-3}\nabla v\cdot\nabla(|\mathrm{II}|^2\varphi)\,\eta - 2\nabla(|\mathrm{II}|^2\varphi)\cdot\nabla\eta\\ &\quad= -2\nabla(|\mathrm{II}|^2\varphi\,\eta)\cdot(\varphi\, v^{-3}\nabla v + \eta^{-1}\nabla\eta) + 2|\mathrm{II}|^2\varphi(\varphi\, v^{-3}\nabla v\cdot\nabla\eta + \eta^{-1}|\nabla\eta|^2)\\ &\quad\le 2|\mathrm{II}|^2\varphi\left(\frac{k}{(1-kv^2)^2}|\nabla v|^2\eta + \left(\frac{(1-kv^2)^2\varphi^2v^{-6}}{4k} + 1\right)\eta^{-1}|\nabla\eta|^2\right)\\ &\qquad - 2\nabla(|\mathrm{II}|^2\varphi\,\eta)\cdot V,\end{aligned}$$

where $V = \varphi\, v^{-3}\nabla v + \eta^{-1}\nabla\eta$. Since we have $(1-kv^2)^2\varphi^2v^{-6} = v^{-2}$ and $\eta^{-1}|\nabla\eta|^2 = 4|\nabla r|^2$, we obtain

$$(\partial_t - \Delta)(|\mathrm{II}|^2\varphi\,\eta) \le 2|\mathrm{II}|^2\varphi\left(\frac{v^{-2}}{k} + 5\right)|\nabla r|^2 - 2\nabla_V(|\mathrm{II}|^2\varphi\,\eta) - 2k(|\mathrm{II}|^2\varphi)^2\,\eta.$$

We conclude that

$$\begin{aligned}(\partial_t - \Delta)(t|\mathrm{II}|^2\varphi\,\eta) \le\ & -2V\cdot\nabla(t|\mathrm{II}|^2\varphi\,\eta) - 2tk(|\mathrm{II}|^2\varphi)^2\,\eta\\ & +2t|\mathrm{II}|^2\varphi\left(\frac{v^{-2}}{k} + 5\right)|\nabla r|^2 + |\mathrm{II}|^2\varphi\,\eta.\end{aligned}$$

Now define

$$\Omega_t(R) = \{X \in \mathcal{M}_t \mid r(X,t) = |X|^2 + 2nt \le R^2\}.$$

Given $\bar t \in (0,T]$, define the "parabolic region"

$$\Omega_{[0,\bar t\,]}(R) = \bigcup_{t\in[0,\bar t\,]} \Omega_t(R) \times \{t\}.$$

Assume that

$$\partial\Omega_{[0,T]}(R) \cap \Omega_{(0,T]}(R) \subset \{(X,t) \in \mathbb{R}^{n+1} \times [0,T] \,:\, |X|^2 + 2nt = R^2\}.$$

Set $\Phi = t|\mathrm{II}|^2\varphi\,\eta$. Then the maximum of Φ on $\Omega_{[0,\bar t\,]}(R)$ is attained at some interior point (X_0,t_0), allowing for $t_0 = \bar t$. By the maximum principle,

$$0 \le -2t_0k(|\mathrm{II}|^2\varphi)^2\,\eta + 2t_0|\mathrm{II}|^2\varphi\left(\frac{v^{-2}}{k} + 5\right)|\nabla r|^2 + |\mathrm{II}|^2\varphi\eta$$

at (X_0,t_0). Multiplying this inequality by $t_0\eta$ yields, at (X_0,t_0),

$$2k\Phi^2 \le \left(2t_0\left(\frac{v^{-2}}{k} + 5\right)|\nabla r|^2 + \eta\right)\Phi.$$

Since $t_0 \le \bar t$, $|\nabla r|^2 \le 4|X|^2 \le 4R^2$, $\eta \le R^4$, and $v \ge 1$, we obtain

$$\sup_{\Omega_{[0,\bar t\,]}(R)} \Phi = \Phi(X_0,t_0) \le \frac{1}{2k}\left(8\bar t\,R^2\left(\frac{1}{k} + 5\right) + R^4\right). \tag{7.28}$$

Now choose

$$k = \frac{1}{2}\inf_{\Omega_{[0,\bar t\,]}(R)}(v^{-2}) = \frac{1}{2}\left(\sup_{\Omega_{[0,\bar t\,]}(R)} v\right)^{-2}.$$

Then $1 - kv^2 \ge \frac{1}{2} > 0$ on $\Omega_{[0,\bar t\,]}(R)$. Since $k \ge 2$, we obtain from (7.28) that

$$\sup_{\Omega_{[0,\bar t\,]}(R)} \Phi = \Phi(X_0,t_0) \le \frac{1}{k^2}\left(14\,\bar t\,R^2 + R^4\right).$$

For $X \in \mathcal{M}_{\bar t}$ such that $r(X,\bar t\,) \le \theta R^2$, we have $\eta(r(X,\bar t\,)) \ge R^4(1-\theta)^2$. Hence for $X \in \Omega_{\bar t}(\theta^{1/2}R)$, $\theta \in (0,1)$, we have

$$\begin{aligned} |\mathrm{II}|^2(X,\bar t\,) &\le \frac{1}{k^2\,\bar t\,\varphi\,\eta}\left(14\,\bar t\,R^2 + R^4\right) \\ &\le \frac{4}{(1-\theta)^2}\left(14\,R^{-2} + \bar t^{-1}\right)\left(\sup_{\Omega_{[0,\bar t\,]}(R)} v\right)^4. \end{aligned}$$

The desired estimate follows. $\square$

The local analogue of Theorem 6.24 provides a local bound for the derivatives of the curvature given a local bound for the curvature.

Theorem 7.14 ([**209**, Corollary 3.5(ii)], [**205**, Proposition 3.22]). *Suppose that $\{\mathcal{M}_t^n\}_{t\in I}$ is an embedded solution to mean curvature flow which is properly defined in $P_r(x_0,t_0)$. If $\sup_{P_r(x_0,t_0)} |\mathrm{II}|^2 \le C_0 r^{-2}$, then*

$$\sup_{P_{\theta r}(x_0,t_0)} |\nabla^m \mathrm{II}|^2 \le C_m r^{-2(m+1)}\,, \tag{7.29}$$

where the constant C_m depends only on n, m, C_0, and θ.

Proof. Set $\psi(t) = \frac{R^2 t}{R^2+t}$, which is asymptotic to t as $t\to 0$. Observe that $\frac{d\psi}{dt}(t) = \frac{R^4}{(R^2+t)^2} \le 1$. Let m be a nonnegative integer. In the following, constants generally depend also on the dimension n, which we shall suppress in our notation. By induction, suppose that there exist constants $I_0, I_1, \ldots, I_{m-1}$ such that

$$\sup_{\Omega_{[0,T]}(R)} \psi^k |\nabla^{k-1}\mathrm{II}|^2 \le I_{k-1}^2 \quad \text{for } 1\le k\le m.$$

Set $\mathrm{D}_k \doteqdot |\nabla^k \mathrm{II}|$, so that

$$\psi^{(k+1)/2}\,\mathrm{D}_k \le I_k \quad \text{for } 0 \le k \le m-1. \tag{7.30}$$

Recall from (6.68) that

$$(\partial_t - \Delta)\,\mathrm{D}_m^2 \le -2\,\mathrm{D}_{m+1}^2 + C_m \sum_{i+j+k=m} \mathrm{D}_i\,\mathrm{D}_j\,\mathrm{D}_k\,\mathrm{D}_m\,.$$

Thus,

$$\begin{aligned}(\partial_t - \Delta)(\psi^{m+1}\,\mathrm{D}_m^2) \;&\le\; -2\psi^{m+1}\,\mathrm{D}_{m+1}^2 + m\psi^m \frac{d\psi}{dt}\,\mathrm{D}_m^2 \\ &\quad + C_m \psi^{m+1} \sum_{i+j+k=m} \mathrm{D}_i\,\mathrm{D}_j\,\mathrm{D}_k\,\mathrm{D}_m\,.\end{aligned} \tag{7.31}$$

Claim 7.15. There exists A_m (depending on n, m, $I_0, \ldots, I_{m-1}$) such that

$$\psi^{m+1} \sum_{i+j+k=m} \mathrm{D}_i\,\mathrm{D}_j\,\mathrm{D}_k\,\mathrm{D}_m \le A_m \sum_{\ell=0}^{m} \psi^\ell\,\mathrm{D}_\ell^2\,.$$

Proof of the claim. We have

$$\psi^{m+1} \sum_{i+j+k=m} \mathrm{D}_i\,\mathrm{D}_j\,\mathrm{D}_k\,\mathrm{D}_m = 3\psi\,\mathrm{D}_0^2\,\psi^m\,\mathrm{D}_m^2 + \psi^{\frac{m}{2}}\,\mathrm{D}_m\,\psi^{\frac{m}{2}+1} \sum_{\substack{i+j+k=m\\ i,j,k\le m-1}} \mathrm{D}_i\,\mathrm{D}_j\,\mathrm{D}_k.$$

On the other hand, $\psi\,\mathrm{D}_0^2 \le I_0^2$ and by (7.30) we have

$$\psi^{\frac{m}{2}+1} \sum_{\substack{i+j+k=m\\ i,j,k\le m-1}} \mathrm{D}_i \mathrm{D}_j \mathrm{D}_k = \sum_{\substack{i+j+k=m\\ i,j,k\le m-1}} \psi^{\frac{i}{2}}\,\mathrm{D}_i\,\psi^{\frac{j}{2}+1}\,\mathrm{D}_j\,\psi^{\frac{k}{2}+1}\,\mathrm{D}_k \le \sum_{\substack{i+j+k=m\\ i,j,k\le m-1}} \psi^{\frac{i}{2}}\,\mathrm{D}_i\,I_j\,I_k.$$

The claim now follows from the Peter–Paul inequality. □

By applying the claim and the estimate $\frac{d\psi}{dt} \le 1$ to (7.31), we obtain

$$(\partial_t - \Delta)(\psi^{m+1}\,\mathrm{D}_m^2) \le -2\psi^{m+1}\,\mathrm{D}_{m+1}^2 + B_m \sum_{\ell=0}^{m} \psi^\ell\,\mathrm{D}_\ell^2,$$

where $B_m = C_m A_m + m$. Let $F = \psi^{m+1}\,\mathrm{D}_m^2 (B + \psi^m\,\mathrm{D}_{m-1}^2)$. Thus,

$$\begin{aligned}(\partial_t - \Delta)F &\le \psi^{m+1}\,\mathrm{D}_m^2 \cdot \left(-2\psi^m\,\mathrm{D}_m^2 + B_{m-1}\sum_{\ell=0}^{m-1}\psi^\ell\,\mathrm{D}_\ell^2\right)\\ &\quad + \left(-2\psi^{m+1}\,\mathrm{D}_{m+1}^2 + B_m \sum_{\ell=0}^{m}\psi^\ell\,\mathrm{D}_\ell^2\right)\cdot(B + \psi^m\,\mathrm{D}_{m-1}^2)\\ &\quad -2\psi^{2m+1}\,\nabla\,\mathrm{D}_m^2 \cdot \nabla\,\mathrm{D}_{m-1}^2\,.\end{aligned}$$

Since $\sum_{\ell=0}^{m-1}\psi^\ell\,\mathrm{D}_\ell^2 \le \psi^{-1}\sum_{\ell=0}^{m-1} I_\ell^2 \doteqdot \psi^{-1} J_m$, we obtain

$$\begin{aligned}(\partial_t - \Delta)F &\le -2\psi^{-1}(\psi^{m+1}\,\mathrm{D}_m^2)^2 + \psi^{-1}(\psi^{m+1}\,\mathrm{D}_m^2)\,B_{m-1}J_m\\ &\quad + \left(-2\psi^{m+1}\,\mathrm{D}_{m+1}^2 + B_m\frac{1}{\psi}(\psi^{m+1}\,\mathrm{D}_m^2) + \frac{1}{\psi}B_m J_m\right)(B + I_{m-1}^2)\\ &\quad -2\psi^{2m+1}\,\nabla\,\mathrm{D}_m^2 \cdot \nabla\,\mathrm{D}_{m-1}^2\,.\end{aligned}$$

Regarding the last term, we have

$$\begin{aligned}-2\psi^{2m+1}\nabla\,\mathrm{D}_m^2 \cdot \nabla\,\mathrm{D}_{m-1}^2 &\le 2\psi^{2m+1}\left|\nabla|\nabla^m \mathrm{II}|^2 \cdot \nabla|\nabla^{m-1}\mathrm{II}|^2\right|\\ &\le 8\psi^{2m+1}\,\mathrm{D}_{m-1}\,\mathrm{D}_m^2\,\mathrm{D}_{m+1}\,.\end{aligned}$$

By the Peter–Paul inequality,

$$\begin{aligned}8\psi^{2m+1}\,\mathrm{D}_{m-1}\,\mathrm{D}_m^2\,\mathrm{D}_{m+1} &\le 2\psi^{m+1}\,\mathrm{D}_{m+1}^2(B + I_{m-1}^2)\\ &\quad + \frac{8}{B + I_{m-1}^2}\psi^{-1}(\psi^m\,\mathrm{D}_{m-1}^2)(\psi^{m+1}\,\mathrm{D}_m^2)^2.\end{aligned}$$

Thus,

$$\begin{aligned}(\partial_t - \Delta)F &\le -2\psi^{-1}(\psi^{m+1}\,\mathrm{D}_m^2)^2 + \psi^{-1}(\psi^{m+1}\,\mathrm{D}_m^2)\,B_{m-1}J_m\\ &\quad + \left(B_m\psi^{-1}(\psi^{m+1}\,\mathrm{D}_m^2) + \psi^{-1}B_m J_m\right)(B + I_{m-1}^2)\\ &\quad + \frac{8 I_{m-1}^2}{B + I_{m-1}^2}\psi^{-1}(\psi^{m+1}\,\mathrm{D}_m^2)^2.\end{aligned}$$

Choosing $B = 1 + 7I_{m-1}^2$, we obtain

$$\begin{aligned}(\partial_t - \Delta)F &\le -\psi^{-1}(\psi^{m+1}\,\mathrm{D}_m^2)^2 + \psi^{-1}B_m J_m(B + I_{m-1}^2)\\ &\quad + \psi^{-1}(\psi^{m+1}\,\mathrm{D}_m^2)\,(B_{m-1}J_m + B_m(B + I_{m-1}^2))\\ &\le -\psi^{-1}\,(B + \psi^m\,\mathrm{D}_{m-1}^2)^{-2}\,F^2 + C\psi^{-1}\,(1 + F).\end{aligned}$$

Since $(B+\psi^m\,\mathrm{D}^2_{m-1})^{-2}\geq (B+I^2_{m-1})^{-2}\doteqdot 2\delta>0$ and by the Peter–Paul inequality, we obtain with a different constant C that

$$(\partial_t-\Delta)F\leq \psi^{-1}(-\delta F^2+C).$$

As before, we localize the calculation and let $\eta=(R^2-r)^2$. Using $(\partial_t-\Delta)\eta=2|\nabla r|^2\leq 8R^2$ and $\eta^{-1}|\nabla\eta|^2\leq 16R^2$, we see that

$$\begin{aligned}(\partial_t-\Delta)(F\eta) &\leq \eta(\partial_t-\Delta)F+F(\partial_t-\Delta)\eta-2\eta^{-1}\nabla(F\eta)\cdot\nabla\eta+2\frac{|\nabla\eta|^2}{\eta}F\\ &\leq \eta\psi^{-1}(-\delta F^2+C)+40R^2F-2\eta^{-1}\nabla(F\eta)\cdot\nabla\eta.\end{aligned}$$

Since the maximum of $F\eta$ on $\Omega_{[0,T]}(R)$ is attained at some interior point (X_0,t_0), we obtain at (X_0,t_0) that $\eta\psi^{-1}(-\delta F^2+C)+40R^2F\geq 0$; that is,

$$\delta(F\eta)^2\leq 40R^2\psi(F\eta)+C\eta^2\leq 40R^4(F\eta)+CR^8,$$

where we used $\psi\leq R^2$ and $\eta\leq R^4$. We conclude that $\sup_{\Omega_{[0,T]}(R)}F\eta\leq CR^4$. Thus, for $X\in\mathcal{M}_t$ such that $r(X,t)\leq\theta R^2$, we have

$$\psi^{m+1}|\nabla^m\mathrm{II}|^2(X,t)\leq C_m(1-\theta)^{-2}. \qquad \square$$

7.6. Proof of Theorem 7.1

We now put the above ingredients together to prove Theorem 7.1 (this is the proof of Theorem 5.1 in [**209**]).

Proof of Theorem 7.1. Assume that $X_0:\mathbb{R}^n\to\mathbb{R}^{n+1}$ is an entire graph given by $X_0(x)=(x,u_0(x))$ and that $u_0\in C^{2,\alpha}(\mathbb{R}^n)$. For each $R>0$, we know by Theorem 7.6 that there exists an instantaneously smooth solution $X_R:\bar{B}_R\times[0,\infty)\to\mathbb{R}^{n+1}$ of mean curvature flow, which is the graph of a function u_R over the ball $\bar{B}_R(0)$ for each $t\geq 0$ and satisfies the initial-boundary condition $u_R=u_0$ on $(B_R\times\{0\})\cup(\partial B_R\times[0,\infty))$.

Fix any compact domain $\Omega\subset\mathbb{R}^n$ and any $T>2$. Choose R_0 so that $\Omega\subset B_{R_0}$. By (7.20), there exists a constant C_n depending only on n such that for $R\geq R_1\doteqdot C_n\sqrt{R_0^2+T}$ the height function of X_R satisfies

$$\sup_{B_{2R_0}\times[0,T]}|u_R|\leq C_0,$$

where C_0 depends only on n,R_0,T, and $\sup_{B_{R_1}}u_0$. The gradient bound (Theorem 7.8) then implies that

$$\sup_{B_{3/2R_0}\times[T^{-1},T]}|\nabla u_R|\leq C_1,$$

where C_1 depends only on n, R_0, T, and $\sup_{B_{2R_0}} |\nabla u_0|$. Theorems 7.13 and 7.14 now imply that

$$\sup_{B_{R_0}\times[2T^{-1},T]} |\nabla^m u_R| \leq C_m\,,$$

where C_m depends only on m, R_0, C_0, and C_1. Since $u_0 \in C^{2,\alpha}$, we actually have a $C^{2,\alpha}$ bound back to time 0 by the parabolic Schauder theory.

So, if we take any $R_k \to \infty$, there exist a sequence of graphical solutions $u_{R_k} : \bar{B}_{R_k} \times [0,\infty) \to \mathbb{R}$ and a function $u \in C^\infty(\mathbb{R}^n \times (0,\infty)) \cap C^{2,\alpha}(\mathbb{R}^n \times [0,\infty))$ such that for any compact domain Ω and time $T > 2$ the sequence u_{R_k} converges to u in $C^\infty(\Omega \times [2T^{-1}, T])$ as $k \to \infty$. The convergence is also in $C^{2,\alpha}(\Omega\times[0,T])$. Thus the limit u defines a graphical solution $X : \mathbb{R}^n \times [0,\infty) \to \mathbb{R}^{n+1}$ of mean curvature flow by $X(x,t) = (x, u(x,t))$ with $X(0) = X_0$. A standard approximation argument implies that there exists a continuous, instantaneously smooth solution for locally Lipschitz continuous initial data. □

7.7. Convergence to self-similarly expanding solutions

In [**208**], Ecker and Huisken study the asymptotic behavior of entire graphs through the rescaling of the initial surfaces that was developed to study type-I singularities using Huisken's monotonicity formula in Chapter 11.

Theorem 7.16 ([**208**, Theorem 5.1]). *Let X_0 be an entire graph with bounded curvature. Assume in addition that $v_0 \leq C$, i.e., $\langle \mathsf{N}, \omega\rangle \geq C^{-1}$, and that X_0 is straight at infinity; i.e.,*

$$|\langle X, \mathsf{N}\rangle| \leq C(1+|X|)^{1-\delta} \tag{7.32}$$

for some $C < \infty$ and $\delta > 0$. Then the solution $\tilde{X}(\tilde{t})$ of the normalized flow $\frac{\partial}{\partial \tilde{t}}\tilde{X} = -\tilde{H}\tilde{\mathsf{N}} - \tilde{X}$, $\tilde{X}(0) = X_0$, exists for all positive time and converges as $\tilde{t} \to \infty$ to a limit $\mathcal{M}_\infty$ that satisfies the expanding self-similar equation $X_\infty^\perp = \vec{H}_\infty$; that is, $\langle X_\infty, \mathsf{N}_\infty\rangle = -H_\infty$.

Proof (sketch). The main idea in the proof is to consider the rescaled solution $\tilde{X}(\tilde{t}) = \frac{1}{\sqrt{2t+1}}X(t)$, $\tilde{t} = \frac{1}{2}\log(2t+1)$, and derive the estimate that for all $\varepsilon < \delta$ there exist constants $\alpha, \beta > 0$ such that

$$\sup_{\tilde{M}_{\tilde{t}}} \frac{\left(\tilde{H} + \langle \tilde{X}, \tilde{\mathsf{N}}\rangle\right)^2 \tilde{v}^2}{\left(1+\alpha|\tilde{X}|^2\right)^{1-\varepsilon}} \leq e^{-\beta\tilde{t}} \sup_{M_0} \frac{(H + \langle X, \mathsf{N}\rangle)^2 v^2}{(1+\alpha|X|^2)^{1-\varepsilon}}.$$

For details, see [**208**]. □

We note that the straightness at infinity condition (7.32) is necessary.

Example 7.17 ([**208**, Proposition 6.1]). Let $\mathcal{M}_0$ be the graph of the function

$$u_0(x) = \begin{cases} |x| \sin \log |x|, & |x| \geq 1, \\ \overline{u}_0(x), & |x| \leq 1, \end{cases}$$

where $\overline{u}_0 : \overline{B}_1 \to \mathbb{R}$ is any smooth function with $\overline{u}_0|_{\partial B_1} \equiv 0$. The resulting solution to mean curvature flow does not converge to a self-similar solution.

7.8. Self-similarly shrinking entire graphs

We observe that the only shrinking self-similar entire graphical solutions with at most polynomial growth are stationary hyperplanes.

Proposition 7.18 (Classification of graphical shrinking self-similar solutions [**208**, Proposition on p. 471]). *Let* $\mathcal{M} = \operatorname{graph} u$ *be an entire graph satisfying the shrinker equation* (*cf.* (6.11)):

$$H = \langle X, \mathsf{N} \rangle . \tag{7.33}$$

If ∇u, $\nabla^2 u$, *and* $\nabla^3 u$ *grow at most polynomially, then* $\mathcal{M}$ *is a hyperplane.*

Proof. Observe that

$$\langle X^\top, \nabla v \rangle = \Delta v - 2v^{-1} |\nabla v|^2 - |\mathrm{II}|^2 v .$$

Multiplying this equation by $e^{-|X|^2/2}$ and integrating by parts yields (recall that $X^\top = \frac{1}{2}\nabla |X|^2$)

$$0 = \int_{\mathcal{M}} \left(2v^{-1} |\nabla v|^2 + |\mathrm{II}|^2 v \right) \mathrm{e}^{-|X|^2/2} d\mu . \tag{7.34}$$

Indeed, from the polynomial growth assumption, v, ∇v, and II have polynomial growth (recall also that $v^{-1} \leq 1$). This justifies the integration by parts formula

$$\int_{\mathcal{M}} \left(\Delta v - \frac{1}{2} \langle \nabla |X|^2, \nabla v \rangle \right) \mathrm{e}^{-|X|^2/2} d\mu = 0 . \tag{7.35}$$

Finally, by (7.34) we conclude that $|\mathrm{II}|^2 \equiv 0$, so that $\mathcal{M}$ is a hyperplane. □

7.9. Notes and commentary

An excellent exposition of the results of this chapter may be found in Klaus Ecker's book [**205**] and in the Ecker–Huisken papers themselves [**208, 209**].

The local Lipschitz condition in Theorem 7.1 was relaxed by Clutterbuck [**174**] and by Colding and Minicozzi [**180**].

G. Drugan and X. H. Nguyen used the Ecker–Huisken interior estimates to construct entire graphical solutions to mean curvature flow in dimensions $n \geq 2$ which behave differently from the heat flow [**203**] (in contrast to the curve shortening flow, where entire graphs converge to solutions of the heat flow with the same initial data [**413**]).

R. Alessandroni and C. Sinestrari have obtained similar results for entire convex surfaces under a generalized class of flows with nonlinear speeds [**6**].

Alex Freire obtained short-time existence for the motion by mean curvature of graphs with free boundary meeting a hyperplane with constant contact angle and proved that the curvature blows up at any finite maximal time [**224**]. Valentina Wheeler obtained similar results in the setting of "half-entire" graphs with free boundary orthogonal to a fixed hyperplane in Euclidean space and also proved convergence to a self-similar solution in this setting [**526**] (see also [**525, 527**]). She also exploited the free boundary mean curvature flow (supported on cylinders and cones) to obtain uniqueness results for free boundary minimal surfaces [**528**].

Figure 7.2. Valentina-Mira Wheeler.

The graphical mean curvature flow with prescribed boundary contact angle over a bounded domain was studied in B. Guan [**252**], where global in time existence and asymptotic convergence results were proved.

7.10. Exercises

Exercise 7.1. Show that if $u : \Omega \times [0,T] \to \mathbb{R}$ is a graphical solution of mean curvature flow and if $\lambda > 0$, then $u_\lambda : \lambda\Omega \times [0, \lambda^2 T] \to \mathbb{R}$ defined by $u_\lambda(x,t) = \lambda u(\frac{x}{\lambda}, \frac{t}{\lambda^2})$ is also a graphical solution of mean curvature flow.

Exercise 7.2. Justify the integration by parts used to obtain (7.35).

Exercise 7.3 ([**208**, Lemma 4.1]). Let $\{\mathcal{M}_t^n\}_{t\in I}$ be a solution to mean curvature flow in $\mathbb{R}^{n+1}$. Given $\omega \in S^n$, set $v = \langle \mathsf{N}, \omega\rangle^{-1}$. Show that

$$(\partial_t - \Delta)\left(|\mathrm{II}|^2 v^2\right) \leq -2\left\langle v^{-1}\nabla v, \nabla\left(|\mathrm{II}|^2 v^2\right)\right\rangle,$$

where the Laplacian, norms, and inner products are all with respect to the induced metric g. Deduce that

$$(\partial_t - \Delta)\left(2t\,|\mathrm{II}|^2\,v^2 + v^2\right) \leq -2\left\langle v^{-1}\nabla v, \nabla\left(2t\,|\mathrm{II}|^2\,v^2 + v^2\right)\right\rangle .$$

See Exercise 10.4 for applications of these heat-type inequalities using the noncompact maximum principle.

Chapter 8

Huisken's Theorem

In this chapter we see that it is possible to completely describe the global in time behavior of the mean curvature flow in the special setting of compact, convex hypersurfaces. The result, due to G. Huisken [**291**], states that the hypersurface remains convex and contracts to a single point at the final time in such a way that, after rescaling by the square root of the remaining time, the resulting family of hypersurfaces converges smoothly to a round sphere.

Theorem 8.1 (Huisken [**291**]). *Let $X : M^n \times [0,T) \to \mathbb{R}^{n+1}$, $n \geq 2$, be a maximal solution to mean curvature flow defined on the maximal time interval $[0,T)$. If $X_0 \doteqdot X(\cdot,0)$ is a convex embedding, then $X_t \doteqdot X(\cdot,t)$ is a convex embedding for all $t > 0$, X_t converges uniformly to a constant $p \in \mathbb{R}^{n+1}$ as $t \to T$, and the rescaled embeddings $\widetilde{X}_t : M^n \to \mathbb{R}^{n+1}$ defined by*

$$\widetilde{X}_t(x) \doteqdot \frac{X_t(x) - p}{\sqrt{2n(T-t)}} \tag{8.1}$$

converge uniformly in the smooth topology to a smooth embedding whose image coincides with the unit sphere S^n.

Figure 8.1. Gerhard Huisken. Author: Gerd Fischer. Source: Archives of the Mathematisches Forschungsinstitut Oberwolfach.

Eventually, we will present *four* different proofs of this result, each of which illustrates important techniques that arise later in other settings. Huisken's argument, presented in the following sections, makes use of strong pointwise pinching and gradient estimates for the second fundamental form, yielding direct, quantitative control on the convergence to a round sphere. The remaining proofs invoke indirect "compactness" arguments to varying degrees: A second argument, which we present at the end of this chapter, relies on a bound for the ratio of circumradius to inradius; the convergence to a round sphere then follows from a straightforward application of the Arzelà–Ascoli theorem and the strong maximum principle. An argument of Hamilton, which we present in Section 11.4, combines very weak control on the solution (namely, preservation of the initial curvature pinching) with a classification of possible "blow-ups" to reach the conclusion but requires a more sophisticated compactness theory (which we develop in Section 11.1). The fourth proof, presented in Section 12.3, makes use of the more recently discovered "noncollapsing" phenomena for mean curvature flow, which provide very strong control over the evolution, and the convergence then follows again via Arzelà–Ascoli. This proof, unlike the previous three proofs, also applies to the curve shortening flow, giving a unified proof of the theorems of Gage–Hamilton and Huisken.

8.1. Pinching is preserved

A uniformly convex hypersurface admits a positive lower bound for its mean curvature. An elementary application of the maximum principle shows that this bound is preserved along the mean curvature flow.

Lemma 8.2 (Nonnegative lower bound for H preserved under mean curvature flow). *Let $X : M^n \times [0,T) \to \mathbb{R}^{n+1}$ be a compact solution of mean curvature flow. If $\min_{M\times\{0\}} H \geq 0$, then*

$$H(x,t) \geq \min_{M\times\{0\}} H \quad \text{for all} \quad (x,t) \in M \times [0,T)\,. \tag{8.2}$$

Proof. Apply the maximum principle to the evolution equation (6.18) for the mean curvature. □

A more careful analysis yields refined estimates for the maximum and minimum of H.

Lemma 8.3. *Let $X : M^n \times [0,T) \to \mathbb{R}^{n+1}$ be a compact solution of mean curvature flow satisfying*

$$\limsup_{t\nearrow T} \max_{M\times\{t\}} H^2 = \infty\,.$$

Then

$$\min_{M\times\{t\}} H \le \frac{\sqrt{n}}{\sqrt{2(T-t)}}\,. \tag{8.3}$$

If, in addition, there is some $C>0$ such that $|\mathrm{II}|^2 \le CH^2$, then

$$\max_{M\times\{t\}} H \ge \frac{1}{\sqrt{2C(T-t)}}\,. \tag{8.4}$$

Proof. Recalling the evolution equation (6.18) for the mean curvature, we estimate

$$\partial_t H \ge \Delta H + \frac{1}{n}H^3 \quad \text{and} \quad \partial_t H \le \Delta H + C\,H^3,$$

with the second inequality holding provided $|\mathrm{II}|^2 \le CH^2$.

Set $\overline{H}(t) \doteqdot \max_{M\times\{t\}} H$ and $\underline{H}(t) \doteqdot \min_{M\times\{t\}} H$. Let $\{t_i\}_{i\in\mathbb{N}}$ be a sequence of times converging to T and satisfying $\lim_{i\to\infty} \overline{H}(t_i) = \infty$. For each $i\in\mathbb{N}$, let $\overline{\varphi}_i$ and $\underline{\varphi}_i$ be the solutions of the ODEs

$$\begin{cases} \overline{\varphi}_i' = C\overline{\varphi}_i^3 & \text{in } [0,t_i]\,,\\ \overline{\varphi}_i(t_i) = \overline{H}(t_i) \end{cases}$$

and

$$\begin{cases} \underline{\varphi}_i' = \frac{1}{n}\underline{\varphi}_i^3 & \text{in } [0,t_i]\,,\\ \underline{\varphi}_i(t_i) = \underline{H}(t_i)\,, \end{cases}$$

respectively. That is,

$$\overline{\varphi}_i(t) = \frac{1}{\sqrt{\overline{H}^{-2}(t_i) + 2C(t_i - t)}} \quad \text{and} \quad \underline{\varphi}_i(t) = \frac{1}{\sqrt{\underline{H}^{-2}(t_i) + \frac{2}{n}(t_i - t)}}\,.$$

By the contrapositive of the ODE comparison principle,

$$\overline{H} \ge \overline{\varphi}_i \quad \text{and} \quad \underline{H} \le \underline{\varphi}_i \quad \text{in} \quad [0,t_i]\,.$$

Taking $i\to\infty$ yields the claims. □

For a convex solution to mean curvature flow, a purely algebraic calculation implies that $|\mathrm{II}|^2 \le H^2$ and hence

$$H_{\min}(t) \lesssim \frac{1}{\sqrt{T-t}} \lesssim H_{\max}(t)\,.$$

The scaling behavior of the mean curvature flow suggests the stronger conclusion

$$H(\cdot,t) \sim \frac{1}{\sqrt{T-t}}\,.$$

We shall soon see that this does indeed hold for convex solutions of the mean curvature flow (although it is not true in general).

Any compact, locally uniformly convex hypersurface is α-**pinched** for some $\alpha > 0$; that is,

$$\frac{\kappa_1}{\kappa_n} \geq \alpha .$$

In particular, $\mathrm{II} \geq \frac{\alpha}{n} Hg$ and $\mathrm{II} \leq \frac{1}{n\alpha} Hg$. A straightforward adaptation of the classical *scalar* maximum principle shows that these *tensor* inequalities are also preserved.

Lemma 8.4 (Pinching is preserved [**291**, Theorem 4.3]). *Let $X : M^n \times [0,T) \to \mathbb{R}^{n+1}$ be a compact, strictly mean convex solution to mean curvature flow.*

(1) *If $\mathrm{II} \geq \alpha Hg$ for some $\alpha \in \mathbb{R}$ at $t = 0$, then the same inequality holds for all $t \in (0,T)$.*

(2) *If $\mathrm{II} \leq CHg$ for some $C \in \mathbb{R}$ at $t = 0$, then the same inequality holds for all $t \in (0,T)$.*

Proof. By (6.15), (6.17), and (6.18), the tensor $S \doteqdot \mathrm{II} - \alpha Hg$ satisfies

$$(\nabla_t - \Delta) S = |\mathrm{II}|^2 S .$$

So the first claim follows from the tensor maximum principle (Theorem 6.14). The second claim is proved similarly. □

8.2. Pinching improves: The roundness estimate

Given an immersed hypersurface $X : M^n \to \mathbb{R}^{n+1}$, recall from Chapter 5 that a point $p \in \mathcal{M}$ is called **umbilic** if the second fundamental form at p is proportional to the metric at p; that is,

$$\mathrm{II}_p = \alpha_p g_p$$

for some $\alpha_p \in \mathbb{R}$. By tracing this, we see that $\alpha_p = \frac{1}{n} H(p)$, so $p \in \mathcal{M}$ is umbilic if and only if the **trace-free second fundamental form**

$$\mathring{\mathrm{II}} \doteqdot \mathrm{II} - \tfrac{1}{n} Hg \tag{8.5}$$

vanishes at p. Another way of saying this is that the norm $|\mathring{\mathrm{II}}|$ of the trace-free second fundamental form is strictly positive at a point p unless p is umbilic. Since the only compact, totally umbilic, connected hypersurfaces of $\mathbb{R}^{n+1}$ are the round spheres (see Exercise 5.5), the norm of the trace-free second fundamental form gives a crude pointwise measure of "**roundness**" for a hypersurface. In fact, since roundness should not depend on scale, we are more interested in the scale-invariant ratio $|\mathring{\mathrm{II}}|^2/H^2$.

It turns out that this measure of roundness is preserved by mean curvature flow.

Proposition 8.5 (Roundness is preserved). *Let $X : M^n \times [0,T) \to \mathbb{R}^{n+1}$ be a compact, mean convex solution of mean curvature flow. Then*

$$\max_{\mathcal{M}_t} \frac{|\mathring{\mathrm{II}}|^2}{H^2} \leq \max_{\mathcal{M}_0} \frac{|\mathring{\mathrm{II}}|^2}{H^2} \quad \textit{for all} \quad t \in [0,T)\,. \tag{8.6}$$

Proof. Recalling (6.17) and (6.18), we observe that

$$\begin{aligned}(\nabla_t - \Delta)\frac{\mathrm{II}}{H} &= \frac{(\nabla_t - \Delta)\,\mathrm{II}}{H} - \frac{\mathrm{II}}{H^2}(\partial_t - \Delta)H + 2\nabla_{\frac{\nabla H}{H}}\frac{\mathrm{II}}{H}\\ &= 2\nabla_{\frac{\nabla H}{H}}\frac{\mathrm{II}}{H}\,.\end{aligned}$$

Thus,

$$\begin{aligned}(\partial_t - \Delta)\frac{|\mathring{\mathrm{II}}|^2}{H^2} &= (\partial_t - \Delta)\frac{|\mathrm{II}|^2}{H^2}\\ &= 2\left\langle (\nabla_t - \Delta)\frac{\mathrm{II}}{H}, \frac{\mathrm{II}}{H}\right\rangle - 2\left|\nabla\frac{\mathrm{II}}{H}\right|^2\\ &= 4\left\langle \nabla_{\frac{\nabla H}{H}}\frac{\mathrm{II}}{H}, \frac{\mathrm{II}}{H}\right\rangle - 2\left|\nabla\frac{\mathrm{II}}{H}\right|^2\\ &= 2\left\langle \nabla\frac{|\mathrm{II}|^2}{H^2}, \frac{\nabla H}{H}\right\rangle - 2\left|\nabla\frac{\mathrm{II}}{H}\right|^2\\ &= 2\left\langle \nabla\frac{|\mathring{\mathrm{II}}|^2}{H^2}, \frac{\nabla H}{H}\right\rangle - 2\left|\nabla\frac{\mathrm{II}}{H}\right|^2.\end{aligned} \tag{8.7}$$

The claim now follows from the maximum principle. □

We expect that diffusion should work to *improve* the roundness ratio. Of course, we cannot hope for too much since, for obvious reasons, a "thin" torus (with bounded roundness) cannot be smoothly deformed to a surface with perfect roundness — the sphere. For convex hypersurfaces, there is no such obstruction.

Theorem 8.6 (Roundness estimate [**291**]). *Let $X : M^n \times [0,T) \to \mathbb{R}^{n+1}$, $n \geq 2$, be a compact, convex solution of mean curvature flow. For any $\varepsilon > 0$ there exists a constant C_ε — which depends only on ε and the initial embedding — such that*

$$|\mathring{\mathrm{II}}|(x,t) \leq \varepsilon H^2(x,t) + C_\varepsilon \quad \textit{for all} \quad (x,t) \in M^n \times [0,T)\,. \tag{8.8}$$

As we shall see in the course of the proof, the constant C_ε can be controlled in terms of ε, n, $\min_{M^n\times\{0\}} \frac{\kappa_1}{H}$, $\mathrm{Area}(\mathcal{M}^n_0)$, $\max_{M^n\times\{0\}} H$, and T (which, by Exercise 8.1, can be controlled in terms of the circumradius or minimum mean curvature of the initial hypersurface).

The roundness estimate implies that solutions become round at a singularity in the sense that, given any sequence of points $(x_k, t_k) \in M \times [0, T)$ satisfying $H(x_k, t_k) \to \infty$ as $k \to \infty$,

$$\frac{|\mathring{\mathrm{II}}|^2}{H^2}(x_k, t_k) \to 0 \quad \text{as} \quad k \to \infty .$$

To prove the theorem, we find constants $\sigma > 0$ and $C < \infty$ such that

$$\frac{|\mathring{\mathrm{II}}|^2 - \varepsilon H^2}{H^2} \leq C H^{-\sigma} . \tag{8.9}$$

The theorem then follows by estimating the right-hand side using Young's inequality. So, given any $\varepsilon > 0$ and any $\sigma \in (0,1)$, define

$$f_{\varepsilon,\sigma} \doteqdot \left(|\mathring{\mathrm{II}}|^2 - \varepsilon H^2 \right) H^{\sigma-2}. \tag{8.10}$$

Then we wish to show, for any $\varepsilon > 0$, that $f_{\varepsilon,\sigma}$ is bounded uniformly in time for some $\sigma > 0$. First, an evolution equation.

Lemma 8.7. *Given any $\varepsilon > 0$ and $\sigma \in (0,1)$, the function $f_{\varepsilon,\sigma}$ defined in (8.10) evolves according to*

$$\begin{aligned}(\partial_t - \Delta) f_{\varepsilon,\sigma} = {} & \sigma \left|\mathrm{II}\right|^2 f_{\varepsilon,\sigma} + 2\left(1-\sigma\right) \left\langle \nabla f_{\varepsilon,\sigma}, \frac{\nabla H}{H} \right\rangle \\ & - 2H^\sigma \left| \nabla \frac{\mathrm{II}}{H} \right|^2 - \sigma\left(1-\sigma\right) f_{\varepsilon,\sigma} \frac{|\nabla H|^2}{H^2}. \end{aligned} \tag{8.11}$$

Proof. The evolution equation for $f_{\varepsilon,\sigma}$ follows from the evolution equations (8.7) for $|\mathring{\mathrm{II}}|^2/H^2$ and (6.18) for H and the "product rule"

$$\begin{aligned}(\partial_t - \Delta)(uv^\alpha) = {} & v^\alpha(\partial_t - \Delta)u + \alpha u v^{\alpha-1}(\partial_t - \Delta)v \\ & - 2\alpha \left\langle \nabla(uv^\alpha), \frac{\nabla v}{v} \right\rangle + \alpha(\alpha+1)uv^\alpha \frac{|\nabla v|^2}{v^2} . \end{aligned} \tag{8.12}$$

□

Unfortunately, Theorem 8.6 does not follow from a direct application of the maximum principle — the reaction term is of the wrong sign wherever $f_{\varepsilon,\sigma} > 0$. We need to make better use of the diffusion term using integral estimates.

First, we need an estimate for the good curvature gradient term (i.e., the *second* to last term in (8.11) — it turns out that the final term, due to its dependence on σ, is not as useful as it appears; we will simply discard it).

Lemma 8.8. *Let $\mathcal{M}^n \hookrightarrow \mathbb{R}^{n+1}$, $n \geq 2$, be a convex hypersurface satisfying $\mathrm{II} \geq \alpha H g$ for some $\alpha > 0$. Then*

$$\left| \nabla \frac{\mathrm{II}}{H} \right|^2 \geq \gamma \frac{|\nabla \mathrm{II}|^2}{H^2}, \tag{8.13}$$

where $\gamma > 0$ depends only on α and n.

Proof. Observe that

$$\left|\nabla \frac{\mathrm{II}}{H}\right|^2 = H^{-4}|H\nabla \mathrm{II} - \nabla H \otimes \mathrm{II}|^2 .$$

Thus, it suffices to prove the purely algebraic inequality

$$|\operatorname{tr}(B)\,T - \operatorname{tr}(T)\otimes B|^2 \geq \gamma(\alpha, n)\,\operatorname{tr}(B)^2\,|T|^2 \tag{8.14}$$

for all symmetric, positive definite $B \in \mathbb{R}^n \otimes \mathbb{R}^n$ satisfying $B \geq \alpha\,\operatorname{tr}(B)\mathrm{I}$ and all totally symmetric $T \in \mathbb{R}^n \otimes \mathbb{R}^n \otimes \mathbb{R}^n$, where I is the standard inner product on $\mathbb{R}^n$ and $\operatorname{tr}(T)$ denotes any choice of the three equivalent traces of T. Moreover, as the estimate is invariant under scaling of either B or T, it suffices to prove (8.14) when $|B| = |T| = 1$ (the inequality is trivial if $T = 0$). So suppose that there is no $\gamma > 0$ for which (8.14) holds on the subset

$$K \doteqdot \{\,(B,T) : B > 0,\ B \geq \alpha\,\operatorname{tr}(B)\,\mathrm{I},\ |B| = |T| = 1\,\}.$$

Then, as K is compact and $\operatorname{tr}(B)^2 \leq n$, there must be some pair $(B,T) \in K$ satisfying

$$\operatorname{tr}(B)\,T = \operatorname{tr}(T) \otimes B\,. \tag{8.15}$$

Note that $\operatorname{tr}(T)$ cannot be the zero vector as, by assumption, T and $\operatorname{tr}(B)$ are nonzero. Thus, tracing over the first and second components, we find that $\operatorname{tr}(T)$ is an eigenvector of B with eigenvalue $\operatorname{tr}(B)$. But this is impossible as B is positive definite. □

In particular, since the inequality $\mathrm{II} \geq \alpha H g$ is preserved under mean curvature flow, we can estimate

$$H^\sigma \left|\nabla \frac{\mathrm{II}}{H}\right|^2 \geq \gamma \frac{|\nabla \mathrm{II}|^2}{H^2} H^\sigma \geq \gamma f_{\varepsilon,\sigma} \frac{|\nabla \mathrm{II}|^2}{H^2}\,,$$

where the constant γ depends only on n and the initial curvature pinching $\alpha \doteqdot \min_{M\times\{0\}} \frac{\kappa_1}{H}$. We can now estimate, wherever $f_{\varepsilon,\sigma} > 0$,

$$(\partial_t - \Delta) f_{\varepsilon,\sigma} \leq \sigma |\mathrm{II}|^2 f_{\varepsilon,\sigma} + 2\,(1-\sigma)\left\langle \nabla f_{\varepsilon,\sigma}, \frac{\nabla H}{H}\right\rangle - 2\gamma f_{\varepsilon,\sigma} \frac{|\nabla \mathrm{II}|^2}{H^2}\,. \tag{8.16}$$

Next, given $p \geq 10$, say, set

$$v = v_{\varepsilon,\sigma,p} \doteqdot (f_{\varepsilon,\sigma})_+^{\frac{p}{2}}, \tag{8.17}$$

where $(f_{\varepsilon,\sigma})_+ \doteqdot \max\{f_{\varepsilon,\sigma}, 0\}$ denotes the nonnegative part of $f_{\varepsilon,\sigma}$. Then v is C^2, so that we may legally calculate

$$2v\nabla v = p(f_{\varepsilon,\sigma})_+^{p-1} \nabla f_{\varepsilon,\sigma}$$

and

$$(\partial_t - \Delta)v^2 = pf_{\varepsilon,\sigma}^{p-1}(\partial_t - \Delta)f_{\varepsilon,\sigma} - p(p-1)f_{\varepsilon,\sigma}^{p-2}|\nabla f_{\varepsilon,\sigma}|^2$$
$$\leq pv^2\left(\sigma|\mathrm{II}|^2 + \frac{4\,(1-\sigma)}{p}\left\langle\frac{\nabla v}{v}, \frac{\nabla H}{H}\right\rangle - 2\gamma\frac{|\nabla \mathrm{II}|^2}{H^2}\right) - \frac{4(p-1)}{p}|\nabla v|^2.$$

Estimating the inner product term, using Young's inequality, by

$$4\,(1-\sigma)\,v^2\left\langle\frac{\nabla v}{v}, \frac{\nabla H}{H}\right\rangle \leq 4v^2\frac{|\nabla v|}{v}\frac{|\nabla H|}{H} \leq v^2\left(\frac{|\nabla v|^2}{v^2} + 4\frac{|\nabla H|^2}{H^2}\right)$$

we arrive at

$$(\partial_t - \Delta)v^2 \leq \sigma p|\mathrm{II}|^2v^2 - 2|\nabla v|^2 - 2\,(\gamma p - 2n)\,v^2\frac{|\nabla \mathrm{II}|^2}{H^2}\,.$$

If we further require that $p > \frac{4n}{\gamma}$, then

$$(\partial_t - \Delta)v^2 \leq \sigma p|\mathrm{II}|^2v^2 - \gamma pv^2\frac{|\nabla \mathrm{II}|^2}{H^2} - 2|\nabla v|^2\,. \tag{8.18}$$

Using Simons's identity, we can control the first term on the right in L^1 by the two good (i.e., negative) terms.

Proposition 8.9. *For each $n \geq 2$ there exists a constant P with the following property: Let $\mathcal{M}^n \hookrightarrow \mathbb{R}^{n+1}$ be a smooth hypersurface and, given $\alpha > 0$, let $u \in W^{2,1}(\mathcal{M})$ be a function with compact support contained in the set $\{x \in \mathcal{M} : \mathrm{II} \geq \alpha Hg\}$. Then*

$$\alpha^4\int_{\mathcal{M}} u^2\,|\mathring{\mathrm{II}}|^2\,d\mu \leq P\int_{\mathcal{M}} u^2\left(\frac{|\nabla \mathrm{II}|^2}{H^2} + \frac{|\nabla \mathrm{II}|}{H}\frac{|\nabla u|}{u}\right)d\mu\,. \tag{8.19}$$

Proof. By Simons's identity (5.50),

$$\nabla_{(i}\nabla_{j)}\mathrm{II}_{kl} - \nabla_{(k}\nabla_{l)}\mathrm{II}_{ij} = \mathrm{C}_{ijkl}\,,$$

where

$$\mathrm{C} \doteqdot \mathrm{II}\otimes\mathrm{II}^2 - \mathrm{II}^2\otimes\mathrm{II}\,.$$

Note that

$$|\mathrm{C}|^2 = \sum_{i,j=1}^n \kappa_i^2\kappa_j^2(\kappa_i - \kappa_j)^2\,.$$

In particular, at any point where $\mathrm{II} \geq \alpha Hg$,

$$|\mathrm{C}|^2 \geq \alpha^4H^4\sum_{i,j=1}^n(\kappa_i - \kappa_j)^2 = 2n\alpha^4H^4|\mathring{\mathrm{II}}|^2\,. \tag{8.20}$$

Thus, integrating by parts,

$$\begin{aligned}
\alpha^4 \int_{\mathcal{M}} u^2 |\mathring{\mathrm{II}}|^2 \, d\mu &\leq \frac{1}{2n} \int_{\mathcal{M}} |\mathrm{C}|^2 H^{-4} u^2 \, d\mu \\
&= \frac{1}{2n} \int_{\mathcal{M}} \frac{u^2}{H^4} \mathrm{C}^{ijkl} \left(\nabla_i \nabla_j \mathrm{II}_{kl} - \nabla_k \nabla_l \mathrm{II}_{ij} \right) d\mu \\
&= \frac{1}{n} \int_{\mathcal{M}} \frac{u^2}{H^4} \mathrm{C}^{ijkl} \nabla_i \nabla_j \mathrm{II}_{kl} \, d\mu \\
&= \frac{1}{n} \int_{\mathcal{M}} \frac{u^2}{H^4} \left(4\mathrm{C}^{ijkl} \frac{\nabla_i H}{H} - 2\mathrm{C}^{ijkl} \frac{\nabla_i u}{u} - \nabla_i \mathrm{C}^{ijkl} \right) \nabla_j \mathrm{II}_{kl} \, d\mu.
\end{aligned}$$

Using the notation of Section 6.8, we observe that

$$\mathrm{C} = \mathrm{II} * \mathrm{II} * \mathrm{II} \quad \text{and} \quad \nabla \mathrm{C} = \mathrm{II} * \mathrm{II} * \nabla \mathrm{II}$$

and hence

$$\begin{aligned}
\alpha^4 \int_{\mathcal{M}} u^2 \, |\mathring{\mathrm{II}}|^2 \, d\mu &\leq \frac{1}{n} \int_{\mathcal{M}} \frac{u^2}{H^4} \left(\mathrm{II} * \frac{\nabla \mathrm{II}}{H} - \mathrm{II} * \frac{\nabla u}{u} - \nabla \mathrm{II} \right) * \mathrm{II} * \mathrm{II} * \nabla \mathrm{II} \, d\mu \\
&\leq P \int_{\mathcal{M}} u^2 \left(\frac{|\nabla \mathrm{II}|^2}{H^2} + \frac{|\nabla \mathrm{II}|}{H} \frac{|\nabla u|}{u} \right) d\mu \,,
\end{aligned}$$

where we estimated $|\mathrm{II}| \leq H$. □

Since $|\mathring{\mathrm{II}}|^2 \geq \varepsilon H^2 \geq \varepsilon |\mathrm{II}|^2$ wherever $f_{\varepsilon,\sigma} > 0$, Proposition 8.9 yields

$$\begin{aligned}
\int_{\mathcal{M}} |\mathrm{II}|^2 \, v^2 \, d\mu &\leq \frac{P}{\varepsilon \alpha^4} \int_{\mathcal{M}} v^2 \left(\frac{|\nabla \mathrm{II}|^2}{H^2} + \frac{|\nabla \mathrm{II}|}{H} \frac{|\nabla v|}{v} \right) d\mu \\
&\leq \frac{P}{\varepsilon \alpha^4} \int_{\mathcal{M}} v^2 \left((1+r) \frac{|\nabla \mathrm{II}|^2}{H^2} + r^{-1} \frac{|\nabla v|^2}{v^2} \right) d\mu
\end{aligned}$$

for any $r > 0$. Choosing $r = p^{\frac{1}{2}}$, we find

$$\begin{aligned}
\frac{d}{dt} \int_{\mathcal{M}} v^2 d\mu &= \int_{\mathcal{M}} \partial_t v^2 \, d\mu - \int_{\mathcal{M}} v^2 H^2 \, d\mu \\
&\leq \sigma p \int_{\mathcal{M}} |\mathrm{II}|^2 \, v^2 \, d\mu - \gamma p \int_{\mathcal{M}} v^2 \frac{|\nabla \mathrm{II}|^2}{H^2} \, d\mu - 2 \int_{\mathcal{M}} |\nabla v|^2 \, d\mu \\
&\leq \left(\frac{\sigma p P}{\varepsilon \alpha^4} (1 + p^{\frac{1}{2}}) - \gamma p \right) \int_{\mathcal{M}} v^2 \frac{|\nabla \mathrm{II}|^2}{H^2} d\mu + \left(\frac{\sigma p^{\frac{1}{2}} P}{\varepsilon \alpha^4} - 2 \right) \int_{\mathcal{M}} |\nabla v|^2 d\mu.
\end{aligned}$$

Choosing p sufficiently large and σ of the order $p^{-\frac{1}{2}}$, we can ensure that the right-hand side is nonpositive.

Lemma 8.10. *There exists $\ell > 0$ depending only on n, α, and ε such that*

$$\frac{d}{dt}\int_{\mathcal{M}} (f_{\varepsilon,\sigma})_+^p \, d\mu \le 0 \tag{8.21}$$

for all $p \ge \ell^{-1}$ and all $\sigma \le \ell p^{-\frac{1}{2}}$.

Integrating yields an L^2 estimate for the function $v \doteqdot f_{\varepsilon,\sigma}^{\frac{p}{2}}$ (at least when p is large and σ is of the order $p^{-\frac{1}{2}}$). To pass from L^2 to L^∞, we make use of an iteration argument originally due to Stampacchia [**330**, Chapter II, Appendix B]. It is based on the following ingenious iteration lemma.

Lemma 8.11 (Stampacchia's lemma). *Let $\varphi : [k_0, \infty) \to \mathbb{R}$ be a nonnegative, nonincreasing function satisfying*

$$\varphi(h) \le \frac{C}{(h-k)^\alpha}\varphi(k)^\beta \quad \text{for all} \quad h > k > k_0 \tag{8.22}$$

and for some constants $C > 0$, $\alpha > 0$, and $\beta > 1$. Then

$$\varphi(k_0 + d) = 0\,, \tag{8.23}$$

where $d^\alpha = C\varphi(k_0)^{\beta-1}2^{\frac{\alpha\beta}{\beta-1}}$.

Proof. Consider the sequence of numbers k_r given by

$$k_r = k_0 + d - \frac{d}{2^r}, \qquad r = 0, 1, 2, \ldots .$$

By assumption,

$$\varphi(k_{r+1}) \le C\frac{2^{(r+1)\alpha}}{d^\alpha}\varphi(k_r)^\beta \quad \text{for all} \quad r = 0, 1, \ldots . \tag{8.24}$$

We will prove by induction that

$$\varphi(k_r) \le \varphi(k_0)2^{-r\mu} \tag{8.25}$$

for all $r \in \mathbb{N}$, where $\mu \doteqdot \frac{\alpha}{\beta-1} > 0$. Clearly (8.25) holds trivially for $r = 0$. Supposing (8.25) holds up to some integer r, we find by (8.24) and the definition of d that

$$\varphi(k_{r+1}) \le C\frac{2^{(r+1)\alpha}}{d^\alpha}\varphi(k_0)^\beta 2^{-r\mu\beta} = \varphi(k_0)2^{-(r+1)\mu}\,.$$

The claim (8.25) follows. Now, by the monotonicity assumption,

$$0 \le \varphi(k_0 + d) \le \varphi(k_r) \qquad \text{for all} \quad r = 0, 1, \ldots .$$

But, by (8.25), $\varphi(k_r) \to 0$ as $r \to \infty$. $\square$

Readers unfamiliar with Stampacchia iteration may wish to attempt Exercise 8.4 before proceeding.

Given any $k \geq k_0 \doteqdot \sup_{\sigma\in(0,1)} \sup_{M\times\{0\}} f_{\varepsilon,\sigma}$, set

$$v_k^2(x,t) \doteqdot \left(f_{\varepsilon,\sigma}(x,t) - k\right)_+^p \quad \text{and} \quad U_k(t) \doteqdot \{x \in M : v_k(x,t) > 0\} .$$

We want to apply Stampacchia's lemma to the function

$$k \mapsto \|U_k\| \doteqdot \int_0^T \int_{U_k(t)} d\mu(\cdot\,,t)\,dt .$$

Note that

$$(h-k)^p \|U_h\| \leq \int_0^T \int_{U_k} v_k^2\,d\mu\,dt . \tag{8.26}$$

We need to estimate the right-hand side in terms of $\|U_k\|$.

First observe that

$$\frac{d}{dt}\int v_k^2\,d\mu + \int |\nabla v_k|^2\,d\mu + \int v_k^2 H^2\,d\mu \leq \sigma p \int_{U_k} f_{\varepsilon,\sigma}^p H^2\,d\mu \tag{8.27}$$

for all $p \geq \min\{4, \frac{2n}{\gamma}\}$ (this is proved by integrating an estimate similar to (8.18)). To exploit the good gradient term, we make use of the Sobolev inequality of Michael and Simon [**400**] (see also [**305**]). For functions on submanifolds, the Sobolev inequality includes an additional term involving the mean curvature.

Theorem 8.12 (Sobolev inequality). *Let M^n be a smoothly immersed submanifold of $\mathbb{R}^{n+k}$ of dimension $n \geq 2$ and codimension $k \geq 1$ and let p be any number satisfying $1 < p < n$. Then every function $u \in W_0^{1,p}(M)$ satisfies*

$$\left(\int_M |u|^{p^*}\,d\mu\right)^{\frac{1}{p^*}} \leq S\left[\int_M \left(|\nabla u|^p + |\vec{H}|^p|u|^p\right)d\mu\right]^{\frac{1}{p}}, \tag{8.28}$$

where $\frac{1}{p^*} \doteqdot \frac{1}{p} - \frac{1}{n}$ and S is a constant which depends only on n and p.

The Sobolev inequality cannot be applied in the critical case $n = p$, which we shall require when $n = 2$. In that case, we instead make use of the following Poincaré inequality, which is obtained from the $p = 1$ case of the Sobolev inequality using Hölder's inequality (albeit with a constant which now depends also on the measure of the support of u).

Corollary 8.13 (Poincaré inequality). *Let M^n be a smoothly immersed submanifold of $\mathbb{R}^{n+k}$ of dimension $n \geq 2$ and codimension $k \geq 1$ and let q be any number satisfying $q \geq 1$. Then every function $u \in W_0^{1,n}(M)$ satisfies*

$$\left(\int_M |u|^{1^* q}\, d\mu\right)^{\frac{1}{1^* q}} \leq S |\operatorname{spt} u|^{\frac{1}{1^* q}} \left[\int_M \left(|\nabla u|^n + |\vec{H}|^n |u|^n\right) d\mu\right]^{\frac{1}{n}}, \tag{8.29}$$

where $\frac{1}{1^*} \doteqdot 1 - \frac{1}{n}$ and S is a constant which depends only on n and q.

Applying the Sobolev inequality (with $p = 2$) when $n \geq 3$ and the Poincaré inequality when $n = 2$ then yields[1]

$$\frac{d}{dt}\int v_k^2\, d\mu + \frac{1}{S}\left(\int v_k^{2^*}\, d\mu\right)^{\frac{2}{2^*}} \leq \sigma p \int_{U_k} f_{\varepsilon,\sigma}^p H^2\, d\mu\,,$$

where, for $n = 2$, we define $2^* \doteqdot 1^* q$ for some $q \geq 1$ and S is a constant which depends only on n when $n \geq 3$ and depends only on $|\mathcal{M}_0|$ and q (which we may take to be fixed) when $n = 2$. Integrating this in time and noting that $U_k(0) = \emptyset$ whenever $k > k_0$, we find

$$\sup_{[0,T)}\left(\int_{U_k} v_k^2\, d\mu\right) + \int_0^T\!\left(\int v_k^{2^*}\, d\mu\right)^{\frac{2}{2^*}} dt \leq S\sigma p \int_0^T\!\int_{U_k} H^2 f_{\varepsilon,\sigma}^p\, d\mu\, dt\,. \tag{8.30}$$

We need to convert the left-hand side into the space-time L^2-norm of v_k. We achieve this using the interpolation inequality for L^p spaces. Recall that this allows us to estimate, for any $\vartheta \in (0,1)$ and any $r, q > 1$,

$$\|f\|_{q_0} \leq \|f\|_q^{\vartheta}\, \|f\|_r^{1-\vartheta}\,,$$

where $\frac{1}{q_0} \doteqdot \frac{\vartheta}{q} + \frac{1-\vartheta}{r}$. Setting $r = 1$, $q = 2^*/2 > 1$, and $\vartheta = 1/(2 - \frac{1}{q}) \in (0,1)$ we find $q_0 = \frac{1}{\vartheta}$ and hence

$$\int_{U_k} v_k^{2q_0}\, d\mu \leq \left(\int_{U_k} v_k^2\, d\mu\right)^{q_0 - 1}\left(\int_{U_k} v_k^{2^*}\, d\mu\right)^{\frac{2}{2^*}}.$$

Young's inequality now yields

$$\begin{aligned}\left(\int_0^T\!\int_{U_k} v_k^{2q_0}\, d\mu\, dt\right)^{\frac{1}{q_0}} &\leq \left(\sup_{[0,T)}\int_{U_k} v_k^2\, d\mu\right)^{\frac{q_0-1}{q_0}}\left(\int_0^T\left(\int_{U_k} v_k^{2^*}\, d\mu\right)^{\frac{2}{2^*}} dt\right)^{\frac{1}{q_0}}\\ &\leq \sup_{[0,T)}\int_{U_k} v_k^2\, d\mu + \int_0^T\left(\int_{U_k} v_k^{2^*}\, d\mu\right)^{\frac{2}{2^*}} dt\,.\end{aligned}$$

[1]We remark that the argument of [**291**] differs slightly from the one presented here, in that it uses the "$p = 1$" Sobolev inequality and the Hölder inequality. The simplification (at least when $n \geq 3$) using the "$p = 2$" inequality was suggested to us by Klaus Ecker.

Returning to (8.30), the Hölder inequality then yields the desired estimate

$$\begin{aligned}(8.31)\qquad \int_0^T\!\!\int_{U_k} v_k^2\,d\mu\,dt &\le \|U_k\|^{1-\frac{1}{q_0}}\left(\int_0^T\!\!\int_{U_k} v_k^{2q_0}\,d\mu\,dt\right)^{\frac{1}{q_0}}\\ &\le S\sigma p\|U_k\|^{1-\frac{1}{q_0}}\int_0^T\!\!\int_{U_k} H^2 f_{\varepsilon,\sigma}^p\,d\mu\,dt\,.\end{aligned}$$

The right-hand side can be bounded in terms of an appropriate power of $\|U_k\|$ using Hölder's inequality. Indeed, for any $r\ge 1$,

$$\begin{aligned}(8.32)\qquad \int_0^T\!\!\int_{U_k} H^2 f_{\varepsilon,\sigma}^p\,d\mu\,dt &\le \|U_k\|^{1-\frac{1}{r}}\left(\int_0^T\!\!\int_{U_k} H^{2r} f_{\varepsilon,\sigma}^{pr}\,d\mu\,dt\right)^{\frac{1}{r}}\\ &= \|U_k\|^{1-\frac{1}{r}}\left(\int_0^T\!\!\int_{U_k} f_{\varepsilon,\sigma'}^{pr}\,d\mu\,dt\right)^{\frac{1}{r}},\end{aligned}$$

where $\sigma' \doteqdot \sigma+\frac{2}{p}$. Let ℓ be as in Lemma 8.10. If $\sigma\le\frac{\ell}{2}(pr)^{-\frac{1}{2}}$ and $p\ge\max\{\ell^{-1},\frac{16r}{\ell^2}\}$, then $pr\ge\ell^{-1}$ and $\sigma'\doteqdot\sigma+\frac{2}{p}\le\ell(pr)^{-\frac{1}{2}}$. Thus we may apply Lemma 8.10 to estimate the right-hand side of (8.32) to obtain

$$(8.33)\qquad \int_0^T\!\!\int_{U_k} H^2 f_{\varepsilon,\sigma}^p\,d\mu\,dt \le C\|U_k\|^{1-\frac{1}{r}},$$

where $C\doteqdot k_0^p\,(T\mu_0(M))^{\frac{1}{r}}$.

Putting together estimates (8.26), (8.31), and (8.33), we arrive at

$$\|U_h\|\le\frac{SC\sigma p}{(h-k)^p}\|U_k\|^\gamma$$

for all $h>k\ge k_1$, where $\gamma\doteqdot 2-\frac{1}{q_0}-\frac{1}{r}$. Now fix any $r>\frac{q_0}{q_0-1}$ (so that $\gamma>1$) and any $p>\max\{\ell^{-1},\frac{16r}{\ell^2}\}$ and choose $\sigma<\frac{\ell}{2}(pr)^{-\frac{1}{2}}$ sufficiently small that $\sigma p\le 1$. Then we may apply Stampacchia's lemma to conclude that $\|U_k\|=0$ for all $k\ge k_1+d$, where $d^p=SC2^{\frac{\gamma p}{\gamma-1}+1}\|U_{k_1}\|^{\gamma-1}$. That is,

$$f_{\varepsilon,\sigma}\le k_1+d\le K\doteqdot k_1+SC2^{\frac{\gamma p}{\gamma-1}+1}(|\mathcal{M}_0|T)^{\gamma-1}\,.$$

Young's inequality then yields

$$|\mathring{\mathrm{II}}|^2\le\varepsilon H^2+KH^{-\sigma}\le 2\varepsilon H^2+C_\varepsilon\,.$$

This completes the proof of the roundness estimate.

8.3. A gradient estimate for the curvature

Next, we use the roundness estimate to derive an estimate for the gradient of the second fundamental form. Roughly speaking, such an estimate implies that the hypersurface looks umbilic in a neighborhood of a singularity and not only at the singular point.

Theorem 8.14 (Gradient estimate [**291**,**303**]). *Let $X : M^n \times [0,T) \to \mathbb{R}^{n+1}$ be a solution of mean curvature flow for $n \geq 2$ with compact, convex initial datum. For any $\varepsilon > 0$ there exists a constant C_ε — which depends only on ε and the initial embedding — such that*

$$|\nabla \mathrm{II}(x,t)| \leq \varepsilon H^2(x,t) + C_\varepsilon \quad \textit{for all} \quad (x,t) \in M \times [0,T)\,. \tag{8.34}$$

Proof. Recalling Lemma 6.22, we may estimate

$$\begin{aligned}(\partial_t - \Delta)|\nabla \mathrm{II}|^2 &= -2|\nabla^2 \mathrm{II}|^2 + \sum_{i+j+k=1} \nabla^i \mathrm{II} * \nabla^j \mathrm{II} * \nabla^k \mathrm{II} * \nabla \mathrm{II} \\ &\leq c(n)|\mathrm{II}|^2|\nabla \mathrm{II}|^2 - 2|\nabla^2 \mathrm{II}|^2\,.\end{aligned}$$

We will control the bad term using the good term in the evolution equation for $|\mathrm{II}|^2$. The following estimate, which exploits the Codazzi equation to improve the constant in the rough estimate $|\nabla H|^2 \leq n|\nabla \mathrm{II}|^2$, is crucial.

Lemma 8.15. *On any smooth hypersurface $X : M^n \to \mathbb{R}^{n+1}$,*

$$|\nabla \mathrm{II}|^2 \geq \tfrac{3}{n+2}|\nabla H|^2. \tag{8.35}$$

Proof. We decompose

$$\nabla \mathrm{II} = E + F$$

into its trace part

$$E_{ijk} \doteqdot \tfrac{1}{n+2}\left(\nabla_i H g_{jk} + \nabla_j H g_{ki} + \nabla_k H g_{ij}\right)$$

and its trace-free part $F_{ijk} \doteqdot \nabla_i \mathrm{II}_{jk} - E_{ijk}$. As E and F are orthogonal (a fact we invite the reader to check) we conclude

$$|\nabla \mathrm{II}|^2 \geq |E|^2 = \tfrac{3}{n+2}|\nabla H|^2\,. \qquad \square$$

By the roundness estimate, given any $\varepsilon > 0$ there is a constant $C_\varepsilon > 0$ such that $|\mathring{\mathrm{II}}|^2 \leq \varepsilon H^2 + C_\varepsilon$. Thus, fixing any $\varepsilon > 0$,

$$G_\varepsilon \doteqdot 2C_\varepsilon + \varepsilon H^2 - |\mathring{\mathrm{II}}|^2 \geq C_\varepsilon > 0\,.$$

Similarly, there is a constant $C_0 > 0$ such that $|\mathring{\mathrm{II}}|^2 \leq \frac{(n-1)}{n(n+2)}H^2 + C_0$, which, setting $\delta \doteqdot \frac{3}{n+2}$, ensures that

$$G_0 \doteqdot 2C_0 + \delta H^2 - |\mathrm{II}|^2 = 2C_0 + \tfrac{2(n-1)}{n(n+2)}H^2 - |\mathring{\mathrm{II}}|^2 \geq C_0 + \tfrac{(n-1)}{n(n+2)}H^2 > 0\,.$$

We compute

$$(\partial_t - \Delta)G_\varepsilon = 2|\mathrm{II}|^2(G_\varepsilon - 2C_\varepsilon) + 2|\nabla \mathrm{II}|^2 - 2\left(\tfrac{1}{n} + \varepsilon\right)|\nabla H|^2 .$$

Since $G_\varepsilon \geq C_\varepsilon$, we can estimate $G_\varepsilon - 2C_\varepsilon \geq -G_\varepsilon$ and, assuming $\varepsilon \leq \varepsilon_0 \doteqdot \frac{n-1}{n(n+2)}$, we can estimate

$$|\nabla \mathrm{II}|^2 - \left(\tfrac{1}{n} + \varepsilon\right)|\nabla H|^2 \geq \frac{\kappa}{2}|\nabla \mathrm{II}|^2 ,$$

where $\kappa \doteqdot \frac{2(n-1)}{3n}$. Thus,

$$(\partial_t - \Delta)G_\varepsilon \geq -2|\mathrm{II}|^2 G_\varepsilon + \kappa|\nabla \mathrm{II}|^2 .$$

Similarly,

$$(\partial_t - \Delta)G_0 \geq -2|\mathrm{II}|^2 G_0 .$$

We seek a bound[2] for the ratio $\frac{|\nabla \mathrm{II}|^2}{G_\varepsilon G_0}$. Note that, at a local spatial maximum of $\frac{|\nabla \mathrm{II}|^2}{G_\varepsilon G_0}$,

$$0 = \nabla_k \frac{|\nabla \mathrm{II}|^2}{G_\varepsilon G_0} = 2\frac{\langle \nabla_k \nabla \mathrm{II}, \nabla \mathrm{II}\rangle}{G_\varepsilon G_0} - \frac{|\nabla \mathrm{II}|^2}{G_\varepsilon G_0}\left(\frac{\nabla_k G_\varepsilon}{G_\varepsilon} + \frac{\nabla_k G_0}{G_0}\right) .$$

In particular,

$$4\frac{|\nabla \mathrm{II}|^2}{G_\varepsilon G_0}\left\langle \frac{\nabla G_\varepsilon}{G_\varepsilon}, \frac{\nabla G_0}{G_0}\right\rangle \leq \frac{|\nabla \mathrm{II}|^2}{G_\varepsilon G_0}\left|\frac{\nabla G_\varepsilon}{G_\varepsilon} + \frac{\nabla G_0}{G_0}\right|^2 \leq 4\frac{|\nabla^2 \mathrm{II}|^2}{G_\varepsilon G_0} .$$

Suppose that $\frac{|\nabla \mathrm{II}|^2}{G_\varepsilon G_0}$ reaches a new interior minimum at (x_0, t_0). Then

$$\begin{aligned}
0 &\leq (\partial_t - \Delta)\frac{|\nabla \mathrm{II}|^2}{G_\varepsilon G_0} \\
&= \frac{(\partial_t - \Delta)|\nabla \mathrm{II}|^2}{G_\varepsilon G_0} - \frac{|\nabla \mathrm{II}|^2}{G_\varepsilon G_0}\left(\frac{(\partial_t - \Delta)G_\varepsilon}{G_\varepsilon} + \frac{(\partial_t - \Delta)G_0}{G_0}\right) \\
&\quad + \frac{2}{G_\varepsilon G_0}\left\langle \nabla \frac{|\nabla \mathrm{II}|^2}{G_\varepsilon G_0}, \nabla(G_\varepsilon G_0)\right\rangle + 2\frac{|\nabla \mathrm{II}|^2}{G_\varepsilon G_0}\left\langle \frac{\nabla G_\varepsilon}{G_\varepsilon}, \frac{\nabla G_0}{G_0}\right\rangle \\
&\leq \frac{|\nabla \mathrm{II}|^2}{G_\varepsilon G_0}\left((c+4)|\mathrm{II}|^2 - \kappa\frac{|\nabla \mathrm{II}|^2}{G_\varepsilon}\right)
\end{aligned}$$

at (x_0, t_0). It follows that

$$\frac{|\nabla \mathrm{II}|^2}{G_\varepsilon G_0} \leq \frac{(c+4)}{\kappa}\frac{|\mathrm{II}|^2}{G_0} \leq \frac{(c+4)}{\kappa}\frac{n(n+2)}{n-1}\frac{|\mathrm{II}|^2}{H^2} \leq \frac{(c+4)}{\kappa}\frac{n(n+2)}{n-1}$$

[2] Eventually, we will want to take ε arbitrarily small, but we will need to keep δ positive and independent of ε, which explains why we do not simply consider $\frac{|\nabla \mathrm{II}|^2}{G_\varepsilon^2}$.

at any local parabolic maximum of $\frac{|\nabla \mathrm{II}|^2}{G_\varepsilon G_0}$. On the other hand, if no such point exists, then

$$\frac{|\nabla \mathrm{II}|^2}{G_\varepsilon G_0} \leq C_0 \doteqdot \max_{M\times\{0\}} \frac{|\nabla \mathrm{II}|^2}{G_\varepsilon G_0}\,.$$

We conclude that

$$\frac{|\nabla \mathrm{II}|^2}{G_\varepsilon G_0} \leq C \doteqdot \max\left\{C_0, \frac{(c+4)}{\kappa}\frac{n(n+2)}{n-1}\right\}.$$

The claim now follows from Young's inequality. □

We will need the following consequence of the gradient estimate.

Corollary 8.16. *Let $X : M^n \times [0,T) \to \mathbb{R}^{n+1}$, where $n \geq 2$, be a maximal solution of mean curvature flow evolving from a compact, convex initial embedding $X_0 : M^n \to \mathbb{R}^{n+1}$. Then*

$$\lim_{t\to T}\frac{H_{\max}}{H_{\min}} = 1 \quad \textit{and} \quad \lim_{t\to T}\operatorname{diam}_t(M) = 0\,, \tag{8.36}$$

where $H_{\max} \doteqdot \max_{M\times\{t\}} H$, $H_{\min} \doteqdot \min_{M\times\{t\}} H$, *and* $\operatorname{diam}_t(M) \doteqdot \max_{x,y\in M} d_{g(t)}(x,y)$ *is the intrinsic diameter of* $(M^n, g(t))$.

Proof. By the gradient estimate (Theorem 8.14), for every $\eta > 0$ there is a constant $C_\eta < \infty$ such that

$$|\nabla H| \leq \frac{1}{2}\eta^2 H^2 + C_\eta\,.$$

Since $H_{\max}(t) \to \infty$ as $t \to T$, there is, for every $\eta > 0$, some point $(x_\eta, t_\eta) \in M \times [0,T)$ such that

$$H_\eta^2 \doteqdot H^2(x_\eta, t_\eta) = H_{\max}^2(t_\eta) \geq 8C_\eta/\eta^2$$

and hence

$$|\nabla H|(x, t_\eta) \leq \eta^2 H(x_\eta, t_\eta)^2$$

for all $x \in M$. Now let γ be a unit speed geodesic (with respect to the metric at time t_η) through $\gamma(0) = x_\eta$. Then, for each $s \leq L \doteqdot \eta^{-1}H_\eta^{-1}$, the mean value theorem provides some $s_0 \in (0,s)$ such that

$$H(\gamma(s), t_\eta) = H_\eta + s\nabla_{\gamma'(s_0)}H(\gamma(s_0), t_\eta) \geq H_\eta(1-\eta)\,. \tag{8.37}$$

Next, recall from the trace of the Gauß equation that

$$\mathrm{Rc} = H\mathrm{II} - \mathrm{II}^2\,.$$

Thus, applying the estimate $\mathrm{II} \geq \frac{\alpha}{n}g$,

$$\mathrm{Rc}(\gamma', \gamma') \geq (n-1)\frac{\alpha^2}{n^2}H^2 \geq (n-1)\frac{\alpha^2}{n^2}H_\eta^2(1-\eta)\,.$$

If $\eta < \frac{1}{2}$, then

$$\mathrm{Rc}(\gamma', \gamma') \geq (n-1)K^2 g\,,$$

where $K \doteqdot \frac{\alpha}{2n} H_\eta$. Choosing further $\eta \leq \frac{\alpha}{2n\pi}$, we obtain $L \geq \pi K^{-1}$. Myers's theorem then implies that every point of M_{t_η} is reached by a geodesic of length at most L and we conclude from (8.37) that

$$H_{\min}(t_\eta) \geq (1-\eta) H_{\max}(t_\eta)\,.$$

Finally, since $H_{\min}$ is nondecreasing,

$$H^2_{\max}(t) \geq (1-\eta)^2 H^2_{\max}(t_\eta) \geq \frac{1}{4} H^2_\eta \quad \text{for all} \quad t \geq t_\eta\,.$$

We leave it to the reader to check that the above arguments then hold for all $t \geq t_\eta$. We conclude that, given any $\eta \leq \min\{\frac{\alpha}{2n\pi}, \frac{1}{2}\}$, there is some time $t_\eta \in [0, T)$ such that

$$\mathrm{diam}_t(M) \leq \frac{1}{\eta H_{\max}(t)} \quad \text{and} \quad H_{\min}(t) \geq (1-\eta) H_{\max}(t)$$

for all $t > t_\eta$. The corollary follows. □

8.4. Huisken's theorem

We now have all the ingredients in place needed to prove Huisken's theorem (Theorem 8.1).

8.4.1. Convergence to a point. By the avoidance principle, the precompact regions $\Omega_t \subset \mathbb{R}^{n+1}$ bounded by the time slices $\mathcal{M}_t$ satisfy $\Omega_{t_2} \subset \Omega_{t_1}$ for all $0 \leq t_1 \leq t_2 < T$. Since $\mathrm{diam}(\mathcal{M}_t) \to 0$ as $t \to T$ (Corollary 8.16), we conclude that $\bigcap_{t \in [0,T)} \Omega_t$ is a single point, p say.

This proves the first part of Theorem 8.1. To complete the proof, we need to obtain appropriate decay estimates for the rescaled solution and its derivatives.

8.4.2. Estimates in C^0 after rescaling. We first show that the images of the rescaled embeddings converge in the Hausdorff topology to the unit sphere.

By Lemma 8.3 and Corollary 8.16, the mean curvature satisfies

$$\sqrt{2n(T-t)}\,\tfrac{1}{n} H(x,t) \to 1 \;\text{ as }\; t \to T \tag{8.38}$$

uniformly in $x \in M$. Integration in time then yields

$$\limsup_{t\to T} \frac{\sup_{x\in M} \|X(x,t) - p\|}{\sqrt{2n(T-t)}} \leq \limsup_{t\to T} \frac{\sup_{x\in M} \int_t^T H(x,s)\,ds}{\sqrt{2n(T-t)}} = 1\,. \tag{8.39}$$

To obtain the reverse inequality, we first combine the roundness estimate (Theorem 8.6) with the uniform curvature blow-up (Corollary 8.16) to deduce

$$\frac{|\mathring{\mathrm{II}}|^2}{H^2}(x,t) \to 0 \text{ as } t \to T$$

uniformly in $x \in M$. It follows that

$$\sqrt{2n(T-t)}\,\frac{\mathrm{II}_{(x,t)}(v,v)}{g_{(x,t)}(v,v)} \to 1 \text{ as } t \to T \tag{8.40}$$

uniformly in $x \in M$ and $v \in T_xM \setminus \{0\}$. In particular,

$$\liminf_{t\to T} \frac{\rho_-(t)}{\sqrt{2n(T-t)}} \geq \liminf_{t\to T} \frac{1}{\sqrt{2n(T-t)}\,\max\limits_{\mathcal{M}\times\{t\}} \kappa_n} = 1\,, \tag{8.41}$$

where

$$\rho_-(t) \doteqdot \sup\{r : B_r(q) \subset \Omega_t \text{ for some } q \in \mathbb{R}^{n+1}\}$$

is the **inradius** of $\mathcal{M}_t$. It follows that

$$\frac{\|p_t - p\|}{\sqrt{2n(T-t)}} \to 0\,,$$

where p_t is the center of the largest ball enclosed by $\mathcal{M}_t$. Indeed, whenever $p_t \neq p$, we can choose $x_t \in \mathcal{M}$ such that p_t lies on the segment joining $X(x_t,t)$ and p, so that

$$\|p - p_t\| = \|X(x_t,t) - p\| - \|X(x_t,t) - p_t\| \leq \|X(x_t,t) - p\| - \rho_-(t) \tag{8.42}$$

and the claim follows from (8.39) and (8.41). Estimating

$$\frac{\|X(x,t) - p\|}{\sqrt{2n(T-t)}} \geq \frac{\|X(x,t) - p_t\|}{\sqrt{2n(T-t)}} - \frac{\|p_t - p\|}{\sqrt{2n(T-t)}} \to 1 \text{ as } t \to T$$

uniformly in $x \in \mathcal{M}$, we may now conclude that

$$\frac{\|X(x,t) - p\|}{\sqrt{2n(T-t)}} \to 1 \text{ as } t \to T$$

uniformly in $x \in M$. This proves Hausdorff convergence of the image hypersurfaces to the unit sphere. To obtain C^0 convergence of the parametrizations, it will suffice to show that the derivative of the rescaled family of embeddings converges.

8.4.3. Estimates in C^1 after rescaling. Recalling the evolution equation (6.15) for the induced metric g we find, for any $x \in M$ and any nonzero $v \in \mathbb{R}^n$, that the rescaled metric satisfies

$$\partial_t \log \frac{g_{(x,t)}(v,v)}{2n(T-t)} = \frac{2n}{2n(T-t)} - 2H(x,t)\frac{\mathrm{II}_{(x,t)}(v,v)}{g_{(x,t)}(v,v)}.$$

Thus, to prove that the rescaled metric has a nondegenerate limit, it suffices to show that

$$-\infty < \int_0^T \left(\frac{1}{2n(T-t)} - \tfrac{1}{n}H(x,t)\frac{\mathrm{II}_{(x,t)}(v,v)}{g_{(x,t)}(v,v)}\right) dt < \infty$$

for all $x \in \mathcal{M}$ and $v \in T_x\mathcal{M} \setminus \{0\}$.

We first consider the lower bound. Recalling that

$$\frac{1}{n}H_{\min} \le \frac{1}{\sqrt{2n(T-t)}},$$

we may estimate

$$\tfrac{1}{n}H\frac{\mathrm{II}(v,v)}{g(v,v)} - \frac{1}{2n(T-t)} \le \tfrac{1}{n}H|\mathring{\mathrm{II}}| + \frac{1}{n^2}\left(H_{\max}^2 - H_{\min}^2\right). \tag{8.43}$$

In order to estimate these terms effectively, we will need to obtain a better decay rate for the terms on the right than given by Corollary 8.16 and (8.40). The following lemma will allow us to bound both terms by an integrable power of the remaining time.

Lemma 8.17. *There are constants $N < \infty$, $C < \infty$, and $\delta > 0$ such that*

$$N\frac{|\mathring{\mathrm{II}}|^2}{H^2} + \frac{|\nabla \mathrm{II}|^2}{H^4} \le C(T-t)^\delta. \tag{8.44}$$

Proof. Recalling Lemma 6.22, we may estimate

$$(\partial_t - \Delta)|\nabla \mathrm{II}|^2 \le (4+c)|\mathrm{II}|^2|\nabla \mathrm{II}|^2 - 2|\nabla^2 \mathrm{II}|^2,$$

where c is a constant which depends only on n. Thus,

$$\begin{aligned}(\partial_t - \Delta)\frac{|\nabla \mathrm{II}|^2}{H^4} &\le c\frac{|\nabla \mathrm{II}|^2}{H^2} - 2\frac{|\nabla^2 \mathrm{II}|^2}{H^4} + 8\left\langle \nabla\frac{|\nabla \mathrm{II}|^2}{H^4}, \frac{\nabla H}{H}\right\rangle \\ &\le c\frac{|\nabla \mathrm{II}|^2}{H^2} - 2\frac{|\nabla^2 \mathrm{II}|^2}{H^4} + 16n\frac{|\nabla^2 \mathrm{II}||\nabla \mathrm{II}|^2}{H^5} \\ &\le c\frac{|\nabla \mathrm{II}|^2}{H^2} - \frac{|\nabla^2 \mathrm{II}|^2}{H^4} + 64n^2\frac{|\nabla \mathrm{II}|^4}{H^6}.\end{aligned}$$

Since $H \sim (T-t)^{-\frac{1}{2}} \to \infty$, the gradient estimate (Theorem 8.14) implies that

$$\frac{|\nabla \mathrm{II}|^2}{H^2} \le H^2$$

and hence

$$(\partial_t - \Delta)\frac{|\nabla \mathrm{II}|^2}{H^4} \le -\frac{|\nabla^2 \mathrm{II}|^2}{H^4} + (c + 64n^2)\frac{|\nabla \mathrm{II}|^2}{H^2}$$

for t sufficiently close to T. By decomposing $\nabla^2 \mathrm{II}$ into certain (orthogonal) symmetric and skew-symmetric parts and recalling (8.20), we can estimate

$$|\nabla^2 \mathrm{II}|^2 \ge \gamma H^4 |\mathring{\mathrm{II}}|^2 , \tag{8.45}$$

where $\gamma > 0$ depends only on n and the initial pinching constant α. Thus,

$$(\partial_t - \Delta)\frac{|\nabla \mathrm{II}|^2}{H^4} \le -\gamma|\mathring{\mathrm{II}}|^2 + c'\frac{|\nabla \mathrm{II}|^2}{H^2}$$

for t sufficiently close to T, where $c' \doteqdot c + 64n^2$.

On the other hand, the roundness estimate (Theorem 8.6) and Lemma 8.8 yield

$$\begin{aligned}(\partial_t - \Delta)\frac{|\mathring{\mathrm{II}}|^2}{H^2} &\le 4\left\langle \nabla \frac{|\mathring{\mathrm{II}}|^2}{H^2}, \frac{\nabla H}{H}\right\rangle - 2\gamma' \frac{|\nabla \mathrm{II}|^2}{H^2} \\ &\le 16n\frac{|\nabla \mathrm{II}|^2}{H^2}\frac{|\mathring{\mathrm{II}}|}{H} - 2\gamma'\frac{|\nabla \mathrm{II}|^2}{H^2} \le -\gamma'\frac{|\nabla \mathrm{II}|^2}{H^2}\end{aligned}$$

for t sufficiently close to T, where $\gamma' > 0$ is a constant which depends only on n and the initial pinching constant α. Setting $N \doteqdot \frac{c'+\delta}{\gamma'}$, where δ is the positive solution of

$$(c' + \delta)\delta = \gamma\gamma' ,$$

we find

$$\begin{aligned}(\partial_t - \Delta)\left(\frac{|\nabla \mathrm{II}|^2}{H^4} + N\frac{|\mathring{\mathrm{II}}|^2}{H^2}\right) &\le -\gamma|\mathring{\mathrm{II}}|^2 - \delta\frac{|\nabla \mathrm{II}|^2}{H^2} \\ &= -\delta H^2\left(N\frac{|\mathring{\mathrm{II}}|^2}{H^2} + \frac{|\nabla \mathrm{II}|^2}{H^4}\right) \\ &\le -\frac{\delta n}{4(T-t)}\left(N\frac{|\mathring{\mathrm{II}}|^2}{H^2} + \frac{|\nabla \mathrm{II}|^2}{H^4}\right)\end{aligned}$$

for t sufficiently close to T. The comparison principle then yields

$$\log\left(\frac{|\nabla \mathrm{II}|^2}{H^4} + N\frac{|\mathring{\mathrm{II}}|^2}{H^2}\right)\Bigg|_{t_0}^{t} \le \frac{\delta n}{4}\log(T-t)\big|_{t_0}^{t}$$

for all $t > t_0$ for t_0 sufficiently close to T. The claim follows. □

Corollary 8.18. *There are constants $C < \infty$ and $\delta > 0$ such that*

$$H_{\max} - H_{\min} \le C(T-t)^{\delta} H_{\max} . \tag{8.46}$$

Proof. This is proved by integrating the gradient bound of the previous lemma along geodesics, as in Corollary 8.16. See Exercise 8.8. □

We conclude that

$$\int_0^T \partial_t \left(\frac{g_{(x,t)}(v,v)}{2n(T-t)} \right) dt > -\infty \,.$$

To obtain the upper bound, observe that

$$H\frac{\mathrm{II}(v,v)}{g(v,v)} - \frac{1}{2(T-t)} \geq -H|\mathring{\mathrm{II}}| - \frac{1}{n}\left(H_{\max}^2 - H_{\min}^2\right) + \frac{1}{n}H_{\max}^2 - \frac{1}{2(T-t)} \,.$$

Lemma 8.19. *There are constants $C < \infty$ and $\delta > 0$ such that*

$$\frac{1}{2n(T-t)} - \frac{1}{n^2}H_{\max}^2 \leq C(T-t)^{\delta-1} \,. \tag{8.47}$$

Proof. By Lemma 8.17, we may estimate

$$(\partial_t - \Delta)H = |\mathrm{II}|^2 H = \left(\frac{|\mathring{\mathrm{II}}|^2}{H^2} + \frac{1}{n} \right) H^3 \leq \left(C(T-t)^\delta + \frac{1}{n} \right) H^3 \,.$$

The ODE comparison principle then yields

$$H_{\max}^{-2}(t) \leq 2\left(\frac{C}{\delta+1}(T-t)^{\delta+1} + \frac{1}{n}(T-t) \right)$$

and hence

$$\frac{1}{n^2}H_{\max}^2(t) - \frac{1}{2n(T-t)} \geq -\frac{\frac{nC}{\delta+1}(T-t)^\delta}{2n(T-t)\left[\frac{nC}{\delta+1}(T-t)^\delta + 1\right]} \,.$$

The claim follows. □

It follows that

$$\int_0^T \partial_t \left(\frac{g_{(x,t)}(v,v)}{2n(T-t)} \right) dt < \infty \,.$$

We can now conclude that the rescaled metrics converge uniformly as $t \to T$ to a nondegenerate limit metric.

8.4.4. Estimates in C^∞ after rescaling.

We have already proved that the rescaled Weingarten tensor converges to the identity map (8.40) and that its derivative decays to zero (Lemma 8.17). The Bernstein estimates (Theorem 6.24) yield bounds for all higher derivatives of the rescaled curvature, so we can obtain decay estimates by interpolation.

Proposition 8.20 (Interpolation inequality for tensor fields). *Let T be a smooth tensor field of rank (p,q) on a compact Riemannian n-manifold (M^n, g) equipped with a metric connection ∇. For every $k \in \mathbb{N}$ there exists a constant $c_k < \infty$, which depends only on n, $p+q$, and k, such that*

$$|\nabla^k T|^2_{C^0} \leq c_k |\nabla^{k-1} T|_{C^0} |\nabla^{k+1} T|_{C^0}. \tag{8.48}$$

Proof. Let T be a tensor field of covariant order p and contravariant order q. It suffices to prove the claim when $k = 1$ since, for other values of k, the claim follows immediately upon replacing T with $\nabla^{k-1} T$.

Consider first the case that $|T|_{C^0} |\nabla^2 T|_{C^0} = 0$. We claim that $|\nabla T|_{C^0} = 0$. This is certainly the case if $|T|_{C^0} = 0$, so suppose that $|\nabla^2 T|_{C^0} = 0$. Consider the function $f : M^n \to \mathbb{R}$ defined by

$$f(x) \doteqdot \frac{1}{2} |T_x|^2.$$

If $|\nabla T|_{C^0} \neq 0$, then we can find $x \in M^n$ and a unit vector $v \in T_x M^n$ such that $\nabla_v T_x \neq 0$. Let $\gamma : \mathbb{R} \to M^n$ be the geodesic with $(\gamma(0), \gamma'(0)) = (x, v)$. Observe that

$$\frac{d}{ds} |\nabla_{\gamma'} T|^2 = 2g(\nabla_{\gamma'} \nabla_{\gamma'} T, \nabla_{\gamma'} T) = 0.$$

So $|\nabla_{\gamma'} T|^2 \equiv |\nabla_v T_x|^2 \neq 0$. A similar calculation yields

$$\frac{d^2}{ds^2} (f \circ \gamma) = |\nabla_{\gamma'} T|^2.$$

It follows that f grows linearly along γ. But this is impossible since M^n is compact (and hence f is bounded).

So we may assume that $|T|_{C^0} |\nabla^2 T|_{C^0} \neq 0$. Fix $x \in M^n$, unit vectors $u, u_1, \ldots, u_p \in T_x M^n$, and unit covectors $\alpha_1, \ldots, \alpha_q \in T_x^* M^n$. Let $\gamma : \mathbb{R} \to M^n$ be the unit speed geodesic with $\gamma'(0) = u$. Extend $u, u_1, \ldots, u_p$ and $\alpha_1, \ldots, \alpha_q$ by parallel translation to form vector and covector fields, respectively, along γ. Define $f : \mathbb{R} \to \mathbb{R}$ by

$$f(s) = T(u_1, \ldots, u_p, \alpha_1, \ldots, \alpha_q)(\gamma(s)).$$

Recalling the interpolation inequality (3.44) for real-valued functions, we obtain at the point x that

$$|\nabla_u T(u_1, \ldots, u_p, \alpha_1, \ldots, \alpha_q)| = f'(0) \leq 2 \sup_{s \in \mathbb{R}} |f(s)|^{1/2} \sup_{s \in \mathbb{R}} |f''(s)|^{1/2}.$$

It follows that

$$|\nabla T|_0^2 \leq 4 |T|_0 |\nabla^2 T|_0,$$

where $|\cdot|_0$ is the operator norm. The claim follows since the operator norm is equivalent to the norm induced by g, with the constant depending only on n and $p+q$. $\square$

Lemma 8.21. *For every $\gamma > 0$ and each $k \in \mathbb{N}$ there exists a constant $C_k < \infty$ such that*

$$(T-t)^{1+k}|\nabla^k \mathrm{II}|^2 \le C_k (T-t)^\gamma . \tag{8.49}$$

Proof. By combining the decay estimate of Lemma 8.17 with the tensor interpolation inequalities of Proposition 8.20 as in the proof of Grayson's theorem (see Lemma 3.27), it suffices to bound the rescaled derivatives $(T-t)^{k+1}|\nabla^k \mathrm{II}|^2$ by some constant, $\widetilde{C}_k$ say, for each $k \ge 2$. We can achieve this using the rapid smoothing estimates (Theorem 6.24) exactly as in the proof of Grayson's theorem (see Lemma 3.25). □

To obtain estimates in C^∞ for the rescaled embeddings, we adapt the proof of Theorem 6.20. We recall that the rescaled embeddings $\widetilde{X}(\cdot,t) : \mathcal{M}^n \to \mathbb{R}^{n+1}$ are defined for $t \in [0,T)$ by

$$\widetilde{X}(\cdot,t) \doteqdot \frac{X(\cdot,t)-p}{\sqrt{2n(T-t)}} .$$

As in the proof of Theorem 6.20, we work with a fixed (time-independent) metric $\langle\cdot,\cdot\rangle$ and connection D (the initial ones, say) on $T\mathcal{M}^n$.

Lemma 8.22. *There exists $\lambda > 0$ such that*

$$\widetilde{g}_t \ge \lambda \widetilde{g}_0$$

for all $t \in [0,T)$, where $\widetilde{g}_t$ is the metric induced by $\widetilde{X}(\cdot,t)$.

There exist $\delta > 0$ and, for each $m \in \mathbb{N}$, $C_m < \infty$ such that

$$|D^m \widetilde{X}| + (T-t)^{1-\delta}|D_t D^m \widetilde{X}| \le C_m . \tag{8.50}$$

Proof. The first claim was proved in Section 8.4.3. So we are free to work with either metric.

Observe that

$$2(T-t)\partial_t \widetilde{X} = \widetilde{X} - \tfrac{1}{n}\widetilde{H}\widetilde{\mathsf{N}} .$$

In particular,

$$\begin{aligned} 2(T-t)D_t\widetilde{X}_* &= \widetilde{X}_* - D\widetilde{H}\otimes\widetilde{\mathsf{N}} - \tfrac{1}{n}\widetilde{H}\widetilde{X}_*\widetilde{\mathrm{L}} \\ &= \left(1-\tfrac{1}{n^2}\widetilde{H}^2\right)\widetilde{X}_* - D\widetilde{H}\otimes\widetilde{\mathsf{N}} - \widetilde{H}\widetilde{X}_*\left(\widetilde{\mathrm{L}} - \tfrac{1}{n}\widetilde{H}\mathrm{I}\right) . \end{aligned}$$

It follows from Lemma 8.17 and Corollary 8.18 that

$$|D_t\widetilde{X}_*| \le C_1(T-t)^{\delta-1}$$

for some $\delta > 0$ and $C_1 < \infty$. Integrating then yields a bound for $|\widetilde{X}_*|$, which proves the claim when $m = 1$. It is not hard to see that the claim follows in general, for if we differentiate $2(T-t)D_t\widetilde{X}_*$ a further m times, then, assuming that the claim holds up to order m, we see that each factor will contain bounded terms multiplied by either a nontrivial derivative of the curvature, the factor $\left(1 - \frac{1}{n^2}\widetilde{H}^2\right)$, or the factor $\left(\widetilde{\mathrm{L}} - \frac{1}{n}\widetilde{H}\mathrm{I}\right)$, each of which is bounded by a constant multiple of $(T-t)^{\delta-1}$. So we obtain

$$|D_t D^{m+1}\widetilde{X}| \le C_1(T-t)^{\delta-1}\,.$$

Integration then yields a bound for $|D^{m+1}\widetilde{X}|$. □

Thus,

$$\partial_t|D^m\widetilde{X}|^2 \le 2C_m^2(T-t)^{\delta-1}\,.$$

Integrating, we find that $|D^m X|$ is Cauchy in t for each $m \in \mathbb{N}$.

We can now conclude that the rescaled embeddings $\widetilde{X}(\cdot,t)$ converge in $C^\infty(M^n,\mathbb{R}^{n+1})$ as $t \to T$ to an embedding whose image coincides with the unit sphere: Choose a finite number of compact sets $K_1,\dots,K_\mu$ whose interiors cover M and charts $\varphi_1 : U_1 \to \mathbb{R}^n,\dots,\varphi_\mu : U_\mu \to \mathbb{R}^n$ with $K_i \subset U_i$ for each $i = 1,\dots,\mu$. Set $\widetilde{X}_i \doteqdot \widetilde{X}\circ\varphi_i$. Since the components of the metric and connection with respect to each coordinate chart are uniformly bounded in each of the corresponding compact sets, equations (6.86) and (6.87) and a simple induction argument imply that $D\widetilde{X}_i$ is Cauchy with respect to t in the space $C^k(K_i,\mathbb{R}^n)$ for each $k \in \mathbb{N}$. Since $C^k(K_i,\mathbb{R}^{n+1})$ is complete, it follows that $D\widetilde{X}_i(\cdot,t)$ converges in $C^\infty(U_i,\mathbb{R}^{n+1})$ for each $i = 1,\dots,\mu$. We can then conclude, as in the proof of Grayson's theorem, that $\widetilde{X}_i(\cdot,t)$ converges in $C^\infty(K_i,\mathbb{R}^{n+1})$ for each $i = 1,\dots,\mu$: Let p_t be the center of an inscribed sphere at each time t. Then $\frac{X_i(\cdot,t)-p_t}{\sqrt{2n(T-t)}}$ converges in $C^\infty(U_i,\mathbb{R}^{n+1})$ for each $i = 1,\dots,\mu$. The arguments of Section 8.4.2 show that rescaled distance $\frac{\|p-p_t\|}{\sqrt{2n(T-t)}}$ goes to zero as $t \to T$, so in fact $\widetilde{X}_i(\cdot,t)$ converges. Pasting the pieces together, we conclude that $\widetilde{X}(\cdot,t)$ converges in the compact-open C^∞ topology to a smooth embedding as $t \to T$. Its image must be the unit sphere by (8.42).

8.5. Regularity of the arrival time

The estimates for the second fundamental form and its derivatives, derived in the previous section, imply that the arrival time of a convex solution to mean curvature flow, a priori only Lipschitz regular, is in fact of class C^2 [**296**, Theorem 6.1].

Theorem 8.23. *The arrival time of a convex, compact solution to mean curvature flow is of class C^2.*

Proof. Let $\{\mathcal{M}_t^n\}_{t\in[0,T)}$ be a solution to mean curvature flow with $\mathcal{M}_0^n$ bounding a bounded, convex region Ω. Recall that the arrival time $u:\overline{\Omega}\to\mathbb{R}$ is defined by

$$u(X)=t \text{ if and only if } X\in\mathcal{M}_t^n\,.$$

Since the hypersurfaces move inward and converge to a single point p as $t\to T$, u is well-defined if we set $u(p)\doteqdot T$. The claim is that $u\in C^2(\Omega)$. Let $X:M^n\times[0,T)\to\mathbb{R}^{n+1}$ be a smooth family of parametrizations $X(\cdot,t)$ of $\mathcal{M}_t^n$. Then

$$u(X(x,t))=t\,. \tag{8.51}$$

Fix a point $q=X(x,t)$ in Ω and local g_t-normal coordinates $\{x^i\}_{i=1}^n$ for M^n about x. Choose the basis $\{e_i\}_{i=1}^{n+1}$ for $\mathbb{R}^{n+1}$ so that $e_{n+1}=\mathsf{N}(x,t)$ and $e_i=\partial_iX(x,t)$ for each $i=1,\ldots,n$. Differentiating (8.51) yields the identities

$$Du\cdot\partial_iX=0 \text{ and } -HDu\cdot\mathsf{N}=1\,. \tag{8.52}$$

Thus,

$$Du=-\frac{\mathsf{N}}{H}\,.$$

Differentiating (8.52) at the point $q=X(x,t)$ then yields

$$D^2u=\begin{pmatrix}-\mathrm{II}/H & \nabla H/H^2\\ \nabla H/H^2 & -\partial_tH/H^3\end{pmatrix}. \tag{8.53}$$

By Lemmas 8.17 and 8.21,

$$\frac{\partial_tH}{H^3}=\frac{\Delta H}{H^3}+\frac{|\mathrm{II}|^2}{H^2}\to\frac{1}{n}$$

as $t\to T$. Thus, by Lemma 8.17,

$$D^2u\to\frac{1}{n}\mathrm{I}$$

as $q\to p$. The claim follows. ☐

In the 1-dimensional setting, u is even C^3 [**334**]. This is not true in higher dimensions, however [**477**].

8.6. Huisken's theorem via width pinching

We shall now present a quite different proof of Huisken's theorem, which appears in [**26**] in the context of a more general class of hypersurface flows. The principal observation is the general fact that pinching of the principal curvatures implies pinching of the widths. Recall that the **width** $w(z)$ of a compact, convex set $\Omega\subset\mathbb{R}^{n+1}$ in a given direction $z\in S^n$ is defined by

$$w(z)\doteqdot\sigma(z)+\sigma(-z)\,, \tag{8.54}$$

where $\sigma : S^n \to \mathbb{R}$ is the **support function** of Ω, defined by

$$\sigma(z) \doteqdot \sup_{X\in\Omega} \langle X, z\rangle \ .$$

Define the **maximum** and **minimum widths** of Ω by

$$w_+ = \max_{z\in S^n} w(z) \ \text{ and } \ w_- = \min_{z\in S^n} w(z)\,, \tag{8.55}$$

respectively. Observe that the maximum width is just the diameter of Ω,

$$w_+ = \max_{X,Y\in\Omega} \|X - Y\| \, .$$

If Ω is open and its boundary $\mathcal{M} \doteqdot \partial\Omega$ is smooth and locally uniformly convex, then

$$\sigma(z) = \left\langle \mathsf{N}^{-1}(z), z\right\rangle \,,$$

where $\mathsf{N} : \partial\Omega \to S^n$ is the Gauß map of $\mathcal{M}$, and hence

$$\sigma(\mathsf{N}(X)) = \langle X, \mathsf{N}(X)\rangle$$

for any $X \in \partial\Omega$. We will abuse notation by defining $\sigma : \mathcal{M} \to \mathbb{R}$ by

$$\sigma(X) = \langle X, \mathsf{N}(X)\rangle \ .$$

Define the maximal width w_+ of $\mathcal{M}$ and minimal width w_- of $\mathcal{M}$ to be those of Ω. We have the following *width pinching estimate.*

Lemma 8.24 (Andrews [**26**]). *Let $\Omega \subset \mathbb{R}^{n+1}$, $n \geq 2$, be a bounded, locally uniformly convex open set with smooth boundary $\mathcal{M} = \partial\Omega$. If the principal curvatures of $\mathcal{M}$ are pointwise pinched, $\kappa_n(x) \leq C\kappa_1(x)$ for all $x \in \mathcal{M}$ for some $C < \infty$, then the widths of $\mathcal{M}$ are pinched by the same constant*:

$$w_+ \leq C w_- \, . \tag{8.56}$$

Proof. We first prove the claim when $n = 2$. Let e be any unit vector in $\mathbb{R}^3$ and set $N \doteqdot \mathsf{N}^{-1}(e)$ and $S \doteqdot \mathsf{N}^{-1}(-e)$, so that $w(e) = \langle N - S, e\rangle$. Define the curves $\Gamma_h \doteqdot \mathcal{M} \cap \{X \cdot e = h\}$ of constant "height", h. Each curve Γ_h is a convex planar curve and hence its tangent vector $\mathsf{T}(h,\cdot)$ and curvature $\kappa(h,\cdot)$ are defined almost everywhere. We claim that

$$\int_{\mathcal{M}} \mathrm{II}(\mathsf{T},\mathsf{T})\, d\mu = \int_{w(S)}^{w(N)} \int_{\Gamma_h} \kappa \, ds_h \, dh = 2\pi w(e)\,, \tag{8.57}$$

where s_h is the arc length parameter of Γ_h. The lemma then follows easily:

$$w_+ \leq \frac{1}{2\pi} \int_{\mathcal{M}} \kappa_n d\mu \leq \frac{C}{2\pi} \int_{\mathcal{M}} \kappa_1 d\mu \leq C w_- \, .$$

To prove (8.57), define $\mathsf{B} \doteqdot \mathsf{T} \times \mathsf{N}$ so that $\{\mathsf{T},\mathsf{B}\}$ is a consistently oriented, almost everywhere defined basis for $T\mathcal{M}$. Since $\nabla h \propto \mathsf{T}^\perp$, the volume form $d\mu$ for $\mathcal{M}$ can be written as

$$d\mu = \phi \, ds \wedge dh$$

for some L^1 function ϕ, where, abusing notation, ds is the 1-form dual to T. Since ∇h is the projection of e onto $T\mathcal{M}$,

$$1 = d\mu(\mathsf{T}, \mathsf{B}) = \phi\, dh(\mathsf{B}) = \phi \left\langle \mathsf{T} \times \mathsf{N}, e \right\rangle = \phi \left\langle e \times \mathsf{T}, \mathsf{N} \right\rangle ,$$

where we used the total antisymmetry of the vector triple product. On the other hand, since the normal vector to Γ_h is equal to $e \times \mathsf{T}$, the Gauß–Weingarten equation implies that

$$\mathrm{II}(\mathsf{T}, \mathsf{T}) = \kappa \left\langle e \times \mathsf{T}, \mathsf{N} \right\rangle .$$

We conclude that

$$\phi\, \mathrm{II}(\mathsf{T}, \mathsf{T}) = \kappa\, .$$

The identity (8.57) follows.

If $n \geq 3$, then the same argument can be applied to the projection of Ω onto any 3-dimensional subspace of $\mathbb{R}^{n+1}$. So we only need to show that, given any 3-dimensional subspace Π of $\mathbb{R}^{n+1}$, the boundary $\Sigma \doteqdot \partial\pi(\Omega)$ of the projection $\pi(\Omega)$ of Ω onto Π satisfies the same pinching estimate as $\mathcal{M}$ almost everywhere. This follows from the fact that the normal is preserved under the projection: Identify S^2 with $\partial\pi(B^{n+1}) = \Pi \cap S^n$. Observe that, for almost every $X \in \mathcal{M}$, $\pi(X) \in \Sigma$ if and only if $\mathsf{N}_{\mathcal{M}}(X) = \mathsf{N}_{\Sigma}(\pi(X))$ (this can be seen by translating a hyperplane with normal $\mathsf{N}(X)$ from infinity until it touches $\mathcal{M}$). That is, $\mathsf{N}_{\Sigma}^{-1} = \pi \circ \mathsf{N}_{\mathcal{M}}^{-1}$. Let $\gamma : I \to S^2$ be a curve parametrized by arc length with $\gamma(0) = \mathsf{N}(X)$ and $\frac{d}{ds}\big|_{s=0} \gamma = v \in T_{\mathsf{N}(X)} S^2 = T_X \Sigma$. If N_{Σ} is smooth at $\pi(X)$ (which is true almost everywhere), then

$$\mathrm{L}_{\Sigma}^{-1}(v) = \frac{d}{ds}\bigg|_{s=0} \mathsf{N}_{\Sigma}^{-1} \circ \gamma = \frac{d}{ds}\bigg|_{s=0} \pi \circ \mathsf{N}_{\mathcal{M}}^{-1} \circ \gamma = d\pi \circ \mathrm{L}_{\mathcal{M}}^{-1}(v)\, .$$

Thus, for any $v \in T_{\mathsf{N}(X)} S^2 = T_X \Sigma$,

$$\left\langle \mathrm{L}_{\mathcal{M}}^{-1}(v), v \right\rangle = \left\langle d\pi \circ \mathrm{L}_{\Sigma}^{-1}(v), v \right\rangle = \left\langle \mathrm{L}_{\Sigma}^{-1}(v), v \right\rangle . \tag{8.58}$$

The claim follows. □

It will be useful to relate the widths of a convex set Ω to its **circumradius** and **inradius**, which we recall are defined by

$$\rho_+ \doteqdot \inf \left\{ r : \Omega \subset B_r(X) \text{ for some } X \in \mathbb{R}^{n+1} \right\} \text{ and} \tag{8.59a}$$

$$\rho_- \doteqdot \sup \left\{ r : B_r(X) \subset \Omega \text{ for some } X \in \mathbb{R}^{n+1} \right\}, \tag{8.59b}$$

respectively. When a smooth hypersurface $\mathcal{M}$ is the boundary of a bounded convex set Ω, we define the circumradius and inradii of $\mathcal{M}$ as those of Ω.

Lemma 8.25. *On any bounded convex body $\Omega \subset \mathbb{R}^{n+1}$,*

$$\rho_+ \leq \tfrac{1}{\sqrt{3}} w_+ \text{ and } \rho_- \geq \tfrac{1}{n+2} w_- . \tag{8.60}$$

Proof. Let S be a sphere of smallest radius which encloses Ω. By translating if necessary, we may arrange that its center is the origin. Choose two points $x, y \in S \cap \Omega$ which maximize the Euclidean distance. The angle between x and y is at least $\frac{2\pi}{3}$ since otherwise we could move S slightly in the direction $x+y$ so that it strictly contains Ω, contradicting the assumption that S has smallest possible radius. Then the distance from x to y is a lower bound for the maximum width w_+ and is at least $\sqrt{3}$ times the radius of S. This proves the first inequality.

Now let S be a sphere of largest radius contained in Ω. By translating if necessary, we may arrange that its center is the origin. We claim that there is a nonempty set of points $P \subset S \cap \partial\Omega$ such that $P \setminus \{z\}$ is linearly independent for any $z \in P$ and such that there is a positive linear combination of the elements of P with value zero. Indeed, if this were not the case, then the convex hull of $S \cap \partial\Omega$ could not contain the origin, and so S could be moved slightly to become properly contained in Ω. Let E be the smallest affine subspace of $\mathbb{R}^{n+1}$ which contains the set P. Note that E has dimension $k-1$, where P has k elements. Let Σ be the simplex $\{y \in E : \langle y, z\rangle \leq \sigma(z) \text{ for all } z \in P\}$, where σ is the support function of Ω. Then Σ contains the projection of Ω onto E since the latter is convex (and hence the intersection of its supporting half-spaces). Hence the minimum width of $\mathcal{M}$ is no greater than the minimum width of Σ, which is the shortest altitude of Σ. This is bounded by the altitude of a regular simplex inscribed by Σ in E, or $k\rho_-$. Since E has dimension at most $n+1$, the second inequality follows. □

Combining Lemmas 8.24 and 8.25, we deduce that a hypersurface with pinched principal curvatures, $\kappa_n(x) \leq C\kappa_1(x)$, satisfies

$$\rho_+ \leq \frac{(n+2)}{\sqrt{3}} C\rho_- . \tag{8.61}$$

Now let $\{\Omega_t\}_{t\in[0,T)}$ be a family of precompact convex open sets whose boundaries $\mathcal{M}_t = \partial\Omega_t$ evolve by mean curvature flow and set

$$C \doteqdot \max_{\mathcal{M}_0} \frac{\kappa_n}{\kappa_1} .$$

We suppose that T is the maximal time, so that $\limsup_{t\nearrow T} \max_{\mathcal{M}_t} H = \infty$.

Lemma 8.26. *The inradius $\rho_-(t)$ of Ω_t goes to zero as $t \to T$. Thus, Ω_t converges in the Hausdorff topology to some point $p \in \mathbb{R}^{n+1}$ as $t \to T$.*

Proof. Suppose, to the contrary, that $\rho_0 \doteqdot \lim_{t\to T} \rho_- > 0$ (the limit exists since ρ_- is nonincreasing). After a translation in space, we may arrange that the sphere of radius ρ_0 centered at the origin lies in Ω_t for all $t \in [0,T)$. For each $t \in [0,T)$, we can represent $\mathcal{M}_t$ as a graph over the sphere. That is, we can find a smooth function $r : S^n \times [0,T) \to (\rho_0, \rho_+(0)]$ such that

$$\mathcal{M}_t = \operatorname{graph} r(\cdot\,,t) = \{r(z,t)\, z : z \in S^n\}\,. \tag{8.62}$$

For a starshaped hypersurface evolving by the mean curvature flow, we have $\partial_t X = -\frac{H}{\langle z,\mathsf{N}\rangle} z$ (since $(\partial_t X)^\perp = -H\mathsf{N}$). Thus,

$$\partial_t r(z) = -\frac{H(z)}{\langle z, \mathsf{N}(z)\rangle}.$$

Recall from (5.74) and (5.76) that

$$g^{ij} = \frac{1}{r^2}\left(\overline{g}^{ij} - \frac{\overline{\nabla}^i r \overline{\nabla}^j r}{r^2 + \left|\overline{\nabla} r\right|^2}\right) \quad \text{and}$$

$$\mathrm{II}_{ij} = -\frac{r}{\sqrt{r^2 + \left|\overline{\nabla} r\right|^2}}\left(\overline{\nabla}_i \overline{\nabla}_j r - 2\frac{\overline{\nabla}_i r \overline{\nabla}_j r}{r} - r\overline{g}_{ij}\right),$$

respectively. Since $H = g^{ij}\,\mathrm{II}_{ij}$ and $\langle z, \mathsf{N}\rangle = \frac{r}{\sqrt{r^2+|\overline{\nabla} r|^2}}$, we conclude that the radial function r satisfies the equation

$$\partial_t r = \frac{1}{r^2}\left(\overline{g}^{ij} - \frac{\overline{\nabla}^i r \overline{\nabla}^j r}{r^2 + \left|\overline{\nabla} r\right|^2}\right)\left(\overline{\nabla}_i \overline{\nabla}_j r - 2\frac{\overline{\nabla}_i r \overline{\nabla}_j r}{r} - r\overline{g}_{ij}\right). \tag{8.63}$$

Since r is uniformly bounded away from zero, equation (8.63) is uniformly parabolic up to time T, so the regularity theory for uniformly parabolic equations implies that the solution can be extended to a larger time interval (see [**340**, Section 5.5]). Perhaps the easiest way to see this is to observe that the inverse $\mathrm{K} \doteqdot r^{-1}$ satisfies the linear parabolic equation

$$\partial_t \mathrm{K} = a^{ij}\left(\partial_{\theta^i}\partial_{\theta^j}\mathrm{K} - \overline{\Gamma}_{ij}{}^k \partial_{\theta^k}\mathrm{K} + \mathrm{K}\,\overline{g}_{ij}\right)$$

in local coordinates $\{\theta^i\}_{i=1}^n$ for S^n, where

$$a^{ij} \doteqdot \frac{1}{r^2}\left(\overline{g}^{ij} - \frac{\overline{\nabla}^i r \overline{\nabla}^j r}{r^2 + \left|\overline{\nabla} r\right|^2}\right).$$

Since any solution of (8.63) gives rise to a starshaped solution of mean curvature flow via (8.62), this contradicts the maximality of the solution.

The remaining claim follows from (8.61). □

By comparing the solution with the evolution of inscribed and circumscribed spheres, we deduce that

$$\rho_-(t) \le \sqrt{2(T-t)} \le \rho_+(t)\,. \tag{8.64}$$

We can now deduce, using an argument of Kai-Seng Chou (a.k.a. Kaising Tso) [**506**], that the curvature growth is of the first type.

Lemma 8.27. *There exists $t_0 \in [0,T)$ such that*

$$\max_{\mathcal{M}_t} H \leq \frac{C^2}{\sqrt{2n(T-t)}} \quad \textit{for all } t \in [t_0, T). \tag{8.65}$$

Proof. Given $t_1 \in [0,T)$, set $\alpha \doteqdot \rho_-(t_1)$. By translating if necessary, we can arrange that the center of a chosen inscribed sphere for Ω_{t_1} is the origin. Then $\sigma(\cdot,t) \geq \rho_-(t) \geq \rho_-(t_1)$ for $t \leq t_1$. Define $q : \mathcal{M} \times [0,t_1] \to \mathbb{R}$ by

$$q \doteqdot \frac{H}{\sigma - \frac{\alpha}{2}}\,.$$

Observe that

$$\begin{aligned}(\partial_t - \Delta)q &= \left\langle \nabla q, \frac{\nabla\sigma}{\sigma - \frac{\alpha}{2}} \right\rangle + \frac{1}{\sigma - \frac{\alpha}{2}}(\partial_t - \Delta)H - \frac{q}{\sigma - \frac{\alpha}{2}}(\partial_t - \Delta)\sigma \\ &= \left\langle \nabla q, \frac{\nabla\sigma}{\sigma - \frac{\alpha}{2}} \right\rangle + |\mathrm{II}|^2 q - \frac{q}{\sigma - \frac{\alpha}{2}}(|\mathrm{II}|^2\sigma - 2H) \\ &= \left\langle \nabla q, \frac{\nabla\sigma}{\sigma - \frac{\alpha}{2}} \right\rangle - \frac{q}{\sigma - \frac{\alpha}{2}}\left(\frac{\alpha}{2}|\mathrm{II}|^2 - 2H\right).\end{aligned}$$

Estimating $|\mathrm{II}|^2 \geq \frac{1}{n}H^2$ and $\sigma - \frac{\alpha}{2} \geq \frac{\alpha}{2}$, we find

$$(\partial_t - \Delta)q \leq \left\langle \nabla q, \frac{\nabla\sigma}{\sigma - \frac{\alpha}{2}} \right\rangle - q^2\left(\frac{\alpha^2}{4n}q - 2\right).$$

Thus, at a point where q achieves a new spatial maximum,

$$0 \leq -q^2\left(\frac{\alpha^2}{4n}q - 2\right), \quad \text{which implies } q \leq \frac{8n}{\alpha^2},$$

and hence

$$\max_{\mathcal{M}\times\{t_1\}} q \leq \max\left\{\max_{\mathcal{M}\times\{0\}} q, \frac{8n}{\alpha^2}\right\}.$$

By Lemma 8.4 we can find some $C < \infty$ such that $\kappa_n \leq C\kappa_1$ and hence, by (8.61), we can estimate $\sigma(\cdot,t_1) \leq \rho_+(t_1) \leq C\rho_-(t_1) = \alpha C$, so that

$$H \leq \left(\sigma - \frac{\alpha}{2}\right)q \leq \frac{C}{2}\max\left\{\alpha \max_{\mathcal{M}\times\{0\}} q, \frac{8n}{\alpha}\right\}$$

at time t_1. By Lemma 8.26, we can find $t_0 \in [0,T)$ such that $\alpha = \rho_-(t_1) \leq \sqrt{8n/\max_{\mathcal{M}\times\{0\}} q}$ if $t_1 > t_0$. In that case,

$$H(\cdot,t_1) \leq \frac{C}{\alpha}.$$

Now, a circumsphere for $\mathcal{M}_{t_1}$ shrinks to a point under mean curvature flow after time $\rho_+(t_1)^2/2n$ and hence

$$T \leq t_1 + \frac{\rho_+(t_1)^2}{2n} \leq t_1 + \frac{\alpha^2 C^2}{2n}.$$

We conclude that

$$H(\cdot, t_1) \leq \frac{C^2}{\sqrt{2n(T-t_1)}}.$$

The claim follows since $t_1 \in [t_0, T)$ can be chosen arbitrarily. □

We now have enough control to obtain convergence of the solution after rescaling. The strong maximum principle implies that the limit is round.

Corollary 8.28. *Set*

$$A_0(t) \doteqdot \max_{\mathcal{M}\times\{t\}} |\mathring{\mathrm{II}}|^2 \quad \textit{and, for each } m \in \mathbb{N}, \quad A_m \doteqdot \max_{\mathcal{M}\times\{t\}} |\nabla^m \mathrm{II}|^2 .$$

Then, for every $m \in \mathbb{N} \cup \{0\}$,

$$\limsup_{t \nearrow T} (T-t)^{-(m+1)} A_m(t) = 0. \tag{8.66}$$

Proof. Suppose, to the contrary, that there is some $\varepsilon > 0$, some $m \in \mathbb{N}\cup\{0\}$, and some sequence of times $t_j \nearrow T$ such that

$$(T-t_j)^{-1} A_m(t_j) \geq \varepsilon \tag{8.67}$$

for all $j \in \mathbb{N}$. Set $\sigma_j \doteqdot \lambda_j^2 t_j$, where $\lambda_j \doteqdot (T-t_j)^{-\frac{1}{2}}$, and consider the sequence $\{\mathcal{M}_t^j\}_{t\in[-\lambda_j^2 t_j, 1), i\in\mathbb{N}}$ of rescaled flows defined by $\mathcal{M}_t^j \doteqdot \partial\Omega_t^j$, where

$$\Omega_t^j \doteqdot \lambda_j \left(\Omega_{\lambda_j^{-2}t + t_j} - p\right) .$$

Observe that $\sigma_j \to \infty$ as $j \to \infty$ and

$$|\mathrm{II}_j|^2(\cdot, t) \leq H_j^2(\cdot, t) = \lambda_j^{-2} H^2\left(\cdot, t_j + \lambda_j^{-2} t\right) \leq \frac{C_0}{1-t},$$

where

$$C_0 \doteqdot \sup_{(x,t)\in\mathcal{M}\times[0,T)} (T-t) H^2(x,t) < \infty .$$

It follows from the arguments in Section 6.8 that, on any compact time interval $J \Subset (-\infty, 0)$, all derivatives of the local parametrizations for the rescaled solutions are bounded uniformly in j and that the induced metrics are uniformly equivalent. Moreover, by Lemma 8.26 and (8.64), the circumradius $\rho_+^j(t)$ of Ω_t^j is bounded by

$$\rho_+^j(t) \leq \frac{(n+2)C}{\sqrt{3}} \sqrt{2(1-t)} .$$

Since each Ω_t^j contains the origin, this implies a C^0 bound on any compact time interval, uniform in j. It follows from the Arzelà–Ascoli theorem and a diagonal subsequence argument that some subsequence of the parametrizations of the rescaled flows converges in the smooth topology to a smooth, compact, convex limit flow, uniformly on compact time intervals. We claim that $\frac{|\mathring{\mathrm{II}}|^2}{H^2}$ is constant on the limit. Recall from (8.7) and Lemma 8.15 that

$$(\partial_t - \Delta)\frac{|\mathring{\mathrm{II}}|^2}{H^2} \leq \left\langle \nabla\frac{|\mathring{\mathrm{II}}|^2}{H^2}, \frac{\nabla H}{H}\right\rangle - \gamma\frac{|\nabla \mathrm{II}|^2}{H^2}, \tag{8.68}$$

where $\gamma > 0$ depends only on n and C. We first show that $\max_{\mathcal{M}\times\{t\}}\frac{|\mathring{\mathrm{II}}|^2}{H^2}$ is constant on the limit. Indeed, given any interval $[a,b]\subset(-\infty,1)$,

$$\max_{\mathcal{M}\times\{b\}}\frac{|\mathring{\mathrm{II}}_j|^2}{H_j^2} - \max_{\mathcal{M}\times\{a\}}\frac{|\mathring{\mathrm{II}}_j|^2}{H_j^2} = \max_{\mathcal{M}}\frac{|\mathring{\mathrm{II}}|^2}{H^2}(\cdot,\lambda_j^{-2}b+t_j) - \max_{\mathcal{M}}\frac{|\mathring{\mathrm{II}}|^2}{H^2}(\cdot,\lambda_j^{-2}a+t_j).$$

By the maximum principle, $\max_{\mathcal{M}_t}\frac{|\mathring{\mathrm{II}}|^2}{H^2}$ is monotone nonincreasing and hence convergent as $t\to T$. Since $\lambda_j\to\infty$, $\lambda_j^{-2}b+t_j$ and $\lambda_j^{-2}a+t_j$ both approach T as $j\to\infty$, and it follows that the right-hand side converges to zero as $j\to\infty$. So $\max_{\mathcal{M}\times\{t\}}\frac{|\mathring{\mathrm{II}}|^2}{H^2}$ is indeed constant on the limit. The strong maximum principle, applied to (8.68), then implies that the ratio $\frac{|\mathring{\mathrm{II}}|^2}{H^2}$ is constant on the limit flow. It then follows from (8.68) that $\nabla\mathrm{II}\equiv 0$ on the limit and we conclude that the limit is a shrinking sphere, contradicting (8.67). □

We can now proceed as in Section 8.4.4 to convert the geometric estimates into estimates for the rescaled embedding and its derivatives and thereby deduce smooth convergence to an embedding of the unit sphere.

8.7. Notes and commentary

There are yet further approaches to Huisken's convergence result. For example, it is possible to bypass the interpolation arguments of Section 8.4.4 by analyzing the volume-preserving mean curvature flow equation: Since the round spheres are stable solutions of the linearized equation (as radial graphs, say), exponential convergence follows directly from well-known results (see [**374**, Theorem 9.1.2]). This approach is taken, for example, by James McCoy in [**395**].

An inspiration for Huisken's theorem is Hamilton's seminal classification of closed 3-manifolds with positive Ricci curvature using Ricci flow [**259**].

8.7.1. Mean curvature flow in the sphere. Recall that the mean curvature flow is well-defined for hypersurfaces in general Riemannian ambient spaces. A particularly nice situation is when the ambientspace is the round sphere, S^{n+1}. In that case, Huisken was able to prove, using similar techniques to those used to obtain Theorem 8.1, that hypersurfaces satisfying the quadratic pinching condition

$$\begin{cases} |\mathrm{II}|^2 < \frac{1}{n-1}H^2 + 2 & \text{if } n \geq 3, \\ |\mathrm{II}|^2 < \frac{3}{4}H^2 + \frac{4}{3} & \text{if } n = 2 \end{cases}$$

either shrink to a round point in finite time or else converge in the smooth topology to a totally geodesic sphere as $t \to \infty$. In case $n \geq 3$, the pinching condition is sharp in the sense that the tori $S^m(r) \times S^{n-m}(s)$ with $r^2 + s^2 = 1$ lie in S^{n+1} and satisfy

$$|\mathrm{II}|^2 - \frac{1}{n-1}H^2 = 2 + \frac{(n-2)r^2}{(n-1)s^2}.$$

Note that the right-hand side can be made arbitrarily close to 2. However, when $n = 2$, the optimal condition should be positive sectional curvature, which is equivalent to

$$|\mathrm{II}|^2 < H^2 + 2.$$

The reason for the stronger pinching condition in this case is purely technical, having to do with obtaining a sign on certain terms in the evolution equation for $|\mathrm{II}|^2 - \alpha H^2$.

It turns out that a different (fully nonlinear) flow for surfaces in S^3 does preserve positive Gaußian curvature and contracts such surfaces to round points [**35**].

8.7.2. Mean curvature flow in Riemannian ambient spaces.

Mean curvature flow of hypersurfaces in Riemannian ambient spaces does not preserve positivity of the second fundamental form. However, it does preserve the stronger convexity condition

$$H\mathrm{II} > nKg + \frac{n^2}{H}Lg, \tag{8.69}$$

where $-K \leq 0$ is a global lower bound for the sectional curvatures of the ambient space and L is a global bound for the norm of the covariant derivative of the Riemann curvature tensor of the ambient space [**293**]. An argument similar in nature to that of Theorem 8.1 can then be used to prove that hypersurfaces satisfying (8.69) contract to points, becoming asymptotically round in the process, so long as the ambient space has sectional curvatures bounded from above and injectivity radius bounded from below (away from zero) [**293**]. Once again, the form of the pinching condition arises from considerations involving the application the maximum principle.

It turns out that a different (fully nonlinear) curvature flow will preserve the pinching condition

$$\mathrm{II} > \sqrt{K}g\,, \tag{8.70}$$

where again $-K \leq 0$ is a lower bound for the sectional curvatures of the ambient space. While the condition (8.70) is slightly more restrictive than (8.69) in case the ambient space is locally symmetric, it is a significant improvement in the general case since it is unaffected by the derivatives of the ambient curvature. An argument similar in nature to that of Section 8.6 can then be used to prove that hypersurfaces satisfying (8.70) contract to points, becoming asymptotically round in the process, so long as the sectional curvatures of the ambient space are also bounded above and the covariant derivative of its Riemann curvature tensor is bounded [**27**] (note that no lower bound on the injectivity radius is required). In particular, a compact hypersurface satisfying (8.70) is diffeomorphic to a sphere and bounds a disk. Combining this with an argument of Gromov (see [**213**]) then yields the following smooth quarter pinching "twisted" sphere theorem [**27**, Theorem 7-7].

Theorem 8.29. *Let (N^n, g), $n \geq 2$, be a compact, simply connected, smooth Riemannian n-manifold with sectional curvatures in the range $(\frac{1}{4}, 1]$. Then N is diffeomorphic to a "twisted" sphere: A manifold obtained by smoothly gluing two smooth n-disks along their boundaries by an orientation-preserving diffeomorphism of S^{n-1}.*

N. Alikakos and A. Freire showed that certain embedded hypersurfaces of Riemannian manifolds that are close to a small geodesic ball converge, as time tends to infinity, to surfaces of constant mean curvature under the volume-preserving mean curvature flow [**14**].

8.7.3. High-codimension mean curvature flow. Since every immersed Riemannian submanifold possesses a mean curvature vector, the mean curvature flow can also be defined for immersed submanifolds, regardless of codimension. It is rather unmanageable in the general case, however (somewhat to be expected given the Nash [isometric] embedding theorem) — the algebraic structure of the evolution equations become vastly more complicated in high codimension and embeddedness is no longer preserved (imagine a space curve with two "linked" segments. These will move closer together for a short time and can eventually collide under the space-curve shortening flow).

One interesting situation where things become manageable is known as the **Lagrangian mean curvature flow**: Suppose that the ambient Riemannian manifold $(N, \langle\cdot,\cdot\rangle)$ admits a compatible complex structure

$J : TN \to TN$. An immersed submanifold $X : M \to N$ is **Lagrangian** if

$$X^*\omega = 0\,, \tag{8.71}$$

where ω is the induced symplectic form:

$$\omega(u,v) \doteqdot \langle Ju, v\rangle\ .$$

If the ambient manifold has vanishing Ricci curvature (e.g., $(N, \langle\cdot,\cdot\rangle)$ is $\mathbb{C}^n$ with its standard Hermitian product) then it turns out that the Lagrangian condition (8.71) is preserved under mean curvature flow. An interesting difference between codimension-one mean curvature flow and Lagrangian mean curvature flow concerns the curvature blow-up rate at a singularity: Whereas it is widely believed that the codimension-one mean curvature flow generically undergoes only type-I singularities, the opposite is believed for Lagrangian mean curvature flow (see Chen and Li [**142**], Groh, Schwarz, Smoczyk, and Zehmisch [**250**], and Neves [**414**,**416**]). Andreas Savas-Halilaj and Knut Smoczyk [**456**] have recently constructed an explicit example of Lagrangian mean curvature flow (with initial submanifold given by an immersed Whitney sphere) which forms a singularity that converges, after rescaling, to a Grim plane. For a more detailed investigation of Lagrangian mean curvature flow, we direct the reader to the surveys by M.-T. Wang [**519**], A. Neves [**415**], and D. Joyce [**322**].

In the general case, some important early results were obtained by Smoczyk (see [**488**]). For instance, an analogue of Huisken's theorem holds when the evolving submanifold has trivial normal bundle.

More recently, Andrews and Baker [**43**, **81**] showed that the quadratic curvature pinching condition

$$|\mathrm{II}|^2 \le c_n |\vec{H}|^2$$

is preserved under codimension $k \ge 1$ mean curvature flow in $\mathbb{R}^{n+k}$, where $c_n \doteqdot \frac{4}{3n}$ if $2 \le n \le 4$ or $c_n \doteqdot \frac{1}{n-1}$ if $n \ge 4$, and that such submanifolds contract to "round points" in an $(n+1)$-dimensional subspace. The pinching condition was improved by Charlie Baker and Huy The Nguyen when $n = k = 2$ [**82**].

8.7.4. Free boundary mean curvature flow.

Axel Stahl obtained short-time existence of solutions to mean curvature flow with free boundary of prescribed contact angle on a fixed "support" hypersurface in Euclidean space. He showed that solutions either exist for all time or develop a curvature singularity [**496**]. He also obtained an analogue of Huisken's theorem in this setting [**495**].

8.8. Exercises

Exercise 8.1. Let $X : M^n \times [0,T) \to \mathbb{R}^{n+1}$ be a compact, mean convex solution of mean curvature flow. Prove that:

(i) $2nT \leq \rho_+(0)$, where $\rho_+(0)$ is the circumradius of $X_0(M^n)$.

(ii) $2nT \leq (\frac{1}{n}H_{\min}(0))^{-2}$, where $H_{\min}(0) \doteqdot \min_{M^n\times\{0\}} H$.

Exercise 8.2. Denote by $\mathrm{S}(n)$ the normed linear space of symmetric $n \times n$-matrices equipped with the Hilbert–Schmidt norm, $\|B\| \doteqdot \sqrt{\operatorname{tr}(B^2)}$. Let $\mathrm{R} \doteqdot \{r\mathrm{I} : r > 0\}$ be the positive ray, where I is the identity matrix. Show that

$$d(B, \mathrm{R}) = \sqrt{\|B\|^2 - \tfrac{1}{n}\operatorname{tr}(B)^2}$$

for any $B \in \mathrm{S}(n)$, where d is the induced distance function.

Exercise 8.3. Prove the product rule (8.12) for smooth u and smooth, positive v. HINT: Consider the function $\log(|u|v^\alpha)$ wherever u is nonzero.

Exercise 8.4 (Stampacchia iteration). Given a bounded open set $\Omega \subset \mathbb{R}^n$, where $n \geq 2$, let $v \in H^1_0(\Omega)$ be a weak subsolution of the equation

$$-\operatorname{div}[A(\nabla v) + V] = 0\,,$$

where $A : \Omega \to \mathrm{GL}(n, \mathbb{R})$ is a family of selfadjoint linear transformations satisfying $A \geq \lambda \mathrm{I}$ for some $\lambda > 0$ and $V : \Omega \to \mathbb{R}^n$ is a vector field satisfying $\left(\int_\Omega |V|^p\,dx\right)^{\frac{1}{p}} \leq \nu$ for some $p > n$ and $\nu > 0$.

Given $k > 0$, set

$$v_k \doteqdot \max\{v - k, 0\} \quad \text{and} \quad A_k \doteqdot \{x \in \Omega : v_k(x) > 0\}\,.$$

(i) Show that

$$\lambda^2 \int_\Omega |\nabla v_k|^2\,dx \leq \int_{A_k} |V|^2\,dx\,.$$

(ii) Assuming $n \geq 3$, use Hölder's inequality and the Sobolev inequality to deduce that

$$\left(\int_\Omega v_k^{2^*}\,dx\right)^{\frac{1}{2^*}} \leq \frac{\nu S}{\lambda}|A_k|^{\frac{1}{2}-\frac{1}{p}}\,,$$

where $\frac{1}{2^*} \doteqdot \frac{1}{2} - \frac{1}{n}$ and S is the Sobolev constant.

(iii) Deduce, for any $h > k > 0$, that

$$(h-k)^{2^*}|A_h| \leq C^{2^*}|A_k|^{2^*\left(\frac{1}{2}-\frac{1}{p}\right)}\,,$$

where $C \doteqdot \frac{S\nu}{\lambda}$.

(iv) Conclude from Stampacchia's lemma that

$$\|v\|_\infty \leq 2^{\frac{n(p-2)}{2(p-n)}}\,C\,|\Omega|^{\frac{1}{n}-\frac{1}{p}}\,.$$

(v) Assuming $n = 2$, derive a bound for $\|v\|_\infty$ using the Poincaré inequality instead of the Sobolev inequality.

Exercise 8.5. Use the Sobolev inequality (Theorem 8.12) and the Hölder inequality to prove the Poincaré-type inequality of Corollary 8.13.

Exercise 8.6. Let $X : M^n \times [0,T) \to \mathbb{R}^{n+1}$ be a convex solution of mean curvature flow satisfying $\kappa_1(\cdot,0) \geq \alpha H(\cdot,0)$ for some $\alpha > 0$.

(1) Verify that the constant C_ε in Theorem 8.6 depends only on n, α, $|\mathcal{M}_0|$, T, $\max_{\mathcal{M}\times\{0\}} H$, and ε.

(2) Set $\rho_+(0) \doteqdot R$ and choose V and Θ so that $|\mathcal{M}_0| \leq VR^n$ and $\max_{\mathcal{M}\times\{0\}} H \leq \Theta R^{-1}$. Show that the constant C_ε can be written as $\mathrm{C}(n,\alpha,V,\Theta,\varepsilon)R^{-2}$.

(3) Show that the same is true if, instead, we set $\min_{\mathcal{M}\times\{0\}} H \doteqdot R^{-1}$.

Exercise 8.7. Fill in the details of the proof of Lemmas 8.15 and 8.17. More precisely, prove that:

(a) $|\nabla \mathrm{II}|^2 \geq \frac{3}{n+2}|\nabla H|^2$ on any hypersurface of $\mathbb{R}^{n+1}$.

(b) $|\nabla^2 \mathrm{II}|^2 \geq \gamma H^4 |\mathring{\mathrm{II}}|^2$ on a compact, convex hypersurface of $\mathbb{R}^{n+1}$ satisfying $\mathrm{II} \geq \alpha H g$, where γ depends only on n and α.

Exercise 8.8. Prove Corollary 8.18.

Exercise 8.9. Let $\{\Omega_t\}_{t\in[0,T)}$ be a family of bounded convex bodies whose boundaries evolve by mean curvature flow. Use the fact that $\rho_-(t) \to 0$ as $t \to T$ to prove that

$$\rho_-(t) \leq \sqrt{2(T-t)} \leq \rho_+(t)\,.$$

Exercise 8.10. Let $X : M^n \times [0,T) \to \mathbb{R}^{n+1}$ be a compact, uniformly convex solution of mean curvature flow. Use the fact that $\mu_t(M^n) \to 0$ as $t \to T$ to prove that

$$\sqrt{2nT} \geq \alpha n \left(\frac{\mu_0(M^n)}{\mathrm{Area}(S^n)}\right)^{\frac{1}{n}},$$

where $\alpha \doteqdot \min_{M^n\times\{0\}} \frac{\kappa_1}{H}$.

Chapter 9

Mean Convex Mean Curvature Flow

In Chapter 8, we saw that a convex hypersurface of $\mathbb{R}^{n+1}$ will shrink to a round point under mean curvature flow when $n \geq 2$. When $n = 1$, this is a consequence of Grayson's theorem (Theorem 3.19) (or, more specifically, the Gage–Hamilton theorem). A key step in the proof was the roundness estimate (Theorem 8.6), which showed that, modulo scaling, curvature pinching becomes "optimal" at the singularity in that the ratio of the norm of the trace-free second fundamental form to the mean curvature goes to zero. Put another way, the distance of the Weingarten curvature from the umbilic ray $\{r\mathrm{I} : r > 0\}$ approaches zero along the solution after rescaling. We will see that analogous pinching phenomena occur in the more general setting of **mean convex** hypersurfaces, that is, hypersurfaces with nonnegative mean curvature. What is not immediately clear, however, is what condition constitutes "optimal" pinching in this case. One might expect, for example, that any embedding of S^n will again be deformed to a round point, as was the case under curve shortening flow. Unfortunately, it turns out that this is rather wishful thinking, so the issue is not entirely topological. To understand what can go wrong, we investigate some explicit examples. Informed by these examples, we obtain optimal pinching estimates and a corresponding estimate for the gradient of the second fundamental form generalizing Theorem 8.14.

9.1. Singularity formation

Two prototypical singularities are the neckpinch and the degenerate neckpinch, both of which can occur for mean curvature flow of embedded mean convex hypersurfaces. In this section we discuss the heuristics in the rotationally symmetric case.

9.1.1. The standard neckpinch. Consider an initial surface which looks like two large spheres connected by a thin cylinder (a "neck") in such a way that the mean curvature in the neck is much larger than in the spheres. We call such a surface a *dumbbell*. We would expect that the larger curvature in the neck causes its radius to go to zero in a short time while the spheres, which have relatively small mean curvature, remain relatively large. If this is the case, then the solution is certainly not contracting to a point. To make this argument rigorous, we can use barriers. We follow the construction in [**205**, Chapter 3] using a 1-sheeted hyperboloid as a barrier. Other barriers, such as Angenent's shrinking torus (doughnut), can be used (see [**68**]).[1]

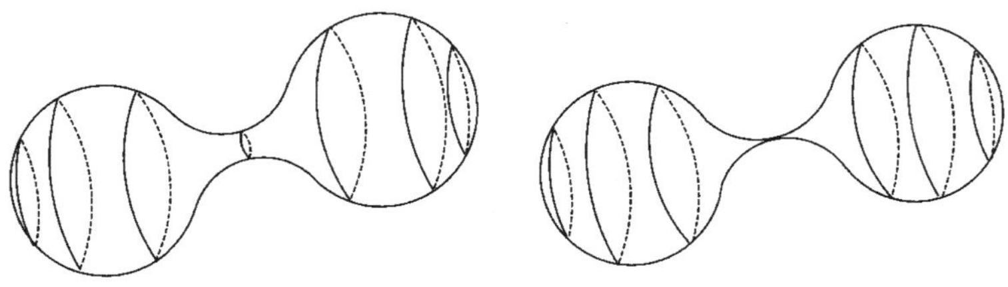

Figure 9.1. Dumbbell neckpinch. L: Earlier time. R: Singular time.

Consider a closed curve $\Gamma \subset \mathbb{R}^2$ given by smoothly joining two circles of radius $R+1$ by a "bridge" of length $2L$ and width ε. More explicitly, consider $\Gamma = \operatorname{graph}(u) \cup \operatorname{graph}(-u)$, where $u : [-L-2R-3, L+2R+3] \to \mathbb{R}$ is a smooth function satisfying

$$u(x) \doteqdot \begin{cases} \varepsilon/2 & \text{if } |x| \leq L\,, \\ \sqrt{(R+1)^2 - (x-L-R-2)^2} & \text{if } x \geq L+2\,, \\ \sqrt{(R+1)^2 - (x+L+R+2)^2} & \text{if } x \leq -L-2\,. \end{cases}$$

Assuming that R is large compared to ε, we can also arrange that u is convex in the interpolating region.

Let $\Sigma^n \subset \mathbb{R}^{n+1}$ be the hypersurface obtained by rotating Γ about the x-axis. The resulting hypersurface is smooth and closed since in a neighborhood of each pole it is simply part of a round sphere.

Given any $R > 0$ there is some $L > 0$ such that the corresponding surface Σ^n lies inside of the 1-sheeted hyperboloid

$$\mathrm{H} \doteqdot \{x \in \mathbb{R}^{n+1} : x_1^2 = x_2^2 + \cdots + x_{n+1}^2 - \varepsilon^2\} \tag{9.1}$$

[1]Roughly the idea is as follows. Consider an initial rotationally symmetric dumbbell with a thin neck. Insert a round sphere inside each of the two bells. Wrap Angenent's shrinking self-similar doughnut around the neck. If the neck is sufficiently thin so that we may take the doughnut to be sufficiently small relative to the size of the two round spheres, then the doughnut shrinks to a point before the spheres shrink to points. By the avoidance principle the neck must pinch before the bells shrink to points.

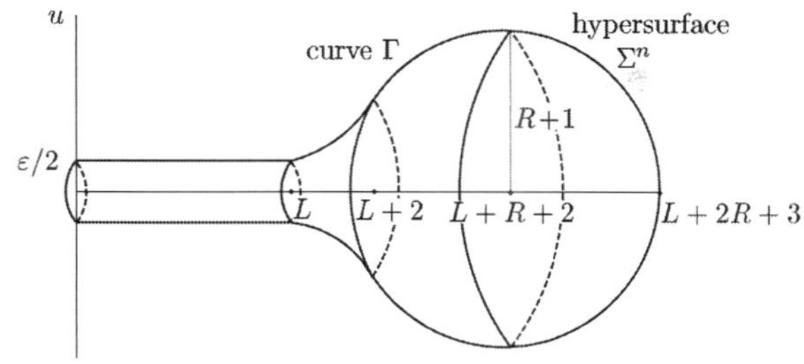

Figure 9.2. The right half of a rotationally symmetric dumbbell.

and outside of the spheres $S^n_R(-(L+R+2)e_1)$ and $S^n_R((L+R+2)e_1)$. Now let $\{\Sigma_t\}_{t\in[0,T)}$ be the solution to mean curvature flow with $\Sigma_0=\Sigma$ and let $\{\mathrm{H}_t\}_{t\in[0,T_\varepsilon)}$ be the family of hyperboloids

$$\mathrm{H}_t \doteqdot \{x\in\mathbb{R}^{n+1} : x_1^2 = x_2^2+\cdots+x_{n+1}^2-\varepsilon^2+2nt\}, \tag{9.2}$$

where $T_\varepsilon \doteqdot \frac{\varepsilon^2}{2n}$ and so that $\mathrm{H}_0=\mathrm{H}$. Recall (see Exercise 6.14) that the family $\{\mathrm{H}_t\}_{t\in[0,T_\varepsilon)}$ is a subsolution to mean curvature flow up to time $T_\varepsilon \doteqdot \frac{\varepsilon^2}{2n}$. Thus, applying the avoidance principle, we conclude that the hypersurface Σ^n_t lies inside the region enclosed by H_t for all $t\in[0,T)\cap[0,T_\varepsilon)$ and outside of the balls

$$B^{n+1}_{\sqrt{R^2-2nt}}(\pm(L+R+2)e_1)$$

for all $t\in[0,T)\cap[0,T_R)$, where $T_R\doteqdot\frac{R^2}{2n}$. But the radius of the "neck" of the hyperboloid H_t tends to zero as $t\to T_\varepsilon$, whereas the radius of the balls is still $\sqrt{R^2-\varepsilon^2}$ at this time. That is, choosing $\varepsilon<R$, we have constructed a solution to mean curvature flow in $\mathbb{R}^{n+1}$ with the topology of S^n which becomes singular before contracting to a point. We shall see in Chapter 11 that, after "zooming in", this singularity looks like a *shrinking cylinder.*

9.1.2. The degenerate neckpinch. Now imagine that, in the example constructed above, we choose one of the "bells" of the dumbbell to have a small radius compared to that of the other. If we choose this radius too close to the radius of the neck, then the inner barrier no longer holds out until the neck has pinched; indeed, we should expect in this situation that one of the bells will sneak through the neck before it pinches off. If the radii are chosen correctly, there should be a borderline case, where the dumbbell almost sneaks through but gets caught in the neck just as it collapses. Such an example, now called a **degenerate neckpinch**, was conjectured by Hamilton and rigorously constructed by Altschuler, Angenent, and Giga [**20**]. Subsequently, Angenent and Velázquez [**75**] constructed, for any odd integer $m\geq 3$, a solution which is rotationally symmetric about the x-axis and becomes singular at the origin at time T, with the maximum of the mean curvature attained at the "tip",

$$p(t) = \left((1+o(1))\sqrt[m]{2(n-1)(T-t)}, 0, \ldots, 0\right),$$

and blowing up like

$$H(p(t)) = \left(\frac{2(n-1)}{n} + o(1)\right)(2(n-1)(T-t))^{\frac{1}{m}-1}.$$

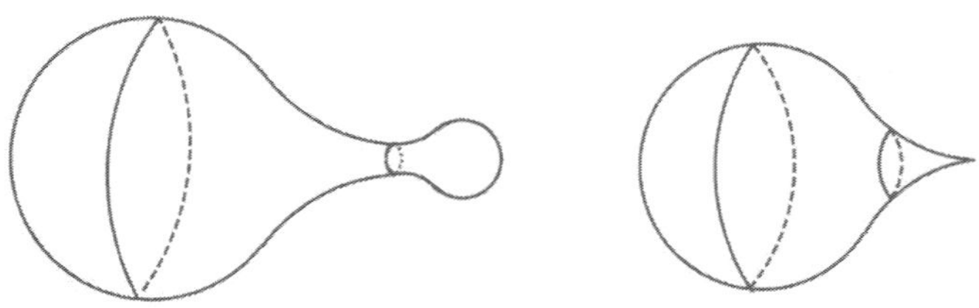

Figure 9.3. Degenerate neckpinch.

We shall see in Chapter 11 that, after "zooming in" near its tip, this singularity looks like a rotationally symmetric convex solution to mean curvature flow which moves purely by translation, called the *bowl soliton.*

9.2. Preserving pinching conditions

Denote by $\mathrm{S}(n)$ the normed linear space of symmetric $n\times n$-matrices equipped with the Hilbert–Schmidt norm. We say that a subset $\Omega \subset \mathrm{S}(n)$ is $\mathrm{SO}(n)$-**invariant** if the orbit $O(P) \doteqdot \{OPO^\top : O \in \mathrm{SO}(n)\}$ is contained in Ω whenever $P \in \Omega$. Recall that the component matrix of the Weingarten tensor with respect to an orthonormal basis is symmetric and that changing to another orthonormal basis changes the component matrix by conjugation through an orthogonal matrix. Thus, by working with orthonormal frames, we can interpret the space $\mathrm{S}(n)$ of symmetric matrices as the space of Weingarten tensors of smooth hypersurfaces of $\mathbb{R}^{n+1}$, and $\mathrm{SO}(n)$-invariant subsets $\Omega \subset \mathrm{S}(n)$ as curvature pinching conditions. Given an $\mathrm{SO}(n)$-invariant subset $\Omega \subset \mathrm{S}(n)$, we shall write $\mathrm{II}_x \in \Omega$ if this is true for the component matrix of II_x with respect to some (and hence any) orthonormal basis at x. Since the mean curvature flow is invariant under parabolic rescaling, it is natural to consider pinching conditions which are scale invariant. Moreover, diffusion systems (including mean curvature flow, as we shall soon see) tend to favor *convex* pinching conditions. These considerations lead to the following definition, which generalizes the notion of *pinched hypersurface* encountered in Chapter 8.

Definition 9.1. A **pinching condition** (for mean curvature flow in $\mathbb{R}^{n+1}$) is an open, convex, $\mathrm{SO}(n)$-invariant cone[2] $\Gamma \subset \mathrm{S}(n)$.

Recall that $\kappa_1 \leq \kappa_2 \leq \cdots \leq \kappa_n$ denotes the ordered principal curvatures of a hypersurface.

Example 9.2. The following inequalities correspond to pinching conditions:

- Mean convexity: $H > 0$.
- (Uniform) local convexity: $\mathrm{II} > 0$ ($\mathrm{II} > \varepsilon H g$, $\varepsilon \geq 0$).
- Quadratic pinching: $|\mathrm{II}|^2 < CH^2$, $C > 0$.
- (Uniform) k-convexity: $\kappa_1 + \cdots + \kappa_k > 0$ ($\kappa_1 + \cdots + \kappa_k > \varepsilon H, \varepsilon \geq 0$).
- Lower pinching: $\kappa_1 > -CH$, $C > 0$.
- Upper pinching: $\kappa_n < CH$, $C > 0$.

We need to recall some convex geometry (see, for example, the book by Rolf Schneider [**461**]). Given a convex subset C of a finite-dimensional normed linear space E, the **signed distance to the boundary of** C is given by

$$d_C(x) \doteqdot \inf_{\ell \in \mathrm{S}C} \ell(x)\,, \tag{9.3}$$

where $\mathrm{S}C$ denotes the set of **supporting affine functionals** for C, that is, the set of affine linear maps $\ell : E \to \mathbb{R}$ satisfying $\|D\ell\| = 1$ and $\ell(x) \geq 0$ for all $x \in C$ with $\ell(x_0) = 0$ at some $x_0 \in \partial C$. We will often say that such a functional ℓ **supports** C **at** x_0. It is a consequence of the Hahn–Banach theorem that the **extremal set** of C at z,

$$\mathrm{S}_z C \doteqdot \{\ell \in \mathrm{S}C : d_C(z) = \ell(z)\}\,, \tag{9.4}$$

is nonempty for every $z \in E$.

Lemma 9.3. *Let E be a finite-dimensional normed linear space and C a convex subset of E.*

1. *$d_C(x)$ is the distance from x to $E \setminus C$ if $x \in C$ and the negative of the distance from x to C if $x \in E \setminus C$.*
2. *If C is a cone, then $\mathrm{S}C \subset E^*$; that is, the supporting affine functionals are linear functionals.*

Proof. See Exercise 9.1. □

It is often more convenient to work at the level of eigenvalues. We say that a subset[3] $Z \subset \mathbb{R}^n$ is **symmetric** (or P_n**-invariant**, where P_n is

[2] A subset C of a linear space E is called a **cone** if the ray $R_x \doteqdot \{rx, r > 0\}$ lies in C whenever $x \in C$.

[3] In practice, Z will actually be a subset of the dual space $(\mathbb{R}^n)^*$ of $\mathbb{R}^n$.

the group of permutations of the set of n elements $\{1, \ldots, n\}$) if the orbit $o(z) \doteqdot \{(z_{\sigma(1)}, \ldots, z_{\sigma(n)}) : \sigma \in P_n\}$ lies in Z whenever $z = (z_1, \ldots, z_n)$ lies in Z. Of course, an $\mathrm{SO}(n)$-invariant subset $\Omega \subset \mathrm{S}(n)$ defines a symmetric subset $Z \subset \mathbb{R}^n$ (and vice versa) via the rule

$$z \in Z \quad \Longleftrightarrow \quad O(\operatorname{diag}(z)) \subset \Omega\,, \tag{9.5}$$

where $O(W)$ denotes the orbit of W under the action of $\mathrm{SO}(n)$ by conjugation.

We claim that convexity and the distance to the boundary are preserved under this identification.

Lemma 9.4. *Suppose that $\Omega \subset \mathrm{S}(n)$ and $Z \subset \mathbb{R}^n$ are related by (9.5).*

(1) *Ω is a convex subset of $\mathrm{S}(n)$ if and only if Z is a convex subset of $\mathbb{R}^n$.*

(2) *If Ω and Z are convex, then*

$$d_Z(z) = d_\Omega(W)$$

whenever $\{z_1, \ldots, z_n\}$ are the eigenvalues of W.

Proof. See Exercise 9.2. □

Henceforth, we shall often conflate an $\mathrm{SO}(n)$-invariant cone $\Gamma \subset \mathrm{S}(n)$ with the corresponding symmetric cone in $\mathbb{R}^n$.

Now let $X : M^n \times I \to \mathbb{R}^{n+1}$ be a solution to mean curvature flow. Then, given a pinching condition $\Gamma \subset \mathrm{S}(n)$, we can form a function $f : M^n \times I \to \mathbb{R}$ by setting

$$f(x,t) \doteqdot d_\Gamma(\mathrm{II}_{(x,t)})\,,$$

where the shorthand $d_\Gamma(\mathrm{II}_{(x,t)})$ really means that d_Γ is evaluated at the component matrix of $\mathrm{II}_{(x,t)}$ with respect to some $g(t)$-orthonormal basis at x. Of course, f is well-defined by the $\mathrm{SO}(n)$-invariance of Γ. We want to understand the evolution of the function f under the mean curvature flow. Note that we have already been introduced to such functions in the previous chapter: When the set Γ is the umbilic ray, $\{r\mathrm{I} : r > 0\}$, we saw that

$$d_\Gamma(W) = \|W - \tfrac{1}{n}\operatorname{tr} W\| = \sqrt{\|W\|^2 - \tfrac{1}{n}(\operatorname{tr} W)^2}\,,$$

which corresponds to $\|\mathring{\mathrm{II}}\|$ when $W = \mathrm{II}$ (see Exercise 8.2). In this case, the function f is smooth and satisfies a nice evolution equation, which we used to show that upper bounds for f/H are preserved (and, in fact, improve). When the set Γ is the **positive cone**, $\Gamma_+ \doteqdot \{W \in \mathrm{S}(n) : W > 0\}$, the function f is simply the smallest principal curvature, κ_1 (see Exercise 9.3). In this case, the function we obtain is not smooth. However, we were still able to deduce (this time using the *tensor* maximum principle) that lower bounds for f/H are preserved. Although in general the function f will not be smooth, we are still able to derive an evolution equation (rather, an inequality) which holds in a suitable weak sense.

Definition 9.5 (Barrier supersolutions). Let $X : M^n \times [0,T) \to \mathbb{R}^{n+1}$ be a smooth 1-parameter family of smooth immersions. Denote by $\mathcal{B}_{r_0}(x_0,t_0)$ the intrinsic ball in M^n of radius r_0 with respect to $g(t_0)$ centered at x_0. A continuous function $f : M^n \times I \to \mathbb{R}$ satisfies

$$(\partial_t - \Delta)f(x,t) \geq F(\nabla f(x,t), f(x,t), x, t)$$

in the **barrier sense** if for every $(x_0,t_0) \in M^n \times I$ there exist $r_0 > 0$ and a smooth function $\psi \in C^\infty(\mathcal{B}_{r_0}(x_0,t_0) \times (t_0 - r_0^2, t_0])$ which satisfies

$$(\partial_t - \Delta)\psi \geq F(\nabla\psi, \psi, \cdot) \quad \text{at} \quad (x_0,t_0)$$

and **supports** f **from above at** (x_0,t_0); that is,

$$\psi(x_0,t_0) = f(x_0,t_0) \quad \text{and} \quad \psi \geq f \ \text{ on } \ \mathcal{B}_{r_0}(x_0,t_0) \times (t_0 - r_0^2, t_0]\,.$$

We refer to f as a **barrier supersolution** of the corresponding equation.

A continuous function is a **barrier subsolution** if the same holds with all inequalities reversed, that is, if at each point it admits a smooth, lower supporting subsolution.

Proposition 9.6. *Let $X : M^n \times I \to \mathbb{R}^{n+1}$ be a solution to mean curvature flow. Given a pinching condition $\Gamma \subset \mathrm{S}(n)$, the function $f : M^n \times I \to \mathbb{R}$ defined by*

$$f(x,t) \doteqdot d_\Gamma(\mathrm{II}_{(x,t)})$$

satisfies

$$(\partial_t - \Delta)f \geq |\mathrm{II}|^2 f + Q(\nabla \mathrm{II}) \tag{9.6}$$

in the barrier sense, where Q is defined with respect to a principal frame (at each point) by

$$Q(\nabla \mathrm{II}) \doteqdot \sup_{\ell \in \mathrm{S}_\kappa\Gamma} \sum_{k=1}^n \sum_{\kappa_i \neq \kappa_j} \frac{\ell^i - \ell^j}{\kappa_j - \kappa_i} (\nabla_k \mathrm{II}_{ij})^2\,, \tag{9.7}$$

where $\kappa = (\kappa_1, \ldots, \kappa_n)$, $\ell = (\ell^1, \ldots, \ell^n) \in (\mathbb{R}^n)^*$, we have identified $\Gamma \subset \mathrm{S}(n)$ with the corresponding symmetric, convex cone in $\mathbb{R}^n$, and $\mathrm{S}_\kappa\Gamma$ is the extremal set of Γ at κ.

Proof. Fix any point $(x_0,t_0) \in M^n \times [0,T)$ and a principal basis $\{e_i^0\}_{i=1}^n$ at (x_0,t_0) and choose some $\ell_0 \in \mathrm{S}_{\kappa_0}\Gamma$, where $\kappa_0 \doteqdot \kappa(x_0,t_0)$. Having fixed an orthonormal basis at (x_0,t_0), we can consider ℓ_0 as a symmetric bilinear form acting on $T_{x_0}M$. That is, $\ell_0 \equiv \ell_0^{ij} e_i^0 \otimes e_j^0$, where $\ell_0^{ij} \doteqdot \ell_0^i \delta_{ij}$. We smoothly extend ℓ_0 to a symmetric bilinear form ℓ defined in a neighborhood of (x_0,t_0)

by setting $\ell \doteqdot \ell_0^{ij} e_i \otimes e_j = \ell_0^i e_i \otimes e_i$, where the local orthonormal frame field $\{e_i\}_{i=1}^n$ is formed by solving first, for some fixed metric connection $\overline{\nabla}$,

$$\text{(9.8)} \qquad \overline{\nabla}_{\gamma'} e_i \equiv 0 \quad \text{with initial condition} \;\; e_i(x_0, t_0) = e_i^0$$

along any $\overline{\nabla}$-geodesic γ emanating from x_0 and then extending in the time direction by solving

$$\text{(9.9)} \quad \nabla_t e_i \equiv 0 \quad \text{with initial condition} \;\; e_i(x, t_0) = e_i(x) \;\; \text{for any} \;\; x \in M^n.$$

Now set

$$\psi \doteqdot \langle \ell, \mathrm{II} \rangle = \ell_0^{ij} \mathrm{II}_{ij} \,.$$

Choosing r_0 sufficiently small, we can arrange that the function ψ is defined on $\mathcal{B}_{r_0}(x_0, t_0) \times (t_0 - r_0^2, t_0]$. Clearly, $\psi(x_0, t_0) = \ell_0(\kappa_0) = f(x_0, t_0)$. Moreover, since the basis $\{e_i\}_{i=1}^n$ remains orthonormal and $\ell_0 \in \mathrm{S}\Gamma$, it is a consequence of the definition (9.3) that $\psi \geq f$ on $\mathcal{B}_{r_0}(x_0, t_0) \times (t_0 - r_0^2, t_0]$. Since ψ is smooth, we can compute

$$\begin{aligned} \text{(9.10)} \qquad (\partial_t - \Delta)\psi &= (\partial_t - \Delta) \langle \ell, \mathrm{II} \rangle \\ &= \langle \ell, (\nabla_t - \Delta)\mathrm{II} \rangle - g^{kl} \left(2\nabla_k \ell^{ij} \nabla_l \mathrm{II}_{ij} + \nabla_k \nabla_l \ell^{ij} \mathrm{II}_{ij} \right) \\ &= |\mathrm{II}|^2 \psi - g^{kl} \left(2\nabla_k \ell^{ij} \nabla_l \mathrm{II}_{ij} + \nabla_k \nabla_l \ell^{ij} \mathrm{II}_{ij} \right) \end{aligned}$$

at (x_0, t_0), where $\nabla_k \ell^{ij} \doteqdot (\nabla_{e_k} \ell)^{ij} = (\nabla_{e_k} (\ell_0^{lm} e_l \otimes e_m))^{ij}$.

Aside. If we now choose $\overline{\nabla} = \nabla$, then $\nabla \ell = 0$ and $\nabla^2 \ell = 0$ at (x_0, t_0) and we have proved that

$$\text{(9.11)} \qquad (\partial_t - \Delta) f \geq |\mathrm{II}|^2 f$$

in the barrier sense. However, by choosing the metric connection $\overline{\nabla}$ more carefully, we can obtain the additional (useful) term in (9.6).

Given any section Λ of $T^*M \otimes T^*M \otimes TM$ satisfying the skew-symmetry condition

$$\text{(9.12)} \qquad g(\Lambda(w, u), v) + g(u, \Lambda(w, v)) = 0 \quad \text{for all} \quad u, v, w \in TM\,,$$

we can define a metric connection $\overline{\nabla}$ (and vice versa) via

$$\overline{\nabla}_u v \doteqdot \nabla_u v - \Lambda(u, v)\,.$$

Define the components of Λ by $\Lambda_{ij}{}^k e_k \doteqdot \Lambda(e_i, e_j)$ and $\Lambda_{ijk} \doteqdot g(\Lambda(e_i, e_j), e_k)$. Then the skew-symmetry becomes $\Lambda_{ijk} + \Lambda_{ikj} = 0$. Observe by (9.8) that

$$\begin{aligned} \gamma' \langle e_i, e_j \rangle &= \langle \nabla_{\gamma'} e_i, e_j \rangle + \langle e_i, \nabla_{\gamma'} e_j \rangle \\ &= \langle \Lambda(\gamma', e_i), e_j \rangle + \langle e_i, \Lambda(\gamma', e_j) \rangle \\ &= 0, \end{aligned}$$

so that $\{e_i\}$ remains orthonormal. Then

$$\begin{aligned}\nabla_k \ell &= \ell_0^{ij}\left(\nabla_{e_k} e_i \otimes e_j + e_i \otimes \nabla_{e_k} e_j\right)\\ &= \ell_0^{ij}\left(\overline{\nabla}_{e_k} e_i + \Lambda_{ki}{}^p e_p\right)\otimes e_j + \ell_0^{ij} e_i \otimes \left(\overline{\nabla}_{e_k} e_j + \Lambda_{kj}{}^q e_q\right),\end{aligned} \tag{9.13}$$

so that, at (x_0, t_0),

$$\nabla_k \ell = \ell_0^{ij}\left(\Lambda_{ki}{}^p e_p \otimes e_j + \Lambda_{kj}{}^q e_i \otimes e_q\right).$$

Taking a second covariant derivative of (9.13) yields, at (x_0, t_0), that

$$\begin{aligned}\nabla_k\nabla_k \ell = \ell_0^{ij}\big(&\nabla_k\Lambda_{ki}{}^p e_p \otimes e_j + \nabla_k \Lambda_{kj}{}^q e_i \otimes e_q + \Lambda_{ki}{}^p \Lambda_{kp}{}^r e_r \otimes e_j\\ &+\Lambda_{ki}{}^p\Lambda_{kj}{}^q e_p \otimes e_q + \Lambda_{kj}{}^q \Lambda_{ki}{}^p e_p \otimes e_q + \Lambda_{kj}{}^q \Lambda_{kq}{}^r e_i \otimes e_r\big)\end{aligned}$$

since $\overline{\nabla}_k \overline{\nabla}_k e_i = \overline{\nabla}_{e_k}(\overline{\nabla}_{e_k} e_i) - \overline{\nabla}_{\overline{\nabla}_{e_k} e_k} e_i = 0$ at (x_0, t_0). That is, componentwise,

$$\nabla_k \ell^{ij} = \ell_0^{pj} \Lambda_{kp}{}^i + \ell_0^{iq} \Lambda_{kq}{}^j$$

and

$$\begin{aligned}\nabla_k \nabla_k \ell^{ij} =& \ell_0^{pj} \nabla_k \Lambda_{kp}{}^i + \ell_0^{iq} \nabla_k \Lambda_{kq}{}^j\\ &+ \ell_0^{rj}\Lambda_{kr}{}^p \Lambda_{kp}{}^i + \ell_0^{pq} \Lambda_{kp}{}^i \Lambda_{kq}{}^j + \ell_0^{pq}\Lambda_{kq}{}^j \Lambda_{kp}{}^i + \ell_0^{ir} \Lambda_{kr}{}^q \Lambda_{kq}{}^j .\end{aligned}$$

Making use of the skew-symmetry of Λ, we find that

$$-2\nabla_k \ell^{ij} \nabla_k \mathrm{II}_{ij} = -4\ell_0^{im} \Lambda_{kmp} \nabla_k \mathrm{II}_{ip}$$

and

$$\begin{aligned} g^{kl}\nabla_k \nabla_l \ell^{ij}\, \mathrm{II}_{ij} =& 2\left(\ell_0^{pi} \nabla_k \Lambda_{kpi} + \ell_0^{ir}\Lambda_{krp}\Lambda_{kpi} + \ell_0^{pq}\Lambda_{kpi}\Lambda_{kqi}\right)\kappa_i\\ =& 2\left(\ell_0^{pr}\Lambda_{kri}\Lambda_{kip}\kappa_p + \ell_0^{pq}\Lambda_{kpi}\Lambda_{kqi}\kappa_i\right)\\ =& 2\ell_0^{pq}\Lambda_{kpi}\Lambda_{kqi}\left(\kappa_i - \kappa_p\right) .\end{aligned}$$

Since $\ell_0^{ij} = \ell_0^i \delta_{ij}$, we obtain, regarding the right-hand side of (9.10),

$$\begin{aligned} Q_{\Lambda,\ell_0}(\nabla \mathrm{II}) \doteqdot & - g^{kl}\left(2\nabla_k \ell^{ij}\, \nabla_l \mathrm{II}_{ij} + \nabla_k \nabla_l \ell^{ij}\, \mathrm{II}_{ij}\right)\\ =& -2\sum_{k,i,p=1}^{n} \ell_0^i \Lambda_{kip}\left[2\nabla_k \mathrm{II}_{ip} + \Lambda_{kip}(\kappa_p - \kappa_i)\right]\\ =& -\sum_{k=1}^{n}\sum_{i\neq j}\left(\ell_0^i - \ell_0^j\right)\Lambda_{kij}\left[2\nabla_k \mathrm{II}_{ij} + \Lambda_{kij}(\kappa_j - \kappa_i)\right]\end{aligned} \tag{9.14}$$

at the point (x_0, t_0).

Choosing

$$\Lambda_{kij} = \begin{cases} -\dfrac{\nabla_k \mathrm{II}_{ij}}{\kappa_j - \kappa_i} & \text{if } \kappa_i \neq \kappa_j ,\\ 0 & \text{if } \kappa_i = \kappa_j \end{cases} \tag{9.15}$$

at (x_0, t_0) yields

$$Q_{\Lambda,\ell_0}(\nabla \mathrm{II}) = \sum_{k=1}^{n} \sum_{\kappa_i \neq \kappa_j} \frac{\ell_0^i - \ell_0^j}{\kappa_j - \kappa_i} (\nabla_k \mathrm{II}_{ij})^2$$

and hence

$$(\partial_t - \Delta)\psi \geq |\mathrm{II}|^2 \psi + \sup_{\ell_0} \sup_{\Lambda} Q_{\Lambda,\ell_0}(\nabla \mathrm{II}) \geq |\mathrm{II}|^2 \psi + Q(\nabla \mathrm{II})$$

at (x_0, t_0). This completes the proof. □

We remark that the choice of Λ in (9.15) *is optimal: When the principal curvatures are all distinct, we can "complete the square" in* (9.14) *to obtain*

$$Q_{\Lambda,\ell_0}(\nabla \mathrm{II}) = \sum_{k=1}^{n} \sum_{i \neq j} (\ell_0^i - \ell_0^j)(\kappa_j - \kappa_i) \left(\frac{(\nabla_k \mathrm{II}_{ij})^2}{(\kappa_j - \kappa_i)^2} - \left[\Lambda_{kij} + \frac{\nabla_k \mathrm{II}_{ij}}{\kappa_j - \kappa_i} \right]^2 \right).$$

We note that the gradient term (9.7) is indeed a "good" term.

Lemma 9.7. *Let $C \subset \mathbb{R}^n$ be a convex, symmetric cone. Then for each $w \in C$ and each $\ell \in \mathrm{S}_w C$,*

$$(\ell^j - \ell^i)(w_j - w_i) \leq 0 \quad \textit{for each} \quad i \textit{ and } j\,. \tag{9.16}$$

Proof. Fix $i, j \in \{1, \dots, n\}$. By the symmetry of C, $\widehat{w} \in C$, where $\widehat{w}$ is obtained from w by interchanging its i-th and j-th components; that is,

$$\widehat{w} \doteqdot w - (w_i - w_j)(e_i - e_j)\,.$$

Since $\ell(w) = d_C(w) = d_C(\widehat{w}) = \inf_{\widehat{\ell} \in \mathrm{S}C} \widehat{\ell}(\widehat{w})$, we obtain

$$(\ell^j - \ell^i)(w_i - w_j) = \sum_{k=1}^{n} \ell^k (\widehat{w}_k - w_k) \geq d_C(\widehat{w}) - d_C(w) = 0\,.$$ □

The maximum principle now implies that pinching conditions (convex, $\mathrm{SO}(n)$-invariant curvature cones) are preserved by mean curvature flow.

Corollary 9.8 (Weak maximum principle for II)**.** *Let $X : M^n \times [0, T) \to \mathbb{R}^{n+1}$ be a compact solution to mean curvature flow and let $\Gamma \subset \mathrm{S}(n)$ be a pinching condition. If $\mathrm{II}_{(x,0)} \in \overline{\Gamma}$ for all $x \in M^n$, then $\mathrm{II}_{(x,t)} \in \overline{\Gamma}$ for all $(x, t) \in M^n \times [0, T)$.*

Proof. We will prove that the signed distance f of the curvature II to the boundary of Γ remains nonnegative under the flow. The argument is essentially that of the classical maximum principle, except that we apply it to the smooth barrier functions ψ rather than to f directly.

Given any $\tau \in [0, T]$ and $\varepsilon > 0$ consider the function $f_{\varepsilon,\tau} \doteqdot f + \varepsilon \mathrm{e}^{(C+1)t}$, where $C \doteqdot \sup_{M^n \times [0,\tau]} |\mathrm{II}|^2$ and f is the signed distance of II to the boundary of Γ. By hypothesis, f_ε is positive at the initial time. We claim that $f_{\varepsilon,\tau}$ remains nonnegative under the flow up to time τ. Since τ and ε are arbitrary, this will suffice to prove the claim. Suppose, to the contrary, that there is some $(x_0, t_0) \in M^n \times (0, \tau]$ such that $f_{\varepsilon,\tau}(x_0, t_0) = 0$. Since $f_{\varepsilon,\tau}$ is strictly positive at the initial time, we may assume that t_0 is the first such time. By Proposition 9.6, there is some $r > 0$ and function $\psi \in C^\infty(\mathcal{B}_r(x_0, t_0) \times (t_0 - r^2, t_0])$ satisfying $\psi \geq f$ on $\mathcal{B}_r(x_0, t_0) \times (t_0 - r^2, t_0]$ with equality at (x_0, t_0) and

$$(\partial_t - \Delta)\psi \geq |\mathrm{II}|^2\psi$$

at (x_0, t_0). The function

$$\psi_{\varepsilon,\tau} \doteqdot \psi + \varepsilon e^{(C+1)t}$$

is therefore positive on $\mathcal{B}_r(x_0, t_0) \times (t_0 - r^2, t_0)$ and zero at (x_0, t_0) so that

$$\begin{aligned} 0 &\geq (\partial_t - \Delta)\psi_{\varepsilon,\tau} \\ &\geq |\mathrm{II}|^2\psi + \varepsilon(C+1)e^{(C+1)t} \\ &= -\varepsilon e^{(C+1)t}|\mathrm{II}|^2 + \varepsilon(C+1)e^{(C+1)t} \\ &\geq \varepsilon e^{(C+1)t} \\ &> 0 \end{aligned}$$

at (x_0, t_0), a contradiction. □

Remark 9.9. Corollary 9.8 can also be obtained directly from Hamilton's tensor maximum principle [**260**, Lemma 8.2].

Applying the strong maximum principle, we can obtain rigidity results for certain pinching conditions (cf. [**260, 382, 531**]). We will need the following lemma.

Lemma 9.10 (Strong maximum principle for barrier supersolutions (cf. [**134**])). *Let $X : M^n \times I \to \mathbb{R}^{n+1}$ be a smooth 1-parameter family of smooth immersions which is properly defined in the extrinsic parabolic cylinder $B_R(p_0) \times (t_0 - R^2, t_0]$ for some $(x_0, t_0) \in M^n \times I$ and $R > 0$, where $p_0 \doteqdot X(x_0, t_0)$. Suppose that $f : M^n \times I \to \mathbb{R}$ is nonnegative and satisfies*

$$(\partial_t - \Delta)f \geq \nabla_b f$$

in the barrier sense for some vector field b. If $f(x_0, t_0) = 0$, then $f \equiv 0$ in the intrinsic parabolic cylinder $\mathcal{P}_r(x_0, t_0) \doteqdot \mathcal{B}_r(x_0, t_0) \times (t_0 - r^2, t_0]$ for some $r > 0$.

Proof of Lemma 9.10. Suppose that the claim does not hold. Then, given any $r > 0$, there is some $(x, t) \in \mathcal{P}_r(x_0, t_0)$ such that $f(x, t) > 0$. By decreasing r if necessary, we may assume that (x, t) is in the parabolic boundary $\partial\mathcal{P}_r(x_0, t_0) \doteqdot \partial\mathcal{B}_r(x_0, t_0) \times (t_0 - r^2, t_0] \cup \mathcal{B}_r(x_0, t_0) \times \{t_0\}$, where

$\partial\mathcal{B}_r(x_0,t_0)$ is the usual boundary of $\mathcal{B}_r(x_0,t_0)$ in M^n. Having chosen r, define

$$V \doteqdot \{(x,t) \in \partial\mathcal{P}_r(x_0,t_0) : f(x,t) = 0\} .$$

We claim that there is a smooth function $h : \overline{\mathcal{P}_r(x_0,t_0)} \to \mathbb{R}$ satisfying

$$\begin{aligned} h > 0 \quad &\text{on} \quad V, \\ h(x_0,t_0) = 0&, \\ (\partial_t - \Delta - \nabla_b)h > 0 \quad &\text{on} \quad \overline{\mathcal{P}_r(x_0,t_0)}, \end{aligned}$$

at least if r is sufficiently small. Indeed, we can find a neighborhood U of V and a smoothly time-dependent coordinate system $(x^1,\ldots,x^n) : \mathcal{P}_{2r}(x_0,t_0) \to \mathbb{R}^n$ satisfying

$$(x^1(x_0,t_0),\ldots,x^n(x_0,t_0)) = 0 \text{ and } U \subset \{(x,t) \in \mathcal{P}_{2r}(x_0,t_0) : x^1(x,t) > 0\} .$$

The function $h \doteqdot 1 - \mathrm{e}^{-\alpha x^1}$ then satisfies the first two of the desired conditions when $\alpha > 0$. Moreover,

$$(\partial_t - \Delta - \nabla_b)h = \alpha \mathrm{e}^{-\alpha x^1}\left((\partial_t - \Delta)x^1 + \alpha|\nabla x^1|^2 - \nabla_b x^1\right) .$$

So we can arrange the third condition by choosing α sufficiently large.

Now let φ be an upper support for f at (x_0,t_0) satisfying

$$(\partial_t - \Delta - \nabla_b)\varphi(x_0,t_0) \geq 0 .$$

Then, provided $\delta > 0$ is sufficiently small, the function $\varphi_\delta \doteqdot \varphi + \delta h$ satisfies

$$\begin{aligned} \varphi_\delta > 0 \quad &\text{on} \quad \partial\mathcal{P}_r(x_0,t_0), \\ \varphi_\delta(x_0,t_0) = 0&, \\ (\partial_t - \Delta - \nabla_b)\varphi_\delta > 0 \quad &\text{on} \quad \overline{\mathcal{P}_r(x_0,t_0)} . \end{aligned}$$

But the first two conditions imply that there is a point $(x',t') \in \mathcal{P}_r(x_0,t_0)$ such that $\varphi_\delta(x',t') = \min_{\mathcal{P}_{r'}(x',t')} \varphi_\delta$ for some $r' > 0$, contradicting the third. ☐

Theorem 9.11 (Rigidity of the positive cone). *Let $X : M^n \times I \to \mathbb{R}^{n+1}$ be a locally convex (i.e., $\mathrm{II} \geq 0$) solution to mean curvature flow which is properly defined in $P_R(p_0,t_0)$ for some $(x_0,t_0) \in M^n \times I$, where $p_0 \doteqdot X(x_0,t_0)$. If $\ker \mathrm{L}_{(x_0,t_0)} \neq \{0\}$, then the solution locally* ***splits isometrically from a plane****; that is, there exist $m \in \{1,\ldots,n\}$ and $r > 0$ such that*

- *$dX(\ker \mathrm{L}_{(x,t)})$ is a constant, m-dimensional linear subspace of $\mathbb{R}^{n+1}$ for each $(x,t) \in \mathcal{P}_r(x_0,t_0)$,*
- *$(\mathcal{P}_r(x_0,t_0), g_t)$ embeds isometrically into $\mathbb{R}^m \times^\perp \Sigma_t^{n-m}$ for each $t \in (t_0 - r^2, t_0]$, where Σ_t^{n-m} is an embedded, m-dimensional submanifold of $\mathbb{R}^{n-m+1}$, and*
- *the family $\{\Sigma_t^{n-m}\}_{t\in(t_0-r^2,t_0]}$ evolves by mean curvature flow.*

Proof. Consider the function $f(x,t) \doteqdot \kappa_1(x,t) = \operatorname{dist}_{\Gamma_+}(\mathrm{II}_{(x,t)})$. By Proposition 9.6, f satisfies

$$(\partial_t - \Delta) f \geq |\mathrm{II}|^2 f + 2 \sup_{\ell \in \mathrm{S}_\kappa \Gamma_+} \sum_{k=1}^{n} \sum_{\kappa_j > \kappa_i} \frac{\ell^i - \ell^j}{\kappa_j - \kappa_i} (\nabla_k \mathrm{II}_{ij})^2$$

in the barrier sense. Since f reaches an interior minimum, Lemma 9.10 implies that $f \equiv 0$ in $\mathcal{P}_r(x_0, t_0)$ for some $r > 0$. In particular, f is smooth in $\mathcal{P}_r(x_0, t_0)$ with

$$0 \equiv \nabla f \quad \text{and} \quad 0 \equiv \sum_{k=1}^{n} \sum_{\kappa_j > \kappa_i} \frac{\ell^i - \ell^j}{\kappa_j - \kappa_i} (\nabla_k \mathrm{II}_{ij})^2$$

for each $\ell \in \mathrm{S}_\kappa \Gamma_+$. The first identity implies that

$$\nabla_w \mathrm{II}(u, v) = 0$$

for all $u, v \in \ker \mathrm{L}_{(x,t)}$ and $w \in T_{(x,t)}M$ if $(x,t) \in \mathcal{P}_r(x_0, t_0)$. Indeed, given $(x,t) \in \mathcal{P}_r(x_0, t_0)$, $U, V \in \Gamma(\ker \mathrm{L})$ with $U_{(x,t)} = u$ and $V_{(x,t)} = v$, and $w \in T_{(x,t)}M$,

$$0 \equiv w(\mathrm{II}(U, V)) = \nabla_w \mathrm{II}(u, v) + \mathrm{II}(\nabla_w U, V) + \mathrm{II}(U, \nabla_w V) = \nabla_w \mathrm{II}(u, v) .$$

Since $e_i \in S_\kappa \Gamma$ whenever $\kappa_i = 0$, the second identity implies that

$$\nabla_w \mathrm{II}(u, v) = 0$$

for all $u \in \ker(\mathrm{L}_{(x,t)})$, all $v \in \left(\ker \mathrm{L}_{(x,t)}\right)^\perp$, and all $w \in T_{(x,t)}M$. By the Codazzi equation, we conclude that

$$\nabla_u \mathrm{II} = 0$$

for all $u \in \ker(\mathrm{L}_{(x,t)})$ when $(x,t) \in \mathcal{P}_r(x_0, t_0)$.

Now observe that $\nabla_k V \in \Gamma(\ker \mathrm{L})$ whenever $V \in \Gamma(\ker \mathrm{L})$. Indeed, given $V \in \Gamma(\ker \mathrm{L})$,

$$0 \equiv \nabla_k(\mathrm{L}(V)) = \nabla_k \mathrm{L}(V) + \mathrm{L}(\nabla_k V) = \mathrm{L}(\nabla_k V) .$$

That is, $\ker \mathrm{L}$ is invariant under parallel translation in space. It follows that $dX(\ker \mathrm{L})$ is parallel in space with respect to the pullback connection ${}^X D$. Indeed, given any $V \in \Gamma(\ker \mathrm{L})$ and any $u \in T_{(x,t)}M$,

$${}^X D_u(dX(V)) = dX(\nabla_u V) - \mathrm{II}(u, v)\mathsf{N} = dX(\nabla_u V) \in dX(\ker \mathrm{L}_{(x,t)}) ,$$

where $v = V_{(x,t)}$.

By the evolution equation (6.17) for L and the parallel invariance of $\ker \mathrm{L}$,

$$\nabla_t \mathrm{L}(V) = \Delta \mathrm{L}(V) + |\mathrm{II}|^2 \mathrm{L}(V) = \Delta \mathrm{L}(V) = 0$$

for any $V \in \ker \mathrm{L}$ so that

$$\mathrm{L}(\nabla_t V) = \nabla_t(\mathrm{L}(V)) - \nabla_t \mathrm{L}(V) = 0\,;$$

that is, $\ker \mathrm{L}$ is also invariant with respect to ∇_t. Since, for any $V \in \Gamma(\ker \mathrm{L})$, we have $\nabla_V H = \mathrm{tr}(\nabla_V \mathrm{II}) \equiv 0$, this implies that

$${}^X\!D_t(dX(V)) = \nabla_V H\,\mathsf{N} + dX(\nabla_t V) = dX(\nabla_t V)\,,$$

so that $dX(\ker \mathrm{L})$ is also parallel in time. We conclude that $dX(\ker \mathrm{L})$ is a constant subspace of $\mathbb{R}^{n+1}$ (after canonically identifying the spaces $T_p\mathbb{R}^{n+1} \cong \mathbb{R}^{n+1}$). In particular, the nullity $m \in \{1, \dots, n\}$ of L is constant in $\mathcal{P}_r(x_0, t_0)$.

Now consider any geodesic $\gamma : I \to M^n$ with respect to the metric at time $t \in (0, t_0]$ with initial data $(\gamma(0), \gamma'(0)) \in \ker \mathrm{L}_{(x,t)}$. Since $\ker \mathrm{L}$ is invariant under parallel translation, $\gamma'(s) \in \ker \mathrm{L}_{(\gamma(s),t)}$ for all s and hence

$${}^X\!D_s\left(dX(\gamma')\right) = dX(\nabla_s \gamma') - \mathrm{II}(\gamma', \gamma')\mathsf{N} = 0\,.$$

That is, $X \circ \gamma$ is geodesic in $\mathbb{R}^{n+1}$.

By Sard's theorem [**401**, Chapter 2], we can find $r' > 0$ (arbitrarily close to r) and x_0' (arbitrarily close to x_0) such that the section $\Sigma_t^{n-m} \doteqdot \mathcal{M}_t \cap \Pi^{n-m+1}$ is a smoothly embedded $(n-m)$-manifold for each $t \in (t_0 - (r')^2, t_0]$, where $\mathcal{M}_t \doteqdot X_t(\mathcal{B}_{r'}(x_0', t_0))$ and $\Pi^{n-m+1} \doteqdot p_0' + [dX(\ker \mathrm{L}_{(x_0',t)})]^\perp$, $p_0' \doteqdot X(x_0', t_0)$. The remaining claims follow. □

In Section 12.4, we will prove a stronger version of the splitting theorem (due to Brian White [**531**, Appendix A]) as well as an analagous rigidity result for the largest principal curvature.

9.3. Pinching improves: Convexity and cylindrical estimates

We want to apply the techniques of Section 8.2 to obtain an optimal pinching result in the setting of mean convex hypersurfaces. Note that, due to the neckpinch examples of Section 9.1, we cannot hope to rule out "cylindrical" singularities in general. We say that a matrix $W \in \mathrm{S}(n)$ is **cylindrical** if it has a null eigenvalue of multiplicity m and a positive eigenvalue of multiplicity $n - m$ for some $m \in \{0, \dots, n-1\}$ and we denote the set of cylindrical points by Cyl. We shall also say that a point x of a hypersurface is **cylindrical** if $\mathrm{II}_x \in \mathrm{Cyl}$. It turns out that these points are the only obstruction to improving curvature pinching, at least among convex curvature conditions.

Given two cones Γ and Γ' in a finite-dimensional normed linear space E, we shall say that Γ' is **compactly contained in** Γ (modulo scaling) and write $\Gamma' \Subset \Gamma$ if the intersection of Γ' with the unit sphere in E is compactly contained in Γ in the usual sense; i.e., $\overline{\Gamma' \cap S} \subset \mathrm{int}\,\Gamma$, where $S \doteqdot \{W \in E : \|W\| = 1\}$. In this section we shall prove the following.

Theorem 9.12 (Convexity estimates (cf. [**301, 302, 531**])). *Let $X : M^n \times [0,T) \to \mathbb{R}^{n+1}$, $n \geq 2$, be a compact, strictly mean convex solution to mean curvature flow. Let $\Gamma_0 \Subset \{W \in \mathrm{S}(n) : \mathrm{tr}(W) > 0\}$ be any pinching condition satisfying $\mathrm{II}_{(x,0)} \in \Gamma_0$ for all $x \in M^n$ and denote by Λ the convex hull of the cylindrical points in Γ_0. Given any $\varepsilon > 0$, there is a constant $C_\varepsilon = C_\varepsilon(\mathcal{M}_0, \varepsilon)$ such that*

$$d_\Lambda(\mathrm{II}_{(x,t)}) \geq -\varepsilon H(x,t) - C_\varepsilon \tag{9.17}$$

for all $(x,t) \in M^n \times [0,T)$. The constant C_ε is of the form $C_\varepsilon = \mathrm{C}_\varepsilon R^{-1}$, where $R \doteqdot \frac{1}{2}\operatorname{diam}(\mathcal{M}_0)$ and C_ε depends only on n, Γ_0, ε, and upper bounds for $R\max_{\mathcal{M}_0} H$ and $R^{-n}|\mathcal{M}_0|$.

The theorem is essentially due to G. Huisken and C. Sinestrari [**301**–**303**] and, independently, B. White [**531**]. We present a streamlined version of the argument of Huisken and Sinestrari, which makes use of integral estimates and Stampacchia iteration, as in the proof of the roundness estimate (Theorem 8.6). White's proof, in contrast, proceeds by contradiction, making use of a blow-up argument and the strong maximum principle for κ_1/H (and κ_n/H), similar to the proof of Corollary 8.28 in Section 8.6 (see the proof of Theorems 12.20 and 12.23 in Chapter 12).

Figure 9.4. Carlo Sinestrari.

Before proving Theorem 9.12, let us note some useful consequences. Observe first that the convexity estimates immediately imply the umbilic estimate. Indeed, if the initial hypersurface is uniformly convex, the only admissible cylindrical points are the umbilic points, so Λ is the umbilic ray $\mathbb{R}_+\mathrm{I}$ and

$$d_{\mathbb{R}_+\mathrm{I}}(\mathrm{II}_{(x,t)}) = -\frac{|\mathring{\mathrm{II}}|}{H}(x,t)\,.$$

We also immediately obtain the following two corollaries.

Corollary 9.13 (Convexity estimate [**301, 302, 531**]). *Let $X : M^n \times [0,T) \to \mathbb{R}^{n+1}$, $n \geq 2$, be a compact, strictly mean convex solution to mean curvature flow. Given any $\varepsilon > 0$ there is a constant $C_\varepsilon = C_\varepsilon(\mathcal{M}_0, \varepsilon)$ such that*

$$\kappa_1(x,t) \geq -\varepsilon H(x,t) - C_\varepsilon \tag{9.18}$$

for all $(x,t) \in M^n \times [0,T)$.

Proof. The convex hull of the cylindrical points allowed by mean convexity is the whole positive cone, Γ_+. The claim follows since, by Exercise 9.3,

$$d_{\Gamma_+}(\mathrm{II}) = \kappa_1 \,. \qquad \square$$

Corollary 9.14 (Cylindrical estimates (cf. [**303**])). *Let $X : M^n \times [0,T) \to \mathbb{R}^{n+1}$, $n \geq 2$, be a compact, strictly $(m+1)$-convex solution to mean curvature flow, $m \in \{0, \ldots, n-2\}$. Given any $\varepsilon > 0$ there is a constant $C_\varepsilon = C_\varepsilon(\mathcal{M}_0, \varepsilon)$ such that*

$$\sum_{i=m+1}^{n} (\kappa_n - \kappa_i)(x,t) \leq \sum_{i=1}^{m} \kappa_i(x,t) + \varepsilon H(x,t) + C_\varepsilon \tag{9.19}$$

for all $(x,t) \in M^n \times [0,T)$.

Proof. The convex hull Λ_m of the cylindrical points allowed by $(m+1)$-convexity is the convex hull of the m-cylindrical points. Since the cone C_m corresponding to the condition

$$\kappa_n \leq \tfrac{1}{n-m} H$$

is convex and contains the m-cylindrical points, it must contain Λ_m. The claim follows since $-d_{C_m}(\mathrm{II})$ is proportional to $\kappa_n - \frac{1}{n-m}H$ (see Exercise 9.4). $\square$

The convexity estimate implies that a mean convex solution is becoming weakly convex at a singularity after rescaling. The cylindrical estimates then imply that an $(m+1)$-convex solution is becoming weakly m-convex and either strictly m-convex or m-cylindrical at a singularity after rescaling.

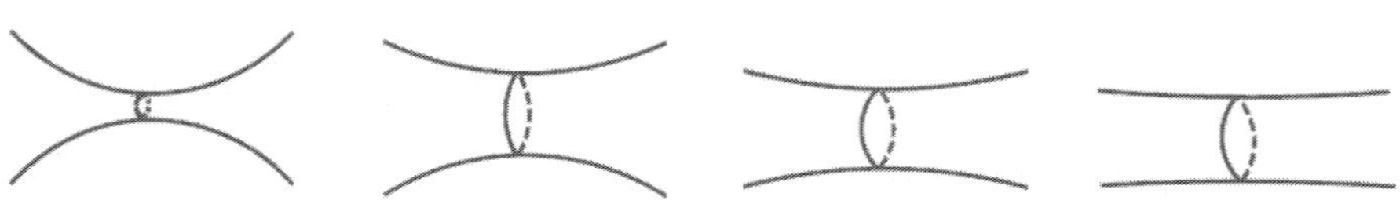

Figure 9.5. A rescaled neckpinch: The amount of nonconvexity limits to zero and the cross section becomes round.

So let $X : M^n \times [0,T) \to \mathbb{R}^{n+1}$ be a solution to mean curvature flow satisfying the hypotheses of Theorem 9.12 and set

$$f(x,t) \doteqdot -d_\Lambda(\mathrm{II}_{(x,t)}).$$

In order to obtain a better gradient term in its evolution equation, we need to modify f slightly. Note that, by Corollary 9.8, there exists some $L = L(\Gamma_0) > 0$ such that $|\mathrm{II}| \le LH$ under the flow.

Lemma 9.15. *Given any $\varepsilon \in (0,1)$, set $f_\varepsilon \doteqdot \max\{f - \varepsilon(LH - |\mathrm{II}|), 0\}$. There exists $\gamma = \gamma(n, \Gamma_0, \varepsilon) > 0$ such that*

$$(\partial_t - \Delta) f_\varepsilon \le f_\varepsilon \left(|\mathrm{II}|^2 - \gamma \frac{|\nabla \mathrm{II}|^2}{H^2} \right) \tag{9.20}$$

in the barrier sense.

Proof. See Exercise 9.6. □

Given any $\varepsilon \in (0,1)$, we seek to bound the function (cf. Section 8.2)

$$f_{\varepsilon,\sigma} \doteqdot f_\varepsilon H^{\sigma - 1} \tag{9.21}$$

for some $\sigma \in (0,1)$. Computing as in Lemma 8.7 using Lemma 9.15 (and the barriers defined in Proposition 9.6), we find that $f_{\varepsilon,\sigma}$ satisfies

$$(\partial_t - \Delta) f_{\varepsilon,\sigma} \le \sigma \left|\mathrm{II}\right|^2 f_{\varepsilon,\sigma} + 2(1-\sigma) \left\langle \nabla f_{\varepsilon,\sigma}, \frac{\nabla H}{H} \right\rangle - 2\gamma f_{\varepsilon,\sigma} \frac{|\nabla \mathrm{II}|^2}{H^2}$$

in the barrier sense for some $\gamma = \gamma(n, \Gamma_0, \varepsilon)$. Given $p \ge \max\{10, 4n\gamma^{-1}\}$, set $v = v_{\varepsilon,\sigma,p} \doteqdot f_{\varepsilon,\sigma}^{\frac{p}{2}}$. Then

$$\begin{aligned}
(\partial_t - \Delta) v^2 &\le p v^2 \left(\sigma |\mathrm{II}|^2 + \frac{4(1-\sigma)}{p} \left\langle \frac{\nabla v}{v}, \frac{\nabla H}{H} \right\rangle - 2\gamma \frac{|\nabla \mathrm{II}|^2}{H^2} \right) \\
&\quad - \frac{4(p-1)}{p} |\nabla v|^2 \\
&\le \sigma p |\mathrm{II}|^2 v^2 - 2|\nabla v|^2 - 2(\gamma p - 2n) v^2 \frac{|\nabla \mathrm{II}|^2}{H^2} \\
&\le \sigma p |\mathrm{II}|^2 v^2 - 2|\nabla v|^2 - \gamma p v^2 \frac{|\nabla \mathrm{II}|^2}{H^2} \qquad (9.22)
\end{aligned}$$

in the barrier sense (cf. (8.18)).

Using Simons's identity, we can control the first term on the right in L^1 by the two good terms (cf. Proposition 8.9). But first we note that (9.22) holds in L^1.

Lemma 9.16. *The function $v^2 \doteqdot f_{\varepsilon,\sigma}^p$ is locally Lipschitz in t and satisfies (9.22) in the* ***distributional sense****. That is,*

$$\frac{d}{dt} \int_M \eta v^2 d\mu \le \int_M v^2 \left[\eta \left(\sigma p |\mathrm{II}|^2 - 2 \frac{|\nabla v|^2}{v^2} - \gamma p \frac{|\nabla \mathrm{II}|^2}{H^2} \right) + (\partial_t + \Delta - H^2)\eta \right] d\mu$$

for all nonnegative test functions $\eta \in C^\infty(M^n \times [0,T))$ and almost every $t \in [0,T)$.

Proof. To prove that v^2 is locally Lipschitz in t it suffices to prove that $f = \operatorname{dist}_\Lambda(\mathrm{II})$ is locally Lipschitz in t. Given $(x,t_0) \in M^n \times [0,T)$ choose $L_0 \in \mathrm{S}_{\mathrm{II}_{(x,t_0)}}\Lambda$ so that $f(x,t_0) = L_0(\mathrm{II}_{(x,t_0)})$. By parallel translation in space and time we can extend L to a local section L of $TM \otimes TM$ defined near (x,t_0). In particular, for t_1 sufficiently close to t_0,

$$\begin{aligned} f(x,t_0) = L(\mathrm{II}_{(x,t_0)}) &= L(\mathrm{II}_{(x,t_1)}) + L(\mathrm{II}_{(x,t_0)}) - L(\mathrm{II}_{(x,t_1)}) \\ &\geq f(x,t_1) - C|t_0 - t_1| . \end{aligned}$$

Since M^n is compact, we can choose the constant C depending only on t_0.

Next, we claim that v^2 is semiconvex. It suffices to prove that $f \doteqdot \operatorname{dist}_\Lambda(\mathrm{II})$ is semiconcave. Given $(x_0,t) \in M^n \times [0,T)$ choose $L_0 \in \mathrm{S}_{\mathrm{II}_{(x_0,t)}}\Lambda$ so that $f(x_0,t) = L_0(\mathrm{II}_{(x_0,t)})$. By parallel translation in space we can extend L to a local section L of $TM \otimes TM$ defined in $\mathcal{B}_r(x_0,t)$ for some $r > 0$. Set $\varphi(x) \doteqdot L(\mathrm{II}_{(x,t)})$ for $x \in \mathcal{B}_r(x_0,t)$. Let $\gamma : I \to M^n$ be a unit speed geodesic in M^n (with respect to the metric at time t) with initial data $(\gamma(0), \gamma'(0)) = (x_0, v)$ and set $u(s) \doteqdot f \circ \gamma$. Then

$$\begin{aligned} u(s) - u(0) &\leq \varphi(\gamma(s)) - \varphi_0(x_0) \\ &= s\nabla_v\varphi(x_0) + \frac{1}{2}s^2\nabla^2_{v,v}\varphi(x_0) + o(s^2) . \end{aligned}$$

Now set $u_\lambda(s) \doteqdot u(s) - \lambda s^2$, where $\lambda = \sup_{\mathcal{B}_{r/2}(x_0,t)} |\nabla^2\varphi|$. Then

$$\begin{aligned} u_\lambda(s) - u_\lambda(0) &\leq s\nabla_v\varphi(x_0) + \frac{1}{2}s^2\nabla^2_{v,v}\varphi(x_0) - \lambda s^2 + o(s^2) \\ &\leq s\nabla_v\varphi(x_0) - \frac{\lambda}{2}s^2 + o(s^2) . \end{aligned}$$

For s sufficiently small, we obtain

$$u_\lambda(s) \leq u_\lambda(0) + s\nabla_v\varphi(x_0) .$$

Thus, at each point, we have found a supporting line lying above the graph of u_λ. This proves that v^2 is semiconvex. Alexandrov's theorem [**8**] now implies that, at each time t, there is a set of measure zero in M away from which f, and hence v^2, admits first and second derivatives. This proves the claim since the almost-everywhere Laplacian is dominated by the distributional Laplacian [**218**, §6.4]. □

In particular,

$$\frac{d}{dt}\int_M v^2\, d\mu \leq \int_M v^2 \left(\sigma p\, |\mathrm{II}|^2 - H^2 - 2\,\frac{|\nabla v|^2}{v^2} - \gamma p\,\frac{|\nabla \mathrm{II}|^2}{H^2} \right) d\mu$$

for almost every $t \in [0,T)$.

To control the remaining term, we need the following Poincaré-type inequality for functions whose support does not contain any cylindrical points.

Proposition 9.17. *Given an integer $n \geq 2$, an $\mathrm{SO}(n)$-invariant cone $\Gamma \Subset \{W \in \mathrm{S}(n) : \mathrm{tr}(W) > 0\} \setminus \mathrm{Cyl}(n)$, and any $\delta > 0$ there exists a constant $P_\delta < \infty$ with the following property. Let $X : M^n \to \mathbb{R}^{n+1}$ be a hypersurface of dimension $n \geq 2$ and let $u \in W^{2,1}(M)$ be a function with compact support contained in the set $\{x \in M : \mathrm{II} \in \Gamma\}$. Then*

$$\int_M u^2 |\mathrm{II}|^2 \, d\mu \leq \delta \int_M |\nabla u|^2 d\mu + P_\delta \int_M u^2 \frac{|\nabla \mathrm{II}|^2}{H^2} \, d\mu \,.$$

Proof. The proof is very similar to the proof of Proposition 8.9. We leave it as an exercise (see Exercise 9.7). □

We are now able to conclude, as in Lemma 8.10, that

$$\frac{d}{dt} \int_M f_{\varepsilon,\sigma}^p \, d\mu \leq 0 \tag{9.23}$$

for p sufficiently large and σ sufficiently small (of order $\sim p^{-\frac{1}{2}}$). Proceeding exactly as in the proof of the roundness estimate (Theorem 8.6) then yields a bound for $f_{\varepsilon,\sigma}$. Theorem 9.12 follows. □

There is actually a simpler characterization of the cones Λ and Γ_0 of Theorem 9.12: Roughly speaking, the largest pinching condition not containing the curvature of the cylinder $\mathbb{R}^{m+1} \times S^{n-m-1}$ is $(m+1)$-convexity,

$$\kappa_1 + \cdots + \kappa_{m+1} > 0,$$

while the smallest pinching condition containing the curvature of $\mathbb{R}^m \times S^{n-m}$ is the intersection of the two natural pinching conditions for the largest and smallest principal curvatures,

$$\kappa_1 \geq 0 \quad \text{and} \quad \kappa_n \leq \tfrac{1}{n-m} H \,.$$

Proposition 9.18. *Given $n \geq 2$ and $m \in \{0, \ldots, n-1\}$, set*

$$W_m^n = \mathrm{diag}(\underbrace{0, \ldots, 0}_{m\text{-times}}, 1, \ldots, 1).$$

(a) *Let Γ_m^n be the largest pinching condition in $\mathrm{S}(n)$ not containing W_{m+1}^n. That is,*

$$\Gamma_m^n \doteqdot \bigcup \Big\{ \Gamma \underset{\substack{\text{open, convex} \\ \mathrm{SO}(n)\text{-invariant}}}{\subset} \mathrm{S}(n) : W_{m+1}^n \notin \Gamma \Big\} .$$

Then

$$\Gamma_m^n = \{W \in \mathrm{S}(n) : \lambda_1(W) + \cdots + \lambda_{m+1}(W) > 0\} ,$$

where $\lambda_1(W) \leq \cdots \leq \lambda_n(W)$ denote the ordered eigenvalues of W.

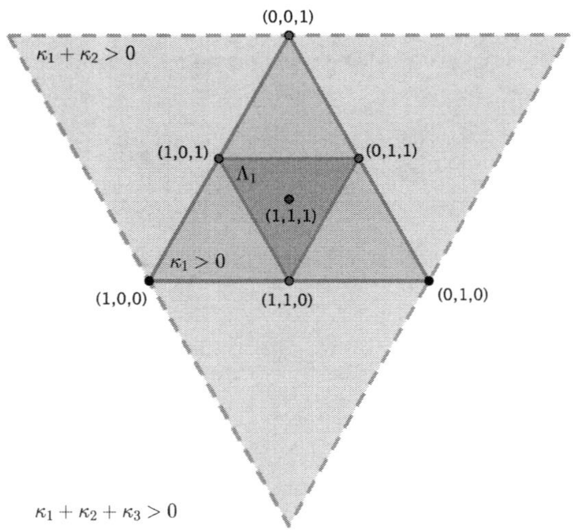

Figure 9.6. Optimal pinching sets in $\mathbb{R}^3$.

(b) *Let* Λ^n_m *be the smallest pinching condition in* $\mathrm{S}(n)$ *containing* W^n_m. *That is,*

$$\Lambda^n_m \doteqdot \bigcap \Big\{ \Gamma \underset{\substack{\text{closed, convex}\\ \mathrm{SO}(n)\text{-invariant}}}{\subset} \mathrm{S}(n) : W^n_m \in \Gamma \Big\}.$$

Then

$$\Lambda^n_m = \overline{\Gamma}^n_+ \cap \overline{\mathrm{K}}^n_m ,$$

where $\overline{\Gamma}^n_+ \doteqdot \{W \in \mathrm{S}(n) : W \geq 0\}$ *is the closure of the positive cone and* $\overline{\mathrm{K}}^n_m \doteqdot \{W \in \mathrm{S}(n) : \lambda_n(W) \leq \frac{1}{n-m}\operatorname{tr}(W)\}$.

In particular, if $\Gamma \subset \mathrm{S}(n)$ *is a pinching condition and* Λ *is the convex hull of the cylindrical points in* Γ, *then there is some* $m \in \{0, \ldots, n-1\}$ *such that* $\Lambda = \Lambda^n_m$ *and* $\Gamma_0 \subset \Gamma^n_m$.

Proof. We shall only prove (b). It is readily verified that $\Lambda^n_m \subset \overline{\Gamma}^n_+ \cap \overline{\mathrm{K}}^n_m$ for each $m \in \{0, \ldots, n-1\}$ and that $\Lambda^n_{n-1} = \overline{\Gamma}^n_+ = \overline{\Gamma}^n_+ \cap \overline{\mathrm{K}}^n_{n-1}$ for each $n \geq 2$. So it remains to prove that $\overline{\Gamma}^n_+ \cap \overline{\mathrm{K}}^n_m \subset \Lambda^n_m$ for every $m \in \{0, \ldots, n-2\}$ for every $n \geq 2$. We will achieve this by induction. Note that the claim is true when $n = 2$. Indeed, Λ^2_0 and $\overline{\mathrm{K}}^2_0$ are both the positive ray $\{(r,r) : r > 0\}$. So fix any $n > 2$ and suppose that $\overline{\Gamma}^{n-1}_+ \cap \overline{\mathrm{K}}^{n-1}_m \subset \Lambda^{n-1}_m$ for every $m \in \{0, \ldots, n-2\}$. It will suffice to show that $\partial(\overline{\Gamma}^n_+ \cap \overline{\mathrm{K}}^n_m) \subset \Lambda^n_m$. So consider a point

$$W \in \partial(\overline{\Gamma}^n_+ \cap \overline{\mathrm{K}}^n_m) = (\partial\overline{\Gamma}^n_+ \cap \overline{\mathrm{K}}^n_m) \cup (\overline{\Gamma}^n_+ \cap \partial\overline{\mathrm{K}}^n_m)$$

and denote its eigenvalues by $w_1 \leq \cdots \leq w_n$. Suppose first that $W \in \partial\overline{\Gamma}^n_+ \cap \overline{\mathrm{K}}^n_m$. Then $w_1 = 0$ and

$$w_n \leq \tfrac{1}{n-m}(w_1 + \cdots + w_n) = \frac{1}{(n-1)-(m-1)}(w_2 + \cdots + w_n)$$

so that, applying the induction hypothesis,

$$W = W|_{e_1^\perp} \in \overline{\Gamma}^{n-1}_+ \cap \overline{\mathrm{K}}^{n-1}_{m-1} \subset \Lambda^{n-1}_{m-1} = \partial\overline{\Gamma}^n_+ \cap \Lambda^n_m \subset \Lambda^n_m\,.$$

Next consider the case $W \in \overline{\Gamma}^n_+ \cap \partial\overline{\mathrm{K}}^n_m$. Then

$$w_{n-1} \leq w_n = \frac{1}{n-(m+1)}(w_1 + \cdots + w_{n-1})$$

and hence, applying the induction hypothesis, $W|_{e_n^\perp} \in \overline{\Gamma}^{n-1}_+ \cap \overline{\mathrm{K}}^{n-1}_m \subset \Lambda^{n-1}_m$. Thus,

$$(w_1, \ldots, w_{n-1}) = \sum_{\sigma \in P_{n-1}} a_\sigma (w^m_{\sigma(1)}, \ldots, w^m_{\sigma(n-1)})$$

for some positive coefficients $\{a_\sigma\}_{\sigma \in P_{n-1}}$, where $w^m_1 = \cdots = w^m_m = 0$ and $w^m_{m+1} = \cdots = w^m_{n-1} = 1$. In particular,

$$w_1 + \cdots + w_{n-1} = (n-(m+1)) \sum_{\sigma \in P_{n-1}} a_\sigma$$

so that

$$(w_1, \ldots, w_n) = \sum_{\sigma \in P_{n-1}} a_\sigma (w^m_{\sigma(1)}, \ldots, w^m_{\sigma(n-1)}, 1)\,.$$

We conclude that $W \in \Lambda^n_m$, which completes the proof. □

So an alternative proof of Theorem 9.12 can be obtained by considering the functions κ_1 and κ_n in place of $f = \mathrm{dist}_\Lambda(\mathrm{II})$.

9.4. A natural class of initial data

In light of the dependency of the constants in the convexity estimates, it is natural to introduce the following class of hypersurfaces.

Definition 9.19. Given integers $n \geq 2$ and $m \in \{0, \ldots, n-2\}$ and positive real numbers α_0, α_1, α_2, and R, the **class** $\mathcal{C}^n_m(R, \alpha_0, \alpha_1, \alpha_2)$ consists of all compact, mean convex hypersurfaces $X : M^n \to \mathbb{R}^{n+1}$ which satisfy

(i) $\kappa_1 + \cdots + \kappa_{m+1} \geq \alpha_0 H$,
(ii) $H \geq \alpha_1 R^{-1}$, and
(iii) $\mu(M^n) \leq \alpha_2 R^n$,

where μ is the induced measure on M^n.

When $m = n-1$, $\mathcal{C}^n_m(R, \alpha_0, \alpha_1, \alpha_2)$ is defined similarly, except that the condition $\kappa_1 + \cdots + \kappa_{m+1} \geq \alpha_0 H$ is replaced by $H \geq \alpha_0 |\mathrm{II}|$.

It will be convenient to use the shorthand notation $\alpha \doteqdot (\alpha_0, \alpha_1, \alpha_2)$ and $\mathcal{C}^n_m(R, \alpha) \doteqdot \mathcal{C}^n_m(R, \alpha_0, \alpha_1, \alpha_2)$.

The constant R plays the role of scaling parameter, making α scale invariant: If $X \in \mathcal{C}^n_m(R, \alpha)$, then $\lambda X \in \mathcal{C}^n_m(\lambda R, \alpha)$. Of course, *every* compact, strictly $(m+1)$-convex immersed hypersurface $X : M^n \to \mathbb{R}^{n+1}$ admits parameters $\alpha_0, \alpha_1, \alpha_2$, and R which place it in the class $\mathcal{C}^n_m(R, \alpha)$. One convenient way of choosing the parameters is to first choose R so that $\max_{M^n} |\mathrm{II}| \le R^{-1}$ and then choose α so that

$$\alpha_0 \le \min_{M^n} \frac{\kappa_1 + \cdots + \kappa_{m+1}}{H}, \quad \alpha_1 \le \min_{M^n} RH, \quad \text{and} \quad \alpha_2 \ge R^{-n}\mu(M^n)$$

in case $m < n-1$ or

$$\alpha_0 \le \min_{M^n} \frac{H}{|\mathrm{II}|}, \quad \alpha_1 \le \min_{M^n} RH, \quad \text{and} \quad \alpha_2 \ge R^{-n}\mu(M^n)$$

in case $m = n-1$.

Observe that each class is preserved under the mean curvature flow: If $X : M^n \times [0,T) \to \mathbb{R}^{n+1}$ satisfies mean curvature flow and $X_0 \in \mathcal{C}^n_m(R, \alpha)$, then $X_t \in \mathcal{C}^n_m(R, \alpha)$ for each $t \in (0,T)$.

We also have the following universal interior estimates.

Lemma 9.20. *Let $X : M^n \times [0,T) \to \mathbb{R}^{n+1}$ be a maximal solution to mean curvature flow with initial condition in the class $\mathcal{C}^n_m(R, \alpha)$. Suppose that $\sup_{M^n \times \{0\}} |\mathrm{II}| \le R^{-1}$. Then*

$$R^2 \le 2T \le n\alpha_1^{-2} R^2 \tag{9.24}$$

and

$$\max_{M^n \times \{R^2/4\}} |\nabla^k \mathrm{II}| \le c_k R^{-(k+1)} \tag{9.25}$$

for every $k \in \mathbb{N}$, where c_k depends only on n and k.

Proof. By a straightforward ODE comparison argument applied to the inequality

$$(\partial_t - \Delta)|\mathrm{II}|^2 \le 2|\mathrm{II}|^4,$$

we obtain

$$\max_{M^n \times \{t\}} |\mathrm{II}|^2 \le \frac{1}{\max_{M^n \times \{0\}} |\mathrm{II}|^{-2} - 2t}.$$

Combined with Exercise 8.1 and Theorem 6.20, this yields (9.24). Moreover,

$$|\mathrm{II}|^2(\cdot, t) \le \frac{2}{R^2} \quad \text{for all } t \le \frac{R^2}{4}. \tag{9.26}$$

So (9.25) follows from Theorem 6.24. $\square$

Let us restate Corollaries 9.13 and 9.14 in terms of these classes.

Corollary 9.21 (Convexity estimate). *Let $X : M^n \times [0,T) \to \mathbb{R}^{n+1}$, $n \geq 2$, be a solution to mean curvature flow with initial condition in the class $\mathcal{C}^n_m(R,\alpha)$. Suppose also that $\max_{M^n\times\{0\}} |\mathrm{II}| \leq R^{-1}$. Given any $\varepsilon > 0$ there is a constant $C_\varepsilon = C_\varepsilon(n,\alpha,\varepsilon)$ such that*

$$\kappa_1(x,t) \geq -\varepsilon H(x,t) - C_\varepsilon R^{-1} \tag{9.27}$$

for all $(x,t) \in M^n \times [0,T)$.

Corollary 9.22 (Cylindrical estimates). *Let $X : M^n \times [0,T) \to \mathbb{R}^{n+1}$, $n \geq 2$, be a solution to mean curvature flow with initial condition in the class $\mathcal{C}^n_m(R,\alpha)$. Suppose also that $\max_{M^n\times\{0\}} |\mathrm{II}| \leq R^{-1}$. Given any $\varepsilon > 0$ there is a constant $C_\varepsilon = C_\varepsilon(n,\alpha,\varepsilon)$ such that*

$$\sum_{i=m+1}^{n} (\kappa_n - \kappa_i)(x,t) \leq \sum_{i=1}^{m} \kappa_i(x,t) + \varepsilon H(x,t) + C_\varepsilon R^{-1} \tag{9.28}$$

for all $(x,t) \in M^n \times [0,T)$.

9.5. A gradient estimate for the curvature

We will use the convexity estimates to obtain a gradient estimate for the mean curvature analogous to the one proved in Section 8.3. First observe that the linear pinching estimate immediately implies the following quadratic pinching estimate.

Proposition 9.23. *Let $X : M^n \times [0,T) \to \mathbb{R}^{n+1}$, $n \geq 2$, be a solution to mean curvature flow with initial condition in the class $\mathcal{C}^n_m(R,\alpha)$ and satisfying $\max_{M^n\times\{0\}} |\mathrm{II}| \leq R^{-1}$. For any $\varepsilon > 0$ there exists $C_\varepsilon = C(n,\alpha,\varepsilon) < \infty$ such that*

$$\left(|\mathrm{II}|^2 - \tfrac{1}{n-m}H^2\right)(x,t) \leq \varepsilon H^2(x,t) + C_\varepsilon R^{-2} \tag{9.29}$$

for all $(x,t) \in M^n \times [0,T)$.

Proceeding as in the proof of the gradient estimate for convex solutions (Theorem 8.14) now yields a curvature gradient estimate for $(m+1)$-convex solutions *but only if m is not too large.*

Theorem 9.24 (Gradient estimate [**291, 303**]). *Let $X : M^n \times [0,T) \to \mathbb{R}^{n+1}$, $n \geq 2$, be a solution to mean curvature flow with initial condition in the class $\mathcal{C}^n_m(R,\alpha)$ and satisfying $\max_{M^n\times\{0\}} |\mathrm{II}| \leq R^{-1}$. If $m < \frac{2(n-1)}{3}$, then there exist constants $c = c(n)$ and, for every $\varepsilon > 0$, $C_\varepsilon = C(n,\alpha,\varepsilon)$ such that*

$$|\nabla \mathrm{II}|^2 \leq c\left(\varepsilon H^2 + \tfrac{1}{n-m}H^2 - |\mathrm{II}|^2\right)H^2 + C_\varepsilon R^{-4} \quad \text{in } M^n \times [R^2/4, T). \tag{9.30}$$

Proof. Observe that

$$\beta \doteqdot \frac{3}{n+2} - \frac{1}{n-m} = \frac{2(n-1)-3m}{(n+2)(n-m)}$$

is positive if $m < \frac{2(n-1)}{3}$. By Proposition 9.23, we can find constants $C_0 = C_0(n,\alpha)$ and, for any $\varepsilon \le \beta/2$, $C_\varepsilon = C_\varepsilon(n,\alpha,\varepsilon)$ such that

$$|\mathrm{II}|^2 - \tfrac{1}{n-m}H^2 \le \beta H^2 + C_0 R^{-2} \quad\text{and}\quad |\mathrm{II}|^2 - \tfrac{1}{n-m}H^2 \le \varepsilon H^2 + C_\varepsilon R^{-2}.$$

Set

$$G_0 \doteqdot 2C_0R^{-2} + \tfrac{3}{n+2}H^2 - |\mathrm{II}|^2 \quad\text{and}\quad G_\varepsilon \doteqdot 2C_\varepsilon R^{-2} + \varepsilon H^2 - \left(|\mathrm{II}|^2 - \tfrac{1}{n-m}H^2\right).$$

Then, computing as in Theorem 8.14, we obtain, at any newly occurring maximum of $\frac{|\nabla \mathrm{II}|^2}{G_0G_\varepsilon}$,

$$\frac{|\nabla \mathrm{II}|^2}{G_0G_\varepsilon} \le c_n\,,$$

where c_n is a constant that depends only on n. By Lemma 9.20,

$$\max_{M^n\times\{R^2/4\}} |\nabla \mathrm{II}|^2 \le c_0 R^{-4}\,,$$

where c_0 depends only on n. Since $G_0G_\varepsilon \ge C_0C_\varepsilon R^{-4}$, we conclude, assuming without loss of generality that $C_0C_\varepsilon \ge 1$, that

$$\max_{M^n\times[R^2/4,T)} \frac{|\nabla \mathrm{II}|^2}{G_0G_\varepsilon} \le \max\{c_n, c_0\}\,.$$

The desired estimate is now obtained by applying Young's inequality. □

Note that the conclusion is not vacuous since, by Lemma 9.20, the maximal existence time of a solution with initial data in the class $\mathcal{C}^n_m(R,\alpha)$ and satisfying $\max_{M^n\times\{0\}} |\mathrm{II}| \le R^{-1}$ is at least $R^2/2$.

Choosing $\varepsilon = 1$, say, yields the simpler estimate

$$|\nabla \mathrm{II}|^2 \le cH^4 + CR^{-4} \quad\text{in}\quad M^n \times [R^2/4, T)\,, \tag{9.31}$$

where C depends on n and α, while c depends only on n.

We remark that some bound for m is necessary since immersed, mean convex solutions can form "cusp-like" singularities. These singularities admit Grim hyperplane *blow-up limits* (see Chapter 11), on which $|\nabla \mathrm{II}|^2/H^4$ is unbounded (see Exercise 9.10). It is not clear whether the given bound is optimal, however.

The following basic lemma illustrates the utility of scale-invariant, pointwise gradient estimates for the curvature (cf. [**303**, Lemma 6.6]).

Lemma 9.25. *Let $X : M^n \to \mathbb{R}^{n+1}$ be a mean convex, properly immersed hypersurface. If*

$$\sup_{M^n} \frac{|\nabla H|}{H^2} \le C < \infty\,,$$

then

$$\frac{H(x)}{2} \le H(y) \le 2H(x) \tag{9.32}$$

for all $y \in \mathcal{B}_{\frac{1}{2CH(x)}}(x)$, the intrinsic ball of radius $\frac{1}{2CH(x)}$ about the point x.

Proof. For any unit speed geodesic $\gamma : [0, s] \to M^n$ joining the points $y = \gamma(0)$ and $x = \gamma(s)$, we have

$$\nabla_{\gamma'} H^{-1} \le C\,.$$

Integrating yields

$$-Cs \le H^{-1}(x) - H^{-1}(y) \le Cs,$$

or, if $s \le \frac{1}{2CH(x)}$,

$$\frac{H(x)}{2} \le \frac{H(x)}{1 + CH(x)s} \le H(y) \le \frac{H(x)}{1 - CH(x)s} \le 2H(x)\,.$$

The claim follows. □

In particular, if we translate and rescale so that the mean curvature is 1 at the origin, then it is bounded by 2 in an intrinsic ball of radius $(2C)^{-1}$ about the origin.

The gradient estimate can be used to bound the first-order terms which arise in the evolution equation for $\nabla^2 \mathrm{II}$. A straightforward maximum principle argument exploiting this observation yields an analagous estimate for $\nabla^2 \mathrm{II}$ (cf. Lemma 8.17).

Theorem 9.26 (Hessian estimate [**291, 303**]). *Let $X : M^n \times [0, T) \to \mathbb{R}^{n+1}$, $n \ge 2$, be a solution to mean curvature flow with initial condition in the class $C^n_m(R, \alpha)$ and satisfying $\max_{M^n \times \{0\}} |\mathrm{II}| \le R^{-1}$. If $m < \frac{2(n-1)}{3}$, then there exist constants $c = c(n)$ and $C = C(n, \alpha)$ such that*

$$|\nabla^2 \mathrm{II}|^2 \le cH^6 + CR^{-6} \;\; \text{in } M^n \times [R^2/4, T)\,. \tag{9.33}$$

Proof. By (6.66), we can estimate

$$(\partial_t - \Delta)|\nabla^2 \mathrm{II}|^2 \le c\left(|\mathrm{II}|^2 |\nabla^2 \mathrm{II}|^2 + |\nabla \mathrm{II}|^4\right) - 2|\nabla^3 \mathrm{II}|^2\,,$$

where c depends only on n. It follows that

$$(\partial_t - \Delta)\frac{|\nabla^2 \mathrm{II}|^2}{H^5} \leq \frac{c}{H^5}\left(|\mathrm{II}|^2|\nabla^2\mathrm{II}|^2 + |\nabla \mathrm{II}|^4\right) - 2\frac{|\nabla^3\mathrm{II}|^2}{H^5}$$
$$- 5|\mathrm{II}|^2\frac{|\nabla^2\mathrm{II}|^2}{H^5} - 30\frac{|\nabla^2\mathrm{II}|^2}{H^7}|\nabla H|^2 + \frac{10}{H^6}\langle \nabla H, \nabla|\nabla^2\mathrm{II}|^2\rangle.$$

The first term on the second line is not helpful since the constant c is greater than 5, so we discard it. We can use the good third-order term on the first line to absorb the final term on the second line since

$$\frac{10}{H^6}\left\langle \nabla H, \nabla|\nabla^2\mathrm{II}|^2\right\rangle \leq \frac{|\nabla^3\mathrm{II}|^2}{H^5} + \frac{100}{H^7}|\nabla H|^2|\nabla^2\mathrm{II}|^2 .$$

Estimating the first-order terms using Theorem 9.24 (cf. (9.31)) then yields

$$(\partial_t - \Delta)\frac{|\nabla^2\mathrm{II}|^2}{H^5} \leq c_1\frac{|\nabla^2\mathrm{II}|^2}{H^3} + C_1R^{-4}\frac{|\nabla^2\mathrm{II}|^2}{H^7} + \frac{c_1H^8 + C_1R^{-8}}{H^5} - \frac{|\nabla^3\mathrm{II}|^2}{H^5},$$

where c_1 depends only on n and C_1 depends only on n and α.

Similarly, we obtain

$$(\partial_t - \Delta)\frac{|\nabla\mathrm{II}|^2}{H^3} \leq \frac{c_2H^8 + C_2R^{-8}}{H^5} - \frac{|\nabla^2\mathrm{II}|^2}{H^3},$$

where c_2 depends only on n and C_2 depends only on n and α, as well as

$$(\partial_t - \Delta)\frac{|\nabla\mathrm{II}|^2}{H^7} \leq \frac{c_3H^8 + C_3R^{-8}}{H^9} - \frac{|\nabla^2\mathrm{II}|^2}{H^7},$$

where c_3 depends only on n and C_3 depends only on n and α.

Setting

$$f \doteqdot \frac{|\nabla^2\mathrm{II}|^2}{H^5} + c_1\frac{|\nabla\mathrm{II}|^2}{H^3} + C_1R^{-4}\frac{|\nabla\mathrm{II}|^2}{H^7}$$

and estimating $H \geq \alpha_1 R^{-1}$, we obtain

$$(\partial_t - \Delta)f \leq c_1(1+c_2)H^3 + \left(\frac{C_1 + c_1C_2}{\alpha_1^5} + C_1\left(\frac{c_3}{\alpha_1} + \frac{C_3}{\alpha_1^9}\right)\right)R^{-3}$$
$$\leq c_4|\mathrm{II}|^2H + C_4R^{-3},$$

where c_4 depends only on n and C_4 depends only on n and α, and hence

$$(\partial_t - \Delta)(f - c_4H) \leq C_4R^{-3}.$$

The maximum principle and Lemma 9.20 then yield

$$\max_{M^n\times\{t\}}(f - c_4H) \leq \max_{M^n\times\{R^2/4\}}(f - c_4H) + C_4R^{-3}(t - R^2/4)$$
$$\leq C_5R^{-1}$$

for all $t \geq R^2/4$, where C_5 depends only on n and α. We conclude that

$$|\nabla^2\mathrm{II}|^2 \leq cH^6 + CR^{-1}H^5 \text{ in } M^n \times [R^2/4, T),$$

where c depends only on n and C depends only on n and α. The claim then follows from Young's inequality. $\square$

This immediately yields an analogous bound for the time derivative of II using the evolution equation for II. Indeed,

$$\begin{aligned}|\nabla_t \mathrm{II}| &= |\Delta \mathrm{II} + |\mathrm{II}|^2 \mathrm{II}| \\ &\le c|\mathrm{II}|^3 + CR^{-3} \text{ in } M^n \times [R^2/4, T),\end{aligned} \tag{9.34}$$

where c depends only on n and C depends only on n and α.

In particular, we obtain the following a priori bounds for ∇H and $\partial_t H$ for flows with initial condition in the class $\mathcal{C}^n_m(R,\alpha)$ when $m < \frac{2(n-1)}{3}$.

Corollary 9.27. *Let $X : M^n \times [0,T) \to \mathbb{R}^{n+1}$, $n \ge 2$, be a solution to mean curvature flow with initial condition in the class $\mathcal{C}^n_m(R,\alpha)$ and satisfying $\max_{M^n\times\{0\}} |\mathrm{II}| \le R^{-1}$. If $m < \frac{2(n-1)}{3}$, then there exist $h_\sharp = h_\sharp(n,\alpha)$ and $c_\sharp = c_\sharp(n,\alpha)$ such that*

$$H(x,t) \ge h_\sharp R^{-1} \quad \textit{implies} \quad \frac{|\nabla H|}{H^2}(x,t) \le c_\sharp \quad \textit{and} \quad \frac{|\partial_t H|}{H^3}(x,t) \le \frac{c_\sharp^2}{2}. \tag{9.35}$$

This is a very useful estimate in light of the following "parabolic" version of Lemma 9.25.

Lemma 9.28. *Let $X : M^n \times [0,T) \to \mathbb{R}^{n+1}$ be a compact, mean convex solution to mean curvature flow satisfying*

$$\frac{|\nabla H|}{H^2} \le C \quad \textit{and} \quad \frac{|\partial_t H|}{H^3} \le \frac{C^2}{2}.$$

Then

$$\frac{H(x,t)}{10} \le H(y,s) \le 10H(x,t) \tag{9.36}$$

for all $(y,s) \in \mathcal{P}_{\frac{1}{10CH(x,t)}}(x,t)$, the intrinsic parabolic cylinder of radius $\frac{1}{10CH(x,t)}$ about the point (x,t).

Proof. Fix $\gamma \in [\frac{1}{2}, 1)$. As in Lemma 9.25, given any $r \le \frac{1-\gamma}{CH(x,t)}$,

$$\gamma H(x,t) \le \frac{H(x,t)}{1 + CH(x,t)r} \le H(y,t) \le \frac{H(x,t)}{1 - CH(x,t)r} \le \gamma^{-1} H(x,t)$$

for all $y \in \mathcal{B}_r(x,t)$, the g_t-intrinsic ball of radius r about the point x. Given $y \in \mathcal{B}_r(x,t)$, set $h(t) \doteqdot H(y,t)$. Then

$$-C^2 \le (h^{-2})'(s) \le C^2.$$

Since $r \le \frac{1-\gamma}{CH(x,t)} \le \frac{\gamma}{CH(x,t)} \le \frac{1}{CH(y,t)}$, integrating between $s \in (t-r^2, t]$ and t yields

$$\frac{H(y,t)}{\sqrt{1-C^2H^2(y,t)r^2}} \le H(y,s) \le \frac{H(y,t)}{\sqrt{1-C^2H^2(y,t)r^2}}$$

and hence

$$\frac{H(x,t)}{\sqrt{\gamma^{-2}+C^2H^2(x,t)r^2}} \le H(y,s) \le \frac{H(x,t)}{\sqrt{\gamma^2-C^2H^2(x,t)r^2}}$$

for all $(y,s) \in \mathcal{P}_r(x,t)$, so long as $r \le \frac{\gamma}{CH(x,t)}$. The claim follows by choosing, say, $\gamma = 1/2$. □

An inductive argument, exploiting estimates for lower-order terms in the evolution equations for higher derivatives of II (cf. Theorem 9.26), can be applied to obtain estimates for spatial derivatives of II to all orders. The evolution equation for II then yields bounds for the mixed space-time derivatives (cf. (9.34)) [**303**, Theorem 6.3 and Corollary 6.4].

Theorem 9.29 (Higher-order estimates [**291**, **303**]). *Let $X : M^n \times [0,T) \to \mathbb{R}^{n+1}$, $n \ge 2$, be a solution to mean curvature flow with initial condition in the class $\mathcal{C}^n_m(R,\alpha)$ and satisfying $\max_{M^n\times\{0\}} |\mathrm{II}| \le R^{-1}$. If $m < \frac{2(n-1)}{3}$, then there exist, for each pair of nonnegative integers k and ℓ, constants $c_{k,\ell} = c_{k,\ell}(k,\ell,n)$ and $C_{k,\ell} = C_{k,\ell}(k,\ell,n,\alpha)$ such that*

$$(9.37) \quad |\nabla_t^k \nabla^\ell \mathrm{II}|^2 \le c_{k,\ell}|\mathrm{II}|^{2+4k+2\ell} + C_{k,\ell}R^{-(2+4k+2\ell)} \text{ in } M^n \times [R^2/4, T).$$

9.6. Notes and commentary

The proof of Theorem 9.11 was inspired by ideas of Lawrence Christopher Evans [**217**], who obtained an analogous result for certain parabolic systems.

Theorem 9.12 made use of various notions from convex geometry.[4] This can be avoided if we are willing to split the result into two: Essentially the same argument can be applied to the smallest and largest principal curvatures κ_1 and κ_n, respectively, in place of the distance function $d_\Lambda(\mathrm{II})$ to obtain Corollaries 9.13 and 9.14 directly.

For the 3-dimensional Ricci flow, an analogue of Corollary 9.13 is the Hamilton–Ivey estimate [**268**, Theorem 24.4] and an analogue of Corollary 9.14 is [**268**, Theorem 24.7] for type-I solutions, which are instrumental in the development of the Hamilton–Perelman–Brendle 3-dimensional singularity classification theorem.

As we have mentioned, an alternative proof of the convexity estimate (Corollary 9.13) was obtained (in the embedded case) by B. White using

[4] An excellent reference on convex geometry is the book by Rolf Schneider [**461**].

a blow-up analysis and the strong maximum principle [**531**]. A similar argument yields the cylindrical estimates (Corollary 9.14). This approach is taken in Chapter 12 to obtain Theorems 12.20 and 12.23 (following R. Haslhofer and B. Kleiner [**276**]).

An important recent development in mean convex mean curvature flow is the discovery and exploitation of noncollapsing phenomena [**41**, **276**, **481**, **529**]. This will be the topic of Chapter 12.

Knut Smoczyk extended the Huisken–Sinestrari convexity estimate to a larger class of initial surfaces (which include mean convex and starshaped surfaces) evolving by mean curvature flow [**487**]. Similar results were obtained by Longzhi Lin using noncollapsing methods [**369**].

As we shall see in Chapter 19, convexity and cylindrical estimates can also be obtained for flows by certain nonlinear functions of curvature [**53**, **56**, **57**, **353**].

Huy The Nguyen obtained certain convexity and cylindrical estimates for the mean curvature flow of hypersurfaces of the sphere satisfying a natural quadratic pinching condition [**417**].

Roundness and convexity estimates for free boundary mean curvature flow have been obtained by Nick Edelen [**211**].

9.7. Exercises

Exercise 9.1. Let E be a finite-dimensional normed linear space and C a convex subset of E.

(1) Show that $d_C(x)$ is the distance from x to $E \setminus C$ if $x \in C$ and the negative of the distance from x to C if $x \in E \setminus C$.

(2) Suppose that C is a cone. Show that $\mathrm{S}C \subset E^*$; that is, the supporting affine functionals are linear functionals.

Exercise 9.2. Suppose that $\Omega \subset \mathrm{S}(n)$ and $Z \subset \mathbb{R}^n$ are related by (9.5).

(1) Show that Ω is a convex subset of $\mathrm{S}(n)$ if and only in Z is a convex subset of $\mathbb{R}^n$.

(2) Suppose that Ω and Z are convex. Show that

$$d_Z(z) = d_\Omega(W)$$

whenever $\{z_1, \ldots, z_n\}$ are the eigenvalues of W.

Exercise 9.3. Show that

$$d_{\Gamma_+}(W) = \min\{w_1, \ldots, w_n\}\,,$$

where $w_1, \ldots, w_n$ denote the eigenvalues of W.

Exercise 9.4. Denote by $\overline{\Lambda}_m \subset \mathrm{S}(n)$ the convex cone

$$\overline{\Lambda}_m \doteqdot \bigcap_{i=1}^{n} \left\{ W : \operatorname{tr}(W) > 0 \;\text{ and }\; w_i \leq \tfrac{1}{n-m} \operatorname{tr}(W) \right\} .$$

Show that

$$-d_{\overline{\Lambda}_m}(W) = \frac{\max\{w_1, \ldots, w_n\} - \frac{1}{n-m} \operatorname{tr}(W)}{\sqrt{n-1+(n-1-m)^2}} ,$$

where $w_1, \ldots, w_n$ denote the eigenvalues of W.

Exercise 9.5. Let (M^n, g) be a Riemannian manifold. Given any $\Lambda^0 \in \mathbb{R}^n \otimes \mathbb{R}^n \otimes \mathbb{R}^n$ satisfying $\Lambda^0_{kij} + \Lambda^0_{kji} = 0$, any $x_0 \in M^n$, and any orthonormal basis $\{e_i^0\}_{i=1}^n$ for $(T_{x_0}M, g_{x_0})$, construct a section $\Lambda \in \Gamma(T^*M \otimes T^*M \otimes TM)$ satisfying

$$g(\Lambda(w,u),v) + g(u, \Lambda(w,v)) = 0$$

and

$$g_{x_0}\left(\Lambda_{x_0}\left(e_k^0, e_i^0\right), e_j^0\right) = \Lambda^0_{kij} .$$

Deduce that the choice in equation (9.15) is indeed possible.

Exercise 9.6. Prove Lemma 9.15.

Exercise 9.7. Prove the Poincaré-type inequality of Proposition 9.17.

Exercise 9.8. Prove that the classes $\mathcal{C}^n_m(R, \alpha)$ are preserved under mean curvature flow.

Exercise 9.9. Prove the quadratic pinching estimate of Proposition 9.23.

Exercise 9.10. Show that

$$\frac{|\nabla \mathrm{II}|}{H^2}(x,t) = |\tan x_1|$$

on the Grim hyperplane, $\Gamma^n_t \doteqdot \{(x, t - \log\cos x_1) : x \in (-\frac{\pi}{2}, \frac{\pi}{2}) \times \mathbb{R}^{n-1}\}$.

Chapter 10

Monotonicity Formulae

In this chapter, we extend to higher dimensions Huisken's monotonicity formula and Hamilton's differential Harnack estimate (first encountered in the context of curve shortening flow in Section 4.3), which, as we shall see in Chapter 11, are fundamental to the understanding of singularity formation.

10.1. Huisken's monotonicity formula

Monotonicity formulae have been crucial to the study of the long-term behavior of curvature flows. We have already encountered entropy monotonicity, a differential Harnack inequality, and the 1-dimensional version of Huisken's monotonicity formula in the context of curve shortening flow. In higher dimensions, we have seen that the mean curvature flow is a gradient flow, with the area functional playing the role of the energy. We have also seen the value of monotone quantities in the proof of Huisken's theorem presented in Section 8.6. In this section, we prove Huisken's monotonicity formula for the Gaußian area and discuss some of its consequences.

If we scale an immersion $X : M \to \mathbb{R}^{n+1}$ by $\lambda > 0$, the area of M with the induced metric scales like $A_\lambda(M) = \lambda^n A(M)$, where A_λ is the area functional of the scaled immersion X_λ and A that of X. The scaling properties of the heat kernel can be exploited to construct a weighted area functional which is invariant under *parabolic* scaling along any solution to mean curvature flow.

Given $(X_0, t_0) \in \mathbb{R}^{n+1} \times \mathbb{R}$, the **Gaußian area** $\Theta_{(X_0,t_0)}$ **based** at (X_0, t_0) of a solution $X : M \times I \to \mathbb{R}^{n+1}$ to mean curvature flow is defined for

$t \in I \cap (-\infty, t_0)$ by

$$\Theta_{(X_0,t_0)}(t) \doteqdot \int_M \Phi_{(X_0,t_0)}(x,t)\, d\mu_t(x)\,, \tag{10.1}$$

where

$$\begin{aligned}\Phi_{(X_0,t_0)}(x,t) &\doteqdot \rho_{(X_0,t_0)}(X(x,t),t)\\ &= (4\pi(t_0-t))^{-\frac{n}{2}}\, \mathrm{e}^{-\frac{|X(x,t)-X_0|^2}{4(t_0-t)}}\end{aligned} \tag{10.2}$$

is the n-dimensional backward heat kernel along X, centered at (X_0, t_0). Since, for an embedded solution $\{\mathcal{M}_t\}_{t\in I}$, the pushforward of the Riemannian measure is the n-dimensional Hausdorff measure $\mathscr{H}^n$, we have

$$\Theta_{(X_0,t_0)}(t) = \int_{\mathcal{M}_t} (4\pi(t_0-t))^{-\frac{n}{2}}\, \mathrm{e}^{-\frac{|X-X_0|^2}{4(t_0-t)}}\, d\mathscr{H}^n(X)\,. \tag{10.3}$$

If we parabolically rescale the solution X about (X_0, t_0) by $\lambda > 0$, to obtain the solution $X_\lambda : M \times \lambda^2(I - t_0) \to \mathbb{R}^{n+1}$ defined by

$$X_\lambda(x,t) \doteqdot \lambda\left(X(x, \lambda^{-2}t + t_0) - X_0\right),$$

then the Gaußian area $\Theta^\lambda_{(0,0)}$ of X_λ is

$$\begin{aligned}\Theta^\lambda_{(0,0)}(t) &= (-4\pi t)^{-\frac{n}{2}} \int_M \mathrm{e}^{-\frac{|X_\lambda(x,t)|^2}{-4t}}\, d\mu^\lambda_t(x)\\ &= \left(-4\pi\lambda^{-2}t\right)^{-\frac{n}{2}} \int_M \mathrm{e}^{-\frac{\left|X(x,\lambda^{-2}t+t_0)-X_0\right|^2}{-4\lambda^{-2}t}}\, d\mu_{\lambda^{-2}t+t_0}(x)\\ &= \Theta_{(X_0,t_0)}(\lambda^{-2}t + t_0)\,.\end{aligned}$$

In particular, $\Theta \doteqdot \Theta_{(0,0)}$ is constant on any shrinking self-similar solution to the flow.

Example 10.1 (Stone [**498**]).

(1) The Gaußian area of the stationary hyperplane $\{\mathbb{R}^n \times \{0\}\}_{t\in(-\infty,0)}$ in $\mathbb{R}^{n+1}$ is

$$\Theta(t) \equiv 1\,.$$

Indeed,

$$\Theta(t) \equiv \Theta(-\tfrac{1}{2}) = (2\pi)^{-\frac{n}{2}} \int_{\mathbb{R}^n\times\{0\}} \mathrm{e}^{-\frac{|X|^2}{2}}\, d\mathscr{H}^n(X) = 1\,.$$

(2) The Gaußian area of the shrinking cylinder

$$\{\mathbb{R}^{n-m} \times S^m_{\sqrt{-2mt}}\}_{t\in(-\infty,0)}\,, \qquad \text{where } m \in \{1, \dots, n\},$$

in $\mathbb{R}^{n+1}$ is

$$\Theta(t) \equiv \left(\frac{m}{2\pi \mathrm{e}}\right)^{\frac{m}{2}} \mathscr{H}^m(S^m_1).$$

Indeed, writing $X = (x, y) \in \mathbb{R}^{n-m} \times S^m_{\sqrt{m}}$, we find

$$\begin{aligned}
\Theta(t) \equiv \Theta(-\tfrac{1}{2}) &= (2\pi)^{-\frac{n}{2}} \int_{\mathbb{R}^{n-m}\times S^m_{\sqrt{m}}} \mathrm{e}^{-\frac{|x|^2}{2}} \mathrm{e}^{-\frac{|y|^2}{2}}\, d\mathscr{H}^n(X)\\
&= (2\pi)^{-\frac{n}{2}} \int_{\mathbb{R}^{n-m}} \mathrm{e}^{-\frac{|x|^2}{2}} \int_{S^m_{\sqrt{m}}} \mathrm{e}^{-\frac{|y|^2}{2}}\, d\mathscr{H}^m(X)\, d\mathscr{H}^{n-m}(X)\\
&= (2\pi)^{-\frac{n}{2}} \mathrm{e}^{-\frac{m}{2}} \mathscr{H}^m(S^m_{\sqrt{m}}) \int_{\mathbb{R}^{n-m}} \mathrm{e}^{-\frac{|x|^2}{2}}\, d\mathscr{H}^{n-m}(X)\\
&= (2\pi)^{-\frac{m}{2}} \mathrm{e}^{-\frac{m}{2}} \mathscr{H}^m(S^m_{\sqrt{m}})\\
&= \left(\frac{m}{2\pi\mathrm{e}}\right)^{\frac{m}{2}} \mathscr{H}^m(S^m_1)\,.
\end{aligned}$$

In particular, the Gaußian area of a nontrivial shrinking cylinder is equal to that of its spherical factor.

Lemma 10.2 (Stone [**498**, Lemma A.4]). *For each $m \in \mathbb{N}$, let*

$$\vartheta_m \doteqdot \left(\frac{m}{2\pi\mathrm{e}}\right)^{\frac{m}{2}} \mathscr{H}^m(S^m_1)$$

be the Gaußian area of the shrinking sphere in $\mathbb{R}^{m+1}$. The sequence $\{\vartheta_m\}_{m\in\mathbb{N}}$ is strictly decreasing and satisfies

$$\lim_{m\to\infty} \vartheta_m = \sqrt{2}\,.$$

Proof. Define a function $f : (0, \infty) \to \mathbb{R}$ by

$$f(x) \doteqdot \mathrm{e}^x x^{-x} \Gamma\left(x + \frac{1}{2}\right),$$

where Γ is the gamma function. Since

$$\mathscr{H}^m(S^m_1) = \frac{2\pi^{\frac{m+1}{2}}}{\Gamma(\frac{m+1}{2})},$$

we have

$$\vartheta_m = \frac{2\sqrt{\pi}}{f(\frac{m}{2})}\,.$$

We claim that f is strictly increasing. Set

$$\phi(x) \doteqdot \log f(x) = \log\left(\Gamma\left(x + \frac{1}{2}\right)\right) + x - x\log x\,.$$

Using the formula [**3**, p. 200]

$$\left(\frac{\Gamma'(t)}{\Gamma(t)}\right)' = \sum_{j=0}^{\infty} (t+j)^{-2},$$

we compute

$$
\begin{aligned}
\phi''(x) &= \sum_{j=0}^{\infty}\left(x+\frac{1}{2}+j\right)^{-2}-\frac{1}{x}\\
&< \sum_{j=0}^{\infty}\frac{1}{(x+j)(x+j+1)}-\frac{1}{x}\\
&= \sum_{j=0}^{\infty}\int_j^{j+1}\frac{1}{(x+t)^2}dt-\frac{1}{x}\\
&= \int_0^{\infty}\frac{1}{(x+t)^2}dt-\frac{1}{x}\\
&= 0\,.
\end{aligned}
$$

Now observe, using the identity $\Gamma(x+1) = x\Gamma(x)$, that

$$
\phi(x+1)-\phi(x) = 1+\log\left(\frac{x+\frac{1}{2}}{x+1}\right)+x\log\left(\frac{x}{x+1}\right).
$$

Since this is positive for all x sufficiently large and ϕ is strictly concave, we conclude that ϕ is strictly increasing. It follows that f is strictly increasing, which proves the first claim.

To obtain the limit, we use the identity [**3**, p. 206]

$$
\Gamma\left(\frac{m+1}{2}\right) = \sqrt{2\pi}\left(\frac{m+1}{2}\right)^{\frac{m}{2}}\mathrm{e}^{-\frac{m+1}{2}}\mathrm{e}^{\zeta\left(\frac{m+1}{2}\right)/6(m+1)}\,,
$$

where ζ is a function satisfying $0 < \zeta < 1$. Thus,

$$
\begin{aligned}
f\left(\frac{m}{2}\right) &= \mathrm{e}^{\frac{m}{2}}\left(\frac{m}{2}\right)^{-\frac{m}{2}}\Gamma\left(\frac{m+1}{2}\right)\\
&= \sqrt{2\pi}\mathrm{e}^{-\frac{1}{2}}\left(1+\frac{1}{m}\right)^{\frac{m}{2}}\mathrm{e}^{\zeta\left(\frac{m+1}{2}\right)/6(m+1)}\,.
\end{aligned}
$$

The claim follows since

$$
\lim_{m\to\infty}\left(1+\frac{1}{m}\right)^m = \mathrm{e}\,. \qquad \square
$$

Huisken's monotonicity formula states that $\Theta_{(X_0,t_0)}$ is monotone decreasing along the mean curvature flow unless the solution is homothetically shrinking about (X_0, t_0).

Theorem 10.3 (Huisken's monotonicity formula I). *Let* $X : M^n \times I \to \mathbb{R}^{n+1}$ *be a* compact *solution to mean curvature flow. Given* $(X_0, t_0) \in \mathbb{R}^{n+1} \times \mathbb{R}$, *the Gaußian area* $\Theta_{(X_0,t_0)}(t)$ *defined by* (10.1) *is nonincreasing*:

$$\frac{d}{dt}\Theta_{(X_0,t_0)}(t) = -\int_M \left(H - \frac{\langle X - X_0, \mathsf{N}\rangle}{2(t_0 - t)}\right)^2 \Phi \, d\mu_t \leq 0 \tag{10.4}$$

in $(-\infty, t_0) \cap I$. *In particular, if* $\Theta_{(X_0,t_0)}$ *is constant in* $I \cap (-\infty, t_0)$, *then* $\{\mathcal{M}_t\}_{t \in I \cap (-\infty, t_0)}$ *is a shrinking self-similar solution about* (X_0, t_0).

Proof. By translating in space-time it suffices to consider the case $(X_0, t_0) = (0, 0)$. First observe that

$$\partial_t \Phi = \left(\frac{n}{-2t} - \frac{\langle X, \partial_t X\rangle}{-2t} - \frac{|X|^2}{4t^2}\right)\Phi\,.$$

Recalling the evolution equation (6.19) for the measure, we obtain

$$\frac{d}{dt}\int_M \Phi \, d\mu = \int_M \left(\frac{n}{-2t} + \frac{H\,\langle X, \mathsf{N}\rangle}{-2t} - \frac{|X|^2}{4t^2} - H^2\right)\Phi \, d\mu\,. \tag{10.5}$$

By the identity (5.32) and the divergence theorem,

$$\begin{aligned}
\int_M H\,\langle X, \mathsf{N}\rangle \,\Phi \, d\mu &= -\int_M \langle X, \Delta X\rangle \,\Phi \, d\mu \\
&= \int_M \left(|\nabla X|^2 \Phi + \frac{1}{2}\left\langle \nabla |X|^2, \nabla \Phi\right\rangle\right) d\mu \\
&= \int_M \left(n - \frac{1}{-2t}\left|X^\top\right|^2\right)\Phi \, d\mu\,.
\end{aligned}$$

Multiplying this by $\frac{1}{-2t}$ and adding the difference (which is zero) into (10.5) yields

$$\begin{aligned}
\frac{d}{dt}\int_M \Phi d\mu &= \int \left(\frac{H\,\langle X, \mathsf{N}\rangle}{-t} - \frac{\langle X, \mathsf{N}\rangle^2}{4t^2} - H^2\right)\Phi \, d\mu \\
&= -\int_M \left(H - \frac{\langle X, \mathsf{N}\rangle}{-2t}\right)^2 \Phi \, d\mu\,.
\end{aligned}$$ □

According to Ecker [**205**, p. 53], "*the monotonicity formula intuitively says that, under mean curvature flow, the area of a hypersurface near any point is nonincreasing on any scale. In fact, it strictly decreases unless the hypersurface is homothetically contracting about this point.*"

10.1.1. Pointwise monotonicity formula. There is a pointwise version of the monotonicity formula which helps shed light on Huisken's derivation (see [**205**, (4.6) on p. 56]). Observe that, on any solution $X : M \times I \to \mathbb{R}^{n+1}$ of mean curvature flow,

$$\begin{aligned} \frac{d}{dt}\int_M uv\, d\mu &= \int_M \left(v\partial_t u + u\partial_t v - uvH^2\right) d\mu \\ &= \int_M \left(v(\partial_t - \Delta)u + u(\partial_t + \Delta - H^2)v\right) d\mu \end{aligned} \tag{10.6}$$

for any pair of smooth functions u and v, at least one of which is compactly supported in M^n at each fixed time. We say that the operator

$$(\partial_t - \Delta)^* \doteqdot -(\partial_t + \Delta - H^2)$$

is the L^2-**dual** of the heat operator $(\partial_t - \Delta)$.

Proposition 10.4 (Pointwise monotonicity formula I). *Let $X : M \times I \to \mathbb{R}^{n+1}$ be a solution to mean curvature flow. For any $(X_0, t_0) \in \mathbb{R}^{n+1} \times \mathbb{R}$,*

$$(\partial_t + \Delta - H^2)\Phi_{(X_0,t_0)} = -\left(H - \frac{\langle X - X_0, \mathsf{N}\rangle}{2(t_0 - t)}\right)^2 \Phi_{(X_0,t_0)} \tag{10.7}$$

for $t \in I \cap (-\infty, t_0)$, where $\Phi_{(X_0,t_0)}$ is defined by (10.2). *In particular,*

$$\frac{d}{dt}\int_M u\,\Phi_{(X_0,t_0)} d\mu = \int_M \left[(\partial_t - \Delta)u - u\left(H - \frac{\langle X - X_0, \mathsf{N}\rangle}{2(t_0 - t)}\right)^2\right] \Phi_{(X_0,t_0)} d\mu$$

for $t \in I \cap (-\infty, t_0)$ and any smooth $u : M \times [0, T) \to \mathbb{R}$ with compact support in M at each $t \in I \cap (-\infty, t_0)$.

Proof. See Exercise 10.1. □

Integrating equation (10.7) and making use of appropriate cutoff functions, we obtain a more general version of Huisken's monotonicity formula which applies to solutions with finite Gaußian area.

Corollary 10.5 (Huisken's monotonicity formula II). *Suppose that $X : M^n \times I \to \mathbb{R}^{n+1}$ is a properly immersed solution to mean curvature flow. Given $(X_0, t_0) \in \mathbb{R}^{n+1} \times \mathbb{R}$, assume that $\Theta_{(X_0,t_0)}(t) < \infty$ for each $t \in I \cap (-\infty, t_0)$. Then*

$$\frac{d}{dt}\Theta_{(X_0,t_0)}(t) = -\int_M \left(H - \frac{\langle X - X_0, \mathsf{N}\rangle}{2(t_0 - t)}\right)^2 \Phi\, d\mu_t \le 0 \tag{10.8}$$

in $(-\infty, t_0) \cap I$. In particular, if $\Theta_{(X_0,t_0)}$ is constant in $I \cap (-\infty, t_0)$, then $\{\mathcal{M}_t\}_{t \in I \cap (-\infty, t_0)}$ is a shrinking self-similar solution about (X_0, t_0).

Proof. We assume, without loss of generality, that $(X_0, t_0) = (0,0)$. We begin by constructing an appropriate cutoff function. Recall that, given any positive numbers r and R with $r < R$, there exists a smooth function $\eta : \mathbb{R}^{n+1} \to [0,1]$ which vanishes outside B_R, is 1 inside B_r, and satisfies

$$(R-r)|D\eta| + (R-r)^2|D^2\eta| \leq C\,,$$

where C may depend on n but depends on neither r nor R. Indeed, setting

$$\rho_\varepsilon(x) \doteqdot \varepsilon^{-n}\rho_1(\varepsilon^{-1}x)\,,$$

where $\rho_1 : \mathbb{R}^n \to \mathbb{R}$ is the standard bump function, defined by

$$\rho_1(x) \doteqdot \begin{cases} \mathrm{e}^{-\frac{1}{1-|x|^2}} \Big/ \int_{B_1} \mathrm{e}^{-\frac{1}{1-|y|^2}} dy & \text{if } x \in B_1\,, \\ 0 & \text{if } x \notin B_1\,, \end{cases}$$

such a function can be constructed by mollifying the characteristic function $\chi_{B_{\frac{R+r}{2}}}$ of the ball $B_{\frac{R+r}{2}}$ against ρ_ε for some $\varepsilon \in (0, \frac{R-r}{2})$.

Set $R = 2r$ and define the function $\phi : M^n \times I \to \mathbb{R}$ by

$$\phi(x,t) \doteqdot \eta(X(x,t))\,.$$

Then ϕ is smooth and satisfies

$$(\partial_t - \Delta)\phi = -\operatorname{tr}_{TM} D^2\eta \leq \frac{Cn}{r^2}\chi_{B_{2r}\setminus B_r}(X)\,.$$

Proposition 10.4 then yields

$$\frac{d}{dt}\int_M \phi\,\Phi\,d\mu \leq \frac{Cn}{r^2}\int_{X_t^{-1}(B_{2r}\setminus B_r)} \Phi\,d\mu - \int_{X_t^{-1}(B_r)} \left(H - \frac{\langle X, \mathsf{N}\rangle}{-2t}\right)^2 \Phi\,d\mu\,.$$

Taking $r \to \infty$ yields the claim. □

The following proposition generates further monotonicity formulae (cf. [**263**], [**297**], and [**205**, pp. 57–58]).

Proposition 10.6 (Pointwise monotonicity formula II). *Let $X : M \times I \to \mathbb{R}^{n+1}$ be a solution to mean curvature flow. Set $\Phi \doteqdot (t_0 - t)^{\frac{1}{2}}\rho$, where $\rho : \mathbb{R}^{n+1} \times (-\infty, t_0) \to \mathbb{R}$ is a positive solution to the backward heat equation,*

$$(\partial_t + \Delta_{\mathbb{R}^{n+1}})\rho = 0\,.$$

Then

$$\begin{aligned}(10.9)\qquad \left(\partial_t + \Delta - H^2\right)\Phi = &- \left(H + D_{\mathsf{N}}\log\rho\right)^2 \Phi \\ &- \left(D^2_{\mathsf{N},\mathsf{N}}\log\rho + \frac{1}{2(t_0 - t)}\right)\Phi\,d\mu\end{aligned}$$

for all $t \in I \cap (-\infty, t_0)$*. In particular,*

$$\frac{d}{dt}\int_M u\,\Phi_{(X_0,t_0)}d\mu = \int_M \left[(\partial_t - \Delta)u - u\,(H + D_{\mathsf{N}}\log\rho)^2\right]\Phi\,d\mu - \int_M u\left(D^2_{\mathsf{N},\mathsf{N}}\log\rho + \frac{1}{2(t_0-t)}\right)\Phi\,d\mu \tag{10.10}$$

for $t \in I \cap (-\infty, t_0)$ *and any smooth* $u : M \times [0,T) \to \mathbb{R}$ *satisfying*

$$\int_M \left(|u| + |\partial_t u| + |\nabla u| + |\nabla^2 u|\right)\Phi_{(X_0,t_0)}\,d\mu_t < \infty$$

at each time $t \in I \cap (-\infty, t_0)$*.*

Proof. See Exercise 10.3. □

Note that the matrix differential Harnack estimate (see Exercise 1.14) implies that

$$D^2_{\mathsf{N},\mathsf{N}}\log\rho + \frac{1}{2(t_0-t)} \geq 0$$

for any positive solution $\rho : \mathbb{R}^{n+1} \times (-\infty, t_0) \to \mathbb{R}$ to the backward heat equation.

10.1.2. A maximum principle for noncompact solutions.

Proposition 10.7. *Let* $X : M \times [t_0, T) \to \mathbb{R}^{n+1}$ *be a solution to mean curvature flow and* u *a smooth subsolution to the induced heat equation on* $M \times [t_0, T)$*. If* u *satisfies*

$$\int_M \left(|u| + |\partial_t u| + |\nabla u| + |\nabla^2 u|\right)\Phi_{(X_0,t_0)}\,d\mu_t < \infty \tag{10.11}$$

at each time $t \in [t_0, T)$*, then*

$$\sup_{M^n\times\{t\}} u \leq \sup_{M^n\times\{t_0\}} u \tag{10.12}$$

for each $t \in [t_0, T)$*.*

Proof. Given $k > \sup_{M^n\times\{t_0\}} u$, set

$$u_k \doteqdot (u-k)^3_+.$$

Then u_k also satisfies (10.11), and we conclude via Proposition 10.6 that

$$\int_{M^n} u\Phi_{0,T}\,d\mu_t \leq \int_{M^n} u\Phi_{0,T}\,d\mu_{t_0} = 0$$

for all $t \in (0,T)$. The claim follows. □

10.1.3. Local area bounds. The monotonicity formula can be used to obtain local area estimates.

Lemma 10.8. *Let $X : M^n \times [0,T) \to \mathbb{R}^{n+1}$ be a compact solution to mean curvature flow. Given $\vartheta \in (0,1)$, $p_0 \in \mathbb{R}^{n+1}$, and $r > 0$, we have*

$$\frac{\mu_t(X_t^{-1}(B_r(p_0)))}{r^n} \leq V \doteqdot \mathrm{e}^{\frac{1}{4}}\vartheta^{-\frac{n}{2}}\frac{\mu_0(M^n)}{T^{\frac{n}{2}}} \tag{10.13}$$

for every $t \in (\vartheta T, T)$.

Proof. Given $t \in (0,T)$ set $t_0 \doteqdot t + r^2$. Then, by the monotonicity formula (Theorem 10.3),

$$\begin{aligned}(4\pi)^{-\frac{n}{2}}\mathrm{e}^{-\frac{1}{4}}\frac{\mu_t(X_t^{-1}(B_r(p_0)))}{r^n} &\leq [4\pi(t_0 - t)]^{-\frac{n}{2}}\int_{X_t^{-1}(B_r(p_0))} \mathrm{e}^{-\frac{|X(x,t)-p_0|^2}{4(t_0-t)}}\,d\mu_t(x)\\ &\leq \Theta_{(p_0,t_0)}(t)\\ &\leq \Theta_{(p_0,t_0)}(0)\\ &\leq (4\pi t_0)^{-\frac{n}{2}}\mu_0(M^n)\\ &\leq (4\pi t)^{-\frac{n}{2}}\mu_0(M^n)\,.\end{aligned}$$

The claim follows. □

10.2. Hamilton's Harnack estimate

In Section 4.3.2, we proved a differential Harnack inequality for curve shortening flow (Theorem 4.10). We now generalize this estimate to higher dimensions. We present Hamilton's proof of the result, which nicely illustrates the use of evolving orthonormal frames in curvature flows (see Section 5.3.2) to simplify computations. Equivalently, one could work with the space-time formalism of Section 5.3.1. The proof illuminates the relationship between self-similar solutions and Harnack estimates. Its derivation is similar to, but less complicated than, that of the matrix Harnack estimate for the Ricci flow that Hamilton proved in [**262**]. In a later chapter (see Section 18.6), we shall discuss Harnack estimates for geometric flows of hypersurfaces with more general speeds. In this section, we take the perspective of the parametrized flow, whereas in Section 18.6 we use the Gauß map parametrization, which results in an easier proof for more general flows (cf. Remark 10.13 below).

10.2.1. The Harnack estimate as the nonnegativity of a quadratic. Recall from (1.16) that the Li–Yau inequality for positive solutions u to the heat equation $\partial_t u = \Delta u$ on a complete Riemannian manifold (M^n, g) with

Figure 10.1. Richard Hamilton.

nonnegative Ricci curvature states that

$$\partial_t \log u + \frac{n}{2t} - |\nabla \log u|^2 \geq 0.$$

Multiplying this inequality by u, we obtain

$$\partial_t u + \frac{n}{2t}u - \frac{1}{u}|\nabla u|^2 \geq 0. \tag{10.14}$$

We may also reformulate this inequality as nonnegativity of a quadratic form on TM. Namely,

$$\partial_t u + \frac{n}{2t}u + 2\langle \nabla u, V\rangle + u\, g(V,V) \geq 0 \tag{10.15}$$

for all tangent vectors $V \in TM$. Indeed, by completing the square, we see that (10.15) is equivalent to

$$\partial_t u + \frac{n}{2t}u - \frac{1}{u}|\nabla u|^2 + \frac{1}{u}|\nabla u + uV|^2 \geq 0$$

for all $V \in TM$. By taking $V = -\frac{1}{u}\nabla u$, which minimizes the quadratic on the left-hand side, we see the equivalence of this inequality with (10.14).

It turns out that for geometric evolution equations it is often convenient (and sometimes necessary) to consider the Harnack expressions in the form of quadratics. With this in mind, we present the following (see Theorem 1.1 in [**269**]).

Theorem 10.9 (Differential Harnack inequality). *Suppose that* $X : M^n \times (0,T) \to \mathbb{R}^{n+1}$, *where* $0 < T \leq \infty$, *is a weakly convex solution to the mean curvature flow which is either* (1) *closed* (*i.e., compact without boundary*) *or*

(2) *complete with bounded second fundamental form on each compact time subinterval. Then on* $M^n \times (0,T)$ *we have*

$$\partial_t H + \frac{1}{2t} H + 2\langle \nabla H, V\rangle + \mathrm{II}(V,V) \geq 0 \tag{10.16}$$

for all $V \in TM$. *In particular,*

$$\partial_t(t^{1/2}H) \geq 0\,. \tag{10.17}$$

Thus, for any fixed $x \in M^n$, $t \mapsto t^{1/2}H(x,t)$ *is a nondecreasing function.*

We will prove Theorem 10.9 in Sections 10.2.3 and 10.2.4. When $\mathcal{M}_t$ is locally uniformly convex, we can phrase the result in the following way.

Corollary 10.10. *If the time slices* $\mathcal{M}_t$ *are locally uniformly convex, then*

$$\partial_t H + \frac{1}{2t} H - \mathrm{II}^{-1}(\nabla H, \nabla H) \geq 0. \tag{10.18}$$

Proof. Since the Weingarten map L is invertible, we may simply take

$$V = -\,\mathrm{L}^{-1}(\nabla H)$$

in (10.16), which then yields (10.18). □

An important consequence of the Harnack inequality states that locally uniformly convex **eternal solutions** (that is, locally uniformly convex solutions defined for $-\infty < t < \infty$) necessarily move by translation if the quantity $\partial_t H - \mathrm{II}^{-1}(\nabla H, \nabla H)$ vanishes somewhere.

Theorem 10.11 (Eternal solutions are translating self-similar solutions). *Let* $\{\mathcal{M}_t\}_{t\in(-\infty,\infty)}$ *be a complete, locally uniformly convex, eternal solution to the mean curvature flow. If*

$$\partial_t H - \mathrm{II}^{-1}(\nabla H, \nabla H) = 0 \tag{10.19}$$

at some point in space and time (for example, if the mean curvature attains its space-time maximum), then the solution is a translating self-similar solution.

We will prove Theorem 10.11 in Section 10.2.5.

Recall that a **translating self-similar solution** is a solution to mean curvature flow $\{\mathcal{M}_t^n\}_{t\in(-\infty,\infty)}$ that evolves only by translation in space; i.e., there exists a vector $e \in \mathbb{R}^{n+1}$ such that

$$\mathcal{M}_t^n = \mathcal{M}_s^n + (t-s)e \tag{10.20}$$

for all $s,t \in (-\infty,\infty)$. Since $\{\mathcal{M}_t^n\}_{t\in(-\infty,\infty)}$ evolves by mean curvature flow, the component of e normal to $\mathcal{M}_t$ must be the mean curvature vector. Thus,

$$H + \langle e, \mathsf{N}\rangle = 0\,. \tag{10.21}$$

Conversely, if $\mathcal{M}^n$ satisfies (10.21), then $\{\mathcal{M}_t^n\}_{t\in(-\infty,\infty)}$ evolves by mean curvature flow, where $\mathcal{M}_t^n \doteqdot \mathcal{M}^n + te$. We will sometimes refer to e as the **bulk velocity** of the translating solution.

There is a nice correspondence between translating self-similar solutions and **gradient solitons**, i.e., solutions whose metrics flow along diffeomorphisms generated by gradient vector fields (see (10.25) below). To see this, consider a translating solution with bulk velocity $e \in \mathbb{R}^{n+1}$. Since the solution satisfies mean curvature flow, e decomposes as

$$e = V - H\mathsf{N}\,, \tag{10.22}$$

where V is tangential to the solution.

Lemma 10.12. *Let $\{\mathcal{M}_t^n\}_{t\in(-\infty,\infty)}$ be a translating self-similar solution to mean curvature flow with bulk velocity $e \in \mathbb{R}^{n+1}$. Then*

$$\nabla H + \mathrm{L}(V) = 0\,, \tag{10.23}$$

and

$$\nabla V = H\,\mathrm{L} \tag{10.24}$$

where $V \doteqdot e^\top$ is the tangential part of e.

Proof. The first identity follows by differentiating the translator equation (10.21). The second follows by differentiating (10.22) and resolving tangential and normal components: In local coordinates, $V = V^i\,\partial_i X$. Differentiating (10.22) and using the fact that $D_i e = 0$, we obtain

$$\begin{aligned} 0 &= (D_iV)^\top + (D_iV)^\perp - D_iH\,\mathsf{N} - HD_i\mathsf{N} \\ &= \nabla_iV - \mathrm{II}(V,\partial_i)\,\mathsf{N} - \nabla_iH\,\mathsf{N} - H\mathrm{II}(\partial_i) \\ &= g^{jk}(\nabla_iV_j - H\mathrm{II}_{ij})\,\partial_kX - (\mathrm{II}_{ij}V^j + \nabla_iH)\,\mathsf{N}\,, \end{aligned}$$

where we used (5.19) and (5.27). Equating the normal and tangential components yields the lemma. □

Equation (10.24) implies that $\nabla_iV_j = \nabla_jV_i$. Assuming that all closed 1-forms on $\mathcal{M}$ are exact, there exists a function f such that $V = \nabla f$. Hence the translating self-similar solution is a gradient soliton:

$$\mathcal{L}_{\partial_t}g = -2H\,\mathrm{II} = -2\nabla^2 f = -2\mathcal{L}_{\nabla f}g\,, \tag{10.25}$$

where $\mathcal{L}$ denotes the Lie derivative.

In the proof of Theorem 10.11, we will see (Lemma 10.20 below) that if the solution is locally uniformly convex, then the converse of Lemma 10.12 holds as well.

Remark 10.13. Recall that a locally uniformly convex solution $X : M^n \times I \to \mathbb{R}^{n+1}$ to mean curvature flow can be parametrized by its Gauß map, that is, by the parametrization $Y : \Gamma_t \times I \to \mathbb{R}^{n+1}$, $\Gamma_t \doteqdot \mathsf{N}(M^n, t)$, defined by

$$Y(\mathsf{N}(x,t),t) = X(x,t)\,.$$

Denoting by H_X the mean curvature with respect to the fixed parametrization X and by H_Y the mean curvature with respect to the Gauß map parametrization Y, we find

$$\begin{aligned}\partial_t H_X &= \partial_t H_Y + dH_Y(\partial_t \mathsf{N})\\ &= \partial_t H_Y + dH_Y(\nabla H_X)\\ &= \partial_t H_Y + \mathrm{II}^{-1}(\nabla H_X, \nabla H_X)\,.\end{aligned}$$

So, for a locally uniformly convex solution, the differential Harnack inequality (10.18) is equivalent to the inequality

$$\partial_t H + \frac{1}{2t} H \geq 0$$

with respect to the Gauß map parametrization.

10.2.2. Saturation by expanding self-similar solutions. To motivate the differential Harnack inequality, we will show that it is saturated by expanding self-similar solutions[1]. Recall that an **expanding self-similar solution** is a solution $\{\mathcal{M}^n_t\}_{t\in(0,\infty)}$ that evolves purely by positive dilations. That is,

$$\mathcal{M}^n_t = \sqrt{\frac{t}{s}}\mathcal{M}^n_s \quad \text{for all } \; s,t \in (0,\infty)\,.$$

Its parametrization vector $X : M^n \times (0,\infty) \to \mathbb{R}^{n+1}$ satisfies

$$H = -\frac{1}{2t}\langle X, \mathsf{N}\rangle\,. \tag{10.26}$$

Lemma 10.14. *Let $\{\mathcal{M}^n_t\}_{t\in(0,\infty)}$ be an expanding self-similar solution of mean curvature flow. Then*

$$X \doteqdot \nabla H + \mathrm{L}(V) = 0\,, \tag{10.27}$$

$$Y \doteqdot \nabla V - H\,\mathrm{L} - \frac{1}{2t}\mathrm{I} = 0\,, \quad \textit{and} \tag{10.28}$$

$$W \doteqdot \nabla_t \mathrm{L} + \nabla_V \mathrm{L} + \frac{1}{2t}\mathrm{L} = 0\,, \tag{10.29}$$

*where V is the **flow vector field**, $V \doteqdot \frac{1}{2t}X^\top$.*

[1]Compare this with the differential Harnack inequality for the heat equation (Proposition 1.12), which is saturated by the fundamental solution.

Proof. The first identity follows by differentiating the expander equation (10.26): Given a tangent vector u,

$$\nabla_u H = \frac{1}{2t}\langle X, \mathrm{L}(u)\rangle = \frac{1}{2t}\left\langle X^\top, \mathrm{L}(u)\right\rangle = \langle \mathrm{L}(V), u\rangle\,.$$

The second identity is left as an exercise (see Exercise 10.5).

To obtain the third, we make use of the evolving orthonormal frame formalism of Section 5.3.2 (alternatively, one can use the space-time formalism of Section 5.3.1). Differentiating equation (10.27), we obtain

$$\nabla_a\nabla_b H + \nabla_a(\mathrm{II}_{bc}\, V_c) = 0.$$

Applying (10.28) and the Codazzi equation (5.47), this implies that

$$\nabla_a\nabla_b H + V_c\nabla_c\,\mathrm{II}_{ab} + H\,\mathrm{II}^2_{ab} + \frac{1}{2t}\,\mathrm{II}_{ab} = 0\,. \tag{10.30}$$

On the other hand, by (6.17) for II as a vector-valued function on the orthonormal frame bundle, we have

$$\nabla_t\,\mathrm{II}_{ab} = \nabla_a\nabla_b H + H\,\mathrm{II}^2_{ab}\,. \tag{10.31}$$

The lemma follows from combining the two equations above. □

Corollary 10.15. *Let $\{\mathcal{M}^n_t\}_{t\in(0,\infty)}$ be an expanding self-similar solution to mean curvature flow. Then*

$$\partial_t H + 2\nabla_V H + \mathrm{II}(V,V) + \frac{1}{2t}H = 0\,, \tag{10.32}$$

where V is the flow vector field.

Proof. Taking the trace of (10.29) yields

$$\partial_t H + \nabla_V H + \frac{1}{2t}H = 0\,. \tag{10.33}$$

On the other hand, by (10.27),

$$\nabla_V H + \mathrm{II}(V,V) = 0\,.$$

□

10.2.3. The Harnack calculation. To derive the Harnack estimate, we shall apply the maximum principle to an evolution equation for the **Harnack quadratic** Q, defined by

$$V \mapsto Q(V) \doteqdot \partial_t H + 2\langle \nabla H, V\rangle + \mathrm{II}(V,V) + \frac{1}{2t}H\,. \tag{10.34}$$

Since convex expanding self-similar solutions saturate the differential Harnack inequality, a necessary condition for the nonnegativity of $Q(V)$ is the vanishing of its first variation at an expanding self-similar solution flowing along V. We compute that this is indeed the case for the choice of Q above:

$$DQ_V(W) = 2\langle \nabla H + \mathrm{II}(V), W\rangle,$$

which by (10.27) vanishes on an expanding self-similar solution flowing along V.

To simplify the Harnack calculation, we introduce, for a given vector field V,

$$X \doteqdot \nabla H + \mathrm{L}(V), \tag{10.35a}$$

$$Y \doteqdot \nabla V - H\,\mathrm{L} - \frac{1}{2t}\mathrm{I}, \quad \text{and} \tag{10.35b}$$

$$W \doteqdot \nabla_t \mathrm{L} + \nabla_V \mathrm{L} + \frac{1}{2t}\mathrm{L}\,. \tag{10.35c}$$

We saw that each of these tensors vanishes on an expanding self-similar solution when V is the flow vector field. We also define the vector field

$$U \doteqdot (\nabla_t - \Delta)V + \mathrm{L}(\nabla H) + \frac{1}{t}V\,. \tag{10.36}$$

We observe that U also vanishes on a locally uniformly convex expanding self-similar solution when V is the flow vector field (though this is not needed for the proof of the Harnack inequality).

Proposition 10.16. *Let $\{\mathcal{M}_t^n\}_{t\in(0,\infty)}$ be a locally uniformly convex expanding self-similar solution. Then*

$$U \doteqdot (\nabla_t - \Delta)V + \mathrm{L}(\nabla H) + \frac{1}{t}V \equiv 0\,, \tag{10.37}$$

where V is the flow vector field.

Proof. We compute in an evolving orthonormal frame. By (10.27), on an expanding self-similar solution,

$$(\nabla_t - \Delta)(\mathrm{II}_{ab}\, V_b) = -(\nabla_t - \Delta)\nabla_a H\,. \tag{10.38}$$

Recall from (6.18) and (6.17) that

$$\partial_t H = \Delta H + |\mathrm{II}|^2\, H \quad \text{and} \quad \nabla_t \mathrm{II}_{ab} = (\Delta\,\mathrm{II})_{ab} + |\mathrm{II}|^2\, \mathrm{II}_{ab}\,,$$

respectively. Using this and the commutator formula (6.26),

$$(\nabla_t - \Delta)\nabla_a f = \nabla_a(\nabla_t - \Delta)f + \mathrm{II}^2_{ab}\nabla_b f\,,$$

we may rewrite (10.38) as

$$\begin{aligned} |\mathrm{II}|^2\, \mathrm{II}_{ab}\, V_b + \mathrm{II}_{ab}(\nabla_t - \Delta)V_b - 2\nabla_c \mathrm{II}_{ab}\nabla_c V_b &= -\nabla_a(\nabla_t - \Delta)H - \mathrm{II}^2_{ab}\nabla_b H \\ &= -\nabla_a(|\mathrm{II}|^2\, H) - \mathrm{II}^2_{ab}\,\nabla_b H\,. \end{aligned}$$

Applying (10.28), (10.27) twice, and the Codazzi equation, yields

$$\begin{aligned}\mathrm{II}_{ab}(\nabla_t - \Delta)V_b &= 2\nabla_c \mathrm{II}_{ab}\, \nabla_c V_b - |\mathrm{II}|^2\, \mathrm{II}_{ab}\, V_b \\ &\quad - 2\,\mathrm{II}_{bc}\, \nabla_a\, \mathrm{II}_{bc}\, H - |\mathrm{II}|^2\, \nabla_a H - \mathrm{II}^2_{ab}\, \nabla_b H \\ &= 2\nabla_c\, \mathrm{II}_{ab}\, H\, \mathrm{II}_{cb} + 2\nabla_c\, \mathrm{II}_{ab}\, \frac{1}{2t} g_{cb} + |\mathrm{II}|^2\, \nabla_a H \\ &\quad - 2\,\mathrm{II}_{bc} \nabla_a\, \mathrm{II}_{bc}\, H - |\mathrm{II}|^2\, \nabla_a H - \mathrm{II}^2_{ab}\, \nabla_b H \\ &= -\,\mathrm{II}^2_{ab}\, \nabla_b H + \frac{1}{t}\nabla_a H \\ &= -\,\mathrm{II}_{ab}\left(\mathrm{II}_{bc}\, \nabla_c H + \frac{1}{t} V_b\right).\end{aligned}$$

Since $\mathrm{II} > 0$, this implies that $U_a = 0$. □

Proposition 10.17. *Let $\{\mathcal{M}_t\}_{t\in I}$ be a convex solution to mean curvature flow. Given any time-dependent vector field V,*

$$(10.39)\qquad \begin{aligned}(\partial_t - \Delta)(Q(V)) = &\left(|\mathrm{II}|^2 - \frac{2}{t}\right) Q(V) + 2X(U) \\ &- 2\,\mathrm{tr}(Y \cdot \mathrm{L} \cdot Y) - 4g(Y, W),\end{aligned}$$

where X, Y, and W are defined by (10.35a)–(10.35c) *and U is defined by* (10.36). *Furthermore, given any point (x_0, t_0) and any $V \in T_{x_0}\mathcal{M}_{t_0}$, V may be extended so that Y and U are both zero at (x_0, t_0). Thus, at any space-time point and under such an extension of V, we have*

$$(10.40)\qquad (\nabla_t - \Delta)Q(V) = \left(|\mathrm{II}|^2 - \frac{2}{t}\right) Q(V).$$

Proof. We shall compute using an evolving orthonormal frame. Observe that $Q(V) = W_{aa} + V_a X_a$. So we need to calculate evolution equations for $V_a X_a$ and W_{aa}. By (6.18) and the commutator rule (6.26), the evolution equation for $\nabla_a H$ may be written as

$$(10.41)\qquad \begin{aligned}(\nabla_t - \Delta)\nabla_a H &= \nabla_a(\nabla_t - \Delta)H + \mathrm{II}^2_{ab}\nabla_b H \\ &= H\nabla_a|\mathrm{II}|^2 + |\mathrm{II}|^2\nabla_a H + \mathrm{II}^2_{ab}\nabla_b H.\end{aligned}$$

Using (6.17) and the product rule, we see that

$$(\nabla_t - \Delta)(\mathrm{II}_{ab}V_b) = |\mathrm{II}|^2\, \mathrm{II}_{ab}V_b + \mathrm{II}_{ab}(\nabla_t - \Delta)V_b - 2\nabla_c \mathrm{II}_{ab}\nabla_c V_b.$$

Summing the two formulae above yields

$$(\nabla_t - \Delta)X_a = |\mathrm{II}|^2\, X_a + H\nabla_a|\mathrm{II}|^2 + \mathrm{II}^2_{ab}\nabla_b H + \mathrm{II}_{ab}(\nabla_t - \Delta)V_b - 2\nabla_c \mathrm{II}_{ab}\nabla_c V_b.$$

Thus, by using the product rule again, we obtain the evolution of V_aX_a:

$$\begin{aligned}(\partial_t-\Delta)(V_aX_a) &= X_a(\nabla_t-\Delta)V_a-2\nabla_cV_a\nabla_cX_a\\ &\quad+V_a\big(|\mathrm{II}|^2X_a+H\nabla_a|\mathrm{II}|^2+\mathrm{II}^2_{ab}\nabla_bH\\ &\qquad+\mathrm{II}_{ab}(\nabla_t-\Delta)V_b-2\nabla_c\mathrm{II}_{ab}\nabla_cV_b\big)\\ &=|\mathrm{II}|^2V_aX_a+(X_b+\mathrm{II}_{ab}V_a)(\nabla_t-\Delta)V_b\\ &\quad+HV_a\nabla_a|\mathrm{II}|^2+\mathrm{II}^2_{ab}V_a\nabla_bH-2(\nabla_cX_b+V_a\nabla_c\mathrm{II}_{ab})\nabla_cV_b\,.\end{aligned} \tag{10.42}$$

We next calculate an evolution equation for W_{aa}. Applying (6.18) and the commutator rule (6.27),

$$(\nabla_t-\Delta)\Delta f=\Delta(\nabla_t-\Delta)f+2H\,\mathrm{II}_{ab}\,\nabla_a\nabla_bf+2\,\mathrm{II}_{ab}\,\nabla_aH\nabla_bf\,,$$

we find that

$$\begin{aligned}(\nabla_t-\Delta)\nabla_tH &= (\nabla_t-\Delta)(\Delta H+|\mathrm{II}|^2H)\\ &=\Delta(\nabla_t-\Delta)H+2H\mathrm{II}_{ab}\nabla_a\nabla_bH+2\mathrm{II}_{ab}\nabla_aH\nabla_bH\\ &\quad+(\nabla_t-\Delta)(|\mathrm{II}|^2H)\\ &=2H\mathrm{II}_{ab}\nabla_a\nabla_bH+2\mathrm{II}_{ab}\nabla_aH\nabla_bH+2H\mathrm{II}_{ab}\nabla_t\mathrm{II}_{ab}+|\mathrm{II}|^2\nabla_tH.\end{aligned}$$

Firstly, by applying (10.31) to this, we obtain

$$(\nabla_t-\Delta)\nabla_tH=|\mathrm{II}|^2\nabla_tH+2H\mathrm{II}_{ab}(2\nabla_t\mathrm{II}_{ab}-H\,\mathrm{II}^2_{ab})+2\mathrm{II}_{ab}\nabla_aH\nabla_bH\,.$$

Secondly, using the product rule and (10.41), we obtain

$$\begin{aligned}(\nabla_t-\Delta)(V_c\nabla_cH) &= \nabla_cH(\nabla_t-\Delta)V_c-2\nabla_cV_a\nabla_c\nabla_aH+V_c(\nabla_t-\Delta)\nabla_cH\\ &=|\mathrm{II}|^2V_c\nabla_cH+\nabla_cH(\nabla_t-\Delta)V_c+2HV_c\mathrm{II}_{ab}\nabla_c\mathrm{II}_{ab}\\ &\quad-2\nabla_cV_a\nabla_c\nabla_aH+V_c\mathrm{II}^2_{cb}\nabla_bH\,.\end{aligned}$$

Thirdly,

$$(\nabla_t-\Delta)\left(\frac{1}{2t}H\right)=\frac{1}{2t}|\mathrm{II}|^2H-\frac{1}{2t^2}H\,.$$

Since

$$W_{aa}=\partial_tH+\nabla_VH+\frac{1}{2t}H,$$

summing the three equations above yields

$$\begin{aligned}(\nabla_t-\Delta)W_{aa} &= |\mathrm{II}|^2W_{aa}+\nabla_cH(\nabla_t-\Delta)V_c+\mathrm{II}^2_{cb}\nabla_bHV_c\\ &\quad+2H\mathrm{II}_{ab}(2\nabla_t\mathrm{II}_{ab}-H\,\mathrm{II}^2_{ab}+V_c\nabla_c\mathrm{II}_{ab})-2\nabla_cV_a\nabla_c\nabla_aH\\ &\quad+2\mathrm{II}_{ab}\nabla_aH\nabla_bH-\frac{1}{2t^2}H\,.\end{aligned} \tag{10.43}$$

Now, putting (10.42) and (10.43) together yields

$$\begin{aligned}(\nabla_t - \Delta)Q &= (\nabla_t - \Delta)(V_a X_a) + (\nabla_t - \Delta)W_{aa} \\ &= |\mathrm{II}|^2 Q + 2X_b(\nabla_t - \Delta)V_b \\ &\quad + 2H\mathrm{II}_{ab}(2\nabla_t \mathrm{II}_{ab} - H\mathrm{II}^2_{ab} + 2V_c\nabla_c\mathrm{II}_{ab}) \\ &\quad - 2\nabla_c V_b(\nabla_c\nabla_b H + \nabla_c X_b + V_a\nabla_c\mathrm{II}_{ab}) \\ &\quad + 2\mathrm{II}^2_{ab}V_a\nabla_b H + 2\mathrm{II}_{ab}\nabla_a H\nabla_b H - \frac{1}{2t^2}H\,.\end{aligned}$$

Substituting $\nabla_c X_b = \nabla_c\nabla_b H + \nabla_c(\mathrm{II}_{ab}V_a)$ and using (10.31) to replace the resulting Hessian term $2\nabla_c\nabla_b H$, we obtain

$$\begin{aligned}(\nabla_t - \Delta)Q = {}& |\mathrm{II}|^2 Q + 2X_b(\nabla_t - \Delta)V_b \\ & + 2H\mathrm{II}_{ab}(2\nabla_t\mathrm{II}_{ab} + 2V_c\nabla_c\mathrm{II}_{ab} - H\mathrm{II}^2_{ab}) \\ & - 2\nabla_c V_b(2\nabla_t\mathrm{II}_{cb} + 2V_a\nabla_a\mathrm{II}_{cb} - 2H\mathrm{II}^2_{cb} + \mathrm{II}_{ab}\nabla_c V_a) \\ & + 2\mathrm{II}^2_{ab}V_a\nabla_b H + 2\mathrm{II}_{ab}\nabla_a H\nabla_b H - \frac{1}{2t^2}H\,.\end{aligned}$$

Since Q vanishes on an expanding self-similar solution to the mean curvature flow, we know that the rest of the terms in the equation above must also. It is reasonable to expect that these terms are sums of terms, each of which vanishes on an expanding self-similar solution. So we proceed to convert these terms into U_b, W_{ab}, and Y_{ab} and see if there is anything left over. We start by converting the "highest-order terms":

$$\begin{aligned}(\nabla_t - \Delta)Q = {}& |\mathrm{II}|^2 Q + 2X_b\left(U_b - \mathrm{II}_{ab}\nabla_a H - \frac{1}{t}V_b\right) \\ & + 2H\mathrm{II}_{ab}\left(2W_{ab} - \frac{1}{t}\mathrm{II}_{ab} - H\mathrm{II}^2_{ab}\right) \\ & - 2\nabla_c V_b\left(2W_{cb} + \mathrm{II}_{ab}Y_{ca} - \frac{1}{2t}\mathrm{II}_{cb} - H\mathrm{II}^2_{cb}\right) \\ & + 2\mathrm{II}^2_{ab}V_a\nabla_b H + 2\mathrm{II}_{ab}\nabla_a H\nabla_b H - \frac{1}{2t^2}H\,.\end{aligned}$$

Next, we look for simplifications and cancellations via regrouping terms:

$$\begin{aligned}(\nabla_t - \Delta)Q = {}& |\mathrm{II}|^2 Q + 2X_b U_b \\ & - 4Y_{ab}W_{ab} - \frac{2}{t}(W_{aa} + X_b V_b) - 2H\mathrm{II}_{ab}\left(\frac{1}{2t}\mathrm{II}_{ab} + H\mathrm{II}^2_{ab}\right) \\ & - 2\nabla_c V_b \mathrm{II}_{ab}Y_{ca} + 2\nabla_a V_b\left(\frac{1}{2t}\mathrm{II}_{ab} + H\mathrm{II}^2_{ab}\right) \\ & + 2\mathrm{II}_{ab}\nabla_a H(\nabla_b H + \mathrm{II}_{bc}V_c - X_b) - \frac{1}{t}g_{ab}\left(\frac{1}{2t}\mathrm{II}_{ab} + H\mathrm{II}^2_{ab}\right).\end{aligned} \tag{10.44}$$

Equation (10.39) now follows from applying to the expression above the following identities:

$$\begin{aligned} Q &= W_{aa} + X_b V_b \,, \\ 2Y_{ab} &= -2H\mathrm{II}_{ab} + 2\nabla_a V_b - \frac{1}{t} g_{ab} \,, \\ 0 &= \nabla_b H + \mathrm{II}_{bc} V_c - X_b \end{aligned}$$

and by then combining terms.

Finally, we show that any vector V at any point (x_0, t_0) may be extended locally in space-time so that Y_{ab} and U_a are both zero at that point. In this case, that equation (10.40) holds at (x_0, t_0) follows immediately from (10.39).

Claim 10.18. For any vector $V \in T_{x_0}M$ and 2-tensor $a \in T^*_{x_0}M \otimes T_{x_0}M$ at x_0, we may extend V in a spatial neighborhood of x_0 so that $\nabla V = a$ at (x_0, t_0).

Proof of the claim. Let $(\mathcal{U}, \{x^i\})$ be geodesic normal coordinates centered at x_0 with respect to $g(t_0)$, so that $\Gamma_{ij}^k(x_0) = 0$. Firstly, extend a to $\mathcal{U}$ by defining $a(x) = a_i{}^j dx^i \otimes \partial_{x^j}$. Secondly, extend V over $\mathcal{U}$ by defining $V^j(x) = V^j + a_i^j x^i$. We then have $\nabla_i V^j(x_0) = \partial_i V^j(x_0) = a_i^j$. This proves the claim. □

By the claim, we may extend any vector V at (x_0, t_0) to a spatial neighborhood so that

$$\nabla_a V_b = H\mathrm{II}_{ab} + \frac{1}{2t} g_{ab} \quad \text{at } (x_0, t_0);$$

that is, $Y_{ab}(x_0, t_0) = 0$. With the extension above, we simply further extend V in a space-time neighborhood so that

$$\nabla_t V_a = \Delta V_a - \mathrm{II}_{ab} \nabla_b H - \frac{1}{t} V_a \tag{10.45}$$

at (x_0, t_0). It immediately follows from (10.45) that $U_a(x_0, t_0) = 0$. This completes the proof of the proposition. □

10.2.4. The Harnack maximum principle argument. We are now in a position to complete the proof of Theorem 10.9. By the local derivative estimates, we may assume without loss of generality that the solution is defined on a closed, finite time-interval $[0, T]$ with both the second fundamental form and its higher derivatives uniformly bounded on $M \times [0, T]$. For example, to obtain the result on $(0, T]$ from this, we consider $[\varepsilon, T]$ and let $\varepsilon \to 0$. That the higher derivatives of the second fundamental form are uniformly bounded on $M \times [\varepsilon, T]$, albeit depending on ε, follows from Theorem 7.14.

Recall that by Proposition 10.17, at any (x_0, t_0) with $t_0 > 0$ and for any tangent vector V at x_0, we may extend V to a space-time neighborhood so that $Q(V)$ satisfies

$$(\partial_t - \Delta)(Q(V)) = \left(|\mathrm{II}|^2 - \frac{2}{t}\right) Q(V)\,. \tag{10.46}$$

This suggests that $Q(V)$ should remain nonnegative. To effect a rigorous maximum principle argument, we perturb $Q(V)$ and consider

$$\hat{Q}(V) \doteqdot Q(V) + \phi + \psi\,|V|^2\,, \tag{10.47}$$

where $\phi : M^n \times (0,T] \to \mathbb{R}$ and $\psi : [0,T] \to \mathbb{R}$ are to be chosen below. We shall assume that ψ is uniformly bounded from below by a positive constant and that $\phi(x,t)$ tends to infinity as $x \to \infty$ uniformly in t.

Observe that the components of the Harnack quadratic $Q(V)$ are all bounded, except for the $\frac{1}{2t}H$ term, which is nonnegative.

Suppose that $\hat{Q}(V)$ is not positive on all of $M^n \times (0,T]$. Then, by the assumptions on ψ and ϕ, there exists a first time $t_0 > 0$ at which $\hat{Q}(V) = 0$ for some $x_0 \in M$ and $V \in T_{x_0}M$. We extend V in a space-time neighborhood of (x_0, t_0) so that $Y_{ab} = 0$ and $U_a = 0$ at (x_0, t_0) and hence so that (10.46) holds at (x_0, t_0). With this local extension of V, we have at (x_0, t_0) that

$$\begin{aligned} 0 \geq (\nabla_t - \Delta)\hat{Q}(V) = &\left(|\mathrm{II}|^2 - \frac{2}{t}\right) Q(V) + (\partial_t - \Delta)\phi + (\partial_t\psi)|V|^2 \\ &+ 2\psi V \cdot (\nabla_t - \Delta)V - 2\psi|\nabla V|^2 \end{aligned}$$

since $\nabla_t g = 0$ and ψ depends only on t. Rewriting the right-hand side using $Y_{ab} = 0$ and $U_a = 0$ again, at (x_0, t_0) we have

$$\begin{aligned} 0 \geq &\left(|\mathrm{II}|^2 - \frac{2}{t}\right) \hat{Q}(V) + \left(\frac{2}{t} - |\mathrm{II}|^2\right) \phi - |\mathrm{II}|^2\psi|V|^2 + (\partial_t - \Delta)\phi \\ &+ (\partial_t\psi)|V|^2 - 2\psi \mathrm{II}(\nabla H, V) - 2\psi\left|H\mathrm{II} + \frac{1}{2t}g\right| \\ \geq\ &(\partial_t - \Delta)\phi + \left(\frac{2}{t} - C\right)\phi + (\partial_t\psi - C\psi)|V|^2 - C\psi|V| - \frac{n}{t^2}\psi - C\psi. \end{aligned} \tag{10.48}$$

Note that we can absorb the term $-C\psi|V|$ by changing the constant coefficients of the terms $-C\psi|V|^2$ and $-C\psi$.

Firstly, we set $\psi(t) = \delta e^{At}$, where $\delta > 0$, so that $\partial_t\psi = A\psi > C\psi$ provided $A > C$. Secondly, set

$$\phi(x,t) = \varepsilon e^{Bt} \frac{1}{t^p} f(x,t)\,,$$

where $\varepsilon, B, p > 0$, $f \geq 1$, and

$$(\partial_t - \Delta) f \geq 2Bf \tag{10.49}$$

(e.g., see Lemma 12.7 in [**163**]). Then

$$(\partial_t - \Delta)\phi \geq \left(2B - \frac{p}{t}\right)\phi .$$

Thus, by (10.48), we have

$$\begin{aligned} 0 &\geq \left(\frac{2-p}{t} + 2B - C\right)\phi - \frac{n}{t^2}\psi - C\psi \\ &\geq \left(\frac{2-p}{t} + 2B - C\right)\varepsilon \mathrm{e}^{Bt}\frac{1}{t^p} - \frac{n}{t^2}\delta \mathrm{e}^{At} - C\delta \mathrm{e}^{At} . \end{aligned}$$

Taking $p = \frac{3}{2}$, $B \geq \max\{A, C\}$, and $\varepsilon = \delta$, we obtain

$$0 \geq \left(\frac{1}{2t} + B\right)\frac{1}{t^{3/2}} - \frac{n}{t^2} - C .$$

Finally, by choosing B large enough depending only on n, A, C, and T (and independent of $\varepsilon = \delta$), we obtain a contradiction. Theorem 10.9 follows from taking $\varepsilon \to 0$.

10.2.5. Convex eternal solutions. We move to the application of the Harnack inequality to locally uniformly convex eternal solutions; namely we prove Theorem 10.11. Throughout this subsection, $X : M^n \times (-\infty, \infty) \to \mathbb{R}^{n+1}$ is a complete, locally uniformly convex solution of mean curvature flow with the property that there exists a space-time point at which $\partial_t H + \mathrm{II}^{-1}(\nabla H, \nabla H) = 0$.

We will work with the Harnack quadratic

$$Z(V) = \nabla_t H + 2\nabla_V H + \mathrm{II}(V, V),$$

which is $Q(V)$ without the $\frac{1}{2t}H$ term, since we are concerned with translating solitons instead of expanding solitons. We correspondingly modify the expressions for X, Y, W, and U: Define

$$X \doteqdot \nabla H + \mathrm{L}(V) , \tag{10.50a}$$

$$Y \doteqdot \nabla V - H\,\mathrm{L} , \tag{10.50b}$$

$$W \doteqdot \nabla_t \mathrm{L} + \nabla_V \mathrm{L} , \tag{10.50c}$$

$$U \doteqdot (\nabla_t - \Delta)V + \mathrm{L}(\nabla H) , \tag{10.50d}$$

which together with Z vanish on translating solitons. These are the same definitions as before, except without the $1/t$ terms. Proposition 10.17 (simply remove the $1/t$ terms) yields

$$(\nabla_t - \Delta)Z(V) = |\mathrm{II}|^2 Z(V) + 2X(U) - 2\,\mathrm{tr}(Y \cdot \mathrm{L} \cdot Y) - 4g(Y, W) . \tag{10.51}$$

We will show that if $\mathrm{II} > 0$ and H attains its maximum at a point (x_0, t_0) in space-time, then there is exactly one vector V at each point (y, t) with $t \le t_0$ at which $Z(V) = 0$. But then $\nabla H + \mathrm{L}(V) = 0$ and $\nabla_t \mathrm{L} + \nabla_V \mathrm{II} = 0$, which will imply that the solution is a translating self-similar solution.

By assumption, the solution is eternal and locally uniformly convex, and by Theorem 10.9, $Z(V) \ge 0$. Since $\mathrm{II} > 0$, at each space-time point there exists at most one vector V such that $Z(V) = 0$. Our first task is to show that there is at least one V at all times $t \le t_0$ so that $Z(V) = 0$. To do so, we employ the strong maximum principle:

Lemma 10.19. *Let $s \in \mathbb{R}$ and let $F : M^n \times [s, \infty) \to \mathbb{R}$ be a nonnegative function satisfying $\partial_t F = \Delta F$. If the Harnack quadratic satisfies $Z(V) \ge F$ for every V at s, then $Z(V) \ge F$ at all times $t > s$ for every V.*

Proof. We argue as in the previous subsection, this time with the quantity

$$\hat{Z}(V) \doteqdot Z(V) - F + \phi + \psi\, |V|^2 ,$$

where $\phi(x, t) > 0$ and $\psi(t) > 0$ are to be chosen below. Then $\hat{Z}(V) > 0$ at time $t = s$. Suppose that there is a first time $t_0 > s$ at which $\hat{Z}(V) = 0$ for some V at some point (x_0, t_0). Extend V locally so that $U_a = Y_{ab} = 0$ at (x_0, t_0). Then

$$(\nabla_t - \Delta) Z(V) = |\mathrm{II}|^2\, Z(V)$$

at (x_0, t_0). Since $\nabla_t \hat{Z} \le 0$ and $\Delta \hat{Z} \ge 0$ at (x_0, t_0), this yields

$$\begin{aligned} 0 \ge (\nabla_t - \Delta) \hat{Z}(V) &= |\mathrm{II}|^2 \left(F - \phi - \psi |V|^2\right) + (\nabla_t - \Delta)\phi + (\nabla_t \psi)\, |V|^2 \\ &\quad + 2\psi V \cdot (\nabla_t - \Delta) V - 2\psi |\nabla V|^2 \\ &= |\mathrm{II}|^2 \left(F - \phi - \psi |V|^2\right) + (\nabla_t - \Delta)\phi + (\nabla_t \psi)\, |V|^2 \\ &\quad - 2\psi \mathrm{II}(\nabla H, V) - 2\psi H^2 |\mathrm{II}|^2 \\ &\ge (\partial_t - \Delta - C_1)\phi + (\partial_t \psi - C_2 \psi)|V|^2 - C_3 \psi \end{aligned}$$

since $F \ge 0$, $|\mathrm{II}| \le C$, and $|\nabla H| \le C$. A contradiction results if ϕ and ψ are chosen so that additionally

$$(\nabla_t - \Delta)\phi \ge 2C_1 \phi, \quad \nabla_t \psi \ge C_2 \psi, \quad \text{and} \quad C_1 \phi > C_3 \psi .$$

Functions ϕ and ψ that satisfy all of these requirements can be chosen (essentially the same as the choices in the proof of Theorem 10.9) so that they are arbitrarily small perturbations. □

Now H assumes its maximum at a point (x_0, t_0), and so $\partial_t H = 0$ and $\nabla H = 0$ at this point. Therefore $Z(V) = \partial_t H + 2\nabla_V H + \mathrm{II}(V, V) = 0$ in the direction $V = \vec{0}$. More generally, we only need to assume that (10.19), i.e., $Z(V) = 0$, holds at (x_0, t_0) for some $V \in T_{x_0} M$. We claim that then Lemma 10.19 implies that $Z(V)$ has a zero vector everywhere prior to t_0, for if not, then there exists a point (x, t) with $t < t_0$ at which $Z(V) > 0$ for every V. Thus we may choose a nonnegative smooth function $F : \mathcal{M}_t \to \mathbb{R}$ with $F > 0$ in a neighborhood of x, with $F = 0$ outside the neighborhood, and such that $Z(V) \geq F$ for all $V \in TM$ at $t = t_0$. If we evolve F by the heat equation, then $F > 0$ for subsequent times by the strong maximum principle for functions, and so by Lemma 10.19 we have that $Z(V) > 0$ after time t. This contradicts $Z(\vec{0}) = 0$ at (x_0, t_0), so at each time prior to t_0, $Z(V)$ must have had a zero vector everywhere.

Now consider a point (x, t) with $t \leq t_0$ and a direction V in which $Z(V) = 0$. In that direction

$$X = \nabla H + \mathrm{L}(V) = 0.$$

Since $\mathrm{II} > 0$, this implies that L is invertible and V is uniquely determined by

$$V = -\mathrm{L}^{-1}(\nabla H).$$

We can show that $W = \nabla_t L + \nabla_V L = 0$ in this direction as well: Extend V in space so that $Y = W \cdot \mathrm{L}^{-1}$ and then in time so that $U = X$. Then

$$(\nabla_t - \Delta) Z = 2\,|X|^2 + 2\,\mathrm{tr}(W \cdot \mathrm{L}^{-1} \cdot W),$$

which implies that both $X = 0$ (which we already knew) and $W = 0$: Otherwise, $\nabla_t Z > 0$, which implies that Z would have been negative at an earlier time. Theorem 10.11 is now a consequence of the following lemma.

Lemma 10.20. *Given a locally uniformly convex solution to the mean curvature flow, if V is a vector field with*

$$\nabla H + \mathrm{L}(V) = 0 \tag{10.52}$$

and

$$\nabla_t \mathrm{II} + \nabla_V \mathrm{II} = 0, \tag{10.53}$$

then the solution is a translating self-similar solution.

Proof. We again compute in an evolving orthonormal frame. Differentiating (10.52), we have

$$\nabla_a \nabla_b H + V_c \nabla_c \mathrm{II}_{ab} + \mathrm{II}_{bc} \nabla_a V_c = 0.$$

By (10.31) and (10.53), this implies that

$$\mathrm{II}_{bc} \nabla_a V_c = H\, \mathrm{II}_{ac}\, \mathrm{II}_{bc},$$

and so since $\mathrm{II} > 0$,

$$\nabla_a V_c = H\, \mathrm{II}_{ac}\,.$$

We now define the vector

$$e = V - H\mathsf{N},$$

which we recognize from the beginning of this section. Differentiating, we find

$$D_i e = g^{jk}(\nabla_i V_j - H\,\mathrm{II}_{ij})D_k X - (\nabla_i H + g^{jk}\mathrm{II}_{ij}V_k)\mathsf{N} = 0\,.$$

So e is constant. Since its normal component is H, the solution must proceed by translation with bulk velocity e. □

10.2.6. Space-time formulation of Hamilton's Harnack estimate. Following ideas developed by Chu and the second author [**161, 162**], Brett Kotschwar developed a "space-time" approach to the Harnack inequality, which we now describe [**337**].

Let $X : M^n \times [0,T) \to \mathbb{R}^{n+1}$ be an embedded mean curvature flow and let $\mathcal{M}_t = X_t(M^n)$. Define its **space-time track** by

$$\overline{\mathcal{M}} = \bigcup_{t\in[0,T)} \mathcal{M}_t \times \{t\} \subset \mathbb{R}^{n+1} \times [0,T). \tag{10.54}$$

A parametrization of $\overline{\mathcal{M}}$ is given by $\overline{X} : M^n \times [0,T) \to \mathbb{R}^{n+1} \times [0,T)$, where

$$\overline{X}(x,t) = (X(x,t),t)\,.$$

Given $N > 0$, define on the ambient space-time $\mathbb{R}^{n+1} \times [0,T)$ the (time-stretched) Euclidean metric

$$g_{\mathbb{E},N} = g_{\mathbb{E}^{n+1}} + N\,dt^2\,.$$

Let ${}^N\overline{g}$ be the Riemannian metric on $\overline{\mathcal{M}}$ induced by $g_{\mathbb{E},N}$.

By defining the time coordinate $x^0 = t$ we may extend any local coordinate system $\{x^i\}_{i=1}^n$ on $U \subset M^n$ to local coordinates $\{x^a\}_{a=0}^n$ on $U \times [0,T)$. Observe that $\partial_i\overline{X} = (\partial_i X, 0)$ for $i \geq 1$ and $\partial_0\overline{X} = (-H\mathsf{N}, 1)$. Equivalently, $\partial_a\overline{X} = (\partial_a X, \delta_{a0})$. Thus, for $0 \leq a, b \leq n$, we have that

$$\begin{aligned} {}^N\overline{g}_{ab} &= g_{\mathbb{E},N}\left(\partial_a\overline{X}, \partial_b\overline{X}\right) \\ &= g_{\mathbb{E}^{n+1}}\left(\partial_a X, \partial_b X\right) + N\delta_{a0}\delta_{b0} \\ &= \begin{cases} g_{ab} & \text{if } a,b \geq 1, \\ 0 & \text{if } a > b = 0 \text{ or } b > a = 0, \\ H^2 + N & \text{if } a = b = 0. \end{cases} \end{aligned} \tag{10.55}$$

A choice of unit normal vector field ${}^N\overline{\mathsf{N}}$ to the space-time track $\overline{\mathcal{M}}$ with respect to the ambient metric $g_{\mathbb{E},N}$ is given by

$${}^N\overline{\mathsf{N}} = \frac{(\mathsf{N}, N^{-1}H)}{\sqrt{1 + N^{-1}H^2}}\,. \tag{10.56}$$

Indeed, one easily checks that this vector is unit and perpendicular to each $\partial_a\overline{X}$, $a \geq 0$, all of which is with respect to $g_{\mathbb{E},N}$.

Let D and ${}^N D$ denote the Levi-Civita connections of the metrics $g_{\mathbb{E}^{n+1}}$ and $g_{\mathbb{E},N}$, respectively. The second fundamental form of the space-time track $\overline{\mathcal{M}}$ with respect to the ambient metric $g_{\mathbb{E},N}$ is given by

$$
\begin{aligned}
{}^N\mathrm{II}_{ab} &= -g_{\mathbb{E},N}\left({}^N D_{\partial_a \overline{X}} \partial_b \overline{X},\, {}^N\overline{\mathsf{N}}\right) \\
&= -\frac{1}{\sqrt{1+N^{-1}H^2}}\, g_{\mathbb{E}^{n+1}}(D_{\partial_a X}\partial_b X, \mathsf{N}) \\
&= \frac{1}{\sqrt{1+N^{-1}H^2}} \begin{cases} \mathrm{II}_{ab} & \text{if } a, b \geq 1, \\ \partial_a H & \text{if } a > b = 0, \\ \partial_t H & \text{if } a = b = 0 \end{cases}
\end{aligned} \tag{10.57}
$$

and ${}^N\mathrm{II}_{ba} ={}^N \mathrm{II}_{ab}$. Thus we obtain a scalar multiple of Hamilton's Harnack quadratic for the mean curvature flow.

These ideas were developed further by Sebastian Helmensdorfer and Peter Topping [**283**], who obtained Hamilton's differential Harnack inequality as a consequence of the preservation of convexity of the **canonical** ($(n+1)$-dimensional) **expanding solution** constructed from a given (n-dimensional) convex solution.

10.2.7. Concavity of the arrival time. Recall, from the proof of Theorem 8.23, that the time-of-arrival function $u : \bigcup_{t\in(0,T)} \mathcal{M}_t^n \to \mathbb{R}$ of a convex solution $\{\mathcal{M}_t^n\}_{t\in[0,T)}$ to mean curvature flow satisfies

$$
Du = -\frac{1}{H}\mathsf{N}
$$

and

$$
D^2 u = \begin{pmatrix} -\mathrm{II}/H & \nabla H/H^2 \\ \nabla H/H^2 & -\partial_t H/H^3 \end{pmatrix}.
$$

The Hessian D^2u looks suspiciously like the quadratic form in Hamilton's differential Harnack inequality.

Theorem 10.21. *Let u be the arrival time of a convex (possibly noncompact) ancient solution $\{\mathcal{M}_t^n\}_{t\in(-\infty,T)}$ to mean curvature flow. Then u is locally concave.*

Proof. Fix a point $p \in \mathcal{M}_t^n$ and any vector $V = V^\top + \alpha\mathsf{N}(p,t) \in T_p\mathbb{R}^{n+1}$. If $\alpha = 0$, then

$$
-HD^2u(V,V) = \mathrm{II}(V^\top, V^\top) \geq 0\,.
$$

Otherwise, we can scale V so that $\alpha = -H$, in which case the differential Harnack inequality yields

$$
-HD^2u(V,V) = \mathrm{II}(V^\top, V^\top) + 2\nabla_{V^\top} H + \partial_t H \geq 0\,. \qquad \square
$$

In fact, it is clear from the proof of Theorem 10.21 that, for a convex ancient solution, the differential Harnack inequality is *equivalent* to local concavity of the arrival time.

A similar argument shows that the Harnack inequality for a general (nonancient) convex solution $\{\mathcal{M}_t^n\}_{t\in[0,T)}$ to mean curvature flow is equivalent to local concavity of the *square root* of the arrival time.

Theorem 10.22. *Let u be the arrival time of a convex (possibly noncompact) solution $\{\mathcal{M}_t^n\}_{t\in[t_0,T)}$ to mean curvature flow. Then $\sqrt{2(u-t_0)}$ is locally concave.*

Let us provide a direct proof of these facts, in the compact case, and thereby a different proof of the differential Harnack inequality (cf. [**521**, Lemmas 4.1 and 4.4]).

Alternative proof of Theorems 10.21 and 10.22. Let $\{\mathcal{M}_t^n\}_{t\in[t_0,T)}$ be a solution to mean curvature flow with $\mathcal{M}_{t_0}^n$ bounded and convex. Since $\mathcal{M}_t^n$ is smooth and uniformly convex for $t>t_0$, we may assume, without loss of generality, that $\mathcal{M}_{t_0}^n$ is smooth and uniformly convex. Denote by $\Omega\subset\mathbb{R}^n$ the convex body bounded by $\mathcal{M}_{t_0}^n$. Since, by Huisken's theorem, the solution shrinks to a point (without loss of generality at time T), the arrival time is well-defined and of class C^1 in $\overline{\Omega}$ and has a unique critical point. Set $w\doteqdot\sqrt{2(u-t_0)}$ and define the concavity function [**335**] $Z:\overline{\Omega}\times\overline{\Omega}\times[0,1]\to\mathbb{R}$ by

$$Z(x,y,r)\doteqdot w(rx+(1-r)y)-rw(x)-(1-r)w(y).$$

This function measures how far the point $\big(rx+(1-r)y, w(rx+(1-r)y)\big)$ in $\overline{\Omega}\times\mathbb{R}$ lies above the line joining the points $(x,w(x))$ and $(y,w(y))$. We need to prove that $Z\geq 0$.

Choose (x_0,y_0,r_0) so that

$$Z(x_0,y_0,r_0)=\min_{\overline{\Omega}\times\overline{\Omega}\times[0,1]} Z.$$

If $r_0=0$ or $r_0=1$, then $Z(r_0,x_0,y_0)=0$, which implies the claim. So we may assume that $r_0\in(0,1)$.

Next, suppose that $x_0\in\partial\Omega$. We claim that Z decreases as x_0 moves towards y_0 (keeping y_0 and z_0 fixed), contradicting minimality of (x_0,y_0,r_0). To see this, set, for small $\varepsilon>0$,

$$x_\varepsilon\doteqdot x_0+\varepsilon(y_0-x_0)\ \text{ and }\ r_\varepsilon\doteqdot\frac{1}{1-\varepsilon}r_0.$$

Then

$$z_\varepsilon\doteqdot r_\varepsilon x_\varepsilon+(1-r_\varepsilon)y_0\equiv z_0,\quad \frac{d}{d\varepsilon}x_\varepsilon=y_0-x_0,\quad\text{and}\quad \frac{d}{d\varepsilon}r_\varepsilon=\frac{r_0}{(1-\varepsilon)^2},$$

and hence

$$\frac{d}{d\varepsilon}Z(x_\varepsilon, y_0, r_\varepsilon) = \frac{r_0}{(1-\varepsilon)^2}\big(w(y_0) - w(x_\varepsilon) - (1-\varepsilon)Dw(x_\varepsilon)\cdot(y_0 - x_0)\big).$$

Since $\mathcal{M}^n_{t_0}$ is smooth and uniformly convex and $Dw = \frac{Du}{\sqrt{2(u-t_0)}}$, the right-hand side goes to $-\infty$ as $\varepsilon \to 0$. This proves the claim (contradicting minimality of (x_0, y_0, r_0)). A similar argument applies if $y_0 \in \partial\Omega$.

So we may assume that $x_0, y_0 \in \Omega$. Let us abuse notation by writing $Z(x,y) \doteqdot Z(x,y,r_0)$. Then (x_0, y_0) is a stationary point of Z and hence, setting $z_0 \doteqdot r_0 x_0 + (1-r_0)y_0$,

$$0 = \partial_{x^i} Z(x_0, y_0) = r_0(w_i(z_0) - w_i(x_0))$$

and

$$0 = \partial_{y^i} Z(x_0, y_0) = (1-r_0)(w_i(z_0) - w_i(y_0))\,.$$

So

$$Dw(z_0) = Dw(x_0) = Dw(y_0) \doteqdot p_0\,. \tag{10.58}$$

Since Dw has a unique critical point, we may assume that $p_0 \neq 0$ (otherwise, $x_0 = y_0 = z_0$ and hence $Z(x_0, y_0) = 0$).

Since (x_0, y_0) is a local minimum,

$$\begin{aligned} 0 &\leq (\partial_{x^i} + \partial_{y^i})(\partial_{x^j} + \partial_{y^j})Z(x_0, y_0) \\ &= w_{ij}(z_0) - r_0 w_{ij}(x_0) - (1-r_0)w_{ij}(y_0)\,. \end{aligned} \tag{10.59}$$

Set

$$a_0 \doteqdot \mathrm{I} - \frac{p_0 \otimes p_0}{|p_0|^2}.$$

Then, since $B \mapsto -(a_0^{ij}B_{ij})^{-1}$ is a concave function, (10.58) and (10.59) yield

$$\begin{aligned} w(z_0) &= -\left(a_0^{ij} w_{ij}(z_0)\right)^{-1} \\ &\geq -\left(a_0^{ij}(r_0 w_{ij}(x_0) + (1-r_0)w_{ij}(y_0))\right)^{-1} \\ &\geq -r_0\left(a_0^{ij} w_{ij}(x_0)\right)^{-1} - (1-r_0)\left(a_0^{ij} w_{ij}(y_0)\right)^{-1} \\ &= r_0 w(x_0) + (1-r_0)w(y_0). \end{aligned}$$

This completes the proof that w is concave.

Now suppose that $\{\mathcal{M}_t^n\}_{t\in(-\infty,t_0)}$ is a convex, compact ancient solution. Then $w \doteqdot \sqrt{2(u-t_0)}$ is concave for any $t_0 < T$ and hence, for points in the convex body Ω_{t_0} bounded by $\mathcal{M}_{t_0}^n$,

$$\begin{aligned} D^2 u &= w D^2 w + Dw \otimes Dw \\ &\leq \frac{Du \otimes Du}{2(u-t_0)}. \end{aligned}$$

Taking $t_0 \to -\infty$, we conclude that u is concave. □

10.3. Notes and commentary

10.3.1. Huisken's monotonicity formula for Brakke flows. One can greatly relax smoothness requirements in Huisken's monotonicity formula. Some of the early results in this direction are by Ilmanen in [**306**], who proved a monotonicity formula for integral Brakke flows by adapting Huisken's proof to varifolds. He used it to prove convergence of a general family of rescalings to a self-similar mean curvature flow, in the weak sense of Brakke.

10.3.2. General monotonicity formulae. A nice framework for the derivation of Huisken's and other monotonicity formulae is given by Ecker et al. in [**210**], based on the following simple observation. Consider a general (intrinsic) flow

$$\partial_t g_{ij} = 2T_{ij}$$

on $(M^n, g(t))$. Then since $\partial_t d\mu = (\operatorname{tr}_g h)\, d\mu$, the conjugate of the heat operator $\partial_t - \Delta$ is

$$-\left(\partial_t + \Delta + \operatorname{tr}_g T\right),$$

and one has

$$\frac{d}{dt}\int_{\mathcal{M}} \phi\psi\, d\mu = \int_{\mathcal{M}} \left\{\psi\left[\left(\partial_t - \Delta\right)\phi\right] + \phi\left[\left(\partial_t + \Delta + \operatorname{tr} T\right)\psi\right]\right\} d\mu.$$

If ϕ solves the heat equation and ψ solves the conjugate heat equation, then the integral is time-independent. So, whenever the two terms on the right-hand side have the same sign, one has a monotonicity formula. For example, to derive Huisken's monotonicity formula, one takes $\phi = 1$ and lets ψ be the backward heat kernel. One calculates that

$$\left(\partial_t + \Delta + \operatorname{tr} T\right)\psi = \left(\partial_t + \Delta - H^2\right)\psi = -\left|\frac{x^\perp}{2\tau} - H\mathsf{N}\right|^2$$

and obtains the monotonicity formula.

This also leads to a general framework for local monotonicity formulae. Using the same ideas as in Ecker's derivation, one calculates that

$$\int_{E_r} \left(\psi_r \left[(\partial_t - \Delta)\,\phi\right] + \Psi^{-1}\left[(\partial_t + \Delta + \operatorname{tr} T)\,\Psi\right]\phi\right) d\mu dt$$
$$= \int_{E_r} \left(|\nabla\psi|^2 - (\operatorname{tr} T)\psi_r\right)\phi d\mu dt - \frac{r}{n}\frac{d}{dr}\int_{E_r} \left(|\nabla\psi|^2 - (\operatorname{tr} T)\psi_r\right)\phi d\mu dt.$$

Defining

$$P_{\phi,\Psi}(r) = \int_{E_r} \left(|\nabla\psi|^2 - (\operatorname{tr} T)\psi_r\right)\phi d\mu dt,$$

one obtains local monotonicity theorems of the form

$$\frac{d}{dr}\left(\frac{P_{\phi,\Psi}(r)}{r^n}\right) = \frac{n}{r^{n+1}}\int_{E_r} \log(r^n\Psi)\left[\left(\frac{\partial}{\partial t} - \Delta\right)\phi\right] d\mu dt$$
$$+ \frac{n}{r^{n+1}}\int_{E_r} \Psi^{-1}\left[\left(\frac{\partial}{\partial t} + \Delta + \operatorname{tr} T\right)\Psi\right]\phi d\mu dt.$$

Justifying each of these steps is nontrivial. We refer the reader to [**210**] for the details.

10.3.3. Huisken's monotonicity formula for mean curvature flow in general ambient spaces.

Huisken's monotonicity formula was generalized by Hamilton [**263**] as follows. Let $X : M^n \times I \to P^{n+k}$ be a family of submanifolds in a Riemannian manifold $\left(P^{n+k}, g\right)$ evolving by the mean curvature flow,

$$\partial_t X = \vec{H}\,,$$

and suppose $u : P^{n+k} \times [0, T) \to \mathbb{R}$ is a positive solution of the backward heat equation,

$$\partial_t u = -\Delta u\,,$$

where Δ is the Laplacian induced by g. Then

$$\frac{d}{dt}\left((T-t)^{\frac{k}{2}}\int_M u d\mu\right) = -(T-t)^{\frac{k}{2}}\int_M \left|\vec{H} - (\nabla \log u)^{\perp}\right|^2 u d\mu$$
$$- (T-t)^{\frac{k}{2}}\int_M \operatorname{tr}^{\perp}\!\left(\nabla^2 \log u + \frac{1}{2(T-t)}g\right) u d\mu,$$

where $d\mu$ is the volume form of the M with its induced metric and $\operatorname{tr}^{\perp}$ denotes the trace restricted to the normal bundle $NM \subset X^*TP$ of M. By Hamilton's matrix Harnack estimate for the heat equation (1.33), if $\left(P^{n+k}, g\right)$ has parallel Ricci tensor and nonnegative sectional curvature, then $\nabla\nabla \log u + \frac{1}{2(T-t)}g \geq 0$ and hence

$$\frac{d}{dt}\left((T-t)^{\frac{k}{2}}\int_M u\, d\mu\right) \leq 0$$

with equality if and only if

$$\vec{H} - (\nabla \log u)^{\perp} = 0 \quad \text{and} \quad \left(\nabla^2 \log u + \frac{1}{2\,(T-t)} g\right)^{\perp} = 0\,.$$

10.3.4. Ecker's local monotonicity formula. As we have seen, Huisken's monotonicity formula for mean curvature flow involves a (global) integral over $\mathcal{M}_t$. In contrast, for the heat equation in Euclidean space one has a local monotonicity formula, as the result of a mean value formula. There is also a local monotonicity formula for minimal hypersurfaces, which are stationary solutions of the mean curvature flow. Motivated by these considerations, Klaus Ecker derived a local monotonicity formula for the mean curvature flow: Given $(p_0, t_0) \in \mathbb{R}^{n+1} \times \mathbb{R}$ and $R > 0$, define the cutoff function $\varphi^R_{(p_0,t_0)} : \mathbb{R}^{n+1} \times \mathbb{R} \to \mathbb{R}$ by

$$\varphi^R_{(p_0,t_0)}(X,t) \doteqdot \left(1 - \frac{|X-p_0|^2 + 2n(t-t_0)}{R^2}\right)^3_+ .$$

Recall also that the backward heat kernel $\rho_{(p_0,t_0)} : \mathbb{R}^{n+1} \times (-\infty, t_0) \to \mathbb{R}$ is defined by

$$\rho_{(p_0,t_0)}(X,t) \doteqdot (4\pi(t_0-t))^{\frac{n}{2}} \mathrm{e}^{-\frac{|X-p_0|^2}{4(t_0-t)}}\,.$$

Theorem 10.23 (Ecker's local monotonicity formula [**204**], [**205**, Proposition 4.17]). *Let $X : M^n \times I \to \mathbb{R}^{n+1}$ be a smooth mean curvature flow which is properly defined in $B_{\sqrt{1+2n}R}(p_0) \times (t_0 - R^2, t_0]$. Then*

$$\frac{d}{dt}\int_M \varphi^R_{(p_0,t_0)}\,\rho_{(p_0,t_0)}\,d\mu \leq -\int_M \left|H - \frac{\langle X - p_0, \mathsf{N}\rangle}{2(t-t_0)}\right|^2 \varphi^R_{(p_0,t_0)}\,\rho_{(p_0,t_0)}\,d\mu\,,$$

where $\varphi^R_{(p_0,t_0)}$ and $\rho_{(p_0,t_0)}$ are pulled back to M along X.

Ecker's local monotonicity formula can be used to localize arguments which rely on Huisken's monotonicity formula. For instance, the argument of Lemma 10.8 yields an improved local area estimate.

Lemma 10.24. *Let $X : M^n \times [0,T) \to \mathbb{R}^{n+1}$ be a solution to mean curvature flow which is properly defined in $B_{\sqrt{1+2n}R}(p_0) \times (t_0 - R^2, t_0]$. Given $\vartheta \in (0,1)$ and $r \in (0,R)$,*

$$\frac{\mu_t(X_t^{-1}(B_r(p_0)))}{r^n} \leq V \doteqdot \mathrm{e}^{\frac{1}{4}}\vartheta^{-\frac{n}{2}} \frac{\mu_0(X_0^{-1}(B_R))}{T^{\frac{n}{2}}} \tag{10.60}$$

for every $t \in (\vartheta T, T)$.

10.3.5. Monotonicity formula for free boundary mean curvature flow. John Buckland proved a free boundary version of Huisken's monotonicity formula using a localized reflection in a neighborhood of the free boundary [**128**].

10.3.6. Smockzyk's Harnack estimate for flows with speed $f(H)$. Knut Smoczyk, following Hamilton's approach, has obtained differential Harncak estimates for curvature flows of the form

$$\partial_t X = -f(H)\mathsf{N}, \tag{10.61}$$

where f satisfies certain structure conditions [**486**].

In addition to investigating for which functions f the flow is parabolic and for which f self-similar solutions exist, he proved the following Harnack estimate.

Theorem 10.25. *Let $f : \Omega \to \mathbb{R}$ be a smooth function satisfying, for all $h \in \Omega$,*

$$f' > 0, \ \frac{f''}{f'}h^2 \geq ah, \ \left(\frac{f''}{f'}h\right)' \leq 0, \ ff''h + ff' - (f')^2 \geq 0,$$

for some $a \in \mathbb{R}$. Then there exists $d > 0$ such that

$$\partial_t f + 2\nabla_V f + \mathrm{II}(V,V) + \frac{1}{d + (a+2)t} f'H \geq 0$$

for all $V \in T\mathcal{M}^n_t$, $t \in [-\frac{d}{a+2}, T)$, along any compact, uniformly convex solution $\{\mathcal{M}_t\}_{t\in[0,T)}$ to (10.61).

Note that Smoczyk's result applies, in particular, to the mean curvature flow, where $f(H) = H$, and to the inverse-mean curvature flow, where $f(H) = -1/H$. The full generality of Hamilton's Harnack estimate is not quite a corollary of this theorem, however, since Smockzyk only considers locally uniformly convex solutions.

10.3.7. Harnack estimates for fully nonlinear flows. A Harnack inequality for the Gauß curvature flow was obtained by the second author, using similar ideas as Hamilton, in [**159**]. See Section 15.4.

A generalization of the Harnack estimate to a large class of flows was obtained, using a somewhat different argument, by the first author in [**29**]. We present the proof of these estimates in Section 18.6.

Harnack inequalities for mean curvature flow in the sphere [**122**, **125**] and in more general Riemannian and Lorentzian ambient spaces [**124**] have been obtained by Paul Bryan, Mohammad Ivaki, and Julian Scheuer.

Figure 10.2. Mohammad Najafi Ivaki.

10.4. Exercises

Exercise 10.1 (Pointwise monotonicity formula, I)**.** Let $X : M\times I \to \mathbb{R}^{n+1}$ be a solution to mean curvature flow. Given $(X_0, t_0) \in \mathbb{R}^{n+1}\times\mathbb{R}$, prove that

$$(\partial_t + \Delta - H^2)\Phi_{(X_0,t_0)} = -\left(H - \frac{\langle X - X_0, \mathsf{N}\rangle}{2(t_0 - t)}\right)^2 \Phi_{(X_0,t_0)} \tag{10.62}$$

for $t \in I \cap (-\infty, t_0)$, where $\Phi_{(X_0,t_0)}$ is the n-dimensional backward heat kernel along X centered at (X_0, t_0). Deduce that

$$\frac{d}{dt}\int_M u\,\Phi_{(X_0,t_0)}d\mu = \int_M \left[(\partial_t - \Delta)u - u\left(H - \frac{\langle X - X_0, \mathsf{N}\rangle}{2(t_0 - t)}\right)^2\right]\Phi_{(X_0,t_0)}d\mu$$

for $t \in I \cap (-\infty, t_0)$ and any smooth $u : M \times [0, T) \to \mathbb{R}$ with compact support in M at each time t.

Exercise 10.2. Given $e \in \mathbb{R}^{n+1}$, show that any compact solution $X : M^n \times I \to \mathbb{R}^{n+1}$ to mean curvature flow satisfies

$$\frac{d}{dt}\int_M \rho\, d\mu \le -\int_M (H + \langle e, \mathsf{N}\rangle)^2 \rho\, d\mu\,,$$

where

$$\rho(x, t) \doteqdot \mathrm{e}^{\langle X(x,t),e\rangle - t}\,.$$

Exercise 10.3 (Pointwise monotonicity formula, II)**.** Let $X : M\times I \to \mathbb{R}^{n+1}$ be a solution to mean curvature flow and let $\rho : \mathbb{R}^{n+1} \times (-\infty, t_0) \to \mathbb{R}$ be a positive solution to the backward heat equation,

$$(\partial_t + \Delta_{\mathbb{R}^{n+1}})\rho = 0\,.$$

Show that

$$\left(\partial_t + \Delta - H^2\right)\Phi = -\left[(H + D_{\mathsf{N}}\log\Phi)^2 + D^2_{\mathsf{N},\mathsf{N}}\log u + \frac{1}{2(t_0 - t)}\right]\Phi$$

for $t \in I \cap (-\infty, t_0)$, where $\Phi(x,t) \doteqdot (t_0 - t)^{\frac{1}{2}}\rho(X(x,t),t)$. Deduce that

$$\begin{aligned}\frac{d}{dt}\int u\,\Phi_{(X_0,t_0)}d\mu &= \int \left[(\partial_t - \Delta)u - u\,(H + D_{\mathsf{N}}\log\rho)^2\right]\Phi\,d\mu \\ &\quad - \int u\left(D^2_{\mathsf{N},\mathsf{N}}\log\rho + \frac{1}{2(t_0 - t)}\right)\Phi\,d\mu \\ &\leq \int \left[(\partial_t - \Delta)u - u\,(H + D_{\mathsf{N}}\log\rho)^2\right]\Phi\,d\mu\end{aligned}$$

for $t \in I \cap (-\infty, t_0)$ and any smooth $u : M \times [0,T) \to \mathbb{R}$ with compact support in M at each time t.

Exercise 10.4. Let $\{\mathcal{M}^n_t\}_{t\in[0,T]}$ be mean curvature flow of entire graphs with bounded gradient and curvature. Use the noncompact maximum principle (similarly as in Proposition 10.7) and Exercise 7.3 to prove that

$$\sup_{\mathcal{M}^n_t}|\mathrm{II}| \leq C\sup_{\mathcal{M}^n_0}|\mathrm{II}| \text{ and } \sup_{\mathcal{M}^n_t}|\mathrm{II}| \leq C/t \text{ for all } t \in [0,T],$$

where C depends only on n and the bound for $v \doteqdot \langle \mathsf{N}, e_{n+1}\rangle^{-1}$. A similar argument yields analogous bounds for the derivatives of II.

Exercise 10.5. Let $\{\mathcal{M}^n_t\}_{t\in(0,\infty)}$ be an expanding self-similar solution of mean curvature flow. Show that

$$\nabla_i V_j - H\,\mathrm{II}_{ij} - \frac{1}{2t}g_{ij} = 0\,,$$

where $V \doteqdot \frac{1}{2t}X^\top$.

Exercise 10.6. Prove Lemma 8.27 using (8.64) and the differential Harnack inequality. HINT: Work in the Gauß map parametrization.

Exercise 10.7. Prove Lemma 8.27 using (8.64) and Theorem 10.22.

Chapter 11

Singularity Analysis

We would like to analyze singularity formation in mean curvature flow by "blowing up" — rescaling about a sequence of space-time points approaching a singularity. This worked well in the convex case (see Section 8.6) since the width pinching estimate allowed us to invoke the Arzelà–Ascoli theorem, yielding uniform subsequential convergence of the blow-up sequence to a limit flow. Combined with monotonicity of the curvature pinching and the strong maximum principle, we were able to characterize the blow-up limit as a round, shrinking sphere. Our goal in this chapter is to extend these ideas to the mean convex setting. To achieve this, we must first develop a notion of *local* uniform convergence of mean curvature flows and prove an associated compactness theorem.

11.1. Local uniform convergence of mean curvature flows

Our main goal is to obtain a robust precompactness theorem for sequences of mean curvature flows which extends the classical Arzelà–Ascoli theorem which we applied in the setting of uniform convergence (cf. Proposition 6.39). We have seen that a bound for the second fundamental form implies bounds for all of its derivatives. These bounds translate into bounds for the derivatives of any parametrization map. So we expect that a family of flows with uniformly bounded second fundamental form is precompact in an appropriate topology, so long as it does not wander off to infinity. This is indeed the case, although we need to be careful about the notion of convergence we work with.

The most basic notion of convergence for subsets of Euclidean space is that of **Hausdorff convergence**.

Definition 11.1. A sequence $\{K_i\}_{i\in\mathbb{N}}$ of compact subsets $K_i \subset \mathbb{R}^n$ converges **in the Hausdorff topology** to a compact set K if for every $\varepsilon > 0$ there exists $i_0 \in \mathbb{N}$ such that

$$d_H(K_i, K) \leq \varepsilon$$

for all $i \geq i_0$, where the **Hausdorff distance** between two compact subsets $X, Y \subset \mathbb{R}^{n+1}$ is defined by

$$d_H(X,Y) \doteqdot \max\left\{\sup_{x\in X}\inf_{y\in Y}|x-y|\,,\, \sup_{y\in Y}\inf_{x\in X}|x-y|\right\}.$$

Hausdorff convergence is readily extended to (noncompact) closed sets.

Definition 11.2. A sequence $\{F_i\}_{i\in\mathbb{N}}$ of closed sets $F_i \subset \mathbb{R}^{n+1}$ converges **locally uniformly in the Hausdorff topology** to a closed set $F \subset \mathbb{R}^{n+1}$ if, given any compact $K \subset \mathbb{R}^{n+1}$, the sequence $\{F_i \cap K\}_{i\in\mathbb{N}}$ of compact sets $F_i \cap K$ converges in the Hausdorff topology to the compact set $F \cap K$.

We would like to similarly extend the notion of uniform convergence in the *smooth topology* of solutions to mean curvature flow to a notion of *local* uniform convergence. A new difficulty that arises in this setting stems from the fact that the topology of the sequence could change in the limit. This is illustrated by the following example.

Example 11.3. Consider the sequence of embeddings $X_k : S^n \to \mathbb{R}^{n+1}$ given by

$$X_k(z) \doteqdot k(z + e_{n+1})\,.$$

Observe that $X_k(S^n)$ is a round sphere of radius k touching the hyperplane $\Pi_{n+1} \doteqdot \{X \in \mathbb{R}^{n+1} : \langle X, e_{n+1}\rangle = 0\}$ from above at the origin. The image sets $X_k(S^n)$ converge locally uniformly in the Hausdorff topology to Π_{n+1}. But we cannot obtain local uniform convergence of the parametrizations in the usual sense since no embedding of S^n into $\mathbb{R}^{n+1}$ maps onto Π_{n+1}.

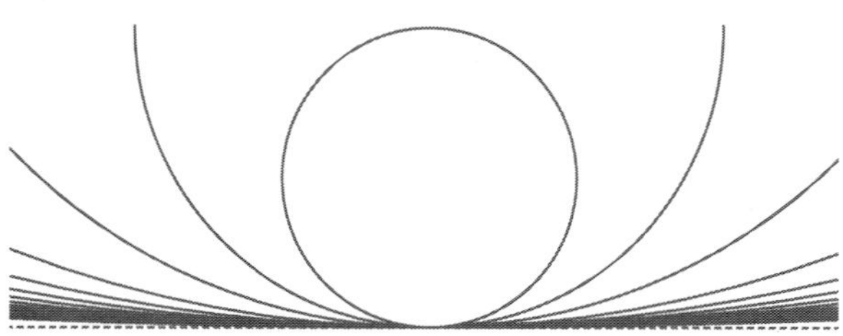

Figure 11.1. Circles of increasing radius converging to a line.

So it would be too restrictive to simply extend the notion of uniform convergence in $C^\infty(M^n, \mathbb{R}^{n+1})$ to *local* uniform convergence in $C^\infty(M^n, \mathbb{R}^{n+1})$. To accommodate possible topological changes in the domain manifold, we shall make use of the compactness theorem for pointed Riemannian manifolds developed by Cheeger, Gromov, and others [**25, 139, 249, 251, 267, 328, 441, 442**], a version of which we now state (see [**404**, §5.1] and [**163**, Chapter 4]). We recall that a **pointed Riemannian manifold** is a triple (M, g, x), where (M, g) is a Riemannian manifold and x is a point of M.

Theorem 11.4 (Compactness theorem for Riemannian manifolds). *Suppose that $\{(M_k, g_k, x_k)\}_{k\in\mathbb{N}}$ is a sequence of pointed, connected Riemannian manifolds satisfying the following conditions*:

(i) *For each $A < \infty$ there exists $k_A \in \mathbb{N}$ such that the geodesic ball $B_A(x_k)$ has compact closure in (M_k, g_k) for all $k \geq k_A$.*

(ii) *For each $m \in \mathbb{N}$ and each $A < \infty$ there exist $C_{m,A} < \infty$ and $k_A \in \mathbb{N}$ such that*
$$\sup_{B_A(x_k)} |{}^k\nabla^m \operatorname{Rm}_k| \leq C_{m,A}$$
for every $k \geq k_A$, where ${}^k\nabla$ and Rm_k denote, respectively, the Levi-Civita connection and Riemann curvature tensor of g_k.

(iii) *There exists $\delta > 0$ such that $\operatorname{inj}_k(x_k) \geq \delta$ for every $k \in \mathbb{N}$, where inj_k denotes the injectivity radius function of (M_k, g_k).*

Then there exist a pointed, complete Riemannian manifold $(M_\infty, g_\infty, x_\infty)$ and a subsequence of $\{(M_k, g_k, x_k)\}_{k\in\mathbb{N}}$ which converges in the (smooth) ***Cheeger–Gromov topology*** *to $(M_\infty, g_\infty, x_\infty)$; that is, after passing to the subsequence, there exist an exhaustion $\{U_k\}_{k\in\mathbb{N}}$ of M_∞ by precompact open sets U_k such that $\overline{U}_k \subset U_{k+1}$ and $x_\infty \in U_1$ and a sequence of diffeomorphisms $\varphi_k : U_k \hookrightarrow V_k \subset M_k$ such that $\varphi_k(x_\infty) = x_k$ for each k and $\varphi_k^* g_k$ converges in the smooth topology to g_∞ uniformly on compact subsets of M_∞. The limit is unique up to pointed isometries.*

Note that the first hypothesis is superfluous if the elements of the sequence are known to be complete Riemannian manifolds. A sequence of Riemannian manifolds satisfying conditions (i)–(iii) is said to have locally uniformly **bounded geometry**.

11.1.1. Hypersurfaces with bounded geometry.

Our first step is to prove precompactness of the space of hypersurfaces with locally uniformly bounded *extrinsic* geometry. We consider two slightly different notions of convergence.

Definition 11.5. Let $X : M^n \to \mathbb{R}^{n+1}$ be a smooth immersion and let $X_k : M_k^n \to \mathbb{R}^{n+1}$ be a sequence of smooth immersions.

(1) The sequence converges **smoothly on compact subsets of** M^n to $X : M^n \to \mathbb{R}^{n+1}$ if there exist an exhaustion $\{U_k\}_{k\in\mathbb{N}}$ of M^n by open sets

U_k and a sequence of smooth embeddings $\varphi_k : U_k \to M_k^n$ such that, for each compact subset $K \subset M^n$, the sequence of immersions $\varphi_k^* X_k|_K : U_k \cap K \to \mathbb{R}^{n+1}$ converges in $C^\infty(K, \mathbb{R}^{n+1})$ to $X|_K : K \to \mathbb{R}^{n+1}$.

(2) The sequence converges **smoothly on compact subsets of** $\mathbb{R}^{n+1}$ to $X : M^n \to \mathbb{R}^{n+1}$ if, in addition, for each $A < \infty$ there exists $k_A \in \mathbb{N}$ such that $\varphi_k^* X_k(U_k) \cap \overline{B}_A = X_k(M_k^n) \cap \overline{B}_A$ for each $k \geq k_A$.

The limit $X : M^n \to \mathbb{R}^{n+1}$ is called **maximal** if, whenever the sequence $\{X_k : M_k^n \to \mathbb{R}^{n+1}\}_{k\in\mathbb{N}}$ converges smoothly on compact subsets of N^n to $Y : N^n \to \mathbb{R}^{n+1}$, there exists an embedding $\iota : N^n \hookrightarrow M^n$ such that $Y = X \circ \iota$.

Given a smoothly immersed hypersurface $\mathcal{M}^n$ of $\mathbb{R}^{n+1}$, we will also say that a sequence $\{\mathcal{M}_j^n\}_{j\in\mathbb{N}}$ of smoothly immersed hypersurfaces $\mathcal{M}_j^n$ of $\mathbb{R}^{n+1}$ converges **locally uniformly in the smooth (compact-open) topology** (respectively, **smoothly on compact subsets of** $\mathbb{R}^{n+1}$) to $\mathcal{M}^n$ if there are parametrizations $X_j : M_j^n \to \mathbb{R}^{n+1}$ for $\mathcal{M}_j^n$ and $X : M^n \to \mathbb{R}^{n+1}$ for $\mathcal{M}^n$ such that the sequence $\{X_j : M_j \to \mathbb{R}^{n+1}\}_{j\in\mathbb{N}}$ converges smoothly on compact subsets of M^n (respectively, smoothly on compact subsets of $\mathbb{R}^{n+1}$) to $X : M^n \to \mathbb{R}^{n+1}$.

An important difference between the two different notions of convergence is that the latter may require the limit to have additional connected components. Consider, for example, a sequence of simple closed curves $\Gamma_k \subset \mathbb{R}^2$ consisting of the two parallel line segments $\{(x, \pm 1) \in \mathbb{R}^2 : |x| \leq k\}$ smoothly capped off by convex arcs. Set $L_\pm \doteqdot \mathbb{R} \times \{\pm 1\}$. Then the sequence $\{\Gamma_k\}_{k\in\mathbb{N}}$ converges locally uniformly in the smooth topology to both L_+ and L_- and also to their union, whereas it converges smoothly on compact subsets of $\mathbb{R}^2$ only to the union $L_+ \cup L_-$. Moreover, even if the sequence converges smoothly on compact subsets of $\mathbb{R}^{n+1}$, the limit may fail to be maximal (consider the sequence of immersions $X_k : \underbrace{\mathbb{R} \sqcup \cdots \sqcup \mathbb{R}}_{k\text{-times}} \to \mathbb{R}^2$ which maps each factor to the x-axis in the canonical way).

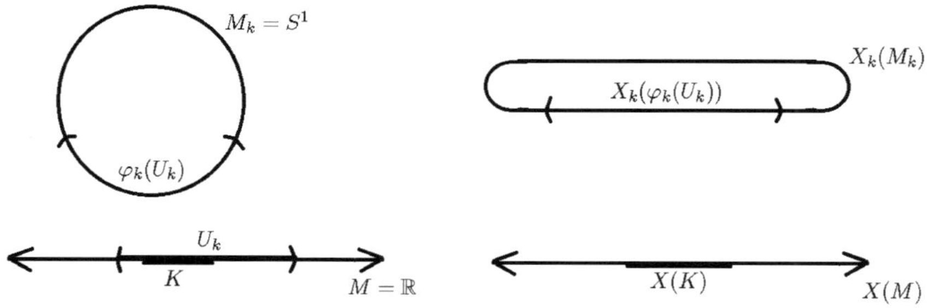

Figure 11.2. Smooth convergence on compact subsets of M.

Theorem 11.6 (Compactness theorem for immersed hypersurfaces I). *Let $X_k : M_k \to \mathbb{R}^{n+1}$ be a sequence of smooth immersions of smooth, connected manifolds M_k of dimension n and let $\{x_k\}_{k\in\mathbb{N}}$ be a sequence of points $x_k \in M_k$. Suppose that the following conditions hold:*

(i) $X_k(x_k) = 0$ *for every* $k \in \mathbb{N}$.

(ii) *For each $A < \infty$ there exists $k_A \in \mathbb{N}$ such that X_k is a proper immersion into B_A for every $k \geq k_A$, where B_A is the ball of radius A about the origin in $\mathbb{R}^{n+1}$.*

(iii) *For every $m \in \mathbb{N} \cup \{0\}$ and $A < \infty$ there exists $C_{m,A} > 0$ such that*
$$\sup_{X_k(x)\in B_A} \left|{}^{X_k}\nabla^m \mathrm{II}^{X_k}(x)\right|_{X_k} \leq C_m$$
for every $k \in \mathbb{N}$, where ${}^{X_k}\nabla$, II^{X_k}, and $|\cdot|_{X_k}$ denote, respectively, the Levi-Civita connection, second fundamental form, and norm induced by X_k.

Then there exist a smooth, connected n-manifold M^n, a smooth, proper immersion $X : M^n \to \mathbb{R}^{n+1}$, and a subsequence of $\{X_k : M_k \to \mathbb{R}^{n+1}\}_{k\in\mathbb{N}}$ which converges smoothly on compact subsets of M^n to $X : M^n \to \mathbb{R}^{n+1}$.

Proof. Our first step is to extract a subsequence of the sequence of pointed Riemannian manifolds $\{(M_k, g_k, x_k)\}_{k\in\mathbb{N}}$ which converges in the Cheeger–Gromov topology, where g_k denotes the metric induced on M_k by X_k.

First observe that, by differentiating the Gauß equation (5.41) and applying an induction argument, the extrinsic curvature (i.e., second fundamental form) derivative bounds yield uniform bounds on the Riemann curvature tensor of g_k and its covariant derivatives. By Lemma 6.19 and Klingenberg's lemma (see, for example, [**138**, Theorem III.2.4]), the injectivity radius of (M_k, g_k) at x_k is bounded from below by a positive constant which depends only on n and the uniform curvature bound. Since the intrinsic distance is bounded from below by the extrinsic distance, the conditions of the compactness theorem for pointed Riemannian manifolds are met, and we obtain a subsequence of $\{(M_k, g_k, x_k)\}_{k\in\mathbb{N}}$ which converges to a pointed Riemannian manifold $(M_\infty, g_\infty, x_\infty)$ in the Cheeger–Gromov topology. That is, passing to the convergent subsequence, there exists an exhaustion $\{U_k\}_{k\in\mathbb{N}}$ of M_∞ with $x_\infty \in U_1$ and a sequence of diffeomorphisms $\{\varphi_k : U_k \to V_k \subset M_k\}_{k\in\mathbb{N}}$ with $\varphi_k(x_\infty) = x_k$ such that $\varphi_k^* g_k$ converges smoothly to g_∞ on each compact set $K \subset M_\infty$.

To complete the proof, we apply the Arzelà–Ascoli theorem to extract a limit immersion. We claim that, for any integer $m \geq 1$ and any compact set $K \subset M_\infty$, the m-th derivative of $\varphi_k^* X_k$ is bounded on K with respect to g_∞, uniformly in k. To see this, first note that the Cheeger–Gromov convergence of the metrics yields the claim for $m = 1$. Using the Gauß equation, uniform bounds for the derivatives of the metrics $\varphi_k^* g_k$ and the assumed uniform bounds for $\sup_{\mathcal{M}_k} |^{X_k}\nabla^m \mathrm{II}^{X_k}|_{X_k}$ can be converted inductively into bounds for the derivatives of $\varphi_k^* X_k$ (cf. Lemmas 2.14 and 6.25). Since $X_k(x_k, 0) = 0$ for all k, we can now apply the Arzelà–Ascoli theorem (as in the proof of Theorem 6.20) and a diagonal subsequence argument to obtain a smooth limit map $X_\infty : M_\infty \to \mathbb{R}^{n+1}$ and a subsequence of $\{\varphi_k^* X_k\}_{k\in\mathbb{N}}$ which converges smoothly on any compact subset of M_∞ to X_∞. By the convergence to g_∞ of the corresponding subsequence of metrics $\varphi_k^* g_k$ induced by $\varphi_k^* X_k$, the limit map X_∞ must be an immersion and g_∞ its induced metric. Finally, since the convergence is uniform on compact subsets and X_k is a proper immersion into B_A for each A when k is sufficiently large, X_∞ is proper. □

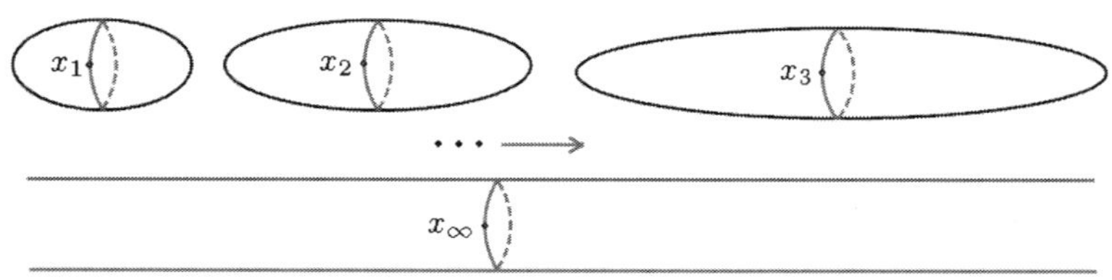

Figure 11.3. A sequence of ovaloids converging to a cylinder.

Remark 11.7. Observe that condition (i) in Theorem 11.6 can be replaced by the following weaker condition:

(i′) There exists $A < \infty$ such that $X_k(x_k) \in \overline{B}_A$ for each $k \in \mathbb{N}$.

The next result provides convergence of a subsequence on compact subsets of $\mathbb{R}^{n+1}$ assuming an additional uniform local area bound.

Theorem 11.8 (Compactness theorem for immersed hypersurfaces II). *Let $X_k : M_k \to \mathbb{R}^{n+1}$ be a sequence of smooth immersions of smooth, connected manifolds M_k of dimension n and let $\{x_k\}_{k\in\mathbb{N}}$ be a sequence of points $x_k \in M_k$. Suppose that the following conditions hold:*

(i) *$X_k(x_k) = 0$ for every $k \in \mathbb{N}$.*

(ii) *For each $A < \infty$ there exists $k_A \in \mathbb{N}$ such that X_k is a proper immersion into B_A for every $k \geq k_A$, where B_A is the ball of radius A about the origin in $\mathbb{R}^{n+1}$.*

(iii) *For every $m \in \mathbb{N} \cup \{0\}$ and $A < \infty$ there exists $C_{m,A} > 0$ such that*

$$\sup_{X_k(x)\in B_A} \left|^{X_k}\nabla^m \mathrm{II}^{X_k}(x)\right|_{X_k} \leq C_m$$

for every $k \in \mathbb{N}$, where ${}^{X_k}\nabla$, II^{X_k}, and $|\cdot|_{X_k}$ denote, respectively, the Levi-Civita connection, second fundamental form, and norm induced by X_k.

(iv) *For every $A < \infty$ there exists C_A such that*

$$\mu_k(X_k^{-1}(B_A)) < C_A$$

for every $k \in \mathbb{N}$, where μ_k is the measure induced by X_k.

Then there exist a smooth n-manifold M^n (with finitely many connected components), a smooth, proper immersion $X : M^n \to \mathbb{R}^{n+1}$, and a subsequence of $\{X_k : M_k \to \mathbb{R}^{n+1}\}_{k\in\mathbb{N}}$ which converges smoothly on compact subsets of $\mathbb{R}^{n+1}$ to $X : M^n \to \mathbb{R}^{n+1}$. Moreover, $X : M^n \to \mathbb{R}^{n+1}$ is a maximal limit.

Proof. The argument of Theorem 11.6 yields a connected manifold M^n, a proper immersion $X : M^n \to \mathbb{R}^{n+1}$, an exhaustion $\{U_k\}_{k\in\mathbb{N}}$ of M^n by open sets U_k, and a sequence of embeddings $\varphi_k : U_k \to M^n_k$ such that, for each compact subset $K \subset M^n$, the sequence of immersions $\varphi_k^* X_k : U_k \to \mathbb{R}^{n+1}$ converges in $C^\infty(K, \mathbb{R}^{n+1})$ to $X|_K : K \to \mathbb{R}^{n+1}$. We will achieve the remaining condition in each ball $\overline{B}_j$ by adding in any missing components which intersect $\overline{B}_j$ and then passing to a subsequence.

Consider first the unit ball $\overline{B}_1$. Suppose that there exists no $k_1 \in \mathbb{N}$ such that $\varphi_k^* X_k(U_k) \cap B_1 = X_k(M^n_k) \cap B_1$ for all $k \geq k_1$. Then, after passing to a subsequence, we can find for each $k \in \mathbb{N}$ a point $x_k \in M_k \setminus \varphi_k(U_k)$ such that $X_k(x_k) \in B_1$. By Theorem 11.6 (note Remark 11.7), we can find a connected manifold N, a proper immersion $Y : N \to \mathbb{R}^{n+1}$, an exhaustion $\{V_k\}_{k\in\mathbb{N}}$ of N, and a sequence of embeddings $\psi_k : V_k \to M_k$ such that the sequence of immersions $\psi_k^* X_k : V_k \to \mathbb{R}^{n+1}$ converges smoothly on each compact subset $K \subset N$ to $Y|_K : K \to \mathbb{R}^{n+1}$. Now we replace M with $M \sqcup N$, U_k with $U_k \sqcup V_k$, and φ_k with the corresponding embedding of $U_k \sqcup V_k$ into M_k and repeat the process. Since each of the additional components intersects $\overline{B}_1$ and has area inside B_2 bounded from below, the local area bound implies that this process must stop after finitely may steps. The resulting data are a manifold M^n with finitely many connected components, an exhaustion $\{U_k\}_{k\in\mathbb{N}}$ of M^n, and embeddings $\varphi_k : U_k \to M_k$ satisfying $\varphi_k^* X_k(U_k) \cap \overline{B}_1 = X_k(M_k) \cap \overline{B}_1$. By a similar argument, if we have a subsequence which converges to a limit $X : M^n \to \mathbb{R}^{n+1}$ locally uniformly in compact subsets of M^n with data satisfying the second convergence criterion on $\overline{B}_j$, then we can add in finitely many more components if there are points in $\overline{B}_{j+1}$ that are in $X_k(M^n_k)$ but not in $\varphi_k^* X_k(U_k)$ to obtain a subsequence satisfying the second convergence criterion on $\overline{B}_{j+1}$. Passing to a "diagonal" subsequence, we obtain a subsequence satisfying both convergence criteria.

A similar argument can be used to arrange that the limit is maximal. □

The arguments presented here are essentially those given by Baker [**81**]. Similar results were proved by Cooper [**190**] and Breuning [**118**] using local graph parametrizations, generalizing a method of Langer [**349**]. It is worth noting that straightforward modifications of the arguments yield similar conclusions for immersed submanifolds of a given Riemannian manifold. In fact, one can even allow the target Riemannian manifold to vary. See [**190**].

11.1.2. Mean curvature flows with bounded curvature. Theorems 11.6 and 11.8 can be used to deduce analogous precompactness criteria for mean curvature flows. These ideas originate in Hamilton's extension of the Cheeger–Gromov compactness theorem to Ricci flows [**267**].

Recall that the parabolic cylinders $P_r(p,t) \subset \mathbb{R}^{n+1} \times \mathbb{R}$ are defined by $P_r(p,t) \doteqdot B_r(p) \times (t-r^2, t]$.

Definition 11.9. A solution $X : M^n \times I \to \mathbb{R}^{n+1}$ to mean curvature flow is **properly defined** in $P_r(p_0, t_0)$ if $(t_0 - r^2, t_0] \subset I$ and $X_t^{-1}(K)$ is compact for each compact subset $K \subset B_r(p_0)$ and each $t \in (t_0 - r^2, t_0]$; $X : M^n \times I \to \mathbb{R}^{n+1}$ is a **proper solution** if X_t is proper for each $t \in I$.

Definition 11.10. Let $X : M^n \times I \to \mathbb{R}^{n+1}$ be a smooth solution to mean curvature flow and let $X_k : M_k^n \times I_k \to \mathbb{R}^{n+1}$ be a sequence of smooth solutions to mean curvature flow.

(1) The sequence converges **smoothly on compact subsets of** $M^n \times I$ to $X : M^n \times I \to \mathbb{R}^{n+1}$ if there is an exhaustion $\{U_k\}_{k\in\mathbb{N}}$ of M^n by open sets U_k, an exhaustion $\{J_k\}_{k\in\mathbb{N}}$ of I by compact intervals $J_k \subset I_k$, and a sequence of smooth embeddings $\varphi_k : U_k \to M_k^n$ such that, for each compact subset $K \subset M^n \times I$, the sequence of maps $\varphi_k^* X_k|_K : (U_k \times J_k) \cap K \to \mathbb{R}^{n+1}$ converges in $C^\infty(K, \mathbb{R}^{n+1})$ to $X|_K : K \to \mathbb{R}^{n+1}$, where $\varphi_k^* X_k(x,t) \doteqdot X_k(\varphi_k(x), t)$.

(2) The sequence converges **smoothly on compact subsets of** $\mathbb{R}^{n+1} \times \mathbb{R}$ to $X : M^n \times I \to \mathbb{R}^{n+1}$ if, in addition, for each $A < \infty$ there exists $k_A \in \mathbb{N}$ such that $\varphi_k^* X_k(U_k \times J_k) \times J_k \cap \overline{P}_A = X_k(M_k^n \times I_k) \times I_k \cap \overline{P}_A$ for $k \geq k_A$.

The limit $X : M^n \times I \to \mathbb{R}^{n+1}$ is called **maximal** if, whenever the sequence $\{X_k : M_k^n \times I_k \to \mathbb{R}^{n+1}\}_{k\in\mathbb{N}}$ converges smoothly on compact subsets of $N^n \times J$ to $Y : N^n \times J \to \mathbb{R}^{n+1}$, then $J \subset I$ and there exists an embedding $\iota : N^n \hookrightarrow M^n$ such that $Y_t = X_t \circ \iota$ for each $t \in J$.

We will also say that a sequence of mean curvature flows $\{\mathcal{M}_{t,j}^n\}_{t\in I_j}$ converges to a mean curvature flow $\{\mathcal{M}_t^n\}_{t\in I}$ **smoothly on compact subsets of** $\{\mathcal{M}_t^n\}_{t\in I}$ (respectively, **smoothly on compact subsets of** $\mathbb{R}^{n+1} \times I$) if there are parametrizations $X_j : M_j^n \times I_j \to \mathbb{R}^{n+1}$ for $\{\mathcal{M}_{t,j}^n\}_{t\in I_j}$ and $X : M^n \times \mathbb{R} \to \mathbb{R}^{n+1}$ for $\{\mathcal{M}_t\}_{t\in I}$ such that the sequence

$\{X_j : M_j \times I_j \to \mathbb{R}^{n+1}\}_{j\in\mathbb{N}}$ converges to $X : M^n \times I \to \mathbb{R}^{n+1}$ smoothly on compact subsets of $M^n \times I$ (respectively, smoothly on compact subsets of $\mathbb{R}^{n+1} \times \mathbb{R}$).

Theorem 11.11 (Compactness theorem for mean curvature flows I). *Let $\{X_k : M_k^n \times I_k \to \mathbb{R}^{n+1}\}_{k\in\mathbb{N}}$ be a sequence of solutions to mean curvature flow and let $\{x_k\}_{k\in\mathbb{N}}$ be a sequence of points $x_k \in M_k^n$. Suppose that:*

(i) $0 \in I_k$ *and* $X_k(x_k, 0) = 0$ *for every* $k \in \mathbb{N}$.

(ii) *For each* $A < \infty$ *there exists* $k_A \in \mathbb{N}$ *such that* X_k *is properly defined in* $B_A \times (-A^2, 0]$ *for every* $k \geq k_A$, *where* B_A *is the ball of radius* A *about the origin in* $\mathbb{R}^{n+1}$.

(iii) *For* $A > 0$ *there exist* $C_{0,A} < \infty$ *and* $k_A \in \mathbb{N}$ *such that for* $k \geq k_A$,

$$\sup_{(X_k(x,t),t)\in B_A\times(-A^2,0]} |\mathrm{II}_k(x,t)| \leq C_{0,A}\,.$$

Then there exist a smooth, connected n-manifold M_∞^n, a smooth, proper solution $X_\infty : M_\infty^n \times (-\infty, 0] \to \mathbb{R}^{n+1}$ to mean curvature flow, and a subsequence of $\{X_k : M_k^n \times I_k \to \mathbb{R}^{n+1}\}_{k\in\mathbb{N}}$ which converges smoothly on compact subsets of $M_\infty^n \times (-\infty, 0]$ to $X_\infty : M_\infty \times (-\infty, 0] \to \mathbb{R}^{n+1}$.

Proof. By the interior estimates of Ecker and Huisken (Theorem 7.14), the local uniform bounds for II_k yield bounds

$$\sup_{B_{A/2}\times(-A^2/4,0]} \left|{}^{X_k}\nabla^m \mathrm{II}_k\right|_{X_k} < C_{m,A}$$

for each $m \in \mathbb{N}$ and $A > 0$ for all k sufficiently large. In particular, we may apply Theorem 11.6 to the sequence of immersions $X_{k,0} \doteqdot X_k(\cdot, 0)$ to extract a smooth, proper immersion $X_{\infty,0} : M_\infty \to \mathbb{R}^{n+1}$ of a smooth manifold M_∞, a point $x_\infty \in M_\infty$, an exhaustion $\{U_k\}_{k\in\mathbb{N}}$ of M_∞ by precompact open sets $U_k \subset M_\infty$ satisfying $\overline{U} \subset U_{k+1}$, and a sequence of smooth diffeomorphisms $\varphi_k : U_k \to V_k \subset M_k$ such that $\varphi_k(x_\infty) = x_k$ for each k and, after passing to a subsequence, $\varphi_k^* X_{k,0}$ converges to $X_{\infty,0}$ smoothly on compact subsets of M_∞. Let us now define the diffeomorphisms

$$\begin{aligned} \psi_k : U_k \times I_k &\to V_k \times I_k\,, \\ (x,t) &\mapsto (\varphi_k(x), t)\,. \end{aligned}$$

As in Section 6.8, we can translate the local uniform curvature derivative estimates for X_k into local uniform bounds for the derivatives of $\psi_k^* X_k$ to all orders in space and time. By the Arzelà–Ascoli theorem and a diagonal subsequence argument, there exists a smooth map $X_\infty : M_\infty \times (-\infty, 0] \to \mathbb{R}^{n+1}$ such that, after passing to a further subsequence, the maps $\psi_k^* X_k$ converge to X_∞ smoothly on compact subsets of $M_\infty \times (-\infty, 0]$. We have seen that uniform bounds for the curvature imply uniform bounds from above and below for the metrics g_k, from which we conclude that $X_\infty(\cdot, t)$ is an immersion for each t. Since the convergence is uniform on compact subsets, $X_\infty(\cdot, t)$ is proper for each t. Finally, since the convergence is smooth, X_∞ satisfies mean curvature flow. □

Using Theorem 11.8, we obtain convergence in the stronger sense if we assume a local area bound at each time.

Combining the local area bound of Lemma 10.8 with Theorem 11.8 we obtain the following compactness result.

Theorem 11.12 (Compactness theorem for mean curvature flows II). *Let $\{X_k : M_k^n \times [-T_k, 0] \to \mathbb{R}^{n+1}\}_{k\in\mathbb{N}}$ be a sequence of compact solutions to mean curvature flow and let $\{x_k\}_{k\in\mathbb{N}}$ be a sequence of points $x_k \in M_k^n$. Suppose that the following conditions are met:*

(i) $X_k(x_k, 0) = 0$ *for every* $k \in \mathbb{N}$.

(ii) *For every $A > 0$ there exist $C_{0,A} < \infty$ and $k_A \in \mathbb{N}$ such that*

$$\sup_{(X_k(x,t),t)\in B_A\times(-A^2,0]} |\mathrm{II}_k(x, t)| \leq C_{0,A}$$

for all $k \geq k_A$.

(iii) *For every $A > 0$ there exists $C_A < \infty$ such that*

$$\mu_{k,-T_k}(X_{k,-T_k}^{-1}(B_A)) \leq C_A$$

for all $k \in \mathbb{N}$, where $\mu_{k,-T_k}$ is the Riemannian measure induced by the initial immersion $X_{k,-T_k} \doteqdot X_k(\cdot, -T_k)$.

Then there exist a smooth n-manifold M_∞^n, a smooth, proper solution $X_\infty : M_\infty^n \times (-\infty, 0] \to \mathbb{R}^{n+1}$ to mean curvature flow, and a subsequence of $\{X_k : M_k^n \times I_k \to \mathbb{R}^{n+1}\}_{k\in\mathbb{N}}$ which converges smoothly on compact subsets of $\mathbb{R}^{n+1} \times (-\infty, 0]$ to $X_\infty : M_\infty^n \times (-\infty, 0] \to \mathbb{R}^{n+1}$.

Proof. Since, by Lemma 10.8, the initial local area bounds imply local area bounds at later times, the theorem follows as in the proof of Theorem 11.11 using Theorem 11.8 instead of Theorem 11.6. □

11.2. Neck detection

We can use the compactness theorem in conjunction with the cylindrical and gradient estimates to prove that 2-convex solutions to mean curvature flow either form "necks" or else become locally uniformly convex in regions of very high curvature. This is a crucial step in the construction of mean curvature flow with surgery (cf. [**303**, Lemma 7.4 and Corollary 7.7]). In fact, we will prove a more general statement.

Remark 11.13. Neck detection in 3-dimensional Ricci flow was pioneered by Richard Hamilton; see [**268**, §24] and [**271**]. A deeper understanding of singularity formation in 3-dimensional Ricci flow was obtained by Perelman in [**438**] for his proof of Thurston's geometrization conjecture [**439**].

Recall from Definition 9.19 that, given $R > 0$ and $\alpha = (\alpha_0, \alpha_1, \alpha_2) \in (0,\infty)^3$, the class $\mathcal{C}^n_m(R,\alpha)$ consists of all compact, mean convex immersed hypersurfaces $X : M^n \to \mathbb{R}^{n+1}$ satisfying

$$\kappa_1 + \cdots + \kappa_{m+1} \geq \alpha_0 H\,, \quad H \geq \alpha_1 R^{-1}\,, \quad \text{and} \quad \mu(M^n) \leq \alpha_2 R^n\,. \tag{11.1}$$

Lemma 11.14 (Neck detection: Curvature necks (cf. [**303**, Lemma 7.4])). *Given an integer $n \geq 3$ and positive constants α_0, α_1, α_2 and $\varepsilon \leq \frac{1}{100}$, there exist $\varepsilon_\sharp > 0$ and $h_\sharp > 0$ with the following property: Let $X : M^n \times [0,T) \to \mathbb{R}^{n+1}$ be a solution to mean curvature flow with initial condition in the class $\mathcal{C}^n_m(R,\alpha)$, with $\max_{M^n\times\{0\}} |\mathrm{II}| \leq R^{-1}$ and $m < \frac{2(n-1)}{3}$. Suppose that $(x_0,t_0) \in M^n \times [0,T)$ is a **high curvature, m-necklike point of quality** $(h_\sharp, \varepsilon_\sharp)$. That is,*

$$H(x_0,t_0) > H_\sharp \doteqdot h_\sharp R^{-1} \quad \textit{and} \quad \frac{\kappa_1 + \cdots + \kappa_m}{H}(x_0,t_0) \leq \varepsilon_\sharp\,. \tag{11.2}$$

*Then (x_0,t_0) lies at the center of a **shrinking curvature m-neck of quality** ε. That is, we then have*

$$A_{r_0,k,\varepsilon}(x_0,t_0) \leq \varepsilon r_0^{k+1}$$

for each $k = 0, \ldots, \lfloor \frac{2}{\varepsilon} \rfloor$, where $r_0 \doteqdot \frac{n-m}{H(x_0,t_0)}$,

$$A_{r,0,\varepsilon}(x,t) \doteqdot \max_{\mathcal{B}_{\varepsilon^{-1}r}(x,t)\times(t-10^4r^2,t]} \sqrt{\sum_{i=1}^{m} \kappa_i^2 + \sum_{j=m+1}^{n} (\kappa_n - \kappa_j)^2},$$

and, for each $k \geq 1$,

$$A_{r,k,\varepsilon}(x,t) \doteqdot \max_{\mathcal{B}_{\varepsilon^{-1}r}(x,t)\times(t-10^4r^2,t]} |\nabla^k \mathrm{II}|\,.$$

Proof. The proof is essentially that of [**303**, Lemma 7.4]. Suppose that the claim does not hold. Then there must exist $n \geq 3$, parameters α_0, α_1, and α_2, some $\varepsilon_0 < \frac{1}{100}$, a sequence $X_j : M^n_j \times [0,T_j) \to \mathbb{R}^{n+1}$ of solutions to mean curvature flow with $X_j(\cdot,0) \in \mathcal{C}^n_m(R_j,\alpha)$, where $R_j^{-1} = \max_{M^n_j\times\{0\}} |\mathrm{II}_j|$ and $m < \frac{2(n-1)}{3}$, and points $(x_j,t_j) \in M^n_j \times [0,T_j)$ such that

$$H_j(x_j,t_j) \geq (n-m) j R_j^{-1} \ \text{ and } \ \frac{\kappa_1^j + \cdots + \kappa_m^j}{H_j}(x_j,t_j) \leq j^{-1}\,,$$

and yet

$$A^j_{k,\varepsilon_0} \geq \varepsilon_0 r_j^{k+1} \tag{11.3}$$

for some $k \leq \lfloor \varepsilon_0^{-1} \rfloor$ for each j, where $r_j \doteqdot \frac{n-m}{H_j(x_j,t_j)}$ and we denote objects defined along X_j using the sub- or superscript j. After passing to a subsequence, we may arrange that (11.3) holds for some fixed $k = k_0 \leq \lfloor \varepsilon_0^{-1} \rfloor$ for all j. After translating the points $(X_j(x_j,t_j), t_j)$ to the space-time origin and parabolically rescaling by r_j, we obtain the sequence of flows $\hat{X}_j : M_j^n \times [-r_j^{-2}t_j, r_j^{-2}(T_j - t_j)) \to \mathbb{R}^{n+1}$ given by

$$\hat{X}_j(x,t) \doteqdot r_j^{-1}(X_j(x, r_j^2 t + t_j) - X_j(x_j,t_j)) . \tag{11.4}$$

Each $\hat{X}_j$ is in the class $\mathcal{C}_m^n(r_j^{-1}R_j, \alpha)$ and satisfies $\hat{X}_j(x_j, 0) = 0$,

$$\hat{H}_j(x_j,0) = (n-m) \quad \text{and} \quad \frac{\hat{\kappa}_1^j + \cdots + \hat{\kappa}_m^j}{\hat{H}_j}(x_j,0) \leq j^{-1}, \tag{11.5}$$

but

$$\hat{A}^j_{k_0,\varepsilon_0} \geq \varepsilon_0 , \tag{11.6}$$

where we denote objects defined along $\hat{X}_j$ using a $\hat{\cdot}$ and the sub- or superscript j. We claim that the new sequence subconverges locally uniformly in the smooth topology to a shrinking m-cylinder solution, in contradiction with (11.6). First note that $R_j^{-2}t_j$ is bounded uniformly away from zero for j sufficiently large. Indeed, by an ODE comparison argument (cf. Lemma 9.20),

$$\max_{M_j^n \times \{t\}} |\mathrm{II}_j|^2 \leq \frac{1}{\max_{M_j^n \times \{0\}} |\mathrm{II}_j|^{-2} - 2t} .$$

Since $|\mathrm{II}_j|^2(x_j,t_j) \geq \frac{1}{n}H^2(x_j,t_j) \geq \frac{j^2}{n}R_j^{-2}$, whereas $\max_{M_j^n \times \{0\}} |\mathrm{II}_j|^2(0) = R_j^{-2}$, we obtain

$$2R_j^{-2}t_j \geq 1 - nj^{-2}$$

as claimed. In particular, $-r_j^{-2}t_j \to -\infty$ as $j \to \infty$. We claim that the mean curvature of $\hat{X}_j$ is uniformly bounded on a parabolic cylinder of uniform radius about $(x_j, 0)$, so long as j is sufficiently large. Indeed, by the gradient estimates of Theorem 9.29, we can find constants $c_\sharp$ (depending only on n) and C (depending only on n and α) such that

$$|\hat{\nabla}_j \hat{H}_j| \leq \frac{c_\sharp}{2}\hat{H}_j^2 + Cj^{-2} \quad \text{and} \quad |\partial_t \hat{H}_j| \leq \frac{c_\sharp^2}{4}\hat{H}_j^3 + Cj^{-3}$$

in $M_j^n \times [-r_j^{-2}t_j + j^{-2}/4, 0]$. Thus, given any $\rho > 0$, we can find $j_0 \in \mathbb{N}$ such that

$$|\hat{\nabla}_j \hat{H}_j| \leq c_\sharp \hat{H}_j^2 \quad \text{and} \quad |\partial_t \hat{H}_j| \leq \frac{c_\sharp^2}{2}\hat{H}_j^3$$

in $M_j^n \times [-\rho, 0]$ for $j \geq j_0$. Lemma 9.28 now implies that

$$\frac{n-m}{10} = \frac{\hat{H}_j(x_j,0)}{10} \leq \hat{H}_j(y,s) \leq 10\hat{H}_j(x_j,0) = 10(n-m)$$

for any $(y,s) \in \mathcal{P}^j_{\frac{1}{10c_\sharp}}(x_j,0)$ for all sufficiently large j. It follows (cf. Section 11.1) that some subsequence of the restricted mean curvature flows $\hat{X}_j|_{\mathcal{P}^j_{\frac{1}{10c_\sharp}}(x_j,0)}$ converges locally uniformly in the smooth topology to a limiting mean curvature flow $\hat{X} : U \times (-\frac{1}{100c_\sharp^2}, 0] \to \mathbb{R}^{n+1}$ (which may not be compact or proper). We claim that the limit flow is part of a shrinking cylinder. We shall denote objects defined along the limit using a $\hat{\cdot}$. Indeed, by the convexity estimate (Corollary 9.13) and the cylindrical estimate (Corollary 9.14), $\hat{X}$ satisfies

$$\hat{\kappa}_1 \geq 0 \quad \text{and} \quad \sum_{j=m+1}^{n} (\hat{\kappa}_n - \hat{\kappa}_j) \leq \sum_{i=1}^{m} \hat{\kappa}_i\,. \tag{11.7}$$

By (11.5), the principal curvatures $\hat{\kappa}_i$ vanish at the origin for $i = 1, \ldots, m$. Thus, by the splitting theorem (Theorem 9.11), $\hat{X}$ splits locally off an m-plane. Since $\hat{\kappa}_{m+1} = \hat{\kappa}_1 + \cdots + \hat{\kappa}_{m+1} \geq \alpha_0 \hat{H}$, (11.7) implies that the cross section of the splitting is umbilic, so the claim follows from Exercise 5.5. We need to extend the convergence to a sufficiently large region. This can be achieved since, a posteriori, the mean curvature could not have increased very much in $\mathcal{P}^j_{\frac{1}{10c_\sharp}}$ (due to the convergence to a shrinking cylinder solution). That is,

$$\hat{H}_j(y,s) \leq 2(n-m)$$

for all $(y,s) \in \mathcal{P}_{\frac{1}{10c_\sharp}}(x_j,0)$ so long as j is sufficiently large. Applying the gradient estimates as before, we obtain uniform bounds for $\hat{H}_j$ on the uniformly larger neighborhood $\mathcal{P}^j_{\frac{2}{10c_\sharp}}(x_j,0)$. Repeating the previous argument, we conclude that a subsequence of the flows $\hat{X}_j|_{\mathcal{P}^j_{\frac{2}{10c_\sharp}}(x_j,0)}$ converge to a part of a shrinking cylinder. After repeating the argument a finite number of times, we obtain convergence of a subsequence of the flows $\hat{X}_j|_{\mathcal{P}^j_{2\varepsilon_0^{-1}}(x_j,0)}$ to a part of a shrinking cylinder. Since the convergence is smooth on compact subsets, this contradicts (11.6). □

Next, we show that it is possible to "integrate" a sufficiently high quality shrinking curvature neck to obtain a region which can be represented as a graph over a shrinking cylinder.

Corollary 11.15 (Neck detection: Geometric necks (cf. [**303**, Propositions 3.4 and 3.5])). *Given an integer $n \geq 3$ and positive constants α_0, α_1, α_2, and $\varepsilon > 0$, there exist $\varepsilon_\sharp > 0$ and $h_\sharp > 0$ with the following property:*

Let $X : M^n \times [0,T) \to \mathbb{R}^{n+1}$ be a solution to mean curvature flow with initial condition in the class $C^n_m(R,\alpha)$, with $\max_{M^n\times\{0\}} |\mathrm{II}| \leq R^{-1}$ and $m < \frac{2(n-1)}{3}$. Suppose that (x_0,t_0) is a high curvature, m-necklike point of quality $(h_\sharp, \varepsilon_\sharp)$ (in the sense of (11.2)). Then (x_0,t_0) lies at the center of a ***shrinking** m**-neck of quality** ε.*

That is, there is a function $u : S^{n-m} \times \mathbb{R}^m \times (-\infty, 0] \to \mathbb{R}$ (with $C^{\lfloor 2\varepsilon^{-1}\rfloor,\lfloor\varepsilon^{-1}\rfloor}$-norm bounded by ε) such that, after translating $(X(x_0,t_0),t_0)$ to the space-time origin, applying a rotation, and parabolically rescaling by $r_0 \doteqdot \frac{n-m}{H(x_0,t_0)}$, the image of the intrinsic parabolic cylinder $\mathcal{B}_{\varepsilon^{-1}}(x_0,0) \times (-1,0]$ lies in the normal graph of u over the shrinking cylinder $\{\mathbb{R}^m \times S^{n-m}_{\sqrt{1-2(n-m)t}}\}_{t\in(-1,0]}$.

Proof. By Lemma 11.14, we can find, for any $\delta > 0$, constants $\varepsilon_\sharp$ and $h_\sharp$ (depending on n, α, and δ) such that any point $(x_0,t_0) \in M^n \times [0,T)$ satisfying

$$H(x_0,t_0) > h_\sharp R^{-1} \quad \text{and} \quad \frac{\kappa_1 + \cdots + \kappa_m}{H}(x_0,t_0) \leq \varepsilon_\sharp$$

lies at the center of a shrinking curvature m-neck of quality δ. Up to a space-time translation and a parabolic rescaling, we may assume that $t_0 = 0$, $X(x_0,0) = 0$, and $H(x_0,0) = n-m$. Then $|\nabla H| \leq c\delta$, and

$$\begin{aligned}|\partial_t H| = |\Delta H + |\mathrm{II}|^2 H| &\leq n^2\delta + \left(\frac{1}{n-m} + c\delta^2\right) H^3 \\ &\leq \tfrac{1}{n-m} H^3 + c\delta\left(1 + \delta H^3\right)\end{aligned}$$

in $\mathcal{B}_{\delta^{-1}}(x_0,0) \times (-10^4, 0]$, where c depends only on n and α_0. Integrating yields

$$\frac{H(x_0,0)}{10} \leq H(y,s) \leq 10H(x_0,0)$$

for $(y,s) \in \mathcal{B}_{(10\delta)^{-1}}(x_0,0) \times (-100,0]$ for $\delta \leq \delta_0(n)$ sufficiently small. Choosing δ_0 even smaller if necessary, we deduce that

$$|\kappa_i| \leq \frac{1}{100} \quad \text{for each } i = 1, \ldots, m,$$

$$|\kappa_j - 1| \leq \frac{1}{100} \quad \text{for each } i = m+1, \ldots, n$$

in $\mathcal{B}_{(10\delta)^{-1}}(x_0,0) \times (-100,0]$. It follows that the spatial tangent bundle to $\mathcal{B}_{(10\delta)^{-1}}(x_0,0) \times (-100,0]$ splits as a direct sum $E^m \otimes F^{n-m}$ where E^m has rank m, F^{n-m} has rank $n-m$, and

$$|\mathrm{II}(u,u)|\leq\frac{1}{100}\quad\text{for each }u\in E^m\text{ with }|u|=1 \tag{11.8a}$$

and

$$|\mathrm{II}(v,v)-1|\leq\frac{1}{100}\quad\text{for each }v\in F^{n-m}\text{ with }|v|=1. \tag{11.8b}$$

Given a principal frame $\{e_i^0\}_{i=1}^n$ at $(x_0,0)$, let $\overline{X}:\mathbb{R}^m\times S^{n-m}\times(-1,0]\to\mathbb{R}^{n+1}$ be the shrinking cylinder whose time 0 slice has radius 1, touches the time 0 slice of X at the origin (with the same orientation), and satisfies $d\overline{X}_{(\overline{x}_0,0)}(T\mathbb{R}^m\oplus\{0\})=dX_{(x_0,0)}(E^m\oplus\{0\})$, where $\overline{X}(\overline{x}_0,0)=0$. Let $\{e_i\}_{i=1}^n$ be the local g_0-orthonormal frame obtained from $\{e_i^0\}_{i=1}^n$ by parallel translation in space along radial g_0-geodesics and set $\mathsf{E}_i\doteqdot dX(e_i)$ for each $i=1,\dots,n$. Then, along a g_0-geodesic $\gamma_\xi(s)=\exp_{(x_0,0)}(s\xi^ie_i^0)$, $\xi\in S^{n-1}$, the frame $\{\mathsf{E}_i,\mathsf{N}\}_{i=1}^n$ satisfies the system

$$\frac{d}{ds}(\mathsf{E}_i\circ\gamma_\xi)=\sum_{p=1}^n\xi^p(\mathrm{II}_{ip}\circ\gamma_\xi)(\mathsf{N}\circ\gamma_\xi)\,,$$

$$\frac{d}{ds}(\mathsf{N}\circ\gamma_\xi)=\sum_{i,p=1}^n\xi^p(\mathrm{II}_{ip}\circ\gamma_\xi)(\mathsf{E}_i\circ\gamma_\xi)\,.$$

Similarly, the frame $\{\overline{\mathsf{E}}_i,\overline{\mathsf{N}}\}_{i=1}^n$, where $\overline{\mathsf{E}}_i\doteqdot d\overline{X}(\overline{e}_i)$, on the shrinking cylinder $\overline{X}$ obtained from the basis $\{\overline{e}_i^0\}_{i=1}^n$ satisfying $d\overline{X}_{(\overline{x}_0,0)}\overline{e}_i=dX_{(x_0,0)}e_i$ by parallel translation along the radial $\overline{g}_0$-geodesic $\overline{\gamma}_\xi(s)\doteqdot\overline{\exp}_{(\overline{x}_0,0)}(s\xi^i\overline{e}_i^0)$ satisfies the system

$$\frac{d}{ds}(\overline{\mathsf{E}}_i\circ\overline{\gamma}_\xi)=\sum_{p=1}^n\xi^p(\overline{\mathrm{II}}_{ip}\circ\overline{\gamma}_\xi)(\overline{\mathsf{N}}\circ\overline{\gamma}_\xi),$$

$$\frac{d}{ds}(\overline{\mathsf{N}}\circ\overline{\gamma}_\xi)=\sum_{i,p=1}^n\xi^p(\overline{\mathrm{II}}_{ip}\circ\overline{\gamma}_\xi)(\overline{\mathsf{E}}_i\circ\overline{\gamma}_\xi)\,.$$

In view of the gradient estimate $|\nabla\mathrm{II}|\leq\delta$, we find

$$|\mathrm{II}_{ij}(\gamma_\xi(s),0)-\mathrm{II}_{ij}(x_0,0)|\leq\delta s$$

for each $i,j=1,\dots,n$. Since

$$|\mathrm{II}_{ij}(x_0,0)-\overline{\mathrm{II}}_{ij}(\overline{x}_0,0)|\leq\delta$$

for each $i,j=1,\dots,n$, we deduce, for any $L>1000$ and $\delta\leq\delta_0(L,n)$, that

(11.9)

$$\max_{\xi\in S^{n-1}}\left\{\sum_{i=1}^n\left|\mathsf{E}_i(\gamma_\xi(s))-\overline{\mathsf{E}}_i(\overline{\gamma}_\xi(s))\right|+\left|\mathsf{N}(\gamma_\xi(s))-\overline{\mathsf{N}}(\overline{\gamma}_\xi(s))\right|\right\}\leq\delta C(\mathrm{e}^{cs}-1)$$

for $x \in \mathcal{B}_L(x_0, 0)$, where c depends only on n and C depends on n and L. The hypersurface $\mathcal{M}_0$ can now be recovered from the frame by integration starting at x_0, which, combined with the higher-derivative estimates, yields

$$(11.10) \qquad \left|(X \circ \gamma_\xi)(s) - (\overline{X} \circ \overline{\gamma}_\xi)(s)\right|_{C^{\lfloor \delta^{-1} \rfloor + 2}} \leq C\delta$$

for $s \leq L$, where C depends on n and L.

To show that $\mathcal{M}_0$ closes up as a cylinder, consider the coordinates $\{y_i\}_{i=1}^{n+1}$ on $\mathbb{R}^{n+1}$ associated with the orthonormal frame $\{\mathsf{E}_i^0\}_{i=1}^{n+1}$, where $\mathsf{E}_i^0 \doteqdot \mathsf{E}_i(x_0)$ for each $i = 1, \dots, n$ and $\mathsf{E}_{n+1}^0 \doteqdot \mathsf{N}(x_0, 0)$. Given $h = (h_1, \dots, h_m) \in \mathbb{R}^m$, define the level set $\Sigma_h^{n-m} \doteqdot \{x \in M^n : \hat{y}(X(x,0)) = h\}$, where $\hat{y} \doteqdot (y_1, \dots, y_m)$. For $s \leq L$, (11.9) implies that

$$\sup_{\xi \in S^{n-1}} \left|\mathsf{E}_i(\gamma_\xi(s)) - \mathsf{E}_i^0\right| \leq c(n, L)\delta$$

in $\mathcal{B}_L(x_0, 0)$. Thus, for $\delta \leq \delta_0(n, L)$ sufficiently small, the implicit function theorem implies that $\Sigma_h^{n-m} \cap \mathcal{B}_L$ is a smooth codimension m submanifold of M^n. Observe that (11.8a), (11.8b), and the derivative bound for II ensure that $\Sigma_h^{n-m} \cap \mathcal{B}_L$ is uniformly convex. Since $n - m > 1 + \frac{m}{2} > 1$, Myers's theorem implies that Σ_h is a smooth embedding of the $(n-m)$-sphere with intrinsic diameter bounded by 10 (cf. the proof of Corollary 8.16). To write $\mathcal{M}_0^n$ as a cylindrical graph, let, for each $x \in \Sigma_h^{n-m}$,

$$\begin{aligned} z_h(x) &\doteqdot X(x,0) - \frac{n-m}{H_h(x,0)} \mathsf{N}_h(x,0) \\ &= (h, y_{m+1}(X(x,0)), \dots, y_{n+1}(X(x,0))) - \frac{n-m}{H_h(x,0)} \mathsf{N}_h(x,0) \end{aligned}$$

be an approximate center for Σ_h^{n-m} as seen from x, where H_h and N_h are the mean curvature and normal, respectively, of Σ_h^{n-m} as a hypersurface in $\mathbb{E}^{n-m+1}$. Observe that

$$|\nabla^h z_h(x)| \leq C(n, L)\delta$$

in view of the bounds for II and its derivatives, where ∇^h is the Levi-Civita connection of Σ_h^{n-m}. It follows, for δ sufficiently small, that the image of Σ_h^{n-m} under $X(\cdot, 0)$ can be written as the graph of a function $u(h, \cdot) : S_1^{n-m}(z_h(x)) \to \mathbb{R}$ over the unit sphere $S_1^{n-m}(z_h(x))$ in $\mathbb{E}^{n-m+1}$ satisfying

$$|u(h, \cdot)|_{C^{\lfloor \delta^{-1} \rfloor + 2}} \leq C\delta,$$

where C depends on n and L. Since

$$|\partial_{h^i} z_h - \mathsf{E}_i| \leq C\delta,$$

we conclude that the image of $\Sigma^n \cap \mathcal{B}_L$ under $X(\cdot, 0)$ can be written as the graph of a function $u : \mathbb{R}^m \times S_1^{n-m}(z_0(x_0)) \to \mathbb{R}$ over the unit cylinder

$\mathbb{R}^m \times S_1^{n-m}(z_0(x_0))$ which satisfies

$$|u|_{C^{\lfloor \delta^{-1} \rfloor + 2}} \le C\delta\,,$$

where C depends on n and L.

Since

$$|\partial_t \left\langle X - z_0, \overline{N} \right\rangle| \le -C\delta t$$

in $\mathcal{B}_L(x_0, 0) \times (-100, 0]$, where the constant C depends on n and L, the graphical parametrization can be extended in time with fixed decay on a fixed time interval. This completes the proof. □

For 2-convex mean curvature flows, the following lemma allows us to deal with singularities where κ_1/H is not necessarily small [**303**, Lemma 6.6 and Theorem 7.14].

Lemma 11.16. *Let $X : M^n \to \mathbb{R}^{n+1}$, $n \ge 2$, be a smooth, connected, immersed, mean convex hypersurface. Suppose that there exist $c_\sharp$ and $H_\sharp$ such that*

$$H(x) \ge H_\sharp \quad \textit{implies} \quad \frac{|\nabla H|}{H^2}(x) \le c_\sharp\,.$$

For every $\eta > 0$, there exist $\alpha > 0$ and $\gamma > 1$ (depending only on $c_\sharp$ and η) with the following property: Suppose that

$$\frac{\kappa_1}{H}(x) \ge \eta \textit{ and } H(x) \ge \gamma H_\sharp\,.$$

Then $H(z) \ge \gamma^{-1} H(x)$ for all $z \in \mathcal{B}_{\frac{\alpha}{H(x)}}(x)$ and either M^n is compact with $\kappa_1 \ge \eta H$ everywhere or there exists a point $y \in \mathcal{B}_{\frac{\alpha}{H(x)}}(x)$ satisfying $\kappa_1(y) \le \eta H(y)$.

Proof. Choose $\gamma > 1$ and suppose that $H(x) \ge \gamma H_\sharp$. We claim that

$$H(x) \ge \gamma H_\sharp \text{ implies } \begin{aligned} H(y) &\ge \frac{H(x)}{1 + c_\sharp d(x,y) H(x)} \\ &\ge \frac{H(x)}{\gamma} \end{aligned} \quad \text{for all } y \in \mathcal{B}_{\frac{\gamma}{c_\sharp H(x)}}(x)\,.$$

Suppose first that the set of points $y \in M^n$ satisfying $H(y) \le H(x)/\gamma$ is nonempty and let $y_0 \in M^n$ be the nearest such point to x. Set $d_0 \doteqdot d(x, y_0) H(x)$ and $\theta_0 \doteqdot \min\{d_0, \frac{\gamma - 1}{c_\sharp}\}$. Assume $y \in \mathcal{B}_{\frac{\theta_0}{H(x)}}(x)$ and let $\omega : [0, d(x,y)] \to M^n$ be a unit speed geodesic joining x to y. Then

$$H(s) \ge \frac{H(x)}{\gamma} \ge H_\sharp\,,$$

where $H(s) \doteqdot H(\omega(s))$, and hence

$$\frac{d}{ds} H \ge -c_\sharp H^2 \text{ for all } s \in [0, d(x,y)]\,.$$

Integrating then yields

$$H(s) \geq \frac{H(x)}{1 + c_\sharp s H(x)} \quad \text{for all } s \in [0, d(x, y)]$$

and we conclude that

$$H(y) \geq \frac{H(x)}{1 + c_\sharp d(x, y) H(x)} \geq \frac{H(x)}{1 + c_\sharp \theta_0} \quad \text{for all } y \in \mathcal{B}_{\frac{\theta_0}{H(x)}}(x)\,. \tag{11.11}$$

If $d_0 < \frac{\gamma - 1}{c_\sharp}$, then (11.11) implies that $H(y_0) > \frac{H(x)}{\gamma}$, in contradiction with the definition of y_0. So in fact $d_0 \geq \frac{\gamma-1}{c_\sharp}$, which implies the assertion.

On the other hand, if $H(y) \geq H(x)/\gamma$ for all $y \in M^n$, then $|\nabla H| \leq c_\sharp H^2$ everywhere. In that case, the assertion follows from the same argument in a more direct way (cf. 9.25). So the claim is proved.

Given $\eta > 0$ and $\alpha > 0$, set $\gamma \doteqdot 1 + c_\sharp \alpha$ and suppose that $x \in M^n$ satisfies $H(x) \geq \gamma H_\sharp$ and $\frac{\kappa_1}{H}(x) > \eta$. Then

$$H(y) \geq \frac{H(x)}{1 + c_\sharp d(x, y) H(x)} \geq \frac{H(x)}{\gamma} \quad \text{for all } y \in \mathcal{B}_{\frac{\alpha}{H(x)}}\,.$$

Suppose that $\frac{\kappa_1}{H}(y) > \eta$ for all $y \in \mathcal{B}_{\frac{\alpha}{H(x)}}$. We will show that α can be chosen sufficiently large that $M^n \subset \mathcal{B}_{\frac{\alpha}{H(x)}}$. It will suffice to show that the Gauß map $\mathrm{G} : \mathcal{B}_{\frac{\alpha}{H(x)}} \to S^n$ is surjective. Given $z \in S^n \setminus \{\pm \mathrm{G}(x)\}$, let ω be the solution to the ODE

$$\begin{cases} \dfrac{d\omega}{ds}(s) = \dfrac{z^\top(s)}{|z^\top(s)|}\,, \\ \omega(0) = X(x)\,, \end{cases}$$

where, denoting $\mathrm{G}(s) \doteqdot \mathrm{G}(\omega(s))$, $z^\top(s) \doteqdot z - \langle z, \mathrm{G}(s)\rangle\, \mathrm{G}(s)$ is the projection of z onto the tangent space at $\omega(s)$. By definition, the solution ω is parametrized by arc length. It can be extended beyond $s = s_0$ whenever $\mathrm{G}(s_0) \neq \pm z$. Observe that

$$\frac{d}{ds} \langle \mathrm{G}, z\rangle = \frac{\mathrm{II}(z^\top, z^\top)}{|z^\top|} \geq \eta H |z^\top| = \eta H \sqrt{1 - \langle \mathrm{G}, z\rangle}\,.$$

That is,

$$\frac{d}{ds} \arcsin \langle \mathrm{G}, z\rangle \geq \eta H\,.$$

If $\omega(s)$ is defined for $s \in [0, \frac{\alpha}{H(x)}]$, then integrating between 0 and $\frac{\alpha}{H(x)}$ yields

$$\begin{aligned}\pi &> \arcsin\left\langle \mathrm{G}(\tfrac{\alpha}{H(x)}), z\right\rangle - \arcsin\langle \mathrm{G}(x), z\rangle \\ &\geq \eta \int_0^{\frac{\alpha}{H(x)}} H(s)\, ds \\ &\geq \eta \int_0^{\frac{\alpha}{H(x)}} \frac{H(x)}{1+c_\sharp s H(x)}\, ds = \frac{\eta}{c_\sharp}\ln(1+c_\sharp\alpha)\,,\end{aligned}$$

where $H(s) \doteqdot H(\omega(s))$. If we choose

$$\alpha \geq \frac{1}{c_\sharp}\left(\mathrm{e}^{\frac{c_\sharp \pi}{\eta}} - 1\right),$$

then this is impossible. So, in that case, there must be some $s_0 \in (0, \frac{\alpha}{H(x)})$ such that either $\mathrm{G}(s) \to 1$ or $\mathrm{G}(s) \to -1$. In fact, the former must be the case since $\frac{d}{ds}\arcsin\langle \mathrm{G}(s), z\rangle \geq 0$. Since $z \in S^n \setminus \{\pm \mathrm{G}(x)\}$ was arbitrary, it follows that $\mathrm{G} : \mathcal{B}_{\frac{\alpha}{H(x)}} \to S^n$ is surjective, as desired. □

Corollary 11.17 (Existence of necks at high curvature scales)**.** *Given an integer $n \geq 3$ and positive constants α_0, α_1, α_2, and $\varepsilon > 0$, there exists $h_\sharp > 0$ with the following property:*

Let $X : M^n \times [0,T) \to \mathbb{R}^{n+1}$ be a solution to mean curvature flow with initial condition in the class $C_1^n(R,\alpha)$, with $\max_{M^n\times\{0\}} |\mathrm{II}| \leq R^{-1}$. If

$$\max_{M^n\times\{t_0\}} H \geq h_\sharp R^{-1},$$

then either $X(\cdot, t_0)$ is a convex embedding (and hence subsequently shrinks to a round point by Huisken's theorem) or there exists $x_0 \in M^n$ such that (x_0, t_0) lies at the center of a shrinking 1-neck of quality ε.

Proof. The claim follows from Lemmas 11.14 and 11.16. □

11.3. The Brakke–White regularity theorem

In this section, we present White's proof of Brakke's regularity theorem for smooth mean curvature flows [**532**], following the treatment of Haslhofer and Kleiner [**276**, Appendix C]. A more comprehensive treatment may be found in Ecker [**205**].

Given $(p,t) \in \mathbb{R}^{n+1}\times\mathbb{R}$, $r > 0$, and t such that $t - r^2 \in I$, the **Gaußian density ratio** $\Theta(p,t,r)$ of a solution to mean curvature flow $X : M^n \times I \to \mathbb{R}^{n+1}$ is defined by

$$\Theta(p,t,r) \doteqdot \Theta_{(p,t)}(t-r^2) = \left(4\pi r^2\right)^{-\frac{n}{2}} \int_M \mathrm{e}^{-\frac{|X-p|^2}{4r^2}}\, d\mu_{t-r^2}\,. \tag{11.12}$$

Figure 11.4. Brian White. Author: Renate Schmid. Source: Archives of the Mathematisches Forschungsinstitut Oberwolfach.

By Huisken's monotonicity formula (10.8), $\Theta(p,t,r)$ is nondecreasing in r. We also set

$$\Theta(p,t) \doteqdot \lim_{r\to 0}\Theta(p,t,r) \quad \text{and} \quad \Theta(\infty) \doteqdot \lim_{r\to\infty}\Theta(0,0,r) \tag{11.13}$$

when the limits exist. Note that $\Theta(p,t)$ is defined if $(t-\varepsilon,t)\subset I$ for some $\varepsilon>0$ and $\Theta(\infty)$ is defined if X is an ancient solution; i.e., X is defined on a time interval of the form $(-\infty,\omega)$, where $\omega\in(-\infty,\infty]$. We call $\Theta(p,t)$ the **Gaußian density** at (p,t) of the solution. For an ancient solution, we call $\Theta(\infty)$ the **asymptotic Gaußian density**.

Recall from Example 10.1 that $\Theta(p,t,r)\equiv 1$ on a stationary hyperplane which passes through p.

Lemma 11.18 ([**532**, Proposition 2.10]). *Let $X : M^n\times I\to\mathbb{R}^{n+1}$ be a properly embedded solution to mean curvature flow with finite Gaußian area. Then $\Theta(\infty)\geq 1$, with equality if and only if $\{\mathcal{M}_t\}_{t\in I}$ is a stationary hyperplane, where $\mathcal{M}_t \doteqdot X(M^n,t)$.*

Proof. We claim that $\Theta(p,s)\geq 1$ for any $s\in I$ and $p\in\mathcal{M}_s$. Consider the rescaled flows $X^\lambda : M\times I_\lambda\to\mathbb{R}^{n+1}$, where $I_\lambda\doteqdot\lambda^2(I-s)$, defined by $X^\lambda(x,t)\doteqdot\lambda(X(x,\lambda^{-2}t+s)-p)$. Denote by Θ_λ the Gaußian density and by μ_t^λ the measure on M induced by X_t^λ. Then

$$\begin{aligned}\Theta_\lambda(0,0,r) &= \left(4\pi r^2\right)^{-\frac{n}{2}}\int_M \mathrm{e}^{-\frac{|X_\lambda(x,-r^2)|^2}{4r^2}}\,d\mu^\lambda_{-r^2}\\ &= \left(4\pi r^2\right)^{-\frac{n}{2}}\int_M \mathrm{e}^{-\frac{\lambda^2|X(\cdot,-\lambda^{-2}r^2+s)-p|^2}{4r^2}}\,d\mu_{-\lambda^{-2}r^2+s}\\ &=\Theta\left(p,s,\frac{r}{\lambda}\right).\end{aligned}$$

Since the flows $\{\mathcal{M}_t^\lambda\}_{t\in I_\lambda}$ converge smoothly on compact subsets of $\mathbb{R}^{n+1}\times\mathbb{R}$ to a stationary hyperplane as $\lambda\to\infty$,

$$\Theta(p,s)=\lim_{\lambda\to\infty}\Theta\left(p,s,\frac{r}{\lambda}\right)=\lim_{\lambda\to\infty}\Theta_\lambda(0,0,r)\geq\Theta_\infty(0,0,r)=1\,.$$

The monotonicity formula (10.8) then implies that $\Theta(\infty)\geq 1$. On the other hand, if $\Theta(\infty)=1$, then, by the rigidity case of the monotonicity formula, $\{\mathcal{M}_t\}_{t\in I}$ is invariant under dilations about (p,t) and hence is a stationary hyperplane. The claim follows. □

Theorem 11.19 (Brakke's regularity theorem for smooth flows). *Given $n\in\mathbb{N}$, there exist constants $\varepsilon>0$ and $C<\infty$ with the following property: If $\{\mathcal{M}_t^n\}_{t\in I}$ is a mean curvature flow of smooth, properly embedded hypersurfaces $\mathcal{M}_t^n\subset\mathbb{R}^{n+1}$ satisfying*

$$\sup_{(p,t)\in P(p_0,t_0,r)}\Theta(p,t,r)<1+\varepsilon$$

for some $r>0$, where Θ is the Gaußian density ratio of $\{\mathcal{M}_t^n\}_{t\in I}$, then

$$\sup_{P(p_0,t_0,r/2)}|\mathrm{II}|\leq Cr^{-1}\,. \tag{11.14}$$

Proof. Observe that, by parabolically rescaling and translating in space and time, it suffices to prove the theorem when $r=1$ and $(p_0,s_0)=(0,0)$. Suppose that the conclusion fails for all $\varepsilon>0$ and $C<\infty$. Then for each $j\in\mathbb{N}$ there must exist some smooth mean curvature flow $\{\mathcal{M}_{j,t}^n\}_{t\in I_j}$ such that

$$\sup_{(p,t)\in P(0,0,1)}\Theta(p,t,1)<1+j^{-1},$$

but

$$|\mathrm{II}|(p_j,t_j)>j$$

for some $(p_j,t_j)\in P(0,0,1/2)$. The idea is to now apply Perelman's point selection method [**438**, §10]. We claim that there is some $(q_j,s_j)\in P(0,0,3/4)$ such that

$$Q_j\doteqdot|\mathrm{II}|(q_j,s_j)>j\ \text{ and }\ \sup_{P(q_j,s_j,j/10Q_j)}|\mathrm{II}|\leq 2Q_j\,. \tag{11.15}$$

Indeed, if $(q_j^0,s_j^0)\doteqdot(p_j,t_j)$ already satisfies (11.15), then it is the point we seek. Otherwise, there is a point $(q_j^1,s_j^1)\in P(q_j^0,s_j^0,j/10Q_j^0)$ such that $Q_j^1\doteqdot|\mathrm{II}|(q_j^1,s_j^1)>2Q_j$. If (q_j^1,s_j^1) satisfies (11.15), then it is the point we seek. Otherwise, there is a point $(q_j^2,s_j^2)\in P(q_j^1,s_j^1,j/10Q_j^1)$ such that $Q_j^2\doteqdot|\mathrm{II}|(q_j^1,s_j^1)>2Q_j^1$, etc. Note that

$$\frac{1}{2}+\frac{j}{10Q_j^0}\left(1+\frac{1}{2}+\frac{1}{4}+\cdots\right)<\frac{3}{4}\,.$$

Since $|\mathrm{II}|$ is finite, the iteration terminates after a finite number of steps, and the final point of the iteration lies in $P(0,0,3/4)$ and satisfies (11.15). Now consider the flows $\{\widehat{\mathcal{M}}_{j,t}\}_{t\in I_j}$ obtained by shifting q_j to the origin and parabolically rescaling by Q_j. This new sequence satisfies $|\widehat{\mathrm{II}}_j|(0,0)=1$ and $\sup_{P(0,0,j/10)}|\widehat{\mathrm{II}}_j|\leq 2$. Thus, by Theorem 11.12, we can pass smoothly to a limit flow satisfying

$$|\mathrm{II}|(0,0)=1 \quad \text{and} \quad \Theta(\cdot,\cdot,1)\equiv 1\,.$$

By the preceding lemma, the latter implies that the limit is a stationary hyperplane, contradicting the former. □

11.4. Huisken's theorem revisited

We now present yet another proof of Huisken's theorem, this time due to Hamilton [**264**]. As in the proof presented in Section 8.6, the idea is to show that a suitable sequence of rescalings about the singularity converges to a shrinking sphere solution. Hamilton achieves this in two steps, combining the preservation of the initial pinching with the compactness provided by Theorem 11.11: First, one uses the preserved pinching of the second fundamental form to deduce that blow-up limits are *compact* solutions of the flow, which rules out type-II singularities. Using the monotonicity formula, it is then shown that the blow-up limit is a shrinking self-similar solution. The claim then follows since, when $n\geq 2$, the shrinking sphere is the only compact, locally uniformly convex shrinking self-similar solution.

11.4.1. Proper hypersurfaces with pinched principal curvatures.

By the Gauß equation, any hypersurface of dimension at least 2 having uniformly positive Weingarten curvature, $\mathrm{II}\geq\varepsilon g$ for some $\varepsilon>0$, has uniformly positive Ricci curvature, $\mathrm{Ric}\geq\varepsilon' g$ for some $\varepsilon'=\varepsilon'(\varepsilon,n)>0$. It follows from Myers's theorem that such a hypersurface is compact. The following theorem, which is harder to prove, gives a scale-invariant version of this result.

Theorem 11.20 (Hamilton [**264**]). *If $\mathcal{M}^n\subset\mathbb{R}^{n+1}$ is a smooth properly embedded hypersurface satisfying $H>0$ and $\mathrm{II}_{ij}\geq\varepsilon H g_{ij}$ for some $\varepsilon>0$, then $\mathcal{M}$ is compact.*

Sketch of the proof. The theorem is proved by contradiction. So suppose that $\mathcal{M}^n$ is noncompact. After rotating $\mathcal{M}$ in $\mathbb{R}^{n+1}=\mathbb{R}^n\times\mathbb{R}$, there exists a convex open subset $\mathcal{U}\subset\mathbb{R}^n$ such that $\mathcal{M}$ is the graph of a smooth, locally uniformly convex function $f:\mathcal{U}\to\mathbb{R}$; i.e., $\nabla^2 f>0$ and

$$\mathcal{M}=\{(x,f(x)):x\in\mathcal{U}\}\,,$$

where $\lim_{x \to \partial \mathcal{U}} f(x) = \infty$. The idea is to show the following:

(1) Let g denote the induced metric on $\mathcal{M}$. Then the conformally equivalent metric $\tilde{g}$ on $\mathcal{M}$ defined by

$$\tilde{g}_{(x,y)} \doteqdot \frac{1}{y^2 (\log y)^2} g_{(x,y)} \doteqdot \varphi g, \tag{11.16}$$

for $(x,y) \in \mathcal{M}$, is a complete metric with finite volume (see Theorem 2.1 in [**264**]).

(2) Since $(\mathcal{M}, g)$ is complete and locally uniformly convex, the Gauß map G is a diffeomorphism onto its image $\mathrm{G}(\mathcal{M}) \subset S^n$, which is contained in the southern hemisphere ($\mathcal{M}$ is a graph). The condition $\mathrm{II}_{ij} \geq \varepsilon H g_{ij}$ for some $\varepsilon > 0$ is equivalent to the Gauß map being **quasiconformal**; that is, there exists $C < \infty$ such that

$$\frac{(\lambda_n)_g (\mathrm{G}^* g_{S^n})}{(\lambda_1)_g (\mathrm{G}^* g_{S^n})} (x,y) \leq C \tag{11.17}$$

for all $(x,y) \in \mathcal{M}$. Here,

$$(\lambda_1)_g (\mathrm{G}^* g_{S^n}) \leq \cdots \leq (\lambda_n)_g (\mathrm{G}^* g_{S^n})$$

denote the eigenvalues of $\mathrm{G}^* g_{S^n}$ with respect to g.

We deduce (11.17) as follows: Recall that (5.78) says

$$(\mathrm{G}^* g_{S^n})_{ij} = \mathrm{II}_{ik} g^{k\ell} \mathrm{II}_{\ell j},$$

so that

$$(\lambda_a)_g (\mathrm{G}^* g_{S^n}) = \kappa_a^2,$$

where $\kappa_1 \leq \cdots \leq \kappa_n$ are the principal curvatures of $\mathcal{M}$. Hence $\mathrm{II}_{ij} \geq \varepsilon H g_{ij}$ implies

$$\frac{(\lambda_n)_g (\mathrm{G}^* g_{S^n})}{(\lambda_1)_g (\mathrm{G}^* g_{S^n})} = \frac{\kappa_n^2}{\kappa_1^2} \leq \frac{(1-\varepsilon)^2}{\varepsilon^2}.$$

Equivalent to (11.17) is

$$\frac{(\lambda_n)_{g_{S^n}} \left(g_{(\mathrm{G}^{-1})^* g} \right)}{(\lambda_1)_{g_{S^n}} \left(g_{(\mathrm{G}^{-1})^* g} \right)} \leq C. \tag{11.18}$$

(3) If $\mathcal{V} \subset S^n$ is a nonempty open subset whose closure is not all of S^n, then any complete metric on $\mathcal{V}$ conformal to g_{S^n} has infinite volume (see Theorem 3.1 in [**264**]).

Since $\tilde{g} = \varphi g$ on $\mathcal{M}$ is complete with finite volume, so is the metric

$$(\mathrm{G}^{-1})^* \tilde{g} = (\varphi \circ \mathrm{G}^{-1}) (\mathrm{G}^{-1})^* g$$

on $\mathrm{G}(\mathcal{M})$. By (11.18), there exists a smooth function ψ on $\mathrm{G}(\mathcal{M})$ such that

$$\frac{1}{C'}\left(\mathrm{G}^{-1}\right)^* g \leq \psi g_{S^n} \leq C'\left(\mathrm{G}^{-1}\right)^* g$$

for some constant $C' \in (0,\infty)$. Hence the metric

$$\left(\varphi \circ \mathrm{G}^{-1}\right)\psi g_{S^n}$$

on $\mathrm{G}(\mathcal{M})$ is complete with finite volume. Since $\mathrm{G}(\mathcal{M})$ is contained in the southern hemisphere, this contradicts (3). □

We need one more lemma.

Lemma 11.21. *The only compact, locally uniformly convex, self-similar shrinking solution to mean curvature flow of dimension at least two is the shrinking sphere.*

Proof. Recall from (8.7) that

$$(\partial_t - \Delta)\frac{|\mathrm{II}|^2}{H^2} = \frac{2}{H}\left\langle \nabla H, \nabla \frac{|\mathrm{II}|^2}{H^2}\right\rangle - 2\left|\nabla \frac{\mathrm{II}}{H}\right|^2 .$$

Since $\frac{|\mathrm{II}|^2}{H^2}$ is scale invariant, up to diffeomorphisms it is constant in time on a self-similar shrinker. In particular, since the time slices are compact, $\max_{M^n\times\{t\}} \frac{|\mathrm{II}|^2}{H^2}$ occurs at an interior space-time point and the strong maximum principle then implies that $\frac{|\mathrm{II}|^2}{H^2}$ is constant in space and time. We conclude that

$$\left|\nabla \frac{\mathrm{II}}{H}\right|^2 \equiv 0 .$$

By the splitting theorem (Theorem 9.11), the solution is uniformly convex for all time since, by compactness, it cannot split off a line. Thus, by compactness, we can find a constant $\alpha > 0$ such that $\mathrm{II} \geq \alpha H g$. Lemma 8.8 then implies that $|\nabla \mathrm{II}|^2 \equiv 0$ and we conclude from Corollary 5.11 that the time slices are all round spheres. The lemma follows. □

Hamilton's proof of Huisken's theorem. Given a compact, uniformly convex hypersurface $X_0 : M^n \to \mathbb{R}^{n+1}$, $n \geq 2$, let $X : M^n \times [0,T) \to \mathbb{R}^{n+1}$ be the maximal solution of mean curvature flow with initial condition X_0. By the compactness of M and by Lemma 8.4, there exists $\alpha > 0$ such that $\mathrm{II} \geq \alpha H g$ on the solution. Recall also, from Lemma 8.3, that

$$\max_{M\times\{t\}} H \geq \frac{1}{\sqrt{2(T-t)}} .$$

Using Theorem 11.20 and Lemma 11.21, we can circumvent the Stampacchia iteration argument of Theorem 8.6 using compactness arguments.

We first prove that the curvature blow-up is of **type-I**. That is,

$$\sup_{M\times[0,T)} \sqrt{T-t}\,H(x,t) < \infty\,.$$

The proof is by contradiction. So suppose that

$$\limsup_{t\to T} \sqrt{T-t}\max_{M\times\{t\}} H = \infty\,.$$

Choose a sequence of times $t_j \in [0, T-j^{-1})$ and a sequence of points $x_j \in M$ such that

$$H^2(x_j,t_j)\left(T-\frac{1}{j}-t_j\right) = \max_{(x,t)\in M\times\left[0,T-\frac{1}{j}\right]} H^2(x,t)\left(T-\frac{1}{j}-t\right).$$

Set $\sigma_j \doteqdot \lambda_j^2 t_j$ and $T_j = \lambda_j^2\left(T-\frac{1}{j}-t_j\right)$, where $\lambda_j \doteqdot H(x_j,t_j)$, and consider the sequence of rescaled solutions $X_j : M\times(-\sigma_j, T_j) \to \mathbb{R}^{n+1}$ defined by

$$X_j(x,t) = \lambda_j\left(X\left(x, t_j+\lambda_j^{-2}t\right) - X(x_j,t_j)\right).$$

Since the solution is of **type-II** (that is, not of type-I), we can pass to some subsequence such that $t_j \to T$, $\lambda_j \to \infty$, $\sigma_j \to \infty$, and $T_j \to \infty$. Observe also that $X_j(x_j, 0) = 0$ and

$$H_j^2(x,t) = \lambda_j^{-2}H^2\left(x, t_j+\lambda_j^{-2}t\right) \le \frac{T-\frac{1}{j}-t_j}{T-\frac{1}{j}-(t_j+\lambda_j^{-2}t)} = \frac{T_j}{T_j-t}.$$

So the curvature is uniformly bounded on any compact time interval. By Theorem 11.11 and a diagonal subsequence argument, the rescaled solutions $X_j : M\times[-\sigma_j, T_j) \to \mathbb{R}^{n+1}$ converge in C^∞, after passing to a subsequence, to a smooth, proper limiting solution $X_\infty : M_\infty\times(-\infty,\infty) \to \mathbb{R}^{n+1}$ uniformly on compact subsets of $M_\infty \times (-\infty, \infty)$. By the avoidance principle, any proper solution defined on $(0,\infty)$ cannot have compact time slices (since such time slices are enclosed by shrinking round spheres which become extinct in finite time). Since the uniform curvature pinching is preserved under the rescaling, this is in contradiction with Theorem 11.20.

So the singularity must be of the first type. That is,

$$C \doteqdot \sup_{(x,t)\in M\times[0,T)} \sqrt{T-t}\,H(x,t) < \infty\,.$$

We claim that

$$\limsup_{t\nearrow T}(T-t)^{-(m+1)}A_m(t) = 0 \tag{11.19}$$

for all $m \in \mathbb{N}\cup\{0\}$, where

$$A_0 \doteqdot \max_{M\times\{t\}} |\mathring{\mathrm{II}}|^2 \quad\text{and, for each } m\in\mathbb{N},\quad A_m \doteqdot \max_{M\times\{t\}} |\nabla^m \mathrm{II}|^2\,.$$

Suppose, to the contrary, that there is some $\varepsilon > 0$, some $m \in \mathbb{N} \cup \{0\}$, and some sequence of times $t_j \nearrow T$ such that

$$(T - t_j)^{-(m+1)} A_m(t_j) \geq \varepsilon \tag{11.20}$$

for all $j \in \mathbb{N}$. Set $\sigma_j \doteqdot \lambda_j^2 t_j$, where $\lambda_j \doteqdot (T - t_j)^{-\frac{1}{2}}$, and consider the sequence of rescaled solutions $X_j : M \times (-\sigma_j, 1) \to \mathbb{R}^{n+1}$ defined by

$$X_j(\cdot, t) = \lambda_j \left(X\left(\cdot, t_j + \lambda_j^{-2} t\right) - X(x_j, t_j) \right),$$

where $x_j \in M^n$ is a point at which $\max_{M \times \{t_j\}} H$ is attained. For each j, $X_j(x_j, 0) = 0$, $\sigma_j \to \infty$, and

$$|\mathrm{II}_j|^2(\cdot, t) \leq H_j^2(\cdot, t) = \lambda_j^{-2} H^2\left(\cdot, t_j + \lambda_j^{-2} t\right) \leq \frac{C^2}{1-t}.$$

By Theorem 11.11 and a diagonal subsequence argument, the rescaled solutions $X_j : M \times [-\sigma_j, 1) \to \mathbb{R}^{n+1}$ converge in the smooth topology, after passing to a subsequence, to a smooth, proper limiting solution $X_\infty : M_\infty \times (-\infty, 1) \to \mathbb{R}^{n+1}$ uniformly on compact subsets of $M_\infty \times (-\infty, 1)$. The mean curvature of the limit is certainly nonnegative. In fact, by (8.4) in Lemma 8.3, the mean curvature must be positive at the origin, and hence positive everywhere by applying the strong maximum principle to the evolution equation (6.18) for H. Since the pinching condition $\mathrm{II} \geq \alpha H g$ is scale invariant, we deduce from Theorem 11.20 that the limit is compact; in particular, $M_\infty \cong M$. We claim that the limit is a shrinker, and hence a shrinking round sphere by Lemma 11.21, contradicting (11.20). Indeed, after passing to a subsequence, we can find $x_\infty \in M$ and $p_\infty \in \mathbb{R}^{n+1}$ such that $x_j \to x_\infty$ and $p_j \doteqdot X(x_\infty, t_j) \to p_\infty$. By the type-I curvature bound,

$$|p_\infty - p_j| \leq \int_{t_j}^{T} H(x_\infty, t)\, dt \leq \int_{t_j}^{T} \frac{C}{\sqrt{T-t}}\, dt = 2C\sqrt{T - t_j} = \frac{2C}{\lambda_j}.$$

Thus, after passing to a subsequence, the sequence of rescaled points $q_j \doteqdot \lambda^j (p_\infty - p_j)$ converges to some point $q_\infty \in \mathbb{R}^{n+1}$. The monotonicity formula now yields

$$\begin{aligned} \int_a^b \!\! \int_M \left(H_j - \frac{\langle X_j - q_j, \mathsf{N}_j \rangle}{2(1-t)} \right)^2 d\mu_j dt &= \Theta^j_{q_j,1}(a) - \Theta^j_{q_j,1}(b) \\ &= \Theta_{p_\infty,T}(\lambda_j^{-2} a + t_j) - \Theta_{p_\infty,T}(\lambda_j^{-2} b + t_j) \end{aligned}$$

for any a and b such that $-\sigma_j < a < b < 1$. But $\lambda_j^{-2} a + t_j$ and $\lambda_j^{-2} b + t_j$ both approach T as $j \to \infty$. Since, by the monotonicity formula, the limit $\lim_{t \to T} \Theta_{p_\infty,T}(t)$ exists, we conclude that the right-hand side goes to zero as $j \to \infty$ for any a and b with $-\infty < a < b < 1$. In particular, the integrand on the left must tend to zero identically on the limit. That is,

$$H_\infty - \frac{\langle X_\infty - q_\infty, \mathsf{N}_\infty \rangle}{2(1-t)} \equiv 0\,.$$

It now follows from Lemma 11.21 that the limit is the shrinking round sphere, contradicting (11.20). This proves (11.19). We can now proceed as in Section 8.4.4 to convert the geometric estimates into estimates for the rescaled embedding and its derivatives and thereby deduce convergence to an embedding of the unit sphere in the smooth topology. □

11.5. The structure of singularities

In the preceding section, we used rescaling methods and the compactness theorem (Theorem 11.11) to show that initially convex hypersurfaces become round near a singularity under mean curvature flow. Similar ideas yield a nice description of the *infinitesimal structure of singularities* more generally, particularly in the mean convex setting. As in the preceding section, we consider two complementary settings:

Definition 11.22. We say that a singular solution to mean curvature flow $X : M^n \times [0,T) \to \mathbb{R}^{n+1}$, where $T < \infty$, develops a **type-I** (a.k.a. **rapidly forming**) **singularity** if

$$\sup_{M^n\times[0,T)} (T-t)\,|\mathrm{II}|^2 < \infty \tag{11.21}$$

or a **type-II** (a.k.a. **slowly forming**) **singularity** if

$$\sup_{M^n\times[0,T)} (T-t)\,|\mathrm{II}|^2 = \infty. \tag{11.22}$$

We have seen that mean curvature flows of convex hypersurfaces develop only type-I singularities. The reason for why type-II singularities are called *slowly* forming is because in this case there exists a sequence of times $t_i \to T$ such that the ratio of the time to blow up $T - t_i$ to that of the maximum curvature scale $(\sup_M |\mathrm{II}|(\cdot, t_i))^{-2}$ tends to infinity. That is, the length of time it takes for the singularity to form from a suitable given time is *longer* than that predicted by the maximum curvature at that time.

To study the infinitesimal structure of singularities, we construct certain sequences of *parabolic rescalings* about a "singular point" and study what happens in the limit. Since the mean curvature flow is invariant under parabolic scaling, the resulting limit (if it exists) is a solution to mean curvature flow. There are multiple ways to perform such rescalings. For example, as in Hamilton's proof of Huisken's theorem in the preceding section, we may choose a sequence of space-time points $(x_j, t_j) \in M^n \times [0,T)$ which approach a singularity, in the sense that

$$\lambda_j \doteqdot |\mathrm{II}|(x_j,t_j) \to \infty \quad \text{as } j \to \infty,$$

and rescale about the ambient point $X(x_j,t_j)$ in such a way that the curvature is normalized at the origin. That is:

Definition 11.23. Given a singular solution to the parametrized mean curvature flow, we form the **curvature normalizing sequence**

$$X_j(x,t) \doteqdot \lambda_j \left(X\left(x, \lambda_j^{-2}t + t_j\right) - X(x_j,t_j)\right) \tag{11.23}$$

about the **blow-up sequence** $\{(x_j,t_j)\}_{j\in\mathbb{N}}$. We call any limit of such a sequence, where convergence is locally uniformly in the smooth topology, a **singularity model**.

Observe that

$$X_j(x_j,0) = 0 \quad \text{and} \quad |\mathrm{II}_j|(x_j,0) = 1\,. \tag{11.24}$$

In particular, a singularity model cannot be a hyperplane since it must contain a point (and hence a neighborhood) with nontrivial second fundamental form.

In order to obtain a singularity model using the compactness theorem (Theorem 11.11) we need to arrange that the curvature is bounded on sufficiently large sets. One way to achieve this is to choose the blow-up sequence in such a way that the curvatures of the rescaled sequence of solutions are bounded globally at earlier times. That is,

$$\sup_{M^n\times[0,t_j]} |\mathrm{II}| \le C|\mathrm{II}|(x_j,t_j)$$

for some constant C independent of j. We call such a blow-up sequence **essential**. Singularity models of essential blow-up sequences are glimpses of the forming singularity at the *highest* curvature scale.

There are important variants of the curvature normalizing approach. Given an *embedded* solution $\{\mathcal{M}_t\}_{t\in I}$ to mean curvature flow in $\mathbb{R}^{n+1}$ and given sequences of ambient space-time points $(p_j,t_j) \in \mathbb{R}^{n+1} \times \mathbb{R}$ and positive numbers (scales) $\lambda_j \to \infty$, we have the sequence of rescaled embedded solutions

$$\lambda_j(\mathcal{M}_{\lambda_j^{-2}t+t_j} - p_j)\,. \tag{11.25}$$

Definition 11.24. Any integral Brakke flow $\{\mu_{t,\infty}\}$ which is the limit in the sense of Brakke flows of such a sequence is called a **limit flow** (cf. the compactness Theorem 6.28). As above, we call (p_j,t_j) a blow-up sequence. If $\{\mu_{t,\infty}\}$ is a smooth flow and the convergence is locally uniformly in the smooth topology, then we also call the limit a limit flow.

Given a mean curvature flow $\{\mathcal{M}_t\}_{t\in I}$ in $\mathbb{R}^{n+1}$, a *fixed* ambient space-time point $(p_0, t_0) \in \mathbb{R}^{n+1} \times \mathbb{R}$, and a sequence of scales $\lambda_j \to \infty$, we can form the sequence of mean curvature flows $\{\mathcal{M}_{t,j}\}_{t\in I_j}$ defined by

$$\text{(11.26)} \quad \mathcal{M}_{t,j} = \lambda_j(\mathcal{M}_{\lambda_j^{-2}t+t_0} - p_0)\,, \quad I_j \doteqdot \{\lambda_j^2(s - t_0) : s \in I \cap (-\infty, t_0]\}\,.$$

For example, if $I = [0, T)$ and $t_0 = T$, then $I_j = [-\lambda_j^2 T, 0)$.

Definition 11.25. If the sequence converges in the sense of Brakke flows to an integral Brakke flow $\{\mu_{t,\infty}\}_{t\in(-\infty,0)}$, then we call the limit flow a **tangent flow** at p_0. So a tangent flow is simply a limit flow where the blow-up sequence $(p_j, t_j) \equiv (p_0, t_0)$ is constant.

A different approach to rescaling, which we have encountered in the statements of Grayson's theorem (Theorem 3.19) and Huisken's theorem (Theorem 8.1), is to study the **normalized flow** of $X : M^n \times I \to \mathbb{R}^{n+1}$ associated to the **type-I normalization** about an ambient space-time point (X_0, t_0), which is the family of immersions $\widetilde{X}_{\tilde{t}}$ defined by

$$\text{(11.27a)} \qquad \widetilde{X}_{\tilde{t}}(x) = \frac{X(x,t) - X_0}{\sqrt{2(t_0 - t)}}\,,$$

$$\text{(11.27b)} \qquad \tilde{t}(t) = -\frac{1}{2}\log(t_0 - t)\,.$$

A type-I normalized solution $\widetilde{X} : M^n \times \left(-\log\sqrt{t_0}, \infty\right) \to \mathbb{R}^{n+1}$ satisfies the type-I normalized flow

$$\text{(11.28)} \qquad \partial_{\tilde{t}}\widetilde{X} = \widetilde{X} - \widetilde{H}\widetilde{\mathsf{N}}\,.$$

Note that in the discussion above, "type-I" refers to the normalization and not to the type of singular solution.

11.5.1. Singularity models for type-I singularities.

We will prove that every type-I singularity in mean convex mean curvature flow is, up to a rigid motion, modeled by either a shrinking round, orthogonal cylinder $\{\mathbb{R}^m \times S^{n-m}_{\sqrt{-2(n-m)t}}\}_{t\in(-\infty,0)}$ or by a shrinking Abresch–Langer cylinder $\{\mathbb{R}^{n-1}\times\Gamma_t\}_{t\in(-\infty,0)}$, where $\{\Gamma_t\}_{t\in(-\infty,0)}$ is one of the self-shrinking Abresch–Langer curves. The idea is a straightforward extension of those used in the convex case:

By the type-I hypothesis, the second fundamental form becomes uniformly bounded after rescaling the flow by the square root of the remaining time, which allows the application of the compactness theorem to obtain an ancient limit solution with bounded curvature and Gaußian area. The monotonicity formula then implies that the limit is a shrinking self-similar solution. For mean convex flows, the convexity estimate and the splitting theorem imply that the limit splits as a product $\{\mathbb{R}^m \times \Sigma_t^{n-m}\}_{t\in(-\infty,0)}$ for some $m \in \{0, \ldots, n-1\}$, where $\{\Sigma_t^{n-m}\}_{t\in(-\infty,0)}$ is a locally uniformly convex shrinking self-similar solution of dimension $n-m$.

Theorem 11.26 (Type-I singularity models). *Let $X : M^n \times [0,T) \to \mathbb{R}^{n+1}$ be a compact solution to mean curvature flow which undergoes a type-I singularity. Given sequences of times $t_j \in [0,T)$ and points $x_j \in M^n$ with $\limsup_{j\to\infty} |\mathrm{II}|(x_j,t_j) \to \infty$, the sequence of rescaled solutions $X_j : M^n \times [-\lambda_j^2 t_j, 0) \to \mathbb{R}^{n+1}$ defined by*

$$X_j(x,t) \doteqdot \lambda_j \left(X(x, \lambda_j^{-2}t + t_j) - X(x_j,t_j) \right) ,$$

where $\lambda_j \doteqdot (T-t_j)^{-\frac{1}{2}}$, converges locally uniformly in the smooth topology, after passing to a subsequence, to a shrinking self-similar solution.

Proof. See the argument given in Hamilton's proof of Huisken's theorem in Section 11.4. □

Given a type-I singular solution, call a sequence (x_j,t_j) **type-I essential** if $(T-t_j)^{1/2}|\mathrm{II}|(x_j,t_j) \geq c$ for some constant $c > 0$. Call (x_j,t_j) **type-I inessential** if $(T-t_j)^{1/2}|\mathrm{II}|(x_j,t_j) \to 0$. A singularity model associated to a type-I essential sequence is necessarily not a hyperplane; otherwise it may be a hyperplane.

By Theorem 11.26, it remains to extend Lemma 11.21.

Lemma 11.27 (Huisken [**297**]). *Let $\{\mathcal{M}_t^n\}_{t\in(-\infty,0)}$, $n \geq 2$, be a locally uniformly convex, self-similarly shrinking solution to mean curvature flow with finite Gaußian density. Suppose that H, $\frac{\mathrm{II}}{H}$, ∇H, and $\nabla \frac{\mathrm{II}}{H}$ are bounded at time -1. Then $\{\mathcal{M}_t^n\}_{t\in(-\infty,0)}$ is the shrinking sphere.*[1]

Proof. Set $u \doteqdot \frac{|\mathrm{II}|^2}{H^2}$. By (8.7),

$$(\partial_t - \Delta)u = 2\left\langle \nabla u, \frac{\nabla H}{H} \right\rangle - 2\left|\nabla \frac{\mathrm{II}}{H}\right|^2 .$$

By (6.10),

$$u\left(\varphi\left(x, -\log\sqrt{-t}\right), t\right) = u_0(x) ,$$

where u_0 is the restriction of u to the time slice $X_0 \doteqdot X(\cdot,-1)$ and φ is the flow of the vector field

$$V \doteqdot -dX_0^{-1}(X_0^\top) = -dX_\varphi^{-1}(X_\varphi^\top) ,$$

where

$$X_\varphi(x,t) \doteqdot X(\varphi(x, -\log\sqrt{-t}), t) = \sqrt{-t}X_0(x) .$$

[1] We will prove a more general version of this result in Theorem 13.14.

It follows that

$$\begin{aligned}0 &\equiv \frac{d}{dt}u(\varphi(x, -\log\sqrt{-t}), t) \\ &= \partial_t u(\varphi(x, -\log\sqrt{-t}), t) + \frac{1}{-2t}\nabla_V u(\varphi(x, -\log\sqrt{-t}), t)\end{aligned}$$

and hence

$$-\Delta u = \frac{1}{-2t}\nabla_V u + 2\left\langle \nabla u, \frac{\nabla H}{H}\right\rangle - 2\left|\nabla \frac{\mathrm{II}}{H}\right|^2 .$$

Multiplying by $H^2 u\Phi$ and integrating by parts yields

$$\begin{aligned}&\int_{\mathcal{M}_t}\left(H^2|\nabla u|^2 + 2uH\left\langle \nabla u, \nabla H\right\rangle + H^2 u\left\langle \nabla u, \frac{\nabla\Phi}{\Phi}\right\rangle\right)\Phi\, d\mu \\ &\quad = -\int_{\mathcal{M}_t} H^2 u\Delta u\Phi\, d\mu \\ &\quad = \int_{\mathcal{M}_t} H^2 u\left(\frac{1}{-2t}\nabla_V u + 2\left\langle \nabla u, \frac{\nabla H}{H}\right\rangle - 2\left|\nabla\frac{\mathrm{II}}{H}\right|^2\right)\Phi\, d\mu.\end{aligned}$$

Since

$$\nabla\Phi = -\frac{1}{-2t}\Phi V ,$$

this yields

$$\int_{\mathcal{M}_t} H^2\left(|\nabla u|^2 + 2u\left|\nabla\frac{\mathrm{II}}{H}\right|^2\right)\Phi\, d\mu = \underbrace{\int_{\mathcal{M}_t} H^2\left\langle \nabla u, \frac{1}{-2t}V - \frac{\nabla\Phi}{\Phi}\right\rangle\Phi\, d\mu}_{=0}$$

and we conclude that

$$|\mathrm{II}|^2 \equiv \alpha H^2 \text{ for some } \alpha > 0 \text{ and } \nabla\frac{\mathrm{II}}{H} \equiv 0 .$$

The second identity implies that

$$0 \equiv H^2\nabla\frac{\mathrm{II}}{H} = H\nabla\mathrm{II} - \nabla H \otimes \mathrm{II} . \tag{11.29}$$

Taking the trace over the first two components then yields

$$H\nabla H = \mathrm{L}(\nabla H) . \tag{11.30}$$

If $\nabla H \equiv 0$, then by (11.29) we have $\nabla\,\mathrm{II} \equiv 0$, and the claim follows from Corollary 5.11. Otherwise, there exists a point where $|\nabla H|^2 \neq 0$. By (11.30), $\nabla H/|\nabla H|$ is a principal direction at this point. Let $\{e_i\}_{i=1}^n$ be a principal orthonormal frame with $e_1 = \nabla H/|\nabla H|$. By (11.29),

$$\nabla_k \mathrm{II}_{ij} = 0, \text{ for } k \geq 2 \text{ and all } i, j \quad \text{and} \tag{11.31a}$$

$$H\nabla_1 \mathrm{II}_{ij} = \nabla_1 H\, \mathrm{II}_{ij} \text{ for all } i, j . \tag{11.31b}$$

By (11.31a), (11.31b), and the Codazzi equation,

$$\mathrm{II}_{ij} = 0 \quad \text{for all } j \geq 2 \text{ and all } i\,.$$

So the only nonvanishing component of II_{ij} is II_{11}. But this contradicts local uniform convexity when $n \geq 2$. □

Corollary 11.28 (Mean convex, type-I singularity models). *Let $X : M^n \times [0,T) \to \mathbb{R}^{n+1}$ be a compact, mean convex solution to mean curvature flow which undergoes a type-I singularity. Given sequences of times $t_j \in [0,T)$ and points $x_j \in M^n$ with $\limsup_{j\to\infty} H(x_j,t_j) \to \infty$, the sequence of rescaled solutions $X_j : M^n \times [-\lambda_j^2 t_j, 0) \to \mathbb{R}^{n+1}$ defined by*

$$X_j(x,t) \doteqdot \lambda_j \left(X(x, \lambda_j^{-2}t + t_j) - X(x_j,t_j) \right) ,$$

where $\lambda_j \doteqdot (T-t_j)^{-\frac{1}{2}}$, converges locally uniformly in the smooth topology, after passing to a subsequence and performing a rigid motion, to one of the following:

(i) *the stationary hyperplane $\{\mathbb{R}^n \times \{0\}\}_{t\in(-\infty,1)}$,*

(ii) *the shrinking sphere $\{S^n_{\sqrt{2n(1-t)}}\}_{t\in(-\infty,1)}$,*

(iii) *a shrinking round orthogonal cylinder*

$$\{\mathbb{R}^m \times S^{n-m}_{\sqrt{2(n-m)(1-t)}}\}_{t\in(-\infty,1)}, \quad \textit{where } m \in \{1,\ldots,n-1\},$$

or

(iv) *a shrinking Abresch–Langer cylinder $\{\mathbb{R}^{n-1} \times \Gamma_t\}_{t\in(-\infty,1)}$.*

Proof. By Theorem 11.26, the sequence subconverges to a self-similarly shrinking solution to mean curvature flow. If the mean curvature vanishes everywhere, then the limit must be a stationary hyperplane since the estimate $|\mathrm{II}| \leq CH$ passes to the limit. Otherwise, by the strong maximum principle, the mean curvature must be positive everywhere. In that case, the convexity estimate (Corollary 9.13) and the splitting theorem (Theorem 9.11) imply that the limit is, up to a rotation, a product $\{\mathbb{R}^m \times \Sigma_t^{n-m}\}_{t\in(-\infty,1)}$ for some $m \in \{0,\ldots,n-1\}$, where $\{\Sigma_t^{n-m}\}_{t\in(-\infty,1)}$ is a locally uniformly convex, $(n-m)$-dimensional self-similarly shrinking solution. The claim now follows from Theorem 4.4 and Lemma 11.27 since the limit inherits bounds for the Gaußian area, H, $\frac{\mathrm{II}}{H}$, ∇H, and $\nabla\frac{\mathrm{II}}{H}$ at time -1. □

Note that the stationary hyperplane can be ruled out if we have a lower bound of the form $H(x_j,t_j) \geq \varepsilon(T-t_j)^{-\frac{1}{2}}$ along the blow-up sequence $\{(x_j,t_j)\}_{j\in\mathbb{N}}$. By Lemma 8.3, such a bound holds, for example, when $H(x_j,t_j) = \max_{M\times\{t_j\}} H$.

11.5.2. The normalized flow about type-I singularities. Recall that for type-I singularities there is a natural way to rescale the mean curvature flow equation. As a special case of (11.27a) and (11.27b), define the type-I normalization

$$\tilde{X}_{\tilde{t}} : \mathcal{M} \to \mathbb{R}^{n+1}$$

by

$$\tilde{X}_{\tilde{t}}(x) \doteqdot (2\tau)^{-1/2} X_t(x), \tag{11.32a}$$

$$\tilde{t}(t) \doteqdot -\frac{1}{2}\log\tau, \tag{11.32b}$$

where $\tau \doteqdot T - t$ and $\tilde{t} \in [-\frac{1}{2}\log T, \infty)$. Then $\mathcal{M}_{\tilde{t}} \doteqdot \tilde{X}_{\tilde{t}}(\mathcal{M}^n)$ evolves by the equation

$$\frac{d\tilde{X}}{d\tilde{t}} = -\tilde{H}\tilde{\mathsf{N}} + \tilde{X}. \tag{11.33}$$

The point is that if $\max_{M_t} |\mathrm{II}|^2 \leq \frac{C_0}{2(T-t)}$, then the rescaled second fundamental form satisfies $\left|\tilde{\mathrm{II}}\right|^2 \leq C_0$, and so it stays bounded as $t \to T$. By the scaling identities

$$d\tilde{\mu} = \frac{1}{(2\tau)^{n/2}} d\mu, \tag{11.34}$$

$$|\tilde{X}|^2 = \frac{|X|^2}{2\tau}, \tag{11.35}$$

$$\tilde{H} = \sqrt{2\tau}H, \tag{11.36}$$

$$\tilde{\mathsf{N}} = \mathsf{N}, \tag{11.37}$$

we have

$$\begin{aligned}\int_{\mathcal{M}_t} \Phi(X,\tau)d\mu &= (4\pi\tau)^{-\frac{n}{2}} \int_{\mathcal{M}_t} \mathrm{e}^{-\frac{|X|^2}{4\tau}} d\mu \\ &= (2\pi)^{-\frac{n}{2}} \int_{\mathcal{M}_t} \mathrm{e}^{-\frac{1}{2}|\tilde{X}|^2} d\tilde{\mu}\end{aligned}$$

and

$$\int_{\mathcal{M}_t} \left(H - \frac{\langle X, \mathsf{N}\rangle}{2\tau}\right)^2 \Phi(X,\tau)d\mu = (2\pi)^{-\frac{n}{2}} \frac{1}{2\tau} \int_{\mathcal{M}_t} \left(\tilde{H} - \langle \tilde{X}, \tilde{\mathsf{N}}\rangle\right)^2 \mathrm{e}^{-\frac{1}{2}|\tilde{X}|^2} d\tilde{\mu}.$$

The two formulae above and $\frac{d}{d\tilde{t}} = 2\tau\frac{d}{dt}$ (since $\frac{d\tilde{t}}{dt} = \frac{1}{2\tau}$) imply that (10.8) becomes the normalized monotonicity formula:

$$\frac{d}{d\tilde{t}} \int_M \mathrm{e}^{-\frac{1}{2}|\tilde{X}|^2} d\tilde{\mu}_{\tilde{t}} = -\int_M \left(\tilde{H} - \langle \tilde{X}, \tilde{\mathsf{N}}\rangle\right)^2 \mathrm{e}^{-\frac{1}{2}|\tilde{X}|^2} d\tilde{\mu}_{\tilde{t}}. \tag{11.38}$$

Let $X : M^n \times [0,T) \to \mathbb{R}^{n+1}$, $T < \infty$, be a solution to mean curvature flow. We say that a point $Y_0 \in \mathbb{R}^{n+1}$ is a **blow-up point** if there exists $x_0 \in \mathcal{M}^n$ such that

$$\lim_{t\to T} X_t(x_0) = Y_0$$

and

$$\sup_{t\in[0,T)} |\mathrm{II}|(x_0,t) = \infty.$$

By translating the solution we may assume the origin $0 = Y_0$ is the blow-up point. We then rescale the solution as in (11.32a) and (11.32b) to get $\tilde{X}_{\tilde{t}}$. Note that we can generalize the notion of blow-up point by either (1) replacing x_0 by x_t or (2) replacing (x_0,t) by a sequence (x_j,t_j).

We first show uniform convergence to a limit map:

Lemma 11.29. *Let $X_t : \mathcal{M} \to \mathbb{R}^{n+1}$ be type-I. Then X_t converges uniformly to a limit map $X_T : \mathcal{M} \to \mathbb{R}^{n+1}$.*

Proof. First we see that for the original solution we have for all $0 \le t_1 < t_2 < T$,

$$\begin{aligned}|X_{t_2}(x) - X_{t_1}(x)| &\le \int_{t_1}^{t_2} |H(x,t)|\,dt \\ &\le C\int_{t_1}^{t_2} (T-t)^{-1/2}\,dt \\ &\le 2C\left((T-t_1)^{1/2} - (T-t_2)^{1/2}\right) \\ &\le 2C(t_2-t_1)^{1/2}.\end{aligned}$$

The claim follows. □

Next we note that at least one point for the rescaled hypersurfaces stays in a bounded region: Taking $x = x_0$, $t_1 = t$, and $t_2 \to T$ and since $\lim_{t_2\to T} X_{t_2}(x_0) = 0$, we have for all $t \in [0,T)$,

$$|X_t(x_0)| \le 2Ct^{1/2}.$$

Hence for all $\tilde{t} \in [-\frac{1}{2}\log T, \infty)$ we have the uniform bound

$$\left|\tilde{X}_{\tilde{t}}(x_0)\right| = (2\tau)^{-1/2}|X_t(x_0)| \le \sqrt{2}C. \tag{11.39}$$

We next obtain the crucial uniform bound on the derivatives of the rescaled hypersurfaces, similarly to Theorem 6.20.

Proposition 11.30. *Given a type-I solution, the second fundamental form of the rescaled solutions* (11.32a) *satisfies*

$$\left|\tilde{\nabla}^m\tilde{\mathrm{II}}\right|^2 \le C_m. \tag{11.40}$$

Proof. For any $x \in \mathcal{M}$ and $\tilde{t} \in [-\frac{1}{2}\log T, \infty)$ we have

$$\left|\widetilde{\mathrm{II}}\right|(x,\tilde{t}) = (2\tau)^{1/2}\left|\mathrm{II}\right|(x,t) \leq \sqrt{2}C,$$

where we used the type-I condition. We also obtain uniform bounds for the higher derivatives of $\widetilde{\mathrm{II}}$. The reason is that from (6.66) we see that the rescaled hypersurfaces $\tilde{X}_{\tilde{t}}$ satisfy

$$\frac{\partial}{\partial \tilde{t}}\left|\tilde{\nabla}^m\widetilde{\mathrm{II}}\right|^2 \leq \tilde{\Delta}\left|\tilde{\nabla}^m\widetilde{\mathrm{II}}\right|^2 - 2\left|\tilde{\nabla}^{m+1}\widetilde{\mathrm{II}}\right|^2 + \sum_{i+j+k=m} \tilde{\nabla}^i\widetilde{\mathrm{II}} * \tilde{\nabla}^j\widetilde{\mathrm{II}} * \tilde{\nabla}^k\widetilde{\mathrm{II}} * \tilde{\nabla}^m\widetilde{\mathrm{II}}$$

since the rescaling only adds the negative term $-2(m-1)\left|\tilde{\nabla}^m\widetilde{\mathrm{II}}\right|^2$ to the equation (see Lemma 2.2 and Proposition 2.3 of [**296**]). We can then apply the Bernstein-type estimates for the higher derivatives of $\widetilde{\mathrm{II}}$ to obtain uniform bounds for $\left|\tilde{\nabla}^m\widetilde{\mathrm{II}}\right|^2$ (since $\left|\widetilde{\mathrm{II}}\right|$ has uniform bounds).

In particular, assume by induction that the $\left|\tilde{\nabla}^\ell\widetilde{\mathrm{II}}\right|^2$ are uniformly bounded for $1 \leq \ell \leq m-1$. We then have for any time $\tilde{t}_0$,

$$\begin{aligned}\frac{\partial}{\partial \tilde{t}}\left((\tilde{t}-\tilde{t}_0)\left|\tilde{\nabla}^m\widetilde{\mathrm{II}}\right|^2\right) &\leq \tilde{\Delta}\left((\tilde{t}-\tilde{t}_0)\left|\tilde{\nabla}^m\widetilde{\mathrm{II}}\right|^2\right) + \left|\tilde{\nabla}^m\widetilde{\mathrm{II}}\right|^2 \\ &\quad + C_1\left(1+\left|\tilde{\nabla}^m\widetilde{\mathrm{II}}\right|^2\right), \\ \frac{\partial}{\partial \tilde{t}}\left|\tilde{\nabla}^{m-1}\widetilde{\mathrm{II}}\right|^2 &\leq \tilde{\Delta}\left|\tilde{\nabla}^{m-1}\widetilde{\mathrm{II}}\right|^2 - 2\left|\tilde{\nabla}^m\widetilde{\mathrm{II}}\right|^2 + C_2,\end{aligned}$$

where C_1 and C_2 depend on n, m, and the bounds for $\left|\tilde{\nabla}^\ell\widetilde{\mathrm{II}}\right|^2$, $1 \leq \ell \leq m-1$. Hence

$$\begin{aligned}&\left(\frac{\partial}{\partial \tilde{t}} - \tilde{\Delta}\right)\left((\tilde{t}-\tilde{t}_0)\left|\tilde{\nabla}^m\widetilde{\mathrm{II}}\right|^2 + (C_1+1)\left|\tilde{\nabla}^{m-1}\widetilde{\mathrm{II}}\right|^2\right) \\ &\qquad \leq C_1 + (C_1+1)\,C_2.\end{aligned}$$

Since $\left|\tilde{\nabla}^{m-1}\widetilde{\mathrm{II}}\right|^2$ is uniformly bounded, we conclude that

$$\begin{aligned}(\tilde{t}-\tilde{t}_0)\left|\tilde{\nabla}^m\widetilde{\mathrm{II}}\right|^2 &\leq (\tilde{t}-\tilde{t}_0)\left|\tilde{\nabla}^m\widetilde{\mathrm{II}}\right|^2 + (C_1+1)\left|\tilde{\nabla}^{m-1}\widetilde{\mathrm{II}}\right|^2 \\ &\leq C_3 + (C_1 + (C_1+1)\,C_2)\,(\tilde{t}-\tilde{t}_0)\end{aligned}$$

for all $\tilde{t} \geq \tilde{t}_0$. Applying this estimate for $\tilde{t} \in \left[\tilde{t}_0, \tilde{t}_0+1\right]$ yields a uniform bound for $\left|\tilde{\nabla}^m\widetilde{\mathrm{II}}\right|^2$ since $\tilde{t}_0$ is arbitrary. □

The following gives a rescaled version of Theorem 11.26.

Theorem 11.31 (Type-I singularity models (rescaled version))**.** *Let $X : M^n \times [0,T) \to \mathbb{R}^{n+1}$ be a compact immersed solution to mean curvature flow. Given any sequence of times $\tilde{t}_j \to \infty$, there is a subsequence along* *which the rescaled immersions* $\tilde{X}_{\tilde{t}_j}$ *converge locally uniformly in the smooth topology to a* ***self-similar shrinker****, that is, a hypersurface satisfying*

$$\tilde{H}_\infty = \langle \tilde{X}_\infty, \tilde{\mathsf{N}}_\infty \rangle. \tag{11.41}$$

Proof. Using (11.39) and Proposition 11.30, Theorem 11.6 implies that a subsequence of the rescaled immersions $\tilde{X}_{\tilde{t}_i}$ converges locally uniformly in the smooth topology to a smoothly immersed limit hypersurface. By the rescaled monotonicity formula (11.38),

$$\int_{s_0}^{\infty} \int_M \left(\tilde{H} - \langle \tilde{X}, \tilde{\mathsf{N}} \rangle \right)^2 \mathrm{e}^{-\frac{1}{2}|\tilde{X}|^2} d\tilde{\mu}_{\tilde{t}} < \infty\,.$$

So the limit is a self-similar shrinker. □

Of course, by the splitting theorem, Theorem 4.4, and Lemma 11.27, the limit must be either a round sphere, a round orthogonal cylinder, or an orthogonal Abresch–Langer cylinder whenever the underlying flow is mean convex (cf. Corollary 11.28).

11.5.3. Singularity models for type-II singularities. We will prove that every type-II singularity in mean convex mean curvature flow admits an associated singularity model which is a product $\{\mathbb{R}^m \times \Sigma_t^{n-m}\}_{t\in(-\infty,\infty)}$ for some $m \in \{0, \ldots, n-1\}$, where $\{\Sigma_t\}_{t\in(-\infty,\infty)}$ is a locally uniformly convex translating solution of dimension $n - m$. The idea of the proof is again a straightforward extension of the convex case:

After rescaling the flow in such a way that the second fundamental form of the rescaled hypersurfaces is uniformly bounded, we may apply the compactness theorem to obtain an eternal limit solution. By the convexity estimate (Theorem 9.12) and the splitting theorem (Theorem 9.11), it will split into a product of a lower-dimensional plane with a nontrivial locally uniformly convex eternal solution. By Hamilton's Harnack inequality (Theorem 10.11), the latter is a translating self-similar solution provided H attains its space-time maximum.

Theorem 11.32 (Type-II singularity models). *Let* $X : M^n \times [0, T) \to \mathbb{R}^{n+1}$ *be a compact, mean convex solution to mean curvature flow which undergoes a type-II singularity at time* T. *Then there are sequences of times* $t_j \in [0, T)$, *points* $x_j \in M^n$, *and scales* $\lambda_j > 0$ *such that the sequence of rescaled solutions* $X_j : M^n \times [-\lambda_j^2 t_j, \lambda_j^2(T - t_j - j^{-1})) \to \mathbb{R}^{n+1}$ *defined by*

$$X_j(x,t) \doteqdot \lambda_j \left(X(x, \lambda_j^{-2} t + t_j) - X(x_j, t_j) \right)$$

converges locally uniformly in the smooth topology, after passing to a subsequence and applying a rigid motion, to a product $\{\mathbb{R}^m \times \Sigma_t^{n-m}\}_{t\in(-\infty,\infty)}$

for some $m \in \{0, \dots, n-1\}$, where $\{\Sigma_t^{n-m}\}_{t\in(-\infty,\infty)}$ is a locally uniformly convex translating self-similar solution.

Proof. As in Hamilton's proof of Huisken's theorem, we choose a sequence of times $t_j \in [0, T)$ and a sequence of points $x_j \in M$ such that

$$H^2(x_j,t_j)\left(T-\frac{1}{j}-t_j\right) = \max_{(x,t)\in M\times\left[0,T-\frac{1}{j}\right]} H^2(x,t)\left(T-\frac{1}{j}-t\right)$$

and set $\sigma_j \doteqdot \lambda_j^2 t_j$ and $T_j = \lambda_j^2\left(T-\frac{1}{j}-t_j\right)$, where $\lambda_j \doteqdot H(x_j,t_j)$. Since the singularity is of type-II, we can pass to some subsequence such that $t_j \to T$, $\lambda_j \to \infty$, $\sigma_j \to \infty$, and $T_j \to \infty$. Observe also that $X_j(x_j, 0) = 0$ and

$$(11.42)\quad H_j^2(x,t) = \lambda_j^{-2}H^2\left(x,t_j+\lambda_j^{-2}t\right) \le \frac{T-\frac{1}{j}-t_j}{T-\frac{1}{j}-(t_j+\lambda_j^{-2}t)} = \frac{T_j}{T_j-t}.$$

So the curvature is uniformly bounded on any compact time interval. By Theorem 11.11 and a diagonal subsequence argument, the rescaled solutions $X_j : M \times [-\sigma_j, T_j) \to \mathbb{R}^{n+1}$ converge in C^∞, after passing to a subsequence, to a smooth, proper limiting solution $X_\infty : M_\infty \times (-\infty,\infty) \to \mathbb{R}^{n+1}$ uniformly on compact subsets of $M_\infty \times (-\infty,\infty)$.

Since the underlying flow is mean convex, the limit satisfies $H \ge 0$. In fact, since $H = 1 > 0$ at the space-time origin, it must be strictly positive everywhere by the strong maximum principle. It then follows from the convexity estimates (Theorem 9.12) that the limit is weakly convex. By the splitting theorem (Theorem 9.11) it splits as a product of a lower- (possibly zero) dimensional plane and a locally uniformly convex solution. By (11.42), $H \le 1$, so the maximum $H = 1$ is attained (at the space-time origin). The rigidity case of Hamilton's Harnack inequality (Theorem 10.11) then implies that the limit is a translating solution to the flow. □

11.5.4. Tangent flows are shrinkers. We have seen that flows undergoing type-I singularities always admit a sequence of rescalings which converge to a self-similarly shrinking solution. Minor adaptations to the proof show that all *tangent flows* about a *type-I* singularity are self-similarly shrinking solutions. In fact, Ilmanen showed that we *always* obtain self-similarly shrinking solutions as tangent flows (independent of type), so long as we are willing to interpret the convergence and limits in the weak sense of Brakke flows.

Theorem 11.33 (Existence of weak tangent flows). *Let $\{\mathcal{M}_t^n\}_{t\in[0,t_0)}$ be a compact solution to mean curvature flow. Given $(p_0, t_0) \in \mathbb{R}^{n+1}\times\mathbb{R}$ and* *a sequence* $\lambda_j \to \infty$, *a subsequence of the mean curvature flows* $\{\mathcal{M}^n_{t,j}\}_{t\in I_j}$ *defined by*

$$\mathcal{M}^n_{t,j} = \lambda_j(\mathcal{M}^n_{\lambda_j^{-2}t+t_0} - p_0)\,, \quad I_j \doteqdot \{\lambda_j^2(s-t_0) : s \in [0,t_0)\} \tag{11.43}$$

converges to a limit integral Brakke flow $\{\mu_t\}_{t\in(-\infty,0)}$ *in the sense of Brakke flows. The limit integral Brakke flow is self-similarly shrinking, in the sense that*

$$\mu_t(K) = \lambda^{-n}\mu_{\lambda^2 t}(\lambda K) \tag{11.44}$$

for all $\lambda > 0$, $t < 0$, *and compact* $K \subset \mathbb{R}^{n+1}$, *and*

$$\vec{H}(x) + \frac{x^\perp}{2} = 0 \tag{11.45}$$

for μ_{-1}*-almost every* $x \in \mathbb{R}^{n+1}$, *where* $\vec{H}$ *is the weak mean curvature vector of* μ_{-1} *and* $\cdot^\perp$ *is the projection onto the* (μ_{-1}*-almost everywhere defined*) *tangent space of* μ_{-1}.

Proof. See Ilmanen [**309**, Lemma 8]. □

This appears to be in conflict with Theorem 11.32, which asserts that the flow should look like a self-similarly *translating* solution near a type-II singularity. The reason for the apparent discrepancy is that the curvature normalizing sequence "follows" the points of highest curvature, which eventually form the "tip" of the translating limit flow, whereas the tangent flow focuses on a fixed ambient space-time point approached by these points. In the latter case, the points of highest curvature are moving too fast to be picked up in the limit of the rescalings based at the fixed ambient space-time point. In the rotationally symmetric degenerate neckpinch example, the curvature normalizing sequence limits to the rotationally symmetric bowl soliton (see Section 13.8), whereas the rescalings about the fixed ambient point only see the "neck" part of the degenerate neckpinch, which is at lower curvature scales, and in the limit forms a shrinking cylinder.

11.5.5. Conjectures on singularity formation in mean curvature flow.

Tom Ilmanen has posed a number of conjectures on singularity formation for mean curvature flow [**309–311**]. In this subsection we discuss some of these as well as conjectures stemming from the work of Brendle, Colding, Huisken, Ilmanen, Minicozzi, Sinestrari, and White. We say that an integral Brakke flow $\{\mu_t\}$ is **smooth** if $\operatorname{spt}\mu_t$ is a smooth hypersurface for each t.

Theorem 11.34 ([**309**, Theorems 1 and 2]). *If* $\{\mathcal{M}_t\}$ *is an embedded mean curvature flow in* $\mathbb{R}^3$, *then any tangent flow* $\{\mu_t\}$ *is smooth. Moreover,* $\mathcal{M}_{t,\infty} \doteqdot \operatorname{spt}\mu_t$ *is a shrinking self-similar solution* (*cf. Theorem* 11.33).

Figure 11.5. Tom Ilmanen. Author: Renate Schmid. Source: Archives of the Mathematisches Forschungsinstitut Oberwolfach.

As a caveat, Ilmanen further states [**309**, p. 4]:

> "The reader should beware that the convergence to the blowup might not be smooth, even when the support of the blowup is smooth; there might, in principle, be small necks and other topological kinks that pinch off and disappear in the limit."

In [**309**, p. 7] he conjectured that this cannot occur.

Conjecture 11.35. *If $\{\mathcal{M}_t\}$ is an embedded mean curvature flow in $\mathbb{R}^3$, then the sequence of rescaled surfaces converges smoothly and with multiplicity one to the tangent flow $\{\mu_t\}$.*

We emphasize that there are two parts to this conjecture, which we call the **smooth convergence conjecture** (that the convergence is smooth) and the **multiplicity one conjecture** (that each tangent flow at the first singular time has multiplicity one). The relation between these two conjectures is explained below.

A natural question is: To what extent are singularities nice in all dimensions? In the following we assume that we have a mean curvature flow of a hypersurface in $\mathbb{R}^{n+1}$.

Conjecture 11.36 (Multiplicity one). *Tangent flows to mean curvature flows of embedded hypersurfaces are necessarily multiplicity one.*

Ilmanen comments: "(1) In higher dimensions, the tools for proving this are weaker, so the evidence for it is weaker. (2) The conjecture is worth making not just for the first singular time, but also for 'embedded' flows (appropriately defined) even with singularities allowed. (3) The conjecture is trivially false for immersed flows: Two sheets of the immersion will develop a momentary point of tangency just before becoming embedded."

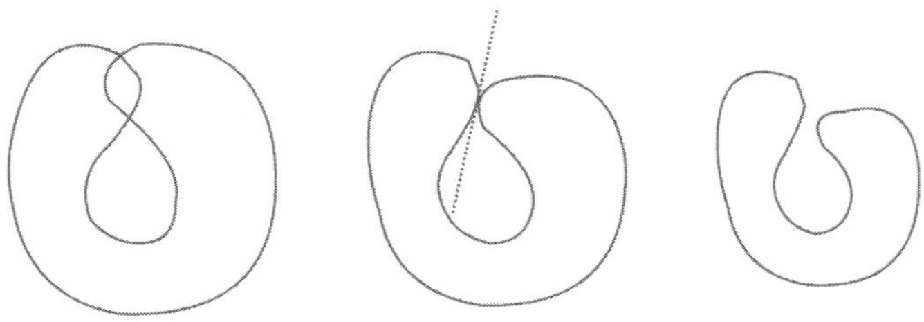

Figure 11.6. A solution to curve shortening flow with a momentary point of tangency.

Problem 11.37 (Smoothness)**.** Are tangent flows of mean curvature flow of embedded hypersurfaces necessarily smooth in sufficiently low dimensions?

Regarding the problem above, Ilmanen has stated the following: "(1) This is a very accessible conjecture for the boundary of the level-set flow in $\mathbb{R}^3$ even past the first singularity. (2) The conjecture fails in $\mathbb{R}^8$ because there are rotationally symmetric examples of Velázquez [**513**] with singularities modeled on area minimizing cones. (3) In $\mathbb{R}^4$ there are (unstable) minimal cones with point singularities. Can such a cone be the tangent flow of a smoothly embedded mean curvature flow? For related work, see Angenent, Ilmanen, and Velázquez [**72**]."

Problem 11.38 (Smooth convergence)**.** Given a tangent flow of an embedded mean curvature flow, does any associated sequence of rescaled surfaces converge smoothly to the tangent flow?

According to Ilmanen, we have the following regarding the above: "(1) If the tangent flow has multiplicity one, then this follows immediately from Brakke's theorem. (2) Smooth convergence in the higher multiplicity case is impossible."

Conjecture 11.39 (Uniqueness for tangent flows)**.** *Tangent flows of immersed smooth mean curvature flows at the first singular time are unique.*

Regarding the problem above, Ilmanen wrote: "(1) This is more a question than a conjecture, since there are some evil examples of nonuniqueness for other problems. (2) You could call this the problem of 'crypto-irregularity', where something is technically smooth, but actually belongs to a family that does not have estimates, so the smoothness is an illusion and does not help. (3) I would hesitate to make a conjecture for uniqueness of MCF [mean curvature flow] tangent flows outside of the smooth world. There I would look for counterexamples instead."

Problem 11.40 (Self-similarity of limit flows)**.** In general, are limit flows of embedded mean curvature flows either self-similar shrinkers, translators (including static minimal hypersurfaces), or expanders?

Ilmanen made the following remarks: "(1) Assume that the mean curvature flow, with singularities and possibly empty, is defined for all nonnegative time. Tangent flows are shrinkers for negative time. Conjecture: Tangent flows are expanders for positive time. (2) Using a monotonicity formula, I proved this in the case of an isolated singularity with a smooth tangent cone (unpublished)."

Problem 11.41 (Uniqueness for limit flows)**.** Is there some sort of duplicity for limit flows of embedded mean curvature flows in low dimensions, having in mind that a "standard" rotationally symmetric degenerate neckpinch has only two associated limit flows: the bowl soliton and its "dimension reduction", the shrinking cylinder?

Problem 11.42 (Canonical neighborhood)**.** Is there a canonical neighborhood theorem for embedded mean curvature flows in low dimensions?

For related work, see Colding, Ilmanen, Minicozzi, and White [**178**] and K. Choi, R. Haslhofer, and O. Hershkovits [**151**].

In each of the above questions, can one obtain an affirmative answer under some assumptions? The multiplicity one conjecture for planar tangent flows is Problem 2 (**nonsqueezing conjecture**) in Ilmanen [**311**]. The uniqueness of multiplicity one tangent flows, without assuming embeddedness, is Problem 3 in [**311**]. The self-similarity conjecture (including expanders) for limit flows is Problem 6 in [**311**].

Colding and Minicozzi [**183**, Theorem 0.2] proved, via a Łojasiewicz-type inequality, the following uniqueness theorem for cylindrical tangent flows.

Figure 11.7. Tobias Colding. Photo courtesy of Bryce Vickmark/MIT.

Theorem 11.43. *If a mean curvature flow of a compact hypersurface has a tangent flow at $p_0 \in \mathbb{R}^{n+1}$ which is a multiplicity one cylinder $S^k \times \mathbb{R}^{n-k}$, then every tangent flow at p_0 must be ambient isometric to this cylinder. That is, all tangent flows at p_0 are exactly the same cylinder in $\mathbb{R}^{n+1}$.*

Following the general scheme of Colding and Minicozzi and taking the advantage of some stronger properties of cones, O. Chodosh and F. Schulze [**147**] proved uniqueness in the case of asymptotically conical tangent flows. For the uniqueness of compact tangent flows, see Schulze [**474**].

Ilmanen (see Problem 4 in [**311**]) conjectured the following.

Conjecture 11.44. *If a mean curvature flow of a compact hypersurface has a tangent flow at $(p_0, t_0) \in \mathbb{R}^{n+1} \times \mathbb{R}$ which is a multiplicity one cylinder $S^k \times \mathbb{R}^{n-k}$, then there exists $\varepsilon > 0$ such that the mean curvature of the solution is positive in the extrinsic parabolic cylinder $P_\varepsilon(p_0, t_0) = B_\varepsilon(p_0) \times (t_0 - \varepsilon^2, t_0]$.*

This conjecture has been proved for $n = 2$ by Choi, Haslhofer, and Hershkovits [**151**].

In [**501**] Ao Sun proved for $n = 2$ that the multiplicity one conjecture for *compact* tangent flows is true in a generic sense.

11.5.6. Mean curvature flow with surgery.

Let us describe some of the ingredients needed for the mean curvature flow with surgery construction of Huisken and Sinestrari. Given positive constants α_0, α_1, α_2, and R, recall that a compact, immersed hypersurface $X : M^n \to \mathbb{R}^{n+1}$, $n \geq 3$, **is in the class** $\mathcal{C}_2^n(R, \alpha)$, where $\alpha \doteqdot (\alpha_0, \alpha_1, \alpha_2)$, if $\kappa_1 + \kappa_2 \geq \alpha_0 H$, $H \geq \alpha_1 R^{-1}$, and $\mu(M^n) \leq \alpha_2 R^n$.

The following is Huisken and Sinestrari's surgery definition in abstract; see pp. 145–146 of [**303**]. A **mean curvature flow with surgeries** is a sequence of smooth mean curvature flows $X_i : M_i^n \times [T_{i-1}, T_i] \to \mathbb{R}^{n+1}$, $1 \leq i \leq N$, where $X_{i+1}(\cdot, T_i) : M_{i+1}^n \to \mathbb{R}^{n+1}$ is defined for each $1 \leq i \leq N-1$ as follows. First, an immersion $\hat{X}_{i+1}(\cdot, T_i) : \hat{M}_i^n \to \mathbb{R}^{n+1}$ is obtained from $X_i(\cdot, T_i) : M_i^n \to \mathbb{R}^{n+1}$ by a "standard surgery" which replaces finitely many necks in M_i^n by pairs of spherical caps. Second, each connected component of $\hat{M}_i^n$ which is recognized as being diffeomorphic to S^n or to $S^{n-1} \times S^1$ is discarded to obtain M_{i+1}^n. Then $X_{i+1}(\cdot, T_i)$ is the restriction of $\hat{X}_{i+1}(\cdot, T_i)$ to M_{i+1}^n.

A mean curvature flow with surgeries **terminates after finitely many steps** at time T_N provided, for M_N^n at time T_N or for $\hat{M}_N^n$ at time T_N (obtained by a standard surgery), each connected component is recognized as being diffeomorphic to S^n or to $S^{n-1} \times S^1$.

Given threshold parameters $H_1 < H_2 < H_3$, we consider mean curvature flows with surgeries satisfying the following properties:

(i) For each $1 \leq i \leq N$, T_i is the first time t in $[T_{i-1}, T_i]$ for which the maximum mean curvature of the hypersurface $X_i(\cdot, t)$ is at least (and thus equal to) H_3.

(ii) The mean curvature of the postsurgery hypersurface $X_{i+1}(\cdot, T_i)$ is at most H_2.

(iii) Each surgery is performed on necks whose end $(n-1)$-spheres lie in the region of the hypersurface where the mean curvature is in the interval $[H_1/2, 2H_1]$.

Huisken and Sinestrari prove that, for any initial hypersurface in a class $\mathcal{C}_2^n(R, \alpha)$, one can construct a mean curvature flow with surgey which terminates after finitely many steps and satisfies the properties (i)–(iii) above with threshold parameters $H_i = h_i R^{-1}$, where h_i depend only on $\alpha_0, \alpha_1, \alpha_2$, and $\varepsilon > 0$ (a sufficiently small parameter which determines the "quality" of the necks on which the surgeries are performed).

11.5.7. Piecewise smooth mean curvature flow. Colding and Minicozzi [**181**] define a **piecewise smooth mean curvature flow** as a sequence of smooth mean curvature flows $X_i : M^n \times [T_{i-1}, T_i] \to \mathbb{R}^{n+1}$, $1 \leq i \leq N$, with the following properties:

(i) $X_{i+1}(\cdot, T_i) : M^n \to \mathbb{R}^{n+1}$ is a normal graph over $X_i(\cdot, T_i) : M^n \to \mathbb{R}^{n+1}$.

(ii) $\operatorname{Area}(X_{i+1}(\cdot, T_i)) = \operatorname{Area}(X_i(\cdot, T_i))$.

(iii) $\lambda(X_{i+1}(\cdot, T_i)) \leq \lambda(X_i(\cdot, T_i))$, where λ is the **entropy** (see (11.49)).

(iv) The final hypersurface $X_N(\cdot, T_N)$ is allowed to be singular.This definition is inspired by the following theorem [**181**, Theorem 0.12].

Theorem 11.45. *Let $\mathcal{M}^n$ be a complete, embedded shrinking self-similar solution to mean curvature flow in $\mathbb{R}^{n+1}$ with polynomial area growth (see Definition* 13.12*). If $\mathcal{M}^n$ is not isometric to a cylinder $S^m \times \mathbb{R}^{n-m}$, then for each $j \in \mathbb{N}$ there exists a function u_j on $\mathcal{M}$ with $\|u_j\|_{C^j} \leq 1/j$ and $\lambda(\mathcal{M}_j^n) < \lambda(\mathcal{M}^n)$, where $\mathcal{M}_j^n \doteqdot \{x + u_j(x)\mathsf{N}(x) : x \in \mathcal{M}^n\}$ is the normal graph of u_j over $\mathcal{M}^n$. Furthermore, if $\mathcal{M}^n$ does not split off a line, then u_j may be chosen to have compact support.*

This result should provide the "invisible hand" to adjust (to effect replacements for) the mean curvature flow solution to avoid noncylindrical shrinking self-similar solutions as blow-ups.

Based on the aforementioned works, one may pose the following.

Conjecture 11.46. *If $\mathcal{M}_0$ is an embedded, compact surface in $\mathbb{R}^3$, then there exists a piecewise smooth mean curvature flow with surgeries (a hybrid of piecewise smooth mean curvature flow and mean curvature flow with surgeries) starting at $\mathcal{M}_0$ which terminates after finitely many steps.*

This would provide another proof of the 3-dimensional Schoenflies theorem. Apparently, the main difficulty is the multiplicity one conjecture and in particular showing that higher multiplicity cylinders cannot occur as tangent flows. For a potential application of mean curvature flow of surfaces to give a new proof of the Smale conjecture, proposed by Shing-Tung Yau, see Hyam Rubinstein [**448**]. For a discussion of all of this, see Choi, Haslhofer, and Hershkovits [**151**, §1.2].

A further result supporting the conjecture is the smooth compactness theorem for 2-dimensional shrinkers of Colding and Minicozzi [**182**].

Theorem 11.47. *Given a nonnegative integer g and a positive constant V, let $\mathcal{S} = \mathcal{S}(g, V)$ denote the space of complete, embedded mean curvature flow shrinkers $\mathcal{M}$ in $\mathbb{R}^3$, where $\mathcal{M}$ is diffeomorphic to a closed surface of genus at most g minus a finite number of points and where $\operatorname{Area}(\mathcal{M} \cap B_r(p)) \leq Vr^2$ for all $p \in \mathbb{R}^3$ and $r > 0$. Then $\mathcal{S}$ is compact in the smooth topology uniformly on compact subsets; that is, for any sequence in $\mathcal{S}$, there exists a subsequence which converges to a complete, embedded mean curvature flow shrinker in C^k on compact sets for each k.*

By the compactness theorem, the entropy drops by at least a fixed amount after each replacement in piecewise smooth mean curvature flow.

For higher dimensions, any program for mean curvature flow with surgery should be more complicated.

Conjecture 11.48. *If $\mathcal{M}_0$ is a compact, embedded hypersurface of $\mathbb{R}^4$, then there exists a piecewise smooth mean curvature flow with surgeries (defined as a hybrid of piecewise smooth mean curvature flow and mean curvature flow with surgeries) starting at $\mathcal{M}_0$. Topologically, the transitions from $\mathcal{M}_i$ to $\mathcal{M}_{i+1}$ are finite collections of 1-surgeries and 2-surgeries.*

Consider the following nefarious example. Let K be a knot in S^3, where S^3 is embedded in $\mathbb{R}^4$ as the unit sphere. Since any loop in $\mathbb{R}^4$ is unknotted, there exists a diffeomorphism $\phi : \mathbb{R}^4 \to \mathbb{R}^4$ with the following properties. The image of the knot $\phi(K)$ is a circle in $\mathbb{R}^2 \times \{(0, 0)\}$ of arbitrarily small radius. A tubular neighborhood N of $\phi(K)$ in $\phi(S^3)$ is the warped product of $S^1 \subset \mathbb{R}^2$ and $D^2 \subset \mathbb{R}^2$, where D^2 is a unit disk and the radii $r = r(|y|)$ of the S^1 is an increasing radial function of $y \in D^2$. So the minimal radius is that of $\phi(K)$. Consider the mean curvature flow with initial hypersurface given by $\phi(S^3)$. Then provided that r is defined suitably, mean curvature flow with surgery should topologically perform a 1-surgery on the knot $\phi(K)$ in $\phi(S^3)$. That is, it should remove a tubular neighborhood $S^1 \times D^2$ of $\phi(K)$ and glue in $D^2 \times S^1$ along the boundary $S^1 \times S^1$. See Figure 11.8.

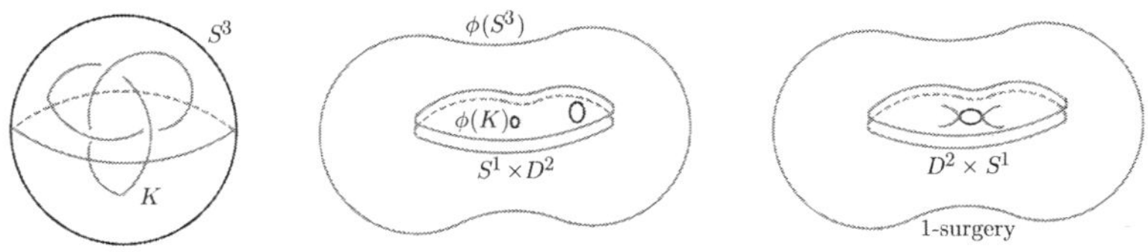

Figure 11.8. Mean curvature flow performing a 1-surgery. L: A knot K in S^3. M: An embedding ϕ of S^3 where $\phi(K)$ is a small circle. R: The embedded 3-manifold after a 1-surgery on $\phi(K)$.

Problem 11.49. Formulate an approach toward the 4-dimensional smooth Schoenflies conjecture via piecewise smooth mean curvature flow with surgery.

Some of the estimates required for the surgery program are discussed in Chapters 9 and 12.

Colding and Minicozzi have suggested approaching the smooth 4-dimensional Poincaré conjecture using mean curvature flow based on the fact that every exotic 4-sphere embeds smoothly in $\mathbb{R}^5$; see [**188**].

11.6. Notes and commentary

11.6.1. A local compactness theorem. Ecker's local monotonicity formula can be used to localize Theorem 11.12.

Theorem 11.50 (Compactness theorem for mean curvature flows III). *Let $\{X_k : M_k^n \times [-T_k, 0] \to \mathbb{R}^{n+1}\}_{k\in\mathbb{N}}$ be a sequence of solutions to mean curvature flow, and let $\{x_k\}_{k\in\mathbb{N}}$ be a sequence of points $x_k \in M_k^n$. Suppose that the following conditions are met:*

(i) $0 \in I_k$ *and* $X_k(x_k, 0) = 0$ *for every* $k \in \mathbb{N}$.

(ii) *There exists an exhaustion $\{U_k\}_{k\in\mathbb{N}}$ of $U \subset \mathbb{R}^{n+1}$ by precompact open sets U_k satisfying $\overline{U}_k \subset U_{k+1}$ such that each X_k is* properly defined *in $U_k \times [-T + \frac{1}{k}, 0]$.*

(iii) *For each k there exist $C_k < \infty$ and $k \in \mathbb{N}$ such that*

$$\sup_{(X_k(x,t),t)\in U_k\times[-T+\frac{1}{k},0]} |\mathrm{II}_k(x,t)| \le C_k\,.$$

(iv) *For every $k > 0$ there exists $V_k < \infty$ such that*

$$\mu_{k,-T+\frac{1}{k}}(X^{-1}_{k,-T+\frac{1}{k}}(U_k)) \leq V_k$$

for all $k \in \mathbb{N}$, where $\mu_{k,t}$ is the Riemannian measure induced by the immersion $X_{k,t} \doteqdot X_k(\cdot, t)$.

There exist a smooth n-manifold M^n_∞, a smooth solution $X_\infty : M^n_\infty \times (-T, 0] \to \mathbb{R}^{n+1}$ to mean curvature flow which is properly defined in $U \times (-T, 0]$, and a subsequence of $\{X_k : M^n_k \times [-T_k, 0] \to \mathbb{R}^{n+1}\}_{k\in\mathbb{N}}$ which converges smoothly on compact subsets of $U \times (-T, 0]$ to $X_\infty : M^n_\infty \times (-T, 0] \to \mathbb{R}^{n+1}$.

11.6.2. A local Brakke–White regularity theorem. The local monotonicity formula can also be used to prove a local version of the Brakke–White regularity theorem. Given a solution $X : M^n \times I \to \mathbb{R}^{n+1}$ to mean curvature flow properly defined in $B_{\sqrt{2n}R}(p_0) \times (t_0 - R^2, t_0]$, set

$$\Theta^R(p, t, r) \doteqdot \int_{\mathcal{M}} \varphi^R_{(p,t)} \rho_{(p,t)} \, d\mu_{t-r^2} \,. \tag{11.46}$$

Replacing Θ by Θ^R in the proof of Brakke's regularity theorem (Theorem 6.29) yields the following local version.

Theorem 11.51 (Brakke's local regularity theorem for smooth flows). *There exist constants $\varepsilon > 0$ and $C < \infty$ with the following property: Let $X : \mathcal{M}^n \times I \to \mathbb{R}^{n+1}$ be a smooth mean curvature flow which is properly defined in $B_{\sqrt{1+2n}R}(p_0) \times (t_0 - R^2, t_0]$ and satisfies*

$$\sup_{(p,t)\in B_r(p_0)\times(t_0-r^2,t_0]} \Theta^R(p, t, r) < 1 + \varepsilon$$

for some $r \in (0, R)$. Then

$$\sup_{B_{r/2}(p_0)\times(t_0-r^2/4,t_0]} |\mathrm{II}| \leq Cr^{-1} \,. \tag{11.47}$$

11.6.3. Curvature pinching and compactness. Curvature pinching results have formed a major topic of study in Riemannian geometry. For example, the Berger–Klingenberg sphere theorem says that if a Riemannian manifold $(\mathcal{M}^n, g)$ has sectional curvatures $1/4 < \operatorname{sect}(g) \leq 1$, then $\mathcal{M}^n$ is homeomorphic to the n-sphere (see [**140**, Chapters 5 and 6]). Later, Huisken [**292**], Margerin [**386**], and Nishikawa [**426**] proved differentiable spherical space form theorems assuming only pointwise curvature pinching by using Hamilton's Ricci flow. In particular they showed that there exist constants $\delta_n > 0$ such that if $(\mathcal{M}^n, g)$ is a closed Riemannian manifold with

$$\frac{\max \operatorname{sect}(x)}{\min \operatorname{sect}(x)} \leq 1 + \delta_n$$

for all $x \in \mathcal{M}$ ($\max\operatorname{sect}(x)$ and $\min\operatorname{sect}(x)$ denote the maximum and minimum sectional curvatures at x, respectively), then $\mathcal{M}^n$ is diffeomorphic to a spherical space form. Differentiable spherical space form theorems culminated in the spectacular works of Böhm and Wilking [**93**] on 2-positive curvature operator and Brendle and Schoen [**117**] on pointwise 1/4-pinched sectional curvature, which also used Ricci flow.

The following analogue of Theorem 11.20 was proved by Chen and Zhu using Ricci flow [**141**].

Theorem 11.52. *If* $(\mathcal{M}^3, g_0)$ *is a* 3*-dimensional complete nonflat Riemannian manifold with bounded nonnegative sectional curvature and* $\mathrm{Rc}_0 \geq \varepsilon R_0 g_0$ *for some* $\varepsilon > 0$, *then* $\mathcal{M}$ *is compact.*

11.6.4. Generic singularities and uniqueness of tangent flows. In their seminal work [**181**], Colding and Minicozzi generalized Huisken's monotonicity result for type-I singularities to generic smooth closed surfaces in $\mathbb{R}^3$. Motivated by the Gaußian area, the authors define the $\mathcal{F}$-functional and related entropy as follows: Let $\mathcal{M} \subset \mathbb{R}^{n+1}$ be a complete, embedded hypersurface and fix $(x_0, t_0) \in \mathbb{R}^{n+1} \times \mathbb{R}_+$. Then define the $\mathcal{F}$**-functional** based at x_0 and at the scale t_0 by

$$\mathcal{F}_{x_0,t_0}(\mathcal{M}) \doteqdot (4\pi t_0)^{-n/2} \int_{\mathcal{M}} \mathrm{e}^{-\frac{|x-x_0|^2}{4t_0}} \, d\mu. \tag{11.48}$$

Given $c \in \mathbb{R}_+$, define the map $c_{x_0} : \mathbb{R}^{n+1} \to \mathbb{R}^{n+1}$ by $c_{x_0}(x) = x_0 + c(x - x_0)$. Observe the scale invariance:

$$\mathcal{F}_{x_0,c^2 t_0}(c_{x_0}(\mathcal{M})) = \mathcal{F}_{x_0,t_0}(\mathcal{M}).$$

Given $v \in \mathbb{R}^{n+1}$, we have the translation invariance:

$$\mathcal{F}_{x_0+v,t_0}(\mathcal{M} + v) = \mathcal{F}_{x_0,t_0}(\mathcal{M}).$$

Given $A \in \mathrm{O}(n+1)$, define $A_{x_0} : \mathbb{R}^{n+1} \to \mathbb{R}^{n+1}$ by $A_{x_0}(x) = x_0 + A(x - x_0)$. We have the rotation invariance:

$$\mathcal{F}_{x_0,t_0}(A_{x_0}(\mathcal{M})) = \mathcal{F}_{x_0,t_0}(\mathcal{M}).$$

A triple $(\mathcal{M}, x_0, t_0)$ is said to be $\mathcal{F}$**-critical** if it is a critical point of $\mathcal{F}$ with respect to variations in $\mathcal{M}$, x_0, and t_0. One shows that $(\mathcal{M}, x_0, t_0)$ is $\mathcal{F}$-critical if and only if $\mathcal{M}$ is a time t_0 slice of a shrinker that becomes extinct at time $t = 0$ at the point x_0. If we just assume that $\mathcal{M}$ is an integer rectifiable varifold, then we also call $\mathcal{M}$ a **weak shrinker**.

The **entropy functional** of a hypersurface is defined to be the supremum of the $\mathcal{F}$-functional over all base points and scales,

$$\lambda(\mathcal{M}) = \sup_{x_0 \in \mathbb{R}^{n+1}, t_0 \in \mathbb{R}_+} \mathcal{F}_{x_0,t_0}(\mathcal{M}). \tag{11.49}$$

Observe that $\lambda(\mathcal{M}) \geq 0$ and that $\lambda(\mathcal{M})$ is invariant under scaling, translating, and rotating $\mathcal{M}$. In particular, if $\{\mathcal{M}_t\}$ is either a shrinking, translating, or expanding self-similar solution, then $\lambda(\mathcal{M}_t)$ is constant in t.

By Huisken's monotonicity formula (10.8), if $\mathcal{M}_t$ is a solution to the mean curvature flow with finite entropy, then $\frac{d}{dt}\mathcal{F}_{x_0,t_0-t}(\mathcal{M}_t) \leq 0$. Thus,

$$\frac{d}{dt}\lambda(\mathcal{M}_t) \leq 0. \tag{11.50}$$

Example 11.53 (Cf. Example 10.1).

(1) For any hyperplane $\mathcal{P}$ in $\mathbb{R}^{n+1}$, $\lambda(\mathcal{P}) = 1$.

(2) If $S^n_{r_0}(x_0)$ is the n-sphere of radius r_0 centered at x_0, then (11.48) yields

$$\mathcal{F}_{x_0,t_0}(S^n_{r_0}(x_0)) = \sigma_n r_0^n (4\pi t_0)^{-n/2} \mathrm{e}^{-\frac{r_0^2}{4t_0}}, \tag{11.51}$$

where σ_n is the area of the unit n-sphere. The supremum of $\mathcal{F}_{x_0,t_0}$ over $t_0 > 0$ is attained at $t_0 = \frac{r_0^2}{2n}$ (which is the same as the time for the mean curvature flow with initial hypersurface $S^n_{r_0}(x_0)$ to become extinct). Moreover, the supremum of $\mathcal{F}_{x,t_0}$ over $x \in \mathbb{R}^{n+1}$ is attained at $x = x_0$ (see Stone [**498**, Lemma 5.2], [**499**]). Thus

$$\lambda(S^n_{r_0}(x_0)) = \sigma_n \left(\frac{n}{2\pi e}\right)^{n/2}. \tag{11.52}$$

(3) If $\mathcal{N}^k$ is a hypersurface in $\mathbb{R}^{k+1}$, then

$$\lambda(\mathcal{N}^k \times \mathbb{R}^{n-k}) = \lambda(\mathcal{N}^k).$$

In particular,

$$\lambda(S^k \times \mathbb{R}^{n-k}) = \sigma_k \left(\frac{k}{2\pi e}\right)^{k/2}.$$

For example, $\lambda(S^2) = \frac{4}{e} \approx 1.4715$ and $\lambda(S^1 \times \mathbb{R}) = \lambda(S^1) = \sqrt{\frac{2\pi}{e}} \approx 1.5203$.

Given an embedded hypersurface $\mathcal{M} \subset \mathbb{R}^{n+1}$, we say that $\{\mathcal{M}_s\}_{s\in(-\varepsilon,\varepsilon)}$ is a **compactly supported variation** of $\mathcal{M}$ if

$$\mathcal{M}_s = \{x + f(x)\mathsf{N}(x) \,|\, x \in \mathcal{M}\}$$

for some $f : \mathcal{M} \to \mathbb{R}$ with compact support. Given (x_0, t_0), a critical point $\mathcal{M}$ for the functional $\mathcal{F}_{x_0,t_0}$ is called **$\mathcal{F}$-stable** if for each compactly supported variation $\{\mathcal{M}_s\}$ of $\mathcal{M}$, there exist variations $\{x_s\}$ of x_0 and $\{t_s\}$ of t_0 such that

$$\left.\frac{d^2}{ds^2}\right| \mathcal{F}_{x_s,t_s}(\mathcal{M}_s) \geq 0.$$

T. Colding and W. P. Minicozzi II proved that smooth, properly embedded self-shrinkers with polynomial volume growth $\mathcal{M}$ which locally minimize the entropy functional and don't split off a line must be $\mathcal{F}$-stable [**181**]. This then implies mean convexity, and so the classification of mean convex self-shrinkers kicks in. An exposition of their approach can be found in [**188**]. These results extend to higher codimension mean curvature flow [**59**].

The **asymptotic entropy** of an ancient solution $\mathcal{M} = \{\mathcal{M}_t\}_{t\in(-\infty,0)}$ is defined to be

$$\lambda_{-\infty}(\mathcal{M}) = \sup_{t\in(-\infty,0)} \lambda(\mathcal{M}_t) = \lim_{t\to-\infty} \lambda(\mathcal{M}_t). \tag{11.53}$$

Denote by $\mathcal{B} = \{\mathcal{B}_t\}$ the bowl soliton and by $\mathcal{O} = \{\mathcal{O}_t\}$ the ancient ovaloid. Then

$$\lambda_{-\infty}(\mathcal{B}) = \lambda_{-\infty}(\mathcal{O}) = \sqrt{2\pi/e}.$$

We say that an embedded surface in $\mathbb{R}^3$ has **subcylindrical entropy** if its entropy λ is at most $\sqrt{2\pi/e}$. We say that a solution $\{\mathcal{M}_t\}_{t\in I}$ to mean curvature flow has subcylindrical entropy if $\lambda(\mathcal{M}_t) \leq \sqrt{2\pi/e}$ for all $t \in I$. If the solution is ancient, then this is equivalent to $\lambda_{-\infty}(\mathcal{M}) \leq \sqrt{2\pi/e}$. Bernstein and Wang [**88**] proved that any 2-dimensional shrinker with subcylindrical entropy is either a static flat plane, a shrinking round sphere, or a shrinking round cylinder. Hershkovits [**284**] proved that any 2-dimensional translator with subcylindrical entropy is either a static flat plane or the bowl translating soliton. Extending these two results, Choi, Haslhofer, and Hershkovits [**151**] proved the following.

Theorem 11.54. *Let $\{\mu_t\}_{t\in(-\infty,T)}$, $T \in (-\infty, \infty]$, be an ancient, unit-regular, cyclic, integral Brakke flow in $\mathbb{R}^3$. If $\{\mu_t\}$ has subcylindrical entropy, then it is either a static flat plane, a shrinking round sphere, a shrinking round cylinder, a translating bowl soliton, or an ancient oval.*

Observe that the definitions (11.12) of the Gaußian density ratio and (11.13) of the Gaußian density carry over to integral Brakke flows. We say an integral Brakke flow is **unit-regular** if for each $(p_0, t_0) \in \mathbb{R}^{n+1} \times \mathbb{R}$ satisfying $\Theta(p_0, t_0) = 1$, there exists $\varepsilon > 0$ such that $\{\operatorname{spt} \mu_t \cap B_\varepsilon(p_0)\}_{t\in[t_0-\varepsilon^2, t_0+\varepsilon^2]}$ is a smooth mean curvature flow. We say that an integral Brakke flow is **cyclic** if the *rectifiable mod 2 flat chain* $[V_t]$ corresponding to V_t satisfies $\partial[V_t] = 0$ for a.e. t; see White [**533**, Definition 4.1].

11.6.5. Singularities in free boundary mean curvature flow.

John Buckland obtained a classification of type-I singularities for hypersurfaces evolving by mean curvature flow with free boundary on an umbilic support hypersurface analogous to the results presented here for mean curvature flow without boundary [**127**].

11.7. Exercises

Exercise 11.1. Prove that the sequence $\{X_j\}_{j\in\mathbb{N}}$ of embeddings $X_j : S^n \to \mathbb{R}^{n+1}$ defined by

$$X_j(z) = j(z + e_{n+1})$$

converges locally uniformly in the smooth topology to the standard embedding of the hyperplane $\mathbb{R}^n \times \{0\}$.

Exercise 11.2. Let $\{X_k : M_k \times I_k \to \mathbb{R}^{n+1}\}_{k\in\mathbb{N}}$ be a sequence of mean curvature flows, $\{x_k\}_{k\in\mathbb{N}}$ a sequence of points $x_k \in M_k$, and $\{t_k\}_{k\in\mathbb{N}}$ a sequence of times $t_k \to T \in \mathbb{R} \cup \{\infty\}$ satisfying the following conditions:

(i) $t_k \in I_k$ and $X_k(x_k, t_k) = 0$ for every $k \in \mathbb{N}$.

(ii) For each $A < \infty$ there exists $k_A \in \mathbb{N}$ such that X_k is properly defined in $B_A \times (-A^2, t_k]$ for every $k \geq k_A$, where B_A is the ball of radius A about the origin in $\mathbb{R}^{n+1}$.

(iii) For every $A > 0$ there exist $C_{0,A} < \infty$ and $k_A \in \mathbb{N}$ such that

$$\sup_{(X_k(x,t),t)\in B_A\times(-A^2,t_k]} |\mathrm{II}_k(x,t)| \leq C_{0,A}$$

for all $k \geq k_A$.

Prove that there exist a solution to mean curvature flow $X : M \times I \to \mathbb{R}^{n+1}$ properly defined in $\mathbb{R}^{n+1} \times (-\infty, T)$ and a subsequence of $\{X_k : M_k \times I_k \to \mathbb{R}^{n+1}\}_{k\in\mathbb{N}}$ that converges smoothly on compact subsets of $M^n \times I$ to X.

Exercise 11.3. Let S^2 be the unit 2-sphere in $\mathbb{R}^3$. Show that

$$\mathcal{F}_{x_0,t_0}(S^2) = \begin{cases} t_0^{-1}\mathrm{e}^{-\frac{1}{4t_0}} & \text{if } x_0 = 0, \\ |x_0|^{-1}\mathrm{e}^{-\frac{1}{4t_0}(1+|x_0|^2)}\left(\mathrm{e}^{\frac{|x_0|}{2t_0}} - \mathrm{e}^{-\frac{|x_0|}{2t_0}}\right) & \text{if } x_0 \neq 0. \end{cases}$$

(The formula for $\mathcal{F}_{0,t_0}(S^2)$ is a special case of (11.51).) Conclude that

$$\sup_{x_0\in\mathbb{R}^3} \mathcal{F}_{x_0,t_0}(S^2) = \mathcal{F}_{0,t_0}(S^2) = t_0^{-1}\mathrm{e}^{-\frac{1}{4t_0}}.$$

Further conclude that $\lambda(S^2) = \mathcal{F}_{0,1/4}(S^2) = 4/\mathrm{e}$, as a special case of (11.52).

Exercise 11.4. Prove that if $\mathcal{M}_t$ is a solution to the mean curvature flow with finite entropy, then (11.50) holds; that is, $\lambda(\mathcal{M}_t)$ is nonincreasing.

Exercise 11.5. Show that the entropy of the Grim Reaper is $\lambda(\Gamma^1) = 2$.

Chapter 12

Noncollapsing

In our proof that the curve shortening and mean curvature flows preserve embeddedness, we made use of the function which gives the extrinsic distance between two points on the hypersurface. For curve shortening flow, we were able to obtain further control over the embeddedness of our solution by considering the chord-arc profile. In this chapter, we introduce a further quantitative measure of embeddedness, the extrinsic ball curvature, and present some powerful applications.

12.1. The inscribed and exscribed curvatures

To simplify notation, we shall often, in this chapter, conflate points of an embedded hypersurface with their image under the embedding.

Figure 12.1. Extrinsic balls: ball interior, upper half-space, ball exterior.

Definition 12.1 (Inscribed and exscribed curvatures [**41, 55, 481**]). Let $\mathcal{M} = \partial\Omega^{n+1} \hookrightarrow \mathbb{R}^{n+1}$, $n \geq 1$, be a smooth, properly embedded boundary of an open set $\Omega^{n+1} \subset \mathbb{R}^{n+1}$, equipped with its outer unit normal field N. Define the **extrinsic ball curvature** $k : (\mathcal{M} \times \mathcal{M}) \setminus \{(x, x) : x \in \mathcal{M}\} \to \mathbb{R}$

by

$$k(x,y) \doteqdot \frac{2\langle x-y,\mathsf{N}(x)\rangle}{\|x-y\|^2}\,, \tag{12.1}$$

the **inscribed curvature**, $\overline{k} : \mathcal{M} \to \mathbb{R}$, by

$$\overline{k}(x) = \sup_{y\in\mathcal{M}\setminus\{x\}} k(x,y) \tag{12.2}$$

and the **exscribed curvature**, $\underline{k} : \mathcal{M} \to \mathbb{R}$, by

$$\underline{k}(x) = \inf_{y\in\mathcal{M}\setminus\{x\}} k(x,y)\,. \tag{12.3}$$

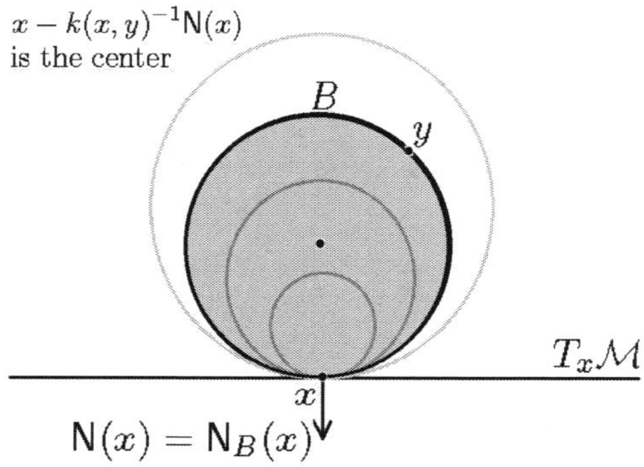

Figure 12.2. B is the unique extrinsic ball passing through y and touching $\mathcal{M}$ at x. Pictured: The $k(x,y) > 0$ case.

For the purposes of the following lemma, which gives a geometric characterization of the ball curvatures, an **extrinsic ball** B will mean either an open ball, an open half-space, or the complement of a closed ball in $\mathbb{R}^{n+1}$ equipped with its outward pointing boundary normal N_B (in particular, the curvature of ∂B is positive if B is a ball and negative if B is a ball complement); see Figure 12.1. An extrinsic ball B **passes through** a point y if $y \in \partial B$ and **touches** a hypersurface $\mathcal{M}$ at $x \in \mathcal{M}$ if $x \in \partial B$ and $\mathsf{N}(x) = \mathsf{N}_B(x)$; see Figure 12.2. In the following, by the curvature of ∂B, we mean its unique principal curvature.

Lemma 12.2 (**[41, 55]**). *Let $\mathcal{M}^n = \partial\Omega^{n+1} \hookrightarrow \mathbb{R}^{n+1}$, $n \geq 1$, be a smooth, proper embedding equipped with its outer (with respect to Ω^{n+1}) unit normal field N. Then $k(x,y)$ is equal to the curvature of the boundary of the unique extrinsic ball which touches the hypersurface at x and passes through the point y. Furthermore, $\overline{k}(x)$ is equal to the curvature of the boundary of the largest extrinsic ball which is contained in Ω and touches $\mathcal{M}$ at x and $\underline{k}(x)$ is equal to the curvature of the boundary of the smallest extrinsic ball which is contained in $\mathbb{R}^{n+1} \setminus \overline{\Omega}$ and touches $\mathcal{M}$ at x.*

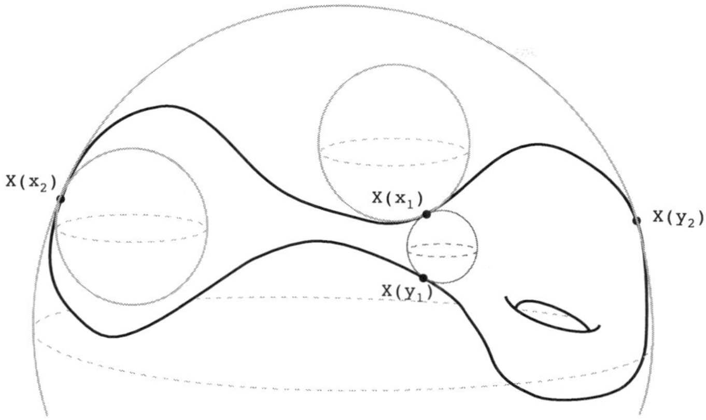

Figure 12.3. Inscribed and exscribed curvatures of an embedded hypersurface $X : \mathcal{M}^2 \hookrightarrow \mathbb{R}^3$. At x_1, $\overline{k}(x_1) = k(x_1, y_1)$ and $\underline{k}(x_1) = \kappa_1(x_1)$ (which is less than zero since the smallest touching region is a ball complement). At x_2, $\overline{k}(x_2) = \kappa_2(x_2)$ and $\underline{k}(x_2) = k(x_2, y_2)$ (which is greater than zero since the smallest touching region is a ball).

Proof of Lemma 12.2. By (12.1), we have the identities

$$\left\| \left(x - k(x,y)^{-1}\mathsf{N}(x) \right) - y \right\|^2 = k(x,y)^{-2} \tag{12.4}$$

when $k(x,y) \neq 0$ and

$$\langle x - y, \mathsf{N}(x) \rangle = 0 \tag{12.5}$$

when $k(x,y) = 0$. In particular, (12.4) implies that y lies on the boundary of the open ball B_y centered at the point $x - k(x,y)^{-1}\mathsf{N}(x)$ with radius $k(x,y)^{-1}$, while (12.5) implies that y lies on the boundary of the open half-space $B_y \doteqdot \{z \in \mathbb{R}^{n+1} \,|\, \langle x - z, \mathsf{N}(x) \rangle > 0\}$. In both cases, we have that $\mathsf{N}_{B_y}(x) = \mathsf{N}(x)$ and that $k(x,y)$ is the curvature of ∂B_y.

Observe that the collection of extrinsic balls touching $\mathcal{M}$ at x is a nested collection with balls of larger curvature inside balls of smaller curvature. Let B be the unique extrinsic ball touching $\mathcal{M}$ at x with curvature $\overline{k}(x)$. Given $k \in \mathbb{R}$, let $B(k)$ denote the extrinsic ball touching $\mathcal{M}$ at x with curvature k. We then have $B = B(\overline{k}(x)) \subset B_y = B(k(x,y))$ for all $y \in \mathcal{M} \backslash \{x\}$.

Suppose that B is not contained in Ω. Since B intersects Ω, this implies that there exists $z \in \mathcal{M} \cap B$. Thus $z \neq x$ and B_z is properly contained in B. Therefore $k(x,z) > \overline{k}(x)$, a contradiction. We conclude that $B \subset \Omega$.

Now suppose that B is not the largest extrinsic ball in Ω and touching $\mathcal{M}$ at x. Then there exists $k < \overline{k}(x)$ such that $B(k)$ is contained in Ω. By the definition of $\overline{k}(x)$, there exists $w \in \mathcal{M} \backslash \{x\}$ such that $k(x,w) > k$. We have $B_w = B(k(x,w)) \subset \Omega$ and $w \in \mathcal{M} \cap B(k)$, contradicting $B(k) \subset \Omega$.

Similarly, one shows that $\underline{k}(x)$ is the curvature of the boundary of the smallest extrinsic ball contained in $\mathbb{R}^{n+1} \setminus \overline{\Omega}$ and touching $\mathcal{M}$ at x. □

Remark 12.3. Denote by D the diagonal of $\mathbb{R}^{n+1} \times \mathbb{R}^{n+1}$ and define

$$\mathrm{k} : S^n \times (\mathbb{R}^{n+1} \times \mathbb{R}^{n+1} \setminus D) \to \mathbb{R}$$

by

$$\mathrm{k}(\mathsf{N}, x, y) = \frac{2\langle \mathsf{N}, x - y\rangle}{\|x - y\|^2}. \tag{12.6}$$

Of course, $k(x, y) = \mathrm{k}(\mathsf{N}(x), x, y)$ in (12.1). If $x - y$ is not a multiple of N, then $\mathrm{k}(\mathsf{N}, x, y)$ is the signed curvature of the unique circle or line C having N as a normal vector and passing through y.[1] Note that C lies in the plane spanned by N and $x - y$. On the other hand, if $x - y$ is a multiple of N, then for any unit vector V perpendicular to N, there exists a unique circle C tangent to V and passing through y. Then $\mathrm{k}(\mathsf{N}, x, y)$ is the signed curvature of any such C. Note that the set of all C comprise an n-dimensional sphere with nonzero constant signed principal curvature $\mathrm{k}(\mathsf{N}, x, y)$, where y is antipodal to x.

Example 12.4.

- (Convex body) It follows immediately from the definition of k that $k \geq 0$ ($k > 0$) if and only if Ω is a (strictly) convex body.
- (Half-space) By (12.5), $\partial\Omega$ satisfies $k \equiv 0$ if and only if Ω is a half-space, $L(p, v) \doteqdot \{z \in \mathbb{R}^{n+1} : \langle z - p, v\rangle > 0\}$ for some $p, v \in \mathbb{R}^{n+1}$.
- (Ball or its complement) By (12.4), $\partial\Omega$ satisfies $k \equiv r^{-1}$ for some $r \in \mathbb{R} \setminus \{0\}$ if and only if Ω is a ball, $B_r^{n+1}(p) = \{z \in \mathbb{R}^{n+1} : \|z - p\|^2 < r^2\}$, or a ball complement. In that case,
$$k \equiv r^{-1} \equiv \frac{1}{n} H .$$
- (Round cylinder) For $0 < m < n$, the cylinders $\partial B_r^{n-m+1}(p) \times \mathbb{R}^m = \{(z_1, z_2) \in \mathbb{R}^{n-m+1} \times \mathbb{R}^m : \|z_1\|^2 = r^2\}$ in $\mathbb{R}^{n+1}$ with their standard orientations satisfy
$$k(x, y) = \frac{1}{r}\left(1 - \frac{\|x_2 - y_2\|^2}{\|x - y\|^2}\right),$$
so that $\overline{k} \equiv \frac{1}{r} \equiv \frac{1}{n-m} H$ and $\underline{k} \equiv 0$. Modulo rigid motions, the converse follows from Exercise 12.9.

[1] If $\mathrm{k}(\mathsf{N}, x, y) \neq 0$, then it is positive if N is the outward unit normal to C and it is negative if N is the inward unit normal to C.

- (When cylinders are round) More generally, in case Ω is a convex body, $\overline{k}$ and $\underline{k}$ are both constant on $\partial\Omega$ if and only if, modulo rigid motions, Ω is a cylindrical region, $B_r^{n-m+1}(p) \times \mathbb{R}^m = \{(z_1, z_2) \in \mathbb{R}^{n-m+1} \times \mathbb{R}^m : \|z_1\|^2 < r^2\}$ for some $0 \leq m \leq n$ (see Exercise 12.9).
- (Grim Reaper) The Grim Reaper,
$$\left\{(x, t - \log\cos x) : x \in \left(-\frac{\pi}{2}, \frac{\pi}{2}\right)\right\}_{t\in\mathbb{R}},$$
satisfies
$$\overline{k}(x,t) = \frac{\sin x}{x} \quad \text{and} \quad \underline{k}(x,t) \equiv 0\,. \tag{12.7}$$
On the other hand,
$$\kappa(x,t) = \cos x$$
and hence
$$\frac{\overline{k}}{\kappa}(x,t) = \frac{\tan x}{x} \to \infty \text{ as } x \to \pm\frac{\pi}{2}\,.$$
- (Paperclip) The paperclip, $\{(x,y) \in \mathbb{R}^2 : \cos x = \mathrm{e}^t \cosh y\}_{t\in\mathbb{R}}$, satisfies
$$\overline{k} = \frac{\tan x}{x\sqrt{\tan^2 x + \tanh^2 y}} \quad \text{and} \quad \underline{k} = \frac{\tanh y}{y\sqrt{\tan^2 x + \tanh^2 y}}\,. \tag{12.8}$$
On the other hand,
$$\kappa = \frac{1}{\sqrt{\tan^2 x + \tanh^2 y}}$$
and hence
$$\frac{\overline{k}}{\kappa} = \frac{\tan x}{x} \to \begin{cases} \infty & \text{as } x \to \pm\frac{\pi}{2}\,, \\ 1 & \text{as } x \to 0 \end{cases} \quad \text{and} \quad \frac{\underline{k}}{\kappa} = \frac{\tanh y}{y} \to \begin{cases} 0 & \text{as } y \to \pm\infty\,, \\ 1 & \text{as } y \to 0\,. \end{cases}$$

Definition 12.5. A smooth, mean convex solution $X : M^n \times I \to \mathbb{R}^{n+1}$ of mean curvature flow is

- **interior noncollapsing** if there exists a constant $K \in \mathbb{R}$ such that
$$\frac{\overline{k}}{H}(x,t) \leq K \quad \text{for all} \quad (x,t) \in M^n \times I\,, \tag{12.9}$$
- **exterior noncollapsing** if there exists a constant $K \in \mathbb{R}$ such that
$$\frac{\underline{k}}{H}(x,t) \geq K \quad \text{for all} \quad (x,t) \in M^n \times I\,, \tag{12.10}$$
- **noncollapsing** if it is both interior and exterior noncollapsing; otherwise, it is (either **interior** or **exterior**) **collapsing**. We refer to noncollapsing with a given constant K as K**-noncollapsing**.

Remark 12.6.

- Nonembedded immersed solutions are necessarily collapsing.
- Any compact, smoothly embedded solution defined on a compact time interval $I \subset \mathbb{R}$ is trivially noncollapsing.
- If M^n is noncompact, then, as demonstrated by the Grim Reaper, a smooth solution $X : M^n \times I \to \mathbb{R}^{n+1}$ can be collapsing, even when I is compact.
- If $I = (-\infty, T]$, then, as demonstrated by the Angenent oval, a smooth solution $X : M^n \times I \to \mathbb{R}^{n+1}$ can be collapsing, even when M^n is compact.
- A priori, smooth embedded solutions defined on a time interval $I = [0, T)$ could be collapsing, even when M^n is compact. The main purpose of this chapter is to prove that this cannot happen for mean convex solutions.

The following lemma shows that k recovers the second fundamental form at the diagonal submanifold, $D \doteqdot \{(x, x) : x \in \mathcal{M}\}$, of $\mathcal{M} \times \mathcal{M}$.

Lemma 12.7. *Let $\mathcal{M} = \partial\Omega^{n+1} \hookrightarrow \mathbb{R}^{n+1}$ be a smooth, properly embedded hypersurface and $\gamma : (-s_0, s_0) \to \mathcal{M}$ any regular curve through $\gamma(0) = x$. Then*

$$\lim_{s \to 0} k(x, \gamma(s)) = \frac{\mathrm{II}_x(v, v)}{g_x(v, v)}, \tag{12.11}$$

where $v = \gamma'(0) \neq 0$. In particular, k is bounded on any smooth, compact embedded hypersurface. Moreover,

$$\overline{k}(x) \geq \limsup_{y \to x} k(x, y) = \kappa_n(x) \tag{12.12}$$

and

$$\underline{k}(x) \leq \liminf_{y \to x} k(x, y) = \kappa_1(x), \tag{12.13}$$

where $\kappa_1 \leq \cdots \leq \kappa_n$ denote the principal curvatures of $\mathcal{M}$.

Proof. By definition,

$$k(x, \gamma(s)) = \frac{2\langle x - \gamma(s), \mathsf{N}(x)\rangle}{\|x - \gamma(s)\|^2}.$$

Defining $d(s) \doteqdot \|x - \gamma(s)\|$ and $f(s) \doteqdot 2\langle x - \gamma(s), \mathsf{N}(x)\rangle$, we find

$$d^2(s) = s^2|\gamma'(0)|^2 + o(s^2) \quad \text{and} \quad f(s) = s^2\mathrm{II}(\gamma'(0), \gamma'(0)) + o(s^2).$$

The claim follows. □

Lemma 12.7 shows that k becomes discontinuous at the diagonal submanifold. On the other hand, it also shows us how to "resolve" this singularity[2]: The diagonal D is a submanifold of $\mathcal{M}\times\mathcal{M}$ of dimension and codimension n. Thus, D does not possess sufficient degrees of freedom to accommodate the $2n-1$ degrees required by the limit, $\lim_{s\to 0} k(x,\gamma(s)) = \mathrm{II}_x(v,v)$ (n for each point x and $n-1$ for each direction v). The idea is to "blow $(\mathcal{M}\times\mathcal{M})\setminus D$ up" near D by attaching the $(2n-1)$-dimensional unit tangent bundle to form a $2n$-dimensional manifold-with-boundary $\widehat{\mathcal{M}}$: As a set, $\widehat{\mathcal{M}}$ is defined as the disjoint union of $(\mathcal{M}\times\mathcal{M})\setminus D$ with the unit tangent bundle $S\mathcal{M} = \{(x,v)\in T\mathcal{M} : |v| = 1\}$. The manifold-with-boundary structure is defined by the atlas generated by all charts for $(\mathcal{M}\times\mathcal{M})\setminus D$ together with the boundary charts $\widehat{Y} : S\mathcal{M}\times(0,r_0)\to\widehat{\mathcal{M}}$ given by $\widehat{Y}(r,z,\vartheta) \doteqdot \big(\exp(rY(z,\vartheta)), \exp(-rY(z,\vartheta))\big)$, where Y is a chart for $S\mathcal{M}$. Note that the normal space $N_{(x,x)}D$ of D at each point (x,x) is the n-dimensional subspace $\{(u,-u) : u\in T_x\mathcal{M}\}$ of $T_{(x,x)}(\mathcal{M}\times\mathcal{M})\cong T_x\mathcal{M}\times T_x\mathcal{M}$. The tubular neighborhood theorem provides some $r_0>0$ such that the exponential map is a smooth diffeomorphism on $\{((x,u),(x,-u))\in T(\mathcal{M}\times\mathcal{M}) :\ 0<|u|<r_0\}$, which ensures that the boundary charts are well-defined. We can now extend k to $\widehat{\mathcal{M}}$ by setting

$$k(x,y) \doteqdot \mathrm{II}_x(y,y) \tag{12.14}$$

for every $(x,y)\in S\mathcal{M} = \partial\widehat{\mathcal{M}}$. We claim that $k\in C^\infty(\widehat{\mathcal{M}})$. Indeed, let

$$\widehat{Y}(r,z,\vartheta) = \big(\exp(rY(z,\vartheta)), \exp(-rY(z,\vartheta))\big)$$

be any boundary chart for $\widehat{\mathcal{M}}$ and write

$$k(r,z,\vartheta) = \frac{N(r,z,\vartheta)}{D(r,z,\vartheta)},$$

where

$$N(r,z,\vartheta) \doteqdot 2\left\langle \exp\big(rY(z,\vartheta)\big) - \exp\big(-rY(z,\vartheta)\big), \mathsf{N}(r,z,\vartheta)\right\rangle$$

and

$$D(r,z,\vartheta) = \left\|\exp\big(rY(z,\vartheta)\big) - \exp\big(-rY(z,\vartheta)\big)\right\|^2.$$

Note that N and D are smooth functions. Moreover, straightforward computations reveal that

$$N(r,z,\vartheta) = 4\mathrm{II}_z(\vartheta,\vartheta)r^2 + o(r^2) \quad\text{and}\quad D(r,z,\vartheta) = 4r^2 + o(r^2).$$

It follows that $d \doteqdot D/r^2$ and $n\doteqdot N/r^2$ are also smooth. The claim follows since $d(0,z,\vartheta) = 4\neq 0$ and $n(0,z,\vartheta)/d(0,z,\vartheta) = \mathrm{II}_z(\vartheta,\vartheta) = k(0,z,\vartheta)$.

[2]We will not actually use this fact in the sequel, so the uninterested reader may safely skip to the following section.

12.2. Differential inequalities for the inscribed and exscribed curvatures

We shall now derive differential inequalities for the inscribed and exscribed curvatures. Although these functions are not in general smooth, we are able to derive differential inequalities for appropriate barrier functions.

Proposition 12.8. *Suppose that $X : M^n \times [0,T) \to \mathbb{R}^{n+1}$ is a solution of mean curvature flow such that $\mathcal{M}_0 = \partial(\Omega^{n+1})$ is properly embedded. Then*

$$(\partial_t - \Delta)\overline{k} \leq |\mathrm{II}|^2 \overline{k} - 2 \sum_{\kappa_i < \overline{k}} \frac{(\nabla_i \overline{k})^2}{\overline{k} - \kappa_i} \tag{12.15}$$

and

$$(\partial_t - \Delta)\underline{k} \geq |\mathrm{II}|^2 \underline{k} + 2 \sum_{\kappa_i > \underline{k}} \frac{(\nabla_i \underline{k})^2}{\kappa_i - \underline{k}} \tag{12.16}$$

in the barrier sense (see Definition 9.5*).*

Before proving Proposition 12.8, we first prove that the identities hold in the "boundary case", i.e., for the largest and smallest principal curvatures. These identities are useful and interesting in their own right and they also offer some insight into the proof of Proposition 12.8 in the "interior case".

Proposition 12.9. *Let $X : M^n \times [0,T) \to \mathbb{R}^{n+1}$ be a solution of mean curvature flow. Then, with respect to a principal frame,*

$$(\partial_t - \Delta)\kappa_n \leq |\mathrm{II}|^2 \kappa_n - 2 \sum_{\kappa_i < \kappa_n} \frac{(\nabla_i \kappa_n)^2}{\kappa_n - \kappa_i} \tag{12.17}$$

and

$$(\partial_t - \Delta)\kappa_1 \geq |\mathrm{II}|^2 \kappa_1 + 2 \sum_{\kappa_i > \kappa_1} \frac{(\nabla_i \kappa_1)^2}{\kappa_i - \kappa_1} \tag{12.18}$$

in the barrier sense of Definition 9.5.

Proof. We will only prove the identity for the largest principal curvature, κ_n, since the proof of the identity for κ_1 is analogous (see Exercise 12.6).

Define the function $K : T\mathcal{M} \setminus \{0\} \times [0,T) \to \mathbb{R}$, where $\{0\}$ denotes the zero section, by

$$K(x,y,t) \doteqdot \frac{\mathrm{II}_{(x,t)}(y,y)}{g_{(x,t)}(y,y)}, \tag{12.19}$$

so that

$$\kappa_n(x,t) = \sup_{y \in T_x\mathcal{M} \setminus \{0\}} K(x,y,t).$$

Note that K is a smooth function. We will use it to define an appropriate barrier for κ_n.

Given any point $(x_0, t_0) \in \mathcal{M}^n \times [0, T)$, let $\{x^i\}_{i=1}^n$ be geodesic normal coordinates for $(\mathcal{M}^n, g_{t_0})$ centered at x_0 and denote by $y^j = dx^j$ the corresponding fiber coordinates. So, as local coordinates on $T\mathcal{M}$, $x^i(x, y) = x^i(x)$ and $y^j(x, y) = dx^j(y)$. Given $y_0 \in T_{x_0}\mathcal{M}^n$, we shall compute the derivatives of K at (x_0, y_0) in these coordinates. Note first that, although $\mathcal{M}$ is Riemannian and there are "natural" metrics on the manifold $T\mathcal{M}$, there is no "canonical" metric (and hence no canonical gradient or Hessian operators) on $T\mathcal{M}$. This is actually a good thing, for it provides us with a lot of freedom in our computation: We will want our elliptic operator on $T\mathcal{M}$ to project to the Laplacian on $\mathcal{M}$ under the natural projection map but otherwise we are free to choose. Indeed, we can even allow it to degenerate on the second factor.

Given a symmetric $n \times n$ matrix Λ (to be determined momentarily), we compute

$$(\partial_{x^k} + \Lambda_l{}^p \partial_{y^p}) K = \frac{1}{g(y,y)} \big([\partial_{x^k} \mathrm{II}_{ij} - K \partial_{x^k} g_{ij}] \, y^i y^j + 2\Lambda_l{}^p [\mathrm{II}_{pj} - K g_{pj}] \, y^j \big) . \tag{12.20}$$

Choose $y_0 \in S\mathcal{M}_{(x_0, t_0)}$ so that

$$K(x_0, y_0, t_0) = \max_{y \in T_{x_0}\mathcal{M} \setminus \{0\}} K(x_0, y, t_0) = \kappa_n(x_0, t_0) .$$

Then

$$\begin{aligned} \mathrm{II}_{(x_0,t_0)}(y_0) &= K(x_0, y_0, t_0) y_0 \quad \text{and} \\ \partial_{x^k} K|_{(x_0, y_0, t_0)} &= \nabla_k \mathrm{II}_{(x_0,t_0)}(y_0, y_0). \end{aligned} \tag{12.21}$$

The relevant components of the Hessian are

$$\begin{aligned} &(\partial_{x^k} + \Lambda_k{}^p \partial_{y^p})(\partial_{x^l} + \Lambda_l{}^q \partial_{y^q}) K \\ &= \frac{1}{g(y,y)} \Big([\partial_{x^k} \partial_{x^l} \mathrm{II}_{ij} - K \partial_{x^k} \partial_{x^l} g_{ij}] \, y^i y^j \\ &\quad - [(\partial_{x^l} + \Lambda_l{}^q \partial_{y^q}) \, K \partial_{x^k} g_{ij} + (\partial_{x^k} + \Lambda_k{}^p \partial_{y^p}) \, K \partial_{x^l} g_{ij}] \, y^i y^j \\ &\quad + 2\Lambda_k{}^p \, (\partial_{x^l} \mathrm{II}_{pj} - K \partial_{x^l} g_{pj}) \, y^j + 2\Lambda_l{}^q \, (\partial_{x^k} \mathrm{II}_{iq} - K \partial_{x^k} g_{iq}) \, y^i \\ &\quad - 2\Lambda_k{}^p \, (\partial_{x^l} + \Lambda_l{}^q \partial_{y^q}) \, K g_{pj} y^j - 2\Lambda_l{}^q \, (\partial_{x^k} + \Lambda_k{}^p \partial_{y^p}) \, K g_{iq} y^i \\ &\qquad + 2\Lambda_k{}^p \Lambda_l{}^q \, [\mathrm{II}_{pq} - K g_{pq}] \Big) . \end{aligned} \tag{12.22}$$

Thus, at the extreme point (x_0, y_0, t_0),

$$\begin{aligned}
-\Delta_\Lambda K \doteqdot &-\delta^{kl}(\partial_{x^k} + \Lambda_k{}^p\partial_{y^p})(\partial_{x^l} + \Lambda_l{}^q\partial_{y^q})K \\
= &-\Delta \mathrm{II}(y,y) \\
&- 2\delta^{kl}\Lambda_k{}^p\left[2\left(\nabla_l \mathrm{II}_{pj} - \nabla_l \mathrm{II}(y,y)\delta_{pj}\right)y^j + \Lambda_l{}^q\left(\mathrm{II}_{pq} - K\delta_{pq}\right)\right] \\
= &|\mathrm{II}|^2\mathrm{II}(y,y) - H\mathrm{II}^2_{ij}y^iy^j - \nabla_y\nabla_y H \\
&- 2\delta^{kl}\Lambda_k{}^p\left[2\left(\nabla_l \mathrm{II}_{pj} - \nabla_l \mathrm{II}(y,y)\delta_{pj}\right)y^j + \Lambda_l{}^q\left(\mathrm{II}_{pq} - K\delta_{pq}\right)\right].
\end{aligned}$$

Finally, observe that the time derivative of K is given by

$$\begin{aligned}
\partial_t K &= \frac{1}{g_{ij}y^iy^j}\left(\partial_t \mathrm{II}_{ij} - K\partial_t g_{ij}\right)y^iy^j \\
&= \frac{1}{g_{ij}y^iy^j}\left(\nabla_t \mathrm{II}_{ij} - 2H\mathrm{II}^2_{ij} + 2KH\mathrm{II}_{ij}\right)y^iy^j.
\end{aligned}$$

Recalling that y_0 is an eigenvector of II at (x_0, t_0), this yields

$$(\partial_t - \Delta_\Lambda)K = |\mathrm{II}|^2 K - 2\sum_{\kappa_i < K}\frac{(\partial_{x^i}K)^2}{K - \kappa_i} - \mathrm{L}_\Lambda \tag{12.23}$$

at the point (x_0, y_0, t_0), where

$$\begin{aligned}
\mathrm{L}_\Lambda \doteqdot &-2\sum_{\kappa_i<K}\frac{(\partial_{x^i}K)^2}{K-\kappa_i} \\
&+ 2\delta^{kl}\Lambda_k{}^p\left[2\left(\nabla_l \mathrm{II}_{pj} - \nabla_l \mathrm{II}(y,y)\delta_{pj}\right)y^j + \Lambda_l{}^q\left(\mathrm{II}_{pq} - K\delta_{pq}\right)\right].
\end{aligned} \tag{12.24}$$

We now introduce our barrier function. Recall that $y_0 \in S\mathcal{M}^n_{(x_0,t_0)}$ is chosen so that $\kappa_n(x_0, t_0) = K(x_0, y_0, t_0)$. Choose coordinates (x^i, y^j) for $T\mathcal{M}$ as before and set

$$\varphi(x,t) \doteqdot \frac{\mathrm{II}^{(x,t)}_{ij}y_0^iy_0^j}{g^{(x,t)}_{ij}y_0^iy_0^j} = K(x, y_0^i\partial_{x^i}, t).$$

Then φ is smooth and supports κ_n from below at the point (x_0, t_0). Moreover, since y_0 is a local maximum of $K(x_0, \cdot, t_0)$, $\partial_{y^i}K = 0$, $\partial_{y^i}\partial_{y^j}K \le 0$ and hence

$$\begin{aligned}
(\partial_t - \Delta)\varphi = (\partial_t - \Delta_\Lambda)\Phi &\le \left(\partial_t - \Delta_\Lambda\right)K \\
&= |\mathrm{II}|^2\varphi - 2\sum_{\kappa_i<\varphi}\frac{(\nabla_i\varphi)^2}{\varphi - \kappa_i} - \mathrm{L}_\Lambda
\end{aligned} \tag{12.25}$$

at (x_0, y_0, t_0), where Φ is the extension of φ to $T\mathcal{M}$ via $\Phi(x,y,t) \doteqdot \varphi(x,t)$.

Aside. At this stage, we can already deduce that

$$(\partial_t - \Delta)\kappa_n \le |\mathrm{II}|^2\kappa_n \tag{12.26}$$

(and the reverse inequality for κ_1) in the barrier sense by choosing $\Lambda = 0$. This is enough for some applications; however, the good gradient term will also prove useful.

So by (12.25) it remains to prove that $\sup_{\Lambda\in \mathrm{S}(n)} \mathrm{L}_\Lambda(x_0, y_0, t_0) \geq 0$. Since L_Λ is quadratic in Λ, we can actually compute the supremum explicitly. Indeed, after rotating our orthonormal frame at x_0 so that $y_0 = e_n$,

$$\begin{aligned}\mathrm{L}_\Lambda &= 2\sum_{k,p=1}^{n} \Lambda_k{}^p\left[2\left(\nabla_k \mathrm{II}_{pn} - \nabla_k \mathrm{II}_{nn}\delta_{pn}\right) - \Lambda_{kp}(\kappa_n - \kappa_p)\right] - 2\sum_{\kappa_i\neq K}\frac{(\partial_{x^i}K)^2}{K-\kappa_i}\\ &= 2\sum_{\substack{k=1\\ \kappa_p<\kappa_n}}^{n} \Lambda_k{}^p\left[2\nabla_k \mathrm{II}_{pn} - \Lambda_{kp}(\kappa_n - \kappa_p)\right] - 2\sum_{\kappa_i\neq K}\frac{(\partial_{x^i}K)^2}{K-\kappa_i}.\end{aligned}$$

Completing the square, we obtain

$$\mathrm{L}_\Lambda = 2\sum_{\substack{k=1\\ \kappa_p<\kappa_n}}^{n}\left[\frac{(\nabla_k \mathrm{II}_{pn})^2}{\kappa_n-\kappa_p} - (\kappa_n-\kappa_p)\left(\frac{\nabla_k \mathrm{II}_{pn}}{\kappa_n-\kappa_p} - \Lambda_{kp}\right)^2\right] - 2\sum_{\kappa_i\neq K}\frac{(\partial_{x^i}K)^2}{K-\kappa_i}.$$

By this, we see that in order to obtain the optimal inequality from (12.25), we should choose

$$\Lambda_{kp} \doteqdot \begin{cases}\dfrac{\nabla_k \mathrm{II}_{pn}}{\kappa_n-\kappa_p} & \text{if} \quad \kappa_p < \kappa_n\,,\\ 0 & \text{if} \quad \kappa_p = \kappa_n\,,\end{cases}$$

which yields

$$\begin{aligned}\mathrm{L}_\Lambda &= 2\sum_{\substack{k=1\\ \kappa_p<\kappa_n}}^{n}\frac{(\nabla_k \mathrm{II}_{pn})^2}{\kappa_n-\kappa_p} - 2\sum_{\kappa_i\neq K}\frac{(\partial_{x^i}K)^2}{K-\kappa_i}\\ &= 2\sum_{\kappa_p<\kappa_n}\frac{(\nabla_p \mathrm{II}_{nn})^2}{\kappa_n-\kappa_p} + 2\sum_{\substack{k<n\\ \kappa_p<\kappa_n}}\frac{(\nabla_k \mathrm{II}_{pn})^2}{\kappa_n-\kappa_p} - 2\sum_{\kappa_i<\kappa_n}\frac{(\partial_{x^i}K)^2}{\kappa_n-\kappa_i}\\ &= 2\sum_{\substack{k<n\\ \kappa_p<\kappa_n}}\frac{(\nabla_k \mathrm{II}_{pn})^2}{\kappa_n-\kappa_p}\\ &\geq 0\,,\end{aligned} \tag{12.27}$$

where the last equality is by $\partial_{x^p}K = \nabla_p \mathrm{II}_{nn}$ from (12.21). This completes the proof. □

Remark 12.10. In fact, we proved that

$$(\partial_t - \Delta)\kappa_n \leq |\mathrm{II}|^2\kappa_n - 2\sum_{k=1}^{n}\sum_{\kappa_i<\kappa_n}\frac{(\nabla_k\,\mathrm{II}_{in})^2}{\kappa_n-\kappa_i} \tag{12.28}$$

and

$$(\partial_t - \Delta)\kappa_1 \geq |\mathrm{II}|^2\kappa_1 + 2\sum_{k=1}^{n}\sum_{\kappa_i>\kappa_1}\frac{(\nabla_k \mathrm{II}_{i1})^2}{\kappa_i - \kappa_1} \tag{12.29}$$

in the barrier sense.

We now move on to the proof of Proposition 12.8, which provides the corresponding heat-type inequalities for the inscribed and exscribed curvatures.

Proof of Proposition 12.8. Once again, we shall only prove the first inequality, as the proof of the second is analogous.

Given a point $(x_0, t_0) \in \mathcal{M}^n \times (0, T)$, there are two situations to consider: the "boundary case", in which $\overline{k}(x_0, t_0) = K(x_0, y_0, t_0)$ for some $y_0 \in T_{x_0}\mathcal{M}^n \setminus \{0\}$, and the "interior case", in which $\overline{k}(x_0, t_0) = k(x_0, y_0, t_0)$ for some $y_0 \in \mathcal{M}^n \setminus \{x_0\}$. The boundary case is dealt with using the barrier constructed in the proof of Proposition 12.9. So assume that $\overline{k}(x_0, t_0) > \kappa_n(x_0, t_0)$ and hence $\overline{k}(x_0, t_0) = k(x_0, y_0, t_0)$ for some $y_0 \in \mathcal{M}^n \setminus \{x_0\}$. We will use k to define a barrier for $\overline{k}$ at such points.

We shall need to compute certain derivatives of k at (x_0, y_0, t_0). To this end, choose normal coordinates $(x^1, \ldots, x^n) : U_{x_0} \to \mathbb{R}^n$ and $(y^1, \ldots, y^n) : U_{y_0} \to \mathbb{R}^n$ centered at x_0 and y_0, respectively, with respect to the metric at time t_0. (So here $\{y^i\}$ has a different meaning than in our earlier discussion.) We can assume that $U_{x_0} \cap U_{y_0} = \emptyset$ so that k is smooth in $U_{x_0} \times U_{y_0} \times (0, t_0]$. As in the previous computations over $\mathcal{M} \times \mathcal{M}$, we define

$$d(x, y, t) \doteqdot \|X(x, t) - X(y, t)\| \,, \quad w(x, y, t) \doteqdot \frac{X(x, t) - X(y, t)}{d} \,,$$

and

$$\partial_i^x(x, y, t) \doteqdot \partial_{x^i}X(x, t) \,, \quad \partial_i^y(x, y, t) \doteqdot \partial_{y^i}X(y, t) \,,$$

and use sub- and superscripts x and y to denote quantities relating to the first and second factors, respectively.

Once again, there is no canonical choice of metric on $\mathcal{M} \times \mathcal{M}$. We are constrained only by the requirement that it project to the metric on $\mathcal{M}$ on the first factor. So choose any $\Lambda \in \mathrm{S}(n)$ and consider

$$(\partial_{x^i} + \Lambda_j{}^p\partial_{y^p})k = \frac{2}{d^2}\left(\langle \partial_i^x - \Lambda_j{}^p\partial_p^y, \mathsf{N}_x - kdw\rangle + \langle dw, \mathrm{II}^x(\partial_{x^i})\rangle\right) .$$

Differentiating once more and considering only the trace components yields

$$
\begin{aligned}
(\partial_{x^j} + \Lambda_j{}^p \partial_{y^p})(\partial_{x^i} + \Lambda_i{}^p \partial_{y^p})k \\
= \frac{2}{d^2}\Big\{ \langle -\mathrm{II}^x{}_{ij}\mathsf{N}_x + \Lambda_i{}^p\Lambda_j{}^q \mathrm{II}^y{}_{pq}\mathsf{N}_y, \mathsf{N}_x - kdw\rangle + \left\langle \partial_i^x - \Lambda_i{}^p\partial_p^y, \mathrm{II}_j^{x\,q}\partial_q^x\right\rangle \\
- \nabla_{\partial_{x^j}+\Lambda_j{}^q\partial_{y^q}} k \left\langle \partial_i^x - \Lambda_i{}^p\partial_p^y, dw\right\rangle - k\left\langle \partial_i^x - \Lambda_i{}^p\partial_p^y, \partial_j^x - \Lambda_j{}^q\partial_q^y\right\rangle \\
+ \left\langle \partial_j^x - \Lambda_j{}^q\partial_q^y, \mathrm{II}_i^{x\,p}\partial_p^x\right\rangle + \left\langle dw, \nabla \mathrm{II}^x_{ij} - (\mathrm{II}^x)^2_{ij}\mathsf{N}_x\right\rangle \\
- (\partial_{x^i} + \Lambda_i{}^p\partial_{y^p})k\left\langle \partial_j^x - \Lambda_j{}^q\partial_q^y, dw\right\rangle \Big\}.
\end{aligned}
$$

We can use the vanishing of the y-derivatives at y_0 to determine the tangent plane to $\mathcal{M}$ at y_0.

Lemma 12.11. *Given a properly embedded hypersurface $\mathcal{M}^n \subset \mathbb{R}^{n+1}$, suppose that $y_0 \in \mathcal{M}^n \setminus \{x_0\}$ is an extremum of $k(x_0, \cdot)$. Then*

$$\mathsf{N}_y = \mathsf{N}_x - dkw \tag{12.30}$$

at (x_0, y_0).

Proof of Lemma 12.11. By Lemma 12.2, there is an inscribed ball $B \subset \Omega$ of radius $1/k$ touching $\partial\Omega$ at x_0 and y_0. The outward normals to B at these points agree with the outward normals to $\partial\Omega$. That is,

$$
\begin{aligned}
\mathsf{N}(y_0) &= k(x_0, y_0)\left(y_0 - \left(x_0 - \frac{1}{k(x_0, y_0)}\mathsf{N}(x_0)\right)\right) \\
&= (\mathsf{N}_x - dkw)|_{(x_0,y_0)}\,.
\end{aligned}
$$

□

Thus, the tangent plane at y_0 is the reflection of the tangent plane at x_0 about the plane orthogonal to $y_0 - x_0$. In particular, we can choose the basis at y_0 to be the reflection of the basis at x_0. That is,

$$\partial_i^y = \partial_i^x - 2\left\langle \partial_i^x, w\right\rangle w \tag{12.31}$$

at (x_0, y_0, t_0).

We can also choose the basis at x_0 so that II^x is diagonal. Then

$$\frac{2}{d^2}\left\langle dw, \partial_{x^i}\right\rangle = \frac{\partial_i^x k}{\kappa_i^x - k}\,.$$

Putting these observations together yields, at the point (x_0, y_0, t_0),

$$\begin{aligned}
-\Delta_\Lambda k \doteqdot & - \delta^{kl}(\partial_{x^k} + \Lambda_k{}^p\partial_{y^q})(\partial_{x^l} + \Lambda_l{}^q\partial_{y^q})k \\
&= |\mathrm{II}_x|^2 k - H_x k^2 + \frac{2}{d^2}\delta^{ij}\left(\mathrm{II}^x_{ij} - \mathrm{II}^y_{ij} - \left\langle dw, \nabla \mathrm{II}^x_{ij}\right\rangle\right) \\
&\quad - \frac{2}{d^2}\delta^{ij}\Big\{\Lambda_i{}^p\Lambda_j{}^q\mathrm{II}^y_{pq} - \mathrm{II}^y_{ij} + 2\left\langle \partial^x_i - \Lambda_i{}^p(\partial^x_p - 2\left\langle\partial^x_p, w\right\rangle w), \mathrm{II}^{xq}_j\partial^x_q\right\rangle \\
&\quad - 2\partial_{x^j}k\left\langle\partial^x_i - \Lambda_i{}^p(\partial^x_p - 2\left\langle\partial^x_p, w\right\rangle w), dw\right\rangle \\
&\quad - k\left\langle\partial^x_i - \Lambda_i{}^p\left(\partial^x_p - 2\left\langle\partial^x_p, w\right\rangle w\right), \partial^x_j - \Lambda_j{}^q\left(\partial^x_q - 2\left\langle\partial^x_q, w\right\rangle w\right)\right\rangle\Big\}.
\end{aligned}$$

To simplify things, we set

$$\mathrm{H} \doteqdot \frac{2}{d^2}\left(H_y - H_x - \langle y - x, \nabla H_x\rangle\right) \tag{12.32}$$

and

$$\begin{aligned}
\mathrm{L}_\Lambda \doteqdot \frac{2}{d^2}\delta^{ij}\Big(&\mathrm{II}^x_{ij} - \mathrm{II}^y_{ij} + (\mathrm{II}^x_{ij} - k\delta_{ij}) \\
&- 2\Lambda_i^p(\mathrm{II}^x_{pj} - k\delta_{pj}) + \Lambda_i{}^p\Lambda_j{}^q(\mathrm{II}^y_{pq} - k\delta_{pq})\Big)
\end{aligned} \tag{12.33}$$

so that, after some rearranging,

(12.34)

$$\begin{aligned}
-\Delta_\Lambda k &= |\mathrm{II}_x|^2 k - H_x k^2 - \mathrm{H} - \mathrm{L}_\Lambda \\
&\quad - \frac{2}{d^2}\delta^{ij}\Big\{4\Lambda_i^p(\mathrm{II}^{xq}_j - k\delta_j{}^q)\langle\partial^x_p, w\rangle\langle\partial^x_q, w\rangle - 2\partial_{x^j}k\langle\partial^x_i + \Lambda_i^p\partial^x_p, dw\rangle\Big\} \\
&= |\mathrm{II}_x|^2 k - H_x k^2 - 2\sum_{i=1}^n \frac{(\partial_{x^i}k)^2}{k - \kappa_i} - \mathrm{H} - \mathrm{L}_\Lambda\,.
\end{aligned}$$

Next, observe that the time derivative of k is given by

$$\partial_t k = \frac{2}{d^2}\big(\langle -H_x\mathsf{N}_x + H_y\mathsf{N}_y, \mathsf{N}_x - k\,dw\rangle + \langle dw, \nabla H_x\rangle\big)$$

so that, applying Lemma 12.11,

$$\partial_t k = \mathrm{H} + k^2 H_x \tag{12.35}$$

at the point (x_0, y_0, t_0).

We conclude that

$$(\partial_t - \Delta_\Lambda)k = |\mathrm{II}_x|^2 k - 2\sum_{i=1}^n \frac{(\partial_{x^i}k)^2}{k - \kappa_i} - \mathrm{L}_\Lambda \tag{12.36}$$

at (x_0, y_0, t_0).

Now consider the barrier function $\varphi : U_{x_0} \times (0, t_0] \to \mathbb{R}$ defined by

$$\varphi(x, t) \doteqdot k(x, y_0, t)\,.$$

Since $y_0 \notin U_{x_0}$, φ is smooth and supports $\overline{k}$ from below at (x_0, t_0). Since y_0 is a local maximum of $k(x_0, \cdot, t_0)$, $\partial_{y^i} k = 0$, $\partial_{y^i}\partial_{y^j} k \leq 0$ and hence

$$(\partial_t - \Delta)\varphi = (\partial_t - \Delta_\Lambda)\varphi \leq (\partial_t - \Delta_\Lambda)k \leq |\mathrm{II}|^2\varphi - 2\sum_{i=1}^{n} \frac{(\nabla_i \varphi)^2}{\varphi - \kappa_i} - \mathrm{L}_\Lambda$$

at (x_0, y_0, t_0). Choosing Λ to be the identity yields $L_\Lambda = 0$ by (12.33). This completes the proof. □

By an argument similar to the one used in Lemma 9.16, the inequalities (12.15) and (12.16) also hold in the distributional sense (cf. [**107**]).

Applying the maximum principle to the inequalities (12.15) and (12.16) yields the following result.

Corollary 12.12 (Compact, embedded, mean convex flows are noncollapsing). *Let $X : M^n \times [0, T) \to \mathbb{R}^{n+1}$, $n \geq 1$, be a compact, embedded, mean convex solution to mean curvature flow. Then*

$$\frac{\overline{k}}{H} \leq \max_{M^n \times \{0\}} \frac{\overline{k}}{H} \quad \text{and} \quad \frac{\underline{k}}{H} \geq \min_{M^n \times \{0\}} \frac{\underline{k}}{H}. \tag{12.37}$$

That is, X is noncollapsing.

Remark 12.13. Corollary 12.12 was first proved in [**41**], although closely related statements were previously obtained (via blow-up analyses) by White [**529**] and Sheng and Wang [**481**]. The argument presented here (namely, Proposition 12.8) was developed in [**41, 55**], where inequalities (12.15) and (12.16) were obtained without the gradient terms. The gradient terms were discovered by Brendle [**107**], where they were exploited to obtain sharp noncollapsing estimates, which we discuss in Section 12.5.1 below (however, they are not needed for the coarse estimates of Corollary 12.12).

Similar ideas were crucial in the proof of the Lawson conjecture on the uniqueness of embedded minimal tori in S^3 by Brendle [**104**]. This argument was subsequently adapted to constant mean curvature tori in S^3 by H. Li and the first author, proving the Pinkall–Sterling conjecture [**58**], and to Weingarten tori by Brendle [**105**].

As an immediate application, we may rule out collapsed singularity models. In particular, we obtain the following.

Corollary 12.14 ((Don't fear) the Reaper).

(1) *Multiplicity two hyperplanes cannot arise as tangent flows to a compact, mean convex, embedded mean curvature flow.*

(2) *The Grim hyperplane cannot arise as a limit flow of a compact, mean convex, embedded mean curvature flow.*

12.2.1. Simons-type differential inequalities. The above computations also yield elliptic "Simons-type" differential inequalities for the interior and exterior ball curvatures for a general hypersurface (recall Simons's formula for the Hessian of the second fundamental form from Proposition 5.10). We first consider the extremal principal curvatures, κ_1 and κ_n.

Similarly to Definition 9.5, given a continuous function $f : M^n \to \mathbb{R}$ and a symmetric 2-tensor a_{ij} on M, we say that $-\nabla_i\nabla_j f \geq a_{ij}$ in the **barrier sense** if for each $x_0 \in M$ there exists a C^2 function ψ defined in a neighborhood U of x_0 satisfying

$$-\nabla_i\nabla_j\psi(x_0) \geq a_{ij}(x_0), \quad \psi(x_0) = f(x_0), \quad \text{and} \quad \psi \geq f \text{ in } U.$$

Similarly, we say that $-\nabla_i\nabla_j f \leq a_{ij}$ in the barrier sense if for each $x_0 \in M$ there exists a C^2 function ψ defined in a neighborhood U of x_0 satisfying

$$-\nabla_i\nabla_j\psi(x_0) \leq a_{ij}(x_0), \quad \psi(x_0) = f(x_0), \quad \text{and} \quad \psi \leq f \text{ in } U.$$

Proposition 12.15. *Let $X : M^n \to \mathbb{R}^{n+1}$ be a smooth hypersurface. Then*

$$(12.38) \quad -\nabla_i\nabla_j\kappa_n \leq \mathrm{II}^2_{ij}\,\kappa_n - \kappa_n^2\,\mathrm{II}_{ij} - \nabla_n\nabla_n\,\mathrm{II}_{ij} - 2\sum_{\kappa_p<\kappa_n}\frac{\nabla_i\,\mathrm{II}_{pn}\,\nabla_j\,\mathrm{II}_{pn}}{\kappa_n-\kappa_p}$$

and

$$(12.39) \quad -\nabla_i\nabla_j\kappa_1 \geq \mathrm{II}^2_{ij}\,\kappa_1 - \mathrm{II}_{ij}\,\kappa_1^2 - \nabla_1\nabla_1\,\mathrm{II}_{ij} + 2\sum_{\kappa_p>\kappa_1}\frac{\nabla_i\,\mathrm{II}_{p1}\,\nabla_j\,\mathrm{II}_{p1}}{\kappa_p-\kappa_1}$$

in the barrier sense. In particular,

$$(12.40) \qquad -\Delta\kappa_n \leq |\mathrm{II}|^2\kappa_n - H\kappa_n^2 - \nabla_n\nabla_n H - 2\sum_{\kappa_i<\kappa_n}\frac{(\nabla_i\kappa_n)^2}{\kappa_n-\kappa_i}$$

and

$$(12.41) \qquad -\Delta\kappa_1 \geq |\mathrm{II}|^2\kappa_1 - H\kappa_1^2 - \nabla_1\nabla_1 H + 2\sum_{\kappa_i>\kappa_1}\frac{(\nabla_i\kappa_1)^2}{\kappa_i-\kappa_1}$$

in the barrier sense.

Proof. In the proof of the heat-type inequality for κ_n (Proposition 12.9) we computed, in equation (12.22), components of the Hessian of the function $K : T\mathcal{M}^n \setminus \{0\} \to \mathbb{R}$ defined by

$$K(x,y) \doteqdot \frac{\mathrm{II}_x(y,y)}{g_x(y,y)}.$$

Evaluating (12.22) at an extreme point (x_0, y_0) of K with $K(x_0, y_0) = \kappa_n$ and $y_0 = e_n$ (the principal direction corresponding to κ_n) yields

$$\begin{aligned}-\nabla_k\nabla_l\varphi \leq &-(\partial_{x^k} + \Lambda_k{}^p\partial_{y^p})(\partial_{x^l} + \Lambda_k{}^q\partial_{y^q})K\\ = &-\nabla_k\nabla_l \mathrm{II}_{nn} + 2\Lambda_k{}^p(\nabla_l K\delta_{pn} - \nabla_l \mathrm{II}_{pn})\\ &+ 2\Lambda_l{}^q(\nabla_k K\delta_{nq} - \nabla_k \mathrm{II}_{nq}) + 2\Lambda_k{}^p\Lambda_l{}^q(K - \kappa_p)\delta_{pq}\,,\end{aligned}$$

where $\varphi \leq \kappa_n$ is the barrier function defined near x_0 by

$$\varphi(x) \doteqdot K(x, y^i\partial_{x^i})\,,$$

where the x^i are geodesic normal coordinates about x_0 with respect to a principal basis at x_0 and the $y^i = dx^i$ are the corresponding fiber coordinates. Choosing

$$\Lambda_{kp} \doteqdot \begin{cases}\dfrac{\nabla_k \mathrm{II}_{pn}}{\kappa_n - \kappa_p} & \text{if} \quad \kappa_p < \kappa_n\,,\\ 0 & \text{otherwise}\end{cases}$$

yields

$$-\nabla_k\nabla_l\varphi \leq -\nabla_k\nabla_l \mathrm{II}_{nn} - 2\sum_{p<n}\frac{\nabla_k \mathrm{II}_{pn}\nabla_l \mathrm{II}_{pn}}{\kappa_n - \kappa_p}\,.$$

Taking the trace and estimating the gradient terms as in (12.27) yields

$$-\Delta\varphi \leq -\Delta \mathrm{II}_{nn} - 2\sum_{k=1}^{n}\sum_{\kappa_i<\kappa_n}\frac{(\nabla_i\varphi)^2}{\kappa_n - \kappa_i}\,.$$

Combined with Simons's equation (5.50), this proves (12.38) and (12.40). The proofs of (12.39) and (12.41) are similar. □

Remark 12.16. We shall see in Section 17.3.3 an application of Proposition 12.15 to the Gauß curvature flow.

Next, we consider the inscribed and exscribed curvatures, $\overline{k}$ and $\underline{k}$. Consider the extension

$$\mathrm{H}(x, y) \doteqdot \begin{cases}\dfrac{2\,(H(y) - H(x) - \langle y - x, \nabla H(x)\rangle)}{d(x, y)^2} & \text{if} \quad y \in \mathcal{M}\setminus\{x\}\,,\\ \nabla_y\nabla_y H(x) & \text{if} \quad y \in S\mathcal{M}_x\end{cases}$$

of the function $\mathrm{H} : (\mathcal{M}\times\mathcal{M})\setminus D \to \mathbb{R}$ given by (12.32). Define the extremal sets

$$\overline{U}_x \doteqdot \begin{cases}\{y \in S\mathcal{M}_x : \overline{k}(x) = \mathrm{II}_x(y, y)\} & \text{if} \quad \overline{k}(x) = \kappa_n(x)\,,\\ \{y \in \mathcal{M}\setminus\{x\} : \overline{k}(x) = k(x, y)\} & \text{if} \quad \overline{k}(x) > \kappa_n(x)\end{cases}$$

and

$$\underline{U}_x \doteqdot \begin{cases}\{y \in S\mathcal{M}_x : \underline{k}(x) = \mathrm{II}_x(y, y)\} & \text{if} \quad \underline{k}(x) = \kappa_1(x)\,,\\ \{y \in \mathcal{M}\setminus\{x\} : \underline{k}(x) = k(x, y)\} & \text{if} \quad \underline{k}(x) < \kappa_1(x)\,.\end{cases}$$

Setting

$$\overline{\mathrm{H}}(x) \doteqdot \sup_{y\in \overline{U}_x} \mathrm{H}(x,y) \quad \text{and} \quad \underline{\mathrm{H}}(x) \doteqdot \sup_{y\in \underline{U}_x} \mathrm{H}(x,y)\,,$$

we obtain the following inequalities.

Proposition 12.17. *Let $\mathcal{M} \hookrightarrow \mathbb{R}^{n+1}$ be a smooth, compact hypersurface. Then*

$$-\Delta\overline{k} \leq |\mathrm{II}|^2\,\overline{k} - H\overline{k}^2 - 2\sum_{i=1}^{n}\frac{(\nabla_i\overline{k})^2}{\overline{k}-\kappa_i} - \overline{\mathrm{H}} \tag{12.42}$$

and

$$-\Delta\underline{k} \geq |\mathrm{II}|^2\,\underline{k} - H\underline{k}^2 + 2\sum_{i=1}^{n}\frac{(\nabla_i\underline{k})^2}{\kappa_i-\underline{k}} - \underline{\mathrm{H}} \tag{12.43}$$

in the barrier sense.

Proof. Fix $x_0 \in \mathcal{M}$. Suppose that $\overline{k}(x_0) > \kappa_n(x_0)$. Then $k(x_0,y_0) = \overline{k}(x_0)$ for some $y_0 \in \mathcal{M}\setminus\{x_0\}$. In the proof of the heat-type inequalities for $\overline{k}$ and $\underline{k}$ (Proposition 12.8) we computed, in equation (12.34), a differential inequality for k at $(x_0,y_0)\in\mathcal{M}\times\mathcal{M}$. Setting $\Lambda = \mathrm{I}$, (12.34) yields

$$\begin{aligned}-\Delta\varphi &\leq -\Delta_\Lambda k\\ &= |\mathrm{II}_x|^2 k - H_x k^2 - 2\sum_{i=1}^{n}\frac{(\partial_{x^i}k)^2}{k-\kappa_i} - \mathrm{H}\,,\end{aligned}$$

where $\varphi \leq \overline{k}$ is the barrier function defined near x_0 by

$$\varphi(x) \doteqdot k(x,y_0)\,.$$

If there is no such $y_0 \in \mathcal{M}\setminus\{x_0\}$, then $\overline{k}(x_0) = \kappa_n(x_0)$ and we apply Proposition 12.15. The arguments for $\underline{k}$ are analogous. □

12.3. The Gage–Hamilton and Huisken theorems via noncollapsing

The noncollapsing property of embedded, mean convex mean curvature flows has numerous useful consequences. In this section, we will see that it yields a particularly efficient proof of Huisken's theorem. In fact, the proof will also apply to the curve shortening flow, providing yet another proof of the Gage–Hamilton theorem. Indeed, recalling Lemma 8.3, exterior noncollapsing immediately implies that a convex solution $\{\mathcal{M}_t\}_{t\in[0,T)}$ shrinks to a point as t approaches the maximal time T since

$$\frac{1}{\rho_+(\mathcal{M}_t)} \geq \max_{\mathcal{M}_t}\underline{k} \gtrsim \max_{\mathcal{M}_t} H \gtrsim (T-t)^{-\frac{1}{2}} \xrightarrow[t\to T]{} \infty\,, \tag{12.44}$$

where $\rho_+(\mathcal{M}_t)$ is the circumradius of $\mathcal{M}_t$.

We also obtain a lower bound for the inradius[3] $\rho_-(\mathcal{M}_t)$ of $\mathcal{M}_t$ from the interior noncollapsing estimate:

$$\frac{1}{\rho_-(\mathcal{M}_t)} \leq \min_{\mathcal{M}_t} \overline{k} \lesssim \min_{\mathcal{M}_t} H \lesssim (T-t)^{-\frac{1}{2}} . \tag{12.45}$$

Comparing circum- and inradii, we find that $H \sim (T-t)^{-\frac{1}{2}}$.

The Arezelà–Ascoli theorem and the strong maximum principle yield decay of the second fundamental form and its derivatives.

Proposition 12.18. *Set*

$$A_0(t) \doteqdot \max_{M^n\times\{t\}} \left(\overline{k} - \tfrac{1}{n}H\right)^2 \quad \textit{and, for each } m \in \mathbb{N}, \quad A_m \doteqdot \max_{M^n\times\{t\}} |\nabla^m \mathrm{II}|^2 .$$

Then

$$\limsup_{t\nearrow T} (T-t)^{-(m+1)} A_m(t) = 0 \tag{12.46}$$

for every $m \in \mathbb{N}\cup\{0\}$.

Proof. Suppose, to the contrary, that there is some $\varepsilon > 0$, some $m \in \mathbb{N}\cup\{0\}$, and some sequence of times $t_j \nearrow T$ such that

$$(T-t_j)^{-1} A_m(t_j) \geq \varepsilon \tag{12.47}$$

for all $j \in \mathbb{N}$. Set $\sigma_j \doteqdot \lambda_j^2 t_j$, where $\lambda_j \doteqdot (T-t_j)^{-\frac{1}{2}}$, and consider the sequence $\{\mathcal{M}^j_t\}_{t\in[-\lambda_j^2 t_j,1),i\in\mathbb{N}}$ of rescaled flows defined by

$$\mathcal{M}^j_t \doteqdot \lambda_j \left(\mathcal{M}_{\lambda_j^{-2}t+t_j} - p\right) .$$

As in the proof of Corollary 8.28, some subsequence of the parametrizations of the rescaled flows converges in the smooth topology to a smooth, compact, convex limit flow, uniformly on compact time intervals. Recall that

$$(\partial_t - \Delta)\frac{\overline{k}}{H} \leq 2\left\langle \nabla \frac{\overline{k}}{H}, \nabla \log H\right\rangle - \frac{2}{H}\sum_{\kappa_i<\overline{k}} \frac{(\nabla_i \overline{k})^2}{\overline{k}-\kappa_i} \tag{12.48}$$

in the barrier sense. Arguing as in Corollary 8.28, the strong maximum principle (see Lemma 9.10) implies that $\frac{\overline{k}}{H}$ is constant on the limit flow. So $\overline{k} \equiv K_0 H$ on the limit flow for some constant $K_0 \geq \frac{1}{n}$. In particular, $\overline{k}$ is smooth. We will show that $K_0 = \frac{1}{n}$. This follows from Exercise 12.5 when $n = 1$, so assume that $n \geq 2$. Fix $t < 0$ and consider the open set $V_t \doteqdot \{x \in \mathcal{M}_t : \overline{k}(x,t) > \kappa_n(x,t)\}$. By (12.48),

$$K_0 \nabla H \equiv \nabla \overline{k} \equiv 0 \quad \text{in} \quad V_t .$$

[3]In fact, equality holds in the first of the inequalities in both (12.44) and (12.45). See Exercise 12.2.

So $\overline{k}$ and H are constant on each connected component of V_t. Since $\mathcal{M}_t$ is convex, the inequality (12.42) of Proposition 12.17 implies that

$$0 \leq -\sum_{i=1}^{n} \overline{k}\kappa_i(\overline{k} - \kappa_i)$$

in V_t. Since $\mathcal{M}_t$ is convex, it follows that V_t is umbilic and hence, by Exercise 5.5, a union of open subsets of round spheres. We claim that V_t is closed in $\mathcal{M}_t$. Suppose, to the contrary, that there is a point $x \in \overline{V}_t \setminus V_t$. Then there is a sequence of points $x_j \in V_t$ such that $x_j \to x$. As $x \in \mathcal{M}_t \setminus V_t$ and V_t is part of a sphere, we conclude that

$$\frac{\overline{k}}{H}(x,t) = \frac{\kappa_n}{H}(x,t) = \lim_{j\to\infty} \frac{\kappa_n}{H}(x_j,t) = \frac{1}{n}.$$

But this can only happen if $\mathcal{M}_t$ itself is a round sphere, in which case $V_t = \emptyset$. So V_t is closed as claimed. By Exercise 12.5, $\mathcal{M}_t \setminus V_t$ is nonempty, so we conclude that V_t is empty. That is, $\kappa_n \equiv \overline{k} \equiv K_0 H$. In particular, κ_n is smooth and $\nabla \kappa_n = K_0 \nabla H$. Consider the endomorphism N of $T\mathcal{M}$ dual to the bilinear form $\mathrm{II} - K_0 H \mathrm{I}$. Observe that

$$\nabla_w \mathrm{II}(u,v) = (\nabla_w \mathrm{II} - K_0 \nabla_w H\, \mathrm{I})(u,v) = 0 \tag{12.49}$$

for any $u, v \in \ker \mathrm{N}$ with $u \perp v$ and any w. By (12.48),

$$\nabla_w \mathrm{II}(u,u) = 0 \tag{12.50}$$

for any $w \in (\ker \mathrm{N})^{\perp}$ and $u \in \ker \mathrm{N}$. Finally, by Exercise 12.4,

$$\nabla_u \mathrm{II}(u,u) = 0 \tag{12.51}$$

whenever $u \in \ker \mathrm{N}$. Combining (12.49), (12.50), and (12.51), we conclude that

$$0 \equiv \nabla_v \mathrm{II}(u,u) \equiv K_0 \nabla_v H$$

for any unit vector $u \in \ker \mathrm{N}$ and any vector v. Proposition 5.12 now implies that $\mathcal{M}_t$ is a round sphere for all t. We conclude that K_0 is indeed equal to $\frac{1}{n}$, which proves the claim. This is in contradiction with (12.47). □

We can now proceed as in Section 8.4.4 to convert the geometric estimates into estimates for the rescaled embedding and its derivatives and thereby deduce convergence in the smooth topology to an embedding of the unit sphere.

12.4. The Haslhofer–Kleiner curvature estimate

In the previous section we have seen how noncollapsing can be applied to efficiently control the formation of singularities in mean curvature flow of convex hypersurfaces. A powerful tool in extending the techniques to (embedded) *mean convex* mean curvature flow is the local curvature estimate of Robert Haslhofer and Bruce Kleiner [**276**, Theorem 1.8].

Figure 12.4. Robert Haslhofer. Photo courtesy of Xiao Xiao.

The estimate is local in nature, so let us recall that the (**extrinsic**) **parabolic cylinder** of radius $r > 0$ centered at $(x, t) \in \mathbb{R}^{n+1} \times \mathbb{R}$ is defined by

$$P_r(x,t) \doteqdot B_r(x) \times (t - r^2, t] \subset \mathbb{R}^{n+1} \times \mathbb{R}\,.$$

Given a mean convex embedded solution of mean curvature flow $\{\partial\Omega_t\}_{t\in[0,T)}$, we will adopt the practice of conflating points $x \in \partial\Omega_t$ with pairs $(x,t) \in \overline{\Omega}_0 \times [0,T)$. Since the flow moves in only the inwards direction, a point $x \in \overline{\Omega}_0$ can lie in at most one time slice, so this should lead to no ambiguity.

Theorem 12.19 (Local curvature estimate [**276**, Theorem 1.8]). *Given $n \in \mathbb{N}$ and $\alpha > 0$ there exist $\sigma \in (0,1)$ and, for each $k \in \mathbb{N} \cup \{0\}$, $C_k < \infty$ with the following property: Let $\{\partial\Omega_t\}_{t\in(t_0-r^2,t_0]}$ be a properly embedded, mean convex mean curvature flow in $\mathbb{R}^{n+1}$. If*

$$-\alpha^{-1} \le \inf_{P_r(x_0,t_0)} \frac{\underline{k}}{H} \quad \text{and} \quad \sup_{P_r(x_0,t_0)} \frac{\overline{k}}{H} \le \alpha^{-1} \tag{12.52}$$

for some $x_0 \in \partial\Omega_{t_0}$ satisfying $H(x_0,t_0) \le r^{-1}$, then

$$\sup_{P_{\sigma r}(x_0,t_0)} |\nabla^k \mathrm{II}| \le C_k r^{-k-1}\,. \tag{12.53}$$

Proof. We need only prove the case $k = 0$, i.e., a local bound for $|\mathrm{II}|$, as the remaining estimates then follow from the interior estimates of Ecker and Huisken (Theorem 7.14). The proof proceeds by contradiction.

So suppose that the claim is false. Then there exist a sequence of scales $r_j > 0$, a sequence of embedded, mean convex mean curvature flows $\{\partial\widehat{\Omega}^j_t\}_{t\in(t_j-r_j^2,t_j]}$ in $\mathbb{R}^{n+1}$, and a sequence of points $x_j \in \partial\widehat{\Omega}^j_{t_j}$ such that $\widehat{H}(x_j,t_j) \le r_j^{-1}$ but $\sup_{P_{r_j}(x_j,t_j)} |\widehat{\mathrm{II}}| \ge jr_j^{-1}$. After translating the points x_j to the origin, rotating so that $\widehat{\mathsf{N}}(x_j,t_j) = e_{n+1}$ and parabolically rescaling by jr_j^{-1}, we obtain a sequence of embedded, mean convex mean curvature flows $\{\partial\Omega^j_t\}_{t\in(-j,0]}$ such that $0 \in \partial\Omega^j_0$ and $\mathsf{N}_j(0,0) = e_{n+1}$ for each j, $H_j(0,0) \to 0$ as $j \to \infty$ but $\sup_{P_1(0,0)} |\mathrm{II}_j| \ge 1$ for each j. We claim that, after passing to a subsequence, the flows $\{\partial\Omega^j_t\}_{t\in(-j,0]}$ converge locally uniformly in the smooth topology to the static hyperplane $\{\partial\{x_{n+1} < 0\}\}_{t\in(-\infty,0]}$, contradicting the fact that $\liminf_{j\to\infty} \sup_{P_1(0,0)} |\mathrm{II}_j| \ge 1$.

First note that, given any $T < 0$ and any compact subset K of the lower half-space $\{x_{n+1} < 0\}$, there exists j so that $K \subset \Omega^j_t$ for all $t > T$. Indeed, this is certainly true when $T = 0$ because $H_j(0,0) \to 0$ and the sequence is uniformly noncollapsing. The claim follows since $\Omega_0 \subset \Omega_t$ for all $t < 0$ by mean convexity. We need to prove the same for the complements. Given $R > 0$ (large) and $d > 0$ (small), consider the ball $B_{R,d} \doteqdot B_R((-R+d)e_{n+1})$ in $\mathbb{R}^{n+1}$ tangent to the hyperplane $\{x_{n+1} = d\}$ at the point de_{n+1}. If we evolve $\partial B_{R,d}$ by mean curvature flow for time $T \doteqdot \frac{2dR+d^2}{2n} \sim dR$, we obtain the boundary of the ball $B_R(-Re_{n+1})$. Since this sphere intersects $\partial\Omega^j_0$ at the origin, we conclude from the avoidance principle that $\partial B_{R,d}$ must intersect $\partial\Omega^j_t$ for each $t \in [-T,0]$ and every j. Fix $t \in [-T,0]$. Since every compact subset of $\{x_{n+1} < 0\}$ lies in Ω^j_t for some j sufficiently large, we can find some $j \in \mathbb{N}$ and some $d_j < d$ so that $\partial B_{R,d_j}$ has first-order contact with $\partial\Omega^j_t$ at some point $x_j \in \partial\Omega^j_t$. The point x_j must therefore satisfy $\langle x_j, e_{n+1}\rangle < d$ and $H_j(x_j,t) \le \frac{n}{R}$. Since $\partial\Omega^j_t \cap B_{R,d}$ has height bounded below by $-\delta \doteqdot -d^2/2R$ for j sufficiently large, we can also assume that $\langle x_j, e_{n+1}\rangle \ge -\delta$ and

$$\|x_j\| \le \sqrt{2R(d+\delta) - (d+\delta)^2 + \delta^2} = \sqrt{2d(R-\delta)}\,.$$

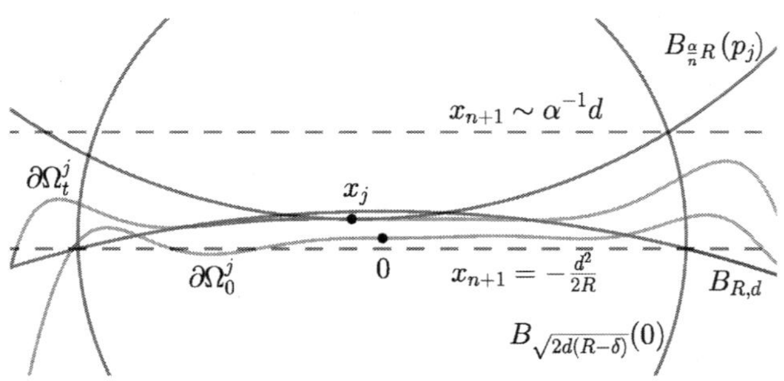

Figure 12.5. Hausdorff convergence to the static hyperplane $\{x_{n+1} = 0\}$.

By the noncollapsing hypothesis, the ball $B_{\frac{\alpha}{n}R}(y_j)$ of radius $\frac{\alpha}{n}R$ centered at the point $y_j \doteqdot x_j + \frac{\alpha}{n}R\mathsf{N}_j(x_j,t)$ lies outside of Ω_t^j (touching it at x_j). But the normal $\mathsf{N}_j(x_j,t)$ coincides with the normal to B_{R,d_j} at x_j, so it must be of the form $\mathsf{N}_j(x_j,t) = \frac{x_j+(R-d)e_{n+1}}{R}$ and hence $\langle \mathsf{N}_j(x_j,t), e_{n+1}\rangle \geq 1 - \frac{d+\delta}{R}$. It follows that Ω_t^j has height bounded by approximately $2n\alpha^{-1}d$ in $B_{\sqrt{2d(R-\delta)}}$ for R sufficiently large. Indeed, any point z in the intersection of $B_{\frac{\alpha}{n}R}(y_j)$ and $B_{\sqrt{2d(R-\delta)}}(0)$ satisfies

$$
\begin{aligned}
\left(\tfrac{\alpha}{n}R\right)^2 &= \|z_j - y_j\|^2 \\
&= \langle z - y_j, e_{n+1}\rangle^2 + \sum_{k=1}^{n} \langle z_j - y_j, e_k\rangle^2 \\
&= 2\langle z, e_{n+1}\rangle \langle y_j, e_{n+1}\rangle + \|y_j\|^2 + \|z\|^2 - 2\sum_{k=1}^{n} \langle z, e_k\rangle \langle y_j, e_k\rangle
\end{aligned}
$$

and hence

$$
\begin{aligned}
2\langle z, e_{n+1}\rangle &= \frac{\|y_j\|^2 - \left(\frac{\alpha}{n}R\right)^2 + \|z\|^2 - 2\sum_{k=1}^{n} \langle z, e_k\rangle \langle y_j, e_k\rangle}{\langle y_j, e_{n+1}\rangle} \\
&\sim \frac{2d(R-\delta)}{\frac{\alpha}{n}R} \leq 2n\alpha^{-1}d\,.
\end{aligned}
$$

To deduce smooth convergence, we apply Brakke's local regularity theorem (Theorem 6.29). It suffices to prove that, given any $\varepsilon > 0$ and any ball $B_r(x) \subset \mathbb{R}^{n+1}$,

$$
|\partial\Omega_t^j \cap B_r(x)| \leq (1+\varepsilon)\omega_n r^n \tag{12.54}
$$

for j sufficiently large. To prove this, observe that, for any $t \in (-j, 0]$ and any closed domain K that agrees with $\partial\Omega_t$ outside of $\overline{B}_r(x)$,

$$
\begin{aligned}
|\partial K \cap B_r(x)| - |\partial\Omega_t^j \cap B_r(x)| &\geq \int_{\partial K \cap B_r(x)} \langle \mathsf{N}, \mathsf{N}_0\rangle - \int_{\partial\Omega_t^j \cap B_r(x)} \langle \mathsf{N}, \mathsf{N}_t\rangle \\
&= \int_{(K\setminus\Omega_t^j)\cap B_r(x)} \operatorname{div} \mathsf{N} \\
&= \int_{(K\setminus\Omega_t^j)\cap B_r(x)} H > 0\,,
\end{aligned}
$$

where N is the vector field on $\Omega_0 \setminus \Omega_t$ induced by the unit normal field of the flow and H is the function on $\Omega_0 \setminus \Omega_t$ induced by the mean curvature of the flow. Setting $K \doteqdot \Omega_t^j \cup (B_r(x) \cap \{x_{n+1} \leq \varepsilon\})$ yields (12.54). □

The local curvature estimate can be used in conjunction with the compactness theory developed in Chapter 11 to obtain versions of the "pinching improves" estimates (Corollaries 9.13 and 9.14) by blow-up arguments along the lines of the one used in Hamilton's proof of Huisken's theorem in Section 11.4, or the proof of Proposition 12.18.

Theorem 12.20 (Haslhofer and Kleiner [**276**] (following White [**531**])). *Given $\alpha > 0$ and $\varepsilon > 0$ there exists $\eta > 0$ with the following property: Let $\{\mathcal{M}_t\}_{t\in I}$ be an embedded, α-noncollapsing mean curvature flow which is properly defined in a parabolic ball $P_{\eta r}(x,t)$ with $t \in I$ and $x \in \mathcal{M}_t$. If $H(x,t) \le r^{-1}$, then*

$$\kappa_1(x,t) \ge -\varepsilon r^{-1}\,. \tag{12.55}$$

We will need the following rigidity result (cf. Theorem 9.11).

Theorem 12.21 (Rigidity of lower pinching, following White [**531**, Appendix A] (cf. [**297**, §5])). *Let $X : M^n \times I \to \mathbb{R}^{n+1}$ be a strictly mean convex solution to mean curvature flow which is properly defined in $P_R(x_0,t_0) \doteqdot B_R(p_0)\times(t_0-r^2,t_0]$, where $p_0 \doteqdot X(x_0,t_0)$. If $\min_{P_R(x_0,t_0)}\frac{\kappa_1}{H}$ attains a nonpositive minimum at (x_0,t_0), then $\kappa_1(x_0,t_0)=0$ and $X : M^n\times I \to \mathbb{R}^{n+1}$ locally splits isometrically off a plane in the direction of* $\ker \mathrm{L}$.

Proof. By Proposition 12.9 (see Remark 12.10) or Proposition 9.6, κ_1/H satisfies

$$(\partial_t - \Delta)\frac{\kappa_1}{H} \ge 2\left\langle \nabla\frac{\kappa_1}{H}, \frac{\nabla H}{H}\right\rangle + \frac{2}{H}\sum_{k=1}^{n}\sum_{\kappa_i>\kappa_1}\frac{(\nabla_k \mathrm{II}_{i1})^2}{\kappa_i-\kappa_1} \tag{12.56}$$

in the barrier sense. Since κ_1/H reaches an interior minimum, Lemma 9.10 implies that $\kappa_1 \equiv K_0 H$ in $\mathcal{P}_r(x_0,t_0)$ for some $K_0 \le 0$ and some $r>0$. Set $\mathrm{N} \doteqdot \mathrm{II} - K_0 H\,\mathrm{I}$. Observe that

$$\nabla_w \mathrm{N}(u,v) = 0 \tag{12.57}$$

for any $u,v \in \ker \mathrm{N}_{(x,t)}$ and any $w \in T_{(x,t)}M$ if $(x,t)\in\mathcal{P}_r(x_0,t_0)$, where we conflate N with its dual (selfadjoint) endomorphism of TM. Indeed, given $U,V \in \Gamma(\ker \mathrm{N})$ with $U_{(x,t)} = u$ and $V_{(x,t)} = v$ and $w \in T_{(x,t)}M$,

$$0 \equiv w\mathrm{N}(U,V) = \nabla_w\,\mathrm{N}(u,v) + \mathrm{N}(\nabla_w U, v) + \mathrm{N}(u, \nabla_w V) = \nabla_w\,\mathrm{N}(u,v)\,.$$

By (12.56),

$$\nabla_w \mathrm{N}(u,v) = \nabla_w\,\mathrm{II}(u,v) = 0 \tag{12.58}$$

for any $u \in \ker \mathrm{N}_{(x,t)}$, $v \in (\ker \mathrm{N}_{(x,t)})^\perp$ and any $w \in T_{(x,t)}M$ if $(x,t) \in \mathcal{P}_r(x_0,t_0)$. Combining (12.57) and (12.58) yields

$$0 = \nabla_w \mathrm{N}(u)$$

for all $u \in \ker(\mathrm{N}_{(x,t)})$ and all $w \in T_{(x,t)}M$ if $(x,t) \in \mathcal{P}_r(x_0,t_0)$. It follows that $\nabla_k V \in \Gamma(\ker \mathrm{N})$ whenever $V \in \Gamma(\ker \mathrm{N})$. Indeed, given $V \in \Gamma(\ker \mathrm{N})$ and $w \in T_{(x,t)}M$,

$$0 \equiv \nabla_w(\mathrm{N}(V)) = \nabla_w \mathrm{N}(v) + \mathrm{N}(\nabla_w V) = \mathrm{N}(\nabla_w V)\,,$$

where $v \doteqdot V_{(x,t)}$. That is, $\ker \mathrm{N}$ is invariant under parallel translation in space. By the Frobenius theorem [**522**, Theorem 1.60], the null spaces of N form an integrable distribution, and the leaves are totally geodesic. So the slices $\mathcal{M}_t$ locally split as a nontrivial, isometric product $H_t \times V_t$ near (x_0,t_0). In particular, the sectional curvature $\kappa_1\kappa_n$ of the 2-plane $e_1 \wedge e_n$ must vanish (since e_1, a null vector of N, is horizontal and e_n, a positive eigenvector of N, is vertical). We conclude that $\kappa_1(x_0,t_0) = 0$. The remaining claim then follows from Theorem 9.11. □

Proof of Theorem 12.20. Let ε_0 be the infimum over all $\varepsilon > 0$ for which the conclusion holds. By the noncollapsing hypothesis, $\varepsilon_0 \leq \alpha^{-1} < \infty$. Suppose, contrary to the claim, that $\varepsilon_0 > 0$. Then, by making use of the invariance properties of the flow, we can find a sequence $\{\mathcal{M}_{t,j}\}_{t\in I_j, j\in\mathbb{N}}$ of α-noncollapsing mean curvature flows which are properly defined in $P_j(0,0)$, with $0 \in I_j$ and $0 \in \mathcal{M}_{0,j}$, and satisfy $H_j(0,0) = 1$ and $\kappa_1^j(0,0) \to \varepsilon_0$. By Theorems 12.19 and 11.11, we can find an embedded mean curvature flow $\{\mathcal{M}_t\}_{t\in(-\infty,0]}$ which is properly defined in $P_\rho(0,0)$ and a subsequence $\{\mathcal{M}_{t,j}\}_{t\in I_j, j\in\mathbb{N}}$ which converges smoothly to $\{\mathcal{M}_t\}_{t\in I}$ on compact subsets of $\{\mathcal{M}_t\}_{t\in I}$, where $\rho > 0$ is the constant from Theorem 12.19.

The limit flow satisfies $H(0,0) = 1$ and $\kappa_1(0,0) = -\varepsilon_0$. By continuity, $H > \frac{1}{2}$ in $P_r(0,0)$ for some $r \in (0,\rho)$. Moreover, by construction, the ratio $\frac{\kappa_1}{H}$ attains a negative minimum at 0. But this contradicts Theorem 12.21. □

Corollary 12.22. *Let $\{\mathcal{M}_t\}_{t\in[0,T)}$ be a compact, embedded, mean convex solution to mean curvature flow. For every $\varepsilon > 0$ there exists a constant $H_\varepsilon = H_\varepsilon(\mathcal{M}_0,\varepsilon) < \infty$ such that*

$$H(x,t) \geq H_\varepsilon \quad \textit{implies} \quad \kappa_1(x,t) \geq -\varepsilon H(x,t)\,. \tag{12.59}$$

Proof. Fix $\varepsilon > 0$ and choose $r_0 = r_0(\mathcal{M}_0,\eta) > 0$ such that $H \leq \frac{1}{2r_0}$ on $\mathcal{M}_t$ for $t \in [0,T/2]$ and $(t-\eta r_0^2, t] \subset (0,t]$ for each $t \in [T/2,T)$, where η is the constant from Theorem 12.20. Then, by Theorem 12.20, the claim holds with $H_\varepsilon \doteqdot r_0^{-1}$. □

By a similar argument, we also recover a version of Corollary 9.14. We will say that a strictly mean convex solution to mean curvature flow is β-$(m+1)$-**convex** if

$$\frac{\kappa_1 + \cdots + \kappa_{m+1}}{H} \geq \beta \tag{12.60}$$

at all points and times.

Theorem 12.23 (Haslhofer and Kleiner [**276**] (following White [**531**])). *Given $\alpha > 0$, $\beta > 0$, and $\varepsilon > 0$ there exists $\eta > 0$ with the following property: Let $\{\mathcal{M}_t\}_{t\in I}$ be an embedded, α-noncollapsing, β-$(m+1)$-convex mean curvature flow which is properly defined in the parabolic ball $P_{\eta r}(x,t)$ with $t \in I$ and $x \in \mathcal{M}_t$. If $H(x,t) \leq r^{-1}$, then*

$$\kappa_n(x,t) - \frac{1}{n-m} H \leq \varepsilon r^{-1}. \tag{12.61}$$

We need the following rigidity result.

Theorem 12.24 (Rigidity of upper pinching). *Let $X : M^n \times I \to \mathbb{R}^{n+1}$ be a strictly mean convex solution to mean curvature flow which is properly defined in $P_R(x_0,t_0) \doteqdot B_R(p_0) \times (t_0 - r^2, t_0]$, where $p_0 \doteqdot X(x_0,t_0)$. If $\max_{P_R(x_0,t_0)} \frac{\kappa_n}{H}$ is attained at (x_0,t_0), then $\frac{\kappa_n}{H}(x_0,t_0) = \frac{1}{n-m}$ for some $m \in \{0,\ldots,n-1\}$ and $X|_{\mathcal{P}_r(x_0,t_0)} : \mathcal{P}_r(x_0,t_0) \to \mathbb{R}^{n+1}$ is part of a round, orthogonal shrinking cylinder for some $r > 0$.*

Proof. By Proposition 12.9 (see Remark 12.10), κ_n/H satisfies

$$(\partial_t - \Delta)\frac{\kappa_n}{H} \leq 2\left\langle \nabla \frac{\kappa_n}{H}, \frac{\nabla H}{H} \right\rangle - \frac{2}{H}\sum_{k=1}^{n}\sum_{\kappa_i<\kappa_n} \frac{(\nabla_k \mathrm{II}_{in})^2}{\kappa_n - \kappa_i} \tag{12.62}$$

in the barrier sense. Since κ_n/H reaches an interior maximum, Lemma 9.10 implies that $\kappa_n \equiv K_0 H$ in $\mathcal{P}_r(x_0,t_0)$ for some $K_0 \geq \frac{1}{n}$ and some $r > 0$. Consider the bilinear form $\mathrm{N} \doteqdot \mathrm{II} - K_0 H \mathrm{I}$. Arguing as in Theorem 12.21, we find that the solution splits locally as an isometric product $H_t \times V_t$ near (x_0,t_0) such that $TH_t = \ker \mathrm{N}$ (including the possibility that V_t is empty). So the leaves H_t are umbilic. On the other hand, the vectors u tangent to V_t must be null eigenvectors of II since the sectional curvature of the 2-plane $u \wedge v$ vanishes for each v tangent to H_t. The splitting then follows as in Theorem 12.21. Roundness of the leaves follows from Exercise 5.5. □

Proof of Theorem 12.23. The proof is similar to the proof of Theorem 12.20 (with Theorem 12.21 replaced by Theorem 12.24). □

Corollary 12.25. *Let $\{\mathcal{M}_t\}_{t\in[0,T)}$ be a compact, embedded, $(m+1)$-convex solution to mean curvature flow. For every $\varepsilon > 0$ there exists a constant*

$H_\varepsilon = H_\varepsilon(\mathcal{M}_0, \varepsilon) < \infty$ *such that* x

$$H(x,t) \geq H_\varepsilon \quad \text{implies} \quad \kappa_n(x,t) - \frac{1}{n-m}H(x,t) \leq \varepsilon H(x,t)\,. \tag{12.63}$$

Proof. See the proof of Corollary 12.22. □

12.5. Notes and commentary

For surveys of the use of "multipoint" maximum principle arguments in geometric analysis and PDE, see the first author [**40**, **42**] and Brendle [**106**].

Related noncollapsing properties for the Ricci flow on closed manifolds were crucial components of Perelman's resolution of the geometrization conjecture [**438**, **439**].

12.5.1. Noncollapsing improves.

By applying a Stampacchia iteration argument not unlike in the proofs of the roundness estimate (Theorem 8.6) and the convexity estimates (Theorem 9.12), it is possible to obtain asymptotically sharp estimates for the noncollapsing ratios at a singularity.

Theorem 12.26. *Let* $X : M^n \times [0,T) \to \mathbb{R}^{n+1}$, $n \geq 1$, *be a compact, embedded,* $(m+1)$*-convex solution of mean curvature flow. For every* $\varepsilon > 0$ *there are constants* $\overline{C}_\varepsilon(n, \mathcal{M}_0, \varepsilon)$ *and* $\underline{C}_\varepsilon(n, \mathcal{M}_0, \varepsilon)$ *such that*

$$\overline{k} \leq \left(\varepsilon + \tfrac{1}{n-m}\right) H + \overline{C}_\varepsilon \quad \text{and} \quad \underline{k} \geq -\varepsilon H - \underline{C}_\varepsilon\,. \tag{12.64}$$

The constant $\overline{C}_\varepsilon$ is of the form $\overline{C}_\varepsilon = \overline{\mathrm{C}}_\varepsilon R^{-1}$, where $R \doteqdot \max_{\mathcal{M}_0} |\mathrm{II}|$ and $\overline{\mathrm{C}}_\varepsilon$ depends only on n, ε, a lower bound for $\min_{\mathcal{M}_0} \frac{\kappa_1+\cdots+\kappa_{m+1}}{H}$, and upper bounds for $\max_{\mathcal{M}_0} \frac{\overline{k}}{H}$, $R^{-1}\operatorname{diam}(\mathcal{M}_0)$ and $R^{-n}\operatorname{Area}(\mathcal{M}_0)$. The constant $\underline{C}_\varepsilon$ is of the form $\underline{C}_\varepsilon = \underline{\mathrm{C}}_\varepsilon R^{-1}$, where $\underline{\mathrm{C}}_\varepsilon$ depends on the same data except that the upper bound for $\max_{\mathcal{M}_0} \frac{\overline{k}}{H}$ is replaced by a lower bound for $\min_{\mathcal{M}_0} \frac{\underline{k}}{H}$.

The theorem was obtained by Brendle in case $m = n-1$ [**107**] and extended to the other cases in [**352**]. A similar statement can be obtained by a blow-up argument as in the proofs of Theorems 12.20 and 12.23. This was observed by Haslhofer and Kleiner [**275**] (see also [**351**]).

The "noncollapsing improves" estimate is crucial to the construction of mean curvature flow with surgery for mean convex surfaces in $\mathbb{R}^3$ by Brendle and Huisken [**114**]. Roughly speaking, it provides a replacement for the cylindrical estimate of Corollary 9.14 (which is vacuous when $n = 2$).

12.5.2. Extensions.

The arguments of this chapter can also be applied to solutions of more general (fully nonlinear) flows when the speed is a degree one homogeneous function of the curvature. Roughly speaking, solutions of flows by convex speeds or inverse-concave speeds are exterior noncollapsing, whereas solutions of flows by concave speeds are interior noncollapsing [**54, 55, 353**]. We shall discuss this further in Chapter 19.

The arguments also work for flows of hypersurfaces of spaceforms, with some modification due to the presence of ambient curvature: The noncollapsing ratios improve exponentially in time if the ambient curvature is positive and they decay exponentially in time if the ambient curvature is negative [**54, 59**]. In fact, the arguments even yield useful information in general Riemannian ambient spaces of bounded geometry [**109**].

Following Haslhofer–Kleiner, Longzhi Lin obtained a local curvature estimate in the setting of starshaped hypersurfaces [**369**].

12.6. Exercises

Exercise 12.1. Identify regions on the surface in Figure 12.3 where the exscribed curvature is negative, zero, and positive, respectively.

Exercise 12.2. For any bounded convex body, prove that

$$\min \overline{k} = \frac{1}{\rho_-} \quad \text{and} \quad \max \underline{k} = \frac{1}{\rho_+},$$

where ρ_+ and ρ_- are its circum- and inradii, respectively.

Exercise 12.3. Prove the identities (12.7) and (12.8) for the Grim Reaper and the Angenent oval, respectively.

Exercise 12.4. ([**104**, Proposition 8]) Let $\Omega^{n+1} \subset \mathbb{R}^{n+1}$ be an open set with smooth boundary, $\mathcal{M} = \partial\Omega$. Suppose that either $\underline{k}(x) = k(x,y) = \kappa_1(x)$ or $\overline{k}(x) = k(x,y) = \kappa_n(x)$ at some boundary point $(x,y) \in S\mathcal{M}$. Prove that

$$\nabla_y \mathrm{II}_x(y,y) = 0\,.$$

HINT: Let $\gamma : (-r_0, r_0) \to \mathcal{M}$ be the geodesic through $\gamma(0) = x$ in the direction of $y = \gamma'(0)$ and consider the function

$$r \mapsto f(r) \doteqdot 2\left\langle x - \gamma(r), \mathsf{N}_x \right\rangle - \mathrm{II}_x(y,y) \left\| x - \gamma(r) \right\|^2 .$$

Exercise 12.5. Let $\Omega^{n+1} \subset \mathbb{R}^{n+1}$ be a smoothly bounded, locally uniformly convex domain with boundary $\mathcal{M} = \partial\Omega$.

(a) Assume that $\sup_{\mathcal{M}} \overline{k}$ is attained at $\overline{x} \in \mathcal{M}$. Show that $\overline{k}(\overline{x}) = \kappa_n(\overline{x})$.

(b) Assume that $\inf_{\mathcal{M}} \underline{k}$ is attained at $\underline{x} \in \mathcal{M}$. Show that $\underline{k}(\underline{x}) = \kappa_1(\underline{x})$.

Exercise 12.6. Prove (12.18) in Proposition 12.9.

Exercise 12.7. Prove Propositions 12.15 and 12.17.

Exercise 12.8. By a similar argument to Lemma 12.7 and the remarks thereafter, show that H defines a C^∞ function on the manifold with boundary $\widehat{\mathcal{M}}$.

Exercise 12.9. Let $\Omega \subset \mathbb{R}^{n+1}$ be a convex body. Use Proposition 12.17 to show that Ω is a cylindrical region,

$$C_r(L) \doteqdot \{z \in \mathbb{R}^{n+1} : d(z, L) < r\}$$

for some affine subspace $L \subsetneq \mathbb{R}^{n+1}$ and some $r > 0$, if and only if $\overline{k}$ and $\underline{k}$ are both constant.

Chapter 13

Self-Similar Solutions

We have seen that self-similarly shrinking and translating solutions model singularity formation under the mean curvature flow. On the other hand, self-similarly *expanding* solutions model the behavior of a solution to mean curvature flow as it flows *out of* a singularity [**71, 175**]; so self-similarly expanding solutions model singularity *de*-formation. They also model the long-time behavior of the topological ends of the evolving hypersurface once all singularities have occurred (see Chapter 7). We shall now look more closely at these special classes of solutions.

13.1. Shrinkers — an introduction

Recall that a **shrinking self-similar solution** $X : M^n \times (-\infty, 0) \to \mathbb{R}^{n+1}$ to mean curvature flow is one which evolves by homothetic contraction:

$$X_t(x) \doteqdot X(x,t) = \sqrt{-t}X_{-1}(\varphi(x,t)) \tag{13.1}$$

for some time-dependent diffeomorphism $\varphi : M^n \times (-\infty, 0) \to M^n$. In particular, the time slices $\mathcal{M}_t = X(M^n, t)$ are all congruent; since they evolve by mean curvature flow equation $\partial_t X = -H\mathsf{N}$, they must satisfy

$$H(x,t) = \frac{1}{-2t}\langle X(x,t), \mathsf{N}(x,t)\rangle \quad \text{for all } (x,t) \in M^n \times (-\infty, 0)\,. \tag{13.2}$$

Conversely, if a hypersurface $X : M^n \to \mathbb{R}^{n+1}$ satisfies

$$H(x) = \frac{1}{2}\langle X(x), \mathsf{N}(x)\rangle \quad \text{for all } x \in M\,, \tag{13.3}$$

then the 1-parameter family of immersed hypersurfaces $\mathcal{M}^n_t \doteqdot \sqrt{-t}\mathcal{M}^n$, where $\mathcal{M}^n = X(M^n)$, evolves by mean curvature flow. A hypersurface satisfying (13.3) is called a (**mean curvature flow**) **shrinker** or **self-shrinker**.

We have already encountered the following examples.

Example 13.1.

- The stationary hyperplane $\{\mathbb{R}^n \times \{0\}\}_{t\in(-\infty,0)}$.
- The shrinking sphere $\{S^n_{\sqrt{-2nt}}\}_{t\in(-\infty,0)}$.
- Shrinking cylinders $\{\mathbb{R}^m \times S^{n-m}_{\sqrt{-2(n-m)t}}\}_{t\in(-\infty,0)}$.
- Abresch–Langer curves $\{\Gamma^{p,q}_t\}_{t\in(-\infty,0)}$ in the plane and their cylinders $\{\mathbb{R}^{n-1} \times \Gamma^{p,q}_t\}_{t\in(-\infty,0)}$.

The only *minimal* shrinkers are the hyperplanes.

Lemma 13.2. *Let $\mathcal{M}$ be a shrinker. If $\mathcal{M}$ is also a minimal hypersurface, then it is a hyperplane passing through the origin.*

Proof. Since $\mathcal{M}$ is minimal, the corresponding shrinking self-similar solution $\{\mathcal{M}_t\}_{t\in(-\infty,0)}$ to mean curvature flow is stationary: $\mathcal{M}_t = \mathcal{M}_s$ for all $t, s < 0$. On the other hand, by (13.1), $\mathcal{M}_t = \sqrt{-t}\mathcal{M}_{-1}$ for all $t < 0$. It follows that $\mathcal{M}_{-1} = \sqrt{-t}\mathcal{M}_{-1}$ for all $t < 0$ and hence $\mathcal{M}$ is a cone. Since $\mathcal{M} = \mathcal{M}_{-1}$ is smooth, it must be a linear subspace. □

13.2. The Gaußian area functional

In Section 10.1 we discussed Huisken's monotonicity formula for the Gaußian area $\Theta_{(X_0,t_0)}$ defined by (10.1). Consider the Gaußian area of a hypersurface $X : M^n \to \mathbb{R}^{n+1}$ centered at the origin and at a fixed scale (time); that is, define

$$\mathcal{F}(X) \doteqdot \int_{\mathcal{M}} \mathrm{e}^{-\frac{|p|^2}{4}}\, d\mu(p)\,. \tag{13.4}$$

In Lemma 5.25 we computed the first variation of the area functional. Similar computations yield the following well-known first variation formula.

Lemma 13.3 (First variation of the Gaußian area)**.** *Let $X_s : M^n \to \mathbb{R}^{n+1}$, $s \in (-s_0, s_0)$, be a normal variation of $X_0 = X : M^n \to \mathbb{R}^{n+1}$ and set $F \doteqdot -\langle \partial_s|_{s=0} X, \mathsf{N}\rangle$. Then*

$$\left.\frac{d}{ds}\right|_{s=0} \mathcal{F}(X_s) = \int_{\mathcal{M}} \left(\frac{1}{2}\langle \mathsf{N}, p\rangle - H\right) F\mathrm{e}^{-\frac{|p|^2}{4}}\, d\mu(p)\,. \tag{13.5}$$

In particular, X is a critical point of $\mathcal{F}$ if and only if X is a shrinker.

Proof. The variation formula for the volume form (5.132) immediately yields

$$\frac{d}{ds}\mathcal{F}(X_s) = \int_{\mathcal{M}} \left(-\frac{1}{2}\langle -F\mathsf{N}, p\rangle - FH\right) \mathrm{e}^{-\frac{|p|^2}{4}}\, d\mu(p)\,.$$ □

Observe that $\mathcal{F}(X)$ is the area of $\mathcal{M}$ with respect the **Gaußian metric**

$$g_{\text{ß}}(p) \doteqdot \mathrm{e}^{-\frac{|p|^2}{2n}} g_{\mathbb{R}^{n+1}}(p) \tag{13.6}$$

(see Exercise 13.2). Thus the lemma says:

Corollary 13.4. *The mean curvature flow shrinkers are exactly the* minimal hypersurfaces *with respect to the ambient Gaußian metric* $g_{ß}$ *on* $\mathbb{R}^{n+1}$.

Note that the Gaußian metric is incomplete, has finite diameter, and has finite volume.

The second variation for the Gaußian area, first utilized by Colding and Minicozzi [**181**], will play an important role in what follows.

Lemma 13.5 (Second variation of the Gaußian area). *Let* $X : M^n \to \mathbb{R}^{n+1}$ *be a mean curvature flow shrinker and let* $X_s : M^n \to \mathbb{R}^{n+1}$, $s \in (-s_0, s_0)$, *be a normal variation of* $X_0 = X$. *Set* $F \doteqdot -\langle \partial_s|_{s=0}X, \mathsf{N}\rangle$. *Then*

$$\left.\frac{d^2}{ds^2}\right|_{s=0} \mathcal{F}(X_s) = -\int_{\mathcal{M}} \Big(\Delta F + |\mathrm{II}|^2 F + \frac{1}{2}F - \frac{1}{2}\langle X, \nabla F\rangle\Big) F \mathrm{e}^{-\frac{|X|^2}{4}} d\mu. \tag{13.7}$$

Proof. Using the identity $\frac{1}{2}\langle X, \mathsf{N}\rangle - H = 0$ and the variations for the unit normal (5.104) and mean curvature, we find

$$\begin{aligned}\frac{d^2}{ds^2}\mathcal{F}(X_s) &= \int_{\mathcal{M}} \Big(\frac{1}{2}\partial_s\langle \mathsf{N}, X\rangle - \partial_s H\Big) F\mathrm{e}^{-\frac{|X|^2}{4}} d\mu \\ &= \int_{\mathcal{M}} \Big(\frac{1}{2}\langle \nabla F, X\rangle - \frac{1}{2}\langle \mathsf{N}, F\mathsf{N}\rangle - (\Delta F + F|\mathrm{II}|^2)\Big) F\mathrm{e}^{-\frac{|X|^2}{4}} d\mu. \ \square\end{aligned}$$

Following [**181**], define the second-order linear elliptic operator

$$LF \doteqdot \Delta F + |\mathrm{II}|^2 F + \frac{1}{2}F - \frac{1}{2}\langle X, \nabla F\rangle . \tag{13.8}$$

Then (13.7) says that

$$\left.\frac{d^2}{ds^2}\right|_{s=0} \mathcal{F}(X_s) = -\int_{\mathcal{M}} F\,LF\,\mathrm{e}^{-\frac{|X|^2}{4}} d\mu . \tag{13.9}$$

We call L the **$\mathcal{F}$-stability operator**. Its higher-order part, given by the operator defined by

$$\mathrm{OU}(F) \doteqdot \Delta F - \frac{1}{2}\langle X, \nabla F\rangle = \Delta F - \frac{1}{2}\langle X^\top, \nabla F\rangle , \tag{13.10}$$

is called the **Ornstein–Uhlenbeck operator**. Note that

$$\mathrm{OU}(F) = \mathrm{e}^{\frac{|X|^2}{4}} \operatorname{div}\left(\mathrm{e}^{-\frac{|X|^2}{4}} \nabla F\right) .$$

Given a function f on $\mathcal{M}$ we define, more generally, the ***f*-Laplacian** by

$$\Delta_f F \doteqdot \Delta F - \langle \nabla f, \nabla F\rangle . \tag{13.11}$$

Then

$$\mathrm{OU}(F) = \Delta_f F\,,$$

where

$$f(p) \doteqdot \frac{|p|^2}{4}$$

since $\nabla f(p) = \frac{1}{2}p^\top$ by (5.13).

The operators L and $\Delta_f = \mathrm{OU}$ are selfadjoint with respect to the induced Gaußian measure $\mathrm{e}^{-f(p)}d\mu(p) = \mathrm{e}^{-\frac{|p|^2}{4}}d\mu(p)$. Namely, if F and G are C^2 functions on an embedded hypersurface $\mathcal{M}$, one of which has compact support, then integration by parts yields

$$\int_{\mathcal{M}} F\Delta_f G\,\mathrm{e}^{-f}d\mu = \int_{\mathcal{M}} \langle \nabla F, \nabla G\rangle\,\mathrm{e}^{-f}d\mu = \int_{\mathcal{M}} G\Delta_f F\,\mathrm{e}^{-f}d\mu\,.$$

The operators L and Δ_f extend to act on tensors in the canonical way. For example, if α is a tensor, then

$$L\,\alpha \doteqdot \Delta\alpha + |\mathrm{II}|^2\alpha + \frac{1}{2}\alpha - \frac{1}{2}\langle p, \nabla\alpha\rangle\,. \tag{13.12}$$

Henceforth, we shall often use the notation

$$dm(p) \doteqdot \mathrm{e}^{-f(p)}d\mu(p) = \mathrm{e}^{-\frac{|p|^2}{4}}d\mu(p) \tag{13.13}$$

for the weighted measure.

13.2.1. Differential identities. Let $X : M^n \times (-\infty, 0) \to \mathbb{R}^{n+1}$ be a shrinker. By differentiating (13.3) and using (5.27), we find

$$\begin{aligned} \nabla_i H &= \frac{1}{2}\langle D_i X, N\rangle + \frac{1}{2}\langle X, D_i N\rangle \\ &= \frac{1}{2}\langle X, \mathrm{II}_{ij}g^{jk}\partial_k\rangle \\ &= \frac{1}{2}\mathrm{II}(X^\top)_i\,. \end{aligned} \tag{13.14}$$

Taking the divergence of this, we obtain

$$\begin{aligned} \Delta H &= \frac{1}{2}\nabla_i \mathrm{II}_{ij}\,(X^\top)_j + \frac{1}{2}\mathrm{II}_{ij}\,\nabla_i(X^\top)_j \\ &= \frac{1}{2}\nabla_j H\,(X^\top)_j + \frac{1}{2}\mathrm{II}_{ij}\,(g_{ij} - \langle X, \mathsf{N}\rangle)\mathrm{II}_{ij} \\ &= \frac{1}{2}\langle X^\top, \nabla H\rangle + \frac{1}{2}H - |\mathrm{II}|^2 H\,. \end{aligned}$$

As an alternate derivation of the formula above, we can use the fact that

$$\partial_t X\,|_{t=-1} = -H\mathsf{N} = -\frac{1}{2}\,\langle X, \mathsf{N}\rangle\,\mathsf{N} = -\frac{1}{2}X + \frac{1}{2}X^\top, \tag{13.15}$$

where $X^\top \in T\mathcal{M}_{-1} = T\mathcal{M}$ is the tangential projection of $X = X_{-1}$. Hence, by (6.18) and the following lemma, we have

$$\frac{1}{2}\left\langle X^\top, \nabla H\right\rangle + \frac{1}{2}H = \partial_t H = \Delta H + |\mathrm{II}|^2 H\,. \tag{13.16}$$

Definition 13.6. We say that a function or tensor expression $Q = Q(X)$ of immersed hypersurfaces $X : M^n \to \mathbb{R}^{n+1}$, which is a pointwise function of X, N, g, II, and their covariant derivatives, has **degree** β if it has the scaling property $Q(\lambda X) = \lambda^\beta Q(X)$ for any $\lambda > 0$.

Lemma 13.7. *Let $X : M^n \times (-\infty, 0) \to \mathbb{R}^{n+1}$ be a shrinking self-similar solution to mean curvature flow and let $Q = Q(X)$ be a tensor expression of degree β. Then, at $t = 0$,*

$$\partial_t Q = L_{X^\top} Q - \beta Q\,, \tag{13.17}$$

where L denotes the Lie derivative.

Proof. We have

$$\partial_t X|_{t=0} = -H\mathsf{N} = -\langle X, \mathsf{N}\rangle \mathsf{N} = -X + X^\top\,.$$

Since for $\partial_t\, Q$ we are discussing a linearization, we may consider the effects of the infinitesimal homotheties $(-X)$ and diffeomorphisms $(+X^\top)$ separately.

So suppose first that $\partial_t X|_{t=0} = -X$. Then

$$\begin{aligned}\partial_t\, Q(X(t))|_{t=0} &= \partial_t\, Q((1-t)X)|_{t=0} \\ &= \partial_t \left((1-t)^\beta Q(X)\right)\Big|_{t=0} \\ &= -\,\beta\, Q(X)\,.\end{aligned}$$

Second, suppose that $\partial_t X|_{t=0} = X^\top$. Let $\mathcal{M} = X(M^n)$ and let φ_t be a 1-parameter family of diffeomorphisms of $\mathcal{M}$, where $\frac{d}{dt}\varphi_t\big|_{t=0} = X^\top$. Since Q is a pointwise expression of X, N, g, II, and their covariant derivatives, we have $Q(\varphi_t^* X) = \varphi_t^*(Q(X))$. Thus

$$\begin{aligned}\partial_t\, Q(X(t))\,|_{t=0} &= \partial_t\, Q(\varphi_t^* X)\,|_{t=0} \\ &= \partial_t\,(\varphi_t^* Q(X))\,|_{t=0} \\ &= L_{X^\top}(Q(X))\,.\end{aligned}$$

The lemma follows from adding the two formulae above. ☐

In terms of the $\mathcal{F}$-stability operator, equation (13.16) becomes [**181**]

$$LH = H\,. \tag{13.18}$$

We will exploit the following Simons-type formula [**181**].

Lemma 13.8. *On a self-shrinker,*

$$L\,\mathrm{II} = \mathrm{II}\,. \tag{13.19}$$

Tracing this, we recover (13.18).

Proof. Under mean curvature flow,

$$\mathcal{L}_{\partial_t}\mathrm{II} = \Delta\mathrm{II} + |\mathrm{II}|^2\mathrm{II} - 2H\mathrm{II}^2\,. \tag{13.20}$$

Since we are on a self-shrinker, by Lemma 13.7 (II has degree 1) we also have (assume that we are at time $t = -1$)

$$\begin{aligned}\partial_t\mathrm{II}_{ij} &= \frac{1}{2}\mathcal{L}_{X^\top}\mathrm{II}_{ij} - \frac{1}{2}\mathrm{II}_{ij}\\ &= \frac{1}{2}\left(\nabla_{X^\top}\mathrm{II}_{ij} + \nabla_i X_k^\top\,\mathrm{II}_{kj} + \nabla_j X_k^\top\,\mathrm{II}_{ik} - \mathrm{II}_{ij}\right),\end{aligned}$$

where $\mathcal{L}$ is the Lie derivative and where we used that for any 2-tensor α and vector field V,

$$(\mathcal{L}_V\alpha)_{ij} = (\nabla_V\alpha)_{ij} + \nabla_i V_k\,\alpha_{kj} + \nabla_j V_k\,\alpha_{ik}\,. \tag{13.21}$$

Since

$$\nabla X^\top = g - \langle X, \mathsf{N}\rangle\,\mathrm{II}\,, \tag{13.22}$$

we obtain

$$\partial_t\mathrm{II} = \frac{1}{2}\nabla_{X^\top}\mathrm{II} + \mathrm{II} - \langle X,\mathsf{N}\rangle\,\mathrm{II}^2 - \frac{1}{2}\mathrm{II}\,.$$

Combining this with (13.20), we conclude that

$$\begin{aligned}0 &= \Delta\mathrm{II} - \frac{1}{2}\nabla_{X^\top}\mathrm{II} + |\mathrm{II}|^2\mathrm{II} - 2H\mathrm{II}^2 + \langle X,\mathsf{N}\rangle\,\mathrm{II}^2 - \frac{1}{2}\mathrm{II}\\ &= L\mathrm{II} - \mathrm{II}.\end{aligned}$$

□

As the reader may have guessed, alternatively one may take the covariant derivative of (13.14) to derive (13.19). Indeed,

$$\begin{aligned}\nabla_i\nabla_j H &= \frac{1}{2}\nabla_j\mathrm{II}_{ik}\,(X^\top)_k + \frac{1}{2}\mathrm{II}_{ik}\,\nabla_j(X^\top)_k\\ &= \nabla_{X^\top}\mathrm{II}_{ij} + \frac{1}{2}\mathrm{II}_{ij} - H\,\mathrm{II}^2_{ij}\,.\end{aligned} \tag{13.23}$$

On the other hand, by the trace of Simons's equation (5.52), we have

$$\nabla_i\nabla_j H = \Delta\mathrm{II}_{ij} + |\mathrm{II}|^2\,\mathrm{II}_{ij} - H\,\mathrm{II}^2_{ij}\,.$$

Now (13.19) follows from this and (13.23).

By Kato's inequality,

$$|\nabla|\mathrm{II}||^2 \leq |\nabla\mathrm{II}|^2\,. \tag{13.24}$$

Lemma 13.9. *If X is a self-shrinker, then*

$$L|\mathrm{II}| = |\mathrm{II}|^{-1}\left(|\nabla \mathrm{II}|^2 - |\nabla|\mathrm{II}||^2\right) + |\mathrm{II}| \geq |\mathrm{II}|\,. \tag{13.25}$$

Proof. We first compute

$$\begin{aligned}
\Delta|\mathrm{II}|^2 &= 2|\nabla \mathrm{II}|^2 + 2\langle \Delta \mathrm{II}, \mathrm{II}\rangle \\
&= 2|\nabla \mathrm{II}|^2 + 2\left\langle \left(-|\mathrm{II}|^2\mathrm{II} + \frac{1}{2}\mathrm{II} + \frac{1}{2}\langle X^\top, \nabla \mathrm{II}\rangle\right), \mathrm{II}\right\rangle \\
&= 2|\nabla \mathrm{II}|^2 - 2|\mathrm{II}|^4 + |\mathrm{II}|^2 + \frac{1}{2}\langle X^\top, \nabla|\mathrm{II}|^2\rangle\,.
\end{aligned} \tag{13.26}$$

Using this, we calculate that

$$\begin{aligned}
\Delta|\mathrm{II}| &= -\frac{1}{4|\mathrm{II}|^3}|\nabla|\mathrm{II}|^2|^2 + \frac{1}{2|\mathrm{II}|}\Delta|\mathrm{II}|^2 \\
&= -\frac{1}{|\mathrm{II}|}|\nabla|\mathrm{II}||^2 + \frac{1}{|\mathrm{II}|}|\nabla \mathrm{II}|^2 - |\mathrm{II}|^3 + \frac{1}{2}|\mathrm{II}| + \frac{1}{2}\langle X^\top, \nabla|\mathrm{II}|\rangle\,.
\end{aligned} \tag{13.27}$$

The lemma now follows from the definition of L and Kato's inequality. □

Remark 13.10. Alternatively, one may use the identity (13.19) to obtain the lemma.

Let ω be a fixed vector in $\mathbb{R}^{n+1}$. By (7.16) (which holds even for nonunit vectors) and by Lemma 13.7 ($\langle \mathsf{N}, \omega\rangle$ has degree 0),

$$\Delta\langle \mathsf{N}, \omega\rangle + |\mathrm{II}|^2\langle \mathsf{N}, \omega\rangle = \partial_t\langle \mathsf{N}, \omega\rangle = \frac{1}{2}\left\langle X^\top, \nabla\langle \mathsf{N}, \omega\rangle\right\rangle.$$

Thus, by the definition of L,

$$L\langle \mathsf{N}, \omega\rangle = \frac{1}{2}\langle \mathsf{N}, \omega\rangle\,. \tag{13.28}$$

This equation also follows easily from the formulae (7.14) and (7.15) for the gradient and Laplacian of $\langle \mathsf{N}, \omega\rangle$, respectively, while using $\nabla H = \frac{1}{2}\mathrm{II}(X^\top)$ to obtain a cancellation.

13.3. Mean convex shrinkers

We have seen only a limited number of examples: the shrinking spheres $S^n_{\sqrt{-2nt}}$ and their products $\mathbb{R}^k \times S^{n-k}_{\sqrt{-2(n-k)t}}$, $k \in \{0, \ldots, n\}$, and the Abresch–Langer curves $\Gamma^{k,l}_{\sqrt{-2t}}$ and their products $\Gamma^{k,l}_{\sqrt{-2t}} \times \mathbb{R}^{n-1}$. In this section we shall show that these are the only mean convex examples. We have already seen that the Abresch–Langer curves are the only convex self-similarly shrinking solutions to curve shortening flow, so it suffices to show that the shrinking cylinders are the only mean convex examples in higher dimensions. We begin with the *compact* case.

Theorem 13.11 (Huisken [**296, 297**]). *If $\mathcal{M}^n$, $n \geq 2$, is a compact, weakly mean convex shrinker, then $\mathcal{M}^n$ is the sphere of radius $\sqrt{2n}$ centered at the origin.*

Proof. By applying the strong maximum principle to the evolution equation (6.18) for H, we conclude that either $H > 0$ or $H \equiv 0$. But the latter is impossible because $\mathcal{M}^n$ is compact, so in fact $\mathcal{M}^n$ is strictly mean convex. It is easy to see from (8.7) that

$$\partial_t \left(\frac{|\mathrm{II}|^2}{H^2} \right) = \Delta \left(\frac{|\mathrm{II}|^2}{H^2} \right) + \frac{2}{H} \left\langle \nabla H, \nabla \left(\frac{|\mathrm{II}|^2}{H^2} \right) \right\rangle - \frac{2}{H^4} \left| H \nabla \mathrm{II} - \nabla H \otimes \mathrm{II} \right|^2 .$$

So, by Lemma 13.7 and the fact that $\frac{|\mathrm{II}|^2}{H^2}$ is scale invariant (i.e., has degree 0), we have the Bochner formula:

$$\frac{1}{2} \left\langle X^\top, \nabla \left(\frac{|\mathrm{II}|^2}{H^2} \right) \right\rangle = \Delta \left(\frac{|\mathrm{II}|^2}{H^2} \right) + \frac{2}{H} \left\langle \nabla H, \nabla \left(\frac{|\mathrm{II}|^2}{H^2} \right) \right\rangle - \frac{2}{H^4} \left| H \nabla \mathrm{II} - \nabla H \otimes \mathrm{II} \right|^2 . \tag{13.29}$$

Since $\mathcal{M}$ is compact, the strong maximum principle yields

$$\frac{|\mathrm{II}|^2}{H^2} \equiv \text{const} , \tag{13.30}$$

which implies that

$$H \nabla \mathrm{II} = \nabla H \otimes \mathrm{II} . \tag{13.31}$$

Taking the trace with respect to the first two components yields

$$H \nabla H = \mathrm{L}(\nabla H) . \tag{13.32}$$

If $|\nabla H|^2 \equiv 0$, then (13.31) implies that $\nabla \mathrm{II} \equiv 0$. It then follows from Corollary 5.11 that the shrinker is a round sphere. (Alternatively, the shrinker equation $H = \frac{1}{2} \langle X, \mathsf{N} \rangle$ and $H \equiv$ const imply the same.) It is easy to see from the shrinker equation that the radius is $\sqrt{2n}$ and the center is the origin.

Otherwise, there exists a point where $\nabla H \neq 0$. By (13.32), $\nabla H / |\nabla H|$ is a principal direction at this point. Let $\{e_i\}_{i=1}^n$ be a principal frame with $e_1 = \nabla H / |\nabla H|$. By (13.31),

$$\nabla_k \mathrm{II}_{ij} = 0 \text{ for } k \geq 2 \text{ and all } i, j \quad \text{and} \tag{13.33a}$$

$$H \nabla_1 \mathrm{II}_{ij} = \nabla_1 H \, \mathrm{II}_{ij} \quad \text{for all } i, j . \tag{13.33b}$$

By (13.33a), (13.33b), and the Codazzi identity,

$$\mathrm{II}_{ij} = 0 \quad \text{for all } j \geq 2 \text{ and all } i .$$

So the only nonvanishing component of II_{ij} is II_{11}. Since $|\mathrm{II}|^2 = H^2$ at such a point and $\frac{|\mathrm{II}|^2}{H^2}$ is constant on $\mathcal{M}$, we conclude that $R = H^2 - |\mathrm{II}|^2 \equiv 0$ on $\mathcal{M}$. Integrating (13.16) then yields

$$2\int_{\mathcal{M}} H^3\, d\mu = 2\int_{\mathcal{M}} |\mathrm{II}|^2 H\, d\mu = \int_{\mathcal{M}} \left(\left\langle X^\top, \nabla H\right\rangle + H\right) d\mu. \tag{13.34}$$

Integrating the first term on the right-hand side by parts yields

$$\int_{\mathcal{M}} \left\langle X^\top, \nabla H\right\rangle d\mu = -\int_{\mathcal{M}} H \operatorname{div}\left(X^\top\right) d\mu\,.$$

Recall from (5.13) that $X^\top = \frac{1}{2}\nabla |X|^2$. Applying (5.39) and $H = \frac{1}{2}\langle X, \mathsf{N}\rangle$ then yields

$$\operatorname{div}\left(X^\top\right) = \frac{1}{2}\Delta |X|^2 = -H\langle X, \mathsf{N}\rangle + n = -2H^2 + n.$$

Thus,

$$\int_{\mathcal{M}} \left\langle X^\top, \nabla H\right\rangle d\mu = \int_{\mathcal{M}} \left(2H^3 - nH\right) d\mu.$$

Combining this with (13.34) yields

$$(1-n)\int_{\mathcal{M}} H d\mu = 0.$$

Since $n \geq 2$ and $H > 0$, this is impossible. □

We now consider the case of *noncompact* shrinkers (see Theorem 5.1 in Huisken [**297**] and Theorem 10.1 in Colding and Minicozzi [**181**]).

Definition 13.12. We say that $\mathcal{M}$ has **polynomial area growth** if there exists a constant C such that $\mathrm{Area}(\mathcal{M} \cap B_r(0)) \leq C\, r^C$ for $r \geq C$.

We need a consequence of Lemma 10.8 for self-shrinking tangent flows.

Lemma 13.13. *Let $X : M^n \times [0, T) \to \mathbb{R}^{n+1}$ be a compact solution to mean curvature flow. If an integral Brakke flow $\{\mu_{t,\infty}\}_{t\in(-\infty,0)}$ is a tangent flow of X at $(p_0, T) \in \mathbb{R}^{n+1} \times \mathbb{R}$, then there exists a constant C such that $\mu_{t,\infty}(B_r(0)) \leq Cr^n$ for all $t \in (-\infty, 0)$ and $r > 0$.*

Proof. Since X is a compact solution, we may take $\vartheta = \frac{1}{2}$ in (10.13) to obtain that for all $r > 0$ and $t \in (\frac{T}{2}, T)$,

$$\mathscr{H}^n(\mathcal{M}_t \cap B_r(p_0)) = \mu_t(X_t^{-1}(B_r(p_0))) \leq V r^n, \tag{13.35}$$

where V depends only on n, $\mu_0(M^n)$, and T. Since $\{\mu_{t,\infty}\}$ is a tangent flow of X at p_0, there exists a sequence $\lambda_j \to \infty$ such that the rescaled solutions $\mathcal{M}_{t,j} = \lambda_j(\mathcal{M}_{\lambda_j^{-2}t+T} - p_0)$ converge in the sense of Brakke flows to

$\{\mu_{t,\infty}\}_{t\in(-\infty,0)}$. By (13.35), we have that

$$\mathscr{H}^n(\mathcal{M}_{t,j} \cap B_r(0)) \leq V r^n$$

for all $r > 0$ and $t \in (-\frac{T}{2}\lambda_j^2, 0)$. Taking the limit as $j \to \infty$ of this inequality, we obtain that $\mu_{t,\infty}(B_r(0)) \leq V r^n$ for all $t \in (-\infty, 0)$ and $r > 0$. □

In particular, any self-shrinker $\mathcal{M}_\infty$ which arises as a *tangent flow* satisfies $\text{Area}(\mathcal{M}_\infty \cap B_r(0)) \leq C r^n$ and hence has polynomial area growth.

The following theorem was proved first in the special case where $|\text{II}|$ is bounded by Huisken [**297**]. The general case was proved by Colding and Minicozzi [**181**].

Figure 13.1. William P. Minicozzi II. Photo courtesy of Bryce Vickmark/MIT.

Theorem 13.14 (Classification of mean convex shrinkers [**181, 296, 297**]). *If $\mathcal{M}^n$, $n \geq 2$, is a noncompact shrinker with $H \geq 0$ and polynomial area growth, then $\mathcal{M}^n$ is, after a rotation, either*

(1) $\mathbb{R}^n \times \{0\}$,

(2) $S^k(\sqrt{2k}) \times \mathbb{R}^{n-k}$, *or*

(3) $\Gamma^{k,\ell} \times \mathbb{R}^{n-1}$, *where $\Gamma^{k,\ell} \subset \mathbb{R}^2$ is one of the Abresch–Langer curves.*

The proof of this theorem will occupy the rest of the section; we follow [**181**]. By Lemma 13.2 and the strong maximum principle, we need only consider the case that $H > 0$. The main idea is the same as in the compact case, namely integrating the Bochner formula (13.29). However, in the noncompact case there are significant technical issues regarding the integrations by parts, which employ the use of cutoff functions. So we begin by deriving some inequalities and estimates which we shall need.

The following algebraic lemma, which may be viewed as a qualitative improvement of Kato's inequality (13.24), is a consequence of the Codazzi equation.

Lemma 13.15. *On any hypersurface,*

$$\left(1+\frac{2}{n+1}\right)|\nabla|\mathrm{II}||^2 \leq |\nabla\mathrm{II}|^2+\frac{2n}{n+1}|\nabla H|^2 \tag{13.36}$$

wherever $|\mathrm{II}|\neq 0$.

Proof. With respect to a principal frame,

$$\begin{aligned}|\mathrm{II}|^2\,|\nabla|\mathrm{II}||^2 &= \frac{1}{4}|\nabla|\mathrm{II}|^2|^2\\ &= \sum_i\left(\sum_j \kappa_j\nabla_i\mathrm{II}_{jj}\right)^2\\ &\leq \sum_i\left(\sum_j\kappa_j^2\sum_j(\nabla_i\mathrm{II}_{jj})^2\right)\\ &= |\mathrm{II}|^2\sum_{i,j}(\nabla_i\mathrm{II}_{jj})^2.\end{aligned}$$

Next, observe that

$$\begin{aligned}|\nabla|\mathrm{II}||^2 &\leq \sum_{i,j}(\nabla_i\mathrm{II}_{jj})^2\\ &= \sum_{i,j\,:\,i\neq j}(\nabla_i\mathrm{II}_{jj})^2+\sum_i(\nabla_i\mathrm{II}_{ii})^2\\ &= \sum_{i,j\,:\,i\neq j}(\nabla_i\mathrm{II}_{jj})^2+\sum_i\left(\nabla_iH-\sum_{j\,:\,i\neq j}\nabla_i\mathrm{II}_{jj}\right)^2\\ &\leq \sum_{i,j\,:\,i\neq j}(\nabla_i\mathrm{II}_{jj})^2+n\sum_i\left((\nabla_iH)^2+\sum_{j\,:\,i\neq j}(\nabla_i\mathrm{II}_{jj})^2\right),\end{aligned}$$

where we used the inequality $(a_1+\cdots+a_n)^2\leq n(a_1^2+\cdots+a_n^2)$ to obtain the last line. Thus

$$|\nabla|\mathrm{II}||^2\leq n|\nabla H|^2+(n+1)\sum_{i,j\,:\,i\neq j}(\nabla_j\mathrm{II}_{ji})^2,$$

where we used the Codazzi equation. By adding $\frac{2}{n+1}$ times this inequality to the inequality $|\nabla|\mathrm{II}||^2 \leq \sum_{i,j}(\nabla_i \mathrm{II}_{jj})^2$, we obtain

$$\begin{aligned}\left(1+\frac{2}{n+1}\right)|\nabla|\mathrm{II}||^2 &\leq \frac{2n}{n+1}|\nabla H|^2 + \sum_{i,j}(\nabla_i \mathrm{II}_{jj})^2 + 2\sum_{i,j\,:\,i\neq j}(\nabla_j \mathrm{II}_{ji})^2 \\ &\leq \frac{2n}{n+1}|\nabla H|^2 + \sum_{i,j,k}(\nabla_i \mathrm{II}_{jk})^2 \\ &= \frac{2n}{n+1}|\nabla H|^2 + |\nabla \mathrm{II}|^2 . \qquad \square\end{aligned}$$

Next we proceed to prove some weighted integral estimates for II and its covariant derivative. Let $\mathcal{M}$ be a shrinking self-similar solution with $H > 0$. Using (13.16), we calculate that

$$\begin{aligned}\Delta_f \log H &= \frac{\Delta_f H}{H} - |\nabla \log H|^2 \\ &= \frac{1}{2} - |\mathrm{II}|^2 - |\nabla \log H|^2 ,\end{aligned} \tag{13.37}$$

where $f(p) = \frac{|p|^2}{4}$. Thus, if $\phi : \mathcal{M} \to \mathbb{R}$ has compact support, then

$$\begin{aligned}\int_{\mathcal{M}} \langle \nabla \phi^2, \nabla \log H \rangle \, dm &= -\int_{\mathcal{M}} \phi^2 \, \Delta_f \log H \, dm \\ &= \int_{\mathcal{M}} \phi^2 \left(-\frac{1}{2} + |\mathrm{II}|^2 + |\nabla \log H|^2\right) dm ,\end{aligned} \tag{13.38}$$

where $dm = \mathrm{e}^{-f} d\mu = \mathrm{e}^{-\frac{|p|^2}{4}} d\mu$. On the other hand,

$$\left|\int_{\mathcal{M}} \langle \nabla \phi^2, \nabla \log H \rangle dm\right| \leq \int_{\mathcal{M}} |\nabla \phi|^2 \, dm + \int_{\mathcal{M}} \phi^2 \, |\nabla \log H|^2 \, dm .$$

Therefore

$$\int_{\mathcal{M}} \phi^2 \, |\mathrm{II}|^2 \, dm \leq \int_{\mathcal{M}} |\nabla \phi|^2 \, dm + \frac{1}{2}\int_{\mathcal{M}} \phi^2 \, dm . \tag{13.39}$$

Let $\xi : [0, \infty) \to \mathbb{R}$ be a smooth, nonnegative, nonincreasing cutoff function with $\xi(s) = 1$ for $s \in [0, 1]$ and $\xi(s) = 0$ for $s \in [2, \infty)$. Taking $\phi(X) = \xi(a|X|)$, where $a \to 0_+$, and by the monotone convergence theorem (and since $|\nabla \phi|^2 \leq a^2(\xi')^2$), we conclude that

$$\int_{\mathcal{M}} |\mathrm{II}|^2 \, dm \leq \frac{1}{2}\int_{\mathcal{M}} dm .$$

We have proved the following.

Lemma 13.16 (See [**181**, Proposition 10.14]). *If $\mathcal{M}$ is a shrinker with $H > 0$ and finite Gaußian area (e.g., polynomial area growth), then $\int_{\mathcal{M}} |\mathrm{II}|^2 \, dm$ is finite.*

A sufficient condition for integration by parts on noncompact hypersurfaces is the following. To ensure the hypotheses of this lemma motivates some of the integral estimates we prove.

Lemma 13.17. *Let $\mathcal{M} \subset \mathbb{R}^{n+1}$ be a complete hypersurface. If $u, v \in C^\infty(\mathcal{M})$ satisfy*

$$\int_{\mathcal{M}} (|u\nabla v| + |\nabla u|\,|\nabla v| + |u\Delta_f v|)\, dm < \infty\,,$$

then

$$\int_{\mathcal{M}} u\,\Delta_f v\, dm = -\int_{\mathcal{M}} \langle \nabla u, \nabla v\rangle\, dm\,. \tag{13.40}$$

Proof. Define the piecewise linear cutoff function

$$\eta_j(x) \doteqdot \begin{cases} 1 & \text{if } |x| \leq j, \\ j+1-|x| & \text{if } j < |x| \leq j+1, \\ 0 & \text{if } |x| > j+1. \end{cases}$$

Then

$$\int_{\mathcal{M}} \eta_j u\,\Delta_f v\, dm = -\int_{\mathcal{M}} \eta_j \langle \nabla u, \nabla v\rangle\, dm - \int_{\mathcal{M}} u \langle \nabla \eta_j, \nabla v\rangle\, dm\,.$$

(1) Since $\int_{\mathcal{M}} |u\Delta_f v| dm < \infty$, we have $\int_{\mathcal{M}} \eta_j u\,\Delta_f v\, dm \to \int_{\mathcal{M}} u\,\Delta_f v\, dm$ by the dominated convergence theorem.

(2) We have $\int_{\mathcal{M}} \eta_j \langle \nabla u, \nabla v\rangle\, dm \to \int_{\mathcal{M}} \langle \nabla u, \nabla v\rangle\, dm$ by the dominated convergence theorem and since $\int_{\mathcal{M}} |\nabla u|\,|\nabla v| dm < \infty$.

(3) By $\int_{\mathcal{M}} |u\nabla v| dm < \infty$ and $|\nabla \eta_j| \to 0$, we have $\int_{\mathcal{M}} u \langle \nabla \eta_j, \nabla v\rangle\, dm \to 0$ by the dominated convergence theorem. The lemma follows. □

Recall that the functions H and $\langle \mathsf{N}, \omega\rangle$ are eigenfunctions of the elliptic operator L defined by (13.8); i.e., $LH - H = 0$ and $L\langle \mathsf{N}, \omega\rangle - \frac{1}{2}\langle \mathsf{N}, \omega\rangle = 0$ for $\omega \in \mathbb{R}^{n+1}$.

The **bottom of the spectrum** of L is defined as the infimum of a **Rayleigh–Ritz quotient**:

(13.41)

$$\lambda_1 = \lambda_1(L) \doteqdot \inf_{u \in C_c^\infty(\mathcal{M})} \frac{\int_{\mathcal{M}} \left(|\nabla u|^2 - \left(|\mathrm{II}|^2 + \frac{1}{2}\right) u^2\right) dm}{\int_{\mathcal{M}} u^2\, dm} \in [-\infty, \infty)\,.$$

Lemma 13.18. *If $\phi \in W^{1,2}(dm)$ and if $g : \mathcal{M} \to \mathbb{R}$ is a positive C^2 function satisfying $Lg + \lambda g = 0$, where $\lambda \in \mathbb{R}$, then*

$$\int_{\mathcal{M}} (2|\mathrm{II}|^2 + |\nabla \log g|^2)\, \phi^2\, dm \leq \int_{\mathcal{M}} \left(4|\nabla \phi|^2 - (1+2\lambda)\phi^2\right)\, dm\,. \tag{13.42}$$

Proof. We compute that

$$\begin{aligned}\Delta_f \log g &= \frac{\Delta_f g}{g} - |\nabla \log g|^2 \\ &= -\left(|\mathrm{II}|^2 + \frac{1}{2} + \lambda + |\nabla \log g|^2\right).\end{aligned}$$

Integrating this against $\eta^2\, dm$, where η is a cutoff function with compact support, we obtain

$$\begin{aligned}\int_{\mathcal{M}} \left(|\mathrm{II}|^2 + \frac{1}{2} + \lambda + |\nabla \log g|^2\right) \eta^2\, dm &= \int_{\mathcal{M}} 2\langle \eta\, \nabla \log g, \nabla \eta\rangle\, dm \\ &\leq \int_{\mathcal{M}} \left(\frac{1}{2}|\nabla \log g|^2\, \eta^2 + 2|\nabla \eta|^2\right) dm.\end{aligned}$$

Thus

$$\int_{\mathcal{M}} \left(2|\mathrm{II}|^2 + |\nabla \log g|^2\right) \eta^2\, dm \leq \int_{\mathcal{M}} \left(4|\nabla \eta|^2 - (1+2\lambda)\eta^2\right) dm.$$

This inequality holds for η in $W^{1,2}(dm)$ with compact support.

Define the approximating sequence

$$\eta_j(x) \doteqdot \begin{cases} \phi(x) & \text{if } |x| \leq j, \\ (j+1-|x|)\,\phi(x) & \text{if } j < |x| \leq j+1, \\ 0 & \text{if } |x| > j+1. \end{cases}$$

The lemma follows from the monotone convergence theorem. □

Lemma 13.19 (See [**181**, Proposition 10.14])**.** *If $\mathcal{M}$ is a shrinker with $H > 0$ and polynomial area growth, then*

$$\int_{\mathcal{M}} |\mathrm{II}|^2\, |X|^2\, dm < \infty. \tag{13.43}$$

Proof. Since $\mathcal{M}$ has polynomial area growth, $\phi(X) \doteqdot |X|$ is in $W^{1,2}(dm)$. Hence, by Lemma 13.18 with $g = H > 0$, we have

$$\int_{\mathcal{M}} \left(2|\mathrm{II}|^2 + |\nabla \log H|^2\right) |X|^2\, dm \leq \int_{\mathcal{M}} \left(4|\nabla |X||^2 + |X|^2\right) dm < \infty. \quad \square$$

Lemma 13.20 (See [**181**, Proposition 10.14])**.** *For a shrinking self-similar solution with $H > 0$ we have that the following two integrals are finite:*

$$\int_{\mathcal{M}} |\mathrm{II}|^4\, dm, \quad \int_{\mathcal{M}} |\nabla \mathrm{II}|^2\, dm.$$

Proof. By (13.39) with $\phi = \eta|\mathrm{II}|$, where $\eta \geq 0$ has compact support, we have

(13.44)

$$\begin{aligned}\int_{\mathcal{M}} \eta^2 \, |\mathrm{II}|^4 \, dm &\leq \int_{\mathcal{M}} \left(\eta^2 \, |\nabla|\mathrm{II}||^2 + |\mathrm{II}|^2 \, |\nabla\eta|^2 + 2\eta \, |\nabla|\mathrm{II}|| \, |\mathrm{II}| \, |\nabla\eta|\right) dm \\ &\quad + \frac{1}{2} \int_{\mathcal{M}} \eta^2 \, |\mathrm{II}|^2 \, dm \\ &\leq \int_{\mathcal{M}} \left((1+\varepsilon)\eta^2 \, |\nabla|\mathrm{II}||^2 + \left(\frac{1}{2}\eta^2 + (1+\varepsilon^{-1})|\nabla\eta|^2\right) |\mathrm{II}|^2\right) dm,\end{aligned}$$

where the second inequality is by the Peter–Paul inequality.

On the other hand, by (13.26) and Lemma 13.15, we have

$$\begin{aligned}(13.45) \quad \frac{1}{2}\Delta_f |\mathrm{II}|^2 &= |\nabla \mathrm{II}|^2 - |\mathrm{II}|^4 + \frac{1}{2}|\mathrm{II}|^2 \\ &\geq \left(1 + \frac{2}{n+1}\right) |\nabla|\mathrm{II}||^2 - \frac{2n}{n+1}|\nabla H|^2 - |\mathrm{II}|^4 + \frac{1}{2}|\mathrm{II}|^2 .\end{aligned}$$

Let $c_n \doteqdot \frac{2}{n+1}$. Multiplying this by $\eta^2 \, dm$ and integrating while discarding the nonnegative $\frac{1}{2}|\mathrm{II}|^2 \, \eta^2$ term, we obtain

$$\int_{\mathcal{M}} \left((1+c_n) \, |\nabla|\mathrm{II}||^2 - nc_n|\nabla H|^2 - |\mathrm{II}|^4\right) \eta^2 \, dm \leq -2\int_{\mathcal{M}} \langle |\mathrm{II}|\nabla\eta, \eta\nabla|\mathrm{II}|\rangle \, dm.$$

Now, by the Peter–Paul inequality,

$$-2\int_{\mathcal{M}} \langle |\mathrm{II}|\nabla\eta, \eta\nabla|\mathrm{II}|\rangle \, dm \leq \int_{\mathcal{M}} \left(\varepsilon|\nabla|\mathrm{II}||^2 \, \eta^2 + \varepsilon^{-1}|\mathrm{II}|^2|\nabla\eta|^2\right) \, dm \, ,$$

where we choose $\varepsilon > 0$ below. Thus

(13.46)

$$(1+c_n-\varepsilon) \int_{\mathcal{M}} \eta^2|\nabla|\mathrm{II}||^2 \, dm \leq \int_{\mathcal{M}} \left((nc_n|\nabla H|^2 + |\mathrm{II}|^4)\eta^2 + \tfrac{1}{\varepsilon}|\mathrm{II}|^2|\nabla\eta|^2\right) dm.$$

By combining (13.44) and (13.46), we obtain

$$\begin{aligned}\int_{\mathcal{M}} \eta^2|\mathrm{II}|^4 \, dm &\leq \frac{1+\varepsilon}{1+c_n-\varepsilon} \int_{\mathcal{M}} \eta^2|\mathrm{II}|^4 \, dm \\ &\quad + \int_{\mathcal{M}} \left(\left(\frac{1}{2}\eta^2 + C|\nabla\eta|^2\right) |\mathrm{II}|^2 + C|\nabla H|^2\right) dm.\end{aligned}$$

Taking $\varepsilon = \frac{c_n}{4} = \frac{1}{2(n+1)}$, so that $1 - \frac{1+\varepsilon}{1+c_n-\varepsilon} = \frac{2c_n}{4+3c_n} > 0$, and assuming $|\eta| \leq 1$ and $|\nabla\eta| \leq 1$, we obtain that

$$\int_{\mathcal{M}} \eta^2 \, |\mathrm{II}|^4 \, dm \leq C \int_{\mathcal{M}} \left(|\mathrm{II}|^2 + |\nabla H|^2\right) dm \, .$$

By (13.14), we have

$$|\nabla H|^2 \le \frac{1}{4}\,|\mathrm{II}|^2\,|X|^2\,. \tag{13.47}$$

Hence, by (13.43) we conclude that $\int_{\mathcal{M}} \eta^2\,|\mathrm{II}|^4\,dm \le C$. By the monotone convergence theorem, we obtain

$$\int_{\mathcal{M}} |\mathrm{II}|^4\,dm < \infty\,.$$

With this, (13.46) yields $\int_{\mathcal{M}} \eta^2\,|\nabla|\mathrm{II}||^2\,dm \le C$, so that by the monotone convergence theorem,

$$\int_{\mathcal{M}} |\nabla|\mathrm{II}||^2\,dm < \infty\,.$$

Now, integrating the first line of (13.45) against $\eta^2\,dm$, we have

$$\begin{aligned}\int_{\mathcal{M}} |\nabla \mathrm{II}|^2\,\eta^2\,dm &= \int_{\mathcal{M}} \left(\frac{1}{2}\Delta_f|\mathrm{II}|^2 + |\mathrm{II}|^4 - \frac{1}{2}|\mathrm{II}|^2\right)\eta^2\,dm \\ &\le \int_{\mathcal{M}} \left(\eta\,\langle\nabla|\mathrm{II}|^2,\nabla\eta\rangle + |\mathrm{II}|^4\eta^2\right)dm\,.\end{aligned}$$

Since $|\eta| \le 1$ and $|\nabla\eta| \le 1$, we have

$$|\eta\,\langle\nabla|\mathrm{II}|^2,\nabla\eta\rangle| \le 2|\mathrm{II}|\,|\nabla|\mathrm{II}||\,|\nabla\eta| \le |\nabla|\mathrm{II}||^2 + |\mathrm{II}|^2\,.$$

Hence

$$\int_{\mathcal{M}} |\nabla \mathrm{II}|^2\,\eta^2\,dm \int_{\mathcal{M}} \left(|\nabla|\mathrm{II}||^2 + |\mathrm{II}|^2 + |\mathrm{II}|^4\right)dm < \infty\,.$$

The lemma follows from the monotone convergence theorem. □

Lemma 13.21. *For a shrinking self-similar solution with $H > 0$, we have*

$$|\nabla \mathrm{II}| = |\nabla|\mathrm{II}|| \quad \textit{and} \quad |\mathrm{II}| = cH, \quad \textit{where } c \in \mathbb{R}_+\,. \tag{13.48}$$

Proof. Since $|\mathrm{II}| \in W^{1,2}(dm)$, we may take $\phi = |\mathrm{II}|$ and $g = H$ in Lemma 13.18 to obtain

$$\int_{\mathcal{M}} |\nabla \log H|^2\,|\mathrm{II}|^2\,dm \le \int_{\mathcal{M}} \left(|\nabla|\mathrm{II}||^2 + |\mathrm{II}|^2\right)dm < \infty\,.$$

Moreover, by (13.37) we have

$$\int_{\mathcal{M}} |\mathrm{II}|^2\,|\Delta_f \log H|\,dm = \int_{\mathcal{M}} |\mathrm{II}|^2\left|\frac{1}{2} - |\mathrm{II}|^2 - |\nabla \log H|^2\right|dm < \infty\,.$$

We also have

$$\int_{\mathcal{M}} |\nabla|\mathrm{II}|^2|\,|\nabla \log H|\,dm \le \int_{\mathcal{M}} |\mathrm{II}|^2\,|\nabla \log H|^2\,dm + \int_{\mathcal{M}} |\nabla|\mathrm{II}||^2\,dm < \infty\,.$$

Because of all of this, we may take $u = |\mathrm{II}|^2$ and $v = \log H$ in Lemma 13.17 to obtain

$$\begin{aligned} (13.49) \quad \int_{\mathcal{M}} \langle \nabla |\mathrm{II}|^2, \nabla \log H \rangle \, dm &= - \int_{\mathcal{M}} |\mathrm{II}|^2 \, \Delta_f \log H \, dm \\ &= \int_{\mathcal{M}} |\mathrm{II}|^2 \left(-\frac{1}{2} + |\mathrm{II}|^2 + |\nabla \log H|^2 \right) dm , \end{aligned}$$

where we used (13.37) for the last equality.

By (13.27), we have

$$\Delta_f |\mathrm{II}| \geq -|\mathrm{II}|^3 + \frac{1}{2} |\mathrm{II}| .$$

It is easy to check that we may use Lemma 13.17 to integrate by parts to get

$$\int_{\mathcal{M}} |\nabla |\mathrm{II}||^2 \, dm = - \int_{\mathcal{M}} |\mathrm{II}| \Delta_f |\mathrm{II}| \, dm \leq \int_{\mathcal{M}} |\mathrm{II}|^2 \left(|\mathrm{II}|^2 - \frac{1}{2} \right) dm .$$

By subtracting this inequality from (13.49), we obtain

$$\int_{\mathcal{M}} \left(-|\nabla |\mathrm{II}||^2 + \langle \nabla |\mathrm{II}|^2, \nabla \log H \rangle - |\mathrm{II}|^2 \, |\nabla \log H|^2 \right) dm \geq 0 .$$

Since the integrand on the left-hand side is the negative of norm squared of a vector, we obtain that the vector is zero:

$$0 = \nabla |\mathrm{II}| - |\mathrm{II}| \, \nabla \log H = |\mathrm{II}| \, \nabla \log \frac{|\mathrm{II}|}{H} .$$

We conclude that $|\mathrm{II}| = cH$ on $\mathcal{M}$ for some $c > 0$. Since $LH = H$, this implies that $L|\mathrm{II}| = |\mathrm{II}|$. Comparing this last equation with (13.25), we obtain

$$|\nabla \mathrm{II}|^2 = |\nabla |\mathrm{II}||^2 . \qquad \square$$

We may now argue as in Theorem 13.11.

Case 1. *$|\nabla H|^2 \equiv 0$ on $\mathcal{M}$.* As before, (13.31) implies $\nabla \mathrm{II} \equiv 0$. By a theorem of Lawson [**356**], $\mathcal{M}$ is a product of a round sphere and a linear subspace. It is easy to see from the shrinker equation that the sphere must be centered at the origin and the subspace must pass through the origin.

Case 2. *There exists a point where $|\nabla H|^2 \neq 0$.* Arguing as before, we find that

$$\mathrm{II} = \mathrm{II}_{11} \, e_1^* \otimes e_1^*$$

at any point for which $\nabla H \neq 0$, where $e_1 \doteqdot \nabla H / |\nabla H|$. Note that this implies that $\mathrm{II}(V, V) = 0$ if and only if $\mathrm{L}(V) = 0$.

Since $H\nabla \mathrm{II} = \nabla H \otimes \mathrm{II}$, the only nonzero component of $\nabla \mathrm{II}$ is $\nabla_1 \mathrm{II}_{11}$. Hence we may express the components of $\nabla\mathrm{II}$ as

$$\nabla_k \mathrm{II}_{ij} = \frac{\nabla_1 \mathrm{II}_{11}}{H} \delta_{k1} \mathrm{II}_{ij} . \tag{13.50}$$

Let $x \in \mathcal{M}$ and $V \in T_x\mathcal{M}$. Let $\gamma : (a,b) \to \mathcal{M}$, $0 \in (a,b)$, be a path with $\gamma(0) = x$. Let $V(t)$ be the parallel translation of V_0 along $\gamma(t)$. Using (13.50), we compute that

$$\begin{aligned}\frac{d}{dt}\left(\mathrm{II}_{ij}V^i(t)V^j(t)\right) &= \langle\dot\gamma(t), e_1\rangle(\nabla_1\mathrm{II}_{ij})V^i(t)V^j(t)\\ &= \langle\dot\gamma(t), e_1\rangle\frac{\nabla_1\mathrm{II}_{11}}{H}\mathrm{II}_{ij}V^i(t)V^j(t) .\end{aligned}$$

Since the function $\langle\dot\gamma(t), e_1\rangle\frac{\nabla_1\mathrm{II}_{11}}{H}$ is bounded, we conclude by ODE comparison that if $\mathrm{II}_{ij}V^i(0)V^j(0) = 0$, then $\mathrm{II}_{ij}V^i(t)V^j(t) = 0$ for all t. Thus, the kernel of II is invariant under parallel translation. Furthermore, we calculate that

$$D_{\dot\gamma(t)}V(t) = \nabla_{\dot\gamma(t)}V(t) - \mathrm{II}(\dot\gamma(t), V(t))\,\mathsf{N}(\gamma(t)) = 0 ,$$

where D is the Euclidean covariant derivative. Hence, we have that $V(t)$ is a constant vector field along $\gamma(t)$.

Since the kernel of II is $(n-1)$-dimensional, there exist an orthonormal set of constant vector fields $\{e_2, \ldots, e_n\}$ globally defined on $\mathcal{M}$ which spans the kernel of II at each point. This implies that $\mathcal{M}$ is the product of an $(n-1)$-dimensional hyperplane (passing through the origin) with a convex immersed self-shrinker curve (in the orthogonal 2-plane). This curve must be an Abresch–Langer shrinker. This completes the proof of Theorem 13.14.

Colding, Ilmanen, and Minicozzi proved that cylinders are rigid in the sense that "a shrinker that is close to a cylinder in a sufficiently large ball must be isometric to the cylinder" [**177**].

Theorem 13.22. *For any integer $n \geq 2$ and positive real numbers λ_0 and C there exists $R = R(n, \lambda_0, C)$ with the following property. Let $\mathcal{M}^n$ be an integer rectifiable varifold in $\mathbb{R}^{n+1}$ that is $\mathcal{F}$-critical; i.e., $\mathcal{M}^n$ is a weak shrinker. Assume that the intersection of $\mathcal{M}^n$ with any compact set is a closed set. If $\lambda(\mathcal{M}) \leq \lambda_0$ and if $\mathcal{M} \cap B_R(0)$ is C^∞ with $H \geq 0$ and $|\mathrm{II}| \leq C$ on $\mathcal{M} \cap B_R(0)$, then $\mathcal{M} = S^k \times \mathbb{R}^{n-k}$ for some $0 \leq k \leq n$. Here $S^0 \doteqdot \{0\}$.*

Although the results above strongly restrict the class of mean convex shrinkers, recent years have seen the construction of a large number of examples which are not mean convex. This activity has been driven in part by the observation that shrinkers are minimal hypersurfaces with respect to the Gaußian metric. We describe some of these results in the notes and commentary at the end of this chapter. A more comprehensive discussion can be found in the survey by G. Drugan, H. Lee, and X. H. Nguyen [**202**].

13.4. Compact embedded self-shrinking surfaces

We have seen that mean convex shrinkers admit a simple classification, particularly in the "noncurve" case. Without the mean convexity hypothesis there appear to be a huge number of examples, however. Recently, there has been some progress towards understanding shrinkers of dimension two in terms of their topology.

Figure 13.2. Simon Brendle. Author: Tatjana Ruf. Photo courtesy of Archives of the Mathematisches Forschungsinstitut Oberwolfach.

13.4.1. Compact, embedded shrinkers of genus 0. We first present Simon Brendle's proof that compact embedded genus 0 shrinkers in $\mathbb{R}^3$ are round spheres.

Theorem 13.23 (Brendle [**108**]). *Let $\mathcal{M}^2$ be a compact shrinking self-similar solution of mean curvature flow embedded in $\mathbb{R}^3$. If $\mathcal{M}^2$ has genus zero, i.e., if it is diffeomorphic to S^2, then $\mathcal{M}^2$ is the round sphere of radius* 2 *centered at the origin.*

Proof. Since $\mathcal{M}^2$ is a compact self-shrinker embedded in $\mathbb{R}^3$, it satisfies, by definition, $H(p) = \frac{1}{2}\langle p, \mathsf{N}(p)\rangle$ for $p \in \mathcal{M}$, where $\mathsf{N}(p)$ is the unit outward normal to $\mathcal{M}$ at p. Suppose that $\mathcal{M}$ is not the round sphere of radius 2 centered at 0. Then, by Theorem 13.11, H must be negative somewhere. On the other hand, the farthest point on $\mathcal{M}$ from the origin has positive mean curvature. So H changes sign and hence we can find a point $p_0 \in \mathcal{M}$ such that

$$0 = 2H(p_0) = \langle p_0, \mathsf{N}(p_0)\rangle. \tag{13.51}$$

Set $\mathsf{N}_0 \doteqdot \mathsf{N}(p_0)$ and consider the coordinate function $f : \mathcal{M} \to \mathbb{R}$ defined by

$$f(p) \doteqdot \langle p, \mathsf{N}_0\rangle$$

and its zero set

$$\mathcal{Z} \doteqdot \{p \in \mathcal{M} \,:\, f(p) = 0\} \subset \mathcal{M}\,. \tag{13.52}$$

Observe that $\mathcal{Z}$ is closed (since f is continuous) and nonempty (since $p_0 \in \mathcal{Z}$).

Note that $\nabla f(p_0) = 0$ since $\nabla f = \mathsf{N}_0^\top$, where $\cdot^\top$ denotes the projection onto the tangent bundle of $\mathcal{M}$. Using $\nabla f = \mathsf{N}_0^\top$ and (5.32), we find that

$$-\Delta f = \langle -\Delta p, \mathsf{N}_0 \rangle = H \langle \mathsf{N}, \mathsf{N}_0 \rangle\,.$$

Since $p = p^\top + \langle p, \mathsf{N} \rangle \mathsf{N}$, we conclude that f satisfies the equation

$$-\Delta f + \frac{1}{2} \langle p, \nabla f \rangle = \frac{1}{2} f\,. \tag{13.53}$$

We can remove the drift term in the equation above for f by defining the function $g : \mathcal{M} \to \mathbb{R}$ by

$$g(p) \doteqdot \mathrm{e}^{-\frac{|p|^2}{8}} f(p)\,.$$

Note that $\mathcal{Z}$ is also the zero set of g. We compute that g satisfies

$$\begin{aligned} -\Delta g = &- \mathrm{e}^{-\frac{|p|^2}{8}} \left(\Delta f - \frac{1}{4} \nabla |p|^2 \cdot \nabla f \right) - f \Delta \left(\mathrm{e}^{-\frac{|p|^2}{8}} \right) \\ = &\,\mathrm{e}^{-\frac{|p|^2}{8}} f \left(\frac{1}{2} - \mathrm{e}^{\frac{|p|^2}{8}} \Delta \left(\mathrm{e}^{-\frac{|p|^2}{8}} \right) \right) \\ = &\, hg\,, \end{aligned} \tag{13.54}$$

where

$$h(p) \doteqdot \frac{1}{2} - \mathrm{e}^{\frac{|p|^2}{8}} \Delta \left(\mathrm{e}^{-\frac{|p|^2}{8}} \right)\,.$$

Since $\mathcal{M}$ is a 2-dimensional smooth manifold and g is a smooth function satisfying $\Delta g + hg = 0$ on $\mathcal{M}$, where $h \in C^\infty(\mathcal{M})$, a result of S. Y. Cheng [**144**, Theorem 2.5] implies that each point $p \in \mathcal{Z}$ admits an open neighborhood $\mathcal{U} \subset \mathcal{M}$ such that $\mathcal{Z} \cap \mathcal{U}$ is the union of m C^2-arcs which intersect transversally at p, where $m \in \mathbb{N}$ is equal to the order of vanishing of f at p (largest $m \in \mathbb{N}$ for which $|f(q)| = O(|p-q|^m)$ as $q \to p$). In fact $\mathcal{Z}$ is the union of a finite number of C^2 immersed loops. Moreover, the singular set

$$\mathcal{Z}_{\text{sing}} \doteqdot \{p \in \mathcal{M} \,:\, f(p) = 0 \text{ and } \nabla f(p) = 0\}$$

is finite and its complement $\mathcal{Z} \backslash \mathcal{Z}_{\text{sing}}$ is a C^2 embedded curve (a disjoint union of open embedded arcs and embedded loops).

Since f and ∇f both vanish at p_0, there exists an open neighborhood $\mathcal{U} \subset \mathcal{M}$ of p_0 such that $\mathcal{Z} \cap \mathcal{U}$ is the union of at least two C^2-arcs which intersect transversally at p_0.

In particular, $\mathcal{Z}$ cannot be a single embedded loop, so we can choose a continuous, piecewise C^2 embedded loop Γ such that $\Gamma \subset \mathcal{Z}$ but $\Gamma \neq \mathcal{Z}$. Since $\mathcal{M}$ has genus 0, the loop Γ bounds a disk in $\mathcal{M}$.

The complement $\mathbb{R}^3 \setminus \mathcal{M}$ has two connected components, exactly one of which is bounded. Denote by Ω the *bounded* component and denote by $\mho$ the *unbounded* component. Of course, $\mathbb{R}^3 = \mathcal{M} \cup \Omega \cup \mho$ is a disjoint union and $\partial\bar{\Omega} = \partial\bar{\mho} = \mathcal{M}$.

Recall from Lemma 13.3 that the shrinking self-similar solutions of mean curvature flow in $(\mathbb{R}^{n+1}, g_{\mathbb{R}^{n+1}})$ are exactly the minimal hypersurfaces in $(\mathbb{R}^{n+1}, g_{\mathrm{ß}})$, where $g_{\mathrm{ß}}(p) = \mathrm{e}^{-\frac{|p|^2}{2n}} g_{\mathbb{R}^{n+1}}(p)$ is the Gaußian metric. By (13.9), a *stable* minimal hypersurface $\mathcal{N}$ in $(\mathbb{R}^{n+1}, g_{\mathrm{ß}})$ satisfies

$$\int_{\mathcal{N}} F\, LF\, dm \leq 0$$

for any compactly supported $F : \mathcal{N} \to \mathbb{R}$, where the stability operator L is defined by (13.8) and $dm(p) = \mathrm{e}^{-\frac{|p|^2}{4}} d\mu(p) = d\mu_{g_{\mathrm{ß}}}(p)$. In the following, it is perhaps good to keep in mind that although $\mathbb{R}^{n+1}$ is noncompact and $g_{\mathrm{ß}}$ is incomplete, $(\mathbb{R}^{n+1}, g_{\mathrm{ß}})$ has finite diameter and volume.

Now consider a general metric that is conformal to the Euclidean metric: $\tilde{g}_{\mathbb{R}^{n+1}} = \mathrm{e}^{2u} g_{\mathbb{R}^{n+1}}$, where u is a smooth function on $\mathbb{R}^{n+1}$. Denote by $\tilde{D}$ the covariant derivative induced by $\tilde{g}_{\mathbb{R}^{n+1}}$. The corresponding connection coefficients are given, in Euclidean coordinates, by

$$\begin{aligned}\tilde{\Gamma}_{ij}^k &= \frac{1}{2}\mathrm{e}^{-2u}\delta^{k\ell}\left(\partial_i(\mathrm{e}^{2u}\delta_{j\ell}) + \partial_j(\mathrm{e}^{2u}\delta_{i\ell}) - \partial_\ell(\mathrm{e}^{2u}\delta_{ij})\right)\\ &= \partial_i u\, \delta_j^k + \partial_j u\, \delta_i^k - \partial^k u\, \delta_{ij}\,.\end{aligned}$$

So the covariant derivative is given by

$$\begin{aligned}\tilde{D}_V W &= D_V W + V^i W^j \tilde{\Gamma}_{ij}^k e_k\\ &= D_V W + V(u) W + W(u) V - \langle V, W\rangle Du\,.\end{aligned}$$

Let $\mathcal{M}$ be a hypersurface in $\mathbb{R}^{n+1}$ and let N be a choice of unit normal field for $\mathcal{M}$ with respect to $g_{\mathbb{R}^{n+1}}$. Then $\tilde{\mathsf{N}} = \mathrm{e}^{-u}\mathsf{N}$ is a unit normal field for $\mathcal{M}$ with respect to $\tilde{g}_{\mathbb{R}^{n+1}}$.

The second fundamental form $\tilde{\mathrm{II}}$ of $\mathcal{M}$ with respect to $\tilde{g}_{\mathbb{R}^{n+1}}$ is then given by

$$\begin{aligned}\tilde{\mathrm{II}}(V, W) &= -\,\tilde{g}_{\mathbb{R}^{n+1}}(\tilde{D}_V W, \tilde{\mathsf{N}})\\ &= -\,\mathrm{e}^u g_{\mathbb{R}^{n+1}}(\tilde{D}_V W, \mathsf{N})\\ &= \mathrm{e}^u\, \mathrm{II}(V, W) + \mathrm{e}^u \langle V, W\rangle\, \langle Du, \mathsf{N}\rangle\,.\end{aligned} \tag{13.55}$$

Since we are interested in the Gaußian metric, we now set $u(p) = -\frac{|p|^2}{4n}$, so that $\tilde{g}_{\mathbb{R}^{n+1}} = g_{\text{ß}}$. If $\mathcal{M} = S^n(r)$, the n-sphere of radius r centered at the origin, then $\mathrm{II}(V,W) = \frac{1}{r}\langle V, W\rangle$ and $\langle Du, \mathsf{N}\rangle = -\frac{r}{2n}$. So the second fundamental form of $S^n(r)$ with respect to $g_{\text{ß}}$ is

$$\tilde{\mathrm{II}}(V,W) = \mathrm{e}^{-\frac{r^2}{4n}}\left(\frac{1}{r} - \frac{r}{2n}\right)\langle V, W\rangle .$$

In particular, if $r > \sqrt{2n}$, then the ball $\bar{B}_r(0)$ has concave boundary $\bar{B}_r(0) = S^n(r)$ with respect to $g_{\text{ß}}$ since $\tilde{\mathrm{II}}(V,W) < 0$.

For a general function u, the second fundamental form of $S^n(r)$ with respect to $\tilde{g}_{\mathbb{R}^{n+1}}$ is

$$\tilde{\mathrm{II}}(V,W) = \mathrm{e}^{u}\left(\frac{1}{r} + \langle Du, \mathsf{N}\rangle\right)\langle V, W\rangle .$$

Let $n = 2$. Given $R > 0$, define a smooth function $\psi_R : [0,\infty) \to [0,1]$ so that $\psi_R = 0$ on the interval $[0,R]$ and $\psi_R'(2R) > R$. Let $u(p) = u_R(p) = -\frac{|p|^2}{8} + \psi_R(|p|)$. For $p \in S^2(2R)$, we have $\langle Du(p), \mathsf{N}(p)\rangle = -\frac{R}{2} + \psi_R'(2R) > \frac{R}{2} > 0$. Thus $S^2(2R)$ has positive second fundamental form. Choose $R_0 \in \mathbb{N}$ large enough so that $\Omega \subset B_{2R}(0)$ for $R \geq R_0$. Assume that $R \geq R_0$. Then $A_R \doteqdot \overline{\mho} \cap \bar{B}_{2R}(0)$ is a domain with smooth boundary $\partial A_R = \Omega \cup S^2(2R)$ which has $\tilde{H} = 0$ for one component and $\tilde{\mathrm{II}} > 0$ for the other component and in particular the boundary ∂A_R is weakly mean convex, all with respect to the ambient metric $g_R \doteqdot \tilde{g}_{\mathbb{R}^{n+1}} = \mathrm{e}^{2u_R} g_{\mathbb{R}^{n+1}}$.

By the work of William Meeks and Shing-Tung Yau [**398**], since A_R has weakly mean convex boundary and since $\Gamma \subset \partial A_R$, there exists a smooth embedded open disk $\mathcal{D}_R$ in A_R bounding Γ which minimizes area with respect to g_R among all branched immersions with boundary Γ. In particular, $\mathcal{D}_R$ is a stable minimal surface with respect to g_R.

Since each disk $\mathcal{D}_R$ bounds Γ, i.e., $\partial\mathcal{D}_R \doteqdot \overline{\mathcal{D}}_R \backslash \mathcal{D}_R = \Gamma$, and since each $\mathcal{D}_R$ is area minimizing with respect to g_R, where $g_R \leq \mathrm{e}^2\, g_{\text{ß}}$ independent of R, we have that

$$\sup_{R \geq R_0} \operatorname{Area}_{g_R}(\mathcal{D}_R) < \infty .$$

Since $g_{\text{ß}} \leq g_R$ for each R, this implies that

$$\sup_{R \geq R_0} \operatorname{Area}_{g_{\text{ß}}}(\mathcal{D}_R) < \infty .$$

Since $\mathcal{D}_R$ is minimal with respect to g_R and since $g_R = g_{\text{ß}}$ on $\bar{B}_R(0)$, we have that the surface $\mathcal{D}_R \cap \bar{B}_R(0)$ is a $g_{\text{ß}}$ minimal surface; i.e., it satisfies $H(p) = \frac{1}{2}\langle p, \mathsf{N}(p)\rangle$. Since $\mathcal{D}_R \cap \bar{B}_R(0)$ is area minimizing with respect to $g_{\text{ß}}$, by the stability inequality (13.9) we have that

$$\int_{\mathcal{D}_R} F\,LF\,dm \leq 0 \tag{13.56}$$

for all $F : \mathcal{D}_R \to \mathbb{R}$ vanishing on $\Gamma \cup (\mathcal{D}_R \backslash B_R(0))$, where $dm = d\mu_{g_{\mathbb{B}}}$.

To facilitate calculating LF, we first observe the following.

Claim 13.24. For any self-shrinker $\mathcal{M}$ and function $F : \mathbb{R}^3 \to \mathbb{R}$, we have

$$\Delta F - \frac{1}{2}\langle p^\top, \nabla F\rangle = \sum_{i=1}^{2} D^2F(e_i, e_i) - \frac{1}{2}\langle p, DF\rangle\,, \tag{13.57}$$

where the F on the left-hand side is the restriction of F to $\mathcal{M}$, where D is the Euclidean covariant derivative, and where $\{e_1, e_2\}$ is any orthonormal frame field for $\mathcal{M}$.[1]

Proof of Claim 13.24. We have

$$\sum_{i=1}^{2}(\nabla^2 F - D^2 F)(e_i, e_i) = \sum_{i=1}^{2} -(\nabla_{e_i} e_i + D_{e_i} e_i)F = -H\,\mathsf{N}(F)$$

and

$$-\frac{1}{2}\langle p^\top, \nabla F\rangle + \frac{1}{2}\langle p, DF\rangle = \frac{1}{2}\langle p, \mathsf{N}\rangle\langle \mathsf{N}, DF\rangle = H\,\mathsf{N}(F)\,, \tag{13.58}$$

where we used the self-shrinker equation to obtain the last equality. The claim follows. □

We shall use the above to prove the following.

Claim 13.25. There exists a constant C, independent of R, such that for all $R \geq R_0$ we have

$$\int_{\mathcal{D}_R \cap \bar{B}_{\sqrt{R}}(0)} |\mathrm{II}|^2(p)\,\langle p, \mathsf{N}_0\rangle^2\,dm(p) \leq \frac{C}{\log R}\int_{\mathcal{D}_R \cap (\bar{B}_R(0)\backslash B_{\sqrt{R}}(0))} dm\,. \tag{13.59}$$

Proof of Claim 13.25. Define a smooth cutoff function $\eta : \mathbb{R} \to [0,1]$ so that $\eta = 1$ on $(-\infty, \frac{1}{2}]$, $\eta = 0$ on $[1, \infty)$, and $-C \leq \eta' \leq 0$ on $\mathbb{R}$. Define $F_R : \mathbb{R}^3 \to \mathbb{R}$ by

$$F_R(p) = \eta\left(\frac{\log|p|}{\log R}\right)\langle p, \mathsf{N}_0\rangle = \eta\left(\frac{\log|p|}{\log R}\right) f(p)\,.$$

Since $\langle p, \mathsf{N}_0\rangle = 0$ on $\Gamma \subset \mathcal{Z}$ and since $\eta\left(\frac{\log|p|}{\log R}\right) = 0$ on $\mathcal{D}_R \backslash B_R(0)$, we have that the restriction of F_R to $\mathcal{D}_R$, which we also denote by F_R, is an admissible test function for the stability inequality (13.56). Note that $F_R(p) = \langle p, \mathsf{N}_0\rangle$ for $p \in \bar{B}_{\sqrt{R}}(0)$.

[1]Since the right-hand side of (13.57) at a point p depends only on $\{e_1(p), e_2(p)\}$, this frame field need not be continuous. Indeed, the first term is simply the trace of D^2F restricted to $T\mathcal{M}$.

By $Df = \mathsf{N}_0$ and $D\log|p| = \frac{p}{|p|^2}$, we have

$$\langle p, DF_R(p)\rangle = F_R + \frac{\eta'}{\log R} f\,. \tag{13.60}$$

We compute that

$$\begin{aligned} D^2F_R = &\frac{\eta'}{\log R}\left(D\log|p| \otimes Df + Df \otimes D\log|p|\right) \\ &+ \frac{f\eta''}{\log^2 R} D\log|p| \otimes D\log|p| + \frac{f\eta'}{\log R} D^2\log|p|\,. \end{aligned}$$

So using in addition that $D^2 f \equiv 0$ and $D^2 \log|p| = \frac{|p|^2 g_{\mathbb{R}^{n+1}} - 2p\otimes p}{|p|^4}$, we have

$$\begin{aligned} D^2F_R = &\frac{\eta'}{|p|^2\log R}\left(p\otimes \mathsf{N}_0 + \mathsf{N}_0 \otimes p\right) + \frac{f\eta''}{|p|^4\log^2 R} p\otimes p \\ &+ \frac{f\eta'}{|p|^4\log R}\left(|p|^2 g_{\mathbb{R}^{n+1}} - 2p\otimes p\right). \end{aligned}$$

From this it is easy to see that

$$|D^2F_R| \le \frac{C}{|p|\log R}\chi_{\sqrt{R}\le|x|\le R}\,,$$

where C is independent of R, and hence that

$$|F_R|\,|D^2F_R| \le \frac{C}{\log R}\chi_{\sqrt{R}\le|x|\le R}\,. \tag{13.61}$$

By (13.57), (13.60), and (13.61), we have

$$\begin{aligned} -F_R\left(\Delta F_R - \frac{1}{2}\langle p^\top, \nabla F_R\rangle\right) &= -F_R\left(\sum_{i=1}^{2} D^2F_R(e_i,e_i) - \frac{1}{2}\langle p, DF_R\rangle\right) \\ &\le \frac{C}{\log R}\chi_{\sqrt{R}\le|x|\le R} + \frac{1}{2}F_R^2 + \frac{1}{2}\frac{\eta\,\eta'}{\log R} f^2 \\ &\le \frac{C}{\log R}\chi_{\sqrt{R}\le|x|\le R} + \frac{1}{2}F_R^2\,, \end{aligned}$$

with the last inequality since $\eta' \le 0$. Hence, by the definition of L, we have

$$-F_R\, LF_R \le \frac{C}{\log R}\chi_{\sqrt{R}\le|x|\le R} - |\mathrm{II}|^2 F_R^2\,.$$

Finally, by the stability inequality (13.56) with $F = F_R$, we obtain

$$
\begin{aligned}
0 &\leq -\int_{\mathcal{D}_R} F_R\, L F_R\, dm \\
&\leq \frac{C}{\log R}\int_{\mathcal{D}_R \cap \{\sqrt{R}\leq |x| \leq R\}} dm - \int_{\mathcal{D}_R} |\mathrm{II}|^2 F_R^2 dm \\
&\leq \frac{C}{\log R}\int_{\mathcal{D}_R \cap \{\sqrt{R}\leq |x| \leq R\}} dm - \int_{\mathcal{D}_R} |\mathrm{II}|^2 f^2 dm\,.
\end{aligned}
$$

This completes the proof of the claim. □

By the claim (13.59), we have

$$
\limsup_{R\to\infty} \int_{\mathcal{D}_R \cap \bar{B}_{\sqrt{R}}(0)} |\mathrm{II}|^2(p)\, \langle p, \mathsf{N}_0\rangle^2\, dm(p) = 0\,. \tag{13.62}
$$

As a special case of [**468**, Theorem 3] by Richard Schoen and Leon Simon, we have the following: Let $k \subset K'$ be open domains with compact closures and smooth boundaries such that $\overline{K} \subset K'$. Let $\mu \in \mathbb{R}_+$. Then there exists a constant C depending only on μ, K, and K' with the following property: If $\mathcal{N}$ is an orientable, C^2 stable minimal surface with respect to g_{B} with $(\overline{\mathcal{N}}\backslash\mathcal{N}) \cap K' = \emptyset$ and $\operatorname{Area}_{g_{\mathrm{B}}}(\mathcal{N}\cap K') \leq \mu$, then

$$
\sup_{\mathcal{N}\cap K} |\tilde{\mathrm{II}}|_{g_{\mathrm{B}}} \leq C\,,
$$

where $\tilde{\mathrm{II}}$ is the second fundamental form of $\mathcal{N}$ with respect to g_{B}. From Schoen and Simon's result, we see that for any compact subset K of $\mathbb{R}^3\backslash\Gamma$,

$$
\limsup_{R\to\infty} \sup_{\mathcal{D}_R\cap K} |\tilde{\mathrm{II}}|_{g_{\mathrm{B}}} < \infty\,.
$$

Note that by this and (13.55), we also obtain that

$$
\limsup_{R\to\infty} \sup_{\mathcal{D}_R\cap K} |\mathrm{II}|_{g_{\mathbb{R}^{n+1}}} < \infty\,.
$$

By a standard compactness theorem for minimal surfaces with bounded second fundamental form (cf. Theorem 11.11), there exists a sequence of radii $R_j \to \infty$ such that the disks $\mathcal{D}_{R_j}$ converge in $C^\infty_{\mathrm{loc}}(\mathbb{R}^3\backslash\Gamma)$ to a smooth g_{B}-minimal surface $\mathcal{D} \subset \mathbb{R}^3\backslash\Gamma$. Since $\mathcal{D}$ is g_{B}-minimal, $\mathcal{D}$ is a self-shrinker satisfying $H(p) = \frac{1}{2}\langle p, \mathsf{N}(p)\rangle$. Let $\partial\mathcal{D} \doteqdot \overline{\mathcal{D}}\backslash\mathcal{D}$ denote the boundary of $\mathcal{D}$. Since $\partial\mathcal{D}_R \cap (\mathbb{R}^3\backslash\Gamma) = \emptyset$, we have $\partial\mathcal{D} \cap (\mathbb{R}^3\backslash\Gamma) = \emptyset$; i.e., $\partial\mathcal{D} \subset \Gamma$.

By (13.62) and the convergence in $C^\infty_{\mathrm{loc}}(\mathbb{R}^3\backslash\Gamma)$, we have $|\mathrm{II}|(p)\langle p, \mathsf{N}_0\rangle = 0$ for all $p \in \mathcal{D}$. From this we can deduce that

$$
\mathrm{II} \equiv 0 \quad \text{on } \mathcal{D}\,.
$$

Indeed, suppose there exists $q \in \mathcal{D}$ such that $|\mathrm{II}|(q) \neq 0$. Then there exists a neighborhood $\mathcal{U}$ of q in $\mathcal{D}$ such that $|\mathrm{II}| > 0$ on $\mathcal{U}$, which implies that

$\langle q, \mathsf{N}_0\rangle = 0$ for $q \in \mathcal{U}$. This says that $\mathcal{U}$ lies in a plane and hence $|\mathrm{II}| \equiv 0$ on $\mathcal{U}$, contradicting $|\mathrm{II}|(q) \neq 0$.

Hence, $\langle p, \mathsf{N}(p)\rangle = H(p) = 0$ on $\mathcal{D}$. Since $\mathcal{D}_R \subset A_R \subset \overline{\mho}$, we have that $\mathcal{D} \subset \overline{\mho}$. Since the g_{B}-minimal surface $\mathcal{M} = \partial\overline{\mho}$ is not totally geodesic (with respect to $g_{\mathbb{R}^{n+1}}$), by the avoidance principle we have that $\mathcal{D} \cap \mathcal{M} = \emptyset$ and hence $\mathcal{D} \subset \mho$. Moreover, since $\mathcal{D}$ is totally geodesic, $\mathcal{D}$ lies in a plane and $\mathsf{N}(p) \doteqdot \mathsf{N}_1$ is a constant vector field.

Claim 13.26. $\partial\mathcal{D} = \Gamma$.

Proof of Claim 13.26. Firstly, recall that $\partial\mathcal{D} \subset \Gamma$. Secondly, suppose that Γ is not a subset of $\overline{\mathcal{D}}$. Then it is easy to construct a 1-form α on $\mathbb{R}^3$ with the following properties:

(1) $\alpha \equiv 0$ in a neighborhood of $\overline{\mathcal{D}}$.

(2) $d\alpha \equiv 0$ in a neighborhood of Γ.

(3) $\int_\Gamma \alpha \neq 0$.

(4) α has compact support.

By (1), (2), (4), and the convergence of $\mathcal{D}_R$ to $\mathcal{D}$ in $C^\infty_{\mathrm{loc}}(\mathbb{R}^3\backslash\Gamma)$, we have that $\lim_{j\to\infty} \int_{\mathcal{D}_{R_j}} d\alpha = 0$. On the other hand, $\int_{\mathcal{D}_{R_j}} d\alpha = \int_{\partial\mathcal{D}_{R_j}} \alpha = \int_\Gamma \alpha \neq 0$ for each j by Stokes's theorem and (3), which is a contradiction. Now, since $\Gamma \cap \mathcal{D} = \emptyset$ and $\Gamma \subset \overline{\mathcal{D}}$, we conclude that $\Gamma \subset \partial\mathcal{D}$. This proves the claim. □

Now, since $\partial\mathcal{D} = \Gamma$ lies in the plane $P \doteqdot \{x \in \mathbb{R}^3 : \langle x, \mathsf{N}_0\rangle = 0\}$ and since $\mathcal{D}$ lies in some plane, we have that $\mathcal{D}$ also lies in the plane P and $\mathsf{N}(p) \equiv \mathsf{N}_0$ on $\mathcal{D}$.

Since $\partial\Omega = \mathcal{M}$ is a g_{B}-minimal surface, the domain $\overline{\Omega}$ is weakly mean convex with respect to g_{B}. Hence we can argue just as above to show that there exists a g_{B}-minimal surface $\mathcal{D}' \subset \Omega$ such that $\partial\mathcal{D}' = \overline{\mathcal{D}}'\backslash\mathcal{D}' = \Gamma$ and such that $\mathrm{II} \equiv 0$ and $\langle p, \mathsf{N}(p)\rangle \equiv 0$ on $\mathcal{D}'$. Moreover, as above, $\mathcal{D}'$ lies in the plane P.

From combining all of the above, we obtain the following. The plane P is the disjoint union of the loop Γ, the surface $\mathcal{D}$, and the disk $\mathcal{D}'$, where $\mathcal{D} \subset \mho$ and $\mathcal{D}' \subset \Omega$. This implies that $\mathcal{Z} = \Gamma$, yielding a contradiction which proves the theorem. □

More generally, Brendle proved the following (see [**108**, Theorem 2]).

Theorem 13.27. *Let $\mathcal{M}^2$ be a properly embedded shrinker in $\mathbb{R}^3$. Suppose that the intersection number of any two loops in $\mathcal{M}^2$ is zero modulo 2. Then $\mathcal{M}^2$ is either a round sphere, a standard cylinder, or a flat plane.*

13.4.2. Compact embedded shrinkers of higher genus. For shrinkers of higher genus, the classification problem appears to be quite difficult (compare the situation with that of minimal surfaces in the 3-sphere). The following result shows that they are, at least, "topologically standard".

Theorem 13.28 (A. Mramor and S. Wang [**408**]). *Every compact, embedded, genus g shrinker in $\mathbb{R}^3$ is ambiently isotopic to a standard embedded genus g surface.*

According to Mramor and Wang, the inspiration for their result is a classical result of Lawson [**357**] (cf. [**227**]) which states that closed minimal surfaces in S^3 (with the round metric) are topologically standard. One reason to expect that shrinkers in $\mathbb{R}^{n+1}$ might share properties with minimal surfaces in S^{n+1} is that shrinkers are minimal with respect to the Gaußian metric $G \doteqdot \mathrm{e}^{-f}\delta$, $f(X) \doteqdot \frac{|X|^2}{4}$, which is f-Ricci positive in the sense of Bakry–Emery.

By the Jordan–Brouwer theorem, a compact, embedded shrinker $\mathcal{M}^2$ separates $\mathbb{R}^3$ into two connected components, a bounded "interior", Ω, and an unbounded "exterior", $\mathbb{R}^3 \setminus \overline{\Omega}$. By an observation of Lawson [**357**], $\mathcal{M}^2$ is a **Heegaard surface**[2] if the universal covers of both components have connected boundaries. On the other hand, by Waldhausen's theorem [**515**], any two Heegaard surfaces in $\mathbb{R}^3$ are ambiently isotopic.[3] Since the standard genus g surface in $\mathbb{R}^3$ is a Heegaard surface, it therefore suffices to prove that the boundaries of the universal covers of Ω and $\mathbb{R}^3 \setminus \overline{\Omega}$ are connected.

Proposition 13.29. *Let $\mathcal{M}^2$ be a compact, embedded shrinker in $\mathbb{R}^3$. Denote by $\Omega \subset \mathbb{R}^3$ the bounded component of $\mathbb{R}^3 \setminus \mathcal{M}^2$. Then the boundaries of the universal covers of $\overline{\Omega}$ and $\mathbb{R}^3 \setminus \Omega$ are connected.*

Proof sketch. Suppose that the claim does not hold. Then the universal cover E of one of the components of $\mathbb{R}^3 \setminus \mathcal{M}^2$ has at least two boundary components. Since shrinkers are minimal surfaces with respect to the Gaußian metric, each boundary component of E is minimal with respect to the lifted Gaußian metric. We claim that the interior of E contains a properly embedded, orientable, least-area minimal surface (with respect to the Gaußian metric). The idea is as follows (for the complete details, see [**408**]): Since E is path connected, we can connect the two boundary components by a simple curve, γ. Let $\{K_i\}_{i\in\mathbb{N}}$ be a compact exhaustion of one of the boundary components of E such that K_1 contains one of the endpoints of γ. The solution Σ_i to the Plateau problem for ∂K_i intersects γ for each i. Since each surface Σ_i is area minimizing, we can find a subsequential limit Gaußian

[2]That is, it divides S^3 (the 1-point compactification of $\mathbb{R}^3$) into two handlebodies.

[3]See Rieck [**443**] and Schleimer [**460**] and the references therein for additional proofs of Waldhausen's theorem.

minimal surface, which is nonempty since $\Sigma_i \cap \gamma \neq 0$ for each i. By the second variation formula and the strong maximum principle this limit will in fact be disjoint from the boundary of E as claimed. This Gaußian minimal surface projects to a shrinker in $\mathbb{R}^3$ which is disjoint from $\mathcal{M}^2$. But this violates the avoidance principle since the corresponding shrinking solutions to mean curvature flow both reach the origin at the same time. □

13.5. Translators — an introduction

Recall that a **translating self-similar** solution to mean curvature flow is one whose image evolves purely by translation. In that case, the time slices are all congruent and satisfy

$$H = -\langle \mathsf{N}, e\rangle\,, \tag{13.63}$$

where $e \in \mathbb{R}^{n+1}$. Conversely, if a hypersurface satisfies (13.63), then the 1-parameter family of translates with velocity e satisfies mean curvature flow. A hypersurface satisfying (13.63) is called a (**mean curvature flow**) **translator** or **self-translator**.

When e is the zero vector, any minimal hypersurface is a solution. We shall consider only the case that e is nonzero, in which case we may eliminate the scaling invariance and isotropy of (13.63) by restricting attention to translating solutions which move with unit speed in the "upwards" direction. That is, *we henceforth assume that* $e = e_{n+1}$.

By Exercise 13.9 below, *mean convex* embedded translators are graphs.

Lemma 13.30. *A properly embedded translator satisfying $H > 0$ is necessarily the graph $\{x + u(x)e_{n+1} : x \in \Omega\}$ of a function $u : \Omega \to \mathbb{R}$, over a domain $\Omega \subset \mathbb{R}^n \times \{0\}$, satisfying*

$$\sqrt{1+|Du|^2}\,\mathrm{div}\left(\frac{Du}{\sqrt{1+|Du|^2}}\right) = 1\,. \tag{13.64}$$

This facilitates the study of embedded, mean convex translators using tools from PDE. We call a properly embedded, mean convex translator **entire** if the corresponding function u is an entire function, that is, if the projection onto $\mathbb{R}^n \times \{0\}$ is surjective.

Note that the avoidance principle implies (by comparing with a shrinking sphere) that there can be no compact translators.

We have already encountered the following examples.

Example 13.31.

- The **vertical hyperplane** $\Pi^n(e) \doteqdot \{X \in \mathbb{R}^{n+1} : \langle X, e\rangle = 0\}$, $e \in \{e_{n+1}\}^{\perp}$.

- The Grim Reaper curve,

$$\Gamma^1 \doteqdot \left\{(x, -\log\cos x) : |x| < \tfrac{\pi}{2}\right\}.$$

- The **Grim hyperplanes** (a.k.a. **Grim Reaper cylinders**)

$$\Gamma^n \doteqdot \left\{(x_1, \ldots, x_n, -\log\cos x_1) : |x_1| < \tfrac{\pi}{2}\right\}.$$

  Further Grim hyperplanes are obtained by rotating Γ^n about the x_{n+1}-axis.

- More generally, if Σ^{n-k} is a translator in $\mathbb{R}^{n-k+1}$, then $\mathbb{R}^k \times \Sigma^{n-k}$ is a translator in $\mathbb{R}^{n+1}$. Further examples are obtained by rotating about the x_{n+1}-axis.

Either by computing directly or by using the fact that $\Sigma^n_t \doteqdot \Sigma^n + te_{n+1}$ is (up to a time-dependent reparametrization) a translating solution to mean curvature flow, we obtain the following identities on a translator Σ^n, some of which are derived in Section 10.2.

Lemma 13.32. *Let Σ^n be a translator in $\mathbb{R}^{n+1}$. Denote by $h : \Sigma \to \mathbb{R}$ the **height function** of Σ^n, defined by*

$$h(X) \doteqdot \langle X, e_{n+1}\rangle\,, \tag{13.65}$$

and set $V \doteqdot \nabla h$. Then

$$V = e_{n+1}^{\top}\,, \tag{13.66}$$

$$-\nabla H = \mathrm{L}(V)\,, \tag{13.67}$$

$$-\nabla V = H\,\mathrm{II}\,, \tag{13.68}$$

$$-(\nabla_V + \Delta)\,\mathrm{II} = |\mathrm{II}|^2\,\mathrm{II}, \quad \textit{and} \tag{13.69}$$

$$-(\nabla_V + \Delta)H = |\mathrm{II}|^2 H\,. \tag{13.70}$$

Proof. The identity (13.66) follows immediately from the definition of V. The next two identities, (13.67) and (13.68), are proved in Lemma 10.12. The ultimate identity (13.70) is the trace of the penultimate one (13.69), which can be obtained either by differentiating (13.67) and applying the Codazzi identity (13.68) and the Simons identity (5.50) or by applying the fact that

$$X(\varphi(x,t),t) + te = X(x,0)$$

and hence

$$\mathrm{II}_{(\varphi(x,t),t)} = \mathrm{II}_{(x,0)}\,,$$

where φ is the flow of V, to the evolution equation for II under mean curvature flow (6.17). □

13.6. The Dirichlet problem for graphical translators

By well-known methods, the Dirichlet problem for the translator equation can be solved over a convex domain so long as a lower barrier exists.

Proposition 13.33. *Let $\Omega \subset \mathbb{R}^n$ be an open, bounded, convex set. Suppose there exists a subsolution $\underline{u} \in C^2(\Omega) \cap C^0(\overline{\Omega})$ to the translator equation (13.64) with $\underline{u} \equiv 0$ on $\partial\Omega$. Then there exists a solution $u \in C^\infty(\Omega) \cap C^0(\overline{\Omega})$ to the Dirichlet problem*

$$\begin{cases} \operatorname{div}\left(\dfrac{Du}{\sqrt{1+|Du|^2}}\right) = \dfrac{1}{\sqrt{1+|Du|^2}} & \text{in } \Omega, \\ u \equiv 0 & \text{on } \partial\Omega . \end{cases} \tag{13.71}$$

It is the unique solution in $C^2(\Omega) \cap C^0(\overline{\Omega})$.

Proof. Since the zero function is an upper barrier, well-known methods for quasilinear elliptic equations (see [**245**, Chapters 13–15]) yield the claims. Briefly, the upper and lower barriers immediately provide an a priori supremum estimate and an a priori boundary gradient estimate. The maximum principle can then be applied to obtain an a priori interior gradient estimate. At this point, the de Giorgi–Nash–Moser theorem and the Schauder theorem yield a priori Hölder estimates for the derivatives of any solution. The desired smooth solution can then be obtained using the method of continuity. Finally, uniqueness is deduced by applying the maximum principle to the linear equation satisfied by the difference between two given solutions. □

Figure 13.3. Ling Xiao.

The following estimates provide compactness criteria for the space of translators. They are proved by geometric measure-theoretical methods, in particular, Allard's regularity theorems [**15, 16**] (see also [**95**]). In low dimensions, analogous (interior) estimates can be obtained by exploiting the stability inequality for graphical translators following R. Schoen [**465**] (see L. Shahriyari [**479**] and Spruck and Xiao [**494**]).

Theorem 13.34. *Given $K > 0$ and $\ell_0 \in \mathbb{N}$ there exists $C < \infty$ with the following property: Let $u : \Omega \to \mathbb{R}$ be a solution to the Dirichlet problem*

$$\begin{cases} \operatorname{div}\left(\dfrac{Du}{\sqrt{1+|Du|^2}}\right) = \dfrac{1}{\sqrt{1+|Du|^2}} & \text{in } \Omega, \\ u = \psi & \text{on } \partial\Omega, \end{cases} \tag{13.72}$$

for some bounded open set $\Omega \subset \mathbb{R}^n$ and smooth function $\psi : \overline{\Omega} \to \mathbb{R}$. Suppose either that $n \leq 6$ or that graph u *is rotationally symmetric with respect to the subspace* $\operatorname{span}\{e_2, \dots, e_n\}$. *If, for some $\alpha \in (0,1)$, $\partial\Omega$ and $\psi|_{\partial\Omega}$ are bounded in $C^{\ell_0+2,\alpha}$ by K, then*

$$\sup_{\operatorname{graph} u} |\nabla^\ell \mathrm{II}| \leq C \tag{13.73}$$

for each $\ell = 0, \dots, \ell_0$.

Proof sketch. Since the mean curvature of a translator is always bounded by 1, classical methods of geometric measure theory can be applied. (Roughly speaking, this is because tangent cones are *minimal* hypersurfaces.) The result follows by standard arguments when $n \leq 6$. In higher dimensions, singular tangent cones are ruled out by the rotational symmetry hypothesis. See [**98**] for the complete details. □

For convex domains, interior estimates follow immediately from the convexity estimate of Spruck and Sun (Theorem 13.41 below).

13.7. Cylindrical translators

In addition to the rather trivial product Grim hyperplane examples, there is also a family of "oblique" Grim hyperplanes Γ^n_θ parametrized by $(\theta, \phi) \in [0, \frac{\pi}{2})$. These are obtained by rotating the "standard" Grim plane Γ^n through the angle $\theta \in [0, \frac{\pi}{2})$ in the plane $\operatorname{span}\{e_n, e_{n+1}\}$ and then scaling by the factor $\sec\theta$. To see that the result is indeed a translator, we need only check that

$$\begin{aligned} -H_\theta = -\cos\theta H &= \cos\theta \, \langle \mathsf{N}, e_{n+1} \rangle \\ &= \langle \cos\theta \mathsf{N} + \sin\theta e_n, e_{n+1} \rangle = \langle \mathsf{N}_\theta, e_{n+1} \rangle \,, \end{aligned}$$

where H_θ and N_θ are the mean curvature and outward unit normal to Γ^n_θ, respectively.

More generally, if Σ^{n-k}, $k \geq 1$, is a translator in $\mathbb{R}^{n-k+1}$, then the hypersurface $\Sigma^n_{\theta,\phi}$ obtained by rotating $\mathbb{R}^k \times \Sigma^{n-k}$ counterclockwise through angle θ in the plane $\phi \wedge e_{n+1}$ and then scaling by $\sec\theta$ is a translator in $\mathbb{R}^{n+1}$ so long as $\phi \in \operatorname{span}\{e_1, \ldots e_k\} \setminus \{0\}$.

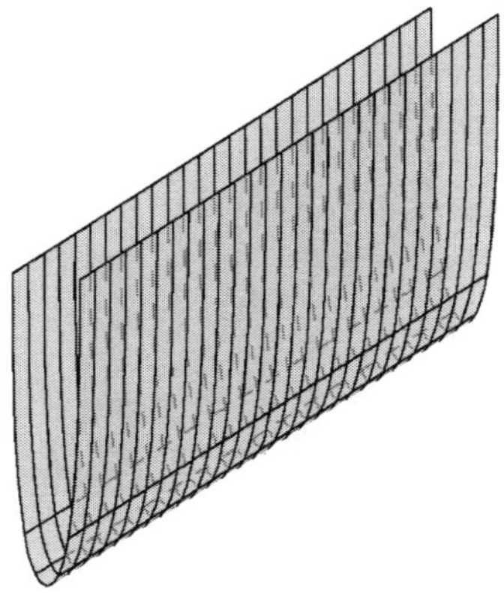

Figure 13.4. An oblique Grim plane ($\theta = \pi/6$). The translation direction is vertical.

Another interesting family, which generalizes the vertical hyperplanes, is the **vertical minimal cylinders**. These are products $\Sigma^n = \Sigma^{n-1} \times \mathbb{R}$, where Σ^{n-1} is a minimal hypersurface of $\mathbb{R}^n$. It turns out that these are the only translators of constant mean curvature.

Theorem 13.35 (Constant mean curvature translators)**.** *Let Σ^n be a translator in $\mathbb{R}^{n+1}$. Suppose that Σ^n has constant mean curvature H_0. Then $H_0 = 0$ and Σ^n is a vertical minimal cylinder. If $n = 2$, then Σ^2 is a vertical plane. More generally, if Σ^n has at most two distinct principal curvatures at each point, then Σ^n is a vertical hyperplane.*

Proof. By (13.70),

$$\langle \mathsf{N}, e_{n+1} \rangle = -H \equiv 0\,,$$

so e_{n+1} is tangential and hence $V \equiv e_{n+1}$. It follows that the integral curves of V are vertical lines. The claim follows. □

13.8. Rotational translators

Another interesting class of examples is the translators of revolution. Note that the axis of revolution is necessarily parallel to the e_{n+1}-axis, unless the translator is a vertical hyperplane (see Exercise 13.3). Parametrizing the solution (locally) as the graph $\{(x, u(|x|)) : r_1 \leq |x| \leq r_2\}$ of a function $u : (r_1, r_2) \to \mathbb{R}$, $0 \leq r_1 < r_2$, the translator equation becomes

$$\frac{u''}{1 + (u')^2} + (n-1)\frac{u'}{r} = 1\,. \tag{13.74}$$

The following result of Clutterbuck, Schnürer, and Schulze yields solutions away from the rotation axis and provides asymptotics.

Proposition 13.36 (Clutterbuck, Schnürer, and Schulze [**176**]). *Given $R > 0$ and $a, b \in \mathbb{R}$, there exists a unique $u \in C^\infty([R,\infty))$ satisfying*

$$\begin{cases} \dfrac{u''}{1+(u')^2} + (n-1)\dfrac{u'}{r} = 1 & \textit{in } [R,\infty), \\ u(R) = a, \quad u'(R) = b\,. \end{cases}$$

Moreover, there exists $A \in \mathbb{R}$ such that

$$u(r) = \frac{r^2}{2(n-1)} - \log r + A + O(r^{-1}) \quad \textit{as } r \to \infty\,. \tag{13.75}$$

Proof. The derivative, $v \doteqdot u'$, satisfies the problem

$$\begin{cases} \dfrac{v'}{1+v^2} + (n-1)\dfrac{v}{r} = 1 & \text{in } [R,\infty), \\ v(R) = b\,. \end{cases}$$

We first show that a solution exists for all $r \geq R$. By the Picard–Lindelöf theorem, the problem admits a unique solution in the interval $[R, R+\delta)$, where $\delta > 0$. If $1-(n-1)\frac{v(R)}{R}$ is negative, then $v'(R) < 0$, so v is decreasing. It follows that $1-(n-1)\frac{v}{r}$ eventually becomes positive and remains so. That is, $1-(n-1)\frac{v(r)}{r}$ is positive for all $r \geq r_0$, say. But then $v' > 0$ for $r > r_0$, which implies that v is also bounded from below. We conclude that v cannot become infinite for finite r, so the solution exists for all $r \geq R$ (by the Picard–Lindelöf theorem).

Next, we study the asymptotic expansion. We claim that, for every $\varepsilon > 0$ and $r_0 > R$, there exists $r_1 > r_0$ such that

$$v(r_1) \geq \frac{r_1}{n-1}(2-\varepsilon)\,.$$

Indeed, if this were not the case, then we could find $\varepsilon > 0$ and $r_0 > 0$ such that $v' \geq \varepsilon(1+v^2)$ for all $r \geq r_0$. But this is impossible since v remains finite for all $r \geq R$. Observe that the function $\zeta(r) \doteqdot (1-\varepsilon)r$ satisfies

$$\zeta' \leq (1+\zeta^2)\left(1-(n-1)\frac{\zeta}{r}\right)$$

for r sufficiently large. Thus, choosing r_1 larger if necessary,

$$(1-\varepsilon)\frac{r}{n-1} \leq v(r) \leq \frac{r}{n-1}$$

for all $r > r_1$. Since ε was arbitrary, we conclude that

$$v(r) = \frac{r}{n-1} + o(r)$$

as $r \to \infty$. Set $w \doteqdot v - \frac{r}{n-1}$. Then

$$w' = -(n-1)\frac{w}{r}\left(1 + \left[w + \frac{r}{n-1}\right]^2\right) - \frac{1}{n-1}.$$

We have shown that w is nonpositive for r sufficiently large and that $w = o(r)$ as $r \to \infty$. We claim that, in fact, $w = o(1)$ as $r \to \infty$. Fix $\varepsilon > 0$. Choose $r_2 \geq 10/\varepsilon$ such that $w(r) \geq -\frac{r}{2(n-1)}$. If $w(r) \leq -\varepsilon$ for $r > r_2$, then

$$w' \geq \frac{(n-1)\varepsilon}{r}\left(1 + \frac{r^2}{4(n-1)^2}\right) - \frac{1}{n-1} > \frac{1}{n-1} > 0\,.$$

We conclude that $w \geq -\varepsilon$ for all $r \geq r_2$. Since ε was arbitrary, we conclude that $w = o(1)$.

Now set $\lambda(r) \doteqdot rw(r)$. We claim that $\lambda(r) \to -1$ as $r \to -\infty$. We have shown that, for any $\mu > 0$, there exists $r_3 > 0$ such that $|\lambda(r)| \leq \mu r$ for all $r \geq r_3$. Fix $\varepsilon > 0$ and assume that $\lambda(r) \geq -1 + \varepsilon$ for some $r \geq r_3$. Then

$$\begin{aligned} \lambda'(r) &= -\frac{r}{n-1}\left(1 + \lambda + 2(n-1)\frac{\lambda^2}{r^2}\right) - (n-2)\frac{\lambda}{r} - (n-1)\frac{\lambda^3}{r^3} \\ &\leq -\frac{r}{n-1}(1+\lambda) + (n-2)\mu + (n-1)\mu^3. \end{aligned}$$

If we require $r_3 \geq \frac{n-1}{\varepsilon} + (n-2)\mu + (n-1)\mu^3 + 1$, then

$$\lambda'(r) \leq -1 < 0\,.$$

We conclude that $\lambda(r) \leq \varepsilon - 1$ for $r \geq r_3$. A similar argument shows that $\lambda(r) \geq -1 - \varepsilon$ for r sufficiently large; we conclude that $\lambda(r) \to -1$ as $r \to \infty$.

Now set $\eta(r) \doteqdot (1 + \lambda(r))r^2$. We claim that $\eta(r) \to (n-1)(n-4)$ as $r \to \infty$. We have shown that, for any $\mu > 0$, there exists $r_4 > 0$ such that $|\eta(r)| \leq \mu r^2$ for all $r \geq r_4$. Fix $\varepsilon > 0$ and suppose that $\eta(r) > (n-1)(n-4) + \varepsilon$ for some $r \geq r_4$. Then

$$\begin{aligned} \eta'(r) &= \frac{r}{n-1}\left((n-1)(n-4) - \eta\right) + \frac{1}{r}\left((n-1) - (n-8)\eta\right) \\ &\quad - \frac{\eta}{r^3}\left(2\eta + 3(n-1)\right) + 3(n-1)\frac{\eta^2}{r^5} - (n-1)\frac{\eta^3}{r^7} \\ &\leq \frac{r}{n-1}\left((n-1)(n-4) - \eta\right) + \frac{n-1}{r} + |n-8|\mu r \\ &\quad - 3(n-1)\frac{\eta}{r^3} + 3(n-1)\frac{\eta^2}{r^5} - (n-1)\frac{\eta^3}{r^7} \\ &\leq \frac{r}{n-1}\left(|n-8|(n-1)\mu - \varepsilon\right) + O(1) \end{aligned}$$

as $r \to \infty$. So r_4 can be chosen sufficiently large, and μ sufficiently small, that $\eta'(r) \leq -1$, say, for any $r > r_4$ satisfying $\eta(r) > (n-4)(n-1) + \varepsilon$. It follows that $\eta(r) \leq (n-4)(n-1) + \varepsilon$ for r sufficiently large and hence $\limsup_{r\to\infty}\eta(r) \le (n-1)(n-4)$. A similar calculation shows that $\liminf_{r\to\infty}\eta(r) \ge (n-1)(n-4)$ and we conclude that $\lim_{r\to\infty}\eta(r) = (n-1)(n-4)$. This completes the proof. □

Corollary 13.37 (Translating catenoids [**176**]). *For every $R>0$ there exists a properly embedded translator of revolution C_R in $\mathbb{R}^{n+1}$, $n\ge 2$, which is the union of the graphs $C_R^\pm \doteqdot \{(x, c_R^\pm(|x|)) : x\in\mathbb{R}^n\setminus B_R\}$ of two functions $c_R^\pm \in C^\infty((R,\infty))\cap C^0([R,\infty))$ satisfying* (13.74) *with*

$$c_R^\pm(R) = 0 \quad and \quad \lim_{r\searrow R}(c_R^\pm)'(r) = \pm\infty$$

and

$$c_R^\pm(r) = \frac{r^2}{2(n-1)} - \log r + A_R^\pm + O(r^{-1}) \quad as \quad r\to\infty\,.$$

Proof. We will obtain the solution locally near $r=R$ by writing the profile curve as a graph over the x_{n+1}-axis and then apply Proposition 13.36 to extend this local solution indefinitely. So consider a hypersurface of revolution

$$\Sigma = \{(x_1,\dots,x_{n+1})\in\mathbb{R}^n\times(-a,a) : x_1^2+\dots+x_n^2 = h^2(x_{n+1})\}\,,$$

where $h\in C^\infty((-a,a))$. That is, Σ is the zero set of the function

$$F(x) \doteqdot x_1^2+\dots+x_n^2 - h^2(x_{n+1})\,.$$

A unit normal field for Σ is thus given by the restriction to Σ of the vector field

$$\mathsf{N} \doteqdot \frac{DF}{|DF|} = \left(\frac{x_i}{h\sqrt{1+(h')^2}}, -\frac{h'}{\sqrt{1+(h')^2}}\right).$$

The second fundamental form is then the restriction to $T\Sigma$ of the bilinear form

$$D\mathsf{N} = \begin{pmatrix} \frac{\delta_{ij}}{h\sqrt{1+(h')^2}} & 0 \\ -x_i\left(\frac{h'}{h^2\sqrt{1+(h')^2}} + \frac{h'h''}{h(1+(h')^2)^{\frac32}}\right) & -\frac{h''}{(1+(h')^2)^{\frac32}} \end{pmatrix}.$$

Since

$$D\mathsf{N}(\mathsf{N},\mathsf{N}) = \frac{1}{h\sqrt{1+(h')^2}}\,,$$

the mean curvature is given by the restriction to Σ of the function

$$\begin{aligned} H = \operatorname{div}_\Sigma \mathsf{N} &= \operatorname{tr}_{\mathbb{R}^{n+1}} D\mathsf{N} - D\mathsf{N}(\mathsf{N},\mathsf{N}) \\ &= \frac{n-1}{h\sqrt{1+(h')^2}} - \frac{h''}{(1+(h')^2)^{\frac32}}\,. \end{aligned}$$

On the other hand,

$$-\langle \mathsf{N}, e_{n+1}\rangle = \frac{h'}{\sqrt{1+(h')^2}}\,.$$

We conclude that Σ satisfies the translator equation if and only if

$$\frac{h''}{1+(h')^2} = \frac{n-1}{h} - h'\,.$$

By the Picard–Lindelöf theorem, the boundary value problem

$$\begin{cases} \dfrac{h''(y)}{1+(h'(y))^2} - \dfrac{n-1}{h(y)} + h'(y) = 0\,, & y \in (-y_0, y_0)\,, \\ h(0) = R\,, \;\; h'(0) = 0 \end{cases}$$

admits a unique solution, so long as y_0 is sufficiently small. Since

$$h''(0) = \frac{n-1}{R} > 0\,,$$

we can arrange, by choosing y_0 smaller if necessary, that h is strictly convex. Returning to the original coordinate system and applying Proposition 13.36 then yields the claims. □

Since (13.74) is singular at $r = 0$, obtaining a solution up to the axis of rotation requires a different approach. This was achieved by Altschuler and Wu [**22**] by solving a sequence of prescribed contact angle problems using a continuity argument. In the proof presented here, we solve the Dirichlet problem for the graphical translator equation near the axis and then extend the solution uniquely using Proposition 13.36.

Theorem 13.38 (The bowl soliton [**22**]). *For each $n \in \mathbb{N}\setminus\{1\}$ there exists a rotationally symmetric solution $u \in C^\infty(\mathbb{R}^n)$ to* (13.64)*. This solution is convex and satisfies*

$$u(x) = \frac{|x|^2}{2(n-1)} - \log|x| + O\left(|x|^{-1}\right) \quad \textit{as} \quad |x| \to \infty\,.$$

Modulo translations, it is the only rotationally symmetric proper solution to (13.64)*.*

Proof. Consider the Dirichlet problem

$$\begin{cases} \operatorname{div}\left(\dfrac{Du}{\sqrt{1+|Du|^2}}\right) = \dfrac{1}{\sqrt{1+|Du|^2}} & \text{in } B_1\,, \\ u = 0 & \text{on } \partial B_1\,. \end{cases}$$

Since the function

$$\underline{u}(x) \doteqdot \frac{|x|^2 - 1}{2(n-1)}$$

is a subsolution, there exists, by Proposition 13.33, a unique solution $u \in C^\infty(B_1) \cap C^0(\overline{B}_1)$. By uniqueness of solutions, u must be rotationally symmetric (otherwise, rotating it about the x_{n+1}-axis would yield a new solution with the same boundary values). Thus, $u(x) = f(|x|)$ for some $f \in C^\infty([0,1)) \cap C^0([0,1])$. Note that $f'(0) = 0$. Taking $r \to 0$ in (13.74) therefore yields

$$f''(0) = \frac{1}{n} > 0\,.$$

Applying Proposition 13.36 with R sufficiently small, we conclude that f extends uniquely to a convex function on $[0,\infty)$ satisfying the claimed asymptotics. □

The unique mean convex, rotationally symmetric translator constructed in Theorem 13.38 is called the **bowl soliton** or, sometimes, the **translating paraboloid**.

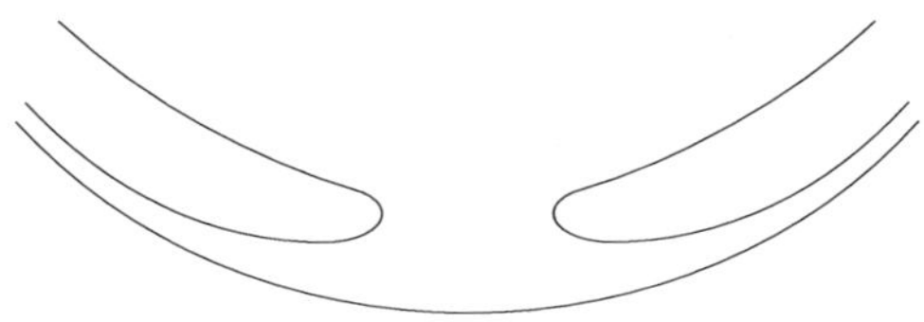

Figure 13.5. The profile curve of the translating catenoid lying above the profile curve of the bowl soliton [**176**].

B. White [**531**, Conjecture 2] conjectured that the Grim hyperplane and the products $\{\mathbb{R}^k \times \mathrm{Bowl}^{n-k}_t\}_{t\in(-\infty,\infty)}$, where $\{\mathrm{Bowl}^{n-k}_t\}_{t\in(-\infty,\infty)}$ is the $(n-k)$-dimensional bowl soliton, are (modulo translations and rotations about the x_{n+1}-axis) the only convex translating solutions to mean curvature flow. Wang has proved that the bowl is the only entire example in $\mathbb{R}^3$ (see Corollary 13.48 below).

On the other hand, Wang constructed convex, nonrotationally symmetric examples in $\mathbb{R}^{n+1}$ for each $n \geq 2$, including entire examples when $n \geq 3$ [**521**]. (We will provide, in Theorem 13.53 below, an explicit construction of a 1-parameter family of examples interpolating between the Grim hyperplane and the bowl.) So White's conjecture turns out to be false. On the other hand, White remarks [**531**, p. 133] that, even if the conjecture is false, it may still hold for limit flows of mean convex mean curvature flow (a conjecture made also by Wang [**521**, p. 1237]). Since a noncollapsing translator is necessarily entire, Corollary 13.48 confirms the conjecture for embedded flows when $n = 2$. It remains unclear whether or not Wang's entire examples arise as singularity models in higher dimensions. Haslhofer [**273**] (see also [**96**]) ruled this out for 2-convex mean curvature flows (cf. Corollary 13.50 below).

13.9. The convexity estimates of Spruck, Sun, and Xiao

Joel Spruck and Ling Xiao proved that mean convex proper translators in $\mathbb{R}^3$ are necessarily convex [**494**]. The result was extended, using the same basic idea, to mean convex translators in $\mathbb{R}^{n+1}$ with at most two principal curvatures in [**98**] and to uniformly 2-convex translators by Spruck and Liming Sun [**493**].

Figure 13.6. Joel Spruck.

Theorem 13.39. *Let $\Sigma^n \subset \mathbb{R}^{n+1}$ be a strictly mean convex, proper translator. Suppose that Σ^n has at most two distinct principal curvatures at each point. Then Σ^n is convex.*

Proof. Denote the principal curvatures by $\kappa \leq \mu$. We need to show that the (open) set $U \doteqdot \{X \in \Sigma : \kappa(X) < 0\}$ is empty. Since $H > 0$, we have $\kappa < 0 < \mu$ in U. It follows that κ is smooth and has constant multiplicity, $m \in \{1, \ldots, n-1\}$ say in U. Recall that

$$-(\nabla_V + \Delta)\,\mathrm{II} = |\mathrm{II}|^2\,\mathrm{II}\,,$$

where $V \doteqdot e_{n+1}^\top$ is the tangential part of e_{n+1}. Computing locally in a principal frame $\{\tau_1, \ldots, \tau_n\}$ with $\kappa_i = \mathrm{II}_{ii} = \kappa$ when $i \leq m$ and $\kappa_i = \mathrm{II}_{ii} = \mu$ when $i \geq m+1$, we obtain by (12.29) that

$$-(\nabla_V + \Delta)\kappa \geq |\mathrm{II}|^2\kappa + 2\sum_{\ell=1}^{n}\sum_{p=m+1}^{n}\frac{(\nabla_\ell \mathrm{II}_{1p})^2}{\mu - \kappa} \quad \text{in} \quad U\,. \tag{13.76}$$

The inequality above holds in the classical sense in U since κ is smooth there. Since the mean curvature satisfies

$$-(\nabla_V + \Delta)H = |\mathrm{II}|^2 H\,,$$

straightforward manipulations then yield

$$-(\nabla_V + \Delta)\frac{\kappa}{\mu} = -(\nabla_V + \Delta)\frac{(n-m)\kappa}{H - m\kappa} \geq \frac{2}{n-m}\frac{H}{\mu^2}\sum_{\ell=1}^{n}\sum_{p=m+1}^{n}\frac{(\nabla_\ell \mathrm{II}_{1p})^2}{\mu - \kappa} + 2\left\langle \nabla\frac{\kappa}{\mu}, \frac{\nabla\mu}{\mu}\right\rangle . \tag{13.77}$$

Suppose that

$$-\varepsilon_0 \doteqdot \inf_{\Sigma}\frac{\kappa}{\mu} < 0\,.$$

(1) If the infimum is attained at some point $X_0 \in \Sigma$, then $\kappa(X_0) < 0$ and the strong maximum principle applied to (13.77) yields $\frac{\kappa}{\mu} \equiv -\varepsilon_0 < 0$. In particular,

$$0 \equiv \nabla_\ell \frac{\kappa}{\mu} = \frac{\nabla_\ell \mathrm{II}_{pp}}{\mu} - \frac{\kappa}{\mu}\frac{\nabla_\ell \mathrm{II}_{qq}}{\mu}$$

when $p \leq m < q$. Since $\mathrm{II}_{ij} \equiv 0$ for $i \neq j$ with respect to our local principal frame, we have whenever $\kappa_i = \kappa_j$ and $i \neq j$

$$0 = \tau_\ell \mathrm{II}_{ij} = \nabla_\ell \mathrm{II}_{ij} + (\kappa_j - \kappa_i)\Gamma_{\ell ij} = \nabla_\ell \mathrm{II}_{ij} \tag{13.78}$$

for each ℓ, where $\Gamma_{\ell ij} \doteqdot \langle \nabla_\ell \tau_i, \tau_j\rangle$. Thus,[4]

$$0 = \nabla_\ell \mathrm{II}_{11} \text{ when } \ell = 2, \ldots, m$$

and

$$0 = \nabla_\ell \mathrm{II}_{nn} \text{ when } \ell = m+1, \ldots, n-1\,.$$

Recalling (13.77), we also find that

$$0 \equiv \sum_{\ell=1}^{n}\sum_{p=m+1}^{n}(\nabla_\ell \mathrm{II}_{1p})^2\,.$$

It follows that the components $\nabla_1 \mathrm{II}_{nn}$, $\nabla_1 \mathrm{II}_{11}$, $\nabla_n \mathrm{II}_{11}$, and $\nabla_n \mathrm{II}_{nn}$ are all identically zero and hence, by the translator equation (13.63),

$$0 \equiv m\nabla_\ell \mathrm{II}_{11} + (n-m)\nabla_\ell \mathrm{II}_{nn} = \nabla_\ell H$$

for each $\ell = 1, \ldots, n$. It follows from Theorem 13.35 that Σ is a vertical hyperplane, contradicting strict mean convexity.

(2) Suppose then that the infimum is not attained. Since $\frac{\kappa}{\mu} \geq -\frac{n-m}{m}$ and the sectional curvatures of Σ are bounded, the Omori–Yau maximum principle (Theorem 1.20) may be applied. This yields a sequence of points $X_i \to \infty$ such that

$$\frac{\kappa}{\mu}(X_i) \to -\varepsilon_0\,, \quad \left|\nabla\frac{\kappa}{\mu}(X_i)\right| \leq \frac{1}{i}, \quad \text{and} \quad -\Delta\frac{\kappa}{\mu}(X_i) \leq \frac{1}{i}\,. \tag{13.79}$$

[4]Here, and elsewhere, we freely make use of the Codazzi equation.

Consider the sequence of translators $\Sigma_i \doteq \Sigma - X_i$. Combining Theorem 13.34 with Theorem 11.11, we find that the sequence of translators Σ_i converges locally uniformly in C^∞, after passing to a subsequence, to a limit translator Σ_∞. Note that, whenever $\kappa < 0 < \mu$,

$$\begin{aligned} \nabla_\ell \frac{m\kappa}{\mu} &= \frac{m\nabla_\ell \mathrm{II}_{11}}{\mu} - \frac{m\kappa}{\mu^2}\nabla_\ell \mathrm{II}_{nn} \\ &= \frac{\nabla_\ell H}{\mu} - m\left(\frac{n-m}{m} + \frac{\kappa}{\mu}\right)\frac{\nabla_\ell \mathrm{II}_{nn}}{\mu}. \end{aligned} \tag{13.80}$$

We claim that

$$\left(\frac{n-m}{m} + \frac{\kappa}{\mu}(X_i)\right)\frac{\nabla_k \mathrm{II}_{nn}}{\mu}(X_i) \to 0 \;\text{ as }\; i \to \infty \tag{13.81}$$

for each $\ell = 1, \ldots, n$. Suppose that this is not the case. Then there exist $i_0 \in \mathbb{N}$ and $\delta_0 > 0$ such that

$$\left(\frac{n-m}{m} + \frac{\kappa}{\mu}(X_i)\right)\frac{|\nabla_k \mathrm{II}_{nn}|}{\mu}(X_i) > \delta_0 \tag{13.82}$$

for all $i > i_0$ and some $k \in \{1, \ldots, n\}$. By (13.79),

$$\left(\frac{\nabla_\ell \mathrm{II}_{11}}{\mu} - \frac{\kappa}{\mu}\frac{\nabla_\ell \mathrm{II}_{nn}}{\mu}\right)(X_i) \to 0$$

for each $\ell = 1, \ldots, n$ as $i \to \infty$ so that, replacing δ_0 and i_0 if necessary,

$$\left(\frac{n-m}{m} + \frac{\kappa}{\mu}(X_i)\right)\frac{|\nabla_k \mathrm{II}_{11}|}{\mu}(X_i) > \delta_0 \tag{13.83}$$

for all $i > i_0$. Moreover, by (13.78),

$$\frac{\nabla_\ell \mathrm{II}_{nn}}{\mu}(X_i) \to 0$$

as $i \to \infty$ for all $\ell = 2, \dots, n-1$. So (13.82) (and hence also (13.83)) holds with $k \in \{1, n\}$. Combining (13.79) and (13.77) we obtain, at the point X_i,

$$
\begin{aligned}
\frac{1}{i} \geq & \frac{1}{n-m}\frac{H}{\mu-\kappa}\sum_{\ell=1}^{n}\sum_{p=m+1}^{n}\frac{(\nabla_\ell \mathrm{II}_{1p})^2}{\mu^2} + \left\langle \nabla\frac{\kappa}{\mu}, \frac{\nabla\mu}{\mu}\right\rangle \\
= & \frac{1}{n-m}\frac{H}{\mu-\kappa}\left(\sum_{p=m+1}^{n}\frac{(\nabla_p\kappa)^2}{\mu^2} + (n-m)\frac{(\nabla_1\mu)^2}{\mu^2}\right) \\
& + \nabla_1\frac{\kappa}{\mu}\frac{\nabla_1\mu}{\mu} + \sum_{\ell=2}^{m}\nabla_\ell\frac{\kappa}{\mu}\frac{\nabla_\ell\mu}{\mu} + \sum_{\ell=m+1}^{n}\nabla_\ell\frac{\kappa}{\mu}\left(\frac{\nabla_\ell\kappa}{\kappa} - \frac{\mu}{\kappa}\nabla_\ell\frac{\kappa}{\mu}\right) \\
= & -\frac{\mu}{\kappa}\sum_{\ell=m+1}^{n}\left(\nabla_\ell\frac{\kappa}{\mu}\right)^2 + \sum_{\ell=2}^{m}\nabla_\ell\frac{\kappa}{\mu}\frac{\nabla_\ell\mu}{\mu} + \nabla_1\frac{\kappa}{\mu}\frac{\nabla_1\mu}{\mu} + \frac{H}{\mu-\kappa}\frac{(\nabla_1\mu)^2}{\mu^2} \\
& + \frac{\mu}{\kappa}\sum_{\ell=m+1}^{n}\nabla_\ell\frac{\kappa}{\mu}\frac{\nabla_\ell\kappa}{\mu} + \frac{1}{n-m}\frac{H}{\mu-\kappa}\sum_{\ell=m+1}^{n}\frac{(\nabla_\ell\kappa)^2}{\mu^2} \\
\geq & -\frac{\mu}{\kappa}\sum_{\ell=m+1}^{n}\left(\nabla_\ell\frac{\kappa}{\mu}\right)^2 + \sum_{\ell=2}^{m}\nabla_\ell\frac{\kappa}{\mu}\frac{\nabla_\ell\mu}{\mu} \\
& + \left(m\frac{\frac{n-m}{m}+\frac{\kappa}{\mu}}{1-\frac{\kappa}{\mu}}\frac{|\nabla_1\mu|}{\mu} - \left|\nabla_1\frac{\kappa}{\mu}\right|\right)\frac{|\nabla_1\mu|}{\mu} \\
& + \sum_{\ell=m+1}^{n}\left(\frac{m}{n-m}\frac{\frac{n-m}{m}+\frac{\kappa}{\mu}}{1-\frac{\kappa}{\mu}}\frac{|\nabla_\ell\kappa|}{\mu} + \frac{\mu}{\kappa}\left|\nabla_\ell\frac{\kappa}{\mu}\right|\right)\frac{|\nabla_\ell\kappa|}{\mu}.
\end{aligned}
$$

Suppose that $k = 1$ in (13.82). If

$$\left(\frac{n-m}{m} + \frac{\kappa}{\mu}(X_i)\right)\frac{|\nabla_n\kappa|}{\mu}(X_i) \not\to 0 \text{ as } i \to \infty,$$

then, taking $i \to \infty$, we find $\frac{|\nabla_1\mu|}{\mu}(X_i) \to 0$ as $i \to \infty$, contradicting (13.82). Otherwise, $\frac{|\nabla_n\kappa|}{\mu}(X_i) \leq \frac{|\nabla_1\mu|}{\mu}(X_i)$ for i sufficiently large and we again obtain $\frac{|\nabla_1\mu|}{\mu}(X_i) \to 0$ as $i \to \infty$, contradicting (13.82). If $k = n$ in (13.82), then we may argue similarly, using (13.83). So (13.81) does indeed hold.

Applying (13.79) and (13.81) to (13.80) yields

$$\frac{\nabla_\ell H}{\mu}(X_i) \to 0$$

for each $\ell = 1, \dots, n$. On the other hand, by the translator equation,

$$\frac{\nabla_\ell H}{\mu} = -\frac{\kappa_\ell\langle\tau_\ell, e_{n+1}\rangle}{\mu}.$$

Since $\frac{\kappa}{\mu}(X_i) \to -\varepsilon_0 \neq 0$, we conclude that $\mathsf{N}(X_i) \to -e_{n+1}$ and hence $H(X_i) \to 1$. Since the infimum of κ/μ is attained at the origin on Σ_∞, we deduce as before that Σ_∞ has constant mean curvature, which must be 1 since $H(X_i) \to 1$. But this contradicts Theorem 13.35. □

The proof of the convexity estimate for 2-convex translators follows the same general idea. It is made easier by the fact that uniform 2-convexity, combined with the translator equation, implies an a priori bound for $|\mathrm{II}|$, but it is complicated by the fact that, although e_1 is smooth wherever $\kappa_1 < 0$, the other principal directions can be singular wherever positive principal curvatures coincide. Spruck and Sun deal with this by exploiting an approximation to κ_1 (introduced by Aarons [**1**] and Heidusch [**281**]) with nice properties. See [**493**] for the proof of the following.

Theorem 13.40. *Let $\Sigma^n \subset \mathbb{R}^{n+1}$ be a strictly mean convex, uniformly 2-convex, proper translator. Then Σ^n is convex.*

Spruck and Sun also discovered a beautiful argument which demonstrates the convexity of solutions to the Dirichlet problem over convex domains.

Theorem 13.41. *Let $\Omega \subset \mathbb{R}^n$ be a bounded open set with smooth, locally uniformly convex boundary $\Gamma \doteqdot \partial\Omega$ and let $u : \overline{\Omega} \to \mathbb{R}$ be a solution to the graphical translator equation over Ω with $u|_\Gamma \equiv 0$. Then* graph u *is locally uniformly convex.*

Proof. Choose a ball $\Omega_0 \subset \Omega_1 \doteqdot \Omega$ and a monotone family $\{\Omega_t\}_{t\in[0,1]}$ of open subsets of Ω whose boundaries $\Gamma_t \doteqdot \partial\Omega_t$ are smooth and locally uniformly convex, and smoothly foliate $\Omega \setminus \Omega_0$. One way to achieve this is to evolve Γ by mean curvature flow until, acccording to Huisken's theorem, it is sufficiently close in the smooth topology to a small, round sphere to be smoothly, monotonically isotoped to an exactly round sphere. Let $u_t : \Omega_t \to \mathbb{R}$ be the solution[5] to the graphical translator equation with $u_t|_{\Gamma_t} \equiv 0$. Since Γ_0 is a round sphere, u_0 is a small piece of the bowl soliton. In particular, $\Sigma_t \doteqdot \operatorname{graph} u_t$ is locally uniformly convex for t sufficiently close to zero.

Denote by $t_* \in (0,1]$ the largest t such that u_t is convex and set $u_* \doteqdot u_{t_*}$. Suppose, contrary to the claim, that $t_* < 1$. Then $\Sigma_* \doteqdot \Sigma_{t_*}$ is weakly locally convex, and hence its second fundamental form II_{Σ_*} is nonnegative definite.

[5]By Proposition 13.33, it suffices to obtain a subsolution for each t. Subsolutions were constructed in terms of the distance-to-the-boundary function by F. Schulze in [**471**, Lemma 3.1]. The important special case of ellipsoidal domains is treated in Exercises 13.11 and 13.12. Continuous dependence on domain then follows from uniqueness of solutions: If $\Omega_i \to \Omega$ uniformly, then a subsequence of the solutions $u_i : \Omega_i \to \mathbb{R}$ converges uniformly in the smooth topology to a solution $u : \Omega \to \mathbb{R}$ since the a priori estimates for u_i depend only on Ω_i and the subsolutions. Since there is only one solution over Ω, we conclude that $u_i \to u$.

In fact, by the splitting theorem (Theorem 9.11) it must be positive definite. On the other hand, since u_t is nonconvex for $t > t_*$ close to t_*, the smallest principal curvature κ_1^* of Σ_* must satisfy

$$\inf_{p\in\Sigma_*} \kappa_1^*(p) = 0\,. \tag{13.84}$$

Since u_* is smooth up to the boundary Γ_* of $\Omega_* \doteqdot \Omega_{t_*}$, there exists a smooth hypersurface $\tilde{\Sigma}_*$ in which the hypersurface-with-boundary $\Sigma_* \cup \Gamma_*$ embeds. By (13.84), there exists a point $p \in \Gamma_* \subset \tilde{\Sigma}_*$ at which the smallest principal curvature $\tilde{\kappa}_1^*$ of $\tilde{\Sigma}_*$ vanishes.

We claim that the multiplicity of $\tilde{\kappa}_1^*$ at p is one. To prove this, we exploit the local uniform convexity of Γ_*: Let $\{e_i\}_{i=1}^n$ be an orthonormal frame for $\tilde{\Sigma}_*$ at p such that $e_1 = \frac{Du_*}{|Du_*|}$ is the outward unit normal to Γ_* and $\{e_2, \ldots, e_n\}$ is a principal frame for Γ_*. Since Γ_* is the zero set of u_*, whereas Σ_* is its graph, we find that the component matrix $[\mathrm{II}_{\tilde{\Sigma}_*}]$ of $\mathrm{II}_{\tilde{\Sigma}_*}$ at p with respect to the frame $\{e_i\}_{i=1}^n$ decomposes as

$$[\mathrm{II}_{\tilde{\Sigma}_*}] = \begin{pmatrix} \frac{D^2u_*(e_1,e_1)}{(1+|Du_*|^2)^{3/2}} & [v] \\ [v]^t & \frac{|Du_*|}{\sqrt{1+|Du_*|^2}}[\mathrm{II}_{\Gamma_*}] \end{pmatrix},$$

where $[\mathrm{II}_{\Gamma_*}]$ is the component matrix of the second fundamental form II_{Γ_*} of Γ_* at p and

$$v \doteqdot \sum_{i=2}^n \frac{D^2u_*(e_i,e_1)}{(1+|Du_*|^2)^{3/2}} e_i\,.$$

Since $[\mathrm{II}_{\Gamma_*}]$ has rank $n-1$, $[\mathrm{II}_{\tilde{\Sigma}_*}]$ has rank at least $n-1$, which implies the claim.

As a consequence, $\tilde{\kappa}_1^*$ is smooth at p and therefore satisfies

$$-(\Delta + \nabla_{e_{n+1}^\top} + |\mathrm{II}|^2)\tilde{\kappa}_1^* = 2\sum_{k=1}^n\sum_{i=2}^n \frac{(\nabla_k \mathrm{II}_{1i})^2}{\kappa_i - \kappa_1} \geq 0$$

in the classical sense. Since $\tilde{\kappa}_1^*$ is nonconstant and attains its minimum over $\overline{\Omega}_*$ at $p \in \partial\Omega_*$, the Hopf boundary point lemma implies that $|\nabla\tilde{\kappa}_1^*(p)| \neq 0$. Thus, by the implicit function theorem, there exists an open subset U of $\tilde{\Sigma}_*$ such that $\Lambda \doteqdot \{q \in U : \tilde{\kappa}_1^*(q) = 0\}$ is a smooth hypersurface. Since Γ_* is locally uniformly convex, Λ must be transverse to Γ_* at p. But this contradicts the fact that Σ_* is locally uniformly convex. ☐

Since, by the translator equation, any *convex* translator satisfies

$$|\mathrm{II}|^2 \leq H^2 = \langle \mathsf{N}, e_{n+1}\rangle^2 \leq 1,$$

Theorems 13.40 and 7.14 immediately yield a priori interior estimates for solutions $u : \Omega \to \mathbb{R}$ to the translator Dirichlet problem over convex domains $\Omega \subset \mathbb{R}^n$ (cf. Theorem 13.34). Namely,

$$\sup_{\Sigma \Subset \operatorname{graph} u} |\nabla^m \mathrm{II}| \leq C_m \tag{13.85}$$

for every $m \in \mathbb{N} \cup \{0\}$, where C_m depends only on n, m, and $\operatorname{dist}(\Sigma, \partial \operatorname{graph} u)$.

13.10. Asymptotics

The following lemma describes the asymptotic geometry of convex translators.

Lemma 13.42 (Asymptotic shape of convex translators). *Let Σ^n be a convex translator in $\mathbb{R}^{n+1}$ and $\{\Sigma^n_t\}_{t\in(-\infty,\infty)}$, where $\Sigma^n_t \doteqdot \Sigma^n + te_{n+1}$, the corresponding translating solution to mean curvature flow.*

(i) (Asymptotic shrinker) *There is a rotation $R \in \mathrm{SO}(n)$ about the x_{n+1}-axis such that the rescaled solutions $\{\lambda R \cdot \Sigma^n_{\lambda^{-2}t}\}_{t\in(-\infty,0)}$ converge locally uniformly in the smooth topology as $\lambda \to 0$ to either*
   - *the shrinking cylinder $\{S^{n-k}_{\sqrt{-2(n-k)t}} \times \mathbb{R}^k\}_{t\in(-\infty,0)}$ for some $k \in \{1, \ldots, n-1\}$ or*
   - *the stationary hyperplane $\{\{0\} \times \mathbb{R}^n\}_{t\in(-\infty,0)}$ of multiplicity either one or two.*

   Moreover, if the limit is the stationary hyperplane of multiplicity one, then Σ^n is a vertical hyperplane.

(ii) (Asymptotic translators) *Given any direction $e \in \partial \mathrm{G}(\Sigma^n)$ in the boundary of the Gauß image $\mathrm{G}(\Sigma^n)$ of Σ^n and any sequence of points $X_j \in \Sigma^n$ with $\mathrm{G}(X_j) \to e$, a subsequence of the translated solutions $\Sigma^n - X_j$ converges locally uniformly in the smooth topology to a convex translator which splits off a line (in the direction $w \doteqdot \lim_{j\to\infty} \frac{X_j}{|X_j|}$).*

Proof. The first claim is a straightforward consequence of a more general result (Lemma 14.10) that we will prove in Section 14.4. It was first proved by X.-J. Wang using a different argument [**521**, Theorem 1.3].

So consider the second claim. Up to a translation, we may assume that $X_1 = 0$. After passing to a subsequence, we can arrange that $w_j \doteqdot X_j/\|X_j\| \to w$ for some $w \in S^n$. Since each Σ^n_j is convex and satisfies the translator equation, the sequence admits the uniform curvature bound

$$|\mathrm{II}_j| \leq H_j = -\langle \mathsf{N}_j, e_{n+1}\rangle \leq 1\,.$$

It follows, after passing to a further subsequence, that the sequence converges locally uniformly in the smooth topology to a convex translator Σ^n_∞. We claim that Σ^n_∞ contains the line

$$L \doteqdot \{sw : s \in \mathbb{R}\}.$$

First note that the closed convex region $\overline{\Omega}$ bounded by Σ^n contains the ray $\{sw : s \geq 0\}$ since it contains each of the segments $\{sw_j : 0 \leq s \leq s_j\}$, where $s_j \doteqdot \|X_j\|$. By convexity, $\overline{\Omega}$ also contains the set $\{rsw + (1-r)X_j : s \geq 0, 0 \leq r \leq 1\}$ for each j. It follows that the closed convex region $\overline{\Omega}_j$ bounded by Σ^n_j contains the set $\{rsw - rs_jw_j : s > 0, 0 \leq r \leq 1\}$ since $s_jw_j = X_j$. In particular, choosing $s = 2s_j$, $\{\vartheta(w - w_j) + \vartheta w : 0 \leq \vartheta \leq s_j\} \subset \overline{\Omega}_j$ and, choosing $s = s_j/2$,

$$\{\vartheta w_j - \vartheta(w - w_j) : -s_j/2 \leq \vartheta \leq 0\} \subset \overline{\Omega}_j.$$

Taking $j \to \infty$, we find $\{sw : s \in \mathbb{R}\} \subset \overline{\Omega}_\infty$. It now follows from convexity of $\overline{\Omega}_\infty$ that $\{sw : s \in \mathbb{R}\} \subset \Sigma^n$. We conclude that κ_1 reaches zero somewhere on Σ^n_∞ and the lemma follows from the splitting theorem (Theorem 9.11). $\square$

We refer to the limit in part (i) as the **blow-down** of Σ^n and the limits in part (ii) as **asymptotic translators**. In some cases, the asymptotic translators can be shown to depend uniquely on $e \in \partial \mathrm{G}(\Sigma^n)$, in which case the convergence is independent of the subsequence.

13.11. X.-J. Wang's dichotomy

A convex translator which is not entire necessarily lies on one side of some vertical hyperplane. The only nonentire translators we have encountered are the vertical hyperplanes and the Grim Reaper and its (oblique) products, which actually lie between *two* vertical hyperplanes. By Lemma 13.42, the blow-down of a nonflat, nonentire translator is a vertical hyperplane of multiplicity two; however, a priori, the distance of the bounding hyperplanes could be going to infinity as $t \to -\infty$ before rescaling. A remarkable result of X.-J. Wang states that this is never the case [**521**, Corollary 2.2] (cf. [**479**]).

Theorem 13.43 (X.-J. Wang's dichotomy for convex translators [**521**, Corollary 2.2]). *Let Σ^n be a convex translator. If its blow-down is a hyperplane of multiplicity two, then Σ^n lies in a vertical slab region (the region between two parallel vertical hyperplanes). In particular, every convex translator is either entire or lies in a vertical slab region.*

The theorem is proved in [**521**, Corollary 2.2]. We will present the proof of a more general result (also due to Wang) in Theorem 14.13.

Figure 13.7. Xu-Jia Wang.

Theorem 13.43 motivates the following definitions.

Definition 13.44. A convex, locally uniformly convex translator is called

- a **bowloid** if it is entire,
- a **flying wing** if it lies in a slab region.

Remark 13.45. The term "flying wing" was coined by R. Hamilton to describe certain translating solutions to both the mean curvature and Ricci flows. An alternative term, coined by T. Ilmanen, is "**Δ-wing**" (or "**delta-wing**"). We note that both of these terms are also used in the literature to describe (nonrotationally symmetric) bowloids.

F. Chini and N. Møller [**146**] have obtained a classification of the *convex hulls* of the projection onto the subspace orthogonal to the translation direction for general translators (see the notes and commentary at the end of this chapter).

Lemma 13.42 and Theorem 13.43 are very powerful tools for analyzing convex translators. (Cf. Corollaries 13.48 and 13.50 below.)

13.12. Rigidity of the bowl soliton

We will use the asymptotics of Lemma 13.42 in conjunction with the following beautiful argument of R. Haslhofer [**273**] to obtain rigidity results for the bowl soliton.

Theorem 13.46 (R. Haslhofer [**273**]). *Let Σ^n, $n \geq 2$, be a convex, locally uniformly convex translator in $\mathbb{R}^{n+1}$. Suppose that the blow-down of the corresponding translating solution $\{\Sigma_t^n\}_{t\in(-\infty,\infty)}$, $\Sigma_t \doteqdot \Sigma^n + te_{n+1}$, is the shrinking cylinder $S^{n-1}_{\sqrt{-2(n-1)t}} \times \mathbb{R}$. Then Σ^n is rotationally symmetric about a vertial axis (and hence, up to a translation, the bowl soliton).*

Proof. First observe that, since its blow-down is the shrinking cylinder, the mean curvature of Σ^n must satisfy

$$H = \sqrt{\frac{n-1}{2h}} + o(h^{-\frac{1}{2}}), \tag{13.86}$$

where $h(X) \doteqdot \langle X, e_{n+1}\rangle$ is the height function.

The rotation and translation fields. Consider the translation and rotation functions $u_p : \Sigma^n \to \mathbb{R}$, $p \in \mathbb{R}^{n+1}$, and $u_J : \Sigma^n \to \mathbb{R}$, $J \in \mathfrak{so}(\mathbb{R}^n \times \{0\})$, respectively, defined by

$$u_p(X) \doteqdot \langle p, \mathsf{N}(X)\rangle, \quad p \in \mathbb{R}^{n+1},$$

and

$$u_J(X) \doteqdot \langle J(X), \mathsf{N}\rangle, \quad J \in \mathfrak{so}(\mathbb{R}^n \times \{0\}).$$

By Exercise 13.7, both families satisfy the Jacobi equation

$$-(\Delta + \nabla_V)u = |\mathrm{II}|^2 u,$$

where $V \doteqdot \nabla h = e_{n+1}^{\top}$.

Since the mean curvature also satisfies the Jacobi equation, we obtain (from the maximum principle) the following weighted estimates for the rotation functions:

$$\max_{\{h\leq z\}} \left|\frac{u_J}{H}\right| \leq \max_{\{h=z\}} \left|\frac{u_J}{H}\right|, \tag{13.87}$$

for any $J \in \mathfrak{so}(\mathbb{R}^n \times \{0\})$.

The rotational defect. Next, we fix a nontrivial rotation vector field $J \in \mathfrak{so}(\mathbb{R}^n \times \{0\}) \hookrightarrow \mathfrak{so}(\mathbb{R}^{n+1})$, say

$$J(X) \doteqdot X^2\partial_1 - X^1\partial_2,$$

and consider the family

$$\mathfrak{R} \doteqdot \{O \cdot J : O \in \mathrm{SO}(\mathbb{R}^{n+1} \times \{0\})\}.$$

Note that $\mathfrak{R}$ is compact. Given $x \in \mathbb{R}^n \times \{0\}$, set

$$R_x \doteqdot R(\cdot - x) \text{ for each } R \in \mathfrak{R}$$

and define

$$\mathfrak{R}_x \doteqdot \{R_x : R \in \mathfrak{R}\}.$$

Consider the "rotational defect"

$$B(z) \doteqdot \inf_{x\in\mathbb{R}^n\times\{0\}} \sup_{R\in\mathfrak{R}_x} \sup_{\{h=z\}} |u_R|. \tag{13.88}$$

Observe that $B(z)$ vanishes only if we can find $x \in \mathbb{R}^n \times \{0\}$ such that the height z slice of Σ^n is rotationally symmetric about the axis $\{x\} \times \mathbb{R}$.

Note that

$$B(z) \doteqdot \inf_{x\in\mathbb{R}^n\times\{0\}} \sup_{R\in\mathfrak{R}} \sup_{\Sigma^n\cap\{h=z\}} |u_R - u_x|\,. \tag{13.89}$$

Indeed, each $R_x \in \mathfrak{R}_x$ is of the form

$$R_x(X) = R(X) - y_{O,x}\,,$$

where

$$y_{O,x} = O(x) + O\cdot J(x) - x\,.$$

The claim follows since, for a dense set of $O \in \mathrm{SO}(n)$, the matrix $O(I+S)-I$ is invertible.

Decay estimate on the shrinking cylinder. We will show in the next step that a nonvanishing rotational defect is in conflict with the following claim.

Claim 13.47 (Decay estimate on the shrinking cylinder)**.** Let u be a solution to the Jacobi equation

$$(\partial_t - \Delta)u = |\mathrm{II}|^2 u$$

on the shrinking cylinder $\{S^{n-1}_{\sqrt{-2(n-1)t}} \times \mathbb{R}\}_{t\in(-\infty,0)}$. Suppose that u is invariant under translations along the axis, that

$$\int_{S^{n-1}_{\sqrt{-2(n-1)t}}\times[-1,1]} u\, d\mu = 0 \ \text{ for all } \ t\in(-1,0),$$

and that

$$|u(\cdot,t)| \le 1 \ \text{ for all } \ t \in (-1,-\tfrac{1}{2})\,.$$

Then there exists $D<\infty$ such that

$$\inf_{p\in\mathbb{R}^n\times\{0\}} \sup_{S^{n-1}_{\sqrt{-2(n-1)t}}\times\mathbb{R}} |u(\cdot,t) - u_p(\cdot,t)| \le D(-t)^{\frac{1}{2}+\frac{1}{n-1}} \tag{13.90}$$

for all $t\in[-\frac{1}{2},0)$.

Proof. Since u is invariant under vertical translations, the Laplacian scales like one over distance squared, and $|\mathrm{II}(\cdot,t)|^2 \equiv \frac{1}{-2t}$, the restriction u_h of u to $S^{n-1}\times\{h\}$ satisfies

$$\partial_t u_h = \frac{1}{-2(n-1)t}\left(\Delta_{S^{n-1}} + n - 1\right) u\,. \tag{13.91}$$

By hypothesis,

$$\int_{S^{n-1}_{\sqrt{-2(n-1)t}}} u_h\, d\mu = 0 \ \text{ for } \ t\in(-1,0) \tag{13.92}$$

and

$$|u_h(\cdot,t)| \le 1 \ \text{ for } \ t\in(-1,-\tfrac{1}{2})\,.$$

Recall that the spectrum $\{\lambda_j\}_{j=0}^{\infty}$ of $-\Delta_{S^{n-1}}$ is given by

$$\lambda_j = j(j+n-2)\,.$$

If $\{\varphi_j\}_{j=0}^{\infty}$ is a corresponding L^2-orthonormal eigenbasis, then we can find functions $\gamma_j : (-\infty, 0) \to \mathbb{R}$ such that

$$u_h(z,t) = \sum_{j=0}^{\infty} \gamma_j(t)\varphi_j(z)\,.$$

Since $\varphi_0 = 1$, (13.92) implies that $\gamma_0 \equiv 0$.

Since u_h satisfies (13.91), we find

$$\gamma_j' = \frac{1}{-2(n-1)t}(-\lambda_j + (n-1))\gamma_j\,,$$

whose solution is easily found to be

$$\gamma_j(t) = \gamma_j(0)(-t)^{\frac{\lambda_j - n+1}{2(n-1)}}\,.$$

In particular,

$$\gamma_1(t) = \gamma_1(0)$$

and

$$\gamma_2(t) = \gamma_2(0)(-t)^{\frac{1}{2}+\frac{1}{n-1}}\,.$$

The claim follows. □

A nonvanishing rotational defect contradicts the decay estimate. Since the blow-down of Σ^n is the shrinking cylinder, we have the estimate

$$B(z) \leq O(z^{\frac{1}{2}})\,. \tag{13.93}$$

We will prove that, given any $z_0 < \infty$, we can find $z > z_0$ such that

$$B(z) = 0\,.$$

Suppose, to the contrary, that there exists $z_0 < \infty$ so that $B(z) > 0$ for all $z \geq z_0$. Fix $\tau \in (0, \frac{1}{2})$ (to be determined momentarily). By (13.93), we can find a sequence of heights $z_j \to \infty$ such that

$$B(z_j) \leq 2\tau^{-\frac{1}{2}} B(\tau z_j)$$

for each j. Indeed, if this were not the case, then we could find $z_0 < \infty$ so that

$$B(z) > 2\tau^{-\frac{1}{2}} B(\tau z)$$

for all $z \geq z_0$ and hence, by induction,

$$B(\tau^{-k} z_0) > 2^k \tau^{-\frac{k}{2}} B(z_0) = 2^k \frac{B(z_0)}{\sqrt{z_0}} \left(\tau^{-k} z_0\right)^{\frac{1}{2}}$$

for all $k \in \mathbb{N}$, contradicting (13.93).

Choose, for every $j \in \mathbb{N}$, a point $x_j \in \mathbb{R}^n \times \{0\}$ so that

$$B(z_j) = \sup_{R \in \mathfrak{R}_{x_j}} \sup_{\{h=z_j\}} |u_R|$$

and a rotation field $R_j \in \mathfrak{R}_{x_j}$ so that

$$B(\tau z_j) = \inf_{x \in \mathbb{R}^n \times \{0\}} \sup_{\{h=\tau z_j\}} |u_{R_j} - u_x|$$

and consider the sequence of functions $\widetilde{u}_j$ defined by

$$\widetilde{u}_j \doteqdot \frac{1}{B(z_j)} u_{R_j} \,.$$

Since $\widetilde{u}_j$ is a constant multiple of u_{R_j}, it satisfies the Jacobi equation. Thus, by (13.86) and (13.87),

$$\sup_{\{h=z\}} |\widetilde{u}_j| \leq 2 \left(\frac{z_j}{z}\right)^{\frac{1}{2}} \tag{13.94}$$

for all $z < z_j$ and j sufficiently large.

If we identify each $\widetilde{u}_j$ with a function on $\{\Sigma_t\}_{t \in (-\infty,\infty)}$, then $\widetilde{u}_j$ satisfies the Jacobi equation

$$(\partial_t - \Delta)\widetilde{u}_j = |\mathrm{II}|^2 \widetilde{u}_j \,.$$

Set $\lambda_j \doteqdot z_j^{-\frac{1}{2}}$ and consider the rescaled flows $\{\widehat{\Sigma}^n_{j,t}\}_{t \in (-\infty,0)}$, where $\widehat{\Sigma}^n_{j,t} \doteqdot \lambda_j \left(\Sigma^n_{\lambda_j^{-2} t} - x_j \right)$.

By hypothesis, the flows $\{\widehat{\Sigma}^n_{j,t}\}_{t \in (-\infty,0)}$ converge smoothly on compact subsets of space-time to the shrinking cylinder $\{S^{n-1}_{\sqrt{-2(n-1)t}} \times \mathbb{R}\}_{t \in (-\infty,0)}$ as $j \to \infty$. The functions $\widetilde{u}_j$ rescale as

$$\widehat{u}_j(X,t) \doteqdot \widetilde{u}_j(\lambda_j^{-1} X + p_j, \lambda_j^{-2} t)$$

and satisfy the Jacobi equation on $\{\widehat{\Sigma}^n_{j,t}\}_{t \in (-\infty,0)}$. By (13.94),

$$\limsup_{j \to \infty} \sup_{t \in [-1+\delta, -\delta]} \sup_{|z| \leq \delta^{-1}} |\widehat{u}_j| < \infty \,.$$

So the sequence $\{\widehat{u}_j\}_{i \in \mathbb{N}}$ converges, after passing to a subsequence, to a function u on the shrinking cylinder which satisfies the corresponding Jacobi equation.

On the one hand, we can estimate

$$\begin{aligned}
\inf_{x\in\mathbb{R}^n\times\{0\}}\sup_{\{h=0\}}|\widehat{u}_j(\cdot,-\tau)-u_x(\cdot,t)|
&=\inf_{x\in\mathbb{R}^n\times\{0\}}\sup_{S^{n-1}_{\sqrt{-2(n-1)t}}\times\{z_j\}}|\widetilde{u}_j(\cdot,-h_j\tau)-u_x(\cdot,t)|\\
&=\inf_{x\in\mathbb{R}^n\times\{0\}}\sup_{S^{n-1}_{\sqrt{-2(n-1)t}}\times\{\tau z_j\}}|\widetilde{u}_j(\cdot,0)-u_x(\cdot,t)|\\
&=\frac{1}{B(h_j)}\inf_{x\in\mathbb{R}^n\times\{0\}}\sup_{S^{n-1}_{\sqrt{-2(n-1)t}}\times\{\tau z_j\}}|u_{r_j}(\cdot,0)-u_x(\cdot,t)|\\
&=\frac{B(\tau h_j)}{B(h_j)}\geq\frac{1}{2}\tau^{\frac{1}{2}},
\end{aligned}$$

which, on taking the limit as $j\to\infty$, yields

$$\tfrac{1}{2}\tau^{\frac{1}{2}}\leq\inf_{x\in\mathbb{R}^n\times\{0\}}\sup_{S^{n-1}_{\sqrt{-2(n-1)t}}\times\mathbb{R}}|u(\cdot,t)-u_x(\cdot,t)|\,. \tag{13.95}$$

On the other hand, the limit u is invariant under vertical translations and satisfies, by (13.94),

$$|u(\cdot,t)|\leq 4\ \text{ for }\ t\in(0,\tfrac{1}{2})\,.$$

Moreover, since the vector fields $R\in\mathfrak{R}$ are divergence free and satisfy $\langle R,V\rangle\equiv 0$, the divergence theorem yields

$$\begin{aligned}
0=\int_{B^n_{\sqrt{-2(n-1)t}}\times[z_1,z_2]}\operatorname{div}_{\mathbb{R}^{n+1}}(R_j)\,d\mathscr{H}^{n+1}&=\int_{S^{n-1}_{\sqrt{-2(n-1)t}}\times[z_1,z_2]}\langle R_j,\mathsf{N}\rangle\,d\mathscr{H}^n\\
&=\int_{S^{n-1}_{\sqrt{-2(n-1)t}}\times[z_1,z_2]}u\,d\mathscr{H}^n\,.
\end{aligned}$$

So u satisfies the decay estimate (13.90). Choosing τ so that $\tau^{-\frac{1}{n-1}}>4D$ results in a contradiction with the estimate (13.95).

Rotational symmetry. We have proved that there exists a sequence of heights z_j with $z_j\to\infty$ as $j\to\infty$ such that

$$B(z_j)=0$$

for each j. For each j, choose $x_j\in\mathbb{R}^n\times\{0\}$ such that

$$\sup_{\{h=z_j\}}|u_{R_{x_j}}|=0\ \text{ for every }\ R_{x_j}\in\mathfrak{R}_{x_j}\,.$$

The estimate (13.87) then implies that

$$u_{R_{x_j}}\equiv 0\ \text{ in }\ \{h\leq z_j\}\ \text{ for every }\ R_{x_j}\in\mathfrak{R}_{x_j}\,.$$

Writing $u_{R_{x_j}} = u_R - u_{y_j}$ for some $R \in \mathfrak{R}$ and some $y \in \mathbb{R}^n \times \{0\}$, we obtain $u_R - u_{y_j} \equiv 0$ in $\{h \leq z_j\}$. It follows that y_j and hence also $x_j \equiv x$ are constant and $u_{R_x} \equiv 0$ on Σ^n for every $R_x \in \mathfrak{R}_x$. It follows that Σ^n is rotationally symmetric about $\{x\} \times \mathbb{R}$. □

Corollary 13.48 (X.-J. Wang [**521**, Theorem 1.1], J. Spruck and L. Xiao [**494**]). *Modulo translation, the bowl soliton is the only mean convex, entire translator Σ^2 in $\mathbb{R}^3$. In particular, it is the only translator which arises as a singularity model for a compact, embedded, mean convex mean curvature flow in $\mathbb{R}^3$.*

Proof. By the Spruck–Xiao convexity estimate (Theorem 13.39), Σ^2 is actually convex. In fact, it must also be locally unformly convex: Otherwise, by the strong maximum principle, it would split off a line; by uniqueness of the Grim Reaper, the result would either be a vertical plane or a Grim plane, neither of which is entire.

Since Σ^2 is entire, Lemma 13.42 and Wang's dichotomy (Theorem 13.43) imply that its blow-down is the shrinking cylinder. The claim now follows from Theorem 13.46. □

Remark 13.49. Corollary 13.48 was proved by X.-J. Wang for convex translators by a different argument: Making use of the fact that the blow-down is the shrinking cylinder, he was able to obtain, by an iteration argument, the estimate

$$|u(x) - u_0(x)| = o(|x|) \text{ as } |x| \to \infty ,$$

where graph u_0 is the bowl soliton whose tip coincides with that of u. A classical theorem of Bernstein then implies that $u - u_0$ is constant, which yields the claim. See [**521**].

The argument can be extended to higher dimensions if we assume that the translator is uniformly 2-convex [**100**].

Corollary 13.50. *Modulo translation, the bowl soliton is the only mean convex, nonplanar translator Σ^n in $\mathbb{R}^{n+1}$, $n \geq 3$, satisfying*

$$\inf_{\Sigma^n} \frac{\kappa_1 + \kappa_2}{H} > 0 . \tag{13.96}$$

In particular, it is the only translator which arises as a singularity model for a compact, 2-convex (immersed) mean curvature flow in $\mathbb{R}^{n+1}$, $n \geq 3$.

Proof. By the Spruck–Sun convexity estimate (Theorem 13.40), Σ^n is convex. We claim that it is entire. Suppose, to the contrary, that this is not the case. Then, by Wang's dichotomy (Theorem 13.43), it lies in a slab. Since Σ^n is locally uniformly convex, there exists $e \in \partial \mathrm{G}(\Sigma)$ with $\langle e, e_{n+1}\rangle > -1$ and $\langle e, e_1\rangle = 0$. By Lemma 13.42(ii), we can find a sequence of points

$X_j \in \Sigma^n$ such that the translates $\Sigma^n - X_j$ converge locally uniformly in the smooth topology to a convex translator which splits off a line parallel to the slab and has normal e at the origin. In particular, the limit cannot be a vertical hyperplane and thus has positive mean curvature. Moreover, since $\langle e, e_{n+1}\rangle < 0$, the cross section of the splitting cannot be compact. But this contradicts the 2-convexity hypothesis (13.96) by Hamilton's compactness theorem (Theorem 11.20). The claim now follows from Haslhofer's result (Theorem 13.46). □

Remark 13.51. Corollary 13.50 was proved under an additional noncollapsing hypothesis (which, by Corollary 12.12, holds for blow-ups of compact, mean convex, embedded mean curvature flows) by Haslhofer [**273**]. It was proved for translators satisfying certain cylindrical and gradient estimates (which, by Corollary 9.14 and Theorem 9.24, hold for blow-ups of compact, 2-convex mean curvature flows of dimension at least 3) in [**96**].

13.13. Flying wings

X.-J. Wang proved that there exist flying wings in $\mathbb{R}^{n+1}$ for each n by taking a limit of a sequence of solutions over bounded convex domains approaching the projection of the slab [**521**, Theorem 1.2]. Since solutions to the Dirichlet problem for the graphical translator equation over convex domains are not necessarily convex, this was achieved by exploiting the Legendre transform and the existence of locally uniformly convex solutions to certain fully nonlinear equations. Unfortunately, this method loses track of the precise geometry of the domain on which the solution is defined and so it remained unclear exactly which slabs admit translators. Note though that there can be no convex, unit speed translator in a slab of width less than π.

Proposition 13.52. *There exists no convex, nonplanar translator contained in a vertical slab in $\mathbb{R}^{n+1}$ of width less than π.*

Proof. Suppose, to the contrary, that there exists a convex, nonplanar translator Σ^n lying in the slab $\Omega_r \doteqdot \{x \in \mathbb{R}^{n+1} : |x_1| < r\}$ for some $r < \frac{\pi}{2}$. By the splitting theorem, we may assume that Σ^n is locally uniformly convex. Indeed, if not, it would split off a line. After rotating and scaling, we would obtain a translator which splits off a *horizontal* line and lies in a *smaller* slab. Since Σ^n is not a vertical hyperplane, we can find a point $p_0 \in \Sigma^n$ with normal $\mathsf{N}_0 \doteqdot \mathsf{N}(p_0)$ parallel to the slab. Then Σ^n lies in the half-slab $\Omega_r \cap \{x \in \mathbb{R}^{n+1} : \langle x - p_0, \mathsf{N}_0\rangle < 0\}$. Since $\cos\theta \doteqdot -\langle \mathsf{N}_0, e_{n+1}\rangle > 0$, there exists h such that the oblique Grim hyperplane $\Gamma^n_\theta - h$ lies below the half-slab. We can now translate $\Gamma^n_\theta - h$ upwards until it makes contact with Σ^n at an interior point. But this contradicts the strong maximum principle. □

The following theorem provides a 1-parameter family of convex translators which interpolate between the Grim hyperplane and the bowl soliton.

Theorem 13.53 (Flying wings [**98, 521**]). *For every $n \geq 2$ and $\theta \in (0, \frac{\pi}{2})$, there exists a strictly convex,* $\mathrm{SO}(1) \times \mathrm{SO}(n-1)$*-invariant translator W^n_θ which lies in $\Pi^{n+1}_\theta \doteqdot \{x \in \mathbb{R}^{n+1} : |x_1| < \frac{\pi}{2}\sec\theta\}$ (and in no smaller slab).*

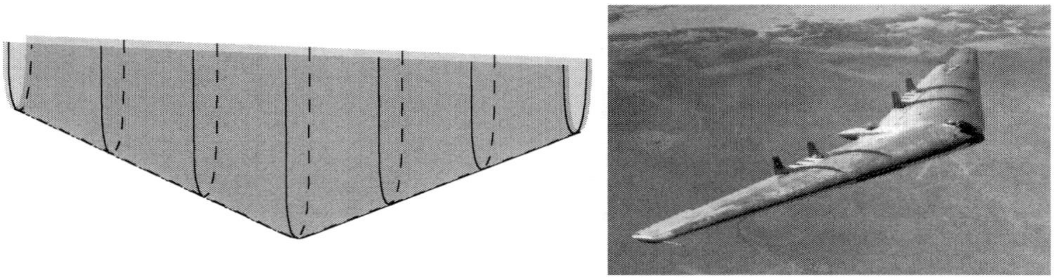

Figure 13.8. L: A "flying wing" translator. R: A Northrop "flying wing" bomber.

Proof of Theorem 13.53. We will obtain the solutions by taking a limit of (translated) solutions to Dirichlet problems

$$\begin{cases} \operatorname{div}\left(\dfrac{Du}{\sqrt{1+|Du|^2}}\right) = \dfrac{1}{\sqrt{1+|Du|^2}} & \text{in } \Omega_R, \\ u = 0 & \text{on } \partial\Omega_R \end{cases} \tag{13.97}$$

over a family of bounded, convex domains Ω_R with $\bigcup_{R>0}\Omega_R = \Pi^n_\theta$. In order to solve (13.97), it will suffice, by Proposition 13.33, to construct lower barriers.

The following subsolution was obtained by modifying the level set flow of the paperclip solution so that it is asymptotic to the correct Grim planes.

Lemma 13.54. *The function $\underline{u} : (-\frac{\pi}{2}\sec\theta, \frac{\pi}{2}\sec\theta) \times \mathbb{R}^{n-1} \to \mathbb{R}$ defined by*

$$\underline{u}(x,y) \doteqdot -\sec^2\theta \log\left(\cos\tfrac{x}{\sec\theta}\right) + \tan^2\theta\log\left(\cosh\tfrac{|y|}{\tan\theta}\right)$$

is a subsolution to the graphical translator equation. In particular, the function $\underline{u}_R : (-\frac{\pi}{2}\sec\theta, \frac{\pi}{2}\sec\theta) \times \mathbb{R}^{n-1} \to \mathbb{R}$ defined by

$$\underline{u}_R(x,y) \doteqdot -\sec^2\theta\log\left(\cos\tfrac{x}{\sec\theta}\right) + \tan^2\theta\log\left(\frac{\cosh\frac{|y|}{\tan\theta}}{\cosh\frac{R}{\tan\theta}}\right)$$

is a subsolution to the Dirichlet problem (13.97) *over the domain*

$$\Omega_R \doteqdot \left\{(x,y) \in (-\tfrac{\pi}{2}\sec\theta, \tfrac{\pi}{2}\sec\theta) \times \mathbb{R}^{n-1} : \underline{u}_R(x,y) < 0\right\}.$$

Proof. The relevant derivatives of $\underline{u}$ are given by

$$D\underline{u} = \left(\sec\theta\tan\left(\tfrac{x}{\sec\theta}\right), \tan\theta\tanh\left(\tfrac{|y|}{\tan\theta}\right)\tfrac{y}{|y|}\right)$$

and

$$D^2\underline{u} = \begin{pmatrix} \sec^2\left(\tfrac{x}{\sec\theta}\right) & \cdots & 0 & \cdots \\ \vdots & & & \\ 0 & & \operatorname{sech}^2\left(\tfrac{|y|}{\tan\theta}\right)\tfrac{y_iy_j}{|y|^2} + \tan\theta\tanh\left(\tfrac{|y|}{\tan\theta}\right)\left(\tfrac{\delta_{ij}}{|y|} - \tfrac{y_iy_j}{|y|^3}\right) & \\ \vdots & & & \end{pmatrix}.$$

So

$$\begin{aligned} 1 + |D\underline{u}|^2 &= 1 + \sec^2\theta\tan^2\left(\tfrac{x}{\sec\theta}\right) + \tan^2\theta\tanh^2\left(\tfrac{|y|}{\tan\theta}\right) \\ &= \sec^2\theta\sec^2\left(\tfrac{x}{\sec\theta}\right) - \tan^2\theta\operatorname{sech}^2\left(\tfrac{|y|}{\tan\theta}\right). \end{aligned}$$

Estimating

$$\Delta\underline{u} \geq \sec^2\left(\tfrac{x}{\sec\theta}\right) + \operatorname{sech}^2\left(\tfrac{|y|}{\tan\theta}\right),$$

we find that

$$\begin{aligned} (1 + |D\underline{u}|^2)^{\frac{3}{2}}H[\underline{u}] &= (1 + |D\underline{u}|^2)\Delta\underline{u} - D^2\underline{u}(D\underline{u}, D\underline{u}) \\ &\geq 1 + |D\underline{u}|^2 + \sec^2\left(\tfrac{x}{\sec\theta}\right)\operatorname{sech}^2\left(\tfrac{|y|}{\tan\theta}\right) \\ &\geq 1 + |D\underline{u}|^2. \end{aligned}$$

□

Proposition 13.33 now yields, for each $R > 0$, a unique solution to the Dirichlet problem over Ω_R.

Corollary 13.55. *For every $R > 0$, the Dirichlet problem* (13.97) *admits a unique smooth solution $u_R \in C^\infty(\overline{\Omega}_R)$.*

By uniqueness of solutions, graph u_R is $\mathrm{SO}(1) \times \mathrm{SO}(n-1)$-invariant. In particular, Theorem 13.34 applies and we deduce, using Theorem 11.6, that the sequence of translators

$$\Sigma_i \doteqdot \operatorname{graph} u_i - \min_{\Omega_i} u_i\, e_{n+1}$$

converges as $i \to \infty$ locally uniformly in the smooth topology, after passing to a subsequence, to a limit translator Σ^n. We claim that Σ^n is proper. It suffices to show that $\min_{\Omega_i} u_i \to -\infty$. Miraculously, rotating the paperclip results in a suitable upper barrier (cf. Theorem 14.27).

Lemma 13.56. *Given $R > 0$, let $\Pi_R \subset \mathbb{R}^{n+1}$ be the surface formed by rotating the time $T \doteqdot -\sec^2\theta\cosh\left(\tfrac{R}{\tan\theta}\right)$ slice of the paperclip of width $\pi\sec\theta$ about the x-axis. That is,*

$$\Pi_R \doteqdot \left\{(x, y, z) \in \mathbb{R} \times \mathbb{R}^{n-1} \times \mathbb{R} : v(x, y, z) = T\right\},$$

where

$$v \doteqdot \sec^2\theta\left[\log\left(\cosh\left(\frac{\sqrt{|y|^2+z^2}}{\sec\theta}\right)\right)-\log\left(\cos\left(\tfrac{x}{\sec\theta}\right)\right)\right].$$

There exists $\varepsilon_0=\varepsilon_0(n,\theta)>0$ *such that the sublevel set*

$$\overline{\Sigma}_{R,\varepsilon}\doteqdot \Pi_R\cap\left\{z\le -R\tfrac{\cos(\theta-\varepsilon)}{\sin\theta}\right\}+R\tfrac{\cos(\theta-\varepsilon)}{\sin\theta}e_{n+1}$$

is a supersolution of the translator equation (13.63) *whenever* $\varepsilon<\varepsilon_0$ *and* $R>R_\varepsilon\doteqdot\frac{2(n-1)}{\varepsilon}$.

Proof. Set $w=(y,z)$. Then

$$Dv=\sec\theta\left(\tan\left(\tfrac{x}{\sec\theta}\right),\tanh\left(\tfrac{|w|}{\sec\theta}\right)\tfrac{w}{|w|}\right)$$

and

$$D^2v=\begin{pmatrix}\sec^2\left(\frac{x}{\sec\theta}\right) & \cdots & 0 & \cdots\\ \vdots & & & \\ 0 & & \operatorname{sech}^2\left(\frac{|w|}{\sec\theta}\right)\frac{w_iw_j}{|w|^2}+\sec\theta\tanh\left(\frac{|w|}{\sec\theta}\right)\left(\frac{\delta_{ij}}{|w|}-\frac{w_iw_j}{|w|^3}\right) & \\ \vdots & & & \end{pmatrix}.$$

It follows, on the one hand, that

$$-\langle \mathsf{N},e_{n+1}\rangle=\left\langle\frac{Dv}{|Dv|},e_{n+1}\right\rangle=\frac{\tanh\left(\frac{\sqrt{|y|^2+z^2}}{\sec\theta}\right)\frac{|z|}{\sqrt{|y|^2+z^2}}}{\sqrt{\tan^2\left(\frac{x}{\sec\theta}\right)+\tanh^2\left(\frac{\sqrt{|y|^2+z^2}}{\sec\theta}\right)}}$$

and, on the other hand, that

$$H=\operatorname{div}\left(\frac{Dv}{|Dv|}\right)=\frac{\frac{1}{\sec\theta}+\frac{n-1}{|w|}\tanh\left(\frac{|w|}{\sec\theta}\right)}{\sqrt{\tan^2\left(\frac{x}{\sec\theta}\right)+\tanh^2\left(\frac{|w|}{\sec\theta}\right)}}.$$

We deduce that Π_R is a supersolution in the region where

$$\frac{|z|-(n-1)}{\sqrt{|y|^2+z^2}}\tanh\left(\frac{\sqrt{|y|^2+z^2}}{\sec\theta}\right)\ge\cos\theta\,.$$

Note that $|y| \le \frac{\sin(\theta-\varepsilon)}{\sin\theta} R$ wherever $|z| \ge \frac{\cos(\theta-\varepsilon)}{\sin\theta} R$. Thus, whenever $R > R_\varepsilon \doteqdot \frac{2(n-1)}{\varepsilon}$ and $z \le -\frac{\cos(\theta-\varepsilon)}{\sin\theta} R$,

$$\begin{aligned}\frac{|z|-(n-1)}{\sqrt{|y|^2+z^2}} &\tanh\left(\frac{\sqrt{|y|^2+z^2}}{\sec\theta}\right)\\ &\ge \left(\cos(\theta-\varepsilon) - \frac{(n-1)}{R}\sin\theta\right)\tanh\left(\frac{\cos(\theta-\varepsilon)}{\tan\theta}R\right)\\ &\ge \cos\theta\left(1+\frac{\varepsilon}{2}\tan\theta + o(\varepsilon)\right)\sqrt{1-4\mathrm{e}^{-\frac{2(n-1)\cos^2\theta\sin\theta}{\varepsilon}}}\,.\end{aligned}$$

This is no less than $\cos\theta$ when $\varepsilon < \varepsilon_0(n,\theta)$. □

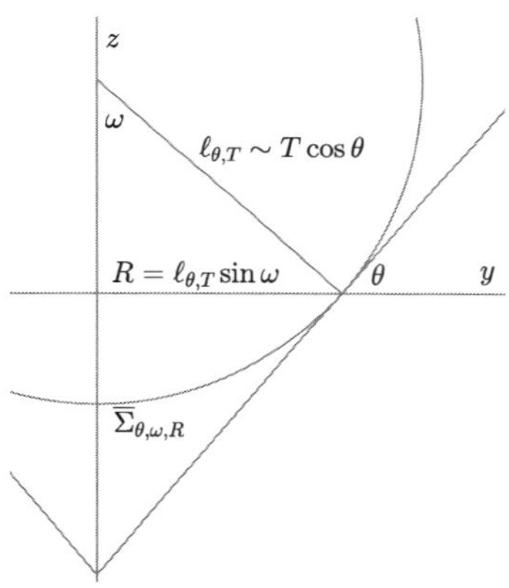

Figure 13.9. Given any $\varepsilon \in (0, \varepsilon_0(n,\theta))$, the portion of Π_R (the rotated time $T = \sec^2\theta\cosh\left(\frac{R}{\tan\theta}\right)$ slice of the paperclip of width $\pi\sec\theta$) lying below height $z = -R\frac{\cos(\theta-\varepsilon)}{\sin\theta}$ is a supersolution of the translator equation when $R > R_\varepsilon \doteqdot \frac{2(n-1)}{\varepsilon}$. The surface $\overline{\Sigma}_{R,\varepsilon}$ is obtained by translating this piece upward so that its boundary lies in $\mathbb{R}^n \times \{0\}$.

Consider the "inner" domain

$$\overline{\Omega}_{R,\varepsilon} \doteqdot \left\{(x,y) \in \Pi^n_\theta : \cos\left(\frac{x}{\sec\theta}\right) < \cosh\left(\frac{\sqrt{|y|^2\sin^2\theta + R^2\cos^2(\theta-\varepsilon)}}{\tan\theta}\right)\Big/\cosh\left(\frac{R}{\tan\theta}\right)\right\}.$$

Note that $\partial\overline{\Omega}_{R,\varepsilon} = \partial\overline{\Sigma}_{R,\varepsilon}$. The following lemma implies that the inner barrier which touches the outer barrier at Re_n lies above it, so long as R is sufficiently large.

Lemma 13.57. *Given any $R > 0$, $\overline{\Omega}_{\rho_\varepsilon,\varepsilon} \subset \Omega_R$, where $\rho_\varepsilon \doteqdot \frac{\sin\theta}{\sin(\theta-\varepsilon)}R$.*

Proof. It suffices to show that the function $f : \mathbb{R}_+ \to \mathbb{R}$ defined by

$$f(\zeta) \doteqdot \frac{\cosh\left(\frac{\sqrt{\zeta^2\sin^2\theta+\rho_\varepsilon^2\cos^2(\theta-\varepsilon)}}{\tan\theta}\right)}{\cosh\left(\frac{\rho_\varepsilon}{\tan\theta}\right)} - \left[\frac{\cosh\left(\frac{\zeta}{\tan\theta}\right)}{\cosh\left(\frac{R}{\tan\theta}\right)}\right]^{\sin^2\theta}$$

is nonpositive. This follows from the log-concavity of the function

$$g(w) \doteqdot \cosh\left(\frac{\sqrt{w}}{\tan\theta}\right).$$

Indeed, given any $s \in (0,1)$ and $w > 0$, the log-concavity of g implies that the function

$$G(z) \doteqdot \frac{g(sz + (1-s)w)}{g(z)^s}$$

is monotone nondecreasing for $z < w$. Since $\zeta < R < \frac{\tan\theta}{\tan(\theta-\varepsilon)}R = \frac{\cos(\theta-\varepsilon)}{\cos\theta}\rho_\varepsilon$, this implies

$$\frac{g(\zeta^2\sin^2\theta + \rho_\varepsilon^2\cos^2(\theta-\varepsilon))}{g(\zeta^2)^{\sin^2\theta}} \le \frac{g(R^2\sin^2\theta + \rho_\varepsilon^2\cos^2(\theta-\varepsilon))}{g(R^2)^{\sin^2\theta}} = \frac{g(\rho_\varepsilon^2)}{g(R^2)^{\sin^2\theta}}.$$

The claim follows. □

Corollary 13.58. *Set* $\varepsilon_R \doteqdot \frac{2(n-1)}{R}$, $\overline{\Sigma}_R \doteqdot \overline{\Sigma}_{\rho_{\varepsilon_R},\varepsilon_R}$, *and* $\overline{\Omega}_R \doteqdot \overline{\Omega}_{\rho_{\varepsilon_R},\varepsilon_R}$. *Then, for* $R > R_0 \doteqdot \frac{2(n-1)}{\varepsilon_0}$, $\overline{\Sigma}_R$ *lies above* $\Sigma_R = \operatorname{graph} u_R$.

We conclude that

$$\min_{\Omega_R} u_R \lesssim -\frac{1-\cos\theta}{\sin\theta}R$$

and hence $\min_{\Omega_i} u_i \to -\infty$ as $i \to -\infty$, as desired.

By the convexity estimate (Theorem 13.39), Σ^n is convex, so it remains only to show that the width of Σ^n did not drop in taking the limit; that is, $\operatorname{proj}_{\mathbb{R}^n}\Sigma^n = \Pi^n_\theta$. Set

$$v(X) \doteqdot 1 - \frac{x(X)}{\frac{\pi}{2}\sec\theta},$$

where $x(X) \doteqdot \langle X, e_1\rangle$. We claim that

$$\inf_{\Sigma\cap\{x>0\}} \frac{H}{v} > 0. \tag{13.98}$$

Since $\inf_\Sigma H = 0$, we conclude that $\sup_\Sigma x = \frac{\pi}{2}\sec\theta$, as desired. To prove (13.98), first observe that

$$-(\Delta + \nabla_V)v = 0$$

and hence

$$-(\Delta+\nabla_V)\frac{H}{v} = |\mathrm{II}|^2\frac{H}{v} + 2\left\langle \nabla\frac{H}{v}, \frac{\nabla v}{v}\right\rangle,$$

where V is the tangential projection of e_{n+1}. The maximum principle then yields

$$\min_{\Sigma_i\cap\{x>0\}} \frac{H}{v} \geq \min\left\{\min_{\partial\Sigma_i\cap\{x>0\}} \frac{H}{v}, \min_{\Sigma_i\cap\{x=0\}} \frac{H}{v}\right\}$$
$$= \min\left\{\min_{\partial\Sigma_i\cap\{x>0\}} \frac{H}{v}, \min_{\Sigma_i\cap\{x=0\}} H\right\}.$$

Since $\underline{\Sigma}^n_R = \operatorname{graph} \underline{u}_R|_{\overline{\Omega}_R}$ lies below $\Sigma^n_R = \operatorname{graph} u_R$ and $\partial\underline{\Sigma}_R = \partial\Omega_R = \partial\Sigma_R$,

$$H[u_R] = -\langle \mathsf{N}_R, e_{n+1}\rangle \geq -\langle \underline{\mathsf{N}}_R, e_{n+1}\rangle \geq \cos\theta\cos(x\cos\theta) \geq \cos\theta\left(1 - \frac{x}{\frac{\pi}{2}\sec\theta}\right) \tag{13.99}$$

on $\partial\Omega_R$, where $\underline{\mathsf{N}}_R$ is the downward pointing unit normal to $\operatorname{graph} \underline{u}_R$. So it remains to show that

$$\liminf_{i\to\infty} \min_{\Sigma_i\cap\{x=0\}} H > 0\,.$$

Suppose, to the contrary, that $\liminf_{i\to\infty} H(X_i) = 0$ along some sequence of points $X_i \in \Sigma_i \cap \{x = 0\}$. Then, by Theorem 13.34, after passing to a subsequence, the translators-with-boundary

$$\widehat{\Sigma}_i \doteqdot \Sigma_i - X_i$$

converge locally uniformly in C^∞ to a translator (possibly with boundary) $\widehat{\Sigma}$ which lies in Π_θ and satisfies $H \geq 0$ with equality at the origin. Since the estimates of Theorem 13.34 hold up to the boundary, (13.99) implies that the origin must be an interior point. The strong maximum principle applied to (13.70) then implies that $H \equiv 0$ on $\widehat{\Sigma}$ and we conclude that $\widehat{\Sigma}$ is either a hyperplane or half-hyperplane. On the other hand, by the reflection symmetry of Σ^n_i across the hyperplane $\{0\} \times \mathbb{R}^{n-1} \times \mathbb{R}$, the normal to the limit (half-) hyperplane is orthogonal to e_1, so $\widehat{\Sigma}$ cannot lie in Π^{n+1}_θ, a contradiction. This completes the proof of Theorem 13.53. □

The following theorem shows that the asymptotic translators of a convex, $\mathrm{SO}(n-1)$-invariant translator which lies in the slab Π^{n+1}_θ (and in no smaller slab) are the vertical planes bounding Π^{n+1}_θ and the oblique Grim hyperplanes which lie in Π^{n+1}_θ (and in no smaller slab).

Theorem 13.59 (Unique asymptotics for flying wings [**98, 494**]). *Given $n \geq 2$ and $\theta \in (0, \frac{\pi}{2})$, let Σ^n be a convex translator which lies in the slab Π^{n+1}_θ (and in no smaller slab). If $n \geq 3$, assume in addition that Σ^n is rotationally symmetric with respect to the subspace* $\operatorname{span}\{e_2, \ldots, e_n\}$. *The* *curve* $\{\sin\omega e_n - \cos\omega e_{n+1} : \omega \in [0,\theta)\}$ *lies in the normal image of* Σ^n *and the translators*

$$\Sigma^n_\omega \doteqdot \Sigma^n_\theta - P(\omega)$$

converge locally uniformly in the smooth topology to the oblique Grim hyperplane Γ^n_θ *as* $\omega \to \theta$, *where* $P(\omega)$ *is the point on* Σ^n_θ *with* $\mathsf{N}(P(\omega)) = \sin\omega e_n - \cos\omega e_{n+1}$.

Proof sketch. The theorem was proved in the 2-dimensional case by Spruck and Xiao [**494**]. We sketch the proof from [**98**], which applies also in the higher-dimensional, symmetric case. The proof makes use of the rotationally symmetric ancient solution constructed in Section 14.8 (see Theorem 14.27) as a barrier.

Define

$$\overline{\omega} \doteqdot \sup\{\omega \in [0,\infty) : \sin\omega e_n - \cos\omega e_{n+1} \in \mathsf{N}(\Sigma)\}\,.$$

Let $\omega_i \in (0,\overline{\omega})$ be a sequence of angles converging to $\overline{\omega}$. By Lemma 13.42, the translators

$$\Sigma_i \doteqdot \Sigma - P(\omega_i)$$

converge locally uniformly in the smooth topology to a limit translator which splits off the line $\{r(\cos\overline{\omega}e_n + \sin\overline{\omega}e_{n+1}) : r \in \mathbb{R}\}$ and lies in a slab parallel to Π_θ. Since, when $n \geq 3$, the limit splits off an additional $n-2$ lines due to the rotational symmetry, the limit must be the oblique Grim hyperplane $\Gamma^n_{\overline{\omega}}$. In fact, since the components of the normal are monotone along the curve $\gamma(\omega) \doteqdot P(\sin\omega e_n - \cos\omega e_{n+1})$, the normal must actually converge (to $\sin\overline{\omega}e_n - \cos\overline{\omega}e_{n+1}$) along γ. It follows that the limit is independent of the subsequence and we conclude that the translators

$$\Sigma_\omega \doteqdot \Sigma - P(\omega)$$

converge locally uniformly in the smooth topology to $\Gamma^n_{\overline{\omega}}$ as $\omega \to \overline{\omega}$. Note that $\overline{\omega} \leq \theta$ since the limit $\Gamma^n_{\overline{\omega}}$ must lie in a slab of width $\frac{\pi}{2}\sec\theta$ (in fact, it must lie in the original slab since it is reflection symmetric). It remains to show that $\overline{\omega} \geq \theta$.

Suppose, to the contrary, that $\overline{\omega} < \theta$. Given $\omega \in [0,\frac{\pi}{2})$, let $\mathrm{P}^\omega_t \doteqdot \sec\omega \mathrm{P}_{\cos^2\omega t}$ be the rotationally symmetric convex ancient solution which lies in the slab Π_ω (and in no smaller slab) and becomes extinct at the origin at time 0 (see Theorem 14.27). The "radius" $\ell_\omega(t)$ of this pancake satisfies (see [**97**])

$$\begin{aligned}(13.100)\quad \ell_\omega(t) \doteqdot \max_{p\in \mathrm{P}^\omega_t}\langle p, e_2\rangle &= \sec\omega\,\ell_0(\cos^2\omega t)\\ &= -t\cos\omega + (n-1)\sec\omega\log(-t) + c + o(1)\end{aligned}$$

as $t \to -\infty$, where the constant c and the remainder term depend on ω and n. Observe that the ray $L_\omega = \{r(\cos\omega e_n + \sin\omega e_{n+1}) : r > 0\}$ is tangent to the circle in the plane $\operatorname{span}\{e_n, e_{n+1}\}$ of radius $-\cos\omega t$ centered at $-te_{n+1}$.

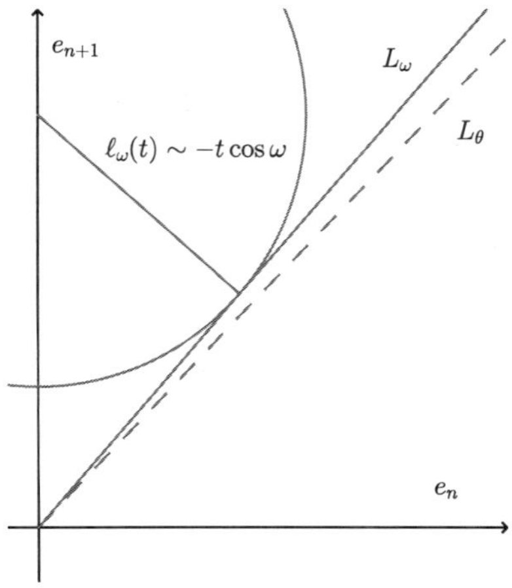

Figure 13.10. If $\overline{\omega} < \theta$, then we can find a pancake that lies above the translator for $-t$ sufficiently large.

Indeed, a point $r(\cos\omega e_n + \sin\omega e_{n+1})$ lies on this circle if and only if

$$\begin{aligned} |r\cos\omega e_n + (r\sin\omega + t)e_{n+1}|^2 &= t^2\cos^2\omega \\ \iff \quad (r + t\sin\omega)^2 &= 0\,. \end{aligned}$$

So there exists a unique point with this property, as claimed. Since, by hypothesis, $\theta < \overline{\omega}$, we conclude from (13.100) that the circle of radius $\ell_\theta(-t)$ lies above the line $L_{\overline{\omega}}$ for $-t$ sufficiently large (cf. Figure 13.10). Thus, for $-t$ sufficiently large, the cross section of the pancake P_t^ω lies above $\Sigma + te_{n+1}$. A straightforward argument exploiting convexity (see Figure 13.11) shows that x_0 and y_0 can be chosen so that the pancake $\mathrm{P}_t^\omega - x_0e_1 - y_0e_n$ lies above $\Sigma_t \doteqdot \Sigma + te_{n+1}$ when $-t$ is sufficiently large (see **[98]** for the details).

Figure 13.11. Linearly interpolating between the "middle" and "edge" regions in the level set $\Sigma_h \doteqdot \{p \in \Sigma : \langle p, e_{n+1}\rangle = h\}$ yields an estimate for the "width" of Σ.

But this contradicts the avoidance principle since $\mathrm{P}^{\omega}_t - x_0 e_1 - y_0 e_n$ contracts to the point $-x_0e_1 - y_0e_n$ as $t \to 0$, which lies below Σ_0. $\square$

Combining the unique asymptotics with the Alexandrov reflection principle[6], it is now easy to show that such a translator is reflection symmetric across the midplane of its defining slab.

Corollary 13.60. *Given $\theta \in (0, \frac{\pi}{2})$, let Σ^n be a strictly convex translator which lies in the slab Π^{n+1}_θ (and in no smaller slab). When $n \geq 3$, assume in addition that Σ^n is rotationally symmetric with respect to the subspace* $\operatorname{span}\{e_2, \ldots, e_n\}$*. Then Σ^n is reflectionally symmetric across the hyperplane $\{0\} \times \mathbb{R}^n$.*

Let us begin by introducing some notation. Given a unit vector $e \in S^n$ and some $\alpha \in \mathbb{R}$, denote by $\mathrm{H}_{e,\alpha}$ the half-space $\{p \in \mathbb{R}^{n+1} : \langle p, e\rangle < \alpha\}$ and by $R_{e,\alpha} \cdot \Sigma \doteqdot \{p - 2(\langle p, e\rangle - \alpha)e : p \in \Sigma\}$ the reflection of Σ across the hyperplane $\partial \mathrm{H}_{e,\alpha}$. We say that Σ **can be reflected strictly about** $\mathrm{H}_{e,\alpha}$ if $(R_{e,\alpha} \cdot \Sigma) \cap \mathrm{H}_{e,\alpha} \subset \Omega \cap \mathrm{H}_{e,\alpha}$.

Lemma 13.61 (Alexandrov reflection principle). *Let Σ^n be a convex translator in $\mathbb{R}^{n+1}$. If $\Sigma^n_h \doteqdot \Sigma^n \cap \{(x, y, z) \in \mathbb{R} \times \mathbb{R}^{n-1} \times \mathbb{R} : z > h\}$ can be reflected strictly about $\mathrm{H}_{e,\alpha}$ for some $e \in \{e_{n+1}\}^\perp$, then Σ^n can be reflected strictly about $\mathrm{H}_{e,\alpha}$.*

Proof. Since the translator equation is invariant under reflections about vertical hyperplanes, this is a consequence of the strong maximum principle and the Hopf boundary point lemma. $\square$

Claim 13.62. For every $\alpha \in (0, \frac{\pi}{4})$ there exists $h_\alpha < \infty$ such that $\Sigma^n_{h_\alpha} \doteqdot \Sigma^n \cap \{(x, y, z) \in \mathbb{R} \times \mathbb{R}^{n-1} \times \mathbb{R} : z > h_\alpha\}$ can be reflected strictly about $\mathrm{H}_\alpha \doteqdot \mathrm{H}_{e_1,\alpha}$.

Proof. Suppose that the claim does not hold. Then there must be some $\alpha \in (0, \frac{\pi}{4})$ and a sequence of heights $h_i \to \infty$ such that $(R_\alpha \cdot \Sigma_{h_i}) \cap \mathrm{H}_\alpha \cap \Sigma_{h_i} \neq \emptyset$. Choose a sequence of points $p_i = x_i e_1 + y_i e_2 \in \Sigma_{h_i}$ whose reflection about the hyperplane H_α satisfies $(2\alpha - x_i)e_1 + y_i e_2 \in (R_\alpha \cdot \Sigma_{h_i}) \cap \Sigma_{h_i} \cap \mathrm{H}_\alpha$ and set $p'_i = x'_i e_1 + y'_i e_2 \doteqdot (2\alpha - x_i)e_1 + y_i e_2$. Without loss of generality, we may assume that $y'_i = y_i \geq 0$. Since $\alpha \leq x_i < \frac{\pi}{2}$, the point p'_i satisfies $\alpha \geq x'_i > -\frac{\pi}{2} + 2\alpha$ so that, after passing to a subsequence, $\lim_{i\to\infty} x'_i \in [-\frac{\pi}{2} + 2\alpha, \alpha]$. But since Σ is convex and converges after translating in the plane $\operatorname{span}\{e_2, e_{n+1}\}$ to the Grim hyperplane Γ^n_θ, we conclude that

$$0 = \lim_{i\to\infty} (x_i + x'_i) = 2\alpha\,.$$

So $\alpha = 0$, a contradiction. $\square$

[6] Alexandrov reflection will be discussed at length in Section 18.8.

Proof of Corollary 13.60. It now follows from Lemma 13.61 that Σ^n can be reflected across H_α for all $\alpha \in (0, \frac{\pi}{2})$. The same argument applies when the half-space H_α is replaced by $-\mathrm{H}_\alpha = \{(x, y, z) \in \mathbb{R} \times \mathbb{R} \times \mathbb{R}^{n-1} : x > -\alpha\}$. Now take $\alpha \to 0$. □

Once the unique asymptotics and reflection symmetry have been established, uniqueness of each of the examples in Theorem 13.53 follows by way of a classical argument of Radó (established in the context of minimal surfaces in $\mathbb{R}^3$). This was observed by Hoffman et al. [**287**].

We first consider the symmetric case and then remove this hypothesis when $n = 2$.

Theorem 13.63. *Given $\theta \in (0, \frac{\pi}{2})$, let Σ^n be a convex, locally uniformly convex translator which lies in the slab Π_θ^{n+1} (and in no smaller slab) and is rotationally symmetric with respect to the subspace* $\mathrm{span}\{e_2, \dots, e_n\}$*. Then Σ^n is a translate of the example constructed in Theorem* 13.53.

Proof. Denote by $\Sigma = \operatorname{graph} u$ the flying wing of width $\pi \sec\theta$ and let $\widehat{\Sigma} = \operatorname{graph} \hat{u}$ be a second convex, locally uniformly convex translator which lies in the slab Π_θ^{n+1} (and in no smaller slab). By Theorem 13.59 and Corollary 13.60, we can assume that $u(0) = \hat{u}(0)$ and $Du(0) = D\hat{u}(0) = 0$. We claim that $Du \equiv D\hat{u}$ on $(-\frac{\pi}{2}\sec\theta, \frac{\pi}{2}\sec\theta) \times \{0\}$.

First note that

$$D_i u = D_i \hat{u} = 0 \quad \text{on} \quad (-\tfrac{\pi}{2}\sec\theta, \tfrac{\pi}{2}\sec\theta) \times \{0\}$$

for each $i = 2, \dots, n$ because of the symmetry hypothesis. So it remains to consider the e_1-direction. Since $D_1 u(x, 0) \to \pm\infty$ as $x \to \pm\frac{\pi}{2}\sec\theta$, given any $x_0 \in (-\frac{\pi}{2}\sec\theta, \frac{\pi}{2}\sec\theta)$ we can find $\hat{x}_0 \in (-\frac{\pi}{2}\sec\theta, \frac{\pi}{2}\sec\theta)$ such that $D_1\hat{u}(\hat{x}_0, 0) = D_1 u(x_0, 0)$. We need to show that $\hat{x}_0 = x_0$. So suppose this is not the case. Define the translated function $\hat{u}_0(x, y) \doteqdot \hat{u}(x + \hat{x}_0 - x_0, y)$. Since u and $\hat{u}_0$ agree to first order at $(x_0, 0)$, a classical argument due to Radó (see [**427**, §II.8]) implies that the nodal set

$$\Gamma_0 \doteqdot \{(x, y) \in I_0 \times \mathbb{R} : u(x, y, 0, \dots, 0) - \hat{u}_0(x, y, 0, \dots, 0) = u(x_0, 0) - \hat{u}(\hat{x}_0, 0)\},$$

where

$$I_0 \doteqdot (-\tfrac{\pi}{2}\sec\theta, \tfrac{\pi}{2}\sec\theta) \cap (-\tfrac{\pi}{2}\sec\theta - \hat{x} + x_0, \tfrac{\pi}{2}\sec\theta - \hat{x} + x_0),$$

consists of at least two proper C^1-arcs which intersect transversally at $(x_0, 0)$ but are otherwise disjoint. Since $\hat{x}_0 \neq x_0$, these arcs do not meet the boundary $\partial I_0 \times \mathbb{R}^{n-1}$ and must therefore extend to infinity in the y-direction. Two of the four ends of these arcs must lie on each side of $\{y = 0\}$. But this is in conflict with the unique asymptotics.

We conclude that $D\hat{u} \equiv Du$ on $(-\frac{\pi}{2}\sec\theta, \frac{\pi}{2}\sec\theta)\times\{0\}$. Integrating then yields $\hat{u} \equiv u$ on $(-\frac{\pi}{2}\sec\theta, \frac{\pi}{2}\sec\theta) \times \{0\}$. The theorem now follows from the Cauchy–Kowalevski theorem. □

We now remove the symmetry hypothesis when $n = 2$.

Theorem 13.64. *Given $\theta \in (0, \frac{\pi}{2})$, let Σ^2 be a convex, locally uniformly convex translator which lies in the slab Π^3_θ (and in no smaller slab). Then Σ^2 is a translate of the example constructed in Theorem* 13.53.

Proof. By translating in the (y, z)-plane, we can arrange that the tip of $\Sigma^2 = \operatorname{graph} u$ is the origin (i.e., $0 \in \Sigma^2$ and $\mathsf{N}(0) = -e_3$). For each $\omega \in (0, \frac{\pi}{2})$, denote by $\Sigma^2_\omega = \operatorname{graph} u_\omega$ the unique bisymmetric flying wing with tip at the origin which lies in the slab Π^3_ω.

Claim 13.65 ([**287**, Lemma 5.5])**.** Given $\varepsilon > 0$ so that $(\theta-\varepsilon, \theta+\varepsilon) \subset (0, \frac{\pi}{2})$,

$$u_{\theta-\varepsilon}(0, y) \leq u(0, y) \leq u_{\theta+\varepsilon}(0, y) \quad \text{for every} \quad y \in \mathbb{R}\,.$$

Proof of Claim 13.65. Since θ and ε are arbitrary, it suffices to prove that $u_{\theta+\varepsilon}(0, y) \leq u(0, y)$. We claim that $u_{\theta+\varepsilon}(0, \cdot) - u(0, \cdot)$ has only one critical point, namely the origin. Suppose, to the contrary, that $u_{\theta+\varepsilon}(0, \cdot) - u(0, \cdot)$ has a second critical point, $y_0 \neq 0$. By Corollary 13.60, the point $-y_0$ is also critical. Radó's argument then implies that the nodal set

$$\Gamma_0 \doteqdot \{u_{\theta+\varepsilon}(0, \cdot) - u(0, \cdot) = u_{\theta+\varepsilon}(0, y_0) - u(0, y_0)\}$$

contains at least two C^1-arcs intersecting transversely at y_0 and a further two intersecting transversely at $-y_0$. At least four of the eight ends of these arcs must reach infinity on the same side of $\{y = 0\}$, contradicting the unique asymptotics.

Thus, since (by the reflection symmetry) $D_1 u_{\theta+\varepsilon}(0, y) = 0 = D_1 u(0, y)$,

$$D_2 u_{\theta+\varepsilon} u(0, y) \neq D_2 u(0, y)$$

for all $y \neq 0$. Since (by the unique asymptotics)

$$\lim_{y\to\infty} D_2 u_{\theta+\varepsilon} u(0, y) > \lim_{y\to\infty} D_2 u(0, y)\,,$$

we conclude that

$$D_2 u_{\theta+\varepsilon} u(0, y) > D_2 u(0, y)$$

for all $y > 0$. Thus,

$$u_{\theta+\varepsilon}(0, y) = \int_0^y D_2 u_{\theta+\varepsilon}(0, \eta)\, d\eta > \int_0^y D_2 u(0, \eta)\, d\eta = u_{\theta-\varepsilon}(0, y)$$

for all $y > 0$. The proof is similar for $y < 0$. □

Claim 13.66 ([**287**, Corollary 5.4])**.** The translators $\Sigma^2_{\theta\pm\varepsilon}$ converge locally uniformly in the smooth topology to Σ^2_θ as $\varepsilon \to 0$.

Proof of Claim 13.66. Let $\{\varepsilon_i\}_{i\in\mathbb{N}}$ be any sequence of positive numbers satisfying $\varepsilon_i \to 0$. Then, after passing to a subsequence, the translators $\Sigma^2_{\theta\pm\varepsilon_i}$ converge locally uniformly in the smooth topology to respective limit translators $\Sigma^2_\pm$. The limits $\Sigma^2_\pm$ are necessarily bisymmetric and lie in the slab Π^3_θ. We claim that they lie in no smaller slab. Indeed, since the translators Σ^2_ω, $\omega \in (0, \frac{\pi}{2})$, are symmetric across the $\{y = 0\}$ plane, the argument used to prove Claim 13.65 shows that

$$u_{\omega_1}(x, 0) \geq u_{\omega_2}(x, 0) \text{ for every } x \in (-\tfrac{\pi}{2}\sec\omega_1, \tfrac{\pi}{2}\sec\omega_1)$$

whenever $\omega_1 < \omega_2$, which implies the claim. By Theorem 13.63, $\Sigma^2_\pm = \Sigma^2_\theta$. The claim follows since the limit is independent of the subsequence. □

We deduce that

$$u(0, y) = u_\theta(0, y) \text{ for all } y \in \mathbb{R}$$

and hence

$$D_2u(0, y) = D_2u_\theta(0, y) \text{ for all } y \in \mathbb{R}\,.$$

Since u and u_θ are reflection symmetric about $\{x = 0\}$, we also have

$$D_1u(0, y) = D_1u_\theta(0, y) = 0 \text{ for all } y \in \mathbb{R}\,.$$

So the claim follows from the Cauchy–Kowalevski theorem. □

Combining these results with Wang's uniqueness theorem for the bowl soliton and the Spruck–Xiao convexity estimate now yields a complete classification of proper, mean convex translators in $\mathbb{R}^3$.

Theorem 13.67 (Classification of mean convex translators in $\mathbb{R}^3$ [**22, 98, 287, 494, 521**]). *Let Σ^2 be a properly embedded, mean convex translator in $\mathbb{R}^3$. Then, up to a translation and a vertical rotation, Σ^2 is either*

- *the vertical plane $\{0\} \times \mathbb{R}^2$,*
- *the Grim plane Γ^2,*
- *an oblique Grim plane Γ^2_θ for some $\theta \in (0, \frac{\pi}{2})$,*
- *one of the flying wings W^2_θ (constructed in Theorem 13.53) for some $\theta \in (0, \frac{\pi}{2})$,*
- *the bowl soliton.*

There are further examples in higher dimensions: In addition to Wang's examples [**521**], Hoffman et al. constructed an $(n-1)$-parameter family of graphical translators in each slab in $\mathbb{R}^{n+1}$ with prescribed principal curvatures (adding to 1) at their "tip" (including, in particular, examples which are not $\mathrm{SO}(1) \times \mathrm{SO}(n-1)$-invariant when $n \geq 3$) [**287**, Theorem 11.1].

Theorem 13.68. *Given $\theta \in (0, \frac{\pi}{2})$ and $0 < k_1 \leq \cdots \leq k_{n-1}$ such that $\sum_{i=1}^{n-1} k_i = 1$, there exists $\mu \in (\frac{k_{n-1}}{1+k_{n-1}}, 1)$ and a convex translator $\Sigma^n \subset \mathbb{R}^{n+1}$ defined in Π_θ^{n+1} which is tangent to $\{e_{n+1}\}^\perp$ at the origin with second fundamental form $\mathrm{II}_0 = (1-\mu)\sum_{i=1}^{n-1} k_i e_i \otimes e_i + \mu e_n \otimes e_n$ at the origin. It is also reflection symmetric across each coordinate hyperplane $\{e_i\}^\perp$. Given $0 < i < j < n$, Σ^n is rotationally symmetric with respect to the subspace $e_i \wedge e_{i+1} \wedge \cdots \wedge e_j$ if and only if $k_i = k_{i+1} = \cdots = k_j$.*

13.14. Bowloids

X.-J. Wang proved that there exist entire, locally uniformly convex translators in $\mathbb{R}^{n+1}$ for each $n \geq 3$ which are not rotationally symmetric [**521**, Theorem 1.2]. The following theorem, due to Hoffman et al. [**287**, Theorem 8.1], provides a full $(n-2)$-parameter family of entire graphical translators in $\mathbb{R}^{n+1}$ with prescribed principal curvatures (adding to 1) at their "tip".

Theorem 13.69. *Given $0 < k_1 \leq \cdots \leq k_{n-1} \leq k_n$ such that $\sum_{i=1}^n k_i = 1$, there exists a convex translator $\Sigma^n \subset \mathbb{R}^{n+1}$ which is tangent to $\{e_{n+1}\}^\perp$ at the origin with second fundamental form $\mathrm{II}_0 = \sum_{i=1}^n k_i e_i \otimes e_i$. It is entire if $k_{n-1} = k_n$. The translator is reflection symmetric across each coordinate hyperplane $\{e_i\}^\perp$. It is rotationally symmetric with respect to the subspace $e_i \wedge e_{i+1} \wedge \cdots \wedge e_j$ if and only if $k_i = k_{i+1} = \cdots = k_j$.*

Proof. By the a priori estimates implied by Exercise 13.12 and by a continuity argument, given any

$$\mathbf{a} = (a_1, \ldots, a_n) \in \Delta^n \doteqdot \{x \in [0,\infty)^n : x_1 + \cdots + x_n = 1\}$$

and $\lambda > 0$, there exist $R > 0$ and a solution $u_{\mathbf{a},\lambda} : \Omega_{\mathbf{a},\lambda} \to \mathbb{R}$ to the Dirichlet problem for the translator equation over the domain $\Omega_{\mathbf{a},\lambda} \doteqdot \{x \in \mathbb{R}^n : \sum_{i=1}^n a_i x_i^2 < R^2\}$ satisfying $u_{\mathbf{a},\lambda}(0) = -\lambda$.

By Theorem 13.41, $u_{\mathbf{a},\lambda}$ is locally uniformly convex. Since $\Omega_{\mathbf{a},\lambda}$ is reflection symmetric across each coordinate hyperplane $\{e_i\}^\perp$, so is $\Sigma_{\mathbf{a},\lambda}$. In particular, the tip (i.e., lowest point) $p_{\mathbf{a},\lambda}$ of $\Sigma_{\mathbf{a},\lambda}$ lies on the x_{n+1}-axis and the coordinate vectors $\{e_i\}_{i=1}^n$ are principal directions at p_0. We shall denote by $\mathbf{k}_{\mathbf{a},\lambda} = (k_1^{\mathbf{a},\lambda}, \ldots, k_n^{\mathbf{a},\lambda})$ the corresponding principal curvature n-tuple. Furthermore, if $a_{i_1} = \cdots = a_{i_k}$ for some subset $\{i_1, \ldots i_k\} \subset \{1, \ldots, n\}$, then $\Omega_{\mathbf{a},\lambda}$ is $\mathrm{SO}(k)$-invariant with respect to the subspace $e_{i_1} \wedge \cdots \wedge e_{i_k}$, and hence so is $\Sigma_{\mathbf{a},\lambda}$.

Claim 13.70. If $a_i < a_j$, then $k_i^{\mathbf{a},\lambda} < k_j^{\mathbf{a},\lambda}$.

Proof of Claim 13.70. Let $J = x_i e_j - x_j e_i$ be the tangent vector field to the counterclockwise rotation in the plane $e_i \wedge e_j$. Then, by Exercise 13.7, the function

$$v(X) \doteqdot \langle J(X), \mathsf{N}(X)\rangle = (1+|Du_{\mathbf{a},\lambda}|^2)^{-\frac{1}{2}}(x_i\partial_j - x_j\partial_i)u_{\mathbf{a},\lambda}$$

satisfies the Jacobi equation:

$$-(\Delta + \nabla_{e_{n+1}^\top} + |\mathrm{II}|^2)v = 0. \tag{13.101}$$

Consider the region $U \doteqdot \Omega_{\mathbf{a},\lambda} \cap \Gamma_{ij}$, where $\Gamma_{ij} \doteqdot \{x \in \mathbb{R}^n : x_i > 0, x_j > 0\}$. Since $\Sigma_{\mathbf{a},\lambda}$ is reflection symmetric across each coordinate hyperplane, we find

$$(x_i\partial_j - x_j\partial_i)u_{\mathbf{a},\lambda} = 0$$

whenever $x \in \Omega_{\mathbf{a},\lambda} \cap \partial\Gamma_{ij}$. If $x \in \partial\Omega_{\mathbf{a},\lambda} \cap \Gamma_{ij}$, we find that

$$(x_i\partial_j - x_j\partial_i)u_{\mathbf{a},\lambda} > 0$$

since $a_i < a_j$. The maximum principle now implies that $u_{\mathbf{a},\lambda}|_U > 0$. Since

$$u(x) - u(p_{\mathbf{a},\lambda}) = \frac{1}{2}\sum_{i=1}^n k_i^{\mathbf{a},\lambda}x_i^2 + O(|x|^3), \tag{13.102}$$

we conclude that

$$0 < (x_i\partial_j - x_j\partial_i)u_{\mathbf{a},\lambda} = x_ix_j(k_j^{\mathbf{a},\lambda} - k_i^{\mathbf{a},\lambda}) + O(|x|^3)$$

in U, which yields the claim. □

Claim 13.71. Given $\mathbf{k} \in \Delta^n$ and $\lambda > 0$ there exists $\mathbf{a} \in \Delta^n$ so that $\mathbf{k}^{\mathbf{a},\lambda} = \mathbf{k}$.

Proof of Claim 13.71. Given $\lambda > 0$ and $n \in \mathbb{N}$, define $F^n_\lambda : \Delta^n \to \Delta^n$ by

$$F^n_\lambda(\mathbf{a}) \doteqdot \mathbf{k}_{\mathbf{a},\lambda}.$$

We need to show that F^n_λ is surjective. Since $\Sigma_{\mathbf{a},\lambda}$ depends continuously on $\mathbf{a}$, F^n_λ is continuous in the interior of Δ^n. We claim that, for any m-dimensional face $\Delta^m \subset \partial\Delta^n$, $F^n_\lambda|_{\Delta^m} = F^m_\lambda$. Briefly, if $\{\mathbf{a}_i\}_{i\in\mathbb{N}}$ is a sequence of points $\mathbf{a}_i \in \Delta^n$ converging to $\mathbf{a} = (a_1, \ldots, a_m, 0, \ldots, 0) \in \Delta^m$, then a subsequence of the sequence of solutions $u_{\mathbf{a_i},\lambda} : \Omega_{\mathbf{a}_i,\lambda} \to \mathbb{R}$ converges locally uniformly in the smooth topology to a limit $u : \Omega_{\mathbf{a},\lambda} \to \mathbb{R}$. Since $\Omega_{\mathbf{a},\lambda} = \Omega_{\hat{\mathbf{a}},\lambda} \times \mathbb{R}^{n-m}$, where $\hat{\mathbf{a}} \doteqdot (a_1, \ldots, a_m)$, and u is convex, the splitting theorem implies that it is constant on the flat factor. Since $\Omega_{\hat{\mathbf{a}},\lambda}$ is compact, uniqueness of solutions implies that $u|_{\Omega_{\hat{\mathbf{a}},\lambda}} = u_{\hat{\mathbf{a}},\lambda}$. We conclude that $F^n_\lambda|_{\Delta^m} = F^m_\lambda$.

We have shown that $F^n_\lambda : \Delta^n \to \Delta^n$ is continuous and maps $\partial\Delta^n$ to $\partial\Delta^n$. Elementary degree theory now implies that F^n_λ is surjective [**255**]. □

Claim 13.72. Suppose that $k_n^{\mathbf{a},\lambda} = \max\{k_1^{\mathbf{a},\lambda}, \ldots, k_n^{\mathbf{a},\lambda}\}$. For each $r \in [0, \frac{R}{\sqrt{a_n}})$,

$$u_{\mathbf{a},\lambda}(re_n) = \max_{x\in\partial B_r\cap\Omega_{\mathbf{a},\lambda}} u_{\mathbf{a},\lambda}(x).$$

Proof of Claim 13.72. By Claim 13.71, $a_n \geq a_j$ for each $j \leq n$. In particular, $B_{R/\sqrt{2a_n}} \subset \overline{\Omega}_{\mathbf{a},\lambda}$. Define

$$r_* \doteqdot \sup\left\{r \in [0, \tfrac{R}{\sqrt{a_n}}] : u_{\mathbf{a},\lambda}(re_n) = \max_{x\in\partial B_r\cap\Omega_{\mathbf{a},\lambda}} u_{\mathbf{a},\lambda}(x)\right\}.$$

The expansion (13.102) (and the fact that $u_{\mathbf{a},\lambda}$ is rotationally invariant in the subspace $e_{i_1}\wedge\cdots\wedge e_{i_k}$ whenever $a_{i_1} = \cdots = a_{i_k}$) implies that $r_* > 0$. Suppose, contrary to the claim, that $r_* < \frac{R}{\sqrt{a_n}}$. Then, by the reflection symmetries of $\Sigma_{\mathbf{a},\lambda}$, for $r > r_*$ sufficiently close to r_*, we can find $x_r \in \Omega_{\mathbf{a},\lambda}\cap\overline{\Gamma}{}^n_+$ not on the e_n-axis such that $u(x_r) = \max_{\partial B_r} u$. But then, since $(x_j\partial_i - x_i\partial_j)|_{x_r}$ is tangent to ∂B_r at x_r, we have that

$$0 = (x_j\partial_i - x_i\partial_j)|_{x_r}u$$

for each i, j. If $(x_r)_n > 0$, then the strong maximum principle applied to (13.101) yields

$$0 \equiv (x_n\partial_i - x_i\partial_n)u$$

for each i such that $(x_r)_i > 0$. It follows that $u_{\mathbf{a},\lambda}$ is rotationally invariant in the subspace $e_{i_1}\wedge\cdots\wedge e_{i_k}\wedge e_n$, where $\{i_1,\ldots,i_k\}$ are the indices corresponding to the nonzero components of x_r. But then

$$\max_{x\in\partial B_r} u(x) = u(x_r) = u(|x_r|e_n).$$

So we may assume that $(x_r)_n = 0$. Taking $r \searrow r_*$, we find $x^* \in \partial B_{r_*}\cap\overline{\Gamma}{}^n_+$ with $(x^*)_n = 0$ and $u(x^*) = \max_{\partial B_{r_*}} u = u(r_*e_n)$. But then there must be some interior point $x_* \in \partial B_{r_*}\cap\Gamma_+$ that *minimizes* u over the arc $\partial B_{r_*}\cap (x^*\wedge e_n)$. Applying the above argument at x_* completes the proof. □

Given any sequence $\lambda_j \to \infty$, choose $\mathbf{a}_j$ so that the principal curvatures at the tip of $\Sigma_{\mathbf{a}_j,\lambda_j}$ are $\mathbf{k}$. By (13.85), a subsequence of the translators $\Sigma_j \doteqdot \Sigma_{\mathbf{a}_j,\lambda_j} - p_{\mathbf{a}_j,\lambda_j}$ converges locally uniformly in the smooth topology to some limit translator $\Sigma_{\mathbf{k}}$ which is tangent to $\{e_{n+1}\}^\perp$ at the origin with second fundamental form $\mathrm{II}_0 = \sum_{i=1}^n k_ie_i\otimes e_i$. It is reflection symmetric across each coordinate hyperplane and, by Claim 13.70, rotationally symmetric with respect to the subspace $e_i\wedge\cdots\wedge e_j$ if $k_i = \cdots = k_j$. In particular, if $k_n = k_{n-1}$, then $\Sigma_{\mathbf{k}}$ is rotationally symmetric with respect to the subspace $e_{n-1}\wedge e_n$. On the other hand, if $\Sigma_{\mathbf{k}}$ is not entire, then Claim 13.72 and Wang's dichotomy (Theorem 13.43) imply that it is defined over a slab domain of the form $\mathbb{R}^n\times(-L, L)$. This completes the proof. □

13.15. Notes and commentary

Self-similar solutions for the Ricci flow are called Ricci solitons. The study of gradient Ricci solitons is fundamental to the understanding of singularities in the Hamilton–Perelman theory of Ricci flow singularity formation.

13.15.1. Classification of shrinkers. Whereas the class of mean convex shrinkers is quite restricted, it turns out that there are a large number of examples which are not mean convex. Let us mention some of the many recent results concerning the construction and classification of shrinkers.

Using the *shooting method* for geodesics, Angenent constructed rotationally symmetric "shrinking doughnuts" in all dimensions $n \geq 2$ — embedded examples with the topology of $S^1 \times S^{n-1}$ [**68**]. As we mentioned in Section 9.1, these examples may be used as barriers to prove the existence of neckpinch singularities. Drugan [**200**] constructed an immersed rotationally symmetric shrinking sphere using similar methods. Building on these results and the work of Kleene and Møller [**333**], Drugan and Kleene [**201**] constructed infinitely many rotationally symmetric, complete, immersed shrinkers in $\mathbb{R}^{n+1}$ with the topological type of S^n, $S^{n-1} \times \mathbb{R}$, and $S^1 \times S^{n-1}$, respectively. McGrath has constructed embedded shrinkers in $\mathbb{R}^{2n}$ with the topology of $S^1 \times S^{n-1} \times S^{n-1}$ and admitting $\mathrm{SO}(n) \times \mathrm{SO}(n)$ symmetry [**397**]. On the other hand, when $n = 2$, Brendle has proved that the only embedded shrinker with the topological type of S^2 is the shrinking round sphere [**108**]. Uniqueness of Angenent's shrinking embedded torus remains open but Mramor and S. Wang have proved that any embedded shrinker in $\mathbb{R}^3$ of genus 1 must at least be unknotted [**408**].

A number of embedded shrinkers in $\mathbb{R}^3$ of large genus have been constructed by *desingularization* arguments (similar to those pioneered by Nicos Kapouleas [**323–325**] and others [**279, 503**] to construct new families of minimal and constant mean curvature surfaces): X. H. Nguyen [**418, 420**] showed that the intersection of the shrinking cylinder and the plane can be "desingularized" by gluing in a "bent" Scherk surface and thereby obtained noncompact embedded examples with two ends and any sufficiently large genus. X. H. Nguyen [**422**] and, independently, Kapouleas, Kleene, and Møller [**326**] were able to construct noncompact embedded examples with one end and sufficiently large genus by desingularizing the intersection of the shrinking sphere and a plane via a similar construction. Compact examples of large genus were constructed by Møller by desingularizing the shrinking round sphere and Angenent's shrinking torus [**402**]. We will briefly describe this method in Section 13.15.3 in the context of translating solutions.

Using *min-max* methods, Ketover has constructed embedded shrinkers in $\mathbb{R}^3$ of genera 3, 5, 7, 11, and 19 which resemble "doublings" of the Platonic solids [**329**]. These shrinkers were conjectured to exist by Chopp based on numerical evidence [**153**]. It is expected that they are compact, although this is not known.

L. Wang has obtained uniqueness results for shrinkers with asymptotically conical or asymptotically cylindrical ends [**517, 518**]. Building on, and

Figure 13.12. Lu Wang. Author: Petra Lein. Photo courtesy of Archives of the Mathematisches Forschungsinstitut Oberwolfach.

proving a conjecture from, the work of Colding, Ilmanen, Minicozzi, and White [**178**], J. Bernstein and L. Wang [**88**] have proved that the round sphere uniquely minimizes the entropy among all nonflat 2-dimensional self-shrinkers.

Cavalcante and Espinar [**136**] proved that the only self-shrinker in $\mathbb{R}^{n+1}$ which lies in the half-space $H \doteqdot \{X \in \mathbb{R}^{n+1} : \langle X, e_{n+1} \rangle \geq 0\}$ is ∂H. They also showed that the only self-shrinker which lies inside the solid cylinder $C \doteqdot \mathbb{R}^m \times B^{n-m+1}_{\sqrt{2(n-m)}}$ is ∂C.

For arbitrary-codimension mean curvature flow of surfaces, Colding and Minicozzi [**186**] obtained entropy and dimension bounds for the space of $\mathcal{F}$-stable shrinking self-similar solutions on closed surfaces of any given genus.

An excellent survey of the various existence and uniqueness results for shrinkers may be found in [**202**].

13.15.2. Open problems related to shrinkers.

Open Problem 13.73. Show that Ketover's shrinkers in $\mathbb{R}^3$ are compact.

Ketover's examples have genera 3, 5, 7, 11, and 19, while the compact examples of Møller have large genera. An obvious question is the following.

Open Problem 13.74. Is there a compact shrinker in $\mathbb{R}^3$ of genus 2?

The following pair of open problems are posed in [**202**].

Open Problem 13.75. Is Angenent's shrinking torus the only closed, embedded self-shrinker in $\mathbb{R}^3$ of genus 1?

Uniqueness is not even known within the class of rotational examples. Embeddedness is a necessary hypothesis due to the immersed examples constructed by Drugan and Kleene [**201**]. By the result of Mramor and S. Wang, an embedded shrinker in $\mathbb{R}^3$ of genus 1 must at least be unknotted [**408**].

Open Problem 13.76. Is the round sphere of radius $\sqrt{6}$ centered at the origin the only embedded shrinker in $\mathbb{R}^4$ with the topology of S^3?

The embeddedness assumption is necessary due to the existence of immersed examples of Drugan [**200**] and Drugan and Kleene [**201**]. Uniqueness is known in the rotational case [**199, 333**].

13.15.3. Classification of translators.

Further examples. Halldorsson constructed a 2-parameter family of immersed planar translators with screw symmetry [**258**]. These examples are sometimes referred to as **translating helicoids**. Due to their screw symmetry, vertical translation is equivalent to rotation about the vertical axis, so these solutions are also rotators.

Graham Smith [**485**] and Dávila, del Pino, and Nguyen [**195**] have independently constructed families of embedded translators of large genus with three ends asymptotic to bowl solitons using desingularization agruments. In Smith's construction, the solutions are obtained by desingularizing the union of three bowl solitons using a Costa–Hoffman–Meeks minimal surface [**485**]. In the construction of Dávila, del Pino, and Nguyen, the solutions are obtained by desingularizing the union of a bowl soliton and a translating catenoid using Scherk minimal surfaces. These examples converge in the Hausdorff sense, with genus tending to infinity, to the union of a bowl soliton and a translating catenoid. Let us give a brief overview of the argument from [**195**].

Figure 13.13. Xuan Hien Nguyen. Photo courtesy of Keo Pierron.

Theorem 13.77 (Dávila, del Pino, and Nguyen [**195**]). *Let P and C be, respectively, a bowl soliton and a translating catenoid in $\mathbb{R}^3$ which intersect transversally. For every sufficiently small $\varepsilon > 0$, there exists a properly* *embedded translator Σ_ε with three ends which lies in an ε-neighborhood of $P \cup C$. The genus $g(\varepsilon)$ of Σ_ε satisfies $g(\varepsilon) \sim \varepsilon^{-1}$ as $\varepsilon \to 0$.*

Proof sketch from [195]. After scaling by ε^{-1}, the problem is equivalent to finding a properly embedded surface $\Sigma_\varepsilon \subset \mathbb{R}^3$ satisfying

$$H = -\varepsilon \langle \mathsf{N}, e_3 \rangle . \tag{13.103}$$

Fix $\varepsilon > 0$. The first step is to construct a surface Σ_{approx} which is almost a solution. This is accomplished by desingularizing the union of $\varepsilon^{-1}P$ and $\varepsilon^{-1}C$ using singly periodic Scherk minimal surfaces. At a large distance from the circle of intersection, Σ_{approx} coincides with $\varepsilon^{-1}P \cup \varepsilon^{-1}C$, and in some neighborhood of the intersection circle, Σ_{approx} is given by a slightly bent singly periodic Scherk surface. The approximating solution Σ_{approx} will not exactly solve (13.103). We seek a solution to (13.103) in the form of a normal graph Σ_ϕ over Σ_{approx}. That is,

$$\Sigma_\phi = \{x + \phi(x)\mathsf{N}(x) : x \in \Sigma\}$$

with $|\phi|$ sufficiently small. The mean curvature H_ϕ and normal N_ϕ of Σ_ϕ are given by

$$H_\phi = H + \Delta\phi + |\mathrm{II}|^2\phi + Q_1$$

and

$$\mathsf{N}_\phi = \mathsf{N} - \nabla\phi + Q_2 ,$$

respectively, where Q_1 and Q_2 are quadratic expressions in ϕ, $\nabla\phi$, and $\nabla^2\phi$. It follows that Σ_ϕ solves (13.103) if and only if ϕ solves

$$\Delta\phi + |\mathrm{II}|^2\phi + \varepsilon \langle \nabla\phi, e_3 \rangle + H - \varepsilon \langle \mathsf{N}, e_3 \rangle = Q , \tag{13.104}$$

where Q is a quadratic expression in x, ϕ, $\nabla\phi$, and $\nabla^2\phi$.

The equation (13.104) is solved by linearizing about $\phi = 0$ and analyzing the linearized operator

$$\phi \mapsto L\phi \doteqdot \Delta\phi + |\mathrm{II}|^2\phi + \varepsilon \langle \nabla\phi, e_3 \rangle . \qquad \square$$

A similar argument was used by X. H. Nguyen to construct a very large class of further **translating trident** examples by desinguarizing (1) the intersection of a Grim plane with a vertical plane [**419**] and (2) any finite family of Grim planes in general position [**421, 423**].

Hoffman, Martín, and White constructed a 2-parameter family of (non-convex) "Scherk-like" translators in $\mathbb{R}^3$ [**288**]. Their construction resembles that of the Scherk minimal surfaces.

These results provide a plethora of examples with a high degree of flexibility, in contrast to the relative order and rigidity we have seen in the mean convex case (at least in low dimensions). On the other hand, F. Chini and N. Møller [**146**] have proved that the convex hull of the projection onto the subspace orthogonal to the translation direction of a proper translator is either a hyperplane, a slab, a half-space, or the whole subspace. We have seen many examples for which the projection is a hyperplane, a slab, or the whole subspace. Wang's dichotomy rules out a half-space when the translator is convex.

13.15.3.1. *Uniqueness results.* An early uniqueness theorem for the bowl soliton was obtained by F. Martín, A. Savas-Halilaj, and K. Smoczyk, who used Alexandrov's method of moving planes to prove that the bowl soliton is the only properly embedded translator with finite genus and a single end that is "smoothly asymptotic" to the bowl [**390**].

F. Martín, J. Pérez-García, A. Savas-Halilaj, and K. Smoczyk proved that a nonflat, connected, properly embedded translator in $\mathbb{R}^3$ which has locally uniformly bounded genus and is C^1-asymptotic to two planes outside a cylinder coincides with the Grim hyperplane [**389**]. The genus bound was removed and the result extended to higher dimensions by E. Gama and F. Martín [**233**].

O. Hershkovits [**284**] proved that any translator in $\mathbb{R}^3$ with entropy no greater than the shrinking cylinder is either a vertical plane or the bowl.

13.15.4. Open problems related to translators. By Theorem 10.11, every convex eternal solution to mean curvature flow which admits a space-time critical point for the mean curvature is necessarily a translating solution.

Conjecture 13.78 (B. White [**531**, Conjecture 1]). *Every nontrivial, convex eternal solution to mean curvature flow is a translating solution.*

By Theorem 14.17, the conjecture is true when $n = 1$. White [**531**, p. 133] also asks whether it is true under the additional assumption of being a limit flow of a mean convex, embedded mean curvature flow.

By Corollaries 9.13, 12.12, and 13.48, the bowl soliton is the only translator which arises as a singularity model in embedded mean convex mean curvature flow in $\mathbb{R}^3$. By Corollaries 9.13 and 13.50, the bowl is the only translator which arises as a singularity model in 2-convex mean curvature flow in $\mathbb{R}^{n+1}$ when $n \geq 3$.

Conjecture 13.79 (B. White [**531**, Conjecture 2]). *Up to rigid motions and parabolic rescaling, the products* $\{\mathbb{R}^k \times \mathrm{Bowl}_t^{n-k}\}_{t\in(-\infty,\infty)}$, $k \geq 1$, *where* $\{\mathrm{Bowl}_t\}_{t\in(-\infty,\infty)}$ *is the bowl soliton, are the only translating solutions which arise as singularity models in mean convex mean curvature flow.*

For progress on this conjecture, see Theorem 14.31 and [**151**, Theorem 1.6].

We have seen that (up to a vertical rotation) locally uniformly convex translators in $\mathbb{R}^3$ are uniquely determined by their Gauß images, $G_\theta = \{e \in S^2 : \langle e, e_2\rangle < \sin\theta\}$ for $\theta \in [0, \frac{\pi}{2}]$. In particular, the bowl soliton is the only convex translator in $\mathbb{R}^3$ whose Gauß image covers the whole southern hemisphere. Wang has conjectured that this is also the case in higher dimensions.

Conjecture 13.80 (X.-J. Wang [**521**, p. 1237]). *Let Σ^n be a convex translator in $\mathbb{R}^{n+1}$. If the Gauß image of Σ^n is the whole southern hemisphere $H^n_- \doteqdot \{e \in S^n : \langle e, e_{n+1}\rangle < 0\}$, then Σ^n is the bowl soliton.*

Open Problem 13.81. Are proper, strictly mean convex translators necessarily convex?

Theorems 13.39, 13.40, and 13.41 provide a partial answer. Wang has conjectured an affirmative answer for *entire* translators.

Conjecture 13.82 (X.-J. Wang [**521**, p. 1237]). *Every entire, strictly mean convex translator is convex.*

A related question is whether entire, strictly (mean) convex translators are necessarily noncollapsing.

Open Problem 13.83. Is every (mean) convex entire translator necessarily noncollapsing?

A positive answer would imply Wang's conjecture by Theorem 12.20.

13.15.5. Expanders. Expanding self-similar solutions *flow out* of a cone at time $t = 0$. They are therefore expected to describe both how a mildly singular hypersurface is smoothed out by mean curvature flow and also how the (weak) flow might resolve a conical singularity that has formed. Expanding self-similar solutions are also prototypical examples of immortal solutions. They appear to be less rigid than their shrinking and translating counterparts and have received less attention.

Since self-expanders are immortal, there can be no closed examples. But there are many asymptotically conical examples: By solving an associated asymptotic Plateau problem (as suggested by Ilmanen [**310**]), Q. Ding constructed a self-similarly expanding solution which evolves out of any given (sufficiently regular) cone in $\mathbb{R}^{n+1}$, $2 \le n \le 6$ [**196**]. Some cones admit multiple asymptotic expanders: Angenent, Chopp, and Ilmanen showed that the rotationally symmetric double cone

$$D_\alpha \doteqdot \{X \in \mathbb{R}^{n+1} : X_{n+1}^2 \sin^2\alpha = (X_1^2 + \cdots + X_n^2)\cos^2\alpha\}$$

admits at least two asymptotic expanders (a "2-sheeted" and a "1-sheeted" example, both rotationally symmetric) when $\alpha \in (\alpha_0, \frac{\pi}{2})$, where $\alpha_0 \sim 66°$. In fact, Helmensdorfer [**282**] proved that D_α, $\alpha \in (\alpha_0, \frac{\pi}{2})$, admits a *second* connected, rotationally symmetric expander. So we do not even have uniqueness within a fixed topological class. L. Wang and J. Bernstein show that this behavior is generic [**89**]: There exists an open set of cones in $\mathbb{R}^3$ which admit at least three asymptotic self-expanders (two of which are connected annuli and one of which consists of a pair of disjoint disks).

F. Fong and P. McGrath proved that any proper convex expander which is asymptotic to a rotationally symmetric cone (with either one end, or two ends with the same rotation axis) is necessarily rotationally symmetric about the axis of the cone [**223**].

13.15.6. Self-similar solutions to other flows. Curvature flows by homogeneous functions of curvature are invariant under appropriate parabolic scaling and hence, in principle, admit self-similarly shrinking and expanding solutions. The behavior of self-similar solutions appears to be highly dependent on the homogeneity (for example, as we shall see in Section 16.1, all ellipsoids are shrinkers for the flow by the $\frac{1}{n+2}$-nd power of the Gauß curvature) as well as the degeneracy of the speed function.

James McCoy extended Huisken's classification of compact, convex mean curvature flow shrinkers (Theorem 13.11) to a large class of contraction flows by 1-homogeneous speeds [**396**].

John Urbas obtained a complete classification of convex *noncompact* shrinking, expanding, and translating solutions to flows by powers of the Gauß curvature [**507, 508**]. In particular, in stark contrast to the mean curvature flow, the graphical translator equation for the Gauß curvature flow admits a (unique) complete solution over every bounded convex domain. Kyeongsu Choi, Panagiota Daskalopoulos, and Kiahm Lee obtained optimal regularity of weakly convex translating solutions to the Gauß curvature flow [**150**]. Uniqueness of the shrinking sphere among convex, *compact*, shrinking self-similar solutions to flows by powers $\alpha > \frac{1}{n+2}$ of the Gauß curvature was proved by Brende, Choi, and Daskalopoulos [**113, 148**] (see Section 17.3).

13.16. Exercises

Exercise 13.1. Let X be a shrinking self-similar solution to the mean curvature flow. Show that (13.2) follows from (13.1) and the mean curvature flow equation. Show also that $\partial_t\varphi_t = -\frac{1}{2t}\left((X_{-1})^*(X_{-1})^\top\right)$.

Exercise 13.2. Show that a hypersurface Σ^n of $\mathbb{R}^{n+1}$ is a shrinker if and only if it is minimal with respect to the Gaußian metric, $G = \mathrm{e}^{-\frac{|X|^2}{2n}}\langle\cdot,\cdot\rangle$.

Exercise 13.3. Suppose that $X : M^n \to \mathbb{R}^{n+1}$ is a translator which is invariant under rotations about the axis $\mathbb{R}e$, $e \in S^n$. Show that $e = e_{n+1}$ unless $X(M^n)$ is a vertical hyperplane.

Exercise 13.4. Given a smooth function $u : (r_1, r_2) \to \mathbb{R}$, $0 \leq r_1 < r_2$, show that the graph $\{(x, u(|x|) : r_1 \leq |x| \leq r_2\}$ satisfies the translator equation (6.8) if and only if

$$\frac{u''}{1+(u')^2} + (n-1)\frac{u'}{r} = 1\,.$$

Exercise 13.5. Show that a hypersurface Σ^n of $\mathbb{R}^{n+1}$ is a translator if and only if it is minimal with respect to the metric $G = \mathrm{e}^{-\langle X, e_{n+1}\rangle}\langle \cdot, \cdot \rangle$.

Exercise 13.6. Let $\{\Sigma^n_\varepsilon\}_{\varepsilon \in (-\varepsilon_0, \varepsilon_0)}$ be a 1-parameter family of translators in $\mathbb{R}^{n+1}$. Set

$$u \doteqdot \langle \partial_\varepsilon|_{\varepsilon=0} X_\varepsilon, \mathsf{N}_0\rangle\ ,$$

where $X_\varepsilon : \Sigma^n_\varepsilon \to \mathbb{R}^{n+1}$ is a parametrization of Σ_ε and N_0 is the unit normal to Σ_0. Prove that u satisfies the Jacobi equation

$$-(\Delta + \nabla_{e^\top_{n+1}})u = |\mathrm{II}|^2 u. \tag{13.105}$$

Exercise 13.7. Let Σ^n be a translator in $\mathbb{R}^{n+1}$. Prove that the functions $u : \Sigma^n \to \mathbb{R}$ given by

(a) $u(X) = u_p(X) \doteqdot \langle p, \mathsf{N}(X)\rangle\,,\ \ p \in \mathbb{R}^{n+1}$, and

(b) $u(X) = u_J(X) \doteqdot \langle J(X), \mathsf{N}(X)\rangle\,,\ \ J \in \mathfrak{so}(\mathbb{R}^n \times \{0\})$,

satisfy the Jacobi equation (13.105).

Exercise 13.8. Let Σ^{n-1} be a translator in $\mathbb{R}^n$. Denote by $R(\theta)$ the counterclockwise rotation by angle $\theta \in (-\frac{\pi}{2}, \frac{\pi}{2})$ in the plane $e_1 \wedge e_{n+1}$ in $\mathbb{R}^{n+1}$ and set $J_\theta = \frac{d}{d\varepsilon}\big|_{\varepsilon=0} R(\theta+\varepsilon)$. Prove that the function $u : \Sigma^n_\theta \to \mathbb{R}$ given by

$$u(X) = u_\theta(X) \doteqdot \langle \tan\theta X + J_\theta(X), \mathsf{N}\rangle$$

solves the Jacobi equation on the translator $\Sigma^n_\theta \doteqdot \sec\theta R(\theta)(\mathbb{R} \times \Sigma^{n-1})$.

Exercise 13.9. Show that a properly embedded, mean convex translator is necessarily the graph of a function $u : \Omega \to \mathbb{R}$, $\Omega \subset \mathbb{R}^n$, satisfying the graphical translator equation (13.64).

Exercise 13.10. Let $u : \Omega \to \mathbb{R}$ solve the translator equation with $u|_{\partial\Omega} \equiv 0$, where Ω is a smoothly bounded, convex set. Suppose that $\Omega \times \mathbb{R}$ is invariant under reflection across a vertical plane $\Pi \subset \mathbb{R}^{n+1}$. Prove, using the Alexandrov reflection method, that $\operatorname{graph} u$ is invariant under reflection across Π.

Exercise 13.11. Given $n \in \mathbb{N} \setminus \{1\}$ and $R > 0$, define

$$\overline{u}_R(x) \doteqdot 0 \quad \text{and} \quad \underline{u}_R(x) \doteqdot \frac{|x|^2 - R^2}{2(n-1)}$$

for each $x \in \overline{B}_R \subset \mathbb{R}^n$. Prove that $\overline{u}_R : \overline{B}_R \to \mathbb{R}$ is a supersolution and $\underline{u}_R : \overline{B}_R \to \mathbb{R}$ a subsolution to the Dirichlet problem (13.71) over $\Omega = B_R$.

Exercise 13.12. Let $A : \mathbb{R}^n \times \mathbb{R}^n \to \mathbb{R}$ be a symmetric, positive definite bilinear form satisfying $\operatorname{tr} A = n$. For each $R > 0$, denote by $\Omega_R \doteqdot \{x \in \mathbb{R}^n : A(x,x) < R^2\}$ the corresponding ellipsoid. Define $\underline{u}_R : \overline{\Omega}_R \to \mathbb{R}$ by

$$\underline{u}_R(x) \doteqdot \frac{A(x,x) - R^2}{2(n-1)} .$$

Prove that $\underline{u}_R$ is a subsolution to the Dirichlet problem (13.71) over $\Omega = \Omega_R$ if R is sufficiently small.

Chapter 14

Ancient Solutions

Recall that a solution to mean curvature flow is called an **ancient solution** if it has existed for all times in the past. We have already encountered a number of ancient solutions, many of which arose as singularity models. While in the cases we have encountered so far we were able to show that the singularity model is actually self-similar (shrinking or translating), this is not guaranteed in general. On the other hand, just as we saw for self-similar solutions, we shall see that ancient solutions tend to exhibit rigidity phenomena. This should be expected since diffusion has had an infinite amount of time to drive the solution towards equilibrium.

Example 14.1.

- Minimal hypersurfaces are stationary solutions and hence exist for all times $t \in \mathbb{R}$. They are necessarily nonconvex and noncompact.
- Shrinkers are defined for all negative times. Examples include the shrinking round sphere, $S^n_{\sqrt{-2nt}}$, and the shrinking round orthogonal cylinders, $S^{n-k}_{\sqrt{-2(n-k)t}} \times \mathbb{R}^k$.
- Translators exist for all times $t \in \mathbb{R}$. Examples include the Grim Reaper, Grim hyperplanes, and the bowl soliton.
- The paperclip and hairclip solutions are nontrivial ancient solutions in the sense that they are not self-similar. The paperclip is compact and convex, while the hairclip is neither.

For compact ancient solutions, we shall always assume that $t = 0$ is the maximal time. Of course, by the avoidance principle, the maximal time for a compact ancient solution is necessarily finite, so this can always be arranged by a translation in time.

14.1. Rigidity of the shrinking sphere

We saw in Section 4.5 that the shrinking circle and the paperclip are the only compact, convex ancient solutions to curve shortening flow. In particular, the shrinking circle is the unique convex solution to the curve shortening flow which admits a uniform bound on its eccentricity. This turns out also to be the case in higher dimensions — the shrinking sphere solution is unique among convex ancient solutions satisfying mild geometric hypotheses. We first prove that it is rigid with respect to uniform pinching and then under other uniform, scale-invariant geometric conditions. The result, due to Huisken and Sinestrari [**304**], is proved by a clever modification of the proof of Huisken's roundness estimate (Theorem 8.6). We remark that a similar result was proved at around the same time by Haslhofer and Hershkovits assuming, in addition, that the solution is noncollapsing [**274**]. We will provide an alternative proof in Section 14.5.

Theorem 14.2 (Huisken and Sinestrari [**304**] (cf. [**274**])). *Given $n \in \mathbb{N} \setminus \{1\}$, let $\{\mathcal{M}_t^n\}_{t\in(-\infty,0)}$ be a compact ancient solution to mean curvature flow satisfying*

$$\liminf_{t\to-\infty} \min_{\mathcal{M}^n\times\{t\}} \frac{\kappa_1}{H} > 0\,.$$

Then $\{\mathcal{M}_t\}_{t\in(-\infty,0)}$ is a shrinking sphere.

Proof. Set

$$\alpha \doteqdot \liminf_{t\to-\infty} \min_{\mathcal{M}^n\times\{t\}} \frac{\kappa_1}{H}\,.$$

By the tensor maximum principle, $\kappa_1 \geq \alpha H$ in $\mathcal{M} \times (-\infty, 0)$. Given any $\varepsilon > 0$, any $\sigma \in (0, 1)$, and any $p \geq 3$, set

$$f_{\varepsilon,\sigma} \doteqdot \left(|\mathring{\mathrm{II}}|^2 - \varepsilon H^2\right) H^{\sigma-2} \quad \text{and} \quad v \doteqdot (f_{\varepsilon,\sigma})_+^{\frac{p}{2}}\,.$$

We will show that there is some $\ell > 0$, which depends only on n, α, and ε, such that

$$\frac{d}{dt}\int_M v^2\,d\mu \leq -\sigma p \int_M v^2|\mathrm{II}|^2\,d\mu \tag{14.1}$$

for all $p > \ell^{-1}$ and $\sigma < \ell p^{-\frac{1}{2}}$. We achieve this by modifying the final step in the proof of Lemma 8.10. Fix any $p > \ell^{-1}$ and $\sigma < \ell p^{-\frac{1}{2}}$ such that $\sigma p > 2n + 1$. Then $\delta \doteqdot \frac{2}{\sigma p+1} < \frac{1}{n+1} < 1$. Thus, noting also that $f_{\varepsilon,\sigma} \leq H^\sigma$, we may estimate

$$\begin{aligned}\int_M v^2\,d\mu &= \int_M v^{2(1-\delta)}v^{2\delta}\,d\mu\\ &\le \int_M v^{2(1-\delta)}H^{\delta\sigma p}\,d\mu\\ &= \int_M (v^2H^2)^{1-\delta}H^{\delta}\,d\mu\\ &\le \left(\int_M v^2H^2\,d\mu\right)^{1-\delta}\left(\int_M H\,d\mu\right)^{\delta}.\end{aligned}$$

Applying (14.1) and recalling the algebraic inequality $n|\mathrm{II}|^2 \ge H^2$, this yields

$$\begin{aligned}\frac{d}{dt}\int_M v^2\,d\mu &\le -(2n+1)\int_M v^2|\mathrm{II}|^2\,d\mu\\ &\le -\int_M v^2H^2\,d\mu\\ &\le -\left(\int_M v^2\,d\mu\right)^{\frac{1}{1-\delta}}\left(\int_M H\,d\mu\right)^{-\frac{\delta}{1-\delta}}\\ &\le -\left(\int_M v^2\,d\mu\right)^{1+\frac{2}{\sigma p-1}}\left(\int_M H\,d\mu\right)^{-\frac{2}{\sigma p-1}}.\end{aligned}$$

Setting

$$\varphi(t) \doteqdot \int_M v^2(x,t)\,d\mu_t(x)\,, \quad \psi(t) \doteqdot \int_M H(x,t)\,d\mu_t(x),$$

and $\beta \doteqdot \frac{2}{\sigma p-1}$, this becomes

$$\frac{d}{dt}\left(\varphi^{-\beta}\right) = -\beta\varphi^{-\beta-1}\frac{d}{dt}\varphi \ge \beta\psi^{-\beta}\,. \tag{14.2}$$

Note that, if $\varphi(s) = 0$ for some $s \in (-\infty, 1)$, then $\varphi(t) = 0$ for all $t \in [s, 1)$ since the inequality $|\mathring{\mathrm{II}}| \le \varepsilon H$ is preserved under the flow. Set

$$\tau_0 \doteqdot \sup\{s \in (-\infty, 0] : \varphi(t) > 0 \text{ for all } t < s\}.$$

Suppose that $\tau_0 > -\infty$. Then (14.2) and Hölder's inequality yield

$$\begin{aligned}\varphi^{-\beta}(\tau) - \varphi^{-\beta}(s) &\ge \beta\int_s^\tau \psi^{-\beta}(t)dt\\ &\ge \beta(\tau-s)^{1+\beta}\left(\int_s^\tau \psi(t)dt\right)^{-\beta}\end{aligned} \tag{14.3}$$

for any $s < \tau < \tau_0$.

Claim 14.3. The solution $\{\mathcal{M}_t\}_{t\in(-\infty,0)}$ has **bounded rescaled volume**:

$$\sup_{t\in(-\infty,0)} \frac{1}{\sqrt{-t}}|\Omega_t|^{\frac{1}{n+1}} < \infty\,, \tag{14.4}$$

where Ω_t is the convex body bounded by $\mathcal{M}_t$.

Proof of Claim 14.3. Estimating

$$K \geq \kappa_1^n \geq \alpha^n H^n\,,$$

we obtain

$$\sup_{t\in(-\infty,0)}\left(\int_M H^n\,d\mu\right)^{\frac{1}{n}} \leq \alpha^{-1} \sup_{t\in(-\infty,0)}\left(\int_M K\,d\mu\right)^{\frac{1}{n}} = \alpha^{-1}\mathrm{Area}(S^n)^{\frac{1}{n}} \doteqdot \Lambda.$$

Hölder's inequality then yields

$$\frac{d}{dt}\mu(\mathcal{M}) = -\int_M H^2 d\mu \geq -\mu(\mathcal{M})^{\frac{2-n}{n}}\left(\int_M H^n\,d\mu\right)^{\frac{2}{n}} \geq -\Lambda^2\mu(\mathcal{M})^{\frac{2-n}{n}}\,.$$

That is,

$$\frac{d}{dt}\mu(\mathcal{M})^{\frac{2}{n}} \geq -\frac{2\Lambda^2}{n}\,,$$

and hence, since Huisken's theorem implies that the area of $\mathcal{M}_t$ goes to 0 as $t \to 0$,

$$\mu_t(\mathcal{M})^{\frac{1}{n}} \leq \frac{\Lambda}{n}\sqrt{-2nt}\,.$$

The isoperimetric inequality

$$|\Omega|^{\frac{1}{n+1}} \leq \frac{\mathrm{Vol}\,(B^{n+1})^{\frac{1}{n+1}}}{\mathrm{Area}\,(\partial B^{n+1})^{\frac{1}{n}}}\mu(\mathcal{M})^{\frac{1}{n}} \tag{14.5}$$

now yields $|\Omega|^{\frac{1}{n+1}} \leq C\sqrt{-t}$, where C is a constant that depends only on n and α. This proves (14.4). □

On the other hand, the enclosed volume $|\Omega_t|$ evolves according to

$$\frac{d}{dt}|\Omega_t| = -\int_M H\,d\mu\,.$$

Integrating and applying (14.4), we conclude from (14.3) that

$$\varphi^\beta(\tau) \leq C\frac{(-s)^{\frac{n+1}{2}\beta}}{(\tau - s)^{1+\beta}}$$

for some $C < \infty$ and all $s < \tau < \tau_0$. But $0 < \frac{n}{2}\beta < 1$, so that, taking $s \to -\infty$, we obtain $\varphi(\tau) = 0$, a contradiction. We are forced to conclude that $\tau_0 = -\infty$ and hence $|\mathring{\mathrm{II}}| \leq \varepsilon H$ in $M \times (-\infty, 0)$. The claim follows since $\varepsilon > 0$ was arbitrary. □

Corollary 14.4 (Huisken and Sinestrari [**304**] (see also Haslhofer and Hershkovits [**274**] and Wang [**521**, Remark 3.1])). *Let* $\{\mathcal{M}^n_t\}_{t\in(-\infty,0)}$, $n \geq 2$, *be a compact, convex ancient solution to mean curvature flow. The following are equivalent*:

(1) $\{\mathcal{M}^n_t\}_{t\in(-\infty,0)}$ *is a shrinking sphere* $\{S^n_{\sqrt{-2nt}}(p)\}_{t\in(-\infty,0)}$, $p \in \mathbb{R}^{n+1}$.

(2) $\{\mathcal{M}^n_t\}_{t\in(-\infty,0)}$ *is uniformly pinched*:
$$\liminf_{t\to-\infty} \min_{M^n\times\{t\}} \frac{\kappa_1}{H} > 0\,.$$

(3) $\{\mathcal{M}^n_t\}_{t\in(-\infty,0)}$ *has bounded rescaled (extrinsic) diameter*:
$$\limsup_{t\to-\infty} \frac{\operatorname{diam}(\mathcal{M}_t)}{\sqrt{-t}} < \infty\,.$$

(4) $\{\mathcal{M}^n_t\}_{t\in(-\infty,0)}$ *has bounded eccentricity*:
$$\limsup_{t\to-\infty} \frac{\rho_+(t)}{\rho_-(t)} < \infty\,,$$
where $\rho_+(t)$ *and* $\rho_-(t)$ *denote, respectively, the circum- and inradii of* $\mathcal{M}^n_t$.

(5) $\{\mathcal{M}^n_t\}_{t\in(-\infty,0)}$ *has type-I curvature decay*:
$$\limsup_{t\to-\infty} \sqrt{-t} \max_{M^n\times\{t\}} H < \infty\,.$$

(6) $\{\mathcal{M}^n_t\}_{t\in(-\infty,0)}$ *has bounded speed ratios*:
$$\limsup_{t\to-\infty} \frac{\max_{M^n\times\{t\}} H}{\min_{M^n\times\{t\}} H} < \infty\,.$$

(7) $\{\mathcal{M}^n_t\}_{t\in(-\infty,0)}$ *satisfies a reverse isoperimetric inequality*:
$$\limsup_{t\to-\infty} \frac{\mu_t(\mathcal{M})^{n+1}}{|\Omega_t|^n} < \infty\,.$$

Proof. We have already proved that conditions (1) and (2) are equivalent. Moreover, it is clear that (1) implies each of the conditions (2)–(7). We will prove that each of the conditions (3)–(7) implies (2).

Let's begin with case (3). We prove (2) by contradiction. So suppose that there is a sequence of space-time points (x_j, t_j) with $t_j \to -\infty$ such that $\frac{\kappa_1}{H}(x_j, t_j) \to 0$ as $j \to \infty$. Set $\lambda_j \doteqdot \frac{1}{\sqrt{-t_j}}$ and consider the sequence of mean curvature flows $X_j : M^n \times (-\infty, 0) \to \mathbb{R}^{n+1}$ defined by
$$X_j(x,t) \doteqdot \lambda_j X\left(x, \lambda_j^{-2} t\right)\,.$$

By the bounded rescaled diameter hypothesis, the diameter of $X_j(M,t)$ is bounded uniformly in j for each t. Next observe that the first inequality of Lemma 8.25 implies that

$$\sqrt{3}\rho_+(\mathcal{M}) \le \operatorname{diam}(\mathcal{M}) \le \operatorname{diam}_I(\mathcal{M}) \le \pi\rho_+(\mathcal{M})$$

for any convex hypersurface $\mathcal{M} \hookrightarrow \mathbb{R}^{n+1}$, where $\operatorname{diam}_I(\mathcal{M})$ is the intrinsic diameter of $\mathcal{M}$. So the Harnack inequality yields

$$\max_{M\times\{t\}} H \le \min_{M\times\{t/2\}} H \exp\left(\frac{\operatorname{diam}_I^2(\mathcal{M}_t)}{-2t}\right) \le \mathrm{e}^{\frac{\pi^2C^2}{4}} \min_{M\times\{t/2\}} H\,,$$

where

$$C \doteqdot \sup_{t\in(-\infty,0)} \frac{\operatorname{diam}(\mathcal{M}_t)}{\sqrt{-t}}\,.$$

Recalling Lemma 8.3, we find

$$\frac{\mathrm{e}^{\frac{\pi^2C^2}{4}}}{2\sqrt{-t}} \le \min_{M\times\{t\}} H \quad\text{and}\quad \max_{M\times\{t\}} H \le \frac{\sqrt{n}\mathrm{e}^{\frac{\pi^2C^2}{4}}}{\sqrt{-t}}\,.$$

Since H and $1/\sqrt{-t}$ scale in the same way, the same bounds hold for the rescaled solutions X_j. We may now conclude from Theorem 11.11 that $X_j\big|_{M^n\times[-2,-1]} : M^n \times [-2,-1] \to \mathbb{R}^{n+1}$ converges uniformly in the smooth topology to a compact, strictly mean convex, weakly convex limit solution $X_\infty : M \times [-2,-1] \to \mathbb{R}^{n+1}$ which satisfies

$$\min_{M\times\{-1\}} \frac{\kappa_1}{H} = 0\,.$$

But the rigidity of the maximum principle for the second fundamental form (Theorem 9.11) then implies that X_∞ splits off a line, in contradiction with compactness.

Case (4) reduces to case (3) since comparison with shrinking spheres implies

$$\rho_-(t) \le \sqrt{-2nt}\,.$$

Integrating the type-I hypothesis of case (5) yields

$$\|X(x,t) - X(x,0)\| \le \int_t^0 H(x,s)\,ds \le 2C\sqrt{-t}\,,$$

where

$$C \doteqdot \sup_{t\in(-\infty,0)} \sqrt{-t}\max_{M\times\{t\}} H\,,$$

which brings us again to case (3).

Case (6) can be reduced to case (5) by way of Lemma 8.3.

The proof is completed by observing that the reverse isoperimetric inequality of case (7) implies a bound for the eccentricity. See [**304**, Lemma 4.4]. □

14.2. A convexity estimate

It is also possible to weaken the convexity hypothesis[1] in Theorem 14.2. The following "convexity estimate" shows that any compact, mean convex ancient solution which admits a scale-invariant lower bound for κ_1 is necessarily convex if it has *bounded rescaled volume* [**352**].

Theorem 14.5 (Convexity estimate for ancient solutions [**352**]). *Let* $X : M^n \times (-\infty, 0) \to \mathbb{R}^{n+1}$, $n \geq 2$, *be a compact, mean convex ancient solution to mean curvature flow satisfying*

$$\limsup_{t\to-\infty} \frac{1}{(-t)^{\frac{n+1}{2}}} \int_t^0 \int_M H(\cdot, s)\, d\mu(s)\, ds < \infty . \tag{14.6}$$

If

$$\liminf_{t\to-\infty} \min_{M^n\times\{t\}} \frac{\kappa_1}{H} > -\infty ,$$

then $\mathcal{M}_t = X(M^n, t)$ *is locally uniformly convex for every* $t \in (-\infty, 0)$.

Proof. The proof is similar to the proof of Theorem 14.2, except that we now take $f_{\varepsilon,\sigma}$ to be the function constructed in (9.21) in the proof of the convexity estimate, making use of the proof of (9.23) in lieu of the arguments of Lemma 8.10. Local uniform convexity follows from the splitting theorem (Theorem 9.11). □

We remark that embeddedness was not required in the proof of Theorem 14.5 (which is why we stated the bounded rescaled volume hypothesis (14.6) without reference to enclosed regions). We also note that the growth rate in (14.6) can be weakened: We leave it to the reader to check that, in fact, *any* polynomial rate,

$$\limsup_{t\to-\infty} \frac{1}{(-t)^r} \int_t^0 \int_M H(\cdot, s)\, d\mu(s)\, ds < \infty , \quad r \in \left[\tfrac{n+1}{2}, \infty\right) ,$$

will suffice (cf. [**353**, Theorem 1.11]). This is satisfied, with $r = n + 1$, by any compact, convex, embedded ancient solution (see Exercise 14.1).

A similar argument yields the following m-convexity estimates (see [**304**, **352**]).

Theorem 14.6. *Let* $X : M^n \times (-\infty, 0) \to \mathbb{R}^{n+1}$, $n \geq 2$, *be a compact, mean convex ancient solution to mean curvature flow satisfying*

$$\limsup_{t\to-\infty} \frac{1}{(-t)^{\frac{n+1}{2}}} \int_t^0 \int_M H(\cdot, s)\, d\mu(s)\, ds < \infty .$$

[1] By Theorem 12.20, every proper *noncollapsing* ancient solution is necessarily convex.

If, for some $m \in \{1, \ldots, n-2\}$,

$$\liminf_{t\to-\infty} \min_{M^n\times\{t\}} \frac{\kappa_1 + \cdots + \kappa_{m+1}}{H} > 0,$$

then $\mathcal{M}_t$ *is strictly* m*-convex for every* $t \in (-\infty, 0)$.

The bounded rescaled volume hypothesis is satisfied, in particular, under each of the conditions in Theorem 14.2.

Proposition 14.7. *Let* $X : M^n \times (-\infty, 0) \to \mathbb{R}^{n+1}$, $n \geq 2$, *be a compact, mean convex, embedded ancient solution to mean curvature flow. If* $\{\mathcal{M}_t\}_{t\in(-\infty,0)}$ *satisfies any of the conditions* (1)–(7) *of Corollary* 14.4, *then it has bounded rescaled volume:*

$$\sup_{t\in(-\infty,0)} \frac{1}{\sqrt{-t}}|\Omega_t|^{\frac{1}{n+1}} < \infty\,, \tag{14.7}$$

where Ω_t *is the precompact region bounded by* $\mathcal{M}_t$.

Proof. The claim is immediate in case (1) and we have already seen in the course of the proof of Theorem 14.2 that it holds in case (2).

The claim follows in case (3) by comparing $|\Omega_t|$ with the volume enclosed by a circumsphere.

The claim follows in case (4) from the comparison principle, which, comparing the solution with shrinking sphere solutions, yields, for all $t < 0$,

$$\rho_-(t) \leq C\sqrt{-t}$$

for some $C < \infty$, reducing the hypothesis to that of case (3).

In case (5), we integrate to obtain

$$\|X(x,t) - X(x,s)\| \leq \int_t^s H(x,\sigma)\,d\sigma \leq C\sqrt{-t}$$

for any $(x,t) \in M \times (-\infty, 0)$ and $s \in (-t, 0)$. Since $\mathcal{M}_s$ is bounded as $s \to 0$, we are back to case (3).

To deal with case (6), we recall that

$$\min_{M^n\times\{t\}} H \leq C\sqrt{\frac{n}{-2t}}$$

for some $C < \infty$, which reduces the claim to case (5).

In the final case, we estimate, using Hölder's inequality,

$$\begin{aligned}|\Omega_t|-|\Omega_0| &= \int_t^0\int_M H(\cdot,s)\,d\mu_s\,ds\\ &\le \left(\int_t^0\int_M H^2(\cdot,s)\,d\mu_s\,ds\right)^{\frac12}\left(\int_t^0\int_M d\mu_s\,ds\right)^{\frac12}\\ &= \big(\mu_t(\mathcal{M}^n)-\mu_0(\mathcal{M}^n)\big)^{\frac12}\left(\int_t^0 \mu_s(\mathcal{M}^n)\,ds\right)^{\frac12}\\ &\le \big(\mu_t(\mathcal{M}^n)-\mu_0(\mathcal{M}^n)\big)^{\frac12}\sqrt{-t}\,\mu_t(\mathcal{M}^n)^{\frac12}.\end{aligned}$$

The reverse isoperimetric inequality now yields

$$|\Omega_t|\le C\sqrt{-t}\,|\Omega_t|^{\frac{n}{n+1}}$$

for $t<-1$, say. Rearranging yields the claim. □

Corollary 14.8 (**[352]** (cf. **[274]**)). *The conclusion of Corollary* 14.4 *holds for compact, embedded,* mean convex *ancient solutions to mean curvature flow in* $\mathbb{R}^{n+1}$*,* $n\ge 2$*, satisfying*

$$\liminf_{t\to-\infty}\min_{M\times\{t\}}\frac{\kappa_1}{H}>-\infty\,.$$

We have seen that the shrinking sphere is rigid under certain geometric hypotheses. In the absence of such hypotheses, however, there exist further convex, compact examples. Indeed, the paperclip solution provides an example for the curve shortening flow. In dimensions $n\ge 2$, there exist a number of compact, convex ancient solutions other than the shrinking sphere. See Sections 14.7 and 14.8 and the notes and commentary at the end of this chapter.

14.3. A gradient estimate for the curvature

Theorem 14.9. *Given integers* $n\ge 2$ *and* $0\le m<\frac{2(n-1)}{3}$*, there exists a constant* $C<\infty$ *with the following property: Let* $X: M\times(-\infty,0)\to\mathbb{R}^{n+1}$ *be a mean convex ancient solution to the mean curvature flow satisfying*

$$\liminf_{t\to-\infty}\min_{M\times\{t\}}\frac{\kappa_1+\cdots+\kappa_{m+1}}{H}>0\,.$$

If

$$\int_{-\infty}^{-1}\min_{M\times\{t\}}|\mathrm{II}|^2\,dt=\infty, \tag{14.8}$$

then

$$|\nabla \mathrm{II}|^2\le CH^4\,. \tag{14.9}$$

Proof. Define

$$G_1 \doteqdot \left(\frac{1}{n-m}+\beta\right)H^2-|A|^2 \quad\text{and}\quad G_2 \doteqdot \frac{3}{n+2}H^2-|A|^2\,,$$

where

$$\beta \doteqdot \frac{1}{2}\left(\frac{3}{n+2}-\frac{1}{n-m}\right)>0\,.$$

By Theorem 14.6,

$$G_2 > G_1 > \beta H^2 > 0\,.$$

As in the proof of Theorem 9.24 (see also Theorem 8.14),

$$(\partial_t-\Delta)\frac{|\nabla \mathrm{II}|^2}{G_1G_2} \le -2\left\langle \frac{\nabla G_1}{G_1},\nabla\frac{|\nabla \mathrm{II}|^2}{G_1G_2}\right\rangle + |\mathrm{II}|^2\frac{|\nabla \mathrm{II}|^2}{G_1G_2}\left(C_n-2\beta^2\frac{n+2}{3n}\frac{|\nabla \mathrm{II}|^2}{G_1G_2}\right),$$

where $C_n>0$ is a constant that depends only on n. Set

$$\phi(t) \doteqdot \max_{M\times\{t\}}\frac{|\nabla \mathrm{II}|^2}{G_1G_2} \quad\text{and}\quad \psi(t)\doteqdot \min_{M\times\{t\}}|\mathrm{II}|^2\,.$$

Then ϕ is Lipschitz and satisfies

$$\phi' \le \frac{2(n+2)\beta^2}{3n}\psi\phi\left(\frac{3n}{2(n+2)\beta^2}C_n-\phi\right) \tag{14.10}$$

at almost every time. We claim that

$$\phi \le \frac{3n}{2(n+2)\beta^2}C_n\,.$$

Suppose, to the contrary, that there exists $t_0\in(-\infty,0)$ such that

$$\phi(t_0) > \frac{3n}{2(n+2)\beta^2}C_n\,.$$

Then by (14.10) we have that ϕ is decreasing in $(-\infty,t_0]$. In particular, there exists $\delta>0$ such that

$$(1-\delta)\phi(t) \ge \frac{3n}{2(n+2)\beta^2}C_n$$

and hence

$$\phi' \le \delta\frac{2(n+2)\beta^2}{3n}\psi\phi^2$$

for all $t<t_0$. But then integrating from t to t_0 yields

$$\frac{1}{\phi(t_0)}-\frac{1}{\phi(t)} \ge \delta\frac{2(n+2)\beta^2}{3n}\int_t^{t_0}\psi(s)\,ds\,.$$

Taking $t\to-\infty$ results in a contradiction to the hypothesis (14.8). □

14.4. Asymptotics

The following lemma describes the asymptotic properties of convex ancient solutions.

Lemma 14.10 (Asymptotic structure of convex ancient solutions). *Suppose that $\{\mathcal{M}_t^n\}_{t\in(-\infty,0)}$ is a convex ancient solution to mean curvature flow in $\mathbb{R}^{n+1}$.*

(i) (Convergence to a round point) *If $\{\mathcal{M}_t^n\}_{t\in(-\infty,0)}$ is compact and 0 is its singular time, then there exists a point $p \in \mathbb{R}^{n+1}$ such that the family of rescaled solutions $\{\lambda(\mathcal{M}^n_{\lambda^{-2}t} - p)\}_{t\in(-\infty,0)}$ converges uniformly in the smooth topology to the shrinking sphere $\{S^n_{\sqrt{-2nt}}\}_{t\in(-\infty,0)}$ as $\lambda \to \infty$.*

(ii) (Asymptotic shrinker) *There is a rotation $R \in \mathrm{SO}(n+1)$ such that the rescaled solutions $\{\lambda R \cdot \mathcal{M}^n_{\lambda^{-2}t}\}_{t\in(-\infty,0)}$ converge locally uniformly in the smooth topology as $\lambda \to 0$ to either*
- *the shrinking sphere $\{S^n_{\sqrt{-2nt}}\}_{t\in(-\infty,0)}$,*
- *the shrinking cylinder $\{\mathbb{R}^k \times S^{n-k}_{\sqrt{-2(n-k)t}}\}_{t\in(-\infty,0)}$ for some $k \in \{1,\dots,n-1\}$, or*
- *the stationary hyperplane $\{\mathbb{R}^n \times \{0\}\}_{t\in(-\infty,0)}$ of multiplicity either one or two.*

Furthermore:

(ii)(a) *If the limit is the shrinking sphere, then $\{\mathcal{M}_t^n\}_{t\in(-\infty,0)}$ is a shrinking sphere.*

(ii)(b) *If the limit is the stationary hyperplane of multiplicity one, then $\{\mathcal{M}_t^n\}_{t\in(-\infty,0)}$ is a stationary hyperplane of multiplicity one.*

(iii) (Asymptotic translators) *Suppose that $|\mathrm{II}|$ is bounded on M_t^n for each t. Let $e \in G \doteqdot \bigcup_{s<0}\bigcap_{t\le s} \mathrm{G}(\mathcal{M}_t^n)$ be a normal direction. Then the translated solutions $\{\mathcal{M}^n_{t+s} - P(e,s)\}_{t\in(-\infty,-s)}$, $s<0$, converge locally uniformly in the smooth topology, along some subsequence of times $s_j \to -\infty$, to a convex translating solution, where $P(\cdot\,,t)$ is the inverse of the Gauß map $\mathrm{G}(\cdot\,,t)$. The bulk velocity v of the limit translator satisfies*

$$\langle v, e\rangle = -\lim_{s\to-\infty} H(P(e,s),s).$$

Proof. (i) This is an immediate consequence of Huisken's theorem.

(ii) This is a consequence of the monotonicity formula[2]: After a finite translation, we can arrange that the solution reaches the origin at time 0. Since the shrinking sphere also reaches the origin at time 0, it must intersect the solution at all negative times by the avoidance principle. Thus,

$$\min_{q\in\mathcal{M}^n_t}|q|\le\sqrt{-2nt}\,.$$

Since the solution is convex and its speed, H, is bounded in any compact subset of $\mathbb{R}^{n+1}\times(-\infty,0)$ after the rescaling, given any sequence $\lambda_i\searrow 0$, we can find, by Theorem 11.12, a subsequence along which $\{\lambda_i\mathcal{M}^n_{\lambda_i^{-2}t}\}_{t\in(-\infty,0)}$ converges smoothly on compact subsets of $\mathbb{R}^{n+1}\times(-\infty,0)$ to a nonempty limit flow.

We claim that the subsequential limit flow is a self-similarly shrinking solution. Indeed, the monotonicity formula (Corollary 10.5) states that

$$\frac{d}{dt}\Theta(t)=-\int_{\mathcal{M}^n_t}\left|\vec{H}(p,t)+\frac{p^\perp}{-2t}\right|^2\Phi(p,t)\,d\mathscr{H}^n(p)\,, \tag{14.11}$$

where Φ is the Gaußian density and Θ the Gaußian area of the solution:

$$\Theta(t)\doteqdot\int_{\mathcal{M}^n_t}\Phi(p,t)\,d\mathscr{H}^n(p)\,,\quad \Phi(p,t)\doteqdot(-4\pi t)^{-\frac{n}{2}}\mathrm{e}^{-\frac{|p|^2}{-4t}}\,.$$

Since $\Theta(t)$ is bounded *uniformly* (see Exercise 14.5), it must converge to some limit as $t\to-\infty$. But then

$$\begin{aligned}-\int_a^b\int_{\lambda\mathcal{M}^n_{\lambda^{-2}t}}\left|\vec{H}(p,t)+\frac{p^\perp}{-2t}\right|^2\Phi(p,t)d\mathscr{H}^n(p)&=\Theta_\lambda(b)-\Theta_\lambda(a)\\&=\Theta(\lambda^{-2}b)-\Theta(\lambda^{-2}a)\\&\to 0\end{aligned}$$

as $\lambda\to 0$ for any $a<b<0$, where Θ_λ is the Gaußian area of the λ-rescaled flow. We conclude that the integrand vanishes identically in the limit and hence any limit of $\{\lambda\mathcal{M}^n_{\lambda^{-2}t}\}_{t\in(-\infty,0)}$ along a sequence of scales $\lambda_i\to 0$ is a self-similarly shrinking solution to mean curvature flow.

By Theorem 13.14, the only convex examples which can arise are the shrinking spheres, the shrinking cylinders, and the stationary hyperplanes of multiplicity either one or two. If the limit flow is the shrinking sphere, then the solution is compact and hence, by part (i), its "blow-up" is also the shrinking sphere. The monotonicity formula then implies that the Gaußian area of the solution is constant and we conclude that it is the shrinking

[2]The argument presented here was shown to us by Theodora Bourni. A similar result was proved by X.-J. Wang using different arguments [**521**]. A similar argument was used by Sonnanburg to prove that the only type-I, noncollapsing ancient solutions in $\mathbb{R}^3$ with polynomial area growth are the shrinking spheres and cylinders [**491**].

sphere. Similarly, if the limit is the hyperplane of multiplicity one, then the solution is the hyperplane of multiplicity one, since the blow-up about any regular point (without loss of generality, the space-time origin) is also a hyperplane of multiplicity one (see Lemma 11.18). Finally, we note that the limit is unique since convexity ensures that the limiting convex region enclosed by any subsequential limit is contained in the limiting convex region enclosed by any other subsequential limit.

(iii) This is a consequence of the differential Harnack inequality (Corollary 10.10 and Theorem 10.11): By the Blaschke selection theorem, we can find a sequence of times $s_j \to -\infty$ such that the flows $\{\mathcal{M}^n_{t+s_j} - P(e, s_j)\}_{t\in(-\infty,-s_j)}$ converge locally uniformly in $C^{0,1}$ to a family of nested convex sets $\{\hat{\mathcal{M}}^n_t\}_{t\in(-\infty,\infty)}$. The Ecker–Huisken interior estimates (see Chapter 7) then imply that the limit $\{\hat{\mathcal{M}}^n_t\}_{t\in(-\infty,\infty)}$ is in fact a smooth mean curvature flow and that the convergence is smooth.

By Theorem 9.11, after a time-independent rotation, $\{\hat{\mathcal{M}}^n_t\}_{t\in(-\infty,\infty)} = \{\mathbb{R}^m \times \Sigma^{n-m}_t\}_{t\in(-\infty,\infty)}$ for some $m \in \{0, \ldots, n\}$ and some $(n-m)$-dimensional convex, locally uniformly convex solution $\{\Sigma^{n-m}_t\}_{t\in(-\infty,\infty)}$ to mean curvature flow in $\mathbb{R}^{n-m+1}$.

By the differential Harnack inequality, the mean curvature of the flow $\{\mathcal{M}^n_t\}_{t\in(-\infty,0)}$ is nondecreasing with respect to the Gauß map parametrization. It follows that the limit

$$H_\infty(z) = \lim_{s\to\infty} H_{\mathcal{M}^n_s}(z)$$

exists for each $z \in G$ and hence the mean curvature of $\{\Sigma^{n-m}_t\}_{t\in(-\infty,\infty)}$ is constant with respect to the Gauß map parametrization:

$$H_{\Sigma^{n-m}_t}(z) = \lim_{j\to\infty} H_{\mathcal{M}^n_{t+s_j}}(z) = \lim_{s\to\infty} H_{\mathcal{M}^n_s}(z)\,.$$

So the rigidity case of the Harnack inequality (Theorem 10.11) implies that $\{\Sigma^{n-m}_t\}_{t\in(-\infty,\infty)}$ moves by translation, with bulk velocity v, say.

Finally,

$$\langle v, e\rangle = \lim_{j\to\infty} \langle \partial_t P(e,t), e\rangle = -\lim_{j\to\infty} H(e,t). \qquad \square$$

Remark 14.11. In case (iii) of Lemma 14.10, the limit will simply be a stationary hyperplane if $\lim_{s\to-\infty} H(P(e,s),s) = 0$.

We refer to the "forwards" limit in part (i) as the **blow-up** of the solution, the "backwards" limit in part (ii) as its **blow-down**, and the translating limits in part (iii) as **asymptotic translators**.

In certain cases, the asymptotic translator can be shown to be independent of the subsequence, so that the convergence holds as $s \to -\infty$.

14.5. X.-J. Wang's dichotomy

Recall from Section 6.9.2 that the arrival time $u : \bigcup_{t\in I}\mathcal{M}^n_t \to \mathbb{R}$ of a mean curvature flow $\{\mathcal{M}^n_t\}_{t\in I}$ of mean convex boundaries $\mathcal{M}^n_t = \partial\Omega_t$ is defined by

$$u(X) = t \iff X \in \mathcal{M}^n_t\,. \tag{14.12}$$

Definition 14.12. An ancient mean curvature flow $\{\mathcal{M}^n_t\}_{t\in(-\infty,T)}$ of mean convex boundaries $\mathcal{M}^n_t = \partial\Omega_t$ is called **entire** if it **sweeps out all of space**, in the sense that $\bigcup_{t\in(-\infty,T)}\mathcal{M}^n_t = \mathbb{R}^{n+1}$. Equivalently, its arrival time is an entire function.

Observe that the shrinking sphere and the bowl soliton are entire ancient solutions, whereas the paperclip and Grim Reaper are not entire (they only sweep out "strip" regions between two parallel lines).

Any convex ancient solution which is not entire necessarily lies at all times inside some stationary half-space. By Lemma 14.10, the blow-down of a nonflat, nonentire ancient solution is a hyperplane of multiplicity two; however, a priori, the distance of the bounding hyperplanes could be going to infinity as $t \to -\infty$ before rescaling. A remarkable result of X.-J. Wang states that this is never the case [**521**, Corollary 2.2] (cf. [**479**]).

Theorem 14.13 (X.-J. Wang's dichotomy for convex ancient solutions [**521**, Corollary 2.2]). *Let $\{\mathcal{M}^n_t\}_{t\in(-\infty,0)}$ be a convex ancient solution to mean curvature flow with $\sup_{\mathcal{M}^n_t}|\mathrm{II}| < \infty$ for each t. If its blow-down is a stationary hyperplane of multiplicity two, then $\{\mathcal{M}^n_t\}_{t\in(-\infty,0)}$ lies in a stationary slab region. In particular, every convex ancient solution with $\sup_{\mathcal{M}^n_t}|\mathrm{II}| < \infty$ for each t is either entire or lies in a stationary slab region.*

Proof. Let $\{\mathcal{M}^n_t\}_{t\in(-\infty,0)}$ be a convex ancient solution which is not entire. After a rotation, we can assume, by Lemma 14.10, that the rescaled flows $\{\lambda\mathcal{M}^n_{\lambda^{-2}t}\}_{t\in(-\infty,0)}$ converge as $\lambda \to 0$ locally uniformly in the smooth topology to the stationary hyperplane $L \doteqdot \{x \in \mathbb{R}^{n+1} : x_1 = 0\}$ with multiplicity two. Since $\mathcal{M}^n_t$ is convex, we may represent each time slice $\mathcal{M}^n_t$ as the union of the graphs of two functions $v^{\pm}(\cdot\,,t)$ over a domain $V_t \subset L$. That is,

$$\mathcal{M}^n_t = \{(v^+(y,t),y) : y \in V_t\} \cup \{(v^-(y,t),y) : y \in V_t\}\,,$$

with $v^+(\cdot\,,t) : V_t \to \mathbb{R}$ concave, $v^- : V_t \to \mathbb{R}$ convex, and v^+ lying above v^-. Thus, the function

$$v \doteqdot v^+ - v^-$$

is positive and concave in its space variable at each time. Note also that $v(p,t) = 0$ for each $p \in \partial V_t$. We need to show that $v(\cdot\,,t)$ remains uniformly bounded as $t \to -\infty$.

Since their graphs move by mean curvature flow, the functions v^+ and v^- satisfy

$$\frac{\partial v^\pm}{\partial t} = \sqrt{1+|Dv^\pm|^2}\,\mathrm{div}\left(\frac{Dv^\pm}{\sqrt{1+|Dv^\pm|^2}}\right). \tag{14.13}$$

Since the blow-down of the solution is the hyperplane $\{X \in \mathbb{R}^{n+1} : \langle X, e_1\rangle = 0\}$ of multiplicity two, there exists, for every $\varepsilon > 0$, some $t_\varepsilon > -\infty$ such that

$$|p| \geq \varepsilon^{-1}\sqrt{-t} \ \text{ for all } \ p \in \partial V_t \tag{14.14a}$$

and

$$v(0,t) \leq \varepsilon\sqrt{-t} \tag{14.14b}$$

for all $t < t_\varepsilon$.

Since the solution is convex, to prove the theorem it suffices, by (14.14a), to show that $v(0,t)$ remains bounded as $t \to -\infty$. We achieve this with the following two claims.

Claim 14.14 ([**521**, Claim 1 in Lemmas 2.1 and 2.2, and Lemma 2.6]). There exist $t_0 < 0$ and $\alpha > 0$ such that

$$|p|v(0,t) \geq -\alpha t \tag{14.15}$$

for each $p \in \partial V_t$ and every $t \leq t_0$.

Proof. We will use equations (14.14a) and (14.14b) to show that the tangent planes to v^+ and v^- at the origin are almost horizontal for $t \ll 0$, which will allow us to estimate $|p|v(0,t)$ by the area in the plane $e_1 \wedge p$ enclosed by $\mathcal{M}_t$ and $\{\langle X, p\rangle = 0\}$. The argument is quite simple when $n = 1$ and the solution is compact. Removing these hypotheses introduces some technical difficulties, but the idea is essentially the same.

Fix $t < t_\varepsilon$ and $p \in \partial V_t$. By rotating about the x_1-axis, we can arrange that $p = de_2$, where $d > 0$. Set

$$q^\pm(t) \doteqdot v^\pm(0,t)e_1\,.$$

By the convexity/concavity of the respective graphs, the segment connecting $(p, v^+(p))$ to $q^+(t)$ lies below the graph of $v^+(\cdot,t)$ and the segment connecting (p, v^-) to $q^-(t)$ lies above the graph of $v^-(\cdot,t)$. Thus, comparing their slopes with the slope of the tangents to $v^\pm(\cdot,t)$ at 0 and applying (14.14a) and (14.14b), we find that

$$\partial_2 v^+(0,t) - \partial_2 v^-(0,t) \geq -\frac{v^+(0,t) - v^-(0,t)}{d} \geq -2\varepsilon^2\,. \tag{14.16}$$

If the ray $\{re_2 : r < 0\}$ does not intersect ∂V_t, then, since $\mathcal{M}_t^n$ is convex,

$$\partial_2 v^+(0,t) - \partial_2 v^-(0,t) \le 0 .$$

Otherwise, applying the same argument to the point $p' \doteqdot \{re_2 : r < 0\} \cap \partial V_t$, we obtain

$$|\partial_2 v^+(0,t) - \partial_2 v^-(0,t)| \le 2\varepsilon^2 .$$

Denote by $\widehat{\mathcal{M}}_t$ the intersection of $\mathcal{M}_t$ with the plane $e_1 \wedge e_2$. Then the inward normal speed $\hat{H}$ of $\widehat{\mathcal{M}}_t$ is given by

$$\hat{H} = \frac{1}{|D\hat{u}|} = H\frac{|Du|}{|D\hat{u}|} ,$$

where $\hat{u}$ is the restriction of the arrival time u to $e_1 \wedge e_2$. On the other hand, since u is locally concave (Theorem 10.21), the curvature $\hat{\kappa}$ of $\widehat{\mathcal{M}}_t$ can be estimated by

$$\begin{aligned}
\hat{\kappa} = & -\frac{1}{|D\hat{u}|}\left(\Delta\hat{u} - \frac{D^2\hat{u}(D\hat{u}, D\hat{u})}{|D\hat{u}|^2}\right) \\
= & -\frac{D^2\hat{u}(\hat{\tau},\hat{\tau})}{|D\hat{u}|} \\
= & -\frac{1}{|D\hat{u}|}\left(\Delta u - \frac{D^2u(Du, Du)}{|Du|^2} - \operatorname{tr}_{\hat{\tau}^\perp} D^2 u\right) \\
\le & \frac{|Du|}{|D\hat{u}|}H ,
\end{aligned}$$

where $\hat{\tau}$ is a unit tangent vector field to $\widehat{\mathcal{M}}_t$ and where $\hat{\tau}^\perp$ is its orthogonal compliment in $T\mathcal{M}_t$. Thus,

$$\hat{H} \ge \hat{\kappa} .$$

We can now estimate

$$\begin{aligned}
\frac{d}{dt}\mathscr{H}^2(\widehat{\Omega}_t \cap \{\langle X, e_2\rangle > 0\}) = & -\int_{\widehat{\mathcal{M}}_t\cap\{\langle X,e_2\rangle>0\}} \hat{H}\, ds \\
\le & -\int_{\widehat{\mathcal{M}}_t\cap\{\langle X,e_2\rangle>0\}} \hat{\kappa}\, ds \\
\le & -\left(\pi - 2\varepsilon^2\right) ,
\end{aligned}$$

where $\widehat{\Omega}_t$ is the convex region in $e_1 \wedge e_2$ bounded by $\widehat{\mathcal{M}}_t$. Integrating this between $t < 4t_\varepsilon$ and t_ε yields, upon choosing $\varepsilon = \frac{\sqrt{\pi}}{2}$,

$$\mathscr{H}^2(\widehat{\Omega}_t^2 \cap \{\langle X, e_2\rangle > 0\}) \ge -\frac{3}{4}(\pi - 2\varepsilon^2)t \ge -\frac{3\pi}{8}t .$$

The convex region $\widehat{\Omega}_t$ lies between the tangent hyperplanes to $v^\pm(\cdot,t)$ at the origin. By (14.16), these tangent hyperplanes intersect the line $\langle X, e_2\rangle = d$

in $e_1 \wedge e$ at two points with distance at most $2(v^+(0,t)-v^-(0,t))$. Comparing $\mathscr{H}^2(\widehat{\Omega}_t^2 \cap \{\langle X, e_2\rangle > 0\})$ with the area of this enclosing trapezium yields

$$\frac{3}{2}v(0,t)\,|p| \geq -\frac{3\pi}{8}t,$$

which completes the proof of the claim. $\square$

For each $k \in \mathbb{N}$, set

$$t_k \doteqdot 2^k t_\varepsilon, \quad v_k \doteqdot v(\cdot, t_k), \quad \text{and} \quad d_k \doteqdot \min_{p \in \partial V_{t_k}} |p|,$$

where $\varepsilon > 0$ is to be determined and t_ε is chosen as in (14.14a) and (14.14b).

Claim 14.15 ([**521**, Claim 2 in Lemmas 2.1 and 2.2, and Lemma 2.7]). There exists $\varepsilon > 0$ such that

$$v_k(0) \leq v_{k-1}(0) + 2^{-\frac{k}{4n}}\sqrt{-t_\varepsilon} \tag{14.17}$$

for all $k \geq 1$.

Proof. Since the arrival time u of $\{\mathcal{M}_t^n\}_{t\in(-\infty,0)}$ is locally concave (Theorem 10.21) and since

$$u(v^\pm(y,t),y) = t$$

for all $t < 0$ and $y \in V_t$, we find that the functions $t \mapsto v^+(y,t)$ and $t \mapsto -v^-(y,t)$, and hence also $t \mapsto v(y,t)$, are concave for each fixed y. Indeed, if $y \in V_{t_0}$, then, for every $t_1, t_2 < t_0$ and every $r \in (0,1)$,

$$\begin{aligned} u(y, v^+(y, rt_1 + (1-r)t_2)) &= rt_1 + (1-r)t_2 \\ &= ru(y, v^+(y,t_1)) + (1-r)u(y, v^+(y,t_2)) \\ &\leq u(ry + (1-r)y, rv^+(y,t_1) + (1-r)v^+(y,t_2)) \\ &= u(y, rv^+(y,t_1) + (1-r)v^+(y,t_2)) \end{aligned}$$

and hence

$$v^+(y, rt_1 + (1-r)t_2) \geq rv^+(y,t_1) + (1-r)v^+(y,t_2)\,.$$

A similar argument proves the claim for $-v^-$.

It follows that

$$\frac{d}{dt}\frac{v(y,t)}{-t} \geq 0 \ \text{ for fixed } \ y \in V_{t_0} \text{ and every } t < t_0\,. \tag{14.18}$$

In particular,

$$\frac{v_k(0)}{-t_k} \leq \frac{v_0(0)}{-t_0}$$

and hence, by (14.14b),

$$v_k(0) \leq 2^k v_0(0) \leq 2^{k+1}\varepsilon\sqrt{-t_\varepsilon}\,.$$

Given $k_0 \geq 8n$ (to be determined momentarily), we choose $\varepsilon = \varepsilon(k_0)$ small enough that $2^{k_0+1}\varepsilon \leq 2^{-\frac{k_0}{4n}}$, so that

$$v_k(0) \leq 2^{-\frac{k_0}{4n}}\sqrt{-t_\varepsilon} \leq \frac{1}{4}\sqrt{-t_\varepsilon} \text{ for all } k \leq k_0. \tag{14.19}$$

In particular, (14.17) holds for each $k \leq k_0$. We will prove the claim by induction on k.

Suppose that (14.17) holds up to some $k \geq k_0$. Throughout the following discussion, k is fixed as such, where k_0 will be chosen depending only on n. Then

$$v_k(0) \leq v_{k_0}(0) + \sqrt{-t_\varepsilon}\sum_{i=k_0+1}^{k} 2^{-\frac{i}{4n}},$$

where the second term on the right-hand side is taken to be zero if $k = k_0$. Since $\sum_{j=1}^{\infty} 2^{-\frac{j}{4n}} < \infty$, we can choose k_0 so that $\sum_{j=k_0+1}^{\infty} 2^{-\frac{j}{4n}} < 1/4$. Applying (14.19), we then obtain

$$v_k(0) \leq \frac{1}{2}\sqrt{-t_\varepsilon}. \tag{14.20}$$

By (14.18),

$$\frac{v_{k+1}(0)}{-t_{k+1}} \leq \frac{v_k(0)}{-t_k}$$

and hence

$$v_{k+1}(0) \leq 2v_k(0) \leq \sqrt{-t_\varepsilon}.$$

Since $t \mapsto v(0,t)$ is decreasing, we conclude that

$$v(0,t) \leq \sqrt{-t_\varepsilon} \text{ for every } t \geq t_{k+1}.$$

Since $d(t) \doteqdot \min_{p\in\partial V_t}|p|$ is monotone decreasing in t, Claim 14.14 yields that there exists $\alpha > 0$ such that

$$d_{k+1} \geq d_k \geq \alpha\frac{-t_k}{v_k(0)} \geq \alpha\frac{-t_k}{\sqrt{-t_\varepsilon}} \tag{14.21a}$$

and

$$d_{k-1} \geq \alpha\frac{-t_{k-1}}{\sqrt{-t_\varepsilon}} = \frac{\alpha}{2}\frac{-t_k}{\sqrt{-t_\varepsilon}}. \tag{14.21b}$$

Define now

$$D_k \doteqdot \left\{y \in \mathbb{R}^n : |y| < \frac{\alpha}{2}\frac{-t_k}{\sqrt{-t_\varepsilon}}\right\}. \tag{14.22}$$

Since $y \mapsto v(y,t)$ is concave for fixed t,

$$v(y+se,t) - v(y,t) \leq sD_e v(y,t)$$

for any $y \in V_t$, any unit vector e, and any s such that $y + se \in V_t$. If $(y,t) \in D_k \times [t_{k+1}, t_k]$ and the ray $\{y + re : r > 0\}$ intersects ∂V_t, we can choose s so that $y + se \in \partial V_t$ and hence

$$D_e v(y,t) \geq -\frac{v(y,t)}{s} \geq -\frac{v(y,t)}{d(t) - |y|}.$$

If, however, the ray $\{y + re : r > 0\}$ does not intersect ∂V_t, then, since $\mathcal{M}_t^n$ is convex, $D_e v(y,t) \geq 0$. Applying the same reasoning with e replaced by $-e$ yields

$$D_e v(y,t) \leq \frac{v(y,t)}{d(t) - |y|}.$$

We conclude the gradient estimate

$$|Dv(y,t)| \leq \frac{v(y,t)}{d(t) - |y|} \tag{14.23}$$

for all $(y,t) \in D_k \times [t_{k+1}, t_k]$.

Since $t \mapsto v(y,t)$ is concave for fixed y, (14.23) implies

$$\begin{aligned} v(y,t) &\leq v(0,t) + |y||Dv(0,t)| \\ &\leq v(0,t)\left(1 + \frac{|y|}{d(t)}\right) \\ &\leq 2v(0,t). \end{aligned}$$

These estimates, (14.21a) and (14.21b), and the concavity of $t \mapsto v(y,t)$ yield, for all $(y,t) \in D_k \times [t_{k+1}, t_k]$,

$$|Dv(y,t)| \leq \frac{v(y,t)}{d_k - |y|} \leq 4\alpha^{-1}\frac{-t_\varepsilon}{-t_k} \tag{14.24}$$

and

$$0 \leq -\partial_t v(y,t) \leq \frac{v(y,t)}{-t} \leq \frac{v_k(y)}{-t_k} \leq \frac{2\sqrt{-t_\varepsilon}}{-t_k}. \tag{14.25}$$

We will use the gradient estimate to bound $\Delta v(y,t)$ in $D_k \times [t_{k+1}, t_k]$ in terms of t_k. Given a positive function $f : \mathbb{N} \to \mathbb{R}$, to be determined later, define

$$\chi = \{(y,t) \in D_k \times [t_{k+1}, t_k] : -\Delta v(y,h) \geq f(k)\}.$$

Recalling (14.24) we find, for any $t \in (t_{k+1}, t_k)$,

$$\begin{aligned}\mathscr{H}^n(\{y \in D_k : (y,t) \in \chi\}) f(k) &\leq -\int_{D_k} \Delta v(\cdot, t)\, d\mathscr{H}^n \\ &\leq \int_{\partial D_k} |Dv(\cdot,t)|\, d\mathscr{H}^{n-1} \\ &\leq \sup_{D_k} |Dv(\cdot,t)| \mathscr{H}^{n-1}(D_k) \\ &\leq C\frac{t_\varepsilon}{t_k}\left(\frac{-t_k}{\sqrt{-t_\varepsilon}}\right)^{n-1} \\ &= C\frac{2^{k(n-1)}}{-t_k}(-t_\varepsilon)^{\frac{n+1}{2}},\end{aligned}$$

where C is a constant which depends only on n. Integrating between t_{k+1} and t_k then yields

$$\mathscr{H}^{n+1}(\chi) \leq C\frac{2^{k(n-1)}}{f(k)}(-t_\varepsilon)^{\frac{n+1}{2}} \cdot \frac{t_k - t_{k+1}}{-t_k} = 2C\frac{2^{k(n-1)}}{f(k)}(-t_\varepsilon)^{\frac{n+1}{2}}.$$

Consider now another positive function $g : \mathbb{N} \to \mathbb{R}$, to be determined later. Since

$$\begin{aligned}\int_{D_k} \mathscr{H}^1(\chi \cap \{(y,t) : y = z\})\, d\mathscr{H}^n(z) &= \mathscr{H}^{n+1}(\chi) \\ &\leq 2C\frac{2^{k(n-1)}}{f(k)}(-t_\varepsilon)^{\frac{n+1}{2}},\end{aligned}$$

there exists $\widehat{D}_k \subset D_k$ with

$$\mathscr{H}^n(\widehat{D}_k) \leq 2C\frac{2^{k(n-1)}}{f(k)g(k)}(-t_\varepsilon)^{\frac{n+1}{2}}$$

such that

$$\mathscr{H}^1(\chi \cap \{(y,t) : y = z\}) \leq g(k) \text{ for all } z \in D_k \setminus \widehat{D}_k. \tag{14.26}$$

Now, for any $y \in D_k \setminus \widehat{D}_k$,

$$\begin{aligned}v_{k+1}(y) - v_k(y) &= -\int_{t_{k+1}}^{t_k} \partial_t v(y,s)\, ds \\ &= -\int_{[t_{k+1},t_k]\setminus I(y)} \partial_t v(y,s)\, ds - \int_{I(y)} \partial_t v(y,s)\, ds,\end{aligned} \tag{14.27}$$

where $I(y) \doteqdot \{t : (y,t) \in \widehat{D}_k\}$. Using (14.25) and (14.26) to bound the first integral on the right-hand side of (14.27) and the graphical mean curvature flow equation (14.13) to bound the second, we find, for any $y \in D_k \setminus \widehat{D}_k$,

$$v_{k+1}(y) - v_k(y) \leq \frac{2\sqrt{-t_\varepsilon}}{-t_k} g(k) - f(k)t_k. \tag{14.28}$$

We now choose $f(k) = \frac{2^{-k(1+\beta/2)}}{\sqrt{-t_\varepsilon}}$ and $g(k) = 2^{k(1-\beta/2)}(-t_\varepsilon)$, where $\beta \in (0,1)$ will be chosen explicitly below. Then

$$\begin{aligned}\mathscr{H}^n(D_k \setminus \widehat{D}_k) &\geq \omega_n \left(\frac{-\pi t_k}{8\sqrt{-t_\varepsilon}}\right)^n - 2^{k\beta+1}C\sqrt{-t_\varepsilon}\left(\frac{-t_k}{\sqrt{-t_\varepsilon}}\right)^{n-1}\\ &= \left(\omega_n \left(\frac{\pi}{8}\right)^n 2^k - 2^{k\beta+1}C\right)\sqrt{-t_\varepsilon}\left(\frac{-t_k}{\sqrt{-t_\varepsilon}}\right)^{n-1},\end{aligned}$$

which is positive and $O(2^{nk})$ if k_0 is chosen sufficiently large (depending on β).

Since

$$\mathscr{H}^n(\widehat{D}_k) \leq 2^{(\beta+n-1)k+1}C(-t_\varepsilon)^{\frac{n}{2}},$$

there is a point $y_0 \in D_k \setminus \widehat{D}_k$ with

$$0 < |y_0| \leq \left(\frac{2C}{\omega_n}\right)^{\frac{1}{n}} 2^{k\left(\frac{\beta-1}{n}+1\right)}\sqrt{-t_\varepsilon} \quad \text{and} \quad v_k(y_0) \leq v_k(0) \tag{14.29}$$

(see Exercise 14.6). Since $y \mapsto v(y, t_{k+1})$ is concave and zero on ∂V_t, we find

$$v_{k+1}(0) \leq \frac{d_{k+1}}{d_{k+1} - |y|} v_{k+1}(y_0).$$

Set $C' \doteqdot \left(\frac{2C}{\omega_n}\right)^{\frac{1}{n}}$. Then (14.21a) implies that

$$\begin{aligned}\frac{d_{k+1}}{d_{k+1} - |y|} &\leq \frac{d_{k+1}}{d_{k+1} - 2^{\left(\frac{\beta-1}{n}+1\right)k}C'\sqrt{-t_\varepsilon}}\\ &= 1 + \frac{2^{\left(\frac{\beta-1}{n}+1\right)k}C'\sqrt{-t_\varepsilon}}{d_{k+1} - 2^{\left(\frac{\beta-1}{n}+1\right)k}C'\sqrt{-t_\varepsilon}}\\ &\leq 1 + 2C' \cdot 2^{\frac{\beta-1}{n}k}\end{aligned}$$

provided k_0 is chosen sufficiently large (depending on β). Thus,

$$v_{k+1}(0) \leq (1 + 2C' \cdot 2^{\frac{\beta-1}{n}k})v_{k+1}(y_0).$$

Applying (14.28), estimating $v_k(y_0) \leq v_k(0)$, and recalling (14.20), we conclude that

$$\begin{aligned}v_{k+1}(0) &\leq (1 + 2C' \cdot 2^{\frac{\beta-1}{n}k})(v_k(0) + 3\sqrt{-t_\varepsilon}2^{-\frac{\beta}{2}k})\\ &\leq v_k(0) + \sqrt{-t_\varepsilon}\left(C' \cdot 2^{\frac{\beta-1}{n}k} + 3 \cdot 2^{-\frac{\beta}{2}k} + 6C' \cdot 2^{\left(\frac{\beta-1}{n}-\frac{\beta}{2}\right)k}\right).\end{aligned}$$

Choosing now $\beta \in (1/2n, 3/4)$, e.g., $\beta = 5/8$, we obtain provided k_0 is sufficiently large that

$$v_{k+1}(0) \leq v_k(0) + \sqrt{-t_\varepsilon}2^{-\frac{k}{4n}}.$$

This completes the proof of the claim. □

We conclude that

$$v(0, 2^{k+1}t_\varepsilon) = v_{k+1}(0) \le v_0(0) + C\sum_{j=1}^{k} 2^{-\frac{j}{4n}}\,,$$

where C is independent of k. Since the sum $\sum_{j=1}^{\infty} 2^{-\frac{j}{4n}}$ is finite and $v(0,t)$ is monotone, we conclude that $v(0,t)$ is bounded uniformly in time.

It now follows from (14.23) that the solution lies in the slab

$$\Pi \doteqdot \{X \in \mathbb{R}^{n+1} : V^- < \langle X, e_1\rangle < V^+\}\,,$$

where

$$V^\pm \doteqdot \lim_{t\to-\infty} v^\pm(0,t) < \infty\,.$$

This completes the proof of Theorem 14.13. □

We note that F. Chini and N. Møller [**145**] have partially classified the *convex hulls* of (nonconvex) ancient solutions to mean curvature flow more generally (see Theorem 14.30 in the notes and commentary at the end of this chapter).

Theorem 14.13 motivates the following definitions.

Definition 14.16. A convex, compact ancient solution to mean curvature flow is called

- an **ancient ovaloid** if it is entire,
- an **ancient pancake**[3] if it lies in a stationary slab region.

We will say more on ancient ovaloids and ancient pancakes in Sections 14.7 and 14.8, respectively.

Lemma 14.10 and Theorem 14.13 are very useful tools for analyzing convex ancient solutions. Let us first observe that they provide a nice proof of Theorem 14.2 and Corollary 14.4 (rigidity of the shrinking sphere) [**100**].

Alternative proof of Theorem 14.2 and Corollary 14.4 (cf. [521]). Let $\{\mathcal{M}_t^n\}_{t\in(-\infty,0)}$ be a compact, convex ancient solution satisfying one of the conditions (2)–(7). By Lemma 14.10(ii)(a), it suffices to prove that the blow-down of $\{\mathcal{M}_t^n\}_{t\in(-\infty,0)}$ is the shrinking sphere. We only need to prove this in cases (2), (3), and (4) since (6) implies (5) (Lemma 8.3) and (7) implies (4) ([**304**, Lemma 4.4]).

Condition (2) (uniform pinching): Since the pinching condition is scale invariant and violated on any shrinking cylinder $\mathbb{R}^k \times S^{n-k}_{\sqrt{-2(n-k)t}}$, $k \in \{1, \ldots, n-1\}$, it suffices, by Lemma 14.10(ii), to rule out the hyperplane of

[3]Cheese lovers may prefer Sigurd Angenent's terminology: **ancient gouda**.

multiplicity two. Suppose, to the contrary, that the blow-down is the hyperplane of multiplicity two. Theorem 14.13 then implies that $\{\mathcal{M}_t^n\}_{t\in(-\infty,0)}$ lies in a slab. By translating parallel to the slab, we obtain, by part (iii) of Lemma 14.10, an asymptotic translator Σ^n whose normal at the origin is parallel to the slab. Thus, it cannot be a hyperplane and hence, by the strong maximum principle, it must satisfy $H > 0$. By the pinching condition, Hamilton's compactness theorem (Theorem 11.20) [**264**] then implies that Σ^n is compact, which is impossible.

Condition (3) (bounded rescaled diameter) implies that the blow-down is compact and hence, by Lemma 14.10(ii)(a), the shrinking sphere.

Condition (4) (bounded eccentricity): This case is similar to case (2) — since the eccentricity condition is scale invariant and violated on any shrinking cylinder $\mathbb{R}^k \times S^{n-k}_{\sqrt{-2(n-k)t}}$, $k \in \{1, \ldots, n-1\}$, it suffices, by Lemma 14.10(ii), to rule out the hyperplane of multiplicity two. Suppose, to the contrary, that the blow-down is the hyperplane of multiplicity two. Theorem 14.13 then implies that $\{\mathcal{M}_t^n\}_{t\in(-\infty,0)}$ lies in a slab. It follows that its inradius is uniformly bounded but its circumradius is not, contradicting the hypothesis.

Condition (5) (type-I curvature decay) implies (by integration) that the displacement is bounded by $\sqrt{-t}$. It follows that the blow-down is compact and the theorem again follows immediately from Lemma 14.10(ii)(a). □

14.6. Ancient solutions to curve shortening flow revisited

By X.-J. Wang's dichotomy and Lemma 14.10, a convex ancient solution to curve shortening flow which is not a shrinking circle necessarily lies in a strip region. By the strong maximum principle, applied to the evolution of the curvature, a nonflat convex ancient solution is locally uniformly convex. We will show that the only convex, locally uniformly convex solutions to curve shortening flow which lie in strip regions are the paperclips and the Grim Reapers. This will prove the following generalization of Theorem 4.25; see [**99**] by T. Bourni, G. Tinaglia, and the fourth author.

Theorem 14.17. *The only convex ancient solutions to curve shortening flow are the stationary lines, the shrinking circles, the paperclips, and the Grim Reapers.*

By the comments above and the invariance properties of the curve shortening flow, it suffices to consider a convex, locally uniformly convex ancient solution $\{\Gamma_t\}_{t\in(-\infty,0)} \subset \mathbb{R}^2$ which lies in the strip $\Pi \doteqdot \{(x, y) : |x| < \pi/2\}$ and in no smaller strip. Denote by θ the turning angle parameter of the

Figure 14.1. Theodora Bourni.

solution (the angle made by the x-axis and its tangent vector with respect to a counterclockwise parametrization). Since the solution is convex and lies in the strip Π, we can arrange (by reflecting across the x-axis if necessary) that the Gauß image $\theta(\Gamma_t)$ is either all of $\mathbb{R}/2\pi\mathbb{Z}$ (compact case) or the interval $(-\frac{\pi}{2}, \frac{\pi}{2})$ (noncompact case) for all $t < 0$. Denote by $\gamma : I \times (-\infty, 0) \to \mathbb{R}^2$, $I \in \{\mathbb{R}/2\pi\mathbb{Z}, (-\frac{\pi}{2}, \frac{\pi}{2})\}$, the turning angle parametrization and set $p_-(t) = \gamma(0,t)$ and, in the compact case, $p_+(t) \doteqdot \gamma(\pi, t)$. By translating in space-time, we can arrange that $y(p_-(0)) = 0$ in the noncompact case and (by the Gage–Hamilton theorem) $\lim_{t\to 0} y(p_\pm(t)) = 0$ in the compact case.

We begin with some basic asymptotics. Denote by $\{\mathrm{G}_t\}_{t\in(-\infty,\infty)}$ the standard Grim Reaper. That is,

$$\mathrm{G}_t \doteqdot \{(\theta, t - \log\cos\theta) : \theta \in (-\tfrac{\pi}{2}, \tfrac{\pi}{2})\}.$$

Lemma 14.18 (Cf. Lemma 14.10). *The translated family $\{\Gamma^\pm_{s,t}\}_{t\in(-\infty,-s)}$ defined by*

$$\Gamma^\pm_{s,t} \doteqdot \Gamma_{t+s} - p_\pm(s)$$

converges locally uniformly in the smooth topology as $s \to -\infty$ to the Grim Reaper $\{\mp r_\pm \mathrm{G}_{r_\pm^{-2}t}\}_{t\in(-\infty,\infty)}$, where

$$r_+ \doteqdot \lim_{s\to-\infty} \kappa^{-1}(\pi, s) \ \ and \ \ r_- \doteqdot \lim_{s\to-\infty} \kappa^{-1}(0, s).$$

Proof. Since the solution is locally uniformly convex, the differential Harnack inequality (4.24) implies that the curvature is nondecreasing in time with respect to the turning angle parametrization. In particular, the limits $\kappa^-_\infty \doteqdot \lim_{s\to-\infty} \kappa(0, s)$ and $\kappa^+_\infty \doteqdot \lim_{s\to-\infty} \kappa(\pi, s)$ exist. Since each translated solution contains the origin at time 0 and the curvature is uniformly bounded on compact subsets of space-time, it follows from Theorem 11.11 that a subsequence of each of the families converges locally uniformly in the smooth topology to a weakly convex eternal limit flow lying in a strip of the same width. By the strong maximum principle, the limit must be locally uniformly convex since its normal at the origin at time 0 is parallel to the strip (which rules out a stationary line as the limit). It follows that $\kappa_\infty^\pm$ is positive. Since the curvature of the limit is constant in time with respect to the turning angle parametrization, the rigidity case of the differential Harnack inequality implies that it moves by translation. Since the Grim Reapers are the only translating solutions to curve shortening flow other than the stationary lines, we conclude that the limit is the Grim Reaper with bulk velocity $v_\pm \doteqdot \kappa_\infty^\pm e_2$. The claim follows since the limit is independent of the subsequence. □

Lemma 14.19. *The solution $\{\Gamma_t\}_{t\in(-\infty,0)}$ sweeps out all of Π; that is, $\bigcup_{t<0}\Omega_t = \Pi$, where Ω_t is the convex region bounded by Γ_t.*

Proof. Since the solution is locally uniformly convex, $\Omega_{t_2} \subset \Omega_{t_1}$ for all $t_1 \le t_2 < 0$. By hypothesis, given any $\delta \in (0,1)$, there exist points $p_1, p_2 \in \Omega \doteqdot \bigcup_{t<0}\Omega_t$ such that

$$x(p_1) = \frac{\pi}{2} - \delta \text{ and } x(p_2) = -\frac{\pi}{2} + \delta\,.$$

By convexity, the segments joining p_i to $p_-(t)$ are also contained in Ω for each t. In the compact case, the segments joining p_i to $p_+(t)$ are contained in Ω for each t, while in the noncompact case, the rays $\{(x(p_i), sy(p_i)) : s > 1\}$ are contained in Ω. Since $y(p_-(t)) \to -\infty$ and, in the compact case, $y(p_+(t)) \to \infty$ as $t \to -\infty$, we conclude that Ω contains the lines $\{(\frac{\pi}{2} - \delta, s) : s \in \mathbb{R}\}$ and $\{(-\frac{\pi}{2} + \delta, s) : s \in \mathbb{R}\}$ and hence, by convexity, also the strip $\Pi_\delta \doteqdot \{(x,y) \in \mathbb{R}^2 : |x| \le \frac{\pi}{2} - \delta\}$. Since $\delta \in (0,1)$ was arbitrary, we conclude that $\Pi = \bigcup_{\delta\in(0,1)}\Pi_\delta \subset \Omega \subset \Pi$, which implies the claim. □

The key to the argument is to show that the asymptotic Grim Reapers are of maximal width.

By the Harnack inequality and Lemma 14.18, the maximal vertical displacements $\ell_\pm(t) \doteqdot \pm y(p_\pm(t))$ satisfy

$$\ell_\pm(t) \ge -\kappa_\infty^\pm t = -r_\pm^{-1} t\,. \tag{14.30}$$

Let $A(t)$ and $B(t)$ be the two points on Γ_t such that (see Figure 14.2)

$$y(A(t)) = y(B(t)) = 0$$

arranged so that

$$x(A(t)) > x(B(t))\,.$$

Let $\theta_A(t)$ and $\theta_B(t)$ be the corresponding turning angles, so that $\gamma(\theta_A(t),t) = A(t)$ and $\gamma(\theta_B(t),t) = B(t)$. In the noncompact case, the area $\mathrm{Area}(t)$ enclosed by Γ_t and the x-axis satisfies

$$-\mathrm{Area}'(t) = \int_{B(t)}^{A(t)} \kappa\, ds = \int_{\theta_B(t)}^{\theta_A(t)} d\theta = \theta_A(t) - \theta_B(t) \leq \pi\,.$$

Since $\mathrm{Area}(0) = 0$, integrating from $-t$ to 0 yields

$$\mathrm{Area}(t) \leq -\pi t\,.$$

Using the displacement estimate (14.30) and the fact that the solution approaches the boundary of the strip Π, we will prove that the enclosed area grows too quickly as $t \to -\infty$ if $r < 1$ (see Figure 14.2).

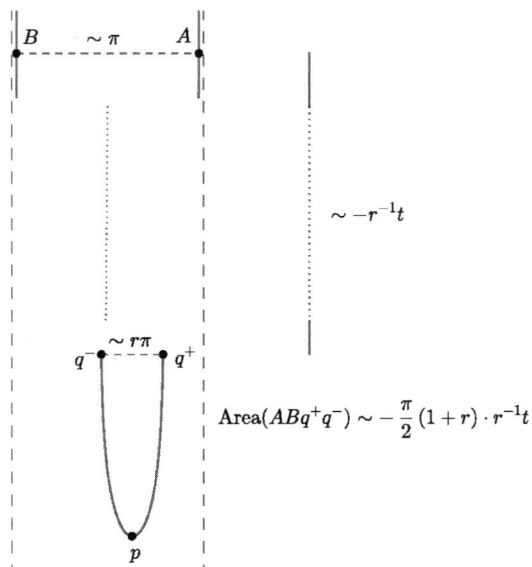

Figure 14.2. The enclosed area grows too quickly if $r < 1$.

Lemma 14.20. *The width of the asymptotic Grim Reaper is maximal:*

$$r_\pm = 1\,.$$

Proof. By Lemma 14.19, for any $\delta \in (0,1)$ we can find $t_\delta < 0$ such that

$$\pi \geq x(A(t)) - x(B(t)) \geq \pi - \delta$$

for every $t < t_\delta$. By Lemma 14.18, choosing t_δ smaller if necessary, we can also find two points $\underline{q}^\pm(t)$ and a constant C_δ with the following properties:

$$y(\underline{q}^+(t)) = y(\underline{q}^-(t))\,,$$

$$\pi r_- - \delta < x(\underline{q}^+(t)) - x(\underline{q}^-(t)) < \pi r_-\,,$$

$$0 < y(\underline{q}^\pm(t)) - y(p_-(t)) < C_\delta\,.$$

Since the enclosed area is bounded below by that of the trapezium with vertices A, B, $\underline{q}^-$, and $\underline{q}^+$, we find, in the noncompact case, that

$$-\pi t \geq \text{Area}(t) \geq \frac{1}{2}\left(\pi r_- - \delta + \pi - \delta\right)\left(-r_-^{-1}t - C_\delta\right)$$

and hence

$$-(\pi(1-r_-) - 2\delta)t \leq (\pi(1+r_-) - 2\delta)r_- C_\delta$$

for all $t < t_\delta$. Taking $t \to -\infty$, we conclude that $(1-r)\pi \leq 2\delta$ for any $\delta > 0$. Taking $\delta \to 0$ then yields the claim.

In the compact case, we can find a further two points $\overline{q}^\pm(t)$ such that

$$\begin{aligned} y(\overline{q}^+(t)) &= y(\overline{q}^-(t)), \\ \pi r_- - \delta < x(\overline{q}^+(t)) &- x(\overline{q}^-(t)) < \pi r_-, \\ 0 < y(p_+(t)) &- y(\overline{q}^\pm(t)) < C_\delta. \end{aligned}$$

Bounding the combined area of the two corresponding trapezia by the area enclosed by Γ_t yields

$$-2\pi t \geq \frac{1}{2}\left(\pi r_- + \pi - 2\delta\right)\left(-r_-^{-1}t - C_\delta\right) + \frac{1}{2}\left(\pi r_+ + \pi - 2\delta\right)\left(-r_+^{-1}t - C_\delta\right)$$

and hence

$$-\left(\pi(r_+^{-1} + r_-^{-1} - 2) - 2\delta(r_+^{-1} + r_-^{-1})\right)t \leq \left(\pi(r_+ + r_-) + 2(\pi - 2\delta)\right)C_\delta$$

for all $t < t_\delta$. Taking $t \to -\infty$, we conclude that $(r_+^{-1} + r_-^{-1} - 2)\pi \leq 2\delta(r_+^{-1} + r_-^{-1})$ for any $\delta > 0$. Taking $\delta \to 0$ then yields the claim. □

The Alexandrov reflection principle of Gulliver and the second author (see [**160, 164**], or Section 18.8) can now be employed to prove that the solution is reflection symmetric about the y-axis (cf. Paul Bryan and Janelle Louie [**126**]).

Lemma 14.21. *Let $\{\Gamma_t\}_{t\in(-\infty,0)}$ be a convex ancient solution which is contained in the strip $\Pi \doteqdot \{(x,y) : |x| < \pi/2\}$ (and in no smaller strip). Then Γ_t is reflection symmetric about the y-axis for all $t < 0$.*

Proof. Set

$$H_\alpha \doteqdot \{(x,y) \in \mathbb{R}^2 : x < \alpha\}$$

and denote by R_α the reflection about ∂H_α. It is a consequence of the convexity of $\{\Gamma_t\}_{t\in(-\infty,0)}$ and the convergence of its "tips" to Grim Reapers that, given any $\alpha \in (0, \pi/2)$, there exists a time t_α such that

$$(R_\alpha \cdot \Gamma_t) \cap (\Gamma_t \cap H_\alpha) \cap \{(x,y) \in \mathbb{R}^2 : \pm y < 0\} = \emptyset$$

for all $t < t_\alpha$. By the Alexandrov reflection principle, this is true for all $t < 0$. Taking $\alpha \searrow 0$ proves the lemma. □

We can now prove that the solution is either the Grim Reaper or the paperclip.

Theorem 14.22. *Let $\{\Gamma_t\}_{t\in(-\infty,0)}$ be a convex ancient solution which is contained in the strip $\Pi \doteqdot \{(x,y) : |x| < \pi/2\}$ and in no smaller strip. Then $\{\Gamma_t\}_{t\in(-\infty,0)}$ is either a Grim Reaper or a paperclip.*

Proof. We will only prove the claim in the noncompact case (we leave the compact case as an exercise). Since, by the Harnack inequality, the curvature of $\{\Gamma_t\}_{t\in(-\infty,0)}$ is nondecreasing in t with respect to the turning angle parametrization, the maximal vertical displacement ℓ of $\{\Gamma_t\}_{t\in(-\infty,0)}$ satisfies

$$\frac{d}{dt}\left(\ell(t)+t\right) \le 0\,.$$

So the limit $\lim_{t\to-\infty}(\ell(t)+t)$ exists in $(0,\infty]$.

Claim 14.23. The asymptotic displacement is equal to that of the Grim Reaper:

$$\lim_{t\to-\infty}(\ell(t)+t) = 0.$$

Proof of the claim. Suppose, contrary to the claim, that $C > 0$ (the case $C < 0$ is ruled out similarly); then we can find t_0 such that

$$\ell(t) > -t \text{ for all } t < t_0 \text{ and } \ell(t_0) = -t_0\,. \tag{14.31}$$

Define the half-spaces H_α and the reflection R_α about ∂H_α as in Lemma 14.21. Given any $\alpha \in (0,\pi/2)$, set

$$\widetilde{\mathrm{G}}_t = (R_\alpha\cdot \mathrm{G}_t)\cap\{(x,y)\in\mathbb{R}^2 : x<0\}\,,$$

where $\{\mathrm{G}_t\}_{t\in(-\infty,\infty)}$ is the Grim Reaper, and

$$\widetilde{\Gamma}_t = \Gamma\cap\{(x,y)\in\mathbb{R}^2 : x<0\}\,.$$

Then, by (14.31), $\partial\widetilde{\mathrm{G}}_t\cap\partial\widetilde{\Gamma}_t = \emptyset$ for all $t < t_0$. Moreover, by convexity and the convergence of the tip to the Grim Reaper, there exists $t_\alpha < t_0$ depending on α such that $\widetilde{\mathrm{G}}_t\cap\widetilde{\Gamma}_t = \emptyset$ for all $t < t_\alpha$. It then follows by the strong maximum principle that $\widetilde{\mathrm{G}}_t\cap\widetilde{\Gamma}_t = \emptyset$ for all $t < t_0$. Letting $\alpha \searrow 0$, we find that Γ_t lies below G_t for all $t < t_0$, contradicting the strong maximum principle since their tips coincide at time $t = t_0$. □

Now consider, for any $\tau > 0$, the solution $\{\Gamma_t^\tau\}_{t\in(-\infty,0)}$ defined by $\Gamma_t^\tau = \Gamma_{t+\tau}$. Since

$$C_\tau \doteqdot \lim_{t\to-\infty}(\ell(t+\tau)+t) > 0\,,$$

we may argue as above to conclude that Γ_t^τ lies above G_t for all $t < 0$. Taking $\tau \to 0$, we find that Γ_t lies above G_t for all $t < 0$. Since the two curves reach the origin at time 0, they intersect for all $t < 0$ by the avoidance principle. The strong maximum principle then implies that the two coincide for all t. $\square$

14.7. Ancient ovaloids

White constructed an ancient ovaloid in $\mathbb{R}^{n+1}$ as the limit as $j \to \infty$ of flows $\{\mathcal{M}^n_{j,t}\}_{t\in[-\alpha_j,\omega_j)}$ of convex, compact initial hypersurfaces $\mathcal{M}^n_{j,-\alpha_j}$ of eccentricity j, translated in time and parabolically rescaled so that the eccentricity is equal to 2, say, at time $t = -1$ [**531**, p. 134]. The limiting ancient solution $\{\mathcal{M}^n_t\}_{t\in(-\infty,T)}$ is convex, compact, nonround (since its eccentricity is 2 at time -1), and entire. Around the same time, Wang [**521**, Theorem 1.2] gave a different construction by taking the limit of a sequence of Dirichlet problems for the level set flow over a sequence of ellipsoidal domains. Haslhofer and Hershkovits extended White's construction to obtain an explicit family of ancient solutions [**274**] and also studied the asymptotics of these solutions using results of Haslhofer and Kleiner [**275**] (see also the formal discussion in [**62**]). These results can be roughly summarized as follows.

Theorem 14.24 (Existence of ancient ovaloids [**62, 274, 521, 531**]). *For each $n \in \mathbb{N}$ and each $m \in \{1, \ldots, n-1\}$ there exists a compact, convex ancient solution $\{\mathcal{O}^{n,m}_t\}_{t\in(-\infty,0)}$ of mean curvature flow in $\mathbb{R}^{n+1}$ which is*

- *uniformly $(m+1)$-convex:* $\displaystyle\liminf_{t\to-\infty} \frac{\kappa_1 + \cdots + \kappa_{m+1}}{H} > 0$,
- $\mathrm{SO}(n+1-m) \times \mathrm{SO}(m)$*-invariant, and*
- *noncollapsing (and hence entire).*

Moreover, $\{\mathcal{O}^{n,m}_t\}_{t\in(-\infty,0)}$ satisfies the following asymptotics:

- Blow-up: $\{\lambda\mathcal{O}^{n,m}_{\lambda^{-2}t}\}_{t\in(-\infty,0)} \to \left\{S^n_{\sqrt{-2nt}}\right\}_{t\in(-\infty,0)}$ *uniformly in the smooth topology as $\lambda \to \infty$.*
- Blow-down: $\{\lambda\mathcal{O}^{n,m}_{\lambda^{-2}t}\}_{t\in(-\infty,0)} \to \left\{S^{n-m}_{\sqrt{-2(n-m)t}} \times \mathbb{R}^m\right\}_{t\in(-\infty,0)}$ *locally uniformly in the smooth topology as $\lambda \to 0$.*
- Asymptotic translators[4]:

$$\left\{\lambda_s\left(\mathcal{O}^{n,m}_{\lambda_s^{-2}t+s} - p_s\right)\right\}_{t\in(-\infty,-\lambda_s^2 s)} \to \{\mathbb{R}^{m-1} \times \mathrm{Bowl}^{n+1-m}_t\}_{t\in(-\infty,\infty)},$$

[4]Note that the asymptotic translators are obtained after rescaling by the curvature at the tip, in contrast to Lemma 14.10.

locally uniformly in the smooth topology as $s \to -\infty$, *where* p_t *is defined by* $\mathsf{N}(p_t, t) = -e_{n+1}$, $\lambda_t \doteqdot H_{\max}(t) = H(p_t, t)$, *and* $\left\{\mathrm{Bowl}_t^k\right\}_{t\in(-\infty,\infty)}$ *is the unit speed, vertically translating bowl soliton in* $\mathbb{R}^{k+1}$.

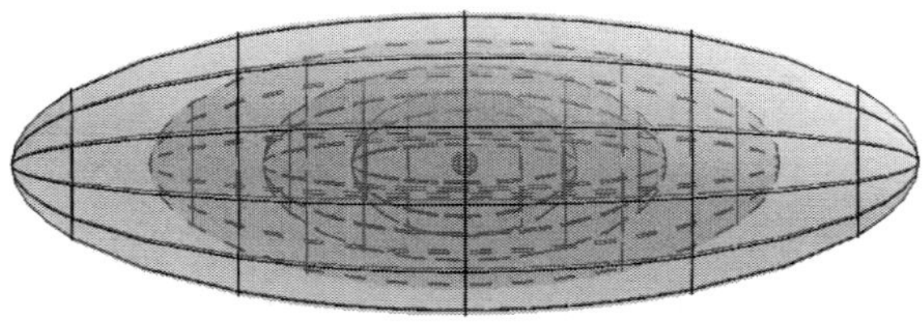

Figure 14.3. Snapshots of the ancient ovaloid $\{\mathcal{O}_t^{2,1}\}_{t\in(-\infty,0)}$ in $\mathbb{R}^3$. It shrinks to a round point as $t \to 0$ and resembles two bowl solitons glued together along a shrinking cylinder as $t \to -\infty$.

Angenent, Daskalopoulos, and Šešum have undertaken a comprehensive study of ancient ovaloids which are uniformly 2-convex and noncollapsing [**63, 70**] (cf. [**111, 112**]). In particular, they derive refined asymptotics for such solutions and, as a consequence, are able to prove that the bisymmetric example described above is unique within this class.

Theorem 14.25 (Uniqueness of 2-convex, noncollapsing ancient ovaloids [**63, 70**]). *Modulo space-time translations, rigid motions, and parabolic scaling, the ancient ovaloid* $\{\mathcal{O}_t^{n,1}\}_{t\in(-\infty,0)}$ *is unique among 2-convex, noncollapsing ancient ovaloids.*

Combining Theorems 14.9 and 14.13 and Lemma 14.10 yields a gradient estimate for ancient ovaloids.

Proposition 14.26. *Let* $\{\mathcal{M}_t^n\}_{t\in(-\infty,0)}$ *be an ancient ovaloid. If*

$$\liminf_{t\to-\infty} \inf_{\mathcal{M}_t^n} \frac{\kappa_1 + \cdots + \kappa_{m+1}}{H} > 0$$

for some $m < \frac{2(n-1)}{3}$, *then*

$$\sup_{t\in(-\infty,0)} \sup_{\mathcal{M}_t^n} \frac{|\nabla A|}{H^2} < \infty \,. \tag{14.32}$$

Proof. Theorem 14.13 and Lemma 14.10 imply that

$$\sqrt{-2t}\min_{\mathcal{M}_t^n} H \sim 1 \tag{14.33}$$

as $t \to -\infty$. The claim then follows from Theorem 14.9. □

A gradient estimate of this form follows for *noncollapsing* ancient ovaloids from the Haslhofer–Kleiner local curvature estimate (Theorem 12.19).

14.8. Ancient pancakes

The existence of ancient pancakes was proved by Wang [**521**, Theorem 1.2], once again by taking the limit of a sequence of solutions to the Dirichlet problem for the level set flow. A different construction was given by Bourni, Tinaglia, and the fourth author by taking the limit of a sequence of mean curvature flows evolving from rotated time slices of the paperclip [**97**].

Theorem 14.27 (Existence of ancient pancakes [**97, 521**]). *For each $n \in \mathbb{N}$ there exists a compact, convex ancient solution $\{\Pi_t^n\}_{t\in(-\infty,0)}$ of mean curvature flow in $\mathbb{R}^{n+1}$ which*

- *is* $\mathrm{SO}(1) \times \mathrm{SO}(n)$*-invariant and*
- *lies at all times in the slab region* $\{x \in \mathbb{R}^{n+1} : |x_1| < \frac{\pi}{2}\}$.

It satisfies the following asymptotics:

- Blow-down[5]: $\{\mathcal{M}_{t+s}\}_{t\in(-\infty,-s)} \to \{\{-\frac{\pi}{2}, \frac{\pi}{2}\} \times \mathbb{R}^n\}_{t\in(-\infty,\infty)}$ *locally uniformly in the smooth topology as* $s \to -\infty$.
- Asymptotic translators: *For each unit vector* $e \in \{e_1\}^\perp$,

$$\{\Pi_{t+s}^n - p_s\}_{t\in(-\infty,-s)} \to \{\Gamma_t^n\}_{t\in(-\infty,\infty)}$$

locally uniformly in the smooth topology as $s \to -\infty$*, where* p_s *is the unique point on* Π_s *satisfying* $\mathsf{N}(p_s, s) = e$ *and* $\{\Gamma_t^n\}_{t\in(-\infty,\infty)}$ *is the Grim hyperplane in the slab which translates with velocity* $-e$.

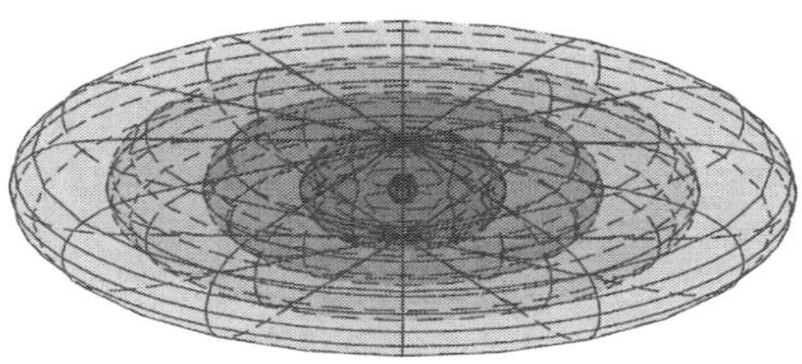

Figure 14.4. The rotationally symmetric ancient pancake in $\mathbb{R}^3$.

Proof (sketch). Existence of the solution is proved by evolving rotated time slices Λ_{-R} of the paperclip solution $\{\Lambda_t\}_{t\in(-\infty,0)}$ and studying the evolution equation

$$\partial_t \gamma_R = -(\kappa_R + (n-1)\lambda_R)\mathsf{N}_R$$

[5]In particular, $\{\lambda\mathcal{M}_{\lambda^{-2}t}\}_{t\in(-\infty,0)}$ converges smoothly on compact subsets of space-time to a hyperplane of multiplicity two as $\lambda \to 0$.

for the profile curve γ_R, where N_R is the outward pointing unit normal to γ_R, κ_R is its curvature, and $\lambda_R \doteqdot -\frac{\cos\theta_R}{y}$, θ_R denoting the angle the tangent vector makes with the x-axis. Let us briefly sketch the ideas involved.

First, we apply the maximum principle to obtain a Sturmian-type lemma: The mean curvature $H_R = \kappa_R + (n-1)\lambda_R$ always has exactly four critical points along γ_R — the maximum occurs when $\theta_R = 0$ and the minimum when $\theta_R = \frac{\pi}{2}$ (note that the double reflection symmetry of Λ_{-R} is preserved under the flow).

The area A_R enclosed by γ_R satisfies

$$-\frac{d}{dt}A_R = \int H_R\,ds_R = \int (\kappa_R + (n-1)\lambda_R)\,ds_R = 2\pi + (n-1)\int \frac{\lambda_R}{\kappa_R}\,d\theta_R\,,$$

where s_R is the arc length along γ_R. By Huisken's theorem, γ_R contracts to a point at the final time, which we take to be zero. So integrating and applying the crude estimate $0 \le \lambda_R \le \kappa_R$ (see [**97**, Lemma 4.1]) yields

$$-2\pi t \le A_R(t) \le -2n\pi t\,. \tag{14.34}$$

Since the area enclosed by the initial curve, Λ_{-R}, is $A_R(-T_R) = 2\pi R$, this gives a uniform estimate for the time interval of existence $[-T_R, 0)$. In particular, $T_R \to \infty$ as $R \to \infty$.

By the compactness theory for mean curvature flow, it remains to obtain a uniform estimate for $H^R_{\max}(t) \doteqdot \max H_R(\cdot\,,t)$ at each time. In fact, it suffices to bound $H^R_{\min}(t) \doteqdot \min H_R(\cdot\,,t)$. Indeed, by convexity, the maximal and minimal displacements ℓ_R and h_R, respectively, satisfy

$$2h_R\ell_R \le A_R \le 4h_R\ell_R$$

and hence

$$-\pi t \le 2h_R\ell_R \le -2n\pi t\,. \tag{14.35}$$

These are related to the largest and smallest curvatures by

$$\frac{dh_R}{dt} = -H^R_{\min} \quad\text{and}\quad \frac{d\ell_R}{dt} = -H^R_{\max}\,. \tag{14.36}$$

An upper bound for $H^R_{\min}$ implies a lower bound for the displacement h_R, which implies an upper bound for ℓ_R via (14.35). An upper bound for $H_{\max}$ can then be obtained from Hamilton's Harnack inequality (see [**97**]).

The required upper bound for $H^R_{\min}$ follows from a simple geometric argument (see Figure 14.5): Since $H^R_{\min}$ occurs when $\theta = \frac{\pi}{2}$, it can be shown that the circle tangent to γ_R at $\theta = \frac{\pi}{2}$ (point p in Figure 14.5) which passes through $\theta = 0$ (point q in Figure 14.5) lies locally inside γ_R at $\theta = \frac{\pi}{2}$. We deduce that

$$H^R_{\min} \le \frac{2nh_R}{\ell_R^2 + h_R^2}\,. \tag{14.37}$$

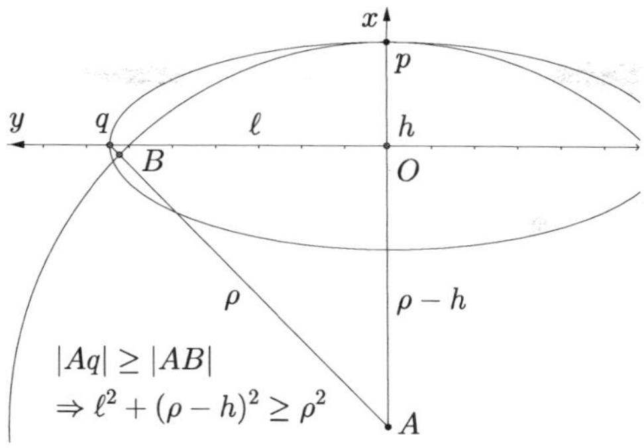

Figure 14.5. Bounding $H^R_{\min}$.

Combining (14.35), (14.36), and (14.37) with Hamilton's Harnack estimate yields a uniform upper bound for $H^R_{\max}$ at each fixed time (see [**97**, Lemma 4.7] for the details).

So we have uniform speed and displacement bounds for γ_R at each fixed time. Since $T_R \to \infty$ as $R \to \infty$, we can extract a compact, weakly convex ancient limit solution along a sequence $R_i \to \infty$. In fact, since the limit solution is compact, the splitting theorem (Theorem 9.11) ensures that it is strictly convex.

The convergence to a "round point" is a consequence of Huisken's theorem. It is also quite easy to show that the "parabolic" region converges to the boundary of the slab (cf. Lemma 14.19). By Lemma 14.10, the "edge" region converges to a Grim hyperplane; however, it is nontrivial to rule out limit Grim hyperplanes which are smaller than the one asymptotic to the boundary of the slab. The argument is similar to the proof of Theorem 14.17 but somewhat more difficult due to the additional rotational curvature term. The key step is a refined estimate for the enclosed area of the limit solution, which is obtained by deriving a better estimate for the integral of the rotational curvature λ. It is then argued (cf. Lemma 14.20) that, if the limit Grim hyperplane is too "thin", then the radius of the pancake, and hence its enclosed area, must actually be quite large, contradicting the aforementioned area estimate. See [**97**, Section 5]. □

Further asymptotics for the maxima and minima of the speed and displacement can be obtained by an iteration argument which makes use of a refined version of the estimate (14.37) (obtained by varying the point B in Figure 14.5). In fact, these asymptotics are actually derived for *any* compact, convex, $\mathrm{SO}(n)$-invariant ancient solution contained in the slab (and no smaller slab). Alexandrov's method of moving planes (cf. Lemma 14.21 and Theorem 14.22) can then be used to obtain the following uniqueness result [**97**, Theorem 1.2].

Theorem 14.28 (Uniqueness of rotationally symmetric ancient pancakes [**97**]). *Modulo space-time translations, rigid motions, and parabolic scaling, the ancient pancake* $\{\Pi_t^n\}_{t\in(-\infty,0)}$ *is unique among ancient pancakes with* $\mathrm{SO}(n)$*-symmetry.*

14.9. Notes and commentary

Three-dimensional singularity models and noncollapsed ancient solutions for the Ricci flow have been completely classified by the combined works of Hamilton [**259, 261, 268**], Perelman [**438, 439**], and Brendle [**110**].

14.9.1. Classification of ancient solutions.

In dimension 1, we have seen that the shrinking circles, paperclips, stationary lines, and Grim Reapers are the only convex ancient solutions [**192**]. In dimensions $n \geq 2$, we have encountered a number of additional examples. Wang's dichotomy divides them into two types — the ancient ovaloids and the ancient pancakes — and Theorems 14.25 and 14.28 provide uniqueness results under additional hypotheses. Without the mean convexity hypothesis, ancient solutions appear to be quite flexible, however.

14.9.1.1. *Ancient trombones.* In addition to the ancient paperclip, the shrinking circle, and the Grim Reaper, we have seen another explicit ancient solution to curve shortening flow: the *ancient sine curve.* The explicit representation of these solutions suggests that they might be unique among ancient solutions, but this turns out not to be true: S. Angenent and Q. You have constructed a plethora of (nonsoliton) ancient solutions to curve shortening flow by gluing together an arbitrary chain of parallel Grim Reapers along their common asymptotes [**76**]. The resulting family is infinite, and each member contains infinitely many free parameters. The family includes nonembedded compact examples and both embedded and nonembedded noncompact examples. Let us briefly describe their construction.

Let $\{x_n : n \in \mathbb{N}\}$ be an increasing sequence of real numbers satisfying

$$\delta \leq x_n - x_{n-1} \leq \delta^{-1}$$

for some $\delta > 0$ and every $n \in \mathbb{N}$. The points x_n are the locations of the asymptotes of the Grim Reaper solutions that will approximate the ancient solution. The velocity of the Grim Reaper between x_n and x_{n+1} is

$$v_n \doteq \frac{\pi}{x_{n+1} - x_n}.$$

Theorem 14.29 (S. Angenent and Q. You [**76**]). *Given any bounded sequence of real numbers* $\{C_n : n \in \mathbb{N}\}$, *there exists an eternal solution* $u : \mathbb{R} \times (-\infty, \infty) \to \mathbb{R}$ *to the graphical curve shortening flow such that*

$$(14.38)\quad (-1)^n u(x,t) = v_n t + C_n + v_n^{-1} \ln \sin v_n\big(x - x_n\big) + o(1) \quad \text{as } t \to -\infty$$

uniformly on compact subsets of the interval (x_n, x_{n+1}) *for each* $n \in \mathbb{N}$.

The ancient sine curve is obtained by setting $x_n = n\pi$ and $C_n = 0$ for all $n \in \mathbb{N}$. The monotonicity assumption on the sequence x_n ensures that the solution is the graph of a function and, in particular, a noncompact, embedded curve. A similar gluing procedure also yields compact, nonembedded ancient solutions for which the sequence x_n is periodic ($x_{n+N} = x_n$ for some $N \in \mathbb{N}$) and thus no longer monotone.

14.9.1.2. *The convex hull of an ancient solution.* Although the ancient trombone examples suggest the existence of a huge number of ancient solutions under general hypotheses, F. Chini and N. Møller have shown that their *convex hulls* are quite restricted (cf. [**289**]).

Theorem 14.30 (Chini and Møller [**145**]). *Let* $X : M^n \times (-\infty, 0) \to \mathbb{R}^{n+1}$ *be a proper ancient solution to mean curvature flow. The convex hull of the swept-out region* $\Omega \doteqdot \bigcup_{t\in(-\infty,0)} \mathcal{M}_t^n$, *where* $\mathcal{M}_t^n \doteqdot X(M^n, t)$, *is either a hyperplane, a slab, a half-space, or* $\mathbb{R}^{n+1}$.

Note that the solution in Theorem 14.30 need be neither compact, mean convex, nor embedded. We have seen examples which sweep out hyperplanes, slabs, and all of $\mathbb{R}^{n+1}$. The half-infinite ancient trombones of Angenent and You provide examples which sweep out half-spaces. Wang's dichotomy (Theorem 14.13) rules out half-spaces when the solution is convex.

Noncompact ancient solutions. By Theorem 14.17, the only *noncompact* convex ancient solutions to curve shortening flow are the straight lines and the Grim Reapers [**99**]. The following partial classification in higher dimensions is due to S. Brendle and K. Choi (cf. [**63, 70**]).

Theorem 14.31 (Brendle and Choi [**111, 112**]). *When* $n \geq 2$, *the bowl solitons are the only noncompact, convex, locally uniformly convex ancient solutions to mean curvature flow in* $\mathbb{R}^{n+1}$ *that are noncollapsing and (when* $n \geq 3$*) uniformly 2-convex.*

A. Mramor and A. Payne have constructed embedded ancient solutions which evolve out of catenoids in any dimension [**407**]. In particular, these provide examples of mean convex ancient solutions that are not convex.

14.9.2. Ancient solutions to further extrinsic curvature flows. Many of the results of this chapter have been extended in various directions. For example, rigidity results for shrinking spheres have been obtained in the context of (1) hypersurface flows by a large class of nonlinear speed functions [**353, 444, 445**], (2) mean curvature flow of hypersurfaces of the sphere (where shrinking spheres are replaced by an ancient family of totally umbilic hyperspheres) [**126, 304**], and (3) mean curvature flow of high codimension submanifolds (in Euclidean space and in the sphere) [**376, 444, 445**]. Convexity estimates for ancient solutions to certain hypersurface flows by nonlinear speeds were obtained in [**353**]. S. Risa [**444**] and, independently, P. Lu and J. Zhou [**372**] have constructed ancient ovaloids evolving by nonlinear curvature flows.

P. Bryan and J. Louie use the differential Harnack inequality and an Alexandrov reflection argument to show that the only closed, embedded, geodesically convex ancient solutions to the curve shortening flow on S^2 are stationary equators or shrinking circles (starting at an equator at time $t = -\infty$ and collapsing to the north pole at time $t = 0$) [**126**]. This method was generalized to obtain uniqueness of shrinking spheres among ancient solutions to a very large class of flows of hypersurfaces in the sphere by P. Bryan, M. Ivaki, and J. Scheuer [**122, 123, 125**] (see also [**124**]).

K. Choi and C. Mantoulidis proved that the equators and shrinking umbilic spheres are the only ancient solutions to codimension k mean curvature flow in the sphere S^{n+k} which have "small" area. In particular, the only ancient solutions to curve shortening flow on S^2 with area uniformly bounded are the equators and the shrinking circles [**152**].

Ben Lambert, Jason Lotay, and Felix Schulze have obtained various classification results for ancient solutions to Lagrangian mean curvature flow in terms of their blow-downs [**348**].

14.9.3. Open problems related to ancient solutions.

Open Problem 14.32. Do there exist noncompact, convex ancient solutions to mean curvature flow which do not evolve by translation (cf. Conjecture 13.78 and Theorems 14.17 and 14.31)?

Open Problem 14.33. Do there exist (strictly mean) convex, type-I ancient solutions to mean curvature flow other than the shrinking cylinders? (Cf. [**491**].)

Open Problem 14.34. Is every (mean) convex entire ancient solution necessarily noncollapsing?

Open Problem 14.35. Is every compact, mean convex ancient solution $X : M^n \times (-\infty, 0) \times \mathbb{R}^{n+1}$, $n \geq 2$, satisfying

$$\liminf_{t \to -\infty} \min_{M^n \times \{t\}} \frac{\kappa_1}{H} > -\infty$$

necessarily convex? Theorem 14.5 gives an affirmative answer for solutions with bounded rescaled volume and the work of Haslhofer and Kleiner [**275**] gives an affirmative answer for noncollapsing solutions.

Open Problem 14.36. Do there exist compact ancient solutions which are mean convex but not convex? The work of Mramor and Payne [**407**] provides noncompact examples.

Conjecture 14.37 (X.-J. Wang [**521**, p. 1237]). *Every entire, mean convex ancient solution to mean curvature flow is convex.*

It is a general principle that finite energy unstable critical points of elliptic energy functionals should give rise to ancient solutions to their gradient flows: Small perturbations in the direction of an unstable mode will flow slowly away from the stationary solution. Since the mean curvature flow is the gradient flow of the area functional, this suggests that each unstable mode of an unstable minimal submanifold should give rise to an ancient solution to mean curvature flow. This is indeed the case for *compact* minimal surfaces (in general Riemannian background spaces) since they have finite area (see the analysis of Choi and Mantoulidis [**152**]). But there are no compact minimal surfaces in Euclidean space. On the other hand, the type-I rescaled mean curvature flow (11.33) is a gradient flow for the Gaußian area. The critical points of the Gaußian area are the self-shrinkers, so each unstable mode of an unstable self-shrinker with finite Gaußian area gives rise to an ancient mean curvature flow[6]. It remains an interesting problem to study the noncompact case (the "ancient catenoids" of Mramor and Payne [**407**] were constructed very explicity).

Open Problem 14.38. Construct:

- General pancakes (Huisken and Sinestrari [**304**]): Ancient pancakes in $\mathbb{R}^3$ asymptotic to Grim planes and a (nonzero, possibly infinite) number of nontrivial flying wing translators.
- General ovailoids (Huisken and Sinestrari [**304**]): Ancient ovaloids in $\mathbb{R}^4$ asymptotic to Bowl planes and a (nonzero, possibly infinite) number of nontrivial bowloid translators.

[6]To avoid trivial examples — rotations, dilations, and space-time translates — one should use the notion of $\mathcal{F}$-stability (see Section 11.6.4).

- An "ancient pineapple ring": A (rotationally symmetric) ancient pancake in $\mathbb{R}^3$ of genus 1.
- An "ancient helicoid": An ancient plane in $\mathbb{R}^3$ with screw-symmetry but which does not evolve by self-similar translation/rotation.

Recall that Bryan and Louie proved that the equators and shrinking hemispheres are the only properly embedded, geodesically convex ancient solutions to curve shortening flow on S^2. K. Choi and C. Mantoulidis [**152**] proved that any other properly embedded ancient solution must have unbounded length as $t \to -\infty$.

Open Problem 14.39 (Huisken)**.** Construct a nontrivial ancient solution to curve shortening flow on S^2 which equally divides its area (and hence exists for all time and converges to an equator as t approaches infinity).

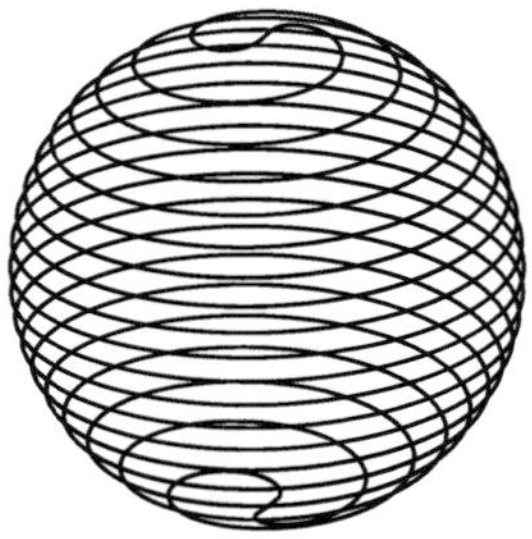

Figure 14.6. A conjectured "ancient spiral" on S^2.

Open Problem 14.40. Does there exist a compact, embedded ancient solution which sweeps out a region whose convex hull is a half-space?

14.10. Exercises

Exercise 14.1. Let $\{\mathcal{M}_t^n\}_{t\in(-\infty,0)}$ be a compact, convex, embedded ancient solution to mean curvature flow. Prove that given any $t_0 < 0$ there exists $K < \infty$ such that

$$|\Omega_t| \leq K(-t)^{n+1}$$

for all $t < t_0$, where Ω_t is the convex body bounded by $\mathcal{M}_t^n$. HINT: Use Hamilton's Harnack inequality to bound the circumradius of Ω_t.

Exercise 14.2. Let $\{\mathcal{M}_t^n\}_{t\in(-\infty,0)}$ be a convex ancient solution to mean curvature flow in $\mathbb{R}^{n+1}$ which lies in a slab. Prove that $\bigcup_{t<0}\Omega_t$ is a slab, where Ω_t is the convex region bounded by $\mathcal{M}_t^n$. HINT: Cf. Lemmas 14.18 and 14.19.

Exercise 14.3. Complete the proof of Theorem 14.22 in the compact case.

Exercise 14.4. Prove that there exists no $\mathrm{SO}(m)\times\mathrm{SO}(n+1-m)$-invariant ancient pancake in a slab perpendicular to e_1 when $m>1$.

Exercise 14.5. Prove that there exists a constant $C=C(n)<\infty$ such that

$$\sup_{k>0}\frac{1}{k^{n/2}}\int_{\mathcal{M}^n}\mathrm{e}^{\frac{-|y|^2}{k}}\,d\mathscr{H}^n(y)<C$$

for any convex hypersurface $\mathcal{M}^n$ of $\mathbb{R}^{n+1}$.

Exercise 14.6. Prove (14.29). HINT: Use convexity and the smallness of the measure of $\widehat{D}_k$ relative to D_k.

Chapter 15

Gauß Curvature Flows

A smooth 1-parameter family $\{\mathcal{M}_t^n\}_{t\in I}$ of hypersurfaces in $\mathbb{R}^{n+1}$ evolves by the **α-Gauß curvature flow** if it admits a smooth 1-parameter family $X : M^n \times I \to \mathbb{R}^{n+1}$ of parametrizations $X_t \doteqdot X(\cdot, t) : M^n \to \mathcal{M}_t^n$, $t \in I$, satisfying

$$\partial_t X = -\operatorname{sign}(\alpha) K^\alpha \mathsf{N}\,, \tag{15.1}$$

where K is the Gauß curvature of the family. When $\alpha = 1$, the flow is called the **Gauß curvature flow**. When $\alpha = -1$, it is called the **inverse-Gauß curvature flow**. In the following three chapters, we shall study the α-Gauß curvature flows by *positive* powers $\alpha > 0$.

William J. Firey proposed the Gauß curvature flow to model the erosion of strictly convex stones as they tumble on a beach [**222**]. To obtain the model, he assumed that the stones were of uniform density, that their wear was isotropic, and that the number of collisions in a region was proportional to the set of normal directions of the region. In this case the rate of wear is proportional to the density per unit surface area of contact directions, which is the Gauß curvature. In Firey's words:

> "Imagine a stone tumbling over a uniformly abrasive plane: the material of the stone has uniform density and is uniform and isotropic as to wear. For reference purposes, fix a Cartesian frame in the stone; relative to this frame, it is the abrasive plane which is mobile. We suppose that for each nonnegative time t up to a certain final time $T < \infty$, the stone, as a point set in Euclidean 3-space, is a convex body $C(t)$ which is smooth in the following sense. At each boundary point of $C(t)$ there is both a tangent plane, whose outer

> normal direction we denote by u, and a Gauß curvature which is a positive, continuously differentiable function $K(t,u)$ in u over the unit sphere Ω of directions...
>
> Contact positions of the abrasive plane with $C(t)$ are distinguished by their outer normal directions. We assume that each direction has equal likelihood of specifying a contact position at each time. More precisely, if σ is any measurable portion of the boundary of $C(t)$, the set of directions, for which the abrasive plane touches $C(t)$ at points in σ, has measure proportional to the measure $m(\omega)$ of the spherical image ω of σ...
>
> The amount of wear on any small part of σ is taken to be proportional to the density per unit surface area on σ of contact points with the abrasive plane in this part of σ. To be more exact, our assumption comes to this: the rate of wear in the direction u is proportional to
>
> $$\lim_{\omega\to u}\frac{m(\omega)}{S(\omega)},$$
>
> where $S(\omega)$ is the area of σ and the limit is taken as ω shrinks to u in a suitable way. This limit is $K(t,u)$. Physically the assumption implies that sharply rounded portions of the stone wear more rapidly than do blunt portions..."

In comparison with mean curvature flow, the Gauß curvature flows have some important differences: Firstly, the Gauß curvature flows are parabolic only for *locally convex* hypersurfaces, so the possible behaviors of solutions are much more limited. Indeed, we will prove a result (due to Kai-Seng Chou [**506**] for the Gauß curvature flow and extended to other α-Gauß curvature flows by the second author [**157**]) that solutions always contract to points; so the neck-pinching and other singularities which were important in mean curvature flow cannot arise here. Secondly, the Gauß curvature flows are *fully nonlinear*, in the sense that their equations of motion cannot be represented by an equation which is linear as a function of the second derivatives. In contrast, the mean curvature flow is *quasilinear* — the equation of motion can be represented by an equation in which the second derivatives appear linearly (but with coefficients which depend on the first derivatives). The added nonlinearity requires some differences in the treatment of regularity of solutions and also leads to some kinds of behavior which do not arise in mean curvature flow (for example, the persistence of "flat sides" in the evolving hypersurfaces). On the other hand, the Gauß curvature flows do have a rich variational structure, and this is exploited in the analysis of their long-time behavior. In particular, this allows the treatment of an entire family of flows (depending on α), while a comparable treatment of flows by powers of mean curvature, for example, is not yet possible.

The Gauß curvature flow can be studied using monotonicity tools that are now familiar to the reader from the curve shortening flow and the mean curvature flow.

15.1. Invariance properties and self-similar solutions

As for the curve shortening and mean curvature flows, the α-Gauß curvature flows are invariant under reparametrizations, time translations, and ambient isometries. On the other hand, since (when $\alpha \neq \frac{1}{n}$) the α-Gauß curvature scales differently to the mean curvature, the flows behave differently under parabolic scaling. Indeed, the correct scaling for the α-Gauß curvature flow is given by

$$X_\varepsilon(x,t) \doteqdot \mathrm{e}^{\varepsilon} X(x, \mathrm{e}^{-(1+n\alpha)\varepsilon} t), \tag{15.2}$$

where $\varepsilon \in \mathbb{R}$. We leave it to the reader to check that $X_\varepsilon : M^n \times I_\varepsilon \to \mathbb{R}^{n+1}$ satisfies the α-Gauß curvature flow whenever $X : M^n \times I \to \mathbb{R}^{n+1}$ does, where $I_\varepsilon \doteqdot \mathrm{e}^{(1+n\alpha)\varepsilon} I$.

So the α-Gauß curvature flow also admits, in principle, self-similarly translating, shrinking, expanding, and rotating solutions. The self-similarly shrinking (and, to a lesser extent, expanding) solutions in particular will play an important role in our analysis of these flows.

15.1.1. Homothetic solutions. As for the curve shortening and mean curvature flows, we call a solution $X : M^n \times I \to \mathbb{R}^{n+1}$ to the α-Gauß curvature flow a **homothetic self-similar solution** if it is invariant under the corresponding parabolic dilation, that is, if there is some vector field $V \in \Gamma(TM)$ such that

$$\mathrm{e}^{\varepsilon} X(\varphi(x,\varepsilon), \mathrm{e}^{-(1+n\alpha)\varepsilon} t) = X(x,t) \tag{15.3}$$

for all $(x,t) \in M^n \times I$ and $\varepsilon \in \mathbb{R}$ such that $t - \varepsilon \in I$, where φ is the flow of V. The solution is called a **shrinking self-similar solution** if $t < 0$ for each $t \in I$ and an **expanding self-similar solution** if $0 < t$ for each $t \in I$.

Fixing any nonzero $t_0 \in I$ and setting $\varepsilon = \frac{1}{1+n\alpha} \log(t/t_0)$, we find that

$$X(x,t) = (t/t_0)^{\frac{1}{1+n\alpha}} X\left(\varphi\big(x, \tfrac{1}{1+n\alpha} \log(t/t_0)\big), t_0\right). \tag{15.4}$$

Differentiating (15.3) with respect to ε at $\varepsilon = 0$, we find that

$$X + dX(V) + (1+n\alpha) t K^{\alpha} \mathsf{N} = 0.$$

In particular,

$$K^{\alpha} = \frac{1}{-(1+n\alpha)t} \langle X, \mathsf{N} \rangle .$$

A hypersurface satisfying

$$K^{\alpha} = \frac{1}{1+n\alpha}\langle X, \mathsf{N}\rangle \tag{15.5}$$

is called an α-**Gauß curvature flow shrinker** and a hypersurface satisfying

$$K^{\alpha} = -\frac{1}{1+n\alpha}\langle X, \mathsf{N}\rangle \tag{15.6}$$

is called an α-**Gauß curvature flow expander**.

If $X_0 : M^n \to \mathbb{R}^{n+1}$ is an α-Gauß curvature flow shrinker, then the family $X : M^n \times (-\infty, 0) \to \mathbb{R}^{n+1}$ defined by

$$X(x,t) \doteqdot (-t)^{\frac{1}{1+n\alpha}} X_0\left(\varphi\big(x, \tfrac{1}{1+n\alpha}\log(-t)\big)\right),$$

where φ is the flow of the vector field $V \doteqdot -dX_0^{-1}(X_0^{\top})$, is a shrinking self-similar solution to the α-Gauß curvature flow. Similarly, if $X_0 : M^n \to \mathbb{R}^{n+1}$ is an α-Gauß curvature flow expander, then the family $X : M^n \times (-\infty, 0) \to \mathbb{R}^{n+1}$ defined by

$$X(x,t) \doteqdot t^{\frac{1}{1+n\alpha}} X_0\left(\varphi\big(x, \tfrac{1}{1+n\alpha}\log(t)\big)\right),$$

where φ is the flow of the vector field $V \doteqdot -dX_0^{-1}(X_0^{\top})$, is an expanding self-similar solution to the α-Gauß curvature flow.

The prototypical example of a shrinking self-similar solution to the α-Gauß curvature flow is the shrinking sphere, $\{S^n_{r_\alpha(t)}\}_{t\in(-\infty,0)}$, where $r_\alpha(t) \doteqdot (-(1+n\alpha)t)^{\frac{1}{1+n\alpha}}$.

15.2. Basic evolution equations

It will be useful to collect some basic evolution equations for flows by powers of the Gauß curvature.

Lemma 15.1. *Let $\{\mathcal{M}^n_t\}_{t\in I}$ be a family of locally uniformly convex hypersurfaces of $\mathbb{R}^{n+1}$ evolving by the α-Gauß curvature flow. Denote by $\mathcal{L}$ the operator*

$$\mathcal{L}\cdot \doteqdot \alpha K^{\alpha}\operatorname{tr}(\mathrm{II}^{-1}\cdot\nabla^2\cdot\,).$$

Then

$$\partial_t \mathsf{N} = \nabla K^{\alpha}, \tag{15.7}$$

$$\begin{aligned}(\partial_t - \mathcal{L})H = {} & K^{\alpha}(\alpha H^2 - (n\alpha-1)|\mathrm{II}|^2) + K^{-\alpha}|\nabla K^{\alpha}|^2 \\ & - \alpha K^{\alpha} g^{ij}(\mathrm{II}^{-1})^{kp}(\mathrm{II}^{-1})^{\ell q}\nabla_i \mathrm{II}_{k\ell}\nabla_j \mathrm{II}_{pq},\end{aligned} \tag{15.8}$$

$$(15.9)\qquad (\nabla_t - \mathcal{L})\mathrm{II}_{ij} = K^\alpha(\alpha H \mathrm{II}_{ij} - (n\alpha - 1)\mathrm{II}^2_{ij}) + K^{-\alpha}\,\nabla_i K^\alpha\,\nabla_j K^\alpha \\ - \alpha K^\alpha (\mathrm{II}^{-1})^{kp}(\mathrm{II}^{-1})^{\ell q}\,\nabla_i \mathrm{II}_{k\ell}\,\nabla_j \mathrm{II}_{pq},$$

and

$$(15.10)\qquad \partial_t K^\alpha = \mathcal{L}K^\alpha + \alpha H K^{2\alpha},$$

where ∇_t is the covariant time derivative (defined either by (5.98) *or* (5.126)*).*

If C is a compact subset of $\mathbb{R}^{n+1}$, then

$$(15.11)\qquad \frac{d}{dt}\,\mathrm{Area}(\mathcal{M}^n_t \cap C) = -\int_{\mathcal{M}^n_t\cap C} H K^\alpha \, d\mu$$

and

$$(15.12)\qquad \frac{d}{dt}\,\mathrm{Vol}(\Omega_t \cap C) = -\int_{\mathcal{M}^n_t\cap C} K^\alpha \, d\mu,$$

where Ω_t is the convex body bounded by $\mathcal{M}^n_t$.

Proof. Applying (5.140) with $F = K^\alpha$, we find

$$(15.13)\qquad \nabla_t \mathrm{II}_{ij} = \nabla_i\nabla_j K^\alpha + K^\alpha \mathrm{II}^2_{ij}.$$

On the other hand,

$$\begin{aligned}\nabla_i\nabla_j K^\alpha &= \nabla_i(\,\alpha K^\alpha (\mathrm{II}^{-1})^{k\ell}\,\nabla_j \mathrm{II}_{k\ell}\,)\\ &= \alpha K^\alpha (\mathrm{II}^{-1})^{k\ell}\nabla_i\nabla_j \mathrm{II}_{k\ell} + \alpha K^\alpha \nabla_i (\mathrm{II}^{-1})^{k\ell}\nabla_j \mathrm{II}_{k\ell}\\ &\quad + \alpha \nabla_i K^\alpha (\mathrm{II}^{-1})^{k\ell}\nabla_j \mathrm{II}_{k\ell}.\end{aligned}$$

Applying Simons's identity (5.53) and the formula

$$\nabla_i (\mathrm{II}^{-1})^{kl} = -(\mathrm{II}^{-1})^{kp}(\mathrm{II}^{-1})^{lq}\nabla_i\,\mathrm{II}_{pq},$$

we arrive at

$$\begin{aligned}\nabla_i\nabla_j K^\alpha &= \mathcal{L}\,\mathrm{II}_{ij} - \alpha K^\alpha (\mathrm{II}^{-1})^{kp}(\mathrm{II}^{-1})^{\ell q}\nabla_i \mathrm{II}_{pq}\nabla_j \mathrm{II}_{k\ell}\\ &\quad + K^{-\alpha}\nabla_i K^\alpha \nabla_j K^\alpha + \alpha K^\alpha (H \mathrm{II}_{ij} - n \mathrm{II}^2_{ij}).\end{aligned}$$

Equation (15.9) now follows from (15.13).

Equation (15.8) is obtained either by taking a trace in (15.9) or by computing directly from (5.141).

Equation (15.10) is readily obtained either from (5.142) or from (5.147) (since $\dot{K}^\alpha = \alpha K^\alpha\,\mathrm{II}^{-1}$).

The final two equations follow from the variational formulae derived in Lemma 5.25. □

We also observe that solutions to Gauß curvature flows obey the avoidance principle.

Theorem 15.2 (The avoidance principle). *Let $X_i : M_i^n \times [0,T) \to \mathbb{R}^{n+1}$, $i = 1, 2$, be two locally uniformly convex solutions to the α-Gauß curvature flow, at least one of which is compact, with $X_1(M_1^n, 0) \cap X_2(M_2^n, 0) = \emptyset$. Then the extrinsic distance*

$$d_{\min}(t) \doteqdot \min_{(x,y)\in M_1^n\times M_2^n} \|X_1(x,t) - X_2(y,t)\| \tag{15.14}$$

is nondecreasing in t. Hence, $X_1(M_1^n, t) \cap X_2(M_2^n, t) = \emptyset$ for $t \in [0,T)$.

The theorem may be proved in the same way as for the curve shortening and mean curvature flows. We omit the proof since a proof is given, in Chapter 18, for a more general class of curvature flows.

15.3. Chou's long-time existence theorem

We now prove K.-S. Chou's long-time existence result for solutions to α-Gauß curvature flows. The argument we present is essentially that of Chou [**506**] who obtained the result for the Gauß curvature flow. The argument is very robust — it was extended to the other α-Gauß curvature flows by the second author [**157**] and to a wide variety of other flows by many authors.

Figure 15.1. Kai-Seng Chou.

Theorem 15.3 (Long-time existence of the α-Gauß curvature flow). *Given $\alpha > 0$ and a smooth uniformly convex hypersurface $\mathcal{M}^n$ of $\mathbb{R}^{n+1}$, there exists a unique maximal smooth solution $\{\mathcal{M}_t^n\}_{t\in[0,T)}$ to the α-Gauß curvature flow* (15.1) *with $\mathcal{M}_0^n = \mathcal{M}^n$; $\mathcal{M}_t^n$ remains uniformly convex on $[0,T)$ and shrinks to a point as $t \to T$. In particular, if $V(t)$ is the volume enclosed by $\mathcal{M}_t^n$, then $\lim_{t\to T} V(t) = 0$.*

Remark 15.4. Here, uniqueness means that any other solution $\{\mathcal{N}_t^n\}_{t\in[0,T')}$ with $\mathcal{N}_0^n = \mathcal{M}^n$ satisfies $\mathcal{N}_t^n = \mathcal{M}_t^n$ for all $t \in [0,T') \cap [0,T)$ and maximal means that we always have $T' \leq T$. By $\mathcal{M}_t$ shrinking to a point as $t \to T$ we mean that there exists a point $\tilde{q} \in \mathbb{R}^{n+1}$ such that $\rho(t) \doteqdot \max_{p \in \mathcal{M}_t} \|p - \tilde{q}\| \to 0$ as $t \to T$. Note that since the speed is positive, the motion is monotone and $\rho(t)$ is monotonically decreasing when $\tilde{q}$ is enclosed by $\mathcal{M}_t$.

The steps required in the proof are as follows: We first show in Section 15.3.1 that the Gauß curvature flows can be rewritten as scalar parabolic equations of Monge–Ampère type for the support function in the Gauß map parametrization. Then in Section 15.3.2 we derive the crucial estimate of Chou, which bounds the speed as long as the inradius remains positive. Then we prove in Section 15.3.3 a time-independent lower bound on the principal curvatures. Combining these allows us to deduce in Section 15.3.4 that the flow remains uniformly parabolic on any time interval where the inradius remains positive. We deduce that the solution has bounds on all derivatives on such a time interval, so the flow can be extended further. It follows that the inradius must approach zero at the final time, and the lower bound on principal curvatures then also implies that the diameter approaches zero, completing the proof of the theorem.

15.3.1. A parabolic Monge–Ampère equation. In this subsection we consider the α-Gauß curvature flow from the point of view of the parabolic Monge–Ampère equation for the support function. Let $\mathcal{M}$ be a uniformly convex hypersurface in $\mathbb{R}^{n+1}$. In Section 5.2.4 we discussed the basics of the support function $\sigma : S^n \to \mathbb{R}$ defined by

$$\sigma(\mathsf{N}) = \langle X, \mathsf{N} \rangle = \left\langle \mathrm{G}^{-1}(\mathsf{N}), \mathsf{N} \right\rangle,$$

where $\mathrm{G} : \mathcal{M} \to S^n$ is the Gauß map defined by (5.28). We shall use the notation introduced in that subsection. By (5.87) we have

$$K^{-1} = \frac{\det\left(\overline{\nabla}^2 \sigma + \sigma \overline{g}\right)}{\det \overline{g}}. \tag{15.15}$$

We will employ the notation

$$A[\sigma] = \overline{\nabla}^2 \sigma + \sigma \overline{g}$$

introduced in Section 5.2.4.

Let $X : M^n \to \mathbb{R}^{n+1}$ be a solution to the α-Gauß curvature flow (15.1). Since the inverse of the Gauß map is a reparametrization of $\mathcal{M}_t = X_t(M^n)$, under the α-Gauß curvature flow we have

$$\partial_\tau \mathrm{G}^{-1}(\mathsf{N}, \tau) = -K^\alpha \circ \mathrm{G}^{-1}(\mathsf{N}, \tau)\, \mathsf{N} + Y(\mathsf{N}, \tau),$$

where Y is a tangential vector field; i.e., $\langle Y, \mathsf{N} \rangle = 0$. (We temporarily switch to τ to distinguish the two parametrizations and we revert to t later.) Thus the evolution of the support function is given by

$$\partial_t \sigma(\mathsf{N}) = \langle \partial_\tau \mathrm{G}^{-1}(\mathsf{N}), \mathsf{N}\rangle = -K^\alpha = -\left(\frac{\det\left(\overline{\nabla}^2\sigma + \sigma\overline{g}\right)}{\det \overline{g}}\right)^{-\alpha}. \tag{15.16}$$

This parabolic Monge–Ampère equation is second order and fully nonlinear.

Since (15.16) is a strictly parabolic equation as long as $A[\sigma] > 0$, we have the following short-time existence result.

Theorem 15.5. *Let $\sigma_0 : S^n \to \mathbb{R}$ be a smooth function with $A[\sigma_0] > 0$. Then there exists $\varepsilon > 0$ and a smooth solution $\sigma : S^n \times [0, \varepsilon) \to \mathbb{R}$ to (15.16) with $\sigma(0) = \sigma_0$ and $A[\sigma] > 0$ on $S^n \times [0, \varepsilon)$.*

Since under the Gauß map reparametrized Gauß curvature flow we have $\partial_\tau \mathsf{N} = 0$ and since $\partial_t \mathsf{N} = \nabla K^\alpha$ under the original parametrized Gauß curvature flow, we obtain

$$0 = \nabla K^\alpha + D_Y \mathsf{N} = \nabla K^\alpha + \mathrm{II}(Y).$$

Thus

$$Y = -\mathrm{II}^{-1}(\nabla K^\alpha). \tag{15.17}$$

So, if φ is a geometric quantity associated to hypersurfaces, then its evolution under the original parametrization $\partial_t\varphi$ and the Gauß map parametrization $\partial_\tau\varphi$ are related by

$$\partial_t\varphi = \partial_\tau\varphi + \langle \mathrm{II}^{-1}(\nabla K^\alpha), \nabla\varphi\rangle. \tag{15.18}$$

Note that

$$g(\mathrm{II}^{-1}(\nabla K^\alpha), \nabla\varphi) = \overline{g}(\overline{\nabla} K^\alpha, \overline{\nabla}\varphi). \tag{15.19}$$

By solving an ODE to make the diffeomorphism change, short-time existence for the parabolic Monge–Ampère equation implies short-time existence for the α-Gauß curvature flow.

Corollary 15.6. *Let $X_0 : M^n \to \mathbb{R}^{n+1}$ be a smooth uniformly convex hypersurface. Then there exists $\varepsilon > 0$ and a smooth solution $X : M^n \times [0, \varepsilon) \to \mathbb{R}^{n+1}$ to (15.1) with $X(0) = X_0$.*

15.3.2. Chou's Gauß curvature estimate. We present Chou's upper bound on the Gauß curvature, which he proved by considering the evolution of

$$\varphi = \frac{K^\alpha}{\sigma - r_0} \tag{15.20}$$

using the Gauß map parametrization, where σ is the support function and r_0 is a positive constant such that $\sigma \geq 2r_0$. The computations are substantially easier in this setting compared to working on the evolving hypersurface (see Exercises 17.2–17.4). Let $A = A[\sigma]$ and write

$$\overline{\mathcal{L}} = \alpha K^{\alpha}(A^{-1})^{ij}\overline{\nabla}_i\overline{\nabla}_j\,. \tag{15.21}$$

By (15.15) we have

$$\partial_\tau \log K = (A^{-1})^{ij}(\overline{\nabla}_i\overline{\nabla}_j K^\alpha + K^\alpha \overline{g}_{ij})$$

and thus

$$\partial_\tau K^\alpha = \overline{\mathcal{L}}K^\alpha + \alpha H K^{2\alpha}\,. \tag{15.22}$$

Using this and the equations $\partial_\tau \sigma = -K^\alpha$ and $\overline{\mathcal{L}}\sigma = n\alpha K^\alpha - \alpha K^\alpha H\sigma$, we calculate that

$$\begin{aligned}
\partial_\tau\left(\frac{K^\alpha}{\sigma - r_0}\right) &= \frac{\overline{\mathcal{L}}K^\alpha + \alpha H K^{2\alpha}}{\sigma - r_0} + \frac{K^{2\alpha}}{(\sigma - r_0)^2}\\
&= \overline{\mathcal{L}}\left(\frac{K^\alpha}{\sigma - r_0}\right) - \frac{2\alpha K^\alpha}{\sigma - r_0}A^{-1}\left(\overline{\nabla}\sigma, \overline{\nabla}\left(\frac{K^\alpha}{\sigma - r_0}\right)\right)\\
&\quad + \frac{\alpha H K^{2\alpha}}{\sigma - r_0} + \frac{K^{2\alpha}}{(\sigma - r_0)^2} + \frac{K^\alpha}{(\sigma - r_0)^2}(n\alpha K^\alpha - \alpha K^\alpha H\sigma)\,.
\end{aligned}$$

That is,

$$\partial_\tau \varphi = \overline{\mathcal{L}}\varphi - \frac{2\alpha K^\alpha}{\sigma - r_0}A^{-1}(\overline{\nabla}\sigma, \overline{\nabla}\varphi) + (1 + n\alpha - r_0\alpha H)\varphi^2\,. \tag{15.23}$$

By the maximum principle, we conclude at a maximum point of φ that

$$H \le \frac{n + \alpha^{-1}}{r_0}\,.$$

By the arithmetic-geometric mean inequality, $K^{1/n} \le H/n$. Hence, at a maximum point of φ we have

$$\varphi \le r_0^{-1}K^\alpha \le r_0^{-1}\left(\frac{H}{n}\right)^{n\alpha} \le r_0^{-1}\left(\frac{1 + (n\alpha)^{-1}}{r_0}\right)^{n\alpha} \doteqdot C\,.$$

We conclude that as long as the solution exists and $\sigma \ge 2r_0$,

$$K^\alpha \le C(\sigma - r_0) \le C\sup_{S^n}\sigma_0\,,$$

where C depends only on n, α, and r_0. This shows that the Gauß curvature K_t of $\mathcal{M}_t$ is bounded above as long as the inradius of $\mathcal{M}_t$ is bounded from below by a positive constant.

15.3.3. Estimate on the radii of curvature. In the next step we prove that the radii of curvatures remain bounded, with a time-independent upper bound. To prove this we derive a differential inequality for the matrix $A[\sigma]$, the eigenvalues of which are the principal radii, and show that we can preserve a bound on the eigenvalues using the maximum principle. We remark that this step (even more than the previous one) is relatively simple in the Gauß map parametrization but rather messy in the parametric flow.

In order to derive an evolution equation for $A = A[\sigma]$ we will need to differentiate the equation (15.15). To convert the result into a parabolic equation for A we will need to commute indices, and this will require the following analogue of the Codazzi identity:

Lemma 15.7. *The covariant derivative of A is totally symmetric:*

$$\overline{\nabla}_i A_{jk} = \overline{\nabla}_j A_{ik}. \tag{15.24}$$

Proof. We compute

$$\begin{aligned}
\overline{\nabla}_i A_{jk} &= \overline{\nabla}_i \left(\overline{\nabla}_j \overline{\nabla}_k \sigma + \overline{g}_{jk}\sigma\right) \\
&= \overline{\nabla}_j \overline{\nabla}_i \overline{\nabla}_k \sigma + \overline{R}_{ijk}{}^p \overline{\nabla}_p \sigma + \overline{g}_{jk} \overline{\nabla}_i \sigma \\
&= \overline{\nabla}_j \overline{\nabla}_i \overline{\nabla}_k \sigma + \overline{g}_{ik} \overline{\nabla}_j \sigma - \overline{g}_{jk} \overline{\nabla}_i \sigma + \overline{g}_{jk} \overline{\nabla}_i \sigma \\
&= \overline{\nabla}_j \left(\overline{\nabla}_i \overline{\nabla}_k \sigma + \overline{g}_{ik}\sigma\right) \\
&= \overline{\nabla}_j A_{ik},
\end{aligned}$$

as required, where we used the expression for the Riemann curvature tensor of the unit sphere: $\overline{R}_{ijkl} = \overline{g}_{ik}\overline{g}_{jl} - \overline{g}_{jk}\overline{g}_{il}$. □

Now we proceed with the computation of the evolution equation for A: Since the metric $\overline{g}$ and connection $\overline{\nabla}$ are not changing in time, we find

$$\begin{aligned}
\partial_t A_{ij} &= -\overline{\nabla}_i \overline{\nabla}_j (\det A)^{-\alpha} - (\det A)^{-\alpha}\, \overline{g}_{ij} \\
&= \overline{\nabla}_i \left(\alpha (\det A)^{-\alpha} \left(A^{-1}\right)^{kl} \overline{\nabla}_j A_{kl}\right) - (\det A)^{-\alpha}\, \overline{g}_{ij} \\
&= \alpha (\det A)^{-\alpha} \left(A^{-1}\right)^{kl} \overline{\nabla}_i \overline{\nabla}_j A_{kl} - \alpha^2 (\det A)^{-2-\alpha}\, \overline{\nabla}_i \det A \overline{\nabla}_j \det A \\
&\quad - \alpha (\det A)^{-\alpha} (A^{-1})^{kp} (A^{-1})^{lq} \overline{\nabla}_i A_{kl} \overline{\nabla}_j A_{pq} - (\det A)^{-\alpha} \overline{g}_{ij}.
\end{aligned} \tag{15.25}$$

To convert this to a parabolic equation, we need to exchange the indices in the second derivative term, using the following analogue of Simons's equation (see (5.49)):

Lemma 15.8. *We have*

$$\overline{\nabla}_{(i} \overline{\nabla}_{j)} A_{kl} = \overline{\nabla}_{(k} \overline{\nabla}_{l)} A_{ij} + \overline{g}_{ij} A_{kl} - \overline{g}_{kl} A_{ij}, \tag{15.26}$$

where $\overline{\nabla}_{(i} \overline{\nabla}_{j)} A_{kl} \doteqdot \frac{1}{2}\left(\overline{\nabla}_i \overline{\nabla}_j A_{kl} + \overline{\nabla}_j \overline{\nabla}_i A_{kl}\right)$.

Proof. Commuting derivatives using Lemma 15.7 and the curvature identity, we find

$$\begin{aligned}\overline{\nabla}_i\overline{\nabla}_j A_{kl} &= \overline{\nabla}_i\overline{\nabla}_k A_{jl}\\ &= \overline{\nabla}_k\overline{\nabla}_i A_{jl} + \overline{R}_{ikj}{}^p A_{pl} + \overline{R}_{ikl}{}^p A_{jp}\\ &= \overline{\nabla}_k\overline{\nabla}_l A_{ij} + \overline{g}_{ij}A_{kl} - \overline{g}_{kj}A_{il} + \overline{g}_{il}A_{jk} - \overline{g}_{kl}A_{ij}.\end{aligned}\tag{15.27}$$

The result follows after symmetrization in k and l (or in i and j). □

Using Lemma 15.8 in equation (15.25), we obtain

$$\partial_t A_{ij} \le \overline{\mathcal{L}}A_{ij} + (n\alpha - 1)(\det A)^{-\alpha}\overline{g}_{ij} - \alpha H(\det A)^{-\alpha}A_{ij},\tag{15.28}$$

where the inequality is in the sense of matrices, and we used the fact that the terms involving $\overline{\nabla}A$ form a negative definite matrix. At a maximum eigenvector v of A, we have $HA(v,v) = r_{\max}(1/r_1 + \cdots + 1/r_n) \ge n$, where $r_1,\ldots,r_n$ are the principal radii of curvature, so we deduce that the right-hand side is negative and hence the maximum eigenvalue is nonincreasing. This gives the estimate

$$A_{ij} \le R_0\overline{g}_{ij}\tag{15.29}$$

on $S^n \times [0,T)$, where $R_0 \doteqdot \sup\{A[\sigma_0]\big|_z(v,v) :\ z\in S^n,\ v\in T_zS^n, |v|=1\}$.

15.3.4. Higher regularity and convergence to a point. We first prove that the inradius of $\mathcal{M}_t$ must converge to zero at the end of the maximal interval of existence of the solution:

By comparison with an enclosing sphere (which shrinks to a point in finite time according to the law $r(t) = \big(r(0)^{1+\alpha} - (1+\alpha)t\big)^{\frac{1}{1+\alpha}}$) we know that the maximal time of existence T must be finite. We suppose, for the sake of obtaining a contradiction, that the inradius does not converge to zero as t approaches T. Since the hypersurface $\mathcal{M}_t$ is contracting, the inradius is decreasing in time, so this implies that there is a positive lower bound for the inradius, or equivalently a positive lower bound on the support function if we choose the origin appropriately.

We show that the Gauß curvature bound $K \le K_0$ and the bound on the principal radii $A \le R_0\overline{g}$ together imply that the principal curvatures are bounded above: To see this, we write

$$\kappa_i = K\prod_{j\neq i}\kappa_j^{-1} = K\prod_{j\neq i} r_j \le K_0 R_0^{n-1}.$$

It follows that the principal radii $r_i(z,t)$ lie in a compact subset of the positive cone $\Gamma_+ = \{(x_1,\ldots,x_n) :\ x_i > 0\}$, and on this subset the function

$F(r_1, \ldots, r_n) = -K^\alpha = -(r_1 \cdots r_n)^{-\alpha}$ is concave and has uniformly positive derivative in the sense that there are constants $0 < \lambda \leq \Lambda < \infty$ such that $\lambda \leq \partial_{r_i} F \leq \Lambda$.

The crucial step in proving that the solution remains smooth is to derive the Hölder continuity of the second derivatives of σ. For this we must appeal to a deep result from the theory of partial differential equations, due to Krylov [**340**]. We use a slightly simplified version of a result from the book by Lieberman [**368**, Corollary 14.9]:

Theorem 15.9. *Let u be a smooth solution on $B_1^n(0)\times[-1,0)$ of an equation of the form*

$$\partial_t u(x,t) = F(D^2 u, x, t)\,, \tag{15.30}$$

where $F:\ \mathrm{S}(n) \times B_1^n(0) \times [-1,0) \to \mathbb{R}$ satisfies the following conditions:

(i) *There exist positive constants λ and Λ such that for any $(x,t) \in B_1^n(0) \times [-1,0)$, any symmetric matrix A, and any positive definite symmetric matrix B,*

$$\lambda\,\mathrm{tr}(B) \leq F(A+B,x,t) - F(A,x,t) \leq \Lambda\,\mathrm{tr}(B)\,.$$

(ii) *For each $(x,t) \in B_1^n(0) \times [-1,0)$, $A \mapsto F(A,x,t)$ is a concave function of $A \in \mathrm{S}(n)$.*

(iii) *For any (x_1,t_1) and (x_2,t_2) in $B_1^n(0) \times [-1,0)$ and any symmetric matrix A,*

$$|F(A,x_2,t_2) - F(A,x_1,t_1)| \leq (b_1 + b_2|A|)\left((x_2-x_1)^2 + |t_2-t_1|\right)^\beta$$

for some positive constants b_1, b_2, and β.

Then there exist positive constants γ (depending only on n, β, λ, and Λ) and C (depending also on b_2) such that

$$|D^2u(y,t) - D^2u(x,s)| + |\partial_t u(y,t) - \partial_t u(x,s)| \leq C(|x-y|^\gamma + |t-s|^{\gamma/2})$$

and

$$|Du(x,t) - Du(x,s)| \leq C|t-s|^{\frac{1+\gamma}{2}},$$

for all (x,s) and (y,t) in $B_{1/2}^n(0) \times [-1/2,0)$.

We will apply this to the Gauß curvature flows, but we cannot do so immediately: First, Theorem 15.9 applies on domains in Euclidean space $\mathbb{R}^n$, while equation (15.16) holds on S^n. Also, equation (15.16) involves the value of σ as well as the second derivatives, and Theorem 15.9 requires that F be defined and uniformly elliptic for all symmetric matrices A, and this is not true for the right-hand side of (15.16).

To overcome these difficulties, we first fix our solution σ of (15.16) on $S^n \times [0,T)$ and assume that we have a positive lower bound on the support function. This implies that there is a constant C such that

$$\text{(15.31)} \qquad \frac{1}{C} \leq \sigma(z,t) \leq C\,, \quad |\overline{\nabla}\sigma|_{\overline{g}}(z,t) \leq C\,, \quad \text{and} \quad \frac{1}{C} \leq r_i(z,t) \leq C$$

for all $(z,t) \in S^n \times [0,T)$, for some $C > 0$. Here the first inequality follows from the lower bound on the support function and the fact that the hypersurfaces are shrinking (so σ is bounded by $\sup\{\sigma_0(z) : z \in S^n\}$); the last inequality is the bounds above and below on principal radii derived above as a consequence of the Chou bound on K and the upper bound on $A[\sigma]$; and the gradient bound (with respect to $\overline{g}$) of the support function is derived as follows: By (5.83), we have $\overline{\nabla}\sigma = X^\top$. Hence

$$|X|^2 = |\overline{\nabla}\sigma|^2_{\overline{g}} + \sigma^2\,.$$

In particular, on $S^n \times [0,T)$ we have the bound

$$\text{(15.32)} \qquad |\overline{\nabla}\sigma|_{\overline{g}}(z,t) \leq \sup_{S^n} |X|(\cdot,t) \leq \sup_{S^n} |X|(\cdot,0)\,.$$

Now fix any $z_0 \in S^n$ and $t_0 \in (0,T)$, and let $\Psi : \mathbb{R}^n \to S^n \setminus \{-z_0\}$ be the stereographic projection (so that $\Psi(0) = z_0$). Then set

$$\tilde{\sigma}(x,t) \doteqdot \sigma(x, t_0(t+1))$$

for $(x,t) \in B_1^n(0) \times [-1,0)$.

The metric $\overline{g}$ and the connection $\overline{\Gamma}_{ij}{}^k$ written in the chart Ψ above are smooth functions on $B_1^n(0)$, and we have

$$A[\sigma]_{ij} = \overline{\nabla}_i\overline{\nabla}_j\sigma + \overline{g}_{ij}\sigma = \partial_i\partial_j\tilde{\sigma} - \overline{\Gamma}_{ij}{}^k\partial_k\tilde{\sigma} + \overline{g}_{ij}\tilde{\sigma}\,.$$

Thus, by (15.16) we have

$$\text{(15.33)} \qquad \partial_t\tilde{\sigma}(x,t) = -t_0\left(\frac{\det\left(\partial_i\partial_j\tilde{\sigma} - \overline{\Gamma}_{ij}{}^k\partial_k\tilde{\sigma} + \overline{g}_{ij}\tilde{\sigma}\right)}{\det\overline{g}}\right)^{-\alpha}.$$

We define $F_0(A,x,t)$ by

$$F_0(A,x,t) = -t_0\left(\frac{\det\left(A_{ij} - M_{ij}(x,t)\right)}{\det\overline{g}(x)}\right)^{-\alpha},$$

for $(x,t) \in B_1^n(0) \times [-1,0)$ and symmetric matrices A in the set

$$\mathcal{C}(x,t) = \left\{A : \frac{1}{C}\mathrm{id} \leq A - M(x,t) \leq C\mathrm{id}\right\},$$

where

$$M_{ij}(x,t) \doteqdot \overline{\Gamma}_{ij}{}^k(x)\partial_k\tilde{\sigma}(x,t) + \overline{g}_{ij}(x)\tilde{\sigma}(x,t).$$

Finally, we define $F(A,x,t)$ as follows:

$$F(A,x,t) = \inf_{B\in\mathcal{C}(x,t)} \left(F_0(B,x,t) + DF_0\big|_{(B,x,t)}(A-B,0,0) \right).$$

Since $F_0(A,x,t)$ is a concave function of $A \in \mathcal{C}(x,t)$ for each (x,t), we have $F(A,x,t) - F_0(A,x,t)$ for each $A \in \mathcal{C}(x,t)$. Since $A = D^2\sigma(x,t)$ lies in $\mathcal{C}$ for each $(x,t) \in B_1^n(0) \times [-1,0)$, we have

$$\partial_t\tilde{\sigma}(x,t) = F_0(D^2\tilde{\sigma}(x,t),x,t) = F(D^2\tilde{\sigma}(x,t),x,t)$$

for $(x,t) \in B_1^n(0) \times [-1,0)$.

With this definition we check the conditions of Theorem 15.9: The uniform ellipiticity (i) of F follows since F is an infimum of linear functions; the concavity (ii) of F in A follows since F is an infimum of linear functions of A each of which is uniformly elliptic; and the regularity (iii) follows since $\overline{g}$ and $\overline{\Gamma}$ are smooth, while $\tilde{\sigma}$ and $\partial\tilde{\sigma}$ are Lipschitz functions.

Applying the theorem, we deduce that $D^2\tilde{\sigma}$ is bounded in $C^{0,\gamma}(B^n_{1/2}(0) \times [-1/2,0))$, and hence $\overline{\nabla}^2\sigma$ is bounded in $C^{0,\gamma}(\psi(B_{1/2}(0)) \times [t_0/2,t_0)$. Since z_0 is arbitrary, $\nabla^2\sigma$ is bounded in $C^{0,\gamma}(S^n \times [t_0/2,t_0))$. This completes the proof of the $C^{2,\gamma}$ estimate.

To complete the argument we need to estimate the higher derivatives of the solution σ. This is technically easier than the $C^{2,\gamma}$ estimate, requiring only the following Schauder estimate (see [**368**, Theorem 4.9]):

Theorem 15.10. *Let u be a smooth solution to*

$$\partial_t u = a^{ij}D_iD_ju + b^iD_iu + cu + f \quad \text{on } B_1(0) \times [-1,0), \tag{15.34}$$

where $\lambda|\xi|^2 \le a^{ij}\xi_i\xi_j \le \Lambda|\xi|^2$ and a^{ij}, b^i, and c have $C^{0,\gamma}(B_1(0)\times[-1,0))$ norms bounded by a constant B. Then there is a constant C determined by λ, Λ, and B such that

$$\|D^2u\|_{C^{0,\gamma}(B_{1/2}(0)\times[-1/2,0))} \le C\left(\sup_{B_1(0)\times[-1,0)}|u| + \|f\|_{C^{0,\gamma}(B_1(0)\times[-1,0))}\right).$$

In order to apply Theorem 15.10 to obtain Hölder continuity of the third derivatives of σ, we define $\tilde{\sigma}$ as before, on a time interval where we already have Hölder bounds for the second derivatives, and differentiate the equation (15.33) to obtain an equation satisfied by partial derivatives of $\tilde{\sigma}$: Writing $A_{ij} = \partial_i\partial_j\tilde{\sigma} - \overline{\Gamma}_{ij}{}^k\partial_k\tilde{\sigma} + \overline{g}_{ij}\tilde{\sigma}$, we find that $\partial_k\tilde{\sigma}$ satisfies (15.34) with

$$\begin{aligned}
a^{ij} &= \alpha t_0(\det\overline{g})^\alpha(\det A)^{-\alpha}(A^{-1})^{ij},\\
b^\ell &= -a^{ij}\overline{\Gamma}_{ij}{}^\ell,\\
c &= a^{ij}\overline{g}_{ij},\\
f &= a^{ij}(\partial_k\overline{\Gamma}_{ij}{}^p\partial_p\tilde{\sigma} + \partial_k\overline{g}_{ij}\tilde{\sigma}) - \alpha t_0(\det\overline{g})^{\alpha-1}(\det A)^{-\alpha}\partial_k\det\overline{g}.
\end{aligned}$$

Since all of these have bounded Hölder norms, the theorem applies to give that the second derivatives of $\partial_k\tilde{\sigma}$ are bounded in $C^{0,\gamma}$, and hence (since k is arbitrary) the third derivatives of σ are Hölder continuous. Higher derivative bounds follow by induction on the order of differentiation in a similar way. We conclude that there are uniform bounds on all derivatives of σ on $S^n \times [a,T)$ for any $a > 0$.

Now we can complete the argument: Since $\partial_t D^\beta\sigma$ is bounded for any multi-index β, we have that $D^\beta\sigma$ converges uniformly as t approaches T, and it follows that $\sigma(\cdot,t)$ converges in C^∞ to a function $\sigma(\cdot,T)$ as t approaches T. Then we can apply the short-time existence theorem to extend the solution σ to $S^n \times [0,T+\delta)$ for some $\delta > 0$, contradicting the assumption that T is the maximal interval of existence.

This contradiction implies that our assumption of a lower bound on the inradius is impossible, so we conclude that the inradius converges to zero as t approaches T. It remains only to prove that the diameter also approaches zero. For this we use the upper bound on principal radii from Section 15.3.3:

Lemma 15.11. *Suppose $\mathcal{M}^n \subset \mathbb{R}^{n+1}$ is a convex hypersurface with principal radii bounded by a constant C. Then*

$$(\mathrm{diam}(\mathcal{M}^n))^2 \leq 8Cr_-(\mathcal{M}^n). \tag{15.35}$$

Proof. Let p and q be points of $\mathcal{M}^n$ at maximal distance, and choose the origin to be at their midpoint. By rotation we can assume that $p = (0, -\frac{D}{2})$ and $q = (0, \frac{D}{2})$ in $\mathbb{R}^n \times \mathbb{R} \simeq \mathbb{R}^{n+1}$, where $D = \mathrm{diam}(\mathcal{M}^n)$. Now for $s > C$ consider the region

$$\Omega_s = \{(x,y) \in \mathbb{R}^n \times \mathbb{R} : \ |x| < \sqrt{s^2 - D^2/4} - \sqrt{s^2 - y^2}, \ |y| < D/2\}.$$

The boundary surface has curvature in the y-direction equal to s everywhere except at the endpoints. As s approaches infinity, this region collapses to the line segment from p to q and so is enclosed by $\mathcal{M}^n$. Therefore it is contained for any sufficiently large s. Also, since Ω_s is conical at p and q for $s < D/2$, Ω_s does not touch $\mathcal{M}^n$ near p or q. Now let $\bar{s}$ be the maximal value of $s \leq D/2$ for which Ω_s is enclosed by $\mathcal{M}^n$. It follows that there is a point of contact between $\mathcal{M}^n$ and the boundary surface of $\Omega_{\bar{s}}$, and at this point we must have that the radius of curvature of $\mathcal{M}^n$ in the y-direction is at least $\bar{s}$.

If $\bar{s} \geq C$, then we have a contradiction to the radius of curvature bound, so we must have $\bar{s} \leq C$.

This leaves two possibilities: First, if $C < D/2$, then we have $\bar{s} = D/2$ and we conclude that $\mathcal{M}^n$ encloses the ball

$$\Omega_{D/2} = \{(x,y) : \ |y|^2 + |x|^2 < D^2/4\}$$

and therefore coincides with this ball since it has diameter D. But then $\mathcal{M}^n$ has principal curvatures equal to $D/2 > C$, contradicting the radius of curvature bound.

The only remaining possibility is that $C \geq D/2$ and $\bar{s} = C$. This proves that the region Ω_C is enclosed by $\mathcal{M}^n$. The result now follows since Ω_C has inradius equal to

$$\frac{D^2}{4(C + \sqrt{C^2 - D^2/4})} \geq \frac{D^2}{8C}.$$

Therefore $r_-(\mathcal{M}^n) \geq r_-(\Omega_C) \geq \frac{D^2}{8C}$, proving the lemma. □

This completes the proof of Chou's theorem: We have proved that the solution continues to exist smoothly until the inradius converges to zero and (by the lemma) that the diameter must also converge to zero.

15.4. Differential Harnack estimates

As usual, we motivate the Harnack form by considering self-similar solutions.

Lemma 15.12. *Let $\{\mathcal{M}^n_t\}_{t\in(0,\infty)}$ be an expanding self-similar solution to the α-Gauß curvature flow. Then*

$$X \doteqdot \nabla K^\alpha + \mathrm{L}(V) = 0, \tag{15.36}$$

$$Y \doteqdot \nabla V - K^\alpha\,\mathrm{L} - \frac{1}{(n\alpha+1)t}\mathrm{I} = 0, \quad \textit{and} \tag{15.37}$$

$$W \doteqdot \nabla_t \mathrm{L} + \nabla_V \mathrm{L} + \frac{1}{(n\alpha+1)t}\mathrm{L} = 0, \tag{15.38}$$

where V is the flow vector field, $V \doteqdot \frac{1}{(n\alpha+1)t}X^\top$.

Proof. The first identity follows by differentiating the expander equation

$$K^\alpha = \frac{1}{(n\alpha+1)t}\langle X, \mathsf{N}\rangle\,.$$

The second follows from differentiating the definition of V,

$$V = \frac{1}{(n\alpha+1)t}(X - \langle X, \mathsf{N}\rangle\,\mathsf{N}),$$

and resolving tangential and normal components. The third follows from differentiating the first, applying the second, and recalling (15.13). □

Following Hamilton's mantra of estimating quantities which are constant on self-similar solutions, we define

$$
\begin{aligned}
P &\doteqdot \nabla_t \mathrm{II} - \nabla_{\mathrm{L}^{-1}(\nabla K^\alpha)} \mathrm{II} \\
&= \nabla^2 K^\alpha - \nabla_{\mathrm{L}^{-1}(\nabla K^\alpha)} \mathrm{II} + K^\alpha \mathrm{II}^2
\end{aligned}
\tag{15.39}
$$

and

$$
\begin{aligned}
p &\doteqdot (\mathrm{II}^{-1})^{ij} P_{ij} \\
&= (\mathrm{II}^{-1})^{ij} \nabla_i \nabla_j K^\alpha - \alpha^{-1} K^{-\alpha} \mathrm{II}^{-1}(\nabla K^\alpha, \nabla K^\alpha) + H K^\alpha \\
&= \alpha^{-1} K^{-\alpha} \left(\partial_t K^\alpha - \mathrm{II}^{-1}(\nabla K^\alpha, \nabla K^\alpha) \right) .
\end{aligned}
\tag{15.40}
$$

Theorem 15.13. *Every compact, locally uniformly convex solution to the α-Gauß curvature flow satisfies*

$$
\partial_t K^\alpha - \mathrm{II}^{-1}(\nabla K^\alpha, \nabla K^\alpha) + \frac{n\alpha}{(n\alpha+1)t} K^\alpha \geq 0 . \tag{15.41}
$$

Proof sketch. Similar calculations as carried out for mean curvature flow yield the evolution equation

$$
\partial_t p = \mathcal{L} p + 2\alpha \, \mathrm{II}^{-1}(\nabla K^\alpha, \nabla p) + |P|^2_{\mathrm{II}} + \alpha p^2 , \tag{15.42}
$$

where

$$
|P|^2_{\mathrm{II}} \doteqdot (\mathrm{II}^{-1})^{ik} (\mathrm{II}^{-1})^{j\ell} P_{ij} P_{k\ell} . \tag{15.43}
$$

Since

$$
|P|^2_{\mathrm{II}} \geq \frac{1}{n} \left((\mathrm{II}^{-1})^{ij} P_{ij} \right)^2 = \frac{1}{n} p^2 , \tag{15.44}
$$

the claim follows from the ODE comparison principle. □

Observe that, in the Gauß map parametrization, (15.41) becomes (cf. Remark 10.13)

$$
\partial_t K^\alpha + \frac{n\alpha}{(n\alpha+1)t} K^\alpha \geq 0 . \tag{15.45}
$$

It turns out that the Harnack quantities admit very simple evolution equations when working in the Gauß map parametrization [**29**] (cf. Section 4.3.2). We will exploit this observation to obtain a generalization of Theorem 15.13 to a large class of geometric evolution equations in Section 18.7.

15.5. Firey's conjecture

Firey proved that the Gauß curvature flow of a bounded, convex hypersurface in $\mathbb{R}^3$ contracts convex hypersurfaces to *round* points, assuming the existence and regularity of solutions of the flow, and that the initial support function is symmetric about the origin [**222**]. He conjectured that the result would hold without the symmetry conditions.

Figure 15.2. William J. Firey. Photo by Patricia Potts.

We have already seen, in Section 15.3, that solutions always exist for a short time and remain smooth and convex until they shrink to a point in finite time. So it remains to prove that a general convex solution becomes spherical as it approaches its final point. The main ingredients are a bound on the difference of the principal curvatures (proved by the maximum principle) and an isoperimetric estimate using the width function [**33**]; the Harnack estimate for the Gauß curvature flow (Theorem 15.13) is also used.

Theorem 15.14 (Fate of the Rolling Stones). *Let $\mathcal{M}_0 = X_0(M^2) \subset \mathbb{R}^3$ be a smooth, uniformly convex surface given by an embedding X_0, and let X be the smooth solution to the Gauß curvature flow:*

$$\partial_t X(x,t) = -K(x,t)\mathsf{N}(x,t), \tag{15.46a}$$

$$X(x,0) = X_0(x) \tag{15.46b}$$

for $t \in [0,T)$, where $T = V(\mathcal{M}_0)/4\pi$. Then the surfaces $\mathcal{M}_t = X_t(M^2)$ remain uniformly convex and converge as $t \to T$ to a point in $\mathbb{R}^3$, which after translating the initial surface we may assume is the origin. After rescaling about this origin, the surfaces converge in C^∞ to the unit 2-sphere:

$$\tilde{X}(x,t) \doteqdot \frac{X(x,t)}{(3(T-t))^{1/3}} \to \tilde{X}_T(x), \tag{15.47}$$

where $\tilde{X}_T$ is a smooth embedding with $\tilde{X}_T(M^2) = S^2(1)$.

Firstly, observe that round solutions are given as follows. Let $S_o^2(r_0)$ be the 2-sphere of radius r_0 centered at a point $o \in \mathbb{R}^3$. Parametrize $S_o^2(r_0)$ by the embedding $X_0 : S^2 \to \mathbb{R}^3$ defined by $X_0(x) = o + r_0 x$. Then the solution to the Gauß curvature flow with this initial data is given by $X_t(x) = o + r(t)x$, where $r(t) = (r_0^3 - 3t)^{1/3}$. Indeed, $\frac{dr}{dt}(t) = -r^{-2}(t)$ and $r(0) = r_0$. From this we see that the singular time is $T = r_0^3/3$ and hence $r(t) = (3(T-t))^{1/3}$. This explains the normalization in (15.47).

As a function of g and II the Gauß curvature is $K = \frac{\det \mathrm{II}_{ij}}{\det g_{ij}} = K(g, \mathrm{II})$, where the dependence of K on g and II is pointwise. The variation of K, in terms of the variations of II and g, is

$$\begin{aligned} \partial_s K &= \frac{\partial K}{\partial \mathrm{II}_{ij}} \partial_s \mathrm{II}_{ij} + \frac{\partial K}{\partial g_{ij}} \partial_s g_{ij} \\ &= K(\mathrm{II}^{-1})^{ij} \partial_s \mathrm{II}_{ij} - K g^{ij} \partial_s g_{ij} \,. \end{aligned} \tag{15.48}$$

We introduce some notation. Define

$$\dot{K}^{ij} = \frac{\partial K}{\partial \mathrm{II}_{ij}} = K(\mathrm{II}^{-1})^{ij} \tag{15.49}$$

and

$$\ddot{K}^{k\ell mn} = \frac{\partial^2 K}{\partial \mathrm{II}_{k\ell} \partial \mathrm{II}_{mn}} = K\Big((\mathrm{II}^{-1})^{k\ell}(\mathrm{II}^{-1})^{mn} - (\mathrm{II}^{-1})^{km}(\mathrm{II}^{-1})^{\ell n}\Big) \,. \tag{15.50}$$

With this notation, the evolution equation (15.8) for the mean curvature may be written as

Lemma 15.15. *For any* $n \geq 2$,

$$\partial_t H = \dot{K}^{k\ell} \nabla_k \nabla_\ell H + g^{ij} \ddot{K}^{k\ell mn} \nabla_i \mathrm{II}_{k\ell} \nabla_j \mathrm{II}_{mn} + K(H^2 - (n-1)\,|\mathrm{II}|^2). \tag{15.51}$$

When $n = 2$, *the identity* $H^2 - |\mathrm{II}|^2 = 2K$ *yields the following equivalent formulation of* (15.51):

$$\partial_t H = \dot{K}^{k\ell} \nabla_k \nabla_\ell H + g^{ij} \ddot{K}^{k\ell mn} \nabla_i \mathrm{II}_{k\ell} \nabla_j \mathrm{II}_{mn} + 2K^2. \tag{15.52}$$

By (5.142), the evolution of the Gauß curvature is given by

$$\partial_t K = \dot{K}^{k\ell} \nabla_k \nabla_\ell K + K^2 H \,. \tag{15.53}$$

The next proposition is the key estimate on the difference of the principal curvatures. This result is surprising since it says that a quantity which measures the difference from roundness and has degree -1 (i.e., is homogeneous of degree 1 in the principal curvatures) is monotonically nonincreasing under the Gauß curvature flow. Provided that the hypersurfaces converge smoothly after rescaling, this shows that the limiting shape is round.

Proposition 15.16. *On any uniformly convex solution $X : M^n \times [0,T) \to \mathbb{R}^3$ to the Gauß curvature flow in $\mathbb{R}^3$,*

$$\max_{M^2\times\{t\}} |\kappa_2 - \kappa_1| \leq \max_{M^2\times\{0\}} |\kappa_2 - \kappa_1| . \tag{15.54}$$

Proof. Define

$$Q \doteqdot H^2 - 4K = (\kappa_2 - \kappa_1)^2 . \tag{15.55}$$

We compute the evolution equation of Q using (15.52) and (15.53):

$$\begin{aligned} \partial_t Q &= 2H\partial_t H - 4\,\partial_t K \\ &= 2H(\dot{K}^{k\ell}\nabla_k\nabla_\ell H + g^{ij}\ddot{K}^{k\ell mn}\nabla_i \mathrm{II}_{k\ell} \nabla_j \mathrm{II}_{mn} + 2K^2) \\ &\quad - 4(\dot{K}^{k\ell}\nabla_k\nabla_\ell K + K^2 H) \\ &= \dot{K}^{k\ell}\nabla_k\nabla_\ell Q - 2\dot{K}^{k\ell}\nabla_k H\nabla_\ell H + 2Hg^{ij}\ddot{K}(\nabla_i \mathrm{II}, \nabla_j \mathrm{II}). \end{aligned} \tag{15.56}$$

To show that $\max_{M^2\times\{t\}} Q$ is nonincreasing, fix $t \in [0,T)$ and let $x_0 \in M$ be a point at which $Q(x_0,t) = \max_{x\in M} Q(x,t)$. Choose coordinates so that $g_{ij}(x_0,t) = \delta_{ij}$ and $\mathrm{II}_{ij}(x_0,t)$ is diagonal.

The first term on the right-hand side of the bottom line of (15.56) is nonpositive, as is the second term, since the solution is uniformly convex.

For the third term on the right-hand side of (15.56), we use $\nabla Q(x_0,t) = 0$ to calculate at (x_0,t) that

$$\begin{aligned} 0 &= \nabla_1 Q \\ &= 2H\nabla_1 H - 4\nabla_1 K \\ &= 2(\kappa_1 + \kappa_2)(\nabla_1 \mathrm{II}_{11} + \nabla_1 \mathrm{II}_{22}) - 4(\kappa_1 \nabla_1 \mathrm{II}_{22} + \kappa_2 \nabla_1 \mathrm{II}_{11}) \\ &= 2(\kappa_2 - \kappa_1)(\nabla_1 \mathrm{II}_{22} - \nabla_1 \mathrm{II}_{11}) , \end{aligned}$$

where $\kappa_i = \mathrm{II}_{ii}$ for $i = 1,2$ are the principal curvatures. If $\kappa_1 = \kappa_2$ at (x_0,t), then $Q(x_0,t) = 0$, and so we assume that $\kappa_1 \neq \kappa_2$. Then $\nabla_1 \mathrm{II}_{11} = \nabla_1 \mathrm{II}_{22}$ and the analogous calculation with $\nabla_2 Q = 0$ gives $\nabla_2 \mathrm{II}_{11} = \nabla_2 \mathrm{II}_{22}$ at (x_0,t). Hence, at the maximum point (x_0,t) we have

$$\begin{aligned} \ddot{K}(\nabla_1 \mathrm{II}, \nabla_1 \mathrm{II}) &= K\left((\mathrm{II}^{-1})^{k\ell}(\mathrm{II}^{-1})^{mn} - (\mathrm{II}^{-1})^{km}(\mathrm{II}^{-1})^{\ell n}\right)\nabla_1\mathrm{II}_{k\ell}\nabla_1\mathrm{II}_{mn} \\ &= K\kappa_k^{-1}\kappa_m^{-1}(\nabla_1\mathrm{II}_{kk}\nabla_1\mathrm{II}_{mm} - (\nabla_1\mathrm{II}_{km})^2) \\ &= 2\nabla_1 \mathrm{II}_{11}\, \nabla_1 \mathrm{II}_{22} - 2(\nabla_1 \mathrm{II}_{12})^2 \\ &= 2(\nabla_1 \mathrm{II}_{11})^2 - 2(\nabla_2 \mathrm{II}_{11})^2 \\ &= 2(\nabla_1 \mathrm{II}_{11})^2 - 2(\nabla_2 \mathrm{II}_{22})^2. \end{aligned}$$

Similarly, one obtains

$$\ddot{K}(\nabla_2 \mathrm{II}, \nabla_2 \mathrm{II}) = 2(\nabla_2 \mathrm{II}_{22})^2 - 2(\nabla_1 \mathrm{II}_{11})^2 .$$

Therefore, at (x_0, t) we have

$$g^{ij}\ddot{K}(\nabla_i \mathrm{II}, \nabla_j \mathrm{II}) = \ddot{K}(\nabla_1 \mathrm{II}, \nabla_1 \mathrm{II}) + \ddot{K}(\nabla_2 \mathrm{II}, \nabla_2 \mathrm{II}) = 0$$

and so the aforementioned third term vanishes. The proposition now follows from applying the maximum principle to (15.56). □

We will next use Proposition 15.16 to derive an isoperimetric estimate. If $\mathcal{M}^n \subset \mathbb{R}^{n+1}$ is a uniformly convex embedded hypersurface, then we can parametrize it by S^n using the Gauß map. Recall that the support function $\sigma : S^n \to \mathbb{R}$ is defined by

$$\sigma(z) \doteqdot \langle z, X \rangle,$$

where $X = X(z) = \mathrm{G}^{-1}(z)$ and $\mathrm{G} : \mathcal{M} \to S^n$ is the Gauß map given by (5.28). The normal to X is z, i.e., $\mathsf{N}(X(z)) = z$, and $\sigma(z)$ is the length of the projection onto z.

Given $z \in S^2$ and a point $x \in M^2$, define $e_z(x,t)$ to be a choice of unit vector at $X(x,t)$ that is tangent to the image under the inverse of the Gauß map of a great circle between $\mathsf{N}(x,t)$ and z. We parametrize S^2 by choosing spherical coordinates $\theta \in [0, \pi]$ and $\phi \in [0, 2\pi)$ so that z has coordinates $(\theta, \phi) = (0,0)$. Then (5.78) and (5.81) imply that

$$\mathrm{II}^{-1}(e_z, e_z) = \partial_\theta^2 \sigma + \sigma. \tag{15.57}$$

We next prove that the difference between a certain (time-dependent) constant and the support function of the surface with respect to a certain (time-dependent) origin is bounded by a constant times the area of the surface. Since the left-hand side of the display below has degree 1 whereas the right-hand side has degree 2, this shows that after rescaling the hypersurfaces converge in C^0 to the unit 2-sphere (see Corollary 15.18).

Proposition 15.17. *Let $\{\mathcal{M}_t\}_{t\in[0,T)}$ be a family of uniformly convex surfaces in $\mathbb{R}^3$ evolving by Gauß curvature flow. Then there exists a point $\tilde{q}(t) \in \mathbb{R}^3$ so that at any point $x \in M^2$,*

$$\left| \langle X(x,t) - \tilde{q}(t), \mathsf{N}(x,t) \rangle - \frac{1}{8\pi} \int_{\mathcal{M}_t} H_t \, d\mu_t \right| \le \frac{C}{4\pi} A(\mathcal{M}_t), \tag{15.58}$$

where $A(\mathcal{M}_t)$ is the surface area of $\mathcal{M}_t$ and $C = \sup_{x\in M^2} |\kappa_1(x,0) - \kappa_2(x,0)|$.

Proof. Fix any $z \in S^2$. We first show that (for any uniformly convex surface $X : M^2 \to \mathcal{M} \subset \mathbb{R}^3$) the *width* can also be represented by the integral

$$w(z) = \frac{1}{2\pi} \int_{M^2} \dot{K}(e_z(x), e_z(x)) \, d\mu(x), \tag{15.59}$$

where $e_z(x)$ is defined the same way as in (15.57) (see Theorem 5.1 in [**26**])[1]. Indeed, parametrize S^2 using θ and ϕ as above, and use (15.57) and integration by parts twice to calculate

$$\begin{aligned}\int_{M^2} K(x)\,\mathrm{II}^{-1}(e_z(x), e_z(x))d\mu(x) &= \int_{S^2} \mathrm{II}^{-1}(e_z(G^{-1}(z')), e_z(G^{-1}(z')))d\bar{\mu}(z') \\ &= \int_0^{2\pi}\int_0^{\pi} (\partial_\theta^2\sigma + \sigma)\sin\theta\, d\theta\, d\phi \\ &= \int_0^{2\pi}\int_0^{\pi} (-\partial_\theta\sigma\, \cos\theta + \sigma\sin\theta)\, d\theta\, d\phi \\ &= \int_0^{2\pi} (\sigma(\pi,\phi) + \sigma(0,\phi))\, d\phi \\ &= 2\pi(\sigma(z) + \sigma(-z))\end{aligned}$$

since $\sigma(z) = \sigma(0,\phi)$ for all ϕ. This proves (15.59).

Next, notice that

$$\begin{aligned}\dot{K}(e_z, e_z) + \dot{K}(e_z^\perp, e_z^\perp) &= K\,\mathrm{II}^{-1}(e_z, e_z) + K\,\mathrm{II}^{-1}(e_z^\perp, e_z^\perp) \\ &= H\,.\end{aligned} \tag{15.60}$$

Then by Proposition 15.16, we have

$$\left|\dot{K}(e_z, e_z) - \dot{K}(e_z^\perp, e_z^\perp)\right| \le C \tag{15.61}$$

since $\dot{K}(e,e)$ is bounded by the principal curvatures κ_1 and κ_2 for any unit vector e. So by (15.59), the difference of the width and a constant is small in the following sense:

$$\left|w(z,t) - \frac{1}{4\pi}\int_{\mathcal{M}_t} H_t\, d\mu_t\right| \le \frac{C}{4\pi}A(\mathcal{M}_t). \tag{15.62}$$

Now let $\mathcal{M}$ be a uniformly convex hypersurface, $p \in \mathcal{M}$, and $z = \mathsf{N}(p)$. The second step of the proof is to show that for the point

$$\tilde{q} = \frac{1}{4\pi}\int_{\mathcal{M}} p\, K(p)\, d\mu(p) = \frac{3}{4\pi}\int_{S^2} z'\,\sigma(z')\, d\mu(z') \tag{15.63}$$

(see Exercise 15.4 for a proof of the second equality in (15.63)), the equality

$$\langle p - \tilde{q}, z\rangle = \frac{1}{2}w(z) + \frac{1}{4\pi}\int_{\mathcal{M}} \dot{K}(e_z, e_z)\langle \mathsf{N}, z\rangle\, d\mu \tag{15.64}$$

[1]The integral representation can be used to prove Lemma 8.24: Indeed, there exist points z_+ and z_- such that $w_{\max} = \sigma(z_+) + \sigma(-z_+)$ and $w_{\min} = \sigma(z_-) + \sigma(-z_-)$. By the pinching hypothesis $\max_i \kappa_i \le C \min_j \kappa_j$, the result follows.

holds. With the spherical coordinates parametrization, we compute that

$$\begin{aligned}\langle\tilde{q},z\rangle &= \frac{3}{4\pi}\int_{S^2}\sigma(z')\langle z',z\rangle\,d\mu(z')\\ &= \frac{3}{4\pi}\int_0^{2\pi}\int_0^{\pi}\sigma(\theta,\phi)\cos(\theta)\sin(\theta)\,d\theta\,d\phi\\ &= \frac{3}{8\pi}\int_0^{2\pi}\int_0^{\pi}\sigma(\theta,\phi)\sin(2\theta)\,d\theta\,d\phi.\end{aligned}$$

By the identity $\partial_\theta^2\sin(2\theta)+\sin(2\theta)=-3\sin(2\theta)$, this is equal to

$$\langle\tilde{q},z\rangle = -\frac{1}{8\pi}\int_0^{2\pi}\int_0^{\pi}\sigma(\theta,\phi)(\partial_\theta^2\sin(2\theta)+\sin(2\theta))\,d\theta\,d\phi.$$

Integrating the expression $\sigma(\theta,\phi)\,\partial_\theta^2\sin(2\theta)$ by parts twice, we obtain

$$\begin{aligned}\langle\tilde{q},z\rangle &= -\frac{1}{8\pi}\int_0^{2\pi}\int_0^{\pi}(\partial_\theta^2\sigma+\sigma)\sin(2\theta)\,d\theta\,d\phi+\frac{1}{2}\sigma(z)-\frac{1}{2}\sigma(-z)\\ &= -\frac{1}{4\pi}\int_{S^2}\mathrm{II}^{-1}(e_z,e_z)\langle z',z\rangle\,d\bar{\mu}(z')+\frac{1}{2}\sigma(z)-\frac{1}{2}\sigma(-z),\end{aligned}\tag{15.65}$$

where we used $d\bar{\mu}(z')=\sin\theta\,d\theta\,d\phi$ and $\langle z',z\rangle=\cos\theta$. Since $z=\mathsf{N}(p)$, we have $\langle p,z\rangle=\sigma(z)$, and so equation (15.65) implies

$$\begin{aligned}\langle p-\tilde{q},z\rangle &= \frac{1}{4\pi}\int_{S^2}\mathrm{II}^{-1}(e_z,e_z)\langle z',z\rangle\,d\bar{\mu}(z')+\frac{1}{2}w(z)\\ &= \frac{1}{4\pi}\int_{M^2}\dot{K}(e_z(x),e_z(x))\,\langle\mathsf{N},z\rangle\,d\mu(x)+\frac{1}{2}w(z),\end{aligned}$$

which is (15.64).

Together with (15.62), this yields

$$\begin{aligned}\left|\langle p-\tilde{q},z\rangle-\frac{1}{8\pi}\int_{\mathcal{M}}H\,d\mu\right| &\le \frac{1}{2}\left|w(z)-\frac{1}{4\pi}\int_{\mathcal{M}}H\,d\mu\right|+\left|\langle p-\tilde{q},z\rangle-\frac{1}{2}w(z)\right|\\ &\le \frac{C}{8\pi}A(\mathcal{M})+\frac{1}{4\pi}\left|\int_{\mathcal{M}}\dot{K}(e_z,e_z)\langle\mathsf{N},z\rangle d\mu\right|.\end{aligned}$$

The expression (15.60) allows us to write

$$\frac{1}{4\pi}\int_{\mathcal{M}}\dot{K}(e_z,e_z)\,d\mu=\frac{1}{8\pi}\int_{\mathcal{M}}\dot{K}(e_z,e_z)\,d\mu+\frac{1}{8\pi}\int_{\mathcal{M}}(H-\dot{K}(e_z^\perp,e_z^\perp))\,d\mu,$$

and so using $\int_{\mathcal{M}} H\mathsf{N}\,d\mu = 0$ and the inequalities $|\langle \mathsf{N}, z\rangle| \le 1$ and (15.61), we conclude that

$$\begin{aligned}\left|\langle p-\tilde{q}, z\rangle - \frac{1}{8\pi}\int_{\mathcal{M}} H\,d\mu\right| &\le \frac{C}{8\pi}A(\mathcal{M}) + \frac{1}{8\pi}\left|\int_{\mathcal{M}} H\langle \mathsf{N}, z\rangle\,d\mu\right| \\ &\quad + \frac{1}{8\pi}\int_{\mathcal{M}}\left|\dot{K}(e_z,e_z) - \dot{K}(e_z^{\perp}, e_z^{\perp})\right| |\langle \mathsf{N}, z\rangle|\,d\mu \\ &\le \frac{C}{4\pi}A(\mathcal{M}).\end{aligned}$$

□

In general, if $\mathcal{M} \subset \mathbb{R}^{n+1}$ is a closed convex hypersurface bounding a compact region Ω^{n+1}, then its n-dimensional surface area is given by the **Crofton formula**:

$$A(\mathcal{M}) = \frac{1}{\omega_n}\int_{S^n} \mathrm{Vol}_n(\pi_{z^\perp}(\Omega^{n+1}))\,d\bar{\mu}(z)\,, \tag{15.66}$$

where ω_n is the volume of the unit Euclidean n-ball, $z^\perp$ is the orthogonal complement of z in $\mathbb{R}^{n+1}$, and $\pi_{z^\perp}$ is the orthogonal projection onto $z^\perp$; see §5.5 in the book by Daniel Klain and Gian-Carlo Rota [**332**]. Call $\mathrm{Vol}_n(\pi_{z^\perp}(\Omega^{n+1})) = \mathrm{Vol}_n(\pi_{z^\perp}(\mathcal{M}))$ the **shadow area** of $\mathcal{M}$ in the direction $z \in S^n$.

If r_+ denotes the circumradius of $\mathcal{M}$, then $\pi_{z^\perp}(\Omega^{n+1})$ is contained in a ball of radius r_+. Hence $\mathrm{Vol}_n(\pi_{z^\perp}(\Omega^{n+1})) \le \omega_n r_+^n$. We conclude that

$$A(\mathcal{M}) \le A(S^n)\,r_+^n\,. \tag{15.67}$$

A consequence of Proposition 15.17 is that at each time t, the hypersurface is bounded by concentric spheres centered at the point $\tilde{q}(t)$:

Corollary 15.18. *The hypersurface $\mathcal{M}_t$ is contained between spheres centered at $\tilde{q}(t)$, with radii $r_-(t) \le r_+(t)$, that satisfy*

$$\frac{r_+(t)}{r_-(t)} \le 1 + C(T-t)^{1/3} \tag{15.68}$$

for $t \in [0,T)$ and some constant C. Of course, the right-hand side converges to 1 as $t \to T$, so this is a C^0 roundness estimate.

Proof. According to Proposition 15.17, $\mathcal{M}_t$ is contained between spheres of radii

$$r_-(t) = \frac{1}{8\pi}\int_{\mathcal{M}_t} H_t\,d\mu_t - \frac{C}{4\pi}A(\mathcal{M}_t)$$

and

$$r_+(t) = \frac{1}{8\pi}\int_{\mathcal{M}_t} H_t\,d\mu_t + \frac{C}{4\pi}A(\mathcal{M}_t),$$

where $A(\mathcal{M}_t)$ is the area of $\mathcal{M}_t$. Since $A(\mathcal{M}_t) \leq 4\pi r_+^2(t)$ by (15.67), we have the inequality

$$\begin{aligned} r_+(t) &\leq \frac{1}{8\pi}\int_{\mathcal{M}_t} H_t\,d\mu_t + Cr_+^2(t) \\ &= r_-(t) + \frac{C}{4\pi}A(\mathcal{M}_t) + Cr_+^2(t) \\ &\leq r_-(t) + 2Cr_+^2(t)\,. \end{aligned}$$

Now assume that t is sufficiently close to T so that $r_+(t) \leq \frac{1}{4C}$. Then the quadratic inequality above yields

$$r_+(t) \leq \frac{1-\sqrt{1-8Cr_-(t)}}{4C} = \frac{2r_-(t)}{1+\sqrt{1-8Cr_-(t)}}. \tag{15.69}$$

We also know that the volume is $V(t) = 4\pi(T-t)$ by the evolution equation

$$\frac{dV}{dt} = -\int_{\mathcal{M}} K\,d\mu = -4\pi,$$

and $V(T) = 0$. Therefore

$$r_-(t) \leq \left(\frac{3V(t)}{4\pi}\right)^{1/3} = (3(T-t))^{1/3},$$

and the corollary follows from the inequality $\frac{2}{1+\sqrt{1-x}} \leq 1+x$, $x \in [0,1]$. □

Convergence to a sphere after rescaling is now a consequence of the following C^2 roundness estimates (curvatures are second derivatives of the embedding).

Proposition 15.19. *There exist constants C_1 and C_2 such that*

$$\left|K(x,t) - (3(T-t))^{-2/3}\right| \leq C_1(T-t)^{-1/2} \quad \textit{for all } (x,t) \in M^2 \times [0,T)$$

and

$$\left|\mathrm{II}(e,e) - (3(T-t))^{-1/3}\right| \leq C_2(T-t)^{-1/6} \quad \textit{for all unit vectors } e \in T\mathcal{M}_t\,.$$

Proof. Fix a point $z \in S^2$ and consider the (unique) points $p(z,t) \in M^2$ satisfying $\mathsf{N}(p(z,t),t) = z$ for each $t \in [0,T)$. In effect, we are using the Gauß reparametrization. From the Gauß curvature flow equation

$$\partial_t X(x,t) = -K(x,t)\mathsf{N}(x,t)\,,$$

if $x_t = p(z,t)$, we have

$$-K(x_t,t) = \left\langle \frac{d}{dt}X(x_t,t), \mathsf{N}(x_t,t)\right\rangle = \frac{d}{dt}\langle X(x_t,t), z\rangle = \frac{d}{dt}\sigma(z),$$

where the second equality holds since $\langle\cdot,\cdot\rangle$ is the Euclidean metric and where σ is the support function.

The idea of the proof is to estimate the average of K over a time interval by controlling the distance traveled by the tangent space in the direction z. Corollary 15.18 allows us to do that. More precisely, fix a time t_0 and translate the solution X_t so that the point $\tilde{q}(t_0)$ found in Proposition 15.17 is the origin. At that time $\mathcal{M}_{t_0}$ lies between the spheres of radii $r_- \doteqdot r_-(t_0)$ and $r_+ \doteqdot r_+(t_0)$. By the avoidance principle and by calculating the evolution of the round sphere of radius r_- under Gauß curvature flow, we know that after a time increment $\tau > 0$, $\mathcal{M}_{t_0+\tau}$ lies outside of the sphere of radius $(r_-^3 - 3\tau)^{1/3}$. Using the inequality

$$(1-3\varepsilon)^{1/3} \geq 1 - \varepsilon - \varepsilon^2$$

with $\varepsilon = \tau / r_-^3$, we see that the distance that the tangent plane moves in the direction z is at most

$$r_+ - (r_-^3 - 3\tau)^{1/3} \leq r_+ - r_- + \tau r_-^{-2} + \tau^2 r_-^{-5}.$$

Therefore

$$\begin{aligned} \text{(15.70)} \qquad \inf_{t_0 \leq t \leq t_0+\tau} K(p(z,t),t) &\leq \frac{1}{\tau}\int_{t_0}^{t_0+\tau} K(p(z,t),t)\,dt \\ &= -\frac{1}{\tau}\int_{t_0}^{t_0+\tau} \frac{d}{dt}(\sigma(p(z,t),t))\,dt \\ &\leq r_-^{-2} + \tau r_-^{-5} + \frac{r_+ - r_-}{\tau}. \end{aligned}$$

We claim that

$$\begin{aligned} \text{(15.71)} \qquad K(p(z,t_0),t_0) &\leq \left(\frac{t_0+\tau}{t_0}\right)^{2/3} \inf_{t_0 \leq t \leq t_0+\tau} K(p(z,t),t) \\ &\leq \left(\frac{t_0+\tau}{t_0}\right)^{2/3} \left(r_-^{-2} + \tau r_-^{-5} + \frac{r_+ - r_-}{\tau}\right) \\ &\leq (3(T-t_0))^{-2/3} + C_1 C^{1/2} (T-t_0)^{-1/2}. \end{aligned}$$

Indeed, the first inequality in (15.71) follows from the Harnack estimate (18.66), which says that

$$\partial_t (t^{2/3} K(p(z,t),t)) \geq 0.$$

The second inequality follows from (15.70). Both of these inequalities hold for $0 < \tau < T - t_0$.

To see the third inequality, we select τ and estimate term by term. Let

$$\tau = r_-^{5/2}(r_+ - r_-)^{1/2},$$

which is possible for t_0 sufficiently close to T since $r_- \leq r_+ \leq C(T-t_0)^{1/2}$. Observe that (15.68) implies

$$r_+(t) - r_-(t) \leq C(T-t)^{5/6}.$$

So, firstly, we have

$$\frac{r_+ - r_-}{\tau} = \tau r_-^{-5} = \frac{(r_+ - r_-)^{1/2}}{r_-^{5/2}} \leq C r_-^{-2} (T - t_0)^{1/6} .$$

Secondly, $\left(\frac{t_0+\tau}{t_0}\right)^{2/3} \leq 1 + C\tau \leq 1 + C(T - t_0)^{5/3}$. Moreover, (15.68) also implies that

$$r_-^{-2} \leq (3(T - t_0))^{-2/3} + C(T - t_0)^{-1/2} .$$

Thus

$$\begin{aligned}
&\left(\frac{t_0 + \tau}{t_0}\right)^{2/3} \left(r_-^{-2} + \tau r_-^{-5} + \frac{r_+ - r_-}{\tau}\right) \\
&\qquad \leq (1 + C(T - t_0)^{5/3}) \left(r_-^{-2}(1 + C(T - t_0)^{1/6}) + C(T - t_0)^{-1/2}\right) \\
&\qquad \leq (3(T - t_0))^{-2/3} + C(T - t_0)^{-1/2} .
\end{aligned}$$

This completes the proof of the third inequality in (15.71). The lower bound for K is found similarly.

Since the mean curvature and the principal curvatures may be rewritten as

$$H^2 = 4K + (\kappa_2 - \kappa_1)^2$$

and

$$\kappa_2 = \frac{1}{2}H + \frac{1}{2}(\kappa_2 - \kappa_1) \text{ and } \kappa_1 = \frac{1}{2}H - \frac{1}{2}(\kappa_2 - \kappa_1),$$

respectively, the bound for $\mathrm{II}(e, e)$ follows immediately from the bound for K and Proposition 15.16. □

To finish the proof of Theorem 15.14, apply the scaling

$$\tilde{X}(x, t) = \frac{X(x, t) - \tilde{q}(t)}{(3(T - t))^{1/3}} .$$

Then the results of Proposition 15.19 become

$$\left|\tilde{K} - 1\right| \leq C_1(T - t)^{1/6}$$

and

$$\left|\tilde{\mathrm{II}}(e, e) - 1\right| \leq C_2(T - t)^{1/6}.$$

This gives C^2 convergence modulo translations, but this is sufficient by Corollary 7.15 of [**26**]. Also, C^∞ convergence follows as in [**506**]. For an outline of the argument, see [**159**]. This completes the proof of Theorem 15.14.

Remark 15.20. With a little more work to bound $|\mathrm{II}|^2$, Firey's conjecture holds also for viscosity solutions (in the sense of Section 6.9.3). Indeed, the bound on $|\mathrm{II}|^2$ gives the required bound on Q, and hence the isoperimetric and convergence results. See [**33**].

15.6. Variational structure and entropy formulae

In this section we investigate some of the special structure of the Gauß curvature flows. This structure leads in particular to a gradient flow structure for the Gauß curvature flows (in the same sense that the mean curvature flow is the steepest descent flow for the area functional). Perhaps more importantly for our later investigations, the Gauß curvature flows also have an associated family of integral quantities which evolve monotonically.

15.6.1. Gradient flow. The mean curvature flow is the gradient flow of the area functional. Here we discuss some ways in which Gauß curvature flows may be considered as gradient flows.

The **mean width** of a uniformly convex hypersurface $\mathcal{M}$ is defined by

$$\mathrm{MW}(\mathcal{M}) \doteqdot \int_{S^n} w(z)\, d\bar{\mu}(z)\,, \tag{15.72}$$

where $w : S^n \to \mathbb{R}$ is the width function given in (8.54) by

$$w(z) = \sigma(z) + \sigma(-z) = \sup\{(x-y)\cdot z :\ x, y \in \mathcal{M}\}, \tag{15.73}$$

where $\sigma : S^n \to \mathbb{R}$ is the support function of $\mathcal{M}$. Recall that the maximum width is the diameter of $\mathcal{M}$. By the invariance of the measure $d\bar{\mu}$ under the antipodal map and since K is the Jacobian determinant of the Gauß map, we have

$$\mathrm{MW}(\mathcal{M}) = 2\int_{S^n} \sigma(z)\, d\bar{\mu}(z) = 2\int_{\mathcal{M}} \langle p, \mathsf{N}(p)\rangle K(p)\, d\mu(p)\,. \tag{15.74}$$

We compute the first variation of the mean width. Let $\{\mathcal{M}_s\}_{s\in(-\varepsilon,\varepsilon)}$ be a smooth 1-parameter family of uniformly convex hypersurfaces with

$$\partial_s \sigma_s|_{s=0} = \bar{v},$$

where the σ_s are the support functions of $\mathcal{M}_s$ and $\bar{v}$ is a function on S^n. We have

$$\partial_s \mathrm{MW}(\mathcal{M}_s)|_{s=0} = 2\int_{S^n} \bar{v}(z)\, d\bar{\mu} = 2\int_{\mathcal{M}_0} \bar{v}(\mathrm{G}_0(p)) K_0(p)\, d\mu_0(p)\,,$$

where G_0 is the Gauß map, K_0 is the Gauß curvature, and $d\mu_0$ is the induced measure on $\mathcal{M}_0$.

Fix $\alpha > 0$ and set $p = 1 + \frac{1}{\alpha}$, so that $\alpha = \frac{1}{p-1}$. Denote $v = \bar{v} \circ \mathrm{G}_0$. By the Hölder inequality,

$$\frac{1}{2}\partial_s \operatorname{MW}(\mathcal{M}_s)|_{s=0} = \int_{\mathcal{M}_0} v K_0 \, d\mu_0$$

$$\geq - \left(\int_{\mathcal{M}_0} |v|^p \, d\mu_0 \right)^{1/p} \left(\int_{\mathcal{M}_0} K_0^{\frac{p}{p-1}} \, d\mu_0 \right)^{(p-1)/p}, \tag{15.75}$$

with equality if and only if $v = CK^{\frac{1}{p-1}}$ for some negative constant C.

Let $\mathcal{U}$ be the space of support functions of smooth uniformly convex hypersurfaces. Then $\sigma_0 \in \mathcal{U}$ and its tangent space $T_{\sigma_0}\mathcal{U}$ is the vector space of smooth real-valued functions on S^n. Let $\|\cdot\|$ be the L^p-norm on $T_{\sigma_0}\mathcal{U}$ defined by

$$\|\bar{v}\| = \left(\int_{\mathcal{M}_0} |\bar{v} \circ \mathrm{G}_0|^p \, d\mu_0 \right)^{1/p}.$$

Then (15.75) says that for any $\bar{v} \in T_{\sigma_0}\mathcal{U}$ we have

$$\bar{v}(\operatorname{MW}) \geq -\|\bar{v}\| \left(\int_{\mathcal{M}_0} K_0^{\frac{p}{p-1}} \, d\mu_0 \right)^{(p-1)/p},$$

with equality if and only if $v = CK_0^\alpha$ for some $C < 0$. Thus, the infinitesimal variation $V \doteqdot -K_0^\alpha \in T_{\sigma_0}\mathcal{U}$ has the property that $W(\operatorname{MW}) \geq V(\operatorname{MW})$ for all $W \in T_{\sigma_0}\mathcal{U}$ with $\|W\| = \|V\|$, with equality only if $W = V$. That is, V is a **negative gradient vector**, or **direction of steepest descent** of MW at σ_0 with respect to the $L^{1+\frac{1}{\alpha}}$-norm. Since the variation $-K_0^\alpha$ of the support function σ_0 corresponds to the α-Gauß curvature flow, we see that the α-Gauß curvature flow is the **steepest descent flow** of the mean width with respect to the $L^{1+\frac{1}{\alpha}}$-norm.

The following proposition shows that the Gauß curvature flow is a gradient flow of the total $(n-1)$-st mean curvature.

Proposition 15.21. *The mean width* MW *is twice the integral of the* $(n-1)$-*st elementary symmetric function* $S_{n-1} = \frac{1}{n}\sum_{1\leq i_1 < \cdots < i_{n-1} \leq n} \kappa_{i_1} \cdots \kappa_{i_{n-1}}$ *of the principal curvatures*:

$$\operatorname{MW}(\mathcal{M}) = \int_{\mathcal{M}} S_{n-1} \, d\mu. \tag{15.76}$$

Proof. Computing in the Gauß map parametrization and recalling (5.88), we find

$$\begin{aligned}\operatorname{MW}(\mathcal{M}) &= 2\int_{S^n} \sigma(z) \, d\bar{\mu}(z) = \tfrac{2}{n} \int_{S^n} (\overline{\Delta}\sigma + n\sigma) \, d\bar{\mu} \\ &= \tfrac{2}{n} \int_{\mathcal{M}} \sum_i \kappa_i^{-1} K \, d\mu = 2 \int_{\mathcal{M}} S_{n-1} \, d\mu. \qquad \square\end{aligned}$$

15.6.2. The Firey entropy. A monotone energy for the Gauß curvature flow was discovered by Firey. It generalizes in a straightforward way to the other α-Gauß curvature flows.

Given a uniformly convex hypersurface $\mathcal{M}^n \subset \mathbb{R}^{n+1}$, define, for each $\alpha > 0$, the **Firey entropy** $\mathcal{E}_\alpha(\mathcal{M}^n)$ of $\mathcal{M}^n$ by

$$\mathcal{E}_\alpha(\mathcal{M}^n) \doteqdot \begin{cases} \left(\dfrac{|B^{n+1}|}{\mathrm{Vol}(\mathcal{M}^n)}\right)^{\frac{1}{n+1}} \exp\left(\dfrac{1}{|S^n|}\displaystyle\int_{S^n} \log\sigma\, d\bar{\mu}\right) & \text{if } \alpha = 1, \\ \left(\dfrac{|B^{n+1}|}{\mathrm{Vol}(\mathcal{M}^n)}\right)^{\frac{1}{n+1}} \left(\dfrac{1}{|S^n|}\displaystyle\int_{S^n} \sigma^{\frac{\alpha-1}{\alpha}}\, d\bar{\mu}\right)^{\frac{\alpha}{\alpha-1}} & \text{if } \alpha \neq 1, \end{cases} \tag{15.77}$$

where $\sigma : S^n \to \mathbb{R}$ is the support function of $\mathcal{M}^n$.

Proposition 15.22. *Let $\{\mathcal{M}_t^n\}_{t\in[0,T)}$ be a family of uniformly convex hypersurfaces evolving by the α-Gauß curvature flow. Suppose that the origin is enclosed by $\mathcal{M}_t^n$ for each $t \in [0,T)$. Then*

$$\frac{d}{dt}\mathcal{E}_\alpha(\mathcal{M}_t^n) \leq 0. \tag{15.78}$$

In fact, the inequality is strict unless $\{\mathcal{M}_t^n\}_{t\in[0,T)}$ is a self-similar shrinking solution to the flow (shrinking to the origin).

Proof. We compute in the case $\alpha \neq 1$, and leave the case $\alpha = 1$ as an exercise (see Exercise 15.5). Then we have

$$\frac{d}{dt}\mathrm{Vol}(\mathcal{M}_t^n) = -\int_{\mathcal{M}_t} K^\alpha\, d\mu = -\int_{S^n} K^{\alpha-1}\, d\bar{\mu},$$

and we also have

$$\frac{d}{dt}\int_{S^n} \sigma^{\frac{\alpha-1}{\alpha}}\, d\bar{\mu} = -\frac{\alpha-1}{\alpha}\int_{S^n} \sigma^{-\frac{1}{\alpha}} K^\alpha\, d\bar{\mu}.$$

Consider the function $f \doteqdot \frac{\sigma}{K^\alpha}$. Note that f is constant (in space) if and only if the solution is shrinking self-similar. We also let $dm \doteqdot K^{\alpha-1}d\bar{\mu}$. Then, by the volume formula (5.40), we have

$$\mathrm{Vol}(\mathcal{M}_t^n) = \frac{1}{n+1}\int_{S^n} f\, dm \quad\text{and}\quad \int_{S^n} \sigma^{\frac{\alpha-1}{\alpha}}\, d\bar{\mu} = \int_{S^n} f^{\frac{\alpha-1}{\alpha}}\, dm,$$

while

$$\int_{S^n} K^{\alpha-1}\, d\bar{\mu} = -\int_{S^n} dm \quad\text{and}\quad \int_{S^n} \sigma^{-\frac{1}{\alpha}} K^\alpha\, d\bar{\mu} = \int_{S^n} f^{-\frac{1}{\alpha}}\, dm.$$

Thus,

$$\begin{aligned}\frac{d}{dt}\log \mathcal{E}_\alpha(\mathcal{M}_t) &= -\frac{\int_{S^n} f^{-\frac{1}{\alpha}}\,dm}{\int_{S^n} f^{\frac{\alpha-1}{\alpha}}\,dm} + \frac{\int_{S^n} dm}{\int_{S^n} f\,dm}\\ &= -\frac{\int_{S^n}\int_{S^n}(f(x)-f(y))\left(f(y)^{-\frac{1}{\alpha}} - f(x)^{-\frac{1}{\alpha}}\right)dm(x)\,dm(y)}{2\int_{S^n} f\,dm\int_{S^n} f^{\frac{\alpha-1}{\alpha}}\,dm}.\end{aligned}$$

The claims follow since the integrand is manifestly nonnegative and vanishes only if $f(x) = f(y)$ for every $x \neq y$. □

The Firey entropy $\mathcal{E}_\alpha$ is central to the proof that rescaled solutions converge to self-similar solutions, which we present in Section 17.2.

15.6.3. The Gaußian entropy.

Define the **Gaußian entropy** $E(\mathcal{M}^n)$ of a uniformly convex hypersurface $\mathcal{M}^n \subset \mathbb{R}^{n+1}$ by

$$E\left(\mathcal{M}^n\right) \doteqdot \left(\frac{\mathrm{Vol}(\mathcal{M}^n)}{|B^{n+1}|}\right)^{\frac{n}{n+1}} \exp\left(\frac{1}{|S^n|}\int_{\mathcal{M}^n} K\log K\,d\mu\right). \tag{15.79}$$

We saw in Section 4.3 that this entropy (introduced by Michael Gage and Richard Hamilton [**228**, equation 4.3.4]) is monotone under the 1-dimensional Gauß curvature flow — the curve shortening flow. It turns out that it is in fact monotone under the Gauß curvature flow in all dimensions.

Proposition 15.23. *The Gaußian entropy $E(\mathcal{M}_t^n)$ of a uniformly convex solution $\{\mathcal{M}_t^n\}_{t\in[0,T)}$ to Gauß curvature flow satisfies*

$$\frac{d}{dt}E(\mathcal{M}_t^n) \leq 0. \tag{15.80}$$

Proof. Under the Gauß curvature flow,

$$\partial_t\left(K\,d\mu\right) = K(\mathrm{II}^{-1})^{ij}\nabla_i\nabla_j K\,d\mu. \tag{15.81}$$

It follows that

$$\begin{aligned}F(t) \doteqdot \frac{d}{dt}\int_{\mathcal{M}_t^n} K\log K\,d\mu_t &= \int_{\mathcal{M}_t}(K(\mathrm{II}^{-1})^{ij}\nabla_i\nabla_j K + HK^2)\,d\mu\\ &\quad + \int_{\mathcal{M}_t}\log K\,K(\mathrm{II}^{-1})^{ij}\nabla_i\nabla_j K\,d\mu\\ &= \int_{\mathcal{M}_t} pK\,d\mu,\end{aligned} \tag{15.82}$$

where p is defined in (15.40) and to obtain the last equality we integrated by parts and applied the identity $\nabla_i(K(\mathrm{II}^{-1})^{ij}) = 0$.

Recalling (15.42) and applying (15.81), we find

$$\frac{dF}{dt} = \int_{\mathcal{M}} \left(|P|^2_{\mathrm{II}} + p^2\right) K d\mu\,, \tag{15.83}$$

where the tensor P is defined in (15.39) and $|P|_{\mathrm{II}}$ is defined by (15.43).

Recalling (15.44) and applying Hölder's inequality, we conclude that

$$\frac{dF}{dt} \geq \frac{n+1}{n}\int_{\mathcal{M}} p^2 K d\mu \geq \frac{n+1}{n}\left(\int_{\mathcal{M}} pK\, d\mu\right)^2 \Big/ \int_{\mathcal{M}} K d\mu = \frac{n+1}{n\sigma_n}F^2,$$

where σ_n is the surface area of the unit n-sphere.

We may assume that T is the singular time. For every $\varepsilon > 0$ consider the solution to the ODE $\frac{d\phi_\varepsilon}{dt} = \frac{n+1}{n\sigma_n}\phi_\varepsilon^2$ with $\lim_{t\to T-\varepsilon}\phi_\varepsilon(t) = +\infty$. We have

$$\phi_\varepsilon(t) = \frac{n\sigma_n}{n+1}\frac{1}{T-\varepsilon-t}.$$

Thus $F(t) \leq \frac{n\sigma_n}{n+1}\frac{1}{T-\varepsilon-t}$ for all $\varepsilon > 0$. Integrating this and taking the limit as $\varepsilon \to 0$, we obtain

$$\int_{\mathcal{M}_t^n} K\log K\, d\mu_t \leq \int_{\mathcal{M}_0^n} K \log K\, d\mu_0 + \frac{n\sigma_n}{n+1}\log\left(\frac{T}{T-t}\right) \tag{15.84}$$

for $t \in (0,T)$. By (15.12), the enclosed volume satisfies

$$\frac{d\,\mathrm{Vol}(\mathcal{M}_t^n)}{dt} = -\int_{\mathcal{M}} K d\mu = -\sigma_n.$$

Thus, by Chou's long-time existence theorem (Theorem 15.3),

$$\mathrm{Vol}(\mathcal{M}_t^n) = \sigma_n\,(T-t)\,. \tag{15.85}$$

Combining (15.84) and (15.85) yields the claim. □

For Ricci flow there is a well-known relation between the **Nash entropy** (or **Boltzmann entropy**) and **Perelman's $\mathcal{W}$-entropy** (for an exposition, see [**163**, §6.1.2]). For the Gauß curvature flow one may think of $E(\mathcal{M})$ defined above as playing the role of the Nash entropy. With this in mind, one may ask if there is a $\mathcal{W}$-entropy for the Gauß curvature flow. In Guo, Philipowski, and Thalmaier [**256**] an entropy is defined by

$$\mathcal{W}(\mathcal{M}_t) = \int_{\mathcal{M}_t} (\tau(t)p - (n+1)\log K - n\log\tau(t))K d\mu\,, \tag{15.86}$$

where $\tau(t) > 0$ is any solution to $\frac{d\tau}{dt} = -(n+1)$.

They proved under the Gauß curvature flow the following entropy formula:

$$\frac{d\mathcal{W}}{dt}(t) = \tau(t)\int_{\mathcal{M}_t}\left(\left|P - \frac{1}{\tau(t)}\mathrm{II}\right|^2_{\mathrm{II}} + \left(p - \frac{n}{\tau(t)}\right)^2\right) K d\mu\,. \tag{15.87}$$

See Exercise 15.7 for the proof. Note that $P - \frac{1}{\tau(t)}\mathrm{II}$ vanishes on shrinking self-similar solutions to the Gauß curvature flow (cf. Lemma 15.12).

15.6.4. The general Gaußian entropies. The Gaußian entropy for $\alpha = 1$ generalizes to other values of α as follows: The α-**Gaußian entropy** of a bounded convex hypersurface $\mathcal{M}^n \subset \mathbb{R}^{n+1}$ is defined by

$$E_\alpha(\mathcal{M}) \doteqdot \begin{cases} \left(\dfrac{\mathrm{Vol}(\mathcal{M}^n)}{|B^{n+1}|}\right)^{\frac{n}{n+1}} \exp\left(\dfrac{1}{|S^n|}\displaystyle\int_{\mathcal{M}^n} K \log K \, d\mu\right) & \text{if } \alpha = 1, \\ \left(\dfrac{\mathrm{Vol}(\mathcal{M}^n)}{|B^{n+1}|}\right)^{\frac{n}{n+1}} \left(\dfrac{1}{|S^n|}\displaystyle\int_{\mathcal{M}^n} K^\alpha d\mu\right)^{\frac{1}{\alpha-1}} & \text{if } \alpha \neq 1. \end{cases} \tag{15.88}$$

The following proposition is a special case of a more general result (Theorem 18.34) which we present in Section 18.8. We present the argument here for Gauß curvature flows, where there is an interesting connection to the Brunn–Minkowski inequality.

Proposition 15.24. *Under the α-Gauß curvature flow of a uniformly convex hypersurface with $\alpha \neq 1$, the normalized Gaußian entropy $\tilde{E}_\alpha(\mathcal{M}_t)$ is a nonincreasing function of time and strictly decreasing unless the solution is a self-similar shrinking solution.*

Proof. We first compute the evolution of the Gaußian entropy under the α-Gauß curvature flow, using the Gauß map parametrization. In this setting the entropies take the form

$$E_\alpha(\mathcal{M}) = \begin{cases} \left(\dfrac{\mathrm{Vol}(\mathcal{M}^n)}{|B^{n+1}|}\right)^{\frac{n}{n+1}} \exp\left(-\dfrac{1}{|S^n|}\displaystyle\int_{S^n} \log\det A \, d\bar{\mu}\right) & \text{if } \alpha = 1, \\ \left(\dfrac{\mathrm{Vol}(\mathcal{M}^n)}{|B^{n+1}|}\right)^{\frac{n}{n+1}} \left(\dfrac{1}{|S^n|}\displaystyle\int_{S^n} (\det A)^{1-\alpha} d\bar{\mu}\right)^{-\frac{1}{1-\alpha}} & \text{if } \alpha \neq 1, \end{cases} \tag{15.89}$$

where

$$A = A[\sigma] \doteqdot \overline{\nabla}^2\sigma + \sigma\bar{g}. \tag{15.90}$$

Since $\bar{g} = (\mathrm{G}^{-1})^*\mathrm{II}$, where G is the Gauß map, (5.40), (15.12), and the change of variables formula yield the identities

$$\mathrm{Vol}(\mathcal{M}_t) = \frac{1}{n+1}\int_{S^n} \sigma \det A \, d\bar{\mu} \quad \text{and} \quad \frac{d}{dt}\mathrm{Vol}(\mathcal{M}_t) = -\int_{S^n} F \det A \, d\bar{\mu},$$

where $F = (\det A)^{-\alpha}$, we obtain

$$\partial_t \log E_\alpha = \frac{\int_{S^n} F \det A \left(A^{-1}\right)^{ij} A[F]_{ij} \, d\bar{\mu}}{\int_{S^n} F \det A \, d\bar{\mu}} - n\frac{\int_{S^n} F \det A \, d\bar{\mu}}{\int_{S^n} \sigma \det A \, d\bar{\mu}}.$$

To show that the entropy is decreasing, we use the **Brunn–Minkowski inequality**, which states that

$$\text{(15.91)} \qquad \mathrm{Vol}(A+B)^{\frac{1}{n+1}} \leq \mathrm{Vol}(A)^{\frac{1}{n+1}} + \mathrm{Vol}(B)^{\frac{1}{n+1}}$$

for any pair of compact, convex bodies A and B, where $A+B$ denotes the **Minkowski sum**,

$$A + B \doteqdot \{a + b : a \in A,\ b \in B\}.$$

Applying this inequality with B replaced by a dilation of B, we find that $\mathrm{Vol}(A+tB)^{\frac{1}{n+1}}$ is a concave function. We conclude that the second derivative with respect to t at $t = 0$ is nonpositive. Since the support function of $A+tB$ is given by $\sigma_A + t\sigma_B$, we have

$$\frac{d}{dt}\mathrm{Vol}(A+tB) = \int_{S^n} F \det A \, d\bar{\mu},$$

where $F = \sigma_B$ and $A = A[\sigma_A]$. Differentiating further gives

$$\frac{d^2}{dt^2}\mathrm{Vol}(A+tB) = \int_{S^n} F \det A \left(A^{-1}\right)^{ij} A[F]_{ij} \, d\bar{\mu}.$$

Thus the Brunn–Minkowski inequality implies that

$$\text{(15.92)} \qquad \int_{S^n} F \det A \left(A^{-1}\right)^{ij} A[F]_{ij} \, d\bar{\mu} \leq n \frac{\left(\int_{S^n} F \det A \, d\bar{\mu}\right)^2}{\int_{S^n} \sigma \det A \, d\bar{\mu}},$$

whenever σ and F are support functions of smooth, uniformly convex regions. But now we observe that this inequality holds also for arbitrary smooth functions F, since then $\tilde{F} = F + C\sigma$ is the support function of a smooth, uniformly convex body for sufficiently large C, and the difference of the left and right sides of the above inequality is unchanged if we replace F by $\tilde{F}$ (see Exercise 15.8).

Comparing (15.92) with the evolution of E_α computed above, we conclude that E_α is nonincreasing under the α-Gauß curvature flow.

To obtain the characterization of the equality case, we argue as follows: Recall that

$$\overline{\nabla}_i \left(\det A \left(A^{-1}\right)^{ij}\right) = 0.$$

Define a symmetric bilinear form B on $H^1(S^n) = W^{1,2}(S^n)$ by

$$\begin{aligned} B(f,g) &= \int_{S^n} f \det A \left(A^{-1}\right)^{ij} A[g]_{ij} \, d\bar{\mu} \\ &= -\int_{S^n} \det A \left(A^{-1}\right)^{ij} \overline{\nabla}_i f \, \overline{\nabla}_j g \, d\bar{\mu} + \int_{S^n} fg \det A \left(A^{-1}\right)^{ij} \bar{g}_{ij} \, d\bar{\mu}. \end{aligned}$$

Associated to B is the elliptic operator

$$\mathcal{L}[f] \doteqdot \sigma \left(A^{-1}\right)^{ij} A[f]_{ij},$$

which satisfies

$$B(f,g) = \langle \mathcal{L}[f], g\rangle_\sigma = \langle f, \mathcal{L}[g]\rangle_\sigma,$$

where $\langle \cdot, \cdot \rangle_\sigma$ is the inner product in $L^2_\sigma \doteqdot L^2\left(\frac{\det A[\sigma]}{\sigma}\, d\bar{\mu}\right)$. Thus there exists an orthonormal basis of eigenfunctions of $\mathcal{L}$ for L^2_σ. The first (and the unique positive) eigenfunction of $\mathcal{L}$ is σ, with eigenvalue n. The inequality (15.92) is equivalent to the statement $B(f,f) \leq 0$ whenever f is orthogonal to σ in L^2_σ. To complete the argument we need the following:

Lemma 15.25. *The null space of $\mathcal{L}$ is the $(n+1)$-dimensional space consisting of the functions $f(z) = \langle p, z\rangle$ for $p \in \mathbb{R}^{n+1}$ (i.e., the restrictions of linear functions to S^n).*

Proof. For f as given in the lemma, we have

$$X(z) \doteqdot fz + \overline{\nabla} f = p,$$

and hence $0 = \partial_i X = A[f]_{ik}\bar{g}^{kl}\partial_l z$, implying that $A[f] = 0$ and hence $\mathcal{L}[f] = 0$.

We must show that these are the only elements of the null space of $\mathcal{L}$. Suppose that f satisfies $\mathcal{L}[f] = 0$, or equivalently $\left(A^{-1}\right)^{ij} A[f]_{ij} = 0$. Then we compute

$$\begin{aligned} 0 &= \int_{S^n} f \det A \left(A^{-1}\right)^{ij} A[f]_{ij}\, d\bar{\mu} \\ &= \int_{S^n} f Q^{ij} A[\sigma]_{ij}\, d\bar{\mu} \\ &= \int_{S^n} \sigma Q^{ij} A[f]_{ij}\, d\bar{\mu} \\ &= \int_{S^n} \sigma \det A \left(A^{-1}\right)^{ik} \left(A^{-1}\right)^{jl} A[f]_{ij} A[f]_{kl}\, d\bar{\mu}, \end{aligned}$$

where we define

$$Q^{ij} = \det A \left(A^{-1}\right)^{ik} \left(A^{-1}\right)^{jl} A[f]_{kl}, \tag{15.93}$$

and use the identity $\overline{\nabla}_i Q^{ij} = 0$ (see Exercise 15.9). It follows that $A[f]_{ij} = 0$ everywhere on S^n and hence that $p = fz + \overline{\nabla} f$ is constant and $f = \langle p, z\rangle$ as claimed. □

Now we can complete the proof of the equality case for (15.92): It follows from the inequality (15.92) and the definition of $\mathcal{L}$ that $\frac{d}{dt}E_\alpha < 0$ unless $F = (\det A)^{-\alpha}$ is a multiple of σ plus an element of the null space of $\mathcal{L}$. By the lemma, this implies that

$$(\det A)^{-\alpha} = c\sigma + \langle z, p\rangle$$

for some $z \in \mathbb{R}^{n+1}$ and $c > 0$. The claim follows. □

15.7. Notes and commentary

The pinching estimate $|\kappa_2 - \kappa_1| \leq C$ derived in the proof of the Firey conjecture was later extended to $K^{\alpha-1}|\kappa_2 - \kappa_1| \leq C$ under the α-Gauß curvature flow for $1/2 < \alpha < 1$ in the case of surfaces ($n = 2$) by Xuzhong Chen and the first author [**46**]. Related ideas have been used for other flows in the 2-dimensional case [**462, 472**] but have not been effective in higher dimensons or for higher α in the case $n = 2$.

The inequality (15.92) is known as Minkowski's second inequality (in the case where F is the support function of a convex body). Without the assumption that σ is smooth and A_{ij} is positive definite, the analysis of the equality cases is much more difficult than in the case treated here (see the discussion in Chapter 7 of [**461**]). The identities proved in Exercises 15.2 and 15.9 are special cases of a more general identity (see Lemma 18.30) reflecting the total symmetry of mixed volumes.

The Firey and Gaußian entropies both have direct analogues for a natural class of anisotropic Gauß curvature flows, where the speed of motion has the form $\mu(\mathsf{N})K^\alpha$ for μ a positive smooth function.

15.8. Exercises

Exercise 15.1. Show that for any $\varepsilon \in \mathbb{R}$, X_ε given by (15.2) satisfies the α-Gauß curvature flow.

Exercise 15.2. Prove that

$$\overline{\nabla}_i(\det A(A^{-1})^{ij}) = 0.$$

Exercise 15.3. Give an alternate derivation of equation (15.51) using the definitions $\dot{K}^{ij} = \frac{\partial K}{\partial \mathrm{II}_{ij}}$ and $\ddot{K}^{k\ell mn} = \frac{\partial^2 K}{\partial \mathrm{II}_{k\ell}\partial \mathrm{II}_{mn}}$.

Exercise 15.4. Prove the second equality in (15.63).

Exercise 15.5. Prove Proposition 15.22 when $\alpha = 1$.

Exercise 15.6. Prove that the Gaußian and Firey entropies are each continuous and monotone nondecreasing with respect to α.

Exercise 15.7. Prove formula (15.86). Hint: Compute the difference of the right side of (15.86) and τ times the right side of (15.83), and write this as the time derivative of a suitable quantity.

Exercise 15.8. Let σ be the support function of a smooth, uniformly convex body, and let F be a smooth function on S^n. Let $\Delta[F]$ be the difference between the left and right sides of (15.92). Show that for any $C \in \mathbb{R}$, $\Delta[F + C\sigma] = \Delta[F]$.

Exercise 15.9. Show that $\overline{\nabla}_i Q^{ij} = 0$, where Q^{ij} is given by the expression (15.93) for an arbitrary smooth function f.

Exercise 15.10. Show that under the α-Gauß curvature flows, the function $t \mapsto \mathrm{Vol}(\mathcal{M}_t)^{\frac{1+n\alpha}{n+1}}$ is concave for $0 < \alpha \le 1$ and convex for $\alpha \ge 1$.

Exercise 15.11. Show that the time of existence $T(A)$ of the solution of Gauß curvature flow with initial hypersurface $M = \partial A$ is locally Lipschitz with respect to Hausdorff distance on the set of open bounded convex sets with smooth, uniformly convex boundary. HINT: Show that if the origin is chosen suitably, there exists C such that $\max\{1-Cd, 0\}A \subset B \subset (1+Cd)A$ whenever B is a convex set within Hausdorff distance d of A. Use the homogeneity of the α-Gauß curvature flow, the comparison theorem, and Chou's theorem to estimate $T(B)$.

Exercise 15.12. We say that a family of open bounded convex regions $\{\Omega_t : 0 \le t < T\}$ forms a **viscosity solution** of α-Gauß curvature flow if it respects comparison with smooth solutions: That is, if $\{M_t = \partial A_t\}_{t\in[t_0,t_1]}$ is any smooth, uniformly convex solution of α-Gauß curvature flow with $[t_0, t_1] \subset [0, T)$ and $A_{t_0} \subset \Omega_{t_0}$ (respectively, $\Omega_{t_0} \subset A_{t_0}$), then $A_t \subset \Omega_t$ for each $t \in [t_0, t_1]$ (respectively, $\Omega_t \subset A_t$ for each $t \in [t_0, t_1]$). By Chou's theorem, for each bounded open convex set A with smooth, uniformly convex boundary there exists a unique solution of α-Gauß curvature flow with initial hypersurface ∂A, which consists of smooth hypersurfaces bounding convex regions which we denote by $G_{\alpha,t}(A)$ for $0 \le t < T(A)$. Given an open bounded convex set Ω_0, let $\mathcal{K}^+(\Omega)$ be the set of open bounded convex regions with smooth, uniformly convex boundary which contain Ω_0. Define

$$A_t = \bigcap_{A\in\mathcal{K}^+(\Omega_0),\ T(A)>t} G_{\alpha,t}(A).$$

Show that A_t is the unique viscosity solution of α-Gauß curvature flow for which A_t approaches Ω_0 in Hausdorff distance as t approaches zero.

Exercise 15.13 (Flat sides persist for $\alpha > 1/n$)**.** Suppose $\alpha > \frac{1}{n}$, and define an open convex subset of $\mathbb{R} \times \mathbb{R}^n$ by

$$A = \{(x, y) \in \mathbb{R} \times \mathbb{R}^n : \ x > 0,\ |y|^2 + 2x - 1 - 2x^{\frac{n\alpha-1}{(n+1)\alpha-1}} < 0\}.$$

Show that there exists $\beta(\alpha, n) > 0$ such that $A_t = (1-\beta t)^{\frac{1}{1+n\alpha}} A$ is a viscosity subsolution of the α-Gauß curvature flow, in the sense that if $M_t = \partial\Omega_t$ is a smooth solution for $t \in [t_0, t_1]$ with $A_{t_0} \subset \Omega_{t_0}$, then $A_t \subset \Omega_t$ for $t \in [t_0, t_1]$. HINT: Show that $K^\alpha \le C\langle X, \mathsf{N}\rangle$ on $M = \partial A$ for some $C > 0$. We say that an open bounded convex region Ω has a flat side if there exist $p \in \mathbb{R}^{n+1}$, $e \in S^n$, and $r > 0$ such that $\Omega \subset \{X : \ \langle X - p, e\rangle > 0\}$ and $B_r(p) \cap \{X : \ \langle X - p, e\rangle = 0\} \subset \bar{\Omega}$. Prove that if Ω_0 has a flat side, then the same is true for Ω_t for small $t > 0$, where Ω_t is the viscosity solution of α-Gauß curvature flow constructed in the previous exercise.

Exercise 15.14 (Translating solutions). Consider a hypersurface M^n given by a graphical embedding of the form $X(x) = (x, u(|x|))$, where u is a smooth function. Show that the Gauß curvature K is given by

$$K = |x|^{1-n}(1 + (u')^2)^{-\frac{n+2}{2}}(u')^{n-1}u''.$$

Show that for any $\alpha > 0$ there exists a convex hypersurface of this form satisfying the equation $K^\alpha = -\langle \mathsf{N}, e_{n+1}\rangle$ by solving the corresponding ordinary differential equation with initial conditions $u = u' = 0$ when $|x| = 0$. Show that the resulting hypersurface is a graph over all of $\mathbb{R}^n$ if $\alpha \leq 1/2$ and it is a graph over a ball $B_R(0)$ if $\alpha > 1/2$. Show that $\lim_{|x|\to R} u(|x|) = \infty$ if $\alpha \leq 1$ and that $\lim_{|x|\to R} u(|x|) = U < \infty$ if $\alpha > 1$. In the latter case show that the union of M with the cylinder $\{(x, y) :\ |x| = R,\ y \geq U\}$ is a complete $C^{2+\frac{1}{\alpha-1}}$ hypersurface which satisfies $K^\alpha = -\langle \mathsf{N}, e_{n+1}\rangle$ everywhere and so generates a translating solution.

Exercise 15.15 (Solutions containing line segments). Let $\alpha > 1$. Let u be the function defined in the previous exercise. Define

$$\begin{aligned}\tilde{M}_t = \{(x, u(|x|) - U + t) :\ |x| < R\} &\cup \{(x, -u(|x|) + U - t) :\ |x| < R\} \\ &\cup \{(x, y) :\ |x| = R,\ t \leq y \leq -t\}.\end{aligned}$$

Show that $\tilde{M}_t$ is a solution of α-Gauß curvature flow for $t \leq 0$. Deduce that if Ω_0 is an open bounded convex set in $\mathbb{R}^{n+1}$ and if $M_0 = \partial\Omega_0$ is $C^{1,1}$ and contains a line segment $\{(1-s)x_- + sx_+ :\ 0 \leq s \leq 1\}$, then the viscosity solution $M_t = \partial\Omega_t$ with initial condition M_0 contains the subsegment $\{(1-s)x_- + sx_+ :\ 1/3 \leq s \leq 2/3\}$ for all sufficiently small $t > 0$.

Exercise 15.16. Let $0 < \alpha \leq 1$. Let Ω_0 be an open bounded convex set with boundary containing the cylinder $\{(x, y) :\ |x| = 1,\ |y| < 1\}$. Show that for any $t > 0$, the viscosity solution Ω_t of the α-Gauß curvature flow is contained in a cylinder $\{|x| \leq R\}$ for some $R < 1$. Hint: Use the translating solutions constructed in Exercise 15.14 as barriers.

Chapter 16

The Affine Normal Flow

Recall that uniformly convex hypersurfaces contract to points under flows by any positive power of the Gauß curvature, in accordance with Chou's theorem (Theorem 15.3). In dimensions 1 and 2, we have also seen that the asymptotic shape is a round circle/sphere when the power is 1 (Theorems 3.19 and 15.14). It is natural to ask whether this happens to be the case more generally; although the answer turns out to be no, this is due to a rich and beautiful phenomenon: As we shall see, the $\frac{1}{n+2}$-Gauß curvature flow is invariant under affine transformations. In particular, any affine transformation of the shrinking sphere solution is also a solution.

Recall that an **ellipsoid** centered at the origin in $\mathbb{R}^{n+1}$ is any set of points $X \in \mathbb{R}^{n+1}$ satisfying $A(X,X) = 1$ for some positive definite, symmetric bilinear form A on $\mathbb{R}^{n+1}$. The volume enclosed by this ellipsoid is equal to $\omega_n(\det A)^{-1/2}$, where ω_n is the volume of the unit n-ball, so we say that the ellipsoid is **normalized** if $\det A = 1$.

Theorem 16.1 (Andrews [**30**]). *Let $\mathcal{M}_0^n$ be a uniformly convex hypersurface in $\mathbb{R}^{n+1}$. The maximal solution $\{\mathcal{M}_t^n\}_{t\in[0,T)}$ to the $\frac{1}{n+2}$-Gauß curvature flow converges to a normalized ellipsoid after reparametrizing and rescaling about the final point.*

Let us be a little more precise about what we mean here: By Chou's theorem (Theorem 15.3) we know that a unique maximal solution exists, parametrized by a smooth 1-parameter family $X : M^n \times [0,T) \to \mathbb{R}^{n+1}$ of embeddings $X(\cdot,t) : M^n \to \mathbb{R}^{n+1}$, and converges to a point $p \in \mathbb{R}^{n+1}$ at the final time. Convergence to an ellipsoid after reparametrizing and rescaling about p means that there exists a smooth 1-parameter family $\varphi : M \times [0,T) \to M^n$ of diffeomorphisms $\varphi(\cdot,t)$ of M^n such that the rescaled embeddings $\hat{X}(\cdot,t) : M^n \to \mathbb{R}^{n+1}$ defined by

$$\hat{X}(x,t) = \frac{X(\varphi(x,t),t) - p}{\left(\frac{2(n+1)}{n+2}(T-t)\right)^{\frac{n+2}{2(n+1)}}} \tag{16.1}$$

converge in $C^\infty(M^n,\mathbb{R}^{n+1})$ to a limiting embedding $\hat{X}(\cdot,T)$ whose image is a normalized ellipsoid centered at the origin.

Since we cannot expect convergence to round points in general, the proof of the convergence theorem requires a rather different set of arguments from those used in previous sections. While for the mean curvature flow and the 2-dimensional Gauß curvature flow (and also the mean curvature-type flows we will study in Chapter 19), the convergence to a sphere can be proved using pointwise control of curvature quantities via the maximum principle, we will obtain convergence in this setting by using the monotonicity of the entropy functional from Proposition 15.24. Our argument is as follows: We first show that the evolution equation is invariant under affine transformations. Exploiting this invariance property, we obtain regularity results and conclude that the solutions, suitably modified by affine transformations, converge to a limit. The monotonicity of the entropy implies that this limit is a self-similar solution, and we complete the argument by using a symmetrization argument to prove that self-similar solutions are ellipsoids, following an argument due to W. Blaschke.

16.1. Affine invariance

The $\frac{1}{n+2}$-Gauß curvature flow is **affine invariant** in the following sense.

Proposition 16.2. *Let $X : M \times [0,T) \to \mathbb{R}^{n+1}$ be a smooth, uniformly convex solution of the $\frac{1}{n+2}$-Gauß curvature flow. Given $L \in \mathrm{SL}(n+1,\mathbb{R})$, there exists a unique smooth family of diffeomorphisms $\varphi : M\times[0,T) \to M$, with $\varphi(x,0) = x$, such that $\tilde{X}(x,t) = L(X(\varphi(x,t),t))$ is also a solution of the α-Gauß curvature flow.*

Proof. Fix a point $x \in M$ and a time $t \in [0,T)$, and choose an orthonormal basis $\{e_1,\dots,e_n\}$ for the tangent space to $\mathcal{M}_t$ at x. Also choose an orthonormal basis $\{\tilde{e}_1,\dots,\tilde{e}_n\}$ for the tangent space to $L(\mathcal{M}_t)$ at the image point. Let N be the unit normal to X at x and $\tilde{\mathsf{N}}$ the unit normal to $L(\mathcal{M}_t)$ at the image point.

With respect to these bases, we can write

$$L(e_i) = P_i^j\tilde{e}_j \quad\text{and}\quad L(\mathsf{N}) = R\tilde{\mathsf{N}} + Q^i\tilde{e}_i \tag{16.2}$$

for some P_i^j, Q^i, and R since L must take a tangent vector to $\mathcal{M}_t$ to a tangent vector to $L(\mathcal{M}_t)$. Since $\det L = 1$, we conclude that $R \det P = 1$.

Working in normal coordinates for $\mathcal{M}_t$ about x such that $\partial_i X = e_i$ at x, we find that

$$\tilde{g}_{ij} \doteqdot \langle \partial_i(L \circ X), \partial_j(L \circ X) \rangle = P_i^k P_j^k, \tag{16.3}$$

so that $\det \tilde{g} = (\det P)^2$. We also have

$$\begin{aligned} -\tilde{\mathrm{II}}_{ij} \tilde{\mathsf{N}} + \tilde{\Gamma}_{ij}^k L(e_k) = \partial_i \partial_j (L \circ X) &= L(\partial_i \partial_j X) \\ &= L(-\mathrm{II}_{ij} \mathsf{N}) = -\mathrm{II}_{ij}(R\tilde{\mathsf{N}} + Q^k \tilde{e}_k). \end{aligned}$$

Equating the $\tilde{\mathsf{N}}$ components, we find that

$$\tilde{\mathrm{II}}_{ij} = R\mathrm{II}_{ij}, \tag{16.4}$$

and taking determinants gives

$$\det \tilde{\mathrm{II}} = R^n \det \mathrm{II}.$$

Combining these expressions yields

$$\tilde{K} = \frac{\det \tilde{\mathrm{II}}}{\det \tilde{g}} = \frac{R^n \det \mathrm{II}}{(\det P)^2} = R^{n+2} K. \tag{16.5}$$

Finally, we compute that $\tilde{X}(x,t) = L(X(\varphi(x,t),t))$ satisfies

$$\begin{aligned} \partial_t \tilde{X} &= L(-K^{\frac{1}{n+2}} \mathsf{N} + \partial_t \varphi^i e_i) \\ &= -K^{\frac{1}{n+2}}(R\tilde{\mathsf{N}} + Q^k \tilde{e}_k) + \partial_t \varphi^i P_i^k \tilde{e}_k \\ &= -\tilde{K}^{\frac{1}{n+2}} \tilde{\mathsf{N}} + (\partial_t \varphi^i P_i^k - K^{\frac{1}{n+2}} Q^k) \tilde{e}_k. \end{aligned}$$

That is, $\tilde{X}$ is again a solution of the $\frac{1}{n+2}$-Gauß curvature flow provided we choose φ so that $\partial_t \varphi^i = K^{\frac{1}{n+2}} Q^k (P^{-1})_k^i$. □

Applying this to the shrinking sphere solution, we obtain the following:

Corollary 16.3. *Let $\mathcal{M}_0$ be the ellipsoid $\{x + p : \ x^T A x = 1\}$, where A is a positive definite symmetric matrix. Then the solution of the affine normal flow with initial data $\mathcal{M}_0$ is given by*

$$\mathcal{M}_t = \left\{ x + p : x^T A x = \left(1 - \tfrac{2(n+1)}{n+2} t\right)^{\frac{n+2}{n+1}} \right\}. \tag{16.6}$$

That is, every ellipsoid is a self-similar shrinking solution of the affine normal flow.

There is a natural explanation for this invariance in the setting of affine differential geometry: There is an affine-invariant vector called the **affine normal vector**, which has the form

$$\mathcal{N} = -K^{\frac{1}{n+2}}\mathsf{N} + a^i\partial_i X, \tag{16.7}$$

where a^i depends on the second fundamental form and the gradient of the Gauß curvature. Thus, up to a tangential term (which merely generates a time-dependent reparametrization), the $\frac{1}{n+2}$-Gauß curvature flow is the same as the motion in the direction of the affine normal vector. We will therefore often refer to the $\frac{1}{n+2}$-Gauß curvature flow as the **affine normal flow** although, strictly speaking, it differs by a time-dependent reparametrization.

It will also be important to use the fact that the entropies associated to this flow are invariant under affine transformations:

Proposition 16.4. *The Firey entropy $\mathcal{E}_{\frac{1}{n+2}}$ defined by* (15.77) *is invariant under linear transformations of $\mathbb{R}^{n+1}$, and the Gaußian entropy $E_{\frac{1}{n+2}}$ defined by* (15.88) *is invariant under affine transformations of $\mathbb{R}^{n+1}$. That is, if $\mathcal{M}$ is a smooth uniformly convex hypersurface enclosing the origin and $L \in \mathrm{GL}(n+1,\mathbb{R})$, then*

$$\mathcal{E}_{\frac{1}{n+2}}[L(\mathcal{M})] = \mathcal{E}_{\frac{1}{n+2}}[\mathcal{M}]; \tag{16.8}$$

and for any $p \in \mathbb{R}^{n+1}$,

$$E_{\frac{1}{n+2}}[L(\mathcal{M})+p] = E_{\frac{1}{n+2}}[\mathcal{M}]. \tag{16.9}$$

Proof. Both entropies are invariant under dilations, and E is invariant under translations since it depends only on curvatures and not directly on the support function. Thus it remains only to check that both entropies are invariant under the action of $\mathrm{SL}(n+1,\mathbb{R})$ transformations.

For the Gaußian entropy we can use the calculations above: If $\tilde{X} = L \circ X$ with $L \in \mathrm{SL}(n+1,\mathbb{R})$, we have $\tilde{K} = R^{n+2}K$; and $\det\tilde{g} = (\det P)^2 \det g = R^{-2}\det g$, so that $d\mu(\tilde{g}) = R^{-1}d\mu(g)$. Since L preserves the enclosed volume, the invariance of the Gaußian entropy is a consequence of the following:

$$\int_{\tilde{\mathcal{M}}} \tilde{K}^{\frac{1}{n+2}}\,d\mu(\tilde{g}) = \int_{\mathcal{M}} \left(RK^{\frac{1}{n+2}}\right)\left(R^{-1}d\mu(g)\right) = \int_{\mathcal{M}} K^{\frac{1}{n+2}}\,d\mu(g).$$

To complete the argument for the Firey entropy, we require the transformation formula for the support function: Writing $X = \sigma \mathsf{N} + a^i e_i$, we find

$$\begin{aligned}
\tilde{\sigma} &= L(X) \cdot \tilde{\mathsf{N}} \\
&= L(\sigma \mathsf{N} + a^i e_i) \cdot \tilde{\mathsf{N}} \\
&= \left(\sigma(R\tilde{\mathsf{N}} + Q^i \tilde{e}_i) + a^i P_i^j \tilde{e}_j\right) \cdot \tilde{\mathsf{N}} \\
&= R\sigma.
\end{aligned}$$

Using this, we conclude that

$$\begin{aligned}
\int_{S^n} \tilde{\sigma}^{-(n+1)}\, d\bar{\mu} &= \int_{\tilde{\mathcal{M}}} \tilde{\sigma}^{-(n+1)} \tilde{K}\, d\mu(\tilde{g}) \\
&= \int_{\mathcal{M}} \left(R^{-(n+1)} \sigma^{-(n+1)}\right) \left(R^{n+2} K\right) \left(R^{-1} d\mu(g)\right) \\
&= \int_{\mathcal{M}} \sigma^{-(n+1)} K\, d\mu(g) \\
&= \int_{S^n} \sigma^{-(n+1)}\, d\bar{\mu}.
\end{aligned}$$

□

These entropies correspond to important invariants in affine differential geometry. Recall that the **polar dual** of a hypersurface $\mathcal{M}$ is

(16.10) $$\mathcal{M}^* = \partial\{x \in \mathbb{R}^{n+1} : \langle x, y\rangle \leq 1 \text{ for all } y \in \mathcal{M}\}.$$

The integral $\int_{S^n} \sigma^{-(n+1)}\, d\bar{\mu}$ appearing in the Firey entropy is the volume of the polar dual of the hypersurface. Given a compact, convex hypersurface $\mathcal{M}$, its **volume product** $P(\mathcal{M})$ is the minimum over all x in the interior of $\mathcal{M}$ of Area($\mathcal{M}$) Area(($\mathcal{M} - x)^*$). Recall also that the **Blaschke–Santaló inequality** says that (see Schneider [**461**] or Gardner [**235**])

(16.11) $$P(\mathcal{M}) \leq P(S^n),$$

with equality if and only if $\mathcal{M}$ is an ellipsoid. The Blaschke–Santaló inequality implies that, provided we translate to minimize the volume of the polar dual, the Firey entropy is minimized by ellipsoids. The integral $\int_{\mathcal{M}} K^{\frac{1}{n+2}}\, d\mu$ appearing in the Gaußian entropy is the **affine surface area**, and the *affine isoperimetric inequality* states that this is maximized by ellipsoids among hypersurfaces with fixed enclosed volume. This implies that the Gaußian entropy is minimized by ellipsoids.

16.2. Affine-renormalized solutions

The result proved above concerning the affine invariance of the affine normal flow allows a very easy proof of convergence of the evolving hypersurfaces after suitable rescaling. A key observation is the following:

Theorem 16.5 (F. John [**320**], 1948). *For any closed convex hypersurface $\mathcal{M}$ there exists a unique ellipsoid E of maximal volume contained in the compact region bounded by $\mathcal{M}$. Furthermore $\mathcal{M}$ is contained in the region bounded by the ellipsoid $\tilde{E}$ given by scaling E by a factor $(n+1)$ about its center. We call E the* ***John ellipsoid*** *of $\mathcal{M}$.*

We use this to perform an affine transformation to keep the geometry controlled, allowing relatively straightforward control of the higher regularity.

Given $\bar{t} \in [0, T)$, we define

$$X_{\bar{t}}(x,t) \doteqdot L_{\bar{t}}(X(\varphi(x,t), \bar{t} + (\det L_{\bar{t}})^{-\frac{2}{n+2}} t)),$$

where $L_{\bar{t}}$ is an affine transformation which takes the John ellipsoid of $\mathcal{M}_{\bar{t}}$ to the unit sphere centered at the origin and φ is a suitably chosen smooth family of diffeomorphisms. Thus the definition of $X_{\bar{t}}$ combines a volume-preserving affine transformation (and an accompanying reparametrization) as in Proposition 16.2 with a rescaling of the form

$$X(x,t) \mapsto rX(x, r^{-\frac{2(n+1)}{n+2}} t).$$

A direct calculation shows that this rescaling takes solutions to solutions, so $X_{\bar{t}}$ is again a solution of the $\frac{1}{n+2}$-Gauß curvature flow. By construction, the John ellipsoid of $X_{\bar{t}}(\cdot, 0)$ is the unit sphere.

From this we conclude (using Chou's theorem) that the interval of existence of the solution $X_{\bar{t}}$ is at least $\frac{n+2}{2(n+1)}$ since this is the interval of existence of the solution starting from the unit sphere, which remains enclosed by $\mathcal{M}_{\bar{t},t} \doteqdot X_{\bar{t}}(M,t)$ by the comparison principle. If we restrict ourselves to the time interval $[0, \frac{n+2}{4(n+1)}]$, then the solution hypersurface is contained between spheres of radii $n+1$ (by John's theorem) and $2^{-\frac{n+2}{2(n+1)}}$ (the smallest radius of the shrinking sphere solution on this time interval).

16.2.1. Gauß curvature bound. Now we use Chou's argument to obtain bounds on the Gauß curvature of the solution $X_{\bar{t}}$, but we must refine the argument slightly in order to obtain bounds that are independent of $\bar{t}$: Let $\sigma_{\bar{t}}$ be the support function of the solution $X_{\bar{t}}$. The last paragraph of the previous subsection shows that

$$\frac{1}{C} \leq \sigma_{\bar{t}}(z,t) \leq C \tag{16.12}$$

on $S^n \times [0, \frac{n+2}{4(n+1)}]$, where C is a constant depending only on n.

Recall the evolution equation (15.23) for $\varphi = \frac{K^\alpha}{\sigma - r_0}$:

$$\partial_\tau \varphi = \overline{\mathcal{L}}\varphi - \frac{2\alpha K^\alpha}{\sigma - r_0} A^{-1}(\overline{\nabla}\sigma, \overline{\nabla}\varphi) + (1 + n\alpha - r_0\alpha H)\varphi^2 .$$

Here we can choose $r_0 = \frac{1}{2C}$, guaranteeing that $\sigma \geq 2r_0$ on $S^n \times [0, \frac{n+2}{4(n+1)}]$. Now we observe that $H \geq nK^{\frac{1}{n}} \geq n(\sigma - r_0)^{\frac{1}{n\alpha}}\varphi^{\frac{1}{n\alpha}} \geq nr_0^{\frac{1}{n\alpha}}\varphi^{\frac{1}{n\alpha}}$, so that

$$\partial_t \varphi \leq \overline{\mathcal{L}}\varphi - \frac{2\alpha K^\alpha}{\sigma - r_0} A^{-1}(\overline{\nabla}\sigma, \overline{\nabla}\varphi) + \left(1 + n\alpha - n\alpha r_0^{1+\frac{1}{n\alpha}}\varphi^{\frac{1}{n\alpha}}\right)\varphi^2.$$

It follows that $\varphi \leq \max\left\{\left(\frac{2(1+n\alpha)}{n\alpha}\right)^{n\alpha} r_0^{-(1+n\alpha)}, \left(\frac{2}{1+n\alpha}\right)^{\frac{n\alpha}{1+n\alpha}} r_0^{-1} t^{-\frac{n\alpha}{1+n\alpha}}\right\}$ since the right-hand side is a supersolution of the same evolution equation. In particular, since $\sigma - r_0$ is uniformly bounded above, this gives a bound of the form $K^\alpha \leq \tilde{C}$ on $S^n \times [\frac{n+2}{8(n+1)}, \frac{n+2}{4(n+1)}]$, where $\tilde{C}$ depends only on n.

We remark that the same argument works for other α-Gauß curvature flows, provided we have upper and lower bounds on the support function, resulting in bounds on the Gauß curvature depending only on n, α, the constant C in the bound (16.12), and the duration of the time interval on which this holds. This will be useful in our later investigations.

16.2.2. Gauß curvature lower bound. The next step in the argument is to derive a lower bound on the Gauß curvature for the solutions $X_{\bar{t}}$, independent of $\bar{t}$. This follows by combining the Harnack inequality with a comparison argument with suitable enclosing spheres.

First, comparison with spheres produces the following result:

Proposition 16.6. *For $\alpha < 1/n$ the following holds for any smooth, strictly convex solution of the α-Gauß curvature flow:*

$$\sigma(z,t) \leq \sigma(z,0) - C(n,\alpha) r_+(\mathcal{M}_0)^{-\frac{2n\alpha}{1-n\alpha}} t^{\frac{1}{1-n\alpha}} \tag{16.13}$$

for all $(z,t) \in S^n \times (0,T)$.

Proof. Fix $z \in S^n$. Then $\mathcal{M}_0$ is enclosed in a hemispherical region obtained by intersecting the sphere of radius $2r_+(\mathcal{M}_0)$ centered at $\bar{x}(z)$ with the half-space $\{y \in \mathbb{R}^{n+1} : \langle y, z\rangle \leq \sigma(z)\}$. For any $\varepsilon < 2r_+(\mathcal{M})$ this hemispherical region is enclosed by the sphere S_ε of radius $(\varepsilon^2 + 4r_+(\mathcal{M})^2)/(2\varepsilon)$ centered at the point $\bar{x}(z) - (4r_+(\mathcal{M})^2 - \varepsilon^2)/(2\varepsilon)z$ (this sphere is chosen to have support function in direction z equal to $\sigma(z) + \varepsilon$). We consider the evolution of these spheres for suitably small ε: Since $\rho \geq \inf_{S^n} \rho$, a sphere of radius r evolves in time t to be contained inside a sphere of radius $\left(r^{1+n\alpha} - (1+n\alpha)t\right)^{1/(1+n\alpha)}$ about the same center, which is enclosed by the sphere of radius $r - \inf \rho r^{-n\alpha} t$. In particular, this applies for each of the spheres S_ε, and by the comparison principle $\mathcal{M}_t$ is also enclosed by this smaller sphere. This implies the inequality

$$\begin{aligned}\sigma(z,t) - \sigma(z,0) &\leq \varepsilon - \left(\frac{\varepsilon^2 + 4r_+(\mathcal{M})^2}{2\varepsilon}\right)^{-n\alpha} t \\ &\leq \varepsilon\left(1 - (4r_+(\mathcal{M})^2)^{-n\alpha}\varepsilon^{n\alpha-1}t\right).\end{aligned}$$

In particular, choosing

$$\varepsilon = \left(\frac{t}{2^{1+2n\alpha} r_+(\mathcal{M}_0)^{2n\alpha}}\right)^{1/(1-n\alpha)},$$

we obtain

$$\sigma(z,t) - \sigma(z,0) \leq -\left(\frac{1}{2^{1+2n\alpha} r_+(\mathcal{M}_0)^{2n\alpha}}\right)^{1/(1-n\alpha)} t^{1/(1-n\alpha)}$$

for $t \leq C(n,\alpha) r_+(\mathcal{M}_0)^{1+n\alpha}$. □

Combining this with the Harnack inequality, we deduce the following lower bound on the Gauß curvature:

Proposition 16.7. *For $\alpha < 1/n$ the following holds for any smooth, strictly convex solution of α-Gauß curvature flow:*

$$K^\alpha \geq C'(n,\alpha) r_+(\mathcal{M}_0)^{-\frac{2n\alpha}{1-n\alpha}} \cdot t^{\frac{n\alpha}{1-n\alpha}} \tag{16.14}$$

for $0 < t < C''(n,\alpha) r_+(\mathcal{M}_0)^{1+n\alpha}$.

Proof. The estimate on the change in the support function can be converted to an estimate on the speed using the Harnack estimate: Applying the estimate on the time interval $[t/2, t]$, we have

$$\begin{aligned}C r_+(\mathcal{M}_0)^{-\frac{2n\alpha}{1+n\alpha}} t^{\frac{1}{1-n\alpha}} &\leq \sigma(z,t/2) - \sigma(z,t) \\ &= \int_{t/2}^t K(z,\tau)^\alpha d\tau \\ &\leq t/2 \sup_{\tau\in[t/2,t]} (K^\alpha).\end{aligned}$$

The Harnack estimate then gives

$$t^{\frac{n\alpha}{n\alpha+1}} K(z,t)^\alpha \geq (t/2)^{\frac{n\alpha}{n\alpha+1}} \sup_{[t/2,t]} K^\alpha$$

and hence

$$K(z,t)^\alpha \geq C' r_+(\mathcal{M}_0)^{-\frac{2n\alpha}{1+n\alpha}} t^{1/(1-n\alpha)-1}.$$ □

In particular, in the case of the solutions $X_{\bar{t}}$ of the affine normal flow, we have uniform control on $r_+(\mathcal{M}_{\bar{t},0})$ and so obtain a bound from below on the Gauß curvature. Combining with the result of the previous section gives the following:

Corollary 16.8. *There exists $C > 0$ depending only on n such that for the solutions $X_{\bar{t}}$ we have*

$$C^{-1} \le K(x,t) \le C \tag{16.15}$$

for all $(x,t) \in M \times [\frac{n+2}{8(n+1)}, \frac{n+2}{4(n+1)}]$ and all $\bar{t} \in [0,T)$.

16.2.3. Radius of curvature bound. The next step is to obtain a lower bound on principal curvatures (equivalently, an upper bound on principal radii) independent of $\bar{t}$. This is similar to the argument used in the proof of Chou's theorem in Section 15.3.3, with some additional reasoning required in order to obtain a bound independent of $\bar{t}$.

Recall the evolution equation (15.28) for the tensor A_{ij}:

$$\partial_t A_{ij} \le \overline{\mathcal{L}} A_{ij} + (n\alpha - 1)(\det A)^{-\alpha}\overline{g}_{ij} - \alpha H (\det A)^{-\alpha} A_{ij}.$$

Applying this on the time interval $[\frac{n+2}{8(n+1)}, \frac{n+2}{4(n+1)}]$ where we have the estimate of Corollary 16.8, we observe that $\det A = r_1 \cdots r_n \ge r_{\max} r_{\min}^{n-1}$, so that $r_{\min}^{-1} \ge (\det A)^{-\frac{1}{n-1}} r_{\max}^{\frac{1}{n-1}}$, and so

$$H = \sum_i r_i^{-1} \ge r_{\min}^{-1} \ge (\det A)^{-\frac{1}{n-1}} r_{\max}^{\frac{1}{n-1}}.$$

Applying the evolution equation at a maximum eigenvector e_1 of A then gives (since $\alpha < 1/n$)

$$\partial_t A_{11} \le -\alpha C^{-\alpha - \frac{1}{n-1}} r_{\max}^{\frac{n}{n-1}}.$$

This gives the estimate

$$r_{\max} \le \left(\tfrac{n-1}{\alpha}\right)^{n-1} C^{1+\frac{\alpha}{n-1}} \left(t - \tfrac{n+2}{8(n+1)}\right)^{-\frac{1}{n-1}} \tag{16.16}$$

since we are applying the estimate on a time interval beginning at $t = \frac{n+2}{8(n+1)}$. In particular, if we restrict to the time interval $\left[\frac{3(n+2)}{16(n+1)}, \frac{n+2}{4(n+1)}\right]$, we have $r_{\max} \le C$, for some C depending only on n.

We remark that the argument (with minor modification) goes through for any α provided that the Gauß curvature is bounded above and below, obtaining a bound on the radius of curvature depending on n, α, the bounds on the Gauß curvature, and the duration of the time interval.

16.2.4. Higher regularity. The result of the previous section also implies a uniform upper bound on principal curvatures since $\kappa_{\max}\kappa_{\min}^{n-1} \le K$ and we have upper bounds on K and lower bounds on $\kappa_{\min} = r_{\max}^{-1}$.

We can now apply the argument of Section 15.3.4 to obtain bounds on $\sigma_{\bar{t}}$ in C^k for every k, independent of $\bar{t}$, on a slightly smaller time interval ($[\frac{7(n+2)}{32(n+1)}, \frac{n+2}{4(n+1)}]$, say).

16.3. Convergence and the limit flow

The previous sections construct a family of functions $\sigma_{\bar{t}} : S^n \times [a,b] \to \mathbb{R}$, where $[a,b] = [\frac{7(n+2)}{32(n+1)}, \frac{n+2}{4(n+1)}]$, for each $\bar{t} \in [0,T)$, with bounds in C^k for each k independent of $\bar{t}$. It follows from the Arzelà–Ascoli theorem and a diagonal sequence argument that there is a sequence $\bar{t}_k$ approaching T along which $\sigma_{\bar{t}_k}$ converges to a limit $\bar{\sigma}$ in the C^∞ topology. Since each $\sigma_{\bar{t}}$ satisfies the evolution equation (15.16), the same is true for $\bar{\sigma}$. That is, the limit $\bar{\sigma}$ is the support function of a smooth, uniformly convex solution $\bar{X} : M \times [a,b] \to \mathbb{R}^{n+1}$ of the affine normal flow.

The invariance of the Gaußian entropy implies, for each $t \in [a,b]$,

$$\begin{aligned}
E_{\frac{1}{n+2}}[\bar{X}(\cdot,t)] &= \lim_{k\to\infty} E_{\frac{1}{n+2}}[X_{\bar{t}_k}(\cdot,t)] \\
&= \lim_{k\to\infty} E_{\frac{1}{n+2}}[L_{\bar{t}_k}(X(\cdot,\bar{t}_k + (\det L_{\bar{t}_k})^{-\frac{2}{n+2}}t))] \\
&= \lim_{k\to\infty} E_{\frac{1}{n+2}}[X(\cdot,\bar{t}_k + (\det L_{\bar{t}_k})^{-\frac{2}{n+2}}t)] \\
&= \lim_{\tau\to T} E_{\frac{1}{n+2}}[X(\cdot,\tau)]\,,
\end{aligned}$$

where we used in the last line the fact that the entropy is monotone and therefore converges to a limit as $t \to T$ and that $\bar{t}_k + (\det L_{\bar{t}_k})^{-\frac{2}{n+2}}t$ is a sequence approaching T as k approaches infinity. Note that the quantity on the last line of the display above is a constant, independent of $t \in [a,b]$.

We conclude that the Gaußian entropy is constant in t on the solution $\bar{X}$. Proposition 15.24 implies that $\bar{X}$ is a self-similar shrinker of the affine normal flow.

16.4. Self-similarly shrinking solutions are ellipsoids

In this section we will prove that the only closed convex hypersurfaces which are self-similar shrinkers of the affine normal flow are the ellipsoids, using a refinement of a symmetrization argument due to Blaschke.

The argument is variational: We will show that self-similar solutions correspond to critical points of the entropy functional and then show by symmetrization that the only critical points are the ellipsoids.

Proposition 16.9. *Let $\mathcal{M}$ be a smooth, uniformly convex closed hypersurface. Then $\mathcal{M}$ is a critical point for the Gaußian entropy functional $E_{\frac{1}{n+2}}$ if and only if $\mathcal{M}$ is a shrinking self-similar solution for the affine normal flow.*

Proof. Since the time derivative of the entropy under the flow is nonzero if the hypersurface is not a shrinking self-similar solution, we conclude that any critical point of the entropy must be a shrinking self-similar solution.

To prove the converse, consider a variation of the support function given by

$$\partial_t \sigma = \eta,$$

where η is a smooth function on S^n. Since the entropy is scale invariant, it suffices to restrict to those η for which $\int_{S^n} \eta \det(A[\sigma])\, d\bar{\mu} = 0$ (these are the variations which preserve the enclosed volume to first order), where $A[\sigma]$ is defined by (15.90). For such variations, the variation in the entropy comes entirely from the variation of the affine surface area

$$S = \int_{\mathcal{M}} K^{\frac{1}{n+2}}\, d\mu = \int_{S^n} (\det A)^{\frac{n+1}{n+2}}\, d\bar{\mu}.$$

We compute

$$\begin{aligned}\frac{d}{dt} S &= \frac{n+1}{n+2} \int_{S^n} (\det A)^{-\frac{1}{n+2}} \det A (A^{-1})^{ij} A[\eta]_{ij}\, d\bar{\mu} \\ &= \frac{n+1}{n+2} \int_{S^n} \eta \det A (A^{-1})^{ij} A[(\det A)^{-\frac{1}{n+2}}]_{ij}\, d\bar{\mu},\end{aligned}$$

where $A[\eta] = \overline{\nabla}^2 \eta + \bar{g}\eta$ and $A[(\det A)^{-\frac{1}{n+2}}] = \overline{\nabla}^2 (\det A)^{-\frac{1}{n+2}} + \bar{g} (\det A)^{-\frac{1}{n+2}}$. Here, to obtain the second equality in the display above, we integrated by parts and used that $\overline{\nabla}_i(\det A(A^{-1})^{ij}) = 0$ (see Exercise 15.2).

In the case where σ is the support function of a shrinking self-similar solution, we have

$$(\det A)^{-\frac{1}{n+2}} = c\sigma + \langle z, p \rangle$$

for some $p \in \mathbb{R}^{n+1}$, so that $A[(\det A)^{-\frac{1}{n+2}}]_{ij} = cA_{ij}$. Therefore

$$\frac{d}{dt} S = c \frac{n+1}{n+2} \int_{S^n} \eta \det A (A^{-1})^{ij} A_{ij}\, d\bar{\mu} = cn \frac{n+1}{n+2} \int_{S^n} \eta \det A\, d\bar{\mu} = 0,$$

proving that σ is a critical point of the entropy. □

Now we prove that the only critical points of the entropy are ellipsoids.

Proposition 16.10. *Suppose $\mathcal{M}$ is a smooth, uniformly convex body which is a critical point of the Gaußian entropy $E_{\frac{1}{n+2}}$. Then for any vector $e \in \mathbb{R}^{n+1}$, there exists a hyperplane Π transverse to e such that $\mathcal{M}$ is invariant under the reflection in Π along lines parallel to e.*

Proof. Fix e, and choose an (orthonormal) basis $\{e_1, \ldots, e_{n+1}\}$ for $\mathbb{R}^{n+1}$ such that $e_{n+1} = e$. Then we can write $\mathcal{M} = \mathcal{M}_+ \cup \mathcal{M}_1 \cup \mathcal{M}_0$, where $\mathcal{M}_+ = \{x \in \mathcal{M} : N \cdot e < 0\}$ and $\mathcal{M}_- = \{x \in \mathcal{M} : N \cdot e > 0\}$ are smooth hypersurfaces and $\mathcal{M}_0$ is a smooth $(n-1)$-dimensional hypersurface of $\mathcal{M}$ consisting of the points where $e \in T_x\mathcal{M}$. Here $\mathcal{M}_0$ is the common boundary of $\mathcal{M}_+$ and $\mathcal{M}_-$.

By construction, $\mathcal{M}_+$ and $\mathcal{M}_-$ are both graphical hypersurfaces over a common domain $\Omega \subset \mathbb{R}^n = \operatorname{span}\{e_1, \ldots, e_n\}$. We write

$$\mathcal{M}_- = \{(x, u_-(x)) : x \in \Omega\} \text{ and } \mathcal{M}_+ = \{(x, -u_+(x)) : x \in \Omega\}$$

and observe that the convexity of $\mathcal{M}$ implies that u_- and u_+ are convex functions on Ω. In these coordinates we can write the affine surface area S as follows:

$$S = \int_{\mathcal{M}_-} K^{\frac{1}{n+2}}\, d\mu + \int_{\mathcal{M}_+} K^{\frac{1}{n+2}}\, d\mu.$$

By Lemma 5.14, we have on $\mathcal{M}_\pm$

$$K = \frac{\det(D^2 u_\pm)}{(1 + |Du_\pm|^2)^{\frac{n+2}{2}}}$$

and

$$d\mu(g) = \sqrt{1 + |Du_\pm|^2}\, dx^n.$$

This gives

$$S = \int_\Omega \left((\det(D^2 u_+))^{\frac{1}{n+2}} + (\det(D^2 u_-))^{\frac{1}{n+2}} \right) dx^n.$$

Now consider the family of smooth convex hypersurfaces defined by

$$\mathcal{M}_s = \{(x, -u_{+,s}(x)) : x \in \bar{\Omega}\} \cup \{(x, u_{-,s}(x)) : x \in \bar{\Omega}\},$$

where we define

$$u_{+,s}(x) = (1-s)u_+(x) + s u_-(x) \text{ and } u_{-,s}(x) = (1-s)u_-(x) + s u_+(x).$$

We observe that the volume enclosed by $\mathcal{M}_s$ is constant since this is given by

$$\int_\Omega (u_{-,s}(x) + u_{+,s}(x))\, dx^n = \int_\Omega (u_+(x) + u_-(x))\, dx^n.$$

If we assume that $\mathcal{M}$ is a critical point for the energy, then $\mathcal{M}$ is also a critical point for S with respect to any volume-preserving smooth variation, so we must have $\frac{d}{ds} S[\mathcal{M}_s]|_{s=0} = 0$.

Next we observe that $S[\mathcal{M}_s]$ is a concave function of s: This follows since for each $x \in \Omega$, $D^2 u_{\pm,s} = (1-s)D^2 u_\pm + s D^2 u_\mp$ is a linear function of s, while the map $A \mapsto (\det A)^{\frac{1}{n+2}}$ is a strictly concave function on the cone of positive definite symmetric matrices. Furthermore, the hypersurface $\mathcal{M}_{1-s}$ is the reflection of the hypersurface $\mathcal{M}_s$ in the hyperplane spanned by

$\{e_1, \ldots, e_n\}$, so we have $S[\mathcal{M}_s] = S[\mathcal{M}_{1-s}]$. It follows that $\frac{d}{ds}S[\mathcal{M}_s]|_{s=0} > 0$, unless $S[\mathcal{M}_s]$ is constant in s, and this occurs only if $D^2u_{\pm,s}(x)$ is constant in s for each x. This in turn implies that u_+ and u_- differ by a linear function, which is equivalent to the statement that $\mathcal{M}$ is invariant under a reflection of the form $(x, y) \mapsto (x, L(x) - y)$, where L is a linear function on $\mathbb{R}^n$. □

Corollary 16.11. *Any smooth, uniformly convex hypersurface which is a critical point of the Gaußian entropy $E_{\frac{1}{n+2}}$ is an ellipsoid.*

Proof. We must show that any hypersurface which has reflection symmetries in all directions, in the sense of the previous proposition, is an ellipsoid.

We observe that the John ellipsoid E_M of a convex hypersurface M is equivariant under affine transformations: We have $E_{L(M)} = L(E_M)$ for any affine transformation L. This is clear from the characterization of the John ellipsoid as the unique enclosed ellipsoid of maximal volume.

Now suppose that M is a critical point of the Gaußian entropy $E_{\frac{1}{n+2}}$ and let E_M be the John ellipsoid of M. Let T be an affine transformation which takes E_M to the unit sphere about the origin. Note that $T(M)$ is still a critical point of the entropy since the entropy is affine invariant. It follows from Proposition 16.10 that for every nonzero vector e, $T(M)$ is invariant under a reflection L_e along lines in the direction e through some transversal plane. But then it follows that

$$L_e(S_1^n(0)) = L_e(E_{T(M)}) = E_{L_e(T(M))} = E_{T(M)} = S_1^n(0),$$

so L_e is an affine transformation preserving the unit sphere and hence an orthogonal transformation. It follows that L_e is the orthogonal reflection in the hyperplane through the origin orthogonal to e. That is, we have proved that $T(M)$ is invariant under every orthogonal reflection and hence under every orthogonal transformation. Therefore $T(M)$ is the unit sphere, and $M = T^{-1}(S_1^n(0))$ is an ellipsoid. □

16.5. Convergence without affine corrections

We have proved that there are a sequence of times t_k approaching T and a sequence of affine transformations L_k such that $L_k(\mathcal{M}_{t_k})$ converges in C^∞ to an ellipsoid. To obtain convergence without using the affine corrections L_k, we first deduce an exponential rate of convergence of the entropy by considering the linearization of the flow about the sphere. To be precise about this, we define a flow which is "normalized" not only by scaling, but by an optimal choice of affine correction.

16.5.1. Center of mass and the Legendre ellipsoid. There are many choices which can be made to normalize the solution with respect to affine transformations. Our choice relies on the *center of mass* and the *Legendre ellipsoid* of the enclosed region.

Let M^n be a compact convex hypersurface, bounding a convex region $\Omega \subset \mathbb{R}^{n+1}$. Then the **center of mass** $\bar{p}(\Omega)$ of Ω is defined by

$$\bar{p}(\Omega) = \frac{1}{|\Omega|} \int_\Omega X d\mathscr{H}^{n+1}(X). \tag{16.17}$$

The center of mass of a region is an affine invariant, in the sense that if L is a $\mathrm{GL}(n+1, \mathbb{R})$ transformation and T is a translation of $\mathbb{R}^{n+1}$, then

$$\bar{p}(L\Omega) = L\left(\bar{p}(\Omega)\right) \quad \text{and} \quad \bar{p}(T\Omega) = T\left(\bar{p}(\Omega)\right).$$

A second useful affine invariant is the **Legendre form**, which is the positive definite quadratic form defined by

$$Z(\Omega)(u,v) \doteqdot \frac{n+3}{|\Omega|} \int_\Omega X \cdot u X \cdot v \, d\mathscr{H}^{n+1}(X). \tag{16.18}$$

This is $\mathrm{GL}(n, \mathbb{R})$-equivariant, in the sense that

$$Z(L\Omega)(u,v) = Z(\Omega)(Lu, Lv)$$

for any $L \in \mathrm{GL}(n+1, \mathbb{R})$. The Legendre form is not invariant under translations, but the modified form

$$\tilde{Z}(\Omega)(u,v) = \frac{n+3}{|\Omega|} \int_\Omega (X - \bar{p}(\Omega)) \cdot u (X - \bar{p}(\Omega)) \cdot v \, d\mathscr{H}^{n+1}(X) \tag{16.19}$$

is invariant under translations.

The modified Legendre form gives rise to the **Legendre ellipsoid**

$$\mathrm{E}(\Omega) \doteqdot \left\{ X + \bar{p}(\Omega) : \ \tilde{Z}(\Omega)(X, X) < 1 \right\}. \tag{16.20}$$

This ellipsoid has a useful equivariance property: For any invertible affine transformation L,

$$\mathrm{E}(L(\Omega)) = L(\mathrm{E}(\Omega)).$$

The normalization in the definition of the Legendre form is chosen so that $\mathrm{E}(B) = B$, where B is the unit ball, and it follows that $\mathrm{E}(E) = E$ for any ellipsoid E.

16.5.2. The affine-normalized affine normal flow ☺. In this subsection we prove that the affine-normalized affine normal flow evolves any uniformly convex embedding of S^n to the unit sphere.

Proposition 16.12. *Let $\tilde{X}_0 : M \to \mathbb{R}^{n+1}$ be a smooth, uniformly convex embedding enclosing a region $\tilde{\Omega}_0$, and let $\tilde{X} : M \times [0,T) \to \mathbb{R}^{n+1}$ be the maximal solution of the $\frac{1}{n+2}$-Gauß curvature flow with initial condition $\tilde{X}_0$. Then there exists a smooth family of linear transformations $T : [0,T) \to \mathrm{GL}(n+1,\mathbb{R})$ (unique up to multiplication on the left by an arbitrary time-independent $\mathrm{SO}(n+1)$ transformation), a unique smooth curve $p : [0,T) \to \mathbb{R}^{n+1}$, and a unique smooth family of diffeomorphisms $\varphi : M \times [0,T) \to M$ with $\varphi(x,0) = x$ for each x, with the following properties.*

(1) *The hypersurface $\mathcal{M}_\tau$ given by the embedding*

$$X(x,\tau(t)) = T(t)\left(\tilde{X}(\varphi(x,t),t) - p(t)\right), \tag{16.21}$$

where $\tau(t) = \int_0^t (|\tilde{\Omega}_s|/|B|)^{-\frac{2}{n+2}}\,ds$, encloses a region Ω_τ with normalizations:

$$\text{volume } |\Omega_\tau| = |B|, \quad \text{center of mass } \bar{p}(\Omega_\tau) = 0, \quad \text{and } \mathring{Z}(\Omega_\tau) = 0, \tag{16.22}$$

where $\mathring{Z}$ is the traceless part of the Legendre form Z.

(2) *The embedding X satisfies the* ***affine-normalized affine normal flow***

$$\partial_\tau X = -K^{\frac{1}{n+2}}\mathsf{N} + b(\tau) + L(\tau)(X), \tag{16.23}$$

where the vector $b(\tau)$ and the linear map $L(\tau)$ are given by the formulae

$$b(\tau) = \frac{1}{|B|}\int_{\mathcal{M}_\tau} K^{\frac{1}{n+2}} X\, d\mathscr{H}^{n+1}$$

and

$$L_{\alpha\beta}(\tau) = \frac{\int_{\mathcal{M}_\tau} K^{\frac{1}{n+2}}\,d\mu}{(n+1)|B|}\delta_{\alpha\beta} + \frac{n+1}{2}\frac{\int_{\mathcal{M}_\tau} K^{\frac{1}{n+2}}\left(X^\alpha X^\beta - \frac{1}{n+1}|X|^2\delta^{\alpha\beta}\right)d\mu}{\int_{\Omega_\tau}|X|^2\,d\mathscr{H}^{n+1}}.$$

(3) *As $\tau \to \infty$, $X(\cdot,\tau)$ converges in C^∞ to a limiting embedding X_∞ which has image equal to the unit sphere with center at the origin, and we have the estimates*

$$\|X(\cdot,\tau) - X_\infty\|_{C^k(M,g_\infty)} \le C(k,\varepsilon)\mathrm{e}^{-\left(\frac{n+4}{n+2}-\varepsilon\right)\tau} \tag{16.24}$$

for each $k \ge 0$ and $\varepsilon > 0$.

Proof. We construct the affine-normalized affine normal flow X as follows. First, let $p(0) = \bar{p}(\Omega_0)$ and let

$$T(0) = \left(\frac{|B|}{|\tilde{\Omega}_0|}\right)^{\frac{1}{n+1}}\left(\det \tilde{Z}(\tilde{\Omega}_0)\right)^{\frac{1}{2(n+1)}}\tilde{Z}(\tilde{\Omega}_0)^{-1/2}$$

(where $\tilde{Z}^{-1/2}$ is the positive definite symmetric matrix with the same eigenvectors as the modified Legendre form $\tilde{Z}$ and eigenvalues λ replaced by

$\lambda^{-1/2}$). By construction, this gives the required conditions for $X(\cdot, 0)$. It follows that if we define $T(t)$ by

$$\frac{d}{d\tau} T_{\alpha\beta} = L_{\alpha\gamma} T_{\gamma\beta} \tag{16.25}$$

and we define $p(t)$ by

$$\frac{d}{d\tau} p(t) = b(\tau), \tag{16.26}$$

then $X(\tau)$ defined as in the proposition satisfies the equation (16.23). It then follows by a direct computation that the volume $|\Omega_\tau|$ is constant:

$$\begin{aligned}
\frac{d}{d\tau}|\Omega_\tau| &= \int_{\partial\Omega_\tau} \left(-K^{\frac{1}{n+2}} + b\cdot \mathsf{N} + L(X)\cdot \mathsf{N}\right) d\mu \\
&= -\int_{\mathcal{M}_\tau} K^{\frac{1}{n+2}}\, d\mu + \int_{\Omega_\tau} \operatorname{div}\left(b + L(X)\right) d\mathscr{H}^{n+1} \\
&= -\int_{\mathcal{M}_\tau} K^{\frac{1}{n+2}}\, d\mu + \operatorname{tr}(L)|\Omega_\tau| \\
&= 0.
\end{aligned}$$

Furthermore, the center of mass of Ω_τ remains at the origin since

$$\begin{aligned}
\frac{d}{d\tau}\int_{\Omega_\tau} X^\alpha\, d\mathscr{H}^{n+1} &= \int_{\mathcal{M}_\tau} X^\alpha \left(-K^{\frac{1}{n+2}} + b\cdot \mathsf{N} + L(X)\cdot \mathsf{N}\right) d\mu \\
&= -\int_{\mathcal{M}_\tau} X^\alpha K^{\frac{1}{n+2}}\, d\mu + \int_{\Omega_\tau} \operatorname{div}\left(X^\alpha b + X^\alpha L(X)\right) d\mathscr{H}^{n+1} \\
&= -\int_{\mathcal{M}_\tau} X^\alpha K^{\frac{1}{n+2}}\, d\mu + b^\alpha|\Omega_\tau| \\
&\quad + \left(L_{\alpha\gamma} + \operatorname{tr}(L)\delta_{\alpha\gamma}\right)\int_{\Omega_\tau} X^\gamma\, d\mathscr{H}^{n+1} \\
&= \left(L_{\alpha\gamma} + \operatorname{tr}(L)\delta_{\alpha\gamma}\right)\int_{\Omega_\tau} X^\gamma\, d\mathscr{H}^{n+1},
\end{aligned}$$

so that the vector $\int_{\Omega_\tau} X\, d\mathscr{H}^{n+1}$ remains zero since it is initially so. Finally, a similar computation shows that the traceless part of the Legendre form $\tilde{Z}(\Omega_\tau)$ remains zero, making use of the facts that the enclosed volume is constant and the center of mass is at the origin.

Next we show how to prove the exponential convergence. For this purpose we consider the evolution of the support function under the evolution equation (16.23). We first need to compute the varation of the support function produced by a 1-parameter family of special linear transformations

$s \mapsto \exp(sL)$: We find that if $X(\cdot, s) = \exp(sL) \circ X(\cdot)$, then

$$\begin{aligned}\partial_s \sigma(z,s)\Big|_{s=0} &= L\bar{X}(z)\cdot z \\ &= \sigma(z) L_{\alpha\beta} z^\alpha z^\beta + \partial_i \sigma(z) \bar{g}^{ij} L_{\alpha\beta} \partial_j z^\alpha z^\beta,\end{aligned}$$

where the Greek indices correspond to any fixed orthonormal basis for $\mathbb{R}^{n+1}$, $\bar{X} = \mathrm{G}^{-1}$ is the inverse of the Gauß map, and we used the representation

$$\bar{X}(z) = \sigma(z) z + \partial_i \sigma(z) \bar{g}^{ij} \partial_j z \tag{16.27}$$

from (5.80). Since $\tilde{\sigma}(t)$ satisfies $\partial_t \tilde{\sigma} = -\tilde{K}^{\frac{1}{n+2}} = -(\det A[\tilde{\sigma}])^{-\frac{1}{n+2}}$, this gives the following evolution equation for the support function $\sigma(\tau)$ of $X(\tau)$:

$$\partial_\tau \sigma = -(\det A[\sigma])^{-\frac{1}{n+2}} + b_\alpha z^\alpha + L_{\alpha\beta}(\sigma z^\alpha + \sigma_i \partial_i z^\alpha) z^\beta \doteqdot G[\sigma], \tag{16.28}$$

where b_α and $L_{\alpha\beta}$ are as in the proposition and so are given in terms of the support function σ by the following formulae:

$$b_\alpha = \frac{n+1}{|S^n|} \int_{S^n} (\det A[\sigma])^{\frac{n+1}{n+2}} \left(\sigma z^\alpha + \sigma_i \bar{g}^{ij} \partial_j z^\alpha\right) d\bar{\mu}(z) \tag{16.29}$$

and

$$\begin{aligned} L_{\alpha\beta} = & \frac{\int_{S^n} (\det A[\sigma])^{\frac{n+1}{n+2}}\, d\bar{\mu}}{|S^n|} \delta_{\alpha\beta} \\ & + \frac{(n+1)(n+3)}{2} \frac{\int_{S^n} (\det A[\sigma])^{\frac{n+1}{n+2}} \left(\bar{X}^\alpha \bar{X}^\beta - \frac{1}{n+1} |\bar{X}|^2 \delta^{\alpha\beta}\right) d\bar{\mu}}{\int_{S^n} \sigma \left(\sigma^2 + |\overline{\nabla} \sigma|^2\right) \det A[\sigma]\, d\bar{\mu}}, \end{aligned} \tag{16.30}$$

where $\bar{X} = \sigma z + \overline{\nabla}_i \sigma \bar{g}^{ij} \partial_j z$.

The analysis of the previous section shows that we can find a sequence of times t_k approaching T (corresponding to a sequence τ_k approaching infinity in equation (16.28)) such that the hypersurfaces $\tilde{\mathcal{M}}_{t_k} = \tilde{X}_{t_k}(M)$ converge after suitable affine transformations to an ellipsoid. It follows that the hypersurfaces $\mathcal{M}_{\tau_k} = X_{\tau_k}(M)$ converge without affine correction to the unit sphere, since this is the unique ellipsoid with center of mass at the origin, which is a spherical Legendre ellipsoid, and with enclosed volume equal to that of the unit ball. This implies that the support functions $\sigma(\cdot, \tau_k)$ approach 1 in $C^\infty(S^n)$ as $k \to \infty$. The regularity estimates also imply that we have uniform bounds on all derivatives, so that there exist constants C_k such that

$$\|\sigma(\cdot, \tau)\|_{C^k(S^n)} \le C_k, \quad \tau \in [0, \infty), \tag{16.31}$$

for each $k \ge 0$.

To proceed toward exponential convergence, we first consider the linearization of the flow (16.28) about the stationary solution $\sigma = 1$. If we have a smooth family of solutions $\sigma(z,\tau,s)$ with $\sigma(z,\tau,0)=1$, then writing

$$\dot{\sigma}(z,\tau)=\left.\frac{\partial}{\partial s}\sigma(z,\tau,s)\right|_{s=0},$$

we find by differentiating (16.28) the following:

$$\frac{\partial}{\partial\tau}\dot{\sigma}=\frac{1}{n+2}(\overline{\Delta}\dot{\sigma}+n\dot{\sigma})+\dot{b}_\alpha z^\alpha+\dot{L}_{\alpha\beta}z^\alpha z^\beta+L_{\alpha\beta}z^\alpha\dot{X}^\beta, \tag{16.32}$$

where "$\cdot$" denotes a derivative with respect to s at $s=0$ and where $L_{\alpha\beta}$ is evaluated at $s=0$. The formulae (16.30) with $\sigma=1$ and (16.27) give $L_{\alpha\beta}=\delta_{\alpha\beta}$ and

$$\dot{X}^\beta=\dot{\sigma}z^\beta+\overline{\nabla}_i\dot{\sigma}\overline{g}^{ij}\partial_j z^\beta,$$

respectively, so that $L_{\alpha\beta}z^\alpha\dot{X}^\beta=z\cdot(\dot{\sigma}z+\overline{\nabla}\dot{\sigma})=\dot{\sigma}$. We also have

$$\dot{b}=\frac{n+1}{|S^n|}\int_{S^n}\left(\dot{\sigma}z+\overline{\nabla}_i\dot{\sigma}\overline{g}^{ij}\partial_j z\right)d\bar{\mu}=\frac{(n+1)^2}{|S^n|}\int_{S^n}\dot{\sigma}z\,d\bar{\mu} \tag{16.33}$$

and

$$\dot{L}_{\alpha\beta}=c_1(n)\int_{S^n}\dot{\sigma}\,d\bar{\mu}\,\delta_{\alpha\beta}+c_2(n)\int_{S^n}\dot{\sigma}z^\alpha z^\beta\,d\bar{\mu} \tag{16.34}$$

for some constants c_1 and c_2.

Furthermore, if the constraints $|\Omega_\tau|=|B|$, $\bar{p}(\Omega_\tau)=0$, and $\mathring{Z}(\Omega_\tau)=0$ all hold for each s, then we have that $\int_{S^n}\dot{\sigma}\,d\bar{\mu}=0$, $\int_{S^n}\dot{\sigma}z^\alpha\,d\bar{\mu}=0$, and $\int_{S^n}\dot{\sigma}z^\alpha z^\beta\,d\bar{\mu}=0$, so that $\dot{b}=0$ and $\dot{L}=0$. This gives the linearized equation

$$\partial_\tau\dot{\sigma}=\frac{1}{n+2}\overline{\Delta}\dot{\sigma}+\frac{2(n+1)}{n+2}\dot{\sigma}.$$

From this we can see that the sphere solution $\sigma=1$ is linearly stable under (16.28): We can express $\dot{\sigma}$ as a spherical harmonics series:

$$\dot{\sigma}=\sum_{k=3}^{\infty}\varphi_k(z,\tau),$$

where each $\varphi_k(\cdot,\tau)$ is a harmonic homogeneous polynomial on $\mathbb{R}^{n+1}$ of degree k (necessarily of degree at least 3 since the three constraints imply that $\dot{\sigma}$ is orthogonal to the spherical harmonics of degree 0, 1, or 2). Then we have $\overline{\Delta}\varphi_k=-k(n-1+k)\varphi_k$, and therefore

$$\frac{d}{d\tau}\varphi_k=-\left(\frac{k(n-1+k)}{n+2}+\frac{2(n+1)}{n+2}\right)\varphi_k=-\frac{(k-2)(n+1+k)}{n+2}\varphi_k,$$

and it follows that $\varphi_k(\cdot,\tau) = \mathrm{e}^{-\frac{(k-2)(n+1+k)}{n+2}\tau}\varphi_k(\cdot,0)$. In particular, since $k \geq 3$, we have $\frac{(k-2)(n+1+k)}{n+2} \geq \frac{n+4}{n+2}$ and hence

$$\|\dot{\sigma}(\cdot,\tau)\|_{L^2}^2 = \sum_{k=3}^{\infty} \|\varphi_k(\cdot,\tau)\|_{L^2}^2 \leq C\mathrm{e}^{-\frac{n+4}{n+2}\tau}$$

for some constant C. This proves the linear stability.

Next we show how to prove that the solution σ of (16.28) converges exponentially to 1 if it is sufficiently close to 1. Recall that we defined $G[\sigma] = \partial_\tau \sigma$ to be the right-hand side of (16.28), and note that G is a smooth map from $C^2(S^n)$ to $C^0(S^n)$ in a $C^2(S^n)$-neighborhood of the function $\sigma = 1$. Then we have by (16.32), (16.33), (16.34), and since $G[1] = 0$, that

$$\begin{aligned}
\frac{d}{d\tau}\int_{S^n}(\sigma-1)^2 d\bar{\mu} &= 2\int_{S^n}(\sigma-1)G[\sigma]\,d\bar{\mu}\\
&\leq 2\int_{S^n}\left((\sigma-1)DG_1(\sigma-1) + C|\sigma-1|\,\|\sigma-1\|_{C^2}^2\right)d\bar{\mu}\\
&\leq 2\int_{S^n}(\sigma-1)\left(\frac{1}{n+2}\overline{\Delta}(\sigma-1) + \frac{2(n+1)}{n+2}(\sigma-1)\right)d\bar{\mu}\\
&\quad + \frac{2(n+1)^2}{|S^n|}\left|\int_{S^n}(\sigma-1)z\,d\bar{\mu}\right|^2 + 2c_1\left(\int_{S^n}(\sigma-1)\,d\bar{\mu}\right)^2\\
&\quad + 2c_2\sum_{\alpha,\beta}\left(\int_{S^n}(\sigma-1)z^\alpha z^\beta\,d\bar{\mu}\right)^2 + C\|\sigma-1\|_{L^2}^{\frac{1}{2}}\|\sigma-1\|_{C^2}^2,
\end{aligned}$$

where DG_1 denotes the derivative of $G[\sigma]$ at $\sigma = 1$. Now we use the constraints to handle some of the terms: The volume constraint $|\Omega_\tau| = |B|$ becomes $\int_{S^n}\sigma\det A[\sigma]\,d\bar{\mu} = |S^n|$, and expanding the integrand about $\sigma = 1$ gives

$$\left|\int_{S^n}\sigma\det A[\sigma]\,d\bar{\mu} - |S^n| - (n+1)\int_{S^n}(\sigma-1)\,d\bar{\mu}\right| \leq C\|\sigma-1\|_{C^2}^2,$$

from which it follows that

$$\left|\int_{S^n}(\sigma-1)\,d\bar{\mu}\right| \leq C\|\sigma-1\|_{C^2}^2. \tag{16.35}$$

Similarly, the constraint on the center of mass implies

$$\left|\int_{S^n}(\sigma-1)z\,d\bar{\mu}\right| \leq C\|\sigma-1\|_{C^2}^2, \tag{16.36}$$

and the constraint of spherical Legendre ellipsoid implies

$$\left|\int_{S^n}(\sigma-1)z^\alpha z^\beta\,d\bar{\mu}\right| \leq C\|\sigma-1\|_{C^2}^2. \tag{16.37}$$

This gives

$$\frac{d}{d\tau}\int_{S^n}(\sigma-1)^2\,d\bar{\mu} \le 2\int_{S^n}(\sigma-1)\left(\frac{1}{n+2}\overline{\Delta}(\sigma-1)+\frac{2(n+1)}{n+2}(\sigma-1)\right)d\bar{\mu}$$
$$+\,C\|\sigma-1\|_{C^2}^{5/2}.$$

To control the terms involving $\|\sigma-1\|_{C^2}$, we use the following interpolation inequality (a special case of the Gagliardo–Nirenberg interpolation inequality):

$$\|f\|_{C^2} \le C(k)\,\|f\|_{C^k}^{\frac{n+2}{n+k}}\,\|f\|_{L^2}^{\frac{k-2}{n+k}} \tag{16.38}$$

for any $k \ge n$. Applying this with large k and using the uniform C^k bounds on σ from (16.31), we deduce that

$$\|\sigma-1\|_{C^2} \le C(\varepsilon)\|\sigma-1\|_{L^2}^{1-\varepsilon} \tag{16.39}$$

for any $\varepsilon>0$. To complete the argument we can again decompose $\sigma-1$ as a sum of spherical harmonics:

$$\sigma-1=\sum_{k=0}^{\infty}\phi_k.$$

A direct computation then gives

$$\begin{aligned}\frac{d}{d\tau}\int_{S^n}(\sigma-1)^2\,d\bar{\mu} &\le -\sum_{k\ge 3}\frac{2(k-2)(n+k+1)}{n+2}\int_{S^n}\phi_k^2\,d\bar{\mu}\\ &\quad+\sum_{k=0}^{2}c_k\int_{S^n}\phi_k^2\,d\bar{\mu}+C\|\sigma-1\|_{L^2}^{5/2-\varepsilon}\\ &\le -\frac{2(n+4)}{n+2}\|\sigma-1\|_{L^2}^2+C\|\sigma-1\|_{L^2}^{5/2-\varepsilon},\end{aligned} \tag{16.40}$$

where we used

$$\|\phi_0\|_{L^2}^2=c\left|\int_{S^n}(\sigma-1)\,d\bar{\mu}\right|^2 \le C\|\sigma-1\|_{C^2}^4$$

and

$$\|\phi_1\|_{L^2}^2=c\left|\int_{S^n}(\sigma-1)z\,d\bar{\mu}\right|^2 \le C\|\sigma-1\|_{C^2}^4$$

from (16.35) and (16.36), respectively, while

$$\|\phi_2\|_{L^2}^2=c\left|\int_{S^n}(\sigma-1)\left(z^\alpha z^\beta-\frac{1}{n+1}\delta^{\alpha\beta}\right)d\bar{\mu}\right|^2 \le C\|\sigma-1\|_{C^2}^4$$

from (16.37), and the inequality (16.39).

It follows from (16.40), while taking $\varepsilon<1/2$, that

$$\|\sigma-1\|_{L^2} \le Ce^{-\frac{n+4}{n+2}\tau} \tag{16.41}$$

for some C, provided $\|\sigma - 1\|_{L^2}$ is initially small enough. In particular, since we have a subsequence of times along which $\sigma - 1$ converges to zero in C^∞, the smallness condition is eventually satisfied and the exponential decay holds as $\tau \to \infty$.

The convergence of higher derivatives (with exponential rate $\frac{n+4}{n+2} - \varepsilon$ for any $\varepsilon > 0$) now follows from the L^2 convergence and the interpolation inequality (16.38). Finally, the convergence of the embeddings $X(\cdot, \tau)$ can be recovered from the convergence of σ. $\square$

Having proved exponential convergence to the unit sphere for the affine-normalized affine normal flow, we can now deduce the convergence of the original flow to an ellipsoid centered at the final point, modulo rescaling and reparametrization:

Corollary 16.13. *Let $X_0 : M \to \mathbb{R}^{n+1}$ be a smooth, uniformly convex embedding and let $X : M \times [0, T) \to \mathbb{R}^{n+1}$ be the maximal solution of the $\frac{1}{n+2}$-Gauß curvature flow with initial data X_0. Then there exists $p \in \mathbb{R}^{n+1}$ and a smooth family of diffeomorphisms $\varphi : M \times [0, T) \to M$ such that*

$$\hat{X}(x, t) \doteqdot \frac{X(\varphi(x, t), t) - p}{\left(\frac{2(n+1)}{n+2}(T - t)\right)^{\frac{n+2}{2(n+1)}}}$$

converges in C^∞ to a limiting embedding $\hat{X}_T$ with image an ellipsoid centered at the origin. Furthermore, we have

$$\|\hat{X}(\cdot, t) - \hat{X}_T\|_{C^k(M, g_T)} \leq C(k, \varepsilon)|T - t|^{\frac{n+4}{2(n+1)} - \varepsilon} \tag{16.42}$$

for any $k \geq 0$ and $\varepsilon > 0$.

Proof. This follows directly since the linear transformations $T(t)$ and the translations $p(t)$ converge exponentially in τ according to (16.25) and (16.26), and we have from the definition of τ that $T - t \sim Ce^{-\frac{2(n+1)}{n+2}\tau}$ as $\tau \to \infty$. $\square$

16.6. Notes and commentary

The treatment of the affine normal flow given here differs from the original treatment in [**30**] and is closer to the argument sketched in [**34**, Section 9]. The original treatment is based on the analysis of affine-geometric invariants under the evolution, essentially following the argument of Calabi in the corresponding elliptic setting of *affine hyperspheres*. In particular, the key step is a decay estimate on the cubic ground form (defined by (16.43a) and (16.43b) below) which implies that the hypersurfaces converge to an ellipsoid. Some of the argument is indicated in Exercises 16.5–16.14 below.

The monotonicity of the Gaußian entropy, together with the convergence to an ellipsoid, implies that the entropy for any convex hypersurface is greater than that for an ellipsoid. This is equivalent to the *affine isoperimetric inequality.* Similarly, the monotonicity of the Firey entropy (optimized over the choice of center) implies the *Blaschke–Santaló inequality.* See the remarks at the end of Section 16.1.

The argument used here to prove that shrinking self-similar solutions of the affine normal flow are ellipsoids, based on Steiner symmetrization of the Gaußian entropy functional, is an adaptation of the argument of Blaschke used to prove the affine isoperimetric inequality: In that argument the symmetrization was used to prove that minimizers of the affine isoperimetric ratio (equivalently, the Gaußian entropy) must be symmetric in any hyperplane, and our argument amounts to the observation that the same must be true for critical points of the affine isoperimetric ratio (equivalently, self-similar solutions of the affine normal flow).

16.7. Exercises

Exercise 16.1. Show that shrinking self-similar solutions of the affine normal flow in the plane have support function satisfying

$$\sigma_{\theta\theta} + \sigma = C\sigma^{-3},$$

where θ is the standard angle parameter on the circle S^1. Prove that the solutions have the form $\sigma^2(\theta) = C_1 + C_2\cos(2(\theta - \theta_0))$ for some C_1, C_2, and θ_0. Check that the image under the embedding $\bar{X}(\theta) = \sigma e^{i\theta} + i\sigma_\theta e^{i\theta}$ is an ellipse centered at the origin.

Exercise 16.2. Show that parabolae are translating self-similar solutions of the affine normal flow in the plane and hyperbolae are expanding self-similar solutions.

Exercise 16.3. Show that a paraboloid of revolution is a translating self-similar solution of the affine normal flow in any dimension.

Exercise 16.4. Show that if A is a positive definite symmetric bilinear form on $\mathbb{R}^n$, then the hypersurface $\left\{(x,y) \in \mathbb{R}^n \times \mathbb{R} : \ y = \frac{1}{2}A(x,x)\right\}$ is a translating self-similar solution of the affine normal flow.

Exercise 16.5. Let $\mathcal{M}$ be a convex hypersurface in $\mathbb{R}^{n+1}$ with support function σ. Define the *affine normal* by $\mathcal{N} = -X\left[(\det A[\sigma])^{-\frac{1}{n+2}}\right]$, where $X[\cdot]$ is defined by (5.85). Show that there is a Riemannian metric $\tilde{g}$ (the *affine metric*), a $(1,1)$-tensor $\mathcal{A}$ (the *affine curvature*), and a $(2,1)$-tensor $\mathcal{C}$

(the *cubic ground form*) such that the following *structural equations* hold:

$$D_U D_V X = \tilde{g}(U,V)\mathcal{N} + DX(\tilde{\nabla}_U V + \mathcal{C}(U,V)), \tag{16.43a}$$

$$D_U \mathcal{N} = -DX(\mathcal{A}(U)) \tag{16.43b}$$

for any vector fields U and V, where $\tilde{\nabla}$ is the Levi-Civita connection of $\tilde{g}$. Verify that these are given by the following expressions:

$$\tilde{g}_{ij} = (\det A)^{\frac{1}{n+2}} A_{ij},$$

$$\mathcal{A}_i^j = (A^{-1})^{jk} A\left[(\det A)^{-\frac{1}{n+2}}\right]_{ik},$$

$$\mathcal{C}_{ijk} = \frac{1}{2}(\det A)^{\frac{1}{n+2}}\left(\overline{\nabla}_i A_{jk} - \frac{A_{ij}\overline{\nabla}_k \det A + A_{ik}\overline{\nabla}_j \det A + A_{jk}\overline{\nabla}_i \det A}{(n+2)\det A}\right),$$

where $\mathcal{C}_{ijk} = \mathcal{C}_{ij}{}^p \tilde{g}_{pk}$. Verify that $\mathcal{A}$ is selfadjoint with respect to $\tilde{g}$ and that $\mathcal{C}$ is totally symmetric and totally traceless with respect to $\tilde{g}$.

Exercise 16.6. Show that the affine normal $\mathcal{N}$ is invariant under special affine transformations: If $X_L = LX + b$, then $\mathcal{N}_L = L(\mathcal{N})$. Deduce from this and the structural equations of the previous exercise that the affine metric $\tilde{g}$, affine curvature $\mathcal{A}$, and cubic ground form $\mathcal{C}$ are all affine invariant. Deduce that the cubic ground form vanishes on any ellipsoid.

Exercise 16.7. By differentiating the structural equations (16.43a) and (16.43b) and commuting derivatives, prove the affine-geometric Gauß and Codazzi equations:

$$\mathcal{R}_{ikjl} = \tilde{g}_{ij}\mathcal{A}_{kl} - \tilde{g}_{jk}\mathcal{A}_{il} + \mathcal{C}_{il}{}^p\mathcal{C}_{jkp} - \mathcal{C}_{ij}{}^p\mathcal{C}_{klp} + \tilde{\nabla}_i\mathcal{C}_{jkl} - \tilde{\nabla}_k\mathcal{C}_{ijl}, \tag{16.44}$$

$$\tilde{\nabla}_i\mathcal{A}_{jk} + \mathcal{A}_j^p\mathcal{C}_{ipk} = \tilde{\nabla}_j\mathcal{A}_{ik} + \mathcal{A}_i^p\mathcal{C}_{jpk}, \tag{16.45}$$

where $\mathcal{R}$ is the Riemann curvature tensor of $\tilde{g}$.

Exercise 16.8. From the affine Gauß identity (16.44) and the symmetries of the Riemann curvature tensor, derive the following Codazzi-type identity for the cubic ground form:

$$\tilde{\nabla}_i\mathcal{C}_{jkl} - \tilde{\nabla}_j\mathcal{C}_{ikl} = \frac{1}{2}\tilde{g}_{jk}\mathcal{A}_{il} + \frac{1}{2}\tilde{g}_{il}\mathcal{A}_{jk} - \frac{1}{2}\tilde{g}_{ik}\mathcal{A}_{jl} - \frac{1}{2}\tilde{g}_{jl}\mathcal{A}_{ik}. \tag{16.46}$$

Exercise 16.9. Suppose that $\mathcal{M}^n$ is a complete connected convex hypersurface for which the cubic ground form $\mathcal{C}$ vanishes. Deduce from the identity (16.46) that the affine curvature $\mathcal{A}$ is a multiple of the identity at each point, so that $\mathcal{A} = \frac{\mathcal{H}}{n}I$, where $\mathcal{H} \doteqdot \operatorname{tr}\mathcal{A}$ is the *affine mean curvature*. Use the affine Codazzi equation (16.45) to deduce that $\mathcal{H}$ is constant. Finally, solve the structural equations along geodesics of $\tilde{g}$ to prove that $\mathcal{M}^n$ is an *ellipsoid* if $\mathcal{H} > 0$, a *paraboloid* if $\mathcal{H} = 0$, or an *elliptic hyperboloid* if $\mathcal{H} < 0$ (i.e., the image under an affine transformation of the sphere $\{x \in \mathbb{R}^{n+1} : |x| = 1\}$, of the paraboloid $\{(x, y) \in \mathbb{R}^n \times \mathbb{R} : \ y = |x|^2\}$, or of the hyperboloid $\{(x, y) \in \mathbb{R}^n \times \mathbb{R} : \ y = \sqrt{1+|x|^2}\}$, respectively).

Exercise 16.10. For a smooth, convex hypersurface $\mathcal{M}$ with support function σ and for a smooth function F, show that

$$X\left[(\det A[\sigma])^{-\frac{1}{n+2}}F\right] = -F\mathcal{N} + \nabla^{\tilde{g}}F = -F\mathcal{N} + \nabla_i F\,\tilde{g}^{ij}\partial_j X, \tag{16.47}$$

and differentiate this to obtain the identity

$$(\det A[\sigma])^{\frac{1}{n+2}} A\left[(\det A[\sigma])^{-\frac{1}{n+2}}F\right]_{ij} = \tilde{\nabla}_i\tilde{\nabla}_j F + \mathcal{C}_{ij}{}^p\nabla_p F + \mathcal{A}_{ij}F. \tag{16.48}$$

Exercise 16.11. By differentiating the expressions derived for $\tilde{g}$ and $\mathcal{A}$ in Exercise 16.5 with respect to t and using the identity (16.48), derive the following evolution equations for the affine metric and affine curvature under the affine normal flow:

$$\partial_t\tilde{g} = -\frac{1}{n+2}\mathcal{H}\,\tilde{g} - \mathcal{A} \tag{16.49}$$

and

$$\partial_t\mathcal{A}_{ij} = \frac{1}{n+2}\tilde{\nabla}_i\tilde{\nabla}_j\mathcal{H} + \frac{1}{n+2}\mathcal{C}_{ij}{}^p\nabla_p\mathcal{H}. \tag{16.50}$$

Exercise 16.12. From the expression for the cubic ground form in Exercise 16.5, derive the following evolution equation under the affine normal flow:

$$\begin{aligned}\partial_t\mathcal{C}_{ijk} = & -\frac{1}{2}\tilde{\nabla}_i\mathcal{A}_{jk} + \frac{1}{2}\mathcal{A}_i^p\mathcal{C}_{pjk} - \frac{\mathcal{H}}{n+2}\mathcal{C}_{ijk} \\ & -\frac{1}{2}\left(\mathcal{A}_i^p\mathcal{C}_{pjk} + \mathcal{A}_j^p\mathcal{C}_{pik} + \mathcal{A}_k^p\mathcal{C}_{pij}\right) \\ & + \frac{1}{2(n+2)}\left(\tilde{g}_{ij}\nabla_k\mathcal{H} + \tilde{g}_{ik}\nabla_j\mathcal{H} + \tilde{g}_{jk}\nabla_i\mathcal{H}\right).\end{aligned} \tag{16.51}$$

Exercise 16.13. Starting from the identity $0 = \tilde{g}^{pq}\tilde{\nabla}_i\tilde{\nabla}_j\mathcal{C}_{pqk}$, apply the Codazzi-type identity (16.46) for the cubic ground form, the affine Gauß equation (16.44), and the Codazzi-type identity (16.45) for the affine curvature to obtain the following Simons-type identity for the cubic ground form:

$$\begin{aligned}\tilde{\Delta}\mathcal{C}_{ijk} = & -\frac{n+2}{2}\left(\tilde{\nabla}_i\mathcal{A}_{jk} - \mathcal{A}_i^p\mathcal{C}_{pjk}\right) + \frac{1}{2}\left(\tilde{g}_{ij}\nabla_k\mathcal{H} + \tilde{g}_{jk}\nabla_i\mathcal{H} + \tilde{g}_{ki}\nabla_j\mathcal{H}\right) \\ & + \frac{\mathcal{H}}{2}\mathcal{C}_{ijk} + \mathcal{C}_{ia}{}^b\mathcal{C}_{pb}{}^a\mathcal{C}_{jk}{}^p + \mathcal{C}_{ja}{}^b\mathcal{C}_{pb}{}^a\mathcal{C}_{ki}{}^p + \mathcal{C}_{ka}{}^b\mathcal{C}_{pb}{}^a\mathcal{C}_{ij}{}^p - 2\mathcal{C}_{ia}{}^b\mathcal{C}_{jb}{}^c\mathcal{C}_{kc}{}^a,\end{aligned}$$

where $\tilde{\Delta}$ is the Laplacian corresponding to the metric $\tilde{g}$.

Exercise 16.14. Deduce from the previous two exercises the following evolution equation:

$$\partial_t \mathcal{C}_{ijk} = \frac{1}{n+2}\tilde{\Delta}\mathcal{C}_{ijk} - \frac{3\mathcal{H}}{n+2}\mathcal{C}_{ijk} - \frac{1}{2}\left(\mathcal{A}_i^p\mathcal{C}_{pjk} + \mathcal{A}_j^p\mathcal{C}_{pki} + \mathcal{A}_k^p\mathcal{C}_{pij}\right)$$

(16.52)

$$+ \frac{1}{n+2}\Big(2\mathcal{C}_{ia}{}^b\mathcal{C}_{jb}{}^c\mathcal{C}_{kc}{}^a - \mathcal{C}_{ia}{}^b\mathcal{C}_{pb}{}^a\mathcal{C}_{jk}{}^p - \mathcal{C}_{ja}{}^b\mathcal{C}_{pb}{}^a\mathcal{C}_{ki}{}^p - \mathcal{C}_{ka}{}^b\mathcal{C}_{pb}{}^a\mathcal{C}_{ij}{}^p\Big).$$

Combine this with (16.49) to show that

$$\begin{aligned}\partial_t|\mathcal{C}|^2 &= \frac{1}{n+2}\tilde{\Delta}|\mathcal{C}|^2 - \frac{2}{n+2}|\tilde{\nabla}\mathcal{C}|^2 + \frac{4}{n+2}\mathcal{C}^{ijk}\mathcal{C}_{ia}{}^b\mathcal{C}_{jb}{}^c\mathcal{C}_{kc}{}^a \\ &\quad - \frac{6}{n+2}\mathcal{C}_p{}^{ab}\mathcal{C}_{ab}{}^q\mathcal{C}_q{}^{cd}\mathcal{C}_{cd}{}^p.\end{aligned} \tag{16.53}$$

Using the inequality $\left|\mathcal{C}_{ij}{}^a\mathcal{C}_{akl} - \mathcal{C}_{ik}{}^a\mathcal{C}_{ajl}\right|^2 \geq 0$, deduce using the maximum principle that the supremum of $|\mathcal{C}|^2$ is nonincreasing in t under the affine normal flow.

Chapter 17

Flows by Superaffine Powers of the Gauß Curvature

We have seen that convex hypersurfaces shrink to points under the α-Gauß curvature flow for any $\alpha > 0$ but that the asymptotic shape is *not* necessarily spherical, at least when $\alpha = \frac{1}{n+2}$. We will see in this chapter that the asymptotic shape is always spherical when $\alpha > \frac{1}{n+2}$: By exploiting the monotonicity of the Gaußian and Firey entropies, we prove that solutions converge, after rescaling, to self-similarly shrinking solutions [**51, 254**]. The argument is then completed by showing that shrinking spheres are the only self-similarly shrinking solutions [**113, 148**].

17.1. Bounds on diameter, speed, and inradius

We will exploit the monotonicity of the Gaußian entropy to obtain upper bounds for diameter and speed as well as a lower bound for the inradius, for any $\alpha > \frac{1}{n+2}$. We begin with the case $\alpha = 1$, following Hamilton [**265**].

17.1.1. The Gauß curvature flow. Recall that the Gaußian entropy $E(\mathcal{M}^n)$ of a convex, uniformly convex hypersurface $\mathcal{M}^n \subset \mathbb{R}^{n+1}$ is defined by

$$E(\mathcal{M}^n) \doteqdot \left(\frac{\mathrm{Vol}(\mathcal{M}^n)}{|B^{n+1}|}\right)^{\frac{n}{n+1}} \exp\left(\frac{1}{|S^n|}\int_{\mathcal{M}^n} K \log K \, d\mu\right).$$

Recall that the shadow area of a convex hypersurface $\mathcal{M}^n$ in the direction $z \in S^n$ is the n-dimensional area of its projection $\pi_{z^\perp}(\mathcal{M}^n)$ onto the subspace $z^\perp$. Define the **least shadow area** $A_\#(\mathcal{M}^n)$ by

$$A_\#(\mathcal{M}^n) \doteqdot \min_{z\in S^n} \mathscr{H}^n(\pi_{z^\perp}(\mathcal{M}^n)). \tag{17.1}$$

The diameter bound follows from three lemmas. The first two allow us to estimate the entropy from below in terms of the least shadow area. The third relates the least shadow area to the diameter.

Lemma 17.1. *Given any projection π of a convex hypersurface $\mathcal{M}^n$ onto a hyperplane P in $\mathbb{R}^{n+1}$ and any subset U of $\mathcal{M}^n$ with associated unit normals N lying in one hemisphere, denote by $A_\pi(U)$ the area of $\pi(U)$ in P and by $A_{\mathrm{G}}(U) \le \frac{1}{2}\sigma_n$ the area of $\mathrm{G}(U)$ in $(S^n, \overline{g})$, where G is the Gauß map. Then*

$$\int_U K \log K \, d\mu \ge A_{\mathrm{G}}(U) \log\left(\frac{A_{\mathrm{G}}(U)}{A_\pi(U)}\right) - C_n, \tag{17.2}$$

where, denoting by S^n_+ the northern hemisphere in S^n and by $\phi(z)$ the angle $z \in S^n_+$ makes with the north pole, we define

$$C_n \doteqdot \int_{S^n_+} \log\sec\phi(z)\, d\bar{\mu}(z), \tag{17.3}$$

which is a finite improper integral.

Proof. Denote by $d\mu_P$ the area element of P. Observe that $d\bar{\mu} = K \sec\phi \, d\mu_P$. By Jensen's inequality,

$$\begin{aligned}\int_U K\log K \, d\mu + C_n &= -\int_{\mathrm{G}(U)} \log\left(\frac{1}{K\sec\phi}\right) d\bar{\mu} \\ &\ge -A_{\mathrm{G}}(U)\log\left(\frac{1}{A_{\mathrm{G}}(U)}\int_{\mathrm{G}(U)}\frac{1}{K\sec\phi}\,d\bar{\mu}\right) \\ &= -A_{\mathrm{G}}(U)\log\left(\frac{A_\pi(U)}{A_{\mathrm{G}}(U)}\right).\end{aligned}$$

□

Lemma 17.2. *Given a convex, uniformly convex hypersurface $\mathcal{M}^n \subset \mathbb{R}^{n+1}$,*

$$\exp\left(\frac{1}{\sigma_n}\int_{\mathcal{M}^n} K \log K \, d\mu\right) \ge \mathrm{e}^{-2\frac{C_n}{\sigma_n}} \frac{\sigma_n}{2A_\#(\mathcal{M}^n)}. \tag{17.4}$$

Proof. Let π be the projection whose shadow has the least area, $A_\#$. The hypersurface $\mathcal{M}^n$ divides into two regions, U_+ and U_-, in such a way that the projection π is one-to-one on each, with image $A_\#$, while the orientation of the projection is opposite on their interiors. Since the images of U_+ and U_- under the Gauß map are two hemispheres, each of area $\sigma_n/2$, the claim follows by applying Lemma 17.1 on both U_+ and U_- and adding. □

It remains to estimate the diameter in terms of the least shadow area.

Lemma 17.3 ([**265**, Lemma 1d4]). *The boundary $\mathcal{M}^n$ of any bounded convex body satisfies*

$$\mathrm{Vol}(\mathcal{M}^n) \geq \frac{1}{n+1}\mathrm{diam}(\mathcal{M}^n)A_\#(\mathcal{M}^n). \tag{17.5}$$

Proof. Choose a line segment L in $\mathcal{M}^n$ with length $\mathrm{diam}(\mathcal{M}^n)$ and let π be the projection onto the subspace orthogonal to L. Denote by $A_L(\mathcal{M}^n) \geq A_\#(\mathcal{M}^n)$ the area of the projection of $\mathcal{M}^n$ under π onto $L^\perp$. We may rotate and translate so that L lies in the e_{n+1}-axis with its midpoint at the origin. Since $\mathcal{M}^n$ is convex, it encloses the two frustra over $\pi(\mathcal{M}^n)$ with respective vertices at the endpoints of L. But these each enclose $(n+1)$-dimensional volume $\frac{1}{n+1}\frac{\mathrm{diam}(\mathcal{M}^n)}{2}A_L(\mathcal{M}^n)$. □

Proposition 17.4 (Scale-invariant diameter bound). *Any maximal, uniformly convex solution $\{\mathcal{M}^n_t\}_{t\in[0,T)}$ to Gauß curvature flow satisfies*

$$\mathrm{diam}(\mathcal{M}^n_t) \leq C(T-t)^{\frac{1}{n+1}}, \tag{17.6}$$

where C depends only on n and $E(\mathcal{M}^n_0)$.

Proof. Combining Lemma 17.2 and the entropy estimate of Proposition 15.23, we find that

$$A_\#(\mathcal{M}^n_t) \geq \sigma_n(n+1)^{\frac{n}{n+1}}\mathrm{e}^{-2\frac{C_n}{\sigma_n}}E(\mathcal{M}^n_0)^{-1}(T-t)^{\frac{n}{n+1}}.$$

The claim now follows from Lemma 17.3 since, by Chou's theorem (Theorem 15.3) and the variation formula (15.12), the volume $\mathrm{Vol}(\mathcal{M}^n_t)$ enclosed by $\mathcal{M}^n_t$ satisfies

$$\mathrm{Vol}(\mathcal{M}^n_t) = \sigma_n(T-t).$$ □

Proposition 17.5 (Scale-invariant speed bound). *Any maximal, uniformly convex solution $\{\mathcal{M}^n_t\}_{t\in[0,T)}$ to the Gauß curvature flow satisfies*

$$K \leq C(T-t)^{-\frac{n}{n+1}} \quad \textit{for } t \in [T/2, T), \tag{17.7}$$

where C depends only on n, $\mathrm{Vol}(\mathcal{M}^n_0)$, and $E(\mathcal{M}^n_0)$.

Proof. By Chou's theorem (Theorem 15.3), we may translate so that the solution contracts to the origin at time T. In the Gauß map parametrization, the differential Harnack inequality for the Gauß curvature flow (Theorem 15.13) becomes

$$\partial_t(t^{\frac{n}{n+1}}K) \geq 0,$$

which yields, for any $(z,t) \in S^n \times [0,T)$ and $s \in [t,T)$,

$$t^{\frac{n}{n+1}}K(z,t) \leq s^{\frac{n}{n+1}}K(z,s).$$

Estimating $s \leq T$ and recalling that $\partial_t \sigma = -K$ yields

$$K(z,t) \leq -\left(\frac{T}{t}\right)^{\frac{n}{n+1}} \partial_t \sigma(z,s).$$

Since the solution contracts to the origin at time T, integrating from $s = t$ to $s = T$ yields

$$K(z,t) \leq \left(\frac{T}{t}\right)^{\frac{n}{n+1}} \frac{\sigma}{T-t} \quad \text{for all } (z,t) \in S^n \times [0,T).$$

Since the origin is enclosed by the solution at all times, we may estimate $\max_{z \in S^n} \sigma(z,t) \leq \operatorname{diam}(\mathcal{M}^n_t)$ for all $t \in [0,T)$. So the claim follows from Proposition 17.4. □

17.1.2. The α-Gauß curvature flow. Recall that, when $\alpha \neq 1$, the Gaußian entropy $E_\alpha(\mathcal{M}^n)$ of a uniformly convex hypersurface $\mathcal{M}^n$ is given by

$$\begin{aligned} E_\alpha(\mathcal{M}^n) &= \left(\frac{\operatorname{Vol}(\mathcal{M}^n)}{|B^{n+1}|}\right)^{\frac{n}{n+1}} \left(\int_{\mathcal{M}^n} K^\alpha \, d\mu\right)^{\frac{1}{\alpha-1}} \\ &= \left(\frac{\operatorname{Vol}(\mathcal{M}^n)}{|B^{n+1}|}\right)^{\frac{n}{n+1}} \left(\int_{S^n} K^{\alpha-1} \, d\bar{\mu}\right)^{\frac{1}{\alpha-1}}. \end{aligned}$$

We begin, as in Section 17.1.1, by seeking a lower bound for the shadow areas in terms of the entropy. Here, a fairly crude estimate will suffice.

Lemma 17.6. *If $\alpha > \frac{1}{n+2}$, then any uniformly convex hypersurface $\mathcal{M}^n \subset \mathbb{R}^{n+1}$ satisfies*

$$E_\alpha(\mathcal{M}^n) \geq C^{-1} \left(\frac{\operatorname{Vol}(\mathcal{M}^n)^{\frac{n}{n+1}}}{A_\#(\mathcal{M}^n)}\right)^\beta, \tag{17.8}$$

where $A_\#(\mathcal{M}^n)$ is the least shadow area of $\mathcal{M}^n$ and where C and β are positive constants which depend only on n and α.

Proof. Given $z_0 \in S^n$, the n-dimensional area $\mathscr{H}^n(\pi_{z_0^\perp}(\mathcal{M}^n))$ of the projection $\pi_{z_0^\perp}(\mathcal{M}^n)$ of $\mathcal{M}^n$ onto the subspace $z_0^\perp$ is given by

$$\mathscr{H}^n(\pi_{z_0^\perp}(\mathcal{M}^n)) = \frac{1}{2}\int_{S^n} |\langle z_0, z\rangle| K(z)^{-1} d\bar{\mu}(z).$$

Hölder's inequality yields, for $\alpha \in (0,1)$,

$$\mathscr{H}^n(\pi_{z_0^\perp}(\mathcal{M}^n)) \geq \frac{1}{2}\left(\int_{S^n} K^{\alpha-1} d\bar{\mu}\right)^{\frac{1}{1-\alpha}} \left(\int_{S^n} |\langle z_0, z\rangle|^{1-\frac{1}{\alpha}} d\bar{\mu}(z)\right)^{\frac{\alpha}{\alpha-1}}.$$

If $\alpha \in (\frac{1}{2}, 1)$, then $1 - \frac{1}{\alpha} > -1$ and hence

$$C \doteqdot \int_{S^n} |\langle z_0, z\rangle|^{1-\frac{1}{\alpha}} d\bar{\mu}(z) < \infty .$$

So we may estimate

$$\mathscr{H}^n(\pi_{z_0^\perp}(\mathcal{M}^n)) \geq \frac{C^{\frac{\alpha}{\alpha-1}}}{2|B^{n+1}|^{\frac{n}{n+1}}} \operatorname{Vol}(\mathcal{M}^n)^{\frac{n}{n+1}} E_\alpha(\mathcal{M}^n)^{-1}. \tag{17.9}$$

The same inequality holds for $\alpha \geq 1$ since, by the Hölder inequality, $E_\alpha(\mathcal{M}^n)$ is nondecreasing in α (see Exercise 15.6).

It remains to consider the case $\alpha \in (\frac{1}{n+2}, \frac{1}{2}]$. We first apply Hölder's inequality to estimate

$$E_\alpha(\mathcal{M}^n) \geq E_{\frac{1}{n+2}}(\mathcal{M}^n)^{\frac{(n+1)(3/4-\alpha)}{4(1-\alpha)(3n+2)}} E_{3/4}(\mathcal{M}^n)^{\frac{\alpha-1/(n+2)}{(1-\alpha)(3-4/(n+2))}} .$$

Since $E_{\frac{1}{n+2}}$ is nonincreasing under the $\frac{1}{n+2}$-Gauß curvature flow, Theorem 16.1 implies that

$$E_{\frac{1}{n+2}}(\mathcal{M}^n) \geq 1 . \tag{17.10}$$

(This inequality is known as the *affine isoperimetric inequality* (see [**92**, §26 and §73] or [**454**] and the notes and commentary at the end of Chapter 16). So the previous bound (17.9) for $E_{3/4}$ yields the desired estimate. □

We next convert this into a bound for the maximum width of $\mathcal{M}^n$.

Lemma 17.7. *Any uniformly convex hypersurface $\mathcal{M}^n \subset \mathbb{R}^{n+1}$ satisfies*

$$w_+(\mathcal{M}^n) \leq 2(n+1) \frac{\operatorname{Vol}(\mathcal{M}^n)}{A_\#(\mathcal{M}^n)}, \tag{17.11}$$

where $w_+(\mathcal{M}^n)$ is its maximum width.

Proof. Let z_0 be the normal direction of a pair of parallel supporting hyperplanes for $\mathcal{M}^n$ of maximal separation. The points of contact of $\mathcal{M}^n$ with these two planes are joined by a segment enclosed by $\mathcal{M}^n$ of length equal to the maximum width of $\mathcal{M}^n$. Choosing the origin to be at the center of this segment, we have $\sigma(z_0) = \sigma(-z_0) = \frac{w_+(\mathcal{M}^n)}{2}$ and $\sigma(z) \geq |\langle z, z_0\rangle| \sigma(z_0)$ for all $z \in S^n$. But then

$$\begin{aligned} \operatorname{Vol}(\mathcal{M}^n) = \frac{1}{n+1}\int_{S^n} \sigma K^{-1} d\bar{\mu} &\geq \frac{1}{n+1}\int_{S^n} \sigma(z_0)|\langle z, z_0\rangle| K(z)^{-1} d\bar{\mu}(z) \\ &= \frac{w_+(\mathcal{M}^n)\mathscr{H}^n(\pi_{z_0^\perp}(\mathcal{M}^n))}{2(n+1)} . \end{aligned}$$

□

Corollary 17.8. *If $\alpha > \frac{1}{n+2}$, then any uniformly convex hypersurface $\mathcal{M}^n \subset \mathbb{R}^{n+1}$ satisfies*

$$\frac{r_+(\mathcal{M}^n)}{r_-(\mathcal{M}^n)} \leq C E_\alpha(\mathcal{M}^n)^{(n+1)\beta}, \tag{17.12}$$

where r_+ and r_- denote the circumradius and inradius, respectively, and C and β are positive constants which depend only on α.

Proof. Note that $\mathrm{Vol}(\mathcal{M}^n) \leq w_-(\mathcal{M}^n) w_+(\mathcal{M}^n)^n$, where $w_-(\mathcal{M}^n)$ is the minimum width of $\mathcal{M}^n$, since $\mathcal{M}^n$ in contained between $n+1$ pairs of parallel planes in any set of orthonormal directions and, in particular, in the case where one of the pairs of planes is at minimal separation, in which case the separation of all the other pairs is bounded by $w_+(\mathcal{M}^n)$. Lemma 17.7 then implies that

$$w_-(\mathcal{M}^n) \geq \mathrm{Vol}(\mathcal{M}^n) w_+(\mathcal{M}^n)^{-n} \geq (2(n+1))^{-n}\,\mathrm{Vol}(\mathcal{M}^n)^{1-n} A_\#(\mathcal{M}^n)^n.$$

Applying Lemma 17.7 again, in conjunction with Lemma 8.25, yields

$$\frac{r_+(\mathcal{M}^n)}{r_-(\mathcal{M}^n)} \leq \frac{n+2}{\sqrt{3}} \frac{w_+(\mathcal{M}^n)}{w_-(\mathcal{M}^n)} \leq (2(n+1))^{n+1} \frac{n+2}{\sqrt{3}} \frac{\mathrm{Vol}(\mathcal{M}^n)^n}{A_\#(\mathcal{M}^n)^{n+1}}.$$

The claim now follows from Lemma 17.6. □

Proposition 17.9 (Upper diameter bound). *Any maximal, uniformly convex solution $\{\mathcal{M}^n_t\}_{t\in[0,T)}$ to the α-Gauß curvature flow satisfies*

$$\mathrm{diam}(\mathcal{M}^n_t) \leq C(T-t)^{\frac{1}{n\alpha+1}}, \tag{17.13}$$

where C depends only on n, α, and $E_\alpha(\mathcal{M}^n_0)$.

Proof. Given $t_0 \in [0,T)$, choose the origin at the center of a ball of radius $r_-(\mathcal{M}^n_{t_0})$ enclosed by $\mathcal{M}^n_{t_0}$. Fix $z \in S^n$ and define, for each $\varepsilon \in (0,1]$, a sphere

$$S_\varepsilon(z) \doteqdot \left\{y \in \mathbb{R}^{n+1} : \left|y - (1-\varepsilon)\overline{X}(z,t_0)\right| = \varepsilon r_-(\mathcal{M}^n_{t_0})\right\},$$

where $\overline{X}(\cdot,t_0) : S^n \to \mathbb{R}^{n+1}$ is the canonical embedding of $\mathcal{M}^n_{t_0}$,

$$\overline{X}(z,t_0) \doteqdot \sigma(z,t_0)z + \overline{\nabla}\sigma(z,t_0).$$

Then $S_\varepsilon(z)$ is contained in the convex hull of $\overline{X}(z,t_0)$ and $B_{r_-(\mathcal{M}^n_{t_0})}(0)$ and hence in the region enclosed by $\mathcal{M}^n_{t_0}$. Since $S_\varepsilon(z)$ shrinks to the point $\overline{X}(z,t_0)$ at time $\frac{\varepsilon^{n\alpha+1}}{n\alpha+1} r_-(\mathcal{M}^n_{t_0})^{n\alpha+1}$, we find

$$\begin{aligned}\sigma(z,t) - \sigma(z,t_0) &\geq -\frac{((n\alpha+1)(t-t_0))^{\frac{1}{n\alpha+1}}}{r_-(\mathcal{M}^n_{t_0})}\sigma(z,t_0)\\ &\geq -((n\alpha+1)(t-t_0))^{\frac{1}{n\alpha+1}}\frac{r_+(\mathcal{M}^n_{t_0})}{r_-(\mathcal{M}^n_{t_0})}.\end{aligned}$$

Since, by Chou's theorem (Theorem 15.3), the solution shrinks to a point at time T, the claim now follows from Corollary 17.8 and the entropy monotonicity (Proposition 15.24). □

The differential Harnack inequality now yields a bound for the speed.

Proposition 17.10 (Upper speed bound). *Any maximal convex, uniformly convex solution $\{\mathcal{M}_t^n\}_{t\in[0,T)}$ to the α-Gauß curvature flow satisfies*

$$K^\alpha \leq C\,(T-t)^{-\frac{n\alpha}{n\alpha+1}} \tag{17.14}$$

for all $t \in [T/2, T)$, where C depends only on n and α.

Proof. The argument is similar to the proof of Proposition 17.4 (and Exercise 10.6). □

To conclude this section, we observe that both the diameter and the inradius are in fact controlled from above and below in terms of the remaining time:

Proposition 17.11. *If $\alpha > \frac{1}{n+2}$ and $\{\mathcal{M}_t\}$ is a smooth solution of the α-Gauß curvature flow, then there exists a constant c depending only on $E_\alpha(\mathcal{M}_0)$, n, and α such that*

$$c^{-1}(T-t)^{\frac{1}{1+n\alpha}} \leq r_-(\mathcal{M}_t) \leq r_+(\mathcal{M}_t) \leq c(T-t)^{\frac{1}{1+n\alpha}}. \tag{17.15}$$

Proof. The bound (17.12) implies that both $r_+(\mathcal{M}_t)$ and $r_-(\mathcal{M}_t)$ are comparable to $\mathrm{Vol}(\mathcal{M}_t)^{\frac{1}{n+1}}$, so it suffices to show that $\mathrm{Vol}(\mathcal{M}_t)$ is comparable to $(T-t)^{\frac{n+1}{1+n\alpha}}$. This follows from the identity

$$\partial_t \mathrm{Vol}(\mathcal{M}_t) = -c(n)E_\alpha(\mathcal{M}_t)^{\alpha-1}\,\mathrm{Vol}(\mathcal{M}_t)^{\frac{n(1-\alpha)}{n+1}},$$

together with the observation that $E_\alpha(\mathcal{M}_t)$ is bounded above and below, and the fact that $\mathrm{Vol}(\mathcal{M}_t)$ approaches zero as t approaches T by Chou's theorem. □

17.2. Convergence to a shrinking self-similar solution

We now complete the proof, due to Guan and Ni [**254**] and the first author, Guan, and Ni [**51**], that, when $\alpha > \frac{1}{n+1}$, convex solutions to the α-Gauß curvature flow converge to self-similar solutions after rescaling.

Theorem 17.12. *If $\alpha > \frac{1}{n+2}$, then any closed strictly convex hypersurface is the initial data for an α-Gauß curvature flow which converges after rescaling to a closed strictly convex shrinking self-similar solution.*

The crucial estimate in the proof is a scale-invariant lower bound on the support function computed about the final point, which we establish by controlling the *entropy point* associated with the Firey entropy. A scale-invariant lower bound on the speed can then be deduced, and this in turn implies scale-invariant upper and lower bounds on the principal curvatures. Higher regularity and convergence to a self-similar solution are straightforward consequences.

Let $\{\mathcal{M}_t^n\}_{0\le t<T}$ be a maximal, smooth, uniformly convex solution of the α-Gauß curvature flow in $\mathbb{R}^{n+1}$ with $\alpha > \frac{1}{n+2}$. By Chou's theorem, the time slices $\mathcal{M}_t^n$ converge to a point as t approaches T, and we choose the origin to be at that point. Define the rescaled hypersurfaces[1]

$$\tilde{\mathcal{M}}_t \doteqdot ((1+n\alpha)(T-t))^{-\frac{1}{1+n\alpha}}\,\mathcal{M}_t. \tag{17.16}$$

By Proposition 17.11, the diameter and inradius of $\tilde{\mathcal{M}}_t$ are comparable to 1 on $[0,T)$. Since each of the hypersurfaces $\tilde{\mathcal{M}}_t$ encloses the origin, the diameter bound implies that the support function $\tilde{\sigma} = (T-t)^{-\frac{1}{n\alpha+1}}\sigma$ of $\tilde{\mathcal{M}}_t$ is bounded and uniformly Lipschitz. Indeed,

$$\tilde{\sigma}^2 + |\overline{\nabla}\tilde{\sigma}|^2 = |\tilde{X}|^2 \le \sup_{t\in[0,T)} \operatorname{diam}(\tilde{\mathcal{M}}_t)\,.$$

We will prove, by reductio ad absurdum, that there exists $a>0$ such that

$$\tilde{\sigma}(z,t) > a \quad \text{for all } z \in S^n \text{ and } t \in [0,T). \tag{17.17}$$

Suppose to the contrary that there exists a sequence (z_k, t_k) such that $\tilde{\sigma}(z_k,t_k)$ converges to zero. Since $K^\alpha > 0$, we may arrange (by passing to a subsequence if necessary) that t_k is an increasing sequence approaching T and that $z_k \to z_0 \in S^n$ as $k\to\infty$. By the uniform Lipschitz bound and the Arzelà–Ascoli theorem (or equivalently, the Blaschke selection theorem) we can assume that $\tilde{\sigma}(t_k,\cdot)$ converges uniformly to the support function $\tilde{\sigma}$ of a convex hypersurface $\tilde{\mathcal{M}} = \partial\tilde{\Omega}$ such that $\tilde{\sigma}(z_0) = 0$ and with the same bounds on diameter and inradius. By construction, $\tilde{\mathcal{M}}$ is contained in the closed half-space $\{X :\ X\cdot z_0 \le 0\}$ and contains the origin.

Now we consider the Firey entropy in more detail: Unlike the Gaußian entropy, the Firey entropy depends on the choice of origin. We make this

[1] Note that we could have chosen to rescale to keep the enclosed volume constant, or various other normalizations, without substantially changing the ensuing argument. The choice we made here amounts to the requirement that the time to extinction of the α-Gauß curvature flow with initial hypersurface $\tilde{\mathcal{M}}_t$ is equal to $\frac{1}{1+n\alpha}$, which equals the time to extinction of the unit sphere.

explicit by defining the entropy relative to a point p:
(17.18)

$$\mathcal{E}_\alpha(\mathcal{M}^n, p) \doteqdot \begin{cases} \left(\dfrac{|B^{n+1}|}{\mathrm{Vol}(\mathcal{M}^n)}\right)^{\frac{1}{n+1}} \exp\left(\dfrac{1}{|S^n|}\displaystyle\int_{S^n} \log(\sigma - p\cdot z)\, d\bar{\mu}\right) & \text{if } \alpha = 1, \\ \left(\dfrac{|B^{n+1}|}{\mathrm{Vol}(\mathcal{M}^n)}\right)^{\frac{1}{n+1}} \left(\dfrac{1}{|S^n|}\displaystyle\int_{S^n} (\sigma - p\cdot z)^{\frac{\alpha-1}{\alpha}}\, d\bar{\mu}\right)^{\frac{\alpha}{\alpha-1}} & \text{if } \alpha \neq 1, \end{cases}$$

where $\sigma : S^n \to \mathbb{R}$ is the support function of $\mathcal{M}^n$ and p is any point in the region enclosed by $\mathcal{M}$. This expression makes sense if $\mathcal{M}$ is the boundary of a bounded open convex region and p is a point in the interior of the region.

Lemma 17.13. *If $\mathcal{M} = \partial\Omega$, where Ω is a bounded open convex region, then the function $p \in \Omega \mapsto \mathcal{E}_\alpha(\mathcal{M}, p)$ attains a maximum at a unique point p_e in $\bar{\Omega}$. We will refer to the point p_e as the* ***entropy point*** *of $\mathcal{M}$.*

Proof. We first establish some properties of the integrals defining the entropy: Define $I_\alpha : \Omega \to \mathbb{R}$ by

$$I_\alpha(p) = \begin{cases} \displaystyle\int_{S^n} \log(\sigma(z) - p\cdot z)\, d\bar{\mu}(z), & \alpha = 1, \\ \displaystyle\int_{S^n} (\sigma(z) - p\cdot z)^{1-\frac{1}{\alpha}}\, d\bar{\mu}(z), & \alpha \neq 1, \end{cases}$$

where σ is the support function of $\mathcal{M}$ and $p \in \Omega$. Then I_α is a smooth function on Ω, and if $\alpha > 1$, then I_α is continuous on $\bar{\Omega}$, while if $\alpha = 1$, then it is continuous on $\bar{\Omega}$ as a map into $\mathbb{R} \cup \{-\infty\}$, and if $0 < \alpha < 1$, then I_α is continuous from $\bar{\Omega}$ to $\mathbb{R} \cup \{\infty\}$. It follows that I_α attains a maximum value in $\bar{\Omega}$ if $\alpha \geq 1$, and it attains a minimum value in $\bar{\Omega}$ if $0 < \alpha < 1$. Thus in all cases, $\mathcal{E}_\alpha(\mathcal{M}, \cdot)$ attains a maximum value in $\bar{\Omega}$.

Uniqueness of the maximum point of $\mathcal{E}_\alpha(\mathcal{M}, \cdot)$ in $\bar{\Omega}$ follows from the concavity properties of I_α: If $\alpha > 1$, then a short computation yields

$$\begin{aligned} D^2 I_\alpha|_p(v, v) &= -\frac{1}{\alpha}\left(1 - \frac{1}{\alpha}\right)\int_{S^n} (\sigma(z) - z\cdot p)^{-\frac{1+\alpha}{\alpha}} (z\cdot v)^2\, d\bar{\mu}(z) \\ &\leq -\frac{1}{\alpha}\left(1 - \frac{1}{\alpha}\right)\mathrm{diam}(\mathcal{M})^{-\frac{1+\alpha}{\alpha}} \frac{|S^n||v|^2}{n+1} < 0 \end{aligned}$$

for $v \in \mathbb{R}^{n+1}$, so that I_α is uniformly concave. The same is true for $\alpha = 1$, while for $0 < \alpha < 1$, I_α is uniformly convex. □

Lemma 17.14. *The entropy point p_e of $\tilde{\mathcal{M}}$ is at the origin.*

Proof. Since $\tilde{\Omega}_{t_k}$ converges to $\tilde{\Omega}$ in Hausdorff distance, for any $p \in \tilde{\Omega}$ we have $p \in \tilde{\Omega}_{t_k}$ for all sufficiently large k. Then for k large and $\ell > k$ and writing $\lambda(t) \doteqdot ((1+n\alpha)(T-t))^{\frac{1}{1+n\alpha}}$, we have

$$\begin{aligned}\mathcal{E}_\alpha\left(\tilde{\mathcal{M}}_{t_k}, \frac{\lambda(t_\ell)}{\lambda(t_k)}p\right) &= \mathcal{E}_\alpha\left(\frac{1}{\lambda(t_k)}\mathcal{M}_{t_k}, \frac{\lambda(t_\ell)}{\lambda(t_k)}p\right)\\ &= \mathcal{E}_\alpha(\mathcal{M}_{t_k}, \lambda(t_\ell)p)\\ &\geq \mathcal{E}_\alpha(\mathcal{M}_{t_\ell}, \lambda(t_\ell)p)\\ &= \mathcal{E}_\alpha(\tilde{\mathcal{M}}_{t_\ell}, p),\end{aligned}$$

where we used the scale invariance of the entropy and the fact that the entropy is decreasing under the flow. Now letting $\ell \to \infty$, the right-hand side converges to $\mathcal{E}_\alpha(\tilde{\mathcal{M}}, p)$, while the left-hand side converges to $\mathcal{E}_\alpha(\tilde{\mathcal{M}}_{t_k}, 0)$. Finally, allowing k to approach infinity gives the inequality

$$\mathcal{E}_\alpha(\tilde{\mathcal{M}}, 0) \geq \mathcal{E}_\alpha(\tilde{\mathcal{M}}, p).$$

Since $p \in \tilde{\Omega}$ is arbitrary, this proves the claim. □

Recall that, by construction, the origin is on the boundary of $\tilde{\Omega}$.

Lemma 17.15. *For any open bounded convex set Ω, the entropy point $p_e(\Omega)$ is in the interior of Ω.*

Proof. We will show that no $p \in \mathcal{M} = \partial\Omega$ maximizes the entropy. Given $p \in \mathcal{M}$, we claim that there exists a supporting hyperplane $H = \{x : x \cdot z_0 = p \cdot z_0\}$ for Ω at p such that $p - sz_0 \in \Omega$ for all sufficiently small s.

To prove the claim, it suffices to show that the intersection $\mathcal{T}(p) \cap (-\mathcal{N}(p))$ is nontrivial, where $\mathcal{T}(p)$ and $\mathcal{N}(p)$ are the tangent and normal cones at p, respectively.[2] If this is not the case, then by the Hyperplane Separation Theorem [**461**, Theorem 1.3.7 and Theorem 1.3.9] there exists $z \in S^n$ such that $z \cdot a \geq 0$ for all $a \in -\mathcal{N}(p)$ and $z \cdot b \leq 0$ for all $b \in \mathcal{T}(p)$. But the last statement implies $z \in \mathcal{N}(p)$, and this contradicts the first statement since we can choose $a = -z$.

Now let $R(z) = z - 2(z \cdot z_0) z_0$ be the reflection in the hyperplane normal to z_0. Write $\sigma_p(z) = \sigma(z) - p \cdot z$. Then for any $z \in S^n$ with $z \cdot z_0 > 0$ there exists $x \in \mathcal{M}$ with $\sigma_p(z) = (x-p) \cdot z$, and then we have

$$\sigma_p(R(z)) \geq (x-p) \cdot R(z) = \sigma_p(z) + 2(z \cdot z_0) z_0 \cdot (p-x).$$

Since $z_0 \in \mathcal{N}(p)$, we conclude that $\sigma_p(R(z)) > \sigma_p(z)$. Furthermore, the inequality is strict when $z = z_0$ (and therefore is strict in a set of positive

[2] Recall that the *normal cone* of Ω at p is the set $\mathcal{N}(p) = \{v \in \mathbb{R}^{n+1} : \sigma(v) = p \cdot v\}$, where $\sigma(v) = \sup\{x \cdot v : x \in \Omega\}$ (thus σ is the support function defined previously when $v \in S^n$ and is positively homogeneous of degree 1). Since σ is a convex function, $\mathcal{N}(p)$ is a closed convex cone, and $\mathcal{N}(p) \cap S^n$ is the set of outward unit normal vectors to all supporting hyperplanes of Ω at p. The *tangent cone* of Ω is the open convex cone defined by $\mathcal{T}(p) = \bigcup_{r>0} r(\Omega - p)$. This consists of those vectors v such that $p + sv \in \Omega$ for all sufficiently small $s > 0$.

measure near z_0). It follows, in case $\alpha > 1$, that

$$\begin{aligned}
\frac{d}{ds}I_\alpha(p - sz_0)\big|_{s=0} &= -\int_{S^n} \sigma_p(z)^{-1/\alpha}\,(z\cdot z_0)\,d\bar\mu(z) \\
&= \int_{\{z:z\cdot z_0>0\}} \left(\sigma_p(z)^{-1/\alpha} - \sigma_p(R(z))^{-1/\alpha}\right) z\cdot z_0\,d\bar\mu(z) \\
&> 0.
\end{aligned}$$

A similar computation holds if $\alpha = 1$ or in the case $0 < \alpha < 1$ in which case the inequality is reversed. Therefore in all cases p is not the entropy point, and the lemma is proved. □

This contradicts our assumption that $\tilde\sigma$ does not have a positive lower bound, so we have established the estimate (17.17) for some $a > 0$.

The next step is to derive a positive lower bound on the Gauß curvature of the hypersurfaces $\tilde{\mathcal{M}}_t$. Since we have positive upper and lower bounds on the support function of the form $a \le \tilde\sigma(z,t) \le R$, this follows directly from the Harnack inequality (15.45), which is equivalent to the inequality $\frac{d}{dt}\left(K^\alpha(z,t)t^{\frac{n\alpha}{1+n\alpha}}\right) \ge 0$: Set $C = (1+\frac{R}{a})^{1+n\alpha}$, and consider t sufficiently close to T that $t' = T - C(T-t) > 0$. This implies

$$\begin{aligned}
\sigma(z,t') - \sigma(z,t) &\ge a(T-t')^{\frac{1}{1+n\alpha}} - R(T-t)^{\frac{1}{1+n\alpha}} \\
&= (C^{\frac{1}{1+n\alpha}}a - R)(T-t)^{\frac{1}{1+n\alpha}} \\
&= a(T-t)^{\frac{1}{1+n\alpha}}.
\end{aligned} \tag{17.19}$$

On the other hand, since $\partial_t\sigma = -K^\alpha$, the differential Harnack inequality (15.45) yields

$$\begin{aligned}
\sigma(z,t') - \sigma(z,t) &= \int_{t'}^{t} K^\alpha(z,s)\,ds \\
&\le K^\alpha(z,t)t^{\frac{n\alpha}{1+n\alpha}}\int_{t'}^{t} s^{-\frac{n\alpha}{1+n\alpha}}\,ds \\
&\le K^\alpha(z,t)\left(\frac{t}{t'}\right)^{\frac{n\alpha}{1+n\alpha}}(t-t') \\
&= (C-1)K^\alpha(z,t)\left(\frac{t}{t'}\right)^{\frac{n\alpha}{1+n\alpha}}(T-t) \\
&= (C-1)\tilde K^\alpha(z,t)\left(\frac{t}{t'}\right)^{\frac{n\alpha}{1+n\alpha}}(T-t)^{\frac{1}{1+n\alpha}}.
\end{aligned} \tag{17.20}$$

Combining (17.19) and (17.20) gives $\tilde K(z,t)^\alpha \ge \frac{a}{C-1}(t'/t)^{n\alpha/(1+n\alpha)}$, and we observe that the last factor can be made as close as desired to 1 by choosing

t sufficiently close to T. This gives a uniform lower bound on $\tilde{K}$ depending only on a, R, α, and n for t sufficiently close to T.

Next, we control the principal curvatures: This follows exactly as in Section 16.2.3, giving a uniform lower bound on the principal curvatures $\tilde{\kappa}_i$ of the hypersurfaces $\tilde{\mathcal{M}}_t$. Upper bounds on $\tilde{\kappa}_i$ follow as in Section 16.2.4. This implies uniform upper and lower bounds on the eigenvalues $A[\tilde{\sigma}]$ on $S^n \times [0, T)$.

Finally, uniform bounds on all higher derivatives of $\tilde{\sigma}$ follow from Theorems 15.9 and 15.10, and convergence to a shrinking self-similar solution along a sequence of times t_k approaching T follows from the monotonicity of either the Gaußian entropy or the Firey entropy, following the argument in Section 16.3.

It follows from the result of [**31**] that $\tilde{\sigma}(\cdot, t)$ converges smoothly to the self-similar limit as $t \to T$, rather than converging only along a subsequence of times. However in view of the result of the next section, which implies that the only self-similar shrinking solution which has interval of existence $1/(1 + n\alpha)$ and final point at the origin is the unit sphere about the origin, the smooth convergence follows by a much easier argument (see Exercise 17.7). This completes the proof of Theorem 17.12.

We remark that, when $\frac{1}{n+2} < \alpha < \frac{1}{n}$, the argument presented here (with the lower bound on Gauß curvature provided by the argument of Proposition 16.7) also applies for anisotropic variants of the Gauß curvature flows, of the form

$$\partial_t X = -\rho(\mathsf{N}) K^\alpha \mathsf{N},$$

where ρ is any smooth positive function on S^n. The conclusion is the same: Solutions converge, after rescaling about the final point to keep the enclosed volume fixed (or by a suitable power of the remaining time), to a self-similar solution of the same flow. The same result holds also when $\alpha = 1/n$, with a slightly different argument required for the lower bound on Gauß curvature [**34**].

17.3. Shrinking self-similar solutions are round

Recall that a **shrinking self-similar solution** $X : M^n \times (-\infty, 0) \to \mathbb{R}^{n+1}$ to the α-Gauß curvature flow is one which evolves by homothetic contraction; that is,

$$X(x, t) = (-t)^{\frac{1}{n\alpha+1}} X_{-1}(\varphi(x, t))$$

for some time-dependent diffeomorphism $\varphi : M^n \times (-\infty, 0) \to M^n$. In particular, the time slices $\mathcal{M}_t = X(M^n, t)$ are all congruent; since they evolve by the α-Gauß curvature flow equation, they must satisfy

$$K^\alpha(x,t) = \frac{1}{-(n\alpha+1)t}\langle X(x,t), \mathsf{N}(x,t)\rangle \tag{17.21}$$

for all $(x,t) \in M^n \times (-\infty, 0)$. Conversely, if a hypersurface $X : M^n \to \mathbb{R}^{n+1}$ satisfies

$$K^\alpha(x) = \langle X(x), \mathsf{N}(x)\rangle \quad \text{for all } x \in M\,, \tag{17.22}$$

then the family of immersed hypersurfaces $\mathcal{M}_t^n \doteqdot (-(n\alpha+1)t)^{\frac{1}{n\alpha+1}}\mathcal{M}^n$, where $\mathcal{M}^n = X(M^n)$, evolves by the α-Gauß curvature flow. We call a hypersurface satisfying (17.22) an α-**Gaußian shrinker**.

In this section we present the proof, due to Choi and Daskalopoulos [**148**] and Brendle, Choi, and Daskalopoulos [**113**], that any uniformly convex shrinking self-similar solution to the α-Gauß curvature flow is a shrinking round sphere when $\alpha > \frac{1}{n+2}$.

Figure 17.1. Kyeongsu Choi.

Theorem 17.16 (Brendle, Choi, and Daskalopoulos). *If $\alpha > \frac{1}{n+2}$ and $n \geq 2$, then any closed strictly convex shrinking self-similar solution to the α-Gauß curvature flow is a round sphere.*

Combined with Theorem 17.12, this proves the convergence of the α-Gauß curvature flow of closed strictly convex hypersurfaces to round points for all powers $\alpha > \frac{1}{n+2}$ and dimensions $n \geq 2$.

Corollary 17.17. *If $\alpha > \frac{1}{n+2}$ and $n \geq 2$, then any compact, locally uniformly convex hypersurface is the initial data for an α-Gauß curvature flow which converges after rescaling to a round sphere.*

For the affine invariant power $\alpha = \frac{1}{n+2}$, we saw that solutions converge to *ellipsoids* after rescaling (Theorem 16.1). A crucial step was to show that shrinking ellipsoids are the only self-similarly shrinking solutions for this power (see Section 16.4). In this section we shall also present an alternate proof of this fact.

Theorem 17.18 (Brendle, Choi, and Daskalopoulos). *If $n \geq 2$, then any closed strictly convex shrinking self-similar solution to the $\frac{1}{n+1}$-Gauß curvature flow is a shrinking ellipsoid.*

The key is to find the right quantities on which to apply the maximum principle.

17.3.1. Preliminary calculations. By differentiating the shrinker equation, we find

$$\frac{1}{2}\nabla|X|^2 = X^\top = \mathrm{II}^{-1}(\nabla K^\alpha). \tag{17.23}$$

By formula (5.23) for $\nabla_i\nabla_j X$,

$$\begin{aligned}\mathcal{L}X &= \alpha K^\alpha(\mathrm{II}^{-1})^{ij}\nabla_i\nabla_j X \\ &= -n\alpha K^\alpha \mathsf{N}.\end{aligned} \tag{17.24}$$

Thus the α-Gauß curvature flow may be written as

$$\partial_t X = \frac{1}{n\alpha}\mathcal{L}X.$$

Note that the case $\alpha = \frac{1}{n}$ is characterized as the case where $\partial_t X = \mathcal{L}X$.

By (17.24) and formula (5.5) for g_{ij}, we compute for any convex hypersurface that

$$\begin{aligned}\frac{1}{2}\mathcal{L}|X|^2 &= \langle \mathcal{L}X, X\rangle + \alpha K^\alpha(\mathrm{II}^{-1})^{ij}\langle\nabla_i X, \nabla_j X\rangle \\ &= \alpha K^\alpha(-n\langle X, \mathsf{N}\rangle + (H_{-1})^{-1}),\end{aligned} \tag{17.25}$$

where $H_{-1} = \frac{H_n}{H_{n-1}}$ is the harmonic mean curvature defined by (5.35).

Suppose that $X(t)$, $t \in [0, \varepsilon)$, is a solution to the α-Gauß curvature flow with $X(0) = X$, where X satisfies $K^\alpha = \langle X, \mathsf{N}\rangle$. We have the following analogue of Lemma 13.7.

Lemma 17.19. *Suppose that $X : M^n \to \mathbb{R}^{n+1}$ is a shrinking self-similar solution to the α-Gauß curvature flow. Let $Q = Q(X)$ be a tensor expression of degree β. Then under the curvature flow we have at $t = 0$,*

$$\partial_t Q = L_{X^\top}Q - \beta Q, \tag{17.26}$$

where $X^\top = \mathrm{II}^{-1}(\nabla K^\alpha)$ and L denotes the Lie derivative.

Since K^α has degree $-n\alpha$, we obtain from Lemma 17.19 that at $t = 0$

$$\partial_t K^\alpha = n\alpha K^\alpha + X^\top \cdot \nabla K^\alpha .$$

Hence the evolution equation (5.142) yields for an α-Gaußian shrinker that

$$\mathcal{L}K^\alpha = X^\top \cdot \nabla K^\alpha + n\alpha K^\alpha - \alpha H K^{2\alpha} . \tag{17.27}$$

Since II has degree 1, by Lemma 17.19 we have that

$$\begin{aligned} \partial_t \mathrm{II}_{ij} &= -\mathrm{II}_{ij} + (L_{X^\top}\mathrm{II})_{ij} \\ &= -\mathrm{II}_{ij} + (\nabla_{X^\top}\mathrm{II})_{ij} + g^{k\ell}(\nabla_i X_k^\top \, \mathrm{II}_{\ell j} + \nabla_j X_k^\top \, \mathrm{II}_{i\ell}) . \end{aligned}$$

On the other hand, by differentiating $(X^\top)_k = (\mathrm{II}^{-1})_k^m \nabla_m K^\alpha$, we obtain

$$\nabla_i X_k^\top = \nabla_i (\mathrm{II}^{-1})_k^m \nabla_m K^\alpha + (\mathrm{II}^{-1})_k^m \nabla_i \nabla_m K^\alpha .$$

Hence

$$\begin{aligned} \partial_t \mathrm{II}_{ij} = &- \mathrm{II}_{ij} + (\nabla_{X^\top}\mathrm{II})_{ij} \\ &+ 2\nabla_i\nabla_j K^\alpha + \nabla_i(\mathrm{II}^{-1})^{m\ell}\nabla_m K^\alpha \, \mathrm{II}_{\ell j} + \nabla_j(\mathrm{II}^{-1})^{m\ell}\nabla_m K^\alpha \, \mathrm{II}_{\ell i} \\ &- \mathrm{II}_{ij} + (\nabla_{X^\top}\mathrm{II})_{ij} + 2\nabla_i\nabla_j K^\alpha - 2(\mathrm{II}^{-1})^{m\ell}\nabla_m K^\alpha \, \nabla_\ell \mathrm{II}_{ij} \\ &- \mathrm{II}_{ij} - (\nabla_{X^\top}\mathrm{II})_{ij} + 2\nabla_i\nabla_j K^\alpha , \end{aligned}$$

where we used the Codazzi equation and (17.23). Combining this with (15.13) yields

$$\partial_t \mathrm{II}_{ij} = \mathrm{II}_{ij} + (\nabla_{X^\top}\mathrm{II})_{ij} - 2K^\alpha \mathrm{II}^2_{ij}. \tag{17.28}$$

By further combining this with (15.9), we conclude for an α-Gaußian shrinker that

$$\begin{aligned} \mathcal{L}\,\mathrm{II}_{ij} = \mathrm{II}_{ij} &+ (\nabla_{X^\top}\mathrm{II})_{ij} - K^{-\alpha}\,\nabla_i K^\alpha\,\nabla_j K^\alpha \\ &+ \alpha K^\alpha (\mathrm{II}^{-1})^{kp}(\mathrm{II}^{-1})^{\ell q}\,\nabla_i \mathrm{II}_{k\ell}\,\nabla_j \mathrm{II}_{pq} \\ &- K^\alpha(\alpha H \mathrm{II}_{ij} - (n\alpha - 1)\mathrm{II}^2_{ij}) . \end{aligned} \tag{17.29}$$

Thus, for an α-Gaußian shrinker,

$$\begin{aligned} \mathcal{L}(\mathrm{II}^{-1})^{ij} = &-(\mathrm{II}^{-1})^{ij} + \nabla_{X^\top}(\mathrm{II}^{-1})^{ij} \\ &+ (\mathrm{II}^{-1})^{ir}(\mathrm{II}^{-1})^{js}\,K^{-\alpha}\,\nabla_r K^\alpha\,\nabla_s K^\alpha \\ &+ \alpha K^\alpha (\mathrm{II}^{-1})^{uv}\,(\mathrm{II}^{-1})^{ir}(\mathrm{II}^{-1})^{jp}(\mathrm{II}^{-1})^{sq}\,\nabla_u \mathrm{II}_{pq}\,\nabla_v \mathrm{II}_{rs} \\ &+ K^\alpha(\alpha H (\mathrm{II}^{-1})^{ij} - (n\alpha - 1)g^{ij}) . \end{aligned} \tag{17.30}$$

Define on $\mathcal{M}$ the contravariant symmetric 2-tensor

$$Z^{ij} \doteqdot K^\alpha(\mathrm{II}^{-1})^{ij} - \frac{n\alpha - 1}{2n\alpha}|X|^2\,g^{ij} . \tag{17.31}$$

Its trace is

$$Z \doteqdot K^\alpha \operatorname{tr}_g(\mathrm{II}^{-1}) - \frac{n\alpha - 1}{2\alpha}|X|^2. \tag{17.32}$$

Note that $\operatorname{tr}_g(\mathrm{II}^{-1}) = g_{ij}(\mathrm{II}^{-1})^{ij} = (H_{-1})^{-1}$ is the reciprocal of the harmonic mean curvature.

Let $\kappa_1 \leq \cdots \leq \kappa_n$ denote the ordered principal curvatures. The maximum eigenvalue of Z^{ij} with respect to g is the function

$$W \doteqdot K^\alpha \kappa_1^{-1} - \frac{n\alpha - 1}{2n\alpha}|X|^2. \tag{17.33}$$

Observe that $W \geq \frac{1}{n} Z$ on $\mathcal{M}$.

Lemma 17.20. *When $\alpha = \frac{1}{n+2}$, i.e., in the affine invariant case, we have that $Z = K^{\frac{1}{n+2}} \operatorname{tr}_g(\mathrm{II}^{-1}) + |X|^2$ is constant on ellipsoids centered at the origin, which are the uniformly convex closed shrinking self-similar solutions of the $\frac{1}{n+2}$-Gauß curvature flow.*

Proof. Recall that an ellipsoid $\mathcal{E}$ centered at the origin in $\mathbb{R}^{n+1}$ is the set of X satisfying $A_{ij}X^iX^j = 1$, where (A_{ij}) is a symmetric, positive-definite $(n+1) \times (n+1)$ matrix.

Claim. On $\mathcal{E}$ we have $Z \equiv \operatorname{tr}(A^{-1})$, i.e., the sum of the squares of the semiaxes of $\mathcal{E}$.

Proof of the claim. We may write $A_{ij} = (L^{-1})_{ik}(L^{-1})_{kj}$, where (L_{ij}) is a symmetric, positive definite matrix. Observe that then $L(S^n) = \mathcal{E}$, so that given $\mathsf{N} \in S^n$, we have $\tilde{X} \doteqdot L(\mathsf{N}) \in \mathcal{E}$ and hence $|\tilde{X}|^2 = |L(\mathsf{N})|^2$. Let $\tilde{g}$ and $\tilde{\mathrm{II}}$ denote the induced metric and second fundamental form of $\mathcal{E}$, respectively. Let $\{e_i\}_{i=1}^n$ be an orthonormal basis for the tangent space to S^n at N. By (16.3) we have $\tilde{g}_{ij} = L(e_i) \cdot L(e_j)$ and by (16.4) we have $(\tilde{\mathrm{II}}^{-1})^{ij} = R^{-1}\tilde{g}^{ij}$, where $R = (\det P)^{-1}$ and P is defined by (16.2). Moreover, by (16.5), we have $\tilde{K} = R^{n+2}K$. Hence

$$\tilde{K}^{\frac{1}{n+2}} \operatorname{tr}_{\tilde{g}}(\tilde{\mathrm{II}}^{-1}) = \sum_{i=1}^n |L(e_i)|^2.$$

Thus

$$\tilde{Z} \doteqdot \tilde{K}^{\frac{1}{n+2}} \operatorname{tr}_{\tilde{g}}(\tilde{\mathrm{II}}^{-1}) + |\tilde{X}|^2 = \sum_{i=1}^n L^2(e_i) \cdot e_i + L^2(\mathsf{N}) \cdot \mathsf{N} = \operatorname{tr}(A^{-1})$$

as claimed. □

Finally, recall from Corollary 16.3, Proposition 16.9, and Corollary 16.11 that the ellipsoids centered at the origin are the uniformly convex closed shrinking self-similar solutions. □

We proceed to calculate $\mathcal{L}Z^{ij}$ and $\mathcal{L}Z$, with the goal of applying the maximum principle. By equations (17.30), (17.27), and (17.25) for $\mathcal{L}(\mathrm{II}^{-1})^{ij}$, $\mathcal{L}K^\alpha$, and $\mathcal{L}|X|^2$, respectively, and by the product rule

$$\begin{aligned}\mathcal{L}(K^\alpha(\mathrm{II}^{-1})^{ij}) = {} & K^\alpha\mathcal{L}(\mathrm{II}^{-1})^{ij} + (\mathrm{II}^{-1})^{ij}\mathcal{L}K^\alpha \\ & + 2\alpha K^\alpha(\mathrm{II}^{-1})^{pq}\nabla_p K^\alpha\nabla_q(\mathrm{II}^{-1})^{ij},\end{aligned}$$

we compute that

$$\begin{aligned}\mathcal{L}\,Z^{ij} = {} & -K^\alpha(\mathrm{II}^{-1})^{ij} + \nabla_{X^\top}(K^\alpha(\mathrm{II}^{-1})^{ij}) \\ & + (\mathrm{II}^{-1})^{ir}(\mathrm{II}^{-1})^{js}\,\nabla_r K^\alpha\,\nabla_s K^\alpha \\ & + \alpha K^{2\alpha}(\mathrm{II}^{-1})^{uv}(\mathrm{II}^{-1})^{ir}(\mathrm{II}^{-1})^{jp}(\mathrm{II}^{-1})^{sq}\nabla_u\mathrm{II}_{pq}\nabla_v\mathrm{II}_{rs} \\ & + 2\alpha K^\alpha(\mathrm{II}^{-1})^{pq}\nabla_p K^\alpha\nabla_q(\mathrm{II}^{-1})^{ij} \\ & + n\alpha(\mathrm{II}^{-1})^{ij}K^\alpha - \frac{n\alpha-1}{n}g^{ij}\,K^\alpha(H_{-1})^{-1}.\end{aligned}$$

Using $\langle X^\top, \nabla|X|^2\rangle = 2|X^\top|^2$ and $X^\top = \mathrm{II}^{-1}(\nabla K^\alpha)$ (by (5.13) and (17.23)) and expressing the right-hand side in terms of Z^{ij} as much as possible, we may rewrite this as

(17.34)

$$\begin{aligned}\mathcal{L}\,Z^{ij} = {} & (n\alpha-1)\left(Z^{ij} - \frac{1}{n}Z\,g^{ij}\right) + (1+2\alpha)(\mathrm{II}^{-1})^{k\ell}\nabla_k K^\alpha\nabla_\ell Z^{ij} \\ & + (\mathrm{II}^{-1})^{ir}(\mathrm{II}^{-1})^{js}\;\nabla_r K^\alpha\,\nabla_s K^\alpha - 2\alpha(\mathrm{II}^{-1})^{ij}(\mathrm{II}^{-1})^{pq}\nabla_p K^\alpha\nabla_q K^\alpha \\ & + (1+2\alpha)\frac{n\alpha-1}{n\alpha}\,\mathrm{II}^{-2}(\nabla K^\alpha,\nabla K^\alpha)\,g^{ij} \\ & + \alpha K^{2\alpha}(\mathrm{II}^{-1})^{uv}\,(\mathrm{II}^{-1})^{ir}(\mathrm{II}^{-1})^{jp}(\mathrm{II}^{-1})^{sq}\,\nabla_u\mathrm{II}_{pq}\,\nabla_v\mathrm{II}_{rs}\,.\end{aligned}$$

Taking the trace of the equation above, i.e., multiplying by g_{ij} and summing, we obtain

$$\begin{aligned}\mathcal{L}\,Z = {} & (1+2\alpha)\mathrm{II}^{-1}(\nabla K^\alpha,\nabla Z) - 2\alpha(H_{-1})^{-1}\,\mathrm{II}^{-1}(\nabla K^\alpha,\nabla K^\alpha) \\ & + \left(2n\alpha + n - 1 - \alpha^{-1}\right)\mathrm{II}^{-2}(\nabla K^\alpha,\nabla K^\alpha) \\ & + \alpha K^{2\alpha}(\mathrm{II}^{-1})^{uv}\,(\mathrm{II}^{-2})^{pr}(\mathrm{II}^{-1})^{sq}\,\nabla_u\mathrm{II}_{pq}\,\nabla_v\mathrm{II}_{rs}\,.\end{aligned} \tag{17.35}$$

In the ensuing subsections, to facilitate applying the maximum principle we delve deeper into the equations above, in particular into the algebra of the right-hand side modulo an appropriate ∇Z^{ij} or ∇Z term.

17.3.2. Powers $\frac{1}{n+2} \leq \alpha \leq \frac{1}{2}$. The idea for enabling the application of the maximum principle to equation (17.35) is to show that the right-hand side terms which are quadratic in the first covariant derivatives of the second fundamental form are in total nonnegative. Here, our leeway is that we may add a term of the form $\langle V, \nabla Z\rangle$, where the vector field V is of our own choosing. It turns out that the right choice is $V = 4\alpha \mathrm{II}^{-1}(\nabla K^\alpha)$, so we may rewrite (17.35) using a principal frame field (so that $g_{ij} = \delta_{ij}$ and $\mathrm{II}_{ij} = \kappa_i \delta_{ij}$) as

$$\begin{aligned}(17.36)\quad \Box Z &\doteqdot \mathcal{L}Z + (2\alpha - 1)\mathrm{II}^{-1}(\nabla K^\alpha, \nabla Z)\\ &= 4\alpha \mathrm{II}^{-1}(\nabla K^\alpha, \nabla Z) + \alpha K^{2\alpha} r_i^2 r_j r_k (\nabla_i \mathrm{II}_{jk})^2\\ &\quad + \alpha K^{2\alpha}\left[-2\alpha^2 r_j r_i + \left(2n\alpha^2 + (n-1)\alpha - 1\right) r_i^2\right] (\nabla_i \log K)^2\\ &\doteqdot B,\end{aligned}$$

where the $r_i = \kappa_i^{-1}$ are the principal radii, we sum over i, j, and k, and we used the Codazzi equation.

We wish to show that $B \geq 0$. Firstly, regarding the term $\langle V, \nabla Z\rangle$, by differentiating (17.32) and using (17.23), we have

$$\begin{aligned}\nabla Z &= K^\alpha \nabla (H_{-1})^{-1} + (H_{-1})^{-1}\nabla K^\alpha - \frac{n\alpha - 1}{\alpha}\mathrm{II}^{-1}(\nabla K^\alpha)\\ &= K^\alpha (\nabla (H_{-1})^{-1} + (\alpha (H_{-1})^{-1} g^{-1} - (n\alpha - 1)\mathrm{II}^{-1})\nabla \log K).\end{aligned}$$

Now,

$$\nabla_i (H_{-1})^{-1} = \nabla_i (g_{jk}(\mathrm{II}^{-1})^{jk}) = -g_{jk}(\mathrm{II}^{-1})^{jp}(\mathrm{II}^{-1})^{kq}\nabla_i \mathrm{II}_{pq}.$$

Hence, with respect to a principal frame field,

$$K^{-\alpha}\nabla_i Z = -\sum_j r_j^2 \nabla_i \mathrm{II}_{jj} + \left(\alpha \sum_j r_j - (n\alpha - 1) r_i\right) \nabla_i \log K.$$

Let D denote the set of (i, j, k) where i, j, and k are distinct. Secondly, regarding the second term of the second line of (17.36), we have

$$\begin{aligned}(17.37)\quad \sum_{i,j,k} r_i^2 r_j r_k (\nabla_i \mathrm{II}_{jk})^2 &= \sum_{(i,j,k)\in D} r_i^2 r_j r_k (\nabla_i \mathrm{II}_{jk})^2\\ &\quad + \sum_{i,j} r_i^2 r_j^2 (\nabla_i \mathrm{II}_{jj})^2 + 2\sum_{i\neq j} r_j^3 r_i (\nabla_i \mathrm{II}_{jj})^2.\end{aligned}$$

Thus the term B, comprising the right-hand side of (17.36) and equaling $\Box Z$, satisfies

$$\begin{aligned}(17.38)\quad &\alpha^{-1}K^{-2\alpha}B\\ &= -4\alpha r_i r_j^2 \nabla_i \log K \nabla_i \mathrm{II}_{jj} + 4\alpha r_i \left(\alpha r_j - (n\alpha - 1) r_i\right) (\nabla_i \log K)^2\\ &\quad + \sum_D r_i^2 r_j r_k (\nabla_i \mathrm{II}_{jk})^2 + \sum_{i,j} r_i^2 r_j^2 (\nabla_i \mathrm{II}_{jj})^2 + 2\sum_{i\neq j} r_j^3 r_i (\nabla_i \mathrm{II}_{jj})^2\\ &\quad + \left(-2\alpha^2 r_i r_j + \left(2n\alpha^2 + (n-1)\alpha - 1\right) r_i^2\right)(\nabla_i \log K)^2,\end{aligned}$$

where we sum over i, j, and k. Since $\sum_D r_i^2 r_j r_k (\nabla_i \mathrm{II}_{jk})^2 \geq 0$ and since $r_i > 0$, in order to prove that $B \geq 0$ it suffices to show for each i that

$$\begin{aligned} J_i \doteqdot & - 4\alpha r_j^2 \nabla_i \log K \nabla_i \mathrm{II}_{jj} + r_i \sum_j r_j^2 (\nabla_i \mathrm{II}_{jj})^2 + 2 \sum_{j\,:\,j\neq i} r_j^3 (\nabla_i \mathrm{II}_{jj})^2 \\ & + \left(2\alpha^2 \textstyle\sum_j r_j + \left(-2n\alpha^2 + (n+3)\alpha - 1\right) r_i\right) (\nabla_i \log K)^2 \end{aligned}$$

is nonnegative. With respect to a principal frame field, we have

$$\nabla_i \log K = (\mathrm{II}^{-1})^{jk} \nabla_i \mathrm{II}_{jk} = \sum_j r_j \nabla_i \mathrm{II}_{jj}\,.$$

Thus, defining $\sigma_j \doteqdot r_j \nabla_i \mathrm{II}_{jj}$, we have

$$\begin{aligned} J_i \doteqdot & - 4\alpha \sum_{j,k} r_j \sigma_k \sigma_j + r_i \sum_j \sigma_j^2 + 2 \sum_{j\,:\,j\neq i} r_j \sigma_j^2 \\ & + \left(2\alpha^2 \sum_j r_j + \left(-2n\alpha^2 + (n+3)\alpha - 1\right) r_i\right) \left(\sum_j \sigma_j\right)^2, \end{aligned} \tag{17.39}$$

which is homogeneous of degree 1 in $r = (r_1, \ldots, r_n)$ and homogeneous of degree 2 in $\sigma = (\sigma_1, \ldots, \sigma_n)$. Without loss of generality, we may assume that $i = n$ in (17.39), so we consider

$$\begin{aligned} J \doteqdot & \; 2\sum_{i=1}^{n-1} r_i \sigma_i^2 - 4\alpha \left(\sum_{i=1}^{n} r_i \sigma_i\right)\left(\sum_{i=1}^{n} \sigma_i\right) + 2\alpha^2 \left(\sum_{i=1}^{n} r_i\right)\left(\sum_{i=1}^{n} \sigma_i\right)^2 \\ & + r_n \sum_{i=1}^{n} \sigma_i^2 + (-2n\alpha^2 + (n+3)\alpha - 1) r_n \left(\sum_{i=1}^{n} \sigma_i\right)^2. \end{aligned} \tag{17.40}$$

The following result may be found in [**113**, Lemma 3]. We present an alternate approach to this algebraic lemma which was graciously provided to us by Greg Anderson. The proof below comprises his notes verbatim.

Lemma 17.21. *If $n \geq 2$, $\frac{1}{n+2} \leq \alpha \leq \frac{1}{2}$, and $r_i > 0$, then $J \geq 0$ and hence $B \geq 0$. Moreover, if $J = 0$, then*

(1) $\sigma_1 = \cdots = \sigma_n = 0$ *or*

(2) *both* $\alpha = \frac{1}{n+2}$ *and* $\sigma_1 = \cdots = \sigma_{n-1} = \frac{1}{3}\sigma_n$,

where $\sigma_j \doteqdot r_j \nabla_n \mathrm{II}_{jj}$. Hence, for a shrinking self-similar solution to the α-Gauß curvature flow, the quantity Z defined by (17.32) *satisfies*

$$\mathcal{L} Z + (2\alpha - 1) \mathrm{II}^{-1} (\nabla K^\alpha, \nabla Z) \geq 0\,. \tag{17.41}$$

Proof. Seemingly regressing rather than progressing we detach some n-th terms from sums on the top line of (17.40) and rearrange arriving at the longer formula

$$\begin{aligned} J &= 2\sum_{i=1}^{n-1} r_i\sigma_i^2 + r_n\sum_{i=1}^{n}\sigma_i^2 \\ &- 4\alpha\left(\sum_{i=1}^{n-1} r_i\sigma_i\right)\left(\sum_{i=1}^{n}\sigma_i\right) + 2\alpha^2\left(\sum_{i=1}^{n-1} r_i\right)\left(\sum_{i=1}^{n}\sigma_i\right)^2 \\ &- 4\alpha r_n\sigma_n\left(\sum_{i=1}^{n}\sigma_i\right) + ((-2n+2)\alpha^2 + (n+3)\alpha - 1)r_n\left(\sum_{i=1}^{n}\sigma_i\right)^2. \end{aligned}$$

This decomposition is the crucial move for the proof. After this we just reap the benefits but the calculations become too difficult to handle with indices.

We move to matrix notation. Let

$$R = \begin{bmatrix} r_1 & & \\ & \ddots & \\ & & r_n \end{bmatrix}, \quad I = I_n = \begin{bmatrix} 1 & & \\ & \ddots & \\ & & 1 \end{bmatrix}, \quad \sigma = \begin{bmatrix} \sigma_1 \\ \vdots \\ \sigma_n \end{bmatrix}, \quad v = \begin{bmatrix} 1 \\ \vdots \\ 1 \end{bmatrix},$$

$$e_1, \ldots, e_n : \text{columns of } I_n.$$

Let H be one-half of the Hessian matrix of J with respect to $\sigma_1, \ldots, \sigma_n$. We have

$$\begin{aligned} H &= 2(I - \alpha vv^T)(I - e_ne_n^T)R(I - e_ne_n^T)(I - \alpha vv^T) \\ &\quad + r_n(I - 2\alpha e_nv^T - 2\alpha ve_n^T + F(\alpha, n)vv^T), \end{aligned} \tag{17.42}$$

where

$$F(\alpha, n) = (-2n+2)\alpha^2 + (n+3)\alpha - 1.$$

It is enough to show that for $\alpha \in \left(\frac{1}{n+2}, \frac{1}{2}\right]$ we have $H > 0$ and otherwise for $\alpha = \frac{1}{n+2}$ we have $H \geq 0$, where the null space of H is spanned by $e_1 + \cdots + e_{n-1} + 3e_n$.

There exists an orthogonal matrix U such that

$$Ue_n = e_n, \quad U(v - e_n) = \sqrt{n-1}\, e_{n-1}, \quad U^2 = I. \tag{17.43}$$

Indeed, one can take U to be a Householder transformation. We now change coordinates using U to simplify H considerably to the point that our problem becomes essentially 2-dimensional. Firstly let

$$L = U(I - e_ne_n^T)(I - \alpha vv^T)U^T = \begin{bmatrix} I_{n-2} & & \\ & 1 - \alpha(n-1) & -\alpha\sqrt{n-1} \\ & 0 & 0 \end{bmatrix}.$$

Secondly let

$$\begin{aligned} M &= U(I - 2\alpha e_n v^T - 2\alpha v e_n^T + F(\alpha,n) v v^T)U^T \\ &= \begin{bmatrix} I_{n-2} & & \\ & 1+(n-1)F(\alpha,n) & (-2\alpha + F(\alpha,n))\sqrt{n-1} \\ & (-2\alpha+F(\alpha,n))\sqrt{n-1} & 1-4\alpha+F(\alpha,n) \end{bmatrix}. \end{aligned}$$

Then we have

$$UHU^T = 2L^T(URU^T)L + r_n M. \tag{17.44}$$

We will use U, L, and M to make quick work of the proof of the lemma.

We consider three cases, namely $\frac{1}{n+2} < \alpha < \frac{1}{2}$, $\alpha = \frac{1}{2}$, and $\alpha = \frac{1}{n+2}$.

Consider the first case $\frac{1}{n+2} < \alpha < \frac{1}{2}$. Since we have

$$\begin{aligned} 1 - 4\alpha + F(\alpha,n) &= (n-1)\alpha(1-2\alpha), \\ \det M &= (n-1)(1-2\alpha)((n+2)\alpha - 1), \end{aligned}$$

by Sylvester's criterion M is positive definite. Obviously $2L^T(URU^T)L$ is positive semidefinite. Thus J is positive definite and the lemma holds in the first case.

Consider the second case $\alpha = \frac{1}{2}$. In this case the matrices

$$-2L = \begin{bmatrix} -2I_{n-2} & & \\ & n-3 & \sqrt{n-1} \\ & 0 & 0 \end{bmatrix}, \quad M = \begin{bmatrix} I_{n-2} & & \\ & n & 0 \\ & 0 & 0 \end{bmatrix}$$

have linearly independent 1-dimensional null (column) spaces. Furthermore, since URU^T is positive definite, the null spaces of L^TURU^TL and L are the same. Thus J is positive definite and the lemma holds in the second case.

Consider the third and last case $\alpha = \frac{1}{n+2}$. We have

$$(n+2)L = \begin{bmatrix} (n+2)I_{n-2} & & \\ & 3 & -\sqrt{n-1} \\ & 0 & 0 \end{bmatrix},$$

$$\frac{(n+2)^2}{3n} M = \begin{bmatrix} \frac{(n+2)^2}{3n} I_{n-2} & & \\ & 3 & -\sqrt{n-1} \\ & -\sqrt{n-1} & \frac{n-1}{3} \end{bmatrix};$$

hence the null spaces of both matrices $L^T(URU^T)L$ and M are spanned by $\sqrt{n-1}e_{n-1} + 3e_n$ and hence the null space of H is spanned by

$$U(\sqrt{n-1}\, e_{n-1} + 3e_n) = e_1 + \cdots + e_{n-1} + 3e_n\,.$$

The proof of the lemma is complete. $\square$

We are now ready to prove the classification of shrinking self-similar solutions for $\frac{1}{n+2} \leq \alpha \leq \frac{1}{2}$. By appplying the strong maximum principle to (17.41), we have that Z is constant on M^n. Thus the right-hand side of (17.36) is zero; that is, both $\sum_D r_j^2 r_k r_\ell (\nabla_j \mathrm{II}_{k\ell})^2 = 0$ and $J_i = 0$ in (17.39), where the r_j are the principal curvatures and $\sigma_j = r_j \nabla_i \mathrm{II}_{jj}$. Thus

$$\nabla_j \mathrm{II}_{k\ell} = 0 \quad \text{for } j, k, \text{ and } \ell \text{ distinct.} \tag{17.45}$$

By Lemma 17.21 and since $J_i = 0$, we have Case 1 or Case 2:

Case 1. $r_j \nabla_i \mathrm{II}_{jj} = 0$ *for all* $j = 1, \ldots, n$.

Case 2. $\alpha = \frac{1}{n+2}$ *and*

$$r_j \nabla_i \mathrm{II}_{jj} = \frac{1}{3} r_i \nabla_i \mathrm{II}_{ii} \quad \textit{for } j \neq i. \tag{17.46}$$

In the first case, since $r_j > 0$ and by the Codazzi equation, we conclude that $\nabla_j \mathrm{II}_{k\ell} = 0$ for all j, k, and ℓ. Therefore $\mathcal{M}$ is a round sphere by Corollary 5.11. Since $\alpha = \frac{1}{n+2}$ in the second case, this completes the proof of the $\frac{1}{n+2} < \alpha \leq \frac{1}{2}$ cases of Theorem 17.16.

Now we consider the second case. By (17.46), we have

$$\nabla_i \log K = \sum_{j=q}^{n} r_j \nabla_i \mathrm{II}_{jj} = \frac{n+2}{3} r_i \nabla_i \mathrm{II}_{ii} = (n+2) r_k \nabla_i \mathrm{II}_{kk} \tag{17.47}$$

for each $k \neq i$. From this we obtain that the cubic ground form vanishes ($\mathcal{C}_{ijk}$ defined below is the pullback by the Gauß map of the 3-tensor defined in Exercise 16.5):

(17.48)
$$2\mathcal{C}_{ijk} \doteqdot K^{-\frac{1}{n+2}} \nabla_k \mathrm{II}_{ij} + \mathrm{II}_{jk} \nabla_i K^{-\frac{1}{n+2}} + \mathrm{II}_{ki} \nabla_j K^{-\frac{1}{n+2}} + \mathrm{II}_{ij} \nabla_k K^{-\frac{1}{n+2}} = 0$$

for all i, j, k. Indeed, we may assume that II is diagonal with respect to our frame. (1) If i, j, k are distinct, then (17.45) implies that $\mathcal{C}_{ijk} = 0$. (2) If $i \neq j$, then by (17.47) we have

$$2\mathcal{C}_{ijj} = K^{-\frac{1}{n+2}} \left(\nabla_j \mathrm{II}_{ij} - \frac{1}{n+2} r_j^{-1} \nabla_i \log K \right) = 0.$$

(3) By using (17.47) again, we obtain

$$2\mathcal{C}_{iii} = K^{-\frac{1}{n+2}} \left(\nabla_i \mathrm{II}_{ii} - \frac{3}{n+2} r_i^{-1} \nabla_i \log K \right) = 0.$$

Finally, by Exercise 16.9, we conclude that $\mathcal{M}$ is an ellipsoid.

17.3.3. Powers $\alpha > \frac{1}{2}$. Instead of considering the scalar function Z as for the previous case $\frac{1}{n+2} \leq \alpha \leq \frac{1}{2}$, here we examine in essence the tensor Z^{ij} defined by (17.31). So we recall the Lipschitz continuous function $W : \mathcal{M} \to \mathbb{R}$ defined by (17.33):

$$W(p) = K^{\alpha}(p)\kappa_1^{-1}(p) - \frac{n\alpha - 1}{2n\alpha}|p|^2 , \tag{17.49}$$

where $\kappa_1 \leq \cdots \leq \kappa_n$ are the ordered principal curvatures. This is the maximum eigenvalue function of the symmetric 2-tensor Z^{ij}.

Let $p_0 \in \mathcal{M}$ be a point where W attains its maximum. Define the function φ by

$$W(p_0) \equiv K^{\alpha}(p)\varphi^{-1}(p) - \frac{n\alpha - 1}{2n\alpha}|p|^2 \quad \text{for all } p \in \mathcal{M} .$$

That is,

$$\varphi(p) = K^{\alpha}(p)\left(W(p_0) + \frac{n\alpha - 1}{2n\alpha}|p|^2\right)^{-1} \tag{17.50}$$

and we see that φ is a well-defined smooth function near p_0. Since $W(p_0) \geq W(p)$ for all p, we have $K^{\alpha}(p)\varphi^{-1}(p) \geq K^{\alpha}(p)\kappa_1^{-1}(p)$, so that

$$\varphi(p) \leq \kappa_1(p) \quad \text{for all } p .$$

Since $\varphi(p_0) = \kappa_1(p_0)$, we have that φ supports κ_1 from below at p_0. Of course,

$$\mathcal{L}\left(K^{\alpha}(p)\varphi^{-1}(p) - \frac{n\alpha - 1}{2n\alpha}|p|^2\right) = \mathcal{L}(W(p_0)) = 0 \tag{17.51}$$

since $W(p_0)$ is a constant. We shall consider this trivial equality from a nontrivial barrier point of view, which is essentially equivalent to applying the maximum principle to either W or Z^{ij}.

Let m denote the multiplicity of the lowest eigenvalue κ_1 of II, so that $\kappa_1 = \cdots = \kappa_m < \kappa_{m+1} \leq \cdots \leq \kappa_n$. By Proposition 12.15, we have at p_0 with respect to the principal basis that both

$$\nabla_i \mathrm{II}_{jk} = \nabla_i \varphi \, \delta_{jk} \quad \text{for } 1 \leq i \leq n \text{ and } 1 \leq j, k \leq m \tag{17.52}$$

and

$$\nabla_i \nabla_i \varphi - \nabla_i \nabla_i \mathrm{II}_{11} \leq -2 \sum_{j>m} (\kappa_j - \kappa_1)^{-1} (\nabla_i \mathrm{II}_{1j})^2 .$$

Multiplying this inequality by $\alpha K^{\alpha} r_i$ and summing yields

$$\mathcal{L}\varphi - (\mathcal{L}\,\mathrm{II})_{11} \leq -2\alpha K^{\alpha} \sum_{i=1}^{n} \sum_{j>m} r_i (\kappa_j - \kappa_1)^{-1} (\nabla_i \mathrm{II}_{1j})^2 . \tag{17.53}$$

At our maximum point p_0 of W we have

$$\nabla_i \varphi = \nabla_i \mathrm{II}_{11} \quad \text{and} \quad \nabla_i \varphi^{-1} = -r_1^2 \nabla_i \mathrm{II}_{11} . \tag{17.54}$$

Thus, by (17.51), we have at p_0

$$(\mathcal{L}Z)^{11} = (\mathcal{L}Z)^{11} - \mathcal{L}\left(K^\alpha\varphi^{-1} - \tfrac{n\alpha-1}{2n\alpha}|p|^2\right) \tag{17.55}$$
$$= K^\alpha((\mathcal{L}(\mathrm{II}^{-1}))^{11} - \mathcal{L}\varphi^{-1})$$
$$= K^\alpha r_1^2(\mathcal{L}\varphi - (\mathcal{L}\mathrm{II})_{11})$$
$$\leq -2\alpha K^{2\alpha}r_1^2\sum_{i=1}^{n}\sum_{j>m} r_i(\kappa_j-\kappa_1)^{-1}(\nabla_i\mathrm{II}_{1j})^2,$$

where we used the product rule

$$\mathcal{L}(AB) = B\mathcal{L}A + A\mathcal{L}B + 2\alpha K^\alpha \mathrm{II}^{-1}(\nabla A, \nabla B)$$

and canceled terms via (17.54) to obtain the second equality. On the other hand, with respect to a principal basis, (17.34) says that

$$(\mathcal{L}Z)^{11} = (n\alpha-1)K^\alpha\left(r_1 - \tfrac{1}{n}\textstyle\sum_i r_i\right) + (1+2\alpha)\, r_i\nabla_i K^\alpha \nabla_i Z^{11} \tag{17.56}$$
$$+ r_1^2\,(\nabla_1 K^\alpha)^2 - 2\alpha r_1\sum_i r_i\,(\nabla_i K^\alpha)^2$$
$$+ (1+2\alpha)\,\frac{n\alpha-1}{n\alpha}\sum_i r_i^2\,(\nabla_i K^\alpha)^2$$
$$+ \alpha K^{2\alpha}r_1^2 r_i r_j\,(\nabla_i\mathrm{II}_{1j})^2.$$

Since at p_0 we have $\nabla_i Z^{11} = 0$, by combining (17.55) and (17.56), we obtain

(17.57)

$$0 \geq (n\alpha-1)K^\alpha\left(r_1 - \tfrac{1}{n}\textstyle\sum_i r_i\right)$$
$$+ r_1^2\,(\nabla_1K^\alpha)^2 + \sum_i\left(-2\alpha r_1 r_i + (1+2\alpha)\,\frac{n\alpha-1}{n\alpha}\,r_i^2\right)(\nabla_iK^\alpha)^2$$
$$+ \alpha K^{2\alpha}r_1^2 r_i r_j\,(\nabla_i\mathrm{II}_{1j})^2 + 2\alpha K^{2\alpha}r_1^2\sum_{i=1}^{n}\sum_{j>m} r_i(\kappa_j-\kappa_1)^{-1}(\nabla_i\mathrm{II}_{1j})^2.$$

Observe that as long as $\alpha > \frac{1}{n}$, the first term on the right-hand side is nonnegative and vanishes if and only if p_0 is an umbilic point. The only possibly negative term on the right-hand side is the second term of the second line. We further take advantage of the vanishing of the "gradient term" at p_0; namely we have

$$0 = \nabla_i Z^{11} = \left(r_1 - \frac{n\alpha-1}{n\alpha}r_i\right)\nabla_i K^\alpha - K^\alpha r_1^2\,\nabla_i\mathrm{II}_{11} \tag{17.58}$$

since

$$\nabla|X|^2 = 2X^\top = 2\mathrm{II}^{-1}(\nabla K^\alpha)$$

by (17.23).

Taking $1 \leq i \leq m$ in (17.58), we have

$$\nabla_i K^\alpha = n\alpha K^\alpha r_1 \nabla_i \mathrm{II}_{11} \, .$$

By (17.52), we have

$$\nabla_i \mathrm{II}_{1k} = \nabla_i \varphi \, \delta_{1k} = 0 \quad \text{for } 1 \leq i \leq n \text{ and } 2 \leq k \leq m \, .$$

Taking $i = 1$, we have

$$\nabla_k \mathrm{II}_{11} = \nabla_1 \mathrm{II}_{1k} = 0 \quad \text{for } 2 \leq k \leq m \, .$$

Hence, by (17.58) we have for $2 \leq k \leq m$ that

$$\nabla_k K^\alpha = n\alpha K^\alpha r_1 \nabla_k \mathrm{II}_{11} = 0 \, . \tag{17.59}$$

Regarding the last line of (17.57), we have

$$\begin{aligned}
\alpha K^{2\alpha} r_1^2 r_i r_j (\nabla_i \mathrm{II}_{1j})^2 &+ 2\alpha K^{2\alpha} r_1^2 \sum_{i=1}^n \sum_{j>m} r_i (\kappa_j - \kappa_1)^{-1} (\nabla_i \mathrm{II}_{1j})^2 \\
&\geq \alpha K^{2\alpha} r_1^4 \, (\nabla_1 \mathrm{II}_{11})^2 + 2\alpha K^{2\alpha} r_1^2 \sum_{j>m} r_1 (\kappa_j - \kappa_1)^{-1} (\nabla_1 \mathrm{II}_{1j})^2 \\
&= \frac{1}{n^2 \alpha} r_1^2 (\nabla_1 K^\alpha)^2 + 2\alpha\kappa_1 \sum_{j>m} (\kappa_j - \kappa_1)^{-1} \left(r_1 - \tfrac{n\alpha - 1}{n\alpha} r_j\right)^2 (\nabla_j K^\alpha)^2.
\end{aligned}$$

Thus, using (17.59), we have

(17.60)

$$\begin{aligned}
0 \geq \; & (n\alpha - 1) K^\alpha \left(r_1 - \tfrac{1}{n} \textstyle\sum_i r_i\right) \\
& + r_1^2 \, (\nabla_1 K^\alpha)^2 + \sum_i \left(-2\alpha r_1 r_i + \tfrac{(1+2\alpha)(n\alpha-1)}{n\alpha} \, r_i^2\right) (\nabla_i K^\alpha)^2 \\
& + \frac{1}{n^2 \alpha} r_1^2 (\nabla_1 K^\alpha)^2 + 2\alpha\kappa_1 \sum_{j>m} (\kappa_j - \kappa_1)^{-1} \left(r_1 - \tfrac{n\alpha-1}{n\alpha} r_j\right)^2 (\nabla_j K^\alpha)^2 \\
= \; & (n\alpha - 1) K^\alpha \left(r_1 - \tfrac{1}{n} \textstyle\sum_i r_i\right) + \tfrac{(n-1)(2n\alpha - 1)}{n^2 \alpha} r_1^2 \, (\nabla_1 K^\alpha)^2 \\
& + \sum_{j>m} \left(-2\alpha r_1 r_j + \tfrac{(1+2\alpha)(n\alpha-1)}{n\alpha} \, r_j^2 + 2\alpha \tfrac{\kappa_1}{\kappa_j - \kappa_1} (r_1 - \tfrac{n\alpha-1}{n\alpha} r_j)^2\right) (\nabla_j K^\alpha)^2.
\end{aligned}$$

Now the coefficient in the last line may be simplified as follows:

$$
\begin{aligned}
&-2\alpha r_1 r_j + \tfrac{(1+2\alpha)(n\alpha-1)}{n\alpha} r_j^2 + 2\alpha \tfrac{\kappa_1}{\kappa_j-\kappa_1}\left(r_1 - \tfrac{n\alpha-1}{n\alpha} r_j\right)^2 \\
&= \tfrac{\kappa_1}{\kappa_j-\kappa_1}\left(-2\alpha r_1(r_1-r_j) + \tfrac{(1+2\alpha)(n\alpha-1)}{n\alpha} r_j(r_1-r_j) + 2\alpha\left(r_1 - \tfrac{n\alpha-1}{n\alpha} r_j\right)^2\right) \\
&= \tfrac{\kappa_1}{\kappa_j-\kappa_1}\left(\left(\tfrac{n\alpha-1}{n\alpha} + \tfrac{2}{n}\right) r_1 r_j - \tfrac{(n+2)(n\alpha-1)}{n^2\alpha} r_j^2\right) \\
&= \tfrac{\kappa_1}{\kappa_j-\kappa_1}\left(\left(\tfrac{n\alpha-1}{n\alpha} + \tfrac{2}{n}\right)(r_1-r_j) r_j + \tfrac{2}{n^2\alpha} r_j^2\right) \\
&= \left(\tfrac{n\alpha-1}{n\alpha} + \tfrac{2}{n} + \tfrac{2}{n^2\alpha}\tfrac{\kappa_1}{\kappa_j-\kappa_1}\right) r_j^2 .
\end{aligned}
$$

Hence (17.60) becomes

$$
\begin{aligned}
(17.61)\qquad 0 \geq (n\alpha-1)K^\alpha\left(r_1 - \tfrac{1}{n}\textstyle\sum_i r_i\right) &+ \tfrac{(n-1)(2n\alpha-1)}{n^2\alpha} r_1^2\left(\nabla_1 K^\alpha\right)^2 \\
&+ \sum_{j>m}\left(\tfrac{n\alpha-1}{n\alpha} + \tfrac{2}{n} + \tfrac{2}{n^2\alpha}\tfrac{\kappa_1}{\kappa_j-\kappa_1}\right) r_j^2 (\nabla_j K^\alpha)^2.
\end{aligned}
$$

Since $\alpha > \frac{1}{n}$, each of the terms on the right-hand side is nonnegative and hence zero and in particular any maximum point p_0 of W is an umbilic point.

At this point we may return to the evolution equation for Z (defined in (17.32)). By (17.36) and (17.37), we may rewrite it as

$$
\begin{aligned}
(17.62)\quad & \mathcal{L}Z - (2\alpha+1)\mathrm{II}^{-1}(\nabla K^\alpha, \nabla Z) \\
&= \alpha K^{2\alpha}\left[-2\alpha^2 r_j r_i + \left(2n\alpha^2 + (n-1)\alpha - 1\right) r_i^2\right] (\nabla_i \log K)^2 \\
&\quad + \alpha K^{2\alpha} r_i^2 r_j r_k (\nabla_i \mathrm{II}_{jk})^2 \\
&= \alpha K^{2\alpha}\left[-2\alpha^2 r_j r_i + \left(2n\alpha^2 + (n-1)\alpha - 1\right) r_i^2\right] (\nabla_i \log K)^2 \\
&\quad + \sum_D r_i^2 r_j r_k (\nabla_i \mathrm{II}_{jk})^2 + \sum_{i,j} r_i^2 r_j^2 (\nabla_i \mathrm{II}_{jj})^2 + 2\sum_{i\neq j} r_j^3 r_i (\nabla_i \mathrm{II}_{jj})^2 .
\end{aligned}
$$

Regarding the sign of the coefficient of the first term of the right-hand side,

$$
c(\alpha) \doteqdot -2\alpha^2 r_j r_i + \left(2n\alpha^2 + (n-1)\alpha - 1\right) r_i^2 ,
$$

which is the only possibly negative term on the right-hand side, suppose that we are at an umbilic point. By rescaling, we may assume that $r_i = 1$ for all i. Then we obtain

$$
c(\alpha) = 2(n-1)\alpha^2 + (n-1)\alpha - 1 .
$$

In particular, $c(\frac{1}{n}) = \frac{n-2}{n^2}$, which is > 0 for $n \geq 3$ and is $= 0$ for $n = 2$. Hence, $c(\alpha) > 0$ for $\alpha > \frac{1}{n}$ and for r_i close enough to each other. As a consequence of continuity, if $\alpha > \frac{1}{n}$ and $n \geq 2$, then near p_0 which is an umbilic point, we have that

$$\mathcal{L}Z - (2\alpha+1)\mathrm{II}^{-1}(\nabla K^{\alpha}, \nabla Z) \geq 0 \tag{17.63}$$

in some neighborhood $\mathcal{U}$ of p_0.

On the other hand, we have for all $p \in \mathcal{U}$,

$$Z(p) \leq nW(p) \leq nW(p_0) = Z(p_0), \tag{17.64}$$

where the last equality is true since p_0 is an umbilic point. Thus p_0 is a local maximum point of Z. Therefore, by applying the strong maximum principle to (17.63) we have that $Z(p) \equiv Z(p_0)$ for $p \in \mathcal{U}$. By (17.64) this implies that $W(p) \equiv W(p_0)$ for $p \in \mathcal{U}$. We obtain that the set of points $p \in \mathcal{M}$ where $W(p) = W(p_0)$ is open, closed (since W is continuous), and nonempty, and hence all of $\mathcal{M}$. That is, W is a constant function. Since each maximum point of W is umbilic, we conclude that $\mathcal{M}$ is totally umbilic and hence a round sphere (see Exercise 5.5). This completes the proof of the $\alpha > \frac{1}{n}$ cases of Theorem 17.16. By combining this with the result of the previous subsection, we see that the proof of Theorem 17.16 is finished.

17.4. Notes and commentary

Theorem 17.12 was first obtained, by Penfei Guan and Lei Ni, for the 1-Gauß curvature flow [**254**]. The argument of Guan and Ni was extended to all powers $\alpha > \frac{1}{n+2}$ by the first author, Guan, and Ni [**51**]. We discussed the affine invariant case, $\alpha = \frac{1}{n+2}$, in Section 16.3. The Firey entropy was exploited by Mohammad Ivaki [**316**] in the context of *expanding* Gauß curvature flows.

The argument of Theorem 17.16 was discovered by Kyeongsu Choi and Toti Daskalopulos [**148**] for powers $\alpha \in (\frac{1}{n}, 1+\frac{1}{n})$ (and, in particular, for the Gauß curvature flow). It was extended to all powers $\alpha > \frac{1}{n+2}$ by Brendle, Choi, and Daskalopoulos [**113**]. We discussed the affine invariant case, $\alpha = \frac{1}{n+2}$, in Section 16.4.

Let us mention some results related to the α-Gauß curvature flow not discussed here. Kevin Olwell [**432**] has constructed translating solutions to the Gauß curvature flow which are graphs of convex functions over convex domains. Vladimir Oliker [**429**], [**430**] showed that immortal strictly convex graphical solutions to the initial-Dirichlet boundary value problem to the Gauß curvature flow over smooth bounded strictly convex domains in $\mathbb{R}^n$ asymptotically limit to zero and after rescaling limit to self-similar solutions. As pointed out in [**429**] the global in time existence of solutions follows with some work from the results in the book of Nicolai Krylov [**340**]. Xiaolong Li and Kui Wang [**367**] extended Oliker's work to the α-Gauß curvature flow for $\alpha > 1/n$. Naoyuki Ishimura [**314**] proved that any positive rotationally symmetric graphical self-similar solution $u : \mathbb{R}^n \to \mathbb{R}$, $u(x) = U(|x|)$, to the Gauß curvature flow with $U(r) \to \infty$ as $r \to \infty$ must be convex and asymptotically linear. Moreover, he showed that for any positive constant K there exists such a solution satisfying $U(r) = Kr + o(r^{-1})$ as $r \to \infty$. Kyeongsu Choi, Panagiota Daskalopoulos, Lami Kim, and Ki-Ahm Lee [**149**] proved the global in time existence of complete strictly convex graphical solutions to the initial value problem for the α-Gauß curvature flow (over compact or noncompact domains). Urbas [**507**] proved the existence of a C^∞ translating self-similar solution to the α-Gauß curvature flow over a bounded strictly convex domain. Choi, Daskalopoulos, and Lee [**150**] proved the existence of a $C^{1,1}_{\text{loc}}$ translating self-similar solution to the α-Gauß curvature flow over a bounded weakly convex domain.

The study of the Gauß curvature flow with so-called **flat sides** is carried out by Richard Hamilton [**266**], Daskalopoulos and Hamilton [**191**], and Daskalopoulos and Lee [**194**].

Claus Gerhardt [**236**] and John Urbas [**510**] proved that expanding flows which are homogeneous of degree $p = -1$ in the principal curvatures asymptotically converge to the shape of round spheres as the hypersurfaces expand to infinity; this includes the speed $K^{-1/n}$, i.e., $\alpha = -1/n$. Urbas [**511**] extended this to degrees $-1 < p < 0$. As a special case of $p = -2 < -1$, Oliver Schnürer [**463**] proved that the inverse Gauß curvature flow (i.e., $\alpha = -1$) expands convex surfaces in $\mathbb{R}^3$ to asymptotically round spheres. More generally, Gerhardt [**239**] extended this in all dimensions to expanding flows of degrees $p < -1$. Regularity for the α-Gauß curvature flow for homogeneity $p = n\alpha \in (-1, 0)$ was studied by Mohammad Ivaki [**315**] by considering the shrinking dual flow.

A C^1 estimate due to Robert Gulliver and the second author [**164**], which is most effective in the case of expanding geometric flows, follows from the Alexandrov reflection principle. James McCoy [**394**] improved this result by removing a Lipschitz continuity assumption for the speed.

As observed by the first author, for any solution to the Gauß curvature flow of a hypersurface in 4-dimensional Euclidean or Minkowski space, its induced metric (provided it is positive definite, i.e., Riemannian) is a solution to the cross curvature flow in [**166**] of Hamilton and the second author. John Buckland [**128**] proved short-time existence for the cross curvature flow when the sectional curvature of the initial metric has a sign.

A logarithmic Gauß curvature flow was used to give an alternate proof of the Minkowski problem by Kaiseng Chou and Xu-Jia Wang [**154**]. Dong-Ho Tsai and the second author [**173**] considered flows by nonhomogeneous functions of the Gauß curvature. Flows by nonhomogeneous functions of the principal curvatures have been considered by Tsai and the second author

[**172**], Liou, Tsai, and the second author [**169**], and Nina Ivochkina, Thomas Nehring, and Friedrich Tomi [**318**].

In [**366**], Qi-Rui Li, Weimin Sheng, and Xu-Jia Wang considered the anisotropic Gauß curvature flow $\partial_t X = f(\mathsf{N})|X|^\alpha K\mathsf{N}$ for a fixed function f and positive number α. They gave a new solution of the classical Alexandrov problem and they solved, assuming some regularity, the dual q-Minkowski problem of Yong Huang, Erwin Lutwak, Deane Yang, and Gaoyong Zhang [**290**].

By adapting Brendle, Choi, and Daskalopoulos's methods, Shanze Gao, Haizhong Li, and Hui Ma [**234**] proved that compact convex shrinking self-similar solutions to the power α of the m-th mean curvature flow, where $\alpha \geq 1/m$, are round spheres.

17.5. Exercises

Exercise 17.1. Prove that C_n in (17.3) is finite. Calculate C_2.

Exercise 17.2. Using the evolution equations derived in Section 15.2, show that under the α-Gauß curvature flow,

$$\partial_t \sigma = \mathcal{L}\sigma - (1 + n\alpha)K^\alpha + \alpha H K^\alpha \sigma \, ,$$

where $\mathcal{L} = \alpha K^\alpha (\mathrm{II}^{-1})^{ij}\nabla_i\nabla_j$ and $\sigma = \langle X, \mathsf{N}\rangle$ is the support function written as a function on the hypersurface.

Exercise 17.3. Using Exercise 17.2 and the evolution equation (5.142), derive the following evolution equation for $\varphi = \frac{K^\alpha}{\sigma - r_0}$ under the α-Gauß curvature flow:

$$(\partial_t - \mathcal{L})\varphi = \frac{2\alpha K^\alpha}{\sigma - r_0}(\mathrm{II}^{-1})^{ij}\nabla_i\varphi\nabla_j\sigma + ((1 + n\alpha) - \alpha H r_0)\,\varphi^2 \, . \tag{17.65}$$

Exercise 17.4. Verify that evolution equations (17.65) and (15.23) are related in the way indicated by the change of parametrization formula (15.18).

Exercise 17.5. We may extend the support function σ from S^n to $\mathbb{R}^{n+1}$ as a function homogeneous of degree 1 and then restrict it to a hyperplane. That is, we first define $\bar{\sigma} : \mathbb{R}^{n+1} - \{\vec{0}\} \to \mathbb{R}$ by

$$\bar{\sigma}(x) = |x|\sigma\left(\frac{x}{|x|}\right) \qquad \text{for } x \in \mathbb{R}^{n+1} - \{\vec{0}\}$$

and then we define $U : \mathbb{R}^n \to \mathbb{R}$ by

$$U(\mathbf{x}) = \bar{\sigma}(\mathbf{x}, -1) \quad \text{for } \mathbf{x} = (x^1, \ldots, x^n) \in \mathbb{R}^n \, .$$

The normal vectors in the southern hemisphere of S^n may be written as

$$\mathsf{N}(\mathbf{x}) = \frac{(\mathbf{x}, -1)}{\sqrt{1 + |\mathbf{x}|^2}} \, .$$

Show that the Gauß curvature at a point in $\mathcal{M}$ with normal vector $\mathsf{N}(\mathbf{x})$ is given by

$$K(\mathsf{N}(\mathbf{x})) = (1+|\mathbf{x}|^2)^{-\frac{n+2}{2}}(\det \partial_i\partial_j U(\mathbf{x}))^{-1}. \tag{17.66}$$

Exercise 17.6. Let X_t be a strictly convex solution to the α-Gauß curvature flow.

(1) Show that

$$\begin{aligned}\partial_t \mathrm{II}_{ij} = \mathcal{L}\mathrm{II}_{ij} &+ \frac{\alpha H^{2n}}{nK^{2-\alpha}}\nabla_i\left(\frac{K}{H^n}\right)\nabla_j\left(\frac{K}{H^n}\right)\\ &- \frac{\alpha K^\alpha}{H^2}(\mathrm{II}^{-1})^{kp}(\mathrm{II}^{-1})^{\ell q}(H\nabla_i \mathrm{II}_{k\ell} - \nabla_i H \mathrm{II}_{k\ell})(H\nabla_j\mathrm{II}_{pq} - \nabla_j H\mathrm{II}_{pq})\\ &+ \alpha\left(\alpha - \frac{1}{n}\right)\frac{1}{K^{2-\alpha}}\nabla_i K\nabla_j K + \alpha K^\alpha\left(H\mathrm{II}_{ij} - \left(n+\frac{1}{\alpha}\right)\mathrm{II}^2_{ij}\right).\end{aligned}$$

(2) Let $f = \frac{K}{H^n}$, which is ($1/n$ times) the ratio of the geometric mean over the arithmetic mean of the principal curvatures to the n-th power. Show that

$$\begin{aligned}\partial_t f = {}& \mathcal{L}f - \alpha\left(1-\frac{1}{n}\right)K^{\alpha-1}\mathrm{II}^{-1}(\nabla K, \nabla f)\\ &+ \frac{\alpha(n+1)K^\alpha}{H}\mathrm{II}^{-1}(\nabla H, \nabla f) - \frac{\alpha H^{n-1}}{K^{1-\alpha}}|\nabla f|^2\\ &+ \frac{\alpha n K^{1+\alpha}}{H^{n+3}}g^{ip}(\mathrm{II}^{-1})^{jq}(\mathrm{II}^{-1})^{kr}(H\nabla_i\mathrm{II}_{jk} - \nabla_i H\mathrm{II}_{jk})(H\nabla_p\mathrm{II}_{qr} - \nabla_p H\mathrm{II}_{qr})\\ &+ \frac{\alpha(\alpha - n^{-1})}{H^n K^{1-\alpha}}\left(\mathrm{II}^{-1} - \frac{n}{H}g^{-1}\right)(\nabla K, \nabla K)\\ &+ \alpha n\left(n - \frac{1}{\alpha}\right)\frac{K^{1+\alpha}}{H^{n+1}}\left(|\mathrm{II}|^2 - \frac{1}{n}H^2\right).\end{aligned}$$

In particular, for a strictly convex solution of the $1/n$-Gauß curvature flow, there is a vector field Y such that

$$\partial_t f \geq \mathcal{L}f + \langle Y, \nabla f\rangle.$$

Thus, when $\alpha = \frac{1}{n}$ we have by the maximum principle that if $K \geq cH^n$ for some $c > 0$ at $t = 0$, then this inequality holds for $t \geq 0$.

Hint: See [**157**].

Exercise 17.7. Suppose that $\sigma : S^n \times [0, T) \to \mathbb{R}$ is the support function of a smooth solution of an α-Gauß curvature flow with $\alpha > \frac{1}{n+2}$ with final point at the origin, so that $\sigma(\cdot, t)$ converges to zero as $t \to T$. By the argument of Section 17.2, the rescaled support function

$$\tilde{\sigma}(z,t) = ((1+n\alpha)(T-t))^{-\frac{1}{1+n\alpha}}\sigma(z,t)$$

is bounded in $C^k(S^n)$ for all k and all $t \in [0, T)$, with bounds independent of t. Using the result of Section 17.3, show that for any sequence of times t_k approaching T, there is a subsequence t'_k such that $\tilde{\sigma}(\cdot, t'_k)$ converges smoothly to 1. Deduce that $\tilde{\sigma}(\cdot, t)$ converges in C^∞ to 1 as $t \to T$.

Chapter 18

Fully Nonlinear Curvature Flows

In this final part, comprised of Chapters 18–20, we consider curvature flows of hypersurfaces from a more general standpoint. We begin by developing a suitable theory of "functions of curvature", which allows us to make sense of a very large class of "curvature flows". We shall then obtain basic results, such as the short-time existence of solutions and the avoidance principle, which apply in the general setting, before considering more specific classes of flows in the final two chapters.

18.1. Introduction

Recall from Section 5.4 that a smooth 1-parameter family $\{\mathcal{M}_t^n\}_{t\in I}$ of immersed hypersurfaces $\mathcal{M}_t^n \subset \mathbb{R}^{n+1}$ **evolves by curvature** if it admits a 1-parameter family $X : M^n \times I \to \mathbb{R}^{n+1}$ of parametrizations $X(\cdot, t) : M^n \to \mathcal{M}_t^n$ which satisfies the **curvature flow**

$$\partial_t X(x,t) = -F(x,t)\mathsf{N}(x,t)\,, \tag{18.1}$$

where F is given by a function f of the principal curvatures.

In order to ensure that (18.1) is well-behaved, we will need to impose certain structure conditions on the speed defining function f. The most important of these are the following.

Conditions 18.1 (Admissibility conditions). A smooth function $f : \Gamma \to \mathbb{R}$, $\Gamma \underset{\text{open}}{\subset} \mathbb{R}^n$, will be deemed an **admissible speed function** for (18.1) if the following conditions are satisfied:

(i) *Symmetry:* f is invariant under permutation of its variables.

(ii) *Parabolicity:* f is monotone increasing in each of its variables.

Symmetry ensures that the flow is well-defined and that the speed depends smoothly on X. Monotonicity ensures that the flow is "parabolic". These conditions are sufficient to obtain short-time existence of solutions given smooth, compact initial data with curvature in Γ (see Theorem 18.18 below). However, it is still possible for solutions to behave quite badly. For example, there are flows by admissible speed functions which admit solutions that lose higher regularity, or on which the speed F blows up while the solution remains regular; see James McCoy, Yu Zheng, and the first author [**61**, §§5–6]. In order to rule this behavior out, we introduce additional conditions. We will, at various points, make use of one or more of the following conditions:

Conditions 18.2 (Auxiliary conditions).

(iii) *Positivity:* f is positive.

(iv) *Homogeneity:* f is homogeneous.

(v) *Surface flows:* $n = 2$.

(vi) *Concavity:* f is concave.

(vii) *Convexity:* f is convex.

(viii) *Inverse-concavity:* f is inverse-concave (see Definition 18.10).

We will discuss the admissibility and auxiliary conditions in greater depth in Section 18.2 below. Although the admissibility conditions (i)–(ii) will always be assumed, the auxiliary conditions (iii)–(viii) will only be assumed as the need arises. The need for concavity conditions (when $n \geq 2$) is related to the deep problem of optimal regularity of solutions to fully nonlinear elliptic and parabolic PDE of second order (see the regularity theory of Craig Evans [**215**] and Nick Krylov [**339**] for concave Hessian equations and the counterexamples of Nikolai Nadirashvili and Serge Vlăduţ [**411**]).

Note that, although we have implicitly assumed that solutions admit a global unit normal field, this is not necessary when f is an odd function. Indeed, in that case

$$f(\vec{\kappa}_{\mathsf{N}})\mathsf{N} = f(-\vec{\kappa}_{-\mathsf{N}})\mathsf{N} = f(\vec{\kappa}_{-\mathsf{N}})(-\mathsf{N}), \tag{18.2}$$

where $\vec{\kappa}_{\pm\mathsf{N}}$ denotes (up to order) the n-tuple of principal curvatures with respect to the choice of normal $\pm\mathsf{N}$. So a local choice of normal field does not affect the vector field $\vec{F} \doteqdot -F\mathsf{N}$, which is therefore globally well-defined.

Equivalently, we can simply work on the orientation cover of M^n since oddness of f implies that F lifts.

The curvature flows (18.1) are **isotropic** — they are invariant under isometries of $\mathbb{R}^{n+1}$. We will also consider more general **anisotropic** curvature flows, for which the speed F also depends on the unit normal at each point. That is,

$$F(x,t) = f(\vec{\kappa}(x,t), \mathsf{N}(x,t)), \tag{18.3}$$

where $f : \Gamma \times S^n \to \mathbb{R}$ is symmetric and monotone increasing in the first factor.

18.2. Symmetric functions and their differentiability properties

Since the principal curvatures are functions of the second fundamental form, so is the flow speed $F = f(\vec{\kappa})$. It will be useful to relate the differentiability properties of F as a function of the second fundamental form to its differentiability properties as a function of the principal curvatures. Let us recall the following definition.

Definition 18.3 (Symmetric functions)**.** A function $q : \mathbb{R}^n \to \mathbb{R}$ is said to be P_n**-invariant** (or simply **symmetric**) if $(z_{\sigma(1)}, \ldots, z_{\sigma(n)}) \in \Omega$ and $q(z_1, \ldots, z_n) = q(z_{\sigma(1)}, \ldots, z_{\sigma(n)})$ for all $\sigma \in P_n$, where P_n is the group of permutations of the set $\{1, \ldots, n\}$.

Let $\mathrm{S}(n)$ denote the set of symmetric $n \times n$ matrices. Then a function $q : \mathrm{S}(n) \to \mathbb{R}$ is $\mathrm{SO}(n)$**-invariant** (or simply **symmetric**) if $\Sigma^{-1} \cdot Z \cdot \Sigma \in O$ and $q(Z) = q(\Sigma^{-1} \cdot Z \cdot \Sigma)$ for all $\Sigma \in \mathrm{SO}(n)$, where $\mathrm{SO}(n)$ is the special orthogonal group of degree n.

The two types of symmetric functions are clearly related: Denote by λ the *eigenvalue map*, that is, the multivalued map which assigns to a symmetric matrix Z the set of n-tuples with components given by its eigenvalues. Then, given any P_n-invariant function q, we obtain an $\mathrm{SO}(n)$-invariant function $\hat{q} : \mathrm{S}(n) \to \mathbb{R}$ by setting $\hat{q}(Z) = q(z)$ for any $z \in \lambda(Z)$. Since q is P_n-invariant, it takes the same value on any choice of $z \in \lambda(Z)$; hence $\hat{q}$ is well-defined. Conversely, we obtain a P_n-invariant function q from any $\mathrm{SO}(n)$-invariant function $\hat{q}$ by setting $q(z) = \hat{q}(Z)$, for any $Z \in \lambda^{-1}([z])$, where $[z]$ is the orbit of z under the P_n-action. Since $\hat{q}$ is $\mathrm{SO}(n)$-invariant, it takes the same value on any two symmetric matrices with equal eigenvalues; hence q is well-defined. Thus every P_n-invariant function gives rise to a canonical $\mathrm{SO}(n)$-invariant function and vice versa. We will henceforth make the notational abuse of using the same letter (q, say) to denote any two functions related in the above way and speak of q either "as a function of matrix variables" or "as a function of eigenvalue variables".

Since the eigenvalues of a symmetric matrix are not smooth at points of multiplicity, we might expect that a symmetric function is less regular with respect to the matrix variables than with respect to the eigenvalue variables. The following theorem shows that the eigenvalue map behaves with respect to smooth, symmetry-preserving compositions as if it were a smooth function.

Theorem 18.4 (Smooth behavior of eigenvalue map [**246**, **475**]). *Let q be a symmetric function. Then q is smooth with respect to the matrix variables if and only if it is smooth with respect to the eigenvalue variables. Moreover, the first and second derivatives are related by the following formulae: For any diagonal matrix Z in the matrix domain of q with eigenvalue n-tuple $z \in \lambda(Z)$,*

$$\dot{q}_Z^{kl} = \begin{cases} \dot{q}_z^k, & k = l, \\ 0, & k \neq l, \end{cases} \tag{18.4}$$

and, if the eigenvalues are all distinct,

$$\ddot{q}_Z^{pq,rs} V_{pq} V_{rs} = \sum_{p,q} \ddot{q}_z^{pq} V_{pp} V_{qq} + \sum_{p \neq q} \frac{\dot{q}_z^p - \dot{q}_z^q}{z_p - z_q} (V_{pq})^2, \tag{18.5}$$

for any $V \in \mathrm{S}(n)$, where we denote

$$\dot{q}_z^i v_i \doteqdot \left.\frac{d}{ds}\right|_{s=0} q(z + sv)\,, \quad \ddot{q}_z^{ij} v_i v_j \doteqdot \left.\frac{d^2}{ds^2}\right|_{s=0} q(z + sv) \tag{18.6}$$

for $z, v \in \mathbb{R}^n$ and

$$\dot{q}_Z^{ij} V_{ij} \doteqdot \left.\frac{d}{ds}\right|_{s=0} q(Z + sV)\,, \quad \ddot{q}_Z^{pq,rs} V_{pq} V_{rs} \doteqdot \left.\frac{d^2}{ds^2}\right|_{s=0} q(Z + sV) \tag{18.7}$$

for $Z, V \in \mathrm{S}(n)$.

Proof. The proof of the "if" implication is due to Georges Glaeser [**246**] and the proof of the "only if" implication is due to Gerald Schwarz [**475**]. Proofs of the relations (18.4) and (18.5) can be found in [**26**, **38**, **236**, **237**]. We present the proof from [**38**].

Let $Z : \mathrm{S}(n) \times \mathbb{R}^n \times \mathrm{SO}(n) \to \mathrm{S}(n)$ be the smooth map defined by

$$Z(A, \lambda, P) = P^t A P - \mathrm{diag}(\lambda)\,.$$

Observe that $Z(A, \lambda, P)$ vanishes if and only if the columns of P are eigenvectors of A, with eigenvalues λ. Note that the tangent space to $\mathrm{SO}(n)$ at

$P \in \mathrm{SO}(n)$ consists of matrices P' of the form $P' = P\Lambda$, where Λ is skew-symmetric. Indeed, given a regular curve $\gamma : I \to \mathrm{SO}(n)$ with $\gamma(0) = P$ and $\gamma'(0) = P'$,

$$0 = \left(\gamma^t \cdot \gamma\right)'(0) = (P')^t P + P^t P' .$$

Setting $\Lambda = P^t P'$, we obtain

$$\Lambda^t + \Lambda = 0 ,$$

which proves the claim.

Thus, the derivative of Z at a point (A, λ, P) with $Z(A, \lambda, P) = 0$ in a direction (A', λ', P') is

$$\begin{aligned} DZ|_{(A,\lambda,P)}(A', \lambda', P')_{ij} &= P_{ik}A'_{kl}P_{lj} + \Lambda_{ki}P_{lk}A_{lp}P_{pj} + P_{ki}A_{kl}P_{lp}\Lambda_{pj} - \lambda'_i\delta_{ij} \\ &= P_{ik}A'_{kl}P_{lj} + \lambda_j\Lambda_{ji} + \lambda_i\Lambda_{ij} - \lambda'_i\delta_{ij} \\ &= P_{ik}A'_{kl}P_{lj} + (\lambda_i - \lambda_j)\Lambda_{ij} - \lambda'_i\delta_{ij} . \end{aligned} \tag{18.8}$$

If λ is simple, i.e., $\lambda_i \neq \lambda_j$ for each $i \neq j$, then the restriction of $DZ|_{(A,\lambda,P)}$ to the last two components has no kernel: If it vanishes, then the diagonal parts imply $\lambda' = 0$, and the off-diagonal parts imply $\Lambda = 0$. The implicit function theorem then implies that the zero set of Z is locally of the form $\{(A, \lambda(A), P(A))\}$, where λ and P are analytic functions of A. If A is diagonal (so $P = I$), then the first derivatives of λ and P can be read off:

$$\lambda'_i = A'_{ii} \text{ and } P'_{ij} = \Lambda_{ij} = -\frac{A'_{ij}}{\lambda_i - \lambda_j} .$$

Equation (18.8) holds everywhere on the zero set of Z, so differentiating it along a curve $(A(t), \lambda(A(t)), P(A(t)))$ with $A'' = A''(0) = 0$ and $P = P(A(0)) = I$ gives

$$0 = P'_{ki}A'_{kj} + A'_{ik}P'_{kj} + (\lambda'_i - \lambda'_j)\Lambda_{ij} + (\lambda_i - \lambda_j)\Lambda'_{ij} = \lambda''_i\delta_{ij} .$$

The second derivative of λ can be read off from the diagonal components:

$$\lambda''_i = \sum_{k=1}^{n}(\Lambda_{ki}A'_{ki} + A'_{ik}\Lambda_{ki}) = -2\sum_{k\neq i}^{n}\frac{(A'_{ik})^2}{\lambda_k - \lambda_i} .$$

The first and second derivatives of f at a diagonal matrix A with distinct eigenvalues λ can now be computed directly:

$$\dot{f}_A^{kl}A'_{kl} = (f \circ A)'(0) = (f \circ \lambda)'(0) = \dot{f}_\lambda^k\lambda'_k = \dot{f}^k A'_{kk} .$$

So

$$\dot{f}_A^{kl} = \dot{f}_\lambda^k\delta^{kl} .$$

Similarly,

$$\begin{aligned}\sum_{p,q,r,s=1}^{n} \ddot{f}_A^{pq,rs} A'_{pq} A'_{rs} &= \sum_{k=1}^{n} (\dot{f}_\lambda^k \lambda'_k)'(0) \\ &= \sum_{k,l=1}^{n} \ddot{f}_\lambda^{kl} \lambda'_k \lambda'_l + \sum_{k=1}^{n} \dot{f}_\lambda^k \lambda''_k \\ &= \sum_{k,l=1}^{n} \ddot{f}_\lambda^{kl} A'_{kk} A'_{ll} - 2 \sum_{k \neq l} \dot{f}_\lambda^k \frac{(A'_{kl})^2}{\lambda_l - \lambda_k} \\ &= \sum_{k,l=1}^{n} \ddot{f}_\lambda^{kl} A'_{kk} A'_{ll} + \sum_{k \neq l} \frac{\dot{f}_\lambda^k - \dot{f}_\lambda^l}{\lambda_l - \lambda_k} (A'_{kl})^2 . \qquad \square\end{aligned}$$

In fact, analogues of Theorem 18.4 hold under much weaker regularity requirements [**86**] (see also [**237**, Chapter 2]). Unless otherwise stated, we will henceforth assume that all symmetric functions are smooth.

18.2.1. Functions of curvature.

Now suppose that we are in possession of a smooth family of hypersurfaces $X : M^n \times I \to \mathbb{R}^{n+1}$ with principal curvature n-tuple $\vec{\kappa} \doteqdot (\kappa_1, \dots, \kappa_n)$ and a symmetric function q. Then we can form the function

$$Q(x,t) \doteqdot q(\vec{\kappa}(x,t)) . \tag{18.9}$$

We shall refer to a function so defined as a **curvature function**.

Using the conventions of Section 5.3.1, denote by End $\subset T^*M \otimes TM$ the bundle of endomorphisms of TM which are selfadjoint with respect to the induced metric g. Note that a choice of (time-dependent) smooth local orthonormal frame $\{e_i\}_{i=1}^n$ provides local identifications $\Phi_{(x,t)} : \mathrm{End}_{(x,t)} \to \mathrm{S}(n)$ of the fibers of End with $\mathrm{S}(n)$ (taking a selfadjoint endomorphism of (TM, g) to its matrix of components with respect to the chosen frame). This allows us to write Q (locally) as $Q = q(\Phi(\mathrm{II}))$, where, as usual, we are using the same letter q to denote the symmetric function as a function of either eigenvalue or matrix variables. We will therefore, by a further slight abuse of notation, usually write $Q = q(\mathrm{II})$. It then follows from Theorem 18.4 that Q is smooth. In fact, we can obtain explicit, invariant formulae for the derivatives of Q in terms of derivatives of q and the covariant derivatives of II: If we choose $\{e_i\}_{i=1}^n$ such that II is diagonalized at a point (x_0, t_0), then the components of $W \doteqdot \Phi(\mathrm{II})$ satisfy

$$\partial_k W_{ij} = \nabla_k \mathrm{II}_{ij} + \Gamma_{kij} (\kappa_j - \kappa_i)$$

at (x_0, t_0), where the $\Gamma_{kij} \doteqdot g(\nabla_k e_i, e_j)$ are the (antisymmetric) connection coefficients. Similarly, we obtain

$$\partial_t W_{ij} = \nabla_t \mathrm{II}_{ij} + \Gamma_{tij}(\kappa_j - \kappa_i)$$

at (x_0, t_0), where $\Gamma_{tij} \doteqdot g(\nabla_t \partial_i, \partial_j)$. Since, by identity (18.4) of Theorem 18.4, $\dot{q}^{ij}$ is diagonal at (x_0, t_0), we obtain the invariant formulae

$$\begin{aligned}
\nabla_i Q &= \dot{Q}^{kl} \nabla_i \mathrm{II}_{kl} = \dot{Q}(\nabla_i \mathrm{II}), \\
\nabla_t Q &= \dot{Q}^{kl} \nabla_t \mathrm{II}_{kl} = \dot{Q}(\nabla_t \mathrm{II}), \\
\nabla_i \nabla_j Q &= \dot{Q}^{kl} \nabla_i \nabla_j \mathrm{II}_{kl} + \ddot{Q}^{pq,rs} \nabla_i \mathrm{II}_{pq} \nabla_j \mathrm{II}_{rs} \\
&= \dot{Q}(\nabla_i \nabla_j \mathrm{II}) + \ddot{Q}(\nabla_i \mathrm{II}, \nabla_j \mathrm{II}),
\end{aligned}$$

etc., where we are denoting the derivatives of Q with respect to the curvature using dots, just as for q; for example, $\dot{Q} \in \Gamma(\mathrm{End}^*)$ is the tensor defined by

$$\begin{aligned}
\dot{Q}_{(x,t)}(\mathrm{A}) &= \left.\frac{d}{ds}\right|_{s=0} q(W + sA) \\
&= \left.\frac{d}{ds}\right|_{s=0} q(\vec{\kappa} + s\alpha),
\end{aligned}$$

where W and A are the component matrices of $\mathrm{II}_{(x,t)}$ and $\mathrm{A} \in \mathrm{End}_{(x,t)}$, respectively, with respect to some local frame, and α denotes the eigenvalue n-tuple of A. In particular, with respect to an orthonormal frame, we have the expressions

$$\begin{aligned}
\dot{Q}^{kl} &= \dot{q}_{\mathrm{II}}^{kl}, \quad \ddot{Q}^{pq,rs} = \ddot{q}_{\mathrm{II}}^{pq,rs}, \\
\dot{Q}^{k} &= \dot{q}_{\vec{\kappa}}^{k}, \quad \ddot{Q}^{pq} = \ddot{q}_{\vec{\kappa}}^{pq},
\end{aligned}$$

and similarly for higher derivatives.

18.2.2. Homogeneity. Let us recall the definition of a homogeneous function.

Definition 18.5. Let E be a linear space. A subset $C \subset E$ is a **cone** if $\lambda z \in C$ whenever $z \in C$ and $\lambda > 0$. Given a cone $C \subset E$, a function $f : C \to \mathbb{R}$ is **homogeneous (of degree** $\alpha \in \mathbb{R}$) if $f(\lambda z) = \lambda^\alpha f(z)$ whenever $z \in C$ and $\lambda > 0$.

We will often use the shorthand notation **α-homogeneous** to describe a function which is homogeneous of degree α.

Clearly, α-homogeneity of a symmetric function with respect to the eigenvalue variables is equivalent to α-homogeneity with respect to the matrix variables. In particular, since the principal curvatures of a hypersurface scale inversely with distance, this gives rise to invariance of (18.1) under an appropriate parabolic rescaling. Indeed, if $f : \Gamma \to \mathbb{R}$, $\Gamma \subset \mathbb{R}^n$, is

α-homogeneous, $\alpha \neq 0$, and $X : M^n \times I \to \mathbb{R}^{n+1}$ is a solution to the corresponding curvature flow, then the rescaled family $X_\lambda : M^n \times I_\lambda \to \mathbb{R}^{n+1}$, $\lambda > 0$, defined by

$$X_\lambda(x,t) \doteqdot \lambda X\left(x, \lambda^{-(1+\alpha)}t\right) , \quad I_\lambda \doteqdot \lambda^{1+\alpha} I, \tag{18.10}$$

is also a solution to the flow.

A very useful property of homogeneous functions is provided by Euler's theorem.

Proposition 18.6 (Euler's theorem for homogeneous functions)**.** *Let E be a finite-dimensional normed linear space and suppose that $q : C \subset E \to \mathbb{R}$ is a smooth, α-homogeneous function. Then*

$$Dq\big|_z(z) = \alpha q \,, \tag{18.11}$$

where $Dq\big|_z$ is the derivative of q at z.

Proof. The proof is a simple computation:

$$Dq\big|_z(z) = \frac{d}{ds}\bigg|_{s=0} q(z+sz) = \frac{d}{ds}\bigg|_{s=0} (1+s)^\alpha q(z) = \alpha q(z) \,. \qquad \square$$

18.2.3. Concavity. The convexity (concavity, respectively) condition requires that f be a convex (concave, respectively) function *with respect to the eigenvalue variables.* A natural question is whether or not this is equivalent to convexity (concavity, respectively) with respect to the matrix variables.

Let us refer to a point $z \in \mathbb{R}^n$ as **simple** if its components are pairwise distinct.

Lemma 18.7 (Cf. [**26, 208**])**.** *Let $q : \Omega \underset{\text{open, convex}}{\subset} \mathbb{R}^n \to \mathbb{R}$ be a smooth, symmetric function. If q is concave, then, for every simple $z \in \Omega$,*

$$\frac{\dot{q}^i_z - \dot{q}^j_z}{z_i - z_j} \leq 0 \quad \text{for each } i \neq j \,. \tag{18.12}$$

Proof. Suppose that q is concave. Then, for any $v \in \mathbb{R}^n$ and $s \geq 0$ such that $z + sv \in \Omega$,

$$0 \geq \frac{d^2}{ds^2} q(z+sv) = \frac{d}{ds}\dot{q}^i(z+sv)v_i \,, \tag{18.13}$$

so that

$$\dot{q}^i(z+sv)v_i \leq \dot{q}^i(z)v_i \,.$$

Setting $v = -(e_i - e_j)$, where e_i is the basis vector in the direction of the i-th coordinate, we obtain

$$\left(\dot{q}^i - \dot{q}^j\right)\big|_z \leq \left(\dot{q}^i - \dot{q}^j\right)\big|_{z - s(e_i - e_j)} \,.$$

We may assume $z_i \geq z_j$. Then there is some $s_0 \geq 0$ such that

$$(z - s_0(e_i - e_j))_i = (z - s_0(e_i - e_j))_j \, .$$

By symmetry and convexity, $z - s_0(e_i - e_j) \in \Omega$ (this point lies on the line joining z and the point obtained from z by switching its i-th and j-th coordinates). Since q is symmetric, $\dot{q}^i = \dot{q}^j$ at this point and the claim follows. □

Remark 18.8. Note that if strict inequality holds in (18.13), then strict inequality also holds in (18.12).

Corollary 18.9. *A smooth, symmetric function* $q : \Omega \underset{\text{open, convex}}{\subset} \mathbb{R}^n \to \mathbb{R}$ *is concave (convex, respectively) with respect to the eigenvalue variables if and only if it is concave (convex, respectively) with respect to the matrix variables.*

Proof. From the identity (18.5) of Theorem 18.4, we have, for any symmetric matrix V,

$$\ddot{q}^{ij,kl} V_{ij} V_{kl} = \ddot{q}^{ij} V_{ii} V_{jj} + 2 \sum_{i>j} \frac{\dot{q}^i - \dot{q}^j}{z_i - z_j} (V_{ij})^2$$

at any diagonal matrix Z with distinct eigenvalues z_i. So suppose that Z is a diagonal matrix with distinct eigenvalues $z_i = \lambda_i(Z)$. Clearly concavity of q at Z with respect to the matrix components implies concavity of q at z with respect to the eigenvalues. The converse follows from Lemma 18.7.

To see that the claim holds at any diagonal matrix Z, we need only observe that this is the limiting case along a sequence $Z^{(k)}$ of diagonal matrices with distinct eigenvalues which limits to Z.

Finally, the general case follows from the invariance of q with respect to similarity transformations. □

18.2.4. Inverse-concavity.

We next consider the inverse-concavity condition (Conditions 18.2(viii)).

Definition 18.10 (Inverse-concavity). Suppose that $q : \Gamma_+^n \to \mathbb{R}$ is a positive, P_n-invariant function, where

$$\Gamma_+^n \doteqdot \{z \in \mathbb{R}^n : z_i > 0 \text{ for each } i\} \tag{18.14}$$

is the positive cone. We say that q is **inverse-concave** if the **dual function** $q_* : \Gamma_+^n \to \mathbb{R}$ defined by

$$q_* \left(z_1^{-1}, \ldots, z_n^{-1}\right) \doteqdot q(z_1, \ldots, z_n)^{-1} \tag{18.15}$$

is concave.

Similarly, let $q : \mathrm{S}_+(n) \to \mathbb{R}$ be a positive $\mathrm{SO}(n)$-invariant function, where $\mathrm{S}_+(n) \doteqdot \{Z \in \mathrm{S}(n) : Z > 0\}$. Then q is **inverse-concave** if the function $q_* : \mathrm{S}_+(n) \to \mathbb{R}$ defined by

$$q_*\left(Z^{-1}\right) \doteqdot q(Z)^{-1} \tag{18.16}$$

is concave.

Note that, by Corollary 18.9, a symmetric function is inverse-concave with respect to the eigenvalue variables if and only if it is inverse-concave with respect to the matrix variables.

The following provides a local characterization of inverse-concavity.

Lemma 18.11. *Let q be a positive symmetric function. Then q is inverse-concave if and only if the quadratic form $Q_Z : \mathrm{S}(n) \times \mathrm{S}(n) \to \mathbb{R}$ defined by*

$$Q_Z(V,V) \doteqdot \ddot{q}_Z(V,V) - \frac{2\dot{q}_Z(V)\dot{q}_Z(V)}{q(Z)} + 2\dot{q}_Z(VZ^{-1}V)$$

is nonnegative definite for all $Z \in \mathrm{S}_+(n)$, where juxtaposition of matrix variables denotes matrix multiplication. Equivalently, q is inverse-concave if and only if the quadratic form $Q : \mathbb{R}^n \times \mathbb{R}^n \to \mathbb{R}$ defined by

$$Q_z(v,v) \doteqdot \ddot{q}_z(v,v) - \frac{2\dot{q}_z(v)\dot{q}_z(v)}{q(z)} + 2\dot{q}_z(vz^{-1}v)$$

is nonnegative definite for all $z \in \Gamma^n_+$ and, in addition,

$$\frac{\dot{q}^i - \dot{q}^j}{z_i - z_j} + \frac{\dot{q}^i}{z_j} + \frac{\dot{q}^i}{z_j} \geq 0$$

for each $i \neq j$ and each simple z, where $z^{-1} \doteqdot (z_1^{-1}, \ldots, z_n^{-1})$ and juxtaposition of eigenvalue variables denotes componentwise multiplication.

Proof. Differentiating $q_*(Z^{-1})$ with respect to $Z \in \mathrm{S}_+(n)$ in the direction of $V \in \mathrm{S}(n)$ yields

$$q(Z)^{-2}\dot{q}_Z(V) = \dot{q}_{*Z^{-1}}(Z^{-1}VZ^{-1}). \tag{18.17}$$

Differentiating once more yields

$$\frac{(\dot{q}_Z(V))^2}{q(Z)^3} - \frac{\ddot{q}_Z(V,V)}{q(Z)^2} = \ddot{q}_{*Z^{-1}}(Z^{-1}VZ^{-1}, Z^{-1}VZ^{-1}) + 2\dot{q}_{*Z^{-1}}(Z^{-1}VZ^{-1}VZ^{-1}).$$

Applying (18.17) yields

$$q(Z)^{-2}\left(\ddot{q}_Z(V,V)-2\frac{(\dot{q}_Z(V))^2}{q(Z)}+2\dot{q}_Z(VZ^{-1}V)\right)\tag{18.18}$$
$$=-\ddot{q}_{*Z^{-1}}(Z^{-1}VZ^{-1},Z^{-1}VZ^{-1})\,.$$

The first claim follows.

For the second claim, we differentiate $q_*(z^{-1})$ with respect to z to obtain, for any $z\in\Gamma^n_+$ and any $v\in\mathbb{R}^n$,

$$\dot{q}_{*z^{-1}}(z^{-1}vz^{-1})=q(z)^{-2}\dot{q}_z(v)\,.\tag{18.19}$$

Differentiating once more and applying (18.19), we obtain

$$-\ddot{q}_{*z^{-1}}(z^{-1}vz^{-1},z^{-1}vz^{-1})=q(z)^{-2}\left(\ddot{q}_z(v,v)-2\frac{(\dot{q}_z(v))^2}{q(v)}+2\dot{q}_z(vz^{-1}v)\right)\,.$$

Next, consider

$$\frac{\dot{q}^i_{*z^{-1}}-\dot{q}^j_{*z^{-1}}}{z_i-z_j}=\frac{1}{q(z)^2(z_i-z_j)}(\dot{q}^i_z z_i^2-\dot{q}^j_z z_j^2)$$
$$=\frac{z_iz_j}{q(z)^2}\left(\frac{\dot{q}^i_z-\dot{q}^j_z}{z_i-z_j}+\frac{\dot{q}^i_z}{z_j}+\frac{\dot{q}^j_z}{z_i}\right)\,.$$

The second claim now follows from Lemma 18.7. □

For 1-homogeneous admissible flow speeds, the local characterization of inverse-concavity is simplified.

Lemma 18.12. *Let $q:\Gamma^n_+\subset\mathbb{R}^n\to\mathbb{R}$ be a monotone increasing, 1-homogeneous symmetric function. Then q is inverse-concave if and only if the quadratic form $\widehat{Q}_Z:\mathrm{S}(n)\times\mathrm{S}(n)\to\mathbb{R}$ defined by*

$$\widehat{Q}_Z(V,V)\doteqdot\ddot{q}_Z(V,V)+2\dot{q}(VZ^{-1}V)$$

is nonnegative definite for every $Z\in\mathrm{S}_+(n)$.

Equivalently, the flow speed q is inverse-concave if and only if the quadratic form $\widehat{Q}_z:\mathbb{R}^n\times\mathbb{R}^n\to\mathbb{R}$ defined by

$$\widehat{Q}_z(v,v)\doteqdot\ddot{q}_z(v,v)+2\dot{q}_z(vz^{-1}v)$$

is nonnegative definite for each $z\in\Gamma^n_+$ and, in addition,

$$\frac{\dot{q}^i_z-\dot{q}^j_z}{z_i-z_j}+\frac{\dot{q}^i_z}{z_j}+\frac{\dot{q}^j_z}{z_i}\geq 0$$

for each $p\neq q$ wherever the eigenvalues z_i are distinct.

Proof. Since q is homogeneous of degree 1, q_* is homogeneous of degree 1. Thus, recalling (18.18), Euler's theorem implies $Q_Z(V, \cdot) = 0$ whenever $V \propto Z$. Thus, Q_Z is nonnegative definite if and only if it is nonnegative definite on the transversal subspace $S_Z \doteqdot \{V \in \mathrm{S}(n) : \dot{q}|_Z(V) = 0\}$. But, given $V \in S_Z$,

$$Q_Z(V,V) = \ddot{q}_Z(V,V) + 2\dot{q}_Z(VZ^{-1}V).$$

This implies the first claim. The second claim follows similarly. □

Lemma 18.13. *Let $q : \Gamma^n_+ \to \mathbb{R}$ be a monotone increasing, 1-homogeneous flow speed. Then q is inverse-concave if and only if the symmetric function $\chi : \Gamma^n_+ \to \mathbb{R}$ defined by*

$$\chi(z^{-1}) = -q(z)$$

satisfies

$$\ddot{\chi} \leq 2\frac{\dot{\chi} \otimes \dot{\chi}}{\chi}.$$

Proof. This follows from Lemma 18.11 since, with respect to the matrix variables,

$$\dot{\chi}_{Z^{-1}}(V) = \dot{q}_Z(ZVZ)$$

and

$$-\ddot{\chi}_{Z^{-1}}(V,V) = \ddot{q}_Z(ZVZ, ZVZ) + 2\dot{q}_Z(ZVZVZ).$$ □

18.3. Examples

We now describe some examples of curvature functions which define admissible flow speeds and discuss subsets of these which satisfy each of the auxiliary conditions.

Recall that the **elementary symmetric polynomials** $S_k : \mathbb{R}^n \to \mathbb{R}$, $k = 0, \ldots, n$, are defined by

$$S_k(z_1, \ldots, z_n) \doteqdot \binom{n}{k}^{-1} \sum_{1 \leq i_1 < \cdots < i_k \leq n} z_{i_1} \cdots z_{i_k}, \text{ for } k = 1, \ldots, n,$$

$$S_0(z_1, \ldots, z_n) \doteqdot 1$$

and the **power means** $P_r : \Gamma^n_+ \to \mathbb{R}$, $r \in \mathbb{R}$, are defined by

$$P_r(z) \doteqdot \left(\frac{1}{n}\sum_{i=1}^n z_i^r\right)^{\frac{1}{r}}, \text{ for } r \neq 0, \tag{18.20a}$$

$$P_0(z) \doteqdot \left(\prod_{i=1}^n z_i\right)^{\frac{1}{n}}. \tag{18.20b}$$

The k**-th mean curvature** of a hypersurface $\mathcal{M}^n \subset \mathbb{R}^{n+1}$ is defined by

$$H_k \doteqdot S_k(\kappa_1, \ldots, \kappa_n),$$

where $\kappa_1, \ldots, \kappa_n$ are the principal curvatures of $\mathcal{M}^n$.

Example 18.14 (1-homogeneous examples)**.** The following symmetric functions define 1-homogeneous admissible flow speeds.

(1) *The curve shortening flow:* The only 1-homogeneous admissible flow speed in one space dimension is, up to normalization, $f(z) = z$. The corresponding flow is, of course, the curve shortening flow.

(2) *The arithmetic mean:*

$$f(z) = S_1(z), \tag{18.21}$$

defines an admissible flow speed on all of $\mathbb{R}^n$. It is positive on the positive mean half-space,

$$\Gamma_1 \doteqdot \{z \in \mathbb{R}^n : z_1 + \cdots + z_n > 0\}.$$

The corresponding curvature function is the (normalized) mean curvature and the corresponding flow is the (normalized) mean curvature flow.

(3) *The geometric mean:*

$$f(z) = P_0(z) = S_n^{\frac{1}{n}}(z) \tag{18.22}$$

defines a concave, inverse-concave admissible flow speed on the positive cone. The corresponding flow speed is the n-th root of the Gauß curvature, studied in [**157, 506**].

(4) *The square root of* S_2 defines a positive admissible flow speed on the cone

$$\Gamma_2 \doteqdot \{z \in \mathbb{R}^n : S_1(z) > 0 \text{ and } S_2(z) > 0\}.$$

It is concave and inverse-concave on Γ_2. Note that Γ_2 contains the positive cone. The corresponding flow speed is the square root of the scalar curvature, studied in [**158**].

(5) *The power means:* The functions

$$f(z) = P_r(z), \quad r \neq 0, \tag{18.23}$$

define admissible flow speeds on the positive cone. They are concave when $r \leq 1$ and convex when $r \geq 1$. They are inverse-concave when $r \geq -1$. Note that Γ_r contains the positive cone. The corresponding curvature functions include (up to normalization) the mean curvature ($r = 1$), the harmonic mean curvature ($r = -1$), and the magnitude of the second fundamental form ($r = 2$).

(6) *Ratios of consecutive elementary symmetric polynomials:* The functions

$$f = \frac{S_k}{S_{k-1}}, \qquad k = 1, \ldots, n, \tag{18.24}$$

define concave admissible speeds on the cone

$$\Gamma_k \doteqdot \{z \in \mathbb{R}^n : S_\ell(z) > 0 \text{ for } \ell \leq k\} \tag{18.25}$$

(see, for example, [**368**, Chapter XV]). These examples are also inverse-concave since $f_* = \frac{S_{n-k+1}}{S_{n-k}}$ is of the same type. Note that the cone Γ_k contains the positive cone Γ_+; in fact, $\Gamma_1 \supset \cdots \supset \Gamma_n = \Gamma_+$ (see, for example, [**301**, Proposition 2.6]). The corresponding curvature functions include (up to normalization) the mean curvature ($n = 1$) and the harmonic mean curvature ($k = n$).

(7) *Roots of the elementary symmetric polynomials:* The functions

$$f = S_k^{\frac{1}{k}}, \quad k = 1, \ldots, n, \tag{18.26}$$

define concave, inverse-concave admissible speeds on the cone Γ_k given by (18.25) (see, for example, [**368**, Chapter XV] or example (10) below). The corresponding curvature functions include (up to normalization) the mean curvature ($n = 1$), the square root of the scalar curvature ($n = 2$), and the n-th root of the Gauß curvature ($k = n$).

(8) *Positive linear combinations:* The functions

$$f = \sum_i \omega_i f_i \quad (\text{such that } \omega_i > 0 \text{ for each } i) \tag{18.27}$$

of (positive) admissible flow speeds $f_i : \Gamma \to \mathbb{R}$ define (positive) admissible flow speeds $f : \Gamma \to \mathbb{R}$.

(9) *Weighted geometric means:* The functions

$$f = \prod_{i=1}^N f_i^{\omega_i} \quad \left(\text{where } \omega_i \geq 0 \text{ for each } i \text{ and } \sum_{i=1}^N \omega_i = 1\right) \tag{18.28}$$

of positive admissible flow speeds $f_i : \Gamma \to \mathbb{R}$ define positive admissible flow speeds $f : \Gamma \to \mathbb{R}$.

(10) *Roots of ratios of elementary symmetric polynomials:* The function

$$f = \left(\frac{S_k}{S_\ell}\right)^{\frac{1}{k-\ell}}, \quad 0 \leq \ell < k \leq n, \tag{18.29}$$

is the geometric mean of $f_i = \frac{S_i}{S_{i-1}}$ for $i = \ell + 1, \ldots, k$ and hence defines a positive admissible speed function on the cone Γ_k defined by (18.25). These examples are concave and inverse-concave.

(11) *Canonical combinations*: A huge class of homogeneous/concave/convex/inverse-concave admissible speeds is obtained by taking homogeneous/concave/convex/inverse-concave combinations of homogeneous/concave/convex/inverse-concave examples. One interesting example is the m-harmonic mean $H_{-1,m} : \Gamma^n_m \to \mathbb{R}$, $m \le n$, defined by

$$H_{-1,m}(z) \doteqdot \left(\sum_{i_1,\dots,i_m \text{ pairwise distinct}} \frac{1}{z_{i_1} + \cdots + z_{i_m}} \right)^{-1}, \tag{18.30}$$

on the cone

$$\Gamma^n_m \doteqdot \{z \in \mathbb{R}^n : z_{i_1} + \cdots + z_{i_m} > 0 \text{ for } i_1, \dots i_m \text{ pairwise distinct}\}.$$

The 2-harmonic mean curvature flow was studied by Brendle and Huisken [**115**].

Example 18.15 (1-homogeneous admissible surface flows)**.** The following symmetric functions define 1-homogeneous admissible speeds for surface flows:

(1) *Admissible speeds:* All of the examples from Example 18.14 (with $n = 2$).

(2) *A general construction* [**57**]*:* Write z_1, z_2 in polar coordinates (r, θ) with angle measured counterclockwise from the positive ray; that is,

$$r = \sqrt{z_1^2 + z_2^2}, \quad \cos\theta = \frac{z_1 + z_2}{\sqrt{2(z_1^2 + z_2^2)}}, \quad \sin\theta = \frac{z_2 - z_1}{\sqrt{2(z_1^2 + z_2^2)}}.$$

Then we can write any 1-homogeneous $f : \Gamma^2 \to \mathbb{R}$ as $f = r\phi(\theta)$ for some $\phi : (-\theta_0, \theta_0) \to \mathbb{R}$. Observe that f is positive and admissible if and only if $\phi > 0$, $\theta_0 < 3\pi/4$, and

$$A(\theta) < \frac{\phi'(\theta)}{\phi(\theta)} < B(\theta),$$

where

$$A(\theta) \doteqdot \begin{cases} -\infty, & \theta \in (-3\pi/4, -\pi/4], \\ \dfrac{\cos\theta - \sin\theta}{\cos\theta + \sin\theta}, & \theta \in (-\pi/4, 3\pi/4), \end{cases}$$

and

$$B(\theta) \doteqdot \begin{cases} \dfrac{\cos\theta + \sin\theta}{\sin\theta - \cos\theta}, & \theta \in (-3\pi/4, \pi/4), \\ +\infty, & \theta \in [\pi/4, 3\pi/4). \end{cases}$$

In particular, given any smooth, odd function $\psi : (-\theta_0, \theta_0) \to \mathbb{R}$, with $0 < \theta_0 \le 3\pi/4$, satisfying $A(\theta) < \psi(\theta) < B(\theta)$, the function

$$f = r \exp\left(\int_0^\theta \psi(\sigma) d\sigma\right) \tag{18.31}$$

is a positive admissible speed function on the cone

$$\Gamma^2_{\theta_0} \doteqdot \{z \in \mathbb{R}^2 : \theta(z) \in (-\theta_0, \theta_0)\} .$$

Example 18.16 (Speeds of other homogeneities)**.** Let f be a positive, 1-homogeneous admissible flow speed.

(1) If $\alpha > 0$, then the power

$$f_\alpha \doteqdot f^\alpha \tag{18.32}$$

defines an α-homogeneous admissible speed. The corresponding flow speeds include the mean curvature, the Gauß curvature, the scalar curvature and their positive powers. Examples in this class were studied in [**5, 60, 61, 272, 462, 473**].

(2) Flows by certain powers of the mean curvature were studied by F. Schulze, who used them to prove certain isoperimetric inequalities [**471**].

(3) If $\alpha < 0$, then the power

$$f_\alpha \doteqdot -f^\alpha \tag{18.33}$$

defines an α-homogeneous admissible speed. The corresponding flow speeds include (minus) the inverse-mean curvature, the inverse-Gauß curvature, the inverse-scalar curvature, and their positive powers. Examples in this class were studied in [**236, 239, 365, 457, 463, 510, 511**].

(4) A particularly important special case of example (3) is the **inverse-mean curvature flow**, where the speed is given by

$$F = -H^{-1} . \tag{18.34}$$

It's importance to general relativity was noted by R. Geroch [**241**], who observed that it increases the Hawking mass of an embedded 2-sphere in an initial data set, providing evidence for the positive mass conjecture[1]. Jang and Wald [**319**] indicated how this might in fact lead to a proof of the (Riemannian) Penrose inequality. A rigorous proof[2] was achieved by Huisken and Ilmanen [**299**], who analyzed the level set formulation of the flow (in order to overcome singularities which occur during the evolution).

[1] Later proved by Schoen and Yau [**469**].

[2] An independent proof of the Riemannian Penrose inequality (also using a geometric flow approach) was given soon thereafter by Hubert Bray [**102**].

Example 18.17 (Inhomogeneous examples).

(1) Alessandroni and Sinestrari have proved convexity estimates for flows of mean convex hypersurfaces by speeds of the form

$$F \doteqdot \frac{H+\gamma}{\log(H+\gamma)}\,, \tag{18.35}$$

where H is the mean curvature and γ is a constant which is no smaller than $\mathrm{e}^{8n(n-1)}$ (n being the dimension of the solution hypersurfaces) [**4**].

(2) Li and Lv studied flows of convex hypersurfaces by inward normal speeds of the form

$$S = \Phi(f)\,, \tag{18.36}$$

where $f : \Gamma_+ \to \mathbb{R}$ is a 1-homogeneous admissible flow speed which is inverse-concave and satisfies $f_*|_{\partial\Gamma_+} \equiv 0$ and $\Phi : \mathbb{R} \to \mathbb{R}$ is a nonhomogeneous function satisfying certain structure conditions [**362**].

(3) K.-S. Chou (a.k.a. K. Tso) and X.-J. Wang studied the **logarithmic Gauß curvature flow**,

$$\partial_t X = -\log K\,\mathsf{N}\,, \tag{18.37}$$

and its anisotropic variants,

$$\partial_t X = -\log\left(\frac{K}{f(\mathsf{N})}\right)\mathsf{N}\,, \tag{18.38}$$

using them to provide a parabolic approach to the **Minkowski problem** of prescribing the Gauß curvature as a function of the unit normal vector for convex hypersurfaces [**154**].

18.4. Short-time existence

Let $X_0 : M^n \to \mathbb{R}^{n+1}$ be a smooth immersion of a compact, orientable[3] n-manifold M^n. By the tubular neighborhood theorem, there exists $\varepsilon_0 > 0$ such that the map $X : M \times (-\varepsilon_0, \varepsilon_0) \to \mathbb{R}^{n+1}$, defined by

$$X(x,h) = X_0(x) + h\mathsf{N}_0(x)\,,$$

is itself an immersion, where N_0 is a choice of unit normal field for X_0. Let $\overline{g}$ denote the metric induced on $\overline{M} \doteqdot M \times (-\varepsilon_0, \varepsilon_0)$ by X. Then, by the Gauß lemma, $\overline{g}$ admits the decomposition $\overline{g} = \overline{g}_h + dh\otimes dh$, where $\overline{g}_h$ is the metric induced on the hypersurface $M \times \{h\}$ by the immersion $X_h \doteqdot X(\cdot, h)$.

[3]The arguments also apply to flows of nonorientable hypersurfaces $X : M \to \mathbb{R}^{n+1}$ — by lifting X to the orientation cover — so long as the speed function satisfies the oddness condition $f(-A) = -f(A)$.

Now consider the time-dependent graph $\mathrm{G}_u : M \times I \to M \times \mathbb{R}$, where

$$\mathrm{G}_u(x,t) = (x, u(x,t)), \tag{18.39}$$

of a smooth function $u : M \times I \to \mathbb{R}$. If u satisfies $\sup_{M\times I} |u| < \varepsilon_0$, then G_u is a time-dependent immersion, with (time-dependent) pullback metric

$$\gamma \doteqdot (\mathrm{G}_u^* \,\overline{g}) = \overline{g}_u + du \otimes du\,.$$

The inverse of γ is

$$\gamma^* = \overline{g}_u^* - \frac{\overline{\mathrm{grad}}_u u \otimes \overline{\mathrm{grad}}_u u}{1 + |du|^2_{\overline{g}_u}}\,,$$

where $\overline{\mathrm{grad}}_u$ is the gradient operator and $|\cdot|_{\overline{g}_u}$ is the norm induced by the metric $\overline{g}_u$ (in coordinates, $\overline{\mathrm{grad}}_u u$ is $\overline{\mathrm{grad}}_u u^i = \overline{g}_u{}^{ik} u_k$, where $u_j = \partial_j u$).

It is also straightforward to compute a unit normal vector to G_u; we find

$$\mathsf{N}_u = \frac{1}{N}\left(\overline{\mathrm{grad}}_u u - \partial_h\right)\,,$$

where

$$N \doteqdot \sqrt{1 + \overline{g}_u\,(du, du)}\,.$$

The second fundamental form of G_u is therefore given by

$$\begin{aligned} \mathrm{II}_{ij} &= -\,\overline{g}\left({}^{\mathrm{G}_u}\overline{D}_i \mathrm{G}_{u*}\partial_j, \mathsf{N}_u\right) \\ &= -\,N^{-1}\overline{g}\left(\overline{D}_{\partial_i}\partial_j + u_{ij}\partial_h, \overline{\mathrm{grad}}_u u - \partial_h\right) \\ &= N^{-1}\left(u_{ij} - \overline{g}_u\left(\overline{D}_i\partial_j, \overline{\mathrm{grad}}_u u\right)\right) \\ &= N^{-1}\overline{D}_i\overline{D}_j u\,, \end{aligned} \tag{18.40}$$

where $\overline{D}$ is the connection induced on $M \times (-\varepsilon_0, \varepsilon_0)$ by X and ${}^{\mathrm{G}_u}\overline{D}$ is the pullback of $\overline{D}$ to $\mathcal{M} \times I$ by G_u. Since the bundle of connections over M is affine, we may rewrite this in terms of the connection ∇^0 induced by the initial immersion X_0 as $\mathrm{II}_{ij} = \nabla^0_i\nabla^0_j u + \sigma_{ij}$, where σ is the tensor defined by

$$\sigma(Y, Z) \doteqdot du\left({}^{\mathrm{G}_u}\overline{D}_Y Z - \nabla^0_X Z\right)\,.$$

Note that σ depends on u only up to first order.

Thus, the components of the Weingarten curvature are

$$\mathrm{L}_i{}^j = \gamma^{jk}\mathrm{II}_{ik} = \left(\overline{g}^{jk} - \frac{\overline{g}^{jp}u_p\overline{g}^{qk}u_q}{N^2}\right)\left(\nabla^0_i\nabla^0_j u + \sigma_{ij}\right)\,.$$

Next, we compose X with G_u to obtain the time-dependent immersion $X(x,t) = X_0(x) + u(x,t)\mathsf{N}_0(x)$. Note first that, since X is an isometric embedding, the curvature of $X \circ \mathrm{G}_u$ agrees with that of G_u. Moreover, the normal part of the time derivative of $X \circ \mathrm{G}_u$ is

$$\langle \partial_t X, \mathsf{N}\rangle = u_t\,\langle \mathsf{N}_0, X_*\mathsf{N}_u\rangle = N^{-1}u_t\,.$$

We conclude that X solves (18.1) (up to a uniquely determined time-dependent reparametrization) if and only if u solves

(18.41)

$$u_t = \sqrt{1+|du|^2_{\overline{g}_u}}\, f\left(\frac{1}{\sqrt{1+|du|^2_{\overline{g}_u}}}\left(\overline{g}^* - \frac{\overline{\mathrm{grad}}_u u \otimes \overline{\mathrm{grad}}_u u}{1+|du|^2_{\overline{g}_u}}\right)\cdot(\nabla^0\nabla^0 u + \sigma)\right),$$

where the symbol "$\cdot$" denotes the tensor contraction $(U\cdot V)_i{}^j = U^{jk}V_{ik}$.

Theorem 18.18 (Short-time existence). *Let $f : \mathbb{R}^n \to \mathbb{R}$ be a smooth, symmetric function which is strictly monotone on the set $\Gamma \subset \mathbb{R}^n$ and let $X_0 : M^n \to \mathbb{R}^{n+1}$ be a smooth immersion of a smooth, compact n-manifold M^n with principal curvatures satisfying $(\kappa^0{}_1(x), \ldots, \kappa^0{}_n(x)) \in \Gamma$ at all points $x \in M^n$. Then there exists $\delta > 0$ and a unique, smooth time-dependent immersion $X : M^n \times [0,\delta) \to \mathbb{R}^{n+1}$ satisfying the initial value problem*

$$\begin{cases} \partial_t X(x,t) = -f(\vec{\kappa}(x,t))\,\mathsf{N}(x,t), & (x,t)\in M\times(0,\delta),\\ X(x,0) = X_0(x), & x\in M\,. \end{cases} \tag{18.42}$$

Proof. The statement of the theorem is equivalent to the existence of a unique solution $u : M \times [0,\delta) \to \mathbb{R}$ to the initial value problem

(18.43)

$$\begin{cases} u_t(x,t) = \hat{f}(\nabla^0\nabla^0 u(x,t), \nabla^0 u(x,t), u(x,t), x, t), & (x,t)\in M\times(0,\delta),\\ u(x,0) = 0, & x\in M\,, \end{cases}$$

where ∇^0 is the connection on M^n induced by X_0 and

$$\hat{f}(r,p,h,x,t) \doteqdot \sqrt{1+|p|^2_{\overline{g}_h}}\, f\left(\frac{1}{\sqrt{1+|p|^2_{\overline{g}_h}}}\left[\overline{g}^*_h - \frac{\overline{g}^*_h p\otimes \overline{g}^*_h p}{1+|p|^2_{\overline{g}_h}}\right][r+\sigma(p,h,x,t)]\right),$$

where we conflate $f : \Gamma \to \mathbb{R}$ with the corresponding $\mathrm{SO}(n)$-invariant function on $\mathrm{S}(n)$. We will show that the initial value problem (18.43) is uniformly parabolic. The claim then follows from well-known machinery of parabolic PDE (which will not be developed here). First note that, since M^n is compact, there is a compact set $\Gamma_0 \subset \Gamma$ such that $\vec{\kappa}_0(M) \subset \Gamma_0$. Then, since $\dot{f}$ is positive definite on Γ,

$$\lambda\,\|\xi\| \le \dot{f}^{ij}\big|_{\Gamma_0}\xi_i\xi_j \le \Lambda\,\|\xi\|$$

for all $\xi \in \mathbb{R}^n$, where $\lambda \doteqdot \min\{\dot{f}^{ij}(z) : z \in \Gamma_0, 1 \le i,j \le n\} > 0$ and $\Lambda \doteqdot \max\{\dot{f}^{ij}(z) : z\in\Gamma_0, 1\le i,j\le n\} < \infty$.

On the other hand, since $u_0 \equiv 0$, a simple computation yields

$$\left.\frac{\partial \hat{f}}{\partial r_{ij}}\right|_{(\nabla^0\nabla^0 u_0(x),\nabla^0 u_0(x),u_0(x),x,0)} = \dot{f}^{ij}\left(\mathrm{II}_{(x,0)}\right) .$$

This proves the required uniform parabolicity and hence the theorem. □

18.5. The avoidance principle

The avoidance principle holds for odd admissible flow speeds. The proof works in much the same way as the one given for mean curvature flow (Theorem 6.16).

Theorem 18.19 (The avoidance principle). *Let $f : \Gamma \subset \mathbb{R}^n \to \mathbb{R}$ be an admissible flow speed with a well-defined odd extension $f(z) \doteqdot -f(-z)$ for $z \in -\Gamma \doteqdot \{-z : z \in \Gamma\}$ and let $X_i : M_i^n \times [0,T) \to \mathbb{R}^{n+1}$, $i = 1,2$, be two solutions to (18.1), at least one of which is compact, with $X_1(M_1^n, 0) \cap X_2(M_2^n, 0) = \emptyset$. Then the least extrinsic distance*

$$d_{\min}(t) \doteqdot \min_{(x,y)\in M_1^n\times M_2^n} \|X_1(x,t) - X_2(y,t)\| \tag{18.44}$$

is nondecreasing in t, and, in particular, $X_1(M_1^n, t) \cap X_2(M_2^n, t) = \emptyset$ for all $t \in [0,T)$.

Proof. Define the extrinsic distance function $d : M_1^n \times M_2^n \times [0,T) \to \mathbb{R}$ by

$$d(x,y,t) \doteqdot \|X_1(x,t) - X_2(y,t)\| . \tag{18.45}$$

We will show that the time derivative of d is nonnegative at a spatial minimum. The claim then follows from the maximum principle (as in the proof of Theorem 6.16, say). To simplify notation, we define

$$w(x,y,t) = \frac{X_1(x,t) - X_2(y,t)}{d(x,y,t)} \tag{18.46}$$

and use super- and subscripts x and y to denote geometric quantities defined on $M_1 \times M_2$ by pulling back those from X_1 and X_2 by the respective projections; for example, $F_x(\xi,\eta,\tau) \doteqdot F(\xi,\tau)$. With this notation in place, we find that d satisfies

$$\partial_t d = \langle w, -F_x \mathsf{N}_x + F_y \mathsf{N}_y \rangle . \tag{18.47}$$

Suppose there is a spatial minimum of d at (x_0, y_0, t_0). Then, at this point, $\nabla^{M_1\times M_2} d = 0$ and $\mathrm{Hess}^{M_1\times M_2} d \geq 0$. Let $\{x^i\}_{i=1}^n$ and $\{y^i\}_{i=1}^n$ be coordinates defined on neighborhoods of x_0 and y_0, respectively, and write $\partial_i^x \doteqdot \frac{\partial X_1}{\partial x^i}$ and $\partial_j^y \doteqdot \frac{\partial X_2}{\partial y^j}$. Then

$$\partial_{x^i} d = \langle \partial_i^x, w\rangle \quad\text{and}\quad \partial_{y^j} d = -\left\langle \partial_j^y, w\right\rangle . \tag{18.48}$$

These vanish at the minimum (x_0, y_0, t_0); that is, $w(x_0, y_0, t_0)$ is orthogonal both to the tangent plane to X_1 at (x_0, t_0) and the tangent plane to X_2 at (y_0, t_0). Now, the assumption that F is odd implies the flow is invariant under change of orientation since the sign of the Weingarten map changes with the orientation of the normal. So we may choose the orientations of M_1 and M_2 such that[4] $\mathsf{N}_x = \mathsf{N}_y = w$ at (x_0, y_0, t_0).

Next, using the vanishing of the gradients (18.48), we obtain, at the point (x_0, y_0, t_0), the identities

$$\nabla_i^{M_1} \nabla_j^{M_1} d = -\,\mathrm{II}_{ij}^x \langle \mathsf{N}_x, w\rangle + \frac{1}{d}\, g_{ij}^x, \tag{18.49}$$

$$\nabla_i^{M_2} \nabla_j^{M_1} d = -\frac{1}{d} \left\langle \partial_j^x, \partial_i^y \right\rangle, \tag{18.50}$$

and

$$\nabla_i^{M_2} \nabla_j^{M_2} d = \mathrm{II}_{ij}^y \langle \mathsf{N}_y, w\rangle + \frac{1}{d}\, g_{ij}^y. \tag{18.51}$$

Thus, for any vector $v \in \mathbb{R}^n$, we find

$$\begin{aligned} 0 &\le v^i v^j \left(\nabla_i^{M_1} \nabla_j^{M_1} d + 2\,\nabla_i^{M_2} \nabla_j^{M_1} d + \nabla_i^{M_2} \nabla_j^{M_2} d\right) \\ &= -\,\mathrm{II}_{ij}^x v^i v^j \langle \mathsf{N}_x, w\rangle + \frac{1}{d} g_{ij}^x v^i v^j + \mathrm{II}_{ij}^y v^i v^j \langle \mathsf{N}_y, w\rangle + \frac{1}{d} g_{ij}^y v^i v^j - \frac{2}{d} v^i v^j \langle \partial_i^x, \partial_j^y\rangle \end{aligned}$$

at the point (x_0, y_0, t_0). We now choose the coordinates $\{x^i\}_{i=1}^n$ and $\{y^i\}_{i=1}^n$ to be normal coordinates centered at x_0 and y_0, respectively. Since the tangent planes of the two hypersurfaces are parallel at x_0 and y_0, we may further assume that $\partial_i^x = \partial_i^y$ for all i at the point (x_0, y_0, t_0). Then, since $g_{ij}^x = g_{ij}^y = \delta_{ij}$ at (x_0, y_0, t_0), we obtain

$$\mathrm{II}_{ij}^x\, v^i v^j \le \mathrm{II}_{ij}^y\, v^i v^j$$

at that point. It follows that $\mathrm{II}_{ij}^x \le \mathrm{II}_{ij}^y$ at any spatial minimum of d. Since the speed is monotone increasing, this implies that $F_x \le F_y$ at such a point. Thus, by (18.47), we obtain

$$\partial_t d = -F_x + F_y \ge 0$$

at any spatial minimum of d. □

Remark 18.20. The assumption that the speed is an odd function of the curvature can be relaxed if we make an additional topological assumption on the hypersurfaces to guarantee the correct orientation. Namely, if we require that $X_1(M_1, 0) = \partial\Omega_1$ and $X_2(M_2, 0) = \partial\Omega_2$ bound domains Ω_1 and Ω_2 satisfying $\Omega_1 \subset \Omega_2 \subset \mathbb{R}^{n+1}$ and the unit normal to M_i points out of Ω_i for $i = 1, 2$, then the above argument goes through unharmed since this guarantees $\mathsf{N}_x = w = \mathsf{N}_y$ at the distance minimizing pair (x_0, y_0). This

[4]See also the remark following the proof.

observation means that we can still compare compact solutions of (18.1) with enclosing spheres, even if F has no odd extension.

The following example shows that the avoidance principle can be violated if the speed function is not odd and the topological assumption of the preceding remarks is not met.

Example 18.21. Observe that the function $F(z) \doteqdot z_1 + z_2 + \frac{1}{2}\sqrt{z_1^2 + z_2^2}$ defines an admissible speed function on the cone $\Gamma = \mathbb{R}^2$. Under the corresponding flow, surfaces with opposite orientation can move closer together (and even cross): Consider the 2-torus T obtained by rotating the circle $\{z \in \mathbb{R}^3 : (z_1 - 1)^2 + z_2^2 = r^2, z_3 = 0\}$, $r < 1$, about the z_2-axis. If we orient T with its inward pointing unit normal, then we obtain

$$F_{\mathrm{T}}(p) = \left(\tfrac{1}{1-r} - \tfrac{1}{r} + \tfrac{1}{2}\sqrt{\tfrac{1}{(1-r)^2} + \tfrac{1}{r^2}}\right)$$

at the point $p = (1 - r, 0, 0)$. On the other hand, the cylinder C obtained by rotating the line $\{z \in \mathbb{R}^3 : z_1 = 1 - r, z_3 = 0\}$ about the z_2-axis satisfies $F_{\mathrm{C}}(p) = -\frac{1}{2(1-r)}$ if we also orient using the inward pointing normal. Note that $F_{\mathrm{T}}(p) \geq \frac{1}{1-r} - \frac{1}{2r}$. Thus, we can achieve $F_{\mathrm{T}}(p) \geq -F_{\mathrm{C}}(p)$ if $\frac{1}{1-r} > \frac{1}{r}$, that is, if $r > \frac{1}{2}$. Since the normals of the two surfaces are pointing in opposite directions at p, this implies that the surfaces will begin to cross.

Embeddedness is preserved under flows by odd admissible speeds (cf. Theorem 6.18).

Theorem 18.22. *Let $f : \Gamma \subset \mathbb{R}^n \to \mathbb{R}$ be an admissible flow speed with a well-defined odd extension $f(z) \doteqdot -f(-z)$ for $z \in -\Gamma \doteqdot \{-z : z \in \Gamma\}$, and let $X : M^n \times [0, T) \to \mathbb{R}^{n+1}$ be a solution to the corresponding curvature flow* (18.1). *Suppose that $X_0 \doteqdot X(\cdot, 0)$ is an embedding. Then $X_t \doteqdot X(\cdot, t)$ is an embedding for each $t \in [0, T)$.*

Proof. We leave the proof as an exercise (cf. Theorem 6.18). □

18.6. Differential Harnack estimates

Let $X : M^n \times [0, T) \to \mathbb{R}^{n+1}$ be a convex solution to the anisotropic curvature flow

$$\begin{aligned} \partial_t X(x,t) &= -\,F(x,t)\mathsf{N}(x,t)\,, \\ F(x,t) &= f\left(\mathrm{II}_{(x,t)}, \mathsf{N}(x,t)\right), \end{aligned} \tag{18.52}$$

where $f : \Gamma_+^n \times S^n \to \mathbb{R}$ is $\mathrm{SO}(n)$-invariant and monotone nondecreasing in its first factor (and, as usual, f is really evaluated at the component matrix of $\mathrm{II}_{(x,t)}$ with respect to a $g_{(x,t)}$-orthonormal basis).

Since the solution is (by hypothesis) convex and locally uniformly convex, the Gauß map is a diffeomorphism for each t, and so we may use its inverse to parametrize the solution. By (5.78) and (5.82),

$$(\mathrm{G}^{-1})^*\mathrm{II}^2 = \overline{g} \quad \text{and} \quad (\mathrm{G}^{-1})^*\mathrm{II} = A[\sigma] \doteqdot \overline{\nabla}^2\sigma + \sigma\,\overline{g}\,, \tag{18.53}$$

where $\overline{g}$ is the standard metric on S^n and the map $\mathrm{G}^{-1} : S^n \times [0,T) \to M^n$ is the inverse of the Gauß map $\mathrm{G} : M^n \times [0,T) \to S^n$; that is, $\mathrm{G}(\mathrm{G}^{-1}(z,t),t) = z$ for each $(z,t) \in S^n \times [0,T)$.

So we may rewrite (18.52) as an evolution equation of the support function:

$$\partial_t\sigma(z,t) = \phi\big(A[\sigma]_{(z,t)}, z\big), \tag{18.54}$$

where ϕ is defined by $\phi(A,z) \doteqdot -f(A^{-1},z)$ for all $(A,z) \in \Gamma^n_+ \times S^n$ and, by our usual abuse of notation, is really evaluated at the component matrix of $A[\sigma]_{(z,t)}$ with respect to a $\overline{g}_z$-orthonormal frame.

Given $z \in S^n$ and $A \in \Gamma^n_+$, the derivatives $\dot{\phi}$ and $\ddot{\phi}$ of ϕ at (A,z) are defined (cf. Section 18.2.1) by

$$\dot{\phi}_{(A,z)}(B) \doteqdot \left.\frac{d}{dr}\right|_{r=0} \phi(A+rB,z) \tag{18.55}$$

and

$$\ddot{\phi}_{(A,z)}(B,B) \doteqdot \left.\frac{d^2}{dr^2}\right|_{r=0} \phi(A+rB,z) \tag{18.56}$$

for $B \in \mathrm{S}(n)$, the set of symmetric $n \times n$ matrices. We also set

$$\Phi(z,t) \doteqdot \phi\big(A[\sigma]_{(z,t)}, z\big) \tag{18.57}$$

and define the tensors $\dot{\Phi}_{(z,t)}$ and $\ddot{\Phi}_{(z,t)}$ by

$$\dot{\Phi}_{(z,t)}(B) \doteqdot \left.\frac{d}{dr}\right|_{r=0} \phi\big(A[\sigma]_{(z,t)} + rB, z\big) \tag{18.58}$$

and

$$\ddot{\Phi}_{(z,t)}(B,B) \doteqdot \left.\frac{d^2}{dr^2}\right|_{r=0} \phi\big(A[\sigma]_{(z,t)} + rB, z\big) \tag{18.59}$$

for $B \in T_zS^n$, where we conflate $A[\sigma]_{(z,t)}$ and B with their respective component matrices with respect to a $\overline{g}_z$-orthonormal basis.

Henceforth, we simply write $A = A[\sigma]$.

Lemma 18.23. *Under the flow* (18.52) (*equivalently,* (18.54)),

$$\partial_t A = \overline{\nabla}^2\Phi + \Phi\,\overline{g}, \tag{18.60}$$

and

$$(\partial_t - \dot{\Phi}^{ij}\overline{\nabla}_i\overline{\nabla}_j)\Phi = \dot{\Phi}^{ij}\overline{g}_{ij}\,\Phi\,. \tag{18.61}$$

Proof. Since $\overline{g}$ and $\overline{\nabla}$ are independent of time, equation (18.60) follows immediately from differentiating (18.54). Equation (18.61) follows immediately from differentiating (18.57) and applying (18.60). □

Define

$$P(z,t) \doteqdot \partial_t \Phi(z,t). \tag{18.62}$$

Proposition 18.24. *Under the flow* (18.52) (*equivalently,* (18.54)),

$$(\partial_t - \dot{\Phi}^{ij}\overline{\nabla}_i\overline{\nabla}_j)P = \dot{\Phi}^{ij}\overline{g}_{ij}P + \ddot{\Phi}^{pq,rs}\partial_t A_{pq}\partial_t A_{rs}. \tag{18.63}$$

Proof. Differentiate the equation (18.61). □

Given a nonzero real number α, a function $\phi : \Gamma_+^n \to \mathbb{R}$ is **α-convex** (respectively, **α-concave**) if $\operatorname{sgn}(\Phi) = \operatorname{sgn}(\alpha)$ and there is a positive convex (resp., concave) function ψ such that $\phi = \operatorname{sgn}(\alpha)\psi^\alpha$. Equivalently,

$$\ddot{\phi} \geq \frac{\alpha - 1}{\alpha\phi}\dot{\phi}\otimes\dot{\phi} \quad \left(\text{respectively, } \ddot{\phi} \leq \frac{\alpha - 1}{\alpha\phi}\dot{\phi}\otimes\dot{\phi}\right).$$

An "anistotropic" function $\phi : \Gamma_+^n \times S^n \to \mathbb{R}$ is **α-convex** (respectively, **α-concave**) if it is **α-convex** (respectively, **α-concave**) with respect to its first variable.

Theorem 18.25 (Andrews). *Let $X : M^n \times [0,T) \to \mathbb{R}^{n+1}$ be a solution to* (18.52) (*equivalently, its support function is a solution to* (18.54)).

(1) *If ϕ is α-convex for some $\alpha > 1$, then*

$$\partial_t \Phi + \frac{\alpha}{(\alpha - 1)t}\Phi \geq 0. \tag{18.64}$$

(2) *If Φ is α-concave for some $\alpha < 1$, then*

$$\partial_t \Phi + \frac{\alpha}{(\alpha - 1)t}\Phi \leq 0. \tag{18.65}$$

(3) *If ϕ is positive and convex, then $t \mapsto \min_{S^n \times \{t\}}(\partial_t \log \Phi)$ is nondecreasing.*

(4) *If ϕ is positive and concave, then $t \mapsto \max_{S^n \times \{t\}}(\partial_t \log \Phi)$ is nonincreasing.*

Proof. We shall prove the convex cases (1) and (3) and we leave the proofs of the concave cases (2) and (4) as an exercise.

(1) Suppose that ϕ is α-convex, where $\alpha > 1$. Define

$$R = t\,\partial_t\Phi + \frac{\alpha}{\alpha - 1}\Phi = t\,P + \frac{\alpha}{\alpha - 1}\Phi.$$

By (18.63) and (18.61),

$$(\partial_t - \dot{\Phi}^{ij}\overline{\nabla}_i\overline{\nabla}_j)R = \dot{\Phi}^{ij}\overline{g}_{ij}R + P + t\ddot{\Phi}^{pq,rs}\partial_t A_{pq}\partial_t A_{rs}.$$

Since Φ is α-convex, where $\alpha > 1$, we have $\Phi > 0$ and

$$\ddot{\Phi}(\partial_t A, \partial_t A) \geq \frac{\alpha - 1}{\alpha\Phi}\left(\dot{\Phi}(\partial_t A)\right)^2 = \frac{\alpha - 1}{\alpha\Phi}P^2,$$

and hence

$$(\partial_t - \dot{\Phi}^{ij}\overline{\nabla}_i\overline{\nabla}_j)R = \left(\dot{\Phi}^{ij}\overline{g}_{ij} + \frac{\alpha - 1}{\alpha}\frac{P}{\Phi}\right)R.$$

Since R is initially positive, the maximum principle implies that it remains so, which proves the claim.

(3) Suppose that ϕ is positive and convex. Define

$$S \doteqdot \partial_t \log \Phi = \frac{P}{\Phi}.$$

By (18.63),

$$(\partial_t - \dot{\Phi}^{ij}\overline{\nabla}_i\overline{\nabla}_j)P \geq \dot{\Phi}^{ij}\overline{g}_{ij}P.$$

Recalling (18.61), we conclude that

$$(\partial_t - \dot{\Phi}^{ij}\overline{\nabla}_i\overline{\nabla}_j)S \geq \frac{2}{\Phi}\dot{\Phi}^{ij}\overline{\nabla}_i\Phi\overline{\nabla}_jS.$$

So the claim follows from the maximum principle. □

Example 18.26. For the α-Gauß curvature flow, where the speed is $-K^\alpha$, we have $\phi(A) = -(\det A)^{-\alpha}$. Here, ϕ is $-(n\alpha)$-concave since the function $A \mapsto (\det A)^{1/n}$ is concave. So Theorem 18.25(2) implies that

$$\partial_t K^\alpha + \frac{n\alpha}{(n\alpha + 1)t}K^\alpha \geq 0, \tag{18.66}$$

where the time derivative is taken with respect to the Gauß map parametrization, recovering Theorem 15.13.

Interpreting Theorem 18.25 in terms of the original parametrization, we obtain the following inequalities.

Corollary 18.27. *Let X_t be a bounded convex solution to* (18.52). *Then, under the original parametrization:*

(1) *If Φ is α-convex, where $\alpha > 1$, then*

$$\partial_t F - \mathrm{II}^{-1}(\nabla F, \nabla F) + \frac{\alpha}{(\alpha - 1)t}F \leq 0.$$

(2) *If Φ is α-concave for some $\alpha < 1$, then*

$$\partial_t F - \mathrm{II}^{-1}(\nabla F, \nabla F) + \frac{\alpha}{(\alpha - 1)t}F \geq 0.$$

(3) *If Φ is convex and positive, then*

$$\min_{S^n \times \{t\}} \left(\partial_t \log|F| - F\,\mathrm{II}^{-1}(\nabla \log|F|, \nabla \log|F|)\right)$$

is a nondecreasing function of t.

(4) *If Φ is concave and positive, then*

$$\max_{S^n\times\{t\}} \left(\partial_t \log|F| - F\,\mathrm{II}^{-1}(\nabla \log|F|, \nabla \log|F|)\right)$$

is a nonincreasing function of t.

In particular, for the α-Gauß curvature flow, we find under the original parametrization that

$$\partial_t K^\alpha - \mathrm{II}^{-1}(\nabla K^\alpha, \nabla K^\alpha) + \frac{n\alpha}{(n\alpha+1)t} K^\alpha \geq 0\,. \tag{18.67}$$

18.7. Entropy estimates

We shall see that a certain special class of curvature flows admit monotone entropy functionals. These flows, and entropies, are constructed in terms of the *mixed volume*, which are most easily defined using the *Minkowski sum*.

Recall that the **Minkowski sum** D_1+D_2 of two convex regions D_1 and D_2 in $\mathbb{R}^{n+1}$ is defined by

$$D_1 + D_2 = \{\vec{x}_1 + \vec{x}_2 \,|\, \vec{x}_1 \in D_1,\ \vec{x}_2 \in D_2\}\,. \tag{18.68}$$

The region $D_1 + D_2$ is convex and its support function σ is given by

$$\sigma(z) = \sigma_1(z) + \sigma_2(z)\,, \tag{18.69}$$

where σ_1 and σ_2 are the support functions of D_1 and D_2, respectively. Note also that dilation distributes over Minkowski addition:

$$\lambda D + \mu D = (\lambda + \mu) D\,.$$

Let $\mathcal{M}^n$ be the boundary of a smooth, bounded, convex region D in $\mathbb{R}^{n+1}$. Since

$$\overline{g} = (\mathrm{G}^{-1})^* \,\mathrm{II}\,,$$

where G is the Gauß map, (5.40) and the change of variables formula imply that the volume of D may be expressed as

$$V(D) = \frac{1}{n+1}\int_{S^n} \sigma \det A[\sigma]\, d\bar{\mu}\,, \tag{18.70}$$

where

$$A[\sigma] \doteqdot \overline{\nabla}^2\sigma + \sigma\overline{g} = (\mathrm{G}^{-1})^*\,\mathrm{II}$$

and the determinant is taken with respect to $\overline{g}$. Let $\{D_\alpha\}_{\alpha=1}^N$ be a finite collection of convex regions in $\mathbb{R}^{n+1}$ with respective support functions σ_α. By (18.70) and (18.69), the volume of $\sum_{\alpha=1}^N \lambda_\alpha D_\alpha$ is a homogeneous polynomial of degree $n+1$ in the variables λ_α, so the **mixed volume** $V(D_{\alpha_0}, \ldots, D_{\alpha_n})$ of $D_{\alpha_0}, \ldots, D_{\alpha_n}$ is uniquely defined by the formula

$$V\left(\sum_{\alpha=1}^{N} \lambda_\alpha D_\alpha\right) = \sum_{1\leq \alpha_0,\ldots,\alpha_n \leq N} V(D_{\alpha_0}, \ldots, D_{\alpha_n})\lambda_{\alpha_0}\cdots\lambda_{\alpha_n}. \tag{18.71}$$

Related to the mixed volume of convex bodies is the **mixed discriminant** of functions $\sigma_1, \ldots, \sigma_n$ on S^n. This is the multilinear operator defined by

$$\mathcal{Q}[\sigma_1, \ldots, \sigma_n] \doteqdot \frac{1}{n!} \sum_{\tau,\upsilon\in S_n} (-1)^{\operatorname{sgn}(\tau)+\operatorname{sgn}(\upsilon)} A[\sigma_1]_{\tau(1)}^{\upsilon(1)} \cdots A[\sigma_n]_{\tau(n)}^{\upsilon(n)}, \tag{18.72}$$

where S_n denotes the symmetric group of permutations of the set $\{1, \ldots, n\}$ and the indices on $A[\sigma_i]$ are raised using the metric $\bar{g}$. Indeed, the mixed volume of $n+1$ convex regions $D_0, \ldots, D_n$ in $\mathbb{R}^{n+1}$ may be expressed as

$$V(D_0, D_1, \ldots, D_n) = \frac{1}{n+1}\int_{S^n} \sigma_0\, \mathcal{Q}[\sigma_1, \ldots, \sigma_n]\, d\bar{\mu}, \tag{18.73}$$

where $\sigma_0, \ldots, \sigma_n$ are the respective support functions of $D_0, \ldots, D_n$.

Observe that the definition (18.72) is independent of the choice of basis $\{e_i\}$ and that

$$\mathcal{Q}[f, \ldots, f] = \det{}_{\bar{g}} A[f].$$

In particular, $\mathcal{Q}[1, \ldots, 1] = 1$.

Example 18.28. For any strictly convex region $D \subset \mathbb{R}^{n+1}$ and any $k = 0, \ldots, n+1$, define the k-th **mean cross-sectional volume** (a.k.a. **quermassintegral**) $V_k(D)$ by

$$V_k(D) \doteqdot V(\underbrace{D, \ldots, D}_{k\text{-times}}, \underbrace{B^n, \ldots, B^n}_{(n+1-k)\text{-times}}),$$

where B^n is the unit ball. By (18.73), this is the volume $\mathscr{H}^{n+1}(D)$ of D when $k = n+1$ and, since V is totally symmetric (a fact we shall prove in Lemma 18.31 below), $\frac{1}{n+1}$ times the area $\mathscr{H}^n(\partial D)$ of its boundary ∂D when $k = n$. When $1 \leq k < n$, $V_k(D)$ is the average $\mathscr{H}^k$ measure of the projections of D onto the k-dimensional subspaces [**461**].

Lemma 18.29. *The mixed discriminant $\mathcal{Q}$ is totally symmetric,*

$$\mathcal{Q}[\sigma_1, \ldots, \sigma_n] = \mathcal{Q}[\sigma_{\tau(1)}, \ldots, \sigma_{\tau(n)}]$$

for any permutation $\tau \in S_n$, and positive definite,

$$\mathcal{Q}[\sigma_1, \ldots, \sigma_n] > 0 \text{ provided } A[\sigma_i] > 0 \text{ for each } i.$$

Proof. The first claim is clear from the definition of $\mathcal{Q}$ in terms of permutations. The second is proved by induction: Given $k \in \mathbb{N}$ and k positive definite, symmetric $k \times k$ matrices $A^1, \dots, A^k$, define

$$Q^k(A^1, \dots, A^k) \doteqdot \sum_{\tau, \upsilon \in S_n} (-1)^{\operatorname{sgn}(\tau)+\operatorname{sgn}(\upsilon)} A^1_{\tau(1)\upsilon(1)} \cdots A^k_{\tau(k)\upsilon(k)}.$$

Then

$$Q^1(A^1) = \sum_{\tau, \upsilon \in S_1} (-1)^{\operatorname{sgn}(\tau)+\operatorname{sgn}(\upsilon)} A^1_{\tau(1)\upsilon(1)} = A^1_{11} > 0.$$

Suppose that $Q^k > 0$ for all choices of k positive definite $k \times k$ matrices. Let $A^1, \dots, A^{k+1}$ be $k+1$ symmetric, positive definite $(k+1) \times (k+1)$ matrices. By rotating the basis for $\mathbb{R}^{k+1}$, we may arrange that A^{k+1} is diagonal, so that

$$(18.74)\quad Q^{k+1}(A^1, \dots, A^{k+1}) = \sum_{\substack{1 \le i \le k+1, \\ \tau, \upsilon \in S_k(i)}} (-1)^{\operatorname{sgn}(\tau)+\operatorname{sgn}(\upsilon)} A^{k+1}_{ii} \prod_{j=1}^{k} A^j_{\tau(j)\upsilon(j)},$$

where $S_k(i)$ denotes the set of permutations from the set $\{1, \dots, k\}$ to the set $\{1, \dots, i-1, i+1, \dots, k+1\}$. But this can be rewritten in terms of the positive definite, symmetric $k \times k$ matrices $A^j|_{e_i^\perp}$ obtained from A^j by eliminating its j-th row and column as

$$Q^{k+1}(A^1, \dots, A^{k+1}) = \sum_{\substack{1 \le i \le k+1, \\ \tau, \upsilon \in S_k(i)}} A^{k+1}_{ii} Q^k(A^1|_{e_i^\perp}, \dots, A^k|_{e_i^\perp}),$$

which is positive by the inductive hypothesis. The claim follows since a choice of local orthonormal frame for TS^n allows us to write $\mathcal{Q}[\sigma_1, \dots, \sigma_n]$ as $Q^n(A[\sigma_1], \dots, A[\sigma_n])$. □

If we fix functions $\sigma_2, \dots, \sigma_n$ on S^n satisfying $A[\sigma_i] > 0$ for each $2 \le i \le n$, then we can define an operator $\mathcal{Q}_1$ on functions f on S^n by

$$\mathcal{Q}_1[f] \doteqdot \mathcal{Q}[f, \sigma_2, \dots, \sigma_n].$$

By (18.74),

$$(18.75)\qquad \mathcal{Q}_1[f] = \dot{\mathcal{Q}}^{ij}(\overline{\nabla}_i \overline{\nabla}_j f + f \overline{g}_{ij}),$$

where the tensor $\mathcal{Q}$ is defined, using an orthonormal frame, by

$$\dot{\mathcal{Q}}^{ij} = \sum_{\substack{1 \le i \le n, \\ \tau, \upsilon \in S_{n-1}(i)}} (-1)^{\operatorname{sgn}(\tau)+\operatorname{sgn}(\upsilon)} \prod_{p=2}^{n} A[\sigma_p]_{\tau(p)\upsilon(p)} \delta^{ij},$$

and now $S_{n-1}(i)$ denotes the set of permutations from the set $\{2, \dots, n\}$ to the set $\{1, \dots, i-1, i+1, \dots, n\}$. We claim that $\dot{\mathcal{Q}}$ is divergence free.

Lemma 18.30. *Given $\sigma_2, \dots, \sigma_n$ such that $A[\sigma_i] > 0$ for each i, the induced operator $\mathcal{Q}_1$ satisfies*

$$\sum_{i=1}^{n} \overline{\nabla}_i \dot{\mathcal{Q}}^{ij} = 0. \tag{18.76}$$

Proof. Setting $\epsilon_{\tau\upsilon} \doteqdot (-1)^{\mathrm{sgn}(\tau)+\mathrm{sgn}(\upsilon)}$, we have

$$\overline{\nabla}_k \dot{\mathcal{Q}}^{ij} = \sum_{\substack{1 \le i \le n, \\ \tau,\upsilon \in S_{n-1}(i)}} \epsilon_{\tau\upsilon} \sum_{q=2}^{n} \prod_{p \notin \{1,q\}} A[\sigma_p]_{\tau(p)\upsilon(p)} \overline{\nabla}_k A[\sigma_q]_{\tau(q)\upsilon(q)} \delta^{ij},$$

so that

$$\begin{aligned}
\overline{\nabla}_i \dot{\mathcal{Q}}^{ij} &= \sum_{\tau,\upsilon \in S_{n-1}(j)} \epsilon_{\tau\upsilon} \sum_{q=2}^{n} \prod_{p \notin \{1,q\}} A[\sigma_p]_{\tau(p)\upsilon(p)} \overline{\nabla}_j A[\sigma_q]_{\tau(q)\upsilon(q)} \\
&= \sum_{\substack{q>1, \tau,\upsilon \in S_n, \\ \tau(1)=j=\upsilon(1)}} \epsilon_{\tau\upsilon} \prod_{p \notin \{1,q\}} A[\sigma_p]_{\tau(p)\upsilon(p)} \overline{\nabla}_{\tau(1)} A[\sigma_q]_{\tau(q)\upsilon(q)} \\
&= \sum_{\substack{q>1, \tau,\upsilon \in S_n, \\ \tau(1)=j=\upsilon(1)}} \frac{\epsilon_{\tau\upsilon}}{2} \prod_{p \notin \{1,q\}} A[\sigma_p]_{\tau(p)\upsilon(p)} \big(\overline{\nabla}_{\tau(1)} A[\sigma_q]_{\tau(q)\upsilon(q)} - \overline{\nabla}_{\tau(q)} A[\sigma_q]_{\tau(1)\upsilon(q)}\big).
\end{aligned}$$

So the claim follows from the total symmetry of $\overline{\nabla} A[\sigma_i]$ (Lemma 15.7). □

By applying Lemma 18.29 to (18.73), we deduce the following.

Lemma 18.31. *The mixed volume V has the following properties: Let D_0, D_0', $D_1, \dots, D_n$ be bounded, convex, locally uniformly convex regions with respective support functions $\sigma_0, \sigma_0', \sigma_1, \dots, \sigma_n$.*

(1) Total symmetry:

$$V(D_0, D_1, \dots, D_n) = V(D_{\tau(0)}, D_{\tau(1)}, \dots, D_{\tau(n)}) \tag{18.77}$$

for any permutation τ of $\{0, 1, \dots, n\}$.

(2) Translation invariance: *For any point $X_0 \in \mathbb{R}^{n+1}$,*

$$V(D_0 + X_0, D_1, \dots, D_n) = V(D_0, D_1, \dots, D_n). \tag{18.78}$$

(3) Positivity:

$$V(D_0, D_1, \dots, D_n) > 0. \tag{18.79}$$

(4) Monotonicity: *If $D_0 \subset D_0'$, then*

$$V(D_0, D_1, \dots, D_n) \le V(D_0', D_1, \dots, D_n). \tag{18.80}$$

Proof. (1) By (18.73) and (18.75), with $\dot{\mathcal{Q}}$ constructed from $\sigma_2, \ldots, \sigma_n$,

$$\begin{aligned} V(D_0, D_1, D_2, \ldots, D_n) &= \int_{S^n} \sigma_0\, \mathcal{Q}[\sigma_1, \sigma_2, \ldots, \sigma_n]\, d\bar{\mu} \\ &= \int_{S^n} \sigma_0\, \dot{\mathcal{Q}}^{ij} A[\sigma_1]_{ij}\, d\bar{\mu} \\ &= \int_{S^n} A[\sigma_0]_{ij}\, \dot{\mathcal{Q}}^{ij} \sigma_1\, d\bar{\mu} \\ &= V(D_1, D_0, D_2, \ldots, D_n), \end{aligned}$$

where for the third equality we integrated by parts twice and applied (18.76). Now (18.77) follows from the total symmetry of $\mathcal{Q}$.

(2) The support function for the region $D_0 + X_0$ is $\sigma + \langle \cdot, X_0\rangle$. But $A[\langle \cdot, X_0\rangle] \equiv 0$.

(3) By part (2), we may choose the origin so that $\sigma_1 > 0$. Then (18.79) follows from the positivity of $\mathcal{Q}$ (see Lemma 18.29).

(4) If we translate a point of D_0 to the origin, then $\sigma(D_0) \leq \sigma(D_0')$. The claim then follows from (18.73) and the positivity of $\mathcal{Q}$. $\square$

The mixed volume satisfies the following remarkable inequality, known as the Alexandrov–Fenchel inequality. For a proof, see [**461**].

Theorem 18.32 (Alexandrov and Fenchel). *If $D_0, D_1, \ldots, D_n$ are bounded convex regions in $\mathbb{R}^{n+1}$, then*

$$V(D_0, D_0, D_2, \ldots, D_n) \cdot V(D_1, D_1, D_2, \ldots, D_n) \leq V^2(D_0, D_1, D_2, \ldots, D_n)\,. \tag{18.81}$$

Example 18.33. The mean cross-sectional volumes $V_k(D)$, $k = 1, \ldots, n$, satisfy

$$V_{k-1}(D) V_{k+1}(D) \leq V_k(D)^2,$$

where $V_0(D) \doteqdot \omega_{n+1}$ (the volume of the unit ball). Iterating yields

$$V_{k-i}(D)^j V_{k+j}(D)^i \leq V_k(D)^{i+j} \tag{18.82}$$

for i, j, k satisfying $0 \leq k - i < k < k + j \leq n + 1$. In particular,

$$\omega_{n+1}^{\ell-k} V_\ell(D)^k \leq V_k(D)^\ell$$

for k, ℓ satisfying $0 < k < \ell \leq n + 1$. When $k = n$ and $\ell = n + 1$, this is the isoperimetric inequality (14.5).

Given $k \in \{1, \ldots, n\}$, let $\sigma_{k+1}, \ldots, \sigma_n$ be support functions of fixed bounded convex regions. Define the associated mixed discriminant by

$$\mathcal{Q}_k[f] = \mathcal{Q}[\underbrace{f, \ldots, f}_{k\text{-times}}, \sigma_{k+1}, \ldots, \sigma_n] \tag{18.83}$$

and the associated mixed volume by

$$V_{k+1}[f] = V[\underbrace{f, \dots, f,}_{(k+1)\text{-times}} \sigma_{k+1}, \dots, \sigma_n]. \tag{18.84}$$

More generally we define, for $0 \leq \ell \leq k+1$ and a given bounded, convex, locally uniformly convex region containing the origin with support function $\upsilon : S^n \to \mathbb{R}$,

$$V_\ell[f] = V[\underbrace{f, \dots, f,}_{\ell\text{-times}} \underbrace{\upsilon, \dots, \upsilon,}_{(k+1-\ell)\text{-times}} \sigma_{k+1}, \dots, \sigma_n]. \tag{18.85}$$

Given $\alpha \in \mathbb{R} \setminus \{0\}$, define

$$Z_\alpha[f] = \begin{cases} \left(\dfrac{1}{V_0} \displaystyle\int_{S^n} \upsilon \mathcal{Q}_k[\upsilon] \left(\dfrac{\mathcal{Q}_k[f]}{\mathcal{Q}_k[\upsilon]} \right)^{1+\alpha} d\mu \right)^{\frac{1}{1+\alpha}} & \text{for } \alpha \neq -1, \\ \exp\left\{ \dfrac{1}{V_0} \displaystyle\int_{S^n} \upsilon \mathcal{Q}_k[\upsilon] \log \left(\dfrac{\mathcal{Q}_k[f]}{\mathcal{Q}_k[\upsilon]} \right) d\mu \right\} & \text{for } \alpha = -1 \end{cases} \tag{18.86}$$

and consider the associated geometric flow corresponding to the evolution equation

$$\partial_t \sigma = \operatorname{sgn}(\alpha) \upsilon(\mathsf{N}) \left(\frac{\mathcal{Q}_k[\sigma]}{\mathcal{Q}_k[\upsilon]} \right)^\alpha. \tag{18.87}$$

Theorem 18.34. *If σ satisfies (18.87), then the entropy defined by*

$$E_\alpha = \left(V_{k+1}^{-\frac{k}{k+1}}[\sigma] \, Z_\alpha[\sigma] \right)^{\operatorname{sgn} \alpha} \tag{18.88}$$

is a nonincreasing function of time and is strictly decreasing unless the solution is self-similar.

Proof. We first show that the Alexandrov–Fenchel inequality implies a useful generalization:

Proposition 18.35. *Suppose that $\sigma_1, \dots, \sigma_n$ are the support functions of smooth, uniformly convex bodies. Then, for any smooth function f on S^n,*

$$\Delta[f] \doteqdot \int_{S^n} f \mathcal{Q}[f, \sigma_2, \dots, \sigma_n] \, d\bar{\mu} - \frac{\left(\int_{S^n} f \mathcal{Q}[\sigma_1, \dots, \sigma_n] \, d\bar{\mu} \right)^2}{\int_{S^n} \sigma_1 \mathcal{Q}[\sigma_1, \dots, \sigma_n] \, d\bar{\mu}} \leq 0, \tag{18.89}$$

with equality if and only if $f(z) = c\sigma_1(z) + \langle z, p \rangle$ for some $c \in \mathbb{R}$ and $p \in \mathbb{R}^{n+1}$.

Proof. In the case where $A[f] > 0$, so that f is the support function of a uniformly convex body, the proposition coincides with the Alexandrov–Fenchel inequality with $\sigma_0 = f$.

In general, if f is a smooth function, then we can choose c sufficiently large so that $A[f + c\sigma_1] > 0$. We then compute directly that

$$\Delta[f + c\sigma_1] = \Delta[f].$$

The left-hand side is nonpositive by the Alexandrov–Fenchel inequality, with equality if and only if $f + c\sigma_1 = \tilde{c}\sigma_1 + \langle z, p\rangle$, so the result follows. □

Next we compute the evolution of the functional in Theorem 18.34 under the flow (18.87). For convenience, we use the notation

$$V_j[f;\sigma] \doteqdot V[\underbrace{f,\ldots,f}_{j\text{-times}},\underbrace{\sigma,\ldots,\sigma}_{(k-j)\text{-times}},\sigma_{k+1},\ldots,\sigma_n].$$

Letting

$$\mathcal{P} \doteqdot V_1[|\partial_t\sigma|;\sigma] = \int_{S^n} v\mathcal{Q}_k[v]\left(\frac{\mathcal{Q}_k[\sigma]}{\mathcal{Q}_k[v]}\right)^{1+\alpha} d\bar{\mu},$$

we have (for $\alpha \neq -1$) that $Z_\alpha = V_0^{-\frac{1}{1+\alpha}}\mathcal{P}^{\frac{1}{1+\alpha}}$ and

$$\partial_t\mathcal{P} = (1+\alpha)k(\operatorname{sgn}\alpha)V_2[|\partial_t\sigma|;\sigma]. \tag{18.90}$$

Then a direct computation gives that the entropy defined by (18.88) evolves according to

$$\partial_t \log E_\alpha = -k\frac{V_1[|\partial_t\sigma|;\sigma]}{V_0[|\partial_t\sigma|;\sigma]} + k\frac{V_2[|\partial_t\sigma|;\sigma]}{V_1[|\partial_t\sigma|;\sigma]} \leq 0$$

since $V_1^2 \geq V_0V_2$ by Proposition 18.35. Furthermore, equality holds if and only if $\partial_t\sigma = c\sigma + \langle z, p\rangle$ for some constant c and some $p \in \mathbb{R}^{n+1}$. This implies that σ is a self-similar solution of the equation (18.87). This completes the proof of Theorem 18.34. □

Example 18.36. When $v \equiv 1$ and $k = n$, we obtain

$$\partial_t\sigma = \operatorname{sgn}(\alpha)\det[A[\sigma]]^\alpha = \operatorname{sgn}(\alpha)K^{-\alpha}.$$

This is the evolution equation satisfied by the support function of a family of hypersurfaces evolving by the $-\alpha$-Gauß curvature flow. The corresponding entropy E_α is the $-\alpha$-Gaußian entropy (15.88).

18.8. Alexandrov reflection

In this section we shall discuss the Alexandrov reflection method as applied to extrinsic geometric flows. We consider curvature flows of the form (18.1), i.e., $X : M^n \times I \to \mathbb{R}^{n+1}$ satisfying

$$\partial_t X = -F\mathsf{N},$$

where $F \doteqdot f([\mathrm{II}]) \doteqdot f(\vec{\kappa})$ and where $f : \mathrm{S}(n) \to \mathbb{R}$ is a *monotone nondecreasing* $\mathrm{SO}(n)$-invariant function, $\mathrm{S}(n)$ is the set of symmetric $n \times n$-matrices, and $[\mathrm{II}] = [\mathrm{II}_{(x,t)}]$ is the component matrix of II with respect to some g_t-orthonormal basis for $T_{(x,t)}M$. The monotonicity condition is equivalent to

$$f(B) \geq f(A) \text{ for all } A, B \in \mathrm{S}(n) \text{ such that } (B - A) \geq 0.$$

This condition corresponds to *weak* parabolicity of (18.1). We do not assume that $F > 0$ (contracting flows). In fact, the results in this section are most effective when $F < 0$ (expanding flows).

We first consider compact strictly convex hypersurfaces from the point of the support function (using the Gauß map parametrization). Then we consider the case of compact embedded hypersurfaces from a more geometric viewpoint.

18.8.1. Alexandrov reflection of convex hypersurfaces. In this subsection we view extrinsic geometric flows through the lens of the support function. As a result, this subsection has a more analytic flavor in contrast to the next subsection which has a more geometric flavor.

Recall from Section 5.2.4 that a bounded, convex, locally uniformly convex hypersurface $\mathcal{M}^n$ in $\mathbb{R}^{n+1}$ may be represented by its support function $\sigma : S^n \to \mathbb{R}$ via the embedding

$$\mathrm{G}^{-1}(z) = \sigma(z)z + \overline{\nabla}\sigma(z),$$

where $\mathrm{G} : \mathcal{M}^n \to S^n$ is the Gauß map. By (5.78) and (5.81), the curvature flow equation (18.1) for convex, locally uniformly convex solutions is equivalent to the equation

$$\partial_t \sigma = \Phi \tag{18.91}$$

for the support function $\sigma : S^n \times I \to \mathbb{R}$, where

$$\Phi \doteqdot \phi(A[\sigma]) \doteqdot -f(A[\sigma]^{-1}) = -f(\mathrm{II}) = -F.$$

Here, $A[\sigma]$ is conflated with its component matrix with respect to a $\overline{g}$-orthonormal frame and II with its component matrix with respect to a g-orthonormal frame (g being the induced metric). Note that ϕ is nondecreasing if and only if f is nondecreasing.

We do not need to assume that ϕ is smooth, but we shall require that it be locally Lipschitz continuous, i.e., that for any constant C there exists a constant L such that if A and B satisfy $|A| \leq C$ and $|B| \leq C$, then $|\phi(B) - \phi(A)| \leq L|B - A|$.

Recall that if we translate $\mathcal{M}$ by a vector $v \in \mathbb{R}^{n+1}$ to obtain a hypersurface $\mathcal{M}'$ with support function σ', then

$$\sigma'(z) = \sigma(z) + \langle z, v \rangle. \tag{18.92}$$

Functions on S^n of the form $z \mapsto \langle z, v\rangle$, i.e., the restrictions of linear functions, shall play an important role in our discussion. Note that the identity

$$A[\sigma'] = \overline{\nabla}^2\sigma' + \sigma'\overline{g} = \overline{\nabla}^2\sigma + \sigma\overline{g} = A[\sigma] \tag{18.93}$$

is clear both geometrically (the second fundamental form of a translated hypersurface is the same as that of the original) and analytically (a standard Hessian property of the first eigenfunctions of the Laplacian on S^n). Thus, if $\sigma(z,t)$ is a solution to (18.91), then $\sigma'(z,t) = \sigma(z,t) + \langle z, v\rangle$ is also a solution to (18.91) (which corresponds to the translation invariance of the flow corresponding to (18.91)).

Given $y \in S^n$, define the **upper and lower hemispheres in the y-direction** by

$$S^n_{y+} \doteqdot \{z \in S^n \mid \langle z, y\rangle \geq 0\} \quad \text{and} \tag{18.94a}$$

$$S^n_{y-} \doteqdot \{z \in S^n \mid \langle z, y\rangle \leq 0\}, \tag{18.94b}$$

respectively. Note that $S^n_{y-} = S^n_{(-y)+}$. Define the **reflection in the y-direction** $\rho_y : S^n \to S^n$ by

$$\rho_y(z) = z - 2\langle z, y\rangle y \ \text{ for } \ z \in S^n. \tag{18.95}$$

We have $\rho_y(S^n_{y+}) = S^n_{y-}$, $\rho_y(S^n_{y-}) = S^n_{y+}$, and ρ_y restricts to the identity on the **equator** $E_y \doteqdot \partial S^n_{y+} = \partial S^n_{y-}$.

Define the restricted linear function $l_y : S^n \to \mathbb{R}$ by

$$l_y(z) = \langle z, y\rangle. \tag{18.96}$$

Observe that l_y is zero on the equator E_y and recall that

$$\overline{\nabla}^2 l_y + l_y\,\overline{g} = 0 \quad \text{on } S^n. \tag{18.97}$$

In particular, $\overline{\Delta} l_y + n l_y = 0$.

Since equation (18.91) is invariant under reflections about hyperplanes passing through the origin and by (18.97), if σ is a solution to (18.91), then

$$\sigma_{y,\lambda} \doteqdot \sigma \circ \rho_y + 2\lambda l_y \tag{18.98}$$

is also a solution to (18.91) for any $y \in S^n$ and $\lambda \in \mathbb{R}$. Observe that if $\mathcal{M}$ is the hypersurface with support function σ, then its reflection $\mathcal{M}_y = \rho_y(\mathcal{M})$ has support function given by $\sigma_y = \sigma \circ \rho_y$. Indeed, let G_y denote the Gauß map of $\mathcal{M}_y$. Then for $z \in S^n$,

$$\sigma_y(z) = \langle \mathrm{G}_y^{-1}(z), z\rangle = \langle \rho_y(\mathrm{G}_y^{-1}(z)), \rho_y(z)\rangle = \langle \mathrm{G}^{-1}(\rho_y(z)), \rho_y(z)\rangle = \sigma(\rho_y(z)).$$

Thus the invariance under reflections in S^n of (18.91) for σ is equivalent to the invariance of curvature flows of $\mathcal{M}$ under reflections in $\mathbb{R}^{n+1}$. That $\sigma_{y,\lambda}$ remains a solution to (18.91) corresponds to the invariance under the composition of a reflection and a translation; this is also discussed in Section 18.8.2 below.

For any smooth function v on S^n, there exists a (suitably large) constant λ such that on any hemisphere S^n_{y+} the reflected function $v \circ \rho_y$ is larger than v minus 2λ times the linear function l_y:

Lemma 18.37. *Let $v : S^n \to \mathbb{R}$ be a smooth function. Then there exists $\lambda \in \mathbb{R}$ such that for any $y \in S^n$,*

$$v_{y,\lambda}(z) \doteqdot v(\rho_y(z)) + 2\lambda l_y(z) \geq v(z) \quad \textit{for all } z \in S^n_{y+}\,. \tag{18.99}$$

Proof. Suppose the lemma is false. Then there exist sequences $\lambda_i \to \infty$, $y_i \in S^n$, and $z_i \in S^n_{y_i+}$ such that

$$v(\rho_{y_i}(z_i)) + 2\lambda_i l_{y_i}(z_i) < v(z_i)\,. \tag{18.100}$$

Clearly, $z_i \in S^{n\circ}_{y_i+} \doteqdot S^n_{y_i+} \backslash \partial S^n_{y_i+}$. Since S^n is compact, by passing to a subsequence, we may assume that $y_i \to y_\infty \in S^n$. By passing to a further subsequence, we may also assume that $z_i \to z_\infty \in S^n_{y_\infty+}$.

Case 1. $z_\infty \in S^{n\circ}_{y_\infty+}$. By the continuity of v and by the convergence of the sequences $\{y_i\}$ and $\{z_i\}$, we have that $v(\rho_{y_i}(z_i)) \to v(\rho_{y_\infty}(z_\infty)) \in \mathbb{R}$ and that $v(z_i) \to v(z_\infty) \in \mathbb{R}$, whereas $\lambda_i l_{y_i}(z_i) \to +\infty$ since $\lambda_i \to \infty$ and since $l_{y_i}(z_i) \to l_{y_\infty}(z_\infty) > 0$ by the assumption $z_\infty \in S^{n\circ}_{y_\infty+}$. This contradicts (18.100) for i sufficiently large.

Case 2. $z_\infty \in \partial S^n_{y_\infty+}$. We then have $\bar{d}(z_i, E_{y_i}) \to 0$ and in particular $z_i \neq y_i$ for i sufficiently large, which we now assume and where $\bar{d}$ denotes the distance with respect to $\overline{g}$. We may rewrite (18.100) as

$$\frac{v(z_i) - v(\rho_{y_i}(z_i))}{l_{y_i}(z_i)} > 2\lambda_i\,. \tag{18.101}$$

Since $l_{y_i}(z_i) \to 0$ and $\frac{\bar{d}(z_i,\rho_{y_i}(z_i))}{l_{y_i}(z_i)} \to 2$, we claim that

$$\frac{v(z_i) - v(\rho_{y_i}(z_i))}{l_{y_i}(z_i)} \to 2\langle \overline{\nabla} v, y_\infty\rangle(z_\infty)\,. \tag{18.102}$$

Since $\lambda_i \to \infty$, we obtain a contradiction to (18.101) for i large enough.

Finally, we argue for (18.102) as follows. Since $z_i \neq y_i$, z_i and $\rho_{y_i}(z_i)$ lie on a unique great circle. Let w_i be the unique point on E_{y_i} closest to z_i (equivalently, closest to $\rho_{y_i}(z_i)$). Note that $w_i \to w_\infty \doteqdot z_\infty$. Then this great circle is parametrized by the unit speed geodesic γ_{y_i} with initial tangent vector $y_i \in T_{w_i}S^n$. Let $\varepsilon_i \doteqdot \frac{1}{2}\bar{d}(z_i, \rho_{y_i}(z_i))$. Then $\varepsilon_i \to 0$. We have

$$\frac{v(z_i) - v(\rho_{y_i}(z_i))}{l_{y_i}(z_i)} = \frac{v(\gamma_{y_i}(\varepsilon_i)) - v(\gamma_{y_i}(-\varepsilon_i))}{2\varepsilon_i}\,\frac{\bar{d}(z_i, \rho_{y_i}(z_i))}{l_{y_i}(z_i)}\,.$$

The claim now follows since

$$\frac{v(\gamma_{y_i}(\varepsilon_i)) - v(\gamma_{y_i}(-\varepsilon_i))}{2\varepsilon_i} = \langle \overline{\nabla} v, y_i \rangle (w_i) + o(1). \qquad \square$$

The constant λ in the lemma above is related to the modulus of continuity, oscillation, and norm of the gradient of the function as follows.

Corollary 18.38. *Suppose that λ is as in Lemma* 18.37. *Then*

$$|v(z_1) - v(z_2)| \leq 2\lambda \sin\left(\frac{\bar{d}(z_1, z_2)}{2}\right) \quad \textit{for all } z_1, z_2 \in S^n, \tag{18.103}$$

$$\max_{z \in S^n} v(z) - \min_{z \in S^n} v(z) \leq 2\lambda, \quad \textit{and} \tag{18.104}$$

$$|\overline{\nabla} v|(z) \leq \lambda \quad \textit{for all } z \in S^n. \tag{18.105}$$

Proof. Let z_1 and z_2 be distinct points in S^n. Define

$$y = \frac{z_2 - z_1}{|z_2 - z_1|} \in S^n.$$

Then $z_2 \in S^n_{y+}$ and $\rho_y(z_2) = z_1$. Thus (18.99) says that

$$v(z_1) + 2\lambda \langle z_2, y \rangle \geq v(z_2).$$

On the other hand,

$$\langle z_2, y \rangle = \frac{|z_2 - z_1|}{2} = \sin\left(\frac{\bar{d}(z_1, z_2)}{2}\right).$$

Now (18.103) follows since z_1 and z_2 are arbitrary distinct points (the case $z_1 = z_2$ is obvious) and also may be switched.

Inequality (18.104) follows immediately from (18.103).

Furthermore, by (18.103) and since $\sin\theta \leq \theta$ for $0 \leq \theta \leq \pi/2$, we have

$$\frac{|v(z_1) - v(z_2)|}{\bar{d}(z_1, z_2)} \leq \lambda \quad \text{for } z_1 \neq z_2.$$

Now (18.105) follows from this Lipschitz estimate since v is C^1. $\square$

The Alexandrov reflection method yields the following monotonicity property.

Theorem 18.39. *Let $\sigma \in C^2(S^n \times [0,T))$ be a solution to* (18.91) *with $\sigma(\cdot, 0) = \sigma_0$, where ϕ is a nondecreasing, locally Lipschitz continuous function. Suppose that $y \in S^n$ and $\lambda \in \mathbb{R}$ are such that*

$$\sigma_0(\rho_y(z)) + 2\lambda l_y(z) \geq \sigma_0(z) \quad \textit{for all } z \in S^n_{y+}. \tag{18.106}$$

Then

$$\sigma(\rho_y(z), t) + 2\lambda l_y(z) \geq \sigma(z, t) \quad \textit{for all } z \in S^n_{y+},\ t \in [0,T). \tag{18.107}$$

Hence there exists λ depending only on σ_0 such that:

(i) (Modulus of continuity estimate)

$$|\sigma(z_1,t)-\sigma(z_2,t)| \leq 2\lambda \sin\left(\frac{\bar{d}(z_1,z_2)}{2}\right) \quad \textit{for } z_1,z_2 \in S^n,\ t \in [0,T). \tag{18.108}$$

(ii) (Oscillation estimate)

$$\max_{z\in S^n} \sigma(z,t) - \min_{z\in S^n} \sigma(z,t) \leq 2\lambda \ \textit{ for all } t \in [0,T). \tag{18.109}$$

(iii) (Gradient estimate)

$$|\overline{\nabla}\sigma|(z,t) \leq \lambda \quad \textit{for all } z \in S^n\,,\ t \in [0,T)\,. \tag{18.110}$$

Proof. Suppose that $y \in S^n$ and $\lambda \in \mathbb{R}$ are such that (18.106) holds. We shall prove (18.107). The rest of the theorem then follows from the previous lemma and its corollary.

Set

$$\sigma_{y,\lambda}(z,t) \doteqdot \sigma(\rho_y(z),t) + 2\lambda l_y(z)\,.$$

Given $A \in \mathbb{R}_+$, to be determined momentarily, define $w : S^n \times [0,T) \to \mathbb{R}$ by

$$w(z,t) = \mathrm{e}^{-At}(\sigma_{y,\lambda}(z,t) - \sigma(z,t))\,.$$

Then by (18.91) and (18.98), we have that

$$\partial_t w = -Aw + \mathrm{e}^{-At}(\Phi_{y,\lambda} - \Phi)\,, \tag{18.111}$$

where

$$\Phi_{y,\lambda} \doteqdot \phi(\overline{\nabla}^2\sigma_{y,\lambda} + \sigma_{y,\lambda}\overline{g})$$

Note that w is zero on $\partial S^n_y \times [0,T)$ and nonnegative on $S^n_y \times \{0\}$.

Suppose for a contradiction that (18.107) is false. Then w is negative somewhere. Since w is continuous, there exists a point $(z_0,t_0) \in S^{n\circ}_y$ such that (see Exercise 18.12)

$$w(z_0,t_0) = \inf_{S^n_y\times[0,t_0]} w < 0\,. \tag{18.112}$$

By the maximum principle, we have at the minimum point (z_0,t_0) that

$$0 \geq \partial_t w \quad \text{and} \quad 0 \leq \overline{\nabla}^2 w = \mathrm{e}^{-At_0}(\overline{\nabla}^2\sigma_{y,\lambda} - \overline{\nabla}^2\sigma)\,.$$

Thus, by (18.111) and since ϕ is nondecreasing, we have at (z_0,t_0) that

$$\begin{aligned}
0 &> A\mathrm{e}^{At}w \\
&\geq \phi(\overline{\nabla}^2\sigma_{y,\lambda} + \sigma_{y,\lambda}\overline{g}) - \phi(\overline{\nabla}^2\sigma + \sigma\overline{g}) \\
&\geq \phi(\overline{\nabla}^2\sigma + \sigma_{y,\lambda}\overline{g}) - \phi(\overline{\nabla}^2\sigma + \sigma\overline{g}) \\
&\geq -L|(\sigma_{y,\lambda} - \sigma)\overline{g}| \\
&= L\sqrt{n}\mathrm{e}^{At}w\,,
\end{aligned}$$

where L is the maximum Lipschitz constant for ϕ on the set of positive definite symmetric matrices with norm at most the maximum of

$$\sup_{S^n\times[0,t_0]} |\overline{\nabla}^2\sigma_{y,\lambda} + \sigma_{y,\lambda}\overline{g}| \quad \text{and} \quad \sup_{S^n\times[0,t_0]} |\overline{\nabla}^2\sigma + \sigma\overline{g}|.$$

Thus, by taking $A > L\sqrt{n}$, we obtain a contradiction. □

In the next subsection, we consider the monotonicity property above from a more geometric point of view.

18.8.2. Alexandrov reflection of embedded hypersurfaces. Let $y \in S^n \subset \mathbb{R}^{n+1}$ be any unit vector. For each $\lambda \in \mathbb{R}$ define the orthogonal hyperplane of signed distance λ from the origin by

$$y_\lambda^\perp \doteqdot \{x \in \mathbb{R}^{n+1} : \langle x, y\rangle = \lambda\}. \tag{18.113}$$

To this hyperplane define the associated closed half-spaces

$$H_{y,\lambda+} \doteqdot \{x \in \mathbb{R}^{n+1} : \langle x, y\rangle \geq \lambda\} \quad \text{and} \quad H_{y,\lambda-} \doteqdot \{x \in \mathbb{R}^{n+1} : \langle x, y\rangle \leq \lambda\}.$$

Their interiors are the open half-spaces

$$H_{y,\lambda+}^\circ = \{x \in \mathbb{R}^{n+1} : \langle x, y\rangle > \lambda\} \quad \text{and} \quad H_{y,\lambda-}^\circ = \{x \in \mathbb{R}^{n+1} : \langle x, y\rangle < \lambda\}.$$

We have that $\partial H_{y,\lambda+} = \partial H_{y,\lambda-} = y_\lambda^\perp = H_{y,\lambda+} \cap H_{y,\lambda-}$ and also that $H_{y,\lambda+}^\circ \cup H_{y,\lambda-}^\circ \cup y_\lambda^\perp = \mathbb{R}^{n+1}$ is a disjoint union. Define the **reflection map** $\rho_{y,\lambda} : \mathbb{R}^{n+1} \to \mathbb{R}^{n+1}$ about the hyperplane $y_\lambda^\perp$ by

$$\rho_{y,\lambda}(x) \doteqdot x - 2(\langle x, y\rangle - \lambda)y \quad \text{for } x \in \mathbb{R}^{n+1}. \tag{18.114}$$

Let $\mathcal{M} \subset \mathbb{R}^{n+1}$ be a compact, embedded, smooth hypersurface. Let $\mathcal{M}_{y,\lambda}$ denote the reflection of $\mathcal{M}$ about the hyperplane $y_\lambda^\perp$; that is,

$$\mathcal{M}_{y,\lambda} \doteqdot \{\rho_{y,\lambda}(x) : x \in \mathcal{M}\}. \tag{18.115}$$

We have that $\mathcal{M}$ is the boundary of a compact, smooth $(n+1)$-submanifold with boundary Ω of $\mathbb{R}^{n+1}$. Let Ω° denote the interior of Ω.

Given $y \in S^n$ and $\lambda \in \mathbb{R}$, we say that $\mathcal{M}$ **reflects at** (y, λ) if both $\mathcal{M}_{y,\lambda} \cap H_{y,\lambda-}^\circ \subset \Omega^\circ \cap H_{y,\lambda-}^\circ$ and $y \notin T_x\mathcal{M}$ for any $x \in \mathcal{M} \cap y_\lambda^\perp$. We say that $\mathcal{M}$ **reflects up to** (y, λ) if $\mathcal{M}$ reflects at $(y, \bar{\lambda})$ for all $\bar{\lambda} \geq \lambda$.

Note that the hypersurface $\mathcal{M} \cap H_{y,\lambda+}^\circ$ reflects to the hypersurface

$$\rho_{y,\lambda}(\mathcal{M} \cap H_{y,\lambda+}^\circ) = \mathcal{M}_{y,\lambda} \cap H_{y,\lambda-}^\circ.$$

However, $\mathcal{M} \cap H_{y,\lambda+}$ is not necessarily a hypersurface with boundary; in particular, this is the case when there are critical points of the function $x \mapsto \langle x, y\rangle$ on $\mathcal{M}$ which lie on $y_\lambda^\perp$.

Observe that if $y \in T_{x_0}\mathcal{M}$ for some $x_0 \in \mathcal{M} \cap y_\lambda^\perp$, then $\Sigma_{y,\lambda} \doteqdot \mathcal{M} \cap y_\lambda^\perp$ is a smooth $(n-1)$-dimensional submanifold of $\mathbb{R}^{n+1}$ near x_0 and we have the orthogonal decomposition $T_{x_0}\mathcal{M} = T_{x_0}\Sigma_{y,\lambda} \oplus \mathbb{R}y$. Hence, in this case

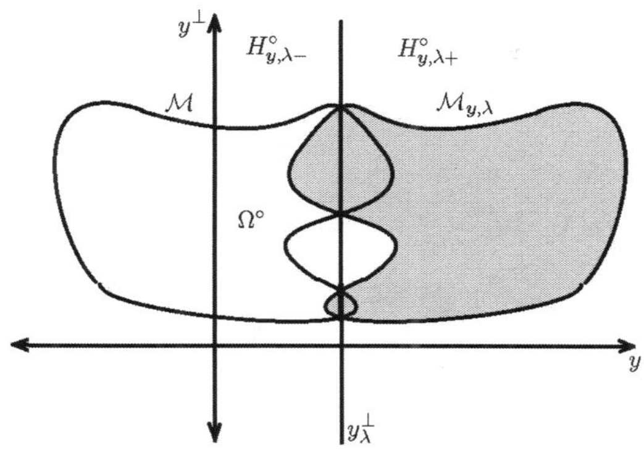

Figure 18.1. An embedded hypersurface and its reflection.

$T_{x_0}\mathcal{M} = T_{x_0}\mathcal{M}_{y,\lambda}$. Conversely, suppose that $T_{x_0}\mathcal{M} = T_{x_0}\mathcal{M}_{y,\lambda}$. Given any $V \in T_{x_0}\mathcal{M}$, we then have

$$(\rho_{y,\lambda})_* V = V - 2\langle V, y\rangle\, y \in T_{x_0}\mathcal{M}\,.$$

Hence $\langle V, y\rangle\, y \in T_{x_0}\mathcal{M}$. Since there exists $V \in T_{x_0}\mathcal{M}$ such that $\langle V, y\rangle \neq 0$, we conclude that $y \in T_{x_0}\mathcal{M}$.

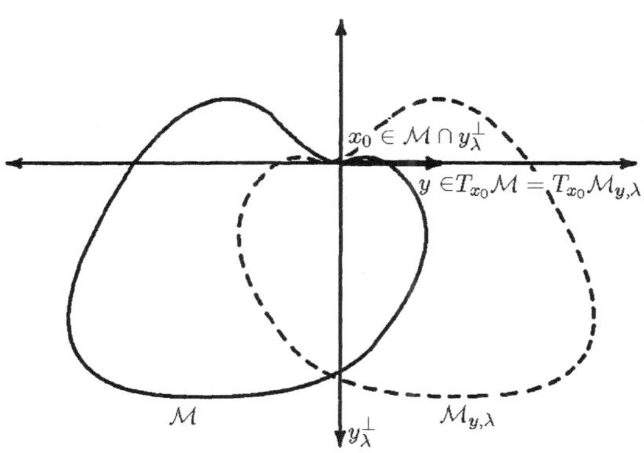

Figure 18.2. The case $y \in T_{x_0}\mathcal{M}$: Tangent space is invariant under reflection.

Thus $\mathcal{M}$ reflecting at (y, λ) is equivalent to that both $T_{x_0}\mathcal{M} \neq T_{x_0}\mathcal{M}_{y,\lambda}$ and

$$\rho_{y,\lambda}(\mathcal{M} \cap H^\circ_{y,\lambda+}) \subset \Omega^\circ\,.$$

That is to say, $T_{x_0}\mathcal{M} \neq T_{x_0}\mathcal{M}_{y,\lambda}$ and the reflection about the hyperplane $y_\lambda^\perp$ of the part of $\mathcal{M}$ in the open half-space $H^\circ_{y,\lambda+}$ lies in the interior of the convex body bounded by $\mathcal{M}$.

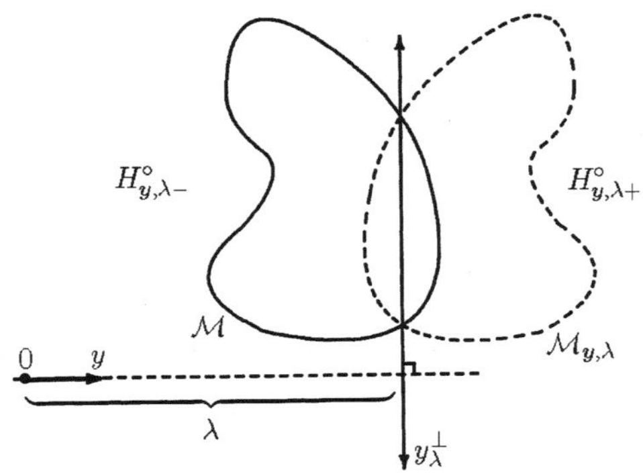

Figure 18.3. $\mathcal{M}$ reflects at (y, λ).

Let $\sigma_{y,\lambda}$ denote the support function in the Gauß map parametrization of $\mathcal{M}_{y,\lambda} = \rho_{y,\lambda}(\mathcal{M})$. We compute that

$$\begin{aligned}\sigma_{y,\lambda} &= \langle \mathrm{G}_{y,\lambda}^{-1}(z), z\rangle \\ &= \langle \rho_y(\mathrm{G}_{y,\lambda}^{-1}(z)), \rho_y(z)\rangle \\ &= \langle \mathrm{G}^{-1}(\rho_y(z)) - 2\lambda, \rho_y(z)\rangle \\ &= \sigma(\rho_y(z)) + 2\lambda\langle z, y\rangle.\end{aligned}$$

This agrees with the choice of definition of $\sigma_{y,\lambda}$ in (18.98).

Theorem 18.40. *Let $X : M^n \times [0, T) \to \mathbb{R}^{n+1}$ be a C^2 solution to (18.1), where $F \doteqdot f(\vec{\kappa})$ is a C^1 function satisfying the strict parabolicity condition*

$$\frac{\partial f}{\partial \kappa_i} > 0 \quad \textit{for all } 1 \leq i \leq n.$$

Suppose that $\mathcal{M}_t = X_t(M^n)$ is embedded for all $t \in [0, T)$. If $\mathcal{M}_0$ reflects at (up to, respectively) (y, λ), then $\mathcal{M}_t$ reflects at (up to, respectively) (y, λ) for all $t \in [0, T)$.

Proof. We shall prove that the property of $\mathcal{M}_t$ reflecting *at* (y, λ) is preserved under the flow. Then the property of $\mathcal{M}_t$ reflecting *up to* (y, λ) being preserved immediately follows.

Suppose for a contradiction that there exists (y, λ) such that the property of $\mathcal{M}_t$ reflecting *at* (y, λ) is not preserved under the flow. We claim that then there exists $t_0 \in (0, T)$ such that:

(1) (*Reflects up to but not including time* t_0) For all $t \in [0, t_0)$ we have that both (a) $\mathcal{M}_{t,y,\lambda} \cap H^\circ_{y,\lambda-} \subset \Omega^\circ_t \cap H^\circ_{y,\lambda-}$ and (b) $y \notin T_x\mathcal{M}_t$ for each $x \in \mathcal{M}_t \cap y_\lambda^\perp$, where $\mathcal{M}_{t,y,\lambda} \doteqdot \{\rho_{y,\lambda}(x) : x \in \mathcal{M}_t\}$.

(2) (*Does not reflect at time* t_0) (i) $\mathcal{M}_{t_0} \cap \mathcal{M}_{t_0,y,\lambda} \cap H^\circ_{y,\lambda-} \neq \emptyset$ or (ii) there exists $x_0 \in \mathcal{M}_{t_0} \cap y_\lambda^\perp$ such that $y \in T_{x_0}\mathcal{M}_{t_0} = T_{x_0}\mathcal{M}_{t_0,y,\lambda}$.

Case (i). Suppose that (2)(i) holds. Then there exists $x_0 \in \mathcal{M}_{t_0} \cap \mathcal{M}_{t_0,y,\lambda} \cap H^{\circ}_{y,\lambda-}$. We obtain a contradiction to the avoidance principle (Theorem 18.19) for geometric flows since $\mathcal{M}_t \cap \mathcal{M}_{t,y,\lambda} \cap H^{\circ}_{y,\lambda-} = \emptyset$ for $t \in [0, t_0)$.

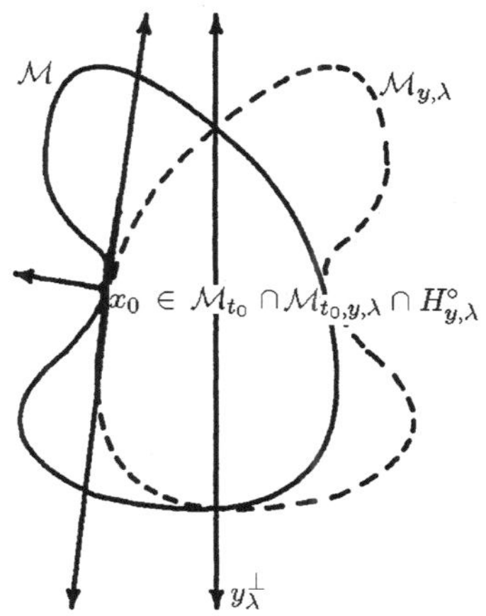

Figure 18.4. Case (i): $\mathcal{M}_{t_0} \cap \mathcal{M}_{t_0,y,\lambda} \cap H^{\circ}_{y,\lambda-} \neq \emptyset$.

Case (ii). Suppose that (2)(ii) holds. Since $T_{x_0}\mathcal{M}_{t_0} = T_{x_0}\mathcal{M}_{t_0,y,\lambda}$, we obtain a contradiction to the Hopf boundary point lemma (Lemma 1.6; cf. [**245**, Chapter 10]). See Figure 18.2. $\square$

Next, we show that the norm of the tangential component of the position vector has a uniform bound.

Corollary 18.41. *Let $X : M^n \times [0,T) \to \mathbb{R}^{n+1}$ be a solution to* (18.1) *satisfying the hypotheses of Theorem* 18.40. *Then there exists a constant λ depending only on $X(0)$ such that*

$$|X^\top|(x,t) = |X - \langle X, \mathsf{N}\rangle \mathsf{N}|(x,t) \leq \lambda \quad \text{for all } x \in M^n,\ t \in [0,T). \tag{18.116}$$

Proof. By Theorem 18.40 there exists $\bar{\lambda} \in \mathbb{R}_+$ such that $\mathcal{M}_t$ reflects at (y,λ) for all $y \in S^n$, $t \in [0,t)$, and $\lambda \geq \bar{\lambda}$. Let $X \in \mathcal{M}_t$. Suppose that $y \in S^n$ satisfies $\langle X, y\rangle \geq \bar{\lambda}$. Let $\lambda \doteqdot \langle X, y\rangle$. Since $\mathcal{M}_t$ reflects at (y,λ) and since $X \in \mathcal{M}_t \cap y_\lambda^\perp$, we have $y \notin T_X\mathcal{M}_t$. Thus, for all unit vectors $z \in T_X\mathcal{M}_t$ we have $\langle X, z\rangle < \bar{\lambda}$. Now suppose that $X^\top \neq 0$. Taking $z = \frac{X^\top}{|X^\top|} \in S^n$, we obtain $|X^\top| < \bar{\lambda}$. $\square$

By (5.83), $\overline{\nabla}\sigma = X^\top$, and we recover the gradient estimate of Theorem 18.39 under a stronger hypothesis.

Corollary 18.42. *Let $X : M^n \times [0,T) \to \mathbb{R}^{n+1}$ be a strictly convex solution to* (18.1) *satisfying the hypotheses of Theorem* 18.40. *Then there exists a*

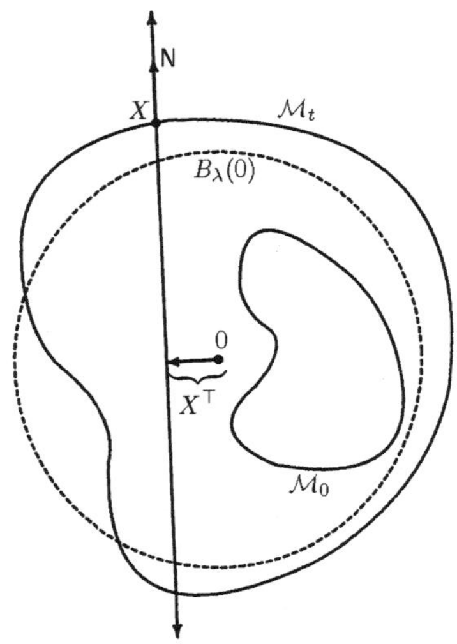

Figure 18.5. Bound for the tangential component of X: $|X^\top| < \lambda$.

constant λ depending only on $X(0)$ such that

$$|\overline{\nabla}\sigma|(z,t) \leq \lambda \quad \text{for all } z \in S^n,\ t \in [0,T)\,, \tag{18.117}$$

where σ denotes the support function in the Gauß parametrization and $\overline{\nabla}$ is the Riemannian covariant derivative of the standard metric $\overline{g}$ on S^n.

We also have a uniform bound for the oscillation of the distance to the origin function on each hypersurface.

Corollary 18.43 (Uniform oscillaton bound). *Let $X : M^n \times [0,T) \to \mathbb{R}^{n+1}$ be a solution to* (18.1) *satisfying the hypotheses of Theorem* 18.40. *Then there exists a constant λ depending only on $\mathcal{M}_0 = X(0)(M^n)$ such that*

$$\max_{x\in M}|X(x,t)| - \min_{x\in M}|X(x,t)| \leq 2\lambda \quad \text{for all } x \in M\,,\ t \in [0,T)\,. \tag{18.118}$$

Thus, for each $t \in [0,T)$, $\mathcal{M}_t$ lies in an annulus of width λ centered at the origin.

Proof. Since $\mathcal{M}_0$ is compact, there exists $\lambda \in \mathbb{R}_+$ such that for all $y \in S^n$ we have that $\mathcal{M}_0$ reflects up to (y,λ). By Theorem 18.40, we have that $\mathcal{M}_t$ reflects up to (y,λ) for all $y \in S^n$ and $t \in [0,T)$. Let $X_1, X_2 \in \mathcal{M}_t$ be such that

$$|X_1| = \min_{X\in\mathcal{M}_t}|X| \quad \text{and} \quad |X_2| = \max_{X\in\mathcal{M}_t}|X|\,.$$

We may assume that $\mathcal{M}_t$ is not a round sphere centered at the origin, so that X_1 and X_2 are distinct points. Define $\bar{y} = \frac{X_2 - X_1}{|X_2 - X_1|} \in S^n$. Since $\mathcal{M}_t$ reflects up to $(\bar{y},\lambda)$, we have

$$d_{\mathbb{R}^{n+1}}(X_2, \bar{y}_\lambda^\perp) \leq d_{\mathbb{R}^{n+1}}(X_1, \bar{y}_\lambda^\perp)\,;$$

that is,

$$\left\langle X_2, \frac{X_2 - X_1}{|X_2 - X_1|} \right\rangle - 2\lambda = \langle X_2, \bar{y} \rangle - 2\lambda \leq -\langle X_1, \bar{y} \rangle + 2\lambda = \left\langle X_1, \frac{X_1 - X_2}{|X_2 - X_1|} \right\rangle + 2\lambda.$$

Thus

$$|X_2|^2 - |X_1|^2 \leq 4\lambda |X_2 - X_1| \leq 8\lambda |X_2| ,$$

which implies

$$|X_2| \leq \frac{|X_1|^2}{|X_2|} + 8\lambda \leq |X_1| + 8\lambda . \qquad \square$$

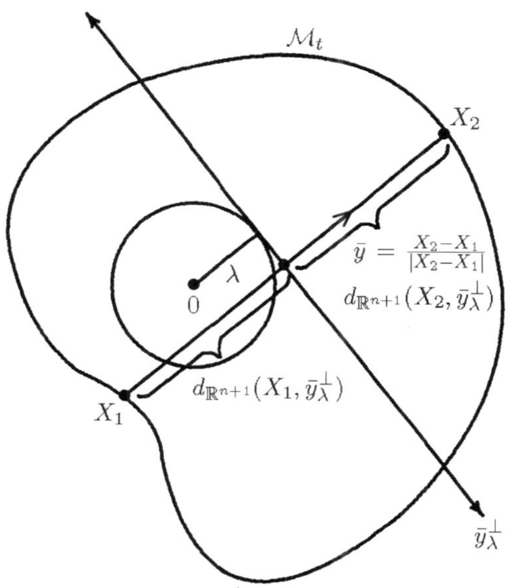

Figure 18.6. $\mathcal{M}_t$ reflects up to $(\bar{y}, \lambda)$.

Under the same hypotheses as above, we have the following.

Corollary 18.44 (Starshaped outside a compact set). *There exists a constant $\lambda \in \mathbb{R}_+$ depending only on $\mathcal{M}_0$ such that for any $t \in [0, T)$ we have that the hypersurface $\mathcal{M}_t \backslash \bar{B}_\lambda(0)$ is starshaped with respect to 0. In particular, if t is such that $\mathcal{M}_t \cap \bar{B}_\lambda(0) = \emptyset$, then $\mathcal{M}_t$ is starshaped with respect to 0.*

Proof. As we saw in the proof of Corollary 18.41, there exists $\lambda \in \mathbb{R}_+$ with the following property. Given any $y \in S^n$, for all $X \in \mathcal{M}_t$ satisfying $\langle X, y \rangle \geq \lambda$ we have $y \notin T_X \mathcal{M}_t$. Thus $\mathcal{M}_t \cap H^\circ_{y,\lambda+}$ is the graph of a function over the hyperplane $y^\perp$. Suppose, for a contradiction, that $\mathcal{M}_t \backslash \bar{B}_\lambda(0)$ is not starshaped with respect to 0. Then there exists $z \in S^n$ and distinct $\lambda_1, \lambda_2 > \lambda$ such that $\lambda_1 z, \lambda_2 z \in \mathcal{M}_t$. This contradicts $\mathcal{M}_t \cap H^\circ_{z,\lambda+}$ being the graph of a function over the hyperplane $z^\perp$. $\square$

Corollary 18.45 (Radial function gradient estimate outside a compact set). *Under the same hypotheses as Corollary* 18.44, *we have that there exists* $\bar{\lambda} \geq \lambda$ *such that the starshaped hypersurface* $\mathcal{M}_t \backslash \bar{B}_{\bar{\lambda}}(0)$ *satisfies the gradient estimate*

$$|\overline{\nabla} r| \leq \bar{\lambda}. \tag{18.119}$$

Proof. By Corollary 18.41, there exists λ such that $|X - \langle X, \mathsf{N}\rangle \mathsf{N}| \leq \lambda$. By (5.75),

$$\mathsf{N}(z) = \frac{rz - \overline{\nabla} r}{(r^2 + |\overline{\nabla} r|^2)^{1/2}}(z).$$

Thus

$$\begin{aligned}
\lambda^2 &\geq |rz - \langle rz, \mathsf{N}(z)\rangle \mathsf{N}(z)|^2 \\
&= \left| rz - \frac{r^2}{r^2 + |\overline{\nabla} r|^2}(rz - \overline{\nabla} r)\right|^2 \\
&= \frac{r^2}{(r^2 + |\overline{\nabla} r|^2)^2}\left||\overline{\nabla} r|^2 z + r\overline{\nabla} r\right|^2 \\
&= \frac{r^2 |\overline{\nabla} r|^2}{r^2 + |\overline{\nabla} r|^2}.
\end{aligned}$$

We conclude that

$$r^2 |\overline{\nabla} r|^2 \leq \lambda^2 (r^2 + |\overline{\nabla} r|^2),$$

which implies that if $r(z) > \lambda$, then

$$|\overline{\nabla} r|^2(z) \leq \frac{\lambda^2 r^2(z)}{r^2(z) - \lambda^2}. \qquad \square$$

18.9. Notes and commentary

For a review of the positive mass theorem and Penrose inequality, see Hubert Bray [**103**].

Our approach to defining curvature functions is not the only one. See, for example, [**237**, Chapter 1] or [**458**].

The approach to short-time existence of solutions, by finding solutions which are graphical over the initial hypersurface, originates in work of Huisken and Polden [**300**] (see also [**81**, **383**, **480**]). A comprehensive treatment is given in [**237**, Chapter 1].

The Alexandrov reflection method was pioneered by Aleksandr D. Alexandrov in a series of brilliant papers [**9**–**13**]. Later important analytic results using this method were obtained by James Serrin [**476**] and by Basilis Gidas, Wei-Ming Ni, and Louis Nirenberg [**242**, **243**]. It is an incredibly robust technique, which has been applied, for example, in the context of the Yamabe problem (by Richard Schoen [**467**]), the Yamabe flow (by Rugang Ye [**539**]), minimal surfaces (by Schoen [**466**] and others), and constant mean curvature surfaces (already in Alexandrov's work, and later by Nick Korevaar, Rob Kusner, and Bruce Solomon [**336**]).

For weakly parabolic extrinsic flows, the Alexandrov reflection method was studied by Robert Gulliver and the second author [**160, 164, 165**]. This includes solutions to level set flows in the viscosity sense; see [**165**]. The Alexandrov reflection method has been more useful in the case of *expanding* flows, although it has also proved useful in obtaining uniqueness results for ancient solutions to the mean curvature flow [**97, 99, 122, 123, 126**] and other contraction flows [**123**]. Some works that use in part this Alexandrov reflection result include Dong-Ho Tsai [**504**], Lii-Perng Liou, Tsai, and the second author [**168**], Tsai and the second author [**171–173**], James McCoy [**394**], Oliver Schnürer [**463**], Yu-Chu Lin and Tsai [**371**], Qi-Rui Li [**365**], and Claus Gerhardt [**239**].

Claus Gerhardt has written an excellent book on curvature flows [**237**]. He considers certain flows in both Riemannian and Lorentzian ambient spaces and describes various applications (e.g., finding closed hypersurfaces of prescribed curvature).

Figure 18.7. Claus Gerhardt.

18.10. Exercises

Exercise 18.1. Let $f : \Gamma^n \to \mathbb{R}$ be an admissible speed function. Show that the associated curvature flow is invariant under space-time translations and ambient rotations. When f is homogeneous of degree $\alpha \neq 0$, show that the associated flow is invariant under the parabolic rescaling (18.10). When f is odd, show that the associated flow is invariant under orientation reversal.

Exercise 18.2. Let $f : \Gamma^n_+ \to \mathbb{R}$ be an admissible speed function. Suppose that f is α-homogeneous for some $\alpha \neq 0$.

(a) Show that $f > 0$ when $\alpha > 0$ and $f < 0$ when $\alpha < 0$.

Define $r_\alpha : I_\alpha \to (0, \infty)$ by

$$r_\alpha(t) \doteqdot \begin{cases} \mathrm{e}^{-\beta t} & \text{if } \alpha = -1, \\ (-(1+\alpha)\beta t)^{\frac{1}{1+\alpha}} & \text{if } \alpha \neq -1, \end{cases}$$

where $\beta \doteqdot f(1, \ldots, 1)$ and

$$I_\alpha \doteqdot \begin{cases} (-\infty, 0) & \text{if } \alpha \in (-\infty, -1) \cup (0, \infty), \\ (-\infty, \infty) & \text{if } \alpha = -1, \\ (0, \infty) & \text{if } \alpha \in (-1, 0). \end{cases}$$

(b) Show that the family $\{S^n_{r_\alpha(t)}\}_{t \in I_\alpha}$ of round spheres $S^n_{r_\alpha(t)}$ satisfies the corresponding curvature flow.

Exercise 18.3. Let $f : \Gamma^n \to \mathbb{R}$ be an admissible speed function and $X_0 : M^n \times [0, T) \to \mathbb{R}^{n+1}$ a solution to the associated curvature flow (18.1). Show that the following functions $u : M \times [0, T) \to \mathbb{R}$ satisfy the linearized flow:

$$\partial_t u = \dot{F}\left(\nabla^2 u + \mathrm{II}^2\, u\right).$$

(a) $u \doteqdot F$.

(b) $u \doteqdot \langle e, \mathsf{N} \rangle$, $e \in \mathbb{R}^{n+1}$.

(c) $u \doteqdot \langle J \cdot X_0, \mathsf{N} \rangle$, $J \in \mathfrak{so}(n+1)$.

(d) (When f is homogeneous of degree $\alpha \neq 0$) $u \doteqdot \langle X_0, \mathsf{N} \rangle + (1+\alpha) t F$.

Exercise 18.4. Prove that the geometric mean is a concave, inverse-concave admissible speed function on the positive cone.

Exercise 18.5. Show that the symmetric function n which gives the norm of a nonzero symmetric matrix,

$$n(A)^2 \doteqdot \operatorname{tr}(A^t A) = \operatorname{tr}(A^2) = \sum_{i=1}^n \lambda_i(A)^2,$$

is **strictly convex in nonradial directions**; that is,

$$\ddot{n}_Z^{ij,kl} V_{ij} V_{kl} > 0$$

for all Z and all $V \in \mathrm{S}(n) \setminus \{\lambda Z : \lambda \in \mathbb{R}\}$.

Exercise 18.6. Prove Theorem 18.22.

Exercise 18.7.

(i) Let g be a positive, α-homogeneous admissible speed, where $\alpha > 0$. Show that $f \doteqdot g^{\frac{1}{\alpha}}$ is a positive, 1-homogeneous admissible speed.

(ii) Let g be a negative, α-homogeneous admissible speed, where $\alpha < 0$. Show that $f \doteq (-g)^{\frac{1}{\alpha}}$ is a positive, 1-homogeneous admissible speed.

Exercise 18.8. Prove parts (2) and (4) of Theorem 18.25.

Exercise 18.9. Given $\mathcal{M}$, $y \in S^n$, and $\lambda \in \mathbb{R}$, define the translated hypersurface

$$\mathcal{M} + \lambda y = \{x + \lambda y \,|\, x \in \mathcal{M}\}.$$

Assume that $\mathcal{M}$ is a uniformly convex, closed, embedded hypersurface. Let $\sigma_{\mathcal{M}} : S^n \to \mathbb{R}$ and $\sigma_{\mathcal{M}+\lambda y} : S^n \to \mathbb{R}$ denote the support functions of $\mathcal{M}$ and $\mathcal{M} + \lambda y$, respectively. Show that

$$\sigma_{\mathcal{M}+\lambda y}(z) = \sigma_{\mathcal{M}}(z) + \lambda l_y(z).$$

Exercise 18.10. Check that $\rho_{y,\lambda}$ defined by (18.114) satisfies $\rho_{y,\lambda}|_{y_\lambda^\perp} = \mathrm{id}_{y_\lambda^\perp}$ and $\rho_{y,\lambda}^2 \doteq \rho_{y,\lambda} \circ \rho_{y,\lambda} = \mathrm{id}_{\mathbb{R}^{n+1}}$.

Exercise 18.11. Recall that $\mathcal{M}_{y,\lambda} = \rho_{y,\lambda}(\mathcal{M})$. Let $\mathrm{G}_{y,\lambda} : \mathcal{M}_{y,\lambda} \to S^n$ denote the Gauß map of $\mathcal{M}_{y,\lambda}$. Show that

$$\mathrm{G}_{y,\lambda} \circ \rho_{y,\lambda} = \rho_y \circ \mathrm{G} : \mathcal{M} \to S^n.$$

Show that

$$\sigma_{\mathcal{M}_{y,\lambda}}(z) = \sigma(\rho_y(z)) + 2\lambda l_y(z) \doteq \sigma_{y,\lambda}(z).$$

Exercise 18.12. Show that (18.112) holds.

Exercise 18.13. Let $v : S^n \times [0, \infty) \to \mathbb{R}$ be a solution to $\partial_t v = \overline{\Delta} v$. Suppose that $\int_{S^n} v(x, 0)\, d\bar{\mu} = 0$. Show that there exists a constant C such that

$$|v(x,t)| \leq C\mathrm{e}^{-nt} \quad \text{for all } (x,t) \in S^n \times [0, \infty).$$

HINT: Consider $u(x,t) = \mathrm{e}^{nt} v(x,t)$.

Exercise 18.14. Let $C > 0$ and $\varepsilon > 0$. Show that there exists $C' = C'(C, \varepsilon)$ with the following property. If $X : M^n \times [0,T) \to \mathbb{R}^{n+1}$ is an embedded solution to a parabolic curvature flow with M^n a closed manifold and $|X_0| \leq C$, then for any $t > 0$ such that $X_t \subset \mathbb{R}^{n+1} \backslash B_{C'}(0)$ we have that

$$\frac{|X_t(x)|}{|X_t(y)|} \leq 1 + \varepsilon$$

for all $x, y \in M^n$.

Exercise 18.15. Suppose that $X : M^n \times [0,T) \to \mathbb{R}^{n+1}$ is an embedded solution to a parabolic curvature flow with M^n a closed manifold and $|X_0| \leq C$. Show that there exists C' such that if $t > 0$ is such that $X_t \subset \mathbb{R}^{n+1} \backslash B_{C'}(0)$, then X_t is starshaped with respect to the origin.

Exercise 18.16 (A nonhomogeneous example). Let $F(\kappa_1, \kappa_2) = \frac{H-2K}{K-H+1}$, where $H = \kappa_1 + \kappa_2$ and $K = \kappa_1\kappa_2$. Show that F is monotone on $\{(\kappa_1, \kappa_2) : 0 < \kappa_i < 1\}$. For any closed convex surface M_0 in $\mathbb{R}^3$ with principal curvatures less than 1, show that there exists a unique solution to the curvature flow with speed F and initial data M_0, which exists on a finite maximal time interval and converges to a sphere of radius 1 as the final time is approached. HINT: Consider the equation satisfied by the surface with support function $\sigma(z,t) - 1$, where $\sigma(z,t)$ is the support function of M_t.

Chapter 19

Flows of Mean Curvature Type

By Euler's theorem for homogeneous functions, flows by admissible speeds of positive homogeneity in the curvature will cause convex hypersurfaces to contract. If the degree of homogeneity is 1, the scaling behavior matches that of the mean curvature flow, so we refer to such flows as **flows of mean curvature type**.

19.1. Convex hypersurfaces contract to round points

The following theorem provides a satisfying extension of Huisken's theorem to a large class of flows of mean curvature type.

Recall that $\Gamma_+^n \doteqdot (0,\infty)^n$ is the positive cone in $\mathbb{R}^n$ and that the dual $f_* : \Gamma_+^n \to \mathbb{R}$ of a function $f : \Gamma_+^n \to \mathbb{R}$ is defined by

$$f_*(z_1^{-1}, \ldots, z_n^{-1}) = f^{-1}(z_1, \ldots, z_n).$$

Theorem 19.1 ([**26,38,39,61,157,158,272**]). *Let $f : \Gamma_+^n \to \mathbb{R}$, $n \geq 2$, be a 1-homogeneous admissible speed function, normalized so that $f(1, \ldots, 1) = 1$ and let $X_0 : M^n \to \mathbb{R}^{n+1}$ be a smooth, locally uniformly convex, compact embedding. Suppose that at least one of the following auxiliary conditions is satisfied:*

(i) $n = 2$.

(ii) *f is convex.*

(iii) *f is concave and extends continuously to $\partial\Gamma_+$ with $f|_{\partial\Gamma_+} \equiv 0$.*

(iv) *f and f_* are concave.*

(v) f_* *is convex.*

(vi) f_* *is concave and extends continuously to* $\partial\Gamma_+$ *with* $f_*|_{\partial\Gamma_+} \equiv 0$.

Then the unique maximal solution $X : M^n \times [0,T) \to \mathbb{R}^{n+1}$ *to*

$$\begin{cases} \partial_t X(x,t) = -f(\mathrm{II}_{(x,t)})\mathsf{N}(x,t) & \text{for } (x,t) \in M^n \times (0,T), \\ X(\cdot,0) = X_0 \end{cases} \tag{19.1}$$

converges uniformly to a constant $p \in \mathbb{R}^{n+1}$ *as* $t \to T$. *The rescaled embeddings* $\widetilde{X}_t : M^n \to \mathbb{R}^{n+1}$ *defined by*

$$\widetilde{X}_t(x) \doteqdot \frac{X(x,t) - p}{\sqrt{2(T-t)}} \tag{19.2}$$

converge uniformly in the smooth topology to a smooth embedding whose image coincides with the unit sphere S^n.

As for the mean curvature flow, there are several possible approaches to the proof. We will follow an argument which is close to that presented in Section 8.6 for the mean curvature flow. The initial step is to control the ratio of principal curvatures at each point, and this is the main point where the proofs of the various cases of the theorem differ. Once the curvature pinching estimate is obtained, the geometric width pinching estimate of Lemma 8.24 is used to prove higher regularity and convergence of rescaled solutions to a limit which is necessarily a sphere. Case (iii) is the simplest and we recommend the reader focus first on this case before moving on to the others.

We remark that the flows to which Theorem 19.1 applies includes the flow by mean curvature (cf. [**291**]), the n-th root of the Gauß curvature (cf. [**157**]), the square root of scalar curvature (cf. [**158**]), the harmonic mean curvature, and the norm of the second fundamental form, as well as many further examples (see Section 18.3).

19.1.1. Preliminaries. Given a uniformly convex hypersurface $X_0 : M^n \to \mathbb{R}^{n+1}$, Theorem 18.18 guarantees the existence of a unique maximal solution $X : M^n \times [0,T) \to \mathbb{R}^{n+1}$ to (19.1), where, by the avoidance principle (Theorem 18.19), $T < \infty$.

Recall that the support function $\sigma : S^n \times [0,T) \to \mathbb{R}$ of the solution is defined by

$$\sigma(z,t) \doteqdot \left\langle \mathrm{G}^{-1}(z,t), z \right\rangle,$$

where $\mathrm{G} : M^n \times [0,T) \to S^n$ is the Gauß map and $\mathrm{G}^{-1}(\cdot,t)$ is the inverse of $\mathrm{G}(\cdot,t)$. Since, by Lemmas 5.20 and 5.21,

$$\mathrm{G}^*\,\mathrm{I} = \mathrm{II}^2 \quad\text{and}\quad \mathrm{G}^*\,\mathrm{II} = A \doteqdot \overline{\nabla}^2\sigma + \sigma\overline{g},$$

where $\overline{g}$ and $\overline{\nabla}$ denote the standard metric and connection on S^n, we find that σ satisfies

$$\partial_t \sigma = \phi(A) \doteq -f_*(A)^{-1} \tag{19.3}$$

where ϕ and f are really evaluated on the component matrix of A with respect to some (and hence any) local orthonormal frame for S^n with respect to its *standard* metric. Recall from Section 5.2.4 that the eigenvalues of A with respect to $\overline{g}$ are the principal radii $\mathbf{r} = (r_1, \ldots, r_n)$ of the hypersurface.

We may also write the solution (at least for a short time) as a radial graph. That is, $\mathcal{M}_t^n = \{\rho(z,t)z : z \in S^n\}$. Then the radial function ρ satisfies

$$\partial_t \rho = -\frac{f(\mathrm{II})}{\langle z, \mathsf{N}(z)\rangle} = -\rho^{-1}\sqrt{\rho^2 + |\overline{\nabla}\rho|^2}\, f(\mathrm{II})$$

and hence, by (5.76) and the 1-homogeneity of f,

$$\partial_t \rho = -f\left(-\overline{\nabla}^2\rho + 2\frac{\overline{\nabla}\rho \otimes \overline{\nabla}\rho}{\rho} + \rho\overline{g}\right), \tag{19.4}$$

where f is evaluated on the component matrix of the argument in an orthonormal frame for TS^n with respect to the *induced* metric.

19.1.2. Preserving pinching. Throughout this section, we consider a fixed 1-homogeneous admissible speed $f : \Gamma_+^n \to \mathbb{R}$ and compact solution $X : M^n \times [0, T) \to \mathbb{R}^{n+1}$ to the corresponding curvature flow (18.1).

The maximum principle immediately implies that lower bounds for the speed are preserved (cf. Lemma 8.2).

Proposition 19.2. *For every* $(x, t) \in M^n \times [0, T)$,

$$F(x,t) \geq \min_{M^n \times \{0\}} F, \tag{19.5}$$

where $F(x,t) = f(\mathrm{II}_{(x,t)})$.

Proof. By (5.147), F satisfies

$$\partial_t F = \dot{F}(\nabla^2 F + \mathrm{II}^2\, F).$$

So the claim follows readily from the maximum principle. □

Since the initial hypersurface is uniformly convex, we can find $\varepsilon_0 > 0$ such that $\mathrm{II} \geq \varepsilon_0\,\mathrm{I}$. The key step in the proof of Theorem 19.1 is to show that the convexity of the solution does not degenerate (cf. Lemma 8.4). To achieve this, we derive a parabolic evolution equation for the second fundamental form.

Lemma 19.3. *The second fundamental form* II *evolves according to*

$$(19.6)\qquad (\nabla_t - \dot F^{kl}\nabla_k\nabla_l)\,\mathrm{II}_{ij} = \ddot F^{pq,rs}\nabla_i\,\mathrm{II}_{pq}\,\nabla_j\,\mathrm{II}_{rs} + \dot F^{kl}\,\mathrm{II}^2_{kl}\,\mathrm{II}_{ij}\,,$$

where $\dot F = \dot f_{\mathrm{II}}$ *and* $\ddot F = \ddot f_{\mathrm{II}}$ *are defined by* (18.7).

Proof. Applying Simons's identity (Proposition 5.10) and the Euler relation $\dot F(\mathrm{II}) = F$ (which follows from the 1-homogeneity of the speed function) to equation (5.140) yields the claim:

$$\begin{aligned}\nabla_t\,\mathrm{II}_{ij} &= \nabla_i\nabla_j F + F\,\mathrm{II}^2_{ij}\\ &= \nabla_i\left(\dot F^{kl}\nabla_j\,\mathrm{II}_{kl}\right) + F\,\mathrm{II}^2_{ij}\\ &= \dot F^{kl}\nabla_i\nabla_j\,\mathrm{II}_{kl} + \ddot F^{pq,rs}\nabla_i\,\mathrm{II}_{pq}\,\nabla_j\,\mathrm{II}_{rs} + F\,\mathrm{II}^2_{ij}\\ &= \dot F^{kl}\left(\nabla_k\nabla_l\,\mathrm{II}_{ij} + \mathrm{II}_{ij}\,\mathrm{II}^2_{kl} - \mathrm{II}_{kl}\,\mathrm{II}^2_{ij}\right) + \ddot F^{pq,rs}\nabla_i\,\mathrm{II}_{pq}\,\nabla_j\,\mathrm{II}_{rs} + F\,\mathrm{II}^2_{ij}\\ &= \dot F^{kl}\nabla_k\nabla_l\,\mathrm{II}_{ij} + \ddot F^{pq,rs}\nabla_i\,\mathrm{II}_{pq}\,\nabla_j\,\mathrm{II}_{rs} + \dot F^{kl}\,\mathrm{II}^2_{kl}\,\mathrm{II}_{ij}\,. \qquad\square\end{aligned}$$

We first consider flows by concave speeds. By Lemma 18.13, the result also applies to flows by inverse-convex speeds.

Proposition 19.4. *If f is concave, then*

$$(19.7)\qquad \left.\frac{|\mathrm{II}|^2}{F^2}\right|_{(x,t)} \le \alpha^{-1} \doteqdot \max_{M^n\times\{0\}}\frac{|\mathrm{II}|^2}{F^2} \quad \textit{for all } (x,t)\in M\times[0,T).$$

In fact, $\max_{M^n\times\{t\}}\frac{|\mathrm{II}|^2}{F^2}$ *is strictly decreasing unless the solution is a shrinking sphere.*

Proof. The evolution equation (19.6) for II yields

$$(\partial_t - \dot F^{ij}\nabla_i\nabla_j)|\mathrm{II}|^2 = \mathrm{II}^{ij}\ddot F^{pq,rq}\nabla_i\mathrm{II}_{pq}\nabla_j\mathrm{II}_{rs} + 2\dot F^{ij}\mathrm{II}^2_{ij}|\mathrm{II}|^2 - 2\dot F^{ij}\nabla_i\mathrm{II}_{kl}\nabla_j\mathrm{II}_{kl}.$$

We also have

$$\partial_t F^2 = \dot F^{ij}\nabla_i\nabla_j F^2 - 2\dot F^{ij}\nabla_i F\nabla_j F + 2F^2\dot F^{ij}\,\mathrm{II}^2_{ij}\,.$$

Combining these, we obtain

$$\begin{aligned}(19.8)\qquad \partial_t\left(\frac{|\mathrm{II}|^2}{F^2}\right) &= \dot F^{ij}\nabla_i\nabla_j\left(\frac{|\mathrm{II}|^2}{F^2}\right) + \frac{4}{F}\dot F^{ij}\nabla_i F\nabla_j\left(\frac{|\mathrm{II}|^2}{F^2}\right)\\ &\quad - \frac{2}{F^2}\dot F^{ij}\nabla_i\,\mathrm{II}_{kl}\,\nabla_j\,\mathrm{II}_{kl} + 2\frac{|\mathrm{II}|^2}{F^4}\dot F^{ij}\nabla_i F\nabla_j F\\ &\quad + \frac{2}{F^2}\,\mathrm{II}_{ij}\,\ddot F^{pq,rs}\nabla_i\,\mathrm{II}_{pq}\,\nabla_j\,\mathrm{II}_{rs}\,.\end{aligned}$$

The last term is nonpositive since F is concave. At a maximum point we have $0 = \nabla\log\left(|\mathrm{II}|^2/F^2\right) = 2\frac{\nabla|\mathrm{II}|}{|\mathrm{II}|} - 2\frac{\nabla F}{F}$, so the terms on the second line become

$$\begin{aligned}
&-\frac{2}{F^2}\left(\dot{F}^{ij}\nabla_i \mathrm{II}_{kl}\nabla_j \mathrm{II}_{kl} - \dot{F}^{ij}\nabla_i|\mathrm{II}|\nabla_j|\mathrm{II}|\right)\\
&\quad = -\frac{2}{F^2}\sum_{i,k,l}\dot{f}^i\left(\nabla_i \mathrm{II}_{kl} - \frac{\mathrm{II}_{pq}}{|\mathrm{II}|^2}\nabla_i \mathrm{II}_{pq}\,\mathrm{II}_{kl}\right)\left(\nabla_i \mathrm{II}_{kl} - \frac{\mathrm{II}_{pq}}{|\mathrm{II}|^2}\nabla_i \mathrm{II}_{pq}\,\mathrm{II}_{kl}\right)\\
&\quad \le 0.
\end{aligned}$$

We conclude from the maximum principle that the maximum value of $\frac{|\mathrm{II}|^2}{F^2}$ is nonincreasing in time. Furthermore, by the strong maximum principle we have that the maximum value is strictly decreasing unless $\frac{|\mathrm{II}|^2}{F^2}$ is constant, in which case all terms on the right-hand side must vanish. In particular, this implies that

$$\nabla_i \mathrm{II}_{kl} - \frac{\mathrm{II}_{pq}}{|\mathrm{II}|^2}\nabla_i \mathrm{II}_{pq}\,\mathrm{II}_{kl} = 0$$

for each i, k, l. Choosing $k = l \neq i$ to be principal directions, we deduce that $\kappa_k \nabla_i \log|\mathrm{II}| = 0$, and hence $\nabla \log|\mathrm{II}| = 0$. So the claim follows from Exercise 19.2. □

Proposition 19.5. *If $n = 2$, then*

$$\frac{\kappa_1}{\kappa_2}(x,t) \ge \alpha \doteqdot \min_{M^n\times\{0\}}\frac{\kappa_1}{\kappa_2}(x,t). \tag{19.9}$$

In fact, $\min_{M^n\times\{t\}} \kappa_1/\kappa_2$ is strictly increasing unless the solution is a shrinking sphere.

Proof. By Exercise 19.1, at an interior critical point of κ_1/κ_2 at which $\kappa_1/\kappa_2 < 1$, both κ_1 and κ_2 are smooth and we find

$$\begin{aligned}
\kappa_2(\partial_t - \dot{F}^{kl}\nabla_k\nabla_l)\frac{\kappa_1}{\kappa_2} &\ge \ddot{F}^{pq,rs}\nabla_1 \mathrm{II}_{pq}\nabla_1 \mathrm{II}_{rs} + 2\dot{F}^k\frac{(\nabla_1 \mathrm{II}_{k2})^2}{\kappa_2-\kappa_1}\\
&\quad - \frac{\kappa_1}{\kappa_2}\left(\ddot{F}^{pq,rs}\nabla_2 \mathrm{II}_{pq}\nabla_2 \mathrm{II}_{rs} - 2\dot{F}^k\frac{(\nabla_2 \mathrm{II}_{k1})^2}{\kappa_2-\kappa_1}\right)\\
&= \ddot{F}^{pq,rs}\left(\nabla_1 \mathrm{II}_{pq}\nabla_1 \mathrm{II}_{rs} - \frac{\kappa_1}{\kappa_2}\nabla_2 \mathrm{II}_{pq}\nabla_2 \mathrm{II}_{rs}\right)\\
&\quad + 2\frac{(1+\kappa_1/\kappa_2)\dot{F}^k}{\kappa_2-\kappa_1}(\nabla_1 \mathrm{II}_{k2})^2.
\end{aligned}$$

Using (18.5), this becomes

$$\begin{aligned}
\kappa_2(\partial_t - \dot{F}^{kl}\nabla_k\nabla_l)\frac{\kappa_1}{\kappa_2} &\ge \ddot{F}^{pq}\left(\nabla_1 \mathrm{II}_{pp}\nabla_1 \mathrm{II}_{qq} - \frac{\kappa_1}{\kappa_2}\nabla_2 \mathrm{II}_{pp}\nabla_2 \mathrm{II}_{qq}\right)\\
&\quad + 2\frac{\dot{F}^2 + \dot{F}^1\kappa_1/\kappa_2}{\kappa_2-\kappa_1}\left((\nabla_2 \mathrm{II}_{11})^2 + (\nabla_1 \mathrm{II}_{22})^2\right).
\end{aligned}$$

The terms on the last line are manifestly nonnegative. Since $\nabla G = 0$, we find

$$\nabla_1 \mathrm{II}_{11} = \frac{\kappa_1}{\kappa_2}\nabla_1 \mathrm{II}_{22} \text{ and } \nabla_2 \mathrm{II}_{11} = \frac{\kappa_1}{\kappa_2}\nabla_2 \mathrm{II}_{22}$$

at the critical point and hence, by Euler's theorem,

$$\begin{aligned}\dot{F}^{pq}\Big(\nabla_1 \mathrm{II}_{pp} \nabla_1 \mathrm{II}_{qq} - \frac{\kappa_1}{\kappa_2}\nabla_2 \mathrm{II}_{pp} \nabla_2 \mathrm{II}_{qq}\Big) \\ &= \frac{1}{\kappa_2^2}(\kappa_1^2 \ddot{F}^{11} + 2\kappa_1\kappa_2\ddot{F}^{12} + \kappa_2^2\ddot{F}^{22})\left((\nabla_1 \mathrm{II}_{22})^2 - \frac{\kappa_1}{\kappa_2}(\nabla_2 \mathrm{II}_{22})^2\right) \\ &= 0\,.\end{aligned}$$

We conclude that

$$(\partial_t - \dot{F}^{kl}\nabla_k\nabla_l)\frac{\kappa_1}{\kappa_2} \geq 0 \tag{19.10}$$

at an interior critical point of κ_1/κ_2. The maximum principle then implies that the minimum of κ_1/κ_2 is nondecreasing. By the strong maximum principle, the minimum is strictly increasing, unless it is constant. Since every compact surface admits an umbilic point, the claim follows from Exercise 5.5. □

To deal with flows by inverse-concave speeds, we exploit a *tensor* maximum principle type argument (cf. Lemma 8.4 and Propositions 9.6 and 12.9). We first show that flows by speeds which are both concave and inverse-concave preserve lower bounds for the ratio κ_1/H and then show that flows by inverse-concave speeds preserve lower bounds for the ratio κ_1/F. Since convex speeds are automatically inverse-concave (see Lemma 18.13), this also yields a pinching estimate for flows by convex speeds. On the other hand, this estimate can be obtained much more directly for flows by convex speeds (see Exercise 19.5).

Proposition 19.6. *If f is concave and inverse-concave, then*

$$\frac{\kappa_1}{H}(x,t) \geq \alpha \doteqdot \min_{M^n\times\{0\}} \frac{\kappa_1}{H}(x,t)\,. \tag{19.11}$$

In fact, the ratio $\min_{M^n\times\{t\}} \frac{\kappa_1}{H}$ is strictly increasing unless the solution is a shrinking sphere.

Proof. Observe that

$$(\partial_t - \dot{F}^{ij}\nabla_i\nabla_j)H = \dot{F}^{ij}\mathrm{II}_{ij}^2 H - \delta^{kl}\ddot{F}^{pq,rs}\nabla_k \mathrm{II}_{pq}\nabla_l \mathrm{II}_{rs}\,.$$

Arguing as in Proposition 12.9 (or Proposition 9.6), we find that

$$(\partial_t - \dot F^{kl}\nabla_k\nabla_l)\kappa_1 \geq \dot F^{kl}\mathrm{II}^2_{kl}\kappa_1 + \ddot F^{pq,rs}\nabla_1\mathrm{II}_{pq}\nabla_1\mathrm{II}_{rs} + 2\sum_{k=1}^n\sum_{\kappa_p>\kappa_1}\frac{\dot F^k}{\kappa_p-\kappa_1}(\nabla_k\mathrm{II}_{1p})^2$$

and hence

$$H(\partial_t - \dot F^{kl}\nabla_k\nabla_l)\frac{\kappa_1}{H} \geq \ddot F^{pq,rs}\Big(\nabla_1\mathrm{II}_{pq}\nabla_1\mathrm{II}_{rs} - \frac{\kappa_1}{H}\sum_{k=1}^n\nabla_k\mathrm{II}_{pq}\nabla_k\mathrm{II}_{rs}\Big) + 2\sum_{k=1}^n\sum_{\kappa_p>\kappa_1}\frac{\dot F^k}{\kappa_p-\kappa_1}(\nabla_k\mathrm{II}_{1p})^2 + \frac{2}{H}\dot F^{ij}\nabla_i\kappa_1\nabla_j\frac{\kappa_1}{H}$$

in the barrier sense.

Suppose that κ_1/H achieves an interior minimum at (x_0,t_0) and set $\varepsilon \doteqdot \frac{\kappa_1}{H}(x_0,t_0)$. Since

$$\nabla_k\mathrm{II}_{ij} = 0 \ \text{ if } \ \kappa_i=\kappa_j, \ \ i\neq j,$$

applying the gradient identities (see Exercise 19.3)

$$\nabla_k\mathrm{II}_{11} = \frac{\varepsilon}{1-\varepsilon}\sum_{i>1}\nabla_k\mathrm{II}_{ii} \ \text{ for each } \ k=1,\dots,n$$

at (x_0,t_0) and recalling (18.5) yields

$$H(\partial_t - \dot F^{kl}\nabla_k\nabla_l)\frac{\kappa_1}{H} \geq Q_1 + \sum_{\kappa_i>\kappa_1}Q_i + \sum_{\kappa_1<\kappa_j<\kappa_k}Q_{1jk} + \sum_{\kappa_1<\kappa_i<\kappa_j<\kappa_k}Q_{ijk}$$

at (x_0,t_0), where

$$Q_1 \doteqdot (1-\varepsilon)\sum_{p,q>1}\left(\ddot F^{pq} + \frac{\varepsilon}{1-\varepsilon}(\ddot F^{p1}+\ddot F^{1q}) + \left(\frac{\varepsilon}{1-\varepsilon}\right)^2\ddot F^{11}\right)\nabla_1\mathrm{II}_{pp}\nabla_1\mathrm{II}_{qq} + 2\sum_{\kappa_p>\kappa_1}\frac{(1-\varepsilon)\dot F^p + \varepsilon\dot F^1}{\kappa_p-\kappa_1}(\nabla_1\mathrm{II}_{pp})^2,$$

$$Q_i \doteqdot -\varepsilon\sum_{p,q>1}\left(\ddot F^{pq} + \frac{\varepsilon}{1-\varepsilon}(\ddot F^{p1}+\ddot F^{1q}) + \left(\frac{\varepsilon}{1-\varepsilon}\right)^2\ddot F^{11}\right)\nabla_i\mathrm{II}_{pp}\nabla_i\mathrm{II}_{qq} + 2\frac{(1-\varepsilon)\dot F^i + \varepsilon\dot F^1}{\kappa_i-\kappa_1}\left(\frac{\varepsilon}{1-\varepsilon}\sum_{p>1}\nabla_i\mathrm{II}_{pp}\right)^2 - 2\varepsilon\sum_{\kappa_p\neq\kappa_i,\kappa_1}\frac{\dot F^p-\dot F^i}{\kappa_p-\kappa_i}(\nabla_i\mathrm{II}_{pp})^2,$$

$$Q_{1jk} \doteqdot 2\left((1-\varepsilon)\frac{\dot F^j-\dot F^k}{\kappa_j-\kappa_k} + \frac{\dot F^j}{\kappa_k-\kappa_1} + \frac{\dot F^k}{\kappa_j-\kappa_1} - \varepsilon\frac{\dot F^j-\dot F^1}{\kappa_j-\kappa_1} - \varepsilon\frac{\dot F^k-\dot F^1}{\kappa_k-\kappa_1}\right)(\nabla_1\mathrm{II}_{jk})^2,$$

and

$$Q_{ijk} \doteqdot -2\varepsilon\left(\frac{\dot F^i-\dot F^j}{\kappa_i-\kappa_j}+\frac{\dot F^j-\dot F^k}{\kappa_j-\kappa_k}+\frac{\dot F^k-\dot F^i}{\kappa_k-\kappa_i}\right)(\nabla_i\mathrm{II}_{jk})^2.$$

We will show that each of these terms is nonnegative. Indeed, Lemma 18.7 implies that $Q_{ijk}\geq 0$ for each $\kappa_i<\kappa_j<\kappa_k$, while Lemmas 18.7 and 18.12 yield $Q_{1jk}\geq 0$ for each $\kappa_1<\kappa_j<\kappa_k$. The nonnegativity of Q_i, for each $\kappa_i>\kappa_1$, follows immediately from the concavity of f since

$$Q_i=-\varepsilon\sum_{p,q=1}^n \ddot F^{pq}\xi_p\xi_q+2\frac{(1-\varepsilon)\dot F^i+\varepsilon\dot F^1}{\kappa_i-\kappa_1}\left(\frac{\varepsilon}{1-\varepsilon}\sum_{p>1}\nabla_i\mathrm{II}_{pp}\right)^2,$$

where

$$\xi\doteqdot\sum_{\kappa_k>\kappa_1}\nabla_1\mathrm{II}_{kk}\left(e_k+\frac{\varepsilon}{1-\varepsilon}e_1\right).$$

To establish the inequality $Q_1\geq 0$, observe that

$$Q_1=(1-\varepsilon)\sum_{\kappa_p>\kappa_q>\kappa_1}\left(\ddot\psi^{pq}+2\frac{\dot\psi^p\delta^{pq}}{\kappa_p-\kappa_1}\right)\nabla_1\mathrm{II}_{pp}\nabla_1\mathrm{II}_{qq},$$

where the function $\psi:\Gamma_+^{n-1}\to\mathbb{R}$ is defined by

$$\psi(z_2,\ldots,z_n)\doteqdot f\left(\frac{\varepsilon}{1-\varepsilon}(z_2+\cdots+z_n),z_2,\ldots,z_n\right).$$

We claim that ψ is inverse-concave. Indeed, since the harmonic mean

$$\phi(z_2,\ldots,z_n)\doteqdot\frac{1-\varepsilon}{\varepsilon}\left(z_2^{-1}+\cdots+z_n^{-1}\right)^{-1}$$

is concave and f is inverse-concave and monotone increasing, we find that

$$\begin{aligned}s\psi_*(z)+(1-s)\psi_*(w)&=sf_*(\phi(z),z)+(1-s)f_*(\phi(w),w)\\&\leq f_*(s\phi(z)+(1-s)\phi(w),sz+(1-s)w)\\&\leq f_*(\phi(sz+(1-s)w),sz+(1-s)w)\\&=\psi_*(sz+(1-s)w)\end{aligned}$$

for any $s\in(0,1)$ and $z,w\in\Gamma_+^{n-1}$. This proves the inverse-concavity of ψ, so the estimate follows from Lemma 18.12.

We have proved that, modulo gradient terms,

$$(\partial_t-\dot F^{kl}\nabla_k\nabla_l)\frac{\kappa_1}{H}\geq 0 \tag{19.12}$$

in the barrier sense. So the maximum principle implies that $\min_{M^n\times\{t\}}\frac{\kappa_1}{H}$ is nondecreasing. In fact, the strong maximum principle (cf. Lemma 9.10) implies that the minimum is strictly increasing unless it is constant, in which case κ_1 is smooth and

$$0 \equiv \nabla_k \frac{\kappa_1}{H}, \quad \text{which implies} \quad \nabla_k \mathrm{II}_{11} = \frac{\kappa_1}{H} \nabla_k H .$$

Since $Q_1 \equiv 0$, we obtain

$$\nabla_1 \mathrm{II}_{pp} \equiv 0$$

whenever $\kappa_p > \kappa_1$.

Since $Q_i \equiv 0$ whenever $\kappa_i > \kappa_1$, we obtain

$$\nabla_i \mathrm{II}_{pp} \equiv 0$$

whenever $\kappa_i > \kappa_1$ and $p > 1$.

Since $\nabla_k \mathrm{II}_{ij} = 0$ for each k whenever $\kappa_i = \kappa_j$, $i \neq j$, we deduce that

$$\left(1 - \frac{\kappa_1}{H}\right) \nabla_k \mathrm{II}_{11} = \frac{\kappa_1}{H} \sum_{p=2}^{n} \nabla_k \mathrm{II}_{pp} = 0 .$$

Since $\kappa_1 < H$, this yields

$$0 = \nabla_k \mathrm{II}_{11} = \frac{\kappa_1}{H} \nabla_k H .$$

We conclude that $\nabla H \equiv 0$, so the claim follows from Proposition 5.12. □

To obtain the pinching estimate for κ_1/F for inverse-concave speeds, we work in the Gauß map parametrization. We need the following evolution equations.

Proposition 19.7. *The speed $\Phi \doteqdot \phi(A) \doteqdot -f_*(A)^{-1}$ satisfies*

$$(\partial_t - \dot{\Phi}^{ij} \overline{\nabla}_i \overline{\nabla}_j) \Phi = \Phi \dot{\Phi}^{ij} \overline{g}_{ij} , \tag{19.13}$$

the inverse-speed $F_* \doteqdot f_*(A)$ satisfies

$$(\partial_t - \dot{\Phi}^{ij} \overline{\nabla}_i \overline{\nabla}_j) F_* = -F_* \dot{\Phi}^{ij} \overline{g}_{ij} - 2 \dot{\Phi}^{ij} \frac{\overline{\nabla}_i F_* \overline{\nabla}_j F_*}{F_*} , \tag{19.14}$$

and the tensor A satisfies

$$(\partial_t - \dot{\Phi}^{ij} \overline{\nabla}_i \overline{\nabla}_j) A_{kl} = \ddot{\Phi}^{pq,rs} \overline{\nabla}_k A_{pq} \overline{\nabla}_l A_{rs} - \dot{\Phi}^{ij} \overline{g}_{ij} A_{kl} , \tag{19.15}$$

where $\dot{\Phi} \doteqdot \dot{\phi}_A$ and $\ddot{\Phi} \doteqdot \ddot{\phi}_A$.

Proof. Since the metric and connection on S^n are fixed, $A = \overline{\nabla}^2 \sigma + \sigma \overline{g}$, and $\partial_t \sigma = \Phi$, we find

$$\partial_t \Phi = \partial_t \phi(A) = \dot{\phi}_A^{kl} \partial_t A = \dot{\phi}_A^{kl} (\overline{\nabla}_i \overline{\nabla}_j \partial_t \sigma + \partial_t \sigma \overline{g}_{ij}) = \dot{\Phi}^{kl} (\overline{\nabla}_i \overline{\nabla}_j \Phi + \Phi \overline{g}_{ij}) ,$$

which proves the first claim. The second follows since $F_* = -\Phi^{-1}$. The third is obtained similarly:

$$\begin{aligned}\partial_t A_{ij} &= \partial_t(\overline{\nabla}_i\overline{\nabla}_j\sigma + \sigma\overline{g}_{ij})\\ &= \overline{\nabla}_i\overline{\nabla}_j\partial_t\sigma + \partial_t\sigma\overline{g}_{ij}\\ &= \overline{\nabla}_i\overline{\nabla}_j\phi(A) + \phi(A)\overline{g}_{ij}\\ &= \dot{\phi}_A^{kl}\overline{\nabla}_i\overline{\nabla}_j A_{kl} + \ddot{\phi}_A^{pq,rs}\overline{\nabla}_i A_{pq}\overline{\nabla}_j A_{rs} + \phi(A)\overline{g}_{ij}\,.\end{aligned}$$

The claim now follows from the Simons identity for A (Lemma 15.8). □

Proposition 19.8. *If f is inverse-concave, then*

$$\frac{\kappa_1}{F}(x,t) \geq \alpha \doteqdot \min_{M^n\times\{0\}}\frac{\kappa_1}{F}(x,t)\,. \tag{19.16}$$

In fact, the ratio $\min_{M^n\times\{t\}}\frac{\kappa_1}{F}$ is strictly increasing unless the solution is a shrinking sphere.

Proof. We will prove that r_1/F_* is nonincreasing, where $r_1 = \kappa_1^{-1}$ is the largest principal radius. Observe that

$$\begin{aligned}(\partial_t - \dot{\Phi}^{ij}\overline{\nabla}_i\overline{\nabla}_j)r_1 &\leq (\partial_t - \dot{\Phi}^{ij}\overline{\nabla}_i\overline{\nabla}_j)A_{11} - 2\sum_{k=1}^n\sum_{r_p<r_1}\frac{\dot{\Phi}^k}{r_1-r_p}(\overline{\nabla}_k A_{1p})^2\\ &= \ddot{\Phi}^{pq,rs}\overline{\nabla}_1 A_{pq}\overline{\nabla}_1 A_{rs} - \dot{\Phi}^{ij}\overline{g}_{ij}r_1 - 2\sum_{k=1}^n\sum_{r_p<r_1}\frac{\dot{\Phi}^k}{r_1-r_p}(\overline{\nabla}_k A_{1p})^2\end{aligned}$$

and hence, modulo gradient terms,

$$\begin{aligned}F_*(\partial_t - \dot{\Phi}^{ij}\overline{\nabla}_i\overline{\nabla}_j)\frac{r_1}{F_*} &= (\partial_t - \dot{\Phi}^{ij}\overline{\nabla}_i\overline{\nabla}_j)r_1 - \frac{r_1}{F_*}(\partial_t - \dot{\Phi}^{ij}\overline{\nabla}_i\overline{\nabla}_j)F_*\\ &\leq \ddot{\Phi}^{pq,rs}\overline{\nabla}_1 A_{pq}\overline{\nabla}_1 A_{rs} - 2\sum_{k=1}^n\sum_{r_p<r_1}\frac{\dot{\Phi}^k}{r_1-r_p}(\overline{\nabla}_k A_{1p})^2\\ &\quad + 2r_1\dot{\Phi}^{ij}\frac{\overline{\nabla}_i F_*\overline{\nabla}_j F_*}{F_*^2}\\ &= F_*^{-2}\ddot{F}_*^{pq,rs}\overline{\nabla}_1 A_{pq}\overline{\nabla}_1 A_{rs} - 2F_*^{-3}(\overline{\nabla}_1 F_*)^2\\ &\quad - \frac{2}{F_*^2}\sum_{k=1}^n\sum_{r_p<r_1}\frac{\dot{F}_*^k}{r_1-r_p}(\overline{\nabla}_k A_{1p})^2 + 2\frac{r_1}{F_*}\dot{F}_*^{ij}\frac{\overline{\nabla}_i F_*\overline{\nabla}_j F_*}{F_*^3}\end{aligned}$$

in the barrier sense.

We need to show that the right-hand side is nonpositive at a local maximum of r_1/F_*. Multiplying by F_*^2 and applying the gradient identities

$$\overline{\nabla}_k A_{11} = \frac{r_1}{F_*}\overline{\nabla}_k F_* \ \text{ for each } k = 1,\ldots,n,$$

and

$$\overline{\nabla}_k A_{ij} = 0 \text{ if } r_i = r_i,\ \ i \neq j,$$

it becomes

$$\begin{aligned}
Q \doteqdot & \sum_{p,q=1}^{n} \ddot{F}_*^{pq} \overline{\nabla}_1 A_{pp} \overline{\nabla}_1 A_{qq} + 2 \sum_{r_q < r_p < r_1} \frac{\dot{F}_*^p - \dot{F}_*^q}{r_p - r_q} (\overline{\nabla}_1 A_{pq})^2 \\
& + 2 \sum_{r_p < r_1} \frac{\dot{F}_*^1 - \dot{F}_*^p}{r_1 - r_p} (\overline{\nabla}_1 A_{p1})^2 - 2 \sum_{k=1}^{n} \sum_{r_p < r_1} \frac{\dot{F}_*^k}{r_1 - r_p} (\overline{\nabla}_k A_{1p})^2 \\
& - \frac{2F_*}{r_1^2} (\overline{\nabla}_1 A_{11})^2 + \frac{2}{r_1} \sum_{k=1}^{n} \dot{F}_*^k (\overline{\nabla}_k A_{11})^2 \\
= & \sum_{r_p, r_q < r_1}^{n} \left(\ddot{F}_*^{pq} - 2 \frac{\dot{F}_*^p \delta^{pq}}{r_1 - r_p} \right) \overline{\nabla}_1 A_{pp} \overline{\nabla}_1 A_{qq} + \ddot{F}_*^{11} (\overline{\nabla}_1 A_{11})^2 \\
& + 2 \sum_{r_q < r_p < r_1} \left(\frac{\dot{F}_*^p - \dot{F}_*^q}{r_p - r_q} - \frac{\dot{F}_*^p}{r_1 - r_q} - \frac{\dot{F}_*^q}{r_1 - r_p} \right) (\overline{\nabla}_1 A_{pq})^2 \\
& + \frac{2}{r_1^2} \left(\dot{F}_*^1 r_1 - F_* \right) (\overline{\nabla}_1 A_{11})^2 + 2 \sum_{r_p < r_1} \left(\frac{1}{r_1} - \frac{1}{r_1 - r_p} \right) \dot{F}_*^p (\overline{\nabla}_p A_{11})^2 .
\end{aligned}$$

The terms on the first and second lines are nonpositive since f_* is concave (recall Lemma 18.7). The final term is clearly nonpositive. Since f_* is 1-homogeneous, the remaining term can be rewritten as

$$\frac{2}{r_1^2} \left(\dot{F}_*^1 r_1 - F_* \right) (\overline{\nabla}_1 A_{11})^2 = -\frac{2}{r_1^2} \sum_{p=2}^{n} \dot{F}_*^p r_p (\overline{\nabla}_1 A_{11})^2 \leq 0 .$$

It follows that $Q \leq 0$. We leave it to the reader to check that $Q < 0$ unless $\overline{\nabla} F_* = 0$. So the claims follow from the strong maximum principle (cf. Lemma 9.10) and Exercise 19.2. □

Corollary 19.9. *In each of the cases* (i)–(vi) *of Theorem* 19.1, *there exists a constant* $\tilde{\alpha}$, *depending only on the dimension* n, *the pinching constant* α, *and the speed function* f, *such that*

$$\frac{\kappa_i(x,t)}{\kappa_j(x,t)} \leq \tilde{\alpha}^{-1} \text{ and } \kappa_i(x,t) \geq \varepsilon_0 \tilde{\alpha} \text{ for all } (x,t) \in M \times [0,T). \tag{19.17}$$

Proof. We leave the proof as an exercise (see Exercise 19.4). □

The curvature pinching implies that the flow is uniformly parabolic, in the sense that we can find $\Lambda < \infty$ (depending on n, f, and the pinching of the principal curvatures) such that

$$\Lambda^{-1} |\xi|^2 \leq \dot{F}_{(x,t)}^{ij} \xi_i \xi_j \leq \Lambda |\xi|^2 \tag{19.18}$$

for all $(x,t) \in M^n \times [0,T)$. This implies that the equations (19.3) (for the support function) and (19.4) (for the radial graph height) are uniformly parabolic and provides bounds for their gradient and Hessian. In cases (ii), (iii), and (iv), we can then obtain a Hölder estimate for $\overline{\nabla}^2\rho$ from Krylov's theorem (Theorem 15.9). In cases (iv), (v), and (vi), a similar argument yields an estimate for $\overline{\nabla}^2\sigma$. In case (i) we can apply the main theorem from [**37**] to bound either $\overline{\nabla}^2\rho$ or $\overline{\nabla}^2\sigma$. Higher-order estimates then follow from the Schauder estimate (Theorem 15.10) and we conclude that the solution exists for all positive times. We omit the details (which can be found in [**26, 61, 157, 158**]) since similar arguments were carried out in detail in Sections 8.6 and 15.3. A further argument (which applies when $n = 2$ or when the speed is either concave or convex) is presented in the following section.

The remainder of the proof of Theorem 19.1 follows the argument of Section 8.6 (cf. Section 15.3). Indeed, the curvature pinching implies pinching of the circum- and inradii by Lemmas 8.24 and 8.25. Representing the solution as a graph over the sphere (or by its support function), we can then use the regularity theory for fully nonlinear PDE, as in Lemma 8.26, to deduce that the inradius goes to zero as the maximal time is approached. Chou's argument (Lemma 8.27) can then be applied to bound the speed by one over the square root of the remaining time. The Arzelà–Ascoli theorem and the strong maximum principle then imply that an appropriate monotone quantity (such as $|\mathrm{II}|^2/F^2$ in case (iii)) becomes optimal at the singular time, which gives control on the curvature pinching and the higher derivatives of the curvature (cf. Corollary 8.28). Interpolation inequalities can then be used to obtain decay estimates for the higher derivatives of the curvature, which imply estimates for the rescaled embedding and its derivatives (cf. Section 8.4.4). Convergence in the smooth topology to an embedding of the unit sphere follows.

19.2. Evolving nonconvex hypersurfaces

Many of the results we have presented for mean convex mean curvature flow extend, in certain cases, to the fully nonlinear setting, with positivity of the speed function playing the role of mean convexity. An argument similar to that presented in Proposition 19.8 shows that flows by positive, concave admissible speeds preserve "pinching from above", while flows by positive, convex admissible speeds preserve "pinching from below". The argument of Proposition 19.5 shows that flows of surfaces by positive admissible speeds preserve any initial pinching.

Proposition 19.10. *Let $f : \Gamma^n \to \mathbb{R}$ be a positive, 1-homogeneous admissible speed function and let $X : M^n \times [0,T) \to \mathbb{R}^{n+1}$ be a compact solution to the corresponding curvature flow* (18.1).

(i) *If $n = 2$, then*

$$\min_{M^2\times\{t\}} \frac{\kappa_1}{\kappa_2} \le \sup_{M^2\times\{0\}} \frac{\kappa_1}{\kappa_2}\,. \tag{19.19}$$

(ii) *If f is concave, then*

$$\max_{M^n\times\{t\}} \frac{\kappa_n}{F} \le \max_{M^n\times\{0\}} \frac{\kappa_n}{F}\,. \tag{19.20}$$

(iii) *If f is convex, then*

$$\min_{M^n\times\{t\}} \frac{\kappa_1}{F} \ge \min_{M^n\times\{0\}} \frac{\kappa_1}{F}\,. \tag{19.21}$$

Proof. We leave the proof as an exercise (see Exercise 19.5). □

A similar argument shows that lower bounds for $(\kappa_1 + \cdots + \kappa_{m+1})/F$ are preserved under flows by positive, convex admissible speeds. In fact, an argument of the form presented in Proposition 9.6 shows that any convex cone of curvatures which contains Γ^n_+ is preserved (see [**350**, §4.2]).

These pinching estimates allow us to extend the long-time existence result (Corollary 19.12) to the nonconvex setting provided that either

(i) $n = 2$,

(ii) f is convex and $\Gamma_+ \subset \Gamma$, or

(iii) f is concave and vanishes on $\partial\Gamma$.

Proposition 19.11. *Let $f : \Gamma^n \to \mathbb{R}$ be a 1-homogeneous admissible speed function satisfying at least one of the conditions* (i)–(iii) *above. Given $C_0 < \infty$, $\varepsilon > 0$, and $m \in \mathbb{N}$, there exists $C_m < \infty$ with the following property: Let $X : M^n \times (0,T] \to \mathbb{R}^{n+1}$ be a smooth solution to the corresponding curvature flow* (18.1). *If $\sup_{M^n\times[T/2,T]} |\mathrm{II}| \le C_0$ and $\inf_{M^n\times[T/2,T]} \frac{\kappa_1}{\kappa_n}(x,t) \ge \varepsilon$, then*

$$\sup_{M^n\times[T/2,T]} |\nabla^m \mathrm{II}| \le C_m\,. \tag{19.22}$$

Proof. The idea is to write the solution locally as an evolving graph (over a tangent hyperplane) and apply known results in the regularity theory for fully nonlinear parabolic PDE.

Fix any $(x_1, t_1) \in M^n \times [T/2, T]$; up to an ambient isometry, we may assume that $X(x_1, t_1)$ is the origin and $\mathsf{N}(x_1, t_1) = e_{n+1}$. The rank theorem implies the existence of neighborhoods U of x_1 and I of t_1, an open set $V \subset \mathbb{R}^n$, a diffeomorphism $\varphi : V \to U$, and a smooth function $u : U \times I \to \mathbb{R}$ such that $X\left(\varphi^{-1}(y), t\right) = (y, u(y,t))$ for all $y \in U$ and $t \in I$. Observe that the Weingarten curvature of X and its derivatives are controlled near (x_1, t_1) by u and its derivatives near (y_1, t_1), where $y_1 = \varphi(x_1)$: We computed, in Section 5.2,

$$\mathrm{II}_{ij} = \frac{u_{ij}}{\sqrt{1 + \|Du\|^2}} \quad \text{and} \quad g^{ij} = \delta^{ij} - \frac{u_i u_j}{1 + \|Du\|^2}\,.$$

Note also that the connection coefficients are given in the graphical coordinates by

$$\Gamma_{ijk} = \left\langle \frac{\partial^2 X}{\partial y^i \partial y^j}, \frac{\partial X}{\partial y^k} \right\rangle = u_{ij} u_k.$$

It follows that, with respect to the graphical coordinates, the components of the m-th covariant derivative of II are given by an expression which depends on $D^k u$ for $1 \leq k \leq m+2$. Since the induced metric depends only on first derivatives of u, we conclude that $|\nabla^m \mathrm{II}|$ is bounded once the derivatives of u up to order $m+2$ are bounded.

Thus, to estimate $|\nabla^m \mathrm{II}|$ at (x_1, t_1), it suffices to obtain estimates for the derivatives of u at (y_1, t_1) up to order $m+2$. We first obtain a local bound for $\|Du\|$ depending only on the bound for $\sup_{M^n \times [T/2, T]} |\mathrm{II}|$: Observe that the Weingarten curvature is given locally by

$$\frac{\partial}{\partial y^i} \left(\frac{u_j}{\sqrt{1 + \|Du\|^2}} \right) = -\,\mathrm{II}_i{}^j\,.$$

Integrating the trace elements, we obtain

$$-\sqrt{C_0}\, y^i \leq \frac{u_i}{\sqrt{1 + \|Du\|^2}} \leq \sqrt{C_0}\, y^i\,.$$

It follows that

$$\frac{\|Du\|^2}{1 + \|Du\|^2} \leq C_0 \, \|y\|^2\,.$$

Thus, $\|Du\|^2 \leq 1$ on $B_{1/\sqrt{2C_0}}(0) \times I$.

We next prove a $C^{2,\alpha}$ estimate for u: A straightforward calculation shows that u satisfies (cf. [**511**])

$$\partial_t u = \hat{f}(D^2 u, Du) \doteqdot f(P D^2 u P)\,, \tag{19.23}$$

where

$$P \doteqdot I - \frac{Du \otimes Du}{\sqrt{1 + \|Du\|^2}\left(1 + \sqrt{1 + \|Du\|^2}\right)}\,.$$

Note that $\hat{f} \equiv \hat{f}(r,p)$ is smooth in each argument and satisfies

$$\left.\frac{\partial \hat{f}}{\partial r_{kl}}\right|_{(D^2u,Du)} = \dot{f}^{pq} P_p{}^k P_q{}^l .$$

It follows from (19.18) that equation (19.23) is uniformly parabolic (with constants depending only on n, C_0, f, and ε). Moreover, since $f(PD^2uP) = \sqrt{1+\|Du\|^2} f(\mathrm{II}) = \sqrt{1+\|Du\|^2} \dot{f}^{kl}\,\mathrm{II}_{kl}$, we see that $\partial_t u$ is bounded by a constant that depends only on n, f, C_0, and ε.

If $n=2$, then a parabolic $C^{2,\alpha}$ estimate (with constants depending only on n, f, C_0, and ε) for u now follows from the main result of [**37**].

Next, observe that

$$\left.\frac{\partial^2 \hat{f}}{\partial r_{kl} \partial r_{mn}}\right|_{(D^2u,Du)} M_{kl}M_{mn} = \ddot{f}^{pq,rs}\big|_{PD^2uP} P_p{}^k P_q{}^l M_{kl} P_r{}^k P_s{}^l M_{kl} .$$

If f is concave, then we obtain a $C^{2,\alpha}$ estimate from the estimate of Krylov [**341**] (see Theorem 15.9) with constants depending only on n, f, C_0, and ε. If f is convex, then we similarly obtain a $C^{2,\alpha}$ estimate for $-u$ since $v \doteqdot -u$ satisfies the equation

$$v_t = \hat{f}_-(D^2v, Dv) ,$$

where $\hat{f}_-(r,p) \doteqdot -\hat{f}(-r,p)$.

Next, we consider the evolution equations for the spatial derivatives of u: Set $v = u_i$. Then v satisfies the equation

$$v_t = a^{ij} v_{ij} + b^i v_i , \tag{19.24}$$

where

$$a^{ij} = \left.\frac{\partial \hat{f}}{\partial r_{ij}}\right|_{(D^2u,Du)} = \dot{f}^{pq} P_p{}^i P_q{}^j$$

and

$$b^i = \left.\frac{\partial \hat{f}}{\partial p_i}\right|_{(D^2u,Du)} = -\frac{1}{1+\|Du\|^2} \dot{f}^{kl} \left(u_{ik}u_l + u_p u_{pk} \delta_{il} - \frac{2u_i u_p u_{pk} u_l}{1+\|Du\|^2} \right) .$$

In particular, the coefficients a^{ij} and b^i are bounded by constants that depend only on bounds for $\|D^2u\|$ and $\|Du\|$ (which we have seen depend only on C_0). Furthermore, we have seen that a^{ij} is uniformly positive definite (with constants depending only on n, f, C_0, and ε). Finally, the $C^{2,\alpha}$ estimate for u implies that the coefficients of (19.24) are Hölder continuous. The Schauder estimate (see Theorem 15.10) now yields $C^{2,\alpha}$ bounds for

v. Higher regularity follows by applying the Schauder estimate inductively: The higher derivatives of u satisfy equations similar to (19.24). Hölder continuity of the m-th spatial derivatives of u permits us to apply Schauder to the evolution equation for the $(m-1)$-st spatial derivatives, yielding Hölder continuity of the $(m+1)$-st derivatives. By induction, this yields bounds on all spatial derivatives of u (which depend only on n, f, C_0, and ε). □

Corollary 19.12 (Long-time existence theorem). *Let $f : \Gamma^n_+ \to \mathbb{R}$ be a 1-homogeneous admissible speed function and let $X : M^n \times [0,T) \to \mathbb{R}^{n+1}$ be a smooth, compact, maximal solution to the corresponding curvature flow* (18.1). *Suppose that at least one of the conditions* (i)–(iii) *in Proposition* 19.10 *is satisfied. Then*

$$\sup_{M^n \times [0,T)} |\mathrm{II}| = \infty \,. \tag{19.25}$$

Proof. The proof can be completed as in Theorem 6.20. □

19.2.1. Convexity and cylindrical estimates.

Arguing similarly as in Theorem 9.12 yields a generalization of the convexity estimate (Corollary 9.13) for convex speeds, or general surface flows.

Theorem 19.13 (Convexity estimate for convex speeds). *Let $f : \Gamma^n \to \mathbb{R}$, $n \geq 2$, be an admissible speed function with $\Gamma^n_+ \subset \Gamma^n$ and let $X : M^n \times [0,T) \to \mathbb{R}^{n+1}$ be a compact solution to the corresponding curvature flow. Suppose that either $n = 2$ or f is convex. For every $\varepsilon > 0$ there exists C_ε (depending on n, f, the initial immersion, and ε) such that*

$$\kappa_1 \geq -\varepsilon F - C_\varepsilon \,. \tag{19.26}$$

Proof. The theorem is proved by a Stampacchia iteration argument along the lines of the proof of Theorem 9.12. See [**56**] and [**57**]. □

Arguing similarly as in Theorem 9.12 yields a generalization of the cylindrical estimates (Corollary 9.14) for flows by convex speeds and certain concave speeds. Recall that the cones $\Gamma^n_k \subset \mathbb{R}^n$ are defined by

$$\Gamma^n_k \doteqdot \{z \in \mathbb{R}^n : z_{\sigma(1)} + \cdots + z_{\sigma(k)} > 0 \text{ for all } \sigma \in P_n\} \,.$$

Theorem 19.14 (Cylindrical estimates). *Let $f : \Gamma^n \to \mathbb{R}$, $n \geq 2$, be an admissible speed function and let $X : M^n \times [0,T) \to \mathbb{R}^{n+1}$ be a compact solution to the corresponding curvature flow. Define*

$$c_m \doteqdot f(\underbrace{0, \ldots, 0}_{m\text{-times}}, 1, \ldots, 1) \,.$$

(i) *Suppose that* $\Gamma^n \subset \Gamma^n_{m+1}$ *and* f *is concave and extends continuously to* $\partial\Gamma^n$ *with* $f|_{\partial\Gamma^n} \equiv 0$. *For every* $\varepsilon > 0$ *there exists* C_ε *(depending only on* n, f, *the initial immersion, and* ε*) such that*

$$\kappa_n - c_m^{-1} F \leq \varepsilon F + C_\varepsilon \,. \tag{19.27}$$

(ii) *Suppose that* f *is convex. For every* $\varepsilon > 0$ *there exists* C_ε *(depending only on* n, f, *the initial immersion, and* ε*) such that*

$$\kappa_1 + \cdots + \kappa_{m+1} - c_m^{-1} F \geq -\varepsilon F - C_\varepsilon \,. \tag{19.28}$$

Proof. The theorem is again proved by a Stampacchia iteration argument. See [**53**] and [**353**]. □

19.2.2. Noncollapsing.

The following generalizes the differential inequalities of Proposition 12.8 to the nonlinear setting.

Proposition 19.15 ([**54, 55**]). *Let* $f : \Gamma^n \to \mathbb{R}$, $n \geq 2$, *be an admissible speed function and let* $X : M^n \times [0,T) \to \mathbb{R}^{n+1}$ *be a compact solution to the corresponding curvature flow* (18.1).

(i) *If* f *is concave, then the inscribed curvature* $\overline{k}$ *satisfies*

$$\partial_t \overline{k} \leq \dot{F}\left(\nabla^2 \overline{k} + \mathrm{II}^2\, \overline{k}\right) \tag{19.29}$$

in the barrier sense (see Definition 9.5*).*

(ii) *If* f *is convex, or* $\Gamma = \Gamma_+$ *and* f *is inverse-concave, then the exscribed curvature* $\underline{k}$ *satisfies*

$$\partial_t \underline{k} \geq \dot{F}\left(\nabla^2 \underline{k} + \mathrm{II}^2\, \underline{k}\right) \tag{19.30}$$

in the barrier sense (see Definition 9.5*).*

Proof. We have already done most of the work in Section 12.2. We begin with the case that f is concave. Fix (x_0, t_0) and suppose that $\overline{k}(x_0,t_0) = \kappa_n(x_0,t_0)$. Then $\overline{k}(x_0,t_0) = \mathrm{II}_{(x_0,t_0)}(y_0,y_0)$ for some $y_0 \in S_{(x_0,t_0)}\mathcal{M}$. Consider the function K defined on $T\mathcal{M} \setminus \{0\} \times [0,T)$ by

$$K(x,y,t) = \frac{\mathrm{II}_{(x,t)}(y,y)}{g_{(x,t)}(y,y)} \,.$$

Let $\{x^i\}_{i=1}^n$ be geodesic normal coordinates for $(\mathcal{M}^n, g_{t_0})$ centered at x_0 and denote by $y^j = dx^j$ the corresponding fiber coordinates. Recalling equations (12.22) and (19.6), we find

$$\partial_t K - \dot{F}^{ij}(\partial_{x^i} + \Lambda_i{}^p \partial_{y^p})(\partial_{x^j} + \Lambda_j{}^q \partial_{y^q}) K = \dot{F}^{ij}\, \mathrm{II}^2_{ij}\, K + L_\Lambda$$

at (x_0, y_0, t_0), where

$$L_\Lambda \doteqdot \ddot{F}^{pq,rs}\nabla_i \mathrm{II}_{pq}\nabla_j \mathrm{II}_{rs}\, y^i y^j + 2\dot{F}^{kl}\Lambda_k{}^p\left[2\left(\nabla_l \mathrm{II}_{jp} - \nabla_l \mathrm{II}(y,y)\delta_{jp}\right)y^j + \Lambda_l{}^q\left(\mathrm{II}_{pq} - K\delta_{pq}\right)\right].$$

Taking $\Lambda = 0$ yields

$$\partial_t K - \dot{F}^{ij}(\partial_{x^i} + \Lambda_i{}^p\partial_{y^p})(\partial_{x^j} + \Lambda_j{}^q\partial_{y^q})K \le \dot{F}^{ij}\,\mathrm{II}^2_{ij}\,K.$$

Set

$$\varphi(x,t) \doteqdot K(x, y_0^i\partial_{x^i}, t_0).$$

Then φ supports $\overline{k}$ from below at (x_0, t_0) and satisfies

$$(\partial_t - \dot{F}^{ij}\nabla_i\nabla_j)\varphi \le \left(\partial_t - \dot{F}^{ij}(\partial_{x^i} + \Lambda_i{}^p\partial_{y^p})(\partial_{x^j} + \Lambda_j{}^q\partial_{y^q})\right)K \le \dot{F}^{ij}\,\mathrm{II}^2_{ij}\,\varphi$$

at (x_0, t_0), which yields the claim.

So suppose that $\overline{k}(x_0, t_0) > \kappa_n(x_0, t_0)$. Then $\overline{k}(x_0, t_0) = k(x_0, y_0, t_0)$ for some $y_0 \in \mathcal{M}\setminus\{x_0\}$. Choose normal coordinates $(x^1, \ldots, x^n) : U_{x_0} \to \mathbb{R}^n$ and $(y^1, \ldots, y^n) : U_{y_0} \to \mathbb{R}^n$ centered at x_0 and y_0, respectively, with respect to the metric at time t_0. We can assume that $U_{x_0} \cap U_{y_0} = \emptyset$, so that k is smooth in $U_{x_0} \times U_{y_0} \times (0, t_0]$. We can also arrange that $\{\partial_{x^i}\}_{i=1}^n$ are principal directions at (x_0, t_0) and, since $\partial_{y^i} k = 0$ at (x_0, y_0, t_0), that

$$\partial_i^y = \partial_i^x - 2\langle \partial_i^x, w\rangle\, w$$

at (x_0, y_0, t_0) (recall Lemma 12.11). Computing as in (12.34) and (12.35), we find

$$(\partial_t - \dot{F}^{ij}_x(\partial_{x^i} + \Lambda_i{}^p\partial_{y^p})(\partial_{x^j} + \Lambda_j{}^q\partial_{y^q}))k = \dot{F}^{ij}\,\mathrm{II}^2_{ij}\,k - 2\sum_{i=1}^n \dot{F}^i\frac{(\partial_{x^i}k)^2}{k - \kappa_i} + L_\Lambda$$

at (x_0, y_0, t_0), where

$$L_\Lambda \doteqdot \frac{2}{d^2}\Big(F_y - F_x + \dot{F}^{ij}_x\left[(k\delta_{ij} - \mathrm{II}^x_{ij}) - 2\Lambda_i{}^p(k\delta_{pj} - \mathrm{II}^y_{pj}) + \Lambda_i{}^p\Lambda_j{}^q(k\delta_{pq} - \mathrm{II}^y_{pq})\right]\Big).$$

Taking Λ to be the identity yields

$$\begin{aligned}
&(\partial_t - \dot{F}^{ij}_x(\partial_{x^i} + \Lambda_i{}^p\partial_{y^p})(\partial_{x^j} + \Lambda_j{}^q\partial_{y^q}))k \\
&\quad \le \dot{F}^{ij}\,\mathrm{II}^2_{ij}\,k + \frac{2}{d^2}\left(F_y - F_x - \dot{F}^{ij}_x\left[\mathrm{II}^y_{ij} - \mathrm{II}^x_{ij}\right]\right) \\
&\quad \le \dot{F}^{ij}\,\mathrm{II}^2_{ij}\,k
\end{aligned}$$

since concavity of f implies that

$$f(B) \le f(A) + \dot{f}_A(B - A)$$

for any $A, B \in \Gamma$. Set

$$\varphi(x,t) \doteqdot k(x, y_0, t).$$

Then φ supports $\overline{k}$ from below at (x_0, t_0) and satisfies

$$(\partial_t - \dot{F}^{ij}\nabla_i\nabla_j)\varphi \leq \big(\partial_t - \dot{F}_x^{ij}(\partial_{x^i} + \Lambda_i{}^p\partial_{y^p})(\partial_{x^j} + \Lambda_j{}^q\partial_{y^q})\big)k \leq \dot{F}^{ij}\,\mathrm{II}_{ij}^2\,k$$

at (x_0, t_0), completing the proof in case f is concave.

In case f is convex, the proof is the same since all inequalities are reversed.

In case f is inverse-concave, we need to work slightly harder. In the "boundary case", $\underline{k}(x_0, t_0) = \kappa_1(x_0, t_0)$, the claim follows from Exercise 12.4 and the following lemma.

Lemma 19.16. *Let $f : \Gamma_+ \subset \mathbb{R}^n \to \mathbb{R}$, $n \geq 1$, be an inverse-concave admissible speed function. Then, for all diagonal $W \in \mathrm{S}(n)$ with positive eigenvalues $\lambda_1 \leq \cdots \leq \lambda_n$ and all symmetric $B \in \mathrm{S}(n)$ with $B_{11} = 0$,*

$$\ddot{f}_W^{pq,rs}B_{pq}B_{rs} + 2\sup_{\Lambda_{i1}=0}\dot{f}_W^k\left[2\Lambda_k{}^pB_{kp} - (\Lambda_k{}^p)^2(\lambda_p - \lambda_1)\right] \geq 0\,. \tag{19.31}$$

Proof. It suffices to prove the claim in the case all λ_i are distinct since any diagonal matrix can be approximated by such W. Set

$$Q \doteqdot \ddot{f}_W^{pq,rs}B_{pq}B_{rs} + 2\sup_{\Lambda_{i1}=0}\dot{f}_W^k\left[2\Lambda_k{}^pB_{kp} - (\Lambda_k{}^p)^2(\lambda_p - \lambda_1)\right]\,.$$

Observe that the supremum occurs when $\Lambda_{kp} = (\lambda_p - \lambda_1)^{-1}B_{kp}$ for $p > 1$. With this choice, we obtain

$$Q = \ddot{f}_W^{pq,rs}B_{pq}B_{rs} + 2\dot{f}_W^{kl}R^{pq}B_{kp}B_{lq}\,,$$

where $R^{pq} \doteqdot (\lambda_p - \lambda_1)^{-1}\delta^{pq}$ for $p, q \neq 1$ and zero otherwise. Therefore, it suffices to prove that

$$0 \leq \left(\ddot{f}_W^{pq,rs} + 2\dot{f}_W^{pr}R^{qs}\right)B_{pq}B_{rs}$$

for any symmetric B with $B_{11} = 0$. Applying Theorem 18.4, we obtain

$$\begin{aligned}
(\ddot{f}_W^{pq,rs} + 2\dot{f}_W^{pr}R^{qs})B_{pq}B_{rs} &= \ddot{f}_W^{pq}B_{pp}B_{qq} + \sum_{p\neq q}\frac{\dot{f}_\lambda^p - \dot{f}_\lambda^q}{\lambda_p - \lambda_q}B_{pq}^2 + 2\sum_{p=1,\,q=2}^{n}\frac{\dot{f}_\lambda^p}{\lambda_q - \lambda_1}B_{pq}^2\\
&= \ddot{f}_\lambda^{pq}B_{pp}B_{qq} + 2\sum_{p>1,\,q>1}\frac{\dot{f}_\lambda^p\delta^{pq}}{\lambda_p - \lambda_1}B_{pp}B_{qq}\\
&\quad + \sum_{p\neq q}\frac{\dot{f}_\lambda^p - \dot{f}_\lambda^q}{\lambda_p - \lambda_q}B_{pq}^2 + 2\sum_{p=2}^{n}\frac{\dot{f}_\lambda^1}{\lambda_p - \lambda_1}B_{p1}^2\\
&\quad + 2\sum_{p>1,\,q>1,\,p\neq q}\frac{\dot{f}_\lambda^p}{\lambda_q - \lambda_1}B_{pq}^2\,.
\end{aligned}$$

We first estimate

$$\begin{aligned}
\ddot f_\lambda^{pq}B_{pp}B_{qq}+2\sum_{p>1,q>1}\frac{\dot f_\lambda^p\delta^{pq}}{\lambda_p-\lambda_1}B_{pp}B_{qq}&\geq \ddot f_\lambda^{pq}B_{pp}B_{qq}+2\sum_{p=2,q=2}\frac{\dot f_\lambda^p}{\lambda_p}\delta^{pq}B_{pp}B_{qq}\\
&=\left(\ddot f_\lambda^{pq}+2\frac{\dot f_\lambda^p}{\lambda_p}\delta^{pq}\right)B_{pp}B_{qq}\\
&\geq 0\,,
\end{aligned}$$

where the final inequality follows from inverse-concavity of f (Lemma 18.12). The remaining terms are

$$\begin{aligned}
&\sum_{p\neq q}\frac{\dot f_\lambda^p-\dot f_\lambda^q}{\lambda_p-\lambda_q}B_{pq}^2+2\sum_{p=2}^n\frac{\dot f_\lambda^1}{\lambda_p-\lambda_1}B_{p1}^2+2\sum_{p>1,\,q>1,\,p\neq q}\frac{\dot f_\lambda^p}{\lambda_q-\lambda_1}B_{pq}^2\\
&=\sum_{p>1,\,q>1,\,p\neq q}\left(\frac{\dot f_\lambda^p-\dot f_\lambda^q}{\lambda_p-\lambda_q}+2\frac{\dot f_\lambda^p}{\lambda_q-\lambda_1}\right)B_{pq}^2+2\sum_{p=2}^n\left(\frac{\dot f_\lambda^p-\dot f_\lambda^1}{\lambda_p-\lambda_1}+\frac{\dot f^1}{\lambda_p-\lambda_1}\right)B_{p1}^2\\
&\geq\sum_{p>1,\,q>1,\,p\neq q}\left(\frac{\dot f_\lambda^p-\dot f_\lambda^q}{\lambda_p-\lambda_q}+\frac{\dot f_\lambda^p}{\lambda_q}+\frac{\dot f_\lambda^q}{\lambda_p}\right)B_{pq}^2+2\sum_{p=2}^n\left(\frac{\dot f_\lambda^p}{\lambda_p-\lambda_1}\right)B_{p1}^2\,.
\end{aligned}$$

The second term is clearly nonnegative. Nonnegativity of the first term follows from inverse-concavity of F (Lemma 18.12). This completes the proof. □

In the "interior" case, $\underline{k}(x_0,t_0)<\kappa_1(x_0,t_0)$, the claim follows from the following lemma.

Lemma 19.17. *Let $f:\Gamma_+\subset\mathbb{R}^n\to\mathbb{R}$, $n\geq 1$, be an inverse-concave admissible speed function. Then, for any $k\in\mathbb{R}$, any diagonal, positive definite $B\in\mathrm{S}(n)$, and any positive definite $A\in\mathrm{S}(n)$ with $k<\min_i\{\lambda_i(A)\}$,*

$$0\leq f(B)-f(A)+\dot f_A^{ij}\sup_\Lambda\Big[(k\delta_{ij}-A_{ij})-2\Lambda_i{}^p(k\delta_{pj}-A_{pj})+\Lambda_i{}^p\Lambda_j{}^q(k\delta_{pq}-B_{pq})\Big]\,.$$

Proof. Since the expression in the square brackets is quadratic in Λ, it is easy to see that the supremum is attained with the choice $\Lambda=(A-kI)\times(B-kI)^{-1}$, where I denotes the identity matrix. Thus, given any positive definite A, we need to show that

$$0\leq Q(B)\doteqdot f(B)-f(A)-\dot f_A^{ij}\Big((A-kI)_{ij}-\big[(A-kI)\cdot(B-kI)^{-1}\cdot(A-kI)\big]_{ij}\Big)\,.$$

Since B is diagonal and the expression $Q(B)$ is invariant under similarity transformations with respect to A, we may diagonalize A to obtain

$$Q(B)\doteqdot f(b)-f(a)-\dot f_a^i\left[(a_i-k)-\frac{(a_i-k)^2}{b_i-k}\right],$$

where we have set $a = \lambda(A)$ and $b = \lambda(B)$. We are led to consider the function q defined on $\Gamma^n_+ \subset \mathbb{R}^n$ by

$$q(z) \doteqdot f(z) - f(a) - \dot{f}^i_a \left[(a_i - k) - \frac{(a_i - k)^2}{z_i - k} \right] .$$

We compute

$$\dot{q}^i = \dot{f}^i - \dot{f}^i_a \frac{(a_i - k)^2}{(z_i - k)^2}$$

and

$$\ddot{q}^{ij} = \ddot{f}^{ij} + 2\dot{f}^i_a \frac{(a_i - k)^2}{(z_i - k)^3} \delta^{ij} = \ddot{f}^{ij} + 2\frac{\dot{f}^i \delta^{ij}}{z_i - k} - 2\frac{\dot{q}^i \delta^{ij}}{z_i - k} .$$

It follows that

$$\ddot{q}^{ij} + 2\frac{\dot{q}^i \delta^{ij}}{z_i - k} = \ddot{f}^{ij} + 2\frac{\dot{f}^i \delta^{ij}}{z_i - k} > \ddot{f}^{ij} + 2\frac{\dot{f}^i \delta^{ij}}{z_i} \geq 0 , \tag{19.32}$$

where the final inequality follows from inverse-concavity of f (Lemma 18.12).

It follows that q has a unique local minimum at the point $z = a$, where it vanishes. □

This completes the proof of Proposition 19.15. □

As a consequence, we obtain the following noncollapsing estimates.

Corollary 19.18. *Let $f : \Gamma^n \to \mathbb{R}$, $n \geq 2$, be an admissible speed function and let $X : M^n \times [0, T) \to \mathbb{R}^{n+1}$ be a compact solution to the corresponding curvature flow.*

(i) *If f is concave, then*

$$\frac{\overline{k}}{F} \leq \max_{\mathcal{M}_0} \frac{\overline{k}}{F} . \tag{19.33}$$

(ii) *If f is convex, or if $\Gamma = \Gamma_+$ and f is inverse-concave, then*

$$\frac{\underline{k}}{F} \geq \min_{\mathcal{M}_0} \frac{\underline{k}}{F} . \tag{19.34}$$

Proceeding as in Section 12.3 now yields an alternate route to Theorem 19.1 for concave, inverse-concave speeds. The 1-sided noncollapsing estimates also simplify the argument for flows by convex speeds, or concave speeds which vanish on the boundary of the positive cone.

19.3. Notes and commentary

By applying a Stampacchia iteration argument, the noncollapsing estimates of Corollary 19.18 can be improved to sharp estimates wherever the curvature is becoming large (cf. Theorem 12.26). See [**116, 353**].

Flows by speeds of positive homogeneity have been studied by several authors (for Gauß curvature flows, see Chapters 15–17). F. Schulze studied the evolution of closed, convex hypersurfaces by positive powers of the mean curvature and proved that solutions contract to points in finite time [**471**]. He also proved that solutions become spherical after rescaling, so long as the principal curvatures are initially sufficiently pinched [**472**]. O. Schnürer showed that closed, convex surfaces in $\mathbb{R}^3$ evolving by $|\mathrm{II}|^2$ or certain integer powers of H contract to round points in finite time [**462**] (see also [**472**, Appendix A]). R. Alessandroni and C. Sinestrari obtained similar results for flows of closed, convex hypersurfaces by powers of the scalar curvature [**5**]: Solutions contract to points in finite time and, when the principal curvatures are initially sufficiently pinched, become round after rescaling. These flows all admit some divergence structure, needed to establish higher regularity of solutions using results for divergence form porous-medium equations. This difficulty is circumvented via a geometric argument in [**60**], where similar results are obtained for flows of sufficiently pinched hypersurfaces by any speed which is a smooth, homogeneous function of the principal curvatures of degree greater than 1. Q. Han [**272**] studied flows by homogeneous speeds $f : \Gamma_+^n \to \mathbb{R}$ of positive degree. He showed that uniformly convex hypersurfaces shrink to points in finite time if the dual function f_* is concave and vanishes on $\partial\Gamma_+$. J. McCoy, Y. Zheng, and the first author [**61**] construct "a surprisingly large family of flows for which such results fail, by a variety of mechanisms: uniformly convex hypersurfaces may become non-convex, and smooth ones may develop curvature singularities; even where this does not occur, non-uniformly convex regions and singular parts in the initial hypersurface may persist, including flat sides, ridges of infinite curvature, or 'cylindrical' regions where some of the principal curvatures vanish; such cylindrical regions may persist even if the speed is positive, and in such cases the hypersurface may even collapse to a line segment or higher-dimensional disc rather than to a point. [They] provide sufficient conditions for these various disasters to occur, and by avoiding these arrive at a class of flows for which arbitrary weakly convex initial hypersurfaces immediately become smooth and uniformly convex and contract to points."

R. Alessandroni and C. Sinestrari obtained convexity estimates for the nonhomogeneous flow in Example 18.17. The nice structure of the speed function allows for a proof using only the maximum principle (avoiding the difficult Stampacchia iteration argument needed in [**301, 302**] and [**56, 57**]).

19.4. Exercises

Exercise 19.1. Let $\mathcal{M}^2$ be a smooth surface in $\mathbb{R}^3$. Suppose that $\kappa_2 > \kappa_1$ at $x \in \mathcal{M}^n$. Show that κ_2 and κ_1 are both smooth at x and satisfy

$$\nabla_k \kappa_1 = \nabla_k \mathrm{II}_{11} \text{ and } \nabla_k \kappa_2 = \nabla_k \mathrm{II}_{22}$$

and

$$\nabla_k \nabla_\ell \kappa_1 = \nabla_k \nabla_\ell \mathrm{II}_{11} + 2\frac{\nabla_k \mathrm{II}_{12} \nabla_\ell \mathrm{II}_{12}}{\kappa_1 - \kappa_2} \text{ and } \nabla_k \kappa_2 = \nabla_k \mathrm{II}_{22} + 2\frac{\nabla_k \mathrm{II}_{12} \nabla_\ell \mathrm{II}_{12}}{\kappa_2 - \kappa_1}$$

at x, where e_1 and e_2 are unit eigenvectors corresponding to κ_1 and κ_2.

Show, furthermore, that $\partial_t \kappa_1 = \nabla_t \mathrm{II}_{11}$ and $\partial_t \kappa_2 = \nabla_t \mathrm{II}_{22}$ at (x,t) if $\mathcal{M}^2$ is the time t slice of a smooth family of surfaces, where ∇_t is the induced covariant time derivative.

Exercise 19.2 (Ecker and Huisken [**207**], the first author, McCoy, and Zheng [**61**]). Let $f : \Gamma_+^n \to \mathbb{R}$, where $n \geq 2$, be an admissible speed function and let $X : M^n \to \mathbb{R}^{n+1}$ be a compact, connected, locally uniformly convex embedding satisfying

$$\nabla F \equiv 0,$$

where $F \doteqdot f(\mathrm{II})$. Show that $X(M^n)$ is a round sphere if:

(a) f is concave. HINT: Show that the mean curvature is constant.

(b) f_* is concave. HINT: Show that the mean radius is constant.

Exercise 19.3. Let $\mathcal{M}^n$ be an immersed hypersurface of $\mathbb{R}^{n+1}$. Suppose that $\kappa_1 - \varepsilon H \geq 0$ on $\mathcal{M}^n$ with equality at x_0. Show that

$$\nabla_k \mathrm{II}_{11} = \frac{\varepsilon}{1-\varepsilon} \sum_{p=2}^{n} \nabla_k \mathrm{II}_{pp}$$

at x_0, where $\{e_i\}_{i=1}^n$ is a principal frame.

Exercise 19.4. Prove Corollary 19.9.

Exercise 19.5. Prove Proposition 19.10.

Exercise 19.6. Let $\alpha > 0$, and let F be an α-homogeneous admissible speed function. Let M_0 be a smooth convex hypersurface enclosing the origin, giving rise to a smooth solution of the flow (18.1). Use the fact that λM_0 encloses M_0 for $\lambda > 1$, the parabolic rescaling (18.10), and the avoidance principle to prove that $\sigma + (1+\alpha)tF \geq 0$.

Exercise 19.7 (Universal lower speed bound for $\alpha < 1$). Let $0 < \alpha < 1$, and let F be an α-homogeneous admissible speed with $F(I) = 1$. Use the argument of Proposition 16.6 to prove that

$$\sigma(z,t) \leq \sigma(z,0) - C(n,\alpha) r_+(\mathcal{M}_0)^{-\frac{2\alpha}{1-\alpha}} t^{\frac{1}{1-\alpha}}$$

for a smooth convex solution of (18.1) with initial hypersurface $\mathcal{M}_0$. Use the previous exercise (and a suitable choice of origin) to deduce that

$$F(z,t) \geq \tilde{C}(n,\alpha) r_+(\mathcal{M}_0)^{-\frac{2\alpha}{1-\alpha}} t^{\frac{\alpha}{1-\alpha}}.$$

Exercise 19.8 (Universal barriers and flat sides for $\alpha > 1$)**.** Let $\alpha > 1$, and let F be an α-homogeneous admissible speed with $F(I) = 1$. Show that the hypersurface $M = \partial\{(x,y) \in \mathbb{R} \times \mathbb{R}^n : x > 0,\ G(x,y) < 0\}$, where $G(x,y) = |y|^2 + 2x - 1 - 2x^{\frac{\alpha-1}{2\alpha-1}}$, satisfies $\langle X, \mathsf{N}\rangle \geq \beta F$, where β depends only on α. Deduce that flat sides persist under the flow (18.1) for $\alpha > 1$ (compare with Exercise 15.13).

Exercise 19.9 (Mean curvature flow of weakly convex hypersurfaces)**.** Let Ω_0 be a bounded open convex subset of $\mathbb{R}^{n+1}$. As in Exercise 15.12, Huisken's theorem implies that there is a unique viscosity solution $\{\mathcal{M}_t = \partial\Omega_t\}$ of mean curvature flow which converges to $\mathcal{M}_0 = \partial\Omega_0$ in Hausdorff distance as t approaches zero, which is obtained as a limit of smooth, uniformly convex solutions $\mathcal{M}_{\varepsilon,t} = \partial\Omega_{\varepsilon,t}$ with $\Omega_{\varepsilon,0}$ approximating Ω_0. By obtaining estimates on $\mathcal{M}_{\varepsilon,t}$ independent of ε, one can deduce that $\mathcal{M}_t$ is smooth and uniformly convex for each $t > 0$, as follows:

(1) Use the avoidance principle to show that there exists $p \in \mathbb{R}^{n+1}$, $t_0 > 0$, $\varepsilon_0 > 0$, and $0 < r_- < r_+$ such that $B_{r_-}(p) \subset \Omega_{\varepsilon,t} \subset B_{r_+}(p)$ for all $t \in [0, t_0]$ and $\varepsilon \in (0, \varepsilon_0)$.

(2) Deduce that for each $z \in S^n$, $t \in [0,t_0]$, and $\varepsilon \in (0,\varepsilon_0)$, $\mathcal{M}_{t,\varepsilon} \cap \left\{X : \langle X-p, z\rangle > 0,\ |X - p\,\langle X-p, z\rangle\, z| < \frac{r_-}{2}\right\}$ is a graph in direction z with gradient bounded by $C(r_-, r_+)$.

(3) Deduce from the estimates of Section 7.5 that $\mathcal{M}_{t,\varepsilon}$ is smooth for $t > 0$, with estimates depending only on t, r_- and r_+, and n (but not ε). Conclude that $\mathcal{M}_t$ is also smooth, with the same estimates.

(4) Apply the strong maximum principle to the evolution equation (6.17) to deduce that the second fundamental form is positive definite for positive times.

Exercise 19.10 (Harmonic mean curvature flow with flat sides)**.** Let $2 \leq k \leq n$. The flat k-disk $D^k = \{(x,y) \in \mathbb{R}^k \times \mathbb{R}^{n+1-k} : y = 0,\ |x| < 1\}$ in $\mathbb{R}^{n+1}$ has support function $\sigma(\cos\theta z, \sin\theta w) = \cos\theta$ for each $z \in S^{k-1}$ and $w \in S^{n-k}$ and $\theta \in [0, \frac{\pi}{2}]$. Show that $A[\sigma](v,v) = \frac{1}{\sigma}\bar{g}(v,v)$ for any v tangent to S^{k-1}, and deduce that the harmonic mean curvature $H_{-1} = \left(\frac{1}{n}\mathrm{Tr}_g A\right)^{-1}$ satisfies $H_{-1} \leq \frac{n}{k-1}\sigma$. Deduce that the shrinking disk $D_t = \sqrt{1 - \frac{2n}{k-1}t}D^k$ is an inner barrier for smooth, strictly convex solutions: If $\{\mathcal{M}_t\}$ is a smooth solution of (18.1) with $F = H_{-1}$ such that $\mathcal{M}_0$ encloses D, then $\mathcal{M}_t$ encloses D_t for each $t \in (0, \frac{k-1}{2n})$. Use this to prove that flat k-dimensional sides persist in viscosity solutions of the harmonic mean curvature flow.

Chapter 20

Flows of Inverse-Mean Curvature Type

By Euler's theorem for homogeneous functions, flows by admissible speeds of *negative* homogeneity in the curvature will cause convex hypersurfaces to *expand*. If the degree of homogeneity is -1, the scaling behavior matches that of the *inverse*-mean curvature flow, so we refer to such flows as **flows of inverse-mean curvature type**.

20.1. Convex hypersurfaces expand to round infinity

The following theorem, which combines results of John Urbas [**510**, **511**], Claus Gerhardt [**236**], and Haizhong Li, Xianfeng Wang, and Yong Wei [**363**], provides a nice analogue of Theorem 19.1 for expansion flows.

Figure 20.1. John Urbas. Photo by Helena Urbas.

Recall that $\Gamma^n_+ \doteqdot (0,\infty)^n$ is the positive cone in $\mathbb{R}^n$ and that the dual $f_* : \Gamma^n_+ \to \mathbb{R}$ of a function $f : \Gamma^n_+ \to \mathbb{R}$ is defined by

$$f_*(z_1^{-1}, \ldots, z_n^{-1}) = f^{-1}(z_1, \ldots, z_n)\,. \tag{20.1}$$

Theorem 20.1 (Urbas [**510, 511**], Gerhardt [**236**], Li, Wang, and Wei [**363**]). *Let $f : \Gamma^n_+ \to \mathbb{R}$ be a 1-homogeneous admissible speed function*[1] *normalized so that $f(1,\ldots,1) = 1$ and let $X_0 : M^n \to \mathbb{R}^{n+1}$ be a smooth, compact, locally uniformly convex embedding. Suppose that at least one of the following auxiliary conditions is satisfied:*

(i) *$n = 2$.*

(ii) *f is convex.*

(iii) *f is concave and extends continuously to $\partial\Gamma_+$ with $f|_{\partial\Gamma_+} \equiv 0$.*

(iv) *f and f_* are concave.*

(v) *f_* is convex.*

(vi) *f_* is concave and extends continuously to $\partial\Gamma_+$ with $f_*|_{\partial\Gamma_+} \equiv 0$.*

Then the unique maximal solution $X : M^n \times [0, T) \to \mathbb{R}^{n+1}$ to

$$\begin{cases} \partial_t X(x,t) = f(\mathrm{II}^{-1}_{(x,t)})\mathsf{N}(x,t) & \text{for } (x,t) \in M^n \times (0,T), \\ X(\cdot, 0) = X_0 \end{cases} \tag{20.2}$$

exists for all time (i.e., $T = \infty$). The rescaled embeddings $\widetilde{X}_t : M^n \to \mathbb{R}^{n+1}$ defined by

$$\widetilde{X}_t(x) \doteqdot \mathrm{e}^{-t} X(x,t) \tag{20.3}$$

converge uniformly in the smooth topology to a smooth embedding whose image is a round sphere centered at the origin.

Note that the speed is given here as a function of the *inverse* of the second fundamental form.

Remark 20.2. There are some important differences between Theorems 19.1 and 20.1:

(1) The argument in the expansion case applies also when the dimension is 1, whereas a different argument is required in the contraction case (excluding the route via noncollapsing; see Chapter 12).

(2) In case (vi) of Theorem 20.1, the condition $f_*|_{\Gamma^n_+} \equiv 0$ is needed in order to preserve convexity of the solution. This can be removed if we are willing to work within the larger class of *starshaped* hypersurfaces: If the 1-homogeneous admissible speed $f_* : \Gamma^n \to \mathbb{R}$

[1] See Conditions 18.1.

is concave and vanishes on $\partial\Gamma^n$, for a general convex cone Γ^n containing Γ^n_+, then starshaped hypersurfaces on which $f_*(\mathrm{II}) > 0$ expand to round infinity under the flow by outward speed $f_*(\mathrm{II})^{-1}$, preserving these conditions [**236, 363, 510**].

(3) For contraction flows, it is crucial that the rescaling be made about the final point, whereas, in the expansion case, the rescaling can be made about any fixed point. For arbitrary centers, however, the solution may never be close to the corresponding expanding sphere (consider two expanding spheres with different centers). Julian Scheuer has proved that one can always find an optimal approximating expanding sphere solution, in the sense that the unscaled time slices will eventually lie in an annulus of any desired width about the corresponding slice of the expanding sphere solution [**457**].

A surprising (explicit) connection between contraction and expansion flows *in the sphere* was discovered by C. Gerhardt [**240**].

20.1.1. Preliminaries. Given a compact, embedded, locally uniformly convex immersion $X_0 : M^n \to \mathbb{R}^{n+1}$, Theorem 18.18 guarantees the existence of a unique maximal solution $X : M^n \times [0,T) \to \mathbb{R}^{n+1}$ to (20.2), where $T \in (0,\infty]$. We assume that the initial hypersurface encloses the origin (which can be arranged by a fixed translation in space).

Recall that the support function $\sigma : S^n \times [0,T) \to \mathbb{R}$ is defined by

$$\sigma(z,t) \doteqdot \left\langle \mathrm{G}^{-1}(z,t), z \right\rangle,$$

where $\mathrm{G} : M^n \times [0,T) \to S^n$ is the Gauß map and $\mathrm{G}^{-1}(\cdot,t)$ is the inverse of $\mathrm{G}(\cdot,t)$. Since, by Lemmas 5.20 and 5.21,

$$\mathrm{G}^*\,\mathrm{I} = \mathrm{II}^2 \quad\text{and}\quad \mathrm{G}^*\,\mathrm{II} = A \doteqdot \overline{\nabla}^2\sigma + \sigma\overline{g},$$

where $\overline{g}$ and $\overline{\nabla}$ denote the standard metric and connection on S^n, we find that σ satisfies

$$\partial_t\sigma = f(A) \doteqdot f([A]), \tag{20.4}$$

where $[A]$ is the component matrix of A with respect to some (and hence any) local orthonormal frame for S^n (with respect to its *standard* metric).

Recall from Section 5.2.4 that the eigenvalues of A with respect to $\overline{g}$ are the principal radii $\mathbf{r} = (r_1, \ldots, r_n)$ of the hypersurface. It will often be convenient to order the principal radii so that $r_1 \geq r_2 \geq \cdots \geq r_n$ (in accordance with our usual ordering convention for the principal curvatures, $\kappa_i = r_i^{-1}$).

We may also write the solution (at least for a short time) as a radial graph, $\mathcal{M}_t^n = \{\rho(z,t)z : z \in S^n\}$. Then the radial function ρ satisfies

$$\partial_t \rho = \frac{f(\mathrm{II}^{-1})}{\langle z, \mathsf{N}(z)\rangle} = \rho^{-1}\sqrt{\rho^2 + |\overline{\nabla}\rho|^2}\, f_*^{-1}(\mathrm{II})$$

and hence, by (5.76) and the 1-homogeneity of f_*,

$$\partial_t \rho = (1+\rho^{-2}|\overline{\nabla}\rho|^2) f_*^{-1}\Big(-\overline{\nabla}^2\rho + 2\rho^{-1}\overline{\nabla}\rho\otimes\overline{\nabla}\rho + \rho\overline{g}\Big), \tag{20.5}$$

where f and f_* are evaluated on the component matrix of the argument in an orthonormal frame for TS^n (with respect to the *induced* metric).

Lemma 20.3. *Let $f : \Gamma_+^n \to \mathbb{R}$ be a 1-homogeneous admissible speed function.*

If f is concave, then

$$\ddot{f}_A^{ij,rs} C_{ij} C_{rs} \leq 0 \ \text{ for any } C \in \mathrm{S}(n), \tag{20.6}$$

$$\dot{f}_A^{ij}\overline{g}_{ij} \geq 1, \tag{20.7}$$

and

$$f(\mathbf{r}) \leq \frac{1}{n}\sum_{i=1}^n r_i, \ \text{ which implies } \ \frac{1}{n}\sum_{i=1}^n (r_1 - r_i) \leq r_1 - f(\mathbf{r}). \tag{20.8}$$

Proof. The first estimate is immediate. To obtain the second, set $\mathbf{1} \doteqdot (1,\ldots,1)$. Then

$$\dot{f}_A^{ij}\overline{g}_{ij} = \sum_{i=1}^n \dot{f}^i|_{\mathbf{r}} \geq \frac{f(\mathbf{r}+s\mathbf{1}) - f(\mathbf{r})}{s} \geq \frac{r_n + s - r_1}{s} \to 1 \text{ as } s \to \infty,$$

where we used the monotonicity and homogeneity of f. The third estimate is left as an exercise (see Exercise 20.1). □

20.1.2. Geometric estimates. Our first step is to obtain an estimate for the support function/radial graph height, which is a straightforward application of the maximum principle. The estimates do not rely on any concavity properties of the speed and hence apply in all of the cases considered.

Proposition 20.4. *For all $(z,t) \in S^n \times [0,T)$, we may estimate*

$$\min_{S^n\times\{0\}} \sigma \leq \mathrm{e}^{-t}\sigma(z,t) \leq \max_{S^n\times\{0\}} \sigma \tag{20.9}$$

and

$$\min_{S^n\times\{0\}} \rho \leq \mathrm{e}^{-t}\rho(z,t) \leq \max_{S^n\times\{0\}} \rho. \tag{20.10}$$

Proof. At an interior spatial maximum of σ, we have $\overline{\nabla}^2\sigma \le 0$, so that

$$\partial_t\sigma = f\left(\overline{\nabla}^2\sigma + \sigma\overline{g}\right) \le f\left(\sigma\overline{g}\right) = \sigma$$

since f is both monotone and 1-homogeneous and normalized so that $f\left(\overline{g}\right) = 1$. The upper bound in (20.9) now follows from the maximum principle. The lower bound follows from considering spatial minimum points.

The estimate for ρ is left as an exercise (Exercise 20.2). □

Observe that the speed $F \doteqdot f(A)$ satisfies the evolution equation

$$(\partial_t - \dot{F}^{ij}\overline{\nabla}_i\overline{\nabla}_j)F = F\dot{F}^{ij}\overline{g}_{ij}, \tag{20.11}$$

where $\dot{F} \doteqdot \dot{f}_A$. Indeed, since the metric and connection on S^n are independent of time,

$$\partial_t f(A) = \partial_t f(\overline{\nabla}^2\sigma + \sigma\overline{g}) = \dot{f}_A^{ij}(\overline{\nabla}_i\overline{\nabla}_j\partial_t\sigma + \partial_t\sigma\overline{g}_{ij}) = \dot{f}_A^{ij}(\overline{\nabla}_i\overline{\nabla}_j f(A) + f(A)\overline{g}_{ij}).$$

Corollary 20.5. *For every $(z,t) \in S^n \times [0,T)$,*

$$\min_{S^n\times\{0\}} \frac{F}{\sigma} \le \frac{F(z,t)}{\sigma(z,t)} \le \max_{S^n\times\{0\}} \frac{F}{\sigma}. \tag{20.12}$$

Proof. Combining the evolution equations (20.4) for the support function σ and (20.11) for the speed F with the identity $F = \dot{F}^{ij}A_{ij}$ coming from Euler's theorem, we find

$$\begin{aligned}\frac{\sigma}{F}(\partial_t - \dot{F}^{ij}\overline{\nabla}_i\overline{\nabla}_j)\frac{F}{\sigma} &= \frac{(\partial_t - \dot{F}^{ij}\overline{\nabla}_i\overline{\nabla}_j)F}{F} - \frac{(\partial_t - \dot{F}^{ij}\overline{\nabla}_i\overline{\nabla}_j)\sigma}{\sigma} + 2\dot{F}^{ij}\frac{\overline{\nabla}_i\frac{F}{\sigma}}{F/\sigma}\frac{\overline{\nabla}_j\sigma}{\sigma}\\ &= \dot{F}^{ij}\overline{g}_{ij} - \frac{F - \dot{f}^{ij}(A_{ij} - \sigma\overline{g}_{ij})}{\sigma} + 2\dot{F}^{ij}\frac{\overline{\nabla}_i\frac{F}{\sigma}}{F/\sigma}\frac{\overline{\nabla}_j\sigma}{\sigma}\\ &= 2\dot{F}^{ij}\frac{\overline{\nabla}_i\frac{F}{\sigma}}{F/\sigma}\frac{\overline{\nabla}_j\sigma}{\sigma}.\end{aligned}$$

So the claim follows from the maximum principle. □

In order to estimate the principal radii, we make use of the following evolution equation for the tensor A.

Lemma 20.6. *The tensor A evolves according to*

$$\begin{aligned}(\partial_t - \dot{F}^{ij}\overline{\nabla}_i\overline{\nabla}_j)A_{k\ell} = \ddot{F}^{pq,rs}\overline{\nabla}_k A_{pq}\overline{\nabla}_\ell A_{rs} + 2F\overline{g}_{k\ell} - \dot{F}^{ij}\overline{g}_{ij}A_{k\ell}\\ + \dot{F}^{ij}\overline{g}_{kj}\overline{\nabla}_i\overline{\nabla}_\ell\sigma - \dot{F}^{ij}\overline{g}_{i\ell}\overline{\nabla}_j\overline{\nabla}_k\sigma,\end{aligned} \tag{20.13}$$

where $\dot{F}^{ij} = \dot{f}_A^{ij}$ and $\ddot{F}^{pq,rs} = \ddot{f}_A^{pq,rs}$ are defined by (18.7).

Proof. Since we are working with a fixed metric and connection on S^n, we may commute ∂_t and $\overline{\nabla}$ to obtain

$$\partial_t\overline{\nabla}_\ell\sigma = \overline{\nabla}_\ell\partial_t\sigma = \overline{\nabla}_\ell f(A) = \dot{f}_A^{ij}\overline{\nabla}_\ell A_{ij}$$

and

$$\partial_t\overline{\nabla}_k\overline{\nabla}_\ell\sigma = \dot{f}_A^{ij}\overline{\nabla}_k\overline{\nabla}_\ell A_{ij} + \ddot{f}_A^{pq,rs}\overline{\nabla}_k A_{pq}\overline{\nabla}_\ell A_{rs}$$

$$= \dot{f}_A^{ij}\overline{\nabla}_k\overline{\nabla}_\ell\overline{\nabla}_i\overline{\nabla}_j\sigma + \dot{f}_A^{ij}\overline{g}_{ij}\overline{\nabla}_k\overline{\nabla}_\ell\sigma + \ddot{f}_A^{pq,rs}\overline{\nabla}_k A_{pq}\overline{\nabla}_\ell A_{rs}. \tag{20.14}$$

Since the Riemann curvature tensor of $\overline{g}$ is given by

$$\overline{R}_{ijk\ell} = \overline{g}_{i\ell}\overline{g}_{jk} - \overline{g}_{ik}\overline{g}_{j\ell},$$

we have the following formulae for commuting covariant derivatives:

$$\begin{aligned}
\overline{\nabla}_k\overline{\nabla}_\ell\overline{\nabla}_i\overline{\nabla}_j\sigma &= \overline{\nabla}_k\overline{\nabla}_i\overline{\nabla}_\ell\overline{\nabla}_j\sigma - \overline{g}_{ij}\overline{\nabla}_k\overline{\nabla}_\ell\sigma + \overline{g}_{\ell j}\overline{\nabla}_k\overline{\nabla}_i\sigma,\\
\overline{\nabla}_k\overline{\nabla}_i\overline{\nabla}_j\overline{\nabla}_\ell\sigma &= \overline{\nabla}_i\overline{\nabla}_k\overline{\nabla}_j\overline{\nabla}_\ell\sigma - \overline{g}_{ij}\overline{\nabla}_k\overline{\nabla}_\ell\sigma + \overline{g}_{kj}\overline{\nabla}_i\overline{\nabla}_\ell\sigma\\
&\quad - \overline{g}_{i\ell}\overline{\nabla}_j\overline{\nabla}_k\sigma + \overline{g}_{k\ell}\overline{\nabla}_j\overline{\nabla}_i\sigma,\\
\overline{\nabla}_i\overline{\nabla}_k\overline{\nabla}_j\overline{\nabla}_\ell\sigma &= \overline{\nabla}_i\overline{\nabla}_j\overline{\nabla}_k\overline{\nabla}_\ell\sigma - \overline{g}_{j\ell}\overline{\nabla}_i\overline{\nabla}_k\sigma + \overline{g}_{k\ell}\overline{\nabla}_i\overline{\nabla}_j\sigma.
\end{aligned}$$

Combining these formulae, we have

$$\begin{aligned}
\overline{\nabla}_k\overline{\nabla}_\ell\overline{\nabla}_i\overline{\nabla}_j\sigma &= \overline{\nabla}_i\overline{\nabla}_j\overline{\nabla}_k\overline{\nabla}_\ell\sigma + 2\overline{g}_{k\ell}\overline{\nabla}_j\overline{\nabla}_i\sigma - 2\overline{g}_{ij}\overline{\nabla}_k\overline{\nabla}_\ell\sigma\\
&\quad + \overline{g}_{kj}\overline{\nabla}_i\overline{\nabla}_\ell\sigma - \overline{g}_{i\ell}\overline{\nabla}_j\overline{\nabla}_k\sigma.
\end{aligned} \tag{20.15}$$

Since $A_{k\ell} = \overline{\nabla}_k\overline{\nabla}_\ell\sigma + \sigma\overline{g}_{k\ell}$, (20.14) and (20.15) yield

$$\begin{aligned}
\partial_t A_{k\ell} &= \dot{f}_A^{ij}\overline{\nabla}_i\overline{\nabla}_j A_{k\ell} + \dot{f}_A^{ij}\overline{g}_{k\ell}\overline{\nabla}_j\overline{\nabla}_i\sigma - \dot{f}_A^{ij}\overline{g}_{ij}\overline{\nabla}_k\overline{\nabla}_\ell\sigma\\
&\quad + \dot{f}_A^{ij}\overline{g}_{kj}\overline{\nabla}_i\overline{\nabla}_\ell\sigma - \dot{f}_A^{ij}\overline{g}_{i\ell}\overline{\nabla}_j\overline{\nabla}_k\sigma\\
&\quad + \ddot{f}_A^{pq,rs}\overline{\nabla}_k A_{pq}\overline{\nabla}_\ell A_{rs} + f(A)\overline{g}_{k\ell}.
\end{aligned} \tag{20.16}$$

Finally, by Euler's theorem for homogeneous functions (Proposition 18.6),

$$f(A) = \dot{f}_A^{ij}A_{ij} = \dot{f}_A^{ij}\overline{\nabla}_i\overline{\nabla}_j\sigma + \dot{f}_A^{ij}\overline{g}_{ij}\sigma. \tag{20.17}$$

Thus (20.13) follows from (20.16). □

If f is concave, this immediately yields an upper bound for the principal radii. The estimate also applies to inverse-convex speeds by Lemma 18.13.

Corollary 20.7. *If f is concave, then the maximum principal radius satisfies*

$$e^{-t}r_1(z,t) \le \max_{S^n\times\{0\}} r_1. \tag{20.18}$$

Proof. Given t_0, suppose that A achieves its spatial maximum eigenvalue at $z_0 \in S^n$ with an associated unit eigenvector $v \in T_{z_0}S^n$. By (20.13) and (20.6) we have at (z_0, t_0) that

$$\partial_t A_{k\ell}v^k v^\ell \le \dot{F}^{ij}(\overline{\nabla}_i\overline{\nabla}_j A_{k\ell})v^k v^\ell + 2F - \dot{F}^{ij}\overline{g}_{ij}A_{k\ell}v^k v^\ell \tag{20.19}$$

because $\dot{F}^{ij}v_j\overline{\nabla}_i\overline{\nabla}_\ell\sigma v^\ell - \dot{F}^{ij}v_i\overline{\nabla}_j\overline{\nabla}_k\sigma v^k = 0$. Now, since v is a maximal unit eigenvector and f is monotone and normalized, we may estimate

$$F = f(A) \le f(A_{k\ell}v^k v^\ell\overline{g}) = A_{k\ell}v^k v^\ell.$$

This and Euler's theorem applied to (20.19) imply at (z_0, t_0) that

$$\partial_t A_{k\ell} v^k v^\ell \le \dot{F}^{ij}(\overline{\nabla}_i \overline{\nabla}_j A_{k\ell}) v^k v^\ell + A_{k\ell} v^k v^\ell .$$

So the estimate follows from the maximum principle. □

When $n = 2$, we can bound the principal radii without requiring any concavity hypothesis (cf. Proposition 19.5).

Corollary 20.8. *If $n = 2$, then, setting $K \doteqdot \max_{S^2 \times \{0\}}(\frac{r_1}{r_2} - 1)$, we have*

$$r_1(z,t) \le (1 + K\mathrm{e}^{-2t}) r_2(z,t) \quad \textit{for all } (z,t) \in S^2 \times [0,T). \tag{20.20}$$

Proof. A similar argument as in Proposition 12.15 yields, wherever $r_1 > r_2$,

$$\begin{aligned}(\partial_t - \dot{F}^{ij}\overline{\nabla}_i\overline{\nabla}_j) r_1 = {} & \ddot{F}^{pq}\overline{\nabla}_1 A_{pp}\overline{\nabla}_1 A_{qq} + 2\frac{\dot{F}^1 - \dot{F}^2}{r_1 - r_2}(\overline{\nabla}_1 A_{12})^2 - 2\dot{F}^k \frac{(\overline{\nabla}_1 A_{k2})^2}{r_1 - r_2} \\ & + 2F - \dot{F}^{ij}\overline{g}_{ij} r_1\end{aligned}$$

and

$$\begin{aligned}(\partial_t - \dot{F}^{ij}\overline{\nabla}_i\overline{\nabla}_j) r_2 = {} & \ddot{F}^{pq}\overline{\nabla}_2 A_{pp}\overline{\nabla}_2 A_{qq} + 2\frac{\dot{F}^1 - \dot{F}^2}{r_1 - r_2}(\overline{\nabla}_2 A_{12})^2 + 2\dot{F}^k \frac{(\overline{\nabla}_2 A_{k1})^2}{r_1 - r_2} \\ & + 2F - \dot{F}^{ij}\overline{g}_{ij} r_2 .\end{aligned}$$

Thus, at a nonzero local maximum of $r_1/r_2 - 1$,

$$\begin{aligned}\frac{r_2}{r_1}(\partial_t - \dot{F}^{ij}\overline{\nabla}_i\overline{\nabla}_j)\left(\frac{r_1}{r_2} - 1\right) &= \frac{1}{r_1}(\partial_t - \dot{F}^{ij}\overline{\nabla}_i\overline{\nabla}_j) r_1 - \frac{1}{r_2}(\partial_t - \dot{F}^{ij}\overline{\nabla}_i\overline{\nabla}_j) r_2 \\ &= \ddot{F}^{pq}\left(\frac{1}{r_1}\overline{\nabla}_1 A_{pp}\overline{\nabla}_1 A_{qq} - \frac{1}{r_2}\overline{\nabla}_2 A_{pp}\overline{\nabla}_2 A_{qq}\right) \\ &\quad + 2\frac{\dot{F}^1 - \dot{F}^2}{r_1 - r_2}\left(\frac{1}{r_1}(\overline{\nabla}_1 A_{12})^2 - \frac{1}{r_2}(\overline{\nabla}_2 A_{12})^2\right) \\ &\quad - 2\left(\frac{1}{r_1} + \frac{1}{r_2}\right)\dot{F}^k \frac{(\overline{\nabla}_2 A_{k1})^2}{r_1 - r_2} + 2F\left(\frac{1}{r_1} - \frac{1}{r_2}\right) \\ &= \ddot{F}^{pq}\left(\frac{1}{r_1}\overline{\nabla}_1 A_{pp}\overline{\nabla}_1 A_{qq} - \frac{1}{r_2}\overline{\nabla}_2 A_{pp}\overline{\nabla}_2 A_{qq}\right) \\ &\quad - 2\frac{\frac{\dot{F}^2}{r_1} + \frac{\dot{F}^1}{r_2}}{r_1 - r_2}\left((\overline{\nabla}_2 A_{11})^2 + (\overline{\nabla}_1 A_{22})^2\right) - 2\frac{f}{r_1}\left(\frac{r_1}{r_2} - 1\right).\end{aligned}$$

We now apply the gradient conditions

$$0 = \frac{r_2}{r_1}\overline{\nabla}_k \frac{r_1}{r_2} = \frac{\overline{\nabla}_k A_{11}}{r_1} - \frac{\overline{\nabla}_k A_{22}}{r_2}$$

and the 1-homogeneity of f to obtain

$$\begin{aligned}\ddot{F}^{pq}\Big(\frac{1}{r_1}\overline{\nabla}_1 A_{pp}\overline{\nabla}_1 A_{qq} - \frac{1}{r_2}\overline{\nabla}_2 A_{pp}\overline{\nabla}_2 A_{qq}\Big) \\ &= \left(\ddot{F}^{11} + 2\frac{r_2}{r_1}\ddot{F}^{12} + \frac{r_2^2}{r_1^2}\ddot{F}^{22}\right)\left(\frac{(\overline{\nabla}_1 A_{11})^2}{r_1} - \frac{(\overline{\nabla}_2 A_{11})^2}{r_2}\right) \\ &= r_1^{-2}(\ddot{F}^{11}r_1^2 + 2\ddot{F}^{12}r_1r_2 + \ddot{F}^{22}r_2^2)\left(\frac{(\overline{\nabla}_1 A_{11})^2}{r_1} - \frac{(\overline{\nabla}_2 A_{11})^2}{r_2}\right) = 0.\end{aligned}$$

The monotonicity of f now yields

$$(\partial_t - \dot{F}^{ij}\overline{\nabla}_i\overline{\nabla}_j)\left(\frac{r_1}{r_2} - 1\right) \le -2\frac{F}{r_2}\left(\frac{r_1}{r_2} - 1\right) \le -2\left(\frac{r_1}{r_2} - 1\right).$$

So the claim follows from the maximum principle. □

For inverse-concave speeds, we will obtain bounds for the principal radii from the evolution equation for the tensor $B \doteqdot A^{-1}$.

Lemma 20.9. *The tensor $B \doteqdot A^{-1}$ evolves according to*

$$\begin{aligned}(\partial_t - \dot{F}^{ij}\overline{\nabla}_i\overline{\nabla}_j)B^{k\ell} &= -2FB^{ki}B^{\ell j}\overline{g}_{ij} + \dot{F}^{ij}\overline{g}_{ij}B^{k\ell} \\ &\quad - B^{ki}B^{\ell j}(\ddot{F}^{pq,rs} + 2\dot{F}^{pr}B^{qs})\overline{\nabla}_i A_{pq}\overline{\nabla}_j A_{rs} \\ &\quad - B^{kp}B^{\ell q}(\dot{F}^{im}\overline{g}_{mp}\overline{\nabla}_i\overline{\nabla}_q\sigma - \dot{F}^{im}\overline{g}_{mq}\overline{\nabla}_i\overline{\nabla}_p\sigma).\end{aligned} \tag{20.21}$$

Proof. Since $A_{kp}B^{p\ell} = \delta_k^\ell$, we find

$$0 = \partial_t(A_{kp}B^{p\ell}) = \partial_t A_{kp}B^{p\ell} + A_{kp}\partial_t B^{p\ell},$$

$$0 = \overline{\nabla}_i(A_{kp}B^{p\ell}) = \overline{\nabla}_i A_{kp}B^{p\ell} + A_{kp}\overline{\nabla}_i B^{p\ell},$$

and

$$\begin{aligned}0 = \overline{\nabla}_i\overline{\nabla}_j(A_{kp}B^{p\ell}) &= \overline{\nabla}_i(\overline{\nabla}_j A_{kp}B^{p\ell} + A_{kp}\overline{\nabla}_j B^{p\ell}) \\ &= \overline{\nabla}_i\overline{\nabla}_j A_{kp}B^{p\ell} + \overline{\nabla}_j A_{kp}\overline{\nabla}_i B^{p\ell} \\ &\quad + \overline{\nabla}_i A_{kp}\overline{\nabla}_j B^{p\ell} + A_{kp}\overline{\nabla}_i\overline{\nabla}_j B^{p\ell},\end{aligned}$$

and hence

$$-\partial_t B^{k\ell} = B^{kp}B^{\ell q}\partial_t A_{pq}, \quad -\overline{\nabla}_i B^{k\ell} = B^{kp}B^{\ell q}\overline{\nabla}_i A_{pq},$$

and

$$-\overline{\nabla}_i\overline{\nabla}_j B^{k\ell} = B^{kp}B^{\ell q}\overline{\nabla}_i\overline{\nabla}_j A_{pq} + B^{kp}\overline{\nabla}_j A_{pq}\overline{\nabla}_i B^{\ell q} + B^{kp}\overline{\nabla}_i A_{pq}\overline{\nabla}_j B^{\ell q}.$$

The claim now follows from the evolution equation (20.13) for A. □

If v is a unit 1-form at a point on S^n, then we obtain from (20.21) that

$$\begin{aligned}\partial_t B(v,v) &= \dot F^{ij}(\overline{\nabla}_i\overline{\nabla}_j B)(v,v) - 2F|B(v)|^2 + \dot F^{ij}\overline{g}_{ij}B(v,v)\\ &\quad - B(v)^p B(v)^q(\ddot F^{ij,rs} + 2\dot F^{ir}B^{js})\overline{\nabla}_p A_{rs}\overline{\nabla}_q A_{ij}\,.\end{aligned} \tag{20.22}$$

If f is inverse-concave, then this immediately yields a lower bound for the principal radii. The estimate also applies to convex speeds by Lemma 18.13.

Corollary 20.10. *If f is inverse-concave, then the minimum principal radius satisfies*

$$\mathrm{e}^{-t}r_n(z,t) \geq \min_{S^n\times\{0\}} r_n\,. \tag{20.23}$$

Proof. By applying Lemma 18.12 to (20.22), we obtain

$$\partial_t B(v,v) \leq \dot F^{ij}(\overline{\nabla}_i\overline{\nabla}_j B)(v,v) - 2F|B(v)|^2 + \dot F^{ij}\overline{g}_{ij}B(v,v)\,. \tag{20.24}$$

Given t_0, suppose that B achieves its spatial maximum eigenvalue at $z_0 \in S^n$ with an associated unit eigenform $v \in T_{z_0}S^n$. Then, since $|B(v)|^2 = B(v,v)^2$ and by $F = \dot F^{ij}A_{ij}$, we have

$$-F|B(v)|^2 + \dot F^{ij}\overline{g}_{ij}B(v,v) = -B(v,v)\dot F^{ij}(A_{ij}B(v,v) - \overline{g}_{ij}) \leq 0\,.$$

Applying this and the monotonicity and normalization of f to (20.24) yields

$$\begin{aligned}\partial_t B(v,v) &\leq \dot F^{ij}(\overline{\nabla}_i\overline{\nabla}_j B)(v,v) - F|B(v)|^2\\ &\leq \dot F^{ij}(\overline{\nabla}_i\overline{\nabla}_j B)(v,v) - B(v,v)\,.\end{aligned} \tag{20.25}$$

In view of $\mathrm{e}^{-t}A \geq \varepsilon\overline{g}$ being equivalent to $\mathrm{e}^{t}B \leq \varepsilon^{-1}\overline{g}$, the estimate now follows from the maximum principle. □

We have proved, in each of the six cases of Theorem 20.1, that the principal radii remain in a compact subset of Γ_+ on any finite interval of time. This implies that the equations (20.4) (for the support function) and (20.5) (for the radial graph height) are uniformly parabolic and provides bounds for their gradient and Hessian. In cases (ii), (iii), and (iv), we can then obtain a Hölder estimate for $\overline{\nabla}^2\sigma$ from Krylov's theorem (Theorem 15.9). In cases (iv), (v), and (vi), a similar argument yields an estimate for $\overline{\nabla}^2\rho$. In case (i) we can apply the main theorem from [**37**] to bound either $\overline{\nabla}^2\sigma$ or $\overline{\nabla}^2\rho$. Higher-order estimates then follow from the Schauder estimate (Theorem 15.10) and we conclude that the solution exists for all positive times. We omit the details (which can be found in [**236, 363, 510, 511**]) since similar arguments were carried out in detail in Sections 8.6 and 15.3.

Proposition 20.11. *The solution exists for all positive times*: $T = \infty$.

20.1.3. Convergence. We now show that the rescaled hypersurfaces converge to spheres. Observe first that, in case (i), Proposition 20.11 and Corollaries 20.5 and 20.8 yield

$$(r_1 - r_2)(z,t) \le K\mathrm{e}^{-t} \tag{20.26}$$

since $r_2 \le f(A)$, where K depends only on the initial embedding.

For concave (or inverse-convex) speeds, we obtain the following decay estimate.

Lemma 20.12. *If f is concave, then*

$$(r_1 - F)(z,t) \le \mathrm{e}^{-t} \max_{S^n\times\{0\}} (r_1 - F) \quad \textit{for all} \quad (z,t) \in S^n \times [0,\infty). \tag{20.27}$$

Proof. Arguing similarly as in Corollary 20.7, we obtain

$$(\partial_t - \dot{F}^{ij}\overline{\nabla}_i\overline{\nabla}_j)(r_1 - F) \le 2F - \dot{F}^{ij}\overline{g}_{ij}r_1 - \dot{F}^{ij}\overline{g}_{ij}F \le -(r_1 - F).$$

So the claim follows from the maximum principle. □

Lemmas 20.3 and 20.12 now yield, in cases (ii), (iii), and (iv),

$$0 \le (r_1 - r_n)(z,t) \le K\mathrm{e}^{-t} \quad \text{for all} \quad (z,t) \in S^n \times [0,\infty), \tag{20.28}$$

where K depends only on n and the initial embedding.

For inverse-concave speeds, we obtain the desired decay estimate by estimating $\kappa_n - F_*$.

Lemma 20.13. *If $n \ge 2$ and f is inverse-concave, then there exist $\delta \in (0,1)$ and $K < \infty$ such that*

$$F - r_n \le K\mathrm{e}^{(1-\delta)t}. \tag{20.29}$$

Proof. Since $F_* = F^{-1}$, the evolution equation (20.11) yields

$$(\partial_t - \dot{F}^{ij}\overline{\nabla}_i\overline{\nabla}_j)F_* = -F_*\dot{F}^{ij}\overline{g}_{ij} - 2\dot{F}^{ij}\frac{\overline{\nabla}_iF\overline{\nabla}_jF}{F^3}.$$

By Lemmas 20.9 and 18.12, κ_n satisfies

$$\begin{aligned}(\partial_t - \dot{F}^{ij}\overline{\nabla}_i\overline{\nabla}_j)\kappa_n \le & -2\kappa_n^2F + \kappa_n\dot{F}^{ij}\overline{g}_{ij} - \kappa_n^2\frac{(\overline{\nabla}_nF)^2}{F} \\ & -2\sum_{k=1}^{n}\sum_{\kappa_p<\kappa_n}\frac{\dot{F}^k}{\kappa_n - \kappa_p}(\overline{\nabla}_kB^{np})^2\end{aligned}$$

in the barrier sense. It follows that

$$
\begin{aligned}
F_*(\partial_t - \dot F^{ij}\overline\nabla_i\overline\nabla_j)\left(\frac{\kappa_n}{F_*}-1\right) \le{}& -2\kappa_n^2F + \dot F^{ij}\overline g_{ij}(\kappa_n+F_*)\\
&+2\dot F^{ij}\overline\nabla_i\frac{\kappa_n}{F_*}\overline\nabla_jF_* + 2\frac{\kappa_n}{F_*}\dot F^{ij}\frac{\overline\nabla_iF\overline\nabla_jF}{F^3}\\
&-2\frac{\kappa_n^2}{F_*^2}\frac{(\overline\nabla_nF)^2}{F^3} - 2\sum_{k=1}^n\sum_{\kappa_p<\kappa_n}\frac{\dot F^k\kappa_n^2\kappa_p^2}{\kappa_n-\kappa_p}(\overline\nabla_kA_{np})^2
\end{aligned}
\tag{20.30}
$$

in the barrier sense. Since the eigenvalues of II/F_* remain in a compact subset of Γ^n_+, the 0-th order terms may be estimated as follows:

$$
\begin{aligned}
-2\kappa_n^2F + \dot F^{ij}\overline g_{ij}(\kappa_n+F_*) &= \sum_{i=1}^n \dot F^i\left(\kappa_n + F_* - 2\frac{\kappa_n^2}{\kappa_i}\right)\\
&= \sum_{i=1}^n \frac{\dot F_*^i\kappa_i}{F_*^2}\left(\kappa_n(\kappa_i-\kappa_n) + \kappa_iF_* - \kappa_n^2\right)\\
&\le -\frac{\kappa_n}{F_*}(\kappa_n - F_*) - \sum_{i=1}^n\frac{\dot F_*^i\kappa_i\kappa_n}{F_*^2}(\kappa_n-\kappa_i)\\
&\le -(1+\delta)(\kappa_n - F_*)
\end{aligned}
$$

for some $\delta>0$ which depends only on the initial curvature pinching.

At a local maximum, we have the gradient conditions

$$
\overline\nabla_kB^{nn} = \frac{\kappa_n}{F_*}\overline\nabla_kF_*\,,\quad \text{which imply}\quad \frac{\overline\nabla_kF}{F} = \kappa_n\overline\nabla_kA_{nn}\,.
$$

So the remaining terms on the right-hand side are (modulo a factor of $2\kappa_n$)

$$
\begin{aligned}
Q \doteqdot{}& \frac{1}{r_n^3}\left(\dot F^nr_n - F\right)(\overline\nabla_nA_{nn})^2 + \sum_{r_p>r_n}\left(\frac{\dot F^p}{r_n^2} - \frac{\dot F^n}{r_p(r_p-r_n)}\right)(\overline\nabla_pA_{nn})^2\\
&-\sum_{r_p>r_n}\frac{\dot F^p}{r_p(r_p-r_n)}(\overline\nabla_nA_{pp})^2 - \sum_{r_q>r_p>r_n}\left(\frac{\dot F^p}{r_p(r_q-r_n)} + \frac{\dot F^q}{r_p(r_p-r_n)}\right)(\overline\nabla_nA_{pq})^2.
\end{aligned}
$$

The final two terms are evidently nonpositive. Nonpositivity of the first term follows from Euler's identity, which yields,

$$
\frac{1}{r_n^3}\left(\dot F^nr_n - F\right)(\overline\nabla_nA_{nn})^2 = -\frac{1}{r_n^3}\sum_{p=1}^{n-1}\dot F^pr_p(\overline\nabla_nA_{nn})^2,
$$

while nonpositivity of the second follows from concavity of f_* via Lemma 18.7 since

$$\sum_{r_p>r_n}\left(\frac{\dot F^p}{r_n^2}-\frac{\dot F^n}{r_p(r_p-r_n)}\right)(\overline{\nabla}_pA_{nn})^2 \le \sum_{r_p>r_n}\left(\frac{\dot F^p}{r_n^2}-\frac{\dot F^n}{r_p^2}\right)(\overline{\nabla}_pA_{nn})^2$$
$$= \sum_{r_p>r_n}\frac{F^2}{r_p^2r_n^2}\left(\dot F_*^p-\dot F_*^n\right)(\overline{\nabla}_pA_{nn})^2 .$$

The ODE comparison principle now yields

$$\kappa_n - F_* \le K\mathrm{e}^{-(1+\delta)t}$$

where $K \doteqdot \max_{S^n\times\{0\}}(\kappa_n - F_*)$. The claim now follows from Proposition 20.4 and Corollary 20.5 since

$$F - r_n = (\kappa_n - F_*)r_nF . \qquad \square$$

We now complete the proof of Theorem 20.1.

Proof of Theorem 20.1. We claim that $\mathrm{e}^{-t}\sigma$ converges in the smooth topology to a constant function as $t\to\infty$.

Since $e^{-t}A$ remains in a compact subset K of Γ_+^n, we deduce, from (20.26) in case (i), from (20.28) in cases (ii), (iii), and (iv), and from Lemma 20.13 in cases (v) and (vi) that $\mathrm{e}^{-t}A$ decays exponentially to a constant multiple of $\overline{g}$. In terms of the principal radii, we have

$$|\mathrm{e}^{-t}\mathbf{r}(z,t) - r_*\mathbf{1}| \le C\mathrm{e}^{-\delta t} \tag{20.31}$$

for some $\delta>0$, $C<\infty$, and $r_*>0$, where $\mathbf{r}\doteqdot(r_1,\dots,r_n)$ and $\mathbf{1}\doteqdot(1,\dots,1)$. In what follows, the particular values of C and δ will be allowed to vary.

By Euler's theorem and the normalization of f, $Df|_{r\mathbf{1}}(\mathbf{1})=1$. Since f is symmetric, we also have $\dot f^i|_{r\mathbf{1}}=\dot f^j|_{r\mathbf{1}}$ for $1\le i,j\le n$, and hence $\dot f^i|_{r\mathbf{1}}=\frac{1}{n}$ for each i. Thus, by the fundamental theorem of calculus,

$$\left|\dot f^i|_{\mathrm{e}^{-t}\mathbf{r}(z,t)}-\tfrac{1}{n}\right| \le \sup_K|D^2f|\left|\mathrm{e}^{-t}\mathbf{r}(z,t)-r_*\mathbf{1}\right| \le C\mathrm{e}^{-\delta t}. \tag{20.32}$$

Applying Euler's theorem, we obtain

$$\left(\tfrac{1}{n}-C\mathrm{e}^{-\delta t}\right)\sum_{i=1}^n\mathrm{e}^{-t}r_i(z,t)\le f(\mathrm{e}^{-t}\mathbf{r}(z,t))\le\left(\tfrac{1}{n}+C\mathrm{e}^{-\delta t}\right)\sum_{i=1}^n\mathrm{e}^{-t}r_i(z,t)$$

and hence

$$\tfrac{1}{n}\mathrm{e}^{-t}\Delta\sigma(z,t)-C\mathrm{e}^{-\delta t}\le\partial_t\left(\mathrm{e}^{-t}\sigma(z,t)\right)\le\tfrac{1}{n}\mathrm{e}^{-t}\Delta\sigma(z,t)+C\mathrm{e}^{-\delta t}. \tag{20.33}$$

Denote by $\overline{\sigma}(t) \doteqdot \frac{1}{|S^n|}\int_{S^n}\sigma(z,t)\,d\bar{\mu}(z)$ the average value of $\sigma(\cdot,t)$. By (20.33),

$$-C\mathrm{e}^{-\delta t} \leq \frac{d}{dt}\left(\mathrm{e}^{-t}\overline{\sigma}(t)\right) \leq C\mathrm{e}^{-\delta t}.$$

It follows that the limit $\sigma_* \doteqdot \lim_{t\to\infty}\mathrm{e}^{-t}\overline{\sigma}(t)$ exists and is positive and that

$$|\mathrm{e}^{-t}\overline{\sigma}(t) - \sigma_*| \leq C\mathrm{e}^{-\delta t}.$$

To see that $\mathrm{e}^{-t}\sigma$ converges to σ_*, we exploit the Poincaré inequality:

$$\frac{d}{dt}\int_{S^n}|\mathrm{e}^{-t}(\sigma-\overline{\sigma})|^2\,d\bar{\mu} \leq -\frac{2}{n}\int_{S^n}|\overline{\nabla}(\mathrm{e}^{-t}\sigma)|^2\,d\bar{\mu} \leq -2\int_{S^n}|\mathrm{e}^{-t}(\sigma-\overline{\sigma})|^2\,d\bar{\mu}.$$

Integrating this and applying the triangle inequality yields

$$\int_{S^n}|\mathrm{e}^{-t}\sigma - \sigma_*|^2\,d\bar{\mu} \leq C\mathrm{e}^{-\delta t}.$$

Since $\overline{\nabla}(\mathrm{e}^{-t}\sigma)$ remains bounded, this proves that $\mathrm{e}^{-t}\sigma$ converges to σ_* in C^0. Since $\overline{\nabla}^2(\mathrm{e}^{-t}\sigma)$ remains bounded, a decay estimate for $\overline{\nabla}(\mathrm{e}^{-t}\sigma)$ follows by interpolation. Decay estimates for the higher derivatives of $\mathrm{e}^{-t}\sigma$ follow similarly, using the bounds for higher derivatives obtained previously. □

20.2. Notes and commentary

The original proof of the convergence result in the case of inverse-concave speeds, due to Gerhardt [**236**] and Urbas [**510**], made use of a decay estimate for the gradient of the radial graph height, in lieu of Lemma 20.13.

Once long-time existence has been established, the convergence result of Theorem 20.1 can also be obtained immediately from the gradient estimate for the support function proved by Gulliver and the second author using Alexandrov reflection [**164**] (which we presented in Theorem 18.39).

Expansion flows have been studied by many authors. The first results were due to Urbas [**511**], who obtained cases (iii) and (iv) of Theorem 20.1 and also proved analogous results for flows of homogeneity $p \in (0,1]$ in the principal radii. Soon after, Urbas [**510**] and Gerhardt [**236**] independently obtained case (vi) of Theorem 20.1. In fact, their results apply more generally to flows of *starshaped* hypersurfaces. Urbas also obtained case (i) of Theorem 20.1 assuming, in addition, that f is concave [**511**]; this hypothesis was later removed by H. Li, X. Wang, and Y. Wei [**363**]. Inverse Gauß curvature flows were studied by Q.-R. Li [**365**] and O. Schnürer [**463**]. C. Gerhardt [**239**] proved an analogue of Theorem 20.1 for expansion flows by speeds of homogeneity $p > 1$ in the principal radii.

For immersed closed convex curves, Tsai [**505**] proved the convergence after rescaling to multiply covered circles for a class of nonhomogeneous expanding flows which generalizes the result for the speeds $\kappa^{-\alpha}$, $\alpha \in (0, 1]$, by Urbas [**511**].

There is a large literature on the expansion of convex hypersurfaces in Riemannian and Lorentzian ambient spaces by curvature functions. See the work of B. Allen [**17**], C. Gerhardt [**237, 238, 240**], T. Mullins [**409**], and J. Scheuer [**459**], for a sample.

Expanding sphere solutions to expansion flows exhibit even stronger rigidity than shrinking sphere solutions to contraction flows — using the Alexandrov reflection method developed by Gulliver and the second author (see Section 18.8), Risa and Sinestrari have proved that, under a very large class of expansion flows, the expanding sphere solutions are the only solutions which evolve out of a single point at the initial time [**446**].

20.2.1. Geometric inequalities. Two nice features of inverse-curvature flows lend themselves to various geometric applications: First is the fact that, in many cases, convexity of the solution hypersurfaces is not necessary. Second is that they often admit useful monotone quantities. One example which we have already mentioned is the proof of the Riemannian Penrose inequality by inverse-mean curvature flow [**299**]. Another important example is the classical Alexandrov–Fenchel inequalities for convex bodies, which were extended to k-convex hypersurfaces using inverse-curvature flows by Pengfei Guan and Junfang Li [**253**]. Similar inequalities were deduced by similar methods for hypersurfaces of hyperbolic space [**516, 523**] and the sphere [**380, 523**]. Ivaki exploited certain anisotropic expansion flows to obtain solutions to certain L^p-Christoffel–Minkowski problems [**317**].

20.2.2. Flows with free boundary. Thomas Marquardt proved an analogue of Theorem 20.1 for inverse-mean curvature flow with free boundary on a cone [**387, 388**].

When the supporting hypersurface is the unit sphere, Ben Lambert and Julian Scheuer proved that the inverse-mean curvature flow evolves convex free boundary hypersurfaces to flat disks [**346**] and used this flow to obtain certain geometric inequalities [**347**].

20.3. Exercises

Exercise 20.1. Prove the estimate (20.8).

Exercise 20.2. Let $\rho : S^n \times [0,T) \to \mathbb{R}$ be a solution to (20.5). Show that

$$\min_{S^n \times \{0\}} \rho \le \mathrm{e}^{-t} \rho(z,t) \le \max_{S^n \times \{0\}} \rho$$

for every $(z,t) \in S^n \times [0,T)$.

Exercise 20.3. Given a positive, 1-homogeneous admissible speed f, show that a compact starshaped hypersurface remains starshaped under the flow

$$\partial_t X = f([\mathrm{II}])^{-1} \mathsf{N}.$$

Bibliography

[1] M. A. S. Aarons, *Mean curvature flow with a forcing term in Minkowski space*, Calc. Var. Partial Differential Equations **25** (2006), no. 2, 205–246, DOI 10.1007/s00526-005-0351-8. MR2188747

[2] U. Abresch and J. Langer, *The normalized curve shortening flow and homothetic solutions*, J. Differential Geom. **23** (1986), no. 2, 175–196. MR845704

[3] L. V. Ahlfors, *Complex analysis: An introduction to the theory of analytic functions of one complex variable*, 3rd ed., International Series in Pure and Applied Mathematics, McGraw-Hill Book Co., New York, 1978. MR510197

[4] R. Alessandroni and C. Sinestrari, *Convexity estimates for a nonhomogeneous mean curvature flow*, Math. Z. **266** (2010), no. 1, 65–82, DOI 10.1007/s00209-009-0554-3. MR2670672

[5] R. Alessandroni and C. Sinestrari, *Evolution of hypersurfaces by powers of the scalar curvature*, Ann. Sc. Norm. Super. Pisa Cl. Sci. (5) **9** (2010), no. 3, 541–571. MR2722655

[6] R. Alessandroni and C. Sinestrari, *Evolution of convex entire graphs by curvature flows*, Geom. Flows **1** (2015), no. 1, 111–125, DOI 10.1515/geofl-2015-0006. MR3420371

[7] J. W. Alexander, *A proof and extension of the Jordan-Brouwer separation theorem*, Trans. Amer. Math. Soc. **23** (1922), no. 4, 333–349, DOI 10.2307/1988883. MR1501206

[8] A. D. Alexandroff, *Almost everywhere existence of the second differential of a convex function and some properties of convex surfaces connected with it* (Russian), Leningrad State Univ. Annals [Uchenye Zapiski] Math. Ser. **6** (1939), 3–35. MR0003051

[9] A. D. Aleksandrov, *Uniqueness theorems for surfaces in the large. I* (Russian), Vestnik Leningrad. Univ. **11** (1956), no. 19, 5–17. MR0086338

[10] A. D. Aleksandrov, *Uniqueness theorems for surfaces in the large. II* (Russian, with English summary), Vestnik Leningrad. Univ. **12** (1957), no. 7, 15–44. MR0102111

[11] A. D. Aleksandrov, *Uniqueness theorems for surfaces in the large. III* (Russian, with English summary), Vestnik Leningrad. Univ. **13** (1958), no. 7, 14–26. MR0102112

[12] A. D. Aleksandrov, *Uniqueness theorems for surfaces in the large. V* (Russian, with English summary), Vestnik Leningrad. Univ. **13** (1958), no. 19, 5–8. MR0102114

[13] A. D. Aleksandrov and Ju. A. Volkov, *Uniqueness theorems for surfaces in the large. IV* (Russian, with English summary), Vestnik Leningrad. Univ. **13** (1958), no. 13, 27–34. MR0102113

[14] N. D. Alikakos and A. Freire, *The normalized mean curvature flow for a small bubble in a Riemannian manifold*, J. Differential Geom. **64** (2003), no. 2, 247–303. MR2029906

[15] W. K. Allard, *On the first variation of a varifold*, Ann. of Math. (2) **95** (1972), 417–491, DOI 10.2307/1970868. MR307015

[16] W. K. Allard, *On the first variation of a varifold: boundary behavior*, Ann. of Math. (2) **101** (1975), 418–446, DOI 10.2307/1970934. MR397520

[17] B. Allen, *IMCF and the stability of the PMT and RPI under L^2 convergence*, Ann. Henri Poincaré **19** (2018), no. 4, 1283–1306, DOI 10.1007/s00023-017-0641-7. MR3775157

[18] D. J. Altschuler, S. J. Altschuler, S. B. Angenent, and L. F. Wu, *The zoo of solitons for curve shortening in $\mathbb{R}^n$*, Nonlinearity **26** (2013), no. 5, 1189–1226, DOI 10.1088/0951-7715/26/5/1189. MR3043378

[19] S. J. Altschuler, *Singularities of the curve shrinking flow for space curves*, J. Differential Geom. **34** (1991), no. 2, 491–514. MR1131441

[20] S. Altschuler, S. B. Angenent, and Y. Giga, *Mean curvature flow through singularities for surfaces of rotation*, J. Geom. Anal. **5** (1995), no. 3, 293–358, DOI 10.1007/BF02921800. MR1360824

[21] S. J. Altschuler and M. A. Grayson, *Shortening space curves and flow through singularities*, J. Differential Geom. **35** (1992), no. 2, 283–298. MR1158337

[22] S. J. Altschuler and L. F. Wu, *Translating surfaces of the non-parametric mean curvature flow with prescribed contact angle*, Calc. Var. Partial Differential Equations **2** (1994), no. 1, 101–111, DOI 10.1007/BF01234317. MR1384396

[23] K. Anada, *Contraction of surfaces by harmonic mean curvature flows and nonuniqueness of their self similar solutions*, Calc. Var. Partial Differential Equations **12** (2001), no. 2, 109–116, DOI 10.1007/PL00009908. MR1821233

[24] K. Anada and T. Ishiwata, *Blow-up rates of solutions of initial-boundary value problems for a quasi-linear parabolic equation*, J. Differential Equations **262** (2017), no. 1, 181–271, DOI 10.1016/j.jde.2016.09.023. MR3567485

[25] M. T. Anderson, *Convergence and rigidity of manifolds under Ricci curvature bounds*, Invent. Math. **102** (1990), no. 2, 429–445, DOI 10.1007/BF01233434. MR1074481

[26] B. Andrews, *Contraction of convex hypersurfaces in Euclidean space*, Calc. Var. Partial Differential Equations **2** (1994), no. 2, 151–171, DOI 10.1007/BF01191340. MR1385524

[27] B. Andrews, *Contraction of convex hypersurfaces in Riemannian spaces*, J. Differential Geom. **39** (1994), no. 2, 407–431. MR1267897

[28] B. Andrews, *Entropy estimates for evolving hypersurfaces*, Comm. Anal. Geom. **2** (1994), no. 1, 53–64, DOI 10.4310/CAG.1994.v2.n1.a3. MR1312677

[29] B. Andrews, *Harnack inequalities for evolving hypersurfaces*, Math. Z. **217** (1994), no. 2, 179–197, DOI 10.1007/BF02571941. MR1296393

[30] B. Andrews, *Contraction of convex hypersurfaces by their affine normal*, J. Differential Geom. **43** (1996), no. 2, 207–230. MR1424425

[31] B. Andrews, *Monotone quantities and unique limits for evolving convex hypersurfaces*, Internat. Math. Res. Notices **20** (1997), 1001–1031, DOI 10.1155/S1073792897000640. MR1486693

[32] B. Andrews, *Evolving convex curves*, Calc. Var. Partial Differential Equations **7** (1998), no. 4, 315–371, DOI 10.1007/s005260050111. MR1660843

[33] B. Andrews, *Gauss curvature flow: the fate of the rolling stones*, Invent. Math. **138** (1999), no. 1, 151–161, DOI 10.1007/s002220050344. MR1714339

[34] B. Andrews, *Motion of hypersurfaces by Gauss curvature*, Pacific J. Math. **195** (2000), no. 1, 1–34, DOI 10.2140/pjm.2000.195.1. MR1781612

[35] B. Andrews, *Positively curved surfaces in the three-sphere*, Proceedings of the International Congress of Mathematicians, Vol. II (Beijing, 2002), Higher Ed. Press, Beijing, 2002, pp. 221–230. MR1957035

[36] B. Andrews, *Classification of limiting shapes for isotropic curve flows*, J. Amer. Math. Soc. **16** (2003), no. 2, 443–459, DOI 10.1090/S0894-0347-02-00415-0. MR1949167

[37] B. Andrews, *Fully nonlinear parabolic equations in two space variables*, arXiv:0402235v1, 2004.

[38] B. Andrews, *Pinching estimates and motion of hypersurfaces by curvature functions*, J. Reine Angew. Math. **608** (2007), 17–33, DOI 10.1515/CRELLE.2007.051. MR2339467

[39] B. Andrews, *Moving surfaces by non-concave curvature functions*, Calc. Var. Partial Differential Equations **39** (2010), no. 3-4, 649–657, DOI 10.1007/s00526-010-0329-z. MR2729317

[40] B. Andrews, *Gradient and oscillation estimates and their applications in geometric PDE*, Fifth International Congress of Chinese Mathematicians. Part 1, 2, AMS/IP Stud. Adv. Math., 51, pt. 1, vol. 2, Amer. Math. Soc., Providence, RI, 2012, pp. 3–19. MR2908056

[41] B. Andrews, *Noncollapsing in mean-convex mean curvature flow*, Geom. Topol. **16** (2012), no. 3, 1413–1418, DOI 10.2140/gt.2012.16.1413. MR2967056

[42] B. Andrews, *Moduli of continuity, isoperimetric profiles, and multi-point estimates in geometric heat equations*, Surveys in differential geometry 2014. Regularity and evolution of nonlinear equations, Surv. Differ. Geom., vol. 19, Int. Press, Somerville, MA, 2015, pp. 1–47, DOI 10.4310/SDG.2014.v19.n1.a1. MR3381494

[43] B. Andrews and C. Baker, *Mean curvature flow of pinched submanifolds to spheres*, J. Differential Geom. **85** (2010), no. 3, 357–395. MR2739807

[44] B. Andrews and P. Bryan, *A comparison theorem for the isoperimetric profile under curve-shortening flow*, Comm. Anal. Geom. **19** (2011), no. 3, 503–539, DOI 10.4310/CAG.2011.v19.n3.a3. MR2843240

[45] B. Andrews and P. Bryan, *Curvature bound for curve shortening flow via distance comparison and a direct proof of Grayson's theorem*, J. Reine Angew. Math. **653** (2011), 179–187, DOI 10.1515/CRELLE.2011.026. MR2794630

[46] B. Andrews and X. Chen, *Surfaces moving by powers of Gauss curvature*, Pure Appl. Math. Q. **8** (2012), no. 4, 825–834, DOI 10.4310/PAMQ.2012.v8.n4.a1. MR2959911

[47] B. Andrews and J. Clutterbuck, *Lipschitz bounds for solutions of quasilinear parabolic equations in one space variable*, J. Differential Equations **246** (2009), no. 11, 4268–4283, DOI 10.1016/j.jde.2009.01.024. MR2517770

[48] B. Andrews and J. Clutterbuck, *Time-interior gradient estimates for quasilinear parabolic equations*, Indiana Univ. Math. J. **58** (2009), no. 1, 351–380, DOI 10.1512/iumj.2009.58.3756. MR2504416

[49] B. Andrews and J. Clutterbuck, *Proof of the fundamental gap conjecture*, J. Amer. Math. Soc. **24** (2011), no. 3, 899–916, DOI 10.1090/S0894-0347-2011-00699-1. MR2784332

[50] B. Andrews and J. Clutterbuck, *Sharp modulus of continuity for parabolic equations on manifolds and lower bounds for the first eigenvalue*, Anal. PDE **6** (2013), no. 5, 1013–1024, DOI 10.2140/apde.2013.6.1013. MR3125548

[51] B. Andrews, P. Guan, and L. Ni, *Flow by powers of the Gauss curvature*, Adv. Math. **299** (2016), 174–201, DOI 10.1016/j.aim.2016.05.008. MR3519467

[52] B. Andrews and C. Hopper, *The Ricci flow in Riemannian geometry: A complete proof of the differentiable 1/4-pinching sphere theorem*, Lecture Notes in Mathematics, vol. 2011, Springer, Heidelberg, 2011. MR2760593

[53] B. Andrews and M. Langford, *Cylindrical estimates for hypersurfaces moving by convex curvature functions*, Anal. PDE **7** (2014), no. 5, 1091–1107, DOI 10.2140/apde.2014.7.1091. MR3265960

[54] B. Andrews and M. Langford, *Two-sided non-collapsing curvature flows*, Ann. Sc. Norm. Super. Pisa Cl. Sci. (5) **15** (2016), 543–560. MR3495438

[55] B. Andrews, M. Langford, and J. McCoy, *Non-collapsing in fully non-linear curvature flows*, Ann. Inst. H. Poincaré Anal. Non Linéaire **30** (2013), no. 1, 23–32, DOI 10.1016/j.anihpc.2012.05.003. MR3011290

[56] B. Andrews, M. Langford, and J. McCoy, *Convexity estimates for hypersurfaces moving by convex curvature functions*, Anal. PDE **7** (2014), no. 2, 407–433, DOI 10.2140/apde.2014.7.407. MR3218814

[57] B. Andrews, M. Langford, and J. McCoy, *Convexity estimates for surfaces moving by curvature functions*, J. Differential Geom. **99** (2015), no. 1, 47–75. MR3299822

[58] B. Andrews and H. Li, *Embedded constant mean curvature tori in the three-sphere*, J. Differential Geom. **99** (2015), no. 2, 169–189. MR3302037

[59] B. Andrews, H. Li, and Y. Wei, *$\mathcal{F}$-stability for self-shrinking solutions to mean curvature flow*, Asian J. Math. **18** (2014), no. 5, 757–777, DOI 10.4310/AJM.2014.v18.n5.a1. MR3287002

[60] B. Andrews and J. McCoy, *Convex hypersurfaces with pinched principal curvatures and flow of convex hypersurfaces by high powers of curvature*, Trans. Amer. Math. Soc. **364** (2012), no. 7, 3427–3447, DOI 10.1090/S0002-9947-2012-05375-X. MR2901219

[61] B. Andrews, J. McCoy, and Y. Zheng, *Contracting convex hypersurfaces by curvature*, Calc. Var. Partial Differential Equations **47** (2013), no. 3-4, 611–665, DOI 10.1007/s00526-012-0530-3. MR3070558

[62] S. Angenent, *Formal asymptotic expansions for symmetric ancient ovals in mean curvature flow*, Netw. Heterog. Media **8** (2013), no. 1, 1–8, DOI 10.3934/nhm.2013.8.1. MR3043925

[63] S. Angenent, P. Daskalopoulos, and N. Sesum, *Unique asymptotics of ancient convex mean curvature flow solutions*, J. Differential Geom. **111** (2019), no. 3, 381–455, DOI 10.4310/jdg/1552442605. MR3934596

[64] S. Angenent, *The zero set of a solution of a parabolic equation*, J. Reine Angew. Math. **390** (1988), 79–96, DOI 10.1515/crll.1988.390.79. MR953678

[65] S. Angenent, *Parabolic equations for curves on surfaces. I. Curves with p-integrable curvature*, Ann. of Math. (2) **132** (1990), no. 3, 451–483, DOI 10.2307/1971426. MR1078266

[66] S. Angenent, *On the formation of singularities in the curve shortening flow*, J. Differential Geom. **33** (1991), no. 3, 601–633. MR1100205

[67] S. Angenent, *Parabolic equations for curves on surfaces. II. Intersections, blow-up and generalized solutions*, Ann. of Math. (2) **133** (1991), no. 1, 171–215, DOI 10.2307/2944327. MR1087347

[68] S. B. Angenent, *Shrinking doughnuts*, Nonlinear diffusion equations and their equilibrium states, 3 (Gregynog, 1989), Progr. Nonlinear Differential Equations Appl., vol. 7, Birkhäuser Boston, Boston, MA, 1992, pp. 21–38. MR1167827

[69] S. B. Angenent, *Curve shortening and the topology of closed geodesics on surfaces*, Ann. of Math. (2) **162** (2005), no. 3, 1187–1241, DOI 10.4007/annals.2005.162.1187. MR2179729

[70] S. Angenent, P. Daskalopoulos, and N. Sesum, *Unique asymptotics of ancient convex mean curvature flow solutions*, J. Differential Geom. **111** (2019), no. 3, 381–455, DOI 10.4310/jdg/1552442605. MR3934596

[71] S. Angenent, T. Ilmanen, and D. L. Chopp, *A computed example of nonuniqueness of mean curvature flow in $\mathbf{R}^3$*, Comm. Partial Differential Equations **20** (1995), no. 11-12, 1937–1958, DOI 10.1080/03605309508821158. MR1361726

[72] S. B. Angenent, T. Ilmanen, and J. J. L. Velázquez, *Fattening from smooth initial data in mean curvature flow*, preprint.

[73] S. Angenent, G. Sapiro, and A. Tannenbaum, *On the affine heat equation for non-convex curves*, J. Amer. Math. Soc. **11** (1998), no. 3, 601–634, DOI 10.1090/S0894-0347-98-00262-8. MR1491538

[74] S. B. Angenent and J. J. L. Velázquez, *Asymptotic shape of cusp singularities in curve shortening*, Duke Math. J. **77** (1995), no. 1, 71–110, DOI 10.1215/S0012-7094-95-07704-7. MR1317628

[75] S. B. Angenent and J. J. L. Velázquez, *Degenerate neckpinches in mean curvature flow*, J. Reine Angew. Math. **482** (1997), 15–66, DOI 10.1515/crll.1997.482.15. MR1427656

[76] S. B. Angenent and Q. You, *Ancient solutions to curve shortening with finite total curvature*, arXiv:1803.01399v1, 2018.

[77] M. Athanassenas, *Volume-preserving mean curvature flow of rotationally symmetric surfaces*, Comment. Math. Helv. **72** (1997), no. 1, 52–66, DOI 10.1007/PL00000366. MR1456315

[78] M. Athanassenas, *Behaviour of singularities of the rotationally symmetric, volume-preserving mean curvature flow*, Calc. Var. Partial Differential Equations **17** (2003), no. 1, 1–16, DOI 10.1007/s00526-002-0098-4. MR1979113

[79] M. Athanassenas and S. Kandanaarachchi, *Convergence of axially symmetric volume-preserving mean curvature flow*, Pacific J. Math. **259** (2012), no. 1, 41–54, DOI 10.2140/pjm.2012.259.41. MR2988482

[80] T. K.-K. Au, *On the saddle point property of Abresch-Langer curves under the curve shortening flow*, Comm. Anal. Geom. **18** (2010), no. 1, 1–21, DOI 10.4310/CAG.2010.v18.n1.a1. MR2660456

[81] C. Baker, *The mean curvature flow of submanifolds of high codimension*, Australian National University, 2011. Thesis (Ph.D.)–Australian National University.

[82] C. Baker and H. T. Nguyen, *Codimension two surfaces pinched by normal curvature evolving by mean curvature flow*, Ann. Inst. H. Poincaré Anal. Non Linéaire **34** (2017), no. 6, 1599–1610, DOI 10.1016/j.anihpc.2016.10.010. MR3712012

[83] P. Baldi, E. Haus, and C. Mantegazza, *Networks self-similarly moving by curvature with two triple junctions*, Atti Accad. Naz. Lincei Rend. Lincei Mat. Appl. **28** (2017), no. 2, 323–338, DOI 10.4171/RLM/765. MR3649351

[84] P. Baldi, E. Haus, and C. Mantegazza, *On the classification of networks self-similarly moving by curvature*, Geom. Flows **2** (2017), no. 1, 125–137, DOI 10.1515/geofl-2017-0006. MR3745455

[85] P. Baldi, E. Haus, and C. Mantegazza, *Non-existence of theta-shaped self-similarly shrinking networks moving by curvature*, Comm. Partial Differential Equations **43** (2018), no. 3, 403–427, DOI 10.1080/03605302.2018.1446162. MR3804202

[86] J. M. Ball, *Differentiability properties of symmetric and isotropic functions*, Duke Math. J. **51** (1984), no. 3, 699–728, DOI 10.1215/S0012-7094-84-05134-2. MR757959

[87] J. Bernstein and T. Mettler, *Characterizing classical minimal surfaces via the entropy differential*, J. Geom. Anal. **27** (2017), no. 3, 2235–2268, DOI 10.1007/s12220-017-9759-6. MR3667429

[88] J. Bernstein and L. Wang, *A topological property of asymptotically conical self-shrinkers of small entropy*, Duke Math. J. **166** (2017), no. 3, 403–435, DOI 10.1215/00127094-3715082. MR3606722

[89] J. Bernstein and L. Wang, *An integer degree for asymptotically conical self-expanders*, arXiv:1807.06494v1, 2018.

[90] M. C. Bertini and C. Sinestrari, *Volume preserving flow by powers of symmetric polynomials in the principal curvatures*, Math. Z. **289** (2018), no. 3-4, 1219–1236, DOI 10.1007/s00209-017-1995-8. MR3830246

[91] M. C. Bertini and C. Sinestrari, *Volume-preserving nonhomogeneous mean curvature flow of convex hypersurfaces*, Ann. Mat. Pura Appl. (4) **197** (2018), no. 4, 1295–1309, DOI 10.1007/s10231-018-0725-0. MR3829571

[92] W. Blaschke, *Vorlesungen über Differentialgeometrie und geometrische Grundlagen von Einsteins Relativitätstheorie II*, 1 ed., vol. 7 of *Grundlehren der mathematischen Wissenschaften*. Julius Springer Verlag, 1923.

[93] C. Böhm and B. Wilking, *Manifolds with positive curvature operators are space forms*, Ann. of Math. (2) **167** (2008), no. 3, 1079–1097, DOI 10.4007/annals.2008.167.1079. MR2415394

[94] W. M. Boothby, *An introduction to differentiable manifolds and Riemannian geometry*, 2nd ed., Pure and Applied Mathematics, vol. 120, Academic Press, Inc., Orlando, FL, 1986. MR861409

[95] T. Bourni, *Allard-type boundary regularity for $C^{1,\alpha}$ boundaries*, Adv. Calc. Var. **9** (2016), no. 2, 143–161, DOI 10.1515/acv-2014-0032. MR3483600

[96] T. Bourni and M. Langford, *Type-II singularities of two-convex immersed mean curvature flow*, previously published in 2 (2016), Geom. Flows **2** (2017), no. 1, 1–17, DOI 10.1515/geofl-2016-0001. MR3562950

[97] T. Bourni, M. Langford, and G. Tinaglia, *A collapsing ancient solution of mean curvature flow in $\mathbb{R}^3$*, arXiv:1705.06981v2, 2017.

[98] T. Bourni, M. Langford, and G. Tinaglia, *On the existence of translating solutions of mean curvature flow in slab regions*, to appear in Anal. PDE, arXiv:1805.05173v3, 2018.

[99] T. Bourni, M. Langford, and G. Tinaglia *Convex ancient solutions to curve shortening flow*, arXiv:1903.02022v1, 2019.

[100] T. Bourni, M. Langford, and G. Tinaglia, *Convex ancient solutions to mean curvature flow*, arXiv:1907.03932v1, 2019.

[101] K. A. Brakke, *The motion of a surface by its mean curvature*, Mathematical Notes, vol. 20, Princeton University Press, Princeton, N.J., 1978. MR485012

[102] H. L. Bray, *Proof of the Riemannian Penrose inequality using the positive mass theorem*, J. Differential Geom. **59** (2001), no. 2, 177–267. MR1908823

[103] H. L. Bray, *On the positive mass, Penrose, and ZAS inequalities in general dimension*, Surveys in geometric analysis and relativity, Adv. Lect. Math. (ALM), vol. 20, Int. Press, Somerville, MA, 2011, pp. 1–27. MR2906919

[104] S. Brendle, *Embedded minimal tori in S^3 and the Lawson conjecture*, Acta Math. **211** (2013), no. 2, 177–190, DOI 10.1007/s11511-013-0101-2. MR3143888

[105] S. Brendle, *Embedded Weingarten tori in S^3*, Adv. Math. **257** (2014), 462–475, DOI 10.1016/j.aim.2014.02.025. MR3187655

[106] S. Brendle, *Two-point functions and their applications in geometry*, Bull. Amer. Math. Soc. (N.S.) **51** (2014), no. 4, 581–596, DOI 10.1090/S0273-0979-2014-01461-2. MR3237760

[107] S. Brendle, *A sharp bound for the inscribed radius under mean curvature flow*, Invent. Math. **202** (2015), no. 1, 217–237, DOI 10.1007/s00222-014-0570-8. MR3402798

[108] S. Brendle, *Embedded self-similar shrinkers of genus 0*, Ann. of Math. (2) **183** (2016), no. 2, 715–728, DOI 10.4007/annals.2016.183.2.6. MR3450486

[109] S. Brendle, *An inscribed radius estimate for mean curvature flow in Riemannian manifolds*, Ann. Sc. Norm. Super. Pisa Cl. Sci. (5) **16** (2016), no. 4, 1447–1472. MR3616339

[110] S. Brendle, *Ancient solutions to the Ricci flow in dimension* 3, arXiv:1811.02559v2, 2018.

[111] S. Brendle and K. Choi, *Uniqueness of convex ancient solutions to mean curvature flow in higher dimensions*, arXiv:1804.00018v2, 2018.

[112] S. Brendle and K. Choi, *Uniqueness of convex ancient solutions to mean curvature flow in $\mathbb{R}^3$*, Invent. Math. **217** (2019), no. 1, 35–76, DOI 10.1007/s00222-019-00859-4. MR3958790

[113] S. Brendle, K. Choi, and P. Daskalopoulos, *Asymptotic behavior of flows by powers of the Gaussian curvature*, Acta Math. **219** (2017), no. 1, 1–16, DOI 10.4310/ACTA.2017.v219.n1.a1. MR3765656

[114] S. Brendle and G. Huisken, *Mean curvature flow with surgery of mean convex surfaces in $\mathbb{R}^3$*, Invent. Math. **203** (2016), no. 2, 615–654, DOI 10.1007/s00222-015-0599-3. MR3455158

[115] S. Brendle and G. Huisken, *A fully nonlinear flow for two-convex hypersurfaces in Riemannian manifolds*, Invent. Math. **210** (2017), no. 2, 559–613, DOI 10.1007/s00222-017-0736-2. MR3714512

[116] S. Brendle and P.-K. Hung, *A sharp inscribed radius estimate for fully nonlinear flows*, Amer. J. Math. **141** (2019), no. 1, 41–53, DOI 10.1353/ajm.2019.0001. MR3904766

[117] S. Brendle and R. Schoen, *Manifolds with 1/4-pinched curvature are space forms*, J. Amer. Math. Soc. **22** (2009), no. 1, 287–307, DOI 10.1090/S0894-0347-08-00613-9. MR2449060

[118] P. Breuning, *Immersions with bounded second fundamental form*, J. Geom. Anal. **25** (2015), no. 2, 1344–1386, DOI 10.1007/s12220-014-9472-7. MR3319975

[119] P. Broadbridge and P. Vassiliou, *The role of symmetry and separation in surface evolution and curve shortening*, SIGMA Symmetry Integrability Geom. Methods Appl. **7** (2011), Paper 052, 19, DOI 10.3842/SIGMA.2011.052. MR2804584

[120] L. Bronsard and F. Reitich, *On three-phase boundary motion and the singular limit of a vector-valued Ginzburg-Landau equation*, Arch. Rational Mech. Anal. **124** (1993), no. 4, 355–379, DOI 10.1007/BF00375607. MR1240580

[121] M. Brown, *A proof of the generalized Schoenflies theorem*, Bull. Amer. Math. Soc. **66** (1960), 74–76, DOI 10.1090/S0002-9904-1960-10400-4. MR117695

[122] P. Bryan and M. Ivaki, *Harnack estimate for mean curvature flow on the sphere*, arXiv:1508.02821v4, 2015.

[123] P. Bryan, M. Ivaki, and J. Scheuer, *On the classification of ancient solutions to curvature flows on the sphere*, arXiv:1604.01694v2, 2016.

[124] P. Bryan, M. Ivaki, and J. Scheuer, *Harnack inequalities for curvature flows in Riemannian and Lorentzian manifolds*, arXiv:1703.07493v1, 2017.

[125] P. Bryan, M. N. Ivaki, and J. Scheuer, *Harnack inequalities for evolving hypersurfaces on the sphere*, Comm. Anal. Geom. **26** (2018), no. 5, 1047–1077, DOI 10.4310/CAG.2018.v26.n5.a2. MR3900479

[126] P. Bryan and J. Louie, *Classification of convex ancient solutions to curve shortening flow on the sphere*, J. Geom. Anal. **26** (2016), no. 2, 858–872, DOI 10.1007/s12220-015-9574-x. MR3472819

[127] J. A. Buckland, *Mean curvature flow with free boundary on smooth hypersurfaces*, J. Reine Angew. Math. **586** (2005), 71–90, DOI 10.1515/crll.2005.2005.586.71. MR2180601

[128] J. A. Buckland, *Short-time existence of solutions to the cross curvature flow on 3-manifolds*, Proc. Amer. Math. Soc. **134** (2006), no. 6, 1803–1807, DOI 10.1090/S0002-9939-05-08204-3. MR2207496

[129] E. Cabezas-Rivas and V. Miquel, *Volume preserving mean curvature flow in the hyperbolic space*, Indiana Univ. Math. J. **56** (2007), no. 5, 2061–2086, DOI 10.1512/iumj.2007.56.3060. MR2359723

[130] E. Cabezas-Rivas and V. Miquel, *Volume-preserving mean curvature flow of revolution hypersurfaces in a rotationally symmetric space*, Math. Z. **261** (2009), no. 3, 489–510, DOI 10.1007/s00209-008-0333-6. MR2471083

[131] E. Cabezas-Rivas and V. Miquel, *Volume preserving mean curvature flow of revolution hypersurfaces between two equidistants*, Calc. Var. Partial Differential Equations **43** (2012), no. 1-2, 185–210, DOI 10.1007/s00526-011-0408-9. MR2886115

[132] E. Cabezas-Rivas and V. Miquel, *Non-preserved curvature conditions under constrained mean curvature flows*, Differential Geom. Appl. **49** (2016), 287–300, DOI 10.1016/j.difgeo.2016.08.006. MR3573835

[133] E. Cabezas-Rivas and C. Sinestrari, *Volume-preserving flow by powers of the mth mean curvature*, Calc. Var. Partial Differential Equations **38** (2010), no. 3-4, 441–469, DOI 10.1007/s00526-009-0294-6. MR2647128

[134] E. Calabi, *An extension of E. Hopf's maximum principle with an application to Riemannian geometry*, Duke Math. J. **25** (1958), 45–56. MR92069

[135] J. R. Cannon, *The one-dimensional heat equation*, with a foreword by Felix E. Browder, Encyclopedia of Mathematics and its Applications, vol. 23, Addison-Wesley Publishing Company, Advanced Book Program, Reading, MA, 1984. MR747979

[136] M. P. Cavalcante and J. M. Espinar, *Halfspace type theorems for self-shrinkers*, Bull. Lond. Math. Soc. **48** (2016), no. 2, 242–250, DOI 10.1112/blms/bdv099. MR3483061

[137] I. Chavel, *Eigenvalues in Riemannian geometry*, including a chapter by Burton Randol; with an appendix by Jozef Dodziuk, Pure and Applied Mathematics, vol. 115, Academic Press, Inc., Orlando, FL, 1984. MR768584

[138] I. Chavel, *Riemannian geometry: A modern introduction*, 2nd ed., Cambridge Studies in Advanced Mathematics, vol. 98, Cambridge University Press, Cambridge, 2006. MR2229062

[139] J. Cheeger, *Finiteness theorems for Riemannian manifolds*, Amer. J. Math. **92** (1970), 61–74, DOI 10.2307/2373498. MR263092

[140] J. Cheeger and D. G. Ebin, *Comparison theorems in Riemannian geometry*, revised reprint of the 1975 original, AMS Chelsea Publishing, Providence, RI, 2008. MR2394158

[141] B.-L. Chen and X.-P. Zhu, *Complete Riemannian manifolds with pointwise pinched curvature*, Invent. Math. **140** (2000), no. 2, 423–452, DOI 10.1007/s002220000061. MR1757002

[142] J. Chen and J. Li, *Singularity of mean curvature flow of Lagrangian submanifolds*, Invent. Math. **156** (2004), no. 1, 25–51, DOI 10.1007/s00222-003-0332-5. MR2047657

[143] Y. G. Chen, Y. Giga, and S. Goto, *Uniqueness and existence of viscosity solutions of generalized mean curvature flow equations*, J. Differential Geom. **33** (1991), no. 3, 749–786. MR1100211

[144] S. Y. Cheng, *Eigenfunctions and nodal sets*, Comment. Math. Helv. **51** (1976), no. 1, 43–55, DOI 10.1007/BF02568142. MR397805

[145] F. Chini and N. M. Møller, *Ancient mean curvature flows and their spacetime tracks*, arXiv:1901.05481v2, 2019.

[146] F Chini and N. M. Møller, *Bi-halfspace and convex hull theorems for translating solitons*, arXiv:1809.01069v2, 2019.

[147] O. Chodosh and F. Schulze, *Uniqueness of asymptotically conical tangent flows*, arXiv:1901.06369v1, 2019.

[148] K. Choi and P. Daskalopoulos, *Uniqueness of closed self-similar solutions to the Gauss curvature flow*, arXiv:1609.05487v1, 2016.

[149] K. Choi, P. Daskalopoulos, L. Kim, and K.-A. Lee, *The evolution of complete non-compact graphs by powers of Gauss curvature*, J. Reine Angew. Math. **757** (2019), 131–158, DOI 10.1515/crelle-2017-0032. MR4036572

[150] K. Choi, P. Daskalopoulos, and K.-A. Lee, *Translating solutions to the Gauss curvature flow with flat sides*, arXiv:1610.07206v2, 2016.

[151] K. Choi, R. Haslhofer, and O. Hershkovits, *Ancient low entropy flows, mean convex neighborhoods, and uniqueness*, arXiv:1810.08467v1, 2018.

[152] K. Choi and C. Mantoulidis, *Ancient gradient flows of elliptic functionals and Morse index*, arXiv:1902.07697v2, 2019.

[153] D. L. Chopp, *Computation of self-similar solutions for mean curvature flow*, Experiment. Math. **3** (1994), no. 1, 1–15. MR1302814

[154] K.-S. Chou and X.-J. Wang, *A logarithmic Gauss curvature flow and the Minkowski problem*, Ann. Inst. H. Poincaré Anal. Non Linéaire **17** (2000), no. 6, 733–751, DOI 10.1016/S0294-1449(00)00053-6. MR1804653

[155] K.-S. Chou and X.-P. Zhu, *A convexity theorem for a class of anisotropic flows of plane curves*, Indiana Univ. Math. J. **48** (1999), no. 1, 139–154, DOI 10.1512/iumj.1999.48.1273. MR1722196

[156] K.-S. Chou and X.-P. Zhu, *The curve shortening problem*, Chapman & Hall/CRC, Boca Raton, FL, 2001. MR1888641

[157] B. Chow, *Deforming convex hypersurfaces by the nth root of the Gaussian curvature*, J. Differential Geom. **22** (1985), no. 1, 117–138. MR826427

[158] B. Chow, *Deforming convex hypersurfaces by the square root of the scalar curvature*, Invent. Math. **87** (1987), no. 1, 63–82, DOI 10.1007/BF01389153. MR862712

[159] B. Chow, *On Harnack's inequality and entropy for the Gaussian curvature flow*, Comm. Pure Appl. Math. **44** (1991), no. 4, 469–483, DOI 10.1002/cpa.3160440405. MR1100812

[160] B. Chow, *Geometric aspects of Aleksandrov reflection and gradient estimates for parabolic equations*, Comm. Anal. Geom. **5** (1997), no. 2, 389–409, DOI 10.4310/CAG.1997.v5.n2.a5. MR1483984

[161] B. Chow and S.-C. Chu, *A geometric interpretation of Hamilton's Harnack inequality for the Ricci flow*, Math. Res. Lett. **2** (1995), no. 6, 701–718, DOI 10.4310/MRL.1995.v2.n6.a4. MR1362964

[162] B. Chow and S.-C. Chu, *Space-time formulation of Harnack inequalities for curvature flows of hypersurfaces*, J. Geom. Anal. **11** (2001), no. 2, 219–231, DOI 10.1007/BF02921963. MR1856176

[163] B. Chow, S.-C. Chu, D. Glickenstein, C. Guenther, J. Isenberg, T. Ivey, D. Knopf, P. Lu, F. Luo, and L. Ni, *The Ricci flow: techniques and applications. Parts I–IV*, Mathematical Surveys and Monographs, vols. 135, 144, 163, 206, American Mathematical Society, Providence, RI, 2007, 2008, 2010, 2015.

[164] B. Chow and R. Gulliver, *Aleksandrov reflection and nonlinear evolution equations. I. The n-sphere and n-ball*, Calc. Var. Partial Differential Equations **4** (1996), no. 3, 249–264, DOI 10.1007/BF01254346. MR1386736

[165] B. Chow and R. Gulliver, *Aleksandrov reflection and geometric evolution of hypersurfaces*, Comm. Anal. Geom. **9** (2001), no. 2, 261–280, DOI 10.4310/CAG.2001.v9.n2.a2. MR1846204

[166] B. Chow and R. S. Hamilton, *The cross curvature flow of 3-manifolds with negative sectional curvature*, Turkish J. Math. **28** (2004), no. 1, 1–10. MR2055396

[167] B. Chow and D. Knopf, *The Ricci flow: an introduction*, Mathematical Surveys and Monographs, vol. 110, American Mathematical Society, Providence, RI, 2004. MR2061425

[168] B. Chow, L.-P. Liou, and D.-H. Tsai, *Expansion of embedded curves with turning angle greater than $-\pi$*, Invent. Math. **123** (1996), no. 3, 415–429, DOI 10.1007/s002220050034. MR1383955

[169] B. Chow, L.-P. Liou, and D.-H. Tsai, *On the nonlinear parabolic equation $\partial_t u = F(\Delta u + nu)$ on S^n*, Comm. Anal. Geom. **4** (1996), no. 3, 415–434, DOI 10.4310/CAG.1996.v4.n3.a3. MR1415750

[170] B. Chow, P. Lu, and L. Ni, *Hamilton's Ricci flow*, Graduate Studies in Mathematics, vol. 77, American Mathematical Society, Providence, RI; Science Press Beijing, New York, 2006. MR2274812

[171] B. Chow and D.-H. Tsai, *Geometric expansion of convex plane curves*, J. Differential Geom. **44** (1996), no. 2, 312–330. MR1425578

[172] B. Chow and D.-H. Tsai, *Expansion of convex hypersurfaces by nonhomogeneous functions of curvature*, Asian J. Math. **1** (1997), no. 4, 769–784, DOI 10.4310/AJM.1997.v1.n4.a7. MR1621575

[173] B. Chow and D.-H. Tsai, *Nonhomogeneous Gauss curvature flows*, Indiana Univ. Math. J. **47** (1998), no. 3, 965–994, DOI 10.1512/iumj.1998.47.1546. MR1665729

[174] J. Clutterbuck, *Parabolic equations with continuous initial data*, PhD thesis, The Australian National University, 2005.

[175] J. Clutterbuck and O. C. Schnürer, *Stability of mean convex cones under mean curvature flow*, Math. Z. **267** (2011), no. 3-4, 535–547, DOI 10.1007/s00209-009-0634-4. MR2776047

[176] J. Clutterbuck, O. C. Schnürer, and F. Schulze, *Stability of translating solutions to mean curvature flow*, Calc. Var. Partial Differential Equations **29** (2007), no. 3, 281–293, DOI 10.1007/s00526-006-0033-1. MR2321890

[177] T. H. Colding, T. Ilmanen, and W. P. Minicozzi II, *Rigidity of generic singularities of mean curvature flow*, Publ. Math. Inst. Hautes Études Sci. **121** (2015), 363–382, DOI 10.1007/s10240-015-0071-3. MR3349836

[178] T. H. Colding, T. Ilmanen, W. P. Minicozzi II, and B. White, *The round sphere minimizes entropy among closed self-shrinkers*, J. Differential Geom. **95** (2013), no. 1, 53–69. MR3128979

[179] T. H. Colding and W. P. Minicozzi II, *Minimal surfaces*, Courant Lecture Notes in Mathematics, vol. 4, New York University, Courant Institute of Mathematical Sciences, New York, 1999. MR1683966

[180] T. H. Colding and W. P. Minicozzi II, *Sharp estimates for mean curvature flow of graphs*, J. Reine Angew. Math. **574** (2004), 187–195, DOI 10.1515/crll.2004.069. MR2099114

[181] T. H. Colding and W. P. Minicozzi II, *Generic mean curvature flow I: generic singularities*, Ann. of Math. (2) **175** (2012), no. 2, 755–833, DOI 10.4007/annals.2012.175.2.7. MR2993752

[182] T. H. Colding and W. P. Minicozzi II, *Smooth compactness of self-shrinkers*, Comment. Math. Helv. **87** (2012), no. 2, 463–475, DOI 10.4171/CMH/260. MR2914856

[183] T. H. Colding and W. P. Minicozzi II, *Uniqueness of blowups and Łojasiewicz inequalities*, Ann. of Math. (2) **182** (2015), no. 1, 221–285, DOI 10.4007/annals.2015.182.1.5. MR3374960

[184] T. H. Colding and W. P. Minicozzi II, *Differentiability of the arrival time*, Comm. Pure Appl. Math. **69** (2016), no. 12, 2349–2363, DOI 10.1002/cpa.21635. MR3570481

[185] T. H. Colding and W. P. Minicozzi II, *Regularity of the level set flow*, Comm. Pure Appl. Math. **71** (2018), no. 4, 814–824, DOI 10.1002/cpa.21703. MR3772402

[186] T. H. Colding and W. P. Minicozzi, II, *Entropy and codimension bounds for generic singularities*, arXiv:1906.07609v2, 2019.

[187] T. H. Colding and W. P. Minicozzi, II, *Optimal bounds for ancient caloric functions*, arXiv:1902.01736v1, 2019.

[188] T. H. Colding, W. P. Minicozzi II, and E. K. Pedersen, *Mean curvature flow*, Bull. Amer. Math. Soc. (N.S.) **52** (2015), no. 2, 297–333, DOI 10.1090/S0273-0979-2015-01468-0. MR3312634

[189] A. A. Cooper, *A characterization of the singular time of the mean curvature flow*, Proc. Amer. Math. Soc. **139** (2011), no. 8, 2933–2942, DOI 10.1090/S0002-9939-2010-10714-1. MR2801634

[190] A. A. Cooper, *Mean curvature flow in higher codimension*, ProQuest LLC, Ann Arbor, MI, 2011. Thesis (Ph.D.)–Michigan State University. MR2941847

[191] P. Daskalopoulos and R. Hamilton, *The free boundary in the Gauss curvature flow with flat sides*, J. Reine Angew. Math. **510** (1999), 187–227, DOI 10.1515/crll.1999.046. MR1696096

[192] P. Daskalopoulos, R. Hamilton, and N. Sesum, *Classification of compact ancient solutions to the curve shortening flow*, J. Differential Geom. **84** (2010), no. 3, 455–464. MR2669361

[193] P. Daskalopoulos and G. Huisken, *Inverse mean curvature evolution of entire graphs*, arXiv:1709.06665v1, 2017.

[194] P. Daskalopoulos and K.-A. Lee, *Worn stones with flat sides all time regularity of the interface*, Invent. Math. **156** (2004), no. 3, 445–493, DOI 10.1007/s00222-003-0328-1. MR2061326

[195] J. Dávila, M. del Pino, and X. H. Nguyen, *Finite topology self-translating surfaces for the mean curvature flow in* $\mathbb{R}^3$, Adv. Math. **320** (2017), 674–729, DOI 10.1016/j.aim.2017.09.014. MR3709119

[196] Q. Ding, *Minimal cones and self-expanding solutions for mean curvature flows*, arXiv:1503.02612v3, 2015.

[197] F. Dittberner, *Curve flows with a global forcing term*, arXiv:1809.08643v1, 2018.

[198] M. P. do Carmo, *Riemannian geometry*, translated from the second Portuguese edition by Francis Flaherty, Mathematics: Theory & Applications, Birkhäuser Boston, Inc., Boston, MA, 1992. MR1138207

[199] G. Drugan, *Self-shrinking Solutions to Mean Curvature Flow*, ProQuest LLC, Ann Arbor, MI, 2014. Thesis (Ph.D.)–University of Washington. MR3271860

[200] G. Drugan, *An immersed S^2 self-shrinker*, Trans. Amer. Math. Soc. **367** (2015), no. 5, 3139–3159, DOI 10.1090/S0002-9947-2014-06051-0. MR3314804

[201] G. Drugan and S. J. Kleene, *Immersed self-shrinkers*, Trans. Amer. Math. Soc. **369** (2017), no. 10, 7213–7250, DOI 10.1090/tran/6907. MR3683108

[202] G. Drugan, H. Lee, and X. H. Nguyen, *A survey of closed self-shrinkers with symmetry*, Results Math. **73** (2018), no. 1, Art. 32, 32, DOI 10.1007/s00025-018-0763-3. MR3763107

[203] G. Drugan and X. H. Nguyen, *Mean curvature flow of entire graphs evolving away from the heat flow*, Proc. Amer. Math. Soc. **145** (2017), no. 2, 861–869, DOI 10.1090/proc/13238. MR3577885

[204] K. Ecker, *A local monotonicity formula for mean curvature flow*, Ann. of Math. (2) **154** (2001), no. 2, 503–525, DOI 10.2307/3062105. MR1865979

[205] K. Ecker, *Regularity theory for mean curvature flow*, Progress in Nonlinear Differential Equations and their Applications, vol. 57, Birkhäuser Boston, Inc., Boston, MA, 2004. MR2024995

[206] K. Ecker, *Local monotonicity formulas for some nonlinear diffusion equations*, Calc. Var. Partial Differential Equations **23** (2005), no. 1, 67–81, DOI 10.1007/s00526-004-0290-9. MR2133662

[207] K. Ecker and G. Huisken, *Immersed hypersurfaces with constant Weingarten curvature*, Math. Ann. **283** (1989), no. 2, 329–332, DOI 10.1007/BF01446438. MR980601

[208] K. Ecker and G. Huisken, *Mean curvature evolution of entire graphs*, Ann. of Math. (2) **130** (1989), no. 3, 453–471, DOI 10.2307/1971452. MR1025164

[209] K. Ecker and G. Huisken, *Interior estimates for hypersurfaces moving by mean curvature*, Invent. Math. **105** (1991), no. 3, 547–569, DOI 10.1007/BF01232278. MR1117150

[210] K. Ecker, D. Knopf, L. Ni, and P. Topping, *Local monotonicity and mean value formulas for evolving Riemannian manifolds*, J. Reine Angew. Math. **616** (2008), 89–130, DOI 10.1515/CRELLE.2008.019. MR2369488

[211] N. Edelen, *Convexity estimates for mean curvature flow with free boundary*, Adv. Math. **294** (2016), 1–36, DOI 10.1016/j.aim.2016.02.026. MR3479560

[212] C. L. Epstein and M. I. Weinstein, *A stable manifold theorem for the curve shortening equation*, Comm. Pure Appl. Math. **40** (1987), no. 1, 119–139, DOI 10.1002/cpa.3160400106. MR865360

[213] J.-H. Eschenburg, *Local convexity and nonnegative curvature—Gromov's proof of the sphere theorem*, Invent. Math. **84** (1986), no. 3, 507–522, DOI 10.1007/BF01388744. MR837525

[214] J. Escher and G. Simonett, *The volume preserving mean curvature flow near spheres*, Proc. Amer. Math. Soc. **126** (1998), no. 9, 2789–2796, DOI 10.1090/S0002-9939-98-04727-3. MR1485470

[215] L. C. Evans, *Classical solutions of fully nonlinear, convex, second-order elliptic equations*, Comm. Pure Appl. Math. **35** (1982), no. 3, 333–363, DOI 10.1002/cpa.3160350303. MR649348

[216] L. C. Evans, *Partial differential equations*, 2nd ed., Graduate Studies in Mathematics, vol. 19, American Mathematical Society, Providence, RI, 2010. MR2597943

[217] L. C. Evans, *A strong maximum principle for parabolic systems in a convex set with arbitrary boundary*, Proc. Amer. Math. Soc. **138** (2010), no. 9, 3179–3185, DOI 10.1090/S0002-9939-2010-10495-1. MR2653943

[218] L. C. Evans and R. F. Gariepy, *Measure theory and fine properties of functions*, revised edition, Textbooks in Mathematics, CRC Press, Boca Raton, FL, 2015. MR3409135

[219] L. C. Evans and J. Spruck, *Motion of level sets by mean curvature. I*, J. Differential Geom. **33** (1991), no. 3, 635–681. MR1100206

[220] L. C. Evans and J. Spruck, *Motion of level sets by mean curvature. III*, J. Geom. Anal. **2** (1992), no. 2, 121–150, DOI 10.1007/BF02921385. MR1151756

[221] H. Federer, *Geometric measure theory*, Die Grundlehren der mathematischen Wissenschaften, Band 153, Springer-Verlag New York Inc., New York, 1969. MR0257325

[222] W. J. Firey, *Shapes of worn stones*, Mathematika **21** (1974), 1–11, DOI 10.1112/S0025579300005714. MR362045

[223] F. T.-H. Fong and P. McGrath, *Rotational symmetry of asymptotically conical mean curvature flow self-expanders*, Comm. Anal. Geom. **27** (2019), no. 3, 599–618, DOI 10.4310/CAG.2019.v27.n3.a3. MR4003004

[224] A. Freire, *Mean curvature motion of graphs with constant contact angle at a free boundary*, Anal. PDE **3** (2010), no. 4, 359–407, DOI 10.2140/apde.2010.3.359. MR2718258

[225] A. Freire, *Mean curvature motion of triple junctions of graphs in two dimensions*, Comm. Partial Differential Equations **35** (2010), no. 2, 302–327, DOI 10.1080/03605300903419775. MR2748626

[226] A. Friedman, *Partial differential equations of parabolic type*, Prentice-Hall, Inc., Englewood Cliffs, N.J., 1964. MR0181836

[227] C. Frohman and W. H. Meeks III, *The topological uniqueness of complete one-ended minimal surfaces and Heegaard surfaces in* $\mathbf{R}^3$, J. Amer. Math. Soc. **10** (1997), no. 3, 495–512, DOI 10.1090/S0894-0347-97-00215-4. MR1443545

[228] M. Gage and R. S. Hamilton, *The heat equation shrinking convex plane curves*, J. Differential Geom. **23** (1986), no. 1, 69–96. MR840401

[229] M. E. Gage, *An isoperimetric inequality with applications to curve shortening*, Duke Math. J. **50** (1983), no. 4, 1225–1229, DOI 10.1215/S0012-7094-83-05052-4. MR726325

[230] M. E. Gage, *Curve shortening makes convex curves circular*, Invent. Math. **76** (1984), no. 2, 357–364, DOI 10.1007/BF01388602. MR742856

[231] M. E. Gage, *Curve shortening on surfaces*, Ann. Sci. École Norm. Sup. (4) **23** (1990), no. 2, 229–256. MR1046497

[232] M. E. Gage, *Deforming curves on convex surfaces to simple closed geodesics*, Indiana Univ. Math. J. **39** (1990), no. 4, 1037–1059, DOI 10.1512/iumj.1990.39.39049. MR1087184

[233] E. Gama and F. Martín *Translating solitons of the mean curvature flow asymptotic to hyperplanes in* $\mathbb{R}^{n+1}$, arXiv:1802.08468v3, 2018.

[234] S. Gao, H. Li, and H. Ma, *Uniqueness of closed self-similar solutions to* σ_k^α*-curvature flow*, NoDEA Nonlinear Differential Equations Appl. **25** (2018), no. 5, Art. 45, 26, DOI 10.1007/s00030-018-0535-5. MR3845754

[235] R. J. Gardner, *Geometric tomography*, 2nd ed., Encyclopedia of Mathematics and its Applications, vol. 58, Cambridge University Press, New York, 2006. MR2251886

[236] C. Gerhardt, *Flow of nonconvex hypersurfaces into spheres*, J. Differential Geom. **32** (1990), no. 1, 299–314. MR1064876

[237] C. Gerhardt, *Curvature problems*, Series in Geometry and Topology, vol. 39, International Press, Somerville, MA, 2006. MR2284727

[238] C. Gerhardt, *Inverse curvature flows in hyperbolic space*, J. Differential Geom. **89** (2011), no. 3, 487–527. MR2879249

[239] C. Gerhardt, *Non-scale-invariant inverse curvature flows in Euclidean space*, Calc. Var. Partial Differential Equations **49** (2014), no. 1-2, 471–489, DOI 10.1007/s00526-012-0589-x. MR3148124

[240] C. Gerhardt, *Curvature flows in the sphere*, J. Differential Geom. **100** (2015), no. 2, 301–347. MR3343834

[241] R. Geroch, *Energy extraction*, Ann. New York Acad. Sci. 224 (1973), 108–117.

[242] B. Gidas, W. M. Ni, and L. Nirenberg, *Symmetry and related properties via the maximum principle*, Comm. Math. Phys. **68** (1979), no. 3, 209–243. MR544879

[243] B. Gidas, W. M. Ni, and L. Nirenberg, *Symmetry of positive solutions of nonlinear elliptic equations in* $\mathbf{R}^n$, Mathematical analysis and applications, Part A, Adv. in Math. Suppl. Stud., vol. 7, Academic Press, New York-London, 1981, pp. 369–402. MR634248

[244] Y. Giga and R. V. Kohn, *Asymptotically self-similar blow-up of semilinear heat equations*, Comm. Pure Appl. Math. **38** (1985), no. 3, 297–319, DOI 10.1002/cpa.3160380304. MR784476

[245] D. Gilbarg and N. S. Trudinger, *Elliptic partial differential equations of second order*, reprint of the 1998 edition, Classics in Mathematics, Springer-Verlag, Berlin, 2001. MR1814364

[246] G. Glaeser, *Fonctions composées différentiables* (French), Ann. of Math. (2) **77** (1963), 193–209, DOI 10.2307/1970204. MR143058

[247] M. A. Grayson, *The heat equation shrinks embedded plane curves to round points*, J. Differential Geom. **26** (1987), no. 2, 285–314. MR906392

[248] M. A. Grayson, *Shortening embedded curves*, Ann. of Math. (2) **129** (1989), no. 1, 71–111, DOI 10.2307/1971486. MR979601

[249] R. E. Greene and H. Wu, *Lipschitz convergence of Riemannian manifolds*, Pacific J. Math. **131** (1988), no. 1, 119–141. MR917868

[250] K. Groh, M. Schwarz, K. Smoczyk, and K. Zehmisch, *Mean curvature flow of monotone Lagrangian submanifolds*, Math. Z. **257** (2007), no. 2, 295–327, DOI 10.1007/s00209-007-0126-3. MR2324804

[251] M. Gromov, *Structures métriques pour les variétés riemanniennes* (French), edited by J. Lafontaine and P. Pansu, Textes Mathématiques [Mathematical Texts], vol. 1, CEDIC, Paris, 1981. MR682063

[252] B. Guan, *Mean curvature motion of nonparametric hypersurfaces with contact angle condition*, Elliptic and parabolic methods in geometry (Minneapolis, MN, 1994), A K Peters, Wellesley, MA, 1996, pp. 47–56. MR1417947

[253] P. Guan and J. Li, *The quermassintegral inequalities for k-convex starshaped domains*, Adv. Math. **221** (2009), no. 5, 1725–1732, DOI 10.1016/j.aim.2009.03.005. MR2522433

[254] P. Guan and L. Ni, *Entropy and a convergence theorem for Gauss curvature flow in high dimension*, J. Eur. Math. Soc. (JEMS) **19** (2017), no. 12, 3735–3761, DOI 10.4171/JEMS/752. MR3730513

[255] V. Guillemin and A. Pollack, *Differential topology*, reprint of the 1974 original, AMS Chelsea Publishing, Providence, RI, 2010. MR2680546

[256] H. Guo, R. Philipowski, and A. Thalmaier, *A note on Chow's entropy functional for the Gauss curvature flow* (English, with English and French summaries), C. R. Math. Acad. Sci. Paris **351** (2013), no. 21-22, 833–835, DOI 10.1016/j.crma.2013.10.003. MR3128971

[257] J. Hadamard, *Sur certaines propriétés des trajectoires en dynamique*, J. Math. Pures Appl. 3 (1897), 331–388.

[258] H. P. Halldorsson, *Self-similar solutions to the curve shortening flow*, Trans. Amer. Math. Soc. **364** (2012), no. 10, 5285–5309, DOI 10.1090/S0002-9947-2012-05632-7. MR2931330

[259] R. S. Hamilton, *Three-manifolds with positive Ricci curvature*, J. Differential Geometry **17** (1982), no. 2, 255–306. MR664497

[260] R. S. Hamilton, *Four-manifolds with positive curvature operator*, J. Differential Geom. **24** (1986), no. 2, 153–179. MR862046

[261] R. S. Hamilton, *Eternal solutions to the Ricci flow*, J. Differential Geom. **38** (1993), no. 1, 1–11. MR1231700

[262] R. S. Hamilton, *The Harnack estimate for the Ricci flow*, J. Differential Geom. **37** (1993), no. 1, 225–243. MR1198607

[263] R. S. Hamilton, *Monotonicity formulas for parabolic flows on manifolds*, Comm. Anal. Geom. **1** (1993), no. 1, 127–137, DOI 10.4310/CAG.1993.v1.n1.a7. MR1230277

[264] R. S. Hamilton, *Convex hypersurfaces with pinched second fundamental form*, Comm. Anal. Geom. **2** (1994), no. 1, 167–172, DOI 10.4310/CAG.1994.v2.n1.a10. MR1312684

[265] R. S. Hamilton, *Remarks on the entropy and Harnack estimates for the Gauss curvature flow*, Comm. Anal. Geom. **2** (1994), no. 1, 155–165, DOI 10.4310/CAG.1994.v2.n1.a9. MR1312683

[266] R. S. Hamilton, *Worn stones with flat sides*, A tribute to Ilya Bakelman (College Station, TX, 1993), Discourses Math. Appl., vol. 3, Texas A & M Univ., College Station, TX, 1994, pp. 69–78. MR1423370

[267] R. S. Hamilton, *A compactness property for solutions of the Ricci flow*, Amer. J. Math. **117** (1995), no. 3, 545–572, DOI 10.2307/2375080. MR1333936

[268] R. S. Hamilton, *The formation of singularities in the Ricci flow*, Surveys in differential geometry, Vol. II (Cambridge, MA, 1993), Int. Press, Cambridge, MA, 1995, pp. 7–136. MR1375255

[269] R. S. Hamilton, *Harnack estimate for the mean curvature flow*, J. Differential Geom. **41** (1995), no. 1, 215–226. MR1316556

[270] R. S. Hamilton, *Isoperimetric estimates for the curve shrinking flow in the plane*, Modern methods in complex analysis (Princeton, NJ, 1992), Ann. of Math. Stud., vol. 137, Princeton Univ. Press, Princeton, NJ, 1995, pp. 201–222, DOI 10.1016/1053-8127(94)00130-3. MR1369140

[271] R. S. Hamilton, *Four-manifolds with positive isotropic curvature*, Comm. Anal. Geom. **5** (1997), no. 1, 1–92, DOI 10.4310/CAG.1997.v5.n1.a1. MR1456308

[272] Q. Han, *Deforming convex hypersurfaces by curvature functions*, Analysis **17** (1997), no. 2-3, 113–127, DOI 10.1524/anly.1997.17.23.113. MR1486359

[273] R. Haslhofer, *Uniqueness of the bowl soliton*, Geom. Topol. **19** (2015), no. 4, 2393–2406, DOI 10.2140/gt.2015.19.2393. MR3375531

[274] R. Haslhofer and O. Hershkovits, *Ancient solutions of the mean curvature flow*, Comm. Anal. Geom. **24** (2016), no. 3, 593–604, DOI 10.4310/CAG.2016.v24.n3.a6. MR3521319

[275] R. Haslhofer and B. Kleiner, *On Brendle's estimate for the inscribed radius under mean curvature flow*, Int. Math. Res. Not. IMRN **15** (2015), 6558–6561, DOI 10.1093/imrn/rnu139. MR3384488

[276] R. Haslhofer and B. Kleiner, *Mean curvature flow of mean convex hypersurfaces*, Comm. Pure Appl. Math. **70** (2017), no. 3, 511–546, DOI 10.1002/cpa.21650. MR3602529

[277] R. Haslhofer and B. Kleiner, *Mean curvature flow with surgery*, Duke Math. J. **166** (2017), no. 9, 1591–1626, DOI 10.1215/00127094-0000008X. MR3662439

[278] A. Hatcher, *Notes on basic 3-manifold topology.* http://pi.math.cornell.edu/~hatcher/3M/3M.pdf, 2000.

[279] L. Hauswirth and F. Pacard, *Higher genus Riemann minimal surfaces*, Invent. Math. **169** (2007), no. 3, 569–620, DOI 10.1007/s00222-007-0056-z. MR2336041

[280] J. Head, *On the mean curvature evolution of two-convex hypersurfaces*, J. Differential Geom. **94** (2013), no. 2, 241–266. MR3080482

[281] M. Heidusch, *Zur Regularität des Inversen Mittleren Krümmungsflusses*, PhD thesis, Eberhard-Karls-Universität Tübingen, 2001.

[282] S. Helmensdorfer, *A model for the behavior of fluid droplets based on mean curvature flow*, SIAM J. Math. Anal. **44** (2012), no. 3, 1359–1371, DOI 10.1137/110824905. MR2982716

[283] S. Helmensdorfer and P. Topping, *The geometry of differential Harnack estimates*, Actes du séminaire de Théorie spectrale et géométrie 30 (2011-2012), 77–89.

[284] O. Hershkovits, *Translators asymptotic to cylinders*, arXiv:1805.10553v1, 2018.

[285] N. J. Hicks, *Notes on differential geometry*, Van Nostrand Mathematical Studies, No. 3, D. Van Nostrand Co., Inc., Princeton, N.J.-Toronto-London, 1965. MR0179691

[286] M. W. Hirsch, *Differential topology*, corrected reprint of the 1976 original, Graduate Texts in Mathematics, vol. 33, Springer-Verlag, New York, 1994. MR1336822

[287] D. Hoffman, T. Ilmanen, F. Martín, and B. White, *Graphical translators for mean curvature flow*, Calc. Var. Partial Differential Equations **58** (2019), no. 4, Art. 117, 29, DOI 10.1007/s00526-019-1560-x. MR3962912

[288] D. Hoffman, F. Martín, and B. White, *Scherk-like translators for mean curvature flow*, arXiv:1903.04617v3, 2018.

[289] D. Hoffman and W. H. Meeks III, *The strong halfspace theorem for minimal surfaces*, Invent. Math. **101** (1990), no. 2, 373–377, DOI 10.1007/BF01231506. MR1062966

[290] Y. Huang, E. Lutwak, D. Yang, and G. Zhang, *Geometric measures in the dual Brunn-Minkowski theory and their associated Minkowski problems*, Acta Math. **216** (2016), no. 2, 325–388, DOI 10.1007/s11511-016-0140-6. MR3573332

[291] G. Huisken, *Flow by mean curvature of convex surfaces into spheres*, J. Differential Geom. **20** (1984), no. 1, 237–266. MR772132

[292] G. Huisken, *Ricci deformation of the metric on a Riemannian manifold*, J. Differential Geom. **21** (1985), no. 1, 47–62. MR806701

[293] G. Huisken, *Contracting convex hypersurfaces in Riemannian manifolds by their mean curvature*, Invent. Math. **84** (1986), no. 3, 463–480, DOI 10.1007/BF01388742. MR837523

[294] G. Huisken, *The volume preserving mean curvature flow*, J. Reine Angew. Math. **382** (1987), 35–48, DOI 10.1515/crll.1987.382.35. MR921165

[295] G. Huisken, *Nonparametric mean curvature evolution with boundary conditions*, J. Differential Equations **77** (1989), no. 2, 369–378, DOI 10.1016/0022-0396(89)90149-6. MR983300

[296] G. Huisken, *Asymptotic behavior for singularities of the mean curvature flow*, J. Differential Geom. **31** (1990), no. 1, 285–299. MR1030675

[297] G. Huisken, *Local and global behaviour of hypersurfaces moving by mean curvature*, Differential geometry: partial differential equations on manifolds (Los Angeles, CA, 1990), Proc. Sympos. Pure Math., vol. 54, Amer. Math. Soc., Providence, RI, 1993, pp. 175–191, DOI 10.1090/pspum/054.1/1216584. MR1216584

[298] G. Huisken, *A distance comparison principle for evolving curves*, Asian J. Math. **2** (1998), no. 1, 127–133, DOI 10.4310/AJM.1998.v2.n1.a2. MR1656553

[299] G. Huisken and T. Ilmanen, *The inverse mean curvature flow and the Riemannian Penrose inequality*, J. Differential Geom. **59** (2001), no. 3, 353–437. MR1916951

[300] G. Huisken and A. Polden, *Geometric evolution equations for hypersurfaces*, Calculus of variations and geometric evolution problems (Cetraro, 1996), Lecture Notes in Math., vol. 1713, Springer, Berlin, 1999, pp. 45–84, DOI 10.1007/BFb0092669. MR1731639

[301] G. Huisken and C. Sinestrari, *Convexity estimates for mean curvature flow and singularities of mean convex surfaces*, Acta Math. **183** (1999), no. 1, 45–70, DOI 10.1007/BF02392946. MR1719551

[302] G. Huisken and C. Sinestrari, *Mean curvature flow singularities for mean convex surfaces*, Calc. Var. Partial Differential Equations **8** (1999), no. 1, 1–14, DOI 10.1007/s005260050113. MR1666878

[303] G. Huisken and C. Sinestrari, *Mean curvature flow with surgeries of two-convex hypersurfaces*, Invent. Math. **175** (2009), no. 1, 137–221, DOI 10.1007/s00222-008-0148-4. MR2461428

[304] G. Huisken and C. Sinestrari, *Convex ancient solutions of the mean curvature flow*, J. Differential Geom. **101** (2015), no. 2, 267–287. MR3399098

[305] J. Hutchinson, *Poincaré-Sobolev and related inequalities for submanifolds of* $\mathbf{R}^N$, Pacific J. Math. **145** (1990), no. 1, 59–69. MR1066398

[306] T. Ilmanen, *Convergence of the Allen-Cahn equation to Brakke's motion by mean curvature*, J. Differential Geom. **38** (1993), no. 2, 417–461. MR1237490

[307] T. Ilmanen, *The level-set flow on a manifold*, Differential geometry: partial differential equations on manifolds (Los Angeles, CA, 1990), Proc. Sympos. Pure Math., vol. 54, Amer. Math. Soc., Providence, RI, 1993, pp. 193–204, DOI 10.1090/pspum/054.1/1216585. MR1216585

[308] T. Ilmanen, *Elliptic regularization and partial regularity for motion by mean curvature*, Mem. Amer. Math. Soc. **108** (1994), no. 520, x+90, DOI 10.1090/memo/0520. MR1196160

[309] T. Ilmanen, *Singularities of mean curvature flow of surfaces*, https://people.math.ethz.ch/∼ilmanen/papers/sing.ps, 1995.

[310] T. Ilmanen, *Lectures on mean curvature flow and related equations*, https://people.math.ethz.ch/∼ilmanen/papers/notes.pdf, 1998.

[311] T. Ilmanen, *Problems in mean curvature flow*, https://people.math.ethz.ch/∼ilmanen/classes/eil03/problems03.pdf, 2003.

[312] T. Ilmanen, A. Neves, and F. Schulze, *On short time existence for the planar network flow*, J. Differential Geom. **111** (2019), no. 1, 39–89, DOI 10.4310/jdg/1547607687. MR3909904

[313] N. Ishimura, *Curvature evolution of plane curves with prescribed opening angle*, Bull. Austral. Math. Soc. **52** (1995), no. 2, 287–296, DOI 10.1017/S0004972700014714. MR1348488

[314] N. Ishimura, *Self-similar solutions for the Gauss curvature evolution of rotationally symmetric surfaces*, Nonlinear Anal. **33** (1998), no. 1, 97–104, DOI 10.1016/S0362-546X(97)00539-7. MR1623054

[315] M. N. Ivaki, *An application of dual convex bodies to the inverse Gauss curvature flow*, Proc. Amer. Math. Soc. **143** (2015), no. 3, 1257–1271, DOI 10.1090/S0002-9939-2014-12314-8. MR3293740

[316] M. N. Ivaki, *Deforming a hypersurface by Gauss curvature and support function*, J. Funct. Anal. **271** (2016), no. 8, 2133–2165, DOI 10.1016/j.jfa.2016.07.003. MR3539348

[317] M. N. Ivaki, *Deforming a hypersurface by principal radii of curvature and support function*, Calc. Var. Partial Differential Equations **58** (2019), no. 1, Art. 1, 18, DOI 10.1007/s00526-018-1462-3. MR3880311

[318] N. M. Ivochkina, Th. Nehring, and F. Tomi, *Evolution of starshaped hypersurfaces by nonhomogeneous curvature functions*, Algebra i Analiz **12** (2000), no. 1, 185–203; English transl., St. Petersburg Math. J. **12** (2001), no. 1, 145–160. MR1758567

[319] P. S. Jang and R. M. Wald, *The positive energy conjecture and the cosmic censor hypothesis*, J. Mathematical Phys. **18** (1977), no. 1, 41–44, DOI 10.1063/1.523134. MR523907

[320] F. John, *Extremum problems with inequalities as subsidiary conditions*, Studies and Essays Presented to R. Courant on His 60th Birthday, January 8, 1948, Interscience Publishers, Inc., New York, N. Y., 1948, pp. 187–204. MR0030135

[321] F. John, *Partial differential equations*, 4th ed., Applied Mathematical Sciences, vol. 1, Springer-Verlag, New York, 1991. MR1185075

[322] D. Joyce, *Conjectures on Bridgeland stability for Fukaya categories of Calabi-Yau manifolds, special Lagrangians, and Lagrangian mean curvature flow*, EMS Surv. Math. Sci. **2** (2015), no. 1, 1–62, DOI 10.4171/EMSS/8. MR3354954

[323] N. Kapouleas, *Complete constant mean curvature surfaces in Euclidean three-space*, Ann. of Math. (2) **131** (1990), no. 2, 239–330, DOI 10.2307/1971494. MR1043269

[324] N. Kapouleas, *Compact constant mean curvature surfaces in Euclidean three-space*, J. Differential Geom. **33** (1991), no. 3, 683–715. MR1100207

[325] N. Kapouleas, *Complete embedded minimal surfaces of finite total curvature*, J. Differential Geom. **47** (1997), no. 1, 95–169. MR1601434

[326] N. Kapouleas, S. J. Kleene, and N. M. Møller, *Mean curvature self-shrinkers of high genus: non-compact examples*, J. Reine Angew. Math. **739** (2018), 1–39, DOI 10.1515/crelle-2015-0050. MR3808256

[327] K. Kasai and Y. Tonegawa, *A general regularity theory for weak mean curvature flow*, Calc. Var. Partial Differential Equations **50** (2014), no. 1-2, 1–68, DOI 10.1007/s00526-013-0626-4. MR3194675

[328] A. Katsuda, *Gromov's convergence theorem and its application*, Nagoya Math. J. **100** (1985), 11–48, DOI 10.1017/S0027763000000209. MR818156

[329] D. Ketover, *Self-shrinking platonic solids*, arXiv:1602.07271v1, 2016.

[330] D. Kinderlehrer and G. Stampacchia, *An introduction to variational inequalities and their applications*, reprint of the 1980 original, Classics in Applied Mathematics, vol. 31, Society for Industrial and Applied Mathematics (SIAM), Philadelphia, PA, 2000. MR1786735

[331] J. R. King, *Emerging areas of mathematical modelling, Science into the next millennium: young scientists give their visions of the future, Part II,* R. Soc. Lond. Philos. Trans. Ser. A Math. Phys. Eng. Sci. **358** (2000), no. 1765, 3–19, DOI 10.1098/rsta.2000.0516. MR1759647

[332] D. A. Klain and G.-C. Rota, *Introduction to geometric probability*, Lezioni Lincee. [Lincei Lectures], Cambridge University Press, Cambridge, 1997. MR1608265

[333] S. Kleene and N. M. Møller, *Self-shrinkers with a rotational symmetry*, Trans. Amer. Math. Soc. **366** (2014), no. 8, 3943–3963, DOI 10.1090/S0002-9947-2014-05721-8. MR3206448

[334] R. V. Kohn and S. Serfaty, *A deterministic-control-based approach to motion by curvature*, Comm. Pure Appl. Math. **59** (2006), no. 3, 344–407, DOI 10.1002/cpa.20101. MR2200259

[335] N. J. Korevaar, *Convex solutions to nonlinear elliptic and parabolic boundary value problems*, Indiana Univ. Math. J. **32** (1983), no. 4, 603–614, DOI 10.1512/iumj.1983.32.32042. MR703287

[336] N. J. Korevaar, R. Kusner, and B. Solomon, *The structure of complete embedded surfaces with constant mean curvature*, J. Differential Geom. **30** (1989), no. 2, 465–503. MR1010168

[337] B. Kotschwar, *Harnack inequalities for evolving convex surfaces from the space-time perspective*, math.la.asu.edu/ kotschwar/pub, 2009.

[338] S. N. Kružkov, *Nonlinear parabolic equations with two independent variables* (Russian), Trudy Moskov. Mat. Obšč. **16** (1967), 329–346. MR0226208

[339] N. V. Krylov, *Boundedly inhomogeneous elliptic and parabolic equations* (Russian), Izv. Akad. Nauk SSSR Ser. Mat. **46** (1982), no. 3, 487–523, 670. MR661144

[340] N. V. Krylov, *Nonlinear elliptic and parabolic equations of the second order*, translated from the Russian by P. L. Buzytsky [P. L. Buzytskiĭ], Mathematics and its Applications (Soviet Series), vol. 7, D. Reidel Publishing Co., Dordrecht, 1987. MR901759

[341] N. V. Krylov, *Lectures on elliptic and parabolic equations in Hölder spaces*, Graduate Studies in Mathematics, vol. 12, American Mathematical Society, Providence, RI, 1996. MR1406091

[342] N. V. Krylov, *Lectures on elliptic and parabolic equations in Sobolev spaces*, Graduate Studies in Mathematics, vol. 96, American Mathematical Society, Providence, RI, 2008. MR2435520

[343] O. A. Ladyženskaja, V. A. Solonnikov, and N. N. Ural′ceva, *Linear and quasilinear equations of parabolic type* (Russian), translated from the Russian by S. Smith, translations of Mathematical Monographs, Vol. 23, American Mathematical Society, Providence, R.I., 1968. MR0241822

[344] A. Lahiri, *A new version of Brakke's local regularity theorem*, arXiv:1601.06710v2, 2016.

[345] A. Lahiri, *Equality of the usual definitions of Brakke flow*, arXiv:1705.08789v1, 2017.

[346] B. Lambert and J. Scheuer, *The inverse mean curvature flow perpendicular to the sphere*, Math. Ann. **364** (2016), no. 3-4, 1069–1093, DOI 10.1007/s00208-015-1248-2. MR3466860

[347] B. Lambert and J. Scheuer, *A geometric inequality for convex free boundary hypersurfaces in the unit ball*, Proc. Amer. Math. Soc. **145** (2017), no. 9, 4009–4020, DOI 10.1090/proc/13516. MR3665052

[348] B. Lambert, J. Lotay, and F. Schulze, *Acient solutions in Lagrangian mean curvature flow*, preprint, arXiv.org:1901.05383.

[349] J. Langer, *A compactness theorem for surfaces with L_p-bounded second fundamental form*, Math. Ann. **270** (1985), no. 2, 223–234, DOI 10.1007/BF01456183. MR771980

[350] M. Langford, *Motion of hypersurfaces by curvature*, Bull. Aust. Math. Soc. **92** (2015), no. 3, 516–517, DOI 10.1017/S000497271500091X. MR3415631

[351] M. Langford, *The optimal interior ball estimate for a k-convex mean curvature flow*, Proc. Amer. Math. Soc. **143** (2015), no. 12, 5395–5398, DOI 10.1090/proc/12624. MR3411154

[352] M. Langford, *A general pinching principle for mean curvature flow and applications*, Calc. Var. Partial Differential Equations **56** (2017), no. 4, Art. 107, 31, DOI 10.1007/s00526-017-1193-x. MR3669776

[353] M. Langford and S. Lynch, *Sharp one-sided curvature estimates for fully nonlinear curvature flows and applications to ancient solutions*, J. Reine. Angew. Math. (to appear).

[354] J. Lauer, *Convergence of mean curvature flows with surgery*, Comm. Anal. Geom. **21** (2013), no. 2, 355–363, DOI 10.4310/CAG.2013.v21.n2.a4. MR3043750

[355] J. Lauer, *A new length estimate for curve shortening flow and low regularity initial data*, Geom. Funct. Anal. **23** (2013), no. 6, 1934–1961, DOI 10.1007/s00039-013-0248-1. MR3132906

[356] H. B. Lawson Jr., *Local rigidity theorems for minimal hypersurfaces*, Ann. of Math. (2) **89** (1969), 187–197, DOI 10.2307/1970816. MR238229

[357] H. B. Lawson Jr., *The unknottedness of minimal embeddings*, Invent. Math. **11** (1970), 183–187, DOI 10.1007/BF01404649. MR287447

[358] N. Q. Le and N. Sesum, *The mean curvature at the first singular time of the mean curvature flow*, Ann. Inst. H. Poincaré Anal. Non Linéaire **27** (2010), no. 6, 1441–1459, DOI 10.1016/j.anihpc.2010.09.002. MR2738327

[359] N. Q. Le and N. Sesum, *Blow-up rate of the mean curvature during the mean curvature flow and a gap theorem for self-shrinkers*, Comm. Anal. Geom. **19** (2011), no. 4, 633–659, DOI 10.4310/CAG.2011.v19.n4.a1. MR2880211

[360] N. Q. Le and N. Sesum, *On the extension of the mean curvature flow*, Math. Z. **267** (2011), no. 3-4, 583–604, DOI 10.1007/s00209-009-0637-1. MR2776050

[361] J. M. Lee, *Riemannian manifolds: An introduction to curvature*, Graduate Texts in Mathematics, vol. 176, Springer-Verlag, New York, 1997. MR1468735

[362] G. Li and Y. Lv, *Contracting convex hypersurfaces in space form by non-homogeneous curvature function*, The Journal of Geometric Analysis (Jan 2019).

[363] H. Li, X. Wang, and Y. Wei, *Surfaces expanding by non-concave curvature functions*, Ann. Global Anal. Geom. **55** (2019), no. 2, 243–279, DOI 10.1007/s10455-018-9625-1. MR3923539

[364] P. Li and S.-T. Yau, *On the parabolic kernel of the Schrödinger operator*, Acta Math. **156** (1986), no. 3-4, 153–201, DOI 10.1007/BF02399203. MR834612

[365] Q.-R. Li, *Surfaces expanding by the power of the Gauss curvature flow*, Proc. Amer. Math. Soc. **138** (2010), no. 11, 4089–4102, DOI 10.1090/S0002-9939-2010-10431-8. MR2679630

[366] Q.-R. Li, W. Sheng, and X.-J. Wang, *Flow by Gauss curvature to the Aleksandrov and dual Minkowski problems*, arXiv:1712.07774v1, 2017.

[367] X. Li and K. Wang, *Nonparametric hypersurfaces moving by powers of Gauss curvature*, Michigan Math. J. **66** (2017), no. 4, 675–682, DOI 10.1307/mmj/1508810813. MR3720319

[368] G. M. Lieberman, *Second order parabolic differential equations*, World Scientific Publishing Co., Inc., River Edge, NJ, 1996. MR1465184

[369] L. Lin, *Mean curvature flow of star-shaped hypersurfaces*, arXiv:1508.01225, 2015.

[370] L. Lin and N. Sesum, *Blow-up of the mean curvature at the first singular time of the mean curvature flow*, Calc. Var. Partial Differential Equations **55** (2016), no. 3, Art. 65, 16, DOI 10.1007/s00526-016-1003-x. MR3509039

[371] Y.-C. Lin and D.-H. Tsai, *Using Aleksandrov reflection to estimate the location of the center of expansion*, Proc. Amer. Math. Soc. **138** (2010), no. 2, 557–565, DOI 10.1090/S0002-9939-09-10155-7. MR2557173

[372] P. Lu and J. Zhou, *Ancient solutions for Andrews' hypersurface flow*, arXiv:1812.04926v1, 2018.

[373] S. L. Lukyanov, E. S. Vitchev, and A. B. Zamolodchikov, *Integrable model of boundary interaction: the paperclip*, Nuclear Phys. B **683** (2004), no. 3, 423–454, DOI 10.1016/j.nuclphysb.2004.02.010. MR2057110

[374] A. Lunardi, *Analytic semigroups and optimal regularity in parabolic problems*, [2013 reprint of the 1995 original] [MR1329547], Modern Birkhäuser Classics, Birkhäuser/Springer Basel AG, Basel, 1995. MR3012216

[375] L. Lusternik and L. Schnirelmann, *Sur le problème de trois géodésiques fermées sur les surfaces de genre 0*, C. R. Acad. Sci. Paris 189 (1929), 269–271.

[376] S. Lynch and H. Nguyen, *Pinched ancient solutions to the high codimension mean curvature flow*, arxiv:1709.09697v1, 2017.

[377] E. Mäder-Baumdicker, *The area preserving curve shortening flow with Neumann free boundary conditions*, Geom. Flows **1** (2015), no. 1, 34–79, DOI 10.1515/geofl-2015-0004. MR3351503

[378] E. Mäder-Baumdicker, *Singularities of the area preserving curve shortening flow with a free boundary condition*, Math. Ann. **371** (2018), no. 3-4, 1429–1448, DOI 10.1007/s00208-017-1637-9. MR3831277

[379] A. Magni, C. Mantegazza, and M. Novaga, *Motion by curvature of planar networks, II*, Ann. Sc. Norm. Super. Pisa Cl. Sci. (5) **15** (2016), 117–144. MR3495423

[380] M. Makowski and J. Scheuer, *Rigidity results, inverse curvature flows and Alexandrov-Fenchel type inequalities in the sphere*, Asian J. Math. **20** (2016), no. 5, 869–892, DOI 10.4310/AJM.2016.v20.n5.a2. MR3622318

[381] C. Mantegazza, *Evolution by curvature of networks of curves in the plane*, joint project with Matteo Novaga and Vincenzo Maria Tortorelli, Variational problems in Riemannian geometry, Progr. Nonlinear Differential Equations Appl., vol. 59, Birkhäuser, Basel, 2004, pp. 95–109. MR2076269

[382] C. Mantegazza, *Lecture notes on mean curvature flow*, Progress in Mathematics, vol. 290, Birkhäuser/Springer Basel AG, Basel, 2011. MR2815949

[383] C. Mantegazza and L. Martinazzi, *A note on quasilinear parabolic equations on manifolds*, Ann. Sc. Norm. Super. Pisa Cl. Sci. (5) **11** (2012), no. 4, 857–874. MR3060703

[384] C. Mantegazza, M. Novaga, and A. Pluda, *Motion by curvature of networks with two triple junctions*, previously published in 2 (2016), Geom. Flows **2** (2017), no. 1, 18–48, DOI 10.1515/geofl-2016-0002. MR3565976

[385] C. Mantegazza, M. Novaga, and V. M. Tortorelli, *Motion by curvature of planar networks*, Ann. Sc. Norm. Super. Pisa Cl. Sci. (5) **3** (2004), no. 2, 235–324. MR2075985

[386] C. Margerin, *Pointwise pinched manifolds are space forms*, Geometric measure theory and the calculus of variations (Arcata, Calif., 1984), Proc. Sympos. Pure Math., vol. 44, Amer. Math. Soc., Providence, RI, 1986, pp. 307–328, DOI 10.1090/pspum/044/840282. MR840282

[387] T. Marquardt, *Inverse mean curvature flow for star-shaped hypersurfaces evolving in a cone*, J. Geom. Anal. **23** (2013), no. 3, 1303–1313, DOI 10.1007/s12220-011-9288-7. MR3078355

[388] T. Marquardt, *Weak solutions of inverse mean curvature flow for hypersurfaces with boundary*, J. Reine Angew. Math. **728** (2017), 237–261, DOI 10.1515/crelle-2014-0116. MR3668996

[389] F. Martín, J. Pérez-García, A. Savas-Halilaj, and K. Smoczyk, *A characterization of the grim reaper cylinder*, J. Reine Angew. Math. **746** (2019), 209–234, DOI 10.1515/crelle-2016-0011. MR3895630

[390] F. Martín, A. Savas-Halilaj, and K. Smoczyk, *On the topology of translating solitons of the mean curvature flow*, Calc. Var. Partial Differential Equations **54** (2015), no. 3, 2853–2882, DOI 10.1007/s00526-015-0886-2. MR3412395

[391] B. Mazur, *On embeddings of spheres*, Bull. Amer. Math. Soc. **65** (1959), 59–65, DOI 10.1090/S0002-9904-1959-10274-3. MR117693

[392] B. C. Mazur, *On embeddings of spheres*, Acta Math. **105** (1961), 1–17, DOI 10.1007/BF02559532. MR125570

[393] R. Mazzeo and M. Saez, *Self-similar expanding solutions for the planar network flow* (English, with English and French summaries), Analytic aspects of problems in Riemannian geometry: elliptic PDEs, solitons and computer imaging, Sémin. Congr., vol. 22, Soc. Math. France, Paris, 2011, pp. 159–173. MR3060453

[394] J. McCoy, *The surface area preserving mean curvature flow*, Asian J. Math. **7** (2003), no. 1, 7–30, DOI 10.4310/AJM.2003.v7.n1.a2. MR2015239

[395] J. A. McCoy, *The mixed volume preserving mean curvature flow*, Math. Z. **246** (2004), no. 1-2, 155–166, DOI 10.1007/s00209-003-0592-1. MR2031450

[396] J. A. McCoy, *Self-similar solutions of fully nonlinear curvature flows*, Ann. Sc. Norm. Super. Pisa Cl. Sci. (5) **10** (2011), no. 2, 317–333. MR2856150

[397] P. McGrath, *Closed mean curvature self-shrinking surfaces of generalized rotational type*, arXiv:1507.00681v1, 2015.

[398] W. W. Meeks III and S. T. Yau, *The existence of embedded minimal surfaces and the problem of uniqueness*, Math. Z. **179** (1982), no. 2, 151–168, DOI 10.1007/BF01214308. MR645492

[399] F. Merle and H. Zaag, *Optimal estimates for blowup rate and behavior for nonlinear heat equations*, Comm. Pure Appl. Math. **51** (1998), no. 2, 139–196, DOI 10.1002/(SICI)1097-0312(199802)51:2⟨139::AID-CPA2⟩3.0.CO;2-C. MR1488298

[400] J. H. Michael and L. M. Simon, *Sobolev and mean-value inequalities on generalized submanifolds of R^n*, Comm. Pure Appl. Math. **26** (1973), 361–379, DOI 10.1002/cpa.3160260305. MR344978

[401] J. W. Milnor, *Topology from the differentiable viewpoint*, based on notes by David W. Weaver; revised reprint of the 1965 original, Princeton Landmarks in Mathematics, Princeton University Press, Princeton, NJ, 1997. MR1487640

[402] N. M. Møller, *Closed self-shrinking surfaces in $\mathbb{R}^3$ via the torus*, arXiv:1111.7318v2, 2011.

[403] F. Morgan, *Geometric measure theory: A beginner's guide*, 5th ed., illustrated by James F. Bredt, Elsevier/Academic Press, Amsterdam, 2016. MR3497381

[404] J. Morgan and G. Tian, *Ricci flow and the Poincaré conjecture*, Clay Mathematics Monographs, vol. 3, American Mathematical Society, Providence, RI; Clay Mathematics Institute, Cambridge, MA, 2007. MR2334563

[405] J. Moser, *A Harnack inequality for parabolic differential equations*, Comm. Pure Appl. Math. **17** (1964), 101–134, DOI 10.1002/cpa.3160170106. MR159139

[406] A. Mramor, *A finiteness theorem via the mean curvature flow with surgery*, J. Geom. Anal. **28** (2018), no. 4, 3348–3372, DOI 10.1007/s12220-017-9962-5. MR3881975

[407] A. Mramor and A. Payne, *Ancient and eternal solutions to mean curvature flow from minimal surfaces*, arXiv:1904.08439v2, 2019.

[408] A. Mramor and S. Wang, *On the topological rigidity of compact self-shrinkers in $\mathbb{R}^3$*, International Mathematics Research Notices (2018), rny050.

[409] T. Mullins, *On the inverse mean curvature flow in warped product manifolds*, arXiv:1610.05234v1, 2016.

[410] W. W. Mullins, *Two-dimensional motion of idealized grain boundaries*, J. Appl. Phys. **27** (1956), 900–904. MR78836

[411] N. Nadirashvili and S. Vlăduţ, *Singular solutions of Hessian elliptic equations in five dimensions* (English, with English and French summaries), J. Math. Pures Appl. (9) **100** (2013), no. 6, 769–784, DOI 10.1016/j.matpur.2013.03.001. MR3125267

[412] K. Nakayama, T. Iizuka, and M. Wadati, *Curve lengthening equation and its solutions*, J. Phys. Soc. Japan **63** (1994), no. 4, 1311–1321, DOI 10.1143/JPSJ.63.1311. MR1280385

[413] M. Nara and M. Taniguchi, *The condition on the stability of stationary lines in a curvature flow in the whole plane*, J. Differential Equations **237** (2007), no. 1, 61–76, DOI 10.1016/j.jde.2007.02.012. MR2327727

[414] A. Neves, *Singularities of Lagrangian mean curvature flow: monotone case*, Math. Res. Lett. **17** (2010), no. 1, 109–126, DOI 10.4310/MRL.2010.v17.n1.a9. MR2592731

[415] A. Neves, *Recent progress on singularities of Lagrangian mean curvature flow*, Surveys in geometric analysis and relativity, Adv. Lect. Math. (ALM), vol. 20, Int. Press, Somerville, MA, 2011, pp. 413–438. MR2906935

[416] A. Neves, *Finite time singularities for Lagrangian mean curvature flow*, Ann. of Math. (2) **177** (2013), no. 3, 1029–1076, DOI 10.4007/annals.2013.177.3.5. MR3034293

[417] H. T. Nguyen, *Convexity and cylindrical estimates for mean curvature flow in the sphere*, Trans. Amer. Math. Soc. **367** (2015), no. 7, 4517–4536, DOI 10.1090/S0002-9947-2015-05927-3. MR3335392

[418] X. H. Nguyen, *Construction of complete embedded self-similar surfaces under mean curvature flow. I*, Trans. Amer. Math. Soc. **361** (2009), no. 4, 1683–1701, DOI 10.1090/S0002-9947-08-04748-X. MR2465812

[419] X. H. Nguyen, *Translating tridents*, Comm. Partial Differential Equations **34** (2009), no. 1-3, 257–280, DOI 10.1080/03605300902768685. MR2512861

[420] X. H. Nguyen, *Construction of complete embedded self-similar surfaces under mean curvature flow. II*, Adv. Differential Equations **15** (2010), no. 5-6, 503–530. MR2643233

[421] X. H. Nguyen, *Complete embedded self-translating surfaces under mean curvature flow*, J. Geom. Anal. **23** (2013), no. 3, 1379–1426, DOI 10.1007/s12220-011-9292-y. MR3078359

[422] X. H. Nguyen, *Construction of complete embedded self-similar surfaces under mean curvature flow, Part III*, Duke Math. J. **163** (2014), no. 11, 2023–2056, DOI 10.1215/00127094-2795108. MR3263027

[423] X. H. Nguyen, *Doubly periodic self-translating surfaces for the mean curvature flow*, Geom. Dedicata **174** (2015), 177–185, DOI 10.1007/s10711-014-0011-2. MR3303047

[424] Y. Ni and M. Zhu, *One-dimensional conformal metric flow*, Adv. Math. **218** (2008), no. 4, 983–1011, DOI 10.1016/j.aim.2008.02.006. MR2419376

[425] C.-H. Nien and D.-H. Tsai, *Convex curves moving translationally in the plane*, J. Differential Equations **225** (2006), no. 2, 605–623, DOI 10.1016/j.jde.2006.03.005. MR2225802

[426] S. Nishikawa, *Deformation of Riemannian metrics and manifolds with bounded curvature ratios*, Geometric measure theory and the calculus of variations (Arcata, Calif., 1984), Proc. Sympos. Pure Math., vol. 44, Amer. Math. Soc., Providence, RI, 1986, pp. 343–352, DOI 10.1090/pspum/044/840284. MR840284

[427] J. C. C. Nitsche, *On new results in the theory of minimal surfaces*, Bull. Amer. Math. Soc. **71** (1965), 195–270, DOI 10.1090/S0002-9904-1965-11276-9. MR173993

[428] J. A. Oaks, *Singularities and self-intersections of curves evolving on surfaces*, Indiana Univ. Math. J. **43** (1994), no. 3, 959–981, DOI 10.1512/iumj.1994.43.43042. MR1305955

[429] V. Oliker, *Evolution of nonparametric surfaces with speed depending on curvature. I. The Gauss curvature case*, Indiana Univ. Math. J. **40** (1991), no. 1, 237–258, DOI 10.1512/iumj.1991.40.40010. MR1101228

[430] V. Oliker, *Self-similar solutions and asymptotic behavior of flows of nonparametric surfaces driven by Gauss or mean curvature*, Differential geometry: partial differential equations on manifolds (Los Angeles, CA, 1990), Proc. Sympos. Pure Math., vol. 54, Amer. Math. Soc., Providence, RI, 1993, pp. 389–402, DOI 10.1090/pspum/054.1/1216597. MR1216597

[431] P. J. Olver, *Applications of Lie groups to differential equations*, 2nd ed., Graduate Texts in Mathematics, vol. 107, Springer-Verlag, New York, 1993. MR1240056

[432] K. D. Olwell, *A family of solitons for the Gauss curvature flow*, ProQuest LLC, Ann Arbor, MI, 1993. Thesis (Ph.D.)–University of California, San Diego. MR2690491

[433] H. Omori, *Isometric immersions of Riemannian manifolds*, J. Math. Soc. Japan **19** (1967), 205–214, DOI 10.2969/jmsj/01920205. MR215259

[434] S. Osher and J. A. Sethian, *Fronts propagating with curvature-dependent speed: algorithms based on Hamilton-Jacobi formulations*, J. Comput. Phys. **79** (1988), no. 1, 12–49, DOI 10.1016/0021-9991(88)90002-2. MR965860

[435] R. Osserman, *The isoperimetric inequality*, Bull. Amer. Math. Soc. **84** (1978), no. 6, 1182–1238, DOI 10.1090/S0002-9904-1978-14553-4. MR500557

[436] R. Osserman, *A survey of minimal surfaces*, 2nd ed., Dover Publications, Inc., New York, 1986. MR852409

[437] L. E. Payne and H. F. Weinberger, *An optimal Poincaré inequality for convex domains*, Arch. Rational Mech. Anal. **5** (1960), 286–292 (1960), DOI 10.1007/BF00252910. MR117419

[438] G. Perelman, *The entropy formula for the Ricci flow and its geometric applications*, arXiv:math.DG/0211159v1, 2002.

[439] G. Perelman, *Ricci flow with surgery on three-manifolds*, arXiv:math.DG/0303109v1, 2003.

[440] G. Perelman, *Finite extinction time for the solutions to the Ricci flow on certain three-manifolds*, arXiv:math/0307245v1, 2003.

[441] S. Peters, *Convergence of Riemannian manifolds*, Compositio Math. **62** (1987), no. 1, 3–16. MR892147

[442] P. Petersen, *Riemannian geometry*, 3rd ed., Graduate Texts in Mathematics, vol. 171, Springer, Cham, 2016. MR3469435

[443] Y. Rieck, *A proof of Waldhausen's uniqueness of splittings of S^3 (after Rubinstein and Scharlemann)*, Workshop on Heegaard Splittings, Geom. Topol. Monogr., vol. 12, Geom. Topol. Publ., Coventry, 2007, pp. 277–284, DOI 10.2140/gtm.2007.12.277. MR2408250

[444] S. Risa, *Ancient solutions of curvature flows*, PhD thesis, Università di Roma, Tor Vergata, 2016.

[445] S. Risa and C. Sinestrari, *Ancient solutions of geometric flows with curvature pinching*, J. Geom. Anal. **29** (2019), no. 2, 1206–1232, DOI 10.1007/s12220-018-0036-0. MR3935256

[446] S. Risa and C. Sinestrari, *Strong spherical rigidity of ancient solutions of expansive curvature flows*, arXiv:1907.12319v1, 2019.

[447] R. T. Rockafellar, *Convex analysis*, reprint of the 1970 original, Princeton Landmarks in Mathematics, Princeton Paperbacks, Princeton University Press, Princeton, NJ, 1997. MR1451876

[448] H. Rubinstein, *Some of Hyam's favourite problems*, Geometry and topology down under, Contemp. Math., vol. 597, Amer. Math. Soc., Providence, RI, 2013, pp. 165–175, DOI 10.1090/conm/597/11778. MR3186672

[449] W. Rudin, *Principles of mathematical analysis*, 3rd ed., International Series in Pure and Applied Mathematics, McGraw-Hill Book Co., New York-Auckland-Düsseldorf, 1976. MR0385023

[450] R. Sacksteder, *On hypersurfaces with no negative sectional curvatures*, Amer. J. Math. **82** (1960), 609–630, DOI 10.2307/2372973. MR116292

[451] M. Sáez and O. C. Schnürer, *Mean curvature flow without singularities*, J. Differential Geom. **97** (2014), no. 3, 545–570. MR3263514

[452] M. Sáez-Trumper, *Relaxation of the flow of triods by curve shortening flow via the vector-valued parabolic Allen-Cahn equation*, J. Reine Angew. Math. **634** (2009), 143–168, DOI 10.1515/CRELLE.2009.071. MR2560408

[453] M. Sáez Trumper, *Uniqueness of self-similar solutions to the network flow in a given topological class*, Comm. Partial Differential Equations **36** (2011), no. 2, 185–204, DOI 10.1080/03605302.2010.539892. MR2763337

[454] L. A. Santaló, *An affine invariant for convex bodies of n-dimensional space* (Spanish), Portugal. Math. **8** (1949), 155–161. MR39293

[455] G. Sapiro and A. Tannenbaum, *On affine plane curve evolution*, J. Funct. Anal. **119** (1994), no. 1, 79–120, DOI 10.1006/jfan.1994.1004. MR1255274

[456] A. Savas-Halilaj and K. Smoczyk, *Lagrangian mean curvature flow of Whitney spheres*, Geom. Topol. **23** (2019), no. 2, 1057–1084, DOI 10.2140/gt.2019.23.1057. MR3939057

[457] J. Scheuer, *Pinching and asymptotical roundness for inverse curvature flows in Euclidean space*, J. Geom. Anal. **26** (2016), no. 3, 2265–2281, DOI 10.1007/s12220-015-9627-1. MR3511477

[458] J. Scheuer, *Isotropic functions revisited*, Arch. Math. (Basel) **110** (2018), no. 6, 591–604, DOI 10.1007/s00013-018-1162-4. MR3803748

[459] J. Scheuer, *Inverse curvature flows in Riemannian warped products*, J. Funct. Anal. **276** (2019), no. 4, 1097–1144, DOI 10.1016/j.jfa.2018.08.021. MR3906301

[460] S. Schleimer, *Waldhausen's theorem*, Workshop on Heegaard Splittings, Geom. Topol. Monogr., vol. 12, Geom. Topol. Publ., Coventry, 2007, pp. 299–317, DOI 10.2140/gtm.2007.12.299. MR2408252

[461] R. Schneider, *Convex bodies: the Brunn-Minkowski theory*, second expanded edition, Encyclopedia of Mathematics and its Applications, vol. 151, Cambridge University Press, Cambridge, 2014. MR3155183

[462] O. C. Schnürer, *Surfaces contracting with speed $|A|^2$*, J. Differential Geom. **71** (2005), no. 3, 347–363. MR2198805

[463] O. C. Schnürer, *Surfaces expanding by the inverse Gauß curvature flow*, J. Reine Angew. Math. **600** (2006), 117–134, DOI 10.1515/CRELLE.2006.088. MR2283800

[464] O. C. Schnürer, A. Azouani, M. Georgi, J. Hell, N. Jangle, A. Koeller, T. Marxen, S. Ritthaler, M. Sáez, F. Schulze, and B. Smith, *Evolution of convex lens-shaped networks under the curve shortening flow*, Trans. Amer. Math. Soc. **363** (2011), no. 5, 2265–2294, DOI 10.1090/S0002-9947-2010-04820-2. MR2763716

[465] R. Schoen, *Estimates for stable minimal surfaces in three-dimensional manifolds*, Seminar on minimal submanifolds, Ann. of Math. Stud., vol. 103, Princeton Univ. Press, Princeton, NJ, 1983, pp. 111–126. MR795231

[466] R. M. Schoen, *Uniqueness, symmetry, and embeddedness of minimal surfaces*, J. Differential Geom. **18** (1983), no. 4, 791–809 (1984). MR730928

[467] R. M. Schoen, *On the number of constant scalar curvature metrics in a conformal class*, Differential geometry, Pitman Monogr. Surveys Pure Appl. Math., vol. 52, Longman Sci. Tech., Harlow, 1991, pp. 311–320. MR1173050

[468] R. Schoen and L. Simon, *Regularity of stable minimal hypersurfaces*, Comm. Pure Appl. Math. **34** (1981), no. 6, 741–797, DOI 10.1002/cpa.3160340603. MR634285

[469] R. Schoen and S. T. Yau, *On the proof of the positive mass conjecture in general relativity*, Comm. Math. Phys. **65** (1979), no. 1, 45–76. MR526976

[470] A. Schoenflies, *Beiträge zur Theorie der Punktmengen. III* (German), Math. Ann. **62** (1906), no. 2, 286–328, DOI 10.1007/BF01449982. MR1511377

[471] F. Schulze, *Evolution of convex hypersurfaces by powers of the mean curvature*, Math. Z. **251** (2005), no. 4, 721–733, DOI 10.1007/s00209-004-0721-5. MR2190140

[472] F. Schulze, *Convexity estimates for flows by powers of the mean curvature*, Ann. Sc. Norm. Super. Pisa Cl. Sci. (5) **5** (2006), no. 2, 261–277. MR2244700

[473] F. Schulze, *Nonlinear evolution by mean curvature and isoperimetric inequalities*, J. Differential Geom. **79** (2008), no. 2, 197–241. MR2420018

[474] F. Schulze, *Uniqueness of compact tangent flows in mean curvature flow*, J. Reine Angew. Math. **690** (2014), 163–172, DOI 10.1515/crelle-2012-0070. MR3200339

[475] G. W. Schwarz, *Smooth functions invariant under the action of a compact Lie group*, Topology **14** (1975), 63–68, DOI 10.1016/0040-9383(75)90036-1. MR370643

[476] J. Serrin, *A symmetry problem in potential theory*, Arch. Rational Mech. Anal. **43** (1971), 304–318, DOI 10.1007/BF00250468. MR333220

[477] N. Sesum, *Rate of convergence of the mean curvature flow*, Comm. Pure Appl. Math. **61** (2008), no. 4, 464–485, DOI 10.1002/cpa.20209. MR2383930

[478] J. A. Sethian, *Numerical algorithms for propagating interfaces: Hamilton-Jacobi equations and conservation laws*, J. Differential Geom. **31** (1990), no. 1, 131–161. MR1030668

[479] L. Shahriyari, *Translating graphs by mean curvature flow*, Geom. Dedicata **175** (2015), 57–64, DOI 10.1007/s10711-014-0028-6. MR3323629

[480] J. J. Sharples, *Linear and quasilinear parabolic equations in Sobolev space*, J. Differential Equations **202** (2004), no. 1, 111–142, DOI 10.1016/j.jde.2004.03.020. MR2060534

[481] W. Sheng and X.-J. Wang, *Singularity profile in the mean curvature flow*, Methods Appl. Anal. **16** (2009), no. 2, 139–155, DOI 10.4310/MAA.2009.v16.n2.a1. MR2563745

[482] L. Simon, *Lectures on geometric measure theory*, Proceedings of the Centre for Mathematical Analysis, Australian National University, vol. 3, Australian National University, Centre for Mathematical Analysis, Canberra, 1983. MR756417

[483] C. Sinestrari, *Convex hypersurfaces evolving by volume preserving curvature flows*, Calc. Var. Partial Differential Equations **54** (2015), no. 2, 1985–1993, DOI 10.1007/s00526-015-0852-z. MR3396440

[484] S. Smale, *A classification of immersions of the two-sphere*, Trans. Amer. Math. Soc. **90** (1958), 281–290, DOI 10.2307/1993205. MR104227

[485] G. Smith, *On complete embedded translating solitons of the mean curvature flow that are of finite genus*, arXiv:1501.04149v2, 2015.

[486] K. Smoczyk, *Harnack inequalities for curvature flows depending on mean curvature*, New York J. Math. **3** (1997), 103–118. MR1480081

[487] K. Smoczyk, *Starshaped hypersurfaces and the mean curvature flow*, Manuscripta Math. **95** (1998), no. 2, 225–236, DOI 10.1007/s002290050025. MR1603325

[488] K. Smoczyk, *Mean curvature flow in higher codimension: introduction and survey*, Global differential geometry, Springer Proc. Math., vol. 17, Springer, Heidelberg, 2012, pp. 231–274, DOI 10.1007/978-3-642-22842-1_9. MR3289845

[489] H. M. Soner, *Motion of a set by the curvature of its boundary*, J. Differential Equations **101** (1993), no. 2, 313–372, DOI 10.1006/jdeq.1993.1015. MR1204331

[490] K. Sonnanburg, *Blow-up continuity for type-I, mean-convex mean curvature flow*, arXiv:1703.02619v1, 2017.

[491] K. Sonnanburg, *A Liouville theorem for mean curvature flow*, arXiv:1711.02261v1, 2017.

[492] P. Souplet and Q. S. Zhang, *Sharp gradient estimate and Yau's Liouville theorem for the heat equation on noncompact manifolds*, Bull. London Math. Soc. **38** (2006), no. 6, 1045–1053, DOI 10.1112/S0024609306018947. MR2285258

[493] J. Spruck and L. Sun, *Convexity of 2-convex translating solitons to the mean curvature flow in* $\mathbb{R}^{n+1}$, arXiv:1910.02195v1, 2019.

[494] J. Spruck and L. Xiao, *Complete translating solitons to the mean curvature flow in* $\mathbb{R}^3$ *with nonnegative mean curvature*, arXiv:1703.01003v3, 2017.

[495] A. Stahl, *Convergence of solutions to the mean curvature flow with a Neumann boundary condition*, Calc. Var. Partial Differential Equations **4** (1996), no. 5, 421–441, DOI 10.1007/BF01246150. MR1402731

[496] A. Stahl, *Regularity estimates for solutions to the mean curvature flow with a Neumann boundary condition*, Calc. Var. Partial Differential Equations **4** (1996), no. 4, 385–407, DOI 10.1007/BF01190825. MR1393271

[497] J. J. Stoker, *Über die Gestalt der positiv gekrümmten offenen Flächen im dreidimensionalen Raume* (German), Compositio Math. **3** (1936), 55–88. MR1556933

[498] A. Stone, *A density function and the structure of singularities of the mean curvature flow*, Calc. Var. Partial Differential Equations **2** (1994), no. 4, 443–480, DOI 10.1007/BF01192093. MR1383918

[499] A. G. Stone, *Singular and boundary behaviour in the mean curvature flow of hypersurfaces*, ProQuest LLC, Ann Arbor, MI, 1994. Thesis (Ph.D.)–Stanford University. MR2691040

[500] C. F. Sturm, *Mémoire sur une classe d'équations à différences partielles*, Collected Works of Charles François Sturm, Birkhäuser Basel, 2009, pp. 505–576.

[501] A. Sun, *Local entropy and generic multiplicity one singularities of mean curvature flow of surfaces*, arXiv:1810.08114v2, 2018.

[502] Y. Tonegawa, *Brakke's mean curvature flow: An introduction*, SpringerBriefs in Mathematics, Springer, Singapore, 2019. MR3930606

[503] M. Traizet, *Construction de surfaces minimales en recollant des surfaces de Scherk* (French, with English and French summaries), Ann. Inst. Fourier (Grenoble) **46** (1996), no. 5, 1385–1442. MR1427131

[504] D.-H. Tsai, *Geometric expansion of starshaped plane curves*, Comm. Anal. Geom. **4** (1996), no. 3, 459–480, DOI 10.4310/CAG.1996.v4.n3.a5. MR1415752

[505] D.-H. Tsai, *Blowup and convergence of expanding immersed convex plane curves*, Comm. Anal. Geom. **8** (2000), no. 4, 761–794, DOI 10.4310/CAG.2000.v8.n4.a3. MR1792373

[506] K. Tso, *Deforming a hypersurface by its Gauss-Kronecker curvature*, Comm. Pure Appl. Math. **38** (1985), no. 6, 867–882, DOI 10.1002/cpa.3160380615. MR812353

[507] J. Urbas, *Complete noncompact self-similar solutions of Gauss curvature flows. I. Positive powers*, Math. Ann. **311** (1998), no. 2, 251–274, DOI 10.1007/s002080050187. MR1625754

[508] J. Urbas, *Complete noncompact self-similar solutions of Gauss curvature flows. II. Negative powers*, Adv. Differential Equations **4** (1999), no. 3, 323–346. MR1671253

[509] J. Urbas, *Convex curves moving homothetically by negative powers of their curvature*, Asian J. Math. **3** (1999), no. 3, 635–656, DOI 10.4310/AJM.1999.v3.n3.a4. MR1793674

[510] J. I. E. Urbas, *On the expansion of starshaped hypersurfaces by symmetric functions of their principal curvatures*, Math. Z. **205** (1990), no. 3, 355–372, DOI 10.1007/BF02571249. MR1082861

[511] J. I. E. Urbas, *An expansion of convex hypersurfaces*, J. Differential Geom. **33** (1991), no. 1, 91–125. MR1085136

[512] J. van Heijenoort, *On locally convex manifolds*, Comm. Pure Appl. Math. **5** (1952), 223–242, DOI 10.1002/cpa.3160050302. MR52131

[513] J. J. L. Velázquez, *Curvature blow-up in perturbations of minimal cones evolving by mean curvature flow*, Ann. Scuola Norm. Sup. Pisa Cl. Sci. (4) **21** (1994), no. 4, 595–628. MR1318773

[514] J. von Neumann, *Discussion: Shape of metal grains*, Metal Interfaces, American Society for Testing Materials, Cleveland, 1952, p. 108.

[515] F. Waldhausen, *Heegaard-Zerlegungen der 3-Sphäre* (German), Topology **7** (1968), 195–203, DOI 10.1016/0040-9383(68)90027-X. MR227992

[516] G. Wang and C. Xia, *Isoperimetric type problems and Alexandrov-Fenchel type inequalities in the hyperbolic space*, Adv. Math. **259** (2014), 532–556, DOI 10.1016/j.aim.2014.01.024. MR3197666

[517] L. Wang, *Uniqueness of self-similar shrinkers with asymptotically conical ends*, J. Amer. Math. Soc. **27** (2014), no. 3, 613–638, DOI 10.1090/S0894-0347-2014-00792-X. MR3194490

[518] L. Wang, *Uniqueness of self-similar shrinkers with asymptotically cylindrical ends*, J. Reine Angew. Math. **715** (2016), 207–230, DOI 10.1515/crelle-2014-0006. MR3507924

[519] M.-T. Wang, *Some recent developments in Lagrangian mean curvature flows*, Surveys in differential geometry. Vol. XII. Geometric flows, Surv. Differ. Geom., vol. 12, Int. Press, Somerville, MA, 2008, pp. 333–347, DOI 10.4310/SDG.2007.v12.n1.a9. MR2488942

[520] X. Wang, *A remark on strong maximum principle for parabolic and elliptic systems*, Proc. Amer. Math. Soc. **109** (1990), no. 2, 343–348, DOI 10.2307/2047994. MR1019284

[521] X.-J. Wang, *Convex solutions to the mean curvature flow*, Ann. of Math. (2) **173** (2011), no. 3, 1185–1239, DOI 10.4007/annals.2011.173.3.1. MR2800714

[522] F. W. Warner, *Foundations of differentiable manifolds and Lie groups*, corrected reprint of the 1971 edition, Graduate Texts in Mathematics, vol. 94, Springer-Verlag, New York-Berlin, 1983. MR722297

[523] Y. Wei and C. Xiong, *Inequalities of Alexandrov-Fenchel type for convex hypersurfaces in hyperbolic space and in the sphere*, Pacific J. Math. **277** (2015), no. 1, 219–239, DOI 10.2140/pjm.2015.277.219. MR3393689

[524] H. F. Weinberger, *Invariant sets for weakly coupled parabolic and elliptic systems* (English, with Italian summary), collection of articles dedicated to Mauro Picone on the occasion of his ninetieth birthday, Rend. Mat. (6) **8** (1975), 295–310. MR397126

[525] G. Wheeler and V.-M. Wheeler, *Mean curvature flow with free boundary outside a hypersphere*, Trans. Amer. Math. Soc. **369** (2017), no. 12, 8319–8342, DOI 10.1090/tran/7305. MR3710626

[526] V. M. Wheeler, *Mean curvature flow of entire graphs in a half-space with a free boundary*, J. Reine Angew. Math. **690** (2014), 115–131, DOI 10.1515/crelle-2012-0028. MR3200337

[527] V.-M. Wheeler, *Non-parametric radially symmetric mean curvature flow with a free boundary*, Math. Z. **276** (2014), no. 1-2, 281–298, DOI 10.1007/s00209-013-1200-7. MR3150205

[528] V.-M. Wheeler, *Mean curvature flow with free boundary in embedded cylinders or cones and uniqueness results for minimal hypersurfaces*, Geom. Dedicata **190** (2017), 157–183, DOI 10.1007/s10711-017-0236-y. MR3704818

[529] B. White, *The size of the singular set in mean curvature flow of mean-convex sets*, J. Amer. Math. Soc. **13** (2000), no. 3, 665–695, DOI 10.1090/S0894-0347-00-00338-6. MR1758759

[530] B. White, *Evolution of curves and surfaces by mean curvature*, Proceedings of the International Congress of Mathematicians, Vol. I (Beijing, 2002), Higher Ed. Press, Beijing, 2002, pp. 525–538. MR1989203

[531] B. White, *The nature of singularities in mean curvature flow of mean-convex sets*, J. Amer. Math. Soc. **16** (2003), no. 1, 123–138, DOI 10.1090/S0894-0347-02-00406-X. MR1937202

[532] B. White, *A local regularity theorem for mean curvature flow*, Ann. of Math. (2) **161** (2005), no. 3, 1487–1519, DOI 10.4007/annals.2005.161.1487. MR2180405

[533] B. White, *Currents and flat chains associated to varifolds, with an application to mean curvature flow*, Duke Math. J. **148** (2009), no. 1, 41–62, DOI 10.1215/00127094-2009-019. MR2515099

[534] H. Whitney, *On regular closed curves in the plane*, Compositio Math. **4** (1937), 276–284. MR1556973

[535] D. V. Widder, *The role of the Appell transformation in the theory of heat conduction*, Trans. Amer. Math. Soc. **109** (1963), 121–134, DOI 10.2307/1993650. MR154068

[536] D. V. Widder, *The heat equation*, Pure and Applied Mathematics, Vol. 67, Academic Press [Harcourt Brace Jovanovich, Publishers], New York-London, 1975. MR0466967

[537] H. Yagisita, *Asymptotic behaviors of star-shaped curves expanding by $V = 1-K$*, Differential Integral Equations **18** (2005), no. 2, 225–232. MR2106103

[538] S. T. Yau, *Harmonic functions on complete Riemannian manifolds*, Comm. Pure Appl. Math. **28** (1975), 201–228, DOI 10.1002/cpa.3160280203. MR431040

[539] R. Ye, *Global existence and convergence of Yamabe flow*, J. Differential Geom. **39** (1994), no. 1, 35–50. MR1258912

Index

2-convex, 214

α-concave/convex, 662
α-Gauß curvature flow, 543
α-homogeneous function, 645
Abresch–Langer curves, 100
admissible speed function, 639
affine
  Codazzi equation, 603
  curvature, 602
  curve shortening flow, 91
  Gauß equation, 603
  invariance, 582, 603
  isoperimetric inequality, 585, 602, 611
  mean curvature, 603
  metric, 602
  normal flow, 584
    affine-normalized, 595
  normal vector, 584, 602
  structural equations, 603
  surface area, 585, 591
Alexandrov reflection, 486, 500, 670, 723, 724
Alexandrov–Fenchel inequality, 668, 724
Allard's regularity theorem, 208
ancient helicoid, 540
ancient ovaloid, 524
ancient pancake, 524
ancient pineapple ring, 540
ancient solution, 31, 105, 503
  entire, 516
Angenent oval, *see also* paperclip solution
Angenent's doughnut, 282, 493, 494
angle parameter, 103
anisotropic
  curvature flow, 641, 724
  curve shortening flow, 91
arc length element, 42
area enclosed by curve, 48
area-normalized mean curvature flow, 183
arrival time, 211, 266, 335, 516
asymptotic entropy, 393
asymptotic Gaußian density, 364
asymptotic translator, 469
avoidance principle
  curve shortening flow, 64
  fully nonlinear flows, 658
  Gauß curvature flows, 548
  mean curvature flow, 192

β-$(m+1)$-convex, 420
ball
  extrinsic, 396
  intrinsic, 287
barrier solutions, 287, 410
  subsolution/supersolution, 287
Bernstein technique, 17, 51, 201
Blaschke–Santaló inequality, 585, 602
blow-down, 469
blow-up point, 378
blow-up sequence, 372
  essential, 372
BÖC, 409
Boltzmann entropy, 19, 574

bounded geometry
  locally uniformly, 347
bowl soliton, 461
  rigidity, 470
bowloid, 470
Brakke flow
  integral, 207
Brakke's regularity theorem, 208, 365
Brakke–White regularity theorem, 208, 365, 390
Breakfast can wait, 533
Brunn–Minkowski inequality, 576

caloric function, 2
canonical vector field, 153
catenoid, 169
center of mass, 594
Cheeger–Gromov convergence, 347
chord-arc profile, 75
Chou's theorem, 548
circumradius, 107, 269
class $\mathcal{C}^n_m(R, \alpha)$, 301
Codazzi equation, 139, 157
  affine, 603
compactness theorem
  for shrinkers, 388
  immersed hypersurfaces, 349, 350
  mean curvature flows, 353, 354
  Riemannian manifolds, 347
concave curvature function, 646
cone, 285
  in a linear space, 645
conformal map, 170
connection, 129
  induced, 154
  Levi-Civita, 130
  metric compatible, 129
  symmetric, 129
conservation of energy, heat, 1
convergence
  in Cheeger–Gromov topology, 347
  in the Hausdorff topology, 346
  in the sense of Brakke flows, 208, 381
  locally uniformly in the Hausdorff topology, 346
  locally uniformly in the smooth topology, 348
  on compact subsets of $M^n \times I$, 352
  on compact subsets of $\mathbb{R}^{n+1} \times \mathbb{R}$, 352
  smoothly on compact subsets of M^n, 347
  smoothly on compact subsets of $\mathbb{R}^{n+1}$, 348
convex body, 148
convex curvature function, 646
convex curve, 49
convex hypersurface, 148
  locally, 148
  locally strictly, 148
  locally uniformly, 149
  strictly, 148
  uniformly, 149
convexity estimates, 295, 303
  for ancient solutions, 509
  for fully nonlinear flows, 702
  for translators, 462, 466
covariant derivative, 129
  Euclidean, 128
Crofton formula, 566
crypto-irregularity, 384
cubic ground form, 603
  vanishing characterization, 603
curvature, 39
  boundary of an extrinsic ball, 396
curvature flow, 164, 639
  1-homogeneous speed, 687, 712
  anisotropic, 641, 724
  isotropic, 641
curvature function, 644
  concave/convex, 646
  inverse-concave, 647
curvature neck, 355
curvature normalizing sequence, 372
curvature tensor
  Riemann, 139
curvature vector, 40
curve shortening flow, 42
  affine, 91
cyclic
  integral Brakke flow, 393
cylindrical estimates, 296, 303, 702
cylindrical point, 294

Δ-wing, 470
Daskalopoulus–Hamilton–Šešum theorem, 122
degenerate neckpinch, 283
delta-wing, 470
diameter, 15
  intrinsic, 258
differential Harnack estimates, 103
directional derivative, 128
Dirichlet boundary condition, 3

Dirichlet energy, 3
Dirichlet problem
  translator, 454
distributional sense
  heat equation subsolution, 297
divergence, 135
dual function, 647

elementary symmetric polynomial, 650
ellipsoid, 581
  normalized, 581
embedded planar curve, 38
embedding, 125
entire
  ancient solution, 516
  function, 144
  graph, 144
  translator, 452
entropy, 607, 669
  asymptotic, 393
  convex curve, 106
  normalized, 107
  subcylindrical, 393
entropy functional, 391
entropy point, 615
error function, 23
eternal solutions, 47, 321
Euler characteristic, 152
Euler's theorem
  for homogeneous functions, 646
evolution equations, 179
evolving orthonormal frame, 158
expander, 177, 546
expanding self-similar solution, 177, 239, 323
exscribed curvature, 396
exterior noncollapsing solution, 399
extremal set, 285
extrinsic ball, 396
extrinsic ball curvature, 395

$\mathcal{F}$-critical, 391
$\mathcal{F}$-functional, 391
$\mathcal{F}$-stable, 392
fattening, 210
Firey conjecture, 560
Firey entropy, 572
  relative to a point, 615
first fundamental form, 126
flying wing, 470, 478
Fourier transform, 5
Fourier's law, 1
Fourier, Jean-Baptiste Joseph, 1
frame bundle, 158
Frenet–Serret equations, 39
function
  symmetric, 641
fundamental solution, 4

Gage–Hamilton theorem, 54, 412
Gauß curvature, 136
Gauß curvature flow, 165, 543
Gauß equation, 139, 157
  affine, 603
Gauß map, 134
Gauß–Bonnet theorem, 152
Gaußian area, 311
Gaußian density, 209, 364
Gaußian density ratio, 363
Gaußian entropy, 573, 575
general linear frame bundle, 160
Giga–Kohn theorem, 21
gouda
  aged, 524
gradient, 136
gradient estimate, 24, 27
  curvature, 256, 303
  graphical mean curvature flow, 229
gradient flow, 3, 32, 44, 182, 570
graph, 175
graph hypersurface, 143
Grayson's theorem, 82
Grim hyperplane, 175, 177, 453
Grim Reaper cylinder (see Grim hyperplane), 453
Grim Reaper solution, 47

hairclip solution, 110
Halldorsson's theorem, 100
Hamilton's Harnack estimate, 319
handle, handlebody, 214
harmonic function, 3
harmonic mean curvature, 136
harmonic mean curvature flow, 165
Harnack estimate
  α-Gauß curvature flow, 559
  curve shortening flow, 105
  fully nonlinear flows, 662
  heat equation, 17
  mean curvature flow, 319
Hausdorff convergence, 345
Hausdorff distance, 346
Hausdorff measure, 137

Hausdorff topology convergence, *see also* Hausdorff convergence
heat ball, 10
heat equation, 2
heat kernel, 12
Heegaard surface, 451
height, 225, 453
helicoid, 169
Hessian estimate
  for mean curvature flow, 305
Hilbert–Schmidt norm, 278
homogeneous function
  of degree α, 645
homothetic self-similar solution, 177
Hopf boundary point lemma, 679
Hopf lemma, 10
Huisken's monotonicity formula, 108, 315, 316
  Hamilton's generalization, 339
Huisken's theorem, 243, 366, 412
hypersurface
  α-pinched, 246
  evolves by curvature, 639
  nonconvex, 698

immersed planar curve, 38
immersion, 125
inner product
  Euclidean, 126
inradius
  of a curve, 107
  of a hypersurface, 260, 269
inscribed curvature, 396
integral Brakke flow, 207
  cyclic, 393
interior noncollapsing solution, 399
interpolation inequality, 85
  Gagliardo–Nirenberg, 600
intrinsic ball, 287
invariant solution, 3, 96, 176
inverse-concave
  characterization, 648
inverse-Gauß curvature flow, 543
inverse-mean curvature flow, 165
isoperimetric inequality, 102, 506
  affine, 611
isotropic
  curvature flow, 641

Jagger
  moves like, 560
John ellipsoid, 586
Jordan–Brouwer theorem, 451

k-convex, 285

L^2-dual
  of the heat operator, 316
Lagrangian mean curvature flow, 276
Laplace transform, 5
Laplacian, 136
least shadow area, 608
Legendre ellipsoid, 594
Legendre form, 594
Leibniz rule, 129
lemma
  festive, 165
length of a curve, 44
level set flow, 209
Li–Yau Harnack inequality, 18
limit flow, 372
linearized flow, 166
locally convex, 148
locally strictly convex, 148
locally uniformly convex, 149
long-time existence theorem
  1-homogeneous speed, 702
  α-Gauß curvature flow, 548
  curve shortening flow, 54
  mean curvature flow, 198

m-th mean curvature, 136
mass of a varifold, 208
maximal solution, 197
maximal subsolution
  of mean curvature flow, 211
maximal time, 198
maximum principle, 8, 189
  strong, 10, 291
  tensor, 191
maximum sustained curvature, 107
mean convex hypersurface, 135, 281
mean convex solution, 211
mean cross-sectional volume, 665
mean curvature, 135
  affine, 603
mean curvature flow, 165, 173
  level set, 209
  normalized, 373
  piecewise smooth, 387
  Riemannian ambient spaces, 275
  subsolution, 221
  unnormalized, 182
  with surgery, 213, 386

mean curvature, k-th, 651
mean value property, 10
mean width, 570
minimal hypersurface, 175
minimum principle, 190
Minkowski problem, 655
Minkowski sum, 576, 664
mixed discriminant, 665
mixed volume, 665
modulus of continuity, 22, 674
more cowbell, 409
multiplicity function, 207
multiplicity one, 208
multiplicity one conjecture, 383
musical isomorphisms, 127

Nash entropy, 19, 574
neck detection
  curvature necks, 355
  geometric necks, 358
necklike point, 355
neckpinch, 282
  degenerate, 283
network flow, 59
Neumann boundary condition, 2
noncollapsing solution, 399
noncollapsing theorem
  fully nonlinear flows, 703
  mean curvature flow, 409
nonconvex hypersurface, 698
normal angle, 41
normal bundle, 130, 153
normal cone, 616
normal projection, 153
normalized mean curvature flow, 182

Omori–Yau theorem, 29, 463
ovaloid
  ancient, 524

P_n-invariant, 164, 641
  subset of $\mathbb{R}^n$, 285
pancake
  ancient, 524
paperclip solution, 113
parabolic cylinder
  extrinsic, 232, 415
  intrinsic, 291
parabolicity condition, 640
parallel translation, 128
Penrose inequality, 654, 682, 724
Perelman's $\mathcal{W}$-entropy, 574
periodic boundary condition, 2
pinched hypersurface, 246
pinching condition
  for mean curvature flow, 285
Poincaré inequality, 14, 254
Poincaré-type inequality, 299
point selection, 365
pointed Riemannian manifold, 347
Poisson transform, 5
polar dual, 585
polynomial area growth, 433
positive cone, 286
positive orientation, 41
power mean, 650
principal basis, frame field, 135
principal curvatures, 135, 155
principal directions, 135
principal radii, 151
projections, 130, 153
proper immersion, 232
proper immersion/embedding, 125
properly defined solution
  in a cylinder, 232
  in a parabolic cylinder, 352
pullback bundle, 126, 153
pullback connection, 130, 153
pullback metric, 153

quasiconformal map, 367

radial function
  of a starshaped hypersurface, 146
Radon measure, 207
rapidly forming singularity, 371
reflection map, 676
reflects
  at/up to a hyperplane, 676
regular point, 145
regular value, 145
restriction bundle, 126
restriction connection, 130
Ricci identity, 143
Riemann curvature tensor, 139
Riemannian manifold, 126
Riemannian measure, 137
Riemannian metric, 126
rotating self-similar solution, 178
rotation index, 89
rotator, 178
round circle, 40
roundness estimate, 246

scale-invariant solution, 21
Scherk's surface, 169
Schoenflies conjecture, 214
second fundamental form, 132, 155
self-similar solution, 6, 176
sequence
  blow-up, 372
  curvature normalizing, 372
shadow area, 566
shadow mean curvature flow, 212
shape operator, 135
short-time existence theorem
  curve shortening flow, 50
  fully nonlinear flows, 657
  mean curvature flow, 186
shrinker, 177
  α-Gauß curvature flow, 546
  weak, 391
shrinking cylinder, 175
shrinking m-neck, 358
shrinking self-similar solution, 177, 425, 618
shrinking sphere, 174
Simons's equation, 141
Simons-type inequality, 410
simple closed curve, 38
simple point in $\mathbb{R}^n$, 646
singular solution, 198
  curve shortening flow, 105
singularity model, 372
singularity types, 371
slowly forming singularity, 371
smiley, 594
smooth convergence conjecture, 383
smoothing, 51, 201
Sobolev inequality
  of Michael and Simon, 253
space-time track, 334
spatial tangent bundle, 153
splitting
  isometric, 292
Stampacchia iteration, 252
starshaped hypersurface, 146
straight at infinity, 223
strictly convex, 148
strictly mean convex hypersurface, 135
strong maximum principle
  for barrier supersolutions, 291
structural equations
  affine, 603
Sturm's theorem, 29
subcylindrical entropy, 393
subsolution
  barrier, 287
  of mean curvature flow, 221
  viscosity, 22
supersolution
  barrier, 287
support
  of a measure, 207
  of a varifold, 207
support function, 48, 150, 268
supporting affine functional, 285
supporting half-space, 148
supporting hyperplane, 148
surgery, 213
symmetric
  function, 641
  subset of $\mathbb{R}^n$, 285
symmetry condition, 640

tangent cone, 616
tangent flow, 373
tangential projection, 153
tensor maximum principle, 191
time function, 153
time-dependent vector field, 153
torus
  n-dimensional, 2
totally umbilic, 170
touches a hypersurface
  extrinsic ball, 396
translating catenoid, 459
translating helicoids, 495
translating paraboloid, *see also* bowl soliton
translating self-similar solutions, 321, 452
translating trident, 496
translator, 177
  asymptotic, 469
  constant mean curvature, 456
  Dirichlet problem, 454
  of revolution, 456
traveling wave solution, heat, 4
tree
  holiday, 165
turning angle, 41
turning tangents theorem, 42
type-I
  singularity, 371
type-II
  normalization, 373

singularity, 371

umbilic, 142, 246
umbilic point, 170
Umlaufsatz, 42
uniformly convex hypersurface, 149
unit normal vector, 39
unit tangent vector, 39
unit-regular, 393
universal covering, 41

varifold, 206
integral, 207
vertical minimal cylinder, 456
viscosity solution, 210
Gauß curvature flow, 579
viscosity subsolution, 22
maximal, 211
to mean curvature flow, 211
viscosity supersolution, 22
volume
enclosed, 138
mean cross-sectional, 668
volume product, 585
volume-normalized mean curvature flow, 183

Waldhausen's theorem, 451
Wang's dichotomy
for ancient solutions, 516
for translators, 469
weak shrinker, 391
weak solutions, 206
weight measure, 207
Weingarten equation, 157
Weingarten tensor, 133, 155
well-posedness, 12
width function, 267
width pinching estimate, 268
Willmore energy, 217

yin-yang spiral, 100

Selected Published Titles in This Series

206 **Ben Andrews, Bennett Chow, Christine Guenther, and Mat Langford,** Extrinsic Geometric Flows, 2020

204 **Sarah J. Witherspoon,** Hochschild Cohomology for Algebras, 2019

203 **Dimitris Koukoulopoulos,** The Distribution of Prime Numbers, 2019

202 **Michael E. Taylor,** Introduction to Complex Analysis, 2019

201 **Dan A. Lee,** Geometric Relativity, 2019

200 **Semyon Dyatlov and Maciej Zworski,** Mathematical Theory of Scattering Resonances, 2019

199 **Weinan E, Tiejun Li, and Eric Vanden-Eijnden,** Applied Stochastic Analysis, 2019

198 **Robert L. Benedetto,** Dynamics in One Non-Archimedean Variable, 2019

197 **Walter Craig,** A Course on Partial Differential Equations, 2018

196 **Martin Stynes and David Stynes,** Convection-Diffusion Problems, 2018

195 **Matthias Beck and Raman Sanyal,** Combinatorial Reciprocity Theorems, 2018

194 **Seth Sullivant,** Algebraic Statistics, 2018

193 **Martin Lorenz,** A Tour of Representation Theory, 2018

192 **Tai-Peng Tsai,** Lectures on Navier-Stokes Equations, 2018

191 **Theo Bühler and Dietmar A. Salamon,** Functional Analysis, 2018

190 **Xiang-dong Hou,** Lectures on Finite Fields, 2018

189 **I. Martin Isaacs,** Characters of Solvable Groups, 2018

188 **Steven Dale Cutkosky,** Introduction to Algebraic Geometry, 2018

187 **John Douglas Moore,** Introduction to Global Analysis, 2017

186 **Bjorn Poonen,** Rational Points on Varieties, 2017

185 **Douglas J. LaFountain and William W. Menasco,** Braid Foliations in Low-Dimensional Topology, 2017

184 **Harm Derksen and Jerzy Weyman,** An Introduction to Quiver Representations, 2017

183 **Timothy J. Ford,** Separable Algebras, 2017

182 **Guido Schneider and Hannes Uecker,** Nonlinear PDEs, 2017

181 **Giovanni Leoni,** A First Course in Sobolev Spaces, Second Edition, 2017

180 **Joseph J. Rotman,** Advanced Modern Algebra: Third Edition, Part 2, 2017

179 **Henri Cohen and Fredrik Strömberg,** Modular Forms, 2017

178 **Jeanne N. Clelland,** From Frenet to Cartan: The Method of Moving Frames, 2017

177 **Jacques Sauloy,** Differential Galois Theory through Riemann-Hilbert Correspondence, 2016

176 **Adam Clay and Dale Rolfsen,** Ordered Groups and Topology, 2016

175 **Thomas A. Ivey and Joseph M. Landsberg,** Cartan for Beginners: Differential Geometry via Moving Frames and Exterior Differential Systems, Second Edition, 2016

174 **Alexander Kirillov Jr.,** Quiver Representations and Quiver Varieties, 2016

173 **Lan Wen,** Differentiable Dynamical Systems, 2016

172 **Jinho Baik, Percy Deift, and Toufic Suidan,** Combinatorics and Random Matrix Theory, 2016

171 **Qing Han,** Nonlinear Elliptic Equations of the Second Order, 2016

170 **Donald Yau,** Colored Operads, 2016

169 **András Vasy,** Partial Differential Equations, 2015

168 **Michael Aizenman and Simone Warzel,** Random Operators, 2015

167 **John C. Neu,** Singular Perturbation in the Physical Sciences, 2015

For a complete list of titles in this series, visit the
AMS Bookstore at **www.ams.org/bookstore/gsmseries/**.